D1740412

PROMOTING CREATIVE TOURISM: CURRENT ISSUES IN TOURISM RESEARCH

PROCEEDINGS OF THE 4TH INTERNATIONAL SEMINAR ON TOURISM (ISOT 2020), NOVEMBER 4–5, 2020, BANDUNG, INDONESIA

Promoting Creative Tourism: Current Issues in Tourism Research

Editors

A.H.G. Kusumah, C.U. Abdullah, D. Turgarini, M. Ruhimat, O. Ridwanudin & Y. Yuniawati
Universitas Pendidikan, Indonesia

CRC Press/Balkema is an imprint of the Taylor & Francis Group, an informa business

Typeset by MPS Limited, Chennai, India

Library of Congress Cataloging-in-Publication Data

Applied for

Published by: CRC Press/Balkema
Schipholweg 107C, 2316 XC Leiden, The Netherlands
e-mail: Pub.NL@taylorandfrancis.com
www.routledge.com – www.taylorandfrancis.com

ISBN: 978-0-367-55862-8 (Hbk)
ISBN: 978-0-367-55864-2 (Pbk)
ISBN: 978-1-003-09548-4 (eBook)
DOI: 10.1201/9781003095484
https://doi.org/10.1201/9781003095484

Promoting Creative Tourism: Current Issues in Tourism Research – Kusumah et al. (Eds)
© 2021 Taylor & Francis Group, London, ISBN 978-0-367-55862-8

Table of contents

Destination management

Hospitality management

Safety and crisis management

Promoting Creative Tourism: Current Issues in Tourism Research – Kusumah et al. (Eds)
© 2021 Taylor & Francis Group, London, ISBN 978-0-367-55862-8

Preface

For the past two decades, creative tourism has been a perennial discussion among researchers, scientists, practitioners, and other related parties within the context of tourism worldwide. As creative tourism offers visitors the opportunity to develop their creative potential through active participation in courses and learning experiences which are characteristic of the holiday destination where they are undertaken, it attracts a variety of tourism sectors to take part.

To this relation, this conference aims to invite academics and professionals in tourism-related fields to share their research and experiences in discussing the current issues in tourism research to promote creative tourism. This macro topic of discussion is then broken down into such important sub-themes as sustainable tourism; ICT and tourism; tourism marketing; halal and sharia tourism; destination management; politics, social phenomena, and humanities in tourism; medical and health tourism; community-based tourism; meeting, incentive, convention, and exhibition; restaurant management and operation; corporate social responsibility (CSR); disruptive innovation in tourism; urban and rural tourism planning and development; marine tourism; tourism and education; tourism, economics, and finance; recreation and sport tourism; culture and indigenous tourism; tourism gastronomy; heritage tourism; film-induced tourism; tourism planning and policy; supply chain management; hospitality management; safety and crisis management; tourism geography; infrastructure and transportation in tourism development; and community resilience and social capital in tourism.

Finally, it is hoped that the conference, as reflected in the variety of papers would allow academics, researchers, as well as practitioners to continue being engaged in the process of redefining creative and sustainable tourism. As tourism is an important part in a nation's development, we need to be constantly involved in the process of reconceptualizing it.

Editorial Team

Promoting Creative Tourism: Current Issues in Tourism Research – Kusumah et al. (Eds)
© 2021 Taylor & Francis Group, London, ISBN 978-0-367-55862-8

Scientific Committee

International Advisory Board

1. Dr. Eunice Tan (Murdoch Uni Singapore, Singapore)
2. Dr. Christina Lam (SIT)
3. Dr. Walanchalee Wattana (Mahidol Univ)
4. Dr. Eka Putra (Uni Korea) Sol International Hospitality Management (SIHOM), Sol International School (SIS), Woosong University
5. Hera Oktadiana Ph.D (JCU)
6. Yong-Jac Choi, Ph.D (Hankuk University of Foreign Studies)
7. Prof. Neil Carr (University of Otago)
8. Dr. Craig Lee (University of Otago)
9. Dr. Manisa Piuchan (Chiang Mai University)
10. Dr. Pipatpong Fakfare (School of Humanities and Tourism Management, Bangkok University, Thailand)
11. Dr. Jiwon Seo (Ewha Woman's University, Seoul, South Korea)
12. Dr. Luk Kiano (Community College of City University Hong Kong, Hongkong)
13. Dr. Ryan Smith (San Francisco State University, USA)
14. Prof. Dr. Elly Malihah M.Si (Universitas Pendidikan Indonesia, Indonesia)
15. Prof. Dr. Darsiharjo M.S (Universitas Pendidikan Indonesia, Indonesia)
16. Dr. Caria Ningsih (Universitas Pendidikan Indonesia, Indonesia)
17. Dr. Erry Sukriah (Universitas Pendidikan Indonesia, Indonesia)
18. Usep Suhud Ph.D (Universitas Negeri Jakarta, Indonesia)
19. Dr. Myrza Rahmanita (STP Trisakti, Indonesia)
20. Dr. Liga Suryadana (STP Enhaii, Indonesia)
21. Dr. Any Ariani Noor S.Pd., M.Sc (Polban, Indonesia)
22. Dr. Marceilla Hidayat BA.Honns M.M (Polban, Indonesia)

Promoting Creative Tourism: Current Issues in Tourism Research – Kusumah et al. (Eds)
© 2021 Taylor & Francis Group, London, ISBN 978-0-367-55862-8

Organizing Committee

Chair

Dr. AH. Galihkusumah, MM (Universitas Pendidikan Indonesia, Indonesia)

Co-Chair

1. Prof. Dr. Elly Malihah, M.Si (Universitas Pendidikan Indonesia, Indonesia)
2. Prof. Dr. Darsiharjo. MS (Universitas Pendidikan Indonesia, Indonesia)
3. Dr. Erry Sukriah, M.SE (Universitas Pendidikan Indonesia, Indonesia)
4. Dr. Caria Ningsih (Universitas Pendidikan Indonesia, Indonesia)
5. Oce Ridwanudin, M.M (Universitas Pendidikan Indonesia, Indonesia)

Speakers

1. Professor Bob McKercher (The Hong Kong Polytechnic University, Hong Kong)
2. Professor Iis Tussyadiah (University of Surrey, United Kingdom)
3. Dr. Dewi Turgarini, MM.Par (Universitas Pendidikan Indonesia, Indonesia)

Co-Host

1. Khrisnamurti (Universitas Negeri Jakarta, Indonesia)
2. Nova Riana (Sekolah Tinggi Ilmu Ekonomi Pariwisata Yapari, Indonesia)
3. Rina Suprina (Trisakti School of Tourism, Indonesia)
4. Sienny Thio (Petra Christian University, Indonesia)

Promoting Creative Tourism: Current Issues in Tourism Research – Kusumah et al. (Eds)

Acknowledgements

The committee would like to express gratitude to all who have been involved in the conference. Our highest appreciation goes to the Rector of Universitas Pendidikan Indonesia for his constant support. The same thankfulness also goes to the Vice Rector of Research, Partnership and Business for his insight that helps the committee to execute the conference.

The committee would also like to thank to the members, reviewers, as well as publication team who have collaborated together to ensure the production of both the conference and the proceeding. Our special thanks is also dedicated to the team from CRC, Routledge who provide a space for the research from our participants to be published and disseminated further.

Finally, our gratitude also goes to all participants who have made our conference successful.

The Conference Committee

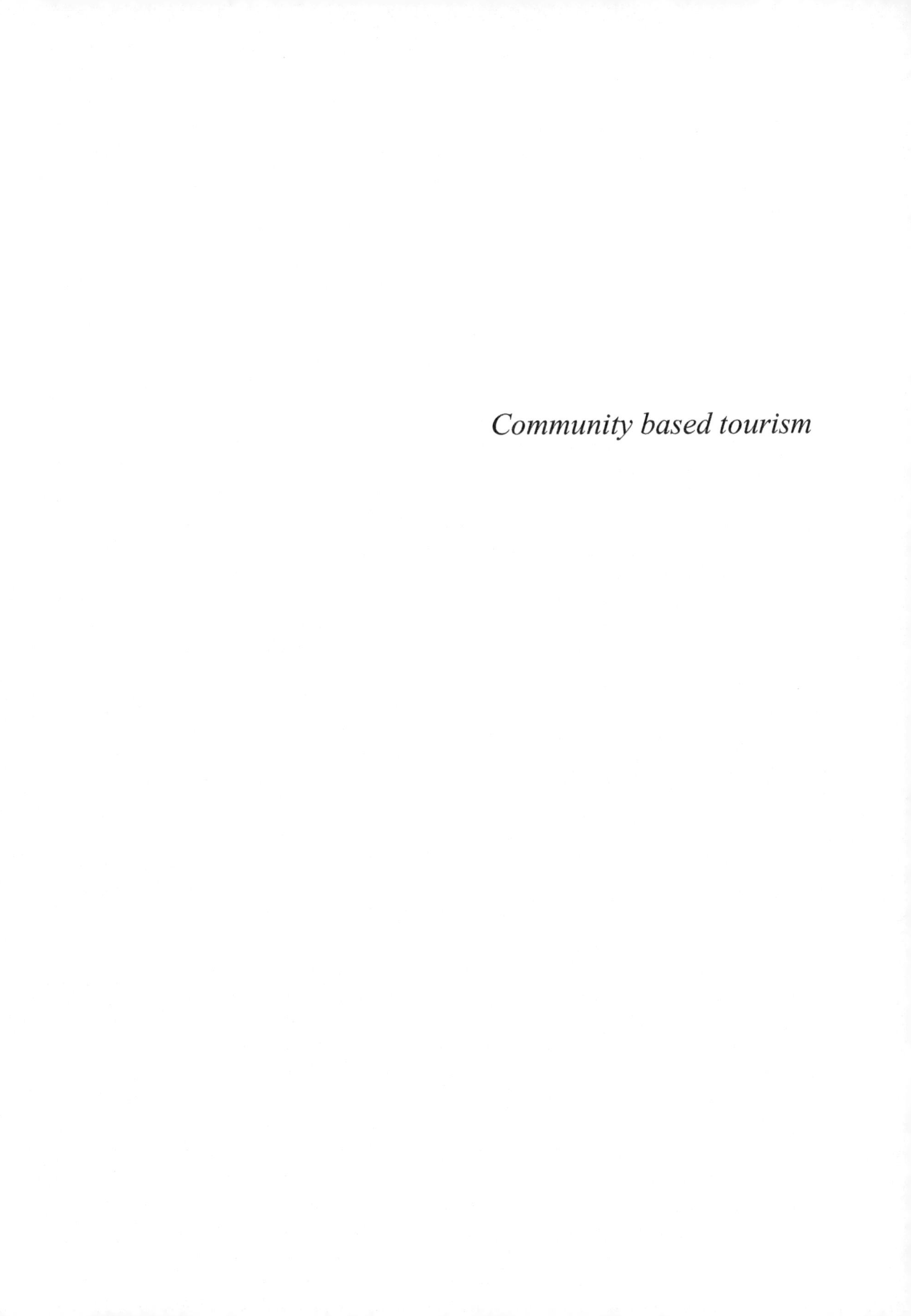

Community based tourism

Promoting Creative Tourism: Current Issues in Tourism Research – Kusumah et al. (Eds)

Border community perception of their local tourist attraction

A. Khosihan, A.R. Pratama & P. Hindayani
Universitas Pendidikan Indonesia, Bandung, Indonesia

ABSTRACT: The growth of worldwide tourism has been followed by the trend of the development of tourism in the lower level. Some of the regions in Indonesia also experience their progress, especially Temajuk, as the territory of Indonesia that borders with Malaysia, which makes the researcher aims to describe the perception of local people in accordance with their tourism attractions obtained. Qualitative research with a more in-depth interview collection data method and focused group discussion (FGD) has been conducted by involving informants from several samplings of the community. The result of this research shows that most of their resident perceive the natural resources as their concern, followed by a cultural perspective, and the sensation or border atmosphere in the last category. Their perceptions of tourism also review their expectation of tourism management for stakeholders seriously. This research aims to support policymakers to map the plan of border tourism.

Keywords: resident's perception, border tourism, attractions

1 INTRODUCTION

Currently, the growth of tourism across the globe creates the tourism sector as one of the possible methods taken by many countries to increase their income. A release of UNWTO as an official organization of the United Nations, in 2019, says that the tourism sector reaches 4% and becomes 1.5 billion of global tourist movement (Lidwina 2020). It absolutely has a big impact as well as the total of the world population. The tourists still dominate their visit to European countries. Meanwhile, Asia-Pacific also experiences the progress of the number of tourists globally.

Nationally, the tourism sector extremely supports the country's income (Nizar 2015). In Indonesia, the growth of tourism obtains well enough evolution in 2019, which the total of foreign exchange gained by tourism reaches around $20 billion. It contributes to the tourism sector as the second-best position in advancing the national income after the export of crude palm oil (CPO). Nowadays, the government of Indonesia makes tourism as one of the development priority sectors called "10 new Bali".

The impact of tourism advancement in Indonesia also affects to un-prioritized regions. A smaller scale of tourism in a specific area is supported by the movement of local tourists that visit several domestic destinations. For instance, one area that borders Indonesia and Malaysia called Sambas. The development of tourism in this place is started to be recognized, especially in Temajuk, a village that borders with Malaysia directly to the land and sea territorial.

The improvement of tourism activity in Temajuk is inseparable from the role of social media that makes stakeholders observe and brand this border area. At present, Temajuk has some facilities and tourist attractions that have not organized well and run without any supporting systems. Therefore, it is very interesting to see the local perception in accordance with the development of tourism in Temajuk.

The research of tourism has been reviewed by several studies, such as the people perception of tourism impact (Gu & Wong 2006), the comparative analysis of Thai people on tourism impact

DOI 10.1201/9781003095484-1

(Mcdowall & Choi 2010), the attitude of local people during tourism development of Masooleh (Zamani-Farahani & Musa 2008), and tourist perception on tourism spot "Batang Dolphin Center" (Murti & Sujali 2013). However, there is a minimum discussion of local 'people's perception in border areas related to the tourist attraction in their area. Then, the researchers asked the main question, "What and how are the perceptions of the Temajuk's community about tourist attractions in their area?".

By understanding the local people perception, we can identify several things, for instance, how the society sees the tourism (May-Chiun et al. 2013), the role of society and community (Eshliki & Kaboudi 2012), and decision making of tourism plan (Moscardo 2011). This article reviews the aspect of the local 'people's perception in border areas specifically. Especially, Temajuk village as the border territory between Indonesia and Malaysia benefits for the plan of theoretical or practical expansion in the upcoming time.

2 LITERATURE REVIEW

Perception, in reality, is a point of view on the object owned by individual circumstances. Related to social construction, perception closely relates to how society perceives the signs or phenomenon occurs. In terms of tourism development, the local 'people's perception on tourism takes a very crucial role that connects with its continuity of the place (Arcana 2016). Perception has two ultimate dimensions; they are positive and negative. Positive perception will be formed if the positive side is visible and perceivable. Meanwhile, it can be negative if it is not as expected (Kurniansah 2016).

In terms of tourism development, the attraction aspect is also a part of tourism components known as 4A; they are the attraction, access, amenity, and ancillary. The attraction is mentioned as a visual attraction or the involvement of the visitors when they visit a certain destination. It can be natural resources, culture, and social (Astuti & Noor 2016), then, besides that there is also an element of special attraction (Sunaryo 2013). Natural attractions can range from beaches, mountains, forests to landscapes. While cultural attractions such as special arts, traditional foods, and other special cultural products, special interests are usually closely related to the combination of nature and the ability to manage areas such as outbound tourism and other special interests, including those related to regional symbols. It also depends on tourism management; therefore, both attraction in the eyes of visitors and the local people must be concerned as well. The perception of a specific tourism attraction will create hope because it relates to understanding experiences and points of view (Huda 2013). When the perception of local people develops to a particular expectation of tourism attraction, this can be called as the process of change.

3 MATERIAL AND METHODS

Qualitative research uses more in-depth interviews collaborated with five informants who represent the local people components and take additional data through FGD to the group of society that consists of mangrove park community officer, representative of village government, and public figure. The main informant of this research is the PO, who acts as the secretary of Temajuk village and covers as of the Headman of Temajuk village, also as the resident who owns a hostel. AR is a citizen of Temajuk village who works as a Primary School teacher and acts as head of youth organization of Temajuk village and moved to Temajuk at the beginning of 2000. Jo is the former chairman of the Travel Aware Community (Pokdarwis) of Temajuk village and acts as the manager of J.lo hostel owned by businessman outside of Temajuk village. NA is the citizen of Temajuk village who is known as one of the pioneers of lodging development in Temajuk village and the owner of the first hostel in Temajuk village. NA is well known by the government of Sambas Regency because acting as the manager of a regional government hostel that is located near his hostel. Wi is the citizen who owns a hostel and acts as a support provider of tour service and restaurant. The hostel constructed by Wi was built personal expenses without businesspeople or private cooperation. And then, in FGD, it is conducted once, attended by eight participants, and produce some essential notes related to management hopes of tourism attraction in Temajuk village. These data collections were fully occupied in February 2019. Analysis data used is triangulation, which is triangulating the results of 'informants' answers during the interviews and FGD, this process begins by examining the answers of one informant and then starting to compare them with other answer of the informants who also appear according to research findings.

4 RESULTS AND DISCUSSION

The perception of local people of Temajuk village towards tourism attraction is various enough. The findings show that the majority of the people concern that natural resource is undoubtedly chosen related to the questions given. As mentioned by informant NA:

> *"It can be declared that the main power of Temajuk village is nautical tourism. In the future, we will probably provide water sports facilities, such as canoe, banana boat, and soon. Hopefully, it can be true". (Interview on February 9, 2019).*

Temajuk village is a coastal region that has a long coastline that faces the west with the view of the Natuna Sea. Nowadays, some other water game facilities have been considered, such as canoe and Jet Ski, which are provided at Camar Bulan Resort. Besides the beach area, the perception of local people is also concerned with protected forest area "Tanjung Dato: as stated by informant Jo:

> *"In Tanjung Dato protected forest', I get the information that there are 25 spots of Rafflesia which only exist in Sumatera island, but in reality, we can find it here." (interview on February 14, 2019)*

Tanjung Dato protected forest area is a protected forest with naturally covered by the hill of rain tropical forest, the local people also can find the clean water sources in this place. Independently, the society builds a dam to hold the water from the mountain then drain it into local 'people's houses through the pipes. Interestingly, this place is shared with Malaysia, wherein the west of the hill is owned by Indonesia, and Malaysia is on the east side of the hill. The natural attraction of forest and other hills in Temajuk village is also in the form of some features, such as a waterfall in several spots. However, this forest is still protected and not organized for tourism yet. Consequently, it is quite difficult to enter the forest or even to enjoy the waterfall view.

Besides that, another perception belongs to Culture Potential, as stated by informant PO:

> *"We have numerous Malay traditions, such as art until custom procession. Moreover, as a coastal Malay society, we have our characteristics, especially when ultimate seasons come, such as jellyfish harvest; this makes our culture potential becomes remarkable". (Interview on February 8, 2019)*

The majority of local people of Temajuk village, more than 95% are Malay and work as a fisherman. From this occupation, they also produce some other stuff, such as processed sell fished products, until traditional food, such as jellyfish soup. The perception of cultural view can be found when some informants answered the jellyfish season could have an impact, such as the tradition of jellyfish harvest in the afternoon, the process of jellyfish until ready to sell, and the creation of jellyfish as merchandises of Temajuk village. It is as stated by informant PO:

> *"We have Malay traditional art, developed into jellyfish dance to brand jellyfish as part of Temajuk village. Besides that, a group of Dzikir can be one of the cultural performances for those who visit Temajuk Village" (interview on February 8, 2019)*

In Sambas Malay society tradition, dzikir is an art show which performs praises to the God and prophet. It is usually sung with a combination of rhythm and poetry and accompanied by drum and maracas. The culture and the 'people's occupation are interesting to discuss. However, they do not make them the potential of tourism attraction.

There are a few societies of Temajuk who concern with another symbolic potential. Most of them are not aware of the beauty and sophisticated culture that becomes a symbolic value of the border region. Along with the blowing up the issue of regional claimed by Malaysia a few years ago due to the lighthouse construction and state boundary peg shifting in Camar Bulan village. There is symbolic attraction from Temajuk village, only one informant who says this potential, namely Ar:

> *"We (youth organization) also involve in Independence Day of the Republic of Indonesia ceremonial. Temajuk is known as a border area, so the sensation of celebrating it is stronger to raise nationalism and persuade this moment for the people or visitors who come". (Interview on February 9, 2019)*

The Independence Day of the Republic of Indonesia ceremonial becomes a valuable moment for all the people of Indonesia, as well as Temajuk society. Celebrating Independence Day in the boundary of the country gives another sensation for visitors who want to enjoy and feel how nationalism is alive. Even though it is covered by any limitations, they do not celebrate a ceremony in the center of the sub-district, because they have their own agenda to do, from the ceremony until traditional games in order to celebrate Independence Day. It can be said that the attraction of Temajuk as a border area can increase the sensation of the holiday with a nationalism atmosphere.

From those perceptions above, they can raise some efforts to conduct. Today, the condition of developing tourism attraction by society only exists in the unit of a personal venture or small group. Then, the problems faced by them are connected to access component and amenity that influence on optimization of the place, such as the role of regional government which is not fully occupied in building up the people, the difficult access and information, travel-aware community and village business entity are malfunctions as well. The result of FGD on 26 February 2019 shows that the society hopes to increase the effort development of tourism attraction among synergy of tourism stake holders, and initiated by the government in the form of socialization of TDUP or programs that connect with tourism expansion in serious way throughout a very serious assistance, strive for BTS tower for communication network and internet connection, and electricity needs within 24 hours to the local houses, access of accommodation gained from the regional, regency, or even province government fund, distribute the recommendation of mangrove management to the group of society in order to elevate their travel aware that basically for the people (people who organize, people who responsible), assistance to the travel aware community or other communities which are from local people in understanding, interpreting, and organizing unbreakable tourism potential with the regional regulation, support maintenance and improvement in several points, the local people are permitted to deliver retribution to the visitors who visit the village, Community Service Program (KKN) are expected to held which brings tourism issue in order to focus on exact field and the local people of Temajuk village can get the advantages directly.

The significant increase in tourists at Temajuk village presented the perception of the development of tourism attraction. This perception can be affected by a clear object and interaction. So, human knows their neighborhood (Listyana 2015), perception is also can be built by the experience of the community (Green 2005). The experience in this context is a process of tourism activities, and the participants happened in the Temajuk village community. It is also caused by the impact that is felt by the community (Green 2005). The interest of the tourist in the jellyfish season to see the fishermen's activities, then it made the community thought to create jellyfish dance as the culture product that describes the harvest season of marine products. The perception of the attraction of tourism gave the local community hope. In this case, the communities' voices are very useful in the expansion of tourism. The communities' voices in tourism development planning will support the growth itself, which is representing a small group for a bigger group (Moscardo 2011). When the communities' voices are heard, hence the support development of synergy numbers will increase. Some studies talked about this positive impact on the life quality of the local community (Andereck & Nyaupane 2011). Villagers are not the type of silent community; they open for any new things, especially in tourism. With all their potential, people want some serious guidance from the related parties to make tourism at Temajuk village becomes better. Due to their limitation at the border area, it makes the aspect of tourism development turns into limited too. Finaly, the perception of temajuk villagers about tourism attractions in their area, not only related to what the attraction is, but also how each attraction should be managed.

5 CONCLUSION

The perception that is formed by the local community about the attraction of tourism is still assumed that nature is the main factor to attract their tourism (mountains and beaches), followed by the culture and other factors (territorial border). This perception is connected to the cooperation and guidance of the government, and the approval of the community to manage the forest area and to support component tourism such as access and amenities. In the end, the community perception on the growth of tourism at Temajuk village has a connection with the intention of making Temajuk village becomes better and more modern. Other studies still can be done by observing the process of community contraction in tourism. It includes the relation between the agent and the structure that affects tourism.

ACKNOWLEDGEMENT

Thanks to the parties involved, especially to all informants of this research who have given an opportunity to do an interview and FGD, the regional government of Temajuk village that provides the research permit, also some colleagues who give access to ease the data findings..

REFERENCES

Andereck, K. L., & Nyaupane, G. P. (2011). Exploring the Nature of Tourism and Quality of Life Perceptions among Residents. *Journal of Travel Research, 50*(3), 248–260. https://doi.org/10.1177/0047287510362918

Arcana, K. T. P. (2016). Persepsi Masyarakat Lokal Terhadap Perkembangan Akomodasi Pariwisata Studi Kasus: Desa Adat Seminyak, Kecamatan Kuta Kabupaten Badung, Bali. *Analisis Pariwisata, 16*(1), 52–60.

Astuti, M. T., & Noor, A. A. (2016). Daya Tarik Morotai Sebagai Destinasi Wisata Sejarah Dan Bahari. *Kepariwisataan Indonesia, 11*(1), 25–46.

Eshliki, S. A., & Kaboudi, M. (2012). Community Perception of Tourism Impacts and Their Participation in Tourism Planning: A Case Study of Ramsar, Iran. *Procedia – Social and Behavioral Sciences, 36*(June 2011), 333–341. https://doi.org/10.1016/j.sbspro.2012.03.037

Green, R. (2005). Community perceptions of environmental and social change and tourism development on the island of Koh Samui, Thailand. *Journal of Environmental Psychology, 25*(1), 37–56. https://doi.org/10.1016/j.jenvp.2004.09.007

Gu, M., & Wong, P. P. (2006). Residents' perception of tourism impacts: A case study of homestay operators in Dachangshan Dao, North-East China. *Tourism Geographies, 8*(3), 253–273. https://doi.org/10.1080/14616680600765222

Huda, A. (2013). *Pengertian Persepsi*. Http://Eprints.Uny.Ac.Id/9686/3/Bab%202.Pdf. https://doi.org/10.1017/CBO 9781107415324.004

Kurniansah, R. (2016). Persepsi Dan Ekspektasi Wisatawan Terhadap Komponen Destinasi Wisata Lakey-Hu'U, Kabupaten Dompu. *Jurnal Master Pariwisata (JUMPA), 3*, 72–91. https://doi.org/10.24843/jumpa.2016.v03.i01.p06

Lidwina, andrea. (2020). Jumlah Wisatawan Global 2019 Hanya Tumbuh 3,8%. In *2020*. https://databoks.katadata.co.id/datapublish/2020/01/24/jumlah-wisatawan-global-2019-hanya-tumbuh-38

Listyana, R. (2015). Persepsi dan Sikap Masyarakat Terhadap Penanggalan Jawa Dalam Penentuan Waktu Pernikahan (Studi Kasus Desa Jonggrang Kecamatan Barat Kabupaten Magetan Tahun 2013). *Jurnal Agastya, 5*, 116–138. https://doi.org/10.3923/ijss.2017.32.38

May-Chiun, L., Peter, S., & Azlan, M. A. (2013). Rural tourism and destination image:community perception in tourism planning. *The Macrotheme Review, 2*(1), 102–118.

Mcdowall, S., & Choi, Y. (2010). A comparative analysis of thailand residents' perception of tourism's impacts. *Journal of Quality Assurance in Hospitality and Tourism, 11*(1), 36–55. https://doi.org/10.1080/15280080903520576

Moscardo, G. (2011). Exploring social representations of tourism planning: Issues for governance. *Journal of Sustainable Tourism, 19*(4–5), 423–436. https://doi.org/10.1080/09669582.2011.558625

Murti, H. C., & Sujali, S. (2013). Persepsi Wisatawan Terhadap Pengembangan Obyek Wisata Batang Dolphin Center. *Jurnal Bumi Indonesia, 2*(2), 260–267.

Nizar, M. A. (2015). Tourism Effect on Economic Growth in Indonesia. *Munich Personal RePEc Archive (MPRA), 65628*. http://mpra.ub.uni-muenchen.de/65628/

Sunaryo, B. 2013. *Kebijakan Pembangunan Destinasi Pariwisata Konsep dan Aplikasinya di Indonesia*. Yogyakarta: Gava Media

Zamani-Farahani, H., & Musa, G. (2008). Residents' attitudes and perception towards tourism development: A case study of Masooleh, Iran. *Tourism Management, 29*(6), 1233–1236. https://doi.org/10.1016/j.tourman.2008.02.008

Promoting Creative Tourism: Current Issues in Tourism Research – Kusumah et al. (Eds)
© 2021 Taylor & Francis Group, London, ISBN 978-0-367-55862-8

Language style and local wisdom in The Travel Documentary Pesona Indonesia: *Tondokku Kondosapata* on TVRI as a medium for tourism promotion

S. Hamidah, N.N. Afidah, I. Kurniawaty, H.T. Abdillah & R.H. Nugraha
Universitas Pendidikan Indonesia, Bandung, Indonesia

ABSTRACT: Language has a very important role in documentary film that is intended as a medium for tourism promotion. The documentary film uses full language to describe information about attractions in a comprehensive manner. The travel documentary Pesona Indonesia: *Tondokku Kondosapata* on TVRI is a documentary film that presents neat cinematography and is accompanied by a narrative with optimal use of language style to support the aesthetic description and documentation of factual information about *Tondok Bakaru* attraction. This study aims to describe (1) language style and (2) local wisdom in the travel documentary Pesona Indonesia: *Tondokku Kondosapata* TVRI's program. The qualitative descriptive method was chosen because the data source was in the form of video films and narrative text of the results of transcription. Based on the results, the analysis obtained the use of language style: simile, metaphor, personification, antithesis, hyperbolic, synecdoche, repetition (anaphora and epizeuxis). Local wisdom in this documentary raises information and documentation of the cultural, behavioral, and philosophical values of the community on these attractions. Utilization of language style and the emergence of local wisdom facts in this documentary strongly support the description and documentation of information about the parts of the tourism object in detail as the working principle of tourism promotion media, namely as a suggestion to communicate tourism to be known by the wider community.

Keywords: language style, documentary film, tourism documentary, tourism promotion media, Pesona Indonesia, *Mamasa* society, *tondok bakaru*

1 INTRODUCTION

Language is a communication tool to convey messages. The medium for conveying messages varies, such as writing, speech, as well as visuals, motion pictures, or films. Messages with interesting and informative language styles are widely used by various professional circles in promoting a work or an object. The goal is that the message conveyed to the public can attract the interest of the community to buy or visit. The use of appropriate language style can attract the attention of the recipient (Zaimar 2002). The tourism industry is one of sectors that also utilizes the attractive and informative language styles for promotional events. The hope is that people who watch or see these promotions decide to visit the attractions that are promoted. Tourism promotion by tourism actors is mostly done in various mass communication media such as social media, online media, and electronic media such as television and radio.

One of the most popular tourist promotion medium today is documentary film. Documentary film is one type of non-fiction film that aims as a media to disseminating cultural and social information that is realist and factual (Fauziah et al. 2018). The documentary film in this study is a film that tells or presents the facts of the tourism potential of *Tondok Bakaru,* which is presented realistically and descriptively. Therefore, documentary film has much potential as a tourism promotion media that optimizes the use of language, especially in conveying the facts about the tourism potential of the filmed tourist attraction. This is because documentary films can be widely accessed not only through conventional media such as television but also through social media such as YouTube as well as online media. The travel documentary *Pesona Indonesia: Tondokku Kondosapata* on TVRI is a documentary film that presents neat cinematography and is accompanied by a narrative with optimal use of language style to support the aesthetic description and documentation of factual information about *Tondok Bakaru* attraction.

Tondok Bakaru is a village located in Mamasa subdistrict, Mamasa District, or nicknamed *Bumi Kondosapata*, West Sulawesi Province, Indonesia. *Tondok Bakaru* is a tourist village that is rich in natural tourism potential and rich in customs and culture. Not surprisingly, *Televisi Republik Indonesia* (TVRI) as the Public Broadcasting Institutions (LPP) elevates the beauty of *Tondok Bakaru* Village in its cultural category program, namely *Pesona Indonesia*. *Pesona Indonesia* "Is a program that will review various kinds of charm that exist in Indonesia. Starting from the natural charm, artistic charm, even culinary charm in Indonesia" (TVRI 2020).

As explained earlier, there are many programs that have similar genre to *Pesona Indonesia* on various television broadcasts and YouTube Channels. DA distinctive feature that distinguishes *Pesona Indonesia* program from similar programs on other television channels is the optimal use of Indonesian language. The optimal use of the Indonesian language in the *Pesona Indonesia* is the influence of 'TVRI's status as a government-owned LPP, namely the functioning of *Bahasa Indonesian* (Indonesian) as the language of national cultural development in accordance with its position as the official language of the state (*Undang-Undang RI Nomor 24 Tahun 2009 Pasal 25 ayat (3)*/Law of Republic Indonesia Number 24 Year 2009 Article 25 paragraph (3)).

One of qualitative elements in language that can be analyzed descriptively is language style. The use of language style in this film is intended to convince or influence the audience that *Bumi Kondosapata* has a beautiful natural landscape with customs and culture that are maintained so that *Tondokku Kondosapata* (*Tondok Bakaru*) is the charm of Indonesia that must be on the list of visits of the audience. The language used by the narrator in *Pesona Indonesia: Tondokku Kondosapata* is full of language style. Tarigan states that the use of language style in a speech or writing is a rhetorical form intended to ensure that it also influences the listener or reader (Sucipto 2018).

In addition to language style, *Pesona Indonesia: Tondokku Kondosapata* program intended as a tourism promotion must contain tourism industry products. According to Bukart and Medlik, one of the products of the tourism industry is tourist attraction (Payangan 2013). This tourist charm must be packaged nicely in order to optimize the function of the film as a medium for tourism promotion. *Pesona Indonesia: Tondokku Kondosapata* program also contains tourist attractions which are categorized as tourism industry products as categorized by Bukart and Medlik (Payangan 2013). The tourist charm is packaged by raising local wisdom in the *Tondok Bakaru* village as a potential tourist destination for tourists to visit.

The documentary film *Pesona Indonesia: Tondokku Kondosapata* visualizes factual information cinematographically about local wisdom packaged in tourist attractions in *Tondok Bakaru* village, including forms of natural attraction, cultural attraction, social attraction, and built attractions (Payangan 2013) by utilizing the language as narration and description of the object. Therefore, this study tries to describe two elements, which are (1) language style and (2) local wisdom in the documentary film *Pesona Indonesia: Tondokku Kondosapata* program on TVRI as a tourism promotion media.

2 LITERATURE REVIEW

2.1 *Language style*

Language style is a technique to use language optimally in a process of delivering messages. The technique of using this language includes the arrangement of words used by the speaker or writer in expressing ideas and experiences to influence and convince the listener or reader (Sucipto 2018). The language style according to Tarigan (Damayanti 2018; Sucipto 2018) is grouped into four groups, namely comparison language style, contradiction language style, interrelated language style, and repetition language style.

One of the definitions of language style in the Linguistic Dictionary (Kridalaksana 2001) is "The use of the richness of language by a person in speaking or writing." Utilization of the wealth of language is an effort to use language optimally in a speech or writing. Optimal use of language in speech or writing, as a medium for conveying messages, with the use of language styles aims to attract the attention of the recipient of the message. As stated by Zaimar (2002), the use of the right language style can attract the attention of the recipient. Furthermore, Tarigan (in Sucipto 2018) explained that the use of language style in a speech or writing aims to influence or convince the recipient. The use of language style is also intended to emphasize the message conveyed (Zaimar 2002). No wonder, if the language style is also used to make the message effective in a tourism promotion media.

2.2 *Tourism promotion*

Promotion is an effort to communicate products and services so that potential customers buy and use these products and services. According to Simamora, promotion is a company effort to influence potential consumers (Handayani & Dedi 2017). In terms of tourism, tourism promotion is a series of efforts or activities to communicate or campaign the attractiveness of tourism products and services to potential consumers or tourists in order to carry out tourist transactions based on these tourist references. As stated by Payangan (2013), the most common function of tourism promotion in the marketing strategy of tourism services is to stimulate transactions (purchase or use of tourism services). According to Payangan (2013), promotional activities are not only a means of communication but are a tool to influence consumers in transactions based on consumer needs.

Local wisdom is an area where tourist visits are part of the tourist attraction. Local wisdom is the cultural wealth of an area, which includes philosophy, values, ethics, and behavior to manage resources (natural, human, and cultural) contextually (Brata 2016; Daniah 2016; Rahmatih et al. 2020). This means that the packaging of tourism industry products in the form of tourist attractions will contain cultural elements as one of the promoted tourism products. These elements of local wisdom influence the forms of tourist attraction, especially cultural attraction, social attraction, and built attractions. In addition, the behavior and cultural ethics that exist in an area where tourists visit will affect the uniqueness of natural attractions as the identity of these tourist visits. As stated by Brata (2016), local wisdom is a cultural identity.

2.3 *Documentary film*

Films are the most comprehensive communication and technology media because films present an attractive audio-visual message and can be widely accessed not only through conventional media such as television but also via YouTube and social media as well as online media. Film as stated in Undang-Undang Republik Indonesia Nomor 33 Tahun 2009 tentang Perfilman (the Law of the Republic of Indonesia Number 33 Year 2009 concerning Film) is one of the cultural arts and mass communication media which is full of cinematographic elements. Film is a marketing tool for postmodern tourism (Juškelyte 2016) because, through the show of attractions in the film, it makes tourists interested to visit the object directly (Gunesch 2017). Film such as virtual brochures offers a smoother way of promotion (Tanskanen 2012). The most common type of film used for tourism promotion is documentary film. This is because the main content of the documentary shows or explores the activity or beauty of a tourist place factually and descriptively through cinematographic processing (Farpember & Akhamd 2012).

Documentary film is one type of non-fiction film that aims as a media in disseminating cultural and social information that is realist and factual (Fauziah et al. 2018). The main element of story building in a documentary film is the delivery of factual references or information and messages from the results of gathering evidence and factual information. Documentary films, besides cinematographic works, (Tambayong 2019) have also proved able to trigger the attention of the audience to the subject of factual information presented in it. This is evidenced by the many television stations with programs in the form of documentaries that raise issues of tourism and culture. As stated by Masdudin (2011), the existence of documentary films has even become one of the shows that is awaited by the public.

3 METHOD

The research method used is a descriptive qualitative research method. This method was chosen because the main data source is in the form of a narrative in the documentary film *Pesona Indonesia: Tondokku Kondosapata*. The documentary film *Pesona Indonesia: Tondokku Kondosapata* is presented in four segments. The narrative in this documentary video is transcribed to produce qualitative data in the form of sentences. There are 43 sentences that take advantage of the language style in this film. The sentences resulting from the transcription were collected and their language style was analyzed based on the theory of language style grouping according to Henry Guntur Tarigan.

Henry Guntur Tarigan (Damayanti 2018; Sucipto 2018) classifies language styles into four groups, namely comparation language style, contradiction language style, interrelated language style, and repetition language style. Sentences have been analyzed, categorized, selected, and grouped, and the results described. Apart from data in the form of sentences, secondary data in this film is in the form of visual content that documents elements of local wisdom. Visual content containing products of the tourism industry, namely tourist attractions, is analyzed and categorized based on the theory of Bukart and Medlik into natural attractions, cultural attractions, social attractions, and built attractions. The first and third segments are forms of natural attraction. The fourth segment is a form of built attractions. The second segment is a form of social and cultural attraction. Furthermore, the results of the grouping are described as a documentation of the philosophy, values, ethics, and behavior that appear in the film as one of the products of the tourism industry that is promoted in this film.

4 RESULTS AND DISCUSSION

4.1 *Language style in "Pesona Indonesia: Tondokku Kondosapata"*

Based on the results of transcription and analysis, there are eight styles of language that the narrator uses to promote the beauty of the *Tondok Bakaru* village in the documentary film *Pesona Indonesia: Tondokku Kondosapata*. These eight language styles are divided into four comparison language styles: simile, metaphor, personification, and antithesis; one contradiction language style: hyperbolic; one interrelated language style: synecdoche; and two repetition language styles: anaphora and epizeuxis.

There are eight narrator's utterances that use the simile language style in *Pesona Indonesia: Tondokku Kondosapata*. Simile is a style of language that compares one thing to another by using comparison particles. The comparison particles in Indonesian include *laksana, ibaratkan, ibaratnya, seperti, semisal,* and other comparison particles (Darheni 2016; Sucipto 2018). Comparison words or particles that appear explicitly in the narrative of *Pesona Indonesia: Tondokku Kondosapata* are *bagaikan, sebagai, seolah, sebagai, seperti, seakan.* In an example of simile style of utterance (1) *Anggrek ibaratnya cinta perlu dirawat diberi perhatian dan juga kesabaran* in the word *ibaratnya* comparison language style that likens orchids to love. Utterance (2) *Deburan air yang terjatuh bagaikan untaian tirai putih,* the word *bagaikan* likens the form of falling water to a string of white curtains.

The next language style is the metaphorical style. There are seven utterances that use the style of the metaphorical language style. Metaphor is a style of language that compares one thing with another implicitly (Darheni 2016; Sucipto, 2018; Zaimar 2002). Implicit because it does not use words of comparison *seperti, ibarat, laksana*, and other comparison words. Utilization of metaphorical language styles such as the word *lautan* in utterance (3) *Lautan awan berarak terbawa angin lembah* describing clouds that swirl in the vast sky like the vast ocean. This description provides visual power to give the impression and reinforce the message of beauty in the *Tondok Bakaru* tourist attraction. Another example, the word *denyut*/pulsation in utterance (4) *Denyut harapan pada setiap insan terhampar seiring naiknya sang surya* is metaphorical because it compares the word *harapan* with a pulsating heart, pulse, or fontanel.

Utilization of the dominant language style in the documentary film *Pesona Indonesia: Tondokku Kondosapata* is language style of personification. Personification is a style of language which likens an inanimate object to being human or as if an inanimate object has a human nature (Darheni 2016; Sucipto 2018). There are 15 utterances that utilize the style of personification in the narrative of the documentary film *Pesona Indonesia: Tondokku Kondosapata*. Examples of the use of personification language style in speech (5) *Embun pagi membasuh tiap helai bunga yang mekar di pagi hari*. *Embun* as if it has human nature. *Embun* (dew) is likened to human actions that can wash away. The next example in utterance (6) *Alam Kondosapata menceritakan betapa baiknya sang maha* pencipta. Alam *Kondosapata* seems to be able to speak so that *Alam,* which is the environment, can tell stories like humans.

In addition to simile, metaphor, and personification styles, there is one more style that is still in the category of comparation language styles, namely the antithesis style used by the cast in the documentary film *Pesona Indonesia: Tondokku Kondosapata*. An antithesis style is a style of language that gives rise to two words that contradict the meaning in one utterance or sentence (Sucipto 2018). *Hidup dan mati adalah keniscayaan.* The word *hidup* is the antonym of the word *mati*, and this is the style of the antithesis.

The contradiction language style in the documentary film *Pesona Indonesia: Tondokku Kondosapata* is in five utterances that utilize the hyperbolic language style. Hyperbolic is a style of language that emphasizes reality that stretched out (Sucipto 2018). The example of utterance (7) *Rumpun bunga memancarkan pesonanya.* The charm of the beauty of the flowers in the *Tondok Bakaru* village is described as if it can radiate like water or light. This description is an example of a description of reality that is stretched out. Hyperbol is also an expressive language style that exceeds reality. In utterance (8) *Menyusuri indahnya setiap jengkal keindahan sang pencipta*, the word *jengkal* to express that beauty is traced in detail, while utterance (9) *Bukit Lenong menjadi saksi akan ide dan kehebatan imajinasi manusia*. The word *saksi* as the title of The Lenong Hill gives an expressive impression to describe the beauty of the arrangement of the Lenong Hill by the local community.

Interrelated language style appears in speech (9) *Berangkat menyusuri setiap puncak, bukit Tanah Kondosapata* and (10) *Senyum penduduknya menyapa dibalik indahnya budaya yang dimilikinya*. The style of language in these two utterances is the style of synecdoche pars prototo. Utterance (9) the word *tanah* (the land) is the expression of a part of the *Kondosapata* area object which shows the whole part of *Tondok Bakaru* village. Likewise, the word *senyum* (the smile) in the speech (10) partial disclosure to show all cultural objects owned by residents of *Tondok Bakaru*.

The style of repetition is a style of language that uses repetition of words that are considered important to provide affirmation (Saptarini 2015). For example, the repetitive style of anaphoric in speech (11) *Rumah adat Mamasa merupakan simbol eksistensi masyarakatnya sarat dengan simbol dan makna kehidupan. Rumah adat Mamasa tak sekadar rumah untuk mengamankan diri dari hewan dan dari gangguan lainnya*. Repetition occurs in the first word of each sentence. However, in speech (12) *Duniaku, dunia kecilku, dan dunia masa depanku*, repetition is done in sequence in the same sentence. This language style is called Epizeuxis style. The style of language that emphasizes the word that is emphasized by repeating it several times in a row (Sucipto 2018).

4.2 *Local wisdom in "Pesona Indonesia: Tondokku Kondosapata"*

In the documentary film *Pesona Indonesia: Tondokku Kondosapata*, the element of local wisdom that exists in the *Tondok Bakaru* village is factual information that is promoted as a tourist attraction. This local wisdom is packaged in the form of natural attraction, cultural attraction, social attraction, and built attractions (Payangan 2013). The documentary film *Pesona Indonesia: Tondokku Kondosapata* is presented in four segments, namely, the first segment documenting agrotourism destinations, the second segment documenting the nature and culture of Mamasa, the third segment documenting pine forest tourist destinations, the Lenong Hill and Liawan waterfalls, the forth segment to documenting the *Makam Tedong-Tedong* (the cemetery) and Rumah Adat Masama (the Mamasa traditional house). From all four segment, in general form of the first segment and third segment are natural attraction. The second segment is social and cultural attraction. Local wisdom is the cultural wealth of an area including philosophy, values, ethics, and behavior to manage resources (natural, human, and cultural) contextually (Brata 2016; Daniah 2016; Rahmatih et al. 2020). Based on the analysis of tourist attraction elements from the four segments in the documentary film *Pesona Indonesia: Tondokku Kondosapata* packaging local wisdom in Mamasa, namely.

4.3 *Tondok bakaru Village is an eco-village*

As stated by Rambalangi et al. (2018) the meaning of the word *kondo* is broad rice fields, and *Sapata'* means one plot; the Mamasa community mostly has farming and gardening livelihoods. This culture is packaged in the form of the use of paddy fields, which at the same time cultivate orchids from the forest around the Kondosapata land. In addition to the use of sapodilla, the *Tondok Bakaru* community also utilizes the pine forest in *Bukit Lenong* as a tourist destination worth visiting. *Bukit Lenong* as the icon of the area is well managed, pine trees still stand tall without being damaged in an effort to beautify the Bukit Lenong. As the narration delivered in metaphorical style likens nature as a friend and personification likens the bend in nature as a human being who can invite.

The waterfall's destination as the next natural attraction is *Serambu Liawan,* which is located in the Sumarorong subdistrict. The documentary film *Pesona Indonesia: Tondokku Kondosapata* visualizes the beauty and cleanliness of *Serambu Liawan*. The beauty and cleanliness of the waterfall Liawan reflects how much the community loves the environment so that it is able to protect the natural surroundings. This has further strengthened Kondosapata residents are residents who live with environmental wisdom.

4.4 *The mamasa community behaves in a friendly manner*

As explained in the first segment, the majority of Kondosapata residents earn a living from farming and gardening, so the common sight in *Tondok Bakaru* is the gathering between farmers at lunch time together in the fields with simple dishes. The scene illustrates the habits of the surrounding community who are happy in togetherness and simplicity regardless of social status. The narrator in the documentary film *Pesona Indonesia: Tondokku Kondosapata* also conveyed this behavior several times, namely in the second segment that was presented by utilizing the language style of synecdoche pars prototo "The smiles of the people greet behind the beauty of their culture. Friendly and unpretentious residents welcome everyone who comes with a smile."

4.5 *Mamasa people have a philosophy of life Kondosapata uai sapolelean*

Kondosapata is a customary land area inhabited by Mamasa, which has a strong philosophy of life (Rambalangi et al. 2018; Wasilah et al. 2013). This is evidenced by the preservation of cultural sites *Makam Tedong-Tedong* in the subdistrict of Balla as a symbol of the spirituality of the Mamasa community. The Mamasa community still preserves historical civilization as a wealth and inheritance of ancestors, which continues to be preserved. In addition to the *Makam Tedong-Tedong* as a distinctive and iconic tourist attraction, *Banua Mamasa* as a traditional house of the Mamasa community is a symbol of cultural existence which is full of symbols and meanings of life. *Banua*

Mamasa has a peculiarity that contains messages and philosophy of life of the Mamasa people, as stated by the narrator with anaphoric language style.

The philosophy held by the people of Kondosapata is "Glorifying the higher degrees, respecting the proper, and loving the lower." The title of the Mamasa region is *Kondosapata uai sapalelean,* philosophically meaning there is an inseparable area. The people live fairly and have the same rights. The consultation system for consensus is still a guideline for the Mamasa community in solving life's problems. Local wisdom is packaged in the documentary film *Pesona Indonesia: Tondokku Kondosapata* as one of the products of the tourist industry, namely tourist attraction. Therefore, describing factual information presented cinematographically about local wisdom in the form of documenting the culture, behavior, and philosophical values of the Mamasa community can be used as a tourism promotion media.

5 CONCLUSION

The documentary film *Pesona Indonesia: Tondokku Kondosapata* has utilized language style optimally in the presentation of factual documents about the object of the film. The use of language style, simile, metaphor, personification, antithesis, hyperbolic, synecdoche, repetition (anaphora and epizeuxis) in the narration of the documentary film *Pesona Indonesia: Tondokku Kondosapata* is intended to support the description and documentation of factual information about the parts of the *Tondok Bakaru* tourist attraction. The form of factual information about the documentary film *Pesona Indonesia: Tondokku Kondosapata* includes natural attractions, cultural attractions, social attractions, and built attractions. These four forms are presented as a means of conveying local wisdom in the form of cultural, behavioral, and philosophical values as a tourist attraction for the *Tondok Bakaru* village and the life of the Mamasa community. Utilization of language style and the emergence of local wisdom facts in this documentary film is a working principle of tourism promotion media, namely as a suggestion to communicate tourism to be known by the wider community as well as offering the tourism industry's products in a subtle way.

REFERENCES

Brata, I.B. 2016. Kearifan Budaya Lokal Perekat Identitas Bangsa. *Jurnal Bakti Saraswati*, 5 (1): 9–18. Bali: LP2M Universitas Mahsaraswati Denpasar. Retrieved from https://jurnal.unmas.ac.id/index.php/Bakti/article/view/226

Damayanti, R. 2018. Diksi dan Gaya Bahasa dalam Media Sosial Instagram. *Jurnal Widyaloka IKIP Widya Darma*, 5 (3): 261–278.

Daniah. 2016. Kearifan Lokal (Local Wisdom) Sebagai Basis Pendidikan Karakter. *Pionir Jurnal Pendidikan,* 5 (2).

Darheni, N. 2016. Stylistic Comparisons in Song Lyrics Cianjuran: Sundanese Ethnic Character Expression in West Java. *Metalingua,* 14 (1): 83–102.

Farpember, T., & Akhamd, F.E. 2012. *Pembuatan Film Dokumenter Karst sebagai Media Pengenalan dan Promosi Wisata Karst di Kabupaten Gunungkidul.* Yogyakarta: Sekolah Tinggi Manajemen Informatika dan Komputer Amikom. Retrieved from http://repository.amikom.ac.id/files/Publikasi%2009.01.2580,%2009.01.2594.pdf

Fauziah, U., Syafwandi, & Ariusmedi. 2018. Eksplorasi Wisata Alam Nagari Batu Bajanjang Melalui Film Dokumenter. *Dekave Jurnal Desain Komunikasi Visual*, 8 (2). Universitas Negeri Padang.

Gunesch, K. 2017. Film and Tourism: Attracting Travelers with Moving Images, Traveling to Destinations. *RSEP International Conferences on Social Issues and Economic Studies*. Prague: Czechia.

Handayani, E., & Dedi, M. 2017. Pengaruh Promosi Wisata Bahari dan Kualitas Pelayanan terhadap Peningkatan Jumlah Kunjungan Wisata di Pelabuhan Muncar Banyuwangi. *Jurnal Wira Ekonomi Mikroskil*, 7 (2): 151–160.

Juškelyte, D. 2016. *Film Induced Tourism: Destination Image Formation and Development. Regional Formation and Development Studies*. Lithuania: Vytautas Magnus University.

Kridalaksana, H. 2001. *Kamus Linguistik*. Jakarta: Gramedia Pustaka Utama.

Masdudin, I. 2011. *Mengenal Dunia Film*. Jakarta: Multi Kreasi Satudelapan.
Payangan, O.R. 2013. *Pemasaran Jasa Pariwisata.* Bogor: IPB Press.
Rahmatih, A.N., Maulyda, M.A., & Syazali, M. 2020. Reflection of Local Wisdom Value on Sains Learning in Elementary School: Literature Review. *Jurnal Pijar MIPA*, 15 (2): 151–156. doi:10.29303/jpm.v15i2.1663
Rambalangi, Sambiran, S., & Kasenda, V. 2018. Eksistensi Lembaga Adat dalam Pembangunan Kecamatan Tawalian Kabupaten Mamasa: Studi Kasus di Kecamatan Tawalian Kabupaten Mamasa Pro.vinsi Sulawesi Barat. *Eksekutif Jurnal Jurusan Ilmu Pemerintahan*, 1 (1). Manado: Universitas Sam Ratulangi.
Saptarini, T. 2015. The Use of Styles Based on Diction and Sentence Structure on The Advertisements of Legislative Election Campaign in 2014. *Metalingua,* 13 (1): 87–102.
Sucipto, M.G. 2018. *Gaya Bahasa Pengetahuan dan Penerapan*. Klaten: Intan Pariwara.
Tambayong, Y. 2019. *Ensiklopedi Seni: Seni Film*. Bandung: Nuasa Cendikia.
Tanskanen, T. 2012. *Film Tourism: Study on How Films Can Be Used*. Laurea Kerava: Laure University of Applied Sciences. Retrieved from https://core.ac.uk/download/pdf/38073247.pdf
TVRI. 2020. *TV Program: Pesona Indonesia*. Retrieved from http://tvri.go.id/tvprogram/detail?id=68
Wasilah, Prijotomo, J., & Rachmawaty, M. 2013. Filosofi Tipologi Bentuk dan Ekspresi Arsitektur. *san121212*. Ref No: (B.1.5).
Zaimar, O.K.S. 2002. Majas dan pembentukannya. *Makara Hubs-Asia,* 6 (2): 45–57. doi: *10.7454/mssh.v6i2.38*

Promoting Creative Tourism: Current Issues in Tourism Research – Kusumah et al. (Eds)
© 2021 Taylor & Francis Group, London, ISBN 978-0-367-55862-8

Community education in developing edutourism values in Geopark Ciletuh

D.S. Logayah, M. Ruhimat & R. Arrasyid
Universitas Pendidikan Indonesia, Bandung, Indonesia

ABSTRACT: The behavior of the community and tourists need to be equipped with the values of education and knowledge to support tourism activities in order to always preserve nature and culture. The potential of the Ciletuh Geopark Palabuhanratu (geodiversity, biodiversity, and culture diversity) becomes the most valuable value after the Ciletuh Geopark becomes part of the Global Geopark Networks (GGN), which was formalized by UNESCO. The purpose of this study is to equip community and tourist education in developing Educational Values in the Ciletuh Palabuhanratu Geopark, Sukabumi Regency. Based on the study of various literatures, the edutourism development model has three dimensions which are the main components in the analysis in research. The first dimension is the dimension of the readiness of local communities in developing edutourism. The second dimension is the dimension of infrastructure readiness and management of edutourism areas, and the third dimension is edutourism tourist satisfaction. Researchers used 126 questionnaire study data sets as 100 community tourists and 11 tourist managers. The results show the value of education is influenced by tourism knowledge, and community motivation, partially tourism knowledge variables, community motivation, partially significantly influence marketing skills, community hospitality and readiness to develop edutourism so that the hypothesis can be proven. The implication of the results of this study is that the public, tourists, and tour managers must raise awareness and motivate people's behavior to care for the environment.

Keywords: edutourism, Ciletuh Geopark, community, values

1 INTRODUCTION

Geopark is an area that has unique geological, archaeological, ecological, and cultural values, where local people are invited to actively contribute in efforts to preserve and improve the function of natural heritage (Andriany et al. 2016). Geopark Ciletuh Palabuhanratu is an area management concept that harmonizes geological, biological, and cultural diversity through the principles of conservation, education, and sustainable development in eight subdistricts in Sukabumi, West Java, Indonesia. The Ciletuh Palabuhanratu Geopark has an area of 126,100 hectares or 1,261 km2. This area covers 74 villages in eight subdistricts, namely, Ciracap, Surade, Ciemas, Waluran, Simpenan, Palabuhanratu, Cikakak, and Cisolok districts. The eight districts are divided into three geoareas, namely Geoarea Ciletuh, Geoarea Simpenan, and Geoarea Cisolok. Ciletuh Palabuhanratu Geopark has several tourist attractions, including Panenjoan Hill, Puncak Darma, Awang Waterfall, Cimarinjung Waterfall, Palangpang Beach, and Kunti Island. The Ciletuh Geopark area itself is the unique and the oldest geological diversity in West Java (Manoharan et al. 2016).

Educational tourism, or better known as edutourism, is a very appropriate tourism activity to manage in the Ciletuh geopark area. Educational tourism or edutourism aims to provide tourists with a real learning experience in a tourist destination (Ratih et al. 2013). The development of edutourism aspects must refer to these goals. However, the first thing to be considered is that the people and communities in the Ciletuh geopark area have to understand the existence of the area better. This can provide a learning experience for tourists, which can be interpreted as experiences that can have

DOI 10.1201/9781003095484-3

a positive impact on the cognitive and affective aspects of society (Ma'rifah & Suryadarma 2015). The positive impact obtained on the cognitive aspect can be seen in the increase in knowledge gained by tourists and of course the community who manages the Ciletuh geopark. In this aspect, the role of the tour guide needs to be considered very carefully. Meanwhile, the positive impact on the affective aspect can be seen in increasing the awareness of the Ciletuh community and tourists toward nature and cultural conservation efforts. In other words, if this happens, the community and tourists will be motivated to contribute to nature and culture conservation efforts (Darsiharjo et al. 2016).

The participation of the local community is an important factor in the development of the Ciletuh Palabuhanratu Geopark (Darsiharjo et al. 2016; Nabila & Yuniningsih 2016). With the support of the local community, the construction or development of the Ciletuh Palabuhanratu Geopark will run smoothly. If the construction or development of the Ciletuh Geopark area does not involve the local community, it will break the relationship between the developer company and the local community. In the absence of a good relationship between the development company and the local community, it might cause some bad things to happen, such as the destruction of the Ciletuh Palabuhanratu Geopark or access blocking at the entrance of the Ciletuh Palabuhanratu Geopark location. Therefore, the relationship between the manager and the local community must be well maintained so that local people can also participate in protecting the Ciletuh Palabuhanratu Geopark area.

The majority of people in the Ciletuh Palabuhanratu Geopark work as fishermen and farmers. The distance of the South Sukabumi area to the center of activity is the reason why formal education is still lacking and only a few jobs are available (Darsiharjo et al. 2016). There are still people who carry out activities that can damage the environment by overexploiting. The potential that exists in the Ciletuh Palabuhanratu Geopark is expected to change the mindset of the community from initially exploiting natural resources by destroying the environment to switching to utilizing them by maintaining and using the potential of natural beauty that puts forward sustainable aspects so that the region becomes a leading tourist area in West Java.

The purpose of writing this article is to provide education to the community around the Ciletuh Geopark to gain cognitive and affective knowledge through the concept of developing edutourism. This is important because the community is the social capital for the existence of the Ciletuh geopark area so that its environmental sustainability can be maintained. Many articles discuss the Ciletuh geopark, but not many researchers have written articles about people's knowledge in managing the Ciletuh geopark area, especially value development.

2 LITERATURE REVIEW

2.1 *Tourism product*

Travel destinations are an important element in the tourism system. A goal is a place where the complexity of tourism activities occurs to fulfil one's needs and desires while traveling. Destinations strive to offer products and services whose characteristic are preferred by tourists (Guo & Chung 2019; Xu et al. 2020). Motivation is the driving factor for tourists, and the pull factor is the factor that influences when, where, and how tourists travel. The choice of destinations is strongly influenced by drive and pull factors (Arcana & Wiweka 2016; Lam et al. 2011). As a tourism provider in the tourism industry, tourist destinations consist of a combination of tourist attractions, accessibility, facilities, and additional services.

2.2 *The importance of community participation in instilling Geopark educational values*

The key to the success of developing and managing a geopark lies in the role and participation of active local people who understand the meaning of the geopark itself (Erfina 2019). Unfortunately, in the Ciletuh Geopark area, there are still some people who do not understand the meaning of a

Table 1. Population and sample of research area by gender.

		All documents		
		Gender		
Subdistrict Name		Male	Women	Amount
1	Ciemas	24,717	23,364	48,081
2	Ciracap	24,435	23,060	47,495
3	Waluran	13,494	12,457	25,951
4	Surade	36,614	35,469	72,083
5	Pelabuhan Ratu	49,247	47,428	96,675
6	Simpenan	24,513	23,768	48,281
7	Cisolok	31,783	30,293	62,076
8	Cikakak	19,205	18,195	37,400
Regency of Sukabumi		224,008	214,043	438,042

Source: Badan Pusat Statistika (2010)

geopark and keep doing rock mining and forests logging that violate the geopark principle, namely sustainable development. Edutourism is one type of alternative tourism that develops between conventional and modern tourism. In general, tourism provides an alternative that is developed based on the principle of sustainability. Educational tourism also makes use of various resources, such as natural resources, cultural resources, and artificial packaging of potential tourist areas (Manoharan et al. 2016; Yerry Yanuar 2018) at Ciletuh Geopark.

2.3 *Edutourism*

Educational system consists of elements of supply and demand that result in educational tourism experiences. Meanwhile, The Canadian Tourism Commission (CTC) (2001) notes two major components of edutourism in terms of supply: the main tourism product and its supporting elements. The main components of edutourism include attractions, natural and human resources (Ojo & Yusof 2019). Meanwhile, supporting services according to your needs include transportation, hotels, and other services. Educational values can have many purposes, such as curiosity about other people such as language and culture; an interest in art, architecture, music, or folklore; inspiring about caring for the environment, nature, plants and animals; or deepen interest in cultural heritage and historical sites.

3 RESEARCH METHOD

This research was conducted in the UNESCO Global Geopark Ciletuh Palabuhanratu, Sukabumi Regency in 2018 until 2020. This geopark area consists of eight subdistricts, and only three subdistricts were taken as the main research samples, namely Ciemas, Ciracap, and Waluran. This is because the number of samples taken is based on the results of the researchers' brief observations with the aim of tourism, especially the condition of the Ciletuh people who are directly related to these attractions. A clearer picture can be seen in Table 1.

The human samples in this study were the people and communities in the UGG Ciletuh Palabuhanratu area. There are two groups of respondents involved in the research, namely the local community totaling 126 people and three representative communities with different roles, namely Panguyuban Alam Pakidulan Sukabumi (PAPSI), Panguyuban Lahan Parahyangan (PALAPAH),

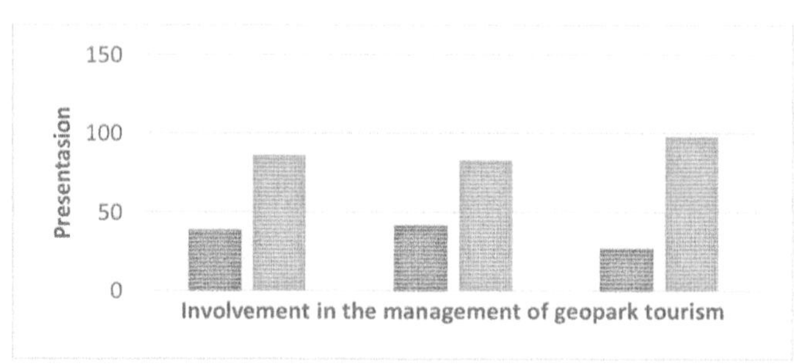

Source: Research Results, 2018

Figure 1. Community involvement in the development of the tourism sector in the UGG Ciletuh Palabuhanratu area.

and the tourism activating group (KOMPEPAR). Data collection was carried out through questionnaires for local communities and interviews for community representatives. There are three variables examined in this study, namely: (1) community involvement in the development of UGG Ciletuh Palabuhanratu; (2) community support for UGG Ciletuh Palabuhanratu edutourism; and (3) the role of the community in developing UGG Ciletuh Palabuhanratu.

4 RESULTS AND DISCUSSION

The implementation of the edutourism concept will present educational values that are packaged in tour packages to create a pleasant educational atmosphere. The community must have an active role in maintaining the geopark area because of social and natural capital for the community to preserve the environment (Santoso et al. 2020). The concept of edutourism is directed at providing services and education for local communities, tourists, and managers of tourist destinations to preserve nature. Even the Ciletuh geopark edutourism tour is one step in the diversification of tourism products in West Java province. In living the UNESCO Global Geopark concept, edutourism is one of the most important elements. UGG Ciletuh Palabuhanratu can sustainably develop the economy through this edutourism activity. Not only education about geology, but also knowledge about conservation and culture will be embedded along with this tourism concept.

The results of field data collection and interviews show that the community tends to play a role in the development of edutourism activities at UGG Ciletuh Palabuhanratu so that the community needs to be equipped with cognitive and affective knowledge about this Ciletuh geopark area. Summary of research data can be seen in Figure 2 and Figure 3.

Figure 1 shows that the number of people who get involved is less than the people who don't. This means that there are still a few local people who contribute to the development of the tourism sector. The development of the tourism sector cannot be done by chance but must be carefully prepared. The role of each element in society must be determined so that tasks and functions do not overlap. Therefore, the data in Figure 2 must be taken seriously so that an increase in community contributions can increase.

Based on the data presented in Figure 1 and Figure 2, the community will not be an inhibiting factor in the development of edutourism. Instead, the community is an important part of managing the Ciletuh geopark area. A small part of the community has participated in the development of UGG Ciletuh Palabuhanratu and also has the desire to contribute to further development efforts. It is felt by many tourists who come and researchers who provide education about the Ciletuh geopark value existence. However, based on the results of field data and interviews, some respondents or the community did not know about the Ciletuh geopark. They think what happens in their environment are just naturally happening; therefore, the geopark is only used for plantation or cultivation. The

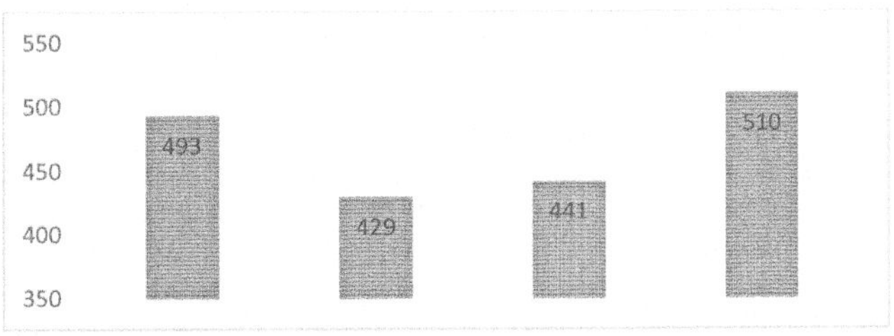

Source: Research Results, 2018

Figure 2. Community support for edutourism in the UGG Ciletuh Palabuhanratu area.

efforts to involve the community also need to be supported by educational programs to contribute the proper procedures. On the other hand, the community also plays a dominant role in the development of the UGG Ciletuh Palabuhanratu area, not only in strengthening their existence but also providing education to the public to be aware of the achieved geopark status. Based on the results of the interview, it is known that the three communities in the Ciletuh Palabuhanratu UGG have different roles. The role of the PAPSI community is to focus on preserving nature through reducing waste and monitoring the exploration of natural resources by the surrounding community. PAPSI is the leading community in the development of the Ciletuh geopark. The PALAPAH community focuses on developing cultural aspects in the geopark area so that it is sustainable and known to many people, whereas the KOMPEPAR community focus is on developing the tourism sector through increasing local community participation in the provision of goods and services related to tourism activities. For now, what is developing is the culinary business, namely the promotion of the liwet hanjeli package.

Based on field findings, the assignment of the three communities above can facilitate the way to provide knowledge and values of edutourism knowledge to the surrounding community regarding the existence of the Ciletuh geopark. Edutourism Geopark is a model of the tourism programs that combines the concept of tourism with education and learning, which of course has educational values in it (Lam et al. 2011; Sulaiman et al. 2019). These educational values incorporate contextual learning in society because tourists must be smart in choosing tours that can provide education. Edutourism Geopark at UGG Ciletuh Palabuhanratu is a tourism program approach that combines the concept of tourism with education and learning about the main elements of UGG, namely, geological, biological, and cultural diversities. By presenting educational values mixed in tourism packages to create an educational atmosphere in a pleasant journey, this concept is directed to provide the first service for schools that have traditionalized study tours as part of their education process so that the purpose and objectives of study tours or field studies for students can achieve optimal goals. In addition, this concept also provides services and education to the community, tourists, and managers to conserve nature (Ojo & Yusof 2019).

In living the UNESCO Global Geopark concept, Edutourism is one of the elements that play a major role in tourism activities. UGG Ciletuh Palabuhanratu can sustainably develop the economy through this Edutourism activity, not only education about geology, but knowledge about conservation and culture will also be embedded along with this tourism concept. The purpose of this edutouirsm tourism is to facilitate a tourism curriculum that contains educational values and learning about the main elements of UGG, namely, geological, biological, and cultural diversities so that it can support learning and provide broader educational insights. In carrying out the concept of Edutourism, it is necessary to build a community synergy that plays an active role, adequate infrastructure, and also tourist cooperation in maintaining the existence of the Ciletuh Geopark (Arcana & Wiweka 2016; Hasanah & Ruhimat 2019).

Learning or education is a change in the behavior of a person due to the experience he gets. So when the person acts, they will learn from the experience. Learning occurs because there is encouragement, awareness, questions, responses, and strengths. There are three suggested sources of learning, namely:

- Cognitive learning, which is defined as a process where people form associations between concepts, such as memorizing lists, and also finish the problem and get an entry. Learning like this abuses the intuition-awakening process where people have the belief to make their data even more transparent. So, cognitive learning is an active process where people have to control the information they get.
- Learning through education, which involves obtaining information from companies through advertising where the merchants and consumers are searching for information themselves.
- Learning through experience, which is gaining knowledge through contact with the product. Learning through experience is generally a more effective way to gain knowledge for consumers.

Talking about learning/education, there are several educational values obtained in the Edutourism Geopark at the UNESCO Global Geopark Ciletuh Palabuhanratu, namely, knowledge relating to geological diversity in Cieltuh Palabuanratu UGG and Ciletuh Palabuhanratu geological history. Educational values that can be circulated to the public regarding the existence of the Ciletuh geopark are geodiversity and biodiversity tourist destinations in the form of tourist destinations that can be visited by tourists, and the community takes part in protecting them such as the Ciletuh Mega amphitheatre or the Panenjoan landscape, which is an area of geological structures in the form of normal faults, which resulted in a large horseshoe-shaped avalanche that occurred at the Miocene age (Hadian et al. 2016). The value of education for the community and tourists is the process of the formation of the area in terms of geography and geology. Meanwhile, the biodiversity geopark Ciletuh offers a mangrove forest conservation area in Mandrajaya Village, which has 18 types of mangrove trees consisting of 12 families (*Alangiaceae, Apocynaceae, Avicenniaceae, Combretaceae, Euphorbiaceae, Fabaceae, Malvaceae, Myrsinaceae, Myrtaceae, Rhizhophoraceae, Sterculiaceae, and Verbenaceae*).

The dominant mangrove species is *Avicennia officinalis*. Tourists are expected to understand that mangroves have an ecological function as a habitat for various fauna. Generally, mangroves have various types of fauna that live in them. The habitat formed in mangrove forests is likely very productive because it can be the habitat for various types of fauna such as various types of birds, vertebrates, and invertebrates. The types of fauna found in the Ciletuh mangrove forest consist of *Periophthalmus gracilis, Episesarma sp., Clibanarius ambonensis, Varanus salvator, Boiga dendrophila, Phyton sp., Egretta sp., Acridotheres sp., Streptopelia chinensis, Prinia sp.,* and *Pycnonotus aurigaster*. The educational value that tourists, communities, and managers will gain is the acquisition of understanding and knowledge about efforts to protect or conserve the existence of mangrove forests so that their existence is maintained in supporting the ecosystem.

5 CONCLUSION

The potential for developing edutourism at Ciletuh Palabuhanratu UGG is very high. It can be seen from the level of participation of local communities and their tendency to contribute more actively. The role of the community is already classified as good, but the role played by the community is known to have not been optimal and well coordinated. Therefore, an edutourism development model in UGG Ciletuh Palabuhanratu is needed in order for both parties to be able to work together without having to depend on the government and other parties, especially when the community began to feel the benefits and feel of the educational values of the Ciletuh geopark.

Based on the research conducted, the researchers proposed two recommendations. First, the development of edutourism requires a long educational process for the public to provide awareness of the potential for geopark tourism. The active role of educational institutions, especially higher educations, is very much needed in realizing a knowledgeable society capable of empowering themselves. Second, the current economic conditions of the community are still inadequate to develop a business in the tourism sector because people have not been able to compete with big entrepreneurs in that field. In this regard, the role of the government is expected to protect local entrepreneurs from foreign investors who intend to dominate the tourism market in UGG Ciletuh Palabuhanratu. This is necessary so that local community economic activities are not taken over by people who are not native to the geopark.

REFERENCES

Andriany, S. S., Fatimah, M. R., & Hardiyono, A. 2016. Geowisata Geopark Ciletuh: Geotrek Mengelilingi Keindahan Mega Amfiteater Ciletuh (The Magical of Ciletuh Amphitheater). *Bulletin of Scientific Contribution, 14*(1), 75–88.

Arcana, K., & Wiweka, K. 2016. *Educational Tourism's Product Strategy at Batur Global Geopark, Kintamani – Bali. 3*(7). https://doi.org/10.2991/atf-16.2016.46

Badan Pusat Statistika. 2010. *BPS dalam Angka*. Retrieved from https://sp2010.bps.go.id/index.php/site/tabel?tid=337&wid=3202000000

Darsiharjo, Supriatna, U., & Saputra, I. M. 2016. Pengembangan Geopark Ciletuh Berbasis Partisipasi Masyarakat sebagai Kawasan Geowisata di Kabupaten Sukabumi. *Jurnal Manajemen Resort Dan Leisure, 13*(1), 55–60. https://doi.org/10.17509/jurel.v13i1.2036

Erfina, I. Y. D. 2019. Social Change in Ciletuh Geopark. *Review of Integrative Business and Economic, 8*(1), 244–259.

Guo, W., & Chung, S. 2019. Using Tourism Carrying Capacity to Strengthen UNESCO Global Geopark Management in Hong Kong. *Geoheritage, 11*(1), 193–205. https://doi.org/10.1007/s12371-017-0262-z

Hadian, M. S. D., Yuliwati, A. K., & Pribadi, K. N. 2016. Increasing community environmental awareness through geodiversity conservation activities at Ciletuh, Sukabumi, West Java. *Journal of Environmental Management and Tourism, 7*(2), 334–340. https://doi.org/10.14505/jemt.v7.2(14).18

Hasanah, S., & Ruhimat, M. 2019. *Edu-Tourism: An Alternative of Tourism Destination Based on Geography Literacy. 259*(Isot 2018), 193–195. https://doi.org/10.2991/isot-18.2019.42

Lam, J. M. S., Ariffin, A. A. M., & Ahmad, H. J. A. 2011. Edutourism: Exploring the push-pull factors in selecting a university. *International Journal of Business and Society, 12*(1), 63–78.

Ma'rifah, D. R., & Suryadarma, I. G. P. 2015. Penyusunan Panduan Edutourism Hutan Wisata Tlogo Nirmolo Guna Memunculkan Karakter Peserta Didik Kelas X [The Preparation of the Tourism Forest Edutourism Guide for Tlogo Nirmolo to Bring Up the Character of Class X Students]. *Jurnal Inovasi Pendidikan IPA, 1*(2), 126–137.

Manoharan, V. K., Khattar, S. K., Paldurai, A., Kim, S. H., & Samal, S. K. 2016. Geowisata Geopark Ciletuh: Geotrek Mengelilingi Keindahan Mega Amfiteater Ciletuh (the Magical of Ciletuh Amphitheater). *Journal of General Virology, 97*(2), 287–292. https://doi.org/10.1099/jgv.0.000350

Nabila, A. R., & Yuniningsih, T. 2016. Analisis Partisipasi Masyarakat dalam Pengembangan Desa Wisata Kandri Kota Semarang. *Journal of Public Policy and Management Review, 5*(3), 1–20.

Ojo, B. Y., & Yusof, R. N. R. 2019. Edu-Tourism Destination Selection Process in an Emerging Economy. *Journal of Tourism Management Research, 6*(1), 45–59. https://doi.org/10.18488/journal.31.2019.61.45.59

Ratih, N., Suryokusumo, B., & Sujudwijono, N. 2013. Perancangan Wisata edukasi Lingkungan Hidup di Batu dengan Penerapan Material Alami. *Jurnal Mahasiswa Jurusan Arsitektur, 1*(1).

Santoso, M. B., Apsari, N. C., & Raharjo, S. T. 2020. *Ciletuh Geopark: Toward the Tourism Industry. 117*(Gcbme 2018), 65–68. https://doi.org/10.2991/aebmr.k.200131.015

Sulaiman, A. I., Chusmeru, C., & Kuncoro, B. 2019. The Educational Tourism (Edutourism) Development Through Community Empowerment Based on Local Wisdom and Food Security. *International Educational Research, 2*(3), p1. https://doi.org/10.30560/ier.v2n3p1

Xu, F., Huang, L., & Whitmarsh, L. 2020. Home and away: cross-contextual consistency in tourists' pro-environmental behavior. *Journal of Sustainable Tourism, 28*(10), 1443–1459. https://doi.org/10.1080/09669582.2020.1741596

Yerry Yanuar. (2018). Prospective Analysis of Sustainable Development Strategy of Geopark Tourism of Ciletuh-Palabuhanratu West Java Indonesia. *International Journal of Current Innovation Research, 4*(2), 969–974. https://doi.org/10.24327/IJCIR

Promoting Creative Tourism: Current Issues in Tourism Research – Kusumah et al. (Eds)
© 2021 Taylor & Francis Group, London, ISBN 978-0-367-55862-8

Building the character of community tourism village in the preservation of Culture *Ngalaksa*

D.M. Nugraha, Supriyono & A. Gumelar
Universitas Pendidikan Indonesia, Bandung, Indonesia

ABSTRACT: Pandemic Covid-19 affects the life order of the world. The state of Indonesia experienced its impact that resulted in the life order of urban and rural communities to change, including the tourism village community character. The challenge of Pandemic Covid-19 is the tourism village community in preserving the culture of the *Ngalaksa* must prepare the character to face the situation of normal new era. This research aims to determine the strategy of tourism village government in building the character of tourism village community in the preservation of cultural *Ngalaksa* in the new era. The research method used is a case study in Rancakalong subdistrict, Sumedang district, with data collection techniques using interviews and observations, and analyzed qualitatively using data reduction. The results of the study gained that efforts to build the character of tourism village community in the preservation of cultural *Ngalaksa* in the new era normal by 1) making policies with a family spirit and the value of local wisdom; and 2) the character of the community owned is a strong character, discipline, cooperation, creative, honest, obedient to the rules, the security of family food, and the preservation of local wisdom.

Keywords: community character, tourist village, culture *Ngalaksa*, new era normal

1 INTRODUCTION

Pandemic Covid-19 affects the life order of the world, and the state of Indonesia experienced its impact that resulted in the life order of urban and rural communities to change, including the tourism village community character. The change of rural community character is influenced by economic sector problems caused by the development of obstacles. To overcome the problem, a business is ideally to build a character that is free from dependency (Sarip & Abdul 2020). One of the characters of the community that must be built is that economic and government character must implement the program of introduction and use of digital technology for UMKM as well as preparation to enter the era of Industrial 4.0 (Pakpahan 2020). Building a community character needs to be done through a tourism conscious program that touches the hearts of the community in creating business opportunities based on local creativity, so it can support the preservation of local cultures (Hariyanto 2017). The state of Indonesia under the Constitution of the Republic of Indonesia year 1945 article 32 paragraph 1 is that "the state furthered the national culture of Indonesia in the midst of world civilization by guaranteeing the freedom of society in maintaining and developing its cultural values." The basis of the Constitution can be a legal basis for the community in building the character of society based on the values of local wisdom. The challenge of Pandemic Covid-19 is that for tourism village community in preserving cultural *Ngalaksa* must prepare character to face the situation of the normal new era. The main problem in this article is 1) the policy issued by the government tourism village of Rancakalong subdistrict to build the character of the community tourism Village of Rancakalong subdistrict in preservation of the culture of the new era; and 2) the character must have by the Community Tourism Village of Rancakalong subdistrict in the preservation of cultural *Ngalaksa* in the new era.

 DOI 10.1201/9781003095484-4

2 METHOD

The study uses a qualitative approach, which is methods for exploring and understanding the meaning by a number of individuals or groups of people considered to be derived from social or humanitarian issues (Creswell 2010). The research method used is a case study in Rancakalong subdistrict Sumedang district. This method is used because the culture of *Ngalaksa* has uniqueness and only exists in the district of Rancakalong which aims to reveal in detail and comprehensively the events of the special events studied (Al Muchtar 2015). Data collection techniques use interviews and observations, and are analyzed qualitatively using data reduction. Researchers conducted an in-depth interview to the community, the officials of the Rancakalong village and the caretaker group of Rancakalong subdistrict Tourism Village to know the efforts to build the community character of tourism village in the preservation of cultural *Ngalaksa* in the new normal era. The study uses a qualitative approach, which is methods for exploring and understanding the meaning by a number of individuals or groups of people considered to be derived from social or humanitarian issues (Creswell 2010). The research method used is a case study in Rancakalong subdistrict Sumedang district. This method is used because the culture of *Ngalaksa* has uniqueness and only exists in the district of Rancakalong, which aims to reveal in detail and comprehensively the events of the special events studied (Al Muchtar 2015).

3 RESULTS AND DISCUSSION

Pandemic Covid-19 has an impact on changes in urban and rural life order, including in tourism village, which means tourism village apparatus must make a policy in terms of preservation of cultural *Ngalaksa*. *Ngalaksa* is the ritual of making food from the harvest of the local community with the aim that the community can express gratitude for all the blessings of God, given the love in the process of farming land accompanied by the music of Tarawangsa and Kacapi for seven days and seven nights (Yulaeliah 2008).

Based on the results of the study, the government of tourism village with the management of Paguyuban Desa tour Kecamatan Rancakalong created a policy to build the character of society in the preservation of cultural *Ngalaksa* in the new normal era by promoting the spirit of family and local wisdom. It is already based on Pancasila as the foundation of the country and the identity of a nation and in accordance with the research results of Kesuma (2016) stating that the culture of *Ngalaksa* is useful as a means of preserving the Sundanese culture and as an event to strengthen the harmony, togetherness, and mutual assistance between citizens, because in preparing the ritual *Ngalaksa* from the beginning to the end of the community from various circles with the village government shoulder to the extent of mutual cooperation for smooth and successful activities.

From the policy that has been taken then the Government ofTourism village with the management of Paguyuban Desa tour of Rancakalong subdistrict conducted various activities in the Corona virus pandemic situations (Covid-19) namely: 1) Increase the faith and pray so that the virus pandemic Corona (Covid-19) soon ends; 2) Socializing enhances the agricultural sector and new business sectors; 3) Provide socialization regarding the importance of togetherness in addressing each problem; 4) To build health protocols by sharing free masks and providing counseling briefly on the importance of maintaining cleanliness in the prevention efforts of Corona virus spreading; 5) The village apparatus provides examples and exemplary related discipline; and 6) Regrow the attitude of mutual cooperation that has become characteristic of the community by advancing and mutual care for others.

From the activities undertaken by the Government ofTourism village along with the caretaker of the Paguyuban Desa Tourism village in the situation of Corona virus pandemic (Covid-19) needs to improve the activities of preparing the community in terms of steadiness, management of agriculture sector, new business, protocol health, and spirit of mutual assistance. This is very important because people become objects and subjects in realizing the success of the district Rancakalong Tourism village in preserving the culture of *Ngalaksa* in Corona virus pandemic situations (Covid-19). So, the key to the success of this tourism village is the readiness of the whole population to open and change, and then it takes a continuous training in the management of tourism village (Susyanti & Latianingsih 2014).

In the activities of "village officials provide examples and the transparency related to discipline" is a very good activity, because in growing the character in the preservation of cultural *Ngalaksa* in Corona virus pandemic situations (Covid-19) it requires the example and the precision of the village apparatus to apply discipline, so that people can emulate and follow it. Village officials, community leaders, and indigenous people need to provide regular and continuous examples and not enough once (Hariyanto 2017). Through preservation of cultural *Ngalaksa,* then expected to complete the Corona virus pandemic (Covid-19), especially in the environment of Rancakalong Tourism village and can generally become a demonstration for people in other areas, because in the preservation of culture *Ngalaksa* there is a content of character education. Thus, the revitalization of the culture of local wisdom should be implemented because the revitalization of local wisdom can respond to the completion of various acute problems faced by nations and countries, such as corruption, poverty, and social gaps (Fajarini 2014).

The village government has socialized to the public about Corona virus pandemic countermeasures (Covid-19) to the neighboring level in the government's way with the health and safety team to socialize the tour to each region directly by using a vehicle. The move was precise by involving various parties in disconnecting the chain of Corona virus spread (Covid-19). Therefore, because the role of various elements is needed, it is certainly to maintain the integrity of the unitary Republic of Indonesia (Nurhalimah 2020).

Condition of the community character tourism impact of the pandemic virus Corona (Covid-19) is a culture of tolerance that has been neglected because of the condition in the limitation of socialization, but can finally understand. This proved that the community already has an awareness in supporting the village government's efforts to countermeasure the virus pandemic Corona (Covid-19) in tourist villages. The characteristic of the Community Tourism Village of Rancakalong subdistrict that has a family spirit and Gotong Royong is very effective in tackling Corona virus pandemic (Covid-19). The collective consciousness of the community in tackling the spread of Corona virus (Covid-19) is perceived to be quite precise as a form of anticipation for the ongoing impact of this pandemic (Jati & Putra 2020). Therefore, the character that the community should have to support the recovery of the village post-pandemic Corona virus (Covid-19) is: 1) Creative economy or a family side business; 2) Resilient or strong; 3) Discipline; 4) Cooperation; 5) Honest; 6) Compliance with the rules; 7) Local wisdom; and 8) Regenerate the attitude of tolerance and the gotong royong that has become characteristic of Rancakalong community. In overcoming the impact of the virus pandemic corona (Covid-19), then the community needs to implement a culture *Ngalaksa* by observing health protocols because in a culture *Ngalaksa* there is a content of character education based on local content. It is in accordance with the opinion (Kesuma 2016) that in the culture of *Ngalaksa* there is a character education that can be taken and built, namely, the character of tolerance, democracy, discipline, daring, hard work, creative, responsibility, religious, caring environment, social care, the spirit of nationality, and love of homeland. Honest character is a very important character to build the character of tourism village community in the preservation of cultural *Ngalaksa* in the new era normal, because the state of Indonesia needs an honest society to be able to release the state of Indonesia from the practice of corruption, collusion, and nepotism (Mulyadi et al. 2019).

4 CONCLUSION

The results of the study gained that efforts to build the character of the tourism village community in the preservation of culture *Ngalaksa* in the new era in the normal way: 1) Create policies with a family spirit and value of local wisdom, by carrying out the faith-enhancing activities, empowering communities from the gross of agriculture, enforcing health protocols, and clean-living attitudes;

and 2) the character of the community owned is a strong character, discipline, cooperation, creative, honest, obedient to the rules, family food security, and preservation of local wisdom.

REFERENCES

Al Muchtar, Suwarma. (2015). *Dasar Penelitian Kualitatif* Bandung: Gelar Pustaka Mandiri.

Creswell, John W. (2010). *Research Design Pendekatan Kualitatif, Kuantitatif, dan Mixed.* Yogyakarta: Pustaka Pelajar.

Fajarini, U. (2014). Peranan kearifan lokal dalam pendidikan karakter. *SOSIO DIDAKTIKA: Social Science Education Journal.* 1(2): 123–130. https://doi.org/10.15408/sd.v1i2.1225

Hariyanto, O.I.B. (2017) Membangun Karakter Sadar Wisata Masyarakat Di Destinasi Melalui Kearifan Lokal Sunda. *Jurnal Pariwisata.* IV(1): 32–39. https://doi.org/10.31311/par.v4i1.1830

Jati, B., & Putra, G.R.A. (2020). Optimalisasi Upaya Pemerintah Dalam Mengatasi Pandemi Covid 19 Sebagai Bentuk Pemenuhan Hak Warga Negara. *SALAM: Jurnal Sosial Dan Budaya Syari.* 7(5): 473–484. https://doi.org/10.15408/sjsbs.v7i5.15316

Kesuma, G.C. (2016). Pendidikan Karakter Berbasis Kearifan Lokal Adat Sunda "Ngalaksa" Tarawangsa Di Rancakalong Jawa Barat. *Al-Tadzkiyyah: Jurnal Pendidikan Islam.* 7(1); 35–44. https://doi.org/10.1017/CBO9781107415324.004

Mulyadi, D., Sapriya, S., & Rahmat, R. (2019). Kajian tentang Penumbuhan Karakter Jujur Peserta Didik sebagai Upaya Pengembangan Dimensi Budaya Kewarganegaraan (Civic Culture) di SMA Alfa Centauri Bandung. *MODELING: Jurnal Program Studi PGMI.* 6(2): 220–232. https://doi.org/10.36835/modeling.v6i2.471

Nurhalimah, N. (2020). Upaya Bela Negara Melalui Sosial Distancing Dan Lockdown Untuk Mengatasi Wabah Covid-19 (Efforts to Defend the Country Through Social Distancing and Lockdown to Overcome the COVID-19 plague). *SSRN Electronic Journal.* 19. https://doi.org/10.2139/ssrn.3576405

Pakpahan, A.K. (2020). Covid-19 dan implikasi bagi usaha mikro, kecil, dan menengah. *JIHI: Jurnal Ilmu Hubungan Internasional.* 20(April): 2–6. https://doi.org/10.26593/jihi.v0i0.3870.59-64

Sarip, A.S., & Abdul, M. (2020). Dampak Covid-19 Terhadap Perekonomian Masyarakat Dan Pembangunan Desa. *Al-Mustashfa: Jurnal Penelitian Hukum Ekonomi Islam.* 21(1): 1–9. http://dx.doi.org/10.24235/jm.v5i1.6732.g3120

Susyanti, D.W., & Latianingsih, N. (2014). Potensi Desa melalui Pariwisata Pedesaan. *Epigram: Jurnal Penelitian dan Pengembangan Humaniora.* 11(1): 65–70. https://doi.org/10.32722/epi.v11i1.666

Yulaeliah, E. (2008). Musik Pengiring dalam Upacara Ngalaksa Masyarakat Rancakalong Sumedang. *Resita: Jurnal Seni Pertunjukan.* 9(1): 31–36. https://doi.org/10.24821/resital.v9i1.447

Promoting Creative Tourism: Current Issues in Tourism Research – Kusumah et al. (Eds)

Development strategies for parenting tourism villages based on digital literacy

N.N. Afidah, D.M. Nugraha, A. Gumelar, P. Hyangsewu & Y.A. Tantowi
Universitas Pendidikan Indonesia, Bandung, Indonesia

ABSTRACT: The development of globalization that occurred in this century affected all countries including Indonesia. All fields experienced significant changes, including education and tourism. Education and tourism are important sectors for the progress of Indonesia, because the quality of society and tourism will affect the development of a country. Therefore, both fields need special attention in developing a long-term program. One form of special attention can be applied in the preparation of development strategies for parenting tourism villages based on digital literacy. Parenting tourism village can be one manifestation of the ideals of the Indonesian people. This is in line with one of the goals of Indonesian, which is to educate the life of the nation. The main problems in this article are (1) the factors that are indicative of the realization of a parenting tourism village based on digital literacy; and (2) development strategies for parenting tourism villages based on digital literacy.

Keywords: parenting tourism village, digital literacy, and development strategy

1 INTRODUCTION

The development of globalization that occurred in this century affected all countries including Indonesia. All fields experienced significant changes, including education and tourism. Education and tourism are important sectors for the progress of Indonesian because the quality of society and tourism will affect the development of a country. Therefore, both fields need special attention in developing a long-term program. One form of special attention can be applied in the preparation of development strategies for parenting tourism villages based on digital literacy. The village is a legal entity in which a group of people in power reside and these communities establish their own government (Kartohadikoesoemo 1984). On the other hand, Widjaja (2003) emphasizes that the village is a legal community unit that has an original structure based on special rights of origin. The foundation of thought regarding village governance is diversity, participation, genuine autonomy, democratization and community empowerment. Parenting tourism village can be one manifestation of the ideals of the Indonesian people. This is in line with one of the goals of the Indonesian state, which is to educate the life of the nation. The main problems in this article are (1) the factors that are indicative of the realization of a parenting tourism village based on digital literacy; and (2) development strategies for parenting tourism villages based on digital literacy.

2 LITERATURE REVIEW

2.1 *The nature of the village*

Understanding Village. The terms of etymology, 'village' comes from the Sanskrit language, 'deca', which means homeland or land of birth. From a geographical perspective, a village is defined as "a group of houses or shops in a country area which is smaller than and town". Village is a legal

 DOI 10.1201/9781003095484-5

community unit that has the authority to manage its household based on the origin and customary rights recognized in the National Government and is located in the regency. In general, the village is interpreted by the community as a place to live a population group that is marked usually with a special accent by villagers whose level of education is relatively low and generally residents' livelihood in the field of agriculture or marine. In the Indonesian General Dictionary, the words village meanings are: (1) an area inhabited by a number of families that have their own government system (headed by the village head), (2) a group of houses that are not in a city (3) hicks or hamlets (in the sense of rural areas or opponents of the city), (4) place, land, area.

Villages based on the provisions of Article 1 of Law Number 6 Year 2014, villages are defined as villages and customary villages or referred to by other names, hereinafter referred to as villages are a legal community unit that has the authority to regulate and manage affairs government, the interests of local communities, then regarding the rights of origin, and traditional rights (Poerwadarminta 2007). "The village is a legal entity in which a group of people in power resides and the community establishes its own government" (Kartohadikoesoemo 1984). The village is an autonomous region which is at the lowest level of the regional autonomy hierarchy in Indonesia, as stated by Nurcholis which says that the village is the lowest administrative unit. One of the affairs of the village government which is the authority of the village is the management of village finances. Village finance is about all the rights and obligations of the village that are valued in money, as well as everything in the form of money and goods that can be owned by the village in relation to the implementation of rights and obligations (Hanif 2011).

"The village is a legal community unit that has an original arrangement based on special original rights. The rationale for village governance is diversity, participation, genuine autonomy, democratization and community empowerment (Widjaja 2003). Based on the review of geography, village is a result of the geographical, social, political, and cultural manifestations that exist in one region and has a reciprocal relationship with other regions. The definition of village according to law is: Government Regulation Number 72 Year 2005 About Village Article 1.7, village or referred to by other names, is a legal community unit that has territorial boundaries authorized to regulate the interests of the local community based on local origins and customs that are recognized and respected in the Republic of Indonesia State Government system.

Law Number 6 of 2014 concerning Villages Article 1, Villages are Hamlets and Customs or what are referred to by other names, hereinafter referred to as Villages, are legal community units that have territorial boundaries authorized to regulate and administer government affairs, the interests of local communities based on initiatives community, original rights, and/or traditional rights recognized and respected in the government system of the Republic of Indonesia. Thus, as a part of the government system of the Republic of Indonesia which is recognized for its autonomy and the Head through the government may be given the assignment of delegation from the government or from the regional government to carry out a particular government.

According to Zakaria in Wahjudin Sumpeno in Candra Kusuma stated that the village is a group of people that lives together or an area, which has a set of regulations that are self-determined, and are in the leadership of the chosen and self-determined. Meanwhile, the government, based on Law Number 72 of 2005 Article 6, states that the Consultative Government in regulating and managing the interests of local communities is based on local origins and customs that are recognized and respected in the system of Government of the Republic of Indonesia.

2.2 *Parenting*

2.2.1 *Definition of child parenting*

All parents would want their children to be people with good personality, healthy mental attitude and good character. Parents are the first personal shaper in a child's life, and must be a good role model for their children. Parents' personalities, attitudes and ways of life are elements of education that will indirectly enter the person of a growing child (Daradjat 1996). Parents have the responsibility of educating their children as well as being called parents as coordinators in the family. And the person called the coordinator must be able to behave proactively. If a child opposes authority, he must be put in order immediately because in the family there are rules and expectations. Children will feel safe even they don't always realize it. So, parents direct their children in accordance with the goal, which is to help children to develop the basics of self-discipline. His parents as individuals and as educators can reveal parenting patterns in developing self-discipline in the relevant circumstances.

2.2.2 *Forms of parenting*

Education in the family provides religious beliefs, cultural values that include moral values and religious rules, in association and views, skills and attitudes that support life in community life. Therefore, to realize this, there are various ways in parenting done by parents. "Parenting patterns are divided into three, namely: (a) parenting with strict rules (parents emphasize all the rules that must be obeyed by children and must not counter parental orders); (b) parenting laisses fire (this parenting usually tends to be applied by modern families, i.e. parenting by means of parents educating their children freely, what children do is allowed by parents, children are considered adults and given freedom in action); and (c) democratic parenting (democratic parenting tends to be applied to modern education among moderate families, and this pattern is more pragmatic, adapting to existing conditions)" (Yusuf 2000).

2.2.3 *Factors that influence parenting*

Several factors influence parental care, namely: (1) the socioeconomic level of parents who come from the middle socioeconomic level is more warm, compared to parents whose socioeconomic level is low; (2) the educational background of parents who are higher in their care practices is seen reading articles more often to see their children's development, whereas parents with low levels of education tend to be authoritarian and treat their children strictly; (3) parental personality greatly influences parenting (Hurlock 2008).

The family environment is the environment that is first and foremost for children. Therefore, the position of the family in developing the child's personality is very dominant. In this case, parents have a very important role in fostering the child's religious nature. The family is a training center for the role of values. In addition, guidance and development of the child's religious nature or soul must be adjusted to the development of his personality from rom birth even more than that from the womb because they are influenced by emotional states or attitudes of parents. The period in the womb is based on the views and observations of psychologists (Yusuf 2001).

2.3 *The nature of digital literacy*

Literacy in English is derived from Latin literacy, which is litera (letters) often interpreted as literacy. If seen from the literal meaning it means someone's ability to read and write. Regularly, people who can read and write are called literates, while people who cannot read and write are called illiterates. Kern (2000) stated, "Literacy as the ability to read and write". In addition, literacy also has the same meaning as learning and understanding the reading sources. On the other hand, "Literacy is a social event involving certain skills, which are needed to convey and obtain information in written form" (Romdhoni 2013).

This is in line with opinion of Kern (2000) which defines literacy more comprehensively, "Literacy is the use of socially, historically, and culturally-situated practices of creating and interpreting meaning through texts. It entails at least a tacit awareness of the relationship between textual conventions and their contexts of use and, ideally, the ability to reflect critically on those relationships. Because its purpose is sensitive, literacy is dynamic-not static-and variable across and within discourse communities and cultures. It draws on a wide range of cognitive abilities, on knowledge of written a spoken language, on knowledge of genres, and on cultural knowledge".

Based on the opinions above, basically it can be explained that literacy is a social event that is equipped with skills to create and interpret meaning through text. Literacy requires a series of abilities to convey and obtain information in written form. On the other hand, "Now literacy is not only related to the ability to read and write texts, because now "text" has expanded its meaning to include "text" in the form of visual, audiovisual, and dimensions of computerization. Therefore, in "text", together cognitive, affective, and intuitive elements emerge. In today's technological era, the context of the intellectual tradition of a society can be said to be literate culture when the community has utilized the information, so they can carry out communication in social and science. Based on the explanation above, it can be understood that literacy is a stage of social behavior, namely the ability of individuals to read, interpret, and analyze the information and knowledge in order to gain the prosperity of life (superior civilization)" (Iriantara 2009).

"Digital literacy is the ability to use technology and information from digital devices effectively and efficiently in various contexts such as academic, career and daily life" (Gilster 1997; Riel et al. 2012). This opinion simplifies digital media which actually consists of various forms of information at once such as sound, text and images. "Digital literacy should be more than the ability to use various digital sources effectively. Digital literacy is also a particular form of thinking" (Eshet 2002). The new understanding of digital literacy is rooted in computer literacy and information literacy. "Computer literacy developed in the 1980 when microcomputers became widely used not only in the business environment but also in the community. Meanwhile, information literacy became widespread in the 1990 as information became more easily compiled, accessed, and disseminated through information technology network" (Bawden 2001).

3 METHOD

The approach used in this research is qualitative, which is a process of research and understanding based on methodologies that investigate social phenomena and human problems. In this approach, the researcher makes a complex picture, examines words, detailed reports from the respondents' views, and conducts studies in natural situations (Creswell 2013). Qualitative methodology is a research procedure that produces descriptive da-ta in the form of written and oral words from people and observed behavior. This is done because the stuff being studied is a modeling of the parenting village, so a comprehensive study is needed regarding the form of its implementation.

The method used is the Naturalistic Inquiry, which is a way of observing and collecting data carried out in a natural setting, meaning that without manipulating the subject under study (as it is nature) (Lincoln & Guba 1985). Therefore, through this approach, it is hoped that information can be obtained in a focused and in-depth manner. The method in naturalistic inquiry is carried out by describing phenomenally related to the stand out information about the formation of a parenting pilot village model.

4 RESULT AND DISCUSSION

4.1 *Factors that indicate the realization of digital literacy-based parenting tourism village*

"Media intelligence in society is very important. Currently the use of digital media in the world has become a lifestyle, which is connected with information technology. Digital literacy is defined as the ability to under-stand and use information in various forms from a very wide variety of sources that are accessed through computer devices" (TIM GLN Kemendikbud 2017). "New understanding of digital literacy is rooted in computer literacy and information literacy" (Bawden 2001). However, the growth of digital media allows a shift in people's behavior. Openness of information on social media is not accompanied by an intelligence of media to analyze existing data and content.

The travel trend has changed from year to year. Many people nowadays prefer to travel in the anti-mainstream places. The fact drives the development of tourism in all corners of Indonesia. The development of tourism is also a concern of the government so that it can be a source of income for the people in the tourist destination. The government also provides adequate infrastructure support for the development of tourist villages so that tourist destinations have high accessibility. Related to this, there are many tourist villages that do require digital literacy-based parenting strategies and the role of parents in preparing children for their day. Parents as educators must first do retrospect and self-introspection by continuing to try to prepare their children in the face of the current digital era and the era going forward. Parents need to make projections by building a commitment or determination to protect the children from the threat of the digital age, but do not preclude the potential benefits they can offer. Therefore, there are several factors to create a parenting tourism village based on digital literation. First, parents need to accompany the children as a digital generation. Children are being late talking due to lack of exercise, too much self-play, too passive, watching too much TV. Parents also have to provide assistance to children as a digital generation. Second, the use of digital media according to age and stages of child development. Parents and children need agreement on the use of digital media, not to protect children, but to provide appropriate opportunities when children are exposed to information from the media, because parents may not always be able to supervise the activities of children. At this stage, parents need to accompany the child according to their age and developmental stages. Third, assisting digital generation as well as providing understanding and knowledge in relation to the page, directing the use of digital devices and media clearly, balancing children's time using digital media with interaction in the real world, lending digital devices as needed, choosing positive programs/applications, as well as accompanying and increasing interactions towards children. In addition, when with children use digital devices wisely, as well as accessing pages that are appropriate to their ages and development stages. Those are some factors that indicate the formation of a digital literacy-based parenting tourism village as well as a supporting factor for the development of a digital literacy-based parenting tourism village.

4.2 *Digital literacy-based parenting village development strategy*

Digitalisation changes the order of new social and cultural systems according to the times. Humans today are faced with a variety of digital information. The challenges are so broad especially whether it is wise or not in managing digital information systems. Surely it is a challenge in digital titration in the community. Digital literacy can be done not only in formal or non-formal education but in an informal way. As in the parenting tourism village that develops digital literacy. To achieve this, a variety of strategies must be done. Based on the research results obtained, the village government has a strategy in developing asus tourism village based on digital literacy with a focus on the family, school and community environment. This is in accordance with the Tricentric of Education that the success of the implementation of education is influenced by the nature of the family, the nature of the school and the nature of youth (community). The three realms must synergize and work well together (Hikmat 2004).

The village government makes assistance programs for families in the implementation of parenting based on digital literacy. This program is very important because it will reduce the dangers posed by social media, such as pornography, hoax news and negative comments. The village government conducts regular training for families in relation to implementing education for their children based on digital literacy. The village government invites and cooperates with experts who understand about implementing digital literacy-based asus patterns, such as working with *Jabar Saber Hoax* which contains ways so as not to be affected by hoaxes in carrying out digital literacy. This is a recommendation that digital literacy should be given at the level of family, school and country (Hurlock 1997).

The strategy of developing a digital literacy-based parenting tourism village is carried out through the literacy movement in the community including: (1) training in the use of responsible internet applications; (2) socialization of the legal basis of using digital media properly and correctly (wisely); (3) providing internet access in public spaces; (4) provision of digital information media; (5) cooperating between tourism villages with experts and practitioners in the field of digital literacy, such as holding community services in the form of digitization training; (6) cooperating with organizations in the community, such as youth groups, PKK, community literacy, and others to improve digital literacy; and (7) involving stakeholders such as local government, business and industry, the media and educational institutions.

The strategies described above are expected to educate the public in utilizing technology and communication wisely and creatively or responsibly. In addition, the increasing number of active participation of communities, institutions or agencies in developing digital literacy-based parenting tourism villages, as well as people understanding the use of the internet and the ITE Law. So, the development of parenting tourism village based on digital literacy can be done through the collaboration of schools, families and communities.

This is in line with what was conveyed by the Ministry of Education and Culture in a book entitled "Digital Literacy Supporting Materials" that community digital literacy can be developed through study groups, PKK, youth clubs, hobby communities, and community organizations. Digital literacy is an important tool for over-coming various social problems, such as pornography and bullying. Digital literacy enables people to access, sort and understand various types of information that can be used to improve the quality of life, such as health, expertise and skills" (TIM GLN Kemendikbud 2017).

Other findings obtained in the study (Silvana & Darmawan 2018) show "the importance of digital literacy programs that have a positive impact on knowledge, understanding and skills in using media, especially social media which is currently often used as a source of information by the audience, especially among youngsters".

In his article Pandapotan (2018) at a national seminar on community service at an open university, entitled "Development of a literacy village model to increase educational motivation and reading interest in the village community, a regency pond Deli Serdang", that the formation of a pool village as a literacy village with literate literacy communities, highly motivated and highly educated have access to books and other sources of knowledge.

It was also strengthened by Raya (2018) in his research entitled "Utilization of Village Application Applications in Agricultural Information Literacy" that the UGM Faculty of Agriculture had developed an application called Desaapps. Desaapps is an agricultural application that aims to provide agricultural counseling in digital form. The existence of Desaapps can be useful for application users to provide information to each other, including interaction and transaction in agriculture. The utilization strategy is to improve application performance, promote village applications, promote expert responsiveness, and optimize the use of Desaapps as a marketing venue. This indicates that digital literacy is applied in agriculture in the community.

5 CONCLUSION

Parenting tourism village can be a manifestation of the ideals of the Indonesian people. In its development, there are several factors that indicate the realization of a digital literacy-based tourism village, including: (1) the need for parents to accompany children as a digital generation; (2) use of digital media that is adapted to the age and stages of child development; (3) means that help the digital generation while providing understanding and knowledge related to pages, clearly directing the use of digital devices and media, balancing children's time using digital media with interactions in the real world, lending needed digital devices, choosing positive programs/applications, also help and enhance children's interactions. In addition, parents can provide understanding to children about the use of digital devices wisely according to the child's age and development stage. Furthermore, the strategy for developing a digital literacy-based parenting tourism village is carried out through a literacy movement in the community, including: (1) training in responsible use of internet applications; (2) socialization of the legal basis for using digital media properly (wisely); (3) providing internet access in public spaces; (4) providing digital information media; (5) cooperation in tourism villages with experts and practitioners in the field of digital literacy, such as organizing community service in the form of digitalization training; (6) collaborating with organizations in the community such as youth community, PKK, community literacy, and others to improve digital literacy; and (7) involving stakeholders such as local government, business and industry, media, and educational institutions. With this strategy, it is hoped that it can educate the public in using technology and communication wisely and creatively or responsibly.

REFERENCES

Bawden, D. 2001. Information and digital literacies: a review of concepts. *Journal of documentation*, 57(2), 218–259.

Creswell, J.W. 2013. *Research design: Qualitative, quantitative, and mixed methods approaches*. (3rd eds.). Thousand Oaks California: SAGE Publications.

Daradjat, Z. 1996. *Metodologi pengajaran agama Islam*. Jakarta: Bumi Aksara.

Eshet, Y. 2002. Digital literacy: A new terminology framework and its application to the design of meaningful technology-based learning environments. *In Proceedings of World Conference on Educational Multimedia*, (pp.493–498).

Gilster. 1997. *Digital literacy*. New York: Wiley.

Hanif, N. 2011. *Pertumbuhan dan penyelenggaraan pemerintahan desa*. Jakarta: Erlangga

Hikmat, H. 2004. *Strategi pemberdayaan masyarakat*. Bandung: Humaniora Utama Press.

Hurlock, E.B. 1997. *Psikologi perkembangan suatu pendekatan sepanjang masa*. (5 eds.). Jakarta: Erlangga.

Hurlock. 2008. *Perkembangan anak*. Jakarta: Erlangga.

Iriantara, Yosal. 2005. *Media relations: Konsep, pendekatan dan praktik*. Bandung: Simbiosa Rekatama Media.

Kartohadikoesoemo, S. 1984. *Desa*. Jakarta: Balai Pustaka.

Kern, R. 2000. *Literacy & language teaching*. Oxford: Oxford University Press.

Lincoln, YS. & Guba, EG. 1985. *Naturalistic Inquiry*. Newbury Park, CA: Sage Publications.

Raya, A.B., Kriska, M., Wastutiningsih, S.P., Cahyaningtyas, M.U., Djitmau, A., & Cahyani, G.F. 2018. Strategi pemanfaatan aplikasi desa apps dalam literasi informasi pertanian. Jurnal Komunikasi Pembangunan, 16(2), 274–285. https://doi.org/10.46937/16201826341.

Riel, J., Christian, S., & Hinson, B. 2012. *Charting digital literacy: A framework for information technology and digital skills education in the community college*. Presentado en Innovations.

Romdhoni, A. 2013. *Alquran dan literasi: Sejarah rancang-bangun ilmu-ilmu keislaman*. Depok: Literatur Nusantara.

Silvana & Darmawan. 2018. Pendidikan literasi digital di kalangan usia muda di Kota Bandung. *Pedagogia*. 16(2).

TIM GLN Kemendikbud. 2017. *Materi pendukung literasi digital*. Jakarta: Kementrian Pendidikan dan Kebudayaan.

Undang-Undang Nomor 72 Tahun 2005 Tentang Desa.

Pandapotan, S. 2018. *Pengembangan model kampung literasi untuk meningkatkan motivasi pendidikan dan minat membaca masyarakat Desa Kolam Kab. Deli Serdang*. In: Seminar Nasional Pengabdian Kepada Masyarakat. Jakarta: Universitas Terbuka.

Peraturan Pemerintah Nomor 6 Tahun 2014 Tentang Desa.

Peraturan Pemerintah Nomor 72 Tahun 2005 Tentang Desa.

Poerwadarminta, W.J.S. 2007. *Kamus Umum Bahasa Indonesia*. Jakarta: Balai Pustaka.

Widjaja, H.A.W. 2003. *Pemerintahan desa/marga*. Jakarta: Raja Grafindo Persada.

Yusuf, S. 2000. *Psikologi perkembangan anak dan remaja*. Bandung: Remaja Rosdakarya.

Yusuf, L.N. 2001. *Psikologi perkembangan anak dan remaja*. Bandung: Rosdakarya.

Promoting Creative Tourism: Current Issues in Tourism Research – Kusumah et al. (Eds)
© 2021 Taylor & Francis Group, London, ISBN 978-0-367-55862-8

Citizenship education in community development in Indonesia: Reflection of a community development Batik Tourism Village

Katiah, A. Dahliyana, Supriyono & V.A. Hadian
Universitas Pendidikan Indonesia, Bandung, Indonesia

ABSTRACT: Tourism becomes a global industry that can be transformed into a governance and management environment with the aim of the development of people sustainably. The inclusion of citizenship education concept in the context of community development aims to provide a position of the nature of citizenship education that Indonesia must enter in the practice of community development. This article is a case study of batik craftsmen in Kalitengah village of Cirebon Regency of West Java, Indonesia. This study shows, one alternative is rural development in suburban areas. Indirectly, the tourism that opened in this area "batik Tourism Village" will help the economic level of batik maker that has been channeling the results of batik to places of sale in the city with the income that is very far from the results obtained by the batik sellers in the shops located in the city.

Keywords: citizenship education, community development, batik, tourism village

1 INTRODUCTION

Citizen participation as the core of citizenship education and community development has a fundamental value orientation in the development process (Koneya 1978; Matarrita-Cascante & Brennan 2012). Citizen participation in community development will help to build healthier social support networks in various environments and strengthen community capacity toward independence (Vasoo & Tiong 1992). However, community development in the context of citizenship education should define the educational aspect of each tradition, so as not to get stuck on the strict competition in society to earn money (Veen 2003). Therefore, community development must be done by creating groups that are able to solve the problem especially in improving the economic ability of the people (Koneya 1978).

In resolving that problem, Matarrita-Cascante and Brennan (2012) said that community development should be given a conceptual definition, an applied perspective, and its sustainability direction by involving policy-making, practitioners, and academics. It is done to minimize the immolation and form of promotion to be offered. Thus, community development is the process of learning citizenship education in the same way as unions to active citizens for low-income groups (Veen 2003).

With regard to the above statement, it is necessary to develop the concept of citizenship education in the context of community development that is able to foster innovation, sustainability, and dedication from citizens involved through action, dialogue, and self-reflection, recognizing the importance of change in worldview (Hochachka 2005; Matarrita-Cascante & Brennan 2012). Therefore, development is more than just the accumulation of wealth, infrastructure, or economic growth. It is about creating space to explore self-relationships enabling self-expression and self-awareness as well as collective compassionate acts (Hochachka 2005).

One of the efforts made is by developing tourist destinations in areas with low citizen income. Agua Blanca's experience shows that much can be achieved through belief in cultural identity, and

DOI 10.1201/9781003095484-6

that community-managed tourism communities and focus on culture can diversify livelihoods that bring tangible economic benefits to local communities (Hudson et al. 2016; Mbaiwa & Sakuze 2009). Mbaiwa and Sakuze (2009) mention that tourism directed at culture can diversify livelihoods, promote local participation in tourism development, alleviate poverty, and contribute to sustainable development. Therefore, villages need to be further empowered to determine the level of tourism growth that is liked in the village as part of governance and environmental management, aiming for a more sustainable economy (Tirasatayapitak et al. 2015; Zhang 2019). Because community development through tourism development brings economic improvement, it inspires awareness of civil rights, democracy, and equality (Xiaoping et al. 2014).

So how is citizenship education in community development in the village Kalitengah Cirebon Regency of West Java Indonesia as a batik-producing village that is surrounded by areas famous for batik shopping destinations such as the Trusmi area? This paper has differences compared to other studies that focus on community empowerment such as Voth (1975), Arches (1999), Bridger and Alter (2006), Hochachka (2005), Hendrickson et al. (2011), Matarrita-Cascante and Brennan (2012), Moore (2002), Sklar et al., (2014). The difference is with research that focuses on community empowerment through tourism such as Hudson et al. (2016), Idziak et al. (2015), Mair and Reid (2007), Mbaiwa and Sakuze (2009), Mokoena (2019), Prasad (2012), Prince (2017), Tirasatayapitak et al. (2015), and Zhang (2019); there is the use of civic education as an empowerment approach for underprivileged communities in villages that produce batik. In addition, there are still studies that are close to this paper such as Koneya (1978), Vasoo and Tiong (1992), and Veen (2003), but not focus on community empowerment by producing batik. Thus, this article is truly original because there is no similar writing yet.

2 METHOD

The approach used in this paper is qualitative, which focuses on the case study method. In obtaining data, it is done through observation techniques and interviews. Both techniques simply represent the retrieval of the data done to find the answer that became the focus of the writing. The research location in this article is Kalitengah village Cirebon Regency West Java, which become one of Indonesian batik producers. This study was done of five people of batik craftsmen, ome head of village, and one researcher, who came from alumni of the School of Arts Education University of Indonesia education. The data obtained is then analyzed through the stages found in qualitative traditions. The data that has been obtained are then analyzed through the stages contained in the qualitative tradition, namely through data collection carried out before, during, and at the end of the research, data reduction is carried out by the process of combining and uniforming all forms of data so that it becomes a writing that will be analyzed. The stages are the results of interviews through verbatim interviews, and observations through appendixes. After that, the data display is performed by processing the data in a matrix, and finally through the stages of drawing conclusions.

3 RESULT AND DISCUSSION

Kalitengah Village of Cirebon Regency of West Java, Indonesia, is included in the first category where the main problem of the village is requiring priority handling to fulfill basic needs such as economy, education, health, infrastructure, and environment. This is because of, in general, the livelihoods of Kalitengah village in general as batik craftsmen and batik entrepreneurs whose results will be distributed to the Trusmi region, which became the location of batik sales in Cirebon Regency.

The social capital/character that is still held strongly by the people of Kalitengah village is a program of "Sapa Warga" of the strengthening of mutual assistance that is still done by the community. In addition, physical capital is one of the old buildings of Dutch heritage that is still very well maintained and beautiful. The building is spread evenly on the pedestrian routes starting from the front part of the village building and ended by the building of special batik craftsmen. It is an important component of many community development strategies (Bridger & Alter 2006) especially the batik-oriented tourism village. However, it is not enough, local involvement in creating thematic villages is much more intensive and has so wide coverage that people cannot talk about facilitating or engaging communities in the development process, but rather on the creation of tourism development by local communities. External assistance from experts may be needed at major stages, especially by providing professional knowledge of market and marketing (Idziak et al. 2015). It is related to the division of territory that must have the clarity of parts, position, and its role in the spread of the specificity of both the product and the character of the community.

Based on the results of observation and interviews, there is a picture that in particular, Kalitengah village has the potential to thrive in terms of batik that is supported by old historical buildings as tourism village of thematic forms. In addition, the central shopping tour to the Trusmi region makes Kalitengah village have another potential to be developed, namely the tourism village of batik producing and batik shopping. In order to implement community development through the development of batik tourism village, there have been many trainings and education that are followed by the community for economic strengthening, family strengthening, and cultural strengthening through the creation of traditional Cirebon batik. In addition, the government has also provided trainings such as batik design and quality batik. Thus, the potential of Kalitengah village, which is located in Cirebon, has potential in terms of batik that can be used as a tourist destination. But it has not yet been run in full because it has not aligned with market needs. In fact, in the development of this tourism village it is very important to conduct empirical research based on market and products (Mokoena 2019). Therefore, it is necessary for practitioners and those who are involved in it to report using research literature and reports on models, guidelines, and community development theories to help them understand and direct attention to complex problems (Moore 2002), so that the empowerment developed is able to create a culture of community-based participation, interconnectedness, and cross-category interdependence to be able to implement greater empowerment (Hendrickson et al. 2011).

Information obtained was from the head of Desa Kalitengah, in general very many batik craftsmen. However, they are the ones who get orders from stores located in the Trusmi area. In fact, based on the results of observation, the results of batik craftsmen are very good and worthy to be used as a high-quality batik-batik with expensive selling price. Students' Alumni of the University of Indonesia Education Program that conducts community research in Kalitengah Village and produces batik in the area of North Sumatra, when interviewed were very sad, that descriptions, and the activities of the batik should be prosperous but they are not prosperous. In one day, the craftsmen benefited from 50,000 to 70,000 rupiah. It means that various trainings and education conducted by experts and governments have not been able to touch the aspects of empowerment in a holistic and integral. This happens because the empowerment process is often intermittent (Arches 1999). Therefore, the empowerment of the community must be understood as the path of education, training, and learning of citizens, especially citizens with a low level of formal education (Veen 2003). In the sense that the approach can be "top down" for social mobilization, social learning, and radical planning model (Mair & Reid 2007) so that the process done in the empowerment takes place continuously.

From the search results, the batik craftsman in Kalitengah Village amounted to 121 people with the number of craftsmen working in the stores in the Trusmi area compared to the private stores. The craftsmen who managed to develop their business by reducing themselves amounted to 28 people. How it relates to community development will greatly depend on participation and local involvement (Matarrita-Cascante & Brennan 2012). Thus, it is necessary to make efforts to maintain grassroots involvement through community-driven programs, and activities that hold the target responsible for improving conditions, giving hope and strength to the public while providing reasons for sustainable organizing, so it is important to meet the challenges and dilemmas in community-based development (Arches 1999). In this regard are government interference and academics. Because the combined locality in the tourist landscape shows the different ways in which localities emerge as cultivated influences and tradable commodities (Prasad 2012).

The small income gained by batik craftsmen in Kalitengah village is because in addition to not yet understanding the concept of market and marketing, they generally have low education level of elementary school to junior high school. It happened because, when one of the members who have been educated and in the middle of the road ran out of funds, they usually invite their children to continue their work as a batik craftsman. The situation is very logical, considering the raw materials for the making of batik is still very rare and do not attempt to be developed or cultivated in the village Kalitengah so the cost for family education is minimal. In addition, to reduce operational costs it is better to involve your own family than to involve others.

Until now, to obtain raw materials of batik coloring, craftsmen who come from Kalitengah village must be ordered from Pekalongan, precisely Ciwaringin village. Although it is in the environment, the process of making it is still long and expensive so that craftsmen prefer to buy finished materials. The situation has been illustrated by Voth (1975), so that the problem of community development will not be far from (1) the ambiguity of the goal, (2) the lack of a clear causal model, (3) inability to control assignments, (4) combinations of weak effects, and (5) political problems. It is because the idealistic and educational expectations of volunteers, counseling, and academics often clash with the short-term practical objectives of the community: there are also cultural differences and experiences between the parties (Prince 2017).

Thus, the concept of citizenship education in community development should have a sustainable side, interdependent planning, and implementation framework that includes direction, strategy, actors, and strategic instruments for the policy of making tourism village, especially batik, by integrating grassroots community throughout the empowerment process by using coalition and interagency collaboration (Arches 1999; Hendrickson et al. 2011). Because, institutions should invest in citizen empowerment and build their involvement in the process of community development (Sklar et al. 2014). In addition, with the coalition and collaboration, (1) development into effective community development practices to maximize resources and utilize various areas of stakeholders' needs and expertise; (2) community connections are crucial for all stakeholders to fully participate and contribute to community development activities; and (3) the coalition that handles consumer outcomes is more committed to community development and is better able to sustain their momentum and direction over time (Guillory et al. 2006).

By emphasizing the analysis of the problem and vision in a better direction, a community can encourage empowerment based on awareness and social transformation, thereby improving population control of communities and organizations as well as a successful signal of citizenship education in the development of an integral community integrating material and interior needs (cultural, spiritual, and psychological health) (Arches 1999; Hochachka 2005).

The form of citizenship education in community development can be done through stages: (1) education should be directed into local leadership training; As an action oriented and on the learning process in the field is supported informally by community workers. This form of education resembles an informal vocational education where seasoned craftsmen train their students. (2) Education as an enhancement of consciousness, which reverses the order of the learning process: In this case it is not an action that leads to education but that education hopefully leads to action by citizens. (3) Education as a service delivery: Here education is a service to the community in the same way as community development can provide other services to the community (Veen 2003). Veen (2003) again stated that it needed a comprehensive approach in the provision of local community development in order to make it possible to offer residents a variety of activities, and for residents to build their own education with the combination and sequence of activities that best fit their own lifelong learning

If it can be realized, the awareness of citizens to engage in public affairs becomes stronger and the pursuit of equitable distribution and focus on the public interest becomes clearer. It is due to (1) a more developed tourism industry resulting in greater demand for the same distribution of the population; (2) as the tourism industry grows, population awareness increases and the ability to participate in public management; (3) with the development, the demands of the population for democracy become clearer; and (4) with the development of the tourism industry, residents pay more attention to the public interest and the development of public spaces, and they gain more from the public spirit (Xiaoping et al. 2014).

4 CONCLUSION

This writing has provided illustrations of conceptual complexity and citizenship education practices needed in community development. During this time, the concept of citizenship education in Indonesia is most dominated by a paradigm that applies to formal educational institutions only. However, in the concept of practice, citizenship education is not limited to formal education alone, and can be done informally like community development. This gives an insight that there is no adequate single paradigm in theoretical citizenship education. The sample study in the field used in this study shows that citizenship education as part of the community development package can produce a positive impact on the community as the focus of empowerment. This has been done by various communities of community development in the world and currently seeks to be implemented in Kalitengah village of Cirebon Regency of West Java, Indonesia, with a focus on empowering people through batik tourism village. The main argument of citizenship education in community development in Kalitengah Village is done to strengthen economic resilience, family resilience, and cultural resilience that can create the disposition of good citizens with participation rate in strong development without eliminating the identity of the region as the biggest batik producer in West Java, Indonesia. The implication of this paper for researchers is that understanding the concept of citizenship education as community empowerment will be more practical and easier to understand than theoretical and classical learning. In practice, the use of a good and humanist approach to the community for the purpose of community empowerment will be easily accepted because they are given awareness of what they are doing. In addition, the implications for society are that the more they understand the concepts and practices of community empowerment well, the level of welfare in the economic field will be easily achieved.

REFERENCES

Arches, J. L. (1999). Challenges and dilemmas in community development. *Journal of Community Practice, 6*(4), 37–55. https://doi.org/10.1300/J125v06n04_03

Bridger, J. C., & Alter, T. R. (2006). Place, Community Development, and Social Capital. *Community Development, 37*(1), 5–18. https://doi.org/10.1080/15575330609490151

Guillory, J. D., Everson, J. M., & Ivester, J. G. (2006). Community development: Lessons learned about coalition building and community connections for stakeholders with disabilities. *Community Development, 37*(3), 83–96. https://doi.org/10.1080/15575330.2006.10383110

Hendrickson, D. J., Lindberg, C., Connelly, S., & Roseland, M. (2011). Pushing the envelope: Market mechanisms for sustainable community development. *Journal of Urbanism, 4*(2), 153–173. https://doi.org/10.1080/17549175.2011.596263

Hochachka, G. (2005). Integrating Interiority in Community Development. *World Futures, 61*(1–2), 110–126. https://doi.org/10.1080/02604020590902399

Hudson, C., Silva, M. I., & McEwan, C. (2016). Tourism and Community: An Ecuadorian Village Builds on its Past. *Public Archaeology, 15*(2–3), 65–86. https://doi.org/10.1080/14655187.2017.1384982

Idziak, W., Majewski, J., & Zmyslony, P. (2015). Community participation in sustainable rural tourism experience creation: a long-term appraisal and lessons from a thematic villages project in Poland. *Journal of Sustainable Tourism, 23*(8–9), 1341–1362. https://doi.org/10.1080/09669582.2015.1019513

Koneya, M. (1978). Citizen participation is not community development. *Community Development Society. Journal, 9*(2), 23–29. https://doi.org/10.1080/15575330.1978.9987065

Mair, H., & Reid, D. G. (2007). Tourism and community development vs. tourism for community development: Conceptualizing planning as power, knowledge, and control. *Leisure/ Loisir, 31*(2), 403–425. https://doi.org/10.1080/14927713.2007.9651389

Matarrita-Cascante, D., & Brennan, M. A. (2012). Conceptualizing community development in the twenty-first century. *Community Development, 43*(3), 293–305. https://doi.org/10.1080/15575330.2011.593267

Mbaiwa, J. E., & Sakuze, L. K. (2009). Cultural tourism and livelihood diversification: The case of Gcwihaba Caves and XaiXai village in the Okavango Delta, Botswana. *Journal ofTourism and Cultural Change, 7*(1), 61–75. https://doi.org/10.1080/14766820902829551

Mokoena, L. G. (2019). Cultural tourism: cultural presentation at the Basotho cultural village, Free State, South Africa. *Journal of Tourism and Cultural Change, 0*(0), 1–21. https://doi.org/10.1080/14766825.2019.1609488

Moore, A. B. (2002). Community Development Practice: Theory in Action. *Journal of the Community Development Society, 33*(1), 20–32. https://doi.org/10.1080/15575330209490140

Prasad, P. (2012). The baba and the patrao: Negotiating localness in the tourist village. *Critical Arts, 26*(3), 353–374. https://doi.org/10.1080/02560046.2012.705461

Prince, S. (2017). Working towards sincere encounters in volunteer tourism: an ethnographic examination of key management issues at a Nordic eco-village. *Journal of Sustainable Tourism, 25*(11), 1617–1632. https://doi.org/10.1080/09669582.2017.1297450

Sklar, S. L., Autry, C. E., & Anderson, S. C. (2014). How park and recreation agencies engage in community development. *World Leisure Journal, 56*(4), 281–299. https://doi.org/10.1080/16078055.2014.958193

Tirasatayapitak, A., Chaiyasain, C., & Beeton, R. J. S. (2015). Can Hybrid Tourism be Sustainable? White Water Rafting in Songpraek Village, Thailand. *Asia Pacific Journal of Tourism Research, 20*(2), 210–222. https://doi.org/10.1080/10941665.2013.877045

Vasoo, S., & Tiong, T. N. (1992). Enhancing citizen participation community development. *Asia Pacific Journal of Social Work and Development, 2*(2), 2–6. https://doi.org/10.1080/21650993.1992.9755605

Veen, R. Van Der. (2003). Community development as citizen education. *International Journal of Lifelong Education, 22*(6), 580–596. https://doi.org/10.1080/0260137032000138149

Voth, D. E. (1975). Problems In Evaluating Community Development. *Journal of the Community Development Society, 6*(1), 147–162. https://doi.org/10.1080/15575330.1975.10878062

Xiaoping, Z., Zhu, H., & Deng, S. (2014). Institutional ethical analysis of resident perceptions of tourism in two Chinese villages. *Tourism Geographies, 16*(5), 785–798. https://doi.org/10.1080/14616688.2014.955875

Zhang, J. (2019). Tourism and environmental subjectivities in the Anthropocene: observations from Niru Village, Southwest China. *Journal of Sustainable Tourism, 27*(4), 488–502. https://doi.org/10.1080/09669582.2018.1435671

Promoting Creative Tourism: Current Issues in Tourism Research – Kusumah et al. (Eds)
© 2021 Taylor & Francis Group, London, ISBN 978-0-367-55862-8

Poverty alleviation in tourism destination: A new village-owned enterprise on the southern coast of West Java

A.W. Handaru, U. Suhud & S. Mukhtar
Universitas Negeri Jakarta, Jakarta, Indonesia

ABSTRACT: This paper aims to discover new small business clusters in the Southern Coast of West Java. Even though the location is gifted with a beautiful beach landscape and rich Sundanese culture, the majority of local people around the beach are poor. A new small business cluster managed by local people may help to reduce poverty and contributes to environmental sustainability. To accomplish the goal, fact and information are retrieved from an online interview session and an online survey. The result of the triangulation of data and observation has been the basis of a new business cluster. In Saganten village, the communities and the government can establish a village-owned enterprise that specifically gathers and fosters fishing groups. This village-owned enterprise also provides fishing boats to sail far. Besides, this village-owned enterprise can also establish a fish auction market and help promote seafood products made by housewives. The new business unit can help the housewives to produce the banana chips in a bigger scale. The main implication of the study is to establish a village-owned enterprise as soon as possible. The village-owned enterprise should apply cooperation, transparency, participation, accountability, and sustainable principles. Some challenges are waiting for the villages to develop the village-owned enterprises. First, improvement of business knowledge of local people; second, the intensive application of information technology and IT infrastructure development; and third, the development of high-quality tourism infrastructure in both locations.

Keywords: beach tourism, poverty alleviation, village-owned enterprise

1 INTRODUCTION

Poverty in a specific tourism destination is a challenging issue in tourism study. For almost three decades, research and discussion of this phenomenon are massive (Knight 2017; Mahadevan et al. 2017; Truong 2014). In some cases, many tourism destinations fail to contribute to local welfare. Some tourism destinations only provide low-level employment in this industry (Ashley et al. 2000). Critics of this situation continue. Some scholars argued that poverty in tourism destinations occurred because of wrong investment and financial plans (Das & Ghosh 2014). Durydiwka and Duda-Gromada (2014) explain that tourism activity should have a connection to the spatial development of tourist destinations. This fact is consistent with Handaru (2018), who mentions that supporting facilities and destination management have significant impact on pro-poor tourism benefits. Other scholars have discovered that there is a connection between poverty in tourism destinations and the extent of private investment (Dung et al. 2018; Morrison & King 2002; Page et al. 2017). They argued that investment is the best plan to start the business within a tourism destination and contributes to local welfare. Poverty in tourism destinations was also observed as a result of ineffective small business management within the location. Getz and Petersen (2005) argued that a small family business within tourism destinations significantly increases family income. In line with them, Thomas et al. (2011), Wanhill (2000), and Morrison et al. (2010) all highlighted the benefit of small and medium tourism enterprises to poverty alleviation. Some empirical studies reveal that small and medium enterprises (SMEs) in the tourism sector can help local people

DOI 10.1201/9781003095484-7

to get wealthier, not only in the countryside but also in the urban tourism area (Ibanescu et al. 2019; Mura & Kljucnikov 2018). Discussion on SMEs in the tourism business is often related to supporting infrastructures like hotel and accommodation, small shops and retails, and cultural event organizer (Domi 2019; Ivanovic et al. 2013; Vlahov 2014;). Unfortunately, the exploration of SMEs owned and managed by local people in a specific village is limited. Most 'SME's studies in tourism are focused on their best practice in terms of innovation and marketing plan. Therefore, the present research focuses on what type of SMEs that are suitable for a specific village in southern West Java. It will be the basis of the action plan taken by village people, local government, and investors. To achieve that, this research is structured as follows: first, the identification of condition in tourism destination; second, the literature review of tourism management and SMEs; third, the application of interview method to get important fact and information; and fourth, the elaboration of the analysis and discussion. The present study closes with the drawing of conclusions and advice for further research.

2 LITERATURE REVIEW

2.1 *Poverty alleviation and pro-poor tourism agenda*

One of the main agendas of the tourism business should consider poverty alleviation in a specific destination. Therefore, tourism activities should give benefits for all stakeholders and local people near the tourist destination. Poverty in tourism destinations still occurs because local people may not have sufficient knowledge and skills to support the tourism industry. Some scholars argued that poverty in specific tourism destinations is happening because the official government, business owner, and local people lack collaboration to create attractive tourism products (Higham & Hinch 2002; Jennings 2007; Saito 2017).

2.2 *One village one product strategy*

Triharini et al. (2014) discover that the one village one product (OVOP) policy was implemented in Indonesia since 2008. The OVOP strategy has successfully reduced poverty in the various village (Rachmawati et al. 2018). Some scholars argued that the OVOP strategy or village-owned SMEs comprise three foundations: local but global, self-reliance and creativity, and human resource development (Niculescu & Tataru 2015; Triharini et al. 2014). The main objective of the OVOP strategy is to reduce poverty in a specific location. Unfortunately, the implementation is challenging and requires collaboration among many institutions (Imani Khoshkhoo & Nadalipour 2016).

2.3 *Marine-based product and attraction*

Marine-based product has much potential to sell and market. Supply from the sea is relatively stable. In the case of the southern beach of West Java, the main product from the sea is seafood. Nevertheless, the marine-based product can be a non-consumption form. The beach also has the potential to market as a tourism destination. Some scholars mention that the beach attraction comprises sea, sand, and sun (3S). Mair and Jago (2010) and Hritz and Ross (2010) argued that the 3S elements must be supported by good tourism infrastructure. Some researchers noticed that 3S attraction can be applied in various forms. Some famous attractions in the beach area are festivals, recreational sports, international sports events, cultural events, beach resorts, and culinary tourism (Gibson 2012; Handaru et al. 2019; Haber & Reichel 2005).

2.4 *Analysis framework*

In order to succeed, Poverty alleviation in tourism destinations must consider three things. First, local people should have self-reliance and creativity to enter the tourism industry. Basic knowledge

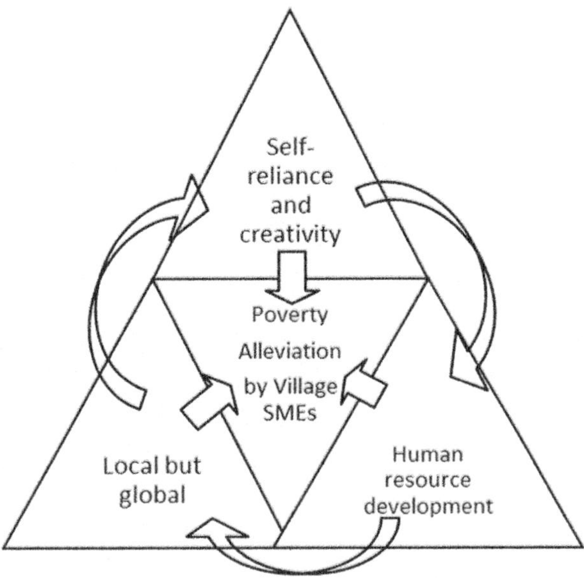

Figure 1. The model of village-owned SMEs and poverty alleviation program.

and skills in tourism are essential to achieve the goal. Second is the human resource development. The development should apply continuously and systematically with the various program. The tourism industry is dynamic and requires highly capable human resources. Third is creating the mindset and principle of local but global. Local people should be proud of their original product and service while maintaining its global quality. The concept of poverty alleviation in this study is presented below.

3 METHOD

The location of the study is in the Saganten Village, South Cianjur, West Java. It is 265 kilometers or six hours driving from Jakarta, and 140 kilometers or four hours driving from Bandung, the capital city of West Java province. Primary information on poverty conditions and SMEs' strategy in location draws from interviews. The interview sessions consist of the interviews with the head of the village, local people near the beach, government officials, and some small business owners in the location. The secondary data on the poverty level were collected from the local government official website. After that, field observation underwent by the assistance from the head of the village. Photos and videos were taken in the location to support qualitative analysis of new SMEs proposals. The triangulation of the new model of OVOP SMEs was done with the assistance of government officials in small focus group discussions.

4 RESULT AND DISCUSSION

4.1 *The Saganten village overview*

Based on observations, it was found that poverty in the village is real. This can be seen from the form of simple houses without fences. Most of the local community houses have the beach as their front yards. Surprisingly, most of the men work as temporary workers and farmers and only a few are fishermen, while most of the women are housewives. The quality of the main road is good, but transportation services are not available. There is low education level and people only rely on

Figure 2. Terrible facilities on the beach. July 9, 2020.

Figure 3. The beach is attractive for recreation and sports. July 10, 2020.

monthly income. The education level is only up to elementary school (SD) and junior high school (SMP). The population in the coastal village is around 151 people. The government has provided financial assistance for village development of nearly 65,000 USD in 2019. However, the funds were not spent for the development of SMEs. Most of the housewives produce "lantak" or banana chips and "bongsang" or small bamboo baskets. Coconut, nuts, and corn thrive in this area.

4.2 *The evaluation of basic and supporting tourism infrastructure*

Field observation discovered several important facts shown in the tables below.

4.3 *Recommendation of village-owned SMEs*

After considering the potential elements in the Saganten Village, the recommendation of village-owned SMEs is to establish the fish business unit and the agricultural processed products business unit. The physical and online promotion of all products can be applied.

Table 1. The evaluation of basic infrastructure in location.

Distance to airport	Bus station	Online transport/ taxi	Electricity	Clean water system	Internet and cellular network	Waste management system	Main road	Hospital/ clinic
243 Kilometers (Soekrno Hatta) or 290 kilometres (Kertajati)	Not available	Not available	Limited, not sufficient for the big tourism industry	Limited, not sufficient for the big tourism industry	Limited, not sufficient for the big tourism industry	Not available	Asphalt, average quality	Not available

Table 2. The assessment of tourism infrastructure in location.

Bank/ATM	Shop/Retail	Hotel	Rental	family recreation areas	Travel office	Restaurant	Market
Not available	Small kiosk available	Not available	Not available	Not available	Not available	Not available	Not available

Table 3. The evaluation of potential product to market.

Farm product	Fish	Handicraft	Food	Hotel	Sports event	Other sea-based product (medicine, cosmetic material)
At the moment, most of the men are farming coconut, corn, beans, and bananas.	The fishermen cannot go far because the boat is small. Only one small boat is available at the baskets. moment. The fish potential is unknown.	Some housewives produce bamboo specialties.	No restaurant at the moment. Traditional food does not have tourists. Some housewives produce banana chips.	The beach is attractive. Beach resort would be appealing for in infrastructures. Unfortunately, other infrastructures are limited.	Possible to be held here. But it will need a lot of refinements	Not available at the moment.

4.3.1 *The fish business unit*

The new village-owned SME should arrange a fishermen group. Then the group is given a modern boat to fish farther on the open sea. Therefore, the cooperation between village-owned SME and the bank is essential. Next, the new business unit also needs to establish a fish market. Besides, it also helps the fishermen to market the fish to other buyers like the restaurant, traditional market, and supermarket in the nearest city.

4.3.2 *The agricultural product business unit*

The new business unit should refine the banana chips distribution and promotion. It can arrange some groups of housewives to produce the banana chips in a bigger scale. The marketing of the chips should reach the nearest city. The product should have a permanent and attractive brand, representing the Saganten Village. The strategy also applies to coconut, corn, and beans.

5 CONCLUSION

The present study aims to discover the new OVOP SMEs to alleviate poverty in the Saganten Village. To achieve the goal, intensive interviews and close observation in locations are complete. After a triangulation of results, the present paper discovers some findings. The recommendation of SMEs in the Saganten Village is the fish business unit and the agricultural product business unit. Future research on poverty alleviation in tourism destinations may address the cultural aspect as a potential tourism attraction.

REFERENCES

Ashley, C., Boyd, C., & Goodwin, H. 2000. Pro-Poor Tourism: Putting Poverty at the Heart of the Tourism Agenda. *Natural Resources Perspective*, 51, 1–6.

Das, J. K., & Ghosh, S. 2014. An Analytical Study on Investment and Financing Scenario of Tourism Industry of West Bengal. *Globsyn Management Journal*, 8(1), 39–52.

Domi, S., Keco, R., Capelleras, J., & Mehmeti, G. 2019. Effects of Innovativeness and Innovation Behavior on Tourism Smes Performance: The Case of Albania. *Economics & Sociology*, 12(3), 67–85.

Dung, H. C., Thanh, P. T. K., Van Oanh, D., & Long, N. H. 2018. Local Government Involvement In Small Tourism Firms Investment: The Case Of Phu Tho Province, Vietnam. *Economics & Sociology*, 11(2), 97–111.

Durydiwka, M., & Duda-Gromada, K. 2014. Influence of Tourism on the Spatial Development of Seaside Resorts: Selected Aspects. *Tourism*, 24(1), 59–65

Getz, D., & Petersen, T. 2005. Growth and profit-oriented entrepreneurship among family business owners in the tourism and hospitality industry. *International journal of hospitality management*, 24(2), 219–242.

Gibson, H. J., Kaplanidou, K., & Kang, S. J. 2012. Small-scale event sport tourism: A case study in sustainable tourism. *Sport management review*, 15(2), 160–170.

Green, B. C., & Chalip, L. 1998. Sport tourism as the celebration of subculture. *Annals of Tourism Research*, 25(2), 275–291.

Haber, S., & Reichel, A. 2005. Identifying performance measures of small ventures—the case of the tourism industry. *Journal of small business management, 43*(3), 257–286.

Handaru, A. W. 2018. Pro-poor tourism: Findings from Bangka Island, Indonesia. *Academy of Strategic Management Journal,* 17(2), 1–10.

Handaru, A. W., Nindito, M., Mukhtar, S and Mardiyati, U. 2019. Beach Attraction: Upcoming Model in Bangka Island, Indonesia. *Academy of Strategic Management Journal*, 18(5).

Higham, J., & Hinch, T. 2002. Tourism, sport and seasons: the challenges and potential of overcoming seasonality in the sport and tourism sectors. *Tourism Management*, 23(2), 175–185.

Hritz, N., & Ross, C. 2010. The perceived impacts of sport tourism: An urban host community perspective. *Journal of Sport Management*, 24(2), 119–138.

Ibanescu, M., Chirita, G. M., Keen, C., & Cisneros, L. 2019. Canadian Smes In The Tourism Sector: A Taxonomy Of Owner-Managers. *Management International*, 23(4), 176–188,195,197,199.

Imani Khoshkhoo, M. H., & Nadalipour, Z. 2016. Tourism Smes And Organizational Learning In A Competitive Environment. *The Learning Organization*, 23(2), 184–200.

Ivanovic, S., Rot, E., & Pretula, M. 2013. Small Enterpreneurship In Hospitality: Croatian Experiences. *Utms Journal Of Economics*, 4(1), 27–35.

Jennings, G. 2007. Water-based tourism, sport, leisure, and recreation experiences. *Water-based tourism, sport, leisure, and recreation experiences* (pp. 19–38). Routledge.

Knight, D. W. 2017. An Institutional Analysis of Local Strategies for Enhancing Pro-Poor Tourism Outcomes in Cuzco, Peru. *Journal of Sustainable Tourism*, 1–18.

Mahadevan, R., Amir, H., & Nugroho, A. 2017. How Pro-Poor And Income Equitable are Tourism Taxation Policies in A Developing Country? Evidence from a Computable General Equilibrium Model. *Journal of Travel Research*, 56(3), 334–346.

Mair, J., & Jago, L. 2010. The development of a conceptual model of greening in the business events tourism sector. *Journal of Sustainable tourism*, 18(1), 77–94.

Morrison, A. J., & King, B. E. 2002. Small tourism businesses and e-commerce: Victorian tourism online. *Tourism and Hospitality Research*, 4(2), 104–115.

Morrison, A., Carlsen, J., & Weber, P. 2010. Small tourism business research change and evolution. *International Journal of Tourism Research*, 12(6), 739–749.
Mura, L., & Kljucnikov, A. 2018. Small Businesses In Rural Tourism And Agro Tourism: Study From Slovakia. *Economics & Sociology*, 11(3), 286–300.
Niculescu, G., & Tataru, D. 2015. Support Of Innovative Smes Focused On Cultural Tourism In Northern Oltenia. A Potential Innovative Project. *Analele Universitatii "Constantin Brancusi" Din Targu Jiu.Serie Litere Si Stiinte Sociale*, (1), 4–10.
Page, S. J., Hartwell, H., Johns, N., Fyall, A., Ladkin, A., & Hemingway, A. 2017. Case study: Wellness, tourism and small business development in a UK coastal resort: Public engagement in practice. *Tourism Management*, 60, 466–477.
Rachmawati, R., Hapsari, S. A., & Cita, A. M. 2018. Virtual Space Utilization In The Digital Smes Kampongs: Implementation Of Smart City And Region. *Human Geographies*, 12(1), 41–53.
Saito, H. 2017. The Role of Intermediaries in Community Capacity Building: Pro-Poor Tourism Perspective. *Academica Turistica-Tourism and Innovation Journal,* 10(1).
Thomas, R., Shaw, G., & Page, S. J. 2011. Understanding small firms in tourism: A perspective on research trends and challenges. *Tourism Management*, 32(5), 963–976.
Triharini, M., Larasati, D., & Susanto, R. 2014. Pendekatan One Village One Product (OVOP) untuk Mengembangkan Potensi Kerajinan Daerah. Studi Kasus: Kerajinan Gerabah di Kecamatan Plered, Kabupaten Purwakarta. *Journal of Visual Art and Design*, 6(1), 29–42.
Truong, V. D. 2014. Pro-Poor Tourism: Looking Backward as We Move Forward. *Tourism Planning & Development*, 11(2), 228–242
Vlahov, A. 2014. Strategic Alliances in The Function of Increasing Competitiveness of Small and Family-Run Hotels in Croatia. *Acta Turistica*, 26(1), 23–54.
Wanhill, S. 2000. Small and medium tourism enterprises. *Annals of Tourism Research*, 27(1), 132–147.

Promoting Creative Tourism: Current Issues in Tourism Research – Kusumah et al. (Eds)
© 2021 Taylor & Francis Group, London, ISBN 978-0-367-55862-8

Readiness and participation of local community for river-based tourism development in Sabak Awor, Muar, Johor

S.N.A. Zuhairi, N.H.A. Rahman, S.A. Abas, S.S.M. Sawari & S.A.A. Latif
International Islamic University Malaysia, Johor, Malaysia

R.M. Wirakusuma
Universitas Pendidikan Indonesia, Bandung, Indonesia

ABSTRACT: The research on river-based tourism in terms of community participation and readiness for river-based tourism development is still limited, especially in the Malaysian context. For instance, this study intends to explore the readiness and participation of the local community for river-based tourism development in Sabak Awor, Muar. A qualitative approach is adopted to this study on the readiness and community participation for river-based tourism development in Sabak Awor village, Muar through the semi-structured interview questions. A total of five residents have participated in the interview. The analyses have yielded into five main findings, namely the low level of knowledge influence local's readiness, tourism resources presents the creation of tourism activities in the area, low involvement in participation in decision-making process and planning, high involvement of the community in tourism operation and management, and benefits gained in terms of economic, infrastructure of the village, and personal development of the community. This study contributes to the planning and development of river-based tourism activities. The findings are beneficial for the management of the key players in the tourism industry to promote the potential of Sabak Awor village as a new tourism destination, specifically for river-based tourism.

Keywords: community readiness, community participation, river based tourism, Muar River, tourism planning and development

1 INTRODUCTION

The tourism industry is considered a pivotal source of economic growth in the world economy. Hence, local community support for tourism is necessary to ensure commercial, socio-cultural, political, and economic sustainability (Hanafiah et al. 2013). River-based tourism provides various tourism activities that are yet to be explored by the local's perspective. According to Prideaux and Cooper (2009), the main tourism resource is the river as it provides a great ambience for recreational opportunities, waterfront landscapes, transportation, and an essential source of living such as water. In Indonesia, the river is one of the recreational resources that has the potential for water sport (Rahmafitria et al. 2017). In addition, Shakiry's (2007) study has supported that River tourism is a worthy market because it could contribute to the local economies. For instance, along the riverbanks, potential jobs could be created especially in tourism activities such as cruise ships, pleasure boats, and parks which include other recreational facilities. It is possible that by implementing tourism projects and activities along the river would potentially give benefits to the locals around the area.

Based on the previous literature, studies that focused on river-based tourism are still limited even though there are a few studies had acknowledged the impact of the Muar river towards the local community (Abu Samah et al. 2013). Precisely, an interesting study of the relationship of the river in Muar that had been conducted by Abu Samah et al. (2013) has revealed there was only a

 DOI 10.1201/9781003095484-8

minority of Muar River's community that still has a strong relationship with Muar River in terms of fulfilling their basic needs and also their source of income. This has shown that the Muar river has less influence on the community. He further explained that the locals agreed to participate in river development despite the low influence of the river (Abu Samah et al. 2013). He noted that the main issue that needs to be addressed is the participation of communities for river-based tourism in rural areas (Abu Samah et al. 2013). A lot of concerns raised in terms of the cooperation from local leaders to be involved with the local communities of Sabak Awor, Muar for the river-based tourism development project. Therefore, this paper aims to explore the local community readiness and participation for river-based tourism development in Sabak Awor Village, Muar, Johor, Malaysia.

2 LITERATURE REVIEW

2.1 *Understanding the river based tourism*

According to Prideaux and Cooper (2009), the rivers are not only one of the major tourism resources, but it also provides spectacular attractions and natural settings, providing recreational opportunities, as well as the waterfront landscapes. It can also be a means of transportation. In this era, there is increasing use of the rivers as tourism resources either as an attraction, a transportation corridor, or as well as the main source of water.

In Malaysia, Nasarudin and Bahar (2013) have discovered that the Pergau River in Kelantan is an example of river-based tourism that own various potentials that can be developed. Among the activities that can aid in its potentials are including the adventure tourism activities. These activities include kayaking, water rafting, river cruise, and other related river-based activities (Kunjuraman et al. 2019). Moreover, the river which is famous for ecotourism activities in Malaysia that have been polluted include Sungai Langat, Sungai Segget, Batang Rajang and Sungai Melaka (Asyraf et al. 2013). Asyraf et al. (2013) stated that rivers are mainly used for transportation purposes in previous years, however, currently, the river is known as ecotourism destination as it offers wide-ranging ecotourism activities dedicated for those who love nature and extreme activities. The activities offered are cruising tours, watching the wildlife and observation, swimming, picnicking, and kayaking. There are two major issues related to river tourism. First, river pollution and bad water quality issues have led to less river-based tourism activities offered. Hence, the polluted rivers will affect people's health, nation economy, marine life, and especially to the environment (Asyraf et al. 2013). Secondly, the study has emphasized that the local community has less awareness of river protection and had contributed to the major reason for river pollution (Asyraf et al. 2013). Therefore, this study is conducted due to these two major issues concerning the river tourism planning and development in Malaysia, especially in the context of the Muar River.

According to Rahman et al. (2020), there are various opportunities for tourism products along the Muar River that include recreational, cultural heritage, community-based tourism, and entertainment. In addition, Muar is still lacking in terms of river-based tourism which this argument has been stated in a previous study by Hamdi et al. (2019) whereby there are two tourism attraction parks in Muar which are Tanjung Emas and Tanjung Ketapang. These two places offered great views and have a unique history, but these two locations are only being used for casual walkway and joggers' lane. In fact, Muar is blessed with its beautiful and wide river that offers a scenic view of mangrove species, and parts with majestic buildings only cater to a simple river cruise activity (Hamdi et al. 2019). Hence, this study will add the body of knowledge in exploring the potential tourism products through discovering the local community readiness and participation in river-based tourism product development for Muar River.

2.2 *Overview of Sabak Awor village in Muar, Johor*

Sabak Awor is recognized as the oldest village in Muar district, Johor, Malaysia. Sabak Awor Village is located along the riverside and can be accessed with vehicles. Based on the report of

Jawatan Kuasa Keselamatan dan Kemajuan Kampung (JKKK Report 2015), the population of Sabak Awor Village is 8890 people in 2015 which had been recorded by the previous headman, Tuan Hj Deris. There is a multiracial community in this village which are, Malay, Chinese, and Indian. There is six areas in Sabak Awor which are Kampung Sabak Awor Dalam, Kampung Sabak Awor Luar, Kampung Jalan Ismail, Kampung Jalan Batu, Kampung Jalan Bakariah, and Kampung Parit Setongkat (JKKK Report 2015).

River-based tourism development in Sabak Awor, Muar is still lacking. However, Sabak Awor village has the potential to be known as river-based tourism attraction because the river has a lot to offers. Sabak Awor village has various tourism activities ranging from agrotourism, nature tourism, and gastronomy tourism (JKKK Report 2015). However, the lack of awareness in developing the village as rural tourism development has contributed to the failure of tourism planning in that area. According to JKKK Report (2015), Sabak Awor has great natural resources to be offered to the tourists such as fireflies watching, fishing tours, boat tours, and the famous "Ikan Bakar" restaurants that managed to attract visitor from outside Muar area to come and experience its fresh seafood. The local community plays an important role to ensure that the tourism planning development in Sabak Awor is successful because the locals will gain benefits from these tourism activities. However, a lot of efforts need to be done in ensuring the sustainability of the tourism attractions and its resources in the village are well maintained. The local community together with stakeholders needs to work together in developing this place as a river-based tourism destination. In fact, Sabak Awor has a variety of amazing natural resources to offer and has huge potentials to be developed as a new tourism attraction. Therefore, this paper intends to explore the readiness and local community participation for river-based tourism development in Sabak Awor village, Muar, Johor, Malaysia.

2.3 *Community participation and readiness models in tourism*

This study has adopted the Thammajinda's (2013) Community Participation Model in which this model was developed from McIntosh, Goeldner, and Ritchie's (1986) study model. Based on these two models, there are three aspects that related to the local community participation. First, the involvement of the local community in decision making. Second, the involvement of the local community in tourism operation and third, the involvement of the local community in benefits to tourism.

Timothy's (2002) study has identified that public participation in decision-making is related to the community involvement in tourism development on deciding the goals and having their voice heard in the organisation and tourism development bodies. Similarly, McIntosh et al. (1986) emphasized that the power of relationship among tourism stakeholders is considerably related to community participation in tourism planning and decision-making.

The second component is community participation in tourism operation and management whereby this component can be explained into two aspects (Ashley & Roe 1998). The first aspect is related to tourism business activities, and the second aspect is related to the management of natural resources and other tourism resources (Ashley & Roe 1998). Based on the study by Ashley and Roe (1998), there is a passive involvement to full participation by the local community. Moreover, there is involvement at the individual's level to the involvement of all community members. For instance, passive participation is portrayed through the local communities as beneficiaries of tourism development and not as influencers. Meanwhile, active participation is regarded as the knowledge in the management of natural resources and tourism development that need to be empowered towards the local community in order for them to make decision in tourism development. Moreover, the full participation is to ensure the stability from the economic, political, and social benefits perspectives.

The third component is the local community participation in terms of readiness that can be explained in three aspects namely, the knowledge, readiness in commitment, and readiness in resources (Ayu 2014). However, people argued that it is important for relevant theoretical knowledge and practical experience by the local community in ensuring the success of plans in tourism development. The local community knowledge is important during the first phase of the planning process. This is supported by Simmons's (1994) study where the low knowledge of tourism amongst local people should be a reason to make the effort in order to encourage the local community to participate in the tourism development process. Kim et al.'s (2014) study in Houay Kaeng Village, Laos has highlighted that among the barrier for the local community's involvement in tourism operation and management is due to having a low level of knowledge and understanding about the tourism development process. This has resulted in low confidence in the local community to involve in tourism development.

In terms of readiness on commitment, a study by Mutambara (2018) concerning Bergville, Kwa-Zulu Natal South Africa as the case study has shown that the rural area is blessed with plenty of natural resources and the natural resources are seen as the tourism products that can boost the quality of life for the local community in this area. Unfortunately, there is no effort from the local authority and its community to develop these natural resources as tourism attractions (Mutambara 2018). This shows that the area is not ready to be developed as a tourism destination because the local community is not ready to fully utilize the tourism resources as the attractions at the place.

Moreover, according to Aref, Redzuan, and Gill (2009), readiness in terms of resources is referred to the ability of the local community to identify potential tourism resources and attractions within their communities as well as their support towards responsible tourism and community development. According to the Report of the World Commission on Environment and Development (2017), the local community has the opportunity of controlling the management of their resources and make decisions for the benefits of the present generation without comprising the organizational structure of the future generation (Donny, Sharon & Nor 2012). Hence, the local community readiness in terms of resources is a vital component to ensure that they will involve in the tourism planning and development process.

3 METHODOLOGY

This study is a qualitative research to explore the readiness and participation of the local community for river-based tourism development in Sabak Awor, Muar. A semi-structured interview question was developed and there were five residents in Sabak Awor Village have participated in this study. This study has implemented the purposeful sampling technique in which the respondents are selected based on different backgrounds and profiles namely, former Head of the village, fishmonger, housewife, unemployed youth, and small business operator. The data collection was conducted from the 17th of October 2019 until the 7th of November 2019.

The semi-structured interview questions were designed based on the components from Thammanjinda's (2013) study. The questions were divided into two sections in which, the first interview section was designed for the local leaders meanwhile, the second section was designed for the local community in Sabak Awor village. Next, the data that has been collected will be classified into themes and sub-themes. The themes are, the readiness of the local, forms of participation, and the impacts of tourism in Sabak Awor village, Muar. Then, the sub-themes are including tourism knowledge, tourism resources, decision-making process and planning, tourism operation and management, benefits, quality of life, employment opportunities, awareness on tourism, household income, accessibility, and infrastructure. The data was transcribed and tabulated to achieve the research objectives.

4 RESULTS AND DISCUSSION

Based on the data, five of the respondents emphasised the potential of the river along with Sabak Awor Village as one of the new tourism attractions in Muar for future tourism development.

4.1 *Aspect one: Readiness*

4.1.1 *Knowledge*

Skills and knowledge are essential components for the local community in order for them to be involved in tourism development. The findings revealed that a low level of knowledge affects the readiness of the local community. First, most of the respondents emphasised the lack of access to tourism information. For instance, one of the respondents from Sabak Awor village stated that ". . . I have heard about community-based eco-tourism development programs and other programs that involve the community in a rural area but I'm not sure what the program is all about. So, I decided not to participate in the end." Another resident has shared that there was no clear information given to them about the village planning. Based on Moscardo (2008), tourism knowledge is important in encouraging the local community to participate in tourism development. The finding from this study has shown that the low level of tourism knowledge has discouraged the local community to participate in tourism development in Sabak Awor village.

4.1.2 *Resources*

Tourism resources are the main initiators for tourists to visit an attraction. Attractive tourist resources with a good ambience and a unique value will initiate more tourist movement and consumption. The findings have shown that there are two tourism resources identified in the context of Sabak Awor village namely, natural resources and anthropogenic tourism resources.

First, a male respondent has listed several natural resources in his statement: ". . . Here, a variety of resources can be found such as a river, types of flora and fauna, and not to forget our symbolic fireflies." This finding also has been supported by another male respondent in which he stated that ". . . One thing that I like about this village is because we have fireflies that not many know. When talking about fireflies they (tourists) will only remember Kuala Selangor. Also, you can find jellyfish in the river while searching for fireflies during the night-time." The findings had reflected Aref et al's (2009) study in which the locals need to identify potential tourism resources and attraction within their areas.

4.1.3 *Commitment*

Based on the transcribed data, the local's commitment to developing Sabak Awor as a river-based tourism destination has been identified. As said by one of the locals, ". . . here already have fireflies tour but the price is high and not many can afford. Since we have locals who can drive their boat especially among the fishmonger why not we offer tourists our services at an affordable price. This can give benefits in terms of increasing their incomes for the community too." Hence, this finding has shown that the locals are committed and ready to upgrade Sabak Awor Village as the new tourism attraction. This finding has supported Ayu's (2014) study on the importance of the local community commitment in the planning and the strategy to develop the area into a tourism destination.

4.2 *Aspect two: Participation*

4.2.1 *Decision making process and planning*

The findings had identified the local community involvement in tourism development and planning in river-based tourism development. One of the respondents has stated that, ". . . I think because I am considered young and most of the committee members are from the older generation are making me and other youngsters feel that we are not eligible also does not have power in joining meetings until we reached their age." Similarly, the issue of eligibility and power is being highlighted by the female respondent which in her statement that ". . . although I have stated before that I don't have time to involve because of being a housewife, actually that is not the major factor. The major reason is because we as the youngsters feel that we are excluded in public engagement activities because our age, and we don't have a power to voice out just like those who are in the committee members." Moscardo (2015) had stressed the issue of exclusion of the locals to participate in tourism planning,

Table 1. Summary of the findings for the local community readiness and participation of River-based Tourism Development in Sabak Awor Muar, Johor.

Aspects	Items	Overall Findings
Readiness	Knowledge	• Low Level of Knowledge ➢ Lack of access to tourism information ➢ Level of English education
	Resources	• Ability to identify tourism resources ➢ Natural resources ➢ Anthropogenic tourism resources ie; folklores, gastronomic.
Participation	Decision-making Process & Planning	• Less Participation from the local community ➢ Tight working schedule and lack of time ➢ Inadequate Opportunities ➢ Insufficient chances ➢ Lack of eligibility and power ➢ The gap between locals and village leaders
	Tourism Operation and Management	• Involvement of the locals in tourism sector ➢ Accommodation ➢ Restaurant ➢ Ikan Bakar (Seafood) ➢ Convenience and Souvenir shops ➢ Boats and speed boat services ➢ Fireflies tour ➢ Fishing tour ➢ Mangrove tour ➢ Sell sea catches
	The benefits of tourism.	• Opportunity for Small Medium Enterprise Sectors (SMEs), • Improvement of the basic infrastructure in the village • Improvement of the social and interpersonal skills • Waste Management ➢ Recycling waste ➢ Management of the river

hence the findings from this study have shown that there is a limited involvement from a different group of the local community in the decision making process in Sabak Awor Village.

4.2.2 *Tourism operation and management*

Based on the data, the local community in Sabak Awor village is involved in the tourism sector either directly or indirectly in tourism development. An interview with the previous village headman, Tuan Hj Deris, has explained that the local community in Sabak Awor village is involved in tourism such as operating accommodation, restaurants, convenience and souvenir shops, boat services, fireflies' tours, and fishing tours. A respondent stated "... Locals here used the source of river as their incomes such as from fishing activities, the sea catches are sold to locals that run 'Ikan Bakar' (Seafood) restaurants." Hence, the finding had shown that the local community is indeed involved in the tourism operation and management for Sabak Awor Village.

4.2.3 *Involvement of locals towards the benefits of tourism*

The findings in relation to the local community involvement in sharing the tourism benefits were identified into three aspects which are; how the tourism industry has improved the Small Medium Enterprise Sectors (SMEs), the basic infrastructure in the village, and the local community skills.

The finding has shown that tourism offers small scale business opportunities compared to other industries. One of the respondents has emphasized on the type of Small Medium Enterprise business in Sabak Awor village in which "... A large number of the local community is involved in the first level of economic activities which are agriculture and fisheries. Most of the villagers are rubber tappers and fisherman, The most important type of businesses are selling fish, sell confectionaries (kuih muih), and open restaurant business such as Seafood restaurant." Interestingly, one respondent shared that his communication skills and tourism knowledge has been improved from the interaction with tourists. Hence, the findings from this study are similar to the study by Hanafiah et al. (2013), in which the local community will gain benefits from participating in tourism development planning.

4.3 *The summary of the findings*

Based on the discussion in previous sections, the local community readiness is viewed into two aspects namely, the knowledge and resources. Meanwhile, the local community participation is viewed into three aspects namely, the decision-making process and planning, tourism operation and management, and benefits of tourism towards the local community. Hence, Table 1 summarises the overall findings of this study.

5 CONCLUSION

This study has highlighted the fundamental approach in exploring the readiness of the local community in river-based tourism in Sabak Awor, Muar, Johor, Malaysia. For the local community readiness, there are two aspects that have been assessed. First, tourism knowledge and second is the tourism resources. Furthermore, this study also highlighted the local community participation in river-based tourism development in terms of the decision-making process and planning, tourism operation and management, and local community participation in the benefits of tourism. Based on the present study, a semi-structured interview had been designed to examine the factors of readiness and participation of the local community. This study has contributed to the body of knowledge in terms of readiness and local participation for river-based tourism development in Sabak Awor village, Muar, Johor, Malaysia. The methodological approach of qualitative data analysis based on transcribing the feedback and responds had a huge influence on this study to explore the readiness and participation of the local community for river-based tourism in Sabak Awor Village, Muar, Johor, Malaysia. Therefore, from this technique, this study had the ability to see community participation in a wider context. This study would have a considerable impact on future planning and development, especially for river-based attraction.

REFERENCES

Aref, F., Redzuan, M., & Gill, S. S. (2009). Community skill & knowledge for tourism development. *European Journal of Social Sciences,* 8(4), 665–671.

Ashley, C., & Roe, D. (1998). Enhancing Community Involvement in Wildlife Tourism: Issues and Challenges. *IIED Wildlife and Development Series,* 11, 1–38.

Asyraf, M. K. M., Nor 'aini, Y., & Suraiyati, R. (2013). Rivers, Lakes, and Swamps: Sustainable Approach towards Ecotourism, (July), 29–31.

Ayu, R. (2014). Sustainable Tourism on Semau Island: Ready or not? *Journal of Tourism & Hospitality*, 03(03). https://doi.org/10.4172/2167-0269.1000133

Donny, S & Mohd Nor, N. (2012). Community-based Tourism (CBT): Local Community Perceptions toward Social and Cultural Impacts

Hamdi, A. E., Maryati, M., & Shafiq Hamdin, M. (2019). The Potential of Nature Tourism at Muar and Tangkak Districts, Johor, Malaysia. *IOP Conference Series: Earth and Environmental Science*, 269(1). https://doi.org/10.1088/17551315/269/1/012008

Hanafiah, M. H., Jamaluddin, M. R., & Zulkifly, M. I. (2013). Local Community Attitude and Support towards Tourism Development in Tioman Island, Malaysia. *Procedia – Social and Behavioral Sciences*, 105, 792–800. https://doi.org/10.1016/j.sbspro.2013.11.082

Jawatan Kuasa Keselamatan dan Kemajuan Kampung (JKKK Report, 2015). Laporan JKKK Kampung Sabak Awor, Muar.

Kim, S., Park, E & Phandanouvong, T. (2014). Barriers to Local Residents' Participation in Community-Based Tourism: Lessons from Houay Kaeng Village in Laos. *SHS Web of Conferences*. 12. 10.1051.

Kunjuraman, V. (2019). River Tourism: A new tourism product for Malaysia, (January).

McIntosh, R. W., Goeldner, C. R., & Ritchie, J. R. B. (1995). *Tourism: Principles, Practices and Philosophies (7 ed.).* New York: John Wiley & Sons.

Moscardo, G. (2008). Building community capacity for tourism development: Conclusions. *Building Community Capacity for Tourism Development.* 172–179.

Mutambara, E. (2018). Critical Resources for the Development of Rural Tourism within the greater Bergville area of Kwa-Zulu Natal South Africa. Retrieved from https://www.ajhtl.com/uploads/7/1/6/3/7163688/article_2_vol_7_5__2018.pdf

Nasarudin, M. H. M., & Bahar, A. M. A. (2013). River Tourism: A Potential in Pergau River, Jeli, Kelantan. *Journal of Tourism, Hospitality and Sports,* 1(2009), 1–17.

Prideaux, B., & Cooper, M. (2009). River tourism. *River Tourism*. https://doi.org/10.9774/gleaf.9781315680088_15

Rahmafitria, F., Wirakusuma, R. M., & Riswandi, A. (2017). Development of Tourism Potential in Watersports Recreation, Santirah River, Pangandaran Regency, Indonesia. *People: International Journal of Social Sciences*, 3(1). Retrieved from https://grdspublishing.org/index.php/people/article/view/442

Rahman, N. H. A., Abas, S. A., Omar, S. R., & Jamaludin, M. I. (2020). Exploring the river-based tourism product for Muar River: A tourism opportunity spectrum (TOS) approach. IOP Conf. *Series: Earth and Environmental Science*, 447(1), 1–7.

Samah, B. A., Sulaiman, M., Shaffril, H. A. M., Hassan, M. S., Othman, M. S., Samah, A. A., & Ramli, S. A. (2011). Relationship to the River: The case of the Muar River community. *American Journal of Environmental Sciences*, 7(4), 362–369. https://doi.org/10.3844/ajessp.2011.362.369

Shakiry, A.S. (2007) River Tourism: Can Iraq Benefit from Europe's experience? Islamic Tourism Prospects, *Islamic Tourism*, Issue 8.

Simmons, D. G. (1994). Community Participation in Tourism Development. *Tourism Management*, 15(2), 98–108. Social. Rural., 40 (2000), pp. 481–496

Thammajinda, R. (2013). Community participation and social capital in tourism planning and management in a Thai context, 300.

Timothy, D. J. (2002). Tourism and Community Development Issues. In R. Sharpley & D. J. Telfer (Eds.), *Tourism and Development: Concepts and Issues*. England: Channel View Publications.

Promoting Creative Tourism: Current Issues in Tourism Research – Kusumah et al. (Eds)

Local community's cultural attitudes towards support for tourism development and conservation in archaeological heritage of the Lenggong Valley

S.A. Abas, M.A. Nur Afiqah, N.H.A. Rahman & S.S.M. Sawari
International Islamic University Malaysia, Johor, Malaysia

G.R. Nurazizah
Universitas Pendidikan Indonesia, Bandung, Indonesia

ABSTRACT: Overdevelopment not only affects the urban landscape but also not acceptable for any tourism region, particularly for the protected area. Local communities are terrified to lose their local authenticity due to the external influence and visitor's disturbance. This research endeavours to explore the cultural attitudes and local community participation in the Archaeological Heritage of the Lenggong Valley. Employing a quantitative research design, a total of 121 survey questionnaires were administered to the residents of Lenggong Valley. The samples were selected using a random sample technique. As a result, the data revealed that cultural attitudes in the host community are positive. Most of them strongly support the conservation of culture and heritage in the Lenggong Valley. Furthermore, the results also showed that the positive benefits outweigh the negative effects of tourism. Further research on the variations in cultural attitudes based on socio-demographic features that have helped confirm theories related to the social impact of tourism, such as the theory of social change, may further explore possible reasons for these communal attitudes.

Keywords: local cultural, attitudes, archaeological heritage, Lenggong Valley.

1 INTRODUCTION

Tourism is considered the fastest-growing sector, and some countries are making this sector a significant contributor to the national economy. For both developed and developing countries, tourism is seen as a vehicle for improving living standards and becoming a side-income, especially for those who are actively involved in providing packages and operating home-based accommodation services, agricultural products, and cultural experience for tourists. The United Nations World Tourism Organization (UNWTO) announced that the foreign tourist arrival increased by up to 4 per cent in 2019, which amounted to 1.5 billion. The year of 2019 was believed as a steady development of tourism, even though a little slower if compared to 2017 due to the global economic downturn and some problems related to xenophobia, dysfunctional politics, and trade tensions (UNWTO 2020). Refers to the upturn in tourism worldwide, tourism growth is becoming a tool of economic, socio-cultural, and environmental change in people's lives, which is more beneficial than others. As one of the tourism types, culture and heritage tourism are globally acquainted with the financial situation, and at the same time, can contribute to sustainable tourism development.

Heritage tourism is described as the development of a distinctive form of tourism that considers and respects other cultures (UNESCO 2005). The heritage values of tourism consist of words, history, and traditions that constitute a pivotal charm to guests. Those values can gain enormous financial advantages for the destinations. Respectively, sustainable management of heritage tourism seeks stability between the protection of heritage assets and the opportunities offered to improve the host community's finances. Heritage tourism encompasses components of living society, history,

DOI 10.1201/9781003095484-9

and the universal history of a destination to be reserved and preserved for the next human race. More specifically, the focus was to meet cultural situations such as scenes, graphic and contemporary arts, and exceptional lifestyles, traditions, values, and occasions. These components provide for potential growth stability, except for their effect on economic development (Khadar et al. 2014). Unlike general tourism, heritage tourism attracts Special Interest Travellers (S.I.T.), which interconnected with urban tourism (Chen & Chen 2009).

Heritage tourism provides a wide range of special interest activities, from visiting memorials to establishing authentic experiences that can serve the needs of cultural and heritage tourism (Caton & Santos 2007; Khadar et al. 2014). Moreover, the previous study stated that tourism support for residents depends vigorously on how the host community sees the effect of improving tourism (Rasoolimanesh et al. 2017). There are various reasons in understanding the attitudes of communities towards the impacts of tourism, for example, negative attitudes among communities may lead to disadvantages in the growth and sustainability of tourist destinations (Almeida-Garcia et al. 2016; Diedrich & Garcia 2009).

Similar with tourism in general, the heritage tourism development also has a positive impact on the lives of the host community, such as increasing monthly salaries, providing job opportunities, a positive change in living standards and open infrastructure, developing accessibility in recreation and entertainment facilities, and promoting and ensuring local culture (Rasoolimanesh et al. 2017; Deery et al. 2012). However, it can also have adverse effects on local communities by increasing living costs and property costs, upsurging overcapacity and traffic congestion, and rising the presence of crime and drugs (Rasoolimanesh et al. 2017).

Achievement in the tourism sector depends on local attractions and host services (Almeida-Garcia et al. 2016; Latkova & Vogt 2012). Thus, improvement of the World Heritage Site (W.H.S.) needs significant support from the locals to gain positive social and monetary effects while addressing the adverse impact. The achievement of heritage tourism measured by the sequence of fascinating heritage resources administration and effective policies that could be started by the locals. The relationship between local communities' attitudes and their involvement in heritage creation became a focus in various studies. Local participation in planning and improvement phases is also a significant necessity for management's improvement (Dyer et al. 2007). Local community attitudes and involvement become a buzzword among scholars and practitioners nowadays in a different tourism setting. However, there is little attention given to look into local community attitudes toward the World Heritage Site.

In 2019, Malaysia reported a definite increase of 2.7% in the first quarter, with a total of 6,696,230 tourists compared to 6,520,218 tourists in the same period in 2018. ASEAN countries have retained the most significant contributor to international tourist arrivals to Malaysia, mainly from the markets of Indonesia, Thailand, and Vietnam (Tourism Malaysia 2020). The National Heritage Department of Malaysia stated that one of the well-known tourist destinations, the Lenggong Valley, is Malaysia's fourth and gazetted as the world's 953rd World Heritage Site by UNESCO on 30th June 2012 (Khadar et al. 2014). The Lenggong Valley's Archeological Heritage inscribed based on criteria that bear unique or at least exceptional evidence of a cultural tradition and civilization.

Lenggong Valley is the famed oldest place of the Paleolithic, Neolithic, and Metal Ages prehistoric times in Southeast Asia. This place was also an early Homo sapiens migration route from Africa to Australia. In this outdoor museum setting, the legends, skeletons, cave drawings, and precious finds, such as gems, pottery, weapons, and tooling, have been found. The valley consists of photographic evidence at the open-air and cave sites along the Perak river that traverse all periods of hominid history outside of Africa.

The Lenggong Valley itself is located 100 miles north of Ipoh in the north part of Perak, Malaysia. This outdoor museum with an open-air setting has been an attraction for tourists until now. Since opened in 2005, under the management of the Center for Global Archeological Research (CGAR), the Lenggong Valley Museum has been kept for maintenance to support tourism growth. In the operational, local community support is needed in order to preserve and sustain the development of the heritage area as a tourist destination. Local community involvement in tourism development is an essential factor that contributes to economic growth and boosts the national Gross Domestic

Product (G.D.P.). The local community is seen as a crucial factor to attract tourists to visit tourism destinations because they can showcase a way of life to the tourists, mainly in terms of cultures, arts and crafts, languages, customs, and other authentic traditions.

However, based on the finding from the previous study, some of the local communities do not participate well in tourism development because some of them have negative perceptions toward development. The negative impacts of tourism activities, from the social and environmental perspectives, are the reason they are not willing to participate and support for tourism development in that destination. Moreover, the local community reluctant to involve due to no power in decisions making and controlling the development process. Many practices in community-based tourism (C.B.T.) unsuccessfully achieved due to the limited involvement of the community in the tourism development process (Tosun 2006). The governments and local authorities should include the local community in decision making, provide initiatives, and incentive for their involvement in tourism development. Thus, in order to prepare the local community to be part of the tourism operators, the governments and local authorities should provide them with training programs and funding. Subsequently, academic enthusiasm for community involvement with regards to W.H.S. preservation, conservation, and tourism development is noteworthy and significant (Rasoolimanesh 2017).

There are three main components involved in the exchange process of tourism development, namely economic, socio-cultural, and environmental impacts (Nunkoo & Ramkissoon 2010). Tourism development leads to cultural exchange and the establishment of recreational chances but also can bring to the rising of crime rates (Dyer et al. 2007). Nowadays, a lot of tourism development give high impacts to the tourism areas, natural resources, and society. For instance, any social activities might be occurred, such as gambling, vandalism, drugs and alcoholic activity, and any criminal activities. Excessive development of tourism is not suitable for any tourism area, especially for the needs of conservation areas. Not only that, but excessive development will also cause and changed the urban landscape. Local communities are terrified of losing their authenticity due to the external influence and disturbance from the visitors. Tourism is often considered responsible for environmental pollution, noise, and congestion (Latkova & Vogt 2012; Nunkoo & Ramkissoon 2010). Nevertheless, this research endeavours to explore the cultural attitudes and local community participation in the Archaeological Heritage of the Lenggong Valley.

2 METHODOLOGY

2.1 *Instruments and measures*

The questionnaire has been developed based on the previous literature review. It was initially developed in English and then translated into the Malay Language. There are two experts from the International Islamic University Malaysia assisted in verifying the translation to ensure both versions were comparable. The questionnaire consists of cultural attitudes (Rasoolimanesh et al. 2017), community involvement (Nicholas et al. 2009), and support for tourism development (Wang & Pfister 2008). A Likert-scale from 1 to 5 was provided, starting respectively from strongly disagree to agree strongly–all of the items tested in a different tourism setting and considered valid. Three local tourism academics invited to comment on the scales to ensure that the measurement could be applied to the local community in the Archaeological Heritage of the Lenggong Valley.

2.2 *Sampling*

To date, the population of the local community who lived around the Archaeological Heritage of the Lenggong Valley was 6,868. Based on the sample size of Krejcie and Morgan (1970), the relevant sample that suitable to collect for this survey is 361. However, according to Israel (1992), the researcher should consider the precision level of ±10%. Thus, an appropriate sample was 98 until 99 respondents for the 6000 number of populations.

The survey was conducted through self-administered questionnaires. Probability sampling was chosen because it is the most straightforward and standard sampling method integrated with survey-based research (Ritchie & Inkari 2006). Thus, it gave the local community of Archaeological Heritage of the Lenggong Valley an equal chance and opportunity to be included in the research. Questionnaires were distributed door-to-door for two weeks. This method has resulted in better response rates than other methods in past studies (Andereck & Nickerson 1997). This approach is one of the compelling techniques as the researchers were able to feel the real situation of data collection and had an informal interview session with the respondents. Also, the door to door approach is appropriate and valuable for the study. It reflected good quality results because the researchers were personally experiencing the data collection process. After two weeks of distributing the questionnaires, only 121 local communities responded to representing the population.

3 RESULTS AND DISCUSSION

This research assessed local communities' cultural attitudes toward tourism growth. The result shows that the cultural perspectives of the local community are positive. Most of them strongly support the conservation of culture and the heritage in the Lenggong Valley. 61.2% of respondents wish to preserve and protect its culture and lifestyle to gain more positive impacts for future generations. Local communities claimed that the Archaeological Heritage of the Lenggong Valley was established as a tourist destination, and the management should reflect and promote its culture and way of life. This finding is therefore supported by Social Exchange Theory (S.E.T.), where cultural pride within the local community might support tourism industry growth (Andereck et al., 2007; Jaafar et al. 2015). Moreover, the host community in Lenggong Valley revealed their optimistic attitudes towards tourism. Hence, the objective of this research to evaluate the cultural attitudes of the Lenggong Valley Archaeological Heritage Site among local communities and support the potential growth of tourism is achieved.

On the other hand, 43% of the local communities participate in the Archaeological Heritage of the Lenggong Valley's tourism growth. In the meantime, the other local group also contributes to the management and the growth process. Local community participation in the Archaeological Heritage of the Lenggong Valley does not live up to standards. This result showed that local communities participating in tourism growth are equal to those not participating in the tourism industry at all. The previous approaches stated that local communities make excessive contributions to development projects without decision-making or influencing the development process in the sense of coercive community engagement (Rasoolimanesh et al. 2016; Tosun 2006).

Moreover, the whole process of tourism development depends on the government and power of the private sector (Rasoolimanesh et al. 2016; Zhang et al. 2013). Typically, local authorities not easily include the local community in tourism planning because they are reluctant to involve, although as the host community, they should be the one who has knowledge of tourism product and services within the area. Based on the Tourism Area Life Cycle by Butler (1980), the tourism sector in the Archaeological Heritage of the Lenggong Valley still in the developing phase. Most of them do not recognize economic benefits, including months of income in the tourism industry. Therefore, the local community relies on its expectations that the proposals are provided or funded before making decisions to accept the outcome (Bronfman et al. 2009).

The results of this research have also revealed positive support for the development of tourism in the local community. Besides, 58.7% of local communities claimed that the tourism industry in the city and the Lenggong Valley should be actively promoted. They intended to promote tourism and realized that the tourism industry becomes an essential part of the community and support new tourist establishments that attract more tourism in the Lenggong Valley. On the other hand, 83.5% of the respondents agreed that the government should support tourism promotion in the Archaeological Heritage of the Lenggong Valley and developed long-term environmental planning, which could excessively influence the tourism business. Moreover, most of the local community, which represent 91.7% of respondents concurred that the effect of tourism activities indirectly

Table 1. A result from cultural attitudes, local involvement, support for future tourism development.

Variables/Items	Percentage (100%) N=121				
	Strongly Disagree	**Disagree**	**Neutral**	**Agree**	**Strongly Agree**
Cultural Attitudes					
Cultural heritage should be preserved for the future generation	1.7	0	0.8	36.4	**61.2**
Culture and heritage tourism should be protected and promoted	1.7	0	1.7	38	**58.7**
Culture and heritage products are essential to be preserved and conserve	1.7	0	2.5	41.3	**54.5**
Local Community Involvement					
Involved in the management	12.4	8.3	19.8	20.7	**38.8**
Involved in the process of planning and development	11.6	11.6	14.9	19	**43**
Have a chance to gives some ideas in the process of development	18.2	24.8	16.5	14.9	**25.6**
Have the opportunity to involved in the conservation project	23.1	20.7	14	14	**28.1**
Support For Future Tourism And Conservation Development					
The tourism industry should be actively encouraged	0.8	1.7	2.5	39.7	**55.4**
Would support the tourism industry as part of the critical sector	0.8	1.7	2.5	45.5	**49.5**
Would support tourism facilities and services to attract more tourist to come	0.8	0.8	5.8	36.4	**56.2**
Tourism activities and conservation project should be actively encouraged	0.8	0	4.1	**47.9**	47.1
The government play essential roles in the support and promote tourism	0.8	3.3	12.4	**47.1**	36.4
Long-term planning on the environment aspect would help the sustainability	0.8	1.7	11.6	**44.6**	41.3
The impact of tourism development was increased the outdoor recreation opportunity among local people	0.8	0.8	6.6	**50.4**	41.3
It is essential to manage the growth of tourism sectors	0.8	0	3.3	43.8	**52**
Would support the tourism sector plays a significant role in the local community	2.5	1.7	6.6	**47.9**	41.3
The future of the Archaeological Heritage of the Lenggong Valley as a sustainable culture and heritage tourism destination is achievable	2.5	0.8	9.1	29.8	**57.9**

could improve the quality of outdoor recreation opportunities. They also felt that the management of growth tourism is essential in the Archaeological Heritage of the Lenggong Valley and that the tourism sector will continue to play a significant role in the community economy.

Furthermore, the results also showed that the positive benefits of tourism in the local community outweigh the adverse effects. The findings can be related to Social Exchange Theory (S.E.T.) research, which is likely to help the local community grow tourism if they believe that definite advantages are more than the drawback (Lee 2013; Stylidis et al. 2014). Therefore, the local community supports Lenggong Valley as a sustainable tourist attraction in the future. This finding is consistent with many previous studies that local communities have continued to promote engagement and protection of the World Heritage Site as well as the growth of tourism in order to increase positive effects and minimize negative impacts (Rasoolimanesh et al. 2017). Thus, in the future, the Archaeological Heritage of the Lenggong Valley will be a tourist destination in Perak because of not only culture and heritage but also ecotourism, as reported in December 2014 by Penang Monthly.

4 CONCLUSION

To ensure the high levels of support for tourism growth and the development of cultural tourism, the government, local authority, and tourism planners should recognize that local communities are not homogeneous. Most of the tourism benefits from heritage and culture niche cannot be seen and earned. Conservation is required where there are benefits to demonstrate either directly or indirectly as a result of the growth of tourism or cultural tourism. Limited advantages, benefits, and techniques are essential to spread tourism's advantages to the general population. Besides, if the local community was unable to gain the benefit, this will lead to inactive participation in tourism as instruments for economic or social growth. More advanced research that considers the community heterogeneous is needed and, more importantly, begins to understand local community positions and, why and how they arise.

Further research on the variations in cultural attitudes based on socio-demographic features that have helped confirm theories related to the social impact of tourism, such as the theory of social change, may further explore possible reasons for these public attitudes. The results highlight specific reasons for cultural attitudes towards the distribution of tourism benefits, which can more than any other variables relate to levels of income and proximity to the tourist center. These residents are more positive about the benefits of tourism. Still, they do not always consider the adverse effects which local community in Archaeological Heritage of the Lenggong Valley itself do not seem to experience at a higher level. Ironically, as compared to earlier research, there were no statistical variations based on jobs in the tourism industry and impact statements. The statistical differences in the cultural or creative sectors, however, were only evident concerning half the perceived impacts on cultural tourism.

Fundamentally, the findings formed a positive and essential connection between the three critical factors of cultural attitude, community engagement, and personal benefits, as well as support for future growth from local communities. The first hypothesis was supported that the positive attitude of the local community leads to more support for future development in tourism. Nevertheless, the lack of local involvement does not contribute negatively to their promotion of future growth itself. While only some of the local community is involved in tourism, they still believe and support the potential growth of tourism that can indirectly contribute to cultural heritage protection and conservation.

In particular, all results relevant to promoting the sustainable growth and restoration of the W.H.S. in Lenggong Valley can be seen. The results and results of this research have specific practical implications for the local community and authorities responsible for the administration of the Archeological Patrimony of Lenggong Valley. Local authorities will also make concerted and transparent attempts to develop inclusive urban communities. As the findings show, however, local communities still have a positive outlook towards tourism, even though they do not see many economic benefits due to a low level of participation in the tourism industry. Therefore, they continue to help the growth and protection of tourism in Archeological Heritage sites in the Lenggong Valley to detect the positive effects of tourism.

REFERENCES

Almeida-García, F., Peláez-Fernández, M.Á., Balbuena-Vázquez, A., & Cortés-Macias, R. 2016. Residents' perceptions of tourism development in Benalmádena (Spain). Tourism Management, 54, 259–274. https://doi.org/10.1016/j.tourman.2015.11.007

Andereck, K.L., Valentine, K.M., Knopf, R.C., & Vogt, C.A. 2005. Residents' perceptions of community tourism impacts. Annals of Tourism Research, 32(4),1056–1076. Retrieved from http://dx.doi.org/10.1016/j.annals.2005.03.001

Andereck, K., Valentine, K., Vogt, C.A., & Knopf, R. 2007. A cross-cultural analysis of tourism and quality of life perceptions. Journal of Sustainable Tourism, 15(5), 483–502.

Andereck, K. L., and N. P. Nickerson (1997). "Community Tourism Attitude Assessment at the Local Level." In The Evolution of Tourism: Adapting to Change, Proceedings of the 28th Annual Travel and Tourism Research Association Conference. Lexington, KY: Travel and Tourism Research Association, pp. 86–100.

Bronfman, N. C., Vazquez, E. L., & Dorantes, G. (2009). An empirical study for the direct and indirect links between trust in regulatory institutions and acceptability of hazards. Safety Science, 47, 686–692.

Butler, R. (1980). "The Concept of a Tourist Area Life Cycle of Evolution: Implications for Management of Resources." Canadian Geographer, 19 (1): 5–12.

Caton, K., & Santos, C. A. (2007). Heritage tourism on route 66: Deconstructing nostalgia. Journal of Travel Research, 45(4), 371–386. https://doi.org/10.1177/0047287507299572

Chen, C.F. & Chen, F.S. 2009. Experience Quality, Perceived Value, Satisfaction, And Behavioral Intentions for Heritage Tourist. Tourism Management, 31, 29– 35.

Diedrich, A. & García, E. 2009. Local perceptions of tourism as indicators of destination decline. Tourism Management, 30, 512–521.

Deery, M., Jago, L., & Fredline, L. 2012. Rethinking social impacts of tourism research: A new research agenda. Tourism Management, 33(1), 64_73.

Dyer, P., Gursoy, D., Sharma, B., & Carter, J. (2007). Structural modelling of resident perceptions of tourism and associated development on the Sunshine Coast, Australia. Tourism Management, 28(2), 409–422.

Israel, Glenn D. 1992. Sampling The Evidence Of Extension Program Impact. Program Evaluation and Organizational Development, IFAS, University of Florida. PEOD-5. October.

Jaafar, M., Rasoolimanesh, S., & Ismail, S. (2015). Perceived socio-cultural impacts of tourism and community participation: A case study of Langkawi Island. Tourism and Hospitality Research. doi:10.1177/1467358415610373

Khadar, N.Z.A., Jaafar, M., & Mohamad, D. 2014. Community Involvement in Tourism Development: A Case Study of Lenggong Valley World Heritage Site. 4th International Conference on Tourism Research, 4(June 2012), 3–9.

Krejcie, R., Morgan, D., 1970. Determining sample size for research activities. Educational and Psychological Measurement 30, 607–610.

Látková, P., & Vogt, C.A. 2012. Residents' attitudes toward existing and future tourism development in rural communities. Journal of Travel Research, 51(1), 50–67. https://doi.org/10.1177/0047287510394193

Latkova P & Vogt CA. (2012). Residents' Attitudes toward Existing and Future Tourism Development in Rural Communities. *Journal of Travel Research*. 51 (1), 50–67doi: https://doi.org/10.1177/0047287510394193

Lee, T.H. (2013). Influence analysis of community resident support for sustainable tourism development. Tourism Management, 34, 37- 46.

Nicholas, L., Thapa, B., & Ko, Y. (2009). Residents' perspectives of a World Heritage Site: The pitons management area, St.Lucia. Annals of Tourism Research, 36(3), 390–412.

Nunkoo, R. 2015. Tourism development and trust in local government. Tourism Management, 46, 623–634. https://doi.org/10.1016/j.tourman.2014.08.016

Nunkoo, R., & Ramkissoon, H. 2010. Residents' satisfaction with community attributes and support for tourism. Journal of Hospitality & Tourism Research, 35(2), 171–190.

Rasoolimanesh, S.M. & Jaafar, M. 2016. Sustainable tourism development and residents' perceptions of World Heritage Site destinations. Asia Pacific Journal Of Tourism Research. Retrieved fromhttp://dx.doi.org/10.1080/10941665.2016.1175491

Rasoolimanesh, S.M., Jaafar, M., Ahmad, A.G. & Barghi, R. 2017. Community participation in World Heritage Site conservation and tourism development, 58. https://doi.org/10.1016/j.tourman.2016.10.016

Rasoolimanesh, S.M., Jaafar, M., Kock, N., & Ahmad, A.G. 2017. The effects of community factors on residents' perceptions toward World Heritage Site inscription and sustainable tourism development. Journal of Sustainable Tourism, 25(2), 198–216. https://doi.org/10.1080/09669582.2016.1195836

Ritchie BW & Inkari M. (2006). Host community attitudes toward tourism and cultural tourism development: the case of the Lewes District, southern England. International Journal of Tourism Research. 8 (1). Doi: https://doi.org/10.1002/jtr.545.

Stylidis, D., Biran, A., Sit, J., & Szivas, E. M. (2014). Residents' support for tourism development: The role of residents' place image and perceived tourism impacts. Tourism Management, 45, 260–274. https://doi.org/10.1016/j.tourman.2014.05.006

Tosun, C. (2006). Expected nature of community participation in tourism development. Tourism Management, 27(3), 493–504.
Tourism Malaysia (2020). Retrieved from https://www.tourism.gov.my/statistics
UNESCO.(2005). "CulturalTourism". Retrieved from http://portal.unesco.org/culture/en/ev.phpurl_id=114 08&url_do=do_printpage &url_section=201.html, Retrieved 01st March 2008.
World Tourism Organization Tourism Market Trends UNWTO. (2020). UNWTO Tourism Highlights, 2017 Edition. Retrieved from mkt.unwto.org/publication/unwto-tourism-highlights
Wang, Y., & Pfister, R. E. (2008). Residents' attitudes toward tourism and perceived personal benefits in a rural community. Journal of Travel Research, 47(1), 84–93.
Zhang, Y., Cole, S.T., & Chancellor, C.H. 2013. Residents' preferences for involvement in tourism development and influences from individual profiles. Tourism Planning & Development Retrieved from http://dx.doi.org/10.1080/21568316.2012.747984.

Promoting Creative Tourism: Current Issues in Tourism Research – Kusumah et al. (Eds)
© 2021 Taylor & Francis Group, London, ISBN 978-0-367-55862-8

Rural tourism in Jakarta (Ecotourism in Pesanggrahan Riverbank)

E. Maryani, Amin, N. Supriatna & M. Ruhimat
Universitas Pendidikan Indonesia, Bandung, Indonesia

ABSTRACT: As a metropolitan city, Jakarta has various environmental problems started from the air pollution, water pollution, up to flood. However, there is an astonishing thing while there is a tourism village on bank of Pesanggrahan River. This study aims to examine the motivation behind the development of this tourism village. It also attempts to analyze the supports done by the society to the development of this village. This study used qualitative design with phenomenology method. In-depth interviews were conducted to 11 respondents. The findings showed that environmental problems became the main prompt in the development of the village. Economy and culture were the motive to build society's environmental awareness, in which the focus was on water and riverbank conservation. Regarding the society's participation, it was promoted by demonstration, example, and informal socialization. To advance the development of tourism village on the Pesangrahan riverbank, regional government support is needed. The government can do socialization through formal school and make a schedule for tourism activity with conservation nuance for the visitors.

Keywords: tourism village, ecotourism

1 INTRODUCTION

Urban areas are well known for frenzied situations, noise, and high buildings. The busyness of the people in this area demands the need for recreation. Recreation is a leisure time that a person has after his primary needs are met (Maryani 2019). It is not only for releasing a fatigue after daily activity, but also for refreshing before re-creating. Outdoor recreation or rural areas sometimes become a choice. Rural areas have their own attraction as a tourist attraction, especially for urban communities (Kastenholz & Lima 2011). It has a variety of attractions and activities, which focus on agriculture, natural environment, and beautiful atmosphere (Lane 2009).

Excessive exploitation on natural resources causes various environmental damage. Flooding is one disaster in which the mitigation requires community participation. Goleman (2009) explains the need for ecological intelligence (knowledge, attitude, and skill) in protecting the environment. Jakarta is a metropolitan city which is often stricken by flood. Thus, environmental awareness for its society needs to be developed. One of the efforts is through making a community which cares about environment (Jayadi et al. 2014).

One of the environmental lovers' community in Jakarta is Kelompok Tani Lingkungan Hidup (KTLH) Sangga Buana, which is led by Babeh Idin. This community has a unique characteristic as it is formed in the middle of metropolitan life with heterogeneous culture and dense population (Maryani 2000). KTLH Sangga Buanasucceeded in growing urban forest along the Pesanggrahan River in Jakarta. It also received many awards from several institutions and events. Because of this achievement, KLTH Sangga Buana has become a tourism village in Jakarta.

The success of KLTH Sangga Buana in building tourism village finally spurred the researchers to investigate the motivation behind the establishment of the tourism village on the Pesangrahan riverbank and the society's participation in developing this village.

DOI 10.1201/9781003095484-10

2 RESEARCH METHOD

This research used qualitative approach with phenomenology method. Cresswell (2009) affirms that qualitative research attempts to explore and understand the meaning of individuals or groups ascribing to a social or human problem. The research process involves emerging questions. The data are typically collected from participants and analyzed inductively from particular to general themes. Meanwhile, phenomenology research is a method which identifies the essence of human experience as a phenomenon as described by participants. It involves studying the small number of subjects through extensive and prolonged engagement to develop patterns and meaning. It examines social situations through three elements (place, actor, and activity) that interact synergistically (Sugiyono 2008).

This research was conducted in tourism village on the Pesangrahan riverbank. In this village, KTLH Sangga Buana developed 40 hectares of urban forest and 80 hectares of agriculture and livestock land. The respondents were 11 people doing activities in tourism village. The respondents were chosen by using snowball sampling. They are a chairman of KTHL Sangga Buana, farmer, breeder, waste management, society, and tourists.

The data were collected through interviews, observation, participation, and documentation. It was then reduced, displayed, and concluded. Data reduction aims to sort out the important data which could be used to answer the research objective. Data display was related to the presentation of the data in the form of figures, pictures, tables, or graph. Meanwhile, data conclusion was carried out through two stages. If the data was considered inadequate, retaking the data in the field was done to fill the missing point.

3 RESULT AND DISCUSSION

3.1 *Driving factor for the development of tourism village*

KTLH Sangga Buana is an environmental lover community that is formed as a response to the increase of environmental damage in Jakarta, especially in Pesangrahan River. Pesangrahan River is one of the rivers that causes flood during a rainy season in Jakarta. Many people litter their garbage into rivers. As a result, a river is so dirt and smells. Most of the people also have a lack of concern for the riverbank. They build temporary houses along a riverbank. This finally motivates H. Chaerudin, known as Babeh/Abah, to educate the society around the river about environment care. Abah often socializes the importance of protecting the environment to the society. He educates the people about clean living, reforestation, waste segregation, water saving, and other environmental behavior. Over time, an environmental lover community is formed with the name Sangga Buana, and Abah is a leader of community.

The leader of KLTH Sangga Buana, on the above case, realizes that the awareness of environmental preservation should be fitted in a human life. In live networks, every human basically has a role, function, and position that is interrelated with the environment (Ruhimat 2019). They also should take responsibility for their nature by preserving it through creative work (Muhtarom Ilyas 2008). Environmental damage is basically the extermination of life. Viewing from their responsibility, humans should stay in a front line in maintaining nature (transcendent) by using two kind of approaches: human-centered and life-centered. Human-centered focuses on respecting each other (social animal), while life-centered deals with the obligation in preserving nature as life survival (Maryani 2018).

KTLH Sangga Buana also realizes that environmental problems do not stand alone. It is connected with the other problems, such as education, economy, and socio-culture. Therefore, it uses an integrative approach in solving the environmental damage in Pesangrahan River. Educational aspect is developed informally by socializing the importance of preserving the environment. Economics aspect is advanced by meeting the needs of life through suitable livelihoods. Economics needs often make people ignore environmental protection aspect. Natural resources are viewed in

Figure 1. Forest city and activity centre of Sangga Buana Tourism Village.

economic sense context and have not yet led to ecological and sustainable sense (Santoso 1999). KLTH Sangga Buana then develops the Pesangrahan riverbank as "working area" by farming and raising livestock as an income source for the society around the area. In cultural aspects, KLTH Sangga Buana not only familiarizes clean life such as reforestation, disposing trash in its place, processing and sorting waste, but also revives the ancestors' proverb in the form of local wisdom. For example: 1) This nature is not a legacy, but the entrustment for our children and grandchildren. This proverb illustrates that we have to inherit the good things for the next generation. They should not over-exploit a nature. Thus, the next generation can enjoy the well-maintained natural resources. 2) A drop of dew in a dessert. This means that we must do something to improve the environment by providing the advantages for environmental sustainability. 3) River is a source of life. This proverb implies that water river is very useful for human life. It can be used for fulfilling the daily needs, washing vehicles, irrigation, fisheries, and tourism object. The implementation of this proverb on life can be seen from annual ceremony named Sedekah Air dan Bumi. In this ceremony, the community cleans the trash from the river, sows the fish seed into the river, and holds the art or sport event such as pencak silat (a traditional marital art). All of these proverbs become a local wisdom that becomes a guidance for society to behave (Maryani 2015).

3.2 *The aim of tourism village development*

Building community's ecological intelligence is the main objective behind the development of tourism village. Goleman (2009) states that ecological intelligence is the ability to understand how nature works. There are three dimensions of ecological intelligence: 1) Geosphere, a dimension that is concerned with the awareness of land, air, water, and climate condition. 2) Biosphere, a dimension related to humans, other species, and plant life. 3) Sociosphere, a dimension that focuses on work environment, living environment, etc.

To achieve the above objective, KLTH Sangga Buana performs four approaches: kinship, integrative, participatory, and modeling. Kinship becomes the main approach to achieve goals. It is reflected in togetherness, mutual cooperation, and responsibility in utilizing the environment. In an integrative approach, KLTH Sangga Buana integrates tourism village with educational, economy, and socio-cultural centers. Sangga Buana becomes non-formal education in preserving the environment. However, at the same time, it is a learning source for formal education as there are many students from all levels of education that come to the village.

KLTH Sangga Buana manages 120-hectare land, in which 40 hectares of land is an urban forest and 80 hectares of land is agriculture and animal husbandry. The agriculture includes rice and vegetables, while livestock consists of fishery, ducks, and chickens. Two hundred people participate in managing agriculture and livestock. The distribution of land is divided by the leader. It is distributed unevenly based on location, land fertility, and the needs of the community. Fisheries and livestock are carried out collectively with mutual cooperation. In this case, the members of KLTH Sangga Buana attempt to give an example to the people who are not aware and have not

Figure 2. Fisheries and livestock activities on the riverbank.

implemented environmental preservation. This makes other people to be more sympathetic and eventually join with the community. Modeling through example, however, has an important role in the socialization and internalization of environmental.

The forest managed by KLTH Sangga Buana can be enjoyed by everyone. People can do recreation, education, or take forest products. Figure 2 shows the pond managed by KLTH Sangga Buana in which the fish seed is bought from visitors' fees. In the harvest time, some of the products will be sold, and the others are shared to the society around the area.

3.3 *Society's participation in the development of tourism village*

Society's participation is very important because tourism village is developed by the society, for the society, and from society. Society's participation can be seen from the idea to make a tourism village, the energy to develop the village, and the finance to build a village. Even though the prior idea of building a village came from H. Chaerudin, known as Abah/Babeh Iddin, there are many people who helped him. Abah Iddin is a pioneer. He is an agent of change who developed the Pesanggrahan riverbank to be urban forest and productive agriculture. Through intensive socialization and modeling, there are so many people who are interested in developing the village. Nowadays, there are more than 200 people involved in handling fisheries and livestock, managing waste, conserving forest, and becoming informal agents for the society.

As informal agent, the community member gradually delivers the idea about environmental preservation. They attract societies sympathetic and prompt them to participate actively in every environmental activity. They also give the explanation and examples of positive results from keeping the clean environment. They also make some activities in which the fund is sourced from individuals who have greatest awareness of the environment. The activities include: 1) picking up the waste from the Pesangrahan river. The waste is then collected in several points to be disposed between organic and inorganics. Organic waste is used as fertilizer, while inorganic waste is sold or utilized into various uses (recycling). 2) Maintaining Sangga Buana urban forest. The community monitors the forest from various activities that can threaten its preservation. The forest can be visited by many people across the area, so that monitoring needs to be done. 3) Participating in livestock or fisheries activities, such as sowing seeds, feeding, cleaning ponds, etc. The fish can be caught, but it should not be poisoned. 4) Participating in "Sedekah Bumi" activity, especially in the form of reforestation. 5) Participating in maintaining arable land. 6) Participating in various activities organized by the Sangga Buana community, such as receiving a visit, socializing program, counseling, and developing arts or sports.

Environmental quality can be interpreted as environmental conditions in relation to quality of life. The higher the quality of life in a particular environment, the higher the quality of environment. The quality of life, however, depends on the degree of meeting the basic needs. The quality of the environment, then, can be interpreted as the degree of fulfilment of the basic needs in certain

Figure 3. Participantion in Tourism Village.

environmental conditions. The higher the degree of meeting the basic needs, the higher the quality of the environment (Soemarwoto 2004).

3.4 *The attraction in tourism village*

Attraction is everything that can encourage someone to a destination. Attraction can be in the form of natural objects, people's lives, and various objects made by humans (Maryani 2019). The main attraction at KTLH Sangga Buana is the Pesanggrahan River, which flows in the city of Jakarta, one of the downstream rivers flowing in South Jakarta. Downstream becomes an accumulation of water that flows in the upstream and middle sections. The flow of water is slow. It forms a wide and gentle river bank (20 to 30 meters) on the left and right of the river. A tourism activity that can be carried out in the downstream is fishing. The Pesanggrahan riverbank covers 40 hectares that is used for forest conservation. Various types of wood and bamboo flourish. The forest is a haven for various types of birds. Thus, the visitors can see not only various trees and wood, but also various species of birds.

The area of riverbanks used for farming is 80 hectares. Activities that can be carried out on the farm are planting and harvesting rice. In the fish pond, visitors can fish and take fish with a tool (at a certain time). In the field, visitors can play various games, learn pencasilat, and watch art performances.

The ceremonies that are used as tradition and attract many tourists are "Sedekah Kali" and "Sedekah Bumi." "Sedekah Kali" is a ceremony in which the community cleans the river from the waste, sows the fish feed, and eats the food prepared by society. The aim of this ceremony is to restore the ecology and hydrology function of the river. "Sedekah Bumi" is a ceremony in which community plants the trees in conservation areas, cleans the waste, and manages the environment to be more beautiful. Waste obtained is disposed. It is then burned by using an environmentally friendly incinerator. The incinerator can produce ash and liquid smoke. Ash residue is usually used as fertilizer or growing media, while the remaining thin liquid smoke that is free from lead content will be processed as fuel.

Edutourism can be given to the visitors visiting KLTH Sangga Buana. Most visitors are students. They can not only learn about environmental conservation but also perform various farming activities that are very rarely done by urban communities. Agricultural activities that are developed are environmentally friendly with the following characteristics:

- do not use synthetic chemical pesticides
- weed, pest, and disease control is carried out by mechanical, biological, and crop rotation or in an integrated manner
- do not use growth regulators and synthetic chemical fertilizers
- maintain and improve soil fertility and productivity through the return of plant residues, manure, and natural sediment from rivers
- reforestation along the riverbanks of Pesanggrahan with various endemic or local trees. For land conservation on riverbanks, bamboo trees are planted. This reforestation result is Sangga Buana urban forest park, which has now become an icon and recreation area.

Figure 4. Activity in Sangga Buana Urban Forest.

Theoretically, education is an important factor in building environmental awareness (Buttel et al. 1978). Economic and social demographic are control variables that have an important impact on environmental awareness (Bozoglu, et al. 2016). Students are young people who have a significant influence on the formation of environmental policies (Condrea & Bostan 2008). The importance of education in changing people's behavior toward the environment has been done by several studies. Education is the most appropriate vehicle in providing knowledge, skills, and attitudes about environmental care. Environmental attitudes and behavior are influenced by environmental education (Bozoglu, et al. 2016). It can create awareness and positive attitudes about environmental issues while curbing the negative role of human actions on the environment (Bradley et al. 1999). Education is also believed to have a strategic role to foster environmental concern, values, morality, and skills that support sustainable development, which in turn can create effective behaviors to participate in realizing environmental sustainability (Maryani 2015).

Edutourism teachers are the agent in transferring knowledge, experience, ethics, and behavior, sourced from nature, community leaders, and people in the destination as members of indigenous peoples. They also should explain, invest, and give examples (Maryani & Ahmad Yani 2018). This is in line with the eco-pedagogical movement which was first developed by Paulo Freire (2001). Eco-pedagogic aims to develop awareness and concern for environmental conservation and balance in the form of a green curriculum. Green curriculum aims to develop the following things: (a) Students' awareness and sensitivity about the importance of environmental sustainability; (b) Students' knowledge, understanding, and experience about the role and function of the environment to sustain life; (c) Students' attitudes to be more responsible in maintaining the environment and preventing environmental damage; (d) Students' skill in solving environmental problems; (e) Students' active participation in finding and carrying out solutions to environmental problems. In this case, tourism village of Sangga Buana can function as a medium and a source of learning for the community in general and students in particular.

4 CONCLUSION

The main uniqueness of the Sangga Buana tourism village is its location, which is in the middle of the city. Urban communities, which are usually individualistic and hedonistic, have successfully changed their orientation to ecocentric. Dissemination, socialization, and internalization of environmentally friendly behavior continue to be rolled out so that they provide benefits not only for environmental sustainability but also for economic, socio-culture, and education, which is the spearhead for morality, environmental attitudes, and behavior. In this case, KLTH Sangga Buana has succeeded in providing not only a concrete example for education, but also a recreation place for urban people.

REFERENCES

Bozoglu, M.; Bilgic, A.; Topuz, B.K. and Ardali, Y. (2016). Factors Affecting the Students' Environmental Awareness, Attitudes and Behaviors in Ondokuz Mayis University, Turkey. *Fresenius Environmental Bulletin*. Volume 25 (4), 1243–1257

Bradley, J. C., Waliczek, T. M. and Zajicek, J. M. (1999). Relationship between environmental knowledge and environmental attitude of high school students. *The Journal of Environmental Education*, Volume 30 (3), 17–21.

Buttel, F. H. & Flinn, W. L. (1978). Social Class and Mass Environmental Beliefs: A Reconsideration. Environment and Behavior, Volume (10), 433–50.

Bruyere, L.B., Wesson, M. and Teel, T. (2011). Incorporating Environmental Education into an Urban After-School Program in New York City. *International Journal of Environmental and Science Education*, Volume 7(2), 327–341.

Condrea, Petru and Bostan, Ionel. (2008) Environmental Issues from an Economic Perpective. *Environmental Engineering and Management Journal*, Volume 7 (6), 843–849.

Creswell, John. W., (2009). *Research Design: Pendekatan Kualitatif, Kuantitatif, dan Mixed*. Yogyakarta: Pustaka Pelajar.

Freire. (2001). *Pendidikan yang Membebaskan (terjemahan)*. Jakarta: Malibas.

Goleman D. (2009). Ecological intelligence: how knowing the hidden impacts of what we buy can change everything (New York: Broadway Bussines).

Jayadi, E. M. dkk., (2014). Local Wisdom Transformation of Wetu Telu Community on Bayan Forest Management, North Lombok, West Nusa Tenggara. *Research on Humanities and Social Sciences Journal*, 4 (20), hlm. 109–118.

Lane, B. (2009). *Rural Tourism: An Overview", in Jamal, T., & Robinson, M., (Eds.) The SAGE Handbook of Tourism Studies*. Londong: Sage Publications.

Kastenholz, E. and Lima. (2011). Integral Rural Tourism Experience From The Tourist's Point Of View – A Qualitative Analysis of Its Nature and Meaning, 19.10.2011 Accepted: 25.11.2011.

Maryani, E. (2019). *Geografi Pariwisata*. Yogyakarta: Ombak.

Maryani, E dan Ahmad Yani. (2018). Local Wisdom Of Kampung Naga In Mitigating Disaster And Its Potencies For Education Tourism Destination.

Maryani, E. (2015). Pendekatan Eco-Pedagogis Dalam Upaya Menumbuhkembangkan Kepedulian Lingkungan, Makalah Seminar Nasional Program Studi Magister Pendidikan IPS Program Pascasarjana Universitas Lumbung Mangkurat Banjarmasin.

Maryani, E. (2000). *Geografi Kota*. Bandung: Pendidikan geografi.

Muhtarom Ilyas, M. (2008). Lingkungan Hidup Dalam Pandangan Islam. *Jurnal Sosial Humaniora*, Volume 1 (2), 154–165.

Ruhimat,M. (2019). *Manusia, Tempat dan Lingkungan*. Yogyakarta: Ombak.

Santoso, Slamet. (1999). *Dinamika Kelompok Sosial*. Jakarta: Bumi Aksara.

Soemarwoto, Otto. (2004). *Ekologi, lingkungan hidup dan pembangunan*. Jakarta: Djambatan.

Sugiyono. (2008). *Metode Penelitian Kuantitatif, Kualitatif dan R&D*. Bandung: Alfabeta.

Promoting Creative Tourism: Current Issues in Tourism Research – Kusumah et al. (Eds)
© 2021 Taylor & Francis Group, London, ISBN 978-0-367-55862-8

Indonesia's spice route tourism

N. Fathiraini, D.P. Novalita, Labibatussolihah & E. Fitriyani
Universitas Pendidikan Indonesia, Bandung, Indonesia

ABSTRACT: As Indonesia Spice Route tourism initiative emerged, there it also raises the question of why the Spice Route tourism being encouraged and promoted vigorously. Indonesia is one of the countries whose tourism is considered to have the potential to accelerate economic development. Tourism development often emphasizes inland tourism rather than maritime-based tourism. An undergoing Indonesia tourism dynamic development in Spice Route tourism will elaborate further. This study utilizes a qualitative descriptive method. The results showed that Spice Route tourism is closely related to the exercise of power where political actor constructed the historical linkage of spice route as well as Indonesia's identity as a maritime state.

Keywords: Historical linkage, maritime identity, politics, Spice Route tourism

1 INTRODUCTION

The study of Indonesia's Spice Route tourism is still understudied and limited to regional scope that is to optimize tourism resources potent in a single region to increase revenue. It sparked by intensifying the emphasis on service provision by leveraging tourism area potential as a result of the implementation of regional autonomy which requires each region to identify and develop its potential, one of which is from tourism. The previous studies elaborated on the development of the potential for tourism areas based on Indonesia's Spice Route tourism in Ternate focused on a strategy in developing Afo clove tourism objects as a tourist destination for spices in Ternate. The Afo clove agro-tourism is assessed based on three components of the tourist area consisting of amenities, accessibility and attractions (Mulae & Said 2019). Other study elaborated the mapping of the Maluku tourism industry in developing effective marketing communication for the Maluku Tourism Office to increase tourists visits by utilizing media strategy development through integrated marketing communication which one of it discussed a Spice Route Festival on 3 islands (Banda, Saparua, Ambon) that has been encouraged as an event attraction (Saimima et al. 2018). Our study aimed to elaborate Indonesia's Spice Route tourism in a broader frame, where the Indonesia's Spice Route tourism destination is not only a single point development of tourism destinations but as a network that is woven into one big narrative of grand national tourism called Indonesia's Spice Route tourism.

The APEC Summit-Beijing in 2014 marked a milestone where thematic tourism introduced to Indonesian tourism. As an initial initiative responded to thematic tourism discourse, the government began to develop the cultural-based thematic tourism in a cultural route form. A cultural route puts forward tourism products that combine several cultural attractions and destinations with certain themes and narratives in planned travel patterns. One of them is the Cheng Ho Maritime Route which highlights Admiral Cheng Ho's journey to the archipelago from 1405–1433 which leaves a cross of cultural traces in Indonesia. In addition, there is also the Wallacea Route which highlights Sir Alfred Russel Wallace's scientific journey, which captured a lot of Indonesia's biodiversity in 1854–1862, which is located between Bali, Lombok, Kalimantan and Sulawesi. The Wallacea Line refers to the Wallacea route which divides the Indonesian archipelago into two parts, namely the

DOI 10.1201/9781003095484-11

western part with typical Asian fauna and the eastern part with typical Australian fauna, then the flora zone that crosses Asia and Australia.

However, one of the significant initiatives in the development of thematic route tourism is the Spice Route (the destination line refers to the Spice Route). The spice route in tourism has expanded its meaning, including in festivals, performances, culinary, traditional knowledge, underwater culture, etc. Thus the Government developing a planning discourse regarding accessibility with flight routes, shipping along spice routes, culinary promotions and other cultures related to spice products. Since 2018, the Ministry of Tourism arranged a Spice Route tourism through culinary aspects, which will be inaugurated in 2020. In addition to culinary, several regions in Indonesia will be developed for spice tourism including Aceh, Banten, Jakarta, Bali, Banjarmasin, Maluku, Gorontalo and Ternate. Not to mention the incessantly promoted tourism events to support the Spice Route tourism discourse were also held, one of which was the Spice Route Exploration Festival on Nyiur Melambai Timur Beli-tung Beach (Kemenpar 2018).

Indonesia's Spice Route tourism promising tourism potential in terms of destination, attraction and culinary. Encouraging the development of cultural routes into tourism products that respond to global trends in tourism on the one hand, as well as preserving local cultural heritage values as part of Indonesian identity is fascinating. This study raises the question of why the maritime cultural trail, in this case, the Spice Route tourism, emerge to be a government priority in tourism development. In contrast to mainstream tourism studies where tourism is defined as a development sector priority built solely by material dimensions that are expected to strengthen the country's economy and make a very large contribution to foreign exchange earnings and create jobs within the country, this study will elaborate on the politics tourism in the development of Indonesian Spice Route tourism.

2 THE POLITICS OF TOURISM

Tourism often considered as defining industry that inscribes and thereby makes places and spaces, thus negates tourism phenomenon which inherently politics. Harold Dwight Lasswell's premise that politics revolves around the various exercises and relations of power that occur across society. As such, politics ought not to be viewed merely as electoral parties' battleground and the subsequent work of government bodies, but rather as a dynamic manifestation that pervades all sorts of human, social, and cultural relationships. In this broader light political action, it is not simply a strong-arm compulsion that suppresses and subjugates; it can also be an empowering force that endows and invests (Hollinshead & Suleman 2017, p.2).

The Government has a role to play in tourism planning which is not limited to economic development and poverty reduction but includes building the nation's image and identity (Chheang 2009). Spice Route tourism is not formed in a political vacuum, but is formed from political actors implementing power. This study aims to elaborate on the connectivity between politics and tourism and to further examine how tourism is inherently a setting for political action.

3 METHOD

This study utilizes a qualitative descriptive method. Descriptive means the study describing data based on objective reality found. Qualitative means that in describing related concept, sentences are used rather than data or statistics.

4 ANALYSIS OF SPICE ROUTE TOURISM

4.1 *The construction of historical linkage*

Indonesia's Spice Route tourism discourse has significantly introduced to the public and strengthened by the Thematic Tourism Guideline (Kemenpar 2018). At this point, in developing Spice Route tourism, the preference rests toward political actor's practices in articulating an idea and construct the historical linkage regarding Indonesia's Spice Route. As in this study, we identified President Joko Widodo construction over Indonesian narrative as the significant and dominant player by accentuating the abundant spices resources as well as maritime trade route on world history.

At several significance states meetings, Joko Widodo as The President of Indonesia many times highlighted Indonesia's glorious past (Nusantara) as a source of spices. For example, in The Belt and Road Forum for International Cooperation (BRF) Summit 2017 that took place in Beijing, Joko Widodo mentioned that Indonesia was tremendously well known for its Spices Paradise. He further revived the memory from the history when the silk route first developed, when Indonesia was known as the spice archipelago or the Spice Islands. (Romadoni 2017). Silk considered the main driving forces in economic history, connected the land route between mainland China and the European continent, meanwhile the spices also have the same reputation that connected five continents and significance in the field economics and world civilization in general. Spice trades activities were possible at that time through the maritime route. Furthermore, spices even became the main trading commodity in the Silk Route. Nusantara's spices were featured as commodities with highly economical values. Nowadays, under Joko Widodo leadership, the spices narrative directed toward an outstanding universal value that sustainable, memorable and meaningful.

In historical records, the Indonesian archipelago was a shipping lane that connects the western and eastern worlds. Nusantara, historically has a strategic position as a fulcrum that created network that connects India, China, and Middle East to European countries. Other than that Nusantara is significant player and acknowledge for spiceries before Europeans expand their trade in Southeast Asian region. The islands are ot only preoccupied as ships transit place from around the world but also developed due to the abundant natural resources of spices. China noted the circumstance where the Tang, Sung, Yuan, Ming dynasties (7–13 century) had established relations with the kingdoms in the archipelago. Refers to the Chinese news, it was often mentioned about spice plant variants gained from the land of the southern sea (Nanhai) (Kemendikbud 2017, p. 4). Spice trade in the archipelago even affected Chinese dynasties authority wealth. In its development, trade relations with the Chinese dynasties expanded towards diplomatic, religious and educational relations (Kemendikbud 2017, p. 5).

As the time went by, spices from the Indonesian archipelago increasingly known across the world. Approximately in the seventh century, vigorously there was a hunt for high-value spices in most of the islands. Cloves in Ternate, Tidore, Halmahera, Seram and Ambon. Fuli (from nutmeg) were widely grown on Run Island in the Banda Islands. Cinnamon, incense, and camphor were grown in Sumatra and Java. Sandalwood is widely produced in the islands of Timor and Sumba, and pepper in Banten, Java, Sumatra and South Kalimantan. Shipping and trade flow came across the ocean, from East Asia, West Asia and South Asia (Beranda Sejarah 2017, p. 5).

Local political power and number of empires at that time traded spices. Significant transformations in Indonesian society occurred when Europeans at their initial departure intended to seek for spices and trading, then turned out to colonizing and building political power. Their political power mobilized to exploit Indonesia's natural resources for the colonial economic benefit, namely to fulfil their motherland interest. These historical facts undoubtedly underlie several points of view where spices were being the catastrophic source in Indonesia.

Joko Widodo has also mentioned firmly about Indonesia's abundant varieties of spices that even European countries were obsessed with and colonized Indonesia as they tried to dominate the spice trades (Setkab 2019). At first, Europeans did not have any knowledge to spiceries, in contrast to the Eastern countries that were more advance in knowledge regarding spices as medicine, religious mean, ceremonial instrument, natural food preservative as well as a seasoning that spread unique aromas and distinctive flavour. Because of the scarcity of spiceries, then it defined to be an exotic and precious commodity for European. A remarkable highlighted the history Nutmeg of Run island was one of the examples which became a valuable spice that made the tension between England and the Dutch. As Nathaniel Courthope, an English explorer, succeeded to reach Run island and secretly persuaded islander due to their interest to monopolize over nutmeg, the Dutch was furious as they also strived to occupy the largest nutmeg provider in the world (Milton 1999).

Along with perspective in doing historical linkage interpretation of Nusantara's spices resources and maritime trade from time to time, it also signifies contemporary legitimacy efforts for Indonesia's domination in history and will be re-manifested through the Spice Route tourism initiative program. Nusantara's spiceries that were once was shaped the prominent transformation in Indonesian history and even have had a profound influence on the map of the world. And now, under the current administration, the historical linkage of Nusantara's spiceries narrative idea developed significantly as part of political exercises.

4.2 *Indonesia's identity as a maritime state*

When Indonesia's government formulating tourism, the consideration it is not merely to accelerate economic revenue and as to find a better solution to overcome wealth distribution gap, yet it also aimed to construct state image and identity. As in this case, the government has put forward the ideational construction (namely identity) over the material one in formulating thematic tourism through promoting Indonesia's Spice Route tourism. To some extent, the government officials cannot be separated from political actor that emerge with the idea to construct and formed state's identity. Under Joko Widodo administration, we can identify the political actor perceives Indonesia identity as a maritime state, or an archipelagic state where livelihood integrates the maritime aspect. This identity as a maritime state then articulated, one of which in the Spice Route tourism.

The construction of maritime identity is closely influenced by culture that has developed in Indonesia, which is not only maritime culture but also agricultural one (Sulistiyono & Rochwulaningsih 2013). Maritime's culture was emerged and developed as a response to its maritime potential and agrarian culture with its feudal character. Several elements of society are trying to revitalize Indonesia's Maritime culture in the development of Modern Indonesia (Cribb and Ford in Sulistiyono & Rochwulaningsih 2013). Along with the revitalization effort, the discourse and symbol contestation between a feudalistic system based on feudalistic agrarian culture and a more democratic system based on utilizing ocean resources as explained by Rokhmin Dahuri, former Minister of Maritime Affairs and Fisheries of Gotong Royong Cabinet also emerged (in Sulistiyono & Rochwulaningsih 2013). And under Joko Widodo administration, most of the policy is directly encouraged to have acknowledgement and should be recognized as one of the most significant maritime nations.

Even though it is known as an archipelago country, its identity construction as a maritime country is often marginalized. The story of the historical legacy of the great kingdoms of the archipelago as a noble maritime empire and the ancestors of the Indonesian nation as a seafaring nation was not enough to build the identity of the Indonesian nation as a maritime nation. Government policy agendas are often prioritized for inward looking. The centralization of resources causes development to only occur on large islands that are not well connected because each island is pursuing its economic growth. The point of economic growth remains based on the development of land potential rather than the mobility of resources carried out by sea between islands. This is a consequence of the development ideology which is oriented towards economic growth. The situation changed when Joko Widodo initiated the formation of an Indonesian maritime identity built from maritime sovereignty. The sovereignty of the sea is considered a source of pride and has even become the identity of many of the great kingdoms in the archipelago, namely Sriwijaya and Majapahit. Under Joko Widodo administration, this preference was based on the idea of changing Indonesia's identity from an agricultural country to a maritime nation. Where Indonesia, which is an archipelago country, makes the sea a strategic force by distributing its resources and for national defense. Furthermore, the sea has become a symbol of Indonesia's great sovereignty (Paskarina 2016, p. 3). If in the past the maritime nation's identity emerged because of the glory at sea, as revealed in the motto of Jalesveva Jayamahe, now the discourse of a maritime state appears because of a downturn. In the past, this triumphant identity was a symbol of resistance to colonialism and imperialism, on the contrary, today it is a symbol of fighting poverty (Paskarina 2016, p. 5).

Joko Widodo became an agent (political actor) who raised the idea to ignite Indonesia's maritime state identity. He bluntly put forward Indonesian identity as "Maritime State" during ASEAN

Summit amongst international audience in Naypyidaw, Myanmar in November 13th 2014. Joko Widodo revived Indonesia's image as a maritime state with the ancient maritime's refrain "Jalesveva Jayamahe" which means "victorious in the sea". His vision toward maritime identity embedded in the long-standing legacy prominently during the Majapahit and Sriwijaya kingdom whereas the culture is strongly related to Indonesian maritime power (Sulistiyono 2016). This identity construction can also be identified in Sukarno's administration. He traced the early maritime power, namely Majapahit and Sriwijaya kingdom then adopted Maritim power as a national image. At that time Sukarno's fundamental introductory regarding the Sanskrit phrase of "Jalesveva Jayamahe" as Joko Widodo also promoted nowadays as a National campaign introduction. To some extent, the reviving spirit of Indonesia's maritime identity directly promoted both in a domestic and international realm.

The strengthen of Indonesia's identity as a maritime state then articulated in the Global Maritime Fulcrum (GMF) where Joko WidodoPresident publicly announced at his initial stage of Presidency. The GMF consist of several maritime concerns such as, first, to revive the Maritime's culture. It means constructing idea and projecting images that articulate from Maritime's identity. Second, to protect and manage maritime resources and sovereignty, ultimately in the fisheries sector. Third, infrastructure development and maritime's connectivity by constructing sea toll, deep seaport, logistic, ship industry as well as Maritime's tourism. Fourth, to develop maritime diplomacy aims to reduce conflict in the ocean. And fifth, to build maritime's security (Kebijakan Nasional 2019). Before Joko Widodo's administration, maritime aspect was un-prioritized. It remains muted over periods while the government heavily focus towards inland development. As Chen (2014, p. 68–69) explained where maritime state where heavily focused on Indonesia build-up not only as an 'archipelagic state' but also 'maritime state'. It means that Indonesia reprioritizing its maritime environment and sea lanes over traditional land-centric focus (Chen 2014, p. 70). Based on Indonesia's maritime identity then it formed national interest towards economics. In which to some extent it is closely related to states' welfare capacity building (economics development).

With this maritime identity, the expected welfare is strongly determined by the owned marine resources. The strategy's arrangement is by intensifying the fishery industry development towards food security, prioritizing infrastructure development along with accessibility to maritime connectivity, such as sea toll roads, deep seaports, logistics, shipping industry, maritime tourism, conducting maritime diplomacy to eliminate sources of conflict at sea as well as building maritime defense forces to maintain shipping safety and maritime security (Paskarina 2016, p. 6). Indonesia's identity as a maritime state thus determines the state's actions to form the sea as a forefront for the welfare of its citizens.

Alongside Joko Widodo, Indonesian Democratic Party of Struggle (PDI-P) chairwoman Megawati Sukarnoputri also had a significant role arising Spice Route Tourism's discourse. At Trisakti Tourism Award in 2019, she encouraged Indonesia's Spice Route tourism development. Furthermore, she highlighted Nusantara's triumph, which has played a globally significant role. She also compared this country to China that is proud about their Silk Road, where Indonesia should also be proud of the Spice Route (Republika 2019). Political actors at this point emphasized the emergence of Indonesia's identity as a maritime power and showcased more subtly through Spice Route tourism. In terms of power, tourism is part of soft power.

5 CONCLUSION

Through study elaboration, we can conclude that Indonesia's Spice Route tourism initiative did not exist in a political vacuum yet the government exercise of power. To put it bluntly, it was more to the state's power politics, ultimately in the social construction of historical linkage and Indonesia's maritime identity conducted by state leader and the ruling party politician. Historical traces and past cultural heritage are rediscovered and interpreted through the cultural route approach and become a strong attraction for Indonesia's tourism potential. The state defines and redefines identity caused by transformation from agricultural into maritime identity. From Indonesia's identity as a maritime country then it defines the government's priorities for economic interests in order to improve people's welfare by encouraging and developing maritime-related sectors prominently in the tourism sector (as referred in the Global Maritime Fulcrum idea). Thus Indonesia's Spice Route tourism's initiative is compatible with the arguments explained above.

However, the tourism planning development policy should be supported by strong commitment, both from the government and the stakeholders. This long term strategy should be envisioned comprehensively, ultimately for the state's image. We could not aside from the government's involvement in developing destination policy as well as promotion in tourism as their significant role in shaping states' identity (national identity) is urgently needed. As benedict Anderson highlight in the imagined community, the development of Indonesia's Spice Route tourism should establish beyond imagination that is in concrete steps. We also find how tourism development planning can be branded as one product which state asserts its stance towards historical linkage and identity. Thus the future study can improve further towards the mapping of Spice's Route tourism strategy.

REFERENCES

Chen, J. 2014. Indonesia's Foreign Policy under Widodo: Continuity or Nuanced Change? Australia: Perth US Asia Centre.

Chheang, V. Spring 2009. State and Tourism Planning: A case study of Cambodia. Turismos [Electronic Version]: An International Multidisciplinary Journal of Tourism, 4, (1), 63–82.

Hollinshead, K., & Suleman, R. 2017. Politics and Tourism. Retrieved from Research Gate: https://www.researchgate.net/publication/318441095_POLITICS_AND_TOURISM

Kemendikbud Republik Indonesia. 2017. Laporan Utama: Ekspedisi Jalur Rempah 2017. Beranda Sejarah, No. 01, 1–79.

Kementerian Pariwisata. 2018. Pengembangan Wisata Tematik Berbasis Budaya. Retrieved from https://www.kemenparekraf.go.id/asset_admin/assets/uploads/media/pdf/media_1569465804_DRAF_2018_-_Pedoman_Wisata_Tematik_Berbasis_Budaya.pdf

Milton, G. 1999. *Nathaniel's Nutmeg: How One Man's Courage Changed the Course of History*. London: Hodder and Stoughton.

Mulae, S.O., & Said, R.M. 2019. Strategi Penilaian Objek Wisata Cengkeh Afo Sebagai Upaya Penguatan Sektor Pariwisata di Ternate. *Jurnal Penelitian Humano. Vol. 10, No. 1.* DOI: 10.33387/hjp.v10i1.1344

Paskarina, C. 2016. Wacana Negara Maritim Dan Reimajinasi Nasionalisme Indonesia. *Jurnal Wacana Politik, Vol. 1 (1), pp. 1–8.* DOI: 10.24198/jwp.v1i1.10542

Republika. 2019, Dec, 23. Megawati Minta Wishnutama Hidupkan Kembali Jalur Rempah. Retrieved from Republika: https://nasional.republika.co.id/berita/q2xbyi414/megawati-minta-wishnutama-hidupkan-kembali-jalur-rempah

Romadoni, A. 2017, May 16. Ajak Peserta KTT Kerja Sama, Joko Widodo Cerita Indonesia Surga Rempah. Retrieved from Liputan 6: https://www.liputan6.com/news/read/2953025/ajak-peserta-ktt-kerja-sama-Joko Widodo-cerita-indonesia-surga-rempah

Saimima, R.M.M., Zpalanzani, A., & Mutiaz, I.R. 2018. Pemetaan Industri Pariwisata Maluku sebagai Landasan Perancangan Strategi Brand 'Baronda Maluku'. *Barista: Jurnal Kajian Bahasa dan Pariwisata, Vol. 5(1), 87–102.* DOI: 10.34013/barista.v5i1.159

Sekretariat Kabinet. 2019, March 19. President Joko Widodo Tells Farmers to Grow Coffee, Spices. Retrieved from Cabinet Secretariat of the Re-public of Indonesia: https://setkab.go.id/en/president-Joko Widodo-tells-farmers-to-grow-coffee-spices/

Sulistiyono, S. T. 2016. Paradigma Maritim dalam Membangun Indonesia: Belajar dari Sejarah. Lembaran Sejarah, Vol. 12, Number 2, 81–108.

Sulistiyono, S. T., & Rochwulaningsih, Y. 2013. Contest for hegemony: The dynamics of inland and maritime cultures relations in the history of Java island, Indonesia. *Journal of Marine and Inland Culture*, 2(2). DOI: 10.1016/j.imic.2013.10.002

Kebijakan Nasional. 2019. Indonesia Poros Maritim Dunia. Retrieved from Portal Informasi Indonesia: https://www.indonesia.go.id/narasi/indonesia-dalam-angka/ekonomi/indonesia-poros-maritim-dunia.

Promoting Creative Tourism: Current Issues in Tourism Research – Kusumah et al. (Eds)

Tarling art: History and tourism potential in Cirebon

A. Mulyana & S. Sartika
Universitas Pendidikan Indonesia, Bandung, Indonesia

ABSTRACT: The study discussed Tarling Cirebon's art, which is one of the cultural tours in Cirebon. The methods used in this study were: a historical method and descriptive qualitative approach by viewing periodization regarding Tarling Cirebon's art development and its tourism potential. This study aimed as a form to promote historical tourism culture regarding Cirebon's art. The study mainly discussed Tarling art emergence background, Tarling performance's style development, Tarling art function development, factors influencing Tarling art, and Tarling arts as a tourism potential in Cirebon. We expect that this study can introduce and revive the glory of Tarling art, as it was in the past both for artists and the younger generation in Cirebon and surrounding areas. Based on a study's result conducted through literature study, in its development, the art continues to experience changes in terms of performance style and its function. Yet, Tarling art can innovate from time to time and still exist until the present day.

Keywords: Cirebon, music, Tarling art

1 INTRODUCTION

West Java has various attractions, such as natural, cultural, and historical tourism. For cultural and religious tourism, one of the cities that one should visit in West Java is Cirebon (Artati 2018). Cirebon has interesting tourist destinations to visit, especially those related to cultural tourism. Cirebon is a multi-ethnic city that combines elements of Sundanese and Javanese, and then mixed it with Chinese cultural elements, which makes Cirebon unique (Morissan 2002). The city is also well known for producing fish and delicious seafood. Not only culinary, Cirebon's unique cultures also attract tourist attention, such as Mask Dance, Gembyung, Berokan, Genjring Burok, Glass Painting and Tarling.

Tarling Art, among Cirebon's art, is interesting to study further. Tarling is a guitar and flute performance which becomes Cirebon special art. It is known for its thick elements of an outside culture. Although contained with external culture elements, Cirebon people admit Tarling as a traditional art that becomes an integral part of Cirebon culture (Supriatin 2012). Tarling performance art combines cultural influence from outside and inside Cirebon. In Tarling performance, there are performances or drama, music and dance that draw directly and indirectly from Javanese traditions and social dance (Cohen & Effendi 1999).

Tarling is one of a regional artw that has unique song characteristics, both in terms of musical composition, song material, and its development. These aspects are interesting to be used as material for study and research, ultimately in understanding its existence in the supporting community. Tarling, as Cirebon's intellectual music masterpiece, has a role in elevating Cirebon's cultural values. In the development, it was predicted that it had already changed in both form and mode of expression. By several Cirebon art observers, Tarling is considered as the music and melodic identities of the Shrimp city (a term for Cirebon city) (Hidayatullah 2015). Thus based on these characteristics, Tarling art can be used as a tourist attraction, especially cultural tourism in Cirebon. In accordance with the background, the hypotheses in this study are:

1) Tarling art can be one of the tourist attractions in Cirebon;
2) How to develop Tarling art into a way of communication for tourists and local communities.

DOI 10.1201/9781003095484-12

2 MATERIAL AND METHOD

Data was processed based on literature study from related various literacy. The researcher gained sources from books, journals, and several internet sources that can be accounted for. This study used a qualitative approach, which this study was written based on descriptive depictions from sources obtained. Furthermore, a historical methodology was conducted to develop examinations.

According to (Sjamsuddin 2007) the historical method is a way of knowing history, because it consists of heuristics, criticism, interpretation and historiography. In the book, (Ismaun et al. 2006) suggested that heuristics are activities related to the searching and gathering historical sources. The activities are also related to using the content study method, namely by using a literature study approach from several sources, by reviewing several findings in the form of collecting books as a reference source and a comparison of sources in this discussion.

3 RESULTS AND DISCUSSION

3.1 *Background of the emergence of Tarling art*

Tarling is a Cirebon regional art, which has the characteristics of playing guitar and flute music instruments, also the music and vocals produced by Pelog (Hidayatullah 2015). Tarling art is a cultural product of the people of Cirebon which appeared around the 1930s (Salim 2015). According to (Jaeni 2014), it was stated that at first, Tarling was only a youth playing form in breaking time after a day of work. At that time, it was at the end of the Japanese occupation, before entering the beginning of the independence revolution. It was contrasted to Salim point of view where Tarling art emerged in the 1930s. The playing only used a guitar to imitate saron beats pattern (melodies). The beats pattern/saron beats (gamelan) which successfully transferred into guitar passages, eventually became a teenagers' habit and played while circling the village at night. They did spontaneously by delivering traditional songs, in which Cirebon gamelan music ensemble's characteristic could accompany with.

Meanwhile, a figure who pioneered flute guitar music form performance in Cirebon was Jayana and his friend, a Chinese descendant named Sin You, who later became known as Pak Barang. From these two figures' creativity in inserting flute instruments, Tarling's seed emerged. A combination result between guitar and flute produces musical colours and patterns that are unique and interesting. Before independence's war, Tarling Barang CS was formed. Barang term itself came from the *bebarang* which means *ngamen* or busking (Jaeni 2014).

As its development around 1950, Tarling's popularity began to shine. Because of Jayana's effort, Tarling was famous outside Cirebon. Jayana's skill with Barang made Tarling the most interesting show. After receiving a good response from the people, Jayana and Barang added some musical instruments to Tarling such as drums, ketuk, kebluk, kecrek, also gong. Thus Tarling becomes public entertainment's medium. Tarling continues to change, as has been the case and observed in several Tarling art/music, from its initial appearance until nowadays. The shift or change is not only about the music material but also in the interests shift or Cirebon's public views regarding Tarling music (Hidayatullah 2015).

In its development process, Tarling always experiences dynamics, if we look at the early history of the emergence of the Cirebon people, they should be proud because Tarling was born through artists' creativity who created Tarling music from people daily habits and introduced it still exists to this day even though the demand is not as much as its initial appearance. Although many versions regarding the beginning of Tarling, some sources say that Tarling emerged during the 1930s.

3.1.1 *The Tarling's function development*

Since its development to the present day, it has the following functions:

- As an entertainment.
  Tarling comes as an ordinary form of entertainment when Cirebon people take a break at night after doing various activities during the day, both as farmers and fishermen. (Salim 2015).

- As a means of aesthetic presentation.
  Aside from entertainment, Tarling also functions as a means of expressing for the Pantura people. The lyrics or stories that played in Tarling expressed Maskumambang that filled with pain. Besides, the lyrics sung in Tarling, especially in classical Tarling are created in the Wangsalan, Parikan form or other expressions which language considered as aesthetic (Supriatin 2012). The aesthetic presentation has a function to meet aesthetic needs, expression and an art appreciation. When it performed, it will raise people's astonishment and joy (Febrianto, tt)
- As an educational medium.
  The messages contained in Tarling's performance are educational besides being expressed through dialogue in prose language form and expressed through dialogue in a poetic language form that is sung (Supriatin 2012).
- As media communication.
  In Tarling's performance, the dialogues are often sung in local languages so that the communication strength to reach rural people who are still illiterate is very large. This factor caused how Tarling's performance arts have long been used as a communication medium by the government, particularly the Cirebon local government, to convey development's messages (Salim 2015).
- Tarling serves as a ritual means.
  Harvest ritual often held until present day, but the musical entertainment program has now shifted from Islamic song into Tarling Cirebonan that follow the times' demand. Cirebon Tarling means "natar eling" which means a demand or introduction. Besides that, Tarling contains another philosophy, namely "yen wis mlatar kudu eling" which means that if you have done something negative then you must repent (Febrianto, tt).
- The function of Tarling that is no less important is to fascinate girls. However, this function occurred at Tarling's initial. At present, after Tarling turned into a kind of performance, that function has changed with other social functions, namely to convey messages openly and decisively (Supriatin 2012).

The Tarling's function at the beginning of its emergence to the present day was not much different, but only in its development underwent several changes. Until nowadays, Tarling's art functions as a means of entertainment, an aesthetic presentation, as a communication and an educational medium. The function that is rarely found now is as a means to attract girls because it has mult-functionality as stated before since its emergence to the present day Tarling is still loved and preserved.

3.1.2 *The development of Tarling performance's style*

At its initial development, Tarling's only Wangsalan and Parikan, but later it turned into a drama in which dialogue was sung. Changes in performance form carried out by Jayana and Maryati, who presented the story-telling drama entitled Saida-Saeni in the Tarling's performance. Then in the next development is a song and comedy period which began around the 1960s, where Tarling continued to experience some changes. Abdul Adjib made short dramas in the form of jokes that told events based on people's daily lives. It was influenced by Bodoran Reog. At that time the Reog performance used a joke separately by a gamelan accompaniment. The famous Reog's comedians at that time named Dawiya and Goyot, were very popular amongst the public. After that Tarling's puppeteers began to imitate them and included jokes as an intermezzo after singing. The performance structure in song and comedy are (1) tetaluan, (2) tayub dance performance, (3) songs interspersed with jokes, and (4) closing (Salim 2015).

Then the theatre's periodization, started in 1965 as stated by Jana at that time, that Tarling became the most preferred performance because of Abdul Adjib's expertise (Tarling famous figure) in packaging Tarling's performance. The drama elements or Masres incorporated into Tarling performance form. A Tarling performance's new pattern since then was introduced by Abdul Adjib as a theatre-style Tarling's performance (Salim 2015).

Subsequent developments occurred changes in personnel, for example, instruments coupled with gongs, drums, and tilts (two large and small bonang that function as rhythm regulators). Initially, the Tarling songs took parikans and wangsalan's form gathered by Sinden into a series of songs.

The lyrics contain a singer's story in a monologue form. For the next performance, an act conveyed in a dialogue form between Sinden and other Tarling actors (Supriatin 2012).

Today, from several aspects of the performance, Tarling experiences a further modern development. This is evidenced by dangdut music nuance using modern music instruments. Both things dominate Tarling's performing arts. Sometimes, Tarling performance depends on the audience's request (celebration's host). Thus, currently, in the Cirebon area, there are two models of Tarling performances. First, is the Tarling performance which features an act and the second one is a performance that is fully presented in musical form (especially, Cirebonan musical Dangdut (Kurnia et al. 2007).

Tarling performance style's development changes from time to time. This was conducted by artists to give new colours and innovations in Tarling's art that adjusted with people's taste. These changes proved audiences' performance preference. Classic Tarling began to erode because it follows time development which combined with modern music that had made Tarling's art still exist to present-day even by reducing the authenticity of classical Tarling music.

3.2 *Factors that influence Tarling art's development*

There are various factors that cause performance art's life and death. It might be because of changes in the political field, some caused by economic issues, changes in the audiences' preference, and some other alienated by another performance form (Soedarsono 1998). This is in line with (Rosala et al. 1999) a statement that the disappearance of some art was caused by several factors, including the death of the creators, the absence of future generations, the lack of enthusiasts of the art, pressured by new types of art that are more favoured by the public, and so on.

A factor that led to Tarling art development was because of the public support's appreciation and based on artists' creativity who was responsive towards development signals within the people itself. Tarling art finally became the most popular performing art in the 1960s to 1970s (Jaeni 2014). This is mainly supported by artists' creativity by combining additional elements in Tarling performance which adjusted to people's taste so that Tarling can survive in competition with other arts.

However, in subsequent developments, Tarling art experienced asetback, which was caused by several factors such as what was mentioned above namely due to the death of Tarling artists, the absence of regenerating successors and urged by a new type of art that was more favored by the public. As a result of the globalization influence that makes many new cultures, new arts and new music types developed in society besides the absence of regeneration of the successor of classical Tarling art make the art less well known at the present day. As stated by Sedyawati (1981) that "the growth and development of art is determined by their supporting people". Therefore, if people lose their interest in traditional arts, it can be assured that traditional arts cannot be able to develop and become scarce and vice versa.

3.3 *Tarling as tourism potential in Cirebon*

By seeing Tarling arts' development which still exists to the present day, Tarling is seen as potential tourism because of its uniqueness. Tarling is a traditional folk art, which in its appearance and development is an unsacred art. The existence of Tarling art is very attached to people around Cirebon because it becomes one of people's cultural identities. In contrast to Keraton art, which is a high-level art and considered having a noble tradition, originality, authenticity, sacred and classy values that came from the cultural elite, as explained by Hauser (in Tjahyodiningrat et al. 2018) that Keraton artworks can be described as a cultural elite as high, rigorous' supporter and uncompromising art, which tend stable because they respect everything that institutionally secure. Unlike the art of Tarling in Cirebon, that has its characteristics and uniqueness.

Since Tarling came from ordinary people, not from the elite's palace, Tarling can be accepted and liked by the people. This can be a value so that Tarling art can exist and enjoy to the present day, though there have been innovations and changes in Tarling art mixed with other music such as Dangdut's Tarling. When discussing art in Cirebon, Tarling is already familiar to the public because it is one of the most popular arts of its time. Its people significantly favored the overall music and songs of Tarling music's development. Cirebon Tarling music becomes the media of people's feelings and it continues to the present day. Music is able to communicate with its people in generating work passion in the social, economic and cultural fields (Jaeni 2014). Although Tarling's art is uncommon for a younger generation, yet this art's connoisseurs still exist. Tarling's art is still known on the national scene. Like the Mask Dance which is already well-known among domestic and foreign tourists. Thus, Tarling art has its potential to become a tourism attraction and an icon for Cirebon's people if artists, government and Cirebon's people are able to develop cooperation to further introduce and revive Tarling's art to be known both domestically and abroad, such as Cirebon Mask Dance. One effort that has been conducted by The Culture and Tourism Office in Cirebon's Regency to introduce Tarling art to the younger generation and to attract tourists' attention by holding Tarling Festival Performance in 2017.

4 CONCLUSION

In its development, Tarling continues to experience changes and to adapt to this time demands. If you see the development of Tarling art, it has its own place and uniqueness in people's hearts. Tarling is the result of the people thinking regarding Pantura region which colours their lives. Tarling art continues to experience changes from time to time both in terms of performance styles and function. The changing style of Tarling performance was a result of artist innovation which packed Tarling as dynamic art. Tarling serves as entertainment that unites the community kinship through art. Besides, Tarling also has a function as a ritual and aesthetic function. Many factors influence Tarling art development, one of which is the artists' creativity, and people's tastes. Even though today Tarling enthusiasts decrease, especially for the younger generation, the local government and artists can play an active role in developing Tarling. It can open the opportunity for Tarling as Cirebon tourism's potential.

REFERENCES

Artati, Y.B. (2018). *Pesona Wisata Jawa Barat*. Surakarta: PT Aksara Sinergi Media.

Cohen, M.I, & Effendi, P. (1999). The Incantation of Semar Smiles: A Tarling Musical Drama by Pepen Effendi. *Asian Theatre Journal*, Vol. 16(2), hlm. 139–193.

Febrianto, A.S. (tt). *Fungsi dan Ciri Khas Kesenian Tarling pada Masyarakat Pantura*. [daring] diakses dari https://id.scribd.com/doc/31/ pada 9 Juni 2020.

Hidayatullah, R. (2015). Seni Tarling dan Perkembangannya di Cirebon. *Jurnal CaLLs*, Vol. 1(1).

Ismaun. (2005). *Pengantar Belajar Sejarah sebagai Ilmu dan Wahana Pendidikan*. Bandung: Historia Utama Press.

Jaeni. (2014). *Kajian Seni Pertunjukan dalam Perspektif Komunikasi Seni*. Bogor: PT Penerbit IPB Press.

Kurnia, dkk (2003). *Deskripsi Kesenian Jawa Barat*. Bandung: Dinas Kebudayaan & Pariwisata Jawa Barat & Pusat Dinamika pembangunan UNPAD.

Morissan. (2002). *Petunjuk Wisata Lengkap Jawa-Bali*. Jakarta: Ghalia Indonesia.

Rosala, dkk. (1999). *Bunga Rampai Sampai Tarian Khas Jawa Barat*. Bandung: Humaniora Utama.

Salim. (2015). Perkembangan dan Eksistensi Musik Tarling Cirebon. *Jurnal Catharsis* IV(1).

Sjamsuddin, H. (2012). *Metodologi Sejarah*. Yogtakarta: Ombak.

Soedarsono, R.M. (1998). *Seni Pertunjukan di Era Globalisasi*. Jakarta: Departemen Pendidikan dan Kebudayaan.

Supriatin, Y.M. (2012). Teks Tarling: Representasi Sastra Liminalitas (Analisis Fungsi dan Nilai-nilai). *Jurnal Meta Sastra*, Vol. 5(1).

Tjahyodiningrat, H., Kasmahidayat, Y., & Haryana, W. (2018). The Cultural Transformation of Seni Tarling in Cirebon. *Advances in Social Science, Education and Humanities Research*.

Promoting Creative Tourism: Current Issues in Tourism Research – Kusumah et al. (Eds)
 ISBN 978-0-367-55862-8

Tourism and spiritual journey from students' perspective and motivation

S.P. Pandia, M.D. Kembara, A. Gumelar & H.T. Abdillah
Universitas Pendidikan Indonesia, Bandung, Indonesia

ABSTRACT: Many kinds of tourism have motivations. One of them is that some people travel based on spiritual reasons. In recent years, many people traveled to Israel, Egypt, Jordan, and several other countries in the Middle East region not for the purpose of taking spiritual journeys or spiritual tours. Currently, there are also many travel agents that offer tours with a variety of package options at a fairly high cost when compared to traveling in Indonesia. The purpose of this study is to determine student interests and also motivation for students who want to take a spiritual journey or spiritual tour. In addition, through this research, students' knowledge about spiritual tours based on their religious faith is also wanted. This research was conducted on 157 students at a university in Indonesia from various departments. Students provide data using a survey through a questionnaire. The results showed that as many as 66 (42.04 percent) of respondents stated very interested, as many as 69 (43.95 percent) expressed interest, which means that as many as 85.99 percent of students expressed interest in doing spiritual tours. It was also found that the majority of respondents knew that a spiritual tour is not an obligation, but based on the data obtained it was found that the main reason for wanting to take a spiritual tour is to look directly at the places mentioned or written of in the Bible so that they can feel and see parts of the history of Christianity in the world.

Keywords: tourism, spiritual tour, motivation, interest

1 INTRODUCTION

In recent years, many people who toured or traveled to Israel, Egypt, Jordan, and several other countries in the Middle East were for spiritual journeys. Such trips are often referred to as spiritual tours. This trip is different from other tourist trips because its main purpose is not to visit natural tourist spots but to visit places related to religion. Information about spiritual tours can also be found easily on the Internet and mass media. In addition, travel agents can easily be found to accommodate travel and offer choices of tourist destinations in Israel.

Although Christianity itself begins in that country, when we look at the contents of the Bible, a trip to Jerusalem, Israel, and other countries is not an order or obligation. However, the fact is that spiritual tours still have an interest and are often promoted to churches. The cost to do a spiritual tour is also quite large when compared to doing tours in the country, so that in general people who do spiritual tourism are people who are already working and have a steady income.

Therefore, in this research, we want to know how students are interested in spiritual tour and also know the motivation or reasons of students who are interested in doing spiritual tourism. We also wants to know the opinions of students on whether the costs required to conduct spiritual tourism are commensurate and whether the costs are the cause of students not interested in doing spiritual tourism. Through this research, data will be analyzed to obtain students' knowledge about spiritual tours referring to the religion faith.

Generally, people travel or travel for a particular destination, and usually the cost is a factor that is considered in determining a tourist destination. This research needs to be done because currently, the second largest number of Christians are in Indonesia, so it is important to explore perceptions

 DOI 10.1201/9781003095484-13

and also the factors that motivate people to take spiritual tours. This research was conducted only for teenagers whose age ranges from 18 to 21 years old. They are students of university. Therefore, in the case of students' interest and motivation toward spiritual tourism, some of the questions that need to be answered are as follows: 1) Student interest toward spiritual tourism; 2) Student motivation toward spiritual tourism; 3) Student knowledge about whether spiritual tourism is obligation or not.

2 LITERATURE REVIEW

2.1 *Spiritual or religious tourism*

Tourism or journeys with religious motivation have a particular importance in many parts of the world. On one hand, primarily due to intrinsic reasons that trigger them, these practices have special significance for the people who carry them out (Drule 2012).

People perceive differently the places they visit or intend to visit, due to two main categories of reasons (Ruback et al. 2008). On the one hand, there are differences of material, social, and symbolic nature between places; on the other hand, distinct perceptions are due to an individual's experience, beliefs, and attitudes (his background) and the traveling motivations.

Although tourism motivation has been extensively addressed in the literature, from sociological, psychological, or anthropological perspective (Cohen 1972; Crompton 1979; Gnoth 1997), data on travel motivations, (e.g., to sacred sites) are quite rare (Meng et al. 2008).

In a first phase, Maslow's hierarchy of needs (1970) has been widely used to study the tourism demand. The first theories and models focused on tourism motivation belong to Stanley Plog and Pearce (Meng 2008); the latter based his studies on the theory proposed by Maslow. Tourism is the combination of needs and desires that affect the propensity to travel in general (Heelas & Woodhead 2005). Several studies present what motivates travelers from different nationalities to visit foreign countries: reasons for Chinese travelers to visit New Zealand (Ryan & Mo 2001), motivation of Chinese tourists visiting Singapore (Kau & Lim 2005), why German and British travelers are visiting Mallorca and Turkey (Kozak 2002), Japanese travelers' reasons for visiting the United States and Canada (Jang et al. 2002), reasons of foreign tourists to visit Jordan (Mohammad & Som 2010).

A large number of researches in the field of tourism motivation are focused on two main categories (Yuan & McDonald 1990; Yuan 1990; Uysal & Jurowski 1993). The first category refers to internal stimuli that push people to travel, to seek experiences that meet their needs and desires like recreation, escape, social interaction, pleasure seeking, fun, etc. The second category includes external perceptions and expectations, such as novelty and benefit expectation (Jurowski 1993, p.844), that attract tourists. However, these motivations are interrelated and have a dynamic evolution depending on situational factors (Correia 2000). A highly relevant aspect is that tourism motivation is a complex concept that significantly influences the decision-making process of a tourist (McCabe 2000).

There have been several interesting ethnographic studies (Fedele 2013; Hall 2006; Heelas & Woodhead 2005; Wilson et al. 2013; Wood 2007; Taylor 2002), among others conducted in various geographical locations focusing on spiritual.

2.2 *Perceptional and teenager customers*

Perception is the process whereby a person accepts, organizes, and interprets incoming information into a separate meaning (Kotler 2008). In its relation with the environment, perception means a process in which individuals use their sense impressions to give meaning to their environment. Factors influencing perception would be personal characteristics and expectations. These factors will lead an individual to think about an item in a certain way despite when it is not suitable with the reality. In response to the question of why people's reality could be different from the actual, it is possibly due to the fact that people do not consider targets in isolation but could instead group them together (Robbins 2008).

3 METHODOLOGY

The purposive sampling method was chosen, with individual subjects representing individual sampling units. The following tools have been used for data collection: surveys through questionnaire. The survey method used was a questionnaire consisting of seven questions.

The study was conducted from 31 May–5 June 2020. The survey was conducted by filling out a questionnaire online. To encourage students to fill in answers in accordance with their thoughts, in this survey students were asked to fill in complete identities such as names and departments in the university. This gave a total of 157 respondents within six days.

The sample size was 157 respondents. Out of 180 questionnaires distributed around, 157 were returned. The response rate was thus at 87.2%. The respondents were students chosen from different departments and faculty of Universitas Pendidikan Indonesia.

The research instrument was systematically designed having structured and unstructured questions. The questionnaire was framed to measure three important constructs which were: a) student interest toward spiritual tourism; b) motivation toward spiritual tourism; c) student knowledge about whether spiritual tourism is obligation or not.

4 RESULT AND DISCUSSION

4.1 *Result*

The sample consists of 157 Christian teenager respondents. Based on the semester, respondents consisted of 52 sixth semester students (33.1 percent) and 105 second semester students (66.9 percent). Based on the activity in the ministry in the church, 86 (54.78 percent) respondents were actively involved in the ministry and the remaining 71 (45.22 percent) students were not involved in the ministry.

The first question is about the level of interest of students to take a spiritual tour to Israel and places written of in the Bible. This question is divided into four levels of answers, which are very interested, interested, not interested, and is very not interested.

From all respondents, it was found that as many as 66 (42.04 percent) respondents stated very interested, as many as 69 (43.95 percent) expressed interest, as many as 20 (12.74 percent) expressed no interest, and lastly as many as 2 (1.27 percent) stated that they were not interested, shown by the graph in Figure 1.

The next question asked by the respondent is to look at the respondent's perspective on the relationship between spiritual tours and maturity of faith. Of respondents, 61 (38.85 percent) said there was a connection, 45 (28.66 percent) respondents said there was no connection, and as many

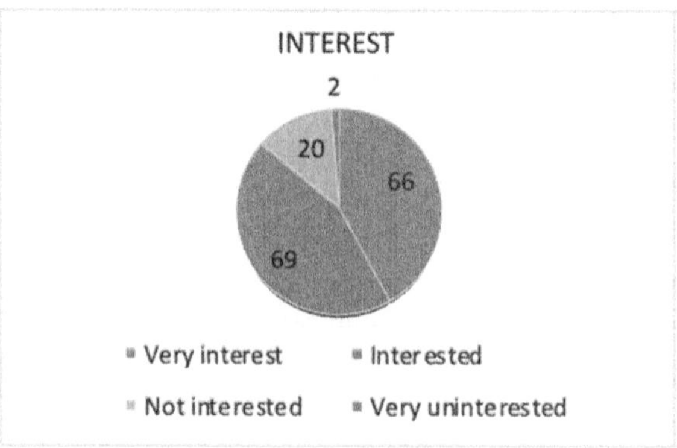

Figure 1. Percentage of student interest in taking a spiritual tour.

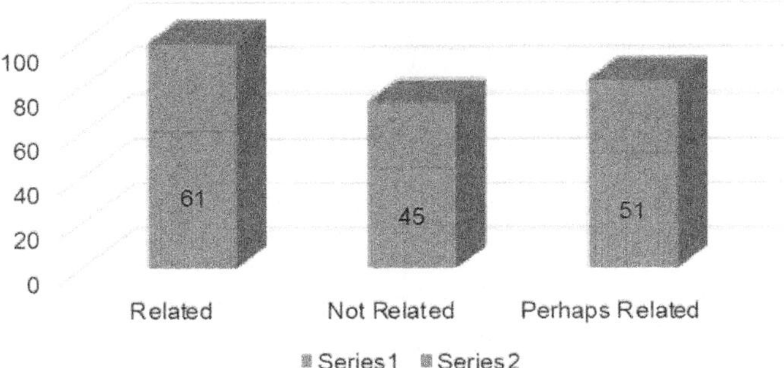

Figure 2. Respondents' perspective on the relationship between spiritual tours and maturity of faith.

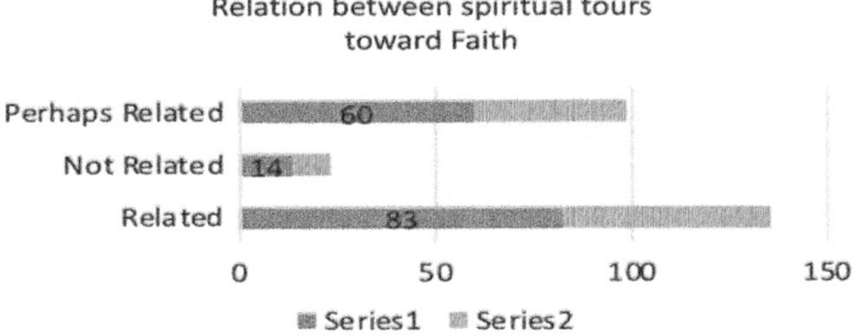

Figure 3. Respondents' perspective on the relationship between spiritual tours toward faith.

as 51 (32.48 percent) said there might be some relation between spiritual tours and the maturity of the faith. The following is displayed in graphical form.

The next question asked to respondents was student's knowledge about whether going on a spiritual tour would increase one's faith. This question has three answer choices: increase, no increase, and maybe increase.

A total of 83 (52.87 percent) stated that spiritual tours relate with the maturity of faith, 14 (8.92 percent) said that spiritual tours did not relate with the maturity of faith, and as many as 60 (38.22 percent) respondents said that spiritual tours perhaps related with the maturity of faith. The result can be seen in Figure 3.

The next question looks at students' perceptions of the costs of going on a spiritual tour. At present, travel agents generally charge a one-time spiritual tour fee of IDR 30,000,000 per person. Students are asked to give an opinion on whether the amount is equivalent or not. As many as 53 (33.76 percent) stated that it was equivalent, 18 (11.46 percent) stated it was not worth it, and as many as 86 (54.78 percent) respondents said it might be equivalent, as shown in Figure 4.

The next question is to look at participants' knowledge of spiritual tours in the Bible. The question asked is whether there are commands in the Bible to tour or travel to Israel.

As many as 35 (22.29 percent) respondents said that spiritual tours might be compulsory, as many as 9 (5.73 percent) said that spiritual tours to Israel were mandatory, and as many as 113 (71.97 percent) respondents said that spiritual tours to Israel were not mandatory. Results are shown in Figure 5.

The next question that was asked was about the student's decision to have money as much as the cost of a spiritual tour, and whether to go on a spiritual tour or use the money for other purposes.

As many as 75 (47.77 percent) respondents said they would go on a spiritual tour, while 82 (52.23 percent) said they would now use the money for other purposes. Results are shown in Figure 6.

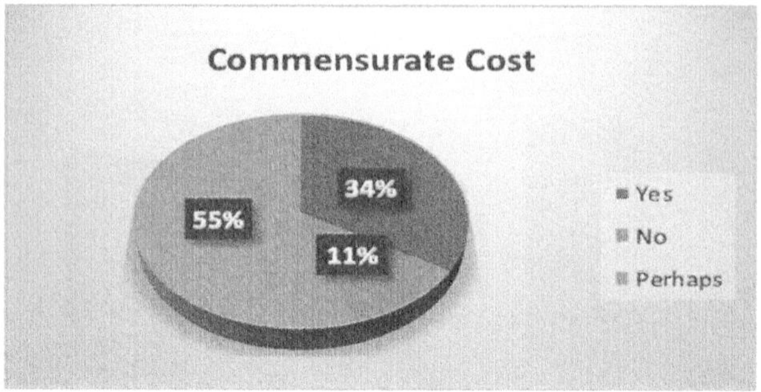

Figure 4. Respondents' perspective about cost.

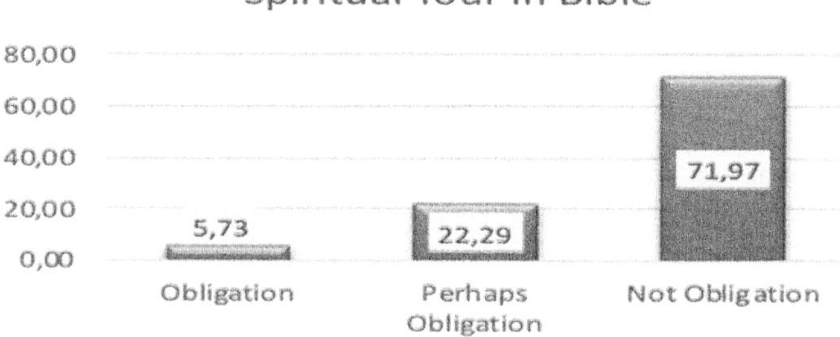

Figure 5. Students' knowledge about spiritual tourism based on the Bible.

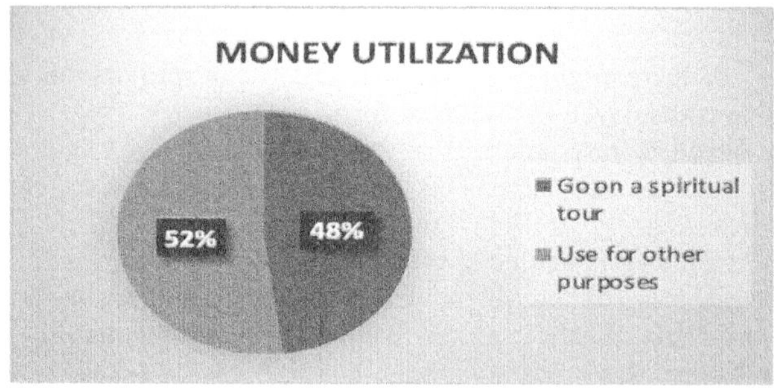

Figure 6. Respondents' decision at this time.

Based on the survey results, it can be seen that the main reason many students who are interested in doing spiritual tours is not to fulfill religious obligations, because most students know that in the Bible, spiritual tours are not an obligation.

And in this study, respondents were also asked the reason why they wanted to take a spiritual tour to Israel, and the following was the answer from the respondents who stated that they were very interested in and were interested in doing a spiritual tour, of 135 respondents.

As many as 39 (24.84 percent) stated that the reason for wanting to take a spiritual tour was because the spiritual tour was the same as tourism and vacation, while 96 (61.15 percent) respondents stated that the main reason why they wanted to take a spiritual tour was to see firsthand the places mentioned or written od in the Bible so that they could can feel and see part of the history of Christianity in the world. Results can be seen in Figure 7.

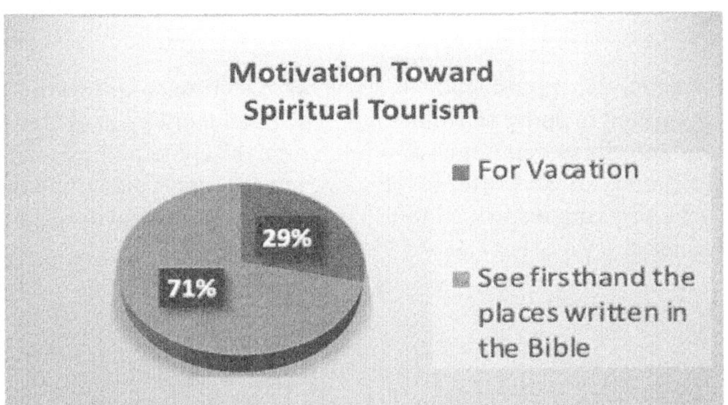

Figure 7. Motivation toward spiritual tourism.

In general, here are some things obtained from the results of this survey

- Student interest toward spiritual tourism
  From all respondents, it was found that the percentage who wanted to take a spiritual tour (interest and very interest) was high at 85.99 percent. It was also found that the majority of respondents considered that taking a spiritual tour would increase one's faith maturity. Another thing that was found from this research is that cost is not a factor that causes students to not be interested in doing a spiritual tour. In addition, it was also found that the majority of respondents (71.97 percent) knew that taking a spiritual tour was not an obligation when referring to the Bible.
- Student motivation toward spiritual tourism.
  Based on the results obtained, it was found that the factors that encourage students interested in taking a spiritual tour can be divided into two categories, namely spiritual reasons and vacation reasons. While respondents who fall into the category of vacation reasons are students who answered that they wanted to go to Israel, Egypt, and Jordan and other holy places, among others who answered wanted to travel abroad, wanted recreation, and wanted a holiday. Respondents who fall into the category of spiritual reasons are students who answered that the main reason for wanting to take a spiritual tour is to look directly at the places mentioned or written of in the Bible so that they can feel and see parts of the history of Christianity in the world.
- Students' knowledge about whether spiritual tourism is an obligation or not
  From the data obtained, it was found that the majority of respondents (71.97 percent) knew that taking a spiritual tour was not an obligation when referring to the Bible. Thus, the main reason students are interested in taking a spiritual tour is not because of religious obligations.

4.2 *Discussion*

The results of this study are also supported by previous research conducted by Alexandra M. Drule (A new perspective of non-religious motivations of visitors to sacred sites: evidence from Romania) which states that for 60% of respondents, the religious motivation is the main reason to visits the sacred sites.

What is interesting to study is the connection between being active in service in the church and interest in taking a spiritual tour.

In addition, another interesting thing to study further is to see whether there is a link between the interest in conducting a spiritual tour if someone has a family or friends who have taken a spiritual tour.

Another thing that can be done is to choose respondents who have already taken a spiritual tour and see an interest in returning to the spiritual tour and whether the spiritual tour can increase the maturity of their faith.

5 CONCLUSION

From this study, it was found that most students knew that a spiritual tour was not an obligation, but students were still interested in doing spiritual tours. The main motivation of students on a spiritual tour is to see firsthand the places mentioned or written of in the Bible so they can feel and see a part of the history of Christianity in the world. In addition, most students also think that visiting places written of in the Bible or taking a spiritual tour has to do with the growth of faith and influences the maturity of one's faith.

REFERENCES

Cohen, E. (1972). Toward sociology of international tourism. *Social Research*: 39.

Correia, A., & Crouch, G.I. (2004). Tourist perceptions of and motivation for visiting Algarve, Portugal. *Tourism Analysis*. 8: 165–9.

Crompton, J. (1979). Motivations of pleasure vacations. *Annals of Tourism Research*. VI(4):408–424.

Drule M. Alexander. (2012) "A new perspective of non-religious motivations of visitors to sacred sites: evidence from Romania", *Elsevier*.

Fedele, A. (2013). *Looking for Mary Magdalene: alternative pilgrimage and ritual creativity at Catholic shrines in France*. Oxford, UK: Oxford University Press

Gnoth, J. (1997). Tourism motivation and expectation formation. *Annals of Tourism Research*. 24, 283–304.

Hall, M. C. (2006). Travel and journeying on the sea of faith: Perspectives from religious humanism. In D. J. Timothy, & D. H. Olsen (Eds.), *Tourism, religion and spiritual journeys*. London, UK: Routledge.

Heelas, P., & Woodhead, L. (2005). *The spiritual revolution: Why religion is giving way to spirituality.* Oxford, UK: Blackwell Publishing

Jang, S.C., Morrison, A.M., & O'Leary, J.T. (2002), Benefit segmentation of Japanese Pleasure Travelers to the USA and Canada; selecting target markets based on the profitability and risk of individual market segments. *Tourism Management*. 23 (2002) 367–378.

Kau, A.K., & Lim, P.S. (2005). Clustering of Chinese tourists to Singapore: an analysis of their motivations, values and satisfaction. *International Journal of Tourism Research*. 7 (4–5). doi: https://doi.org/10.1002/jtr.537

Kozak, M. (2002). Comparative assessment of tourist satisfaction with destinations across 2 nationalities. *Tourism Management*. 22(4):391–401.

McCabe, A.S. (2000). Tourism motivation process. *Annals of Tourism Research*. 27(4):1049–1052

Meng, F., Tepanon, Y., & Uysal, M. (2008). Measuring tourist satisfaction by attribute and motivation: The case of a nature-based resort. *Journal of Vacation Marketing*. 14(1):41–56.

Mohammad, B.A.A.M.A-H., & Som, A.P.M. (2010). An analysis of push and pull travel motivations of foreign tourists to Jordan. *International Journal of Business and Management*. 5(12), 41–50.

P. Kotler and G. Armstrong. (2008). *Prinsip-Prinsip Pemasaran, 1st ed.* Jakarta: Erlangga

Pearce, P. L. (1988). *The Ulysses Factor: Evaluating Visitors in Tourist Settings*. New York: Springer-Verlag

Robbins, P. Stephen, and A. Timothy. (2008). *Perilaku Organisasi.* Jakarta: PT. Salemba Empat

Ruback, Pandey & Kohli. (2008). Evaluations of a sacred place: Role and religious belief at the Magh Mela. *Journal of Environmental Psychology* 28(2):174–184. doi: 10.1016/j.jenvp.2007.10.007.

Ryan, C., & Mo, X. (2001). Chinese visitors to New Zealand – demographics and perceptions. *Journal of Vacation Marketing*. 8(1): 13–27.

Taylor, C. (2002). *Varieties of religion today: William James revisited.* Cambridge, Massachusetts, USA: Harvard University Press

Uysal, M., & Jurowski, C. (1993). An empirical testing of the push and pull factors of tourist motivations. *Annals of Tourism Research*. 21(4): 844–846.

Wilson, G. B., McIntosh, A. J., & Zahra, A. L. (2013). Tourism and spirituality: A phenomenological analysis. *Annals of Tourism Research*. 42, 150–168.

Wood, M. (2007). *Possession, power and the new age*. Belfast, UK: Ashgate

Yuan, S., & McDonald, C. (1990). Motivational determinants of international pleasure time. *Journal of Travel Research*. 29 (1): 42–44

Promoting Creative Tourism: Current Issues in Tourism Research – Kusumah et al. (Eds)

Cultural tourism: Commercialization or preservation?

E. Malihah, S. Komariah, N.F. Utami & E. Prakarsa
Universitas Pendidikan Indonesia, Bandung, Indonesia

ABSTRACT: For tourists, cultural tourism is an attraction for traveling, while for the community it can be a source of income and empowerment because cultural heritage is still preserved. However, the question is whether the culture is really preserved by the community as a cultural heritage or for commercial purposes. This study used a qualitative approach with participatory observation methods and in-depth interviews with traditional leaders and the Sasak Sade community. The results of this study indicate that people must continue to maintain the cultural heritage given by their ancestors because this culture provides economic benefits for the Sade community.

Keywords: cultural tourism, cultural commercialization, cultural preservation

1 INTRODUCTION

Culture is often killed by a noble, holy, and full of sacred society (Yunita & Sugiarti 2019). Culture is formed through a long process, from the start of its emergence, then to adapt or disappear instead of a new culture (Roseta 2020). So when talking about the history of the birth of culture, it is as old as the history of the development of society, because society and culture are two elements that cannot be separated from each other (Hadi & Bogor 2019). Culture has a very wide environment in people's lives since culture can be in the form of creativity, taste, human effort, the environment, life experiences to the glory of life in people's lives (Kian et al. 2019). The development of technology and information also has a big impact on the existence of culture. The dominance of foreign culture and popular culture in Indonesian society makes the original cultural identity fade and weaken. The strong flow of information that is unstoppable and the weakening of indigenous cultural values of Indonesian people cause other indigenous cultures to be left behind (Matondang et al. 2017). Protecting culture in this globalization and modernization era is a very difficult challenge. Therefore, people's perceptions about culture are only interpreted as something that is unfounded, absurd, beyond reason, and does not have any material benefits.

But in reality, the increasingly rapid flow of information and technology can also be seen as having benefits to the existence of culture (Ergashev & Farxodjonova 2020). The existence of cultural tourism provides a new opportunity that culture is not only in the form of sacred dogmas that do not have a reasonable foundation, but are able to provide economic value for the people who continue to guard it (Bakas et al. 2020) The better people maintain the cultural heritage of their ancestors, the higher the culture provides economic value. The presence of cultural tourism has expanded the benefits of culture itself, even exceeding economic values, such as preserving nature, environment and resources, strengthening national identity and unity, to strengthen humanity (Taff et al. 2019). As a tourist activity that cannot be separated from tourist attraction (cultural attraction), cultural tourism has an even greater impact, because in cultural tourism there is a complex integration between attractions, accommodation, and tourist facilities that are integrated with the structure and procedures for community life. Cultural tourism also functions to develop social capital as the main resource in the process (Harianto et al. 2020). Advances in information technology also provide wider reach in promoting cultural tourism. In Indonesia, cultural tourism that is already well known

DOI 10.1201/9781003095484-14

to the international level with the promotion of media is increasingly popular (Goeltom et al. 2020), such as Tanah Lot Temple in Bali, Borobudur Temple in Central Java, Prambanan Temple in Sleman, Uluwatu Temple in Bali, and Sasak Sade Tribe in Lombok.

Based on this background, it is important to conduct research to illustrate how this cultural tourism is commercialized and how the efforts of indigenous people who have unique cultural characteristics maintain and preserve their culture. Through these fundamental questions, it will provide a new pattern which certainly can contribute to the development of economic and social and cultural tourism.

2 LITERATURE REVIEW

Village tourism is one of the many types of tourism that continues to be developed today. Village tourism is closely related to community-based tourism because society is the main element or the most influential actor in the development of tourism itself. The tourism village was developed based on the principles of community-based tourism development (Kristiana et al. 2019). Various tourism village potentials such as physical potential (natural and environmental) and non-physical potential (community and culture) are social capital and important indicators in the development of tourism villages (Shah 2019).

Other important components in the development of a tourist village are attractions, accessibility, and facilities (Abdulhaji & Yusuf 2013). These three components are very important in developing a tourism village in order to create a sustainable tourism village area. (Cerutti & Piva 2020). The development of tourism villages has a huge impact on the community and the environment. The existence of a tourist village increases public awareness of nature conservation. The existence of a tourist village also affirms the social values of the people of an area, and increases the capacity of the community and brings an increase in the community's economy due to the opening of these economic opportunities (Kristiana et al. 2019; Amir et al. 2017). The commercialization of the tourism village as one of the tourism activities increases the economic income of a community. The increase in income also directly increases the people's standard of living for the better. (Amir et al. 2017). In addition, the existence of a tourist village also indirectly expands the interaction of people from a variety of different identities. In this interaction, the process of mutual understanding of each other's values and cultures occurs. This is directly a process of national integration (Eusébio et al. 2016).

Many researchers have conducted research on tourist villages, be it about the community, customs, or local wisdom. However, this study focuses on cultural heritage, which is used as a tourist attraction, which certainly has economic value for the local community.

3 METHOD

The approach used in this study is a qualitative approach using participatory observation and in-depth interviews. The researcher made observations during one month in the traditional Sasak Sade village mingling with the community so that he knew the life of the Sasak Sade indigenous people. Besides, the researchers participated in activities from morning to night, from tourism activities to quiet tourist visits. Actually, the Sasak Sade tourist village has no operating hours because they think the place they live in is not like a museum so tourists can come to the Sasak Sade tourist village at any time. In-depth interviews were conducted with traditional leaders as leaders of indigenous peoples and those who know the history of the traditional Sasak Sade culture. Besides, the researchers interviewed other traditional leaders such as the head of the neighborhood association, the head of the community of Sasak Sade tourism actors, the Sasak Sade community who became the tour guide, government stakeholders (such as village officials), and heads of tourism offices. From various informants interviewed, the researcher conducted data triangulation and member checks to ensure the validity of the data from the informants after the data was collected and then carried out data analysis.

4 RESULT AND DISCUSSION

Sade is located in Rembitan Village, Central Lombok Regency, West Nusa Tenggara Province (NTB). Sade Hamlet is one of the many tourist destinations in NTB. The geographical location of NTB, which lies on the Banda Aceh-Kupang national transportation route, brings considerable economic benefits. Sade Village is a tourist village that has a unique cultural tradition. The villagers of Sade still preserve these features in the midst of increasingly modern times. The various cultural identity features that are owned by the villagers of Sade make the village a tourist village whose existence has penetrated to foreign countries. One of the cultural features possessed by Sade Village is the structure of the building which is still maintained in its authenticity. The structure of the building was inherited by the ancestors of the Sade Citizens to this day. The reason why the Sade people still maintain the shape of the building is because they assume that the shape of the building is also a priceless inheritance and of course must be continually preserved. In addition to its unique building structure, Sade villagers also have other features in their social activities, like having an art show, having values in terms of marriage, and having the habit of weaving, and so forth.

In carrying out their social life, Sade people have a belief in manners called awiq-awiq. The manners are a set of value mechanisms that are believed to be together so as to create a harmonious life of the Sade villagers. In the beginning, Sade villagers were ordinary adat citizens who continued to maintain the authenticity of the culture and traditions of their ancestors. However, the growing development of tourism in the NTB region made Sade Village one of the Tourism Villages that also attracts many visitors. This was also supported by transportation access to Sade Village since Sade Village is passed by tourists who want to travel to Tanjung Aan Beach or Kuta Beach, so indirectly, Sade Village is passed by all tourists. Accessibility is indeed an important thing in a tourism process (Abdulhaji & Yusuf 2013) because the development of tourism processes that do not pay attention to accessibility will not properly develop. Accessibility has major implications for the course of the tourism process (Cerutti & Piva 2020). In carrying out a tourism activity, Sade villagers offer various privileges of their cultural values to tourists, like inviting tourists to get around the village to see the uniqueness of building houses and community activities. Guided by a Sade-native local guide, tourists get a history of the founding of Sade Village and all its sacred values. What was done by the Sade villagers indicated various principles of community-based tourism through community participation activities, nature, social and cultural conservation, and local economy enhancement (Kristiana et al. 2019). This is implicitly a positive impact of a cultural tourism activity (Shah 2019).

In addition to the above tourism activities, Sade villagers also invite tourists to try weaving cloth by learning to weave cloth into its own uniqueness. In this process, many interactions occur as the process of mutual understanding with each other. Interaction in the process of cultural tourism is very important, so it can be an effort to integrate national identity (Eusébio et al. 2016). Sade villagers also display various unique handmade creations such as fabrics, accessories, and woven bamboo in front of their houses. With the aim of commercialization, the craft has different prices. In this case, an increase in economic income is one of the potentials of community-based tourism which directly increases the standard of living of the people (Amir et al. 2017). In carrying out these various tourist activities, the villagers of Sade do not provide a certain rate. Sade villagers only provide donation boxes for tourists who visit. Through the box, tourists can give a sum of money, which will then be managed for the sustainability of tourism and the development of cultural tourism in the Sade Village.

Indirectly, it turns out Sade villagers feel the existence of economic values when they maintain their culture and then manage it into a tourism activity. An increase in the standard of living and income is a positive impact felt by them. The various positive impacts that they feel are manifestations of their determination to remain consistent in maintaining cultural values in the midst of intense social change.

5 CONCLUSION

Based on the explanation above, the Sade villagers still maintain the cultural values of their ancestors' heritage. Various cultural values such as building architecture, art attractions, and so on, remain preserved in the midst of increasingly modern times. Various cultures in the Sade Village later gave birth to the activity of a tourist village. These activities have an economic impact on the villagers of Sade. There has been an increase in the standard of living of Sade villagers as a result of ongoing tourism activities. This resulted in the Sade citizens loving the culture that their ancestors had inherited.

REFERENCES

Abdulhaji, S., & Yusuf, I. S. H. (2013). Pengaruh Atraksi, Aksesibilitas, dan Fasilitas terhadap Citra Objek Wisata Danau Tolire Besar di Kota Ternate. *Journal of Petrology, 369*(1), 1689–1699. https://doi.org/10.1017/CBO9781107415324.004

Amir, S., Osman, M. M., Bachok, S., Ibrahim, M., & Zen, I. (2017). Community-based tourism in Melaka UNESCO world heritage area: A success in food and beverage sector? *Planning Malaysia, 15*(1), 89–108. https://doi.org/10.21837/pmjournal.v15.i6.225

Bakas, D., Kostis, P., & Petrakis, P. (2020). Culture and labour productivity: An empirical investigation. *Economic Modelling, 85*, 233–243. https://doi.org/10.1016/j.econmod.2019.05.020

Cerutti, S., & Piva, E. (2020). Accessibility and sustainable tourism: a kaleidoscope of issues and perspectives. Introduction. *Journal of Research and Didactics in Geography*, 111–113. https://doi.org/10.4458/3099-10

Ergashev, I., & Farxodjonova, N. (2020). Integration of national culture in the process of globalization. *Journal of Critical Reviews, 7*(2), 477–479. https://doi.org/10.31838/jcr.07.02.90

Eusébio, C., Carneiro, M. J., & Caldeira, A. (2016). A structural equation model of tourism activities, social interaction and the impact of tourism on youth tourists' QOL. *International Journal of Tourism Policy, 6*(2), 85–108. https://doi.org/10.1504/IJTP.2016.077966

Goeltom, A. D. L., Wibowo, L. A., Hurriyati, R., & Gaffar, V. (2020). How the Internet Affects the Current Tourism Marketing Theory and Practice. *3rd Global Conference On Business, Management, and Entrepreneurship (GCBME 2018) How, 117*(Gcbme 2018), 53–56. https://doi.org/10.2991/aebmr.k.200131.012

Hadi, M. T., & Bogor, S. M. (2019). Bahasa sebagai pendidikan budaya dan karakter bangsa. *Jurnal Salaka, 1*, 1–4.

Harianto, S. P., Masruri, N. W., Winarno, G. D., Tsani, M. K., & Santoso, P. J. T. (2020). Development strategy for ecotourism management based on feasibility analysis of tourist attraction objects and perception of visitors and local communities. *Biodiversitas, 21*(2), 689–698. https://doi.org/10.13057/biodiv/d210235

Kian, D. A., Rayawulan, R. M., Mberu, Y., & Lily, B. B. (2019). Makna Ruang Dalam Budaya Masyarakat Sikka. *Jurnal Arsitektur KOMPOSISI, 12*(2), 105. https://doi.org/10.24002/jars.v12i2.2045

Kristiana, Y., Pakpahan, R., & Mulyono, S. T. (2019). Pengembangan Pariwisata Berbasis Masyarakat Di Kawasan Seberang Kota Jambi (Sekoja). *Prosiding Konferensi Nasional Pengabdian Kepada Masyarakat Dan Corporate Social Responsibility (PKM-CSR), 2*, 1047–1053. https://doi.org/10.37695/pkmcsr.v2i0.274

Matondang, A., Lubis, Y. A., & Suharyanto, A. (2017). Eksistensi Budaya Lokal Dalam Usaha Pembangunan Karater Siswa Smp Kota Padang Sidimpuan. *Anthropos: Jurnal Antropologi Sosial Dan Budaya (Journal of Social and AN Anthropology), 3*(2), 103. https://doi.org/10.24114/antro.v3i2.8306

Roseta, C. I. (2020). *Dakwah Antarbudaya: Perubahan Sosial udaya Pada Proses Islamisasi Jawa Abad XV. 01*(02), 163–186.

Syah, D. P. (2019). Pengembangan Pariwisata Berbasis Masyarakat (Community Based Tourism). *Eprints.Ums.Ac.Id, 1*(1), 1–13. https://doi.org/10.1017/CBO9781107415324.004

Taff, B. D., Benfield, J., Miller, Z. D., D'antonio, A., & Schwartz, F. (2019). The role of tourism impacts on cultural ecosystem services. *Environments – MDPI, 6*(4), 1–12. https://doi.org/10.3390/environments6040043

Yunita, G. F. R., & Sugiarti. (2019). Kajian Mitos dalam Novel Aroma Karsa Karya Dewi Lestari Perspektif Ekologi Budaya. *Lensa: Kajian Kebahasaan, Kesusastraan, Dan Budaya, 9*(2), 156. https://doi.org/10.26714/lensa.9.2.2019.156-173

Destination management

Promoting Creative Tourism: Current Issues in Tourism Research – Kusumah et al. (Eds)
© 2021 Taylor & Francis Group, London, ISBN 978-0-367-55862-8

The crucial attributes for culinary tourism destination based on tourists' perception

T. Abdullah
University of Otago, Dunedin, New Zealand

Gitasiswhara
Universitas Pendidikan Indonesia, Bandung, Indonesia

R.S. Nugraha
Bandung Institute of Tourism, Bandung, Indonesia

ABSTRACT: Culinary tourism has economic and socio-cultural benefits. Besides providing employment, this tourism activity plays a role in preserving traditional food and beverage of a region. This study was conducted in West Java Province to assess tourists' perceptions of culinary tourism in this province and to identify some attributes considered crucial for tourists when visiting culinary tourism destination. This study used surveys to collect the data, which were gathered from 550 participants. These participants were tourists who visited various districts and cities in West Java, Indonesia. The findings show that some attributes in culinary tourist destinations were considered more crucial than others with the most crucial attribute is the accessibility and public transportation.

Keywords: culinary tourism attributes, tourist's perception, tourist destination, tourism activities

1 INTRODUCTION

Tourism activities continue to grow rapidly, along with the advancement of technology, and all of tourism stakeholders try to take advantage of this improvement, including tourists. Nowadays, tourists could easily access valuable information to plan their trips. Planning a long journey was once considered confusing and frightening for some people; now, it is more comfortable to be done by anyone. The information about tourist destinations, transportation, travel documents, hotels, and restaurants can be found easily on the internet. Tourists can become tour planners for themselves and gain active experience when traveling (López-Guzmán et al. 2014).

When planning their trip, tourists can be driven by many interests. Tourism attractions and activities mostly are the primary purposes for tourists to visit a destination, and the availability of amenities are needed to support it. One of the essential amenities is the foodservice industry. Food has always been a crucial element in supporting tourism activities. However, currently, there are tourists whose main purpose when visiting an area is to enjoy or to taste its local food (López-Guzmán & Sánchez-Cañizares 2012).

The desire to travel to a place to experience unique and authentic food emerged as a new phenomenon in the tourism industry (Smith & Costello 2009). This type of tourism activity is often called culinary tourism. In this type of tourism, people are looking for an interesting, unique and memorable dining experience. Based on the results of a study by Björk and Kauppinen-Räisänen (2014), this experience is formed by consuming local and authentic cuisine which represent the culture of the destination. Therefore, the existence of local and traditional food becomes crucial as the main attraction of culinary tourism activities. Moreover, food is an important part of the culture of an area and it manifests intangible cultural heritage (Updhyay & Sharma 2014). Culinary tourism also shares some economic benefits and one of them is opening opportunities for local people to start small businesses of food and beverage (Sotiriadis 2015). Hence, it does not only provide job opportunities, but also allowing local people to create more various foods through their creativity and later could enhance the attraction of a culinary tourism destination. Culinary tourism is a great potential both as a tourism product and as a marketing tool for tourist destinations. Culinary tourism activities can be a way to promote a destination because food is related with the place and identity (Everett 2008).

This study was conducted to assess tourists' perceptions of a culinary tourism destination and to identify some attributes considered crucial for tourists when visiting culinary tourism destination. Many studies have identified tourism destination attributes. For instance, some literature focused on general tourism destination attributes (e.g., Buhalis 2000; Chahal & Devi 2015; Davis & Sternquist 1987; Mill 2010; Scott et al. 1978; Yüksel & Yüksel 2001). Some others discussed destination attributes in a more specific type of tourism, such as cultural tourism product (Cave et al. 2007), nature-based resort destination attributes (Meng et al. 2008), and ecotourism product attributes (Abdurahman et al. 2016). There were several studies concerning the attributes related to gastronomy or food, and also attributes of the foodservice establishment (Adhikari, 2014; Burusnukul et al. 2011; Sanchez-Cañizares & Castillo-Canalejo 2015). However, the study on culinary tourism attributes is still under-researched. This study was conducted to explore the attributes of culinary tourism considered important by tourists.

2 LITERATURE REVIEW

Attractions and activities often become the main references for tourists when visiting a destination. However, Björk and Kauppinen-Räisänen (2014) argue that food can be one of the evaluation criteria when choosing a travel destination. According to López-Guzmán, Di-Clemente, and Hernández-Mogollón (2014), nowadays, food can even be considered as one of the primary motivations for tourists to visit a destination. Travelers can be attracted to visit a destination to get a dining experience in a restaurant with Michelin Stars predicate, or also in a local restaurant that specializes in providing local cuisine (Björk & Kauppinen-Räisänen 2014).

A type of tourism where the consumption of local food (including drinks) or observing and studying food production (from agriculture or cooking schools) become its motivation and activities can be called as culinary tourism (Ignatov & Smith 2006). There are several other terms which have similar meaning with culinary tourism, namely cuisine tourism, food tourism, gourmet tourism, gastronomy tourism, and gastronomic tourism (López-Guzmán & Sánchez-Cañizares 2012). According to Updhyay and Sharma (2014), all of them are part of food tourism. However, Ignatov and Smith (2006) stated that the most appropriate term is culinary tourism.

Culinary tourism is one of the subcategories of cultural tourism (Updhyay & Sharma 2014). Like other forms of culture, food has a high selling value. Therefore, local food is a valuable resource in tourism and can be the marketing tools (Björk & Kauppinen-Räisänen 2016). Other studies indicate that gastronomy is one of the critical factors which is play an important role in developing and promoting tourism (López-Guzmán & Sánchez-Cañizares 2012; Smith & Costello 2009).

As explained previously, local food has a significant role in tourism, especially culinary tourism. Björk and Kauppinen-Räisänen (2016) explained that local food is food served at a destination, including special food that originates from that place, and food prepared using local raw materials. In daily life, the term of local food is sometimes used interchangeably with traditional food although there is a slightly different meaning between them. Traditional food is a product that is often consumed in or related to certain celebrations and/or seasons. The culture is taught from one generation to another. It is made in a certain way according to a specific gastronomic heritage, processed naturally, distinguished and known for their sensory properties, and related to some local regions, states or countries (Vanhonacker et al. 2006). Concerning Indonesian local and traditional food, Holzen (2006) described that the characteristics of Indonesian food as steamed

Table 1. Questionnaires distribution points.

Distribution Points	Area Coverage	Number of Questionnaires Distributed
1	Bandung and Cimahi	50
2	Bekasi and Bogor	50
3	Majalengka and Kuningan	50
4	Cirebon and Indramayu	50
5	Sukabumi	50
6	Cianjur	50
7	Purwakarta, Subang, and Karawang	50
8	Ciamis and Banjar	50
9	Tasikmalaya	50
10	Garut	50
11	Sumedang	50
	Total	550

rice accompanied by some savory side dishes such as vegetables, fish or maybe meat and poultry dishes with hot chili seasonings or sambal on the side, and also beans, crispy crackers, and fried shallots sprinkled on top to give a contrasting crunchy taste. Holzen states that this illustration is most appropriate for food from Sumatra, Java, and Bali.

López-Guzmán and Sánchez-Cañizares (2012) stated that if a tourist destination wants to promote its place using its local food, some key points need to be considered, namely; 1) the culinary resources must be differ-ent from other regions, 2) having the local cuisine recognized by tourists, 3) having a lot of food and beverage service businesses so tourists can find the local food easily. According to Ignatov and Smith (2006), the supply-sides of culinary tourism are facilities, activities, events, and organizations. This supply-side needs to be available so that culinary tourism activities can be alive.

3 RESEARCH METHOD

This study was conducted in West Java Province, Indonesia. As one of the provinces in Indonesia, it has many tourist attractions and also has potentials in culinary tourism. There are many varieties of local food in this province, and its traditional food is Sundanese cuisine. The administrative division of West Java Province consists of 18 regencies and 9 cities. The population in this study was tourists who visited various regions in West Java Province. There were 550 participants in this study. In order to collect the data, questionnaires were distributed in 11 different places within this province, as depicted in Table 1. This study employed 24 surveyors who live in throughout all of these districts and cities. They were divided into 11 groups according to these geographical locations. Some districts are merged as one point of distributions because their areas are close to each other. Hence, the sampling procedure used in this study was cluster sampling method. The surveyors were firstly briefed before the went to distribute the questionnaires. Afterward, they distributed the questionnaires at some tourist attractions in those districts or cities.

Two questions in this survey used five Likert scale answers, namely the viability level of the region to be developed as a culinary tourism destination and the tourists' willingness level to enjoy traditional food. The data were then calculated using simple statistic calculation, namely mean score and standard deviation. After the average scores were determined, they were classified according to the categories illustrated in Table 2.

This survey also included two open-ended questions; the first one was to explore tourists' motivation when visiting the tourist destination, and the second was to identify attributes considered

Table 2. Likert scale categories.

Average Score Range	Viability Level	Willingness Level
4.2–5	Highly viable	Willing
3.4–4.2	Viable	Somewhat willing
2.6–3.4	Neutral	Neutral
1.8–2.6	Not viable	Not really willing
1–1.8	Highly not viable	Not willing

Table 3. The characteristics of participants.

Characteristics	Frequency	Per cent
Gender		
Male	301	54.7
Female	249	45.3
Total	550	100
Age		
17–20	158	28.73
21–30	283	51.45
31–40	68	12.36
>40	41	7.45
Total	550	100
Occupation		
Private employees	91	16.55
Students	307	55.82
Civil servants	53	9.64
Entrepreneurs	76	13.82
Others	23	4.18
Total	550	100

essential for a culinary tourism destination. The answers from participants later were coded to be used for analysis.

4 RESULTS AND DISCUSSION

As illustrated in Table 3, the number of male participants was higher than women, but all in all, the percentage of these two genders was almost even. Based on the participants' age, the majority of them were 20–30 years old, and most of them were students.

Table 4 depicts tourists' perception for the viability level of the area in West Java Province to be developed as culinary tourism destinations, and the willingness level of tourists to consume traditional food. These levels were measured by five Likert scales and have been categorized according to the classification in Table 2.

Based on Table 4, although participants believed that all of the regions in this province were viable to be developed as culinary tourism destinations, Sukabumi was the area assumed to be 'highly viable'. Most of the regions (if not all) have their unique local food, and domestic tourists, especially those who are from other areas within West Java, recognize this food. Hence, local people might not be the only ones who know precisely the availability of the most well known local and traditional cuisine. Domestic tourists who live in the same province might also be aware of the availability of this food; they also know where to buy this food. On average, participants

Table 4. Tourists' perception of West Java province as culinary tourism destination.

	Viability Level			Willingness Level		
Area	Mean	SD	Category	Mean	SD	Category
Bandung and Cimahi	4.18	0.48	Viable	4.28	0.5	Willing
Bekasi and Bogor	4.02	0.59	Viable	4	0.86	Somewhat willing
Majalengka and Kuningan	3.48	0.5	Viable	3.48	0.65	Somewhat willing
Cirebon and Indramayu	4	0.4	Viable	3.72	0.61	Somewhat willing
Sukabumi	4.6	0.61	Highly Viable	3.8	0.78	Somewhat willing
Cianjur	3.94	0.65	Viable	3.9	0.71	Somewhat willing
Purwakarta, Subang, and Karawang	3.84	0.55	Viable	4.1	0.65	Somewhat willing
Ciamis and Banjar	3.88	0.52	Viable	3.8	0.61	Somewhat willing
Tasikmalaya	3.86	0.53	Viable	3.6	0.76	Somewhat willing
Garut	4	0.7	Viable	3.66	0.98	Somewhat willing
Sumedang	3.76	0.62	Viable	3.9	0.58	Somewhat willing

Table 5. Tourists' motivation when visiting West Java province.

Rank	Motivation	Frequency
1	Recreation and Leisure	216
2	Visiting Friends and Relatives (VFR)	192
3	Culinary Tourism	48
4	Business trip	16
5	Educational, Historical and Cultural Tourism	11
6	Transit only	11
7	Shopping Tourism	11
8	Visiting or Celebrating Events	8
9	Sport Tourism	3
10	Others	34
Total		550

were willing to try and consume traditional food. However, participants who were surveyed in Bandung and Cimahi had a higher willingness level to enjoy the traditional cuisine. Most of the participants' responses, which were 'somewhat willing' is in line with their travel motivations, as can be examined in Table 5, which indicates that most of the participants did not travel to the destination for culinary reasons.

As depicted in Table 5, the number of tourists who went to these areas in West Java for culinary tourism pur-poses was only 48 people. Most of the tourists visited various areas in West Java Province to do recreation and leisure activities ($n = 216$), and also to visit their friends and relatives ($n = 192$). Given the fact that their willingness to consume traditional food mostly were only at the level of 'somewhat willing' and most of their travel motivation were not to enjoy local food, hence culinary tourism activities still were not popular for tourists in West Java Province.

The findings depicted in Table 6 were based on participants' responses to an open-ended question. They were asked what the most important thing that a culinary tourist destination should have. Most of the participants believed that the ease of accessibility and the availability of comfortable and reliable public transportation was the most important thing. The accessibility was considered crucial due to the actual condition in West Java Province (e.g., congestion, damaged road). The problems of public transportations in this province also contributed to the assessment of participants in this matter. Since some tourists might not have a private car or motorcycle, the availability of fast, comfortable, safe, and affordable public transportation is necessary for a culinary tourism destination. Hence, tourists could easily reach the culinary tourism destination. These findings

Table 6. Attributes for culinary tourism destination which are considered important by tourists.

Rank	Attributes	Frequency
1	Accessibility and public transportation	175
2	Centralized culinary tourism area	161
3	Cleanliness	41
4	Accommodations and restaurants	40
5	Public facilities (e.g., parking area, ATM)	25
6	Promotion	21
7	High quality and various traditional foods	17
8	Food souvenir shop	10
9	Culinary event	7
10	Availability of Wi-Fi internet / communication network	6
11	Online transportation services	4
12	The existence of food trucks/street food vendors	6
13	Scenery	1
14	No idea	36
Total		550

emphasized that, even though a tourist destination owns an outstanding core product (i.e., attraction or activities), if it is not supported by good accessibility, it would create unpleasant experiences for tourists.

The second important thing was the existence of a centralized culinary tourism area, this could be in the form of a food market. In this market, many vendors may provide a variety of local and traditional food and beverage. However, this food market cannot stand alone, it has to be supported with other attributes such as a high level of cleanliness and decent public facilities (e.g., ATMs, toilets, internet connection, network, and adequate parking space) and good accessibility.

In Table 6, it is also noticeable that a small number of participants felt that a culinary tourist destination needs to have high quality and variety of traditional foods ($n = 17$). The rank of this attribute might not be the highest, but it is believed as a vital aspect of culinary tourism. Unexpectedly, participants were more focused on the existence of food service providers, which was included in some attributes, namely centralized culinary tourism areas, accommodations and restaurants, food souvenir shops, culinary events, and the food trucks/street food vendors. Hence, it can be argued that the number of food service providers in a culinary tourist destination is more crucial than the availability of high quality and variety of local or traditional food.

5 CONCLUSION

The results show that to enjoy culinary products was not the primary motivation for the majority of tourists who came to West Java Province. In this study, we argue that the high number of food service providers in a culinary tourist destination is more crucial than the availability of high quality traditional or local food. These providers are more preferred if they are centralized in an area such as the food market. In our survey, accessibility and transportation were the most vital aspects in the development of culinary tourism. As a developing country, the acceleration of the development of infrastructure and public transportation in Indonesia is necessary to strengthen culinary tourism development. The results of this study can become a useful reference, especially for the local authority in developing their districts or cities to be culinary tourism destinations. Some attributes uncovered in this study should be prioritized as they are things that are considered essential by tourists for culinary tourism destinations. This study has some limitations, hence it is highly suggested for future studies to include more participants and to implement different methods to understand more about the culinary tourism attributes.

REFERENCES

Abdurahman, A. Z. A., Ali, J. K., Khedif, L. Y. B., Bohari, Z., Ahmad, J. A., & Kibat, S. A. (2016). Ecotourism Product Attributes and Tourist Attractions: UiTM Undergraduate Studies. *Procedia – Social and Behavioral Sciences, 224*, 360–367. https://doi.org/10.1016/j.sbspro.2016.05.388

Adhikari, A. (2014). Differentiating Subjective and Objective Attributes of Experience Products to Estimate Willingness to Pay Price Premium. *Journal of Travel Research*, 1–11. https://doi.org/10.1177/0047287514532366

Björk, P., & Kauppinen-Räisänen, H. (2014). Culinary-gastronomic tourism – a search for local food experiences. *Nutrition & Food Science, 44*(4), 294–309. https://doi.org/10.1108/NFS-12-2013-0142

Björk, P., & Kauppinen-Räisänen, H. (2016). Local food: a source for destination attraction. *International Journal of Contemporary Hospitality Management, 28*(1), 177–194. https://doi.org/10.1108/IJCHM-05-2014-0214

Buhalis, D. (2000). Marketing the competitive destination of the future, *Tourism Management, 21*(1), 97–116

Burusnukul, P., Binkley, M., & Sukalakamala, P. (2011). Understanding tourists' patronage of Thailand foodservice establishments An exploratory decisional attribute approach. *British Food Journal, 113*(8), 965981. https://doi.org/10.1108/00070701111153733

Cave, J., Ryan, C., & Panakera, C. (2007). Cultural Tourism Product: Pacific Island Migrant Perspectives in New Zealand. *Journal of Travel Research, 45*(May), 435–443. https://doi.org/10.1177/0047287506295908

Chahal, H., & Devi, A. (2015). Destination Attributes and Destination Image Relationship in Volatile Tourist Destination: Role of Perceived Risk. *Metamorphosis*, 14(2), 1–19.

Davis, B. D., & Sternquist, B. (1987). Appealing to the Elusive Tourist: An Attribute Cluster Strategy. *Journal of Travel Research, 25*, 25–31. https://doi.org/10.1177/004728758702500405

Everett, S. (2008). Beyond the visual gaze: The pursuit of an embodied experience through food tourism. *Tourist Studies, 8*(3), 337–358. https://doi.org/10.1177/1468797608100594

Holzen, H. V., & Arsana, L. (2006). *The Food of Indonesia, Delicious Recipes from Bali, Java and the Spice Islands*. Hong Kong: Tuttle Publishing.

Ignatov, E., & Smith, S. (2006). Segmenting Canadian Culinary Tourists. *Current Issues in Tourism, 9*(3), 235–255. https://doi.org/10.2167/cit/229.0

López-Guzmán, T., Di-Clemente, E., & Hernández-Mogollón, J. M. (2014). Culinary tourists in the Spanish region of Extremadura, Spain. *Wine Economics and Policy, 3*(1), 10–18. https://doi.org/10.1016/j.wep.2014.02.002

López-Guzmán, T., & Sánchez-Cañizares, S. (2012). Culinary tourism in Córdoba (Spain). *British Food Journal, 114*(2), 168–179. https://doi.org/10.1108/00070701211202368

Meng, F., Tepanon, Y., & Uysal, M. (2008). Measuring tourist satisfaction by attribute and motivation: The case of a nature-based resort. *Journal of Vacation Marketing, 14*(1), 41–56. https://doi.org/10.1177/1356766707084218

Mill, R. C. (2010). *Tourism, The International Business*. Global Text Project

Sanchez-Cañizares, S., & Castillo-Canalejo, A. M. (2015). A comparative study of tourist attitudes towards culinary tourism in Spain and Slovenia. *British Food Journal, 117*(9). https://doi.org/10.1108/BFJ-01-2015-0008

Scott, D. R., Schewl, C. D., & Frederick, D. G. (1978). A Multi-Brand/Multi-Attribute Model of Tourist State Choice. *Journal of Travel Research, 17*(Summer), 23–29. https://doi.org/10.1177/004728757801700105

Smith, S., & Costello, C. (2009). Culinary tourism: Satisfaction with a culinary event utilizing importance-performance grid analysis. *Journal of Vacation Marketing, 15*(2), 99–110. https://doi.org/10.1177/1356766708100818

Sotiriadis, M. D. (2015). Culinary tourism assets and events: suggesting a strategic planning tool. *International Journal of Contemporary Hospitality Management, 27*(6), 1214–1232. https://doi.org/10.1108/IJCHM-11-2013-0519

Updhyay, Y., & Sharma, D. (2014). Culinary preferences of foreign tourists in India. *Journal of Vacation Marketing, 20*(1), 29–39. https://doi.org/10.1177/1356766713486143

Vanhonacker, F., Verbeke, W., Guerrero, L., Claret, A., Contel, M., Scalvedi, L., … Hersleth, M. (2006). How European Consumers Define the Concept of Traditional Food: Evidence From a Survey in Six Countries Filiep. *Agribusiness, 26*(4), 453–476. https://doi.org/10.1002/agr

Yüksel, A., & Yüksel, F. (2001). Comparative performance analysis: Tourists' perceptions of Turkey relative to other tourist destinations. *Journal of Vacation Marketing, 7*(4), 333–355.

Promoting Creative Tourism: Current Issues in Tourism Research – Kusumah et al. (Eds)
© 2021 Taylor & Francis Group, London, ISBN 978-0-367-55862-8

Border tourism in Indonesia's outer islands: The case of Sebatik Island

S.R.P. Wulung & A.K. Yuliawati
Universitas Pendidikan Indonesia, Bandung, Indonesia

M.S.D. Hadian
Universitas Padjajaran, Bandung, Indonesia

ABSTRACT: Sebatik Island is one of the main priorities of Indonesia's outer island that plays a role in sustainable regional development through the tourism sector in supporting the improvement of law and security supervision. The effort to develop Sebatik Island as a border tourism destination requires comprehensive tourism planning according to the existing environmental, socio-cultural, and government policies. This study aims to identify and map the potential distribution of tourist attractions in the border region of Sebatik Island. The existing conditions of tourism activities also identified to determine the existence of Sebatik Island in the tourism destination life cycle. The research approach uses qualitative methods with primary data collections through observation and interviews, while secondary data was obtained from previous research and local government policy. The analysis method uses descriptive qualitative and map analyses. The results found that tourism activities on Sebatik Island are only available in five tourist attractions, while seven other tourist attractions have potential as border tourist attractions. The condition of Sebatik Island is at the stage of exploration and euphoria that requires further tourism planning to avoid the negative impacts caused by increased tourist visits. The development of border tourism on Sebatik Island helps the local community and conserves the natural environment.

Keywords: border tourism, destination management, Sebatik Island, tourism planning

1 INTRODUCTION

Tourism plays a role as an alternative sector in optimizing natural and cultural resources in border areas. The border area becomes an attractive tourism destination by offering various tourist attractions with the theme of natural phenomena (protected areas and landscapes), histories and cultures (warfare and people's lifestyles), and economic or artificial factors (shopping and entertainment centers) (Gelbman 2008). Tourism development is an appropriate strategy for developing border areas in overcoming social, economic, natural environment, and political cooperation (Hall 2009; Prokkola 2007; Timothy 2002). It can be achieved through integrated governance among stakeholders and planning processes in utilizing resources for tourism to attain sustainable regional development and minimize the negative impacts in the border region. (Blasco et al. 2014; Stoffelen & Vanneste 2017).

Borders and boundaries become very important for human insight, thoughts, and perceptions (Mayer et al. 2019a). Borders are an idea to distinguish something as a prerequisite to be recognized from one another rather than a sign or dividing line between two regions or spaces (Mayer-Tasch 2013). A border is identified in tangible and intangible lines through the process of limiting or creating certain boundaries that function in identifying something different (Mayer et al. 2019b; Sofield 2006). Borders are not only related to geographical, but social, economic, political, and virtual aspects that play a role in defining a border (Newman 1998). Borders are identified in two perspectives, namely based on lines (broad and narrow, natural and artificial, visible and invisible,

DOI 10.1201/9781003095484-16

and linear and non-linear) and based on entities/unity (territory, culture, ethnic, language, religion social, economic, climate, vegetation, and geology) (Donec 2014). Borders as tourist attractions are classified based on five types, including material (mountains, rivers, sea, and artificial), territorial (administrative units such as interstate, state, and province), functional (political, administrative boundary, and economic), sociocultural (religion, race, culture, and history), and mental and cognitive (individual perception) (Mayer et al. 2019b).

Border regions affect the tourism system, including travel activities, infrastructure development, marketing, the tourism industry, and tourism development (Gelbman 2010). Border tourism has focused on the different types, scales, scopes, and functions of borders in supporting the development of the economic, socio-cultural, and environmental aspects (Blasco et al. 2014). The border regions as a tourist attraction are reviewed based on two main perspectives, including 1) a border as a tourist attraction, including indicators of demarcation (dividing boundary), fences, walls and border towers; and 2) borders as tourism destinations, such as tourist activities, local community culture, and the natural environment in the border region (Timothy 2002, 1995; Timothy et al. 2016).

Border tourism destinations are created before or after the establishment of the border regions. The study of field conditions and local government policies are needed in designing the spatial of the border region as a tourism destination. There are several schemes for the establishment of border tourism destinations, including 1) antecedents, border regions existed before tourism was developed; 2) subsequent, border regions are established after tourism destinations are developed; 3) transformation, changes in the function of border regions into tourism destinations; and 4) relict, increased tourism development in the border areas and tend to lose their function as a barrier) (Hartshorne 1936; Wachowiak 2016a, 2016b). Border tourism destinations indirectly support sustainable regional development in the border region. There are four models of developing border tourism destinations (Wieckowski 2018), including separated or coexisting space (interaction between border regions is very closed), connected space (starting to interact with one another with limited intensity), open space (openness between policies and procedures in the border region toward tourism activities in increasing tourist visits), and integrated space (a partnership between the border regions of the two parties in developing the border region as a tourism destination). Furthermore, the relationship between the border regions and tourism can be identified based on the border as territorial divides, the borders as a tourism destination, and the borders as modifiers of tourism destinations (Mayer et al. 2019b; Timothy 2002). In general, the border tourism destination model is similar to the concept of tourism area life cycle (exploration, involvement, development, consolidation, and stagnation) and irridex (euphoria, apathy, annoyance, and antagonism) (Butler 1980, 2019; Doxey 1975).

Previous research explained that tourism in the border region positively contributes to economic (Christaller 1964; Diener & Hagen 2012), socio-cultural (Sofield 2006; Stoffelen et al. 2017), and the natural environment (Faby 2016; Horváth & Csüllög 2013; Mayer et al. 2019c). The development of border regions as tourism destinations contributes to improving the economy in 23 priority areas in South Africa (Rogerson 2015); the welfare of the people on the border between Indonesia, Malaysia, and Singapore (Hampton 2010); and the natural environment preservation in the border regions of Germany and Poland (Mayer et al. 2019a). Border tourism can foster a sense of nationalism and strengthen brotherhood and create peace (Gelbman 2010, 2008).

Indonesia's border regions have a diversity of natural resources (abiotic and biotic) and culture that can support sustainable development. Sebatik Island is one of the outermost and foremost islands that has a strategic role in Indonesia's defense and security. Sebatik Island is directly bordered by the state of Sabah, Malaysia, of the northern part of the island. Potential resources on Sebatik Island include geological diversity (Asmoro 2016; Hidayat et al. 1995a, 1995b), biodiversity (Ardiansyah et al. 2012; Hidayanto et al. 2016; Mulyo et al. 2017), and cultural diversity (Fallis 2013; Siburian 2012; Siregar 2008; Sudiar 2017). These resources have the potential for Sebatik Island to develop its territory through tourism planning. Previous studies explained that Sebatik Island has the potential to be developed as ecotourism and agrotourism (Ardiansyah et al. 2012; Sudiar 2017), border tourism (Siregar 2008), and geotourism (Wulung et al. 2019). Integrating tourism planning among stakeholders and making diversity in resources as an attraction make Sebatik Island possibly a border tourism destination. Resource for tourism includes the natural environment (abiotic and biotic) and manufacturing (cultural) supported by governance in realizing sustainable regional development (Schejbal 2015). This study aims to identify and map the potential of tourist attractions in the border region of Sebatik Island and classify them based on abiotic, biotic, and cultural resources. Also identified are the existing conditions of tourism activities in positioning the existence of Sebatik Island in the tourism area life cycle, thus supporting the development of Sebatik Island as a border tourism destination.

2 METHOD

This study uses a qualitative approach that aims to gather actual and detailed information in describing symptoms, identifying problems and field conditions, and making comparisons and evaluations. The study was conducted for four months from September to December 2018 on Sebatik Island, Nunukan Regency, North Kalimantan Province, Indonesia. Primary data was obtained through observation, interviews, and positioning of spatial elements based on satellite using global positioning system technology. Observation activity aims to get data and information about the condition of natural resources (abiotic and biotic) and culture on Sebatik Island. Secondary data were obtained through literature studies on local government policy documents and previous research.

The unit of analysis in assessing the potential of a border tourist attraction on Sebatik Island uses five types of borders, namely, material, territorial, functional, sociocultural, and mental and cognitive cognitive (Mayer et al. 2019b). Meanwhile, to determine the position of Sebatik Island as a border tourism destination using an analysis unit of the concept of tourism area life cycle (exploration, involvement, development, consolidation, and stagnation) (Butler 1980, 2019) and irridex (euphoria, apathy, annoyance, and antagonism) (Doxey 1975). Data and information obtained are then carried out by the process of input and data management (coordinate, digitization, and editing), plot, and data conversion using geographic information system software (Esri ArcGIS). The analytical method used is qualitative descriptive analysis and map analysis.

3 RESULTS

Sebatik Island is one of the small and outermost islands, which is the main priority of its development because it borders directly with neighboring countries. Administratively, Sebatik Island is divided into two regional territories. The northern part of Sebatik Island is part of the Sabah State territory, Malaysia, while the southern part is the sovereign territory of Indonesia and is part of the Nunukan Regency, North Kalimantan Province. Accessibility to Sebatik Island is reached by sea transportation from Nunukan City or Tarakan City. In general, the Sebatik community has social institutions as a result of the integration of various group interests, especially Sulawesi ethnic and Tidung tribes. Four sectors have the potential to be developed on Sebatik Island, including the agriculture, plantation, fisheries, and tourism sectors. The tourism sector has the opportunity to contribute to the welfare of the local community through its independent involvement in tourism activities. Also, the tourism sector has the potential to preserve the natural environment on Sebatik Island due to the exploitation of oil palm, which damages the natural environment. The development of the tourism sector is bound to preserve the natural environment and prosper the people of Sebatik Island.

Pre-existing tourism activities on Sebatik Island are indicated by tourist visits to five tourist attractions, namely Batu Lamampu Beach, Losari Beach, Bukit Keramat, Aji Kuning Village, and Garuda Perkasa Monument. The absence of a tourism development policy on Sebatik Island has an impact on the unintegrated elements of tourism destinations, which is indicated by the inadequate condition of accessibility to tourist attractions. The development of tourism resources can integrate the development of the Sebatik Island border region and create tourism competitiveness. Potential tourism resources in the Sebatik Island border region are classified based on natural resources

Table 1. Potential border tourism attractions on Sebatik Island.

Resources	Tourist Atttraction	Material	Teritorial	Functional	Socio-cultural	Cognitive
Abiotic	1. Batu Lamampu Beach	–	–	X	X	–
	2. Losari Beach	X	X	X	–	X
	3. Sajau Stone Formation	X	X	X	–	X
	4. Sei Pancang River	X	X	X	–	–
	5. Keramat Hill	X	X	X	–	X
Biotic	6. Mangrove	–	–	X	X	–
	7. Oil Palm Plantations	X	X	–	–	–
	8. Setabu Marine Conservation Area	–	–	X	X	–
Culture	9. Bugis Village	–	–	–	X	X
	10. Aji Kuning Village	X	X	X	X	X
	11. Seaweed cultivation	–	–	X	X	–
	12. Garuda Perkasa Monument	X	X	–	X	X

(a)

(b)

(c)

(d)

Figure 1. Sites marking the border between Indonesia and Malaysia in the Keramat Hill Area (a), Aji Kuning TNI-AD Guard Posts (b), Sajau stone formations (c), and mangrove forest areas (d).

(abiotic and biotic) and cultural resources and are identified based on five types of border tourist attractions. There are 12 diversity of tourism resources that support the development of Sebatik Island as a border tourism destination, 5 of which have been identified as tourist arrivals (Tabel 1).

Some features and landscapes that are easily seen and recognized on the border of Sebatik Island become an attraction for tourists, such as in the Keramat Hill Area (Fig 1a). Local people believe that Keramat Hill witnessed a confrontation between Indonesia and Malaysia 1962–1966. In Keramat Hill, some features have the potential to become a tourist attraction in the form of signs/sites/monuments. This feature means the direction or border area between Indonesia and Malaysia. In front of the feature, there is a unique sign, namely the RI-MLY Border Security Task Force with the symbol of the Dayak Indigenous people's shield and the flags of the two countries.

Another feature is at the intersection of Aji Kuning Village (Fig 1b). The existence of the Aji Kuning TNI-AD Guard Post, a flagpole that reads "Steady the Indonesian homeland on the boundary" and a Indonesian flag that stands upright on the Indonesian homeland that indicates the area is the boundary between Indonesia and Sabah, Malaysia. These conditions indicate peace and harmony that exist between the two countries. On the other side, there is a house on the right side of the Aji Kuning TNI-AD Guard Post that has its own story and uniqueness to be interpreted for

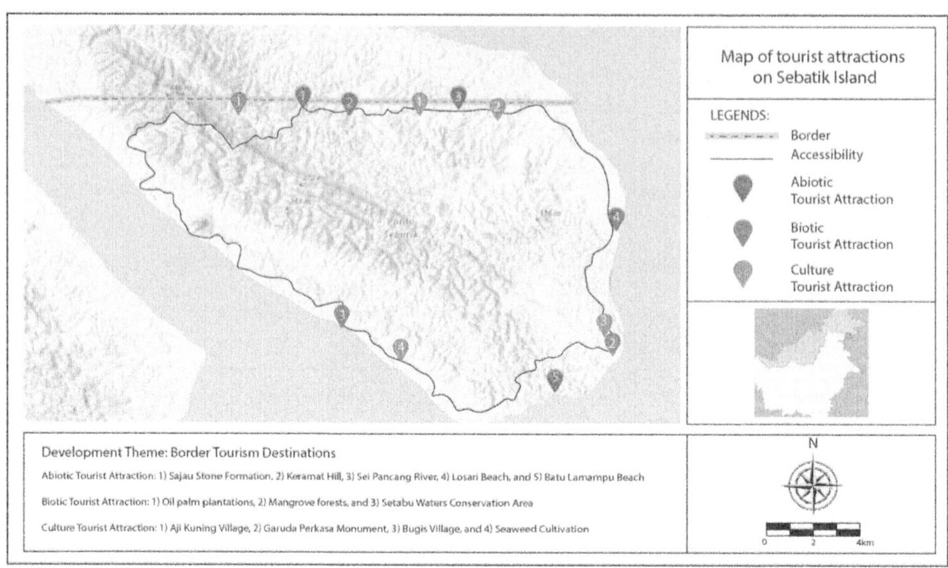

Figure 2. Map of the potential distribution of border tourist attractions on Sebatik Island.

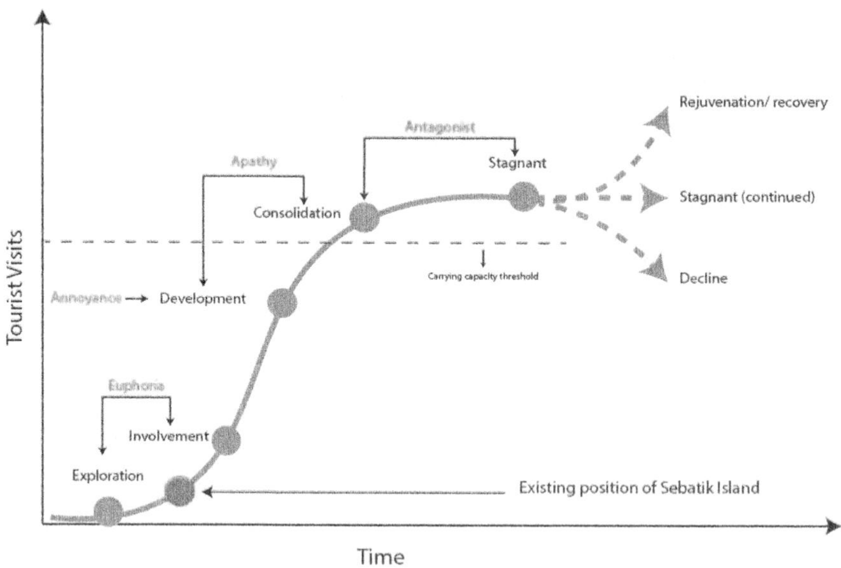

Figure 3. The existing condition of Sebatik Island in the tourism area life cycle adopted from Butler (1980, 2019; and Doxey, 1975).

tourists. The house is owned by Indonesian citizens, the southern part of the house (living rooms) is in Indonesian territory, and the northern part (kitchen) is in Sabah territory.

Landscape features such as the Sajau Stone Formation (Fig. 1c), Losari Beach, Mangrove (Fig. 1d), Keramat Hill, and oil palm plantations are natural features that have a direct view of the border between Indonesia and Sabah, Malaysia. Overall tourism attractions on Sebatik Island can be seen on the map of the distribution of tourist attractions (Fig. 2).

The development of Sebatik Island as a tourism destination depends on the spatial policy of the local government to the national. Sebatik Island is classified as an antecedent because it is indicated by the low number of tourist visits, and only in certain areas (Aji Kuning Village and Losari Beach). Sebatik Island's position is between the exploration and involvement stages in the Butler's Tourism Area Life Cycle, which is identified based on several factors, including 1) not as a tourism destination, both in local policies and in the minds of tourists; 2) low tourist arrivals; 3) there is no spatial pattern of tourist movements; 4) unique and natural tourist attractions (nature and culture); 5) ownership of tourist attractions tends to be owned by local people; 6) no accommodation for tourists; 7) for tourists who have a typical allocentric/tourist interest; 8) the economic contribution of tourism is not significant; 9) low pressure on the environment; and 10) the behavior of local people tends to be pre-euphoric and euphoric toward the presence of tourists.

Sebatik Island is at an exploratory stage requiring the support of stakeholders in planning as a border tourism destination to avoid negative impacts such as environmental degradation, cultural modification, and land ownership by parties outside Sebatik Island. The development of Sebatik Island as a border tourism destination refers to two main models. The first model is separated or coexisting space; this model can be seen in the Bukit Keramat area and the Sajau Stone Formation where there are no community settlements and interactions between the border areas are very closed. Land use in this area is intended for oil palm plantations. The second model is connected space; this model is seen in Aji Kuning Village through the boundary between the two countries. Aji Kuning Village has community interaction with each other, although with limited intensity. That is because it has a fairly good entity through the unity of race, ethnicity, and religion.

4 CONCLUSION

Sebatik Island as one of the foremost and outermost islands in Indonesia has natural and cultural resources that can potentially support sustainable regional development through the development of the tourism sector. The existing condition of Sebatik Island based on the perspective of tourism planning is at the stage between exploration and involvement. This stage has the potential to be developed because of the low intensity of the negative impacts due to tourism activities. The characteristics of local communities who are euphoric toward tourist arrivals have the potential to develop human resources in developing Sebatik Island as a border tourist destination. The tourism sector has a role in increasing the participation of the people of Sebatik Island through their involvement in the development of border tourism destinations and helping to preserve the natural environment. Development of tourism potential requires integration between sectors in government with a view to sustainable development on Sebatik Island.

The findings of this study have implications for designing spatial planning in Sebatik Island as a tourism destination and supporting the Nunukan Regency Government in formulating a tourism development master plan. This research provides knowledge about the development of border areas through tourism, given the limited research on border tourism in Indonesia, especially on the outer islands. The study of border tourism on Sebatik Island led to further research on the literacy of local communities about tourism through the concept of community-based tourism. It helped the local community to participate in tourism management on Sebatik Island. Also, there was need for integrated tourism marketing by utilizing information communication technology as an effort to increase tourist visits to Sebatik Island as other further research.

REFERENCES

Ardiansyah, W.I., Pribadi, R., Soenardjo, W., 2012. Struktur dan Komposisi Vegetasi Mangrove di Kawasan Pesisir Pulau Sebatik, Kabupaten Nunukan, Kalimantan Timur. J. Mar. Res. 1, 203–215.

Asmoro, P., 2016. Gunung {Api} {Purba} {Pulau} {Nunukan}, {Kabupaten} {Nunukan}, {Provinsi} {Kalimantan} {Utara}, in: Seminar {Nasional} {Aplikasi} {Sains} Dan {Teknologi} ({SNAST}), 1. Institut Sains Teknologi AKPRIND Yogyakarta, Yogyakarta, pp. 70–84.

Blasco, D., Guia, J., Prats, L., 2014. Emergence of governance in cross-border destinations. Ann. Tour. Res. 49, 159–173. https://doi.org/10.1016/j.annals.2014.09.002

Butler, R.W., 1980. the Concept of a Tourist Area Cycle of Evolution: Implications for Management of Resources. Can. Geogr. / Le Géographe Can. 24, 5–12. https://doi.org/10.1111/j.1541-0064.1980.tb00970.x

Butler, R.W., 2019. The Tourism Area Life Cycle, Vol.2, The Tourism Area Life Cycle, Vol.2, Aspects of tourism. Channel View Publications, Clevedon; Buffalo. https://doi.org/10.21832/9781845410308

Christaller, W., 1964. Some considerations of tourism location in Europe: The peripheral regions-underdeveloped countries-recreation areas. Pap. Reg. Sci. Assoc. 12, 95–105. https://doi.org/10.1007/BF01941243

Diener, A.C., Hagen, J., 2012. Borders: a very short introduction. Oxford University Press, New York.

Donec, P., 2014. Die {GRENZE}: eine konzeptanalytische {Skizze} der {Limologie}. Würzbg. Königshausen Neumann.

Doxey, G. V, 1975. A causation theory of visitor-resident irritants, methodology and research inferences: The impact of tourism, in: Sixth Annual Conference Proceedings of the Travel Research Association. pp. 195–198.

Faby, H., 2016. Tourism policy tools applied by the European Union to support cross-bordered tourism, in: Tourism and Borders: Contemporary Issues, Policies and International Research. Ashgate, Hampshire, pp. 19–30. https://doi.org/10.4324/9781315550787-11

Fallis, A., 2013. Kebanggaan Masyarakat Sebatik Terhadap Bahasa Indonesia, Bahasa Daerah, Dan Bahasa Asing. J. Chem. Inf. Model. 5, 125–138. https://doi.org/10.1017/CBO9781107415324.004

Gelbman, A., 2010. Border tourism attractions as a space for presenting and symbolizing peace, in: Tourism, Progress and Peace. CABI Publ, Oxfordshire Cambridge MA, pp. 83–98. https://doi.org/10.1079/9781845936778.0083

Gelbman, A., 2008. Border tourism in israel: Conflict, peace, fear and hope. Tour. Geogr. 10, 193–213. https://doi.org/10.1080/14616680802000022

Hall, C.M., 2009. North-south perspectives on tourism, regional development and peripheral areas., in: Tourism in Peripheries: Perspectives from the Far North and South. CABI, Wallingford, pp. 19–37. https://doi.org/10.1079/9781845931773.0019

Hampton, M.P., 2010. Enclaves and ethnic ties: The local impacts of Singaporean cross-border tourism in Malaysia and Indonesia. Singap. J. Trop. Geogr. 31, 239–253. https://doi.org/10.1111/j.1467-9493.2010.00393.x

Hartshorne, R., 1936. Suggestions on the Terminology of Political Boundaries. Ann. Assoc. Am. Geogr. 26, 56–57.

Hidayanto, M., S., S., Yahya, S., Amien, L.I., 2016. Analisis Keberlanjutan Perkebunan Kakao Rakyat di Kawasan Perbatasan Pulau Sebatik, Kabupaten Nunukan, Provinsi Kalimantan Timur. J. Agro Ekon. 27, 213. https://doi.org/10.21082/jae.v27n2.2009.213-229

Hidayat, A., Amiruddin, Satrianas, D., 1995a. Geologi {Lembar} {Tarakan} dan {Sebatik}, {Kalimantan}.

Hidayat, A., Amiruddin, Satrianas, D., 1995b. Geologi {Lembar} {Tarakan} dan {Sebatik}, {Kalimantan}. Pusat Penelitian dan Pengembangan Geologi, Bandung.

Horváth, G., Csüllög, G., 2013. A new slovakian-hungarian cross-border geopark in Central Europe - possibility for promoting better connections between the two countries. Eur. Countrys. 5, 146–162. https://doi.org/10.2478/euco-2013-0010

Mayer-Tasch, P.C., 2013. Raum und Grenze, Raum und Grenze. Springer Fachmedien Wiesbaden, Wiesbaden. https://doi.org/10.1007/978-3-658-03015-5

Mayer, M., Zbaraszewski, W., Pienkowski, D., Gach, G., Gernert, J., 2019a. Cross-Border Tourism in Protected Areas Along the Polish-German Border: A Synthesis 335–361. https://doi.org/10.1007/978-3-030-05961-3_11

Mayer, M., Zbaraszewski, W., Pienkowski, D., Gach, G., Gernert, J., 2019b. Borders, (Protected Area) Tourism and Prejudices: Theoretical and Conceptual Insights 19–64. https://doi.org/10.1007/978-3-030-05961-3_2

Mayer, M., Zbaraszewski, W., Pienkowski, D., Gach, G., Gernert, J., 2019c. Cross-Border Tourism to Protected Areas in Poland and Germany: Methodology 129–158. https://doi.org/10.1007/978-3-030-05961-3_6

Mulyo, J.H., Irham, Jumeri, Perwitasari, H., Rohmah, F., Rosyid, A.H.A., 2017. Studi {Kelayakan} {Komoditas} {Usahatani} {Daerah} {Perbatasan} {Pulau} {Sebatik} {Kalimantan} {Utara}. Purwokerto, pp. 688–694.

Newman, D., 1998. Geopolitics Renaissant: Territory, sovereignty and the world political map. Geopolitics 3, 1–16. https://doi.org/10.1080/14650049808407604

Prokkola, E.K., 2007. Cross-border Regionalization and Tourism Development at the Swedish-Finnish Border: "Destination Arctic Circle." Scand. J. Hosp. Tour. 7, 120–138. https://doi.org/10.1080/15022250701226022

Rogerson, C.M., 2015. Tourism and regional development: The case of South Africa's distressed areas. Dev. South. Afr. 32, 277–291. https://doi.org/10.1080/0376835X.2015.1010713

Schejbal, C., 2015. Diversity as a General Basis of Tourism – System Approach/ Diverzita Jako Základ Turizmu – Systémový Prístup. Geosci. Eng. 61, 18–25. https://doi.org/10.1515/gse-2015-0009

Siburian, R., 2012. Pulau Sebatik: Kawasan Perbatasan Indonesia Beraroma Malaysia. J. Masy. Budaya 14, 53–76. http://dx.doi.org/10.14203/jmb.v14i1.87

Siregar, C., 2008. Analisis Potensi Daerah Pulau-Pulau Terpencil Dalam Rangka Meningkatkan Ketahanan, Keamanan Nasional, Dan Keutuhan Wilayah Nkri Di Nunukanâkalimantan Timur. J. Sosioteknologi 7, 345-368–368.

Sofield, T.H.B., 2006. Border tourism and border communities: An overview. Tour. Geogr. 8, 102–121. https://doi.org/10.1080/14616680600585489

Stoffelen, A., Ioannides, D., Vanneste, D., 2017. Obstacles to achieving cross-border tourism governance: A multi-scalar approach focusing on the German-Czech borderlands. Ann. Tour. Res. 64, 126–138. https://doi.org/10.1016/j.annals.2017.03.003

Stoffelen, A., Vanneste, D., 2017. Tourism and cross-border regional development: insights in European contexts. Eur. Plan. Stud. 25, 1013–1033. https://doi.org/10.1080/09654313.2017.1291585

Sudiar, S., 2017. Kebijakan Pembangunan Perbatasan Dan Kesejahteraan Masyarakat Di Wilayah Perbatasan Pulau Sebatik, Indonesia. J. Paradig. 1, 389–401.

Timothy, D.J., 2002. Tourism and Political Boundaries, 1st ed, Tourism and Political Boundaries. Routledge. https://doi.org/10.4324/9780203214480

Timothy, D.J., 1995. Political boundaries and tourism: borders as tourist attractions. Tour. Manag. 16, 525–532. https://doi.org/10.1016/0261-5177(95)00070-5

Timothy, D.J., Saarinen, J., Viken, A., 2016. Editorial: Tourism issues and international borders in the Nordic Region. Scand. J. Hosp. Tour. 16, 1–13. https://doi.org/10.1080/15022250.2016.1244504

Wachowiak, H., 2016a. Tourism and borders: Contemporary issues, policies and international research, Tourism and Borders: Contemporary Issues, Policies and International Research, New directions in tourism analysis. Ashgate, Aldershot, England; Burlington, VT. https://doi.org/10.4324/9781315550787

Wachowiak, H., 2016b. Tourism and borders: Contemporary issues, policies and international research, Tourism and Borders: Contemporary Issues, Policies and International Research. ROUTLEDGE, Place of publication not identified. https://doi.org/10.4324/9781315550787

Wieckowski, M., 2018. From Periphery and the Doubled National Trails to the Cross-Border Thematic Trails: New Cross-Border Tourism in Poland, in: Müller, D.K., Wieckowski, M. (Eds.), Tourism in {Transitions}. Springer International Publishing, Cham, pp. 173–186. https://doi.org/10.1007/978-3-319-64325-0_10

Wulung, S.R.P., Yuliawati, A.K., Hadian, M.S.D., 2019. Geotourism Potential Analysis of North Kalimantan, in: 3rd International Seminar on Tourism (ISOT 2018), Advances in {Social} {Science}, {Education} and {Humanities} {Research}. Atlantis Press, Bandung, pp. 283–287. https://doi.org/10.2991/isot-18.2019.63

Promoting Creative Tourism: Current Issues in Tourism Research – Kusumah et al. (Eds)
© 2021 Taylor & Francis Group, London, ISBN 978-0-367-55862-8

Visitor satisfaction: The mediating role of crowding perception on environmental characteristics and other visitors' behavior

N.A. Zidany, G.R. Nurazizah & F. Rahmafitria
Universitas Pendidikan Indonesia, Bandung, Indonesia

M.H.Y. Johari
Universiti Teknologi MARA, Melaka, Malaysia

ABSTRACT: The view regarding crowding perceptions and its effect on visitor satisfaction has caused a debate among researchers. Therefore, the purpose of this study was to examine the impact of environmental characteristics and visitor behavior toward crowding perception and its implications for visitor satisfaction. Data was collected by distributing questionnaires to 174 respondents during February 2020. The collected data was then processed using path analysis by examining the direct and indirect effects of environmental characteristics (X1) and visitor behavior (X2) on crowding perceptions (Y) and their implications for visitor satisfaction (Z). The results show that, partially, crowding impressions are influenced by the surrounding characteristics and the other visitors' behavior. The direct impact is indicated by environmental characteristics, while other visitors' behavior requires the mediator variable, which is crowding perception, in influencing visitor satisfaction. Furthermore, the perception of distress also has a positive effect on visitor satisfaction. This research contributes theoretically and methodologically to the literature on the understanding of distress perspective.

Keywords: distress perception, surrounding characteristic, visitor disturbance

1 INTRODUCTION

A large number of visitors in the tourist area is considered as one of the undesirable factors that can reduce the level of visitor satisfaction (Alegre & Garau 2010). When overcrowding occurs in an area, individuals can feel the decrease in tour satisfaction because they feel exhausted because of unwanted interactions or too close contact with others (Sanz-Blas et al. 2019). Crowding is seen as a situation when others limit choices or impede one's ability to achieve individual goals. However, in the context of tourism, visitor satisfaction has an important role that can determine the future actions of visitors, whether to make return visits, suggest to others, or not go back, even tell bad promotion about the place (Dodds & Holmes 2019; Rasoolimanesh et al. 2016). Furthermore, this condition can determine tourism sustainability, which encourages the creation of new and enjoyable experiences for visitors (Kusumoarto et al. 2017).

Regarding crowding perception, this perspective is considered a subjective evaluation of the density level in a particular environment (Zehrer & Raich 2016). Many researches debate on crowding perceptions and their effects on visitor satisfaction have presented mixed findings. Some researches show significant adverse effects of crowding on tourist satisfaction (Zehrer & Raich 2016; Jurado et al. 2013). However, some prove the weak influence (Kalisch & Klaphake 2007) or even show no significant effect on tourist satisfaction (Li et al. 2017). Besides considering location density, crowding perception considers other factors such as personal characteristics of visitors, surrounding characteristics, and others' visitor behavior (Li et al. 2017). Thus, it is essential to

 DOI 10.1201/9781003095484-17

verify the crowding perception of visitor satisfaction through an intermediary variable such as environmental characteristics and other visitors' behavior variables.

This research was conducted at one of the beaches in Indonesia, Ancol Beach, one of the popular tourist areas in Jakarta as the national capital. This beach has the highest number of visits than the other seven leading tourist attractions in the city, reaching 17.5 million visitors (Jakarta Provincial Tourism Office 2019). Visits to this beach were increasing, especially on weekends, school holidays, or year-end holidays. The high interest to visit made the opportunity to cause a crowd at once a time in the same location.

The density that occurs due to the high number of visits caused a variety of tourist responses related to the satisfaction or the decision to revisit Ancol Beach as the major tourist destination. Some visitors expressed a high level of satisfaction despite the density during the visit, but some gave an unpleasant review of overcrowding experience (Google Reviews 2020). This phenomenon presented the anomaly of visitors' behavior and their perception of crowding, still felt satisfied with the beauty and activity even though the condition is very crowded. It might happen because satisfaction also evaluation process of the whole tour, including the quality of activities, the easy access, natural beauty, or the service provided (Munandar et al. 2020)

Based on this description, researchers were interested in researching crowding perception by adding environmental factors and other visitors' behavior factors then analyzing how these perceptions affect visitor satisfaction in beach tourism. Therefore, the study aimed to examine the effect of crowding perceptions on visitor satisfaction through the intermediary variable of environmental characteristics and other visitors' behavior at Ancol Beach.

2 LITERATURE REVIEW

Satisfaction has become one of the most frequently researched variables in the context of tourism because visitor satisfaction becomes an essential consideration in the development of long-term tourism areas. Visitor satisfaction is a post-visitor evaluation of tourist areas (Chiu et al. 2016). This satisfaction is a manifestation of the overall level of pleasure felt by visitors from the results of meeting expectations and travel needs through perceived experience (Jalilvand et al. 2014). However, this opinion is refuted by Hunt (1983 in Chiu et al. 2016) who said that satisfaction is not only about the pleasure of the travel experience but also an evaluation of the whole experience.

From some of these opinions, it can be said that visitor satisfaction is an evaluation of visitor experience after visiting an area compared to their expectations before visiting the area. The indicators in measuring satisfaction can be seen from some factors. The factors are the pleasure over the visit decision (Yuksel et al. 2010), the overall satisfaction along the tourist area (Huang & Hsu 2009), the fulfillment of expectations after coming to the destination (Tjiptono 2011), and including the hope of the other visitors' number can be found in the area. A large number of visitors in the tourist area can be undesirable and reduce visitor satisfaction (Alegre & Garau 2010).

Kalisch and Klaphake (2007) stated that an increase in visitors' numbers may not always result in dissatisfaction with visitors. However, this finding was refuted by Silva and Ferreira (2013). They proposed that an increase in the number of visitors would affect visitors' comfort in the setting of coastal locations. The contradiction between those findings needs further research in terms of the perception of crowding and visitor satisfaction.

Based on the explanation, this study's framework is drawn into a chart with the following hypotheses:

H1: environmental characteristics affect the perception of crowding.
H2: behaviors of other visitors affect the perception of crowding.
H3: environmental characteristics affect visitor satisfaction.
H4: behaviors of other visitors affect visitor satisfaction.
H5: perception of crowding affects visitor satisfaction.
H6: environmental characteristics affect visitor satisfaction mediated by the perception of crowding
H7: behaviors of other visitor affect visitor satisfaction mediated by the perception of crowding

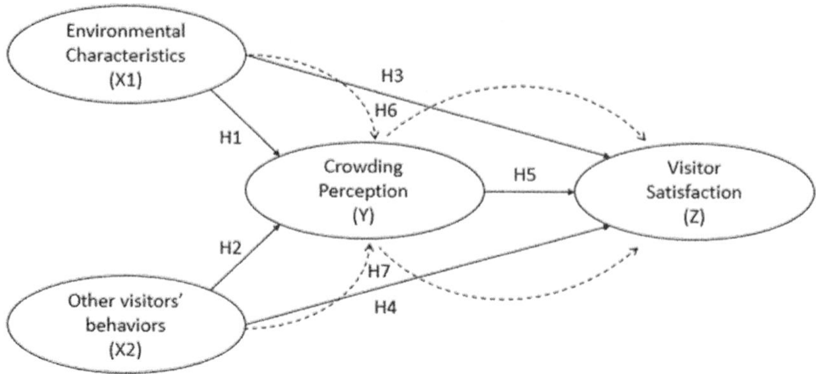

Figure 1. Research framework.

3 RESEARCH METHOD

This study uses descriptive methods to determine the crowding perception and visitor satisfaction in the Ancol Beach tourist area and inferential methods to assess the effect of environmental characteristics and other visitors' behavior on the crowding perception and its implications for visitor satisfaction. The closed-ended questionnaire was distributed to 200 visitors to Ancol Beach incidentally in January 2019 with minimum age criteria of 17 years. Of the 200 surveys collected, only 174 respondents filled out completely and thoroughly.

Data collected from the questionnaire was divided into four parts. The first and two parts are respondent characteristics (age, gender, education level, and occupation) and environmental characteristics (Neuts & Nijkamp 2012; Rasoolimanesh et al. 2016) & behavior of other visitors (Neuts & Nijkamp 2012). The third and fourth parts are crowding perception (Jurado et al. 2013; Kılıçarslan & Caber 2018) and visitor satisfaction (Yuksel et al. 2010).

Data analysis in the study used path analysis with the help of SPSS version 20. In testing the primary hypothesis of decision making based on the calculated t_{value} compared with the t_{table} (1.96). If the t_{value} is higher than the t_{table} with the sig. value less than 0.05, the hypothesis proposed is considered acceptable.

4 RESULTS AND DISCUSSION

Survey results presented that late teens (17–25 years) dominate visits as much as 82%, originating from the province of West Java (50%) with a high school educational background (50%), and with consumption in tourist attractions as much as IDR 100,000 - 300,000 (52%). Furthermore, visitors were considered being loyal since they made repeat visits more than three times (67%).

The partial hypothesis test of environmental characteristics (X1) to the crowding perception (Y) shows the results of t_{value} 2.417 with sig. 0.017 (t_{table} 1.96). So it can be concluded that H0 is rejected, and H1 is accepted, there is a direct positive effect of environmental characteristics variables on the crowding perception. Furthermore, the test result on the behavior of other visitor variables (X2) on the crowding perception also shows a significant positive effect with t_{value} of 9.283 and sig. 0,000 (see Table 1), H2 is accepted.

Excellent environmental characteristics make visitors feel at ease in the beach area. This study's results support the findings of Mehranian and Marzuki (2018), who revealed that sometimes visitors feel crowded due to the uneven distribution of facilities. Thus, decentralization of facilities needs to be done to avoid crowd concentration in only one area (Avenzora et al. 2013). Having to wait a long time to use facilities can also trigger a feeling of crowding. Supported by Rasoolimanesh et al.

Table 1. Partial regression test results X1 & X2 towards crowding perception.

Variable	t_{value}	B	Beta	sig.
(Constant)	2.609	1.713		.010
Environmental Characteristic (X1)	2.417	.093	.151	.017
Behavior of other visitors (X2)	9.283	.229	.581	.000

Table 2. Partial regression test results X1, X2, & Y towards visitor satisfaction.

Variable	t_{value}	B	Beta	Sig.
(Constant)	3.547	3.090		.001
Environmental characteristic (X1)	3.713	.189	.245	.000
Behavior of other visitors (X2)	1.808	.071	.144	.072
Crowding perception (Y)	4.770	.474	.378	.000

(2015), the availability and quality of facilities can influence crowding perceptions in recreation areas. Lower marking than visitors' expectations regarding environmental conditions will affect their perceptions of distress in the field. The findings show that visitors have high scores for the availability and quality of facilities such as restaurants, toilets, and parking lots as their expectation. Thus, crowding perceptions are also low scores because the facilities can accommodate the needs and activities of visitors at the same time and not cause long queues.

Furthermore, findings related to other visitor behavior variables (X2) support the theory of overload stimulus. The approach reveals crowding can be influenced by interactions with other visitors, mainly if unwanted interactions occur. So the research findings show that other tourists' behavior has a significant positive correlation with the perception of distress. The more disturbing actions encountered, the higher the perception of distress.

The testing of H3, H4, and H5 in this study was conducted to analyze the effect of environmental characteristics, other visitor behavior, and crowding perceptions on partial visitor satisfaction using multiple linear regression analysis (see Table 2).

The test results show that both environmental characteristics and crowding perception influence visitor satisfaction. Meanwhile, the behavior of other visitors does not have a significant effect on satisfaction. The availability and quality of facilities have a positive influence on visitor satisfaction. When visitors feel that the tourist area has been able to meet their needs and expectations, the feeling of pleasure will arise as one indicator in measuring satisfaction.

On the other hand, the behavior of other visitors does not directly affect satisfaction as opposed to the findings of Wu (2007) that interaction between visitors positively influences satisfaction. Of course, this direct effect is not created because other visitors' behavior requires mediation from other variables to influence satisfaction. The finding could also occur because the behavior of other visitors is not a significant consideration in assessing the satisfaction of recreation. Although they find unpleasant action, this is not a factor that directly decreases satisfaction because other factors influence satisfaction, such as availability and quality of facilities or services provided by the area (Dodds & Holmes 2019).

Regarding perceptions of distress and their relationship with satisfaction, Kim et al. (2016) stated that behavioral responses tend to be reversed with a high level of enthusiasm. But when the stimulation of interactions exceeds the optimal point, visitors will feel overwhelmed meeting large numbers of people in an area at the same time. The results found that respondents' responses regarding the number of visits were still within reasonable limits at the same time. Also, visitors feel comfortable with the number of visitors when they travel on Ancol Beach, which implies the level of stimulation at Ancol Beach still does not exceed the optimal point limit. The opinion of visitors can also be seen from the point of private space. A lack of privacy can lead to unsatisfactory

Table 3. Mediation effect of crowding perception test results.

Hypothesis	Path Coefficient
Environmental characteristic → Crowding perception → Visitor satisfaction	$0.151 \times 0.378 = 0.057$
Behavior of other visitors → Crowding perception → Visitor satisfaction	$0.581 \times 0.378 = 0.219$

visit experiences. However, during holidays, visitors' privacy and individualism were lost while pleasure and happiness arose from sharing with others (Pons et al. 2006), so they had no problem with the density and remained satisfied. After testing the direct effects, the crowding perception (Y) mediation effect test is conducted to measure the indirect impact of X1 and X2 on visitor satisfaction. The results of the two hypothesis tests are shown in Table 3.

The test results show that environmental characteristics (X1) have a direct effect on visitor satisfaction of 0.275 (see Beta value in Table 2) with a lower indirect effect of 0.057. So the role of crowding perception does not mediate these two variables. Unlike the case of X1, the behavior of other visitors (X2) has an indirect effect on satisfaction (0.219), which is higher than the direct one. Crowding perception plays a full role in the relationship between the behavior of other visitors and visitor satisfaction. That is because the behavior of other visitors cannot directly affect visitor satisfaction.

In this study, the crowding perception does not mediate the effect of environmental characteristics on visitor satisfaction. Even without the crowding perception, surrounding aspects can still affect visitor satisfaction. When visitors assess that the availability and quality of facilities are in line with expectations, it can directly affect their satisfaction. Conversely, the level of satisfaction decreases while visitors determine the facilities cannot accommodate the needs. This assessment can be separated from the number of people they met in the visit.

On the other hand, the crowding perception fully mediates the influence of other visitor behavior on visitor satisfaction. The expression of other visitors cannot influence visitor satisfaction without going through the assessment of distress. When there is density, there must be an increase in stimulation in an area. Eroglu et al. (2005) stated that everyone has a certain optimal point regarding the occurred density. The stimulation can be formed from the interaction between visitors, both desired and unwanted. When visitors make too close contact or encounter unpleasant behavior, visitors can assess the area has exceeded the optimal limit of acceptance of stimulation, causing the increased perception of their distress towards a tourist area.

5 CONCLUSION

Environmental characteristics have a direct influence on the perception of distress and also on visitor satisfaction. However, the behavior of other visitors only has a direct impact on the crowding, not on visitor satisfaction. If seen from the mediating role of the perceived crowding, then the opposite happens. Environmental characteristics do not have an indirect effect on visitor satisfaction, whereas other visitors' behavior has an indirect impact on visitor satisfaction. It happened because other visitor behavior or disturbance of visitors only affects crowding perception, while that perceived crowd directly impacts satisfaction. In contrast, the surrounding characteristics can standalone to affect satisfaction even though the level of crowding is right.

REFERENCES

Alegre, J. & Garau, J. 2010. Tourist satisfaction and dissatisfaction. *Annals of tourism research*, 37(1): 52–73.

Avenzora, R., Dahlan, E.N., Sunarminto, T., Nurazizah, G.R., Utari, W.D. & Utari, A.V. 2013. Ecological and psychological carrying capacity of ecotourism activities. In F. Teguh & R. Avenzora (eds), *Ecotourism and Sustainable Tourism Development in Indonesia: The Potentials, Lessons, and Best Practices*: 505–538. Jakarta: Ministry of Tourism and Creative Economy of Republic of Indonesia.

Chiu, W., Zeng, S. & Cheng, P. 2016. The influence of destination image and tourist satisfaction on tourist loyalty: a case study of Chinese tourists in Korea. *International Journal of Culture, Tourism and Hospitality Research*.

Dodds R. & Holmes, M. 2019. Beach tourists; what factors satisfy them and drive them to return. *Ocean & Coastal Management* 168:158–166.

Eroglu, S., Machleit, K. & Barr, T. 2005. Perceived retail crowding and shopping satisfaction: The role of shopping values. *Journal of Business Research* 58(8): 1146–1153.

Google Review. (2020). Ulasan Pantai Ancol. [Online]. Accessed from https://tinyurl.com/r8xxw2p

Huang, J. & Hsu, C. 2009. Interaction among fellow cruise passengers: Diverse experiences and impacts. *Journal of Travel & Tourism Marketing* 26(5–6):547–567.

Jalilvand, M., Poll, J., Vosta, L. & Nafchali, J. 2014. The effect of marketing constructs and tourists' satisfaction on loyalty to a sport destination. A structural equation model and analysis. *Education, Business and Society: Contemporary Middle Eastern Issues*.

Jurado, E., Morales, & Damian. 2013. Carrying capacity model applied in coastal destinations. *Annals of Tourism Research* 43:1–19.

Kalisch, D. & Klaphake, A. 2007. Visitors Satisfaction and Perception of Crowding in a German National Park: a case study on the island of Hallig Hooge. *Forest Snow and Landscape Research* 81(1/2):109–122.

Kılıçarslan, D. & Caber, M. 2018. The Impacts of Perceived Crowding, and Atmospherics on Visitor Satisfaction at Cultural Heritage Sites: A Comparison of Turkish and British Visitors to Topkapi Palace, Istanbul. *Journal of Tourism and Services* 9(17).

Kim, D., Lee, C. & Sirgy, M. 2016. Examining the differential impact of human crowding versus spatial crowding on visitor satisfaction at a festival. *Journal Travel Tourism* 33(3):293–312.

Kusumoarto, A., Gunawan, A. & Nurazizah, G. R. 2017. Landscape Potential Analysis for Ecotourism Destination in the Resort Ii Salak Mountain, Halimun-Salak National Park. In *IOP Conference Series: Earth and Environmental Science* 9(1):12–29.

Li, L, Zhang, J., Nian, S. & Zhang, H. 2017. Tourists' perceptions of crowding, attractiveness, and satisfaction: a second-order structural model. *Asia Pacific Journal of Tourism Research* 22(12):1250–1260.

Mehranian, H. & Marzuki, A. 2018. Beach Users' Perceptions Toward Beach Quality and Crowding: A Case of Cenang Beach, Langkawi Island, Malaysia. In *Sea Level Rise and Coastal Infrastructure*.

Munandar, A., Gunawan, A., Nurazizah, G.R. & Kusumoarto, A. 2020. Analysis on Willingness to Pay at the Ecotourism Destination, Cigamea Waterfall, Halimun-Salak National Reserve. *Journal of Environmental Management and Tourism* XI(Spring), 2(42):408–421.

Neuts, B. & Nijkamp, P. 2012. Tourist Crowding Perception and Acceptability in Cities - An Applied Modelling Study on Bruges. *Annals of Tourism Research* 39(4):2133–2153.

Pons, F., Laroche, M. & Mourali, M. 2006. Consumer Reactions to Crowded Retail Settings: Cross-Cultural Differences between North America and the Middle East. *Tourism Geographies: An International Journal of Tourism Space, Place and Environment* 23(7):555–572.

Rasoolimanesh, S., Jaafar, M., Marzuki, A. & Abdullah, S. 2016. Tourist's perceptions of crowding at recreational Tourist's perceptions of crowding at recreational. *An International Journal of Tourism and Hospitality Research* 28(1):41–51.

Rasoolimanesh, S., Jaafar, M., Marzuki, A. & Mohamad, D. 2016. How Visitor and Environmental Characteristics Influence Perceived Crowding. *Asia Pacific Journal of Tourism Research* 21(9):952–967.

Sanz-Blas, S., Buzova, D. & Schlesinger, W. 2019. The Sustainability of Cruise Tourism Onshore: The Impact of Crowding on Visitors' Satisfaction. *Sustainability* 11(6):1510.

Silva, S. & Ferreira, J. 2013. Beach Carrying Capacity: The physical and social analysis at Costa de Caparica Portugal. *Journal of Coastal Research* 65:1039–1044.

Tjiptono, F. 2011. *Service Management Mewujudkan Layanan Prima Edisi 2*. Yogyakarta: Andi.

Wu, C.H. 2007. The impact of customer-to-customer interaction and customer homogeneity on customer satisfaction in tourism service—The service encounter prospective. *Tourism Management* 28(6):1518–1528.

Yuksel, A., Yuksel, F. & Bilim, Y. 2010. Destination attachment: Effects on customer satisfaction and cognitive, affective and conative loyalty. *Tourism Management* 31(2):274–284.

Zehrer, A. & Raich, F. 2016. The impact of perceived crowding on customer satisfaction. *Journal of Hospitality and Tourism* 29:88–98.

Promoting Creative Tourism: Current Issues in Tourism Research – Kusumah et al. (Eds)
© 2021 Taylor & Francis Group, London, ISBN 978-0-367-55862-8

Culinary tourism planning and development: A case in Gebang Mekar Cirebon

E. Fitriyani, I.I. Pratiwi & A. Suwandi
Universitas Pendidikan Indonesia

ABSTRACT: This research is as a result of field studies from researchers in studying the planning of new culinary tourism destinations in Gebang Mekar Village, Cirebon. It was conducted through an interview and observation approach with the community, village officials, policy makers, Cirebon tourism office, and tourists around Cirebon. From interviews and observations, researchers found hope from the community, tourists, and the direction of the Cirebon tourism department's policy to develop a new culinary destination in the Village of Gebang Mekar. Seeing the potential of nature, culture, and human resources could become a new culinary tourism destination in Cirebon, researchers also reviewed planning strategies through attractions, accessibility, and facilities. The results of this study are expected to provide input for policymakers in implementing development strategies and planning for new tourist destinations in Gebang Mekar, Cirebon.

Keywords: Tourism Planning, Culinary Tourism Destination, Culinary Tourism, Tourism Planning Development

1 INTRODUCTION

Communication and sustainability-based planning must co-exist in the second domain to have a strong role in the interests of stakeholders and planning in the field of tourism. Knowledge of tourism planning is useful to be contextualized and developed by considering generic planning theory, but it is still very focused on problems and shocks (Rahmafitria et al. 2020). In this study, focusing on the origin of the development of tourism planning that is taking place in Indonesia about the broader field of planning theory. In general, tourism planning can maximize the benefits of tourism for the local community and economy and promote existing resources (Adu-Ampong 2019; Gibson 2009). The idea is, tourism as an important part of the economy and that people live long to appear in the planning literature. As an economic activity among many plans, explicit attention to the needs of several stakeholders to ensure successful tourism is slow to develop (Dredge & Jamal 2015).

The study in this study revealed that the need for tourism planning was only highlighted as the results of inadequate planning. It was to be detrimental to the community and/or the environment. As this is the realization of the potential destructive power of unlimited tourism to emerging, planning for this sector begins to take on many generic characteristics of planning theory (Rahmafitria et al. 2020). In particular, there has been a greater effort to communicate with various stakeholders and the use of sustainability approaches and approaches to defining a new era. Now in contemporary times, tourism planners can become intermediaries for the promotion of social justice in local communities and the natural resource protecting communities which are the basis of a good deal of tourism (Lohmann & Panosso Netto 2017).

Planning can be taken as a set of ideas and principles that are sought to control the spatial distribution of human activities over time (Stylidis et al. 2018). The results of this study are directed at the development of ongoing tourism planning in Indonesia about the broader field of planning

 DOI 10.1201/9781003095484-18

Table 1. Data visit tourists Cirebon regency, 2014–2018.

No	Year	Number of Tourists
1	2014	163.578
2	2015	588.515
3	2016	639.136
4	2017	713.591
5	2018	1,443.069

Source: Tourism Goverment Cirebon regency, 2019.

theory. In general tourism, planning can maximize the benefits of tourism for local people and the economy and promote the use of health resources (World Tourism Organization 2004).

Cirebon Regency is one of the tourist destinations much in demand by tourists, with tourist visits statistics from year to year have increased. Seen in Table 1, the statistics of tourist visits Disparbud Cirebon Regency in 2014–2018 with the number of tourists occurred in 2018 was 1,443,069 while in 2017 the number of tourists was 713,591.

The growth based on statistical data above illustrates the potential opportunities for developing new destinations in Cirebon Regency. In the Great Plan for the Development of World Class Destinations in the Province of West Java, there is a plan for developing destinations using a priority scale divided by nine priority tourist areas. The historical tourism strategic area and the palace in Cirebon and its surroundings get second priority after the strategic area of Bandung's creative tourism and surrounding areas *(Perencanaan & Daerah, n.d.).* In line with the great plan for developing world-class destinations in West Java Province, the Cirebon District Government seeks to improve tourism through the Regional Medium-Term Development Plan (RPJMD) which has set tourism goals and objectives, namely increasing the tourism competitiveness of Cirebon Regency with 3 main programs (RIPPARDA KAB. CIREBON 2019–2025): Tourism marketing development program, tourism destination development program, and partnership development program.

Cirebon Regency has various types of tourism, namely: (1) religious tourism, (2) artificial tourism, (3) culinary tourism, (4) cultural tourism, (5) nature tourism, (6) shopping tourism, (7) and tourism batik. So with this uniformity, Cirebon district is considered having an attractive tourism potential to be developed. However, if examined further, the number of tourist visits above cannot be absorbed evenly in Cirebon Regency. The distribution of many tourist objects in the West and North regions of the Cirebon Regency, while in the East and South regions is still lacking. This will also impact on the economic and social conditions of the people of Cirebon Regency that have not been evenly distributed. Therefore, it is necessary to develop new tourist destinations, especially in the eastern area of Cirebon district, with the hope that it can help improve the welfare of local communities through the study of unique and attractive tourism potential in the area that can be utilized in the development of Indonesian tourism.

For Indonesia, the coastal area has a strategic meaning because it is a transitional area between terrestrial and marine ecosystems, and has very rich natural resource potential and environmental services. This resource wealth attracts various parties to utilize their resources and various agencies to regulate their utilization (Hidayat & Marcellia 2011). One of the coastal areas that has tourism potential in the Cirebon district is Gebang Mekar Village. Based on data from the Village Medium Term Development Plan (RJPMDes) for 2018–2023, Gebang Mekar Village has a diverse culture, namely (1) Barikan, (2) Community Commitments, (3) "Nadran" Ocean Ruwatan, (4) Ngidung, and (5) Alms of the Earth. Traditional culture is the basic capital of development that underlies the development that will be carried out, the cultural heritage is of great value is an asset to attract developing cultural tourism in the Gebang Mekar Village. In addition to culture, Gebang Mekar Village can be a potential for natural tourism supported by natural conditions in the coastal area and seafood culinary tourism by looking at the basic potential of the people of Gebang Mekar Village.

Based on this potential, this study tries to explore the formulation of ideas, or concepts in developing new tourist destinations in Gebang Mekar Village with approaches under the Cirebon Regency Tourism Development Master Plan (RIPPARDA, Cirebon Regency), namely:

1. Sustainable tourism development approach
2. The concept of ecotourism approach
3. Community empowerment approach
4. The supply and demand suitability approach
5. The collaborative management development approach

2 METHOD

The research method used a qualitative approach to dig up data and formulate the concept of developing new culinary tourism destinations in the village of Muara Gebang, Cirebon Regency. A qualitative approach was also used to obtain a broader descriptive picture of the observed phenomenon (Moleong 1995). The qualitative approach is seen as being able to explore the meaning of the phenomenon in more depth (Creswell 1994). The phenomenon in question is regarding the potential of the coastal area of Gebang Mekar Village, Cirebon Regency.

Data collection was done by direct observation of the area to obtain descriptive data. This research was supported by data from the source (key person) to provide an assessment of the variables on the observed object (judgment value). As for the parties who become a key person in determining the value of judgment are; Kuwu Gebang Mekar, Chairman of Youth Organization, Chair of the Village Consultative Body (BPD), Head of Tourism Destination Office of Cirebon Regency Tourism Office, Head of Tourism, Gebang Mekar Residents, Cirebon Regency Travelers, and Gebang Mekar.

3 ANALYSIS AND DISCUSSION

The tourism sector can be an alternative sector for development in the region (Nandi 2016). Tourism can also impact economic activity, whether micro, meso, or macro (Sowwam et al. 2018). Micro impacts include an increase in community income or variations in economic activity. The increase in PAD, employment opening is an impact on the scope of the meso. The macro impact is economic growth and regional GRDP. Besides, other alternative solutions to improve the welfare of rural communities especially in the economic sector are to develop rural tourism sectors based on the utilization of local potential, both natural potential and cultural diversity (Handayani 2016). Tourism in its development has become the focus of world attention because it experiences a big increase in this decade. Therefore, tourism is a strategic policy and a national alternative that has been carried out by various countries around the world. Tourism as an industry has a very large role in terms of Economy, Socio-culture, and the Environment (Atiko et al. 2016).

A tourist destination is a place visited with significant time during one's trip compared to other places traversed during the trip (Khotimah & Wilopo 2017). Tourism destination classification is as follows: 1. Natural resource destinations such as climate, beaches, forests; 2. Destinations for cultural resources such as historic sites, museums, theaters, and local communities; 3. Recreational facilities such as amusement parks; 4. Events such as annual village events (Nadran), night markets, and so on (Pendit 2002). Tourism planning applies the basic concepts of planning in general with adjustments to the characteristics of the tourism system, to achieve tourism development goals (Mason 2012)

A destination must have various facilities needed by tourists so that a tourist's visit can be fulfilled and feel comfortable. Various needs of tourists include transportation facilities, accommodation, travel agencies, attractions (culture, recreation, and entertainment), food service, and souvenir items (World Tourism Organization 2004). The availability of various necessary facilities will make tourists feel comfortable so that more tourists visit (Lozano-Oyola et al. 2012).

Culinary tourism has emerged as a central aspect of every tourist experience. This includes cultural practices, landscapes, seas, local history, values, and cultural heritage. Food functions as a link between us and our inheritance, and the surrounding people. It is a diverse and dynamic channel for sharing stories, forming relationships, and building communities. By combining travel with the experience of eating and drinking, food tourism offers a "place impression" for both locals and tourists (Lockwood & Long 2008). Indonesia is undoubtedly in the culinary field. The diverse tribes and cultures give rise to its characteristics in each region. Similarly, the characteristics are also in the culinary field. Not only on various types, but also Indonesian culinary delights are recognized by the world with the recognition of fried rice and *rendang* as the most delicious food in the world. (Cohen & Avieli 2004).

From interviews and field observations, we can see support from various parties for the development of new culinary tourism destinations in Gebang Mekar Village. The community and village officials responded positively to the construction of new destinations in the Gebang Mekar Village, with the development of a well-managed culinary tourism destination that is expected to overcome the problem of social inequality in the Gebang Mekar Village

3.1 *Market analysis*

Current (existing) market conditions in the tourist district of Cirebon Regency as follows:

a. Analysis of geographical aspects, the majority of tourists visiting the district. Cirebon originated from Java Island with their motivation being for holidays with the interest of shopping tourism, nature tourism, cultural tourism, and culinary.
b. Demographic aspects, in the village of Gebang Mekar most visitors are 68% male and the remaining women. This indicates that men have a greater tendency for nature tourism. The average age of respondents visiting Cirebon District aged 20-30 years with various activities, such as fishing in Gebang Mekar, batik shopping in Trusmi, culinary specialties of Cirebon, visiting the palace (historical places), etc.
c. Psychographic aspects, tourists have a purpose of traveling to Cirebon Regency to enjoy cultural, natural, shopping, and culinary tourism and become one of the fun destinations for visitors to the Gebang Mekar Village to do activities such as fishing or enjoy the beauty of the sunset with a view of Mount Ceremai
d. Behavioral aspects, tourists get information about tourist destinations in Cirebon Regency from family/relatives who have visited several tourist destinations in Cirebon Regency.

From a physical aspect, Gebang Village is growing, which includes tourist destinations in the form of natural resources with cultural and culinary potential. It is stated in Indonesia Law of Tourism Number 10 year 2009, there are at least 3 elements of tourism products or destinations, namely: Tourist Attractions, Amenities, and Accessibilities commonly abbreviated as 3A tourism. The development of tourist destinations at the level of the site is at least taking into account these 3 key points (Undang-Undang Republik Indonesia 2009).

3.2 *Attraction*

According to Pendit (2002), the tourist attraction is defined as everything interesting and valuable to visit and see. The tourist attractions are mentioned more specifically into three types, namely: natural tourism attractions, cultural tourism attractions, and artificial attractions. The attraction of nature tourism is everything that has a uniqueness, beauty, authenticity, and value in the form of natural diversity created by God Almighty. The appeal of cultural tourism is everything that has uniqueness, beauty, authenticity, and value in the form of a diversity of cultural products of the world (*Undang-undang Nomor 10 Tahun 2009 Tentang kepariwisataan*). Natural attractions can be in the form of mountains, beaches, river forests, and so on. While cultural appeal can be in the form of ideas, ideals, values or norms, patterns of activity or patterns of human activity in society, as well as objects of human work (Koenjaningrat 2005).

3.3 *Amenities*

Baud Bovy & Lawson (1998) in his book "Tourism and Recreation handbook of Planning and Design" said that amenities are all forms of facilities that provide services for tourists for all their needs while staying or visiting a tourist destination, for example, hotels, motels, restaurants, bars, shopping centers, souvenir shop, and others. To meet the needs of travel, facilities should be provided, starting from fulfilling the needs since leaving from the tourist residence, while in the tourism destination, and when tourists return to their original places (Suryadana 2015).

3.4 *Accessibility*

Good accessibility will determine whether or not the location is easy to reach. The road network is also one that influences the smoothness of public services, which is very important (Sumarabawa 2013). Accessibility is very important in developing tourist destinations because it guarantees affordability, as well as effectiveness and efficiency for tourist visits.

3.5 *Strategy for developing new culinary tourism destinations*

Strategies for developing new tourism destinations in Gebang Mekar Village, through 3A with the sustainable tourism development approach, the concept of ecotourism, community empowerment, the suitability of supply and demand and the development of collaborative management:

Table 2. Strategy for developing new culinary tourism destinations.

Attraction	
Food Court Seafood	Increasing the production of fish caught by the fishermen of Gebang Mekar Village (culinary tourism)
Selfie Corner (Sunset view Ceremai Mountain)	As an attraction for the new Destination of Gebang Village (Gebang Mekar icon)
Fishing spot	Special interest tourism activities (fishing)
Cultural Studio	As an area to watch the culture of Gebang Mekar Village (Cultural Tourism)
Fish auction	Educate tourists how to sell fish catches
Amenities	
Floating mosque	A place of worship for Muslim travelers (halal tourism)
Parkir area	Parking for 2-wheeled vehicles, 4, tourism bus
Kids' Corner	Children's playground
Restaurant (Food court)	Seafood culinary
Charger area	As a supporter of digital tourism
Free Wifi access connection	As a supporter of digital tourism
Difabel friendly	A tourist destination that everyone can enjoy
Toilet/restroom	Made as comfortable as possible for tourists (instagramable)
Nursing room	Made as comfortable as possible for nursing mothers
Shopping area	Place of purchase souvenir of the community of Gebang Mekar Village and surrounding areas
Accessibility	
Sign	Directions before entering the destination
Access	Good road access
Transportation	Access to transportation for tourists who do not bring private vehicles

Source: Researcher, 2020.

4 CONCLUSION

Based on the results of the field study, interviews with relevant stakeholders, it can be concluded that the enthusiasm of the community, tourists, and the local government to hold new destinations in the village of Gebang Mekar is very high. By adopting five approaches, which are, sustainable tourism development approaches, the concept of ecotourism, community empowerment, the suitability of supply and demand and the development of collaborative management it is hoped that this new destination will bring many benefits, such as:

1. Improving the quality and quantity of fish caught in the village of Gebang Mekar through culinary tourism alliances
2. Improving Community Welfare in Gebang Mekar Village and Cirebon District
3. Preserving the Culture of Gebang Mekar Village
4. Increase tourist visits district. Cirebon, specifically the Gebang Mekar Village

Researchers hope that this study can be taken into consideration for policymakers in the development and planning of new culinary tourism destinations in Gebang Mekar, and also for further researchers are expected to conduct further studies of planning strategies from different perspectives.

REFERENCES

Adu-Ampong, E. A. (2019). Historical Trajectories of Tourism Development Policies and Planning in Ghana, 1957–2017. *Tourism Planning & Development, 16*(2), 124–141. https://doi.org/10.1080/21568316.2018.1537002

Atiko, G., Sudrajat, R. H., & Nasionalita, K. (2016). Analisis Strategi Promosi Pariwisata Melalui Media Sosial Oleh Kementrian Pariwisata RI. *E-Proceeding of Management*.

Baiquni, M. 2004. *Manajemen Strategis*. Buku Ajar Pusat Studi Kajian Pariwisata Sekolah Pascasarjana Universitas Gajah Mada.

Baud-Bovy, M., & Lawson, F. (1998). *Tourism and Recreation: Handbook of Planning and Design*. Butterworth-Heinemann Ltd.

Cohen, E., & Avieli, N. (2004). Food in tourism - Attraction and impediment. *Annals of Tourism Research*. https://doi.org/10.1016/j.annals.2004.02.003

Creswell, John W. 1994. *Research Design–Qualitative, Quantitative, and Mixed Method*. London: SAGE Publications.

Dredge, D., & Jamal, T. (2015). Progress in tourism planning and policy: A post-structural perspective on knowledge production. *Tourism Management, 51*, 285–297. https://doi.org/10.1016/j.tourman.2015.06.002

Gibson, C. (2009). Geographies of tourism: Critical research on capitalism and local livelihoods. *Progress in Human Geography, 33*(4), 527–534. https://doi.org/10.1177/0309132508099797

Handayani, S. N. (2016). *Strategi Kemenpar Gaet Milenial Majukan Pariwisata*. Www.Swa.Co.Id.

Hunger, J. D. dan T. L. Wheelen. 2001. *Strategic Management*. Fifth Editions.

Kasus, S., Pangandaran, P., Ciamis, K., & Barat, J. (2011). *STRATEGI PERENCANAAN DAN PENGEMBANGAN OBJEK WISATA (STUDI KASUS PANTAI PANGANDARAN KABUPATEN CIAMIS JAWA BARAT) Marceilla Hidayat Politeknik Negeri Bandung. I*(1), 33–44.

Khotimah, K., & Wilopo, W. (2017). STRATEGI PENGEMBANGAN DESTINASI PARIWISATA BUDAYA (Studi Kasus pada Kawasan Situs Trowulan sebagai Pariwisata Budaya Unggulan di Kabupaten Mojokerto). *Jurnal Administrasi Bisnis S1 Universitas Brawijaya*.

Koenjaraningrat. (2005). *Pengantar Antropologi II, Pokok-pokok Etnograf* (2nd ed.). Jakarta: Rineka Cipta.

Lockwood, Y. R., & Long, L. M. (2008). Culinary Tourism. *Journal of American Folklore*. https://doi.org/10.2307/20487614

Lohmann, G., & Panosso Netto, A. (2017). *Tourism theory: Concepts, models and systems*. CAB International.

Lozano-Oyola, M., Blancas, F. J., González, M., & Caballero, R. (2012). Sustainable tourism indicators as planning tools in cultural destinations. *Ecological Indicators*. https://doi.org/10.1016/j.ecolind.2012.01.014

Mason, P. (2012). Tourism impacts, planning and management. In *Tourism Impacts, Planning and Management*. https://doi.org/10.4324/9780080481418

Moleong, Lexy. 1995. *Metode Penelitian*. Bandung: Remaja Rosda Karya.

Nandi, N. (2016). PARIWISATA DAN PENGEMBANGAN SUMBERDAYA MANUSIA. *Jurnal Geografi Gea*. https://doi.org/10.17509/gea.v8i1.1689

Pendit, N. S. (2002). Ilmu Pariwisata. *Pariwisata*.

Pitana, I. (2009). *Pengantar Ilmu Pariwisata*. Yogyakarta: andi.

Pitana, I. G., & Putu, G. (2009). *Sosiologi Pariwisata*. Yogyakarta: Andi

Rahmafitria, F., Pearce, P. L., Oktadiana, H., & Putro, H. P. H. (2020). Tourism planning and planning theory: Historical roots and contemporary alignment. *Tourism Management Perspectives, 35*, 100703. https://doi.org/10.1016/j.tmp.2020.100703

Rencana Induk Pembangunan Pariwisata Daerah Kabupaten Cirebon 2019–2025. (2018). Badan Perencanaan Pembangunan dan Pengembangan Daerah Kabupaten Cirebon.

Sumarabawa, I. G. A. dkk. (2013). Ketersediaan Aksesibilitas Serta Sarana dan Prasarana Pendukung Bagi Wisatawan Di Daerah Wisata Pantai Pasir Putih, Desa Prasi, Kecamatan Karangasem. *Jurnal Pendidikan Geografi, 3*(1), 1–14. Retrieved from ejournal.undiksha.ac.id/index.php/JJPG/ article/download/1220/1084, Diakses 03 Juni 2017

Suryadana, Moh Liga dan Vanny Octavia.(2015). Pengantar Pemasaran Pariwisata. Bandung: Alfabeta.

Sowwam, M., Riyanto, Anindita, D., Riyadi, S. A., & Qibthiyyah, R. M. (2018). Kajian Dampak Sektor Pariwisata Terhadap Perekonomian Indonesia. *Kementrian Pariwisata Republik Indonesia*.

Stylidis, D., Weidenfeld, A., & Wall, G. (2018). Tourism planning. In *Tourism Policy and Planning Implementation*. https://doi.org/10.4324/9781315162928-2

Undang-Undang Republik Indonesia, N. 10 T. 2009. (2009). Undang-undang No 10 Tahun 2009 tentang Kepariwisataan. *Bifurcations*. https://doi.org/10.7202/1016404ar

World Tourism Organization. (2004). Indicators of Sustainable Development for Tourism Destinations. In *Tourism's potential as a sustainable development strategy. Proceedings from the 2004 WTO tourism policy forum at the George Washington University, Washington, DC, USA, 18–20 October 2004.*

Promoting Creative Tourism: Current Issues in Tourism Research – Kusumah et al. (Eds)
© 2021 Taylor & Francis Group, London, ISBN 978-0-367-55862-8

Tour guides' multilingualism in the city of Bandung, Indonesia: What does the policy say?

C.U. Abdullah & S.R.P. Wulung
Universitas Pendidikan Indonesia, Bandung, Indonesia

ABSTRACT: Language proficiency and multilingualism play an important role in the tourism industry in order for people to market their product of tourism and for international tourists to communicate during their exploration within a certain destination. Moreover, policies regarding tour guides' self-development, including one related to multilingualism, are also inevitably important so that tour guides have clear rules and regulation as well as a legal basis for their self-development. The aim of this study is to analyze the policy and condition of multilingualism among tour guides in the city of Bandung, Indonesia. This qualitative study collected the data through documents of policies in relation to tour guides and relevant references from previous studies. The collected data were then analyzed using descriptive and content analyses. It has been found that Bandung, the capital city of West Java province, Indonesia, does not have any legal document concerning tour guides, let alone their self-development, including language training. In the meantime, other provinces, such as Bali, Central Java, and East Java, had a policy on that. What West Java province has had is a policy regarding the standardized services and principles of tourism in West Java preserving the noble values of Sundanese culture. Therefore, it is recommended that the local government of Bandung and West Java start discussing the making of policy concerning tour guides, including their language development, either two languages (bilingual) or more than two languages (multilingual).

1 INTRODUCTION

Language diversity is both an advantage and disadvantage in the tourism industry. Of all the official international languages recognized by the United Nations, English is one of the most predominantly used in comparison with the other ones (Arsky & Cherny 1997; Granville 2003; Granville et al. 1998); however, this assumption is questionably challenged in the tourism industry (Manaliyo 2009). One of the alternative solutions to this language glorification is by having a multilingual and multicultural program in the tourism industry through the existing tourism human resources (Jafari and Way 1994; Pek et al. 2019). In global tourism, the number of tourism service providers recognizing language diversity is significantly increasing as they market their products in a variety of languages (Lituchy & Barra 2008).

One of the most common tourism practical services is a visit to a destination, either domestic or international. Tourism industry, particularly in the industrial revolution 4.0, plays an important role in the visit as people involved in the industry are required to master two languages (bilingual) or even more (multilingual). Within the context of the tourism industry, multilingualism is a primary commodity since human interaction is actually the main marketing factor (Morin 2012). For the last couple of years, sociolinguistics and linguistic anthropologists have answered a central question on the globalization impact on language use and its standardization. Therefore, a special attention has been paid to the importance of language and communication to standardize communication patterns, language variety, and identity (Cameron 2000).

DOI 10.1201/9781003095484-19

Bilingualism is defined as an ability to speak two languages in daily lives and bilingual is an individual having the ability in general (Ellis 2015, 2008, 1996; Fabbro 1999). Bloomfield also defined bilingualism as an ability of using more than one language as a means of communication within different contexts (Bloomfield 1933). In the tourism context, a tour guide who speaks a foreign language is automatically bilingual as they master both their native language and a foreign language. In the meantime, Franceschini (2009) defines multilingualism as a capacity of a particular society, group, institution, or individual to use more than one language. According to McArthur et al. (2018), multilingualism is an ability to use three languages or more, both simultaneously through code-mixing and code-switching and partially in different contexts. However, Kemp (2009) avoids the exact number of languages in defining multilingualism for social, cultural, and economic reasons. To this relation, Cummins (2007), highlights multilingualism exclusively as the presence of several different languages in a given space, regardless of who the users are. Both multilingualism and bilingualism have the same principle: to be able to apply the "when" and the "where" of the code-mixing and code-switching, as Duchêne (2009) points out, that multilingual society is a group of people with an ability to switch languages they master in different types of activities properly. Nowadays, the function of language as a means of communication has shifted from monolingual to bilingual and multilingual as are willing and able to master and use two languages or more. This situation, where the society is starting to be bilingual and multilingual, demands the people to be able to select the proper use of language in accordance with the situation they are having (Wahyudi & Widhiasih 2016).

Linguistic abilities have been proven to be a valuable asset in tourism industry (Jafari & Way 1994; Marshall 1996). A study surveying hotels in the USE on how to handle complaints from the visitors showed that an effective response is acquired when reached through a multilingual and multicultural approach (Jafari and Way 1994). In addition, a study conducted in 2000 revealed that hotels in Spain agree to the idea that multilingualism is one of the keys to increase customers' satisfaction. This result is in line with that of a study in 2003 proving that language training placed the third most frequently asked training following computer training (Agut et al. 2003).

Even though linguistic skills are necessary in the tourism industry, it does not mean that language is the only important factor in the industry. Language-centered tasks are identified as jobs requiring their people to use a foreign language as a general skill. Such professions are usually teachers, translators, and interpreters (Manaliyo 2009). In addition, language-related tasks are jobs in which foreign language mastery is supporting skills (Bluford 1994). People with excellent language skills usually take a chance to work within the tourism industry to enhance their skills even though their job is a part time or short-term work (Highley 1997). This is the reason why a good deal of tourism industry holders are recruiting employees with industrial knowledge as well as foreign language skills. Consequently, so many people complete their qualification with foreign language mastery to win their job interview (Manaliyo 2009).

There have been a number of studies discussing multilingualism in tourism, one of which is conducted by Sindik and Božinovic (2013) identifying the difference between third year and fourth year students in the importance of foreign language mastery for the tourism industry. The main result of the study revealed that older students had a tendency to acquire more foreign languages in comparison to the younger ones. Meanwhile, Menike and Pathmalatha (2015), who conducted a study of 50 bachelors in tourism-related majors in the University of Sabargamuwa, Sri Lanka, found that there are such factors as lack of textbooks; workbooks, study hours, and motivation are the main challenges of mastering foreign languages. Another study by Suhaimi and Abdullah (2017) proved that multilingualism plays an important role in the tourism sector.

In the tourism sector, the selection and usage of language by defining which language to use by who is a crucial thing, particularly in relation to giving good services to the tourists (Duchêne 2009). Since the use of a language is attached to its culture, tour guides mastering a foreign language commonly have no problem dealing with its culture. Therefore, linguistic and cultural obstacles are intertwined. Fortunately, tourists are fully aware of this issue, especially when they consider which destinations to visit, prepare to-do and to-bring lists for their vacation, imagine their interaction within the desired destinations, and their quality of experiences (Cohen & Cooper 1986). In this context, multilingualism in the tourism industry is important since it is closely related to both impressions and experiences of the tourists. Therefore, the use of proper language in the tourism industry is very important in giving good services to international tourists. For tourists to have better understanding in communication, the tourism industry is required to perform both verbal and non-verbal standardized bilingualism and multilingualism. Therefore, this study aims to identify the policy made by the local government regarding tour guides, including their multilingualism training for self-development, working in the city of Bandung, West Java, Indonesia.

2 METHOD

The approach used in this study was a qualitative approach, which aims to explore and deeply understand the meanings of a certain individual or group considered a social issue, with a series of research procedures involving research questions. In this study, the focus was the tourism in the city of Bandung; therefore, this study belonged to a descriptive study elaborating a fact systematically, factually, and accurately. The studied phenomenon in this study was multilingualism of tour guides in Bandung, particularly related to the policies made by the local government. Thus, the data collected secondary data acquired from policies and previously related studies from relevant resources. The data were then analyzed through a content/document analysis.

3 RESULTS AND DISCUSSION

The city of Bandung comprises five administrative areas, namely, *Kabupaten Bandung, Kabupaten Bandung Barat, Kabupaten Sumedang, Kota Bandung,* and *Kota Cimahi*, where *Kabupaten* is equal to a district in a rural area, yet *Kota* is equal to a district in an urban area. For the policy regarding multilingualism for tour guides, all the five areas do not have any, either a policy regarding tour guides in general, let alone one in relation to multilingualism of tour guides for their self-development. Moreover, at the provincial level, West Java also does not have any policy regarding the issue. There was a policy related to tourism made in 2015 on a slogan *Someah Hade Ka Semah*, which slightly addressed multilingualism. However, it was just a governor's regulation. In the meantime, there are other provinces in Indonesia which have created a policy regarding tour guides and their multilingualism, including Bali, East Java, and Central Java.

Policies in any sector can give both benefits and drawbacks; they can either increase the quality of a country or reduce or even sometimes damage it. A collaborative study between experts in Australia and China has successfully shown how political movement is sometimes beneficial yet disadvantages the trade between two countries, for instance the United States and China (Cathy et al. 2020). On the other hand, the implementation of poverty alleviation tourism policy (PATP) in China has been able to reduce its poverty rate dramatically (Qin et al. 2019). To maintain sustainable tourism, policies are one of the key factors as every implementing phase in the industry should refer to them. A study in Austria has proven that policies at European countries have supported sustainable tourism, even though an evaluation and improvement of tourism in rural areas are necessary (Wanner et al. 2020). Not only should the policies related to sustainable tourism pay attention to the people involved in the industry, but ones paying attention to animals are also important. A study investigating 123 tourism policies in 73 countries has found that those policies have considered animals as a tourism stakeholder; their existence and thrive are considered particularly when they give economic benefits to the industry (Sheppard & Fennell 2019). In order for the government to be able to make a good quality policy on sustainable tourism, the government and the industry should work well with the researchers (Jamal & Alejandra 2017).

Unfortunately, policies discussing the rights and duties of tour guides are rarely found. In China, for instance, some tour guides are suffering from a zero-fare tour as some experts are interested in discussing the literature and policies related to the issue (Hazel & Mcgehee 2016). As a matter of fact, tour guides play an important key of sustainable tourism; some existing literature even consider tour guides as the key to sustainable tourism (Carmody 2013) as they are ones who can interpret the needs and wants of the tourists (Ababneh 2017) in an interesting way, one of which is in storytelling (Bryon 2012). Therefore, policies regarding multilingualism training let alone regarding their rights and self-development are extremely crucial so that they have job security as well as sustainable self-development. This way, their roles in the tourism industry can be optimal.

4 CONCLUSION

A policy regarding tour guides and their self-development, multilingualism included, is necessary so that tour guides have a legal basis for their job security and improvement. This is such a crucial issue as multilingualism plays an important role in the tourism industry. Tour guides who are multilingual are automatically multicultural; thus, they are preferred by the tourists in comparison with tour guides who are monolingual or bilingual. It is recommended that the government, tourism stakeholders, academics, and researchers work collaboratively to create policies discussing the rights of tour guides in order to maintain sustainable tourism. It is also recommended that further studies investigate the needs and wants of multilingualism training for tour guides, particularly in the city of Bandung, Indonesia.

REFERENCES

Ababneh, A., 2017. Tour guides and heritage interpretation: guides 'interpretation of the past at the archaeological site of Jarash , Jordan. J. Herit. Tour. 0, 1–16.

Agut, S., Grau, R., Peiró, J.M., 2003. Competency needs among managers from Spanish hotels and restaurants and their training demands. Int. J. Hosp. Manag. 22, 281–295.

Arsky, J.M., Cherny, A.I., 1997. The ethno-cultural, linguistic and ethical problems of the "infosphere." Int. Inf. Libr. Rev. 29, 251–260.

Bloomfield, L., 1933. Language history: from Language. Holt, Rinehart and Winston.

Bluford, V., 1994. Working with foreign languages. OCCUP. Outlook Q. 38, 25–27.

Bryon, J., 2012. Tour Guides as Storytellers – From Selling to Sharing Tour Guides as Storytellers – From Selling to Sharing. Scand. J. Hosp. Tour. 12, 37–41.

Cameron, D., 2000. Styling the worker: Gender and the commodification of language in the globalized service economy. J. Socioling. 4, 323–347.

Carmody, J., 2013. Intensive tour guide training in regional Australia: an analysis of the Savannah Guides organisation and professional development schools. J. Sustain. Tour. 21, 37–41.

Cathy, X., Bao, J., Qu, M., 2020. Can tourism be a policy tool to moderate trade balance? Ann. Tour. Res. 1–3.

Cohen, E., Cooper, R.L., 1986. Language and tourism. Ann. Tour. Res. 13, 533–563.

Cummins, J., 2007. Rethinking monolingual instructional strategies in multilingual classrooms. Can. J. Appl. Linguist. 10, 221–240.

Duchêne, A., 2009. Marketing, management and performance: Multilingualism as commodity in a tourism call centre. Lang. Policy 8, 27–50.

Ellis, R., 1996. The Study of Second Language Acquisition. Oxford Univ. Press, Tesl Canada Journa 13, 53–80.

Ellis, R., 2008. The Study of Second Language Acquisition, 2nd ed. Oxford University Press.

Ellis, R., 2015. Understanding Second Language Acquisition 2nd Edition – Oxford Applied Linguistics, 2nd ed. Oxford University Press.

Fabbro, F., 1999. The Neurolinguistics of bilingualism. Psychology Press LTD Hove.

Franceschini, R., 2009. The Genesis and Development of Research in Multilingualism: Perspectives for Future Research. In: The Exploration of Multilingualism, Aila Applied Linguistics Series. John Benjamis Publishing Company, Amsterdam, pp. 27–63.

Granville, S., 2003. Contests over meaning in a south african classroom: Introducing critical language awareness in a climate of social change and cultural diversity. Lang. Educ. 17, 20.

Granville, S., Janks, H., Mphahlele, M., Reed, Y., Watson, P., Joseph, M., Ramani, E., 1998. English with or without g(U)ilt: A position paper on language in education policy for south Africa. Lang. Educ. 12, 254–272.

Hazel, Y.X., Mcgehee, N.G., 2016. Tour guides under zero-fare mode: evidence from China. Curr. Issues Tour. 3500, 1–25.
Highley, J., 1997. Language barrier leads to resignations. Hotel Motel Manag. 212, 20–21.
Jafari, J., Way, W., 1994. Multicultural Strategics in Tourism. Cornell Hotel Restaur. Adm. Q. 35, 72–79.
Jamal, T., Alejandra, B., 2017. Tourism governance and policy: Whither justice? Tour. Manag. Perspect. 2015–2018.
Kemp, C., 2009. Learning, transfer, and creativity in multilingual language learning: A dynamic systems approach. In: Proceeding of Third Language Acquisition and Multilingualism. Free University of Bozen-Bolzano.
Lituchy, T.R., Barra, R.A., 2008. International issues of the design and usage of websites for e-commerce: Hotel and airline examples. J. Eng. Technol. Manag. – JET-M 25, 93–111.
Manaliyo, J.C., 2009. Tourism and multilingualism in Cape Town: language practices and policy. Chicago.
Marshall, A., 1996. Industry must learn to hurdle language barrier. Hotel Motel Manag. 211, 17.
McArthur, T., Lam-McArthur, J., Fontaine, L., McArthur, R. (Eds.), 2018. The Oxford companion to the English language, Second edi. ed, Oxford quick reference. Oxford University Press, Oxford, United Kiingdom; New York, NY.
Menike, H., Pathmalatha, K., 2015. Developing ForeignLanguageCompetencies of TourismIndustry Oriented Undergraduates in Sri Lanka. Tour. Leis. Glob. Chang. 2, 74–87.
Morin, F., 2012. Sennett, Richard, Le travail sans qualités: les conséquences humaines de la flexibilité, Relations industrielles. Albin Michel.
Pek, L.S., Mee, R.W.M., Nadarajan, N.T.M., Mohamad, A.R., Alias, Z., Ismail, M.R., 2019. Tourists' Perceptions on Multilingualism use among Tourism Employees at Major Attractions in Kuala Selangor. Int. J. Acad. Res. Bus. Soc. Sci. 9, 914–919.
Qin, D., Xu, H., Chung, Y., 2019. Perceived impacts of the poverty alleviation tourism policy on the poor in China. J. Hosp. Tour. Manag. 41, 41–50.
Sheppard, V.A., Fennell, D.A., 2019. Progress in tourism public sector policy: Toward an ethic for non-human animals. Tour. Manag. 73, 134–142.
Sindik, J., Božinovic, N., 2013. Importance of foreign languages for a career in tourism as perceived by students in different years of study. Tranzicija 15, 16–28.
Suhaimi, N.I.B., Abdullah, A.T.H. bin, 2017. The Role of Multilingualism in Enhancing Tourism Sector in Malaysia. Int. J. Acad. Res. Bus. Soc. Sci. 7, Pages 816–832.
Wahyudi, N.D., Widhiasih, L.K.S., 2016. KEANEKABAHASAAN (MULTILINGUALISME) DALAM VIDEO PROMOSI DESTINASI PARIWISATA JEGEG BAGUS DENPASAR. In: Prosiding Seminar Nasional Hasil Penelitian. Universitas Mahasaraswati Press, Denpasar.
Wanner, A., Seier, G., Pr, U., 2020. Policies related to sustainable tourism – An assessment and comparison of European policies, frameworks and plans. J. Outdoor Recreat. Tour. 29, 1–12.

Promoting Creative Tourism: Current Issues in Tourism Research – Kusumah et al. (Eds)
© 2021 Taylor & Francis Group, London, ISBN 978-0-367-55862-8

Tourist preferences of activities in the tourist village

S. Marhanah & E. Sukriah
Universitas Pendidikan Indonesia, Bandung, Indonesia

ABSTRACT: The Potential of Free Tourism Village is the uniqueness of the urban villages, such as community activities, culture, typical village arts, culinary, waterfalls, and natural beauty such as rice fields and plantations. However, based on the results of interviews with several managers of Tourism Villages, there are still many local tourists visiting around the Tourism Village and who mostly enjoy the view of the waterfall. Many tourists do not know Cibeusi Village is being developed into a leading tourist village in Subang Regency. The purpose of this study is to recommend preferences for tourism planning in the Tourism Village. The results of this study indicate that tourist preferences for tourism activities in the Tourism Village are to take pictures in exciting places and Instagramable photo spots available in the village that they consider unique and different from other tourist destinations and to enjoy the beauty of the village environment.

Keywords: visit preference, tourism village, tourism activities.

1 INTRODUCTION

At present, many regions are competing to develop their areas into tourist destinations. Since the community is aware of the existence of tourism activities, it will have an impact on improving the economy. Currently, the reality of tourist destinations is everywhere, especially now that the government has implemented a regional autonomy program. Therefore, the tourism industry is one alternative that can be utilized as a source of local revenue (Suwena & Widyatmaja 2017). Subang Regency is one of the districts in West Java that has a lot of tourism potential. In addition, Subang Regency is adjacent to the capital of West Java Province, Bandung. In terms of geomorphological conditions, Subang Regency has varied landscapes, starting from the mountains, highlands, and lowlands. Cibeusi Village, Ciater District, is one of 15 tourism villages in the Subang Regency. The potentials of Cibeusi Village are waterfalls, culture, and culinary. However, over time, Cibeusi Tourism Village is still not developed. Tourists who visit are still limited to local tourists or tourists who live not far from Cibeusi Tourism Village. How to package the right tourist village that can become superior and famous by domestic and foreign tourists still has to be learned. The desire of tourists for a product and service varies from one tourist to another. Therefore, a tourist destination must meet the wants and needs to increase tourist satisfaction when visiting. Before traveling to a place, tourists in general, will consider several factors such as prices, tourist attractions, facilities, and locations that suit tourists' desires. Knowing tourists' desires for decisions in choosing tourist attractions is very important for the tourism industry. It can provide benefits for the providers to prepare or pay attention to what types of activities are desired by tourists. Knowing tourist preferences is also very helpful in creating new products by what is wanted and needed by tourists during a tour to a Tourism Village. Therefore, it is necessary to identify the preferences of tourists for activities in the Tourism Village and further analyze the market share of tourists in the Tourism Village.

 DOI 10.1201/9781003095484-20

2 LITERATURE REVIEW

Under Law Number 10 of 2009 concerning Tourism, tourism is a variety of tourism activities and various facilities and services provided by the community, entrepreneurs, government, and local governments. According to Assauri (2014), a person's preferences are related to the behavior of consumers or buyers in the process of selecting products to be purchased, which is the buying process. The purchase process needs to be studied to determine why someone chooses and buys and becomes more pleased with certain brand products. Usually, tours will visit tourist destinations because of the uniqueness of a place that is an identity.

Various efforts are carried out by the organization of tourism to match market demand, so it is necessary to find out the tourist data needed including the demographic characteristics of the tourists, namely, gender, age, place of origin, occupation (Kusmayadi & Sugiarto 2000); tourist behavior, namely, the type of attraction preferred, the type of accommodation desired, the time required to stay, affordability (Wardiyanta 2006); and tourist psychology, namely, understanding, perception, behavior, the attitude of tourists toward tourism activities, and social and economic conditions (Joaqui & Jaume 2010).

Travel motivation can be an interpreter as a factor that makes someone want to travel (Joseph 2013). Second, previous research has found that tourist motivations could differ depending on the type of regional tourism destinations, even if they in the same country (Huang et al. 2018). Other tourist motivations obtained from the Hong Kong government distinguish motivation into eight (Education Bureau of Hong Kong 2013), namely

- vacation, associated with relaxation and sight-seeing;
- business, related to business interests;
- culture, associated with the addition of insights related to tourist attractions;
- ecotourism, related to enjoying nature;
- learning, compared to knowledge due to academic interests;
- religious, related to ethical travel;health, similar to medical needs;
- visiting friends and relatives

Different Motivational Tourism Activities will determine different types of activities. The Indonesian Ministry of Tourism (2016) categorizes tourism activities into nine categories: 1) marine tourism including surfing, sunbathing, parasailing, cruising, sailing/yachting, fishing (in the sea), diving, and snorkeling; 2) eco-tourism including research/education related to the environment, mountain biking, safari and bird watching, river-going, saving endangered animals, ecoriding, exploring and rallying, farming/gardening; 3) adventure tours include camping, trekking, cave cruising, hunting, rafting, kayaking, river/lake cruise, bungee jumping, sky diving; 4) historical/religious tourism including pilgrimage, visiting shrines, museums, cultural, religious heritage, memorials, traces of civilization; 5) art and culinary tours including folklore performances, theater, pop culture shows, traditional ceremonies, carnivals, traditional arts, dinner, buying local food, tasting unique food; 6) city and rural tourism includes shopping, visiting friends, enjoying night entertainment, staying in traditional villages, visiting traditional markets, darmabakti tours, philanthropic and corporate social responsibility programs, photography and architectural visits, live-in programs; 7) MICE tours include business meetings and forums, incentives, conferences, trade show exhibitions, show business, concerts, films; 8) sports/health tourism including spa and fitness, yoga and meditation, medical and health tourism, participating in international sporting events, watching sporting events, training camps, visiting extreme sports locations, participating in extreme sports, studying and exercising extreme sports; and 9) integrated tourism objects including ecopark, recreation and outbound park, zoo, botanical garden, playground for teenagers and children, pensioners/silver tourism, wedding and honeymoon programs, resorts and islands on the beach, resorts in mountains, dark tourism.

Heritage tourism, creative industries, multiculturalism, and the local way of life are indicators of tourist attractions (Uroševic 2012). Motivation is the internal force that moves individuals to act, and it responds to the tension created by human's needs (Schiffman et al. 2012). Tourist motivation is also a determining factor in the decline and increase in tourist visits. The push and pull factor is a travel motivation theory that is widely used by researchers. Push factors refer to encouragement for individuals to decide to travel from home, whereas pull factors attract individuals (Khuong & Ha 2014). Hamari et al. (2015) conclude that the intrinsic ones are a strong determinant of attitude, but that the extrinsic ones are those that mostly influence the continued use of this type of tourism. Also, there are other opinions, namely Sayangbatti and Baiquni (2013), which state that there are two factors that can influence the final decision of tourists in visiting a tourist destination: 1) internal factors, factors originating from tourists; and 2) external factors, factors that arise from outside the tourists. It could be that the push factor is an internal factor, and the pull factor is an external factor.

3 METHOD

This research is a quantitative descriptive, intended to get a picture or description of tourist preferences for activities in the Tourism Village. The research method uses a survey approach that takes samples from a heterogeneous to look for information factually. Furthermore, the selected respondents will fill out a questionnaire to get the required information in the study—the data analysis technique of this research is quantitative descriptive analysis. Quantitative descriptive analysis was carried out by grouping variable data based on each group from the original. The data is random and easy to interpret informally. Thus, it is hoped that it can provide special characteristics regarding tourist preferences for the Tourism Village.

4 RESULT AND DISCUSSION

Tourist preferences in Tourism Village based on the characteristics of tourist attractions are elements of measurable and immeasurable tourism offers. The tourism offer in question is a Tourism Village, which can provide pleasure and satisfaction for tourists. The analysis of tourism preferences for Tourism Villages aims to determine which Tourist Villages are most in demand and become an attraction for future visitors. Characteristics of visitors to tour packages in Rahmawati's (2010) study include name, age, gender, income level, education level, hometown, marital status, and type of work. In Dwiputra's (2013) research, the characteristics of tourists consist of gender, age group, educational background, area of origin, marital status, family cycle, occupation, monthly income.

Based on the research of types of tourists who have visited the Tourism Village, it shows that most visitors to this Tourism Village are tourists from West Java, which is 73.4%. Tourism Village attracts relatively few visitors who come from outside West Java. Thus, the manager needs to review and develop aspects that can attract preferences for tourists, for example, the type of tourism offered, the tourism facilities provided, tourist village marketing, ease of access to tours, comfort, security guarantees, culinary, and tourist information that can be understood by the traveler.

This study tries find out the comparison between female and male visitors, which obtained data that 65.1% is the female group and 34.9% is male group. The data shows that there are more female visitors than male. Therefore, the Tourism Village is a destination that is much sought after by both male and female visitors. Thus, in its development, it is necessary to consider things that are intended in general, not based on gender differences.

Age of visitors in the group was 18 to 40 years by 80% seen more compared to other age groups. In general, this group is a productive age with good health, thus, the people prefer to travel. In contrast, the age group that visited the least number of tourist attractions was the age group of more than 65 years, only 2%, which consists of parents and the elderly.

Conditions in this age group, in general, have significantly declined both in terms of health and productivity. Based on data from tourists visiting the Tourism Village, more tourists are from young age groups, so then attention should be given priority to the tourist facilities favored by the group.

Based on educational background, visitors are dominated by tertiary education levels of 89% and followed by high school education levels by 11% of the total respondents. It can be a factor in consideration of demand. The level of education will affect the activities, services, and quality of facilities in the Tourism Village.

Tourism Village is dominated by tourists with teenagers, which is 48% of the total visitors. The rest is the cycle of tourists, new couples, couples with toddlers, couples with children, and couples with teenagers. These types of tourists usually come with groups of friends, as teenage is a time when people prefer to look for new experiences and vacation with their friends, including visiting tourist attractions.

The majority of visitors to the Tourism Village–based type of work are students, as many as 38%, and about 24% of employees. It is consistent with the previous discussion that the largest group of visitors is aged 18 to 40 years. Students are those who are in a period that is still full of energy and tend to like looking for new experiences. In carrying out their activities, they also prefer to be with their friends, not least visiting the Tourism Village. Thus, the tour manager needs to pay attention to events and attractions that are of interest and are attractive to young people.

In line with the type of work, data shows that tourists who come to this Tourism Village are groups of students and employees who have small expenses. Data shows the monthly income of tourists are dominated by spending around Rp. 1,000,000–2,000,000 per month, which is as much as 50.91% of total visitors. Based on these conditions, it can be a basis for consideration in determining the price of tourist facilities, souvenirs, and culinary to be affordable for the majority group.

Based on the basic pattern of tourism travel, some factors are included in the criteria definition of people who travel. The types of characteristics of tourist travel patterns consist of tourist destinations, tourist motivations, tourist information, ways of traveling, travel companions, length of travel, frequency of tours in the past one year, willingness to return, costs per day, and vehicles to ride to the attractions (Dwiputra 2013).

The study of tourist destinations is to show the purpose of tourists in visiting the Tourism Village. The survey shows that the majority of tourists visiting are on vacation, with the percentage of 63%. This is supported by data showing that the majority of tourists are students and employees. This group of tourists goes on a vacation by visiting the Tourism Village, amidst daily routine activities.

Many motivations influence human behavior because it is the primary driver of the tour process. Judging from the motive of tourists visiting, the Tourism Village is dominated by the natural beauty factor that is equal to 44%. Cibeusi Tourism Village is in a hilly area with clean and cold air and green nature with trees. The next tourist motivation is due to the availability of various facilities such as the existence of a variety of cultures, various culinary delights, or other reasons.

Tourists from various sources of information can obtain the existence of a tourist destination. Data collected from respondents show that they got the information from three sources: from friends or relatives, 31%; the Internet, 48%; and the remaining 20% from other sources such as brochures and television.

Many tour packages have sprung up, which are offered by tour bureaus, with all its conveniences. However, based on the way tourism came to the Tourism Village, the majority of respondents traveled independently, 83%. In this case, tourists visit freely by their time, desires, and needs. As many as 17% of the respondents choose to use tour packages in traveling. Thus, the manager needs to keep in mind the factors required by independent tourists and tourists using tour packages.

Based on travel companion data, tourist arrivals with friends/relatives are most often found in the Tourism Village, 46%. In the previous analysis, this tourist group generally consists of students and employees who prefer to come with their friends. The remaining largest group is tourists who come with family, which is 54%. This group is tourists who come with their children, wives, husbands, or big families.

As for the time required by visitors for tourism activities in the Tourism Village, the majority is 68% for more than one day. Moreover, the remaining 38% visited for one day. It happens because the majority of tourists are aiming for a vacation and motivation to enjoy the beauty of nature.

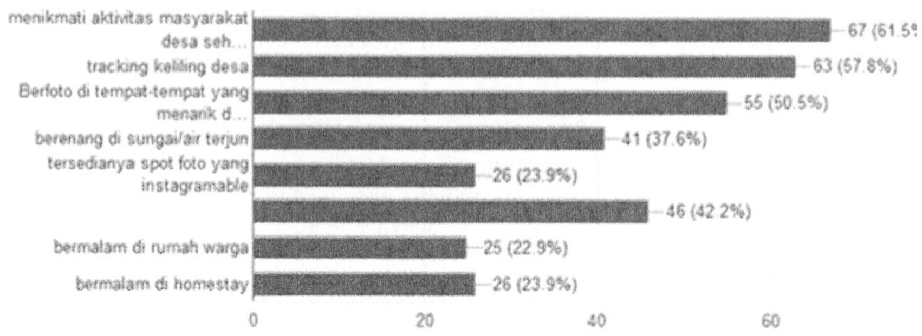

Figure 1. Activity type graph based on traveler preference.

Data number of tourist visits in the past year show that the majority of tourists have visited the Tourism Village once about 61%. Some tourists have visited the Tourism Village more than once, as much as 39%, namely, two times, three times, and more than three times.

Based on the survey results obtained, the majority of respondents, by 93%, expressed their willingness to return to visit the Tourism Village. Furthermore, the remaining 7% said they did not plan to come back to visit, Desa Wisata. This condition can be interpreted that the Tourism Village is considered a beautiful tourist destination so that tourists have the desire to visit again.

Costs incurred by the majority of tourists in the Tourism Village in one day is Rp. 50,000–Rp.100,000 as much as 50%, more than Rp. 150.000 as much as 41%, and the remaining 9% is less than Rp. 50,000 These costs are all expenses, including entrance fees, meals, and souvenirs. These costs are still affordable and relatively inexpensive when compared to the cost of traveling to other tourist destinations. It is related to the previous analysis that the majority of visitors are students and young employees.

The majority of tourists use private vehicles to go to the Tourism Village, which is 62%. In general, private vehicles used by tourists are cars and motorcycles. Most other tourists, 38%, use rental cars for tourism.

Based on the survey results obtained by the activity desired by tourists, 23% of tourists want to take pictures in exciting places in the village and available Instagramable photo spots, 19% want to enjoy community activities in the town, 18% of tourists want to do tracking around the city, 14% of tourists want to spend the night at a resident's home or homestay, 13% of tourists want to interact with villagers, and 12% of tourists want to swim in a river/waterfall. Thus, tourists prefer to do activities to see the scenery and take pictures in exciting places that are considered different from other sites.

5 CONCLUSION

The majority of tourists from West Java who visit the Tourism Village are female. Most of the tourists are aged 18 to 40 years with a college education background. Most tourists are students and employees with monthly salaries starting from Rp. 1,000,000 to Rp. 2,000,000. Most tourists aim for a vacation to see natural scenery, and they get this information from the Internet. The majority of tourists travel independently, with friends or relatives, and visit for more than one day. The majority of tourists have visited tourist attractions in the past year and plan to return. The cost incurred by tourists per day is less than Rp. 50,000–Rp. 100,000 during tourist activities. Most of the vehicles used to get to the tourist areas are private vehicles. For tourist preferences for tourism activities, most respondents want to take pictures at interesting places in the village and available Instagramable photo spots that are considered unique and different from other tourist destinations.

REFERENCES

Asauri, S. 2014. Manajemen Pemasaran Dasar, Konsep dan Strategi. Jakarta: Rajawali Pers.

Dwiputra, R. 2013. Preferensi Wisatawan Terhadap Sarana Wisata di Taman Wisata Alam Erupsi Merapi. Jurnal Perencanaan Wilayah dan Kota, 24(1), 35–48.

Education Bureau of Hong Kong. 2013. Tourism and Hospitality Studies: Introduction to Tourism. Hong Kong: The Government of the Hong Kong Special Administrative Region

Hamari et al., 2015 J. Hamari, M. Sjöklint, A. UkkonenThe sharing economy: why people participate in collaborative consumption J. Assoc. Inf. Sci. Technol., 3 (2) (2015).

Huang, Y., Wu, J., Shi, W. 2018. The impact of font choice on web pages: Relationship with willingness to pay and tourism motivation. Tourism Management, 66, pp. 191–199.

Joaqui and Jaume. 2010. Tourist Satisfaction and Dissatisfaction. Annals of Tourism Research, 37(1), 52–73.

Joseph, P. 2013. Tourism Principles, Policies & Practices. Pondicherry: Directorate of Distance Education, Pondicherry University.

Khuong, M.N. & Ha, H.T. 2014. The Influences of Push and Pull Factors on the International Leisure Tourists' Return Intention to Ho Chi Minh City, Vietnam -A Mediation Analysis of Destination Satisfaction. International Journal of Trade, Economics and Finance, 5(6), 490–496.

Kusmayadi & Sugiarto, E. 2000. Metodologi Penelitian dalam Bidang Kepariwisataan. Jakarta: Sun.

Rahmawati, R. 2010. Analisis Preferensi Pengunjung Terhadap Paket Wisata Kusuma Agrowisata Kota Batu Jawa Timur. Skripsi. Departemen Agribisnis Fakultas Ekonomi dan Manajemen Institut Pertanian Bogor.

Republic of Indonesia. 2009. Undang-undang Republik Indonesia Nomor 10 Tahun 2009 tentang Kepariwisataan. Jakarta: Sekretariat Negara RI.

Schiffman, L.G., Kanuk, L.L., Hansen, H. 2012. Consumer Behaviour: a European Outlook Financial Times: Prentice Hall.

Sayangbatti, D.P. & Baiquni, M. 2013. Motivasi dan Persepsi Wisatawan tentang Daya Tarik Destinasi terhadap Minat Kunjungan Kembali di Kota Wisata Batu. Jurnal Nasional Pariwisata, 126–136.

Suwena, I.K dan Widyatmaja, I.G.N. 2017. Pengetahuan Dasar Ilmu Pariwisata, Denpasar Bali: Pustaka Larasan dan Fakultas Pariwisata Universitas Udayana Bali.

Uroševic, N. 2012. Cultural Identity and Cultural Tourism -between the Local and the Global (A Case Study of Pula, Croatia). Singidunum Journal, 9(1), 67–76.

Undang-Undang No. 10 Tahun 2009 tentang Kepariwisataan

Wardiyanta. 2006. Metode Penelitian Pariwisata. Yogyakarta: Andi.

Promoting Creative Tourism: Current Issues in Tourism Research – Kusumah et al. (Eds)
© 2021 Taylor & Francis Group, London, ISBN 978-0-367-55862-8

The influence of Bandung City image as a fashion city on tourist satisfaction

P. Supriatin, S. Marhanah & Rosita
Universitas Pendidikan Indonesia, Bandung, Indonesia

ABSTRACT: The purpose of this study is to analyze the influence of the image of Bandung City as the city of fashion on tourist satisfaction in Bandung City. This type of research is a quantitative approach with data analysis using descriptive analysis and path analysis. The destination image variable as a fashion city has a significant effect on tourist satisfaction variable. From the results of the study, it is recommended that the local government of Bandung City and tourism businesses maintain the Image of Bandung City and increase the intensity of tourist visits by promoting all types of fashion and optimize the manner in the diverse Bandung City.

Keywords: destination image, tourist satisfaction

1 INTRODUCTION

In this millennial era, the style of clothing for young people in big cities is one of the crucial things that affects the demand for fashion products themselves. Fashion has become a necessity for most people. Because of the increasingly advanced times, styles cannot be separated from young people or teenagers because of their need to find their identity and existence in their social circles. The city of Bandung, which is one of the major cities in Indonesia and famous for its iconic fashion image or Paris Van Java, has helped boost both foreign and domestic tourists in the development of the clothing industry in Bandung. Bandung is a city that is well known as a city with enormous creative industry potential, especially in the fashion world. The current phenomenon is the number of distro outlets, clothing, and factory outlets in Bandung. The more clothing and factory outlets circulating, the tighter the competition. This image is able to attract the attention of tourists who come to Bandung City. Based on data from www.kompas.com, 2019, it was found that the creative industry in Bandung was very developed. The highest percentage of the creative industry is fashion with a rate of 43.71% because style is one type of business that is profitable for entrepreneurs. From this research and the facts about the opinions of tourists about the image of Bandung as a city of fashion, it is hoped that the development of tourism in the city of Bandung can be in harmony with what is needed by tourists, so that tourists can get satisfaction and plan to repeat their visit or at least recommend Bandung as a destination which is worth visiting. In addition, the analysis was also conducted to find out what factors most influence tourist satisfaction from the destination image attribute.

2 LITERATURE REVIEW

Destination tourism has a significant impact on a country's economic development, especially in terms of job creation rates (Jia et al. 2016). However, as Bianchi, et al. (2014) point out, enhancing the positive perception of destination branding in this type of destination is markedly tricky. Through place branding —understood as applying product brand management to the goal—DMOs develop strategies to add value to the brands associated with given tourist destinations. Destination image

 DOI 10.1201/9781003095484-21

consists of functional characteristics regarding tangible aspects such as price level, accommodation, facilities, attractions of a goal, and psychological attributes regarding more intangible elements such as friendliness and safety (Echtner & Ritchie 2003). There are three dimensions of destination image, according to Baloglu and McCleary (Hailin Qu 2011), namely, mental destination image (cognitive destination image), unique destination image (unique destination image), and affective destination image. Cognitive destination image is an assessment of perceptions about the principle of a destination, to produce information about knowledge in memory. Mental imagery consists of "quality of experience, tourist attractions, environment and infrastructure, entertainment/outdoor activities, and cultural traditions."

According to Gunn (2002), image formation before the trip is the most critical phase in selecting tourist destinations. The image perceived after the occurrence of a tourist visit will also affect consumer satisfaction and intensity to make a return visit in the future, depending on the tourist destination in providing experiences that suit the needs and image that tourists have about the tourist area.

Satisfaction can be interpreted as "efforts to fulfill something" or "make something adequate" (Tjiptono, Marketing Strategy 2007). According to Kotler, happiness is decided by the hopes and perceptions of tourists. If the expectations are higher than reality, then tourists will not feel satisfied and not happy. On the contrary, if their expectations are smaller, and the truth is greater, then tourists will contact very satisfied (Kotler 2009).

"Customer satisfaction, especially in the service sector, is imperative for the company to remain successful. The difference occurs because of a gap between customer expectations and the reality (return) of the service received; the gap exists as a result of not meeting the expectations of customers" according to Astuti (2012). Satisfaction is the expectation and assessment of perceptions about the quality of services or products. It is a difference between expectations and the reality that is received by tourists; it means there is a gap. Tourist satisfaction is considered very important because it can cause an increase in the number of tourist visits. Satisfied tourists feel that what they expect and the reality they get is what they want (Setiady et al. 2015). Assaker et al. (2015) suggested that tourist satisfaction has a positive influence on tourist loyalty. Satisfaction will determine tourist's' reliability through their desire to visit again and recommend to others to see. Furthermore, Chiu, et al. (2016) suggested that tourist satisfaction can significantly influence tourist loyalty.

3 METHOD

The research method used in this study is descriptive verification research with a quantitative approach. The technique used in this research is the path analysis technique. Path analysis is carried out to determine the magnitude of influence between variables both directly and indirectly. You can see in the following section the hypothetical model that illustrates the importance of destination images on satisfaction and their impact on loyalty.

4 RESULTS AND DISCUSSION

The path analysis of destination image variables on tourist satisfaction variables are shown in Table 1.

The destination image variable significantly influences tourist satisfaction. The study shows that destination image affects tourist satisfaction, with a probability level (Sig) = 0,000 or Sig ≤ 0.05 and t arithmetic obtained at 8.502. because t arithmetic is more significant than t table (t table with a 5% significance of 1.984), which means that there is a considerable influence of the destination image (X) on tourist satisfaction (Y).

Referring to the trustor knowledge of tourists about the environment and infrastructure, this includes that the quality of the road to Bandung City is considered acceptable, accessibility to the city of Bandung deemed to be easy to go, environmental cleanliness, and public facilities

Table 1. T-test in endogen and exogen variable.

Endogen Variable	Exogen Variable	Beta	tCount	Prob	Note
Destination Image	Satisfaction	0.652	8.502	0.000	Sig
Destination Image	Loyalty	0.564	5.262	0.000	Sig
Satisfaction	Loyalty	0.549	5.461	0.000	Sig

are considered acceptable. The next indicator is the attractiveness of fashion stores, and fashion uniqueness. According to Allameh et al. (2015), destination imagery is an interactive system of thoughts, opinions, feelings, visualizations, and intentions toward a tourist destination that leads to decision making about a tourist destination. Furthermore, Akroush et al. (2016) state that destination images are perceptions proposed by potential visitors to tourist destinations. The development of an excellent tourist destination, of course, can make tourists satisfied and then be able to return to visit other destinations, according to Hanif et al. (2016).

Discussion of tourist satisfaction level in the variable of tourist satisfaction shows that the highest score is on the dimension of interest in visiting again, with a score of 441. The size of interest in seeing again is the highest value because tourists are satisfied to visit Bandung and plan to revisit Bandung because of fashion in Bandung city by expectations or expectations that tourists think about Bandung as a city of fashion. This research is supported by findings of Coban, (2012), which prove that satisfaction positively influences behavior after visiting by revisiting the tourist destination or recommending it to the closest person.

Discussion of the effect of imagery on tourist satisfaction study found that destination image influences tourist satisfaction with a probability level (Sig) = 0,000, or Sig $\leq$ 0.05 and t arithmetic obtained at 8.502. Because t arithmetic is more magnificent than t table (t table with a 5% significance of 1.984), it means that there is a significant influence of the destination image (X) on tourist satisfaction (Y). The results of this study are reinforced by Coban (2012), which explains that there is a positive influence on the image variable (X) on the satisfaction variable (Y). Particularly in cognitive imagery indicators, there are generally six indicators related to tourists' perception of tourists about destinations. The influence of destination image (X) on tourist satisfaction (Y) is also aided by tourist information or insights about the city of Bandung. On average, some respondents get information about Bandung as a city of fashion. This information motivates tourists to visit Bandung for its reasons concerned with fashion shopping tourism.

5 CONCLUSION

After researching stages of observation, interviews, and analysis in Bandung City, the authors conclude the results of this study into the following points: The image of Bandung City, according to respondents as a whole, is in the first category. Between the two destination image subvariables, the cognitive image has the highest score. Mental image refers to the belief or knowledge of tourists about the attraction in Bandung City. Tourist satisfaction, according to respondents, was assessed as being overall in the Very Satisfied category. The variable image of Bandung as a fashion city influences tourist satisfaction. Therefore, what has already existed in Bandung can be maintained and optimized.

REFERENCES

Akroush, M. N., Jraisat, L. E., Kurdieh, D. J., & Qatu, L. T. 2016. Tourism service quality and destination loyalty–the mediating role of destination image from international tourists' perspectives. Tourism Review, 71(1), 18–44. https://doi.org/10.1108/TR-11-2014-0057.

Allameh, S. M., Khazaei Pool, J., Jaberi, A., Salehzadeh, R., & Asadi, H. 2015. Factors influencing sport tourists' revisit intentions: The role and effect of destination image, perceived quality, perceived value, and satisfaction. Asia Pacific Journal of Marketing and Logistics, 27(2), 191–207.

Assaker, G., Hallak, R., Assaf, A. G., & Assad, T. 2015. Validating a structural model of destination image, satisfaction, and loyalty across gender and age: Multigroup analysis with PLS-SEM. Tourism Analysis, 20(6), 577–591.

Astuti, W. W. 2012. Analisis Kepuasan Pelanggan Mengenai Kualitas Pelayanan Service Excellent Komputer Semarang. Management Analysis Journal.

A. Beerli, J.D. 2004. MartínTourists' characteristics and the perceived image of tourist destinations: A quantitative analysis. A case study of Lanzarote, Spain *Tourism Management*, 25 (5), pp. 623–636.

Bianchi, C. N., Morri, C., & Pronzato, R. (2014). The other side of rarity: recent habitat expansion and increased abundance of the horny sponge Ircinia retidermata (Demospongiae: Dictyoceratida) in the southeast Aegean. *Italian Journal of Zoology*.

Chiu, W., Zeng, S., & Cheng, P. S.-T. (2016). The influence of destination image and tourist satisfaction on tourist loyalty: a case study of Chinese tourists in Korea. *International Journal of Culture, Tourism and Hospitality Research*.

Coban, S. 2012The effects of the image of the destination on tourist satisfaction and loyalty: The case of Cappadocia. European Journal of Social Sciences, 222–232.

Echtner, Charlotte M. and Ritchie, J. R. Brent. The meaning and measurement of destination image: [Reprint of original article published in v.2, no.2, 1991: 2–12.] [online]. Journal of Tourism Studies, Vol. 14, No. 1, May 2003: 37–48. Availability: <https://search.informit.com.au/documentSummary;dn=200305723;res=IELAPA> ISSN: 1035-4662. [cited 16 Sep 20].

Gunn, C., 2002. Tourism Planning, Fourth Edition, Basics Concept Cases (4th ed.). New York: Routledge.

Fichtner C. M & Ritchie, J. R. B. 2003. The Meaning and Measurement of Destination Image. The Journal of Tourism Studies. 14 (1): 37–48.

Hailin Qu, L. H. 2011. A model of destination branding: Integrating the concepts of the branding and destination image. Tourism Management, 465–467.

Hanif, A, Kusumawati, A, Mawardi K, 2016. Pengaruh Citra Destinasi terhadap Kepuasan Wisatawan serta dampaknya terhadap Loyalitas Wisatawan, Jurnal Ilmiah Administrasi Bisnis.

Jia, J., Zhang, L., Liu, Z., Xiao, X., & Chou, K.-C. (2016). pSumo-CD: predicting sumoylation sites in proteins with covariance discriminant algorithm by incorporating sequence-coupled effects into general. *Bioinformatics, Volume 32, Issue 20*, 3133–3141.

Kotler, P., & Keller, K. L. 2010. Manajemen Pemasaran. Jakarta: Erlangga.

C.-H. Liu, S.-F. Chou. 2016. Tourism strategy development and facilitation of integrative processes among brand equity, marketing, and motivation Tourism Management, 54, pp. 298–308.

Setiady, T., Sukriah, E., & Rosita. 2015. Pengaruh Servicescepe Terhadap Kepuasan Pengunjung di Floating Market Lembang. Jurnal Managemet Resort dan Leisure, 22.

Syarifuddin. 2018. "Nilai Citra Kota Dari Sudut Pandang Wisatawan. Studi Tentang Citra Kota Bandung Dampaknya Terhadap Kunjungan Ulang.

Tjiptono, F. 2007. Strategi Pemasaran. Yogyakarta: Andi Offset.

Promoting Creative Tourism: Current Issues in Tourism Research – Kusumah et al. (Eds)
© 2021 Taylor & Francis Group, London, ISBN 978-0-367-55862-8

The influence of destination image on revisit intention in Olele Marine Park

M.N. Della & N. Wildan
Universitas Pendidikan Indonesia, Sumedang, Indonesia

O. Sukirman
Universitas Pendidikan Indonesia, Bandung, Indonesia

ABSTRACT: Olele Marine Park in Bone Bolango regency was the focus of this research. Destination image was one of the factors to develop tourism activity in Olele Marine Park. Tourist destination should be creating a positive image to attract and make the tourist come back to spending their leisure time in Olele Marine Park. The study aims were: 1) identify destination image and revisit intention, 2) analyze the influence of destination image on revisit intention in Olele Marine Park. The methods used were descriptive and verification approach by using path analysis. The survey conducted on 180 respondents (tourists who visited Olele Marine Park). The result of the research shows that: 1) Olele Marine Park has a positive image and 2) destination image significantly influenced to revisit intention in Olele Marine Park, which means that the hypothesis proposed in this study was accepted. The manager of Olele Marine Park should increase the authenticity as the unique point to creating positive image; adequate infrastructure and environmental hygiene should be improved to preserve the image of Olele coastal.

Keywords: destination image, revisit intention, Olele Marine Park.

1 INTRODUCTION

Gorontalo has many natural attractions, one of which is Olele Marine Park located in Olele Village, Kabila Bone Regency, Bone Bolango Regency. Olele Marine Park is white sand and the visibility of the water is still good, and the coral reefs are still well preserved because the people around Olele Marine Park are very concerned about the natural wealth in their environment (Asdhiana, 2013). It is known that the visit of domestic tourists to Olele Marine Park in 2018 was 11,714 while foreign tourists numbered 1,631; in 2019 domestic tourists were 17,390 while foreign tourists were 1,228 BPS (2018). This shows that there has been an increase in tourist visits in the last two years. Olele Marine Park is popular for snorkeling and diving activities where there are many dive spots with different characteristics from each other, namely Jin Cave, Traffic Circle, Honeycomb, and Muck Dive. Even the Olele Marine Park has endemic coral sponges that cannot be found in any waters. There is uniqueness in coral motifs that resemble the paintings of one of the artists from Spain, Salvador Dali. The painting was named L'enigma del Desiderio. Therefore, the Petrossian Lignosa sponge is known as Salvador Dali Rizal (2020). The uniqueness and originality of the Olele Marine Park is one of the factors in creating a positive tourist destination image. Qu et al. (2011) suggested that "Destination images consist of cognitive images, unique images, and affective images." A positive destination image will give tourists the confidence to make a return visit to the tourism destination. A good image will certainly attract the interest of tourists to come to visit; on the contrary, bad imaging will make tourism destinations worse off (Indira et al., 2013). Managers and stakeholders of the Olele Marine Park consistently maintain a positive image of their tourist destinations. The author is very interested in examining how much tourists have the intention to return to the Olele Marine Park after seeing and feeling the uniqueness of the attraction of the

 DOI 10.1201/9781003095484-22

Olele Marine Park. According to Chen and Tsai (2007), Lee et al. (2005), and Moutinho (1987) in Chen et al. (2010), there are three consecutive decision-making stages in tourism: pre-visitation (destination choice prior to travel), during-visitation (on-site experience and evaluation), and post-visitation (experience evaluation and future behavioral intention). A study of Samsudin and Worang (2016) mentions that destination image and tourist satisfaction are the factors that affect revisit intention at Bunaken National Park. Therefore, the authors will verify previous research and prove the results of these studies. After the authors describe Olele Marine Park's issue, the formulation of the problem is: 1) how is the image of the tourist destination and revisit intention in Olele Marine Park?; 2) how does the influence of tourist destination image on revisit affect intention in Olele Marine Park?

2 LITERATURE REVIEW

2.1 *Concept of destination image*

Chen et al. (2010) describes destination image as an interactive system of objective knowledge, subjective impressions, prejudice, imaginations, and emotional thoughts toward a destination held by individuals or groups. Tasci (2007) in Rajesh (2013) mentions that destination image is an interactive system of thoughts, opinions, feelings, visualizations, and intentions toward a destination. Kim and Richardson (2003) explain that destination image is a totality of impressions, beliefs, ideas, expectations, and feelings accumulated toward a place over time. Alcaniz et al. (2009) define tourism destination image as the overall perception of a destination, the representation in the tourist's mind of what he/she knows and feels about it. It consists of all that the destination evokes in the individual; any idea, belief, feeling, or attitude that tourists associate with the place. Collectively, destination image could be defined as an idea that is felt by tourists before, while traveling, even after going on a tour of the tourist attraction, facilities and experiences that are suggested to form a positive or negative impression. In this regard, authors such as Baloglu and Brinberg (1997); Chen (2001); Hong et al. (2006); and Walmsley and Young (1998) in Qu et al. (2011) consider that the image is a construct formed from the tourist's rational and emotional interpretations, consisting of two interrelated components: 1) a cognitive or perceptual component, also known as the designative component: beliefs and knowledge about the perceived attributes of the destination; and 2) an affective component: the individual's feelings toward the destination. Agustina (2018) stated a good destination is a destination that must be able to provide tourist facilities that are appropriate to the needs of visitors to provide convenience and meet their needs during the visit.

2.2 *Components of destination image*

Gunn (1972) in Lin (2012) established that the formation of destination image could be separated into organic (i.e., newspapers, magazines, the opinions of friends) and induced (i.e., promotional advertisements, commercial promotions). More recently, Echtner and Ritchie (1993) in Lin (2012) developed an alternative framework for understanding destination image. First, they proposed that destination image included attribute-based and holistic components. The attribute-based component referred to the perception of individual destination features and the holistic component regarded the mental imagery of destination. Second, they argued that the attribute-based and holistic components possessed functional (measurable) and psychological (abstract) characteristics. And, third, they suggested that destination image included attributes common to all destinations as well as attributes unique to specific types of destinations. Destination image is also viewed as an attitudinal construct consisting of cognitive and affective evaluations (Baloglu & McCleary, 1999 in Qu et al., 2011). Although it is argued that cognitive and affective image components are hierarchically correlated to form a destination image (Cai 2002a, b; Gartner 1993, c; Woodside & Lysonski 1989; in Qu et al., 2011), it is still possible that each cognitive and affective brand image component would have unique contributions to the overall image formation. The separate treatment of cognitive and affective components is necessary to examine their unique effects on consumers' attitude structure and future behaviors (Baloglu & Brinberg 1997a, b; Russel 1980, c; Russel & Pratt 1980, d; Russel & Snodgrass 1987, e; Russel, Ward, & Pratt 1981) in Qu et al. (2011). Therefore, this research will focused to measuring cognitive image, affective image, and unique image as the component of destination image. The following dimensions:

- Cognitive image in Olele Marine Park consists of quality of expriences gained by tourists when visiting the destination, the attractiveness of tourist destination, the environment and infrastructure, entertainment /outdoor activities that are usually held to satisfy the needs and desires of the tourists, and the cultural traditions or customs.
- Affective image in Olele Marine Park consists of pleasantness; this is the feeling of comfort by tourists; the relaxed feeling that is felt of tourists, and tourist feel excited about the tourist attraction.
- Unique image in Olele Marine Park consists of the uniqueness of the natural environment in as a superior destination and local attractions that characterize these destinations.

2.3 *Revisit intention*

First-timers' revisit intentions may be influenced mainly by destination performance as a whole because of their initial stay, while repeaters' intentions may be influenced largely by promotional efforts to recall their positive memory and by disseminated information on new attractions (Um et al. 2006a, b; Cole & Scott 2004) in Chen et al. (2010) revisit intention refer to tourist willingness or plans to visit the same destination again. According to Agustina (2018), revisit intention is a feeling of wanting to revisit a tourist destination in the future. Travelers who have an interest in revisiting can be marked by a willingness to review the same destination in the future and recommend destinations to others. To Pratminingsih et al. (2014) in Pangemanan (2014), understanding revisit intention is one the fundamental issue for destination managers because repeat visitors could provide more revenue and minimize the costs. The concept of revisit intention comes from behavioral intention. A behavioral intention can be defined as an intention for planning to perform a certain behavior. It is said that when people have a stronger intention to engage in a behavior, they are more likely to perform the behavior in the tourism and recreation sectors. This takes the form of a repurchase at tourism service or recreational service or a revisit of a destination or visitor attraction.

This research focused to measuring dimensions through Lin (2012). There are also two dimensions: 1) Intention to recommend (desire to recommend to others) and 2) Intention to revisit (desire to return to visit).

2.4 *Destination image and revisit intention*

A study of Samsudin and Worang (2016) mentions that destination image and tourist satisfaction are the factors that affects revisit intention at Bunaken National Park. The result shows that destination image and tourist satisfaction have positive significant effect on revisit intention at Bunaken National Park partially. Destination image is feelings, impressions, and opinions of tourist to the destination. The good destination will attract tourists to revisit the destination. Besides, result of a study from Purnama and Wardi (2019) are that destination image has a significant influence on revisit intention to the most beautiful village in the world (Nagari Tuo Pariangan). The meaning is, if the destination image has a good and interesting image, then a number of tourists visit the most beautiful village in the world (Nagari Tuo Pariangan). The result of Agustina's (2018) study is that destination image has a significant effect on the revisit intention to Mount Batur. According to visitors who have visited, Mount Batur is one of the best tourist attractions in Bali, that it is a must-visit attraction. This means that the better the destination image of Mount Batur, the more tourists who will visit will return to Mount Batur. However, if the image of the destination is bad, tourists don't want to visit. Similarly, the hypotheses of this research are:

H1: Destination image significantly influenced revisit intention to Olele Marine Park
H2: Cognitive image significantly influenced revisit intention to Olele Marine Park

Table 1. Demographic of tourist ($n = 180$).

Variables	Percent (%)	Variables	Percent (%)
Gender		**Purpose**	
Male	57	Leisure	44.9
Female	43	Business	30.2
Age		Study/Research	18.4
17–27	38.2	Others	6.5
28–38	30.1	**Visit duration**	
39–49	25.6	1–3 days	74.1
>50	6.1	4–6 days	13.5
Place of origin		7–10 days	8.6
Gorontalo	20.7	>10 days	3.8
Manado	22.3	**Travel expenditure (IDR)**	
Bitung	13.2	100.000–300.000	39.4
Makassar	11.9	300.000–500.000	31.8
Jakarta	10.6	>500.000	28.8
Bandung	8.2		
Surabaya	6.0		
Medan	4.1		
Jayapura	3.0		

H3: Affective image significantly influenced revisit intention to Olele Marine Park
H4: Unique image significantly influenced revisit intention to Olele Marine Park

3 METHOD

The type of this research was quantitative research using descriptive and verification approaches and also path analysis techniques. The object of research was conducted at Olele Marine Park. Data collection techniques include distributing questionnaires to 180 tourists who have visited Olele Marine Park. The questionnaire was administered personally to the respondents. The tourist profile is displayed in Table 1.

4 FINDINGS

4.1 *Tourist profile*

Questionnaires were distributed to 180 domestic tourists. The most who visited to Olele Marine Park were male tourists, amounting to 57%, while women were 43%. The age range most predominantly was aged 17 to 27 years, which is 38.2%; ages 28 to 38, which is 30.1%; 39 to 49, which is 25.6%; and the last age above 50 years is 6.1%. Most of the domestic tourists visiting were from Manado, namely 22.3%; the second was from Gorontalo, and the least tourists were from Jayapura, with 3.0%. Tourists visiting to spend their leisure time; vacation was by 44.9%, while those aiming for work or business was by 30.2%, and those whose main purpose was for learning was 18.4%; the last, for other purposes, was by 6.5%. The longest tourists enjoy the beauty of the Olele Marine Park for 1 to 3 days, amounting to 74.1%; for 4 to 6 days for 13.5%; while for 7 to 10 days, the purpose is mostly for research or study, is 8.6%; and at the least 3.8% for more than 10 days. For tourist expenditure while in the Marine Park Olele the most by IDR. 100.000 to IDR.200.000, which has a percentage of 39.4%; the cost of IDR. 200.000 to IDR. 300,000 was the percentage 31.8%; while the least percentage is 28.8%, which costs more than IDR. 300,000. The longer tourists are in Olele Marine Park, the more spending on renting snorkeling equipment, diving, and other facilities.

Table 2. ANOVA output (F test).

Model		Sum of Squares	df	Mean Square	F	Sig.
	Regression	9360.886	1	3157.922	92.968	0.001[b]
1	Residual	6021.079	179	30.171		
	Total	15,381.965	180			

a. Dependent Variable: Y
b. Predictors: (Constant), X.3, X.2, X.1

Besides, around Olele Marine Park there are homestays and citizens around offering various types of local food from Gorontalo, and certainly it impacts on spending and helps the economy as well as empowers of the people living in Olele Marine Park.

4.2 *Overview of destination image in Olele Marine Park*

Based on statements from tourists through the distribution of questionnaires, the unique image is the highest dimension chosen by domestic tourists visiting Olele Marine Park, where the percentage is 45.8%. The uniqueness of the Olele marine park lies in its natural and well-preserved environment, especially the uniqueness of a coral reef named Goa Jin. In addition to being presented with beautiful coral reefs, hundreds of small colorful fish make the underwater scenery even more beautiful, and there are some favorite diving spots by tourists such as coral pole, beehive, or traffic jam. Some tourists said that Olele has more varied patterns and models of reefs compared to other marine parks in Indonesia. Affective image as the second highest appraisal rating is that 29.6% state that they are comfortable and happy while in the Olele Marine Park, and certainly very eager to enjoy the beauty of the park underwater. Cognitive image obtained the lowest rating of 24.6%. This is due to the lack of cleanliness around the Olele Marine Park; trash bins provided are inadequate and also insufficient availability of public facilities such as toilets and places of worship, whereas these supporting facilities are needed by tourists to create a good impression of the Olele Marine Park in terms of amenities.

4.3 *The influence of destination image on revisit intention in Olele Marine Park (Simultaneous)*

Results of the processed data are shown in Table 2:

Table 2 shows that the significance test value is 0.001, which means that destination image influences revisit intention simultaneously, which consists of cognitive images, affective images, and unique images and can be interpreted that hypothesis 1 (H1) is accepted. This is in line with Samsudin and Worang (2016), where the results of this study stated that destination image effects revisit intention because destination image is feelings, impressions, and opinions of tourists to the destination. The good destination will attract tourists to visit the destination again. If the image of the destination is better, there will also be more tourists who intend to return to the Olele Marine Park because of their own experience and uniqueness.

4.4 *The influence of destination image on revisit intention in Olele Marine Park (Partial)*

Data shows that there is a partial effect between destination image on revisit intention:

- There is a significant influence between cognitive image on revisit intention with a significance value of $0.004 < 0.050$ then H2 is accepted.
- There is a significant influence between affective image on revisit intention with a significance value of $0.002 < 0.050$ then H3 is accepted.
- There is a significant influence between unique image on revisit intention with a significance value of $0.000 < 0.050$ then H4 is accepted.

Table 3. t Test output.

Model	Unstandardized Coefficients B	Std. Error	Standardized Coefficients Beta	t	Sig.
(Constant)	5.722	2.582		2.547	0.010
X_1	0.939	0.424	0.208	2.078	0.004
X_2	1.489	0.376	0.286	6.898	0.002
X_3	2.371	0.794	0.497	3.924	0.000

a. Dependent Variable: Y

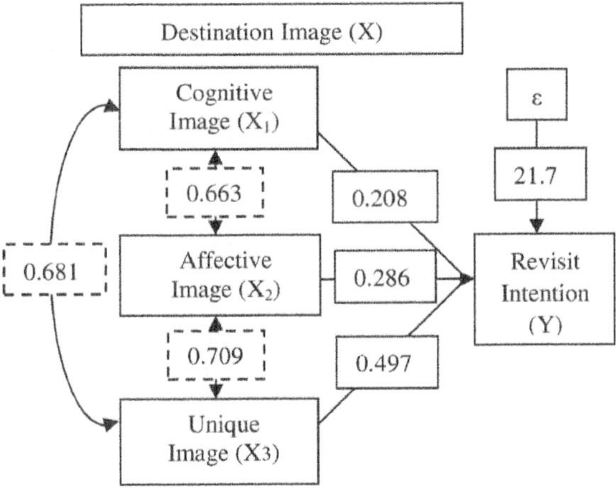

Figure 1. Path diagram of destination image on revisit intention.

Table 4. Test result of path coefficients, direct, and indirect the influence of destination image on revisit intention.

X_1	Direct Effect on Y	Indirect Effect X_1	X_2	X_3	R^2Y $X_{1.1}Y$ $X_{1.2}Y$ $X_{1.3}Y$	Total Effect %
$X_{1.1}$	0.043	–	0.039	0.070	0.153	15.3
$X_{1.2}$	0.082	0.039	–	0.101	0.222	22.2
$X_{1.3}$	0.247	0.070	0.101	–	0.418	41.8
R^2					0.793	79.3

The influence of destination image on revisit intention is shown in Figure 1.

Furthermore, the calculation to determine the direct and indirect influence of the dimensions of the destination image on revisit intention are in Table 4.

Based on Table 4, the results show that the influence of the total destination image on revisit intention to Olele Marine Park is 0.793 or 79.3% while the path coefficient of other variables outside the destination image variable is 21.7%. This shows that X_1, X_2, and X_3 affect the decision of visiting 79.3% and the remaining 21.7% are influenced by other factors. The results of this study are strengthened by findings from Purnama and Wardi (2019) that destination image has a significant influence on revisit intention; it means if the destination image has a good and interesting image, it will increase the number of tourist visits. While the result of the Agustina (2018) study is that destination image has a significant effect on the revisit intention. a tourist destination must have values and characteristics especially having a good image, where the better the image of the destination the more tourists are interested in visiting.

5 CONCLUSION

In this study, destination image used dimension of cognitive image, affective image, and unique image. Destination image has a good rating of the revisit intention to Olele Marine Park. Cognitive image has an influence on the decision to visit Olele Marine Park, environmental cleanliness, availability of public facilities, security, tourist activities, and quality of experience are important in building a good image of tourist destinations in the minds of consumers. Affective image has an influence on the decision to visit Olele Marine Park. During the tour, tourists feel comfortable, feel the spirit to enjoy the attraction of its natural attractions, and feel happy with the tourist activities offered, especially underwater activities where tourists can see the beauty of marine life. Unique image has an influence on the decision to visit Olele Marine Park, and most tourists consider that Olele Marine Park has still maintained the naturalness and uniqueness of its corals; besides that, the Olele Marine Park offers local food of Gorontalo if tourists want to enjoy Gorontalo's cuisine. Thus, tourists can enjoy the natural tourism and culinary attractions of Gorontalo. Even most tourists are willing to recommend Olele Marine Park.

In improving the image of tourist destinations, collaboration among the community, government, and academia is needed to perceive positive destination images in the minds of tourists, especially in this case the government who manages Olele Marine Park should increase the authenticity as the unique point to creating positive images; adequate infrastructure and environmental hygiene should be improved to preserve the image of Olele Marine Park as the popular destination and diving center in Gorontalo.

REFERENCES

Agustina, N. K. W. (2018). The Influence Of Destination Images On Revisit Intention In Mount Batur. *Jbhost, 04*(2), 157–168.

Alcaniz, E. B., Garcıa, I. S., & Blas, S. S. (2009). The functional-psychological continuum in the cognitive image of a destination: A confirmatory analysis. *Tourism Management, 30*, 715–723. https://doi.org/10.1016/j.tourman.2008.10.020

Asdhiana, I. M. (2013). *Olala, Indahnya Taman Laut Olele...* Kompas. https://travel.kompas.com/read/2013/05/02/11273921/olala.indahnya.taman.laut.olele.?page=allhttps://genpi.id/taman-laut-di-pantai-olele-gorontalo/

BPS. (2018). *Kabupaten Bone Bolango Dalam Angka 2020.*

Chen, N., Funk, D. C., Chen, N., & Funk, D. C. (2010). *Exploring Destination Image, Experience and Revisit Intention: A Comparison of Sport and Non-Sport Tourist Perceptions Exploring Destination Image, Experience and Revisit Intention: A Comparison of Sport and Non-Sport Tourist Perceptions. December 2014*, 37–41. https://doi.org/10.1080/14775085.2010.513148

Indira, D., Ismanto, S.U. & Santoso, M. B. (2013). Pencitraan Bandung Sebagai Daerah Tujuan Wisata: Model Menemukenali Ikon Bandung Masa Kini. *Sosiohumaniora, 15*(1), 45–54.

Kim, H., & Richardson, S. L. (2003). *Motion picture impacts. 30*(1), 216–237. https://doi.org/10.1016/S0160-7383(02)00062-2

Lin, C. H. (2012). Effects of Cuisine Experience, Psychological Well-Being, And SelfHealth Perception on the Revisit Intention of Hot Springs Tourist. *Journal of Hospitality & Tourism Research*, 1–22.

Pangemanan, P. & S. S. (2014). The Effect Of Destination Image And Tourist Satisfaction On Intention To Revisit In Lembeh Hill Resort. *Jurnal EMBA, 2*(3), 49–57.

Pratminingsih, S. A., Rudatin, C. L., & Rimenta, T. (2014). *Roles of Motivation and Destination Image in Predicting Tourist Revisit Intention: A Case of Bandung – Indonesia. 5*(1). https://doi.org/10.7763/IJIMT.2014.V5.479

Purnama, W., & Wardi, Y. (2019). *The Influence of Destination Image, Tourists Satisfaction, and Tourists Experience toward Revisit Intention to The Most Beautiful Village in The World (Nagari Tuo Pariangan). 01*, 18–25.

Qu, H., Hyunjung, L., & Hyunjung, H. (2011). A model of destination branding: Integrating the concepts of the branding and destination image. *Tourism Management, 32*(3), 465–476. https://doi.org/10.1016/j.tourman.2010.03.014

Rajesh, R. (2013). Impact of Tourist Perceptions, Destination Image and Tourist Satisfaction on Destination Loyalty: A Conceptual Model. *Revista de Turismo y Patrimonio Cultural, 11*, 67–78.

Rizal. (2020). *Keindahan Tersembunyi di Kedalaman Taman Laut Olele*. IndonesiaKaya. https://www.indonesiakaya.com/jelajah-indonesia/detail/keindahan-tersembunyi-di-kedalaman-taman-laut-olele

Samsudin, A., & Worang, F. G. (2016). *Analysing The Effects Of Destination Image And Tourist Satisfaction On Revisit Intention In Case Bunaken National Park. 16*(04), 23–34.

Tasci, A. D. A. (2007). Assessment of factors influencing destination image using a multiple regression model. *Tourism Review, 62*(2), 23–30.

Promoting Creative Tourism: Current Issues in Tourism Research – Kusumah et al. (Eds)

Topeng Pedalangan as a tourist attraction in Gunungkidul Regency Special Region of Yogyakarta

Kuswarsantyo
Universitas Negeri Yogyakarta, Yogyakarta, Indonesia

ABSTRACT: This is a qualitative study aimed at, first, describing the results of the packaging of the *Topeng Pedalangan* performing arts in Bobung Village, Putat, Pathuk, Gunungkidul, Special Region of Yogyakarta. Second, it seeks to find its impact on tourist attractions. Initially, the art of the *Topeng Pedalangan* was unknown because the village was more focused on mask handicrafts used for the dance, but after some time, the idea to turn the art of *Topeng Pedalangan* into a staged performance for the tourist attraction developed. Assistance methods used to develop the tourist attraction include (1) making inventory and casting supporters who are mostly craftsmen (not dancers), (2) conducting assimilation by involving several dance students from the Department of Dance Education of the Faculty of Languages and Arts, Universitas Negeri Yogyakarta in the training process, and (3) managing the *Topeng Pedalangan* community to carry out an independent performance. The results of the *Topeng Pedalangan* packaging criteria for tourist attractions cover (1) the packaging is adapted to the supporters' capabilities; (2) the packaging model is creatively managed for dynamical changes; (3) The duration of the packaging is compacted; (4) the attraction is highlighted by characters who can communicate with the spectators; and (5) utilizing information technology for publication. The implications for the packaging of *Topeng Pedalangan* include (1) tourist visits in Bobung likely begin to increase, (2) the performing arts can grow the economy of the surrounding community, and (3) young generations can enjoy the arts.

Keywords: *Topeng Pedalangan*, tourist attraction, *Gunungkidul*

1 INTRODUCTION

Desa Wisata Bobung or Bobung tourism village is located in Putat Hamlet, Patuk District, Gunungkidul Regency, Special Region of Yogyakarta. This village has grown its tremendous potential for arts and crafts, especially of wooden batik masks, which have been known to be remarkably unique to the area—having upturned eyes and nose. The shape is extraordinarily distinctive since it is similar to the characters of Wayang Purwa, whose eyes are upturned and the nose is sharp, besides the batik motif that underlies the coloring of the mask, adding to the beauty of the mask. For years this area has been finally developed as a center for wooden batik handicrafts, not only masks but also various other forms of craftwork. This finally allows Putat's local residents, who have all become farmers since the mid-1980s, to partly shift to be craftsmen. Currently, the wooden batik craftwork from Bobung has penetrated the global market.

Another potential available at this tourism village is the art of *Wayang Topeng* (mask puppet) performance, which depicts the story of *Panji*. *Panji* is a story that was adopted during the era of *Kameswara I*, the King of *Daha Kediri*. The story of *Panji* tells an arranged marriage between a prince from *Koripan* (Kediri), namely Raden Inu Kertopati, and Princess Galuh Condro Kirono from Jenggala. The matchmaking of Raden Inu Kertopati and Galuh Condro Kirono, unfortunately, had to take a challenging course through prolonged and great obstacles to finally be realized.

DOI 10.1201/9781003095484-23

Figure 1. A *Topeng Pedalangan* performance in Bobung tourism village. (Photo by Kuswarsantyo 2018).

2 WAYANG TOPENG PEDALANGAN

Wayang *Topeng Pedalangan* (WTP)—special performing arts by puppeteers from the Special Region of Yogyakarta—is generally known as a traditional drama performance, telling the story of *Panji* in which all of the dancers wear a wooden mask on their face. The mask worn by each dancer represents certain figures in the *Panji* story, for example, Prabu Klana Sewanana, *Panji* Inukartapati (Asmarabangun), The King of Jenggala—Prabu Lembu Amiluhur, Dewi Sekartaji, Dewi Ragil Kuning, Raden Gunungsari, and Punakawan figures such as Bancak, Doyok, Regol, Sembunglangu, and so on. The performing arts are traditionally carried out by puppeteer artists and their close relatives. Thus, the word *Pedalangan* has two indicators or meanings: first, the shows are danced or supported by puppeteers and their close relatives, and second, these contributors perform a dance style known as the *Pedalangan* style. In the history of its development, WTP has experienced ups and downs with the development of the era behind. Among the great puppeteer family, WTP has become media to sustain and strengthen the familial ties of the puppeteer artists and their relatives in a larger community called the *trah dhalang*, while the creative spirit of WTP lies in the elements of dance, musical instruments (karawitan), puppeteers, mask crafts, and the myths involved.

According to a study conducted by the Culture and Tourism Office of Sleman Regency, there is only one type of WTP in Sleman. This art is uncommon and threatened with extinction. However, WTP has its own uniqueness and deserves to be preserved. Performance as an art revitalization effort by the mask puppet studio Gondo Wasitan, Ngajeg, Tirtomartani was held on Friday (October 31, 2014) at the Village Hall of Tirtomartani Kalasan, performing the story of "Bancak Nagih Janji." In Gunungkidul Regency, there exists a WTP community in Bobung, Pathuk, named Kenaka Laras. Until now, both Gondo Wasitan and Kenaka Laras are relatively active in holding the performances and exercises, as said by Mbah Sagiman, a respondent in this study. In addition, these arts groups' activities have just been documented by the local state-owned television station TVRI in Central Jakarta on March 20, 2018, followed by a French television station on April 12, 2018, in Bobung.

This documentation offers many benefits for the preservation of mask puppet performing arts in the region, likewise its social effects in the future. Revitalization activities are intended to increase the roles and functions of ancient cultural elements that are still enlivened in society in a new context. Such activity maintains the arts' authenticity with the essence of bringing back the arts susceptible to extinction so that their state is maintained in a way, at least by documenting them textually and visually. This is a vital effort of the local government to preserve and revive the arts from vanishing. However, the problems currently being faced include how to package or present mask puppet performances attractively in harmony with the concept of art in tourism, so as to increase the tourist destinations of the region.

3 ATTRACTION CONCEPTS FOR TOURISTS

Tours cover all activities related to tourists in society (Soekadijo 1997: 2). The word "tours" refers to the meaning of travel, while tourism refers to the meaning of travel plus elements of services or business entities. This is in line with the notion that tourism contains aspects of travel, business entities, and government functions (Ardika 1993). Tours and tourism are both modern industries that currently grow rapidly. The world's attention to tourism has been really constant, indicated by the progress in the fields of infrastructure and superstructure development in both rich and developing countries. Tourism as a foreign-exchange earning commodity has almost become a mainstay of every country in increasing the community's gross domestic product (GDP) through profits earned from local tourists and foreign tourists.

On account of this, the tourism potential in Bobung tourism village can be developed by engaging the components of tourism officials at the village, subdistrict, and district levels. The synergy among government agencies is important to support the efforts of the community members who have tourism prospects, such as *Topeng Pedalangan* performing arts in Bobung.

3.1 *The packaging of Topeng Pedalangan performances in Bobung tourism village*

Performing arts and tourism are two activities that have a very strong relationship. Performing arts in the context of the tourism industry have become a remarkably crucial and riveting tourist attraction, especially when they are associated with cultural tourism activities. In Bobung, there are two focuses that actually support each other. The first is the mask craft-center, and the latter is the mask puppet performances currently being initiated. Performing arts in various tourist destinations in Indonesia have developed and are widely packaged for tourist consumption, which are held in stages, theater, and even in open areas in the courtyard of a typical rural environment. Likewise, the handmade wooden masks from Bobung can be a supporting factor for more tourist visits. Therefore, this potential can be developed in the future.

The development of traditional arts as a tourist attraction, of course, really depends on the quality of the artists and the management of the performing arts and tours. Opportunities for traditional performance business as packaging for unique and attractive tourism art should still refer to the original values of the art genre. Without the intention of plagiarizing the packaging of hula performing arts at the Polynesian Cultural Center of Hawaii and Barong and Kecak dance performances in Bali, the essential issue would be how to develop the concept of packaging art characteristics and kitsch art into actual, unique, and original forms of packaging for tourism art performances from the products of local wisdom that are capable to compete in global tourism.

The criteria for packaging *Topeng Pedalangan* as a tourist attraction, based on the results of the community discussion, would be the following: (1) The packaging is designed according to the ability of the supporting parties. (2) The packaging model is created dynamically to avoid monotony. (3) The duration of the packaging is compressed. (4) The highlights are on the characters or dancers who can communicate with the spectators. (5) Besides, it utilizes information technology for the publication matters. These five results illustrate that the seriousness of the community in Bobung is really great. This potential support makes the improvement of the quality of *Topeng Pedalangan* possible from time to time.

From the point of view of arts, the development of the tourism industry has actually encouraged the growth of artists' creativity to develop their creations so as to attract visitors or tourists. In the case of performing arts, this creativity should be able to be manifested in captivating, attractive choreographic packaging, and present the complete messages and stories for tourists within their time-limited visits.

According to J. Maquet, this kind of packaging is "art by metamorphosis" (art that has undergone a change in shape), or art of acculturation, or pseudo-traditional art, or tourism art. This means that adaptations to the tastes and needs of holiday-makers will be specially created for tourists as a "tourism community." This is in accordance with Adolph S. Tomars' view about the relationship between the class or community system and the art style that develops in it (Soedarsono 1999).

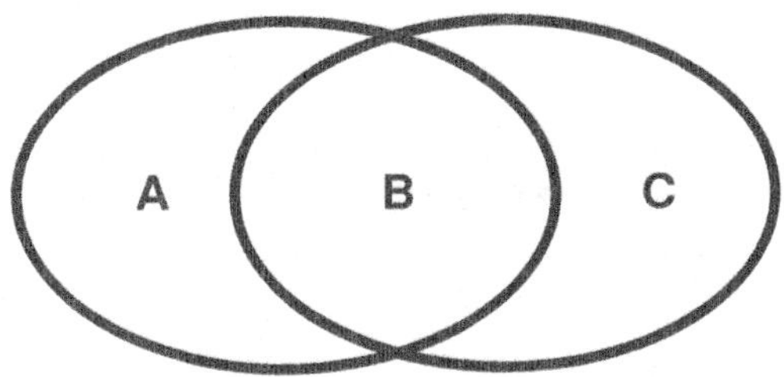

A: Tourism Community/Class
B: Packaging of Tourism Performing Arts
C: Art style

Figure 2. The class and art system. Adapted from Adolph S. Tomars.

Therefore, the art style born is considered an individual and collective expression as the manifestation of local culture. When the art of tourism as a form of local wisdom is aimed at the interests of the creative economy of the community, it can be a means of growing the creative industry.

In relation to the market, performing arts pay attention to the needs of society. This is in line with Riantiarno (1993), who suggested that any show cannot force anyone to buy goods of their needs. In the world of performing arts, people's tastes or desires often go unnoticed, even though knowledge and control of people's tastes are very important to determine the market. As an effort to respond to and master people's tastes, creativity is inevitably required so that these performing arts can always find new things, which in turn can influence the market. Problems that seem essential to be studied, therefore, include (1) the relationship between performing arts and tourism, and (2) tourism performing arts as creative industries.

3.2 *The impacts of the performances on tourists*

The effects of performances on tourists in their sociocultural life, according to Clare A. Gunn (Salim 1991), are structured in five main directions, namely, accommodation, attraction/creation, consumption, information, and transportation. Tourists who come to certain areas will need adequate accommodation as a place to live as they have in their country/region of origin. For the needs of this residence, the establishment of hotels as tourist residences, entertainment venues, souvenir shops, and so on, emerges. Thus, there will be a very intensive and complex interaction there. The segment of creation/attraction suggests that tourists also need entertainment, souvenirs, or keepsakes that become the main characteristics of the areas visited. The provided goods or entertainment packages in this area, hence, are all tourist-oriented and tourism products. Consequently, handicrafts and the packaging of performances will be developed in society.

Besides, information about sociocultural conditions of the area and objects of the visit is very important for tourists. Therefore, the mastery of foreign languages and services to travelers from the time they get off the airplane to return to their country are imperative for tour guides. Direct contact in these activities will have a certain effect on society. The arrival of foreign or local holidaymakers, apart from requiring an accommodation, also requires consumption. They often want some food or beverages being the main icon of the region, locally produced with ingredients specific to the area. In addition, transportation is also a need for them after arriving at their destination, which is generally arranged by the travel agency through the tour guide, then came various types of travel and business agencies, such as vehicle rental and so on.

These five areas of needs, when viewed from the point of view of the economy, benefit other fields, such as foreign exchange reserves, improvement of infrastructure, local products utilization, equitable employment opportunities, and so forth. The advantages in the sociocultural field include the expansion of education, mutual understanding and respect, tolerance, and the decrease of the gap that is racial or related to social status. Meanwhile, negative influences in the sociocultural field include commercialization of art, culture, or religion; gambling; prostitution; and drug-related

Figure 3. The community's enthusiasm in watching *Wayang Topeng Pedalangan*. (Photo by Kuswarsantyo 2018).

crimes (Boedihardjo 1991). With cultural objects as a commodity, there might be an imitation, decline, or reproduction on a large scale so that the quality of those provided in the market decreases. Sometimes people may even find that, in performing arts, the spectacle is packaged and modified in such a way that it eliminates its artistic element (Salim 1991).

Tourism art, the packaged art specifically intended for tourism, is an art form produced by imitating the original. It is often misunderstood as this being cheap art and of low quality. Of course, such interpretation is wrong. Imitative art does not mean an unqualified one. It is cheap in the sense that it is affordable for tourists yet still adheres to good quality. This is because tourism art essentially should be media of information and have such an appeal worth selling.

The era of the tourism industry has indirectly brought a positive atmosphere for traditional performing arts and provided opportunities for artists to create more works of art as a manifestation of their participation. Such a situation, similarly, creates good opportunities for star-rated hotels and big restaurants. One of the tips for bringing in more tourists, therefore, is by presenting quality traditional performing arts.

In view of the tourism industry, in this case, tourism art, it is necessary to look at and consider how to package tourism art appropriately, as any mistake in this area may cause fatal consequences. In relation to this, Soedarsono (1992/1993) suggests that tourism art has five characteristics, namely, (1) imitating the original version, (2) shorter than the original, (3) full of variation, (4) leaving its magical and sacred values, and (5) inexpensive for tourists' rate. Referring to this view, the form or format in packaging traditional performing arts into tourism art can be appropriately developed.

The format as proposed by Soedarsono deserves to be studied and adapted to the setting of Indonesian tourism. It necessarily considers the needs, in terms of performances that reflect the culture demanded by either local or foreign trippers; one of the performing arts studios that refers to Soedarsono's is Sampan Bujana Sentra (Hadi 2001). It is a studio and restaurant capable of serving a mix of dances, music, and vocals, and dinner menus with Indonesian specialties. Sampan, the owner, hopes that such a mixed presentation can attract more holiday-makers. The tourism performances packaged or served include Indang dance from Malay, Saman dance from Aceh, a Sundanese Kecapi show from West Java, Belibis dance from Bali, Blantek mask dance from Betawi, Rampak Kendang, Jaipingan dance, an Angklung music show, and singing with the artists for all tourists. Each performance takes approximately two hours every day, whether tourists are visiting the restaurant or not.

4 IMPLICATIONS FOR TOURIST VISITS IN BOBUNG

The implications of *Topeng Pedalangan* performances in Bobung include (1) tourist visits in Bobung begin to increase; (2) *Topeng Pedalangan* performances can improve the economy of the surrounding community; (3) young generations began to enjoy the arts. The tourism linkages and services cover tourism motives, tourism needs, tourist attractions, and tourism services dealing with social dimensions, especially those referring to the improvement of services with full of economic implications. Economic factors play a significant role and even become a goal. As stated by Soekadijo (1997), tourism is a very complex social phenomenon that concerns the whole human being and has various aspects, namely, sociological, psychological, economic, ecological, and so on. The one receiving the most attention and is almost the only aspect that is considered important is the economic aspect. With the dominance of the economic aspect, it has fairly substantial impacts on changes in the structure of society.

As an integrated tourism village, Bobung necessitates support to make tourism packaging on performances more meaningful, especially through various production efforts of the community for the availability of souvenirs, souvenir-making courses, traditional music courses, traditional dance courses, and so on. In short, all of these include providing various tourism services and facilities, so that visitors feel satisfied and the experiences from Bobung could make great impressions and memories. Likewise, these supporting activities can increase their duration of stay, resulting in the mobility of money and positive impacts on improving the economy of the local people.

5 CONCLUSIONS

Any potential extent in this tourism village would provide hope for the development of other supporting sectors. The village development then will have implications for the economic aspect, where the village community would benefit from the *Topeng Pedalangan* performances. To support this, determination of and cooperation among stakeholders are inevitably required to achieve the agreed goals for the progress of the village. By helping and supporting each other, Bobung tourism village with best-selling *Topeng Pedalangan* performances will be increasingly known to the public. Promoting and conserving arts to youngsters that have long been somewhat serious issues should be resolved by providing opportunities for them to get involved in, appreciate, and embrace the arts, as well as collaborating with others to develop their existing potential to the maximum extent.

The development of tourism art management, especially tourist attractions, of course, depends on the quality of the artists and the management system for tourism performing arts. Opportunities for traditional performance business as a unique and attractive tourism package should still refer to the original values of the art referred to. Support outside of packaged tourism performances has become more meaningful, especially in relation to the productive activities of the community, such as the souvenir-making industry, souvenir-making courses, traditional music courses, traditional dance courses, and so forth.

REFERENCES

Boedihardjo, 1991. *Komersialisasikan Budaya*. Jakarta: Sinar Harapan.

Gunn, Claire A., 1991. *Hubungan Pariwisata dan Pengaruhnya pada Kehidupan Sosial Budaya*. Jakarta: Pustaka Pelajar.

Kayam, Umar. 1983. *"Ngesti Pandowo: Suatu Persoalan Kitsch di Negara Berkembang"* in Sedyawati, Edi & Damono, Sapardi Djoko (Eds). *Seni Dalam Masyarakat Indonesia, Bunga Rampai*. Jakarta: Gramedia.

Hersapandi. 2014. *Seni Wisata Sarana Menumbuhkan Industri Kreatif* in *Greged Joged Jogja*. Yogyakarta: Bale seni Condroradono.

Riantiarno. 1993. *Pasar Seni Pertunjukan*. Jakarta: IKJ Press.

Salim. 1991. *Seni Pertunjukan Tontonan Wisata*. Bandung: CV. Rineka.

Simatupang, Togar W. (2008), "*Industri Kreatif Indonesia*", an internet article published on 15 May 2008.

Soedarsono, R.M. 1992/1993. *Seni Pertunjukan Wisata di Indonesia*. Yogyakarta: Gama Press.

Soedarsono, R.M. 1999. *Seni Pertunjukan Indonesia dan Pariwisata*. Bandung: Masyarakat Seni Pertunjukan Indonesia.

Soekadijo. 1997. *Pariwisata dalam Masyarakat*. Bandung: Bina Taruna.

Promoting Creative Tourism: Current Issues in Tourism Research – Kusumah et al. (Eds)
© 2021 Taylor & Francis Group, London, ISBN 978-0-367-55862-8

Millennial volunteer tourist motivation in West Java Province, Indonesia

D.D. Utami, I. Ramadhani, A.P. Ramdhani & N.T. Murtiani
Sekolah Tinggi Pariwisata Bandung, Bandung, Indonesia

ABSTRACT: Volunteer tourism becomes a growing trend in the world. Interestingly, it tends to be dominated by the millennial generation with various motivations. However, those motivations range from shallow to deep motivation (Haslebacher, et al. 2018; Proyrungroj 2017; Silló 2019). Therefore, this study aims to determine the shallow and deep motivation of millennial generation who participate in the volunteer tourism program in West Java Province. This research uses a qualitative method with content analysis. The data used obtained from research interviews to 18 interviewees by using purposive and snowball sampling techniques. This research found that the shallow motivations are to take a vacation, to build a relationship, to renew a job resume, to develop self-ability, and to get an adventure. Meanwhile, the deep motivations are to give a contribution, to make a change, to feel authenticity, and to transmit goodness.

Keywords: volunteer tourism, volunteer tourist motivation, millennial volunteer tourist

1 INTRODUCTION

Nowadays, world tourism trends continue to develop. Carvache-franco et al. (2019) said that it cannot be denied that new travel trends have begun to emerge and began to shift from mass tourism with 3 S (sea, sand, sun) to various types of alternative tourism. One of the alternative tourism that has become popular and applied is related to social activities, namely volunteer tourism (Booking.com 2018; Pompurová, et al. 2018). Volunteer tourism is a tourist activity that aims to get a tourist experience in which there are social activities during tourism activities (Kazandzhieva 2014). In addition, volunteer tourism activities provide opportunities for tourists to do tourism activities and participate in voluntary activities at the same time (Phelan 2015; Silló 2019). Tourists who undertake voluntary tourism activities are known as volunteer tourists (Wearing 2001). According to Chen & Chen (2011), volunteer tourist is someone who dedicates time, money, and energy to an area that is not his place of origin to get various forms of experience. Moreover, volunteer tourists tend to want a vacation that gives a positive impact, interacts with residents, and changes the situation of a particular community for the better (Han, et al. 2019; Phelan 2015).

Some volunteer tourism program providers in West Java Province are Social Travel and 1000 Teachers. Based on interviews conducted with Social Traveling and 1000 Teachers, the majority of volunteer tourists participating in volunteer tourism activities are millennials. This is in line with the statement of Carvache-franco et al. (2019) and Hallmann & Zehrer (2016) that volunteer tourists who participate in volunteer tourism activities tend to be young people under 35 years old. However, volunteer tourists who take part in volunteer tourism programs have different motivations from one person to another. According to Haslebacher et al. (2018), volunteer tourist motivation is in the shallow to deep range. The shallow motivation is a voluntary tourist who participates with the primary motivation for self-interest compared to the community in the destination area (Coghlan & Fennell 2011; Haslebacher et al. 2018), while deep motivation is a voluntary tourist who participates with the primary motivation for the interests of the community in the destination

 DOI 10.1201/9781003095484-24

area and further contributes to activities for the benefit of the community (Haslebacher et al. 2018; Wright 2013).

If managed properly, volunteer tourism can be an effective program for tourism development and community development, as these activities are often carried out in places that are not mass tourism (Milne, et al. 2018). Based on this, the purpose of this study is to determine the shallow and deep motivation of millennial volunteer tourists who participate in the volunteer tourism program in West Java Province, so that the results of this study can be used as a basis for decision making in non-government organizations (NGOs), foundations, travel agents and communities in making programs that are tailored to volunteer tourist motivation. Furthermore, both volunteer tourists and destination areas will get satisfaction because volunteer tourists get programs that are following their motivation, as well as destination areas will be more sustainable and developed.

2 LITERATURE REVIEW

Volunteer tourists have different motivations from other tourists (Godfrey et al. 2019). Proyrungroj (2017) said that volunteer tourist motivation is multidimensional and generally consists of two dimensions, namely personal interests and altruistic motivation. While Suanpang, et al. (2018) stated that volunteer tourist motivation consists of several factors, namely Increase self-worth, Education, Meeting minded people, Altruism, Adventure/Challenge, and Wanderlust. Meanwhile, Kontogeorgopoulos (2017) revealed that the main motivation of tourists to take part in volunteer tourism programs is to take advantage of opportunities to travel, see new people and places, make friends, learn new things, personal well-being, enter CV experiences, develop a career, want to do kindness, the desire to get the authenticity and uniqueness of culture and look for opportunities to change one's life.

Another opinion from Wright (2013) revealed that the motivation to participate in volunteer tourism is categorized into two mindsets namely volunteering and vacation. While Silló (2019) concluded that volunteer tourists usually have a double motivation because they want to do something useful for the local community and also want to achieve personal development. Whereas Polus & Bidder (2016) divides the main motivations of volunteer tourists into the desire to feel the local culture, make a difference for those who are less fortunate, build relationships with individuals who have similar interests, look for educational opportunities, bond with local children, learn about the environment, involved in conservation work and develops new skills and abilities.

Based on the above explanation, the motivation of volunteer tourists is very diverse, but these motivations can be categorized into shallow and deep motivation. This is based on the statement of Coghlan & Fennell (2011) regarding shallow motivation, which is the motivation of volunteer tourists to participate in activities to help others, but the greatest motivation is to benefit and promote themselves. Furthermore, shallow motivation (egoistic) is a motivation that is based on personal interests such as self-development, improving portfolio, career advancement, improving social status based on experience, thinking about the benefits found for themselves rather than the needs of the local community, as well as being grateful for the circumstances is owned (Haslebacher et al. 2018).

On the other hand, some aspects of motivation, namely education, meeting minded people, feeling and participating in a culture, giving and making differences can be included in the category of deep motivation. This is in line with Wright (2013) statement regarding deep motivation, which is the motivation of volunteer tourists to feel and participate in a culture, giving, and making a difference for the local community. In addition, deep motivation (Altruistic) is a motivation that is based on the interests of the intended community such as contributing more to culture and making changes, thinking about the needs of the local community rather than its benefits, and looking for authentic experiences.

Based on the explanation above, the motivation of volunteer tourists is in the range of shallow and deep motivation, which is expected to explore the dimensions of the millennial volunteer tourist motivation who take part in volunteer tourism programs in West Java.

3 RESEARCH METHOD

3.1 *Research design*

In this study, a qualitative research method was used to determine and analyze the motivation of millennial volunteer tourists in West Java Province. This phenomenon described and analyzed through a specific data collection process that allows for the development of volunteer tourism in West Java Province.

3.2 *Participants*

Researchers determine research participants using non-probability sampling techniques. Moreover, the type of technique used is purposive sampling and snowball sampling. In the initial stages, each prospective participant is identified to find out who is considered capable of providing information to be interviewed. Then, the selected participants recommend other participants as subjects who can provide the information needed. The process lasted until the information needed in this study was fulfilled. The volunteer tourist criteria who participated in this study came from outside West Java Province, participated in volunteer tourism activities in West Java Province, is a millennial generation with an age range of 20 years to 38 years, has participated in volunteer tourism activities in West Java Province with duration of more than 24 hours.

Furthermore, interviews conducted with relevant stakeholders such as the government officer of West Java Province Tourism and Culture Department, the manager of Social Traveling, and the manager of 1000 Teachers.

3.3 *Data collection*

The structured interviews were conducted with the participants. The interview guide used as a reference and guide to determine the motivation of millennial volunteer tourists based on shallow and deep motivation in West Java Province that is academically and practically useful. Interviews were conducted through chat or call to the participants via Whatsapp application because of a COVID-19 pandemic that requires Physical Distancing.

3.4 *Data analysis*

The type of data analysis used in qualitative content analysis with stages in the form of data collection methods, sampling strategies, and selection of suitable units of Analysis (Hsieh & Shannon 2005).

4 FINDINGS AND ANALYSIS

Based on the results of the study, the age range of volunteer tourists who become participants is in the age range of 20 to 27 years, the last education degree (S1), the profession as a private employee, and student. The majority of participant is woman and domiciled in the Province of DKI Jakarta.

4.1 *Shallow motivation*

In this study, it is known that shallow motivation volunteer tourists in West Java Province is very diverse, one of which is a vacation, which means thinking about interests and seeking pleasure for them. In this case, participants stated that they have the motivation to take a vacation because they want to take a break from their routine and use their free time to enjoy the natural beauty of the destination that can make the mind fresh. This is in line with Tomazos & Butler (2010) which states that volunteer tourists have the motivation to take a break from their routine life and make a vacation that is beneficial rather than just lying on the beach.

In addition, participants said that they have the drive to build relationships with individuals who have similar interests, do things that are beneficial to others, interact with new individuals to gain insight, increase self-confidence, want to have the ability to adapt to various individuals who have different backgrounds and characters, want to get new relations from various work and education circles. This is consistent with Haslebacher et al. (2018) that shallow volunteer tourists are tourists who are more focused on their personal needs rather than the needs of the local community, where one of the motivations is to build relationships with other individuals who have similar interests.

Furthermore, based on the results of the interview, it is known that volunteer tourists have the motivation to add the experience of social activities on their work resumes as an added value, this is because many companies judge a person not only from a diploma but also from the experience of social activities. In addition, volunteer tourists want to develop their ability to recognize human characters, increase their ability to respect fellow human beings, develop soft skills, socialize, and be more open-minded. This is in line with the statement of Proyrungroj (2017) that one of the volunteer tourist motivations is to develop their abilities. In addition, the volunteer tourist wants to increase the ability to interact with new individuals, develop self-confidence, improve public speaking skills, and improve soft skills such as leadership. More than that, most of the participants said that they had the motivation to improve skills that could be useful in their daily lives and the future, such as the ability to work together, respect others, be tolerant, respect nature and the environment and improve discipline by following the rules and rundowns that the committee has set. This is in accordance with the statement of Wearing & Grabowski (2011) that volunteer tourists have shallow motivation, which is the motivation for renewing job resumes and self-development. In addition, volunteer tourists want to implement the knowledge that has been mastered such as photography skills that are used to capture moments when volunteer tourism takes place, and increase gratitude for the lives they already have.

Meanwhile, the motivation of volunteer tourists was also found, namely exploring the environment that is different from the usual place or comfort zone so that it can feel something new. That way, they feel the challenges of survival during volunteer tourism activities and explore many things from various perspectives. In addition, participants also revealed that one of their motivations was to practice patience, manage emotions under any conditions, and learn to become better individuals. This is a motivation for volunteer tourists to get adventure, for example, is getting a challenge to control the situation and adapt to new things, such as teaching elementary school children in the destination area, etc.

4.2 *Deep motivation*

Based on the research results of the deep motivation category, it was found that millennial volunteer tourist who participated in volunteer tourism activities in West Java Province wanted to become a volunteer because it was triggered by the desire to help and share happiness with others. Participants become volunteers because they sincerely want to help and express their gratitude by helping the community. This is in accordance with Wright (2013) which states that volunteer activities are described as a group of people who want to help others or who have altruistic motives where volunteer tourists are more concerned with others than themselves.

In the interview results, participants pointed out that being a volunteer is an obligation for him as an individual to carry out social activities. The opportunity to do activities like this is also a place for participants to be able to do things from the heart and without coercion from anyone. In addition, participants want to be useful by helping others. This sense of caring and sharing is indeed a principle of volunteer tourist life before taking part in volunteer tourism activities. After providing benefits to the local community, participants gained many life values that make satisfaction and peace of mind. This is consistent with Domingues & Nojd (2013) which states that volunteer tourists get satisfaction when contributing to create change for local communities, especially those who are less fortunate.

Moreover, participants want to give encouragement for learning, give motivation to pursue dreams, give knowledge for elementary school children and kindergartens in the destination area, conduct counseling, or assist in health screening programs. They want to give contribution with the skills they have mastered so that they can provide change for the better.

Furthermore, volunteer tourism activities are very different from other tourism programs. Sin (2009) states that volunteer tourists must stay with residents and experience cultural authenticity by being, doing, touching, and seeing. In other words, volunteer activities make tourists more familiar with local culture through social interactions that can help the community. In this study, Social Traveling (SocTrav) created a volunteer program to help rural communities called Jelajah Desa. Every participant who joined the program had the opportunity to help local communities' plant tree seedlings and clean the village environment. From here, participants indirectly get new insights such as farming. This process is a symbiosis of mutualism that is mutually beneficial between volunteer tourists and local communities. In addition, volunteer tourists do teaching activities in areas that are less educated. This condition requires them to get to know and learn an everyday language in the area so that they can easily transfer knowledge and take a sense of community trust. In this case, there was a statement from the participants that the motivation to participate in the volunteer tourism program was to feel authenticity.

Subsequently, participants said that they wanted to spread goodness. According to the participant from the 1000 Teachers organization, participants were motivated to transmit goodness by giving their time and energy to teach the community in the destination. Before carrying out the program, Volunteer tourist 1000 Teachers are given materials on how to teach appropriate as well as what lessons must be presented to their students so that the learning process can run smoothly and effectively.

Based on this, volunteer tourism indirectly helps the government in overcoming the social problems of the region by bringing real work solutions. This proves the statement of Guttentag (2009) which states that volunteer tourism has become a promising sector that can benefit the tourists and the local community. This is in accordance with Alexander, (2012) who stated that volunteer tourism is beneficial for the government because it can give sustainable solutions to the problems that are being faced by the area. So it is necessary to make a volunteer tourism program that is adjusted to the motivation of volunteer tourists and the condition of the destination.

5 CONCLUSIONS

From the results of the study, it was found that the millennial volunteer tourists who participated in the volunteer tourism program in West Java province had a variety of deep motivation and shallow motivation. The shallow motivation possessed by millennial volunteer tourists who take part in volunteer tourism activities in West Java province consists of vacationing, intending to optimize their free time; build relationships with individuals who have similar interests; renew job resume; develop self-abilities; and get a challenge or adventure. In addition, the deep motivation possessed by millennial volunteer tourists who participate in volunteer tourism activities in West Java province is to contribute to others through their time, money, and talents; make a change; feel authenticity or culture; and transmit goodness.

6 IMPLICATIONS

Research on volunteer tourism in Indonesia tends to be rarely so that similar studies need to be done in order to give more understanding of this phenomenon. Further research needs to be done about volunteer tourists in other regions in Indonesia so that it is more extensive.

7 SUGGESTIONS

Based on the results of research and discussion, the suggestions obtained for the NGOs, foundations, travel agents, and the community in making volunteer tourism programs are adjusted to the motivation of volunteer tourists.

- In responding to shallow motivation of millennial volunteer tourists, volunteer tourism organizers can make a memorable traveling agenda by choosing interesting, unique, and insightful tourist destinations. The destination that can be chosen is the creative center or production of the local community such as a workshop, factory, or handicraft production. The activities carried out in these places can fulfill the desire of voluntary tourist shallow motivation to develop their abilities, explore and feel a new challenge.
- In responding to deep motivation of millennial volunteer tourists, volunteer tourism organizers can collaborate with the programs planned by the local community in accordance with the specified period of time. For example, a volunteer program for teaching and a volunteer program for health checks are carried out in the same village every month. So that this can accommodate volunteer tourist deep motivations who want to contribute, make changes, and spread goodness.

REFERENCES

Alexander, Z. (2012). International Volunteer Tourism Experience in South Africa: An Investigation Into the Impact on the Tourist. *Journal of Hospitality Marketing and Management, 21*(7), 779–799. https://doi.org/10.1080/19368623.2012.637287

Booking.com. (2018). Booking.com Ungkap 8 Prediksi Travel Untuk 2019. Retrieved January 8, 2020, from https://news.booking.com/bookingcom-ungkap-8-prediksi-travel-untuk-2019/

Carvache-franco, M., Carvache-franco, W., Contreras-Moscol, D., Andrade-alcivar, L., & Carvache-franco, O. (2019). Motivations and Satisfaction of Volunteer Tourism for the Development of a Destination. *GeoJournal of Tourism and Geosites, 26*(3), 714–725. https://doi.org/10.30892/gtg.26303-391

Chen, L. J., & Chen, J. S. (2011). The motivations and expectations of international volunteer tourists: A case study of "Chinese Village Traditions." *Tourism Management, 32*(2), 435–442. https://doi.org/10.1016/j.tourman.2010.01.009

Coghlan, A., & Fennell, D. (2011). Myth or substance: An examination of altruism as the basis of volunteer tourism. *Annals of Leisure Research, 12*(3–4), 377–402. https://doi.org/10.1080/11745398.2009.9686830

Domingues, I. A., & Nojd, P. (2013). *Volunteer Tourism-Who does it benefit? Linnaeus University*. https://doi.org/10.1080/02508281.2003.11081411

Godfrey, J., Wearing, S. L., Schulenkorf, N., & Grabowski, S. (2019). The 'volunteer tourist gaze': commercial volunteer tourists' interactions with, and perceptions of, the host community in Cusco, Peru. *Current Issues in Tourism, 0*(0), 1–17. https://doi.org/10.1080/13683500.2019.1657811

Guttentag, D. A. (2009). The possible negative impacts of volunteer tourism. *International Journal of Tourism Research, 11*(6), 537–551. https://doi.org/10.1002/jtr.727

Hallmann, K., & Zehrer, A. (2016). How do Perceived Benefits and Costs Predict Volunteers' Satisfaction? *Voluntas, 27*(2), 746–767. https://doi.org/10.1007/s11266-015-9579-x

Han, H., Meng, B., Chua, B. L., Ryu, H. B., & Kim, W. (2019). International volunteer tourism and youth travelers–an emerging tourism trend. *Journal of Travel and Tourism Marketing, 36*(5), 549–562. https://doi.org/10.1080/10548408.2019.1590293

Haslebacher, C., Varga, P., & Murphy, H. C. (2018). *Examining the Motivations of Volunteer Tourists: Insights From Images Posted on Social Media*. Research Gate.

Hsieh, H. F., & Shannon, S. E. (2005). Three approaches to qualitative content analysis. *Qualitative Health Research, 15*(9), 1277–1288. https://doi.org/10.1177/1049732305276687

Kazandzhieva, V. (2014). Volunteer tourism in bulgaria. *Research Gate*, (September).

Kontogeorgopoulos, N. (2017). Finding oneself while discovering others: An existential perspective on volunteer tourism in Thailand. *Annals of Tourism Research, 65*, 1–12. https://doi.org/10.1016/j.annals.2017.04.006

Milne, S., Thorburn, E., Hermann, I., Hopkins, R., & Moscoso, F. (2018). *Voluntourism Best Practices: Promoting Inclusive Community-Based Sustainable Tourism Initiatives*. Auckland.

Phelan, K. V. (2015). Examining the Voluntourist Experience in Botswana. *Worldwide Hospitality and Tourism Themes, 7*(2), 127–140. http://dx.doi.org/10.1108/MRR-09-2015-0216

Polus, R. C., & Bidder, C. (2016). Volunteer Tourists' Motivation and Satisfaction: A Case of Batu Puteh Village Kinabatangan Borneo. *Procedia – Social and Behavioral Sciences, 224*(August 2015), 308–316. https://doi.org/10.1016/j.sbspro.2016.05.490

Pompurová, K., Marceková, R., Šebová, L., Sokolová, J., & Žofaj, M. (2018). Volunteer tourism as a sustainable form of tourism-The case of organized events. *Sustainability (Switzerland), 10*(5). https://doi.org/10.3390/su10051468

Proyrungroj, R. (2017). Orphan Volunteer Tourism in Thailand: Volunteer Tourists' Motivations And On-Site Experiences. *Journal of Hospitality and Tourism Research, 41*(5), 560–584. https://doi.org/10.1177/1096348014525639

Silló, Á. (2019). International Volunteers as Strangers in Szeklerland. *Acta Universitatis Sapientiae, Social Analysis, 8*(1), 5–21. https://doi.org/10.2478/aussoc-2018-0001

Sin, H. L. (2009). Volunteer Tourism-"Involve Me and I Will Learn"? *Annals of Tourism Research, 36*(3), 480–501. https://doi.org/10.1016/j.annals.2009.03.001

Suanpang, P., Srisuksai, N., & Tansutichon, P. (2018). The Demand of Voluntourism in a Developing Country. *Journal of Service Science and Management, 11*(03), 333–342. https://doi.org/10.4236/jssm.2018.113023

Tomazos, K., & Butler, R. (2010). The volunteer tourist as "hero." *Current Issues in Tourism, 13*(4), 363–380. https://doi.org/10.1080/13683500903038863

Wearing, S. (2001). *Volunteer Tourism; Experiences that Make a Difference.* (J. Neil & M. Jackson, Eds.) (1st ed.). New York: CABI Publishing.

Wearing, S., & Grabowski, S. (2011). International Volunteer Tourism: One Mechanism for Development Turismo de Voluntariado Internacional: Um mecanismo para o Desenvolvimento International Volunteer Tourism: One Mechanism for Development, (January 2011).

Wright, H. (2013). Volunteer tourism and its (mis)perceptions: A comparative analysis of tourist/host perceptions. *Tourism and Hospitality Research, 13*(4), 239–250. https://doi.org/10.1177/1467358414527984

Promoting Creative Tourism: Current Issues in Tourism Research – Kusumah et al. (Eds)
© 2021 Taylor & Francis Group, London, ISBN 978-0-367-55862-8

Sacred tombs as attraction of tourism village

R. Fedrina, Khrisnamurti, R. Darmawan & U. Suhud
Universitas Negeri Jakarta, Jakarta, Indonesia

ABSTRACT: The development of rural or tourism village in Indonesia has skyrocketed in the past years. With the support of the government, this type of tourism might become the solution for community sustainability tourism. Cisaat village is one of the villages in west Java, Indonesia, with a strong prospective to grow become a tourist attraction. This study aimed to explore the potential factors that can be developed in Cisaat Village, especially tourism village attraction. A qualitative approach was employed to capture the uniqueness of the village. The data collected from observation and interviews at Cisaat Village and analyze using the 3A's tourism (attraction, access, amenities). The result shows that Cisaat Village not only has strong potential to become invigorating village tourism but also has potent imagery about religious activities linked to a sacred heritage tomb. It could be potentially developed to be more interesting for visitors and groups that have not been optimally done by the villagers.

Keywords: tourism village, 3 A's of tourism, religious tourism

1 INTRODUCTION

One of the religious tourism activities is a pilgrimage. One of the new forms of pilgrimage tourism is tourism secularism, which is pilgrimage tourism that is not based on religious motives or general beliefs, but because it is motivated by spiritual ties, admiration and popularism (Tempo 2012). Pilgrimage tourism is a phenomenon and activity that has long been carried out by some Muslims. In Indonesia, pilgrims make short visits (in hours) or quite long (tens of days) to a location or Hajj point (Republika 2019). This pilgrimage point can be a holy grave and its petilasan. Petilasan are ancient relics (generally history), palaces, cemeteries, and so on. This term is taken from the Javanese language, namely a place that has been visited or occupied by someone (which is important). This is what attracts pilgrims or visitors. The existing attraction needs to be developed in tourist villages that have sacred tombs and petilasan. According to Yoeti (1997), a tourist attraction will develop well if it has three things that attract tourists, namely something that can be enjoyed visually (something to see), something that must be done (something to do) and something that can be purchased to enjoy or as a souvenir (something to Buy). Research on religious tourism has been conducted by researchers (Nolan & Nolan. 1992; Shackley 2006; Latifundia 2016). However, there is little unexplored research that develops the appeal of sacred tombs and petilasan, especially in tourist villages. A research on how to build a tourist village that must be done in order to remain attractive and unique has not been done much. Several studies have focused on agriculture and biology. This research tries to analyze how the attraction of the holy graves and petilasan if it is developed as a tourist village attraction in the form of a sacred tomb tourism object which is done in Cisaat Tourism Village, Subang, West Java.

DOI 10.1201/9781003095484-25

2 LITERATURE REVIEW

2.1 *Religious tourism*

Religious groups use religious heritage sites as pedagogical tools to create an authentic 'sense of place' to allow for meditation and to encourage reflection (Shackley 2006). Different spaces can be viewed as sacred or secular by different individuals, persons, groups, and communities. These spaces can also be subject to historical negotiations and contested meanings (Korstanje & Howie 2020). As such, heritage and religious sites and experiences were inevitably entwined (Timothy & Olsen 2001). Pilgrimage or visiting sacred tomb is an activity done by the living people to commemorate the deceased's services by visiting the tomb of the deceased person and praying for the deceased person to be forgiven for his sins (Mirta 2013). Religious tourism or can be referred to as pilgrims are not new things recognized in the Indonesian tourism industry. The emergence of religious tourism has long been developing in Indonesia. International tourism trends have predicted the growing type of psychic-spiritual tourism, namely the emergence of a group of tourist groups that tend to mental and spiritual.

Religious tourism can be referred to as pilgrimage tourism (Zajma et al. 2011). Aulet and Dolors (2018) mentioned religious tourism as a type of tourism primarily motivated by religion (whether in combination with other motivations or not), which has a religious place as a destination and may or may not linked to participation in ceremonies and religious activities.

2.2 *Tourism village*

Wiendu (1993) states that tourism village is a form of integration among attractions, accommodations, and supporting facilities presented in a community life structure that blends with the prevailing ordinances and traditions. Yoeti (2002) argues that the success of a tourist area is very dependent on the 3A that stands for attraction, accessibilities, and amenities. The concept of tourism destination attractiveness has received much attention in tourism literature (Buhalis 2000; Ferrario 1979; Formica & Uysal 2006; Krešic & Prebežac 2011; Tasci 2007). The studies argue that destination attractiveness is the driving force of tourism and without which tourism would be almost nonexistent.

3 METHOD

This method of research was qualitative. As for the analysis unit, there were tourism driving groups, village apparatus, villagers, and visitors. The collection of data in this period was done by an interview or informal speaking technique guided by the instrument or interview guidelines related to 3A of tourism components. Moreover, researchers documented photographs, videos, and supporting instruments to complement the data needs of the research. The observation was done descriptively to explore the meaning of the sacred tombs for the visitors based on the activities undertaken, and the meanings understood.

4 FINDINGS AND DISCUSSION

4.1 *Attractiveness of Cisaat Village*

Cisaat Village has 4 hamlets, 6 RW, 28 RT and 10 villages, namely: hamlet 1 Cisaat covering Kp. Cisaat, Kp. Kabakan Pasir and Kp. Cerelek, hamlet 2 Cilimus covering Kp. Cilimus, Kp. Koleberes and Kp. Palasari Housing, hamlet 3 Cigangsing includes Kp. Cigangsing and Kp. Gunungnutug, hamlet 4 Jagarnaek covering Kp. Jagarnaek and Kp. Cikanyere. As for the attractions include:

4.1.1 *Culture*

Cisaat Village has a culture and art that are rarely owned by other villages including: (a) in the month of Maulud, a visiting joint ceremony is held at places that are considered sacred/ Syeh's tombs and it is held with the implementation of a joint celebration/feast in the sacred place; (b) In ancient times, most of the Cisaat people had traditional beliefs, namely that in certain months they liked to hold traditional ceremonies. For example, in the Muharom Month they hold the ceremony of Ruatan Bumi and Bubur Sura. Also the Hajat Babarit (Joint Thanksgiving held at Bale Lembur) is held every three months (Panyawalaan). (c) The belief of the Cisaat residents is that every time before the Khajatan, both marriage and circumcision, the prospective bride is required to visit the sacred tombs. The day before the Khajatan is carried out, the host must put up offerings (kucingan) either at home, tarub corners, wells and sacred places, the host also entrusts the elders as Goni (People who are considered powerful).

4.1.2 *Religious tourism*

Whereas in Cisaat Village, almost all hamlets have many tombs of Syeh (sacred people), the developer of Islam and Petilasan sites, especially prominent figures of the spread of Islam in the Sagalaherang Wetan area, to be precise in Padukuhan Cisaat, and many other Islamic developer figures.

4.1.3 *Recreation/nature tourism*

There are many recreational/natural tourism places in the Cisaat Village area that can be offered, both managed by the private sector and by the Village Government, such as: (a) Cisaat Village has a recreation area that is managed by the private sector, namely CHR (Ciater Riung Rangga), which is a recreation area that provides Natural Panorama, Horse Riding, Relaxing Walks, Swimming Pools, Camping Ground, Homestay, etc.; (b) Recreation Sports that offers a natural panoramic view of the mountains surrounded by thousands of hectares of cold and beautiful tea plantations as a means of Tea Walk, Fun bike, Road Rest and as a relief to see the charm of beautiful and comfortable mountain scenery.

4.1.4 *Agricultural tourism*

Agro Cisaat Village is a fertile agricultural area divided into 3 functions of agricultural land, namely rice fields, plantations and fields, where this village is one of the bases for vegetable farming and tea plantations, because it allows vegetable and tea plantations to be suitable to be planted here in temperatures ranging 9 to 18 degrees Celcius, especially vegetables in the form of cabbage, broccoli, mustard greens, tomatoes and fruits with low trees such as pineapples, strawberries, and grapes. In addition to agricultural cultivation in Cisaat Village, it is also one of the areas that has the potential to be positive as an area for animal husbandry such as goats/sheeps, purebred chickens, beef cattle and especially dairy cows because it is one of the villages that produces milk for dairy cows and is a center for dairy farming in the Subang Regency

4.2 *Sacred tombs as attraction*

Cisaat Village has a beautiful view. The rice fields are still stretching out, and the village atmosphere is beautiful. The current trend shows visits from cyclists and pedestrians appearing in this village. According to informants, the village government has determined the village's attraction, including cultural, religious, recreational/natural, and agro attractions. In this determination, it does not show the main focus in developing village attraction. Attraction is an important thing to become a potential market motivation to come to this village. According to the village head, apart from the attraction itself, it is also necessary to develop activities that can be carried out in tourist villages to attract visitors.

Pilgrimage tourism has not been touched by the village government, even though repeated visits and rituals occur during this pilgrimage. The pilgrimage is carried out to honor the ancestors and carry out spiritual rituals in Ciputihan/Cikahuripan (namely a place for the descendants of Prabu Siliwangi as a place for solitude and a place for contemplation) and sacred tombs scattered in several hamlets in Cisaat village. Pilgrims have visited this sacred tomb for years. According to the community's belief, these sacred groves are the descendants of Prabu Siliwangi, who contributed to the spread of Islam in the land of Sundanese. This pilgrimage tour has not been touched because there are no planned activities related to this tour. For this, research is needed to determine what activities can be developed, so that pilgrimage and cultural tourism become more attractive to visitors. To control the existence of local culture, it is necessary to consider activities, not mass tourism in nature but the form of small groups such as special interest tourism. Nature tourism is currently the main attraction. Beautiful village conditions and beautiful landscapes provide an exciting experience for visitors. The sacred tombs and the burial or anything that becomes the attraction in the village of Cisaat can be said as a tourism product. The main product is central to other products. The determination of the Government of Cisaat village as a cultural tourism village was just right. These sacred tombs were part of the cultural activities of Cisaat village and its surroundings.

There was some sacred tomb in Cisaat Village. Based on the interview, the history of Cisaat village originated from the Sagalaherang area which is the center of government in the area of Subang at the time his reign was headed by Eyang Suwargi (Rd Arya Wangsa Gofarana).

Sagalaherang government evidenced by the number of officials of the Sagalaherang government in his time living in Cisaat Pedukuhan, including his descendants buried in the village of Cisaat. The next tomb was considered holy and sacred by the inhabitants and visitors. Visiting activities made by visitors or pilgrims are part of religious tourism. Based on the concept above, religious tourism is a spiritual journey that can group to sites that are considered religious or sacred and related to death. Cisaat Village has several sacred tombs of saints who have been widely engaged in spreading Islam's religion during its development.

Furthermore, these tombs have to do with the descendants of Prabu Siliwangi (The King of Pajajaran). Therefore, this sacred is highly respected by the village community, and the hereditary has known its existence for pilgrims. These tombs were guarded and cared for by Kuncens. These sacred tombs scattered in the village of Cisaat, the sacred area/tombs in Cisaat village there were five areas of sacred tombs namely:

- Area of Keramat Ciputihan/Cikahuripan (this area of Petilasan)
  Keramat Ciputihan/Cikahuripan is a historical burial is not a grave as a description of the elders of Cisaat in Sunda.
- Area of Keramat Girang tomb or Karamat Ageung Cisaat (Karamat Girang)
  Keramat Girang location on the outskirts of tea plantation between Kp. Cisaat, Cikaputihan with Kp. Jagarnaek. Approximately 4 Km from the provincial Road to the west of the turns Mariuk complex, Eyang Bao, and Eyang Rd. Pringganata
- Central Sacred Tomb Area
  Sacred Tomb of the location +150 M from Keramat Girang to the North: Rd. Ngabei Madamadija, RD. Ngabei Yuda Anggaprana
- Lower tomb Area
  The tomb of Keramat Hilir location +50 M from the sacred central to the North, eyang Rd. Haji zaenal Mustafa
- Area Keramat Giri Waas (Gunung Cinta)
  Maqom Keramat Giri Waas (Mount Cinta) is located at the top of Mount Cinta in the midst of tea plantation is approximately 1 Km from the provincial road conspiracy to the West Mariuk complex or 300 M from Kampung Cilimus village Cisaat northwards.
- Area Keramat Makam Eyang Santri
  Maqom Keramat Eyang Santri location is located in the middle of the village (Kp. Kolepas Cisaat Village) approximately 1.5 km from the provincial road conspiracy to West Mariuk complex or +2.5 km from the location of Keramat Cikaputihan and Keramat Girang Cisaat

- Another sacred tomb located around Cisaat village: eyang dalu di Palasari at Mount Gedogan, eyang Jamiludin in Cijolang, eyang Rd. Merta kusumah di Sinapeul/Cieuyeub Babakan mount. Eyang Tebol in Curugrendeng, eyang Rd. Priatna di Cibuganggeureung.

The majority of the people of Cisaat Karuhun leave only genealogy, the story told during his lifetime recorded in the history of Sagalaherang and Cianjur, written evidence of the historical figure about them recorded in Babon Rundayan Darmapala (Via Panjalu) around the centuries of XVIII and XIX M, Babon Rundayan Talaga; Galuh Banjar Sari; Panjalu, Babon Rundayan; and also the genealogy of the ancestors of Cisaat Sagalaherang in Karawang Ragion based on the chronicles of Sagalaherang history. The number of visitors who came to the tomb of Kramat in 2019 was about 14,000 people. This number is partially recorded in the guestbook and most are not recorded in the guest book because the pilgrims/visitors many were directly visit the location, especially in the location where special events were held and did not provide a guest book.

The highest visited was occurred on:

- The 8th–29th of Rabiul Awal/Mulud when usually visitors come from various corners for pilgrimage to multiply the blessings and great intercessory in order to glorify the birth of Kangjeng Rosululloh SAW by means of pilgrimage praying and reading Sholawat in the Auliya maqam.
- The 27th & 29th of Rabiul Awal/Mulud when Hajat Maulidan Keramat is usually held. It is carried out once a year, usually organized in the area of central Keramat, when thousands of visitors come from each district to attend the event as well as pilgrimage.
- Every Friday night, especially the Friday Kliwon night, visitors are doing Zikr Pilgrimage and praying.

4.3 *Developing tourism village*

Each religion has a sanctuary that is distributed (Fawaid 2010). The holy place is not only confined to the place of worship but also the sites which are holy graded or worth of history according to their own religious beliefs. The Sacred tomb of Cisaat Village is no exception to this. In almost every religion, visiting the sacred place became the tradition of its adherents, which included Islam. In Islam, in addition to mosques, the tombs of religious denominations (guardians) are often the objects of tourists/visitors. The pilgrimage was also a religious visit. Not to mention other historical sites such as contemplating places, temples, museums, and educational institutions. Culturally, someone is considered less complete in their religious traditions if they never visit holy places such as the sacred tombs of guardians. Even in certain territories, visiting sacred tombs becomes an annual obligation. Besides, economically, the travel on pilgrimage would give a huge economic impact for the visited area. Many souvenirs and merchants in the area of pilgrimage places are a real indicator of how the pilgrimage economic value is quite high. Not to mention the travel companies growing in line with the increasing passion for religious tourism among the community. The research of Latifundia (2016) states that narrative of the society or the key caretaker visited the sites of the Petilasan (memory path of the saints) and the pilgrim because the location of the sites once had been visited by the religious leaders or certain characters who are charismatic. The Research in Cisaat village shows that the village has a petilasan that is Cikahuripan where Islamic propagator people once stayed there to contemplate or deliberate sacred tomb.

The place is always visited by people from various regions with the motivation to get a blessing from Allah. Such tourist activities goes into particular interest. One of the Special interest forms is religious tourism. Based on the interview, it was stated that the Ministry of Culture and Tourism directly gave the determination of the tourist village for Cisaat village in 2010 by raising the cultural aspect as a characteristic of the village without a clear explanation why the establishment of culture is being the specialty of this tourist village. For the Cisaat village community, the culture in question is unclear. As Sundanese people, they feel that their Sundanese culture is the same as other Sundanese people. Pilgrimage tourism has not been touched by the village government, even though repeated visits and rituals occur during this pilgrimage. The pilgrimage is carried out to honor the ancestors and carry out spiritual rituals in Cikahuripan (a descendant of Prabu Siliwangi as a place for seclusion and a place of contemplation) and sacred tombs scattered in several hamlets in Cisaat village. Pilgrims have visited this sacred tomb for years. According to the people's belief, sacred tombs are descendants of King Prabu Siliwangi, who contributed to the spread of Islam at the Sundanese level. This pilgrimage tour has not been touched because there are no plans of activities that can be done related to this tour. For this matter, research is needed to find out what activities can be developed, so that pilgrimage and cultural tourism becomes more attractive to visitors. To control the existence of local culture, it is necessary to consider activities that are not mass tourism in nature but special interest tourism. Nature tourism is currently the main attraction. Beautiful village conditions and beautiful landscapes provide an exciting experience for visitors. Research on how the development of a tourist village must be carried out to remain attractive and have its uniqueness has not been done much. The author is looking for some research in the village of Cisaat which some of them focus on agriculture and biology.

The development of the attraction of this village was explored based on interviews with head village and villagers. It is stated that the village had determined the attractions of the village, including cultural, religious, recreational/natural, and agro attractions. Furthermore, after the interview of this research, the main attraction of this village is to raise the sacred tomb and its pond which is called Ciputihan/ Cikahuripan. This consideration is based on information obtained that the tomb is not just a tomb but has a historical value which is valuable, namely the history of Sundanese and the spread of Islam by Prabu Siliwangi.

To develop these attraction potentials to become a tourist village that relies on the strongest of village's potential resources and visitor preferences who visit the villages. In this research, the researcher refers to the tourism component of 3A formula: (a) accessibility, this village is close to Bandung and Ciater tourism object (Ciater Springwater), and very accessible from the capital city of Jakarta. In terms of access to information, Cisaat village has not described much; (b) attraction, the attraction that becomes the focus for development is the sacred tomb, because this is its peculiarity, the attraction of beautiful natural scenery is an advantage for Cisaat village, and this view is owned by the surrounding villages and visits to this tomb are recorded in the book. Village office guests counted about 14 thousand visitors in 2019; (C) amenities, the village already offers homestays for visitors who want to stay. However, the development of better facilities, especially homestays, needs to be improved, especially cleanliness. The attraction of this village is the existence of sacred tombs and the Saint's journey (descendants of King Siliwangi) or who have the merit of spreading Islam in the region of Subang, Purwakarta, Cianjur, Sukabumi and Limbangan. Based on this formula, it is necessary to develop an attraction that is easier to implement, namely the development of the sacred tomb attractions. This attraction does not require luxurious facilities because what pilgrims required is permission to visit the tomb and Ciputihan. The current tourism awareness activists group is not active because the management is already busy with its business, an interview with one of its members stated that it requires a commitment from the members to develop tourism in this village. In terms of policy, the RPJM has not specified details for the development of tourism villages. All of these will be parameters to be primary data surveyed to be compiled then analyzed.

5 CONCLUSION AND IMPLICATION

This paper contributes to the religious tourism literature using a qualitative approach. The sacred tomb has a unique historical and cultural value. The results showed a strong relationship between belief and faith in the voluntary journey of visitors to the sacred tomb in the hope of getting blessings. In addition, this study also examines a specific time of visit which is the highest because it is related to the sacred dates of certain important months of their journey. Spirituality has created a simple tour package. Strategies need to be reshaped and refocused to accommodate and build relationships with potential future travelers.

Religious tourism in Cisaat Tourism Village is essential. Tourism Village is a form of integration between attractions, accommodation and supporting facilities that are presented in an order of community life that is integrated with prevailing procedures and traditions. The sacred tomb market segment is different from other markets. They accept the conditions of facilities in the village. For them, it is a spiritual journey. Further research on this tourist attraction needs to be analyzed such as developments of direction, scenarios, priorities, technique, required resources, managing institutions/organizations, funding and stages of the implementation period.

REFERENCES

Aulet, Silvia &Vidal, Dolors. (2018). *Tourism and religion: sacred spaces as transmitters of heritage values. Journal Church, Communication and Culture. 3*, pg 237–259.

Beth A. Handwerger, M., Barbara A. Blodi, M., Suresh R. Chandra, M., Timothy W. Olsen, M., & Thomas S. Stevens, M. (2001). Treatment of Submacular Hemorrhage With Low-Dose Intravitreal Tissue Plasminogen Activator Injection and Pneumatic Displacement. *Arch Ophthalmol. 2001;119(1)*, 28–32.

Buhalis, D. (2000). Marketing the competitive destination of the future. Tourism Management, 21, 97–116.

Cracolici, M. F. and Nijkamp, P. *"The attractiveness and competitiveness of tourist destinations: a study of Southern Italianregions," Tourism Management, 30*, (3), pp. 336–344, 2009.

Fawaid, Akhmad. (2010). *Mengunjungi Tempat Suci; Ragam Motivasi Wisata Religious*.ejournal. satinpamekasan.ac.id

Ferrario, F. F. (1979). The Evaluation of Tourist Resources: an Applied Methodology. *Journal of Travel Research.*

Formica, S., & Uysal, M. (2006). Destination attractiveness based on supply and demand evaluations: An analytic framework. Journal of Travel Research, 44, 418–430.

Korstanje, Maximiliano E and Howie, Luke. (2020). *Pilgrimages to Terror: The Role of Heritage in Dark Sites. Dark Tourism and Pilgrimage*.CABI.UK

Krešic, D., & Prebežac, D. (2011). Index of destination attractiveness as a tool for destination attractiveness assessment. *Tourism: An International Interdisciplinary Journal, Vol. 59 No. 4.*

Latifundia, Effie. (2016). Kapata Arkeologi. *Situs Makam-Makam Kuna di Kabupaten Kuningan Bagian Timur: Kaitannya dengan Religi. Kapata Arkeologi 12* (1). Kemdikbud. Jakarta

Mirta, Irmasari. (2013). *Makna Ritual Ziarah Kubur Angku Keramat Junjung Sirih Oleh Masyarakat Nagari Paninggahan*. Jurnal Prodi Pendidikan Sosiologi Antropologi. UNP. Padang

Nolan, M. L., & Nolan, S. (1992). Religious sites as tourism attractions in Europe. *Annals ofTourism Research*, 68–78.

Republika.(2019, February 8). www.republika.co.id.Retrieved August 7,2020, https://republika.co.id/berita/pmk7y9440/mengopt imalkan-peluang-ekonomi-wisata-ziarah

Sesotyaningtyasa, Mega & Manaf, Asnawi.(2015). *Analysis of Sustainable Tourism Village Development at Kutoharjo Village, Kendal Regency of Central Java*. Procedia – Social and Behavioral Sciences 184, pp 273–280.

Shackley,myra.(2006). *Empty Bottle at Sacred Sites: Religious retailing at Ireland's National Shrines. Tourism, Religion and Spiritual Journeys*. Routledge.Canada

Tasci, A., & Gartner, W. (2007). Destination Image and Its Functional Relationships. *Journal ofTravel Research.*

Tempo.(2012, February 24). www.tempo.co.id.Retrieved August 7, 2020, from https://travel.tempo.co/read/437474/apa-itu-wisata-ziarah

Wiendu, Noeryantie. 1993. Concept, *Perspektive and Challenges, Makalah bagian dari laporan Konferensi Internasional.Yogyakarta*. Gadjah Mada Press

Yoeti, O.A. (2002). *Perencanaan dan Pengembangan Pariwisata*. Jakarta: PT Pradaya Paramita.

Promoting Creative Tourism: Current Issues in Tourism Research – Kusumah et al. (Eds)
 ISBN 978-0-367-55862-8

Chinese tourists' perception on Bali Tour Package

H. Utami, R. Darmawan, R. Wardhani & U. Suhud
Universitas Negeri Jakarta, Jakarta, Indonesia

ABSTRACT: To attract more foreign tourists, the Indonesian government offered a free visa for citizens of some countries, including China. As a result, there has been a surge in the number of Chinese tourists visiting Indonesia. This study aims to explore the perceptions of Chinese tourists toward the price and quality of the tour packages on Bali island. This study uses a qualitative method. Data were collected in China using an online survey to the China tourists and also document study. Data were analyzed using descriptive statistical analysis qualitative method. The analysis used in this study is a quantitative analysis, which is commonly used is a statistical analysis and is described in making conclusions. China tourists choose Bali as their destination for holiday because they offered an interesting price from the travel agents. The price of this package became the attraction to Bali. The travel agent then applies the monopolistic practice of only bringing tourists to shop at designated places, which of course has a partner relationship with the travel agents in China. The result of this study proved China tourists are not fully satisfied with the quality of packages offered.

Keywords: perception, Chinese tourists, tour package prices, Bali island

1 INTRODUCTION

Bali traditional culture is unique and can be a magnet for visitors. Dances, paintings, and sculptures are famous Balinese culture. The artworks are part of traditional life and religion in Bali, namely Hinduism. This was also stated in previous study, about Characteristics, Motivation and Activities of Asian Tourist in Ubud Village, that the attraction they want to see in Ubud was the culture (Pratama et al. 2016). The Balinese Hindu community has a caste division within its community as does the Hindu community in India. Bali is one of the provinces and is located in the central part of Indonesia.

The total area of Bali province is 5,633 square kilometers. The highest mountain in Bali is Mount Agung (3,142 meters above sea level) located in Karang Asem Regency. For these attractions, Bali becomes a popular destination in the world. In Asia, Bali is the most favorite destination for holiday. The results of the provincial tourism department data of the Province of Bali then recorded that in the last five years, tourist arrivals from China increased rapidly. with the latest data reached 5 million people in Bali. However, the tourists were dominated by Chinese, and followed by Australians and other Asians (Bali Tourism & Research Center 2020).

Since the travel agent applies the monopolistic practice of only bringing tourists to shop at designated places, which of course has a partner relationship with the travel agents in China, the price of goods offered was higher than the prices offered by local residents. There are even some visits to the industries that have no connection with tourist destinations on vacation. This practice quickly spread among Chinese tourists, giving rise to negative perceptions of Bali, residents, tourism agents, local governments, and Indonesia.

The China tourists' origins are dominantly from Guangzhou, a sub-province which is one of the five National City Centers of China. Guangzhou is considered one of the most prosperous cities because of the rapid growth of industrialization (Foreign Affairs Office of Guangzhou Municipal

 DOI 10.1201/9781003095484-26

People's Government 2019). This study would be a state-of-the-art on Indonesia's tourism development for strategic marketing planning, cultural, and natural attractions development as a support to tourism sustainable development.

2 LITERATURE REVIEW

2.1 *Perception*

Perception is the process through which people choose, organize, and interpret information to form a meaningful picture of the world (Kotler 1997. The process of our perception is influenced by internal and external factors. Internal factors include motives, values, interests, attitudes, past experiences, and expectations, while external factors include motion, intensity, size, novelty, and salience. It is generally accepted that there are three steps in the process of perception (Kenyon & Sen 2015).

In this study, price is the most important factor in consumer perception. Price and quality must be taken into account in this research. Consumer perceptions of Bali tour packages are also influenced by the quality of tour packages including the price of services and products and products obtained by tourists. The previous study stated that Chinese tourists are interested in Bali culture with the very cheap expenses around Rp 201.000–300.000 in a day (Pratama et al. 2016).

There are several definitions of prices, namely, price is the monetary value of the product or service on the market; price is the value of money that customers must exchange to get a product or service; price is also a marker of the value of a product or service for someone, and different customers will give different values for the same goods or services. The price set by a company for one product or service can be seen from several dimensions, namely, affordable or not, price between price and quality/taste, price competition, and price match with portion (quantity) (Ghanimata dan Kamal 2012).

2.2 *Tour packages*

Tour packages are service products, intangible products. Tourists who buy tour package products are essentially more of a buying expectation, relying that the implementation of tour package trips will be in line with expectations. Tour package as a product is formed through the process of fusion of components of other tour packages, such as transportation, hotels, restaurants, tourist attractions, and others.

The previous study stated characteristics of Asian tourists by way of visiting Bali; it was with a group, since most of them are first timers. Components in the packaging are joined together and related to each other and become one of the references for companies in making tour packages. Packaging is one of the ways in marketing to approach potential customers, namely, tourists who want to travel but don't have the time and opportunity to arrange and plan their vacation time. Existing components in the packaging will determine the quality of a tour package, for that tour package is made as well as possible so that it can influence tourists to use it (Nuriata 2014).

3 RESEARCH METHODS

This research used qualitative and quantitative approaches. The total number of participants was 105 participants, who were buyers of Bali tour packages from existing travel agents in Guangzhou, China. Data collection techniques in this study were documentary studies, observations, questionnaires, and also interviews with Chinese tourists who visited Bali in the period July–September 2019. Data collection also worked with several travel agents in Guangzhou, China.

The People's Republic of China (PRC) is the country with the largest population in the world. The total population in the world has a percentage of 18.5% of the world's population. This means

Table 1. The participants' origin.

Origin	Freq	%
Guangzho	49	46,7%
Sekitar Guangzho	29	27,6%
beijing	3	2,6%
Fujian	16	15,2%
Shanghai	8	7,6%
Total	105	100%

Table 2. Range of participants' income.

Monthly Income (Rp)	Freq	%
<2500 yuan	0	0
2500 yuan–5000 yuan	37	35,24%
5000 yuan–10.000 yuan	44	42%
>10.000 yuan	24	22,85%
Total	105	100 %

the Chinese population has great potential as tourists visiting several tourist destinations around the world (Wikipedia 2019). Based on the data from the Central Statistics Agency (BPS) in Bali, arrivals of foreign tourists to Bali in April 2019 were recorded at 476,327 visits. About 476,160 foreign tourists came through the airport and 167 through the port (Tirto.id 2019).

Based on data obtained in 2018, more than 480,000 aircraft landed at or departed from Guangzhou Baiyun International Airport, up to 2.6%. The airport holds 70 million passengers and processes 1.9 million tons of postal goods, both of which experienced a 6% increase. From January to November, the Port of Guangzhou handled 570 million tons of cargo and 20 million TEU, up to 5.2% and 7.5%, respectively, compared to the previous year. This shows the growth rate of tourists through the airport has increased (chinadaily.com.cn 2019).

Meanwhile, the Ministry of Tourism, through the Deputy of the Ministry of Tourism's International Tourism Development, I Gde Pitana, since 2015 stated that China is still the main target of pro-motion because of its enormous potential with the growth of foreign tourists reaching 11% per year, even to Indonesia whose growth is up to 20% (kompas.com 2015).

The analysis used in this study is a quantitative analysis which is commonly used is a statistical analysis and is described in making conclusions. This technique is called descriptive statistics. Descriptive statistical analysis is a statistic used to analyze data by describing or elaborating data that has been collected as it is without intending to make conclusions that apply to the public or generalizations. This analysis is only in the form of accumulation of basic data in the form of a description only in the sense of not seeking or explaining relations, testing hypotheses, making predictions, or drawing conclusions.

4 STUDY RESULTS

4.1 *Participants*

After the questionnaire was distributed to 105 participants, it was found that the number of men was 39 or 37.1% and female participants were 66 or 62.9%.

It can be seen that the Chinese people who went on vacation in Bali are actually the upper middle-class people who had high purchasing power, so that it can be understood if the quality of the prices and the services and products that they get actually did not satisfy them. They argue that low prices are indeed still attracting their attention. The data above shows that the average respondent's income is 5,000–10,000 yuan, which states that the Guangzhou community had fairly good income level and fairly good purchasing power. Education Level Participants, the characteristics of tourists based on the level of education, the most tourists with a level of education from undergraduate to doctoral was 48 people or 45.7% and the least were tourists who did not have an educational background, in this case housewives above 45 years who went with their husbands or children for vacation and children under 5 years old who traveled with their parents.

4.2 *Participant's perceptions*

Participants' perception was of the price of Bali tour packages that are sold by travel agents in their original place (China). It can be seen that most of the Chinese tourists are of the opinion that the tour packages being sold are cheap. It is proven by the number of participants at 63.8% or at 67 people.

The most characteristics of tourists were tourists with jobs as special professions (lecturers, teachers, photographers, services, etc.) as many as 33.4% and civil servants by 22%. Most of the age is dominated by tourists aged 25 to 55 years. This age category was included the group of adults and they have permanent jobs, such as work as employees in a company or entrepreneur so they could travel to Bali. Furthermore, students also occurred with a very significant number of 22%. This category was classified as young, unmarried tourists where they traveled with friends or their spouses and even by school groups who were trying to find new experiences.

Perception about the attractiveness of Bali tour packages was more interesting when compared to domestic tour packages in China. Different perceptions regarding the ease of getting Bali tour packages in travel agents in China emerged from 105 participants who were sampled in this study, this is indicated by 46 people or 43.81% who answered strongly agree, 37 people or 35.24% who answered agreed, 15 people or 14.29% who answered doubtfully, and the remaining 7 people or 6.67% who answered disagree.

Furthermore, participants who were sampled in this study had different perceptions about the price of Bali tour packages that were very cheap. This was indicated by 9 people or 8.57% who answered strongly agreed, 45 people or 42.56% who answered agreed, 34 people or 32.38% who answered doubtfully, and the remaining 15 people or 14.29% who answered disagreed and 2 people or 1.90% who answered strongly disagree.

The participants who were sampled in this study had different perceptions that to buy a Bali tour package, participants had to save for less than one year. This is indicated by 5 people or 4.76% who answered strongly agree, 22 people or 20.95% who answered agreed, 26 people or 24.76% who answered doubtful, and the remaining 26 people or 24.76% who answered disagreed and 26 people or 24.76% who answered strongly disagree.

Based on answers about whether or not the price of a Bali tour package is affordable for participants, that can be seen that the actual score (total score) obtained from participants' answers to the affordability of Bali tour package prices was 2,203 and the highest possible score was $5 \times 6 \times 105 = 3150$. Based on these results, the tendency variable where the maximum score for each questionnaire is 5 or 100% and the minimum score is 1 or 20%. The distance between adjacent scores was one fifth of the difference between the maximum value with a minimum value or equal to 16% of the maximum value of 100%. Based on this, the percentage score intervals obtained for each category followed:

Thus, for the variable perception of affordable or not the price of a Bali tour package for Chinese tourists, the actual score (total score obtained) is 2,203 and the highest possible score (ideal score) is $5 \times 6 \times 105 = 3150$ percentage of the actual score compared to the ideal score that is 69.93%. It can be seen that the percentage of scores obtained is quite good criteria, regarding the suitability between price and quality/taste.

Of the 105 participants who were sampled in this study, they had different perceptions that aviation transportation facilities were in accordance with the Bali tour packages offered. This is indicated by 22 people or 20.95% who answered strongly agree, 50 people or 47.62% who answered

Table 3. Classification of percentage interval.

No	Classification	
	Score	Category
1	88.00–100	Very Good
2	71.00–87.99	Good
3	54.00–70.99	Fair
4	37.00–53.99	Not Good
5	20.00–36.99	Bad

Table 4. Distribution of respondents' answers about the suitability of price and quality.

	1		2		3		4		5		6		ΣSigma	
Score	F	%	F	%	F	%	F	%	F	%	F	%	ΣF	ΣF×S
1	1	0,95	9	8,57	11	10,48	1	0,95	2	1,9	6	5,71	30	30
2	2	01.90	40	38,1	42	40	10	9,52	5	4,76	17	16,19	116	232
3	30	28,57	17	16,19	16	15,24	30	28,57	22	20,95	41	39,05	156	468
4	50	47,62	32	30,48	30	28,57	44	41,9	38	36,19	34	32,38	228	912
5	22	20,95	7	6,67	6	5,71	20	19,05	38	36,19	7	6,67	100	500
Total	105	98	105	100	105	100	105	99,04	105	99,99	105	100	630	2142

agreed, 30 people or 28.57%who answered doubtfully, and the remaining 2 people or 1.90% who answered disagreed and 1 person or 0.95% who answered strongly disagree.

Regarding whether the attractions visited in the itinerary are in accordance with the hopes and desires, this is indicated by 6 people or 5.71% who answered strongly agree, 30 people or 28.57% who answered agreed, 16 people or 15.24% who answered doubtful, 42 people or 40.00% who answered disagree, and 11 people or 10.48% who answered strongly disagree.

Participants had different perceptions that the dining facilities included in the Bali tour package matched the participants' tastes. This is indicated by 20 people or 19.05% who answered strongly agree, 44 people or 41.90% who answered agreed, 30 people or 28.57% who answered doubtfully, 10 people or 9.52% who answered disagreed, and 11 people or 0.95% who answered strongly disagree.

The participants had different perceptions that the tour guide who accompanied the tourists was very interesting and provided the information they needed. This is indicated by 38 people or 36.19% who answered strongly agree, 38 people or 36.19% who answered agreed, 22 people or 20.95% who answered doubtfully, 5 people or 4.76% who answered disagreed, and 2 people or 1.90% who answered strongly disagree.

Accommodation during the tour in Bali in accordance with the expectations of participants is indicated by 7 people or 6.67% who answered strongly agree, 34 people or 32.38% who answered agreed, 41 people or 39.05% who answered doubtful, 17 people or 16, 19% who answered disagree, and 6 people or 5.71% who answered strongly disagree.

Based on the respondent's answer to the Conformity between price and quality/taste, the respondent's answer can be distributed as follows: the actual score (total score) obtained from participants' answers to price competition is 2,142 and the highest possible score is $5 \times 6 \times 105 = 3150$. The actual percentage score compared to the ideal score is 68%. The percentage of scores obtained is quite good.

For price competition, the 105 participants who were sampled in this study have different perceptions of the price of Bali tour packages cheaper than domestic China tourism. This is indicated

Table 5. Distribution of participants' answers about price competition.

	Statement									
	1		2		3		4		Σ	
Score	F	%	F	%	F	%	F	%	ΣF	ΣFxS
1	2	1,9	1	0,95	1	0,95	2	1,9	6	6
2	8	7,62	11	10.48	21	20	24	22,86	64	128
3	16	15,24	21	20	14	13,33	36	34,29	87	261
4	42	40	46	43,81	56	53,33	38	36,19	182	728
5	37	35,24	26	24,76	13	12,38	5	4,76	81	405
Total	105	100	105	100	105	100	105	100	420	1528

by 37 people or 35.24% who answered strongly agree, 42 people or 40.00% who answered agreed, 16 people or 15.24% who answered doubtfully, 8 people or 7.62% who answered disagreed, and 2 people or 1.90% who answered strongly disagree.

Bali tour packages are cheaper when compared to several destinations in the Southeast Asian country from 105 participants who were sampled in this study, indicated by 13 people or 12.38% who answered strongly agree, 56 people or 53.33% who answered agreed, 14 people or 13.33% who answered doubtfully, 21 people or 20.00% who answered disagreed, and 1 person or 0.95% who answered strongly disagreed.

Bali tour packages are cheaper when compared to some other destinations in Indonesia, from 105, indicated that 5 people or 4.76% who answered strongly agree, 38 people or 36.19% who answered agreed, 36 people or 34.29% who answered doubtful, 24 people or 22.86% who answered disagree, and 2 people or 1.90% who answered strongly disagree.

Based on answers about price competition according to participants, it can be seen that the actual score of (total score) obtained from participants' answers to price competition was 1,528 and the highest possible score was $5 \times 4 \times 105 = 2100$. The actual score percentage compared to the ideal score was 72,76%, so the percentage of scores obtained were in good criteria.

Regarding the suitability between price and portion/service, that the number of days in the tour package you bought was in accordance with the price you paid, then the 105 participants who were sampled in this study had different perceptions about the number of days in the tour package that the respondent bought according to the price you pay. This was indicated by 12 people or 11.43% who answered strongly agree, 54 people or 51.43% who answered agreed, 30 people or 28.57% who answered doubtfully, 7 people or 6.67% who answered disagreed, and 2 people or 1.90% who answered strongly disagree. Whereas the length of your stay during your stay in Bali is in accordance with the price you paid when you purchased a tour package in Bali, indicated by 5 people or 4.76% who answered strongly agreed, 35 people or 33.33% who answered agreed, 32 people or 30.48% who answered doubtfully, 28 people or 26.67% who answered disagree, and 5 people or 4.76% who answered strongly disagree.

For transportation facilities that were used according to the price you pay, indicated by 11 people or 10.48% who answered strongly agree, 58 people or 55.24% who answered agreed, 26 people or 24.76% who answered doubtful, 7 people or 6.67% who answered disagree, and 3 people or 2.86% who answered strongly disagree.

The number of attractions you visit in accordance with the price offered in the tour package was indicated by 3 people or 2.86% who answered strongly agree, 44 people or 41.90% who answered agreed, 42 people or 40.00% who answered doubtful, 12 people or 11.43% who answered disagreed, and 4 people or 3.81% who answered strongly disagree. As for the amount of consumption facilities provided in accordance with the price of the tour package bought from the 105 participants who were sampled in this study, it was shown that 3 people or 2.86% who answered strongly agree, 44 people or 41.90% who answered agreed, 38 people or 3619% who answered doubt, 13 people or 12.38% who answered disagree, and 7 people or 6.67% who answered strongly disagree, based on answers about conformity between price and portion/service.

According to participants, the distribution of 'participants' answers, it can be seen that the actual score (total score) obtained from 'participants' answers to price competition is 1,769 and the highest possible score was $5 \times 5 \times 105 = 2625$. The actual percentage score compared to the ideal score was 67.39%. The percentage of scores obtained was in the criteria of quite well.

5 CONCLUSION

This study aimed to explore the perceptions of Chinese tourists toward the price and quality of the tour packages on the island of Bali. Based on the above research, tourists from Guangzhou had the perception that travel packages sold by travel agents in China offered cheap prices. The price of this package included the price of the aircraft, which usually is chartered flight, which adjusted to the capacity or quota of passengers. The service was also tailored from the menu to other types of services that passengers received. In the package price, hotel price also adjusted in terms of service, room size, and classification. The price of a cheap package was not entirely in line with the expectations of tourists in terms of products and services. Many argued that a lot of attractions that were not included in the package but that were only taken to souvenir shops and souvenir shopping. Meanwhile, if they wanted to visit other attractions, then they had to pay more or be included in the chosen tour.

This study will give the contribution to the Indonesia's government to explore the extent of perception of Chinese tourists, especially tourists from Guangzhou about the price of Bali tour packages, and furthermore, they can enhance the image of Bali as a leading international tourist destination in Indonesia.

The limitation of this study was having the tourists involved as participants and collected the data resources, since China tourists mostly had difficulties in understanding English and the China government regulation for the platform for technology.

The future study should involve the local community as a host and the tourism local business industries. Lots of local people in Bali can speak Mandarin. Thus, the community will also be helped by the economy of these Guangzhou tourists.

REFERENCES

Cooper, C. And Jackson, S. L. 1997. *Destination Life Cycle: The Isle of The man Case Study*. (ed. Lesly, France): *T*he Earthscan Reader in Sustainable Tourism. UK: Earthscan Publication Limited.

Fennel, David A. 2003. *Ecotourism Policy and Planning*. UK: Cromwell Press.

Ghanimata, F., & Kamal, M. (2012). ANALISIS PENGARUH HARGA, KUALITAS PRODUK, DAN LOKASI TERHADAP KEPUTUSAN PEMBELIAN (Studi pada Pembeli Produk Bandeng Juwana Elrina Semarang).

Gortazar, L. and Martin, C. (199). *Tourism and Sustainable Development: From Theory to Practice – The Island Experience*. Canary Island: Internastional Scientific Council for Island Development (INSULA)

Ismayanti. 2010. *Pengantar Pariwisata (Introduction to Tourism)*. Jakarta: PT Gramedia Widisarana Indonesia.

Kenyon, George N, Kabir C. Sen. 2015. *The Perception of Quality, Mapping Product and Service Quality to Consumer Peceptions*, Lamar University Beaumont, TX USA.

Kotler, Philip (1997), *Manajemen Pemasaran: Analisis Perencanaan, Implementasi dan Pengendalian, Edisi pertama (Marketing Management: Planning, Implementation and Control Analysis, First Edition)*, Jakarta: Erlangga.

Kusumaningrum, Dian. 2009. *Tesis: Persepsi Wisatawan Nusantara Terhadap Daya Tarik Wisata Di Kota Palembang (Thesis: Perception of Opening the Archipelago Against Tourism Attractions in the City of Palembang)*. Magister Kajian Pariwisata. Universitas Gadjah Mada.

Mathieson, A& Wall, G. 1982. *Tourism: Economic, physical and social impacts*, Longman Scientific and *Techical*, Essex.

Mill, Robert Christie, *The Tourism International Business*, Jakarta: Raja Grafika Persada, 2000.
Nuriata. 2014. *Perencanaan dan Pelaksanaan Perjalanan Wisata (Planning and Implementation of Tours and Travels)*. Bandung: Alfabeta
Paturusi, Syamsul Alam. 2008. *Perencanaan Kawasan Pariwisata (Tourism Sites Planning)*. Denpasar: Press UNUD.
Pratama, Mananda, Sudiarta. *Karakteristik, Motivasi Dan Aktivitas Wisatawan Asia Di Kelurahan Ubud. (Characteristics, Motivation and Activities of Asian Tourists in Ubud Village)*, Fakultas Pariwisata. Faculty of Tourism, UNUD Jurnal IPTA Vo. 4 No. 1, 2016, ISSN: 2338-8633
Yoeti, Oka. Edisi Revisi 2006, *Pengantar Ilmu Pariwisata (Tourism Introduction)*, Penerbit Angkasa, Bandung.

Site:
Foreign Affairs Office of Guangzhou Municipal People's Government, *Guangzhou, place for travel, business, and living*, Chinadaily.com, 2019, http://www.eguangzhou.gov.cn/2019-02/28/c_65096.htm
Gong Fu, Zhong, 2019, *Guangzhou, Britannica.com*, https://www.britannica.com/contributor/Zhong-Gong-fu/3312
Kompas.com, 2015, *Wisatawan China Masih Pasar Utama Indonesia, (Chinese Tourists Still Indonesia's Main Market)*, https://travel.kompas.com/read/2015/10/25/082500727.
'Purnamasari, Dinda, 2018, *Wisatawan Cina Makin Gemar Melancong ke Indonesia' (Chinese tourists are increasingly fond of traveling to Indonesia)*, https://tirto.id/wisatawan-cina-makin-gemar-melancong-ke-indonesia-cDZM.
Subadra, PhD, I Nengah, Bali Tourism and Research Cente https://www.balitourismdirectory.com/tourism-studies/bali-tourism-statistics.html
Tirto.id. (2019). Retrieved from Tirto.id

Promoting Creative Tourism: Current Issues in Tourism Research – Kusumah et al. (Eds)
© 2021 Taylor & Francis Group, London, ISBN 978-0-367-55862-8

Transformational leadership, perceived organizational support, and workplace spirituality on employee engagement of restaurant employees in Surabaya

D.C. Widjaja, R.S.T. Putri & D.E. Febrianto
Petra Christian University, Surabaya, Indonesia

ABSTRACT: Employee engagement has been discussed and researched extensively by many researchers since it is believed to provide a company with competitive advantage. An employee who is engaged will be more likely to give his best to the company he belongs to. As the restaurant business has been one of the fastest growing industry around the world, the competition among the restaurant players becomes more intense from time to time. In that case, having employees who are engaged becomes very crucial for a restaurant to gain competitive advantage. There are many factors that can drive employee engagement. Among the many factors, there are three major factors which are interrelated to one another; they are transformational leadership, perceived organizational support, and workplace spirituality. This study is intended to determine the impacts of transformational leadership, perceived organizational support, and workplace spirituality on creating employee engagement. The study is done in a stand-alone restaurant, Madame Chang in Surabaya. This study used a survey method which involved 45 employees altogether as respondents. The measurement scale used was a 7-point Likert Scale. The findings of this study showed that transformational leadership has an indirect effect on employee engagement. In this case, transformational leadership will lead to employee engagement when employees perceive that there is organizational support. Similarly, transformational leadership which creates workplace spirituality is more likely to result in more engaged employees.

Keywords: transformational leadership, perceived organizational support, workplace spirituality, engagement, restaurant employees

1 INTRODUCTION

The restaurant industry has been growing exponentially. The number of restaurants in Surabaya from 2013 to 2016 has shown a significant growth. In a three-year period (2013–2016), the number of restaurants doubled. from 391 to 790 restaurants. (Badan Pusat Statistik Provinsi Jawa Timur, n.d.). One of the keys to success in running a restaurant business and gaining competitive advantage relies in the human resources who deliver the service to the customers. The quality of the service delivered depends on how the service is delivered by the service provider, in this case, the employees. Employees who are engaged with their job and the organization will perform their job well. Engaged employees will have high work intensity and enthusiasm in performing their job (Saks 2011). The report from Gallup's Global Workplace Analytics on South East Asia Nations' Employee Engagement showed that 8% eof mployees in Indonesia are engaged, 77% are not engaged, and 15% are actively disengaged (Ratanjee & Emond 2013). From this report, it indicates that most of the employees in Indonesia are not engaged in their jobs and organizations. Therefore, employee engagement becomes a crucial research topic to discuss further in this study.

People work not only with their hands but also with their hearts. When people work with their hearts, they can find meaningfulness and purpose. With this kind of fulfilment, the workplace

 DOI 10.1201/9781003095484-27

can be a place where people are able to reveal their whole being (Pfeffer 2010; Petchsawang & Duchon 2009). Based on a survey among 41 big companies in Indonesia, 61% agreed that spirituality at work is vital for their companies (Prakoso et al. 2018). In this case, workplace spirituality can lead to more engaged employees. Moreover, other previous studies showed that perceived organizational support has a positive relationship with employee engagement as well as workplace spirituality (Biswas & Bhatnagar 2013; Saks 2006). When employees are aware that the management is willing to lend a hand, they personally feel appreciated, cared for, and acknowledged. As a result, employees find deeper meaning and purpose at work (Chinomona 2011). In this case, leaders can instill in the minds and hearts of the employees toward workplace spirituality and perceived organizational support. Leaders can be role models and enthuse followers to change and perform effectively. Previous studies showed that transformational leadership has a positive effect on employee engagement (Ghadi et al. 2013; Widjaja et al. 2016). The practice of transformational leadership by the leaders in organizations promotes more engaged employees. The transformational leadership also contributes in the emergence of workplace spirituality by creating meaningfulness in employees' jobs and workplaces, communicating with whole group members in order to make the employees engaged with the company's values and sharing the organization's vision and mission with the employees (Mydin, et al. 2018). Moreover, transformational leaders are able to instill the feeling of being cared about in each subordinate and to provide them with necessary support. A previous study by Anggraini (2019) has showed that transformational leadership has a positive influence on perceived organizational support.

Madame Chang is a restaurant with a healthy food concept, where food and drinks are free of monosodium glutamate, additives, and preservatives and are made with high-quality, always fresh, and natural ingredients. The vision of Madame Chang is promoting the concept of eating to be healthy, not eating to be alive. The mission is to develop a restaurant that promotes the importance of a healthy lifestyle, which is reflected in the slogan: "what you eat, is your life." In accomplishing its vision and mission, the leaders in Madame Chang practice leadership that focuses on building the awareness of the employees on the vision and mission by becoming a role model for the employees, encouraging employees to present their ideas and solve problems with new perspectives, and rewards employees who complete their work very well. Interestingly, most of the employees in Madame Change relatively show enthusiasm and pride in their work and willingness to work overtime in order to help each other, which indicates employee engagement. Therefore, the purpose of this study is to investigate the impact of transformational leadership, perceived organizational support, and workplace spirituality on creating employee engagement in the Madame Chang restaurant. The previous studies have investigated the relationships among transformational leadership, perceived organizational support, workplace spirituality, and employee engagement separately. This study is the first study to investigate the relationship among the variables comprehensively. The findings of this study will benefit the relevant industry, especially the restaurant industry, on how to improve employee engagement effectively by practicing transformational leadership.

2 LITERATURE REVIEW

2.1 *Tranformational leadership*

Transformational leaders emphasize the importance of employees' focus and make joint efforts to inspire and motivate employees to achieve common goals, stimulate intellectual development and provide opportunities for meaningful work, show interest in employee personal and professional development, and show behavior which is consistent with organizational values (Bass 1985). Organizations with leaders with transformational leadership styles will be able to provide innovation, inspiration, motivation, and vision that leads employees to achieve better individual achievement and company goals (Anggiani 2019). This leadership focuses on increasing follower involvement with organizational goals (Bass & Avolio 1994) which leads to employee engagement in the long run. The dimensions of transformational leadership are as follows (Bass & Riggio 2006):

- Idealized influence: leaders become an example or role model for followers. Thus, leaders will be admired, respected, and trusted by followers who want to emulate leaders.
- Inspirational motivation: leaders provide meaning and challenges, which motivate and inspire the followers. In this case, the leader fosters team spirit, enthusiasm, and optimism in followers.
- Intellectual stimulation: leaders grow the innovation and creativity of followers by refraining from existing problems and approaching old problems in new ways.
- Individualized consideration: leaders pay attention to each follower's needs for achievement and growth by acting as a coach or mentor.

2.2 *Perceived organizational support*

Perceived organizational support is a form of organizational support which is about the extent to which the organization appreciates and assesses employee contributions, pays attention to employee welfare, listens to employee complaints, pays attention to employee life, and also considers the goals to be achieved in addition that can be trusted as an effort to treat employees fairly. Perceived organizational support is also considered as a guarantee that assistance will be provided by the company when employees need it, to support employee performance and matters relating to difficult situations (Rhoades & Eisenberger 2002). Perceived organizational support can be interpreted as an appreciation of employee contributions, listening to criticisms and suggestions complained of by workers, appreciating the performance or achievement of workers, and meeting the needs of workers (Rhoades et al. 2001). The dimensions of perceived organizational support are:

- Fairness
  Procedural justice concerns the methods used to determine how to distribute resources among employees.
- Supervisor support
  Supervisor support is the extent to which management assesses employee contributions and is aware of the employee's welfare.
- Organizational rewards and job conditions
  Human resource practices show recognition of employee's contributions, which may include recognition, pay, promotions, job security, autonomy, role stressors, and training.

2.3 *Workplace spirituality*

Dehler and Welsh (2003) describe spirituality at work as an in-depth study of a person or point to something bigger than a person. Research conducted by Tepper (2003) defines spirituality as a person's inner feelings when he has a strong motivation to find meaning and purpose of existence. It is an expression that is driven by one's desire to find meaning and purpose in life, one's circumstances have a high meaning, living in honesty to produce positive attitudes and relationships, and trust in relationships with others and the desire to get out of your comfort zone in order to channel contributions to society as a whole (Neck & Milliman 1994). Based on fundamental research conducted by Duchon & Plowman (2005); Milliman et al. (2003) found the dimensions of the workplace spirituality as:

- Sense of Community
  Harmonious feelings obtained by a person in working through workers' togetherness in the organization, which creates a sense of kinship.
- Meaningful Work
  A person's feelings where work has a meaning in a person's life.
- Value Alignment
  Feelings where a person felt the job which had been done has a greater purpose beyond thoughts and someone needs to contribute to the organization.

2.4 *Employee engagement*

Engagement is associated with focusing the employee on an organization's success where the employee works. According to Kahn (1990), engagement is the self-utilization of an employee in an organization to play a role in work; in its involvement, a person employs and expresses themselves physically, cognitively, and feeling while performing a work role. The dimensions of employee engagement are (Milliman et al. 2018):

- Physical-energetic (vigor)
  An experience where someone has the resilience to work at a high level of energy and mentality and able to deal with problems competently.
- Emotional (dedication)
  A person's feelings when participating in the jobs; where someone feels the value of meaningfulness, enthusiasm, and inspiration.
- Cognitive (absorption)
  A person's high concentration in order to work without realizing time elapse.

2.5 *Transformational leadership, perceived organizational support, workplace spirituality and employee engagement*

Leaders who practice transformational leadership are those who attempt to transform how followers perceive their work so that they are able to find meaning in their work. The research conducted by Mydin, et al. (2018) showed that transformational leaders play a transformational role in enhancing workplace spirituality. According to Giacalone and Jurkiewicz (2003), transfromational leaders will make employees view work as a calling and workplace as a place where employees gain greater identity and meaning. Therefore, the following hypothesis was proposed:

Hypothesis 1: Transformational leadership has a positive and significant influence on workplace spirituality.

Based on the research by Mansor et al. (2017) which stated, a transformational leader has a positive influence on followers and could change employees' future views from negative to positive. Moreover, transformational leaders care about followers by motivating them to stay energized and engage in the organization's main goals and missions to create optimism and enthusiasm in followers which means engagement (Datche & Mukulu 2015). Therefore, the following hypothesis was proposed:

Hypothesis 2: Transformational leadership has a positive and significant influence on employee engagement.

Research conducted by Anggiani (2019) shows managers with transformational leadership styles can encourage the establishment of perceived organizational support appreciating employee achievement, utilizing authority for positive results, providing support for employees to give ideas, becoming a great listener for employees, and supporting employee development and promotion. According to Stinglhamber et al. (2015), transformational leaders will guide and consider the needs of followers and provide a supportive environment in order for followers to be able to develop. Therefore, the following hypothesis was proposed:

Hypothesis 3: Transformational leadership has a positive and significant influence on perceived organizational support.

Besides that, based on the research's result by Krishnakumar and Neck (2002), employees who felt the meaning and purpose of the work will feel complete, where feeling completed will increase engagement to the job. When the two variables are integrated, the employee will believe in the purpose and meaning of the work which they have been done and create a deep relationship with the organization thereby increasing employees' engagement to the organization Therefore, the following hypothesis was proposed:

Hypothesis 4: Workplace spirituality has a positive and significant influence on employee engagement.

Saks (2006) stated that perceived organizational support is one of the variables which are able to influence the development of employee engagement. The results of research conducted by Ahmadi, Tavakoli, and Heidary (2014) showed a correlation between perceived organizational support and employee engagement. The higher the level of perceived organizational support, the higher the level of employee engagement. Therefore, the following hypothesis was proposed:

Hypothesis 5: Perceived organizational support has a positive and significant influence on employee engagement.

Based on the results of Chinomona's research (2011), organizational support has a positive and significant relationship to workplace spirituality. As a result, employees will show a high level of workplace spirituality when they get the support and attention given by the organization Therefore, the following hypothesis was proposed:

Hypothesis 6: Perceived organizational support has a positive and significant influence on workplace spirituality.

3 METHODS

The participants for this study were employees of the Madame Chang Restaurant who were eligible to have worked for at least six months and were permanent employees, with a minimum age of 18 years. Out of 47 questionnaires distributed, there were 45 questionnaires ready to be processed, and 2 questionnaires could not be used because the answers were not fully filled. From 45 completed questionnaires, employees were female (100%) with an age limit dominated by the age of 18–28 years, as many as 22 people (49%). It is also known, employees who work in the service and kitchen departments are 18 people each (40%), with the majority of the work positions of employees are 35 non-managerial employees (78%), and the employee's service period is more than equal to four years for as many as 15 people (33%). To assess employee engagement, we used the 10 items selected from Milliman et al. (2018). To assess transformational leadership, we used 12 items from Vera & Crossan (2004) to create a measure of transformational leadership. To measure employee perceptions of workplace spirituality, we used 12 items from Milliman et al. (2018). To measure employee perceptions of perceived organizational support, we used six items from Rhoades et al. (2001). We used the Likert method of scoring in which participants responded using a 7-point Likert scale (1 = "strongly disagree", 7 = "strongly agree"). The study used Partial Least Square (PLS) to analyze the data and test the proposed research hypotheses.

4 RESULTS AND DISCUSSION

4.1 *Path analysis*

The results of the path analysis as seen in Figure 1 showed that all the loading factor values have good validity as they are all >0.5 while the Average Variance Extracted (AVE) is greater than 0.5. Therefore, it can be concluded that the model has good convergent validity. Moreover, the discriminant validity is related to the principle which the measures of different constructs should not be highly correlated. The discriminant validity was also good as the cross loading which has a value of greater than 0.5 and the value of the indicators of these variables for the indicator has the greatest

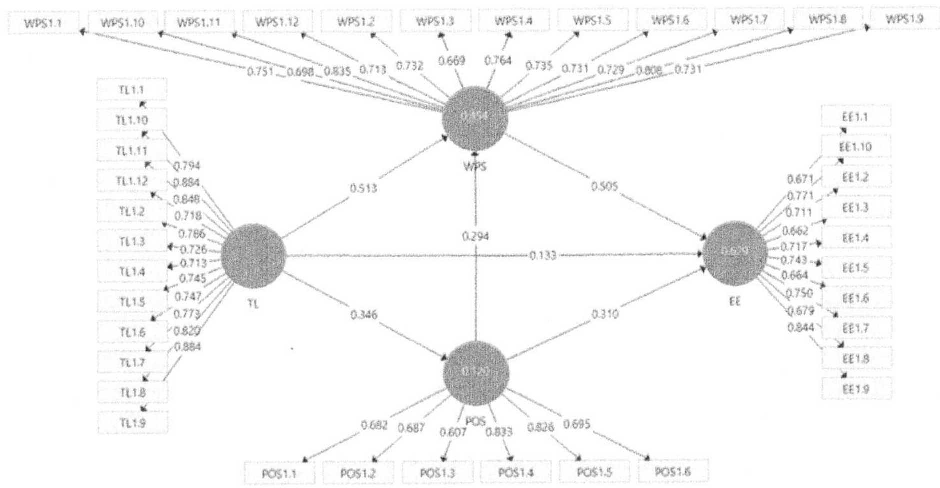

Figure 1. Path analysis.

Table 1. Mean, STDEV, T-statistic.

	Original Sample (O)	Sample Mean (M)	Standard Deviation (STDEV)	T-Statistics (|O|STDEV|)	Remarks
TL→WPS	0,513	0,526	0,126	4,070	Positive & significant
TL→EE	0,133	0,127	0,171	0,781	Positive & insignificant
TL→POS	0,346	0,378	0,134	2,594	Positive & significant
WPS→EE	0,505	0,521	0,138	3,659	Positive & significant
POS→EE	0,310	0,306	0,108	2,871	Positive & significant
POS→WPS	0,294	0,300	0,144	2,045	Positive & significant

value, compared with the value of indicators with other variable while the composite reliability was also good as the Cronbach's Alpha and Composite Reliability's values were above 0.7. The Q-square value was 0.83231248 (83.23%) which shows the structural model structured to explain transformational leadership variables, perceived organizational support, workplace spirituality, and employee engagement at the Madame Chang Restaurant is proven to be good

4.2 *Hypothesis testing*

Hypothesis testing in PLS is done by t-test analysis. The hypothesis can be accepted if the t-value >1.96. In Smart PLS 3.0, the t-test is done by doing the bootstrapping process. Bootstrapping evaluation results obtained Mean, STDEV, and T-statistics are as follows in Table 1.

From Table 1, it can be concluded that all hypotheses were accepted except hypothesis 2 where there was no direct effect of transformational leadership on employee engagement.

From Table 2, it can be seen that there were two significant indirect effects where transformational leadership can impact employee engagement by means of impacting perceived organizational support or workplace spirituality. However, the impact will be stronger when the mediating variable is workplace spirituality. In this case, when leaders give a new perspective on things that employees do not understand and gives personal attention to employees who do not get much attention, employees will feel that the organization is willing to help when employees are in need. As a result, the employees will pay full attention to the work as they become more engaged.

Table 2. Indirect effects.

	Original Sample (O)	Sample Mean (M)	Standard Deviation (STDEV)	T-Statistics (|O|STDEV|)	Remarks
TL→POS→EE	0,107	0,112	0,050	2,135	Positive & significant
TL→WPS→EE	0,259	0,277	0,108	2,409	Positive & significant

5 CONCLUSION AND IMPLICATION

This research demonstrated support for a model which displays the direct effects as well as the indirect effects among the variables. This study found a positive and significant influence of transformational leadership on workplace spirituality as well on perceived organizational support at the Madame Chang Restaurant in Surabaya. In contrast, transformational leadership has a positive but not significant influence on employee engagement at the Madame Chang Restaurant in Surabaya. However, the effect of transformational leadership on employee engagement will be accomplised when there is a mediating role of perceived organizational support or workplace spirituality. As a result, the most effective mediation is through workplace spirituality. On the other hand, we also found workplace spirituality has a positive and significant influence on employee engagement on Madame Chang Restaurant Surabaya's employees which is shown from how the employees perceived values in working resulting in engagement. In addition, we found a positive and significant influence of perceived organizational support on employee engagement as well on workplace spirituality at the Madame Chang Restaurant in Surabaya.

The findings of the study have contributed to the extensive study on employee engagement where leaders especially in the restaurant industry are recommended to practice transformational leadership as an attempt in improving employee engagement. In order to improve employee engagement, leaders are supposed to offer new perspectives to followers and provide personal attention to employees so that they will experience more support from the organization and believe in the organization's conscience. As a result, employees will be more engaged with their work and the organization by giving their full attention at their work assignment.

REFERENCES

Ahmadi, S. A., Tavakoli, S., & Heidary, P. P. 2014. Perceived organizational support and employee engagement. International Journal of Information Technology and Management Studies, 1(1): 54–66.

Anggiani, S. 2019. Effect of transformational leadership on employee creativity: Perceived organizational support mediator (study empiric at five-star hotels in Jakarta). International Journal of Social Sciences, 4(3): 1862–1875.

Badan Pusat Statistik Provinsi Jawa Timur. (n.d.). Jumlah restoran/rumah makan menurut kabupaten/kota di Provinsi Jawa Timur 2013-2016. Retrieved June 18, 2019, from https://jatim.bps.go.id/dynamictable/2017/10/17/137/jumlah-restoran-rumah-ma%20kan-menurut-kabupaten-kota-di-provinsi-jawa-timur-2013–2016.html

Bass, B. M. 1985. Leadership and performance beyond expectations. New York: The Free Press.

Bass, B. M. & Avolio, B. J. 1994. Improving organizational effectiveness through transformational leadership. Thousand Oaks, CA: Sage Publications.

Bass, B. M., & Riggio, E. R. 2006. Transformational leadership. Mahwah, NJ: Lawrence Erlbaum Associates.

Biswas, S. & Bhatnagar, J. 2013. Mediator analysis of employee engagement: Role of perceived organizational support, P-O fit, organizational commitment and job satisfaction. Journal for Decision Makers, 38(1): 27–40.

Chinomona, R. 2011. The impact of organizational support on work spirituality, organizational citizenship behaviour and job performance: The case of Zimbabwe's small and medium enterprises (SME) sector. Journal of Business Management, 6(36): 10003–10014.

Datche, A. E. & Mukulu, E. 2015. The effects of transformational leadership on employee engagement: A survey of civil service in Kenya. Issues in Business Management and Economics, 3(1): 9–16.

Dehler, G. & Welsh, M. 2003. The experience of work: Spirituality and the new workplace. In R.A. Giacalone & C. L. Jurkiewicz (eds). Handbook of workplace spirituality and organizational performance. Armonk, NY: M. E. Sharpe.

Duchon, D. & Plowman, D. A. 2005. Nurturing the spirit at work: Impact on work unit performance. The Leadership Quarterly, 16(5): 807–833.

Ghadi, M. Y., Fernando, M. & Caputi, P. 2013. Transformational leadership and work engagement: The mediating effect of meaning in work. Leadership and Organizational Development Journal, 34(6):532–550

Giacalone, R. A. & Jurkiewicz, C. L. 2010. Handbook of workplace spirituality and organizational performance. Armonk, NY: M. E. Sharpe.

Kahn, W. A. 1990. Psychological conditions of personal engagement and disengagement at work. The Academy of Management Journal, 33(4): 692–724.

Krishnakumar, S. & Neck, C. 2002. The "what", "why" and "how" of spirituality in the workplace. Journal of Managerial Psychology, 17(3): 153–164.

Mansor, Z.D., Mun, C.P., Farhana, B.S.N., Nasuha, W.A., & Tarmizi, W.M. 2017. Influence of transformational leadership style on employee engagement among Generation Y. International Journal of Economics and Management Engineering, 11(1): 161–165.

Milliman, J., Czaplewski, A.J., & Ferguson, J. 2003. Workplace spirituality and employee work attitudes: An exploratory empirical assessment. Journal of Organizational Change Management, 16(4): 426–447.

Milliman, J., Gatling, A., & Kim, J. 2018. The effect of workplace spirituality on hospitality employee engagement, intention to stay, and service delivery. Journal of Hospitality and Tourism Management, 35: 56–65.

Mydin, et al. 2018. Synergising organisational magnificent ambience: The role of transformational leaders to cherish workplace spirituality. Global Business and Management Research: An International Journal, 10(1): 344–355.

Neck, C. & Milliman, J. F. 1994. Thought self-leadership: Finding spiritual fulfilment in organizational life. Journal of Managerial Psychology, 9(6): 9–16.

Petchsawang, P. & Duchon, D. 2009. Measuring workplace spirituality in an Asian context. Human Resource Development International, 12(4): 459–468.

Pfeffer, J. 2003. Business and spirit: Management practices that sustain values. In Giacalone, R.A. & Jurkiewicz, C.L. (eds). The Handbook of Workplace Spirituality and Organizational Performance. Armonk, NY: M.E. Sharpe.

Pfeffer, J. 2010. Building sustainable organizations: The human factor. The Academy of Management Perspectives, 24(1): 34–45.

Prakoso, A. R., Susilo, H., & Aini, E. K. 2018. Pengaruh spiritualitas di tempat kerja (workplace spirituality) terhadap komitmen organisasional (studi pada karyawan PT. Bank Bri Syariah kantor cabang Malang Soekarno Hatta). Jurnal Administrasi Bisnis, 65(1):1–8.

Ratanjee, V. & Emond, L. 2013. Why Indonesia must engage younger workers. Retrieved June 17, 2019, from https://news.gallup.com/businessjournal/166280/why-indonesia-engage-younger-workers.aspx

Rhoades, L. & Eisenberger, R. 2002. Perceived organizational support: A review of the literature. Journal of Applied Psychology, 87(4): 698–714.

Rhoades, L., Eisenberger, R., & Armeli, S. 2001. Affective commitment to the organization: The contribution of perceived organizational support. Journal of Applied Psychology, 86(5): 825–836.

Saks, A. M. 2006. Antecedents and consequences of employee engagement. Journal of Managerial Psychology, 21(7): 600–619.

Saks, A. M. 2011. Workplace spirituality and employee engagement. Journal of Management, Spirituality & Religion, 8(4): 317–340.

Stinglhamber, F., Marique, G., Caesens, G., Hanin, D., & Zanet, F.D. 2015. The influence of transformational leadership on followers' affective commitment. Career Development International, 20(6): 583–603.

Tepper, B. J. 2003. Organizational citizenship behavior and the spiritual employee. In R. A. Giacalone & C. L. Jurkiewicz (eds). Handbook of workplace spirituality and organizational performance. Armonk, NY: M.E. Sharpe.

Widjaja, D. C., Binuko, M. C., Wibawati, L. 2016. Analisa pengaruh transformational leadership terhadap employee engagement dengan kepuasan karyawan sebagai variabel intervening di Artotel Hotel Surabaya. Jurnal Hospitality dan Manajemen Jasa, 4(1):190–208.

Promoting Creative Tourism: Current Issues in Tourism Research – Kusumah et al. (Eds)
 ISBN 978-0-367-55862-8

Destination personality of Labuan Bajo, Indonesia: Local and foreign tourists' perspectives

C.G. Chandra & S. Thio
Petra Christian University, Surabaya, Indonesia

ABSTRACT: The purpose of this study is to identify the personality of Labuan Bajo and then compare the personality perceptions between local and foreign visitors. Labuan Bajo is one of the priority destinations designated by the Indonesian government located in East Nusa Tenggara. This paper adopted the brand personality construct from Aaker (1997), which has five basic dimensions, namely sincerity, excitement, competence, sophistication, and ruggedness. A total of 200 valid responses were obtained in this study and then analyzed using descriptive statistic. Independent t-test was also employed to unravel significant differences in local and foreign perceptions about the destination. The findings revealed that both local and foreign respondents perceived Labuan Bajo as ruggedness, sincerity, and excitement. Local visitors more portrayed the destination with ruggedness personality, while foreign visitors were on sincerity. The dimension personality of competence was perceived to be the least by the two cohorts. Among the five dimensions of destination personality, the two groups were significantly different particularly in their perception of sincerity, excitement, and sophistication. The results of the study may assist destination providers building appropriate branding and positioning strategies for Labuan Bajo.

Keywords: destination personality, local tourists, foreign tourists, Labuan Bajo

1 INTRODUCTION

Tourism is one of the growing industries in Indonesia in recent years and has become the country's source of foreign-currency income. Indonesia has an extraordinary natural resource; thus, the government has been confident to set a target to attract 20 million tourists by 2019 from almost 14 million in 2017 (Ollivaud & Haxton 2019). Indonesian government has prioritized four destinations to be "New Bali," and one of them is Labuan Bajo. Labuan Bajo is the capital city of West Manggarai Regency with a strategic geographical position in the west of Flores Island, East Nusa Tenggara Province. The city of Labuan Bajo itself is surrounded by small islands with sea waters and coastal views, which is very popular by tourists. One of the advantages of Labuan Bajo is the existence of the Komodo National Park, which was inaugurated as a finalist of the Seven New Natural Wonders by UNESCO in 1986. This is because the largest Komodo in the world is only found in the Komodo National Park, surrounded by remarkable natural beauty (Remmer 2017).

As one of the prioritized destinations in Indonesia, government should put some attempts to attract more visitors by strengthening its destination branding to create a positive image of Labuan Bajo. Brands can be considered as a reflection of the quality and popularity of a product or service so that destination branding reflects the quality and popularity of a tourist destination (Bilim & Bilim 2014). Dickinger & Lalicic (2015) stated that destination personality is a form of branding strategy, which is a concept originated from the development of the brand personality theory proposed by Aaker (1997), which is a set of human characteristics associated with a brand. According to Aaker (1997), brand personality can be described into five main characters, namely, sincerity,

 DOI 10.1201/9781003095484-28

excitement, competence, sophistication, and ruggedness. The assessment of the destination personality is strongly influenced by the character of individuals (Lee et al. 2015). Individual perceptions in assessing the destination personality are mainly affected by how the local government develops the concept of personality of a tourist site.

Pong and Noor (2015) define destination personality as a set of human characteristics associated with a destination. Destination personality is also defined as a series of characters chosen to communicate with tourists about destinations (Suleman et al. 2016). Destinations can be linked to human nature and personality and are considered as independent factors for tourists (Huong & Huy 2014). Aaker (1997) suggests that personality traits can be directly linked to destinations through several factors, such as infrastructure, hotels, hotel and restaurant staff, and residents. Destination personalities can be built indirectly through marketing programs such as communication strategies, pricing strategies, and infrastructure development. Destination personality is a metaphor that is appropriate to understand perceptions of someone who will visit a tourist destination and to uniquely manage the identity of the destination. In addition, identifying perception of destination personality may emotionally connect tourists to the destination (Kim & Stepchenkova 2017).

Destination personality is a character that is assessed subjectively by an individual or a group that has a similar background. Stylidis et al. (2014) in their study added that in the assessment of destination personality, attributes and images of a tourist destination may differ according to domestic and foreign tourists. Ayyildiz and Turna (2013) with a similar study also showed that there were differences in the assessment of the attributes and images of tourist destinations from tourists from different countries. Lee et al. (2015) also examined the destination personalities of various tourist destinations in various countries. The results showed that even though the tourist destinations studied were similar (natural attractions and city views), the results can be different. Of the three selected countries (China, the United States, and France), all have different personality characteristics. Dickinger and Lalicic (2015) in their study also examined the destination personality of tourist destinations in Vienna, Austria. The results of the study found that the most prominent character of the destination personality in Vienna, Austria, is the dimension of sincerity.

Even though several studies have been conducted to identify destination personality of several tourist destinations, such as Istanbul (Unurlu & Küçükkancabas 2013), Vienna (Dickinger & Lalicic 2015), China, the United States, France (Lee et al. 2015), Bunaken National Park and Wakatobi (Suleman et al. 2016), most studies focused on destinations in big and popular cities. Thus, this study attempted to identify destination personality of Labuan Bajo as one of the prioritized destinations selected by the Indonesia government. Local and foreign tourists' perspectives are also examined as highlighted by Stylidis et al. (2014) and Ayyildiz and Turna (2013), that domestic and foreign visitors might have different views on how they perceive personality of a destination. The result of this study may provide the local government as well as destination marketers a better understanding how tourists portray Labuan Bajo and then provide marketing strategies to attract more local and foreign tourists to visit Labuan Bajo. As stated by Sahin and Baloglu (2011), the identifying personality of a destination from the perception of visitors will help destination providers capture personality traits to be highlighted into their marketing efforts to enhance tourists' experience and increase their satisfaction.

2 METHOD

This study adopted brand personality construct from Aaker (1997) with five main dimensions, namely, sincerity, excitement, competence, sophistication, and ruggedness to identify destination personality of Labuan Bajo using 42 indicators on a five-Likert scale. Respondents were asked to indicate the extent to which each item within the five dimensions of destination personality is perceived to be suitable for Labuan Bajo ranging from 1 (totally disagree) to 5 (totally agree). As all items in the questionnaires were adopted from Aaker (1997) and then translated into Indonesia, a pilot test was conducted with a small group of 30 to evaluate the reliability and internal consistency of destination personality attributes. Cronbach alpha value is generally accepted with a scale above 0.7 (DeVellis 2003). The results revealed that the alpha coefficients for all personality traits were ranging from 0.723 to 0.787 above the minimum value of 0.7. Therefore, all the dimensions of destination personality developed by Aaker (1997) were applicable to the Indonesia context as the results revealed a satisfactory level of internal consistency.

Quantitative study with descriptive and comparative approach were employed in this study using non-probability sampling design with convenience method. Domestic and foreign visitors who were visiting Labuan Bajo were chosen for sampling purposes. A total of 200 self-administered questionnaires were collected in November 2019 at SkyBajo Hotel in Labuan Bajo with 100 questionnaires for each domestic and foreign visitor. Mean and standard deviation were utilized to portray the data dispersion and to identify central tendency of each personality perceived by the two cohorts. The result of descriptive statistics from this study can be used to identify the main personality of Labuan Bajo perceived by local and foreign travelers. Moreover, statistical comparison was also employed in this study using independent sample t-test for continuous variables to examine whether there are any significant mean differences of personality perceived by local and international tourists who were visiting Labuan Bajo.

3 RESULTS AND DISCUSSION

3.1 *Profile of the respondents*

The profile of the respondents can be seen in Table 1. The gender of the respondents was 41% male and 59% female for domestic respondents and for foreign respondents were 47% male and 53% female. The majority of both respondents were 28 to 37 years old (55% for locals and 51% for foreigners), followed by the age range of 17–27 years old (36% for locals and 32% for foreigners). Most of the local respondents were students (54%) and private employees/entrepreneurs (63%) for foreign respondents with the educational level of undergraduate degree (46% for locals and 45% for foreigners). Some of the domestic respondents were from Jakarta (40%), and the international respondents mostly were from Europe (65%), who came to Labuan Bajo often with their friends/colleagues. More than half of the two group of respondents were first-time visitors, which accounted for 56% and 88% for locals and foreigner, respectively.

3.2 *Destination personality of Labuan Bajo*

Table 2 shows the findings of means and standard deviation of destination personality perceived by both local and foreign visitors. Among five dimensions of personality, ruggedness (overall Mean = 3.93) was perceived the most by domestic respondents to describe Labuan Bajo, followed by sincerity (3.75), excitement (3.75), sophistication (2.72), and competence (1.88), meaning that Indonesian tourists consider Labuan Bajo as a tough and masculine personality with attractive and genuine natural sceneries. Meanwhile, foreign tourists portray Labuan Bajo more on its sincerity (overall Mean=4.03), followed by ruggedness (4.01), excitement (3.97), sophistication (2.55), and competence (1.80). Sincerity personality was acknowledged the most by foreigners to best describe Labuan Bajo with the perception that the destination was more family-oriented with its real and original natural landscape. Even though the two cohorts had slightly different tendencies in describing Labuan Bajo, overall, they portrayed the destination similarly with the three highest personality of Labuan Bajo as ruggedness, sincerity, and excitement. In other words, Labuan Bajo can be personalized as a person who has strong and attractive outdoorsy activities with beautiful natural landscape. It is unsurprising because the biggest attraction of Labuan Bajo is Komodo National Park, which was declared as a world heritage site by UNESCO in 1991 (UNESCO 2011). In addition, unique breathtaking sceneries are also found in Labuan Bajo, such as amazing 360-degree incredible views from Padar Island, which is the top place to see Labuan Bajo, with four small islands, trekking on Komodo and Rinca Islands, which are

Table 1. Profile of the respondents.

Variable	Locals (n = 100)		Foreigners (n = 100)	
	n	%	n	%
Age range				
17–27 years	36	36	32	32
28–37 years	55	55	51	51
38–48 years	7	7	11	11
49–58 years	2	2	6	6
Gender				
Male	41	41	47	47
Female	59	59	53	53
Occupation				
Students	54	54	19	19
Entrepreneurs	44	44	31	31
Employees	21	21	32	32
Professional (teachers, doctors, lawyers, etc.)	2	2	18	18
Others	2	2	0	0
Educational Level				
Junior/Senior high school	38	38	22	22
Diploma degree	15	15	18	18
Bachelor's degree	46	46	45	45
Post-doctoral degree	2	2	15	15
Origin				
East Java	20	20		
Central Java	9	9		
West Java	7	7		
Jakarta	40	40		
Outside Java	24	24		
Asia			15	15
Europe			65	65
America			2	2
Australia			18	18

home to the world's largest lizard, and pink beach as one of the seven pink beaches in the world (trip101, 2020).

Personality dimension of competence was ranked the lowest among the five dimensions by both locals and foreigners. This might be due to lack of awareness of the local community in disposing of waste, which has been one of the problems encountered by Labuan Bajo until now (Agmasari 2018). Labuan Bajo also has several issues such as lack of a clean water supply, inadequate facilities and services, and waste problems, which is unsurprising to be perceived negative by visitors. These problems must be acknowledged by the stakeholders of Labuan Bajo including local government and proper solutions and actions must be taken seriously to promote Labuan Bajo as one of the prioritized destinations.

The Independent sample t-test was deployed to compare the mean score and to identify whether there is a significant difference with the two groups of tourists (locals versus foreigners) regarding their perception about the personality of Labuan Bajo. As appealed in Table 3, it is interesting to find out that all the mean scores of foreign travelers were higher than domestic travelers, suggesting that foreign travelers perceived that Labuan Bajo has stronger personality compared to domestic travelers. From the results, there were significant differences between the two groups in the dimension personality of sincerity, excitement, and sophistication. But there were no significant differences in their perception about ruggedness and competence personality. Meaning that the two cohorts acknowledge that Labuan Bajo offers rugged and masculine activities such as trekking through the

Table 2. Mean and standard deviation of destination personality of domestic and foreign tourists.

	Locals		Foreigners		
Dimension	Mean	StDev	Mean	StDev	Cronbach's Alpha
Ruggedness	**3.93**	**0.44**	**4.01**	**0.40**	0.787
Tough	4.28	0.72	4.24	0.71	
Rugged	4.22	0.67	4.36	0.71	
Outdoorsy	4.20	0.68	4.35	0.73	
Masculine	4.11	0.63	4.44	0.74	
Western	2.84	0.37	2.68	0.73	
Sincerity	**3.75**	**0.36**	**4.03**	**0.28**	0.723
Honest	4.28	0.72	4.24	0.71	
Down-to-earth	4.23	0.69	4.32	0.72	
Sincere	4.22	0.67	4.36	0.71	
Small-town	4.18	0.75	4.19	0.76	
Cheerful	4.13	0.72	4.27	0.77	
Family-oriented	4.11	0.63	4.44	0.74	
Real	4.07	0.67	4.36	0.78	
Original	4.05	0.79	4.32	0.82	
Friendly	3.86	0.77	4.00	0.87	
Wholesome	3.72	0.78	3.84	0.88	
Sentimental	2.14	0.77	2.00	0.87	
Excitement	**3.75**	**0.36**	**3.97**	**0.43**	0.730
Exciting	4.22	0.67	4.36	0.71	
Daring	3.91	0.66	4.28	0.86	
Young	3.83	0.75	4.03	0.89	
Unique	3.82	0.78	3.95	0.88	
Up-to-date	3.82	0.78	3.95	0.88	
Imaginative	3.79	0.65	4.19	0.91	
Spirited	3.75	0.77	3.90	0.89	
Independent	3.72	0.78	3.84	0.88	
Cool	3.68	0.72	3.91	0.92	
Trendy	3.48	0.59	3.81	0.96	
Contemporary	3.22	0.56	3.44	1.12	
Sophistication	**2.72**	**0.40**	**2.55**	**0.52**	0.752
Charming	3.87	0.67	4.23	0.88	
Good looking	3.70	0.74	3.89	0.90	
Glamorous	2.47	0.66	1.94	1.31	
Upper class	2.35	0.52	1.70	1.03	
Feminine	2.09	0.66	1.72	0.86	
Smooth	1.81	0.76	1.82	0.75	
Competence	**1.88**	**0.40**	**1.80**	**0.51**	0.729
Intelligent	2.30	0.46	1.60	0.92	
Leader	2.23	0.42	1.46	0.84	
Confident	2.22	0.59	1.71	0.91	
Secure	1.81	0.76	1.82	0.75	
Successful	1.81	0.91	2.21	0.65	
Reliable	1.68	0.72	1.77	0.69	
Technical	1.64	0.71	1.78	0.67	
Corporate	1.64	0.78	1.93	0.67	
Hard working	1.56	0.74	1.89	0.63	

forests of Komodo, swimming with mantas, and snorkeling at the pink beach. Additionally, both local and foreign visitors agreed that Labuan Bajo has a low-level competence as perceived by the tourist. As stated by Remmer (2017) that infrastructure in Labuan Bajo is considered poor with the

Table 3. Independent sample t-test results of destination personality.

Dimension	Tourists	Mean	t-value	Sig. (2-tailed)
Ruggedness	Locals	3.93	−1.395	0.164
	Foreigners	4.01		
Sincerity	Locals	3.75	−2.658	0.008*
	Foreigners	4.03		
Excitement	Locals	3.75	−3.905	0.000**
	Foreigners	3.97		
Sophistication	Locals	2.72	2.484	0.014**
	Foreigners	2.55		
Competence	Locals	1.88	1.251	0.212
	Foreigners	1.80		

**represents significant levels <0.01

common problem of power shortages. The water pipes installed by the local government do not function well, and compared to other urban areas in Flores, the roads in Labuan Bajo are not in good conditions. Waste problems also contribute to a negative perception of the destination. These issues should be put into account

4 CONCLUSION

The findings of this study may provide useful insights for destination marketers to have better understanding about Labuan Bajo's personality from the views of domestic and foreign travelers. Among the five dimensions of destination personality, ruggedness, sincerity, and excitement had the highest score in representing Labuan Bajo. Local visitors were more likely to perceive Labuan Bajo with ruggedness personality, while foreign visitors were higher on sincerity. The personality dimension of competence was perceived to be the lowest by the two group visitors. Thus, the local government needs to put more emphasis on the improvement of infrastructure such as roads and transportation in order to improve tourists' experiences during their stay in Labuan Bajo.

From the study, it can be concluded that domestic and foreign tourists have difference perceptions on how they portray the personality of Labuan Bajo. Both cohorts perceived Labuan Bajo significantly with difference in the dimension of sincerity, excitement, and sophistication. Destination personality is crucial to build positive image of a destination. Thus, destination marketers and providers need to put more attempts on how they promote and build positioning strategies for both groups in order to attract more tourists to visit to Labuan Bajo. For local tourists, it might be interesting to offer some activities which are challenging and fun with more outdoorsy attractions, as for foreign visitors are more family-friendly and attractive natural attractions. This paper only discusses the destination personality of Labuan Bajo using a quantitative approach; therefore, qualitative method needs to be considered to explore deeper why and how visitors perceive the personality of Labuan Bajo. Other destinations prioritized by the Indonesian government such as Borobudur, Mandalika, and Toba Lake may be worth to be investigated in order to assist tourism agencies to create a better positive destination image.

REFERENCES

Aaker, J. L. 1997. Dimensions of brand personality. *Journal of Marketing Research*, 34(3): 347–356.

Agmasari. S. March 2018. Sampah, Sumber Masalah di Labuan Bajo (trash, the source of problems in Labuan Bajo). Available at: https://travel.kompas.com/read/2018/03/28/120400627/sampah-sumber-masalah-di-labuan-bajo.

Ayyildiz, H., & Turna, G. B. 2013. Perceived image of Spain and Germany as a tourist destination for Dutch travelers. *Journal of Economics, Business and Management,* 1(1): 85–89.
Bilim, Y. B., & Bilim, M. B. 2014. Does a destination have personality? Personality and image issues of a destination. *Athens Journal of Tourism*, 1(2): 121–134.
DeVellis, R.F. 2003. *Scale development: Theory and applications* (2nd Edn). Thousand Oaks, California: Sage.
Dickinger, A., & Lalicic, L. 2015. An analysis of destination brand personality and emotions: A comparison study. *Information Technology & Tourism*, 15(4): 317–340.
Huong, P. T. L., & Huy, N. M. 2014. A study of destination brand personality for Đà Nẵng tourism. *Journal of Economics Development*, 221: 144–160.
Kim, H., & Stepchenkova, S. 2017. Understanding destination personality through visitors' experience: A cross-cultural perspective. *Journal of Destination Marketing & Management*, 6(4): 416–425.
Lee, J., Soutar, G., & Quintal, V. 2016. Destination personality: Cross-country comparisons. *Journal of Economics, Finance and Administrative Science*, 21: 25–29.
Olliveaud, P., & Haxton, P. 2019. Making the most of tourism in Indonesia to promote sustainable regional development. *OECD Economics Department Working Papers*, 1535: 1–41.
Pong, K.S. & Noor, S. M. 2015. The influence of destination personality on brand image evaluation among archaeological tourists. *Malaysian Journal of Communication*, 31 (1): 133–152.
Remmer, S. 2017. Tourism Impacts in Labuan Bajo. Swisscontact WISATA. Available at: https://www.swisscontact.org/fileadmin/user_upload/HEAD_OFFICE/Pictures/Tourismus_Landing_page/Labuan_Bajo_Impact_Assessment.pdf
Sahin, S. & Baloglu, S. 2011. Brand personality and destination image of Istanbul. *Anatolia-an International Journal of Tourism and Hospitality Research*, 22(1): 69–88.
Stylidis, D., Sit, J., & Biran, A. 2014. An exploratory study of residents' perception of place image. *Journal of Travel Research*, 55(5): 659–674.
Suleman, N. L., Rufaidah, P., & Ariawaty, R. N. 2016. The influence of destination personality and perceived value on destination image in National Park Bunaken and Wakatobi. *International Journal of Scientific & Technology Research*, 5(07): 327–337.
Trip101. 2020. The 14 Things To Do In Labuan Bajo, Flores Island – Updated 2020. Available at: https://trip101.com/article/things-to-do-in-labuan-bajo-flores-island
UNESCO (July, 2011), Ecological sciences for sustainable development. Available at: http://www.unesco.org/new/en/natural-sciences/environment/ecological-sciences/biosphere-reserves/asia-and-the-pacific/indonesia/komodo
Unurlu, Ç., & Küçükkancabas, S. (2013). The effects of destination personality items on destination brand image. *International Conference on Eurasian Economies*, 83–88.

Promoting Creative Tourism: Current Issues in Tourism Research – Kusumah et al. (Eds)

Development of tourist visitor management system in Tajur Kahuripan traditional tourism village

A. Agoes & I.N. Agustiani
Sekolah Tinggi Ilmu Ekonomi Pariwisata Yapari, Bandung, Indonesia

ABSTRACT: Tajur Kahuripan Village began to get tourists from different regions by maintaining the indigenous Sundanese cultural traditions, especially its traditional housing. Tourism has since begun to thrive in this area. Since the inhabitants are indigenous farmers, however, no one has any experience and skill in the tourism sector. Consequently, the tourist management system is not optimized to necessarily ideal level. There is no influx of visits to the village as tourists arrive. Similarly, the flow of reception and placement of guests to any homestay has not yet been developed. Those will contribute to a lack of valuable experience on the tourists' part. The goal of this research is to analyze the model of visitor management as to what should be implemented in the Tajur Kahuripan Village, so that it corresponds to the characteristics of the rural tourism offered. Furthermore, this study is intended to build a model of visitor management for Kampung Tajur tourists in order to achieve an optimal visitor experience. The research method used is a qualitative method and a model development method. The targeted output is the concept of visitor management models for tourists coming to Tajur Kahuripan Village that best suits the characteristic of the village. This visitor management model is expected to enhance the tourists' experience coming to visit this village.

Keywords: Tourist visitor management, Tajur Kahuripan, Traditional Tourism Village, tourism.

1 INTRODUCTION

1.1 *Kampung Tajur in brief*

Kampung Tajur in general is an ordinary village, but the inhabitants agreed to preserve the Sundanese tradition, especially in terms of agricultural activities. The distinctive feature of this village is the maintenance of traditional Sundanese-style houses that are sustainable and truly authentically inhabited by their inhabitants. The daily work of Kampung Tajur residents is as a farmer. However, when this village gets attention to become a tourist attraction, the farmers begin to get used to organizing the tourist arrivals. Even the village government also formed a tourism driving group or Kompepar (Kelompok Penggerak Pariwisata). Farmers already tried to manage visitors facilitated with a locally formed organization active in tourism (Kompepar). The main tourism activity conducted by the people of Kampung Tajur is to rent their traditional houses as a homestay. There are 42 traditional houses ready to serve as a homestay during tourist visit. Not only that, but the host of each traditional house acts as a guide to the guests staying at their homestay. They also will accompany the guests to do local activities such as planting rice, cultivating the fields, trekking, and others (Kemala & Agoes 2020). By far, there are possibly half a dozen farmers who already ventured to be a tour guide or tour organizer.

1.2 *Visitor management system in Kampung Tajur*

A visit to a traditional village would have a high selling value if the visit will provide valuable memories for the arriving tourists. There are occasions when the visit flow modeling and the facilities provided would also be an essential part that can leverage on tourist experience. In addition, the activities offered do need to be well organized so that tourists visiting Kampung Tajur can get more experience. Not only limited to ordinary experience, but high-quality experience is expected to be accomplished. Therefore, visitors to Kampung Tajur should be managed properly in order to give higher quality experience.

Despite these farmers' limited knowledge and competence, quite a few tourist groups have visited the Kampung Tajur. They also seemed very pleased with the visit's viable management. Even so, there are tourist standards that do not seem to have been completely met. The aspects of participation, peace of mind, and education were not fully addressed, in keeping with the visitor experience concept. The visitor experience can therefore be further optimized. The visitor management system still needs to be optimized according to Experience Quality that they wished to achieve.

In examining aspects of the flow of services in tourism, the approach can be used through the concept of service in hospitality. This is because hospitality and tourism are two industrial sectors that have the same platform, especially in the service sector. In fact, the idea of hospitality and tourism also have widely overlapped in the concept of the tourist industry. This can be seen from the relationship between hospitality and tourism, as well as the travel industry as adopted from Piboonrungroj and Pizam (Dragan et al. 2015). For these reasons, an approach from the perspective of hospitality is also used in this study.

In developing the visitor management system in Kampung Tajur, a service model adapted from the Guest Experience Process Model of Services (Pijls et al. 2011) will be used. It's just that the emphasis will be on three aspects of Experience Quality, namely, aspects of involvement, peace of mind, and aspects of education. The visitor's experience is divided into three phases, namely, 1) pre-experience that includes information retrieval to the reservation process; 2) experience, i.e., when arriving at Kampung Tajur and getting services and experience of activities there; 3) post-experience, which will see how tourists are expected to get what they are looking for.

1.3 *Improving the quality of experience*

Visitor management system is intended to improve the quality of experience of the tourists (Daniel 2002; Shackley 2003) coming to Kampung Tajur. Experience quality is measured among others through involvement factors, peace of mind, and educational experience. Involvement can mean the involvement of activities offered by a tourist attraction. Peace of mind can be associated with comfort for the services provided, as well as educational experience related to the material knowledge gained from the visits made (Chen & Chen 2010; Zehrer 2009). The things that need to be managed are, first of all, the structure of the Kampung Tajur itself in its context as a place of attraction for visits. The things that need to be planned include the layout of the area including the functions of the building as needed (reception, waiting room, etc.), then the route or path of visitors when entering the area, and the complete facilities such as signs, toilets, parking lots, bins, etc., as needed (Swarbrooke 2002). In addition to the flow of services to visitors, what needs to be managed is also the experience of visitors. Current experiences have been recognized as special and different economic offers, other than service offerings and product offerings. In its offer, experience offers events or activities that will be memorable for the participants. If in previous economic times the things offered were goods or services, then in the economic era this experience offered was a memory (Pine & Gilmore 2013). Thus, it becomes important to be able to manage tourist visits to the Kampung Tajur in order to build valuable experiences for these tourists. In addition, also from the service aspect, it is also necessary to manage the flow of services of tourists who visit there.

It is also important to note that according to Pizam, what is most important in providing visitors with memorable memories is the quality of their experience, not the quality of service (Loureiro 2014). This can definitely also be a consideration in the development of a Visitor Management Program in Kampung Tajur. Where service aspects are seen to have flaws, particularly given the human resources aspects that support them, another factor that needs to be considered is that natural interaction between tourists and local people in the village community will play a key role in creating a high-quality tourist experience (Kastenholz et al. 2018). In view of the condition of traditional villages that are characterized by small and medium-sized enterprises, it is also important to incorporate a service design in managing the tourists. It is important for enhancing the quality of tourism experience. The goal of the service flow design process is to create a memorable experience for tourists (Zehrer 2009).

1.4 *Research objective*

A number of studies exploring this field, namely the management of visitors to a traditional village, could be used as a reference. One of them is a Kampung Cikadu analysis. In this study, it can be seen that the guests who come to the village are mostly from the luxury resort community in the Tanjung Lesung district. Therefore, the handling of visitors is tailored to the character of the guests who come here. Although retaining the traditional character and places of attraction visited, the facilities given need to be customized to the luxury characteristics of the Tanjung Lesung Resort guests (Agoes et al. 2020). In addition, there is work in the Dago Pojok Village, where visitors are limited to researchers and students. In this village, therefore, the management of visitors is very simple, namely with previous correspondence. A visit to Dago Pojok Village does not have a specific route as it is adjusted to the individual needs of each visitor (Agoes 2015). Some data and information are also obtained from previous research in Kampung Tajur.

This research objective is to establish a single method or model to handle visitors coming to Kampung Tajur as tourists. The program should pay attention to both service design and design experience. Through this study, the focus is on enhancing the quality of the experience of tourists visiting Kampung Tajur throughout the aspect of visitor management. However, this is a preliminary research that needs to be further developed into a working model of a visitor management system in Kampung Tajur.

2 METHOD

The research method applied here is by examining the extent to which the managers of tourist visits in this village have applied visitor management concepts that pay attention to the quality of tourist experiences. In addition, the highlighted aspect is the management of the service. To obtain these data, interviews were conducted with three main informants, namely, from the tourism driving group (Kompepar). These data are then processed by being reduced, sorted, and reviewed by researchers. The analytical tool used is from the researchers themselves based on their understanding and experience. Then triangulation was carried out to experts as well as other literature references. Furthermore, the results of the study will offer an ideal management model to be applied in Kampung Tajur so that the tourist experience can be further improved.

The preparation of this model may not necessarily be an ideal single model, but rather it is referred to as a reference model in which tourism activists in Kampung Tajur can use it as a guidance. Of course, the flow of visits and the flow of services further need to be developed and evaluated according to the needs and development of tourists visiting there. However, with the existence of this guidance, it is expected that the managers of tourist visits in Kampung Tajur can be more focused in building the tourist experience so that what is expected can be achieved.

3 RESEARCH FINDINGS AND DISCUSSION

3.1 *Research findings*

First of all, it was found that the flow of tourist visits is currently taking place from the stage of pre-experience, experience, until after-experience has been run by the manager. Tourists who usually come are families or school children (represented by teachers) and the majority is a group. At the pre-experience stage, potential tourists are searching for information and making reservations. The reservation is currently received by one of the residents of Kampung Tajur, who serves as Chairman of Kompepar. The Kompepar Chair then processed the reservation data by confirming to the families whose traditional homes were used as homestays. Following that, the Chairman of Kompepar will hold a meeting with the heads of the families whose homes have been reserved as homestays. A guest activity plan during their stay in Kampung Tajur will be explained in this briefing. In addition to homestay host operators, a number of workers will also be designated from Kompepar to help coordinate tourist arrivals. Many villagers who are also able to handle tourists visits would be assigned as the tour guide.

On the arrival day, tourists will be directed to assemble at a local government-owned house where remarks are made. Then there are also simple shows that are performed in other locations, namely, in the field near the government-owned house. But often this field is also used as a parking lot. Then the place of activities will become increasingly limited. This is also caused by the lack of supporting facilities for proper reception.

The way of handling incoming tourists which is currently applied by Kampung Tajur residents is still not seen as a cohesive concept of the narrative, such that the three aspects of interaction consistency have not been entirely accomplished. Each activity is still treated as a separate element of daily life. So, the whole activity still feels artificial.

By the aspect of tourist service, the issue of peace of mind has not yet been met. Visitors are often faced with a reception that has not been carefully arranged. There is always uncertainty when visitors first move in and are welcomed by the people. No specific procedures are set by default. Even the guest greeter did not have adequate details about what facilities are offered while guests stay there. And, sometimes, visitors don't know what they will expect when they stay.

The visit also still relies on group reservations so it cannot accept walk-in guest visits. So, if the parking facilities, receptionist, and interpretive service in this village are improved, then even walking in, the guests will be able to have a memorable experience. However, while the services have been improved, there have also been other obstacles. Namely, a shortage of human resources that can be available to welcome visitors at any time, since current human resources depend on the participation of the community who work in the field or at the farm. Therefore, it is necessary to immediately set up a working schedule to assign the community to the picket as a receptionist.

Another issue is that, since there are no visiting packages ready to be sold, any time a visit takes place, the villagers must arrange activities according to the wishes of the group of visitors. Therefore, the entire plot may not be comprehended. That's because visitors don't know what Kampung Tajur can offer, so it can provide a memorable experience. When there are packages of visits planned in such a way, it is expected that the flow of the visit will be able to create the experience of the visitors. When carrying out rice planting activities, it is also important to have a predetermined arrangement so that tourists can follow activities with an interesting narrative context. When you're just trying to plant rice, it's definitely not going to be that interesting, so there's no story you want to create. Herein lies one of the shortcomings in the arrangement of tourist visits to Kampung Tajur.

The key offer itself, which resides in a traditional house, seems to be adequate both in terms of service and experience. Nonetheless, improvements should also be made in terms of services. For example, with regard to the information provided. Often the receptionist's description is difficult for visitors to grasp, because it's' not obvious what's appropriate while they visit. Similarly, from the experience point of view, simple homely activities characteristic of the countryside could be added to bring memories to the visitors who live there.

3.2 *Discussion*

In reality, at the beginning of its growth, tourism in the Kampung Tajur started with the Pasanggrahan Village itself. At present, the starting point for the growth of tourism in the Pasanggrahan Village (Kampung Pasanggrahan) has developed to become more modern. The site of the Village Office is also located here. Although tourism is currently more developed in the Tajur Kampung area, actually

Pasanggrahan Village can also be utilized to become a buffer zone or a supporting area as a facility provider in Kampung Tajur that is experiencing shortages. Some of them are, for example, parking facilities and reception facilities. When this can be established, the area of activity in Kampung Tajur will be very accommodating and not disrupted by parking lots. Likewise, the receptionist will be more comfortable in a more developed area, namely in the Pasanggrahan Kampung area. Even then it is necessary for tourists to walk far enough to reach Kampung Tajur. Yet this walking activity is intended even more to be an attraction and enrich the tourist experience. In addition, the entrance to Kampung Tajur is so small that it is difficult for cars to drive through. As an option, you might be given motorbike services to visitors who have trouble walking too far.

For so many types of visitors on the basis of group reservations, then in Kampung Tajur it is considered that they do not yet need to provide a 'ticket center' because any visitor who arrives has already made a reservation in advance. Nevertheless, a special facility in the form of a 'receptionist' which is guarded by picket officers must definitely be provided for its further development. The role of this receptionist is, among other things, to better accommodate more walk-in guests. It is also presumed that two reception points be made, namely, in the Pasanggrahan Village sector, where they can be used as a parking lot at the same time, and in the Kampung Tajur itself.

Indeed, from the point of view of service, it has been perceived to be sufficiently satisfactory considering the characteristics of the village which is a traditional village. Also, tourists would often have unforgettable memories of the quality of their experience and not of the nature of the service. To carefully obtain that elements of service quality at the present level can still be ruled out without compromising the standards of good service from the point of view of tourism (hospitality). For this reason, the priority in developing a visitor management system in Kampung Tajur is to improve the experience management for the tourists, while the next step is to improve the service quality.

Tourist activities are considered quite interesting in the current handling by farmers. Tourists are encouraged to engage in the everyday lives of rural communities. So far as interest is concerned, the handling of tourists is generally good enough. Tourists feel involved in the daily life of the farming community. Some families have even started to take part in activities since waking up, cooking in the traditional kitchen, going to the rice fields, and bathing in the buffalo. Yet arguably, from another perspective, namely the aspect of education, tourists have not been provided with quality information. In reality, the experience gained can still be improved in quality. That is because tour guides do not yet have strong interpretive skills, so that their role is just as the organizer of a program that takes visitors from one event to another. This results in an event after an event not being incorporated as a cohesive story that gives tourists insight. Even the dramaturgy elements have not been included in the Kampung Tajur Visitor Management. As a result, the quality of the tourist experience is not optimum when visiting this village.

The next aspect is from the "Peace of Mind" viewpoint, where visitors are supposed to have peace of mind. That is, of course, related to the quality of service aspects. Visitors would feel refreshed if the delivery of services can be better coordinated by taking visitors' needs into account. For example, it is important to apply reception procedures that pay attention to the elements of comfort, safety, as well as elements of surprise that can offer excitement to tourists. In fact, regardless of all these elements, the reception of tourists has been smooth so far. However, the most important thing in the tourism industry and hospitality practice is a memorable tourist experience. Thus, merely smooth reception of the arriving guests is not something that has saleable value. It is important to pay attention to those additional elements stated earlier.

4 CONCLUSIONS AND SUGGESTIONS

Based on the explanation above, the proposition of this research is, first, from the aspect of service quality, it is necessary to arrange standard procedures immediately from pre-experience, experience, and also post-experience. This procedure is important so that the reception process no longer depends on just one person (i.e., the Kompepar Chair). Instead, some personnel in the village can take part in the activities of these tourist visits. Standard procedures for this service aspect must be prepared with due regard for the elements of peace of mind, in particular in terms of tourist comfort and safety. In addition, surprise elements can also be designed for the tourists who come, probably using the concept normally applied at a resort. For example, providing a welcome scarf, giving a welcome drink, etc. Entertainment elements must also be designed in such a way that they look more natural and do not feel like a segregated program. Developing a tourist visit flow must be in the form of a single whole narrative, beginning with the arrival of tourists, major events, until the return of tourists. That can be achieved by creating a series of events to complement each other, supported by additional services to bring more value to visitors.

The second proposition is from the aspect of the quality of experience. It is important to immediately design a visiting package that can be selected by the tourists. This visiting package should have considered the aspects of tourists' experience quality. The important thing to notice is that there is an aspect of education that needs to be incorporated into every set of activities. The sequence of activities must also consider how they can be conducted. For example, in the activities of going to the rice fields, guides must wear traditional Sundanese clothes that are usually worn to go to the fields. Even if appropriate, the residents of Kampung Tajur should wear traditional clothes when there are tourists visiting. Tourists can also have the option of wearing traditional clothing during activities in this Kampung Tajur. This would increase the sense of authenticity in the experience of the tourist, so that the quality of the experience can be enhanced. Complete citizen participation can be the key to improving the quality of tourists' experiences in order to create unforgettable memories.

Kampung Tajur is a traditional village in which tourists would understand the limited service and quality of experience they have offered, but when Kampung Tajur decides to engage in tourism, it is unavoidable that aspects of tourism services will have to be implemented by this village. The professional reception process must be established and the activities of the quality program must be developed. If the reception of the tourists is mediocre, then the tourists will not have the amazing experience they've been dreaming of. Since the village is not a tourist spot in nature, then additional preparations have to be made. It might sound ironic, but well-designed rural activities are supposed to make visitors feel more authenticity.

REFERENCES

Agoes, A. (2015). Pengembangan Produk Pariwisata Perdesaan Di Kampung Dago Pojok Bandung. *Jurnal Manajemen Resort Dan Leisure,* 12(1). https://doi.org/10.17509/JUREL.V12I1.1049.

Agoes, A., Edison, E., & Kemala, Z. (2019). Designing Rural Tour Program in Connecting Tourism Village To Resort Tourists, Tanjung Lesung, Banten Indonesia. *ASEAN Journal on Hospitality and Tourism*, 17(1), 12. https://doi.org/10.5614/ajht.2019.17.1.2.

Chen, C. F., & Chen, F. S. (2010). Experience quality, perceived value, satisfaction and behavioral intentions for heritage tourists. *Tourism Management*, 31(1), 29–35. https://doi.org/10.1016/j.tourman.2009.02.008.

Daniel, T. C. (2002). Modelling Visitor Flow from the Visitor Perspective: The Psychology of Landscape Navigation. In A. Arnberger, C. Brandenburg, & A. Muhar (Eds.), *Monitoring and Management of Visitor Flows in Recreational and Protected Areas* (pp. 159–165).

Dragan, D., Kramberger, T., & Topolšek, D. (2015). Supply Chain Integration and Firm Performance in the Tourism Sector. (January), 11–13. Retrieved from http://iclst.fl.uni-mb.si/.

Kastenholz, E., Carneiro, M. J., Marques, C. P., & Loureiro, S. M. C. (2018). The dimensions of rural tourism experience: impacts on arousal, memory, and satisfaction. *Journal of Travel and Tourism Marketing*, 35(2), 189–201. https://doi.org/10.1080/10548408.2017.1350617.

Kemala, Z., & Agoes, A. (2020). Buku Panduan Interpretasi Bilingual Kampung Tajur (Cetakan Pe). Retrieved from https://books.google.co.id/books?hl=en&lr=&id=t4DsDwAAQBAJ&oi=fnd&pg=PP1&ots=M_JvUnUqXe&sig=M5CHUGl6lDqZLtDIREEVX_HxMvA&redir_esc=y#v=onepage&q&f=false.

Loureiro, S. M. C. (2014). The role of the rural tourism experience economy in place attachment and behavioral intentions. *International Journal of Hospitality Management*, 40, 1–9. https://doi.org/10.1016/j.ijhm.2014.02.010.

Pine, B. J., & Gilmore, J. H. (2013). *The experience economy: Past, present and future*. Handbook on the Experience Economy, (January 2013), 21–44. https://doi.org/10.4337/9781781004227.00007.

Pijls, Schreiber, & Marle. (2011). *Capturing The Guest Experience In Hotels. Phase Two: Exploratory Study On The Sensory Characteristics Of A Comfortable And Inviting Ambience*. Netherlands: Saxion University of Applied Sciences.

Shackley, M. (Ed.). (2003). *Visitor Management (Case Studies from World Heritage Sites) (Digital Pr)*. Burlington, MA: Butterworth-Heinemam.

Swarbrooke, J. (2002). *The Development and Management of Visitor Attractions* (Second Edi; J. Swarbrooke, Ed.). https://doi.org/10.1017/CBO9781107415324.004.

Zehrer, A. (2009). Service experience and service design: Concepts and application in tourism SMEs. Managing Service Quality, 19(3), 332–349. https://doi.org/10.1108/09604520910955339.

Promoting Creative Tourism: Current Issues in Tourism Research – Kusumah et al. (Eds)
 ISBN 978-0-367-55862-8

Projected destination image on Instagram amidst a pandemic: A visual content analysis of Indonesian National DMO

W.N. Wan Noordin
Universiti Teknologi Mara, Shah Alam, Malaysia

V. Sukmayadi & R.M. Wirakusuma
Universitas Pendidikan Indonesia, Bandung, Indonesia

ABSTRACT: The Covid-19 pandemic that hit globally in 2020 has profoundly impacted the tourism industry. Destination marketing organizations (DMO) have adapted their strategy by shifting more into visual content on social media as part of their survival mode. This paper attempts to explore how the Indonesian official DMO constructed its social media (Instagram) posts to promote tourism destinations while coping with the effects of the crisis. By employing visual content analysis, the authors analyzed selected Instagram posts from the Indonesian DMO account within the period of February–May 2020. Then, the social media engagement rates of the overall posts are measured to see how far the projected images can resonate with the audience. It is expected that this paper could contribute as one of the bases in evaluating the visual communication strategy of presenting the desired destination images as part of tourism resilience.

Keywords: Covid-19, destination image, Instagram, social media, visual content analysis

1 INTRODUCTION

In 2020, more than 1 billion people used Instagram and shared an average of 95 million photos and videos a day (Newberry 2020). Instagram has taken the image sharing niche by storm, and the social media platform is influencing the tourism world in the way of magnifying tourism destinations to affect our travel decisions. The increased online time spent by social media users is directing their beliefs, ideas, and impressions of destinations into the so-called "ideal destinations" and new trends constructed by the published contents they consume on social media (Nixon et al. 2017; Xiang & Gretzel 2010).

The current shift toward a more visual and digital-based promotion means that destination marketing organizations (DMO) and tourism operators need to rethink their promoting strategy. They need to add their promotion arsenal by generating destination images that can be influencing consumers' travel wish list and destination image (Stepchenkova & Zhan 2013).

In carrying the mission to mediate the projected destination images, Instagram has been used extensively by official DMOs in many countries, including Indonesia. Numerous studies have shown the effectiveness of Instagram as a tourism marketing tool. Aside from the fact that people engage with Instagram 10 times more than with Facebook, Instagram is also known to be more effective in strengthening the destination branding and in influencing 'consumers' visiting intention (Fatanti & Suyadnya 2015; Gumpo et al. 2020; Miller 2017; Shuqair & Cragg 2017).

However, due to the outbreak of Covid-19 in early 2020, the tourism industry faced one of its greatest challenges. The pandemic has adversely impacted almost all sectors of the global economy, with travel and tourism sectors among the hardest hit by the pandemic (Becker 2020). As reported by the United Nations World Tourism Organization (UNWTO), the COVID-19 pandemic has resulted in a contraction of the global tourism sector by 20% to 40%, and 80% to 90% of a drop

 DOI 10.1201/9781003095484-30

in international air traffic (Coke-Hamilton 2020). As an illustration, by mid-April, the number of visiting tourists to Japan and Indonesia had dramatically dropped by 90% to 99.9% year over year (CNBC Indonesia 2020; Nikkei 2020).

Although many economic sectors are expected to recover after the lifting of restrictive measures, the pandemic is believed to have a longer-lasting impact on the tourism industry. As estimated by the United Nations Conference on Trade and Development (2020), "This hampered recovery is largely due to reduced consumer confidence and the likelihood of longer restrictions on the international movement of people."

For this reason, it is not surprising that Instagram has been used by DMO's as one of the strategies to adapt to the situation and to help the destination brand stay current and connected amidst the pandemic. Studies have shown that social media engagement can support the destination brand during the time of crisis by reducing adverse public perception so that it can lessen the damaging impact of the crisis (Ryschka et al. 2016; Schroeder et al. 2013).

Thus, it becomes significant to explore how a DMO is constructing and maximizing the destination images on Instagram to maintain its destination brand in a time of crisis. This paper focuses on analyzing and identifying the destination images constructed by the Indonesian official DMO as part of the coping strategy in facing the pandemic.

2 MATERIAL AND METHODS

2.1 *Research design and data collection*

A visual content analysis method is adopted in collecting and analyzing the data. The case of Indonesia is chosen since the country's tourism industry was crippled by the pandemic and became one of the heavily affected countries in the world (CNBC Indonesia 2020; Schlagwein 2020). The purposive sample in this paper consists of selected Instagram posts from the official Instagram account of the Ministry of Tourism and Creative Economy, Indonesia (@indtravel) as the nation's official DMO. The authors selected all Instagram posts published by @indtravel account between February and May 2020, where the main phase of the physical distancing movement occurred globally.

2.2 *Data analysis*

In analyzing the data, the authors employed the visual analytical framework developed by Hunter (2012). In the visual content analysis, coding becomes the essential step in analyzing the data. With visual objects, coding means attaching a set of descriptive labels or categories to the photos and language features (Hao et al. 2016). The coding categories are observable in sample images and can emerge in the composition of images and written captions in each Instagram post. Based on this adopted analysis framework, the authors performed five coding stages.

The first step is the visual data collecting process from the Indonesian official DMO account. Secondly, the authors sorted the gathered visual data to identify the structure and type of the Instagram post's visual representation. In the third step, categories are generated based on the dominant denotative elements. The next step is identifying the frequency and implication of the denotative as well as the connotative elements found on the Instagram posts. Finally, the authors analyzed the social media engagement rates by measuring the total approval actions and the total account followers to get the applause rate percentage.

In ensuring the trustworthiness of the study, the authors established the inter-rater reliability (IRR) method by involving multiple coders to rate the 'authors' coding results. In this paper, the authors used the inter-rater measurement as formulated by Miles and Huberman (1994):

$$\text{reliability} = \frac{\text{number of agreements}}{\text{number of agreements} + \text{disagreements}}$$

Figure 1. Inter-rater reliability formula.

As suggested by Miles and Huberman (1994), an IRR of 80% agreement between coders on 95% of the resulted codes is considered as sufficient agreement to mitigate the interpretative bias. In this study, the percentage of inter-rater reliability (IRR) for the coded categories agreement was 94%, which means it is considered suitable and can be used for further analysis, as described in the following section.

3 FINDINGS AND DISCUSSION

3.1 *Projected image construction*

In capturing and projecting the Indonesian tourism, the Indonesian DMO established its Instagram account (@indtravel) in and achieved more than 624.000 followers. The account described its official Indonesia representative status as the main account profile and attached its primary hashtag (#wonderfulindonesia) along with the official website. The Instagram account was established in 2012, and by mid-2020 had uploaded 2,691 posts. For the purpose of the study, the authors managed to gather 20 Instagram posts during the time of physical and social distancing. As the effects of the Covid-19 continue to be felt across the global economy, business managers in the hospitality industry are working hard to adapt. It is known that to keep posting on social media could serve as one of the strategies to maintain brand awareness during a time of crisis (Ramakhrisnan 2020).

In exploring how the @indtravel account is staying active during the crisis, the authors analyzed the visual elements using the visual analytical adaptation of Hunter (2012). The sampled posts were analyzed based on their denotative and connotative elements. As described by Sukyadi (2014), denotative elements are visible verbal meanings that appear explicitly in the form of elements of images, writings, or captions. While connotative elements are defined as text that carries cultural or emotional associations, in addition to their literal meanings, the extracted denotative and connotative elements from the data can be seen in Table 1.

3.2 *Interpretation and discussion*

As described in Table 1, the visual analysis found that all of the postings were delivered in English, and the most frequent visual content representation was the Indonesian natural landscape delivered in a simple or dynamic style of photography. Bird-eye view, eye level, and high angle were detected as the top three photographic angles that are used by the @indtravel content creator. Photo subjects can be dramatically emphasized simply by which angle we place our camera. By having a bird-eye view and high camera angle, it can be seen that the content creator attempted to alter what seems as typical landscapes from the ground and turn it into more artistic images when seen from above. As described by Setiadi (2017), taking pictures from this point of view can take viewers to see a unique perspective and make them feel as if they are superior to the subject and able to see the whole thing.

The DMO account posts are dominated by the country's natural wonder-based themes as the projected image, where Bali and East Nusa Tenggara Province are framed as the face of Indonesian tourism (89%). The constructing of an image is always based on a certain social context that mediates its impact (Rose 2016). While Bali is a globally known tourism hub, Indonesia, which comprises more than 17,000 islands, has many more tourism destinations with different features and offerings yet to be discovered. The social context in the images represents the Indonesian government's efforts in developing and expanding its tourism destinations to other islands beyond Bali to be completed in 2024 (Antara & Sumarniasih 2017; Sinaga 2017). Subsequently, the images posted by the DMO account were projected to support the government program in pushing the tourism growth by highlighting East Nusa Tenggara Province as one of the undiscovered gems of Indonesian tourism.

Furthermore, the data in Table 1 showed that during the Covid-19 pandemic, two hashtags (#StayAtHome, and # TravelTommorrow) were significantly used alongside the main national featured

Table 1. The Indonesian DMO projected images on Instagram.

Instagram Visual Structure	Category (Denotative Elements)	Frequency (%)
Feed Format	Photo	89%
	Video	11%
Feed Content	Simple Landscape Photography	33%
	Dynamic Landscape Photography	39.5%
	Wildlife Photography	5.5%
	People/human Photography	11%
	Video collages	11%
Photographic Angle	Eye Level	22%
	High Level	22%
	Low Level	11%
	Frog Eye Level	11%
	BirdEye View	34%
Geotag	Bali Province	50%
	East Nusa Tenggara Province	39%
	East Java Province	5.5%
	Kalimantan Province	5.5%
Hashtags (multiple hashtags are found in each of the Instagram posts)	#wonderfulIndonesia	100%
	#StayAtHome	67%
	#ThoughfulIndonesia	22%
	#TravelTommorrow	67%
Emojis	Camera	72%
	Sunshine	17%
	No emoji	11%
Caption Language	English	100%
	Connotative Elements	
Caption Tone (combined tones occurred in each post)	Informative	100%
	Conversational	88%
	Optimistic	65%
	Persuasive	65%
Caption Theme	The Natural Beauty of Indonesia	80%
	Endemic Animals	10%
	Tourism Activities	10%

hashtag of #wonderfulIndonesia. Hashtags are not something new in the digital era. Initially, they are used as a medium to sort and retrieve information on the web (Hanadi & Bashaer 2018). Hashtags then evolve as a medium to annotate images as a branding strategy (Giannoulakis & Tsapatsoulis 2016; Harris 2017). In regard to tourism, hashtag use is not only applicable to assist its marketing purpose but also improving tourists' travel satisfaction, since they assumed that hashtagging is an ethical way to appreciate a tourism destination (Krisna et al. 2019).

The shared projected destination images by the Indonesian DMO act as a promoting product (Hunter 2012). In covering this, the authors analyzed how the Instagram image posting strategy affected its viewer's engagement, as described in the following line graph.

The graph indicated that the average social media engagement during the pandemic was 1.55%. The data also showed that video-based Instagram posts resulted in the highest engagement rate (March 23, April 17, and May 10, 2020) for the DMO account.

In the social media industry standard, an engagement rate on Instagram between 1% and 3% is generally good. A higher than 3% engagement rate is a better indication that the viewers are very engaged with the published content (Mee 2019). Based on the measurement, the DMO content creator has managed to maintain social media engagement during the pandemic time. In fact, the account has a better engagement rate compared to January–February 2020 average rate (1.38%).

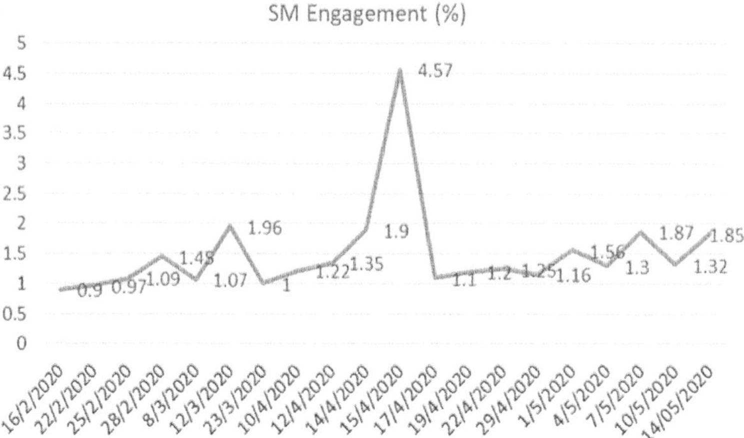

Figure 2. @indtravel social media engagement.

In the current situation (travel and social limitation), the constructed image projection appears to be a sufficient strategy to maintain tourism brand awareness, at least to the online audience.

4 CONCLUSION

Based on the visual analysis, it was found that natural beauty was the prominent theme of the constructed images. The Indonesian DMO account highlighted the Eastern Islands of Indonesia as the projected destination image of Indonesian tourism.

The image construction played a sufficient role in maintaining the DMO social media engagement rate in order to stay active in promoting Indonesian tourism awareness. Further studies can be focused on interviews of destination planners on their creative process of designing the online projected destination image. Otherwise, a further survey can be conducted to see the viewers' future travel decisions after seeing the promotional features published on social media as a form of post-pandemic travel wish lists. It is expected that the analysis in this paper could contribute as one of the bases in evaluating the visual communication strategy and can be used by destination planners in developing promotional destination images as part of tourism resilience during a time of crisis.

REFERENCES

Antara, M., & Sumarniasih, M. S. (2017). Role of Tourism in the Economy of Bali and Indonesia. *Journal of Tourism and Hospitality Management*. https://doi.org/10.15640/jthm.v5n2a4

Becker, E. (2020). *How hard will the coronavirus hit the travel industry?* National Geographic. https://www.nationalgeographic.com/travel/2020/04/how-coronavirus-is-impacting-the-travel-industry/

CNBC Indonesia. (2020, March 28). ASITA: Akibat Corona, Kunjungan Wisatawan Turun Hingga 90%. *CNBC Indonesia*. https://www.cnbcindonesia.com/market/20200326173709-19-147765/asita-akibat-corona-kunjungan-wisatawan-turun-hingga-90

Coke-Hamilton, P. (2020). *Impact of COVID-19 on tourism in small island developing states*. United Nations Conference on Trade and Development. https://unctad.org/en/pages/newsdetails.aspx?OriginalVersionID=2341

Fatanti, M. N., & Suyadnya, I. W. (2015). Beyond User Gaze: How Instagram Creates Tourism Destination Brand? *Procedia – Social and Behavioral Sciences, 211*(September), 1089–1095. https://doi.org/10.1016/j.sbspro.2015.11.145

Giannoulakis, S., & Tsapatsoulis, N. (2016). Evaluating the descriptive power of Instagram hashtags. *Journal of Innovation in Digital Ecosystems, 3*(2), 114–129. https://doi.org/10.1016/j.jides.2016.10.001

Gumpo, C. I. V., Chuchu, T., Maziriri, E. T., & Madinga, N. W. (2020). Examining the usage of Instagram as a source of information for young consumers when determining tourist destinations. *SA Journal of Information Management, 22*(1). https://doi.org/10.4102/sajim.v22i1.1136

Hanadi, B., & Bashaer, A. (2018). Use of hashtags to retrieve information on the web. *The Electronic Library, 36*(2), 286–304. https://doi.org/10.1108/EL-01-2017-0011

Hao, X., Wu, B., Morrison, A. M., & Wang, F. (2016). Worth thousands of words? Visual content analysis and photo interpretation of an outdoor tourism spectacular performance in Yangshuo-Guilin, China. *Anatolia, 27*(2), 201–213. https://doi.org/10.1080/13032917.2015.1082921

Harris, M. (2017). *Marketing with Instagram, the Fastest Growing Social Platform!* The Medium Well. http://mediumwell.com/marketing-instagram/

Hunter, W. C. (2012). Projected Destination Image: A Visual Analysis of Seoul. *An International Journal of Tourism Space, Place, and Environment, 143*, 419–443. https://doi.org/10.1080/14616688.2011.613407

Krisna, D. F., Handayani, P. W., & Azzahro, F. (2019). The antecedents of hashtags and geotag use in smart tourism: a case study in Indonesia. *Asia Pacific Journal of Tourism Research, 24*(12), 1141–1154. https://doi.org/10.1080/10941665.2019.1665559

Mee, G. (2019). *What is a Good Engagement Rate on Instagram?* Scrunch.Com. https://www.scrunch.com/blog/what-is-a-good-engagement-rate-on-instagram

Miles, M., & Huberman, M. (1994). Data management and analysis methods. *Handbook of Qualitative Research*.

Miller, C. (2017). *How Instagram Is Changing Travel*. National Geographic Society. https://www.nationalgeographic.com/travel/travel-interests/arts-and-culture/how-instagram-is-changing-travel/

Newberry, C. (2020). *37 Instagram Stats That Matter to Marketers in 2020*. Hootsuite. https://blog.hootsuite.com/instagram-statistics/

Nikkei. (2020, April 21). Foreign visitors to Japan fall 99.9% amid pandemic. *Asia.Nikkei.Com*. https://asia.nikkei.com/Spotlight/Coronavirus/Foreign-visitors-to-Japan-fall-99.9-amid-pandemic

Nixon, L., Popova, A., & Onder, I. (2017). How Instagram Influences Visual Destination Image – a Case Study of Jordan and Costa Rica 2 Study Methodology. *E-Review of Tourism Research, 8*(Research Notes).

Ramakhrisnan, V. (2020). *Why It's Important for Brands to Keep Posting on Social Media During COVID-19*. Falcon.IO. https://www.falcon.io/insights-hub/case-stories/cs-social-media-strategy/why-its-important-for-brands-to-keep-posting-on-social-media-during-covid-19/

Rose, G. (2016). *Visual methodologies: An introduction to the interpretation of visual materials* (4th ed.). Sage.

Ryschka, A. M., Domke-Damonte, D. J., Keels, J. K., & Nagel, R. (2016). The Effect of Social Media on Reputation During a Crisis Event in the Cruise Line Industry. *International Journal of Hospitality and Tourism Administration, 17*(2), 198–221. https://doi.org/10.1080/15256480.2015.1130671

Schlagwein, F. (2020, March 3). Coronavirus hits global tourism hard. *Deutsche Welle (DW)*. https://www.dw.com/en/coronavirus-hits-global-tourism-hard/a-52619138

Schroeder, A., Pennington-Gray, L., Donohoe, H., & Kiousis, S. (2013). Using Social Media in Times of Crisis. *Journal of Travel and Tourism Marketing, 30*(1–2), 126–143. https://doi.org/10.1080/10548408.2013.751271

Setiadi, T. (2017). *Dasar Fotografi Cara Cepat Memahami Fotografi* (P. Christian (ed.)). CV Andi Offset. https://books.google.co.id/books?id=81NLDwAAQBAJ

Shuqair, S., & Cragg, P. (2017). the Immediate Impact of Instagram Posts on Changing the Viewers' Perceptions Towards Travel Destinations. *Asia Pacific Journal of Advanced Business and Social Studies, 3*(2). https://doi.org/10.25275/apjabssv3i2bus1

Sinaga, D. (2017, May 2). Jokowi Promotes "Ten New Balis" to Investors in Hong Kong. *Jakarta Globe*. https://jakartaglobe.id/context/jokowi-promotes-ten-new-balis-to-investors-in-hong-kong

Stepchenkova, S., & Zhan, F. (2013). Visual destination images of Peru: Comparative content analysis of DMO and user-generated photography. *Tourism Management, 36*, 590–601. https://doi.org/10.1016/j.tourman.2012.08.006.

Sukyadi, didi. (2014). Dampak Pemikiran Saussure bagi Perkembangan Linguistik dan DIsiplin Ilmu Lainya. *PAROLE: Journal of Linguistics and Education, 3*(2). https://ejournal.undip.ac.id/index.php/parole/article/view/5208.

Xiang, Z., & Gretzel, U. (2010). Role of social media in online travel information search. *Tourism Management, 31*(2), 179–188. https://doi.org/10.1016/j.tourman.2009.02.016

Promoting Creative Tourism: Current Issues in Tourism Research – Kusumah et al. (Eds)
© 2021 Taylor & Francis Group, London, ISBN 978-0-367-55862-8

Analysing the factors affecting the purchasing decision of Malaysian batik products

U.H. Simin & N.H.A. Rahman
International Islamic University Malaysia, Johor, Malaysia

ABSTRACT: Malaysian batik is a highly purchased handicraft in comparison to other handicraft products. There has been a scarcity of previous studies in addressing the factors affecting the purchasing decision over the traditional textile, which is batik. Hence, this study aims to analyse the factors that affect the purchasing decision over Malaysian batik products. This study has employed quantitative research by distributing questionnaires consisting of 35 questions, which were categorized into four sections. This study also employed the purposive sampling technique with a total of 62 respondents who completed the questionnaires. The study found that six factors affect the purchasing decision over Malaysian batik products, which are: the quality factor, price factor, availability factor, packaging factor, reference factor, and brand factor. Besides, other factors have been identified from the open-ended question, namely the promotion and awareness, design and color, identity and authenticity, and usability. The study suggested exploring the promotion and awareness factor, design and color factor, identity and authenticity factor, and usability factor that affect the purchasing decisions over Malaysian batik products. The findings of the study would benefit the batik industry by enhancing the use of Malaysian batik and increasing the sales value of Malaysian batik.

Keywords: handicraft, batik products, Malaysian Batik, purchasing decision, quantitative research

INTRODUCTION

Traditional batik is one of the oldest arts of Malaysian culture, and symbolically, the motifs often represent many things (Sidek 2018). Like consumer goods, Malaysian batik may not have a strong identity and character in comparison to Indonesian batik, but it still possesses a big market in Malaysia (Mahdzar et al. 2013). Malaysian Handicraft Development Corporation (MHDC) stated that batik was highlighted as the main contributor to the overall craft sales value in Malaysia (MHDC 2016). The batik sales of a subsidiary company of MHDC, Kraf Holding Sdn Bhd increased from RM 913,778 in 2009 to RM 239, 784, 324.65 in 2016.

Although the sales value of Malaysian batik products is increasing, Annuar (2019), in his article in *Malay Mail* reported that the revival of batik popularity among Malaysians is not for Malaysian batik, but Indonesian batik. There are a few reasons for this; the main one among them is affordability. Amy Blair, the founder of Batik Boutique, explained that price point is still a matter of concern for Malaysians, and Indonesia's mass-printed batik is far more affordable than the East Coast's hand-printed version (Annuar 2019). The report indicated that the price factor contributed to the purchasing decision over Malaysian batik products. Nevertheless, price is not the only factor that consumers consider before purchasing any products (Pesol et al. 2016).

The study concerning the factors that affected the purchasing decision over Malaysian batik products has been under research in prior literature (Fei 2013; Sabijono 2013). The studies specific for Malaysian batik appear to be essential and worth of investigation as the batik industries in Indonesia and Malaysia are different (Leigh 2002). According to Pesol et al. (2016), through

 DOI 10.1201/9781003095484-31

the practical aspect, batik producers or marketers can understand the demand of Malaysian batik consumers when the related factors that affect purchasing decisions over Malaysian batik products are discovered. It would contribute to an in-depth understanding of Malaysian batik consumers, then will provide innovative and creative planning as well as strategic promotional and marketing activities to enhance the use of Malaysian batik and increase its sales value. Approximately 90.5 per-cent of entrepreneurs think that the high consumption of batik can have a positive impact on firm sales (Akhir et al. 2015). Therefore, this study intends to identify the related factors that will affect the purchasing decisions over Malaysian batik products and determine the most influential factors that affect said purchasing decisions.

1 LITERATURE REVIEW

1.1 *Factors affecting the purchasing decision of Malaysian batik products*

This present study adopted the theoretical framework from Ujianto and Abdulrachman (2006). The study incorporated six factors, namely the reference factor, quality factor, price factor, brand factor, packaging factor, and availability factor.

The reference group is an individual or group of people who significantly influence one's behaviour (Sumarwan 2014). Ujianto and Abdulrachman (2006) found that the reference factor was the most influential factor that affects the purchasing decision over batik sarong among consumers in East Java, Indonesia. The study is in line with Mandey and Kawung (2016), which argued that reference factors had a significant effect on the purchasing decision over batik products in Manado, North Sulawesi.

Next, quality is defined as an evaluation of the excellence and superiority of the product (Nguyen & Gizaw 2014). The study by Nurfikriyadi (2016) emphasized that product design and product quality are the significant factors that will affect the purchasing decision over batik products. Furthermore, Rahadi et al. (2016) indicated that quality factors had a significant effect on consumers' batik preferences with pattern and design coronated as the preferable factors in determining the quality of batik products in Indonesia.

Meanwhile, price is the amount of money charged for a product or a service; the sum of the values that customers exchange for the benefits of having or using a product or service (Kotler & Armstrong 2016). Tiningrum (2014) indicated that, aside from product, place, and promotion, the price had a significant influence on purchasing decision towards Small and Medium Enterprises (SME) in Surakarta City. Moreover, Kusumodewi (2016) mentioned that aside from lifestyle and product quality, the price had a significant influence on the purchasing decisions over batik in Mirota, Surabaya. The study is in line with Mandey and Kawung's (2016) study, in which the price factors had a significant effect on the purchasing decision over batik products in Manado, North Sulawesi.

Furthermore, a brand is described as a logo, style, word, name, or any other attribute that distinguishes the services or goods of a seller from other sellers' (Yuliaty 2015). A study by Mubarok (2018) found that brand image has a positive and significant effect on purchasing decisions. Besides, research conducted by Mandey and Kawung (2016) showed that brand factors had a significant positive impact on the purchasing decision over batik products in Manado. Also, Siregar and Widiastuti (2019) mentioned that the brand factor had a significant positive relationship on the purchasing decision over batik products in Danar Hadi Surakarta, Indonesia.

Then, packaging design has become an important marketing factor for various consumer goods. It has a vital role in communicating product benefits to the customer. A study by Ujianto and Abdulrachman (2006) had stated that packaging design affects the purchasing decision over batik sarong in Manado, Indonesia. Besides, Hamdar, Khalil et al. (2018) had also discussed a similar result, in which the research has indicated that there was a significant relationship between the product packaging and the purchasing decision.

Finally, the availability factor refers to the extent of consumer attitudes towards the availability of existing products (Ujianto & Abdulrachman 2006). The study by Ujianto and Abdulrachman (2006) has indicated that there is a significant positive relationship between availability factor and purchasing decision. Besides, a study by Mandey and Kawung (2016) has shown that there is a significant effect on the availability of products towards the consumer purchasing decision. The result has indicated that consumers will choose to buy batik products from famous and well-known shops/markets.

1.2 *The importance of the Malaysian batik for the tourism industry in Malaysia*

As an "intangible cultural heritage of humanity," Malaysian batik is believed to be an attractive heritage product that can be offered to tourists from all over the world, especially those who are interested in culture and the art of society in a country (Akhir et al. 2018). Malaysia has now been successfully promoted internationally by the increasing popularity of batik known as one of the most famous textile crafts. It is not only demanded by batik lovers, but also from international tourists, who buy batiks for gifts and souvenirs. (Pesol et al. 2016). This is because foreign tourists see batik as a fabric that is full of colors and very different from their textiles, which ultimately gives them an interest in buying. Batik is also listed as the Top 10 souvenirs purchased by tourists, where 23 out of 94 tourists prefer to buy batik items, apart from accessories and keychains (Hamden et al. 2015). Hence, the growth in the batik industry can offer attractive tourism products that will further boost the country's tourism industry and contribute to the Malaysian economic growth.

2 METHOD

This study is a quantitative research design that used descriptive analysis. Data for this study have been collected using a purposive sampling technique for domestic tourists who had purchased Malaysian batik products as the target population for the study. Previously, from 1st October until 15th October 2019, a pretest of 30 surveys was conducted to detect possible errors on the questionnaire. This study managed to obtain results from 62 respondents with the implementation of sample size calculation.

Each item for the factors had applied the five-point Likert Scale as an indicator of measurement ranging from "Strongly Disagree" (1) to "Strongly Agree" (6). Then, the responses received were classified into either negative, moderate, or positive. Besides, these data were used to rank the factors that affect the purchasing decision over Malaysian batik.

For the open-ended question, there was only one question provided, where the consumers could state their comments and suggestions on Malaysian batik products. The data from the open-ended section were computed using Microsoft Excel and had been grouped according to their likeness. Then, the result of the data was triangulated with the data stated in the consumers' background, purchase information, and the factors influencing the purchasing decision over Malaysian batik products.

3 RESULTS

Based on the data collected from 62 respondents, quality factor, price factor, availability factor, and packaging factor had a positive impact on the purchasing decision over Malaysian batik products. In contrast, the reference factor and brand factor were moderate. The reference factor indicated a mean score of 3.25, with *'choose Malaysian batik products that are recommended by family members'* contributed to the highest value of reference factor with a mean score of 3.98. Levy and Lee (2004) mentioned that since the family is a fundamental decision-making unit, the interaction between family members is likely to be more significant than those of smaller groups, such as friends or colleagues. Likewise, the consumers were moderately affected by the brand factor, with

Table 1. The result of the open-ended data.

Categories	Frequency	Percent (%)
Promotion and awareness	17	28
Price	12	20
Design and color	8	13
Usability	8	13
Authenticity and identity	6	10
Quality	5	8
Availability	4	7
Total	60	100

a mean score value of 3.13. The highest mean value for the brand factor was *'consider the brand before purchasing Malaysian batik products'*, with a mean score value of 3.27.

Otherwise, the quality factor had the highest mean score value, which is 4.53, making the factor as the most influential factor among others. The data indicated that consumers require a product to have high durability, neat, and strongly sewn to ensure that the product can be used for a longer time. Besides, the consumers were also concerned about the color of Malaysian batik and the durability of said color. They prefer Malaysian batik products with the right color combination and a color that does not fade and wear out quickly. Besides, the tendency to choose Malaysian batik products that are comfortable upon use gave the idea that the consumers do not want to experience discomfort from the material of the products, for example, feeling hot when wearing them. It also explains why smoothness and thickness were considered because the product's material will have an impact on comfort.

Next, the price factor had positively affected the purchasing decision over Malaysian batik products with a total mean score of 4.23. The highest mean score was *'compare prices before purchasing Malaysian batik products'* with a mean score value of 4.65. The data indicated that the price factor is very crucial in determining the purchasing decision over batik. Customers would perceive a different price as a measure of value and quality (Mandey & Kawung 2016). For the availability factor, the results show a mean value of 3.80. The average number of respondents had agreed that they would *'choose to buy Malaysian batik products in a shop or market that has a complete selection'* with the mean score value of 4.15.

Then, the packaging factor indicated a mean score value of 3.63 with the highest mean score for *'choose a Malaysian batik product that has a nice and beautiful packaging design'* with a mean score value of 3.71. Gogoi (2013) mentioned that customers always think that purchasing products that have simple packaging and those that are not so popular will be a high risk because the quality of said products cannot be trusted.

The open-ended data emphasized that other factors can be studied concerning the factors that affect the purchasing decision over Malaysian batik products. From 62 respondents, there are only 60 respondents who answered the open-ended section. Table 1 indicates that seven categories frequently appear in the open-ended data, including promotion and awareness, price, design and color, usability, authenticity and identity, quality, and availability.

Most of the respondents emphasized highly on the promotion and awareness of Malaysian batik products. A female respondent from Perak (R = 27, age = 23), stated that:

> *"Batik should be highly promoted and advertised as one of Malaysia's tradition and cultural pride."*

Additionally, promotion and awareness were also emphasized by a female respondent from Selangor (R = 37, age = 23), who mentioned that:

> *"Batik nowadays become trending, especially during Hari Raya. But I didn't have much information regarding the differences between batik Malaysia and Indonesia. So somehow, I didn't know whether the batik that I buy coming from which country."*

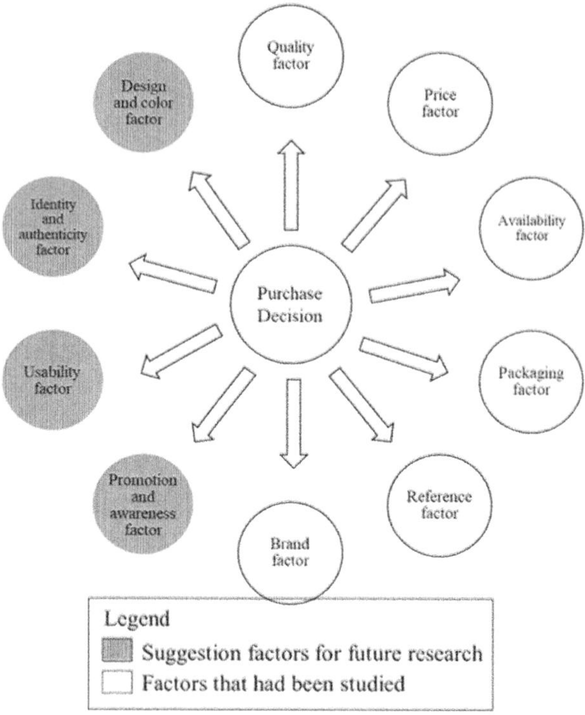

Figure 1. Purchasing decision's factors.

Furthermore, the information from the open-ended section found that the consumers were also concerned with the design and color of Malaysian batik products. A female respondent from Selangor (R = 12, Age 24) stated that:

> *"Selling more batik small flower pattern and pastel colors."*

Meanwhile, another respondent highlighted the design of Malaysian batik products. A female respondent from Kedah (R = 16, Age = 27) suggested that:

> *"Batik industry or designers should always come up with attractive, eye-catching, up-to-date designs that can attract customers to buy batik from time to time as batik as we expect will not be outdated as time passes."*

The data previously mentioned that the consumers were looking towards the pattern and motif of Malaysian batik products before they purchase. This data from the open-ended section proved that consumers are particularly concerned about the pattern and motif of Malaysian batik products and suggested the color and pattern that they want. Besides, a respondent (R = 16, Age = 27) had emphasized on *'up-to-date design'*. As on average, the age range of respondents is between 21 to 24 (66.1%, N = 41), they demand something up-to-date and in line with the current fashion trend.

Moreover, the consumers were also concerned about the usability of Malaysian batik product, as a male respondent from Terengganu (R = 31, Age = 23) mentioned that:

> *"Today, the teacher and some government workers wear batik on Thursday. It is better if all the sectors in Malaysia can wear batik every Thursday, including university students and staff."*

Furthermore, the authenticity and identity of Malaysian batik were also discussed in the open-ended data. A female respondent from Melaka (R = 8, Age = 24) mentioned that:

"It is good if we can keep the authenticity of the batik product. Even trend changing fast, but batik product always gets the demand."

For that, another conceptual framework can be constructed based on the data gained through the open-ended. Figure 1 shows the suggested conceptual framework.

4 CONCLUSION

In conclusion, the research had adopted a study by Ujianto and Abdulrachman (2006) who included six factors in their research, and the factors are; quality factor, price factor, availability factor, packaging factor, reference factor, and brand factor. The average number of respondents had positively agreed that all factors had influenced their purchasing decision over Malaysian batik products. However, the finding from the open-ended questions showed that there are four other factors mentioned by consumers, namely, promotion and awareness, design and color, usability, and identity and authenticity. Therefore, other factors need to be considered and studied for the related factors may and will affect the purchasing decision over Malaysian batik in the Malaysian context.

Based on the data, the most influential factor that affected the purchasing decision over Malaysian batik products is the quality factor. On the price factor, the respondents have also been affected positively by the idea that the item should be worth the quality for certain prices. The study is in line with Djumarno and Djamaluddin (2017), which indicated that better product quality will maintain a high level of customer satisfaction. Hence, this will encourage customers to make future purchases. However, the result is different from the findings in Ujianto and Abdulrachman's (2006) study, in which the most influential factor was the reference factor as the consumers were specifically influenced by a famous officer, idols, and scholars in determining their purchasing decision over batik products in Manado. Therefore, the data emphasized that different populations have different factors and preferences in deciding the choice of batik products that they purchase.

Hence, the findings of the study will provide a better understanding of the critical factors involved in the purchasing decisions over Malaysian batik products, which would benefit batik producers or marketers by enhancing the use of Malaysian batik and increasing its sales value.

REFERENCES

Akhir, N. H. M., Ismail, N. W., Said, R., Ranjanee, S., & Kaliappan, P. (2015). Traditional craftsmanship: The origin, culture, and challenges of the batik industry in Malaysia. In Islamic perspectives relating to business, arts,

Akhir, N. H. M., Ismail, N. W., & Utit, C. (2018). Malaysian batik industry contribution analysis using direct and indirect effects of input-output techniques. *International Journal of Business & Society*, 19(1).

Annuar (2019, October 22). What would it take to make Malaysians wear (our own) batik again? malaymail.com

Djumarno, S. A., & Djamaluddin, S. (2017). Effect of Product Quality and Price on Customer Loyalty through Customer Satisfaction.*International Journal of Business and Management Invention (IJBMI), 7*(8), 12-20.

Fei, Y. S. (2013). The relationship between product personality, attitude, and purchase intention towards Malaysian batik. Kota Kinabalu: University Malaysia Sabah.

Gogoi, B. (2013), the study of antecedents of purchase intention and its effect on brand loyalty of private label brand of apparel, International Journal of Sales & Marketing, 3(2): 73–86.

Hamdar, B., C., Khalil, A., Bissani, M., Kalaydjian, N. (2018). Economic assessment of the impact of packaging design on consumption.

Hamden, H. N. (2015). Perception of tourist's on souvenir purchase intention and perceive authenticity in Kuala Lumpur. Shah Alam: Universiti Teknologi Mara.

Kotler. P., & Amstrong, G. (2016). *Principle of marketing (Sixteenth edition Global version)*. England: Pearson Education. Inc.

Kusumodewi, G. (2016). Pengaruh gaya hidup, kualitas produk, dan harga terhadap keputusan pembelian kain batik Mirota Surabaya. *Jurnal Pendidikan Tata Niaga (JPTN), 3*(3).

Leigh, B. (2002). Batik and pewter: Symbols of Malaysian pianissimo. *Sojourn*, 17, 94–109.

Levy, D. S., & Kwai-Choi Lee, C. (2004). The influence of family members on housing purchase decisions. *Journal of Property Investment & Finance*, 22(4), 320–338.

Malaysian Handicraft Annual Report 2016. Malaysia Handicraft Development Corporation. www.kraftangan.gov.my.

Mahdzar, S. S. S., Jin, C. P., & Safari, H. Development of Historical Culture Tourism Industry through Batik Art to Attract the Local and Tourist.

Manan, D. I. A., & Jan, N. M. (2010). Do resources contribute to firms' performances? Exploring batik industry in Malaysia. *International Review of Business Research Papers,6*(3), 189–204.

Mandey, N., & Kawung, D. (2016). Purchased interests to establish of batik products in Manado, North Sulawesi.

Mubarok, M. M. (2018). The effect of brand image and consumer attitudes on the decision to purchase batik Jetis Sidoarjo mediated by intent to buy.*Journal of Economics, Business, & Accountancy Ventura, 21*(1), 105–116.

Nguyen, T. H., & Gizaw, A. (2014). Factors that influence consumer purchasing decision of Private Label Food Product: A case study of ICA Basic.

Nurfikriyadi, (2016). *Proposed marketing strategy for batik nation (a batik fashion startup company).* Master Thesis. School of Business Management. Institut Teknologi Bandung.

Pesol, N. F., Mustapha, N. A., Ismail, S. S., & Yusoff, N. M. (2016). The attributes of Malaysian batik towards tourist purchase decision.*Tourism, Leisure and Global Change, 3*, 105–116.

Rahadi, R. A., Rahmawati, D., Windasari, N. A., & Belgiawan, P. F. (2020). The Analysis of Consumers' Preferences for Batik Products in Indonesia.*Review of Integrative Businessand Economics Research, 9*, 278–287.

Sabijono, G. (2013). Analysis of Manado consumer preference in buying batik apparel product. Jurnal EMBA: *Jurnal Riset Ekonomi, Manajemen, Bisnis dan Akuntansi*, 1(4), 414–421.

Sidek, H. A. (2018). Contextualizing the metaphor of Malay batik motifs in graphic design or visual communication.

Siregar, J. S., & Widiastuti, E. (2019). Pengaruh brand image, kualitas produk, dan harga terhadap keputusan pembelian produk batik di Danar Hadi Surakarta. *Surakarta Management Journal, 1*(1), 21–29.

Sumarwan, U. (2014). Perilaku konsumen teori dan penerapannya dalam pemasaran Bogor Ghalia Indonesia.

Tiningrum, E. (2014). Pengaruh bauran pemasaran terhadap keputusan pembelian batik di Usaha Kecil Menengah Batik Surakarta. Edisi Pebruari, Vol. 1, No. 2.

Ujianto, U., & Abdurachman, A. (2004). Analisis faktor-faktor yang menimbulkan kecenderungan minat beli konsumen sarung (studi perilaku konsumen sarung di Jawa Timur). *Jurnal Manajemen dan Kewirausahaan,6*(1), 34-53.

Yuliaty, A. (2015). *The dominant factor of batik purchase toward young adults in Jakarta* (Doctoral dissertation, President University).

Promoting Creative Tourism: Current Issues in Tourism Research – Kusumah et al. (Eds)
 ISBN 978-0-367-55862-8

Indonesian mythology as touristic attractiveness: The story of the Queen of the Southern Sea of Java Island

M.V. Frolova
Lomonosov Moscow State University, Moscow, Russia

M.W. Rizkyanfi & N.S. Wulan
Universitas Pendidikan Indonesia, Bandung, Indonesia

ABSTRACT: Java is a land of unique cultural traditions that continually attracts visitors with its ancient temples, court dances, shadow-puppet theatre, and unique cuisine. Due to the Javanese world's cultural diversity, the island may be fascinating for curious tourists because of its rich mythology, among the characters of which is the goddess of the Southern Sea–*Nyai Roro Kidul*. Unlike other female deities, the cult of which developed in pre-Islamic Java, only *Nyai Roro Kidul* organically fit into the Indo-Muslim synthesis of Javanese culture 16th and 17th centuries. This study aims to analyze the queen of the Southern Sea of Java Island's story as touristic attractiveness. The research method used in this study is descriptive with a qualitative approach. The result shows that the vivid cultural tradition contributes to popularizing the image of the sea goddess: folk stories provided by informants, Internet-lore, films, and literary works by some modern Indonesian writers. The article presents the facts discovered by anthropologists and shows that the myth is associated with particular locations on the map of Java, including beaches and hotels. The selection of mythological material is made by the purpose of the study–to increase the tourist potential of Java and make it even more attractive for internal and external tourism.

Keywords: Indonesian mythology, *Nyai Roro Kidul*, the Queen of the Southern Sea of Java Island, touristic attractiveness

1 INTRODUCTION

This paper deals with the mythological character of the goddess of the South Sea *Nyai Roro Kidul* in folklore and modern Indonesian literature in its relation to the touristic attraction to the Java Island. The focus of the study is the cultural impact on the tourism of the famous "Queen of the Southern Sea," which links exclusively to the ethno-cultural area of the Java world (West, Central and East Java, Madura, Bali, Lombok). The basic set of these mythological cycles is formed from the ancient animistic beliefs and rituals of the archipelago and is so firmly entrenched in culture that it goes through folklore, and is preserved as imagined cultural object until this modern stage of global and economic development in Indonesia and the value of the cultural heritage in the touristic sphere. The goal is to select, analyze, define, and estimate the folklore-mythological complex about the sea goddess. The complex demonstrates the two types of cyclization in the provinces of West Java (*Sunda*) and Central Java (Java). The paper suggests several narratives about *Nyai Roro Kidul*, which can be used by the tourist guides Java connected with the goddess myth. The main folklore material used for the research contains the stories of informants mentioned in articles by ethnographers and anthropologists, stories about the sea goddess fragmented in medieval Javanese literature (mostly in mytho-historical chronicles of the 17th century), modern interpretations of myths in popular culture, including popular editions of stories about encounters with spirits, movies, and the Internet-lore (blogs and vlogs about *Nyai Roro Kidul*). The authors of the scientific studies are most often

DOI 10.1201/9781003095484-32

Dutch anthropologists, whose researches consider the image of the goddess as a unique local deity. R. E. Jordaan in the article *The Mystery of Nyai Lara Kidul, Goddess of the Southern Ocean* (1984) analyzes the mismatch of a small amount of information and myths about *Nyai* and her immense popularity, and also makes the emphasis on the comparison with other well-known female deities venerated in ancient times like Dewi Durga and Dewi Sri. In the article by Jordaan *Târâ and Nyai Roro Kidul. Images of the Devine Feminine in Java* (1997) it is suggested that all three goddesses (*Durga, Tara,* and *Nyai*) are the incarnation of one female deity: such conclusion seems doubtful. The article also discusses the options of the original myths about the goddess. Despite the fact that the emergence and the development of the basic myth dates back from the 16th century, R. E. Jordaan writes about proximity and similarity *Nyai Roro Kidul* with some chthonic deities who share common typological traits across the region. An article by C. Brakel *Sadhang-pangan for the Goddess. Offerings to Sang Hyang Bathari Durga and Nyai Roro Kidul* (1997) points out the lack of iconography of the sea goddess in the Javanese shadow puppet theatre *wayang*, while the rest of the gods mainly of Indian origin have their own "shadow." The facelessness and/or many faces of the sea goddess indicates the preservation of the sacredness of her cult, which is proved by the detailed descriptions of the sending offerings on water ceremony (*Labuhan*) and the court dance *Bedhoyo Ketawang*. An article by R. Wessing *Princess from Sunda: Some Aspects of Nyai Roro Kidul Kidul* (1997) examines the origins of the image of *Nyai Roro Kidul*, the mythological roots of this deity in South, East, and Southeast Asia, as well as its functions of transferring power, preserving the fertility of the land and the well-being of the kingdom. Wessing's other article, *Nyai Roro Kidul in Puger: Local Applications of a Myth* (1997), explores the position of *Nyai Roro Kidul* in the modern local pantheon of spirits on the southern coast of East Java. Unlike *the court cycle* (the agreement between the sultan and the goddess), this article examines the folk aspect of the myth. It is noted how exactly this goddess' actions serve as popular explanation of storms, floods, and other dangers that locals and fishermen face. The paper also lists the spirits in the service of *Nyai Roro Kidul*. As an anthropologist, Wessing pays special attention to the ritual side. Australian and American scholars B. J. Smith and M. Woodward came closest to the original interpretation of the myth of the transfer of power to the sultans through the agreement with the goddess of the sea. Their work *Magico-spiritual Power, Female Sexuality and Ritual Sex in Muslim Java: Unveiling the Kesekten of Magical Women* (2015) highlights the main contradiction between the rituals associated with the sea goddess and the official Muslim doctrine, strengthening its position in Java and throughout Indonesia over the past 20 years. In Russian Indonesian studies, the image of the goddess was considered in an article by M. V. Frolova, *Javanese Mythology in the Modern Indonesian Novel by Budi Sarjono "Nyai Roro Kidul–the Goddess of the Southern sea"* (2012) and in her PhD dissertation *Animistic Symbolism in Traditional and Modern Indonesian Literature* (2016). The image of the Goddess is the main theme of the master's dissertation by A. I. Lunyova (scientific supervisor, M. V. Frolova), *The Image of the Goddess of the South Sea Nyai Roro Kidul in Folklore and Modern Literature of Indonesia* (2020). The main literary material was three contemporary works: the novel by Budi Sardjono, *Her Majesty* (*Sang Nyai* 2011), the novel by Ayu Utami, *The Number of Fu* (Bilangan Fu 2008), and the short story by Intan Paramaditha, *The Queen* (Sang Ratu 2005). For the analysis of literary sources were used biographical, mytho-poetic, and comparative-typological methods. The originality of these particular studies lies in the literary aspect, which helps to illuminate the problem from a completely different angle and present new discoveries.

The further narratives about Nyai are presented below; each of them or combined can be used by tourist guides to boost the attractiveness of the Java island for those tourists who share the same passion in mythology with the author of this very paper.

2 METHOD

The research method used in this study is a descriptive study of qualitative research. The data were collected from relevant literatures: folk stories provided by informants, Internet-lore, films, and literary works by some modern Indonesian writers.

3 RESULT AND DISCUSSION

3.1 *The religious and historical background*

In ancient times, Java, like Sumatra or Kalimantan, was under the strong Indian cultural and religious influence. Javanese Buddhist and Hindu dominions arose and fell apart, and the magnificent temples like Borobudur and Prambanan remain the silent witnesses of these times. The world-famous Balinese culture developed on the basis of pre-Islamic Javanese, since between the 15th and 16th centuries many Javanese aristocrats fled to Bali after the fall of the Majapahit Empire (1527) in order to avoid conversion to Islam and the influence of the "revived" Javanese kingdom of Mataram, which was already Muslim (1588–1681). Today, the centers of Javanese culture are two famous cities of the former Mataram: Yogyakarta and Solo. Yogyakarta holds the status of a Special District due to its fame as the "cultural capital" of Indonesia. The main time of Islamization was the turn of the 15th-16th centuries, however, "the oldest evidence of the presence of Muslims on the island dates back to the XIth century: in the town of Leran in East Java, there is a Muslim grave with a tombstone dated 1082" (van der Molen 2012). In general, Islam in Java can be divided into "traditional" and "syncretic," for one part of society, and "pure" for another (*santri*). Despite the fact that most Javanese are Muslims, because of successive waves of cultural and religious influence (Hinduism, Buddhism, Islam, and Christianity), Javanese have developed the special syncretic complex of beliefs. Under these conditions, Javanese folklore preserves its entire mystical and religious complex of beliefs about spirits, ghosts, and local gods. Together with rituals, these beliefs were called *kejawen* (from the root morpheme *Jawa*/*Jawi* – "Java"). Remaining Muslims, many Javanese to this day follow the mysticism of *kejawen*, which also harmonizes well with the pan-Indonesian ideology *Pancasila*, according to which Islam is not the state religion. *Kejawen* as a traditional "Javanese religion" (*agama Jawa*) is a mixture of Indo-Buddhist, Tantric, Sufi, and ancient animistic influences.

3.2 *Two mythological cycles: Sunda and Java*

Among the mythological characters in the Java world, only one occupies a higher status than the spirits, demons, and the walking dead (*hantu*). The goddess, the mistress of the sea, is known all over the island. The Sundanese know her as the fairy princess of Pajajaran; Central and East Java talk about her as the patroness of the sultans. By "she" I mean the goddess of the South Sea, *Nyai Roro Kidul* (Javanese: Lady of the South), one of the most multifaceted characters of Javanese mysticism. "Nyai" means "mistress," "Roro" is interpreted in two ways–"virgin" in Old Javanese and "grief, sorrow" in New Javanese. "Kidul" in Javanese means "south; southern" (Frolova 2012). In a paradoxical way, the *Nyai Roro Kidul* cult developed at the same time with the spread of Islam. The causes and consequences of this phenomenon are no less interesting for research than its artistic embodiment in the folklore of Java. The myths dedicated to this goddess are multilayered and formed at different times. The traditional culture of Java knows two main myths; the first is about the origin of the goddess from the Sundanese (West-Javanese) princess who threw herself into the sea, and the second is about her intimate relationship with the Javanese prince Senopati (1586–1601), the founder of the Mataram kingdom in Central Java in the 16th century, which is mentioned in the chronicles of Mataram *Babad Tanah Jawi* (Javanese: *Chronicles of the Javanese Land*, the common name for mythological and historical manuscripts). This is one of several sources for reconstructing the spread of Islam in Indonesia (Ricklefs 1991). Another source that mentions Nyai is the manuscript *Serat Centhini* (Javanese: *The Book of Centhini*). This is a 12-volume collection of Javanese stories and religious, ethical teachings. The events described in the *Serat Centhini* occurred in the 1630s, when Sultan Agung of Mataram besieged the city of Giri Kedaton in Gresik, East Java (Santoso 2005). Sultan Agung (1593-1645), like his predecessor Senopati, is said to have mystical encounters with *Nyai Roro Kidul*. "Sultan Agung has sexual relations with, and marries Ratu Kidul, but only with the permission of the Muslim Saint Sunan Kalijaga, who, according to the tradition, played a major role in the conversion of Java to Islam" (Smith & Woodward 2015).

3.3 *Sundanese fairy tale*

The Sundanese (West Java) have oral and written circulation of the narrative, in which *Nyai Roro Kidul* appears under the name of Dewi Kadita and is the daughter of the ruler of the West Javanese Kingdom of Pajajaran (1482–1579). Elements of the myth of the transformation of a princess into a powerful goddess differ. The most common is the more recent variant: because of the curse of the envious stepmother/half-sister, the beautiful princess fell ill with leprosy. The disease forced her to leave the palace (opposition stepmother- stepdaughter indicates a degradation of the myth to fairy tales). Or the princess has rejected too many suitors, and the father chased her away. Anyway, after meditation and ascetic practices in the forest for some time, she jumped off the cliff into the Southern sea (the Indian ocean), where she was crowned by the water spirits and became a great Queen of South sea (Wessing 1997).

3.4 *Nyai and power of Sakti*

The Javanese version of the *Nyai Roro Kidul* myth originates from the 16th century, and as perhaps the most famous one in Java still circulates in oral transmission. The main structural element of the myth is the meeting of the goddess with Senopati, the founder of Mataram, which is described in detail in the *Babad Tanah Jawi* chronicles. The future ruler, eager to create a new state, was meditating on the South coast of Java. His meditation effort was so strong that the ocean began to boil and foam. This alarmed the Queen of spirits, and she came to the surface to find out what it was. Senopati told Nyai about his desires, and she promised to help him. Then chronicles describe the journey of Senopati in the underwater palace of the goddess, where he spent three days and three nights, after which he returned to the surface, trained to rule the state, and was able to lead the army of spirits entrusted to him by the goddess. The successful journey to the underworld (underground or underwater kingdom) marks the initiation and obtaining a higher status. In exchange for power over Java, the goddess asked Senopati and all his descendants to marry her. Despite the fact that Senopati was a Muslim, as it is repeatedly mentioned in the text, he agreed on such conditions.

In the Javanese culture, the visible and invisible worlds are closely intertwined; they are inseparable and cannot exist without each other. Therefore, close contacts of these two worlds, as well as human contacts with divine beings, are natural. Therefore, in the traditional religious consciousness of the Javanese there is no conflict between the official religion and *takhayul* (from Arabic "superstition"), which form the basis of Javanese mysticism. Superstitions are well preserved through folklore.

3.5 *Nyi Blorong*

However, the goddess is associated not only with her function of transmitting power. The origins of the *Nyai Roro Kidul* veneration are not entirely clear. Among scholars it is accepted that the most ancient and common motif in folklore of Austronesian peoples is as follows: the founder of the new state marries the princess found in the bamboo or who appeared from the sea foam. W. E. Maxwell in his paper *Two Malay Myths: The Princess of the Foam and the Raja of the Bamboo* (1881) came to the conclusion that the two most common mythological motifs of Nusantara (the princess of the sea foam and the prince found in the trunk of bamboo) have a completely different origin. The myth of foam-born princess has Indian roots, as it is mentioned only in the regions that were under strong cultural influence of India. The myth of the birth of the bamboo prince (later version) and the first man (originally) is sporadically distributed in a wide area from the Andaman Islands to Melanesia, covering the Malay Peninsula, Sulawesi, and Papua, and is also known in some Philippine areas.

In the mythology of the Dayak people (Kalimantan), the upper world is ruled by a male god in the form of the hornbill, and the female ruler of the lower world manifests herself in a snake. The binary oppositions (male-female, bird-snake, up-down) are also reflected in the structure and decoration of Batak iconic vehicle that embodies the unity of the Universe: the nose of the ship was a bird's head (the sky, the upper world). In the cosmology of the peoples of Indonesia, the feminine, the forces of chaos, and the ocean were often represented in the form of a snake. The serpentine forms of the "modern" goddess *Nyai Roro Kidul* received their collective image in the form of a double character Nya Blorong.

There are several different versions about the place of Nyi Blorong in the local mythological system. If you ask Indonesians about their first association with *Nyai Roro Kidul*, they will surely answer that the main thing is not to wear clothes of green color. Both mythological females, *Nyai Roro Kidul* and Nyi Blorong, love the green color and are often depicted in green clothes. *Serat Centhini* mentions that Nyi Blorong is the daughter of Ratu Angin-Angin (the text says that she is the Queen of South Sea). She married the giant snake named Jaka Lelong. During the full moon, she appears in the guise of a beautiful girl, but with the waning moon gradually takes the ghostly form of a huge snake (*ular siluman*). Because of her enthusiasm for meditation (*bertapa*), she earned the title of favorite of *Nyai Roro Kidul*. Sometimes she is even believed to be the daughter of the goddess and the head commander of the troops of the underwater spirit world.

She was also instructed to lead people astray, to force them to make a deal with her. The main function of this character is the granting of wealth to those who dare to perform a bloody ritual *pesugihan* and give Nyi Blorong *tumbal* (sacrifice) in exchange for gold (Kisah Tanah Jawa 2019).

3.6 *Snake goddess*

The given facts allow to carry out the assumption that Nyi Blorong and *Nyai Roro Kidul* are different versions of the same character, each of which perform separate functions.

We can conclude that the myths about the patron of the Javanese sultans exhibit more ancient layers of this genesis. Most likely, in ancient times existed the cult of the chthonic female deity, associated with the sea, the moon, and snakes. The Khmer and the Cham ocean deities were personified as a snake (Brakel 1997). As generally throughout South and Southeast Asia, in Java there was a belief in a close relationship between the rulers and the chthonic creatures, often in the form of the serpent, ensuring their right to rule, and the relationship between snakes, agriculture, and fertility. Indirect evidence to that is in the famous Srivijayan inscription Telaga Batu decorated with seven snakeheads. The name of the ruler Jayanasa was transcribed sometimes as Jayanaga– "the Victorious Serpent." In the inscriptions of Telaga Batu state traitors had to be not just killed (*niwunuh*), but eaten (*nimakan*) by snakes. Wealth of Sriwijaya may also be associated with the special protection of the "snake tribe"–chief warden of Sumatra gold (Parnikel 2003).

3.7 *Internet-lore, desacralization, and touristic attractiveness*

Unlike traditional folklore, most often modern folklore does not bear any didactic load and has become a part of the popular culture. The best illustrations of Indonesian post-folklore can be current anecdotes, and modern urban tales. In modern society with many sources of fundamental knowledge the rudiments of traditional folklore genres like stories about meeting spirits have become entertainment. Urban Yogyakarta folklore tells us that by the command of the sultans there is made a special tunnel leading from the palace to the coast. Ghost-lore evolved oral and written cycles among children and teenagers and students, taxi drivers and rickshaws (including ojek and becak). The transition of post-folklore to its newest stage of the Internet-lore is happening before our eyes. Internet memes, horror films of local production and common Internet blogs, video blogs and YouTube channels about ghost stories are related to this kind of myth desacralization. Narratives about the sea goddess are almost always present in this segment of the Internet, and the narrator goes to her "places of power" and/or interviews local people and "witnesses." The goal of creating an Internet-lore about the goddess Nyai Roro Kidul usually includes the ideas either to persuade the viewer/listener in the idea that she really exists or to reassure the recipient of authenticity and truth about the spoken myth. In this kind of content, there is always objection from Sharia toward general concepts of the supernatural. Most often, faith in Nyai is condemned as "village prejudice" and the mythological construct itself is defined as shirk, polytheism.

Despite that, the goddess figure is still intriguing and the idea of transferring power to the Javanese rules–not only sultans, but high-ranking politicians like presidents–still exists. Bloggers talk about marriage of Nyai with the Indonesian president; YouTube is full of clips, "testifying" meetings with her.

Last but not least, some beaches and hotels across Java and Bali have special rooms for Nyai. For example, Hotels on Sanur beach in Bali, Hotel Inna Grand Bali Beach, always keeps two rooms free: room 327 and room 2401. Opened in 1966, the hotel has repeatedly hosted the first President of the Republic, Soekarno, who always stayed in room 327, which is decorated with a portrait of the Queen of the South Sea. The second room, 2401, is always empty because it can be occupied only by the goddess herself.

Nyai Roro Kidul is inextricably linked with Javanese culture. The vestiges of her cult that survived in the 21st century always put the Queen of the South Sea in the minds of Indonesians to rank above the rest of the heroes of the local "horror stories," Usually movies about the goddess belong to the genre of horror. Unfamiliar with the real descriptions of her conduct from the chronicles or traditional folklore, contemporary images of the Sea goddess switched to her iconic image from the 1980s movies. The influence of cinema on the formation of the image of Nyai Roro Kidul in the public consciousness cannot be underestimated. She is a regular character in numerous films. The new "Portrait of Nyai Roro Kidul" (Lukisan Nyai Roro Kidul) was released in 2019. However, new films are more likely the homage to the old movies of the 1980s starring the "Horror Queen" Suzzanna. The goddess is often referred to as the queen (Ratu) and occupies a significant position in the Javanese cultural paradigm. Mass culture embraces her image: games for mobile phones (Mobile Legends), children's fairy tales, old horror films, and newspaper headlines like "Not only officials, but also Nyai Roro Kidul and Nyi Blorong were invited to the inauguration of President Jokowi" (2019). To this day in some coastal cities in Java, traditional prohibitions on the green color of clothes are preserved, because this is the favourite color of the goddess, and in some hotels she is allocated special eternally empty rooms, draped in green, with offerings constantly renewed. For some, the goddess was and remains a symbol of Javanese traditional self-identity, and for some it is just another fairy tale. Knowing and sharing such information might be interesting and helpful to increase the tourist potential of Java and make it even more attractive for internal and external tourism.

4 CONCLUSION

Indonesia is a country full of mythology. Most of these stories are believed to date. The story of *Nyai Roro Kidul* is a part of this mythology. Stories, places, rituals, and the surrounding community's beliefs make the story a part of everyday life. The story of the Queen of the Southern Sea of Java Island can be a tourist attraction, both for domestic and foreign tourists. From a destination attractiveness perspective, as long as destination are popular, tourists will visit it (Vengesayi et al. 2009). The story of *Nyai Roro Kidul* can make the tourist attractiveness of the southern sea become more popular. Tourists who like mythology or mystical stories can explore places that are inseparable from the mystical tale of *Nyai Roro Kidul* for seeing related rituals and other cultures related to the queen of the Southern Sea.

REFERENCES

Alice. Will the Real 'Queen of the South Sea' Please Stand Up? 2018. URL: https://www.vice.com/en_asia/article/43ezjw/will-the-real-queen-of-the-south-sea-please-stand-up

Brakel, C. Sandhang-pangan for the goddess: Offerings to Sang Hyang Bathari Durga and Nyai Lara Kidul //*Asian Folklore Studies,* 1997. Vol. 56, pp. 253–83.

Frolova, M.V. Javanese mythology in the modern Indonesian novel by Budi Sarjono "*Nyai Roro Kidul* – the goddess of the Southern sea" // *Vestnik SPbSU. Asian and African Studies*, 2012. No. 4. pp. 76–81. (in Russian)

Frolova, M.V. Animistic symbolism in traditional and modern Indonesian literature (PhD dissertation, Moscow State University, 2016). (in Russian).
Haryo, T. Tak Hanya Pejabat, Nyi Roro Kidul dan Nyi Blorong Juga Diundang ke Pelantikan Jokowi! 2019. URL: https://keepo.me/news/tak-hanya-pejabat-nyi-roro-kidul-dan-nyi-blorong-juga-diundang-ke-pelantikan-jokowi/
Jordaan, R.E. Târâ and Nyai Lara Kidul: Images of the Divine Feminine in Java // *Asian Folklore Studies,* 1997. Vol. 56. pp. 285–312.
Jordaan, R.E. The Mystery of Nyai Lara Kidul, Goddess of the Southern Ocean // *Archipel*, 1984. Vol. 28, pp. 99–116.
Kisah Tanah Jawa / *@kisahtanahjawa & Dapoer Tjerita (Mada Ziadan dan Bonaventura D. Genta)* – Jakarta: Gagas Media, 2019.
Lunyova, A.I. The image of the goddess of the South Sea *Nyai Roro Kidul* in folklore and modern literature of Indonesia (Masters' dissertation, Moscow State University, 2020). (in Russian)
Maxwell W.E. Two Malay myths: the princess of the foam and the raja of the bamboo // *Journal of the Royal Asiatic Society,* 1881, vol.13 no.4, pp. 498–523.
Parnickel, B.B. Srivijayan epigraphic and the problem of old Malay Buddhist literature // *Buddhism and literature*. Moscow: IMLI RAN, 2003. pp. 264–292. (in Russian)
Ricklefs, M.C. A History of Modern Indonesia since 1300, 2nd Edition, 1991.
Santoso, S., Pringgoharjono, K. The Centhini Story: The Javanese Journey of Life – Jakarta: Gramedia, 2005.
Smith, B.J., Woodward, M. Magico-spiritual power, female sexuality and ritual sex in Muslim Java: Unveiling the *kesekten* of magical women // *The Australian Journal of Anthropology*. 2015. pp. 1–16.
van der Molen, W. Twelve Centuries of Javanese Literature. St. Petersburg: St. Petersburg State University; 2016. (In Russ.)
Vengesayi, Sebastian & Mavondo, Felix & Reisinger, Yvette. Tourism Destination Attractiveness: Attractions, Facilities, and People as Predictors. Tourism Analysis. (2009). 14. 621–636. 10.3727/108354209X12597959359211.
Vinogradova, L.N. Folk demonology and mythical ritual tradition of the Slavic peoples. Moscow: Indrik, 2000. (In Russ.)
Wessing R. A Princess from Sunda: Some Aspects of *Nyai Roro Kidul* // *Asian Folklore Studies,* 1997, Vol. 56. pp. 317–353.
Wessing, R. *Nyai Roro Kidul* in Puger: Local Applications of a Myth // *Archipel,* 1997. Vol.53, pp. 97–120.

Promoting Creative Tourism: Current Issues in Tourism Research – Kusumah et al. (Eds)
 ISBN 978-0-367-55862-8

The role of psychographic factors in predicting volunteer tourists' stage of readiness: A case of Australia

U. Suhud & A.W. Handaru
Universitas Negeri Jakarta, Jakarta, Indonesia

M. Allan
University of Jordan, Amman, Jordan

B. Wiratama
Universitas Negeri Semarang, Semarang, Indonesia

ABSTRACT: Volunteer tourism is a tourism product that combines volunteer and tourism activities. This study aims at measuring the stage of readiness for a sample of tourists, volunteers, and volunteer tourists to undertake volunteer tourism experience by using psychological factors as predictor variables. We chose lifestyle values, sensation-seeking personality, and social class as representatives of psychographic factors. Data were collected in Australia involving tourists, volunteers, and volunteer tourists who are Australian citizenship. They are chosen using the convenience sampling method. Data obtained using exploratory and confirmatory factor analysis, also using the structural equation model. The main findings of the study indicated that lifestyle value and sensation seeking personality significantly influence the stage of readiness to join volunteer tourism activities.

Keywords: volunteer tourism, sensation seeking personality, lifestyle value, social class, stage of readiness

1 INTRODUCTION

Volunteer tourism is a type of tourism product that continues to receive much attention from volunteers, tourists, volunteering project managers, travel agents, and researchers. This product continues to flourish, both in developed and developing countries. Volunteer tourism is defined by Suhud (2014, pp. 316–317) as:

> *"An intersection of volunteer and tourism concepts and a combination of volunteer and tourism activates in a travel destination (nationally or internationally), which requires motivated (to give) participants to spend a day or kore and pay their own costs (such as transport, accommodation, and meals) and financially contribute to the project itself (for instance, through humanitarian aid, education, health, construction, religion, and conversation), through a sender/host organisation or directly to the needy."*

Volunteer tourism is considered as an alternative tourism product so that only attracts certain people with certain psychographic factors (Wearing 2001). Psychographic factors are used by marketers to create segmentation along with demographic, geographic, and behaviour factors (Kotler & Keller 2013). Lifestyle, sensation seeking, and social class are considered as one's psychographic aspects (Bashar 2020; Demby 2011; Galloway 2002).

In the study of volunteer tourism, particularly in a quantitative approach, there are limited variables to be used to predict behavioural intention, namely attitude, subjective norm, self-efficacy, and motivation (Lee 2011). According to Suhud (2014), intention is part of stage of readiness. There are also minimal studies in the context of tourism exploring influencing factors of stage of readiness.

 DOI 10.1201/9781003095484-33

This study aims to measure the impact of lifestyle, sensation seeking, and social class on stage of readiness to be involved in volunteer tourism. One of the main variables used in this current study is sensation seeking personality and sensation seeking has been connected to tourism, volunteerism, and volunteer tourism (Abdulrahman, 2014; Suhud, 2015; Xu 2010; Zuckerman et al. 1967).

2 LITERATURE REVIEW

2.1 *Stage of readiness*

Stage of readiness is a range of one's behaviour starting from an unawareness to an action, from precontemplation to maintenance (Norcross, Krebs and Prochaska 2011). As defined by Holt and Vardaman (2013, p. 9), stage of readiness is defined as 'the degree to which those involved are individually and collectively primed motivated and technically capable of executing the change'. Suhud (2014, p. 316) states that stage of readiness in the volunteer tourism context contains "awareness", "acceptance of the concept of volunteer tourism", "interest", "desire", and "action". Furthermore, Sugiyama and Andree (2010) presents the model of attention-interest-search-action-share and this model tents to be stage of readiness. In the study of Suhud and Allan (2019), travel intention is part of stage of intention. Prior studies (Suhud 2014; Suhud & Willson 2016) claim that stage of readiness can be influenced by travel motivation, travel constraint, and stage of intention. In the current study, stage of readiness is predicted by lifestyle value, sensation seeking personality, and social class.

2.2 *Psychographic factors*

In this study, we choose lifestyle value, sensation seeking, and social class as psychographic factors to measure stage of readiness to be involved in volunteer tourism.

2.2.1 *Lifestyle value*

Overall, lifestyle is defined as "a pattern of behaviour under constrained resources which conforms to the orientations an individual has towards three major 'life decisions' he or she must make: formation of a household (of any type), participation in the labour force, and orientation towards leisure (Salomon & Ben-Akiva 1983). Lifestyle is used by Qing et al. (2012) to measure intention to purchase fresh fruit. In their study, lifestyle dimensions include risk takers, experiencers, and traditionalists. Based on their findings, risk takers and traditionalist have a significant impact on purchase intention. Furthermore, Lowongan (2015) employs lifestyle and personality to examine purchase decision for a fast food restaurant in Indonesia. This author mentions that lifestyle and personality have an essential role to affect purchase decision.

Considering the previous studies mentioned above, the following hypotheses have been formulated:

H_1 – Lifestyle value will significantly affect stage of readiness.

2.2.2 *Sensation seeking*

Zuckerman et al. (1972, p. 308) define a sensation seeker as "a person who needs varied, novel, and complex sensation and experiences to maintain an optional level of arousal". Sensation seeking is a dimension of personality (Ball & Zuckerman, 1990). To measure this personality, a sensation seeking inventory (the sensation seeking scale – SSS) was for the first time developed by Zuckerman et al. (1964). The sensation-seeking scale established by Zuckerman et al. (1964) consisted of 64 items including four dimensions: experience seeking, boredom susceptibility, thrill and adventure seeking, and disinhibition. This scale was modified several times by scholars in different contexts. For example, Zuckerman et al. (1978) modify it into the sensation-seeking scale form V (SSS-V), followed by Arnett (1994) who changes it into the Arnett inventory of sensation seeking (AISS). Furthermore, Hoyle et al. (2002) developed the brief sensation-seeking scale (BSSS). In the study of Suhud (2014) in volunteer tourism, sensation seeking had two dimensions including excitement seeking and new experience seeking. The scale discussed above is aimed to measure level of sensation seeking personality of sensation seekers.

In the tourism and leisure context, some researchers have explored the impact of sensation seeking personality on stage of readiness. For example, Pizam et al. (2001) assess the impact of sensation seeking on tourist behaviour. Lepp and Gibson (2008) investigate the impact of sensation seeking on destination choice. Furthermore, Li and Tsai (2013) examine the impact of sensation seeking on international tourism choices. In the same vein, Galloway et al. (2008) investigate behaviour of wine tourists in some wineries in Australia. They further claim that sensation seeking significantly influenced tourist behaviour, and destination choice, respectively. Furthermore, in the study of Sharifpour et al. (2013), personality factor is represented by sensation seeking and it is used to examine decision behaviour. This variable is linked to decision behaviour which consists of willingness to travel and need to search for information dimensions. They document a significant impact of sensation seeking on willingness to travel.

The following hypothesis is based on the results of the previous studies discussed above.

H_2 – Sensation seeking will significantly affect stage of readiness.

2.2.3 *Social class*

Social class can be measured using subjective and objective approaches. In measuring social class with a subjective approach, a person is asked to access himself based on three questions relating to the current occupation, annual income, and educational level (Warner et al. 1949). Meanwhile, in measuring social classes with an objective approach, a person is asked to access himself with a more complicated inventory, ranging from hobbies to political views. In many fields of study, social classes are rarely discussed because of social changes that no longer adhere to the feudalism system. Although some previous research shows that social class does not have a prominent role in consumer buying behaviour, however, including social class is also an exciting thing as it is also used for segmentation (Shavitt et al. 2016). Hugstad et al. (1987) measure the relationship between social class and perceived risk. To avoid risk, consumers seek information through various sources of information including newspapers and word-of-mouth communications. They postulate that social class has a significant role on information search. Seeking information is part of stage of readiness (Sugiyama & Andree, 2010). Furthermore, Coleman (1983) investigates consumers in the USA. He classifies social class into upper class, middle class, working class, and lower class and essential for marketers. According to him, different social classes lead to different purchase behaviour.

This is a hypothesis to be tested.

H_3 – Social class will significantly affect stage of readiness.

3 METHODS

3.1 *Sample*

To collect the data, the authors developed an online questionnaire using the Qualtrics survey web service. Potential respondents were attracted using personal and group communication techniques. A total of 551 respondents were recruited and completed the survey.

3.2 *Measures*

All variable employed in this currents study were adapted from prior studies. Lifestyle value was measured using five items adapted from Clary et al. (1998). In addition, sensation seeking personality was measured by eight items adapted from Wymer et al. (2010). A seven-point Likert-type scale was applied ranging from 1 for extremely disagree to 7 for extremely agree. Furthermore, an item to measure stage of readiness was adapted from Prochaska and Norcross (2001).

Table 1. Results of the validation and reliability test.

	Variables and indicators	Factor loadings	Cronbach's alpha
	Lifestyle value		**0.887**
Ls2	I feel compassion towards people in need	0.933	
Ls3	I feel it is important to help others	0.892	
Ls1	I am concerned about those less fortunate than myself	0.875	
Ls4	I am motivated to do something for a cause that is important to me	0.767	
	Sensation seeking – excitement seeking		**0.706**
Ss4	I like wild parties	0.830	
Ss7	I would like to tray bungee jumping	0.802	
Ss6	I prefer friends who are excitingly unpredictable	0.750	
Ss8	I would love to have new and exciting experiences, even if they are illegal	0.550	
	New experience seeking		**0.699**
Ss3	I like to do challenging things	0.877	
Ss1	I like exploring strange places	0.872	
Ss2	I get restless when I spend too much time at home	0.631	
	Social class		**0.766**
Sc2	Current occupation	0.919	
Sc3	Annual income	0.852	
Sc1	Educational level	0.706	

3.3 *Data analysis*

Quantitative data from this study were processed in four stages. The first one was to test the validity of the data using exploratory factor analysis. With this test, we could see whether each variable has dimensions or not. Besides, we could also select valid indicators with factor loadings of 0.5 or more (Hair et al. 2019). The second stage was the data reliability test (Hair et al. 2019). This test examined constructs and only considers constructs that have a Cronbach's alpha value of 0.7 or more. The third stage was the confirmatory factor analysis. This test was to measure the interconnection between one variable and another, including ascertaining which indicators are the most valid. The last one was testing the hypothesis by using a structural equation model. We selected four criteria to ensure that the model we tested was appropriate, including probability, CMIN/DF, CFI, and RMSEA with scores of 0.05 or more, less than 2.0, 0.95 or more, and 0.08 or less, respectively (Hu & Bentler, 1995, 1999; Schermelleh-Engel et al. 2003; Tabachnick et al. 2007).

4 RESULTS

4.1 *Participants*

This study surveyed 311 Australian participants consisting of 158 males (50.8%) and 153 females (49.2%). Most of them were employed for wages (200 participants, 64%) and married/de factor (210 participants, 67.5%).

4.2 *Data validation and reliability tests*

Table 1 presents results of the validation and reliability tests. The EFA calculation produces three variables including lifestyle value, sensation seeking personality, and social class. There were three variables which are tested using exploratory factor analysis, namely lifestyle, sensation seeking, and social class. Lifestyle and social class did not have dimensions while sensation seeking had two dimensions, excitement seeking with a Cronbach's alpha value of 0.706 and new experiences

Table 2. Summary of the hypotheses testing results.

	Paths			C.R.	P	Results
H_1	Lifestyle value	◇	Stage of readiness	3.830	***	Accepted
H_2	Sensation seeking personality	◇	Stage of readiness	2.104	0.035	Accepted
H_3	Social class	◇	Stage of readiness	0.529	0.597	Rejected

seeking with a Cronbach's alpha value of 0.699. Lifestyle had a Cronbach's alpha value of 0.887 and social class had a Cronbach's alpha value of 0.766. All constructs were considered reliable with a Cronbach's alpha score of 0.7 and larger.

A structural model of hypotheses was tested, and it gained a fitness with a probability score of 0.080 and CMIN/DF score of 1.471. Furthermore, it had a CFI score of 0.989, and RMSEA score of 0.030. Table 2 shows results summary of the hypothesis tests. The first hypotheses owned a C.R. score of 3.830 and the second hypotheses obtained a C.R. score of 2.104 indicating significances. On the other hand, the third hypothesis had a C.R. score of 0.529 showing that social class had an insignificant affect on stage of readiness.

As shown in Table 2, the first and second hypotheses had a C.R. score of 3.830 and 2.104 respectively indicating significances. Therefore, these two hypotheses were accepted. On the other hand, the third hypothesis gained a C.R. score less than 2.0 indicating insignificance.

4.3 *Discussion*

As depicted in the literature, little attention has been paid to psychographic factors to predict stage of readiness in different tourism contexts. Thus, this study explores the role of psychographic factors in predicting volunteer tourists' stage of readiness for study cohort in Australia. The findings of the study demonstrate that lifestyle and personality significantly influenced stage of readiness with C.R. scores of 3.830 and 2.104, respectively as predicted by prior studies (Galloway et al. 2008; Lepp & Gibson 2008; Pizam et al. 2001; Qing et al. 2012). Oppositely, social class insignificantly affected stage of readiness with a C.R. score of 0.529. Bukhari et al. (2020) examine factors to influence purchase behaviour in the context of western imported food products in Pakistan. In their study, lifestyle, personality, and social class were linked to purchase intention and purchase intention is linked to purchase behaviour. They found that lifestyle insignificantly affected purchase intention. They also found that social class had nothing to do with purchase intention. On the other hand, they showed that personality had a significant effect on purchase intention. This research as well as previous studies (Bukhari et al. 2020) demonstrate that social class does not have a significant effect on the stage of readiness. These findings suggest that in this modern society social class may indeed not have an important role in determining consumer buying behaviour. However, such studies could be continued to find out on the types of tourism products and groups of participants as to what social class can have a role in forming the stage of readiness.

5 CONCLUSION

This study aimed to measure the impact of psychographic factors on stage of readiness to be involved in volunteer tourism. The psychographic factors included in this current study were lifestyle value, sensation seeking personality, and social class. As a result, lifestyle value and sensation seeking personality significantly affected stage of readiness to be involved in volunteer tourism.

The results of this study help tourism destinations managers and authorities, especially, volunteer tourism products to be able to use a psychographic approach in segmenting and targeting potential tourists. Attention to lifestyle and personality is also essential, apart from looking at motivation and ability to pay rates. In this case, the social class did not take a role in predicting the readiness of the participants to engage in volunteer tourism. This study only involved participants with Australian nationality. Future research can measure participants from other nationalities and countries.

REFERENCES

Abdulrahman, S. (2014) 'The influence of corporate social responsibility on total assets of qouted conglomerates in Nigeria', *Journal of Business Administration and Management Sciences Research*, 3(1), pp. 12–21.

Arnett, J. (1994) 'Sensation seeking: A new conceptualization and a new scale', *Personality and individual differences*. Elsevier, 16(2), pp. 289–296.

Ball, S. A. and Zuckerman, M. (1990) 'Sensation seeking, Eysenck's personality dimensions and reinforcement sensitivity in concept formation', *Personality and Individual differences*. Elsevier, 11(4), pp. 343–353.

Bashar, A. (2020) 'A study of impact of psychographics on impulse buying behaviour with mediating role of brand loyalty: A conceptual framework', *NOLEGEIN-Journal of Consumer Behavior & Market Research*, pp. 1–7.

Bukhari, F. *et al.* (2020) 'Motives and role of religiosity towards consumer purchase behavior in western imported food products', *Sustainability*. Multidisciplinary Digital Publishing Institute, 12(1), p. 356.

Clary, E. G. *et al.* (1998) 'Understanding and assessing the motivations of volunteers: A functional approach.', *Journal of personality and social psychology*. American Psychological Association, 74(6), p. 1516.

Coleman, R. P. (1983) 'The continuing significance of social class to marketing', *Journal of consumer research*. The University of Chicago Press, 10(3), pp. 265–280.

Demby, E. (2011) *Psychographics and from whence it came*. Marketing Classics Press.

Galloway, G. (2002) 'Psychographic segmentation of park visitor markets: evidence for the utility of sensation seeking', *Tourism management*. Elsevier, 23(6), pp. 581–596.

Galloway, G. *et al.* (2008) 'Sensation seeking and the prediction of attitudes and behaviours of wine tourists', *Tourism management*. Elsevier, 29(5), pp. 950–966.

Hair, J. F. *et al.* (2019) *Multivariate data analysis*. 8th edn. Hampshire, UK: Cengage Learning.

Holt, D. T. and Vardaman, J. M. (2013) 'Toward a comprehensive understanding of readiness for change: The case for an expanded conceptualization', *Journal of change management*. Taylor & Francis, 13(1), pp. 9–18.

Hoyle, R. H. *et al.* (2002) 'Reliability and validity of a brief measure of sensation seeking', *Personality and individual differences*. Elsevier, 32(3), pp. 401–414.

Hu, L. and Bentler, P. M. (1995) 'Structural equation modeling: Concepts, issues, and applications', in Hoyle, R. H. (ed.) *Evaluating model fit*. London: Sage, pp. 76–99.

Hu, L. and Bentler, P. M. (1999) 'Cutoff criteria for fit indexes in covariance structure analysis: Conventional criteria versus new alternatives', *Structural equation modeling: A multidisciplinary journal*. Taylor & Francis, 6(1), pp. 1–55.

Hugstad, P., Taylor, J. W. and Bruce, G. D. (1987) 'The effects of social class and perceived risk on consumer information search', *Journal of Consumer Marketing*. MCB UP Ltd.

Kotler, P. and Keller, K. L. (2013) 'Marketing management (814 p.)', *Praha: Grada*.

Lee, S. J. (2011) 'Volunteer tourists' intended Participation: Using the Revised Theory of Planned Behavior'. University Libraries, Virginia Polytechnic Institute and State University.

Lepp, A. and Gibson, H. (2008) 'Sensation seeking and tourism: Tourist role, perception of risk and destination choice', *Tourism Management*. Elsevier, 29(4), pp. 740–750.

Li, C.-Y. and Tsai, B.-K. (2013) 'Impact of extraversion and sensation seeking on international tourism choices', *Social Behavior and Personality: an international journal*. Scientific Journal Publishers, 41(2), pp. 327–333.

Lowongan, E. N. (2015) 'Influence analysis of psychographic factors on consumer purchasing decision in Mcdonald's Manado', *Jurnal Berkala Ilmiah Efisiensi*, 15(4), pp. 401–412.

Norcross, J. C., Krebs, P. M. and Prochaska, J. O. (2011) 'Stages of change', *Journal of clinical psychology*. Wiley Online Library, 67(2), pp. 143–154.

Pizam, A., Reichel, A. and Uriely, N. (2001) 'Sensation seeking and tourist behavior', *Journal of Hospitality & Leisure Marketing*. Taylor & Francis, 9(3-4), pp. 17–33.

Prochaska, J. O. and Norcross, J. C. (2001) 'Stages of change.', *Psychotherapy: theory, research, practice, training*. Division of Psychotherapy (29), American Psychological Association, 38(4), p. 443.

Qing, P., Lobo, A. and Chongguang, L. (2012) 'The impact of lifestyle and ethnocentrism on consumers' purchase intentions of fresh fruit in China', *Journal of Consumer Marketing*. Emerald Group Publishing Limited.

Salomon, I. and Ben-Akiva, M. (1983) 'The use of the life-style concept in travel demand models', *Environment and Planning A*. SAGE Publications Sage UK: London, England, 15(5), pp. 623–638.
Schermelleh-Engel, K., Moosbrugger, H. and Müller, H. (2003) 'Evaluating the fit of structural equation models: Tests of significance and descriptive goodness-of-fit measures', *Methods of psychological research online*. Citeseer, 8(2), pp. 23–74.
Sharifpour, M., Walters, G. and Ritchie, B. W. (2013) 'The mediating role of sensation seeking on the relationship between risk perceptions and travel behavior', *Tourism Analysis*. Cognizant Communication Corporation, 18(5), pp. 543–557.
Shavitt, S., Jiang, D. and Cho, H. (2016) 'Stratification and segmentation: Social class in consumer behavior', *Journal of Consumer Psychology*. Elsevier, 26(4), pp. 583–593.
Sugiyama, K. and Andree, T. (2010) *The Dentsu way: Secrets of cross switch marketing from the world's most innovative advertising agency*. McGraw Hill Professional.
Suhud, U. (2014) *A moment to give, no moment to take: A mixed-methods study on volunteer tourism marketing*. Lap Lambert Academic Publishing.
Suhud, U. (2015) 'A study to examine the role of environmental motivation and sensation seeking personality to predict behavioral intention in volunteer tourism', *International Journal of Research*, 4(1), pp. 17–29.
Suhud, U. and Allan, M. (2019) 'Exploring the impact of travel motivation and constraint on stage of readiness in the context of volcano tourism', *Geoheritage*. Springer, 11(3), pp. 927–934. Available at: https://link.springer.com/article/10.1007/s12371-018-00340-3.
Suhud, U. and Willson, G. (2016) 'The impact of attitude, subjective norm, and motivation on the intention of young female hosts to marry with a Middle Eastern tourist: A projective technique relating to Halal sex tourism in Indonesia', in *Heritage, Culture and Society: Research agenda and best practices in the hospitality and tourism industry - Proceedings of the 3rd International Hospitality and Tourism Conference, IHTC 2016 and 2nd International Seminar on Tourism, ISOT 2016*.
Tabachnick, B. G., Fidell, L. S. and Ullman, J. B. (2007) *Using multivariate statistics*. Pearson Boston, MA.
Warner, W. L., Meeker, M. and Eells, K. (1949) 'Social class in America; a manual of procedure for the measurement of social status.' Science Research Associates.
Wearing, S. (2001) *Volunteer tourism: Experiences that make a difference*. Cabi.
Wymer Jr, W. W., Self, D. R. and Findley, C. S. (2010) 'Sensation seekers as a target market for volunteer tourism', *Services Marketing Quarterly*. Taylor & Francis, 31(3), pp. 348–362.
Xu, S. (2010) 'Motivations and sensation seeking behind recreational storm chasers in the United States'. University of Missouri–Columbia.
Zuckerman, M. *et al.* (1964) 'Development of a sensation-seeking scale.', *Journal of consulting psychology*. American Psychological Association, 28(6), p. 477.
Zuckerman, M. *et al.* (1972) 'What is the sensation seeker? Personality trait and experience correlates of the Sensation-Seeking Scales.', *Journal of consulting and clinical psychology*. American Psychological Association, 39(2), p. 308.
Zuckerman, M., Eysenck, S. B. and Eysenck, H. J. (1978) 'Sensation seeking in England and America: cross-cultural, age, and sex comparisons.', *Journal of consulting and clinical psychology*. American Psychological Association, 46(1), p. 139.
Zuckerman, M., Schultz, D. P. and Hopkins, T. R. (1967) 'Sensation seeking and volunteering for sensory deprivation and hypnosis experiments.', *Journal of Consulting Psychology*. American Psychological Association, 31(4), p. 358.

Promoting Creative Tourism: Current Issues in Tourism Research – Kusumah et al. (Eds)
© 2021 Taylor & Francis Group, London, ISBN 978-0-367-55862-8

Turkish destination image and attitude toward Turkish television drama

U. Suhud & A.W. Handaru
Universitas Negeri Jakarta, Jakarta, Indonesia

M. Allan
University of Jordan, Amman, Jordan

B. Wiratama
Universitas Negeri Semarang, Semarang, Indonesia

ABSTRACT: It is axiomatic that television dramas and movies play a vital role in promoting the tourism destinations to different tourists. More specifically, Turkish dramas have attracted a large portion of Indonesian audiences. Thus, this study aims to explore the destination image of Turkey and the attitude of the audience towards Turkish television dramas. The study was conducted in Jakarta, Indonesia and in a mixed-method approaches. For the qualitative study, data were collected using a group discussion, whereas, for the quantitative study, data were gathered using an online survey involving 170 participants. Data were analysed using content analysis for qualitative data and using exploratory factor analysis for quantitative data. As a result, there are 36 indicators of Turkish destination image, and there are 21 indicators of attitude towards Turkish television dramas.

Keywords: Turkish drama, Turkish destination image, Turkish tourism, movie tourism

1 INTRODUCTION

Non-subscription television stations in Indonesia, almost broadcast some drama series from various countries on daily basis, including India, Korea, the Philippines, and Turkey. However, not only through television, viewers stream through the internet to watch their favourite dramas. Even before this, one of the most popular ways was to watch on DVD before the tradition of watching DVDs was eroded by internet streaming. The audiences of these foreign dramas are not only adults, but also teenagers and children. One of the benefits of watching foreign drama or movies is "the incidental acquisition of a foreign language by children and adolescents" (2010, p. 65). However, this benefit may not be felt by the audience when dubbing is enforced. Another benefit of watching foreign dramas or movies are the creation of an image destination for the country where the drama series taken place (Suhud & Willson 2018). Accordingly, the benefits of watching foreign dramas and movies, according to Bray (2019), that they contain natural and contextual language, allowing viewers to learn foreign cultures, increase motivation, and stimulate thoughts and discussions. Other researchers uncover the role of film and include tv drama as a marketing tool to create and popularise a tourism destination (Vagionis & Loumioti 2011).

Wen et al. (2018) revealed the impact of movies and television on Chinese Tourists perception toward international tourism destinations. Further, the participants were asked to determine which countries they wanted to visit after watching movies or TV dramas. Reportedly, the participants chose South Korea, Thailand, United States, France, Japan, the United Kingdom, and Italy. This current study focuses on Turkish drama and destination image of Turkey. Furthermore, Chiang and Yeh (2011) measured influencing factors of residents' perceptions and attitudes toward film induced tourism, whereas (Thurstone 1931) assessed the impact of film on attitude.

DOI 10.1201/9781003095484-34

The main purpose of this study is to explore destination image of Turkey and attitude viewers toward Turkish drama. According to Sönmez and Sirakaya (2002), Turkey image as a world tourism destination is still weak, so it had an impact on the number of foreign tourist visits, even though it is located in a crowded Mediterranean region of international tourists. Suhud and Willson (2018) interplayed attitude toward Indian drama and destination image of India. They found that attitude toward Indian drama significantly affected destination image and vice versa. Other studies (Herstanti et al. 2014; Kim et al. 2012; Nafisah & Suhud 2016; Suhud et al. 2017) mentioned that attitude and destination image are important to lead tourists to have intention to visit and revisit a destination as well as satisfaction.

It is evident that there is a lack of attention on attitude towards TV drama, movies, film, and cinema. Research on attitudes towards films, cinema, and television dramas seems to be scant. Thurstone (1930) developed a scale about attitudes toward film, whilst Phillips and Noble (2007) focussed on attitude towards the cinema as medium for advertisement. In the same vein, Man and Lewis (1977) researched on attitude towards films. Additionally, Chiang and Yeh (2011) studied attitude toward film induced tourism. Also, Panda and Kanungo (1962) explored different attitudes towards motion pictures. However, there are many studies relating to attitude towards advertisement. Many advertisements are made using a film approach and many films are promoted through advertisements before or during airing in theatres and on television. Therefore, this study will be guided by the pertinent literature relating to attitude towards advertisement. On the other side, we found some studies (Atadil et al. 2017; Sönmez & Sirakaya, 2002) that employ destination image in the context of Turkey.

2 LITERATURE REVIEW

2.1 *Destination image*

Echtner and Ritchie (1991) claim that attribute of destination images are a spectrum between two directions. They call both functional and psychological. The functional direction represents physical attributes and they are measurable, for example, views, costs, climate, and activities. In comparison, the psychological direction leads to abstract attributes, for example, service quality, atmosphere, and hospitality. Some studies have different approaches in learning destination images, namely by grouping attributes or indicators of destination images into affective and cognitive images. Sönmez and Sirakaya (2002) reveal that indicators for affective images may include local attractions and hospitality, socioeconomic and cultural distance, lack of natural attractions and tourist services, comfort/safety and tourist facilitation, outdoor recreation opportunities, and perceived value of vacation. Furthermore, indicators of cognitive image include safe and hospitable environment, general mood and vacation atmosphere, relaxing effect, and authenticity of experience.

Destination image of Turkey (Atadil et al. 2017; Yilmaz et al., 2009) or parts of Turkey for example, Alanya (Artuğer, Çetinsöz & Kılıç 2013), Ankara (Dündar & Güçer 2015), Amasra (Aksoy & Kiyci 2011), Istanbul (Sahin & Baloglu 2011), and Safranbolu (Basaran 2016), have been widely studied. In general, each destination has a different characteristic from other destinations. For this reason, the indicators of each destination will have different indicators. This research is unique because it was conducted outside Turkey, whereas previous studies on the destination image of Turkey were carried out in Turkey.

2.2 *Attitude towards television drama*

A person's attitude towards something or someone in the tourism context, can be favourable and unfavourable (Suhud 2014; Suhud & Willson 2017). According to Chiang and Yeh (2011), attitude can be influenced by socio-demographic and tourism benefit sought. So far, there are limited studies focussing on attitude toward television drama. One of example is conducted by Suhud and Willson (2018). These scholars explore viewers' attitude toward Indian television drama and link to destination image of India. They document a significant impact of attitude on destination image. Other studies look at attitude of viewers toward films or motion pictures.

Thurstone (1931) writes an experimental study of the impact of film on children's social attitudes. This research involved children who were invited to watch a film in a theatre. Tests carried out before and after the film was observed. There were 13 items of social activities tested, including tramp, beggar, speeder, drunkard, gambler, and petty thief. Based on Thurstone reports, although small, social attitudes of children can be formed due to watching films. Another study involving children is conducted by Panda and Kanungo (1962) in the Indian setting. They revealed that there were differences in attitudes between school students and college students with the film. Attitude differences also occur between male and female students, even in Indian students and students from western countries. Many marketers advertise their products in the theatre before the film begins. Phillips and Noble (2007) tests the attitude of the audience towards advertisements that are shown in the theatre before the movie. According to these scholars, the attitude of the audience can result in the audience arriving late into the theatre, going to another theatre, or borrowing films from rental shops.

3 MATERIAL AND METHODS

3.1 *Sample*

In this study, participants were selected using the convenient sampling method. Prospective participants were asked their willingness to be involved in a survey. Once they agreed, they were given a link to an online survey. In total, there were 170 participants involved in the survey.

3.2 *Measures*

To obtain more specific indicators about Turkey, a discussion was held, which involved students from the Business Education study program at a university in Jakarta, Indonesia. They were asked about their perceptions of Turkey as a tourism destination. We picked 36 indicators of attitude toward Turkish drama from the discussion. Further, these indicators were adapted and included in the questionnaire for the quantitative study. The same activity was also carried out to provide indicators of attitudes towards Turkish dramas. Initially, we used the indicators of Thurstone (1930) as a reference to guide the discussion. As a result, we obtained 21 indicators to measure attitudes toward Turkish dramas.

3.3 *Data analysis*

This study was conducted in two phases. Data of the qualitative study were analysed using content analysis. Furthermore, data of the quantitative study were processed in two stages. First stage was to analyse the data using exploratory factor analysis (EFA). This method aims to validate data and to test whether each variable has a dimension. In operating EFA, we chose direct *oblimin* with a minimum loading value of 0.5 (Hair et al. 2019). Second stage was to examine the reliability of the data. Each construct must have a Cronbach's alpha value of at least 0.7 to be considered reliable (Hair et al. 2019).

4 RESULTS

4.1 *Participants*

There were 170 participants from the current study survey, consisting of 119 women (70%) and 51 men (30%). Most participants (78.8%) were less than 20 years old and unmarried (91.8%). Besides, they were unemployed (74.7%). When they were asked about the religion they embraced, 98.2%

Table 1. Exploratory Factor Analysis (EFA) results of Turkish destination image.

Indicators	Factor loadings	Cronbach's alpha
D19 Turkey has a good reputation	0.966	0.995
D30 Tourist attractions in Turkey offer relaxation	0.964	
D23 Turkey provides excellent quality services for tourists	0.961	
D27 Turkey is suitable for family holidays	0.959	
D29 Turkey has complete lodging facilities	0.958	
D32 Sightseeing places in Turkey are interesting for selfies	0.956	
D5 Turkey has buildings with attractive architectural designs	0.956	
D13 Turkey offers halal tourism	0.952	
D22 Turkey has typical food and drinks	0.949	
D26 Turkey offers adventure	0.948	
D17 Turkey prepares information facilities regarding the country's tourism	0.947	
D34 Turkey has a distinctive culture	0.945	
D35 Visiting Turkey can have the opportunity to gain knowledge	0.942	
D31 Turkey has fascinating islands	0.942	
D8 Turkey holds many festivals	0.941	
D11 Tourist attractions in Turkey are easily accessible	0.940	
D36 Turkey has a robust Islamic history	0.939	
D20 Turkey has unique traditional clothing	0.936	
D10 Tourists visit tourist attractions in	0.934	
D7 Turkey has complete shopping facilities	0.934	
D16 Foreign tourists visiting Turkey are friendly	0.933	
D25 The atmosphere in tourist attractions in Turkey is generally familiar to tourists	0.930	
D9 The tourist spots in Turkey are clean	0.926	
D14 Country Turkey is friendly	0.922	
D1 Turkey has an attractive natural landscape	0.919	
D28 Turkey has an exciting nightlife	0.914	
D18 Major cities in Turkey are densely populated	0.909	
D15 Turkish residents are friendly	0.909	
D2 Turkey has a pleasant climate	0.908	
D12 Tourist attractions in Turkey are safe	0.906	
D6 Turkey has interesting historical sites	0.899	
D24 Political conditions in Turkey are stable	0.892	
D4 Turkey has an excellent transportation infrastructure	0.888	
D3 Turkey has national parks	0.862	
D33 Turkey has ethnic and cultural diversity	0.822	
D21 Vacation costs to Turkey are affordable	0.798	

of participants marked Islam as their religion. Of the total, 161 participants (94.7%) said they had watched Turkish drama while nine other participants (5.3%) said they had never watched Turkish drama before. Fifty-two participants (30.6%) reported that they watched Turkish drama in this week, whereas 118 participants (69%) did not watch such dramas. Regarding the *frequency* of watching, 10 participants (5.9%) indicated that they watched a Turkish drama every day. Eighteen participants (10.6%) stated watching this drama 3-4 times a week, then nine participants (5.3%) showed 1-2 times a week, 133 participants (78.2%) reported uncertain/sometimes. Eleven participants (5.5%) indicated that they had visited Turkey. While 159 other participants demonstrated that they had never visited Turkey before. When asked if they have the intention to visit Turkey? Twenty-five participants (14.7%) stated that they had never thought about visiting Turkey at all. Besides, 32 participants (18.8%) indicated that they had the intention to visit Turkey in the next five years, followed by 11 participants (6.5%) who intended to visit Turkey in the next three years. Furthermore,

Table 2. Exploratory Factor Analysis (EFA) results of attitude toward Turkish drama.

Indicators	Factor loadings	Cronbach's alpha
A20 Turkish dramas are made with proper drama making techniques	0.926	0.985
A14 Turkish dramas have an exciting conflict	0.924	
A17 The stories in Turkish dramas are rational	0.920	
A13 Turkish dramas are touching the heart	0.916	
A16 Turkish dramas have interesting storylines	0.914	
A9 Actors/actresses play a proper role	0.914	
A21 Costumes worn by attractive actresses/actors	0.907	
A2 Turkish dramas are exciting	0.903	
A1 In general, Turkish dramas are interesting	0.902	
A6 Turkish dramas are romantic	0.902	
A12 Turkish dramas are inspiring	0.889	
A4 Actors and actresses in Turkish dramas have attractive faces	0.883	
A11 Turkish dramas are not boring	0.882	
A18 The stories in Turkish dramas are not like dramas from other countries	0.881	
A19 The life featured in a Turkish drama is not like the life that took place in Indonesia	0.875	
A10 Turkish dramas are not the wording	0.867	
A5 Turkish dramas have exotic shooting locations	0.864	
A8 Turkish dramas have a moral message	0.852	
A15 Turkish dramas have elements of humour	0.838	
A7 Turkish dramas are dramatic	0.753	
A3 The stories of Turkish dramas are unpredictable.	0.744	

13 participants (7.6%) indicated that they plan to visit Turkey in the next year. At the same time, 89 other participants (52.4%) stated that they were uncertain about their re-visitation.

4.2 *Destination image of Turkey*

Destination image of Turkey formed no dimension. It consisted of 36 items with factor loadings ranging from 0.798 to 0.966 with a Cronbach's alpha score of 0.995 indicating a significance (as seen in Table 1).

4.3 *Attitude toward Turkish drama*

The construct of attitude toward Turkish drama) shaped no dimension and survived 21 indicators with factor loadings ranging from 0.744 to 0.926. A Cronbach's alpha score of 0.985 indicated that this variable was reliable (as seen in Table 2).

4.4 *Discussion*

This study investigates the attitude toward Turkish television dramas for a sample of Indonesian respondents.

The EFA results for the destination image of Turkey, consisted of indicators that sequentially were ordered from psychological to functional. These findings were in line with the conclusions submitted by Echtner and Ritchie (1991). However, they revealed that indicators that correlate with functional will have a higher value than indicators that relate to psychological. In the current study, the results showed the reverse, that the abstract indicators which had a more top value loading. Hence, the results of the current study were more in agreement with the results presented by Suhud and Willson (2018). In their research, the destination image of India formed a spectrum from psychological indicators to functional indicators.

In the studies of Panda and Kanungo (1962), and Ching and Yeh (2011), demographic factors were taken into account to distinguish attitudes and become predictors of attitudes, respectively. However, we only consider demographics as a profile of participants. The findings of the current research are different from the findings from previous studies, for example, showing that attitudes can be formed to be favourable and unfavourable (Suhud & Willson 2017).

5 CONCLUSION

Several previous studies have concentrated on the destination image of Turkey. However, research conducted outside Turkey is still very limited in the literature. Also, research on attitudes toward Turkish television drama is still undertaken rarely. For this reason, this study aims to explore indicators to test destination image and attitudes towards Turkish television dramas. This study brings findings of 36 indicators relating to the destination image of Turkey and 21 indicators relating to attitudes towards Turkish television dramas. Both of the variables performed no dimensions.

The results of this current study suggest that tourism managers, policymakers, planners and authorities can use television dramas as a medium to promote tourism destinations, especially for overseas tourists. One of the weaknesses of this research is that the profile of respondents is mainly restricted to the young audiences. However, in many other studies, decision-makers for family holidays can be stemmed from young children in the family. For this reason, future research can test decision-making abroad, especially Turkey, by including television drama as one of the main attributes.

REFERENCES

Aksoy, R. and Kiyci, S. (2011) 'A destination image as a type of image and measuring destination image in tourism (Amasra case)', *European Journal of Social Sciences*, 20(3), pp. 478–488.

Artuğer, S., Çetinsöz, B. C. and Kılıç, İ. (2013) 'The effect of destination image on destination loyalty: An application in Alanya', *European Journal of Business and Management*, 5(13), pp. 124–136.

Atadil, H. A., Sirakaya-Turk, E. and Altintas, V. (2017) 'An analysis of destination image for emerging markets of Turkey', *Journal of vacation marketing*. Sage Publications Sage UK: London, England, 23(1), pp. 37-54.

Basaran, U. (2016) 'Examining the relationships of cognitive, affective, and conative destination image: A research on Safranbolu, Turkey', *International Business Research*. Canadian Center of Science and Education, 9(5), pp. 164–179.

Bray, E. (2019) 'Using movies in the foreign language classroom: The movie journal approach', *Language Teacher*, 43, p. 9.

Chiang, Y.-J. and Yeh, S.-S. (2011) 'The examination of factors influencing residents perceptions and attitudes toward film induced tourism', *African Journal of Business Management*. Academic Journals, 5(13), pp. 5371–5377.

Dündar, Y. and Güçer, E. (2015) 'The impact of socio-demographics on tourism destination image: a study in Ankara Turkey', *International Journal of Economics, Commerce and Management*. Citeseer, 3(2), pp. 1–22.

Echtner, C. M. and Ritchie, J. R. B. (1991) 'The meaning and measurement of destination image', *Journal of tourism studies*, 2(2), pp. 2–12.

Hair, J. F. *et al.* (2019) *Multivariate data analysis*. 8th edn. Hampshire, UK: Cengage Learning.

Herstanti, G., Suhud, U. and Wibowo, S. F. (2014) 'Three modified models to predict intention of Indonesian tourists to revisit Sydney', *European Journal of Business and Management*, 6(25), pp. 2222–2839.

Kim, S. *et al.* (2012) 'Does a food-themed TV drama affect perceptions of national image and intention to visit a country? An empirical study of Korea TV drama', *Journal of Travel & Tourism Marketing*. Taylor & Francis, 29(4), pp. 313–326.

Kuppens, A. H. (2010) 'Incidental foreign language acquisition from media exposure', *Learning, Media and Technology*. Taylor & Francis, 35(1), pp. 65–85.

Man, J. B. and Lewis, J. M. (1977) 'College students' attitudes toward movies', *Journal of Popular Film*. Taylor & Francis, 6(2), pp. 126–139.

Nafisah, E. and Suhud, U. (2016) 'Who would return to Malioboro? A structural model of factors to influence tourists' revisit', in *International Conference on Education For Economics, Business, and Finance (ICEEBF)*, pp. 28–35. Available at: http://iceebf.um.ac.id/wp-content/uploads/2017/06/4.ELMIA-NAFISAH.pdf.

Panda, K. C. and Kanungo, R. N. (1962) 'A study of Indian students' attitude towards the motion pictures', *The Journal of Social Psychology*. Taylor & Francis, 57(1), pp. 23–31.

Phillips, J. and Noble, S. M. (2007) 'Simply captivating: Understanding consumers' attitudes toward the cinema as an advertising medium', *Journal of Advertising*. Taylor & Francis, 36(1), pp. 81–94.

Sahin, S. and Baloglu, S. (2011) 'Brand personality and destination image of Istanbul', *Anatolia-An International Journal of Tourism and Hospitality Research*. Taylor & Francis, 22(01), pp. 69–88.

Sönmez, S. and Sirakaya, E. (2002) 'A distorted destination image? The case of Turkey', *Journal of travel research*. Sage Publications Sage CA: Thousand Oaks, CA, 41(2), pp. 185–196.

Suhud, U. (2014) *A moment to give, no moment to take: A mixed-methods study on volunteer tourism marketing*. Lap Lambert Academic Publishing.

Suhud, U., Rohyati and Willson, G. (2017) 'Destination image and place attachment on car free day events revisit intention: A gender perspective', in *Proceedings of the 29th International Business Information Management Association Conference – Education Excellence and Innovation Management through Vision 2020: From Regional Development Sustainability to Global Economic Growth*. Available at: https://ibima.org/accepted-paper/destination-image-and-place-attachment-on-car-free-day-events-revisit-intention-a-gender-perspective/.

Suhud, U. and Willson, G. (2017) 'Favourable and unfavourable attitudes of young female residents toward middle eastern male tourists', *Advanced Science Letters*, 23(1). doi: 10.1166/asl.2017.7204.

Suhud, U. and Willson, G. (2018) 'The image of India as a travel destination and the attitude of viewers towards Indian TV dramas', *African Journal of Hospitality, Tourism and Leisure*, 7(3). Available at: https://ro.ecu.edu.au/cgi/viewcontent.cgi?article=5544&context=ecuworkspost2013.

Thurstone, L. L. (1930) 'A scale for measuring attitude toward the movies', *The Journal of Educational Research*. Taylor & Francis, 22(2), pp. 89–94.

Thurstone, L. L. (1931) 'Influence of motion pictures on children's attitudes', *The Journal of Social Psychology*. Taylor & Francis, 2(3), pp. 291–305.

Vagionis, N. and Loumioti, M. (2011) 'Movies as a tool of modern tourist marketing', *Tourismos: an international multidisciplinary journal of tourism*, 6(2), pp. 353–362.

Wen, H. *et al.* (2018) 'Influence of movies and television on Chinese tourists perception toward international tourism destinations', *Tourism management perspectives*. Elsevier, 28, pp. 211–219.

Yilmaz, Yusuf *et al.* (2009) 'Destination image: A comparative study on pre and post trip image variations', *Journal of Hospitality Marketing & Management*. Taylor & Francis, 18(5), pp. 461–479.

Promoting Creative Tourism: Current Issues in Tourism Research – Kusumah et al. (Eds)
© 2021 Taylor & Francis Group, London, ISBN 978-0-367-55862-8

Why do countries allow dark tourism? A review study

S. Barua & E.D. Putra
Woosong University, Daejeon, South Korea

ABSTRACT: Dark tourism is a type of tourism that involves a visit to places connected with death, misery, suffering, atrocities, and crimes. Although there are thousands of dark tourism sites in the world that are categorized into war-related, disaster, cemeteries, holocaust, the number of researches showing the critical purposes of promoting dark tourism sites by the countries is missing. This study probes into the available research material from the journals to search for reasons dark tourism is allowed against the three possible situations of educational purpose, for generating revenue or for motivation and sympathy. This review paper conducted qualitatively using a secondary source of data of published articles from the 2000s onwards using a keyword of dark tourism. As a result, 15 research papers were randomly selected, and the findings showed that the main two reasons countries allowed dark tourism were for educational purposes and revenue generation while motivation or sympathy is subject to the type of site. Despite the limited data collected and analyzed, this study would contribute to dark tourism's study as it highlights the purpose of dark tourism.

Keywords: dark tourism, tourism, countries.

1 INTRODUCTION

Dark tourism is widely referred to as visiting places related to death, disaster, and tragedy (Foley & Lennon, 1997). The most common examples of dark tourism sites are the concentration camps of Auschwitz, Ground Zero of New York, and the dungeons in the Robben Island of South Africa. The common trait of these sites is suffering and death, and they receive large numbers of tourists due to coverage by the media (Virgili et al. 2018).

The connection between tourism and death and disaster has now become a recent research trend in tourism studies. However, most of the research work has focused on the visitors' motivation, inspiration, needs, sensitivity, and requirements visiting the places (Tang 2014), and there is little work about why countries allow such tourism (Light 2017).

The supply side of dark tourism is majorly related to marketing and commercialization. The research into dark tourism, when it started highlighting the economic aspect, the problems of the commodification of tourism sites gave rise to ethical debates about the use of death and pain for financial gains. However, very few research work has discussed the supply side of dark tourism and has shed some light on the commodification of the places of death and disaster (Seaton 1996).

Tourism helps countries generate revenues from tourists, but the question is the utilization and commodification of tragedies and miseries to earn rather than to help the affected communities. The dead are commodified to gain economic benefits as they are presented as products for tourists in dark tourism (Young & Light 2016). As a result, several studies have made ethical debates about the use of sites of deaths and misery for financial gains. However, for some products, there is a need to generate revenues for sustainability and long-term maintenance and integration of history to meet the tourist demand like museums (Strange & Kempa 2003). The main question under scrutiny in this study is, why do countries allow dark tourism? This study's objectives are to highlight the significance of the factors that make countries allow and promote dark tourism. The three main

 DOI 10.1201/9781003095484-35

objectives of this study are: 1)To critically study how dark tourism is allowed and promoted for educational purpose; 2) To analyze if countries promote dark tourism sites for generation revenues; 3) To critically evaluate that dark tourism sites be the source of motivation for the people of countries or win sympathy by wrong interpretation of events.

2 LITERATURE REVIEW

2.1 *Dark tourism*

Dark tourism is the form of tourism which has its existence and history from the dark ages. Due to recent development in studies and specialization in the academic fields, special interest tourism was highlighted by scholars. Dark tourism is also regarded as a type of special interest tourism where the tourist has an interest in the death and disaster-related sites and feelings of grief, misery, and pain (Topsakal & Ekici 2017). The product of dark tourism gained the attention of researchers in the early 1990s and was previously known as "black spot," "thanatourism," "atrocity tourism," or "morbid tourism." The black spot was given to the graves, and such sites where celebrities or masses met death accidentally (Rojek 1993). Thanatourism refers to the visit to the places where the visitor's prime motivation (wholly or partially) is to have a chance to actually or symbolically encounter death. The death may be violent or not but may be activated by specific features of the person who is the focal object (Seaton 1996). The atrocity tourism was related to the acts of violence and atrocity like the holocaust (Fonseca et al. 2016), and morbid tourism was defined as the travel to places with accidents and violent deaths (Blom 2000).

As an umbrella term associated with the tourism of sites related to death and misery, the topic has attracted many studies (Dann 1994). For example, Rojek (1993) probed into the increasing interest of tourists to visit graveyards and cemeteries having labels of celebrities buried there. Lennon and Foley (2000) also wrote a monograph describing their ideas about the visit to places of death and macabre, where they viewed dark tourism as a form of cultural tourism with distinctive features. According to the monograph, the term dark tourism creates doubts and anxiety as it is defined as the visit to the sites having incidents of disaster, atrocity, and death.

2.2 *The dark tourism spectrum*

Only a few researchers have studied the supply side of dark tourism. Some of these sites are actual, while the countries have intentionally created others for entertainment and monetary purposes. Stone (2010) designed a spectrum to categorize dark tourism into sections according to their lightness and darkness. This degree is determined by the various attributes on opposite panels (e.g., education and entertainment) (Light 2017; Virgili 2018). The lighter sites have less political influence and ideology and are developed for commercial purposes and entertainment.

On the other hand, the darkest sites are associated with higher administrative supervision and ideological affiliations. They are the actual sites of incidents of deaths and disasters, such as concentration camps (Liyanage et al. 2015). However, this spectrum lacks any description of experience and motivations (Biran et al. 2011), which are vital role players in tourism. Despite this shortcoming, this study will include the spectrum as it explains the supply side of dark tourism.

3 MATERIAL AND METHODS

The purpose of this paper was to qualitatively review dark tourism research in the hospitality and tourism literature by using a keyword of dark tourism. From many articles available and by using different aspects of the dark tourism sites, 15 research papers on the subject of dark tourism from the 2000s onwards were randomly selected to study the reasons why countries allow dark tourism. As a result, the following articles of different aspects of dark tourism were collected: one

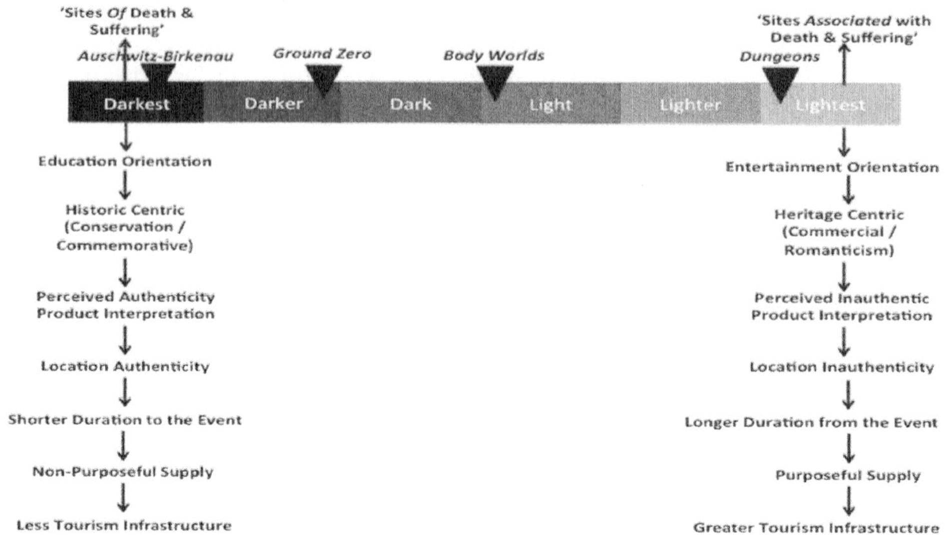

Figure 1. Dark tourism spectrum (Stone, 2010, p. 174).

article about the natural disaster (earthquake), one article about mass murder, two articles about battlefield tourism, one article about the holocaust, one article about dungeon tourism, one article about cemeteries tourism, and eight articles about definitions, concepts, theories, and typologies.

From the list of selected aspects, one article was chosen from 2003, 2006, 2010, and 2015, two articles from 2014, 2016, 2017, and 2018 and three articles from 2012. Then, the articles' data were compiled in the form of a table (Table 1) with descriptions of three main factors of study from the perspective of a country al-lowing dark tourism.

4 FINDINGS AND DISCUSSION

The data analysis of the 15 research articles (Table 1) shows that the primary motivation for countries to pro-mote dark tourism sites is for education and knowledge building about the history and cultures and to keep a record of the incidents related to the sites. Also, the finding shows the country's motivation to open up such touristic sites only if the sites are related to nationalism, national identity, ideology, commemoration, and remembrance.

However, the commercial aspect of dark tourism from a country's perspective is missing in the existing literature. Even though it shows that the commercial use of dark tourism sites is growing, the countries still do not show much interest in commodifying dark sites to generate revenues. The commodification of dark sites is at an early stage and faces ethical debates and challenges. According to Minic (2012), the dark sites present a set of complex moral and ethical standards for the governments and managers. The development of dark tourism on a commercial scale is also facing the challenges of heritage, political ideology, sacredness, and media coverage. It needs policy measures and substantial administrative implications to be beneficial for both tourists and the communities.

Furthermore, the study of the current work and the theoretical framework of the spectrum of dark tourism presented by Stone (2006) reveals that tourist sites falling in the ranks of lighter shades in the spectrum are used for entertainment and commercial activity. Although it generates revenues for the country, no research study found focused on this avenue of financial gains in the national economies. Moreover, no study found focused on seeing it at the leadership or governance level. The analysis of the data collected further makes it clear that each of the dark tourism sites

Table 1. Randomly selected articles of dark tourism.

Year	Title	Author(s)	Journal	Purpose		
				Education	Economy	Motivation
2018	"'From the Flames to the 'Light": 100 years of the commodification of the dark tourist site around the Verdun battlefield	Sandrine Virgili, Hélène Delacour, Frédéric Bornarel, Sébastien Liarte	Annals of Tourism Research	Brief reference to education and knowledge	In-depth discussion of the commodification process and economic potential for the country by creating the battlefield as a tourist attraction	No reference to the motivation
2018	Dark tourism and affect: framing places of death and disaster	Anna Claudia Martini & Dorina Maria Buda	Current Issues in Tourism	Made references to the visit of dark tourism sites for educational purposes and knowledge	Not discussed the economy of dark tourism from a state perspective but a general point.	No reference to a country's motivation using dark tourism for national identity
2017	Progress in dark tourism and thanatourism research: An uneasy relationship with heritage tourism	Duncan Light	Tourism Management	A detailed study about the education and knowledge gained from the dark tourism sites	A detailed study of the commodification of dark tourism places by alteration, revitalization or creation of infrastructure to generate revenue but from the perspective of commercial activity	Briefly discussed how different types of dark sites according to their nature are used for creating a positive image or national identity
2017	Terrorism and tourism in France: the limitations of dark tourism	Hugues Seraphin	Worldwide Hospitality and Tourism Themes	Dark tourism is majorly used for educational purpose	Dark tourism sites are commercialized to get economic benefits	No reference to the motivation
2016	Dark Tourism: Concepts, Typologies and Sites	Claudia Seabra, Carla Silva	Journal of Tourism Research & Hospitality	Discussion of different dark tourism sites and the central role is educational according to the nature of the site	Only a brief reference to the contribution of dark tourism in the economy	Discussion about the supply of dark tourism sites but not motivation from supply-side

(continued)

Table 1. Randomly selected articles of dark tourism.

				Purpose		
2016	Interrogating spaces of and for the dead as "'alternative 'space'': cemeteries, corpses and sites of dark tourism.	Craig Young, Duncan Light	International Review of Social Research	Discussed the cemeteries are subject of educational studies	No reference to the economic benefit	Discussed the motivation behind national recognition of cemeteries as the dark site only to commemorate the sacrifices and honor the dead
2015	Dark destinations-Visitor reflections from a holocaust memorial site	Liyanage, Sherry, Coca-Stefaniak, Andres and Powell, Raymond	International Journal of Tourism Cities	Discussed that holocaust sites mainly serve the educational purpose for nations	Although, tourists visit these sites but countries do not promote for gaining economic benefits	The motivation behind opening such sites for tourism is national identity and adhering to ideologies
2014	Dark touristic perception: Motivation, experience and benefits interpreted from the visit to seismic memorial sites in Sichuan province	Yong Tang	Journal of Mountain Science	The purpose of visiting the seismic site is satisfying curiosity, education and benefit drawn is knowledge (Model on pg. 1338)	Discusses the economic power of dark tourism sites	The data analysis and study reveal the motivations behind visiting as Education Commemoration
2014	Dark Tourism as a Type of Special Interest Tourism: Dark Tourism Potential of Turkey	Yunus Topsakal, Remziye Ekici	Journal of Management Studies	No Reference	No Reference	Discussed the promotion of dark tourism sites to create a positive image of destinations
2012	Benefits of visiting a "'dark' tourism" site: The case of the Jeju April 3rd Peace Park, Korea	Eun-Jung Kang, Noel Scott, Timothy Jeong Yeol Lee, Roy Ballantyne	Tourism Management	The state's key purpose is to educate the citizens about the tragedy by allowing this dark tourist site	No discussion about economic intentions	The motivation of education, commemoration, nationalism, ideology and sympathy

(continued)

Table 1. Randomly selected articles of dark tourism.

				Purpose		
2012	Dark Tourism and Significant Other Deat Towards A Model Of Mortality Mediation	Philip R. Stone	Annals of Tourism Research	Dark tourism sites are promoted for education	Discussed the entertainment side of dark tourism but not from a national point of view	A brief reference to the motivation of Commemoration
2012	Development Of "Dark Tourism" in the Contemporary Society	Natalija Miniæ	Journal Of The Geographical Institute Jovan Cvijic, SASA	Most discussion about the educational aspect of dark tourism	A brief reference to the contribution and role of dark tourism in the economy	No discussion why states promote dark tourism, from motivation aspect
2010	Motivations for War-related Tourism: A Case of DMZ Visitors in Korea	James D. Bigley, Choong-Ki Lee, Jinhyung Chon & Yooshik Yoon	Tourism Geographies	Discussed and analyzed the purpose of the opening which is majorly education (shown in the results)	Briefly discussed the role of dark tourism in the national economies	The national motivation for opening this site is education and remembrance, national identity
2006	Philosophical and methodological praxis in dark tourism: Controversy, contention and the evolving paradigm	A. Craig Wight	Journal of Vacation Marketing	A brief reference to the educational aspect of dark tourism sites	Discuss commercial and economic aspect only from a research point of view	No reference to the motivation
2003	Shades of Dark Tourism Alcatraz and Robben Island	Carolyn Strange Michael Kempa	Shades of Dark Tourism Alcatraz and Robben Island	Not much reference to the education	Discussed commercial aspects of dark sites of dungeons but not from the perspective of a country's interest in earning revenue	There is no national motive except entertainment and heritage to preserve these sites

has its specificity and association. The darkest sites like battlefield and holocaust are associated with education, commemoration, nationalism, identity, ideology, and sympathy, but they are not preferred for revenue generation. However, in some cases, like the Verdun battlefield of France, the study narrates the complete process of commodification and reaping the site's economic potential (Virgili et al. 2018).

The study explains the objectives of the research adequately. It actively supports the first objective that countries do allow dark tourism for educational purposes. The second objective, if the countries promote dark tourism for income generation, is not supported by the study, as it depends on the nature of the tourist site, and there is a lack of evidence from the existing research. Several studies support the third objective that sites related to nationalism and ideologies are visited with the motivation of commemoration and remembrance. The interpretation of events is mostly authentic, according to the Stone (2006), *Dark Tourism Spectrum*.

The study made use of the 15 research articles from the existing literature on dark tourism, covering different aspects explained in the data section. The articles' critical analysis shows that out of 15, 11 articles discuss dark tourism used for educational purposes. Seven articles made a brushing review about the commodification and the commercial use of the dark tourism sites from the perspective of income generation. Still, only two studies carried out a detailed discussion about the government's involvement in the commercialization of the Verdun Battlefield's dark tourism site for its economic potential (Virgili et al. 2018). The national motivation behind allowing dark tourism places is discussed by the seven articles, which mostly fall in the darkest shades of the spectrum. These researches are about Korean Jeju Peace Park, Seismic memorial sites of Sichuan Province, Demilitarized Zones of Korea, Special Interest Tourism sites of Turkey, holocaust memorial sites, and the cemeteries.

5 CONCLUSION

The study results endorse the dark tourism spectrum in light of data analysis using a qualitative approach. The data analysis of the 15 research articles of recent times shows that the countries open and promote dark tour-ism for education and knowledge. It also suggests that for some dark sites, the countries show motivation, and these sites are strongly related to their national identity, ideology, commemoration, and reverence.

The research study has covered the key points raised in the problem statement, but it has certain limitations. This study's main limitation is the availability of the data due to the keywords used to search available published articles. The articles were also randomly selected. Hence, future research should apply more keywords using the type of dark tourism, according to Fonseca et al. (2016). Furthermore, future research can be conducted using qualitative and quantitative data by conducting interviews and surveys to get a clearer picture of the approach of countries toward dark tourism.

REFERENCES

Bigley, J., Lee, C., Chon, J., & Yoon, Y. (2010). Motivations for War-related Tourism: A Case of DMZ Visi-tors in Korea. *Tourism Geographies*, 12(3), 371–394. https://doi.org/10.1080/14616688.2010.494687

Biran, A., Poria, Y., & Oren, G. (2011). Sought experiences at (dark) heritage sites.*Annals of Tourism Re-search*, 38(3), 820e841.

Blom T (2000) Morbid tourism - A postmortem market niche with an example from Althorp. *Nor J Geogr* 54: 29–36.

Dann, G. M. S. (1994). *Tourism: The nostalgia industry of the future. In W. Theobald(Ed.), Global tourism: The next decade (pp. 55e67).* Oxford: ButterworthHeinemann.

Foley, M., & Lennon, J. (1996). JFK and dark tourism: A fascination with assassination. *International Journal of Heritage Studies*, 2(4), 198–211.

Fonseca, A., Seabra, C., & Silva, C. (2016). http://www.scitechnol.com/peer-review/issues-of-croatian-touristic-identity-in-modern-touristic-trends-VJGq.php?article_id=4907. Retrieved April 11 2020, from http://dx.doi.org/10.4172/2324-8807.S2-002.

Kang, E., Scott, N., Lee, T., & Ballantyne, R. (2012). Benefits of visiting a 'dark tourism' site: The case of the Jeju April 3rd Peace Park, Korea. *Tourism Management*, 33(2), 257–265. https://doi.org/10.1016/j.tourman.2011.03.004

Lennon, J. J., & Foley, M. (2000). *Dark Tourism: The Attraction of Death and Disaster*.London: Continuum.

Light, D. (2017). Progress in dark tourism and thanatourism research: An uneasy relationship with heritage tourism. *Tourism Management*, 61, 275–301. https://doi.org/10.1016/j.tourman.2017.01.011

Liyanage, S., Coca-Stefaniak, J., & Powell, R. (2015). Dark destinations – visitor reflections from a holocaust memorial site. *International Journal ofTourism Cities*, 1(4), 282–298. https://doi.org/10.1108/ijtc-08-2015-0019

Logan W (2009) Remembering places of pain and shame. In R. Logan, K.Reeves (Eds.), *Places of pain and shame: Dealing with "difficult heritage" (pp. 1–14).* London: Routledge.

Martini, A., & Buda, D. (2018). Dark tourism and affect: framing places of death and disaster. *Current Issues In Tourism*, 23(6), 679–692. https://doi.org/10.1080/13683500.2018.1518972

Minic, N. (2012). Development of "dark" tourism in contemporary society. *Journal Of The Geographical In-stitute Jovan Cvijic*, SASA, 62(3), 81–103. https://doi.org/10.2298/ijgi1203081m

Rojek C (1993) *Ways of escape: Modern transformations in leisure and travel.* Basingstoke, Australia: Macmillan.

Rojek, C. (1997). Indexing, dragging and the social construction of tourist sights. InC. Rojek, & J. Urry (Eds.), *Touring cultures: Transformations of travel and theory (pp. 52–74)*. London: Routledge.

Seaton, A. (1996). Guided by the dark: From thanatopsis to thanatourism. *International Journal of Heritage Studies*, 2(4), 234–244.

Seraphin, H. (2017). Terrorism and tourism in France: the limitations of dark tourism. *Worldwide Hospitality And Tourism Themes*, 9(2), 187–195. https://doi.org/10.1108/whatt-09-2016-0044

Smith VL (1996) War and its Tourist Attractions. In A. Pizam, Mansfeld Y(Eds.), *Tourism, Crime and Security Issues*. (pp. 247-264) Chichester: Wiley, 18.

Smith VL (1998) War and tourism: An American eth-nography. *Ann Tourism Res* 25: 202–227.

Strange, C., & Kempa, M. (2003). Shades of dark tourism. *Annals of Tourism Research*, 30(2), 386–405. https://doi.org/10.1016/s0160-7383(02)00102-0

Stone, P. R. (2006). A dark tourism spectrum: Towards a typology of death and macabre related tourist sites, attractions, and exhibitions. *Tourism: An Interdisciplinary International Journal*, 54(2), 145–160.

Stone, P. R. (2010). *Death, dying and dark tourism in contemporary society: A theoretical and empirical analysis*. Unpublished doctoral dissertation, University of Central Lancashire (UCLan), United Kingdom.

Stone, P. (2012). Dark tourism and significant other death. *Annals of Tourism Research*, 39(3), 1565–1587. https://doi.org/10.1016/j.annals.2012.04.007

Stone, P. (2016). Enlightening the 'dark' in dark tourism. *Interpretation Journal*, 21(2), 22e24.

Tang, Y. (2014). Dark touristic perception: Motivation, experience and benefits interpreted from the visit to seismic memorial sites in Sichuan province. *Journal of Mountain Science*, 11(5), 1326–1341. https://doi.org/10.1007/s11629-013-2857-4

Topsakal, Yusuf Ekici, Remziye. (2014). Dark Tourism as a Type of Special Interest Tourism: Dark Tourism Potential of Turkey. *AkademikTurizmveYönetimAraştırmalarıDergisi*. 1.

Virgili, S., Delacour, H., Bornarel, F., & Liarte, S. (2018). 'From the Flames to the Light': 100 years of the commodification of the dark tourist site around the Verdun battlefield. *Annals of Tourism Research*, 68, 61–72. https://doi.org/10.1016/j.annals.2017.11.005

Wight, A. (2006). Philosophical and methodological praxes in dark tourism: Controversy, con-tention and the evolving paradigm. *Journal of Vacation Marketing*, 12(2), 119–129. https://doi.org/10.1177/1356766706062151

Young, C., & Light, D. (2016). Interrogating spaces of and for the dead as 'alternative space': ceme-teries, corpses and sites of Dark Tourism. *International Review of Social Research*, 6(2), 61–72. https://doi.org/10.1515/irsr-2016-0009

Promoting Creative Tourism: Current Issues in Tourism Research – Kusumah et al. (Eds)
© 2021 Taylor & Francis Group, London, ISBN 978-0-367-55862-8

Understanding the motivations and preference on ecotourism development: The case of Gunung Leuser National Park, Indonesia

Amrullah, A. Rachmatullah, Nurbaeti, F. Asmaniati & S.P. Djati
Trisakti School of Tourism, Jakarta, Indonesia

ABSTRACT: This study aimed to analyze the orientation of the stakeholders in the development of ecotourism in Gunung Leuser National Park. The framework of the approach used in this research was phenomenology, which was then enriched with the data collection techniques of study documentation, observation, and close-ended questionnaire. The analytical method utilized was One Score One Indicator Scoring System, which was an analysis model that was used through developing elaboration of questionnaires in collecting data and evaluating various variables that had been determined by researchers. The results of the study revealed that various actors (communities, government and tourist) stated high scores or were meaningful both for the development of ecotourism in Gunung Leuser National Park area. Data on motivation and ecotourism reference showed high scores on the distribution of economic, ecological, and socio-cultural benefits. The high economic orientation of the community and government was an important determinant in maintaining the ecological and socio-cultural order, so that it made positive energy to be developed in the development of ecotourism as a whole and integrated. Considering the number of objective approaches made, the synthesis initiated in this study was to optimize several perspectives including: 1) Ecotourism Planning Perspective; 2) Ecotourism Political and Regional Policy Perspective; 3) Collaboration and Partnership Management Perspective.

Keywords: ecotourism, motivation, preference, Gunung Leuser National Park, One Score One Indicator Scoring System

1 INTRODUCTION

Since the last three decades, the acceptance of ecotourism terminology started from the shifting of the anthropocentrism paradigm to biocentrism to ecocentrism. Understanding ecosophy mandated into every regulation and development policy (widely through sustainable development) and sustainable tourism development/ecotourism throughout the world and including Indonesia is a manifestation of the “back to nature” movement to bring public concern to environmental sustainability and/or socio-environmental responsibility. Simply stated, Avenzora (2008) asserts that the paradigm shift in the tourism sector can be seen from two fundamental reasons, namely internal dynamics and external dynamics. Internally, the change is caused by a natural shift of trend, while externally it is caused by a change as a result of the political-pressure of worldwide environmental movements (Avenzora 2008). Mowforth and Munt (1998) explained that there were three main issues that caused the shift, namely, uneven and unjust development, power relations, and globalization. The concept of ecotourism as part of sustainable development is basically intended to improve the welfare and quality of human life, with efforts to meet the needs of human life across generations (Directorate General of Conservation of Natural Resources and Ecosystems/DG KSDAE 2018).

Various ecotourism principles raised by scientists around the world in general are that ecotourism must be able to contribute in minimizing global warming and alleviating other environmental issues, as well as fostering the socio-economic dignity and status of the community. Santiago and

 DOI 10.1201/9781003095484-36

Libosada (1994) explain that "environmentally sound tourism sustainability is implemented in a given ecosystem to yield equitable social and economic benefits and to enhance the conservation of natural and cultural resources." In the economic context, the development of ecotourism has contributed at least 5.5% of Australian gross domestic product, while Malaysia, with its 20 ecotourism destinations, is able to earn foreign exchange of around 13.4 billion ringgit per year (or equivalent to 35 trillion rupiah) from the tourism services sector (Nugroho 2011). In terms of ecology, Nugroho (2011) said that Brazil as the first megabiodiversity country in the world has been able to control the economic benefits of ecotourism without having to damage the Amazon forest environment, which is rich in exotic flora and fauna and other biodiversity. In the context of the theory about the benefits of ecotourism on socio-culture, Perdue et al. (1990) have confirmed that almost all studies have reported a positive relationship between economic benefits and attitudes toward tourism development. Gursoy and Rutherford (2004) said that in addition to tourism, a tool to reduce economic problems such as unemployment, people in fact also see tourism as providing social benefits and cultural preservation.

Indonesia as a megabiodiversity country has a "myriad" of ecotourism resources and destinations, both stored within conservation areas and outside conservation areas. One of the conservation areas that has very high ecotourism resources is the Gunung Leuser National Park (GLNP) area. MacKinnon and MacKinnon (1986) state that the Leuser ecosystem is a habitat for important animal species on the Sunda mainland, where there are 65% of Sumatran mammals (129 species of mammals from 205 species), and habitat for 380 bird species. In addition to being home to key fauna, GLNP also has 4,000 species of flora (including 3 species from 15 species of Rafflessia parasites), and there are many medicinal plants (Brimacombe & Elliot 1996). In addition, the abundance of available resources has also brought the GLNP region to the status of Tropical Rainforest Heritage of Sumatra and World Heritage Site by UNESCO in 2004. Long before that, GLNP was also established as a Biosphere Reserve in 1981 and ASEAN Heritage Park in 1984 by UNESCO. The Leuser Ecosystem, which is surrounded by biological richness, makes this conservation area dubbed the largest and richest tropical sanctuary on earth (Directorate of Utilization of Conservation Forest Environmental Services 2016). In order to realize the development of ecotourism in conservation areas, the harmony and balance of various major interests become necessary to promote the distribution of economic benefits, ecological interests, and socio-cultural interests. The Directorate General of KSDHE (2018) noted that in 2018 more than 7.88 million people visited the conservation area, both for recreational purposes, activities in the wild, education and development of science, research, and others. The visitors to the conservation area consist of 7.37 million domestic visitors and 511,017 foreign visitors. From an economic standpoint, the visit produced non-tax state revenue of Rp. 167,833,158,335. While specifically for GLNP itself, the economic value received in 2018 is Rp. 1,738,194,000 (Directorate General of KSDAE 2018).

Although the development of ecotourism in the GLNP area and in other conservation areas in general is impressive and fantastic in terms of the distribution of benefits, in reality the inherent dynamics of management often hear the dynamics of conflicts of interest that actually cover and hinder the optimum value of potential resources and existing added value. The dynamics of orientation between actors in the development of ecotourism in conservation areas often lead to controversy that makes it a long discourse that seems "just walking in place." If it is simplified, it is strongly suspected that the problem is always trapped on the issue of inequality and unfair distribution of benefits, and even trapped in the issue of "collusive oligopoly" which will always cause social welfare among the lower classes and material loss of other resources. Considering various strategic issues regarding the benefits of sustainable tourism development as well as the policy mandate given to the development of ecotourism in the GLNP region, the development must be managed comprehensively and systemically, and wisely to remain consistent in developing the seven pillars of ecotourism development. In its concept of implementation, all actors must be brave and consistent in resisting the appetite of their neo-capitalism not to fall into "velocity of money," which is "blindly," but must be able to maintain the "rhythm" and harmonization of the sustainability of the ecological pillar and the socio-cultural pillar, and four other pillars raised by Avenzora

(2008) including the pillars of satisfaction, the pillars of experience, the pillars of memory, and the pillars of education. For this, harmonization and consolidation are carried out consistently by all development actors, be it the government, the private sector, non-government organizations (NGOs), and the community itself. Polarization of these perceptions has the potential to be an obstacle in the development of ecotourism; so, to overcome this, we need a harmonious perception among ecotourism development stakeholders (Winarno et al. 2015). The study is intended to understand each stakeholder orientation in ecotourism development in the GLNP region, which is the basis of overall ecotourism planning.

2 METHOD

2.1 *Time and research location*

The study was conducted from September to October 2018 in Gunung Leuser National Park (GLNP) Administrative Region of Ketambe Village, South East Aceh Region, Aceh Province. The location was chosen with the consideration that Ketambe Village was an administrative village located outside the conservation area.

2.2 *Research approach, research instrument and sampling technique*

The framework of the approach used was phenomenology, in which researchers described phenomena that occurred in the field based on the experience and cognitive understanding of researchers (Altinay & Paraskevas 2008). The data collection techniques used were: 1) study documentation, 2) observation; 3) questionnaire instrument (close-ended questionnaire).

The research instrument used was a closed questionnaire (close-ended questionnaire) with a Likert scale guide range of 1–7 scale (modification of the 1–5 Likert scale), reasoning the character of the Indonesian people who articulated a very detailed value (Avenzora 2008). Data obtained from the questionnaire instrument then analyzed using the One Score One Indicator Scoring System method, which was an analysis model that was utilized through the development of a questionnaire elaboration in collecting data and evaluating various variables that had been determined by researchers (Avenzora 2008).

The sampling technique used in the study was incidental sampling, in which anyone who incidentally (incidentally) met the researcher could be used as a sample, if the person was suitable as a source of data (Sugiyono 2012). Roscoe (1982, in Sugiyono 2010) stated that if the sample is divided into categories, then the number of sample members in each category is at least 30 respondents, so the total respondents taken were 90 respondents (30 communities, 30 governments, and 30 tourists).

2.3 *Method of analysis*

Various qualitative data would be processed and presented in tabulated descriptive manner, while quantitative data would be processed using basic descriptive statistical techniques in the form of frequency distribution. Frequency distribution indicated the number and percentage of respondents and the object of study included in the existing category to provide initial information about the respondent or object of study. Thus, the calculation of frequency distribution could be calculated based on the arithmetic mean or mode. The analysis of One Score One Indicator Scoring System, which was an analysis model used through the development of a series of questionnaire elaboration in collecting data and evaluating various variables that had been determined by researchers (Avenzora 2008; Avenzora et al. 2013). This method was utilized to minimize subjectivity and simplify the various components of statements and/or questions arranged in the form of questionnaire, which was then analyzed descriptively and qualitatively as material for consideration to achieve optimum results.

Then to understand differentiation between actors, there were two important issues examined in the analysis of polarization, namely polarization of direction and rating scale. Direction polarization occurred when scores among actors were divided into two dimensions, namely, scores below 4 (3, 2, and 1) and scores above 4 (5, 6, and 7); whereas rating scale polarization occurred when there was absolute score differentiation even though it was in the same dimension (Rachmatullah 2018).

3 RESULTS AND DISCUSSION

3.1 *Dynamics of motivation and preference on ecotourism*

Economic Motivation of Ecotourism. In the context of economic motivation, the data shows that there is no difference in attitude scale as both actors agree (Score 6) over the motivation of ecotourism economics (Figure 5A). Both actors are aware that ecotourism development is not only able to open jobs and increase income but also able to stimulate local infrastructure development in order to increase the distribution of people's economic benefits in their environment. According to Drumm and Moore (2002), aside from ecotourism, it is considered capable of minimizing the impact of ecological conditions; ecotourism is also considered capable of making important contributions in improving the economy of local communities. Nugroho (2011) noted that from the visits of 273 tourists to national parks in the United States, 10 billion dollars of tourist expenditure was obtained, 200 thousand workers and 22 billion dollars entered into economic output each year. In Indonesia alone, according to the Ministry of Tourism Pocket Book (2016), the contribution of the tourism sector to the national GDP in 2014 has reached 9% or as much as Rp. 946.09 trillion, while foreign exchange from the tourism sector in 2014 reached Rp 120 trillion and contributed to employment opportunities of 11 million people (Anggraini 2017).

Motivation for Ecotourism Ecology. In general, the data shows that there is no significant discrepancy between actors regarding ecotourism ecological motivation. In the diagram (Figure 5B), it appears that the community and the government/manager agree (Score 6) of the overall indicators of ecotourism motivation. Both actors believe that ecotourism development can improve the natural mind and psychomotor of the community in protecting the Gunung Leuser NP area. In addition, with the ecotourism downstream industry, the community will not only participate in preserving environmental sustainability but will also actively participate in maintaining the integrity of the function and aesthetics of all ecological elements in the conservation space and its residential environment. Garrod (2003) explains that ecotourism is also something that is put forward as a strategy to help social and economic issues in the local community, as well as an effective and relevant tool for environmental conservation.

Socio-Cultural Motivation of Ecotourism. In many cases, the data shows no signs of polarization of direction among stakeholders on the socio-cultural motivation of ecotourism (Figure 5C). Overall, both the community and the government gave a score of 6 or good significance on the socio-cultural motivation of ecotourism. This can be interpreted that there is a conceptual harmony in looking at the socio-cultural benefits of ecotourism. Alignment of cultural motivation is allegedly due to the degradation of the cultural fabric of society in various regions due to the strong development of technology and information; so, we need an alternative development instrument (ecotourism) which is believed to be able to revitalize the cultural order (socio-cultural) of society to become more entrenched. Stronza (2007) states that ecotourism is not only an economic "tool" for conservation, but also to strengthen new cultural values and social relations. Weaver (2002) explains that "ecotourism is a form of tourism that fosters learning experiences and appreciation of the natural environment, or some components thereof, within its associated cultural context." It has the appearance (in concert with best practice) of being environmentally and socio-culturally sustainable, preferably in a way that enhances the natural and cultural resource base of the destination and promotes the viability of the operation (Weaver 2002).

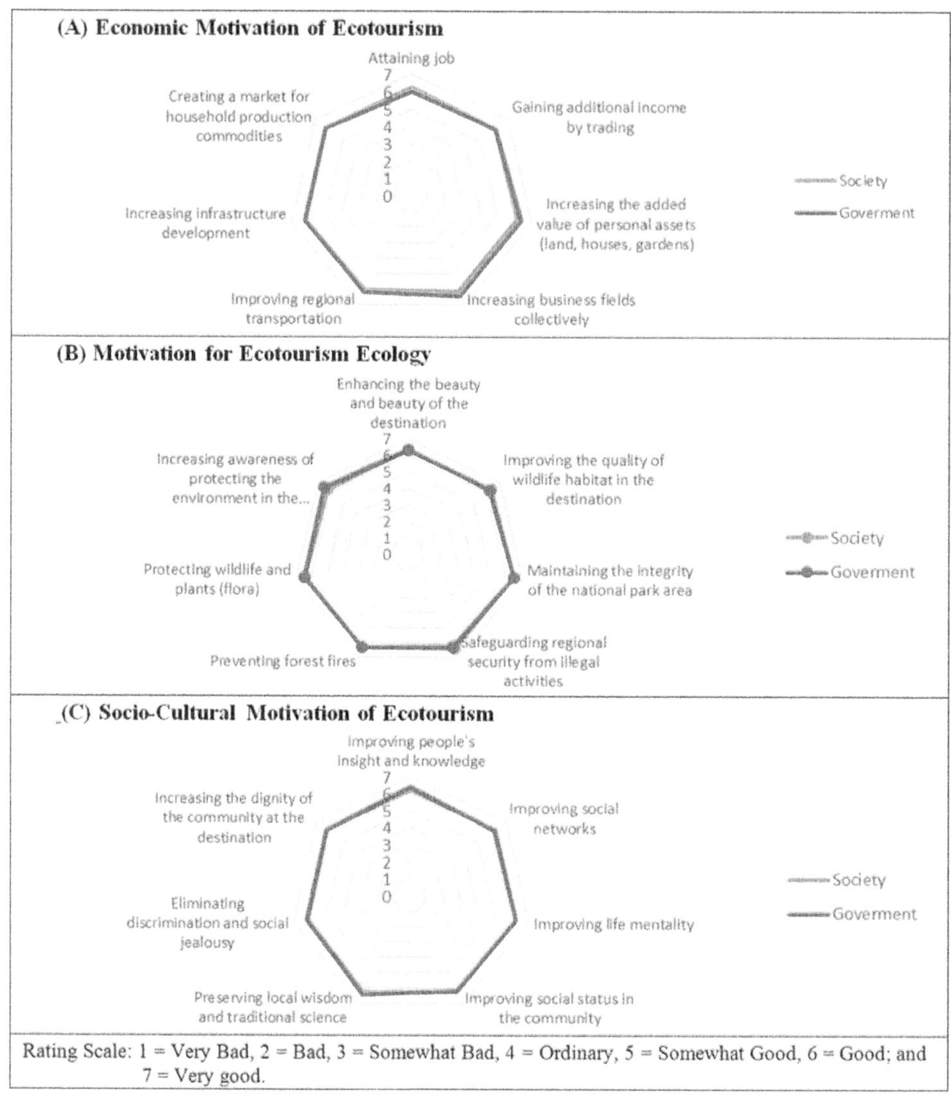

Figure 1. Ecotourism, economic, ecological and socio-cultural motivation.

Tourist's Push and Pull Motivation. In terms of push motivation, available data (Figure 4) shows that the value of tourist motivation in the GLNP region in general is only relatively high (score 5), whereas the highest indicator declared by tourists (score 6) is learning and/or understanding something new to increase knowledge. While in terms of pull motivation, pull data shows that the average value of tourist attraction is to produce a rather high meaning (score 5). It can be understood that the Ketambe GLNP area is still relatively unfamiliar compared to the Tangkahan GLNP ecotourism area, thus causing no optimum pull motivation value. In several studies of tourist motivation, the average primary motivation of a tourist is for the purpose of relaxation and physical refreshment and mind (Abbas, 2000; Fandeli 2002; Reindrawati 2010). The motivation is still the main push factor, while the pull factor is still dominated by natural and cultural ecotourism resources, such as local life styles and eco-activities (Chan et al. 2007; Ros & Iso-Ahola 1991).

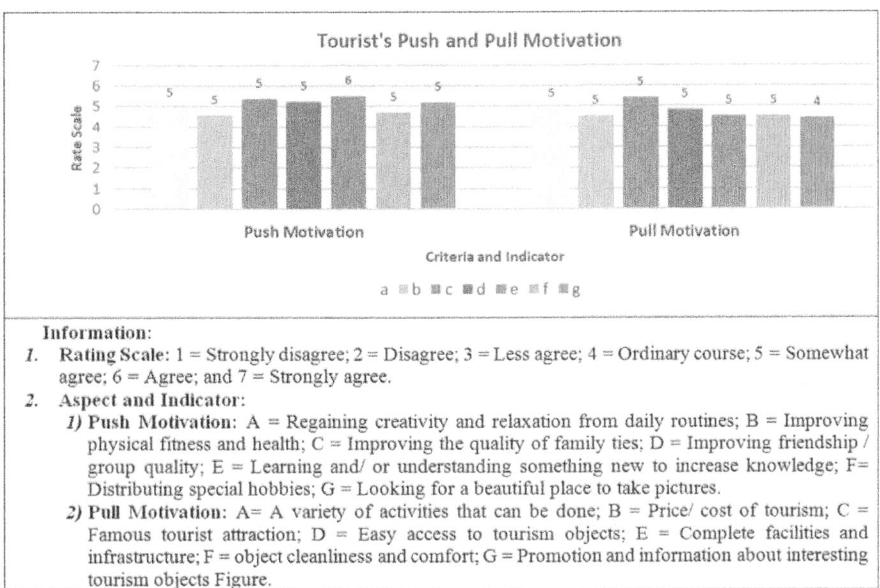

Figure 2. Tourist's push and pull motivation.

Preference on Service Facility. In various criteria, many things prove that stakeholders emphasize the importance of various service facilities in the overall development of ecotourism in the GLNP region. The community and the two other actors prove the high preference (Score 6) for the construction of various service facilities. The high expectations of the community indicate their strong desire and understanding that the various physical developments of the service facilities are developed in order to increase tourist visits and comfort. Good quality facilities and services that pay attention to stakeholder preferences will also have a good impact on the sustainability of ecotourism in an area. Development of physical facilities as an increase in added value in an ecotourism area should adopt the concept of eco-design that is not only superior in aesthetic quality but also in terms of meeting the needs of visitors and adaptive to environmental resources and local socio-cultural life (Pratiekto 2013).

Preference on Infrastructure. Research data shows the high preference of each actor over infrastructure development preferences. The community gives the highest score (Score 7) on infrastructure development preferences. For the community, infrastructure development such as improving main roads and footpaths, electricity and telecommunications networks, and parking space circulation are not only able to make tourists prioritize the principle of length of stay but also can stimulate the community itself to provide good service to tourists. As for the government itself, the average value (Score 6) given by them also shows the strong understanding and desire of the government itself to carry out various infrastructure developments in the GLNP region. The government also said that the distribution of infrastructure development is important in addition to building human resource capacity, which is basically equally fighting for the prosperity of the people. According to Nugroho (2011), infrastructure functions not only as an access method or for economic benefits from the viewpoint of the interests of visitors but also social benefits while supporting the values of environmental conservation.

Preference on Accommodation. Based on the results of the study, the data shows that the score indicated by the community is higher (Score 6) compared to the other two actors. The community considers that homestay development can directly stimulate the economic growth of the community when compared to those who only work in the hotel or private cottages. The average score of 5

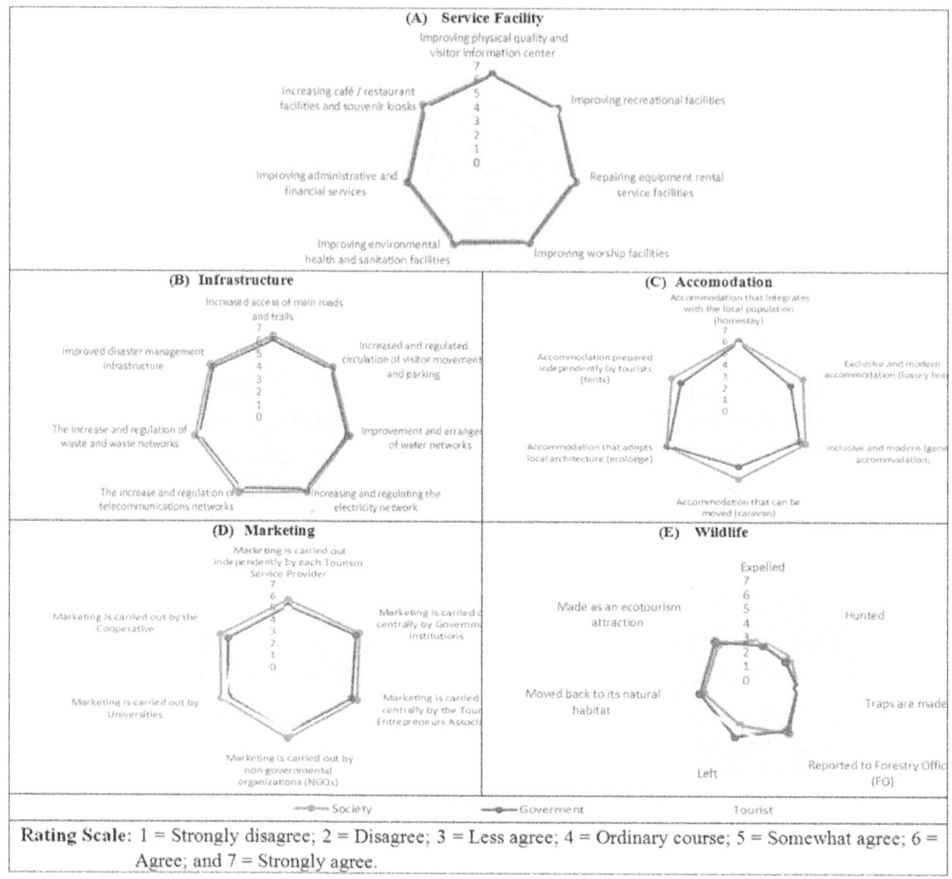

Figure 3. Ecotourism preferences from service facilities, infrastructure, accommodation, marketing, and wildlife.

means somewhat agreed by the government and tourists for the development of various accommodations in the GLNP region. For the government itself, the concept and mechanism of homestay development, in principle, must uphold the justice and prosperity of the people where in each item the accommodation indicator is adaptable to the socio-cultural spirit of the surrounding community. As for tourists themselves, the development of a decent homestay also needs to focus on the criteria of cleanliness, safety, and comfort. According to Ming Gu and Poh Wong (2006), "homestay is not a new concept and its popularity is catching up with many tourists opting to stay in homes and experience the local culture first-hand, rather than checking into more expensive hotels. Tourists eat whatever homestay operators cook for themselves and homestay operators treat tourists more like members of the family. Locally owned and operated, homestays constitute a suitable small-scale tourist accommodation form for the local community to participate in tourism" (Wall and Long, 1996).

Preference on Marketing. The data shows that the preference of stakeholders in the overall ecotourism marketing preference is positive (Score 6). This indicates the seriousness of the actors to distribute information on overall ecotourism resources to various parties in order to bring in tourists. If examined, the high preference is in the local community groups where the preference scores expressed produce arithmetic mean which is rather high (Score 5). For the people themselves, whoever does marketing or a growing number of social groups marketing GLNP ecotourism resources are the same and even considered better because the distribution of information is more colorful and widespread. As for the government itself, the overall score that produces a score of 5 (somewhat agree) is to have meaning including: 1) the government wants ecotourism marketing to be carried out in a participatory manner by all actors; 2) they want marketing carried out simultaneously and integrated so that the information can be carried out objectivity. In this case, it can be said that the government has really realized the importance of ecotourism marketing and promotion, which should be done in an integrated manner to create a climate of sustainable tourism. In the context of visitors, they actually do not think partially who is more entitled to market ecotourism information, but what is more important for them is the content or essence of marketing carried out by various parties is not impressed, "beautiful news from the form," so as not to cause disillusionment to visitors alone. The average value of 5 that was given also emphasized the importance of infographic material that not only contained cheap tour packages but also emphasized the variant of existing ecotourism resource attractions, especially various eco-cultural-tourism attractions. With this cultural ecotourism variant, the tourism sector is not only able to boost the motivation of potential tourists but also can open up opportunities for various other economic indicators to prosper the local community.

Preference on Wildlife. The results of the study show that there is no polarization of stakeholder directions for the preferences of wildlife management and its ecosystem. In this case, the community stated that they totally disagreed (Score 3) on the preferences of wildlife and its ecosystem. When viewed in the first, second, and third indicators, the data shows that the community actually does not agree if wildlife in the forest/GLNP is hunted, evicted, or commercialized. This shows that there is a deep understanding from the local community that various elements of the forest are worth keeping and preserving as stated in the slogan of their ancestors. The same thing was shown by the manager/government, which stated that the hunting of wild fauna in the GLNP area was strongly opposed and violated the law. However, if these wild animals are used as ecotourism attractions, the government in this case is positive or has agreed to become a sustainable environmental services business.

In the context of visitors, the data shows a mean average score (Score 4) over the management preferences of wildlife and its ecosystem. The data shows tourists who oppose various types of hunting and commercialization of protected wildlife. The highest score stated by tourists is on indicators a and b, which in both indicators produce a score of 7 and a score of 6. Visitors think that for wildlife that are outside their natural habitat, it is not natural for them to be hunted or trapped, but they should be moved back to GLNP through various animal management techniques. Then in terms of attractions, visitors agree (Score 6) if wildlife is used as an ecotourism attraction. For visitors, with the attraction of wildlife observation, it can automatically provide a positive value in enhancing the visitor's experience and cognitive space and they take part in various pro movements.

3.2 *Discussion*

In an effort to harmonize the polarization of stakeholder motivation and preferences, the basic idea that must be carried out is to apply collaborative management and sectoral integration. This can be realized in the focused group discussion (FGD) and deliberation activities involving all relevant stakeholders, namely, the local community, government, and universities and non-governmental organizations as well as several travel bureaus. As a form of simplification in the strategic plan, the strategy used is the SWOT analysis; which results in the conclusion of an aggressive strategy. Like IFA and EFA, an aggressive strategy can be interpreted by optimizing two important aspects, namely, 1) political aspects and ecotourism territory policies, and 2) ecotourism planning aspects.

Political and Policy Aspects of the Ecotourism Region. In the domain of ecotourism, political terminology and policy/regulation can be interpreted as an effort to decide cases/issues wisely and fairly for all elements. Through politics, a behavior in interacting to achieve various goals can be carefully formed and various conflicts of interest can be arranged to reach mutual agreement as a form of justice. Meanwhile, Prittwitz (1994) concludes politics as "als Kunst, als Prozess der Selbstaufhebung des Staates, als Interessenkonflikt, als Regelungprozess im Verhältnis gesellschaftlicher Inputs und Outputs bzw; als kybernetischer Steurungsprozess oder als emanzipativer Kommunikationsvorgangg." Avenzora (2013) describes that tourism development is somehow based on various policies needed to ensure the efficient and effective functioning of each role of each stakeholder, in line with the fact that tourism is multisectoral so that the role of government (as a political component) becomes very important.

Nasution (2015) in his research found that in various criteria, the contents of the material substance of all ecotourism policies and regulations in Indonesia only resulted in a score of 5 or rather good. The suboptimal value can be said to not have clear comprehensive and structured objectives so that the dynamics of substance contained and/or contained in the law are still partial. This can be interpreted as follows: 1) the government has not been optimal in understanding the nature of the concept of ecotourism so that in the preparation of regulations there are ambiguities of concepts and meaning; 2) there is a tendency for ecotourism development with a political approach orientation; 3) low political will and objectivity of the drafters relating to ecotourism, which is strongly suspected due to the intervention of various business institutions. Considering existing political and policy dynamics, the orientation of ecotourism development in the GLNP area should be carried out integrally and systemically, and implementing scientific planning oriented. In addition, the government should be a regulator, both from the international scale, national scale, regional scale, up to the individual scale consistently coordinating thoroughly in determining various development priorities as well as the high motivation of the community in participating in ecotourism development in the GLNP region. Hall (1994) also reminded that the most important aspect in discussing regional involvement in tourism politics is the balance of power between the central government and regional governments.

Aspects of Ecotourism Planning. The planning aspect becomes very important not only because the ecotourism sector is multisectoral and multidisciplinary but also because of its characteristic and unique aspects in creating ecotourism resources. In planning for ecotourism to be carried out, in fact, many approaches can be chosen as Gold (1980) outlines as can be done using the approach: 1) demand; 2) resources; 3) use of space; 4) behavior. However, in reality, many practitioners only use the supply approach and demand approach, which actually is classified as dogmatic and prescriptive for a practitioner. But for an academic, the two approaches actually need to be enriched again given the complexity of ecotourism development planning, namely by paying attention to the policy approach, behavior approach, integrated approach, and a priory approach.

In the context of decision making in a planning process, Nelson (1999) reminds that one of the common mistakes made by stakeholders in planning is in the form of "their ignorance of what exactly the way the planning and decision-making process works." Avenzora (2003) reminded that it is necessary for all parties to realize that there are significant differences between the workings of practitioners, government, and academics. Furthermore, in general it is said that the planning process carried out by academics (wissenschaftliches plannung) will elaborate various objectives through a structured process, while the government has a tendency to plan based on political interests (political approach); whereas practitioners tend to put their problem priorities on the basis of the business institutions they own (leitsbild plannung). Considering the various dynamics of existing ecotourism planning, the most logical approach to simplifying "overlapping" ecotourism planning schemes is to adopt an integrated planning approach, which consists of an initial phase, an analysis phase, a synthesis phase, a planning phase, and an implementation phase. The preparation of concepts and strategies for ecotourism planning can be poured into the Master Plan, Site Plan, Detail Plan, to the Visitor Management Plan and the Strategic Plan for Destination Management.

Aspects of Collaboration and Partnership Management. In the practice of collaborative and partnership management, every actor who has an interest in the utilization of GLNP resources and ecotourism should work collectively and consistently adhere to the regulations and orientation of the existing grand plans. All actors such as local governments, GLNP managers, universities, NGOs, private parties, and local communities will play their respective roles, both as regulators, drafters, initiators, collaborators, and implementers protected by the legal umbrella. In order to support the big vision of conservation, there are currently regulations that strengthen investors/private parties to develop a Business Permit for the Provision of Natural Tourism Facilities (IUPSWA) in the GLNP conservation area, Natural Tourism Business Permit (IPPA) regulated in Minister of Environment and Forestry Regulation No. P. 8 / MENLHK / SETJEN / KUM.1 / 3/2019 concerning Nature Tourism Exploitation in Wildlife Reserves, National Parks, Grand Forest Parks, and Nature Tourism Parks. With this regulation, it can basically be interpreted that there is a political will that is quite serious from the relevant Ministries to optimize the benefits of sustainable development and conservation trilogy. In addition, the policy not only opens opportunities to strengthen collaborative and partnership practices to achieve benefit distribution but also fosters the spirit of conservation of all parties to jointly manage common pool resources. With efforts to strengthen collaboration and partnerships in the management of ecotourism services in the GLNP region, it will also indirectly be useful in minimizing and eliminating illegal wildlife, illegal poaching, and overfishing, which in essence optimize the role and function of stakeholders as khalifatullah on the GNLP Earth will become increasingly meaningful.

4 CONCLUSION

In general, the results of the study show that the community and government fully support the various processes of tourism development in the GLNP buffer zone of Ketambe Village. Stakeholders are aware that in addition to ecotourism being able to increase the distribution of economic benefits such as opening new jobs, increasing people's income or added value, as well as increasing demand for goods and services, the community also believes that tourism development is able to safeguard various ecological aspects and preserve the existing socio-cultural values. The phenomenon of direction polarization and polarization of the scale of attitudes that occur between actors in several indicators of ecotourism development in the GLNP region are classified as normative so that they are not to be used as a fundamental obstacle, but rather must be encouraged and made as a strong motivation to engage consistently in integrated and comprehensive ecotourism development.

Considering IFA and EFA, the strategic synthesis that must be carried out aggressively is to optimize two fundamental aspects, namely, 1) Political Aspects and Regional Ecotourism Policies; 2) Aspects of Ecotourism Planning; 3) Aspects of Collaboration and Partnership Management. In the political and policy aspects, the government should be a regulator, both from the international scale, national scale, and regional scale to the individual scale consistently coordinating thoroughly in determining various development priorities as well as the high motivation of the community in participating in ecotourism development in the GLNP region. While in the aspect of ecotourism planning, it is better for practitioners or academics as tourism experts to always pay attention to the policy approach, behavior approach, integrated approach, and a priory approach. The preparation of concepts and strategies for ecotourism planning can be poured into the Master Plan, Site Plan, Detail Plan, to the Visitor Management Plan, and the Strategic Plan for Destination Management. Finally, various regulations relating to the exploitation of natural tourism in conservation areas can be interpreted as political will in strengthening the practice of collaboration and partnership, in order to achieve the distribution of benefits and foster the spirit of conservation of all parties to jointly manage common property resources (common pool resources).

REFERENCES

Abbas, R. 2000. Prospek Penerapan Ekoturisme Pada Taman Nasional Gunung Rinjani di Nusa Tenggara Barat. Bogor: Institut Pertanian Bogor.

Altinay, L. & Paraskevas, A. 2008. *Planning Research in Hospitality and Tourism*. Oxford; Burlington, Mass: Butterworth-Heinemann.

Anggraini, Dewitri. 2017. Analisis Hubungan Komplementer Dan Kompetisi Antar Destinasi Pariwisata (Studi Kasus: 10 Destinasi Pariwisata Prioritas Di Indonesia). Tesis MPKP FEB UI.

Avenzora, R. 2008. *Ekoturisme-Teori dan Praktek*. Bogor: BRR NAD-Nias.

Avenzora, R. 2013. *Ekoturisme Teori dan Implikasi*. Di dalam: Dadursman D, Avenzora R, editor. Avenzora, R. "Ekoturisme; Teori dan Implikasi," In Darusman, D., Avenzora, R., (Eds.), *Pembangunan Ekowisata Pada Kawasan Hutan Produksi;* Potensi dan Pemikiran. Program Studi Pasca Sarjana Manajemen Ekowisata dan Jasa Lingkungan, Fakultas Kehutanan, Institut Pertanian Bogor, 61-95. Boniface.

J. Brimacombe, S. Elliot, 1996, Medical Plants in Gunung Leuser National Park. In van Schaik C, and Supriatna J (eds.), *Leuser Santuary*. Yayasan Bina Sains Hayati, Indonesia, pp. 330-335.

Chan, J.K.L. Baum, T. 2007. Motivation Factors of Ecotourist in Ecolodge Accomodation: The Push amd Pull Factors. *Asia Pasific Journal of Tourism Research*, 12(4), 349–364.

Direktorat Pemanfaatan Jasa Lingkungan Hutan Konservasi. 2016. Pariwisata Alam 51 Taman Nasional Indonesia; Kepingan Surga di Khatulistiwa. Bogor: Direktorat Jenderal Konservasi Sumber Daya Alam dan Ekosistem, Kementerian Lingkungan Hidup dan Kehutanan Republik Indonesia.

Direktorat Jenderal Konservasi Sumber Daya Alam dan Ekosistem. 2019. Statistik Ditjen KSDAE Tahun 2018. Jakarta: Sekretariat Direktorat Jenderal Konservasi Sumber Daya Alam dan Ekosistem, Kementerian Lingkungan Hidup dan Kehutanan Republik Indonesia.

Drumm, A. & Moore, A. 2002. *Ecotourism Development: An Introduction to Ecotourism Planning*. The Nature Concervancy, Arlington, Virginia, USA.

Fandeli, C. 2002. *Perencanaan Kepariwisataan Alam*. Yogyakarta: Universitas Gajah Mada.

Garrod, B. 2003. Local Participation in the Planning and Management of Ecotourism: A Revised Model Approach. *Journal of Ecotourism*, Vol 2 (1): 33–53.

Gold, S.M. 1980. *Recreation Planning and Design*. New York: MC-Growth Hill Book Company.

Gu, Ming. & Wong, P.P. 2006. Residents' Perception ofTourism Impacts: A Case Study of Homestay Operators in Dachangshan Dao, North-East China. *Tourism Geographies: An International Journal ofTourism Space, Place and Enveronment*, 8:3, 253–273.

Gursoy, D. & Rutherford, D.G. 2004. Host Attitudes toward Tourism: An Improved Structural Model. *Annals of Tourism Research*, 31(3): 495–516.

Hall, C.M. 1994. *Tourism and Politics: Policy, Power and Place*. New York: Wiley and Sons.

Kementrian Pariwisata. 2016. *Buku Saku Kementerian Pariwisata*. Jakarta: Kementerian Pariwisata Republik Indonesia

Mowforth, M. & Munt I. 1998. *Tourism and Sustainability: New Tourism in the Third World*. London: Routledge.

Nasution, H. 2017. Analisis Kebijakan dan Peraturan Perundang-Undangan Ekowsata di Indonesia. Media Konservasi Vol 23 No 1: 9–17.

Nelson, J, G. 1999. The Spread of Ecotourism: Some Planning Implication. In Nelson, J.G., Butler, R & Wall, G (eds), Tourism and Sustainable Development: A Civic Approach. Heritage Resources Centre. Join Publication No. 2. University of Waterloo.

Nugroho, I. 2011. *Ekowisata dan Pembangunan Berkelanjutan*. Yogyakarta: Pustaka Pelajar.

Peraturan Menteri Lingkungan Hidup dan Kehutanan No. P. 8/ MENLHK/ SETJEN/ KUM.1/3/2019 tentang Pengusahaan Pariwisata Alam di Suaka Margasatwa, Taman Nasional, Taman Hutan Raya dan Taman Wisata Alam.

Perdue, R. Long, T. & Allen, L. 1990. Resident Support for Tourism Development. *Annals ofTourism Research*, 17(4): 586–599.

Pratiekto. 2013. Studi Permintaan Rekreasi Dan Strategi Pengembangan Ekowisata Spiritual Di TN Ujung Kulon. Bogor: Institut Pertanian Bogor.

Prittwitz, V.V. 1994. *Politikanalyse (Uni-Taschenbücher)*. German Edition. Uni-Taschenbücher German: VS Verlag für Sozialwissenschaften.

Rachmatullah, A. 2017. Polarisasi Orientasi Pemanfaatan Lahan untuk Pembangunan Ekowisata di Ranah Minang Sumatera Barat. Bogor: Institut Pertanian Bogor.

Reindrawati, D. 2010. Motivasi Ekoturis dalam Pariwisata Berbasis Alam (Ekoturism): Studi kasus di Wana Wisata Coban Rondo, Malang. *Jurnal Masyarakat Kebudayaan dan Politik*, Vol 21, No. 2:187–192. Surabaya: Universitas Airlangga.

Ross Dunn, E.L. & Iso-Ahola, S.E. 1991. Sightseeing Tourist' Motivation and Satisfaction. *Annals of Tourism Research*, 18(2), 226–237.

Stronza, A. 2007. The Economic Promise of Ecotourism for Conservation. *Journal of Ecotourism*. 6(3): 210–230.

Santiago, F. & C. Libosada. 1997. *Ecotourism Development in the Philippines*. Manila.

Sugiyono. 2010. *Metode Penelitian Pendidikan Pendekatan Kuantitatif, kualitatif, dan R&D*. Bandung: Alfabeta.

Sugiyono. 2012. *Metode Penelitian Kombinasi (Mixed Methods)*. Bandung: Alfabeta.

Wall, G. & Long, V. 1996. *Balinese homestays: an indigenous response to tourism opportunities, in: R. Butler & T. Hinch (Eds) Tourism and Indigenous Peoples*, pp. 27–48 London: International Thomson Business Press.

Weaver, D. 2002. *Ecotourism*. John Willey & Sons, Milton, Australia.

Winarno, G.D. Avenzora, R. Basuni, S. & Bismark, M. 2015. The Alignment of Perceptions, Motivations and Preferences amongst Stake Holders on Wild Elephant Ecotourism Development in Bukit Barisan Selatan National Park, Lampung Province–Indonesia. *International Journal of Multidisciplinary Research and Development*, 2(5): 277–288.

Wiratno. 2013. *Dari Penebang Hutan Liar ke Konservasi Leuser; Tangkahan dan Pengembangan Ekowisata Leuser*. Medan: YOSL-OIC & UNESCO Jakarta, UNEP-GRASP, Spain-UNEP life web.

Promoting Creative Tourism: Current Issues in Tourism Research – Kusumah et al. (Eds)
© 2021 Taylor & Francis Group, London, ISBN 978-0-367-55862-8

Sequential exploratory mixed methods and scale development: Investigating transformational tourism readiness

J.K. Sabharwal
James Cook University Singapore, Singapore

S. Goh
Auckland University of Technology, Auckland, New Zealand

K. Thirumaran
James Cook University Singapore, Singapore

ABSTRACT: This paper presents two vignettes that might interest tourism scholars in their future research endeavors. Firstly, a journey in multidisciplinary research is described to expound on the successful collaboration between scholars of different backgrounds premised on being open to ideas from non-tourism perspectives, and learning how to hybridize an explanation or method is helpful to the scientific goals. Secondly, we review sequential exploratory mixed methods and scale development as a contribution to the tourism discipline. Utilizing the qualitative and quantitative data collection and analysis in a sequence of phases, the project aimed to develop the scale to measure travel suppliers' readiness to provide transformational tourism services. The significance of this work rests on sharing insights to working in a multidisciplinary team and proposes a conceptual framework to crafting and validating findings using scaling and sequential phases combined with qualitative methods all with the aid of existing data reading software.

Keywords: multidisciplinary, transformational tourism, travel suppliers, sequential exploratory design, mixed methods

1 INTRODUCTION: DISCOVERING AND DEVELOPING KNOWLEDGE

In a quest to contribute knowledge in transformational tourism, key literature on the topic was evaluated, missing links were identified, and an instrument was developed. This paper presents the actual effort that went into the study of travel suppliers' readiness to participate in the transformational tourism service provision. The paper presents insights to scholars and prospective researchers on how a small team with varied specialization and yet common interests converged to help ordinary individuals to transform their state of mind in a beneficial way through travel experiences. In this epistemological exercise of drawing benefits to travelers and industry, the mixed methods of sequential exploration were employed and measured tool was developed. In subsequent sections of this paper we share our collaborative journey and the conceptual framework of investigation using sequential exploratory mixed methods.

2 MULTIDISCIPLINARY COLLABORATION JOURNEY

Interdisciplinary collaboration among the researchers from two or more disciplines leads the way to a more pragmatic research paradigm. This research on transformational tourism brings about researchers from the hospitality, tourism, events, and psychology disciplines; each bearing their

 DOI 10.1201/9781003095484-37

pre-existing research paradigms (post-positivism, social constructionism, and interpretive) in the design of the project. Different disciplines bring about different jargons, experience, research tools, and methods (Bracken & Oughton 2006) and research traditions (Jacobs & Frickel 2009). However, a mixed method design (using both quantitative and qualitative methods) is the pragmatic approach (Creswell & Plano Clark 2018) that this paper undertakes to incorporate the strengths of different research paradigms. The pragmatic approach has given opportunity for triangulation to verify both quantitative and qualitative data. It is the nature of interdisciplinary research to gain different "mental models, conceptual frameworks and methods (Romero-Lankao et al. 2013)." This research collaboration has resulted in a meaningful discussion that prompted the way forward in the development of the existing project. The outcome was a conceptual framework derived from the different paradigmatic approaches of the researchers that is also applicable for future interdisciplinary collaboration in transformational tourism.

This journey of discovery would not be feasible had we not as three specialists from different disciplinary areas converged to deliberate, debate, and derive at an element that requires investigation. Hence, the supply side of an epistemological need emerged.

In an attempt to cogitate about what is the gap in the literature where an epistemological proposition can be presented and validated, we set on a journey to collate existing writings on transformational tourism. It was observed that existing literature focused primarily on the demand side and the tourists' behavior. There was a lack of epistemology account of the supply side and their level of participation. The significance of literature survey cannot be emphasized enough, in order to become a scholar entrepreneur to identify an opportunity to discover and make a difference to the literature.

3 SEQUENTIAL MIXED METHOD DESIGN

There is a paucity of standardized tools that can be used in the tourism studies to collect data in an objective manner with defendable outcomes. The last few decades have seen a surge of mixed method study designs (Timans et al. 2019). Greene (2007) referred to it as "multiple ways of seeing and hearing, multiple ways of making sense of the social world, and multiple standpoints on what is important and to be valued and cherished" (p.20). Qualitative and quantitative methodologies have their inherent strengths and weaknesses.

As stated above, the qualitative and the quantitative methods are combined in mixed methods to explore, collect, analyze, and interpret the research outcomes. The two methods can either be used concurrently or in a sequential manner. While concurrent collection of data using the two methods can be time efficient, the researchers can face issues if the qualitative and quantitative outcomes do not match and the results contradict each other (Creswell & Plano Clark 2018). On the other hand, collecting data in a sequence or in a phased manner helps researchers to develop the critical content after exploring the unknown and subsequently use that knowledge to elaborate the research results (Creswell & Plano Clark 2018).

A two-phase mixed method design can be either explanatory or exploratory (Creswell et al. 2003). In the sequential explanatory design (Figure 1), quantitative phase is followed by qualitative phase (Creswell et al. 2003). Quantitative (numeric) data is analyzed in the first phase and the qualitative or the text data collected in the second phase is used to explain and elaborate the outcomes of the first phase (Creswell & Plano Clark 2007).

Using the qualitative data in the second stage allows for better understanding of the research problem. In other words, it helps the researcher understand the process (reasons) of an outcome rather than only the product (numbers), which the quantitative data will reveal.

The exploratory sequential mixed design focuses first on exploring a topic before finalizing the variables to be measured. The researcher starts with investigation into what is known about a construct or a variable. Generally, this approach helps when investigating variables where not much is known or where there is no universal agreement (Mihas 2019). The procedure involves first gathering the qualitative data, and the outcomes are used to direct the quantitative phase (Creswell

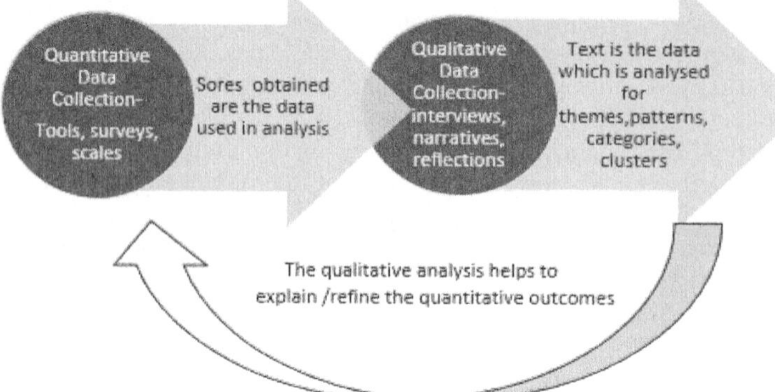

Note: The design starts with quantitative data analysis followed by qualitative data analysis

Figure 1. The flow of sequential explanatory design.

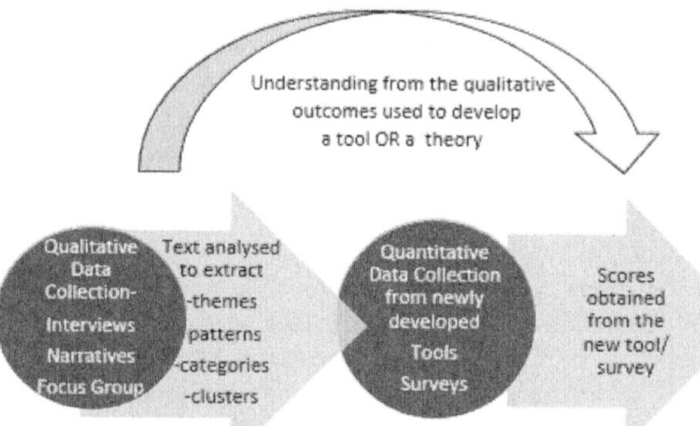

Note: The design starts with qualitative data analysis followed by quantitative data analysis

Figure 2. The flow of sequential exploratory design.

& Plano Clark 2007). This sequential design can focus on either theory development or instrument development (Creswell & Plano Clark 2018). Interviews or ethnographic observations can be used to create measurement instruments (Brownlee et al. 2015; Crede & Borrego 2013) or to help with "the more rigorous (quantitative) investigation" (D'Souza & Yiridoe 2014). The qualitative phase in this instance helps researchers understand underlying hidden processes by providing detailed information about setting or context. The analysis focuses on extracting meaningful quotations/codes to develop larger themes (Mihas 2019). This phase facilitates the collection of data, especially when measures do not exist and researchers need an in-depth understanding of concepts (Klassen et al. 2012).

Ideally, one would expect both phases in the sequential stage to get equal emphasis—in fact, it is in the hands of the researcher as to what phase will get priority. For instance, more resources could be allocated to the qualitative phase as compared to quantitative phase and the design would be labeled QUAL→quant (Creswell & Plano Clark 2018). If on the other hand the qualitative phase involves less resource allocation and the quantitative stage involves large scale survey administration and detailed statistical analysis, as happens in the instrument development variant, this design would be labeled as qual→QUANT (Creswell & Plano Clark 2018).

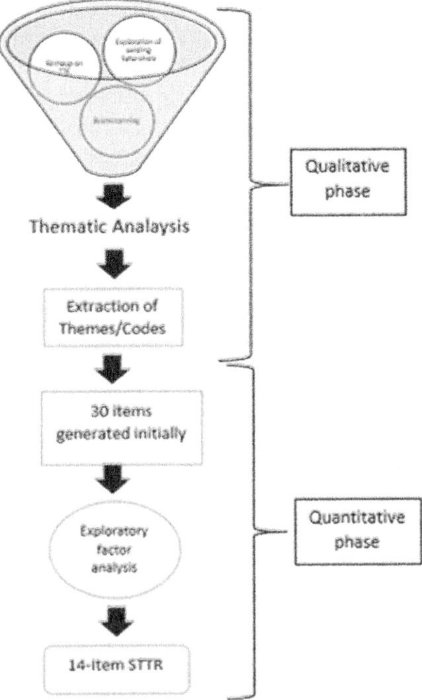

Note: TTE = Transformational Tourism Experience; STTR = Scale of Transformational Tourism Readiness

Figure 3. Flow of the process for scale development in the study.

4 IMPLEMENTING MIXED METHOD DESIGN FOR DEVELOPING SCALE FOR TRANSFORMATIONAL TOURISM READINESS LITERATURE REVIEW

Our study design was shaped by lack of literature addressing transformational tourism from the tourism suppliers' perspectives. Due to the paucity of validated, standardized tool measuring the willingness and readiness in travel suppliers for transformational tourism, we aimed to develop a standard measure to check for the same.

The following hypotheses were tested during the process of scale development to establish the robustness of the scale.

H1-Awareness about transformational tourism is not universal.

H2-Awareness of transformational tourism would result in leading to higher score on STTR.

H3-There would be no difference in scores on STTR for participants from Singapore and New Zealand.

The study utilized qual→QUANT variant of the sequential exploratory design to develop the new scale. The qualitative phase involved a two-pronged approach.

1. In order to operationalize "transformational tourism," existing literature and the travel websites were explored to extract the definitions and meaning.
2. The research team on the project independently responded to the trigger "How would you define a transformative tourism experience?" This was an open-ended exercise, and no limitations were imposed on them regarding what they could or could not write.

Once the research team had made their submissions, the team members analyzed the write-ups and the extracted content from the websites and the extant literature on transformational tourism. An inductive approach to thematic analysis was used to examine responses to this open-ended question. After familiarizing themselves with the data, the researchers coded the responses and recognized patterns in order to identify keywords and themes. Effort was made to establish inter-rater reliability in that two persons coded each work and the third person acted as a mediator if no consensus was reached on a code. The coding did not employ a specific qualitative tradition and was generic in nature. As the study was exploring the factors that would indicate readiness for transformational tourism, the researchers had a set of a priori codes as well. These codes included wellness, potential, profit, expertise, positivity, resistance, and growth.

Scale of transformational tourism readiness (STTR)

Following the themes extracted from the qualitative phase, we developed a reflective measure to evaluate readiness/support for transformational tourism among travel suppliers on a 5-point Likert scale. More than three items were generated for each of the identified themes, so that we had the flexibility in dropping some items during the exploratory factor analysis (EFA) if needed. After a brainstorming session, a total of 30 items were shortlisted for the survey (e.g. "We are ready to offer transformational experience," "We have the resources to offer transformational experiences," "Transformational tourism has added to my/ our company's profitability"). Items for the survey were written from the suppliers' perspective and encompassed themes like awareness, profit, customer need, potential, and collaboration.

Apart from the survey questions, the participants were also asked to seek their definition of transformational tourism and requested them to elaborate the transformation experience, if any, that they had provided to the clients.

4.1 *Demographic and procedure*

The study was approved by the Human Research Ethics Committee from James Cook University, Australia. The survey was uploaded on an online survey platform, Qualtrics, for all participants to access. Participants were recruited using purposive and snowball sampling and through the crowdsourcing site Qualtrics. The demographics form sought information on specific sectors the participants worked in, services that they were providing, number of years in the profession, nature of clientele, and platforms used to attract clients.

The participants in the survey included business operators and suppliers from Singapore and New Zealand. A total of 376 surveys were returned, out of which 22 responses were rejected for satisficing (Krosnick 1991) or, in other words, not being engaged with the survey. Five sets of data were removed due to missing responses or being identified as outliers. The final sample included 349 participants with 174 participants from Singapore and 175 participants from New Zealand.

Once the participants accessed the Qualtrics link, they were directed to the information sheet, which provided the participants with a general background of the study. After providing informed consent, the participants proceeded to the survey on transformational experience. Participants recruited through the Qualtrics platform were paid cash incentives for participation. Those who participated outside of the platform were not paid any incentives for participating in the survey. The completion of the study took approximately 20 minutes. The participants were first asked to complete a demographic form. This was followed by an open-ended question seeking from participants their definition of transformational tourism. They were then administered the STTR in which the order of items was randomized for presentation to counter the "order effect" (Krosnick & Alwin 1987; Salkind 2010). Upon completion of the study, the participants were thanked for his/her participation.

The results were recorded, and the data was analyzed through IBM SPSS Statistics software version 25, and Monte Carlo PCA for Parallel Analysis version 2.3 (Watkins 2008). NVivo 12 was used to analyze the qualitative data. Microsoft Excel 2016 was used for data cleaning. An exploratory factor analysis was conducted for responses on the STTR.

4.2 *Study outcomes*

To examine the factorial validity of the STTR, a total of three EFAs were performed. The Principal Axis Factoring and Promax with Kaiser Normalizations were employed as the extraction and rotation method, respectively. In order to determine the number of factors to be extracted, we used parallel analysis (Horn 1965). While factor loadings of 0.7 or greater are considered as practically significant, factor loadings of 0.45 or greater can be considered as adequate indicators for that factor (Hair et al. 2009; Tabachnick et al. 2007). Items with factor loadings below .45 were suppressed in the SPSS during the analysis.

The preliminary analysis of data was conducted to check if assumptions were met and if the data was suitable for EFA. The Kaiser–Meyer–Olkin (KMO) measure of Sampling Adequacy was .94, which indicated that the factorability of the matrix could be considered marvelous (Kaiser & Rice 1974). The Bartlett's Test of Sphericity was significant indicating that it was appropriate to factor analyze the matrix. The diagonal elements of the anti-image correlation matrix were all observed to be greater than .50, as recommended by Field (2018). The correlation matrix showed that the inter-correlations among items were mostly above the minimum recommended value of .30 (Williams et al. 2010).

The first round of EFA was performed on 30 items, and four underlying factors were identified with eigenvalues more than 1. Parallel analysis was performed with 1,000 replications (Horn 1965) in order to determine the number of factors to be retained. Results from the parallel analysis indicated that three factors should be retained. However, a closer examination of the pattern matrix table showed that only four and two items loaded on the fourth and the fifth factors, respectively. As it is highly recommended for each factor to contain at least five items (Costello & Osborne 2005), two factors were manually extracted before the second EFA was performed on the 24 items.

The second EFA showed that all negatively worded items loaded on the second factor only, a phenomenon known as the method effect (DiStefano & Motl 2006). Method effects generally occurr when "any characteristic of a measurement process or instrument contributes variance to scores beyond what is attributable to the construct of interest" (e.g., Sechrest et al. 2000). The indifferent response pattern for positively and negatively worded items could be attributed to the response style rather than to the intended construct, in this case "transformational tourism." Therefore, all negatively keyed items were removed. The final EFA was run, which gave a one-factor solution and giving14 items which accounted for 62.50% of the variance in the factor analysis (Table 1).

Internal consistency

The new scale had good internal consistency with a Cronbach alpha of .95.

H1- Awareness of transformational tourism.

The participants' inputs on what they understood of the transformational tourism was coded as 1 and 0 to categorize them into whether they understood what transformational tourism was or not, respectively. Of the 349 participants, only 77 participants (22%) had some understanding of what transformational tourism was thus supporting the H1.

H2- Relationship between awareness of transformational tourism and the score on STTR.

Scores on the 14 items were added to give a final awareness/readiness score to each participant. An independent t-test was computed to see if there was any difference in the STTR score for the participants from the two categories of awareness (absent or present). Assumption testing showed that the homogeneity of variance was violated so equal variance was not assumed. The t-test (Table 2) revealed that there was a significant difference in the score for STTR for participants from the two groups ($t\ (157.32) = -2.07$, $p = .04$). Participants from the group with some understanding of transformational tourism scored significantly higher ($M=53.61$, $SD=8.34$) as compared to participants from the group with no understanding of the transformational tourism ($M=51.19$, $SD=11.00$). Thus, the scale was able to distinguish between those who were aware and those who were not aware of the transformational tourism.

Table 1. Exploratory factor analysis of the items of the STTR.

Factor 1: **Readiness toward transformational tourism** ($\alpha = .95$)	Factor Loading
Our company is aware of transformational tourism.	.74
There is a need for transformational services.	.80
We are ready to offer transformational experience.	.81
Transformational tourism is spiritual.	.63
Our customers will react favorably if we offer transformational experience.	.82
We have the resources to offer transformational experiences.	.83
It is easy to add transformational experiences to existing tour packages.	.80
After offering transformational tourism packages, we have tapped into new customer base.	.83
We are open to the idea of including transformational tourism to the existing services offered by us.	.78
Transformational tourism has added to my/our company's profitability.	.84
Transformational tourism has the potential to teach people whatever they wish.	.80
Transformational tourism experience is the future of tourism industry.	.83
Transformational tourism opens up business opportunities to collaborate with smallscale tourism establishments, B&B etc.	.78
Transformational tourism can add to richness to already offered experiences by the travel and tourism sector.	.77

Note: Extraction Method: Principal Component Analysis.

Table 2. t-test results showing difference in awareness of transformational tourism and score on STTR.

Awareness	n	Mean	SD	T
Yes	77	53.61	8.34	−2.07*
No	272	51.19	11.00	

*Note:** $= p < .05$.

Table 3. t-test results showing the difference in the score on STTR by nationality.

Country	n	Mean	SD	T
Singapore	174	54.35	10.11	4.79***
New Zealand	175	49.12	10.29	

*Note.**** $= p < .01$.

H3- Difference in the score on STTR for participants from Singapore and New Zealand

An independent t-test was computed to see if there was any difference in the STTR score for the participants from the two countries (Table 3). Assumption testing showed that the assumption of homogeneity of variance satisfied so equal variance was assumed. The t-test revealed that there was a significant difference in the score for STTR for participants from the two countries ($t\ (347) = 4.79,\ p < .01$). Participants from Singapore scored significantly higher ($M = 54.35$, $SD = 10.11$) as compared to participants from New Zealand ($M = 49.12$, $SD = 10.29$). Thus, the scale was able to differentiate in the levels of transformational tourism readiness for the operators from the two countries.

5 CONCLUDING REMARKS

Also known as the "third methodological orientation" (Teddlie & Tashakkori 2008), mixed methods research draws on the strengths of both qualitative and quantitative research. It provides a wider choice to the researcher in their search for finding answers, which perhaps cannot be answered only by either qualitative or quantitative research (Creswell & Plano Clark 2018). Mixed research designs also add depth to the quality of information researchers can extract from their studies.

The project emerged out of the need to take a first step to standardize how transformational tourism experience is operationalized by the academics and the industry. Majority of existing definitions equate transformational tourism to self-actualization. Therein lies the potential issue. As one moves up the need hierarchy one starts shedding the excess—needs, relationships, desires, possessions. The path to self-actualization is lonely and is driven by intrinsic motivation. The role of the operator in providing transformational experiences then has to be non-intrusive, giving the client space to explore the experiences that might be transformational. Therefore, awareness about the transformational experience among the operators is a must if we want them to be in the role of facilitators for such experiences. This scale is a first step in that direction.

In a multidisciplinary approach to tourism studies, there is always a sense of learning, borrowing, and even extending knowledge in the way discoveries and scientific studies are conducted. However, the challenges posed can be two levels. The first level can be at the human relations between researchers and the second level alludes to the multidisciplinary effort and study outcome. Therefore, the importance of securing a good and cooperative relationship between collaborators is as essential as the outcome that reflects a study based on solid multidisciplinary foundations, whether theoretical or methodological. The mixed method approach acknowledges these and gives researchers an alternative to combine the two approaches to design studies in a way to maximize the outcomes from a research project.

REFERENCES

Bracken, L.J. & Oughton, E.A. (2006). 'What do you mean? The importance of language in developing interdisciplinary research. *Transactions of the Institute of British Geographers*, *31*(3), 371–382.

Brownlee, M., Hallo, J., Jodice, L., Moore, D., Powell, R., Wright, B. (2015). Place attachment and marine recreationists' attitudes toward offshore wind energy development. *Journal of Leisure Research*, *47*, 263–284.

Byrne, B. M. (2013). *Structural equation modeling with Mplus: Basic concepts, applications, and programming*. Routledge.

Comrey, A. L., & Lee, H. B. (2013). *A first course in factor analysis*. Psychology press.

Costello, A. B., & Osborne, J. (2005). Best practices in exploratory factor analysis: Four recommendations for getting the most from your analysis. *Practical Assessment, Research & Evaluation*, *10*(7), 1–9.

Crede, E., & Borrego, M. (2013). From ethnography to items: A mixed methods approach to developing a survey to examine graduate engineering student retention. *Journal of Mixed Methods Research*, *7*(1), 62–80.

Creswell, J. W., & Plano Clark, V. L. (2007). Chapter: 4 Choosing a mixed methods design. In J. W. Creswell & V. L. Plano Clark (Eds.), *Designing and conducting mixed methods research* (pp. 58–88). Thousand Oaks, Calif: SAGE Publications.

Creswell, J. W., & Plano Clark, V. L. (2018). Designing and conducting mixed methods research. Thousand Oaks, CA: Sage Publications.

Creswell, J. W., Plano Clark, V. L., Gutmann, M., & Hanson, W. (2003). Advanced mixed methods research designs. In A. Tashakkori & C. Teddlie (Eds.), *Handbook of mixed methods in social and behavioral research* (pp. 209–240). Thousand Oaks, CA: Sage

DiStefano, C., & Motl, R. W. (2006). Further investigating method effects associated with negatively worded items on self-report surveys. *Structural Equation Modeling: A Multidisciplinary Journal*, *13*(3), 440–464. DOI: 10.1207/s15328007sem1303_6

D'Souza, C., Yiridoe, E. (2014). Social acceptance of wind energy development and planning in rural communities of Australia: A consumer analysis. *Energy Policy*, *74*, 262–270.

Field, A. (2018). *Discovering statistics using IBM SPSS Statistics* (5th ed.). London, UK: Sage Publications Limited.

Greene, J. C. (2008). Is mixed methods social inquiry a distinctive methodology? *Journal of mixed methods research*, *2*(1), 7–22.

Hair, J. F., Black, W. C., Babin, B. J., & Anderson, R. E. (2009). *Multivariate data analysis* (7th ed.). United Kingdom: Prentice Hall.

Horn, J. L. (1965). A rationale and test for the number of factors in factor analysis. *Psychometrika*, *30*, 179–185.

Jacobs, J. A., & Frickel, S. (2009). Interdisciplinarity: A critical assessment. *Annual review of Sociology*, *35*, 43–65. https://doi.org/10.1146/annurev-soc-070308-115954

Kaiser, H.F. and Rice, J. (1974) Little Jiffy, Mark Iv. *Educational and Psychological Measurement*, *34*, 111–117. https://doi.org/10.1177/001316447403400115

Klassen, A.C., Creswell, J., Plano Clark, V.L., Smith, K.C., & Meissner, H.I. (2012). Best practices in mixed methods for quality of life research. *Quality of Life Research*, *21*, 377–380. https://doi.org/10.1007/s11136-012-0122-x

Krosnick, J. A. (1991). Response strategies for coping with the cognitive demands of attitude measures in surveys. *Applied cognitive psychology*, *5*(3), 213–236.

Krosnick, J. A., & Alwin, D. F. (1987). An evaluation of a cognitive theory of response-order effects in survey measurement. *Public Opinion Quarterly*, *51*(2), 201–219.

Mihas, P., & Odum Institute. (2019). *Learn to use an exploratory sequential mixed method design for instrument development.* London, United Kingdom: SAGE Publications, Ltd. doi: 10.4135/9781526496454

Phoenix, C., Osborne, N. J., Redshaw, C., Moran, R., Stahl-Timmins, W., Depledge, M. H. & Wheeler, B. W. (2013). Paradigmatic approaches to studying environment and human health:(Forgotten) implications for interdisciplinary research. *Environmental science & policy*, *25*, 218–228.

Romero-Lankao, P., Borbor-Cordova, M., Abrutsky, R., Günther, G., Behrentz, E., & Dawidowsky, L. (2013). ADAPTE: A tale of diverse teams coming together to do issue-driven interdisciplinary research. *Environmental science & policy*, *26*, 29–39.

Salkind, N. J. (2010). *Encyclopedia of research design* (Vol 1). SAGE. doi: 10.4135/9781412961288

Sechrest L., Davis M. F., Stickle T. R., McKnight P. E. (2000). "Understanding 'method' variance," In Leonard Bickman (ed.), *Research Design: Donald Campbell's Legacy* (pp. 63–87.) Thousand Oaks, CA: Sage Publications.

Tabachnick, B. G., Fidell, L. S., & Ullman, J. B. (2007). *Using multivariate statistics* (Vol. 5, pp. 481–498). Boston, MA: Pearson.

Teddlie, C. & Tashakkori A. (2008). Foundations of mixed methods research: Integrating quantitative and qualitative techniques in the social and behavioral sciences. Thousand Oaks, CA: SAGE Publications.

Timans, R., Wouters, P. & Heilbron, J.(2019). Mixed methods research: what it is and what it could be. Theory and Society, 48, 193–216. https://doi.org/10.1007/s11186-019-09345-5

Watkins, M. (2008). Monte Carlo for PCA parallel analysis (Version 2.3) [Computer software]. Retrieved June 23, 2020 from www.softpedia.com/get/Others/HomeEducation/Monte-Carlo-PCA-for-ParallelAnalysis.shtml

Webster, C., Leigh, J., & Ivanov, S. (2013). *Future tourism: Political, social and economic challenges.* Abingdon, Oxon: New York. doi:10.4324/9780203125038

Williams, B., Onsman, A., & Brown, T. (2010). Exploratory factor analysis: A five-step guide for novices. *Australasian journal of paramedicine*, *8*(3).1–13. DOI: 10.33151/ajp.8.3.93

Promoting Creative Tourism: Current Issues in Tourism Research – Kusumah et al. (Eds)

Economic benefits of selected resorts in Dasmariñas City, Cavite: Basis for a proposed economic strategy

A.R.D. Movido, M.J.L. Tapawan, Q.A.E. Lucero & J.U. Tabuyo
De La Salle University, Cavite, Philippines

ABSTRACT: The City of Dasmariñas had a lot of well-known resorts in the area in which it provides relaxation and recreation that cater to different guests. The study explores the economic benefits of the resorts in the City of Dasmariñas through knowing the different factors on economic benefits such as direct impact, indirect impact, and induced effect. The study had four selected subject resorts such as Tubigan Resort, Saniya Resort, Volets, and Palmas Del Sol. The study employs a quantitative descriptive type of research. Quota sampling used was approximately 100 respondents of the resort employees. Survey questionnaire was used as the instrumentation. An informed consent letter to the City of Dasmariñas was also given to make sure that they are also aware of the study. The study uses percentage, frequency, weighted mean using the 4-point Likert scale, and ANOVA. The results show that most of the workers in the resort industry in the city of Dasmariñas are 18- to 29-year-old males with educational attainments of college graduate and a monthly income of 10,000 to 19,999 pesos. The most economic benefit based on direct impact is that resort industry generates jobs, while on indirect impact was resort industry activity in the area helps creation of jobs from other industries, and induced effect resort industry increases the income of a person needed in purchasing goods and services. Overall, the respondents agree that the resort industry had an economic benefit on direct impact, indirect impact, and induced effect, and there is a significant difference between the profile of the respondents such as age and gender and induced impact economic benefit.

Keywords: Direct Impact, Economic Benefits, Hotel Resort, Indirect Impact, and Induced Effect

1 INTRODUCTION

As tourism grows in an area, it can provide a lot of benefits to the local community. Tourism can provide direct jobs to the community, such as tour guides or hotel housekeeping. Indirect employment is generated through other industries such as agriculture, food production, and retail. Therefore, tourism is a good source of economic income to community and creates a lot of benefits that the locals will extract from it.

The tourism, on the other hand, is a source of economic benefit by providing attractions, important accommodation, and recreation tourism, while hotel only the most purpose is for lodging at a resort that combines different features that are necessary for most tourists, such as recreational activities and entertainment. A resort usually has the features of having a food and drink establishment, activities such as swimming and sport activities, entertainment such as a live bands disco, and even relaxation such as spas and shopping. A resort combines all the wants of the tourist while offering an all-around service to the guest (Nair 2017).

With the rise in number of the resorts globally to cater to the growing demand when it comes to travel, the major global trends when it comes to the resort focus on sustainability, and there is a conception that a resort as a large establishment generates a lot of consumption of different resources from energy requirement, water consumption, to waste management. There is a growing trend with the resorts to go for sustainability and tracking their carbon footprint. Economically,

DOI 10.1201/9781003095484-38

the resorts contribute a lot to an economic activity of an area. However, by applying sustainability, the resorts could run more efficiently and reduce their operating cost in which a major trend also happens when it comes to resort establishment, which is the urban resort, because this resort is located strategically to urban centers so that it can reduce carbon emissions by traveling to resorts located far away from the city and enjoy the same amenities without going farther away (Canter 2020). An example of a known resort that applies sustainability and the urban resort trend is the STAR Sydney, which is located in the Sydney, Australia, and their commitment is to reduce their carbon footprint by reducing plastic use and recycling materials.

In the Philippines, there has been a growing trend when it comes to Urban Resort in which there has been a continuous construction of resorts in the Entertainment City under the development of PAGCOR in which different resorts were developed and owned by different investors. Strategically located inside the Metro Manila near the reclaimed lands in the Manila Bay, the establishment of a resort and entertainment complex, such as Solaire, OKADA, City of Dreams, etc., creates an enormous economic activity in the area providing jobs and attracting investors and tourists to visit the area (entertainmentcityphilippines.com).

Economic Growth is usually connected with economic benefits. Economic rate growth has positive effects in the workforce in an area in a broader sense. If there is a sustained economic growth, then there is an economic benefit that will happen. For example, if an area has a lot of industries like agriculture, tourism, and service, then there will be lots of income that will be generated in an area. Thus, it will create more jobs because the concept of the economic growth and economic benefit has been linked together as it was in a domino effect. If we recognized other factors, such as the social and the environment, then the term of sustainability can be coined (Bautista & Castillo 2014).

Different theories on economic growth have been established for a lot of years; choosing a viable economic growth theory as a conceptualization means that the economy usually evolves over time. The improvement on the sustainability, particularly the economic theory, describes the economy as we know it today. The theory of unified growth defines that endogenous growth theories are consistent with the entire process of development, and in particular the transition from the epoch of Malthusian stagnation that had characterized most of the process of development to the contemporary era of sustained economic growth (Xianmeng & Feng 2014).

Unified growth theory seeks to overcome limitations of these approaches by presenting a coherent, single framework that captures the Malthusian era, the transition to higher growth, and the modern growth state. The process of moving from one state to the next originates within unified growth models, with the seeds of the transition growing during the Malthusian era. As such, the Malthusian state is not equilibrium, but a dynamic process leading to its own end. The benchmark for unified growth models is that they capture the patterns in income, technology, and population through these various states and generate the transition between them (Khan 2014).

The hospitality industry is one of the economic sectors related to the contribution of income to a region. Tourism is always done at leisure as tourism has become a popular global leisure activity and tourists need accommodation, and this is where the hotel industry comes into play. The hotel industry has an important role in the economy because it provides employment in an area and improves social welfare because these hotels are part of an area as well as the heritage it gives (Mowji 2014).

Visitors spending generates income for the local community and can lead to the alleviation of poverty in countries which are heavily reliant on tourism. It can be observed that the economic benefit of tourism can be direct (jobs), indirect (connection with other industries), and induced (other sources of income) (Sharma 2015).

With the increase of the number of tourists, there is an increased demand for hospitality industry in the area, which results in increased production in local industries, as the indirect effects of the hospitality industry (Boita et al. 2015). The economic impact of the hospitality industry is much greater, since many inputs are needed to produce tourism and leisure services, spanning the whole range of farm, agricultural food, and industrial production, including the production of capital goods as well as construction and public works (Vellas 2014).

The economic impact of the hospitality industry can be divided into the direct impact, indirect impact, and induced effect. Direct impact deals more with the direct benefit within the hospitality industry such as an increase in sales and demand for these industries. Indirect impact deals more with the intermediate consumption for the production of goods and services in the tourism sector. These are goods and services that tourism companies purchase from their suppliers, forming the hospitality supply chain. Induced effect concerns expenditure by the employees from wages paid by companies in direct contact with tourists; it also includes the consumption of companies that have benefited directly or indirectly from initial expenditure in the hospitality industry (Shaaban et al. 2014).

When it comes to the potential economic benefits of establishing resorts, the usual direct impact of establishing a resort was the job that it generates, in which these jobs are connected to the induced effect in which an employee has an income they can use to buy goods and services, and at the same time affecting other industries indirectly. An example of this is the indirect effect on agriculture where there is an increase in demand for the food supply service sector. As more people can take advantage of the service, more services can be added to expand the service sector. It is certainly a domino effect when dealing with the economic impact of establishing a resort (Mitchell et al. 2015).

Hospitality industry is a broad category of fields within the service industry that includes lodging, event planning, theme parks, transportation, cruise lines, and additional fields within the tourism industry. The hospitality industry is a several-billion-dollar industry that mostly depends on the availability of leisure time and disposable income. A hospitality unit, such as a restaurant, hotel, or even an amusement park consists of multiple groups such as facility maintenance and direct operations (Boita et al. 2015).

Resorts are considered an important economic component to any community. In some cases, cities and municipalities will offer incentives to stimulate resort development. These incentives can be in the form of tax breaks, favorable land leases, or assistance with financing. Prior to establishing incentives for hotel development, a city or municipality should conduct a study to estimate the economic benefits that result from the development. This assessment should evaluate the economic benefits of the initial investment both in the short term and over the long term. The economic impact to local and regional economies from hotel development is typically separated into four categories: direct, fiscal, indirect, and induced (Suzuki 2014).

The economic benefit of tourism as tourism can provide income to a lot of people in the area, specifically the locals. It can gradually change the way the people make their lives. It can be from an ordinary perspective that their local community that was once much more intensive on other industries can economically benefit with the tourism (Sergui & Florin 2014).

Hospitality industries are a key part of tourism as they provide accommodation to the tourists visiting the area, an integral part of a tourism economic activity. Hotels and resorts are considered an important economic component to any community. Providing jobs to the local people of the area and other business possibilities, it also provides income to the local government by paying taxes and helping in alleviating the class of income in an area. A hotel's impact to the local community is often more than just job creation and additional tax revenue; it can also drive other industries such as agriculture and food production as it consumes these for hotel guests (Suzuki 2014).

However, the research gap of the study was the application on the economic benefits of different tourism establishments because on the concept the economic benefits of the tourism establishment can be divided into four such as the direct impact, indirect impact, induced effect, and fiscal impact. However, the researchers will not use the fiscal impact because it is complicated if applied to hotels; therefore, the researchers opted to use the three economic benefit metrics: the direct impact, indirect impact, and induced effect.

The city of Dasmariñas has borders with the City of Bacoor, Imus, General Trias, and the Municipality of Silang Cavite. The area topography is classified as central hilly and has a population of approximately 700,000. The City of Dasmariñas is an urban area with lots of commercial business and an industrial area which contributes to the local economy of the area. The city also has lots of resorts that cater to the recreation of the locals living in the area (Cavite Demographic Profile 2017).

The setting of the research is Saniya Resort located in the Salawag Dasmariñas Cavite. The Saniya Resort is a famous resort in the city due to the affordable rates of the resort. The rates were 200 pesos during the summer season and 150 pesos during the regular season. The night swimming cost 220 pesos during the summer and 180 pesos during regular season. The Saniya Resort and Hotel also boasts cottages rentals depending on the size, but the prize range was from 500 pesos for a small hut to 2,500 for huts with air-condition rooms. The Saniya Resort and Hotel had also hotel rooms; the price range was from 2,500 up to 6,500 for a family room with an included breakfast.

The Tubigan resort, which is located in Paliparan III Dasmariñas Cavite, was a famous resort in the City of Dasmariñas due to the affordable rates of the resort. The rates were 250 pesos during the summer and 200 pesos during the regular season. The night swimming cost 280 pesos during the summer season and 200 pesos during regular season. The Tubigan Garden resort also boasts cottags rentals depending on the size, but the price range was from 500 pesos for a small hut to 2,500 for huts with air-condition rooms. The Tubigan Garden resort also had hotel rooms; the price range was from 2,950 up to 5,250 for a family room with an included breakfast.

Other famous resorts that can be found in the city of Dasmariñas are the Volet's Resort and the Palmas resort, both located in the downtown of the city of Dasmariñas. The Volet's Hotel and Resort is one of the oldest resorts in the city of Dasmariñas. The resort is famous for its location, which is conveniently located along the highway and near the major malls of the city. The resort is very famous for its wave pool and the tall slides in which guests will sure to have fun. The resort also boasts accommodation rooms ranging from 1,600 pesos on their single rooms to 4,500 on the family suite. The range of their swimming rates ranges from 170 for kids, 230 both daytime and nighttime swimming, and 250 for overnight swimming.

Meanwhile, the Palmas del Sol resort and hotel is located in Marilag village near the town center. The resort boasts a restaurant and a bar to cater to guests. Although the resort does not have a famous feature, the lush green space of the resort is perfect for relaxation and serenity, in which the guest will surely love. The rates of their swimming are very affordable and range from 100 pesos to 150 pesos.

The scope of the study is limited within the area of Dasmariñas City, specifically in the area where the resorts are situated. The study was participated by employees of various resorts in the City of Dasmariñas. The total number of participants was subjected to answer questionnaires and surveys identifying the economic benefits. The time frame of study is from November 2019 to March 2020.

The proposed study generally aimed to evaluate the economic benefits of resorts in the City of Dasmariñas in perception of the resort employees. Specifically, the study sought to answer the following objectives:

1. What are the profile of the respondents in terms of:
   a. Age
   b. Gender
   c. Educational Attainment
   d. Monthly Income
2. How does the respondent assess the economic benefits of the resorts in the City of Dasmariñas based on:
   a. Direct Impact
   b. Indirect Impact
   c. Induced Effect
3. Is there significant difference between the profile variable of the respondents and their assessment of the economic benefits of resorts in Dasmariñas City?
4. Based on the findings, how can the strategy plan to create an enhanced resort economic be proposed?

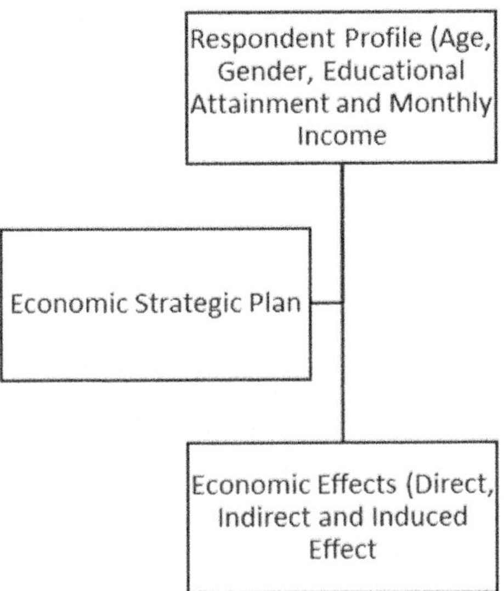

Figure 1. Conceptual framework shows the variables that will be done to get the data and be able to design the output of the study.

2 CONCEPTUAL FRAMEWORK

The figure above shows the conceptual framework of the research that shows profile variables of the respondents, which includes age, gender, educational attainment, income, and employment status. The study uses the concept of Suzuki (2014) on the economic benefit of hotels and resorts which includes the direct impact, indirect impact, and induced effect. These variables of the study are really important as they provide the primary data needed to process the study and to create the output of the study, which is the sustainable plan.

3 METHODOLOGY

The study employs a descriptive type of research. The use of the descriptive type of research in the study is necessary since there was no data to be manipulated. Aside from that, the descriptive research design allows the study to describe the characteristics of the population area and the phenomenon of the study economic benefits of resort in the area in the perception of the resort employees of various resorts in the City of Dasmariñas, Cavite.

The study uses a quota sampling to group the different respondents in the study, the resort employees, and the locals of the town. A quota sample is a type of non-probability sample in which the researcher selects people according to some fixed numbers. Units are selected into a sample based on pre-specified characteristics, so that the total sample has the same distribution of characteristics assumed to exist in the population being studied.

The researchers opted to use quota sampling for the respondents of resort employees. The researchers opted to use 100 participants for the resort employees. The 100 resort employees were also subjected to answer the survey questionnaire; likewise, the city of Dasmariñas has four well-known resorts in the area: the Tubigan Garden Resort in Paliparan, Saniya Resort in Salawag, Volet's Resort, and Palmas Del Sol Resort in the City of Dasmariñas. The employees are divided into four groups with 25 employees each of the four mentioned resorts that equal to 100 employees.

The instrumentation that was used in the study is a survey questionnaire using the descriptive statistics to identify the weighted mean and to give data on the assessment of the economic benefits of resorts in the City of Dasmariñas Cavite.

Table 1. Age of the respondents.

Age	Frequency	Percentage
18 to 29	57	57
30 to 39	28	28
40 to 49	10	10
50 to 59	5	5
60 and above	0	0
Total	**100**	**100**

The survey and questionnaires that were used in the study are based on the direct impact, indirect impact, and induced effect based on economic impact analysis to analyze the perception of both the local residents and the hotel respondents alongside the construction of the Likert scale in order to use the descriptive statistics in the study and so forth.

Before the research was conducted, an informed consent assessment form was filled up for the ethic review committee to review. A consent to participate to research was given to the participants, so that they are aware of the study itself. A legal letter to the City of Dasmariñas was also given to make sure that they are also aware of the study. The data that was gathered was treated as confidential.

The first and second objectives, the researchers distributed the survey questionnaire in a span of two weeks to the employees of each resort. One resort per week, the preferable days for the visit of the site was on weekdays. The survey questionnaire was then listed and was given to the statistician. Lastly, for the third objectives the researchers opted to avail a statistician to compute for the significant difference between the profile of the respondents and their assessment of economic benefits.

For the first objective, percentage frequency was used in analyzing the demographic profile of the respondents participated in the study. This method was useful in expressing the relative frequency of survey responses and other data.

For the second objective, weighted mean was used in the study to interpret the data on economic benefit the respondents answer on the survey. Computing the weighted mean in the study is really important since this was compared to the standard weighted mean. The results then were looked up at the standard weighted mean to interpret it in the scales, whether a category is excellent or if it needs improvement. The interpretation is as follows: 1–1.49 Strongly Disagree, 1.5–2.49 Disagree, 2.5–3.49 Agree, and 3.5–4.00 Strongly Agree.

The third objective uses ANOVA analysis, which was used to assess the relationship between the profile variable of the respondents and their assessment of the economic benefits. This statistical analysis is really important in testing the significance of categorical variables.

4 RESULTS

Table 1 shows the result of the age of the respondents. The result shows that most of the respondents were age 18 to 29 with 57 (57%) respondents followed by respondents with an age group of 30 to 39 with 28 (28%) respondents, while the lowest number of respondents was 50 years old and above with 5 respondents (5%).

The probable reason as to why there are many young adults with the respondents could probably be attributed to there being a lot of young adults working in the hospitality industry, and the hospitality industry is a famous sector among young adults.

According to PSA (2017), the hospitality industry and tourism sector is one of the fastest growing industries today. The hospitality industry grows so fast that there are more professionals needed in this industry. Therefore, it is safe to assume there are more young adults in this industry because of the need for more people since the industry is expanding.

Table 2. Gender of the respondents.

Gender	Frequency	Percentage
Male	52	52
Female	48	48
Total	**100**	**100**

Table 3. Educational attainment of the respondents.

Educational Attainment	Frequency	Percentage
Elementary	0	0
High School	3	3
Vocational	11	11
College Level	23	23
College Graduate	62	62
Post-Graduate	1	1
Total	**100**	**100**

Table 4. Monthly income of the respondents.

Monthly Income	Frequency	Percentage
P10,000 and below	16	16
P10,001 to P19,999	78	78
P20,000 to P29,999	6	6
Total	**100**	**100**

Table 2 shows the result of the gender of the respondents, which shows there are more males with 52 (52%) of respondents as compared with females with 48 (48%) respondents. The probable reason as to why there are more males as compared to females would be probably by chance, but if we are going to look at the data there be not much difference with the number of the respondents.

According to PSA (2017), the number of female workers in the hospitality and tourism sector is much higher compared to males; therefore, the explanation could be that the number of respondents' gender have been by chance because of the sampling technique the researchers have use.

Table 3 shows the result of the educational attainment of the respondents. The result shows that most of the respondents are college graduates with 62 (62%) followed by college level respondents with 23 (23%) respondents, while the lowest in the result was in post-graduate with 1 (1%) respondent.

The probable result was that working in the hospitality sector usually requires to have a Bachelor's degree in either hospitality management or tourism management; however, college level background is still accepted. This result reflects as to why there are more college graduates than other groups of the respondents.

According to PSA (2017) the workers in the hospitality industry are not required to have a college degree, but there are more college degree holders working in the hospitality and tourism sector. Therefore, the result reflects as to why there are more college graduates in the respondent profile.

Table 4 shows the result on the monthly income of the respondents. The result shows that most of the respondents' monthly income ranges from 10,001 to 19,999 with 78 (78%) respondents, while the lowest number of the respondents can be found in respondents with a monthly income of 20,000 to 29,999 pesos with 6 (6%) respondents.

Table 5. Direct impact assessment of the respondents.

Direct Impact	Mean	Interpretation	Rank
1. Resort Industry Generates Jobs.	3.78	Strongly Agree	1
2. It helps the local government to generate income.	3.16	Agree	2
3. Resort industry improves the economy growth in an area.	3.08	Agree	4
4. Resort industry improves the infrastructures of an area.	2.90	Agree	5
5. Resort industry improves the economic status in an area	3.13	Agree	3
6. Resort Industry increases tax revenues	2.81	Agree	6
Overall	**3.14**	**Agree**	

The result of the study can be explained by the monthly income of the most workers in the tourism and hospitality sector, in which the range could be from 10,001 to 19,999 pesos per month.

The average worker salary in the hospitality sector in the Philippines was 16,000 pesos as according to the report from PSA (2017); therefore, in observing the result it can be safe to say that the result reflects as to why there are more respondents with a salary of 10,001 to 19,999.

Table 5 shows the result of the respondents' assessment with direct impact. The result shows that the respondents agree the most with direct impact with the benefit that the establishment of resorts generate jobs with a mean of 3.78 and interpreted as strongly agree. While the lowest mean can be found with the resort industry increases tax revenues with a mean of 2.81 and interpreted as agree, the overall mean of 3.14 suggests that the respondents have agreed that the establishment of the resort industry provides a direct economic impact.

The probable reason as to why the respondents felt that the direct impact of establishing resort industry in an area was that it generates jobs is because this is the easiest way the respondent can feel the impact of the hospitality industry because the respondents were working in this industry.

According to Suzuki (2016), an important aspect when it comes to the economic impact of the resort industry was to provide job in the area, and this is the first directly affected area that could be felt in the jobs that were created with the establishment. This explains the result as to why the respondents felt that the greater direct impact of establishing hotels in the first place is to provide jobs.

In relation with the result the lowest mean can be found on the hotel and resort industry increases the tax revenue in the area, the result shows that this is the least felt direct impact of establishing resort. The probable reason was that tax revenues could probably be hard to understand since there are a lot of tax revenues the City of Dasmariñas gathers from different industry.

According to Moji (2016), the resort industry certainly contributes to the economy of an area by paying taxes. However, the breakdown of taxes an area has differs from different industry that made the economy of the area, and the city of Dasmariñas could probably earn from the tax of the resort and hotel, but the difference here is there are other industries in Dasmariñas such as manufacturing and servicing sectors that contribute much bigger tax revenues.

The result on the respondent's assessment with indirect impact shows that the highest mean can be found in the resort industry activity in the area helps creation of jobs from other industries with a mean of 3.65 and interpreted as strongly agree while the lowest mean can be found in the resort industry activity in an area becomes a source of livelihood for the residents with a mean of 2.87 and interpreted as agree. The overall score of 3.16 and interpreted as agree suggests that the indirect impact based on the assessment of the respondents shows that most of them agree that the resort industry had indirect benefits in the area.

The result above possibly reflects that was the reason the respondents felt that the indirect impact economic benefit are the jobs it could create from other industries, which means that the resort industry could provide an additional benefit by also creating jobs from other industries.

Table 6. Indirect impact assessment of the respondents.

Indirect Impact	Mean	Interpretation	Rank
1. Resort industry increases the economic activity other industries such as agriculture and other supplies.	3.17	Agree	3
2. Resort industry activity in an area helps creation of jobs from other industries.	3.65	Strongly Agree	1
3. Resort industry activity in an area helps in creating other business establishments.	3.22	Agree	2
4. Resort industry activity in an area becomes a source of livelihood for the local residents.	2.87	Agree	6
5. Resort industry activity improves the trade in commerce	2.98	Agree	5
6. Resort Industry improves investments	3.05	Agree	4
Overall	**3.16**	**Agree**	

Table 7. Induced effect assessment of the respondents.

Induced Effect	Mean	Interpretation	Rank
1. Resort industry increases the transactions of goods.	2.79	Agree	6
2. Resort industry increases the transactions of services.	3.17	Agree	3
3. Resort industry increases the income of a person needed in purchasing goods and services.	3.57	Strongly Agree	1
4. Resort industry improves the welfare of a person.	3.20	Agree	2
5. Resort industry increases the quality of life in the area	2.94	Agree	5
6. Resort industry improves the income and standard of living	3.09	Agree	4
Overall	**3.13**	**Agree**	

According to Sergui and Florin (2018), with the establishment of the hospitality industry other sectors and industries are also affected, thus creating a domino effect in which the resort will provide accommodation, and other sectors such as agriculture and other industries are also affected by providing and supplying resources to a resort. This means that the establishment of resort industry in the city of Dasmariñas also creates jobs in other sectors by providing supplies and services during the operation of the hotel and resort.

In relation with the result, the lowest mean can be found in the resort industry activity in an area that becomes a source of livelihood for the residents. The result, although still interpreted as agree, the probable reason to this was that the respondents did not come to think that the hotel and resort industry could become a source of livelihood for the residents. A probable reason to this was due to a lack of partnership with the local community and the hotel and resorts in the City of Dasmariñas.

A probable reason as to why this has the lowest mean could be the awareness of the respondents according to Vellas (2014). The direct impact is easier to realize than the indirect because this is the first one that will realize the indirect effect. On the other hand it can be much harder to realize since there is a need to study the concept of the domino effect and in relation with the economic growth ability of an area, and in addition to this was the possible partnership that can form with the owners of the resort and the local community.

Table 7 shows the result on the induced effect economic benefit of hospitality industry. The result shows that the highest mean that can be found in the resort industry increases the income of a person needed in purchasing goods and services with a mean of 3.57 and interpreted as strongly agree. While the lowest mean that can be found in the resort industry increases the transaction of goods with a mean of 2.79, which is also interpreted as agree, the overall result of 3.13 suggests that the respondents agree that the establishment of resort industry had an induced effect in their area.

Table 8. Significant difference between the age of the respondents and their assessment of economic benefits.

Age	p-value	Interpretation	Decision
Direct Impact	0.1571	Not Significant	Accept
Indirect Impact	0.0117	Significant	Reject
Induced Effect	0.3590	Not Significant	Accept

Table 9. Significant difference between the gender of the respondents and their assessment of economic benefits.

Gender	p-value	Interpretation	Decision
Direct Impact	0.7100	Not Significant	Accept
Indirect Impact	0.400	Significant	Reject
Induced Effect	0.6500	Not Significant	Accept

The result of Table 7 shows that the induced effect that the respondents felt the most was the increase of the income of a person needed in purchasing goods and services. The probable reason as to why this is the result is because of the likelihood that the respondent's jobs where in this resort since the hospitality industry provides a job and this creates a livelihood to them, eventually the induced effect would be an increase of the income of a person that an employee could have the ability to buy more goods and to avail of the services and the transaction because of the money they get from having a job in the company.

According to Suzuki (2016), the resort industry induced effect is the accessory and the connection with the other benefit is not only economic in establishing hotels such as a social benefit. The respondents have felt that the induced effect of the resort industry in the area is through the increased income to buy goods and avail services and transactions because of the eventual realization since with working in the resort comes with salary, this will increase their spending capacity since they have now a salary that they will access buying goods and other stuffs such as services to improve their situations.

In relation with the result of the study, the respondents felt that the resort industry induced effect is less of the increase of the transaction of goods; a probable reason as to why is because the transaction is much harder to realize because transaction is on general a bigger concept and not an intrapersonal one.

According to Sharma (2015), the transaction of goods is an indicator of economic growth, but not in the basis of a personal experience since the transaction is not directly felt because the economic situations can vary such as slow growth or fast growth. Then the transaction in response with the assessment of the respondents cannot quantify easily; therefore, this explains why the respondents do not feel that the resort increase transactions because one must understand the concept of economic growth in order to understand that the transactions of a business is growing too.

Table 8 shows the result on the significant difference on the age of the respondents and their assessment of economic benefits. The result shows that age is not a factor when it comes to direct impact and induced effect because all the p-value was more than the significance level of 0.05. However, there is a significant difference between the indirect impact and the age of the respondents because the p-value was below the level of significance of 0.05.

The result shows that age is a factor when it comes to indirect impact economic benefit because of the respondents 50 to 59 years old have a higher assessment of indirect impact as compared with the other age group.

Accoding to Vellas (2014), the direct impact is easier to realize than the indirect because this is the first one that will realize the indirect effect. On the other hand, it can be much harder to realize; therefore, the older age group since they have more knowledge of the economic benefit is more aware of the indirect impact as compared with the other age groups.

Table 10. Significant difference between the educational attainment of the respondents and their assessment of economic benefits.

Educational Attainment	p-value	Interpretation	Decision
Direct Impact	0.4755	Not Significant	Accept
Indirect Impact	0.9522	Not Significant	Accept
Induced Effect	0.3331	Not Significant	Accept

Table 11. Significant difference between the monthly income of the respondents and their assessment of economic benefits.

Monthly Income	p-value	Interpretation	Decision
Direct Impact	0.2391	Not Significant	Accept
Indirect Impact	0.4130	Not Significant	Accept
Induced Effect	0.2263	Not Significant	Accept

Table 9 shows the result of the significant difference on the gender of the respondents and their assessment of economic benefits. The result shows that gender is not a factor when it comes to direct impact and induced effect because all the p-value was more than the significance level of 0.05. However, there is a significant difference between the indirect impact and the gender of the respondents because the p-value was below the level of significance of 0.05.

The result shows that gender is a factor when it comes to indirect impact; economic benefit because the male respondents have a higher assessment of indirect impact as compared with the females.

The result shows that gender is a factor when it comes to economic benefit. According to McNay (2017), gender could become an attributing factor in realizing an economic potential because males tends to play the role of the worker and provider that they are much aware of the economic potential of a business idea. This could be said the same when it comes to the hotel and resort industry based on the result of the study.

The table above shows the result on the significant difference on the educational attainment of the respondents and their assessment of economic benefits of resort; the result shows that educational attainment is not a factor when it comes to the economic benefit such as direct impact, indirect impact and induced effect because all the p-value was more than the significance level of 0.05.

The result reflects there is no significant difference with the educational attainment of the respondents and economic benefits factors such as direct impact, indirect impact and induced effect. A probable reason for the result was that educational attainment is not a factor when it comes to economic benefits. Different educational attainment groups have the same assessment of direct impact, indirect impact and induced effect.

The probable reason to this could be the awareness of the respondents when it comes to economic benefits. According to Negu (2018), knowing the economic effect could be much harder than it thought since in knowing the economic benefits and effects, there be a need to have good knowledge on how an economy works in relation with the respondents. Although most of them have an educational attainment of college graduate, their knowledge about how economy works could be limited which explains as to why there is no significant difference whatsoever.

Table 11 shows the result of the significant difference on the monthly income of the respondents and their assessment of economic benefits of selected resorts; the result shows that monthly income is not a factor when it comes to the economic benefit such as direct impact, indirect impact, and induced effect because all the p-value was more than the significance level of 0.05.

The result reflects there is no significant difference with the monthly income of the respondents and economic benefits factors such as direct impact, indirect impact, and induced effect. A probable reason for the result was that monthly income is not a factor when it comes to economic benefits. Different educational attainment groups have the same assessment of direct impact, indirect impact, and induced effect.

According to Sergui and Florin (2018), when it comes to the relation of monthly income and economic benefits, there is certainly a crucial relationship to them because a higher monthly income could also suggest higher economic benefits, but the result of the study shows otherwise. The probable reason to this was that most of the respondents had a monthly income range of 10,000 to 19,999 pesos while the maximum income range of the respondents is 20,000 to 29,999, which means that the income range of these respondents are saturated so that it could not produce a significant result.

5 CONCLUSION

Based on the result of the profile of the respondents, the result shows that most of the respondents are age 18 to 29 with 57 (57%) respondents. The result of the gender of the respondents shows that most workers in the hotel and resort industry in Dasmariñas are male with 58 (52%) respondents as compared with female respondents with 48 (48%) respondents. The result of the educational attainment of the respondents shows that most of them had an educational attainment of college graduate with 62 (62%) respondents and a monthly income range of 10,000 to 19,999 with 72 (72%) respondents. The result of the economic benefits of the respondents suggest that in direct impact the highest mean can be found in the resort industry generates jobs with a mean of 3.78 and interpreted as strongly agree, while the lowest mean can be found with the resort industry increases tax revenues, with a mean of 2.81 and interpreted as agree. The overall mean of 3.14 suggests that the respondents have agreed on the statements in direct impact. The result on indirect impact shows that the highest mean can be found in the resort industry activity in the area helps creation of jobs from other industries with a mean of 3.65 and interpreted as strongly agree, while the lowest mean can be found in the resort industry activity in an area becomes a source of livelihood for the residents with a mean of 2.87 and interpreted as agree. The overall score of 3.16 and interpreted as agree means that the respondents agree on the statements of indirect impact. Lastly, the result on induced effect suggests that the highest mean can be found in the resort industry increases the income of a person needed in purchasing goods and services with a mean of 3.57 and interpreted as strongly agree. The lowest mean can be found in the resort industry increases the transaction of goods with a mean of 2.79, which is also interpreted as agree. The overall result of 3.13 suggests that the respondents agree that the establishment of the resort industry had an induced effect in their area. The result of the significant relationship and profile of the respondents shows there is significant difference between the age and gender of the respondents and induced effect, while the other factors do not have a significant difference with the profile of the respondents.

Therefore, based on the findings, the researchers have concluded that most of the workers in the resort industry in the city of Dasmariñas are aged 18 to 29, male with an educational attainment of college graduate, and a monthly income of 10,000 to 19,999 pesos. The most economic benefit based on direct impact is resort industry generates jobs while on indirect impact was resort industry activity in the area that helps the creation of jobs from other industries. Meanwhile, the induced effect, resort industry increases the income of a person needed in purchasing goods and services. Overall, the respondents agree that the resort industry had an economic benefit on direct impact, indirect impact, and induced effect, and there is a significant difference between the profile of the respondents such as age and gender and induced impact economic benefit.

With the conclusion of the study, the researchers have recommended the following to improve the economic benefit of the resort industry in the city of Dasmariñas Cavite based on the lowest mean on each of the factor, such as direct impact, indirect impact, and induced effect.

1. The local government of City of Dasmariñas should provide a copy of the annual report of the tax revenue from the resort industry to the management of each resort.
2. The resort should prioritize the local residents to be the employee of the resort.
3. A seminar on the economic benefit of the resort industry and the advantage of an increased transaction should be held.

REFERENCES

Aaron McNay (2017), Input-Output Models and Economic Impact Analysis: What they can and cannot tell us. Montana Economy at a Glance April 2017.

Aydin Sari, Murat Sari and Cătălin Popescu (2014). Effects of Various Economic Factors on Turkish Imports Economic Insights – Trends and Challenges Vol.III (LXVI) No. 1/2014 1–9.

Bautista, Joseph and Castillo, Jennifer Economic Pillar of Sustainable Development. DLSU-D Sustainable Development Report, 2014.

Boita, Marius, Constantin, Emilia Grigore, Daniela Boita (2015). Tourism and its role in the Economy. Current Context. Annals of Eftimie Murgu University Resita, Fascicle II, Economic Studies. 2015, pp. 188–194. 7p.

Canter, Julie (2020), 6 resort trends in 2020. Retrieved from: https://constructedground.com/6-resort-trends-in-2020/.

Drum, Kevin (2013) The Economy Is Pretty Hard to Understand These Days retrieved from: http://www.motherjones.com/kevin-drum/2013/09/economy-fed-monetary-policy.

Feng Sun (2014). The Dual Political Effects of Foreign Direct Investment in Developing Countries, The Journal of Developing Areas, Volume 48, Number 1, Winter 2014, pp. 107–125.

Khan, Dr. Muhammad Tariq (2015). Effects of Education and Training on "Human Capital - And Effects of Human Capital on Economic Activity (A Literature Based Research) International Journal of Information, Business and Management, Vol. 6, No.3, 2015.

Mitchell, Jonathan Mitchell, Xavier Font & ShiNa Li (2015) What is the impact of hotels on local economic development? Applying value chain analysis to individual businesses, Anatolia, 26:3, 347–358, DOI: 10.1080/13032917.2014.947299 Xavier Font & ShiNa Li (2015) What is the impact of hotels on local economic development? Applying value chain analysis to individual businesses, Anatolia, 26:3, 347–358, DOI: 10.1080/13032917.2014.947299.

Mowji, Mukesh (2016). Entire economy profits when hospitality industry prospers. Hotel & Motel Management. 5/15/2006, Vol. 221 Issue 9, pp. 19–19. 1/2p.

Nair, PK (2017). What is the difference between a resort and a hotel? Retrieved from: https://www.quora.com/What-is-the-difference-between-a-resort-and-a-hotel.

Negu Olimpia (2018), Human Capital: Cause and Effect of the Economic Growth an Emperical Analysis, Western University of Arad, Romania, 2018.

Neil Irwin (2014) Why is it so hard for average people to understand the economy? Retrieved from: https://www.youtube.com/watch?v=dvQSCdXOyzA.

Sergui Rusu and Florin Isac, (2018) Hospitality Indsutry Perspective and Prospects Aurel Vlaicu University of Arad, Romania. Lucari Stienfica (4th Edition).

Shaaban, Ingy A., Ramzy, Yasmine H. And Sharabassy, Azza A. (2015). Tourism as a Tool for Economic Development in Poor Countries: The Case of Comoro Islands. African Journal of Business & Economic Research. 2015, Vol. 8 Issue 1, pp. 127–145. 19p. 4 Graphs, 1 Map.

Sharma, Anupama (2015). Hospitality Industry: A Financial Boom for Indian Economy. Wealth: International Journal of Money, Banking & Finance; Jan-Jun2015, Vol. 2 Issue 1, pp. 54–61.

Suzuki, Alan (2016). The Economic Impact of Hotel Development http://www.pinnacle-advisory.com/economic-impact-of-hotel-development.html.

Vellas, John (2014). Economic Impact of Hospitality Industry on a larger scale and categorization of impacts. Journal of Hospitality and Economics. Vol. 7 Issue 1, pp. 66–80.

Xianming Fang and Yu Jiang (2014). The Promoting Effect of Financial Development on Economic Growth: Evidence from China. Emerging Markets Finance & Trade / January–February 2014, Vol. 50, Supplement 1, pp. 34–50.

Promoting Creative Tourism: Current Issues in Tourism Research – Kusumah et al. (Eds)
© 2021 Taylor & Francis Group, London, ISBN 978-0-367-55862-8

Tourism development and the well-being of local people: Findings from Lembang, West Java, Indonesia

E. Sukriah
Universitas Pendidikan Indonesia, Bandung, Indonesia

ABSTRACT: The high number of tourists equally demands the availability of tourism infrastructure. Tourism requires a high usage of land to ensure tourists' satisfaction. On the other hand, tourism should consider the surrounding eenvironmental sustainability especially for the local community. Hence, it is interesting to see how the local community perceives the development of tourism infrastructure in their region. The paper attempts to investigate the economic impact of tourism development in Lembang, West Java, Indonesia, to the local community and how the limited land in tourism destinations affect the local economy. In gathering the data, the authors have distributed Questionnaires to 400 households in the research site. The present paper uses a difference-in-differences (DD) analysis to investigate the economic impacts felt by the community. The results of data analysis have shown that the tourism have altered the community's economics condition. Farmers are the parties who feel the most impact of tourism in Lembang.

Keywords: local community, economic impact of tourism, difference-in-differences analysis

1 INTRODUCTION

Tourism is becoming one of the increasingly popular sectors utilized in development strategies in developing countries (Bunten 2010; Eusebio et al. 2018; Hampton & Jeyacheya 2015; Rico-Amoros et al. 2009; Wattanakuljarus & Coxhead 2008), including Indonesia. The central and regional governments make various improvements on tourist destination management and access to tourist destinations. The Bandung City Government has also carried out various tourism developments in an effort to achieve these national targets. The annual commemoration of the Asian-African Conference (KAA), the empowerment of Sundanese cultural tourism products, and the preservation of heritage buildings are some examples of local government strategies in attracting foreign tourists. The operation of Husein Sastranegara Airport as an international airport, with the slogan "*Stunning of Bandung, Where The Wonder of West Java Begins*" makes Bandung the main gateway to other tourist destinations in West Java.

Due to its close access to Bandung city as the capital of the province, the district of Lembang has been developing as a nationally known tourism area. Demographically, Lembang is located at an altitude of between 500 m–1,800 m above sea level with a temperature ranging from 19°C to 24°C and in a tropical country like Indonesia, the temperature is considered to be fairly low temperature and comfortable for the visiting travelers. At first, Lembang was a residential area where most of the land is used for housings and agricultural lands as economic support. Lembang region, which was initially an agricultural and residential area, has now turned into a tourist destination with a variety of existing tourist facilities. A large number of tourists and potential tourism destinations has attracted many investors. The rapid popularity and high number of tourists have increased the interests and demands in the provision of various tourist facilities, such as accommodation facilities, restaurants, rest areas, toilets, and other related urban tourism facilities (Ciolac et al. 2017).

 DOI 10.1201/9781003095484-39

The issue of tourism development continues to invite debates among experts. It is recognised that investment can encourage the acceleration of tourist destinations development, but these developments are allegedly not linearly and automatically make a positive contribution to tourist destinations. Numerous Scholars have critically examined the negative impact of tourism development in developing countries, both on aspects of social and economic life and the environment (Rogerson 2008; Wilson 2008).

Tourism is one of the main drivers of environmental change around the world. One of the continuously discussed issues within the realm of tourism is the availability of land. Thus far, there are not many in-depth discussions on this particular issue. Tourism requires high usage of land to ensure tourists' satisfaction. On the other hand, tourism should consider the surrounding environmental sustainability for the local community. The high number of tourists equally demands the availability of tourism infrastructure. Hence, it is interesting to see how the local community perceive the development of tourism infrastructure in their region. Therefore, the aim of this study is to analyze the economic impact of tourism development in Lembang to the local community, especially to community income and expenditures.

2 LITERATURE REVIEW

Tourism is a non-factory - based industry that requires a large workforce to meet the needs of tourists (Harrison 2002). Tourism development provides new employment opportunities for the community (Besculides et al. 2002; Iancu & Stoica 2010). The high absorptive capacity of tourism could give more benefits to people's income and increases their welfare. Furthermore, a land capitalisation for tourism contributes to overall community welfare due to the high absorption of local resources in the industry (Bunten 2010).

In general, local people who have related skills to tourism businesses are absorbed in various means of accommodation, such as camps and lodging (Mbaiwa 2003). This type of workers is a group that benefits from the availability of many new jobs due to changes in land use for tourism. In terms of the skills and workloads carried out by the local communities, most of them are work as supporting staff who are not able to control the business. As supporting staff, they are employed as service personnel, cleaners, kitchen helpers, cooks, carers, housekeeping workers, and a few serve as professional guides or as assistant managers. Agricultural products, plantations, to handicrafts produced by local people will be absorbed into tourism. In addition, transportation facilities, wholesalers, and retail traders become a place of absorption of local labor.

Besides that, tourism purposes can create a suitable environment for local residents' businesses (Garcia et al. 2015). Local businesses are increasingly developing along with the growth of tourism, such as the emergence of local traders, in the form of shops, boutiques, food stalls, places to eat, homestays, or printing business. Moreover, the tourism industry provides new jobs for the surrounding community even though on a small scale, such as artisans, pearl farmers, graphic designers, painters, tailors, local distributors, motorcycle rental owners, and other related business. In other words, the business movement of local traders will affect upstream businesses that are also locally owned.

In international tourism destinations, tourism business ownership is dominated by foreigners and cooperation between foreigners and local residents. Tourism businesses which are 100% owned by local residents are very few, namely only 1% (Mbaiwa 2003). The type of business owned by foreigners is different from the business owned by local residents; this difference is clearly seen. The most foreign-controlled businesses are accommodation and transportation facilities. Their presence only aims to serve the needs of foreign tourists and upper class local tourists. The loss of local control over land in its area has a long-term negative impact on local communities. Not a few rules that apply in the community are violated by investors (Picard 2006), and this is the biggest loss for the local community.

In terms of local infrastructure development, tourism has a positive impact on the development of regional infrastructures, such as road construction and street lighting (Garcia et al. 2015).

Appropriate road conditions will be well maintained not only along the main road to the tourist destination but also to the bus terminal, airports, hotels, restaurants and other tourist destinations (Mbaiwa 2003). Well maintained road conditions will give benefits to the surrounding community since there will be a higher intensity of mobility.

Tourism has an impact on rising land prices, housing construction, and other property values (Korca 1996; Madrigal 1993). Limited amount of land causes an increase in land prices. Local people who own land with high tourism potential are groups that benefit from tourism. In addition to obtaining high land prices, landowners also benefit from increased land rent or building rental prices.

Unfortunately, expensive land and housing prices reduce the ability of many local people to be able to buy their first home, especially local youth (Garcia et al. 2015). Large investors want the location of land that provides great benefits for him. Therefore, lands with strategic locations and high tourism resources will be controlled by them. Local people are only able to buy land that is far from the center of business activities. Changes in land prices will ultimately lead to changes in ownership and function (Zoomers 2011). Land use is driven by activities that have more benefit to their owners. Relinquishment of land ownership is not uncommon due to government coercion of its citizens in the interests of investors (Mbaiwa 2003).

3 METHOD

The paper attempts to investigate the economic impact of tourism development in Lembang to the local community. In collecting the Data, the authors distributed questionnaires to 400 households in the region who have lived there for more than ten years and aged over 30 years. Initially, in referring to Slovin formula, the number of samples is 397 households with 5% margin of error (Tejada & Punzala 2012). For the study purpose, the sample size is rounded to 400 households. Questionnaires were distributed to both regions, namely areas where tourism activities exist and areas where tourism activities do not exist. The proportional stratified random sampling was used in this study, and the samples are taken to represent the studied groups with the same portion (Soehartono 2004; Sugiyono 2012).

In analysing the data, the authors used an impact evaluation analysis to see the effects felt by the community due to tourism development. The method used to analyse the impact evaluation in this study is the difference-in-differences (DD) method as coined by Gertler et al. (2011). The DD analysis compares the communities' income, expenditure or consumption of people in two different regions. The analysis compared areas where there are tourism activities with areas where there are less to none tourism activities. Furthermore, the analyses also covered the differences in the communities' income, expenditure or consumption of the local community before and after the tourism development.

4 RESULTS AND DISCUSSION

4.1 *Community income*

The development of tourism in Lembang encourages changes in land use from vacant, unproductive land and plantation land to various forms of tourism businesses, ranging from attractions, hotels or lodging, restaurants, to parking lots. This causes the loss of income sources for farmers whose land had been altered.

In addition, the livelihoods of people living around tourism activities are changing compared to ten years ago. Figure 1 shows the changes in community income sources at site of the study.

Figure 1 shows that in the last ten years, the livelihoods of people living in Lembang have changed. About 58% of the livelihoods of the people around tourism experienced a change, the rest of the community continued to live the same income. Livelihoods in the agricultural sector have

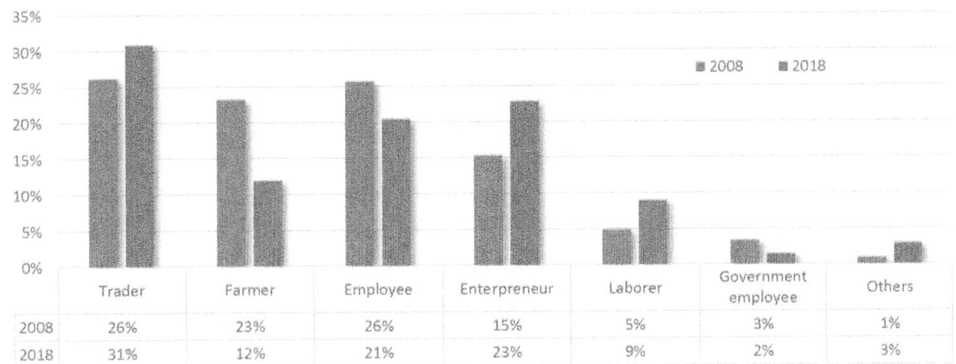

Figure 1. Livelihoods of the population in Lembang.
Source: Primary Data, 2018.

decreased quite sharply from 23% in 2008 to 12% in 2018. On the other hand, livelihoods such as entrepreneurship, trading, and other labors have increased. One of the causes of the decline for the farmers working in the community is the agricultural lands have been turned into tourism facilities. Farmers who lose their land turn into sellers by opening small stalls in front of their terraces for a living. There are also those who do other labors, such as construction laborers, vegetable porters, and even unemployed.

For local sellers who serve the surrounding community, such as meatball sellers, satay sellers, fruit sellers, grocery stores, building material shops, and repair shops, congestion caused by tourism activities reduces their sales turnover. This is because few local people go outside when the holiday mass arrives. They prefer to spend their time at home. On the other hand, traffic jams do not encourage tourists to shop at their stores. The local merchant businesses are micro businesses whose shop does not have parking lot; therefore, tourists are reluctant to stop by at their stores.

The cause is also the same as experienced by sellers, namely traffic jams due to land turned into tourist attractions. There is one public transportation route that directly connects Bandung City with Lembang. This route is commonly used by people living in Lembang who want to go to the city of Bandung. This transportation route is quite strategic because it passes through Bandung City Station so that it can deliver passengers to and from Bandung City Station.

Most tourists who come to Lembang drive in privats cars, motorbikes, and buses (for groups of passengers). Not many tourists take public transportation for a vacation in Lembang. On long holidays, long traffic jams reduce their income because fewer people use public transportation services. This results in a decrease in their income every day. Prior tourism development took place, the average angkot drivers' income was at least IDR 300,000 per day, but now their income is only IDR 120,000 per day at the highest.

Among the people of Lembang, there are also groups who in advantage of the tourism. This group gains an additional income. They are motorcycle taxi drivers, impromptu parking services, and impromptu sellers on the roadside. This group makes their business only as a side during a long holiday period. From this side business, they get additional income averagely ranging from IDR 100,000 to IDR 500,000 per day.

Among the tourists visiting there are also foreign and domestic tourists from outside the island who need fast transportation to get to the Husain Sastranegara Airport. In a long traffic jam, motorcycle taxi drivers and impromptu motorcycle taxi drivers is a quick and appropriate solution for tourists who are in a hurry. They claim not to set certain motorcycle taxi rates to tourists, but will generally get more than usual fees.

Next is the community group that provides vehicle parking for tourists. The large number of four-wheeled vehicles owned by tourists coming to Lembang requires a large parking area at a tourist attraction. Unfortunately, the parking lot provided at a tourist attraction cannot accommodate tourists' buses and cars. This condition is an opportunity for the community to open parking lots on large and empty lands so that an impromptu parking can be utilized at several points around the tourist attraction. The parking fee is very high, which is IDR 10,000 for a motorcycle, IDR 20,000 for a car and IDR 30,000 for a bus. This parking fee is not an official rate. Parking rates are determined based on the agreement of fellow impromptu parking attendants. Income from the impromptu parking business is divided equally for land guardian or for landowners.

Land that has been turned into a place of tourism generally has limited parking space so that another parking lot is needed to accommodate the incoming tourist vehicles. Unproductive empty land or a house terrace becomes an alternative for the surrounding community to be used as a parking lot. Every day on weekends and long holidays, each landowner is able to accommodate 8 to 20 four-wheeled vehicles on large tracts of land. On the terrace, it can accommodate 8 to 20 two-wheeled vehicles. Each four-wheeled vehicle is charged a tariff of Rp. 20,000, while a two-wheeled vehicle is subject to a tariff of Rp. 5,000. Thus, the community will get additional income of IDR 40,000 to IDR 400,000 per day, per parking location.

The next group is impromptu sellers who sell their products, directly to tourists. As in various places in the archipelago, traffic jams will invite hawkers to sell in traffic. This also happened on the streets of Lembang. The hawkers sell the typical snacks of Bandung, such as fried snacks, drinks, grilled sausages, *seblak*, dumplings, *batagor*, and so forth.

4.2 *Community expenditures*

In general, Lembang's community spend more on daily needs among other things such as, school fees, transportation costs, and motorcycle installments. Prices of goods for daily necessities in Lembang are almost the same. The difference is between the price of the products offered on the main street of Lembang and from those Lembang markets. The prices of goods on the market are relatively cheaper compared to those sold on Lembang Main Street.

There are a number of additional expenses arising after tourism development in Lembang, e. g. the cost of motorcycle installments and the cost of home renovations. Most people in Lembang now own motorbikes on installments. Since tourism id present, people living in the tourism area have felt limited space. They can no longer go to other places freely. Public transport is also difficult to board during the holidays. Therefore, many of them buy a motorcycle in installments.

Next is the additional cost of home renovation that must be expended after tourism developmernt. People who live next to a tourist attraction feel a very big impact after a change in land use. Some parts of the house suffered damage, such as cracked walls and broken floors. They have done some repairs to damaged parts of the house, but the damage reoccurs. Nowadays, people who live next to a tourism business feel uneasy. They are worried that the high walls beside the house will collapse onto them, especially during the rainy season and strong winds.

In addition, there are additional expenses that appear only during the long holidays and weekends. After the land is turned into a tourism business, traffic jams always occur. This is what causes the community to add new expenses, such as the cost of riding a motorcycle taxi and the increase of fuel costs allocation.

5 CONCLUSION

Tourism has a negative impact on the economic conditions of the people of Lembang. The communities experiencing the most negative impacts are those who depend on agricultural land. After land capitalization for tourism in Lembang, people's expenditure increases. There are additional expenses following Lembang's becoming a tourist destination, among others, the cost of motor vehicle installments, fuel cost allocation, and transportation costs. Land capitalization for tourism has an impact on increasing the price of land owned by residents. For landowners, rapid price increases encourage them to sell their land, hoping to be able to buy more land in other locations where land prices are still cheap. However, for low-income people, land capitalization for tourism reduces their purchasing power to own residential land. This is because the increase in land price and the size of the land offered is enormous, reducing the purchasing power of the community.

REFERENCES

Besculides, A., Lee, M. E., & McCormick, P. J. (2002). Residents' Perception of The Cultural Benefits of Tourism. *Annals of Tourism Research, 29*(2), 303–319.

Bunten, A. C. (2010). More Like Ourselves: Indigenous Capitalism Through Tourism. *The American Indian Quarterly, 34*(3), 285–311.

Ciolac, R., Rujescu, C., Constantinescu, S., Adamov, T., Dragoi, M.C., Lile, R. (2017). Management of a Tourist Village Establishment in Mountainous Area Through Analysis of Costs and Incomes. *Sustainability*, 9, 875, 1–18.

Eusebio, C., Vieira, A.L., Lima, S. (2018). Place Attachment, Host-Tourist Interactions, And Residents' Attitudes Towards Tourism Development: The Case of Boa Vista Island in Cape Verde, *Journal of Sustainable Tourism*, 1–20.

Hampton, M.P., dan Jeyacheya, J. (2015). Power, Ownership and Tourism in Small Islands: Evidence from Indonesia. *World Development, 70*, 481–495.

Harrison, D. (2002). Tourism, Capitalism And Development in Less Developed Countries. Dalam L. Sklair, *Capitalism & Development*. London dan New York: Routledge.

Iancu, F.-C., Stoica, I.-V. (2010). Tourism Capitalizationof Industrial Heritage Elements: A Strategic Direction of Sustainable Development. Case Study: The Petrosani Depression. *GeoJournal of Tourism and Geosites, 5*(1), 62–70.

Garcia, F. A., Vazquez, A. B., & Macias, R. C. (2015). Resident's Attitudes towards The Impacts of Tourism. *Tourism Management Perspectives, 13*, 33–40.

Gertler, P. J., Martinez, S., Premand, P., Rawlings, L., & Vermeersch, C. M. (2011). *Impact Evaluation in Practice.* Washington DC: The International Bank Reconstruction and Depelopment/The World Bank.

Korca, P. (1996). Resident Attitudes Toward Tourism Impacts. *Annals of Tourism Research, 23*(3), 695–726.

Madrigal, R. (1993). A Tale of Tourism in Two Cities. *Annals of Tourism Research, 20*, 336–353.

Mbaiwa, J. E. (2003). The Socio-Economic and Environmental Impact of Tourism Development on The Okavango Delta, Nort-Western Botswana. *Journal of Arid Environments, 54*, 447–467.

Picard, M. (2006). *Bali: Pariwisata Budaya dan Budaya Pariwisata.* Jakarta: Kepustakaan Populer Gramedia dengan Forum Jakarta-Paris Ecole Francaise d'Extreme-Orient.

Rico-Amoros, A., Olcina-Cantos, J., & Sauri, D. (2009). Tourist Land Use Patterns and Water Demand: Evidence from The Western Mediterranean. *Land Use Policy, 26*, 493–501.

Rogerson, C. M. (2008). Developing Small Tourism Businesses in Southem Africa. *Botswana Notes and Records, 39*, 23–34.

Soehartono, I. (2004). *Metode Penelitian Sosial: Suatu Teknik Penelitian Bidang Kesejahteraan Sosial dan Ilmu Sosial Lainnya*, Bandung: PT Remaja Rosdakarya Offset.

Sugiyono. (2012). *Metode Penelitian Kuantitatif, Kualitatif, Dan R&D.* Bandung: Alfabeta.

Tejada, J. J., & Punzalan, J. R. B. (2012). On the Misuse of Slovin's Formula. *The Philippine Statistician*, 61(1), 129–136.

Wattanakuljarus, A., & Coxhead, I. (2008). Is Tourism-based development goog for the poor? A General Equilibrium Analysisi for Thailand. *Journal of Policy Modeling* (30), 929–955.

Wilson, T. D. (2008). Introduction: The Impacts of Tourism in Latin America. *Latin American Perspectives, 35*(3), 3–20.

Zoomers, A. (2011). Introduction: Rushing for Land: Equitable and Sustainable Development in Africa, Asia and Latin America. *Development, 54*(1), 12–20.

Promoting Creative Tourism: Current Issues in Tourism Research – Kusumah et al. (Eds)
© 2021 Taylor & Francis Group, London, ISBN 978-0-367-55862-8

The effect of individual and destination accessibility on willingness to visit: Nature-based tourism destination

I. Wirajaya, F. Rahmafitria & G.R. Nurazizah
Universitas Pendidikan Indonesia, Bandung, Indonesia

A. Jamin
Universiti Teknologi MARA, Melaka, Malaysia

ABSTRACT: Several studies showed that low perception of individual accessibility negatively influences tourists' willingness to visit, while high perception of destination accessibility positively influences tourists' willingness to visit. The purpose of this study was to measure the influence of individual accessibility and destination accessibility on tourists' revisit intention to nature-based tourism destination. This research design method used Non-correlational Experimental Design. A total of 114 respondents were obtained using convenience and snowball sampling technique to collect the sample, which was then analyzed using multiple linear regression. The result of this study showed that high perception of individual accessibility did not have an influence on tourists' willingness to visit, while high perception of destination accessibility had an influence on tourists' willingness to visit. Simultaneously, both variables have little influence toward tourists' willingness to visit.

Keywords: Individual accessibility, destination accessibility, willingness to visit, nature-based destination

1 INTRODUCTION

Infrastructure development is a critical component in increasing destination accessibility (Rahmafitria et al. 2019). Seetanah (2011) explained that infrastructure development has a positive effect on tourists' visit. Elements of infrastructure development such as train rail, airport, highway, and public transportation have a direct effect on tourism destination, development, and accessibility (Ritchie & Crouch 2010). In accordance with Scott and Gosling (2016), in the past 40 years, massive expansion of road, sea, and air transportation mode, decrease in fuel price, and time efficiency have made it possible for tourists in a massive scale to access destinations.

Several researches suggest that destination accessibility is not one of the primary factors that influences tourist to visit a destination (Mayo & Jarvis 1981; in McGuire et al. 1986; Blazey 1987; Dellaert et al. 2013; Gilbert & Hudson 2000; Pennington-Gray & Kerstetter 2002; Nyaupane & Andereck 2007; Lattman et al. 2016). Crawford et al. (1991) explained that an individual's capability to access a destination is inhibited by constraints, which lower their individual accessibility level. Therefore, destination accessibility is not one of the factor in a tourist's decision making. This research was conducted by employing individual accessibility factor (Dellaert et al. 1998; Fleischer & Pizam 2002; Nyaupene & Andereck 2007) and destination accessibility factor (Lohmann & Kaim 1999; Awaritefe 2004; Gossling et al. 2006; Crouch 2011; Ramseook-Munhurrun et al. 2015; Eby & Molnar 2002; Qiu et al. 2018) as the deciding factor in decision to visit a destination. Researching the thought process of tourist decision making is important to understand how a destination develops and competes with other destination (Huang & Hsu 2009; Jani & Han 2014).

 DOI 10.1201/9781003095484-40

Table 1. Variables operationalization.

Variables	Indicator
Individual Accessiblity	Health condition Psychological condition Opportunity Economic factor Social factor
Destination Accessibility	Driving attributes
	Distance Route Mode of transportation Trip safetiness Congestion Trip scenery Trip facilites
	Destination attributes
	Cost Attraction Climate Destination safetiness Destination facilities
Willingness to visit	Willingness to visit

2 LITERATURE STUDY

2.1 *Individual and destination accessibility*

Manstead and Eekelen (1998) explained that belief in one's capability is divided into two factors: internal and external. It has been identified that several internal and external factors can be utilized to measure individual accessibility: health condition, psychological condition, opportunity, economic factor, and social factor. The researches mentioned below reveal that different demographics have different levels of capabilities (Dellaert et al. 1998; Blazey 1987; McGuire et al. 1986; Nyaupene & Andereck 2007).

Destination accessibility is closely associated with transportation elements that include distance, time, cost, and infrastructure (Prideaux 2000). Handy (2005) explained that destination accessibility can be measured from attractiveness factor, which reflects the quality of attribute of a destination. Those attractiveness factors include cost/value, destination safety, destination attraction, climate, mode of transportation, and route (Lohmann & Kaim 1999; Awaritefe 2004; Gossling et al. 2006; Prideaux & Carson 2010; Crouch 2011; Ramseook-Munhurrun et al. 2015). Rahmafitria et al. (2020) has mentioned the role of perceived accessibility on destination choices. Table 1 shows indicators of individual and destination accessibility.

2.2 *Willingness to visit*

Willingness to visit is the result of a series of evaluations and calculation of cost and benefit before visiting a destination (Rahmafitria et al. 2016). One of the strongest indicators to measure a tourist's intention to visit is destination attraction and readiness (Huang & Hsu 2009; Rahmafitria et al. 2016). Rahmafitria et al. (2016) explained that tourists' willingness to visit is based on overall evaluation of their experience toward a destination, which includes their satisfaction toward destination attraction and readiness, which in turn could lead the tourist to revisit the destination again.

Table 2. Tourist visits to Ujung Genteng.

Year	Number of tourist
2011	24,123
2012	24,335
2013	26,768
2014	29,444
2015	32,391

Data source: Nuraeni (2017).

2.3 *Hypotheses formulation*

Hypothesis 1 was formulated from extensive literature review and a theory by Doran and Pomfret (2019). Doran and Pomfret (2019) explained that an individual with high self-efficacy has positive perception toward their own action. In theory, it means that they are confident they can finish a given task or activity. In the context of this research, if tourists have high self-efficacy, they will likely have a positive perception about their own level of individual accessibility to visit a destination. Therefore, the following hypothesis was formulated:

H_1: Individuals with high self-efficacy have a positive perception about their own individual accessibility. Therefore, it has an effect toward tourist's willingness to visit.

Hypothesis 2 was formulated from extensive literature review and a theory by Qiu et al. (2018). In theory, a tourist's positive perception toward destination accessibility is a strong indicator of a tourist's willingness to visit. Conversely, tourists are unwilling to revisit a destination if they have negative perception toward a destination. Based on the explanation above, the following hypothesis was formulated.

H_2: Positive perception toward destination accessibility influences a tourist's willingness to visit.

This research would also like to test if there were a simultaneous effect between Individual and destination accessibility toward a tourist's willingness to visit. Therefore, the following hypothesis was formulated:

H_3: Individual and destination accessibility affects a tourist's willingness to visit simultaneously.

3 METHODS

This research was conducted at Ujung Genteng Tourism Area, West Java, Indonesia. This location was chosen because of its characteristics, which it is rural and difficult to reach, yet the number of tourist visitation was increasing, as shown in Table 2, which depicts the number of tourist visits from 2011 to 2015. This location can describe the complexity of individual and destination accessibility and its effect on willingness to visit.

This research used Non-experimental Correlational Design (Creswell 2014). The questionnaire consists of 10 questions to measure individual accessibility, 18 questions to measure destination accessibility, and 4 question to measure willingness to visit, all using Likert scales. Samples of this research were tourists who have visited Ujung Genteng at least once or more, which were obtained using convenience sampling method. A total of 114 respondents' data were collected from the questionnaire, which was distributed to respondents from October 5, 2019 to November 6, 2019. Likert scales data were first converted into interval scales using a *successive interval* method. Converted data were then analyzed using multiple linear regression with SPSS.

4 FINDINGS AND DISCUSSION

Findings from this research give several contributions to the conception of individual and destination accessibility as a factor which contribute toward decision making. Based on 114 demographics

Table 3. Hypotheses summary t-test.

Hypothesis	t-value	t-table	Significance
H_1	1.110	1.981	0.269
H_2	4.672	1.981	0.000

Table 4. Result of f-test.

Model	
Independent variable	Individual accessibility Destination accessibility
Dependent variable	Willingness to visit
F-value	19.950
F-table	2.69
Significace	0.000

profiles collected, the majority of respondents are student. Respondents are in the age group between 17–35 years old, which according to Chiu et al. (2015) can be classified as *youth travelers*, a type of tourist who has a high purchasing power. Typically, *youth travelers* have more opportunity to travel because of more leisure time (Chiu et al. 2015). The majority of the respondents come from West Java. Therefore, distance is not an issue. McKercher (2008) explains that it has been recognized that distance affects demand. Tourist visitation will increase the shorter the distance to destination, and vice versa.

Table 3 depicts the result of t-test for individual and destination accessibility. Individual accessibility indicator, which consists of health condition, psychological condition, opportunity, economic factor, and social factors, shows positive results among the respondents. Nevertheless, individual accessibility does not have an effect toward willingness to visit. The result, shown in Table 3, shows significance value of 0.269>0.05 and t-value of 1.110<1.98118, which indicates that the hypothesis is rejected. It was first assumed that lower perception of individual accessibility negatively affects willingness to visit based on the findings of Awaritefe (2004) and Nyaupene and Andereck (2007), but it turns out that high perception of inidivdual accessibility does not have an effect toward tourists' willingness to visit. Thus, it could be intepreted that tourists do not put their high perception of individual accessibility into their decision-making factor when deciding to visit a destination, but when tourists feel like their individual accessibility perception are low, they consider it as a deciding factor when visiting a destination.

Destination accessibility indicator, which consists of driving attribute and destination attribute, shows positive results among the respondents. The result of destination accessibility, shown in Table 3, shows significance value of 0.000<0.05 and t-value of 4.672>1.98118, which indicates that the hypothesis is accepted. Thus, it confirmed the findings of Qiu et al. (2018) that destination accessibility is the main driving force of tourists' decision-making factors when deciding to visit a destination. Furthermore, Qiu et al. (2018) explain that when it came to satisfaction, destination attribute had a significant effect on tourists' satisfaction, whereas driving attribute did not have a significant effect on tourists' satisfaction. Therefore, stakeholders should prioritise the development of destination attribute to be able to accomodate tourists' needs, because a higher satisfaction level means a higher chance that tourists will come to visit and revisit a destination.

The f-test, shown in Table 4, was conducted in order to tell if individual and destination accessibility simultaneously have an effect on tourists' willingness to visit. Simultaneously, individual and destination accessibility has an effect toward willingness to visit as the result shows that significance value of 0.000<0.05 and f-value of 19.950>2.69.

The R-squared test result, shown in Table 5, shows that each dimension of individual and destination accessibility could explain variance of data from dependent variable with a value of 26.4%, the

Table 5. Result of coefficient of determination.

Model	
Independent variable	Individual accessibility Destination accessibility
Depedent variable	Willingness to visit
R	0.514
R-squared	0.264
Adjusted R-squared	0.251

rest of the variance (73.6%) could be explained with different variables, which were not included. The result indicates that simultaneously both independent variables were considered weak to explain the variance of data of dependent variable (Sugiyono 2012). It should be noted that decision making is a complex and a multifaceted aspect of tourists' lives. Thus, the complexity of decision making cannot be reduced into a single factor, whereas there are multiple factors that came into consideration in the decision-making process.

5 CONCLUSION

This research contributes toward the development of individual and destination accessibility as a concept, whereas destination accessibility is an important factor, but not a primary one in the development of nature-based tourism destinations. The results of R-squared shows that both individual and destination accessibility is not the primary factor of decision making. Nevertheless, the significance of destination accessibility from the test results indicates that it is one of the factors that should be considered in developing a destination.

The significance of destination accessibility toward willingness to visit shows the level of importance of a destination to accommodate tourists' needs. Primarily in terms of accommodation, affordability, and quality of facility. Those elements are an important aspect in an effort to entice tourists to visit. A positive or negative assessment given by tourists is important for the growth of tourism destination.

REFERENCES

Albayrak, T., & Caber, M. (2016). Destination attribute effects on rock climbing tourist satisfaction: an Asymmetric Impact-Performance Analysis. *Tourism Geographies, 18*(3), 280–296.

Awaritefe, O. (2004). Motivation and Other Considerations in Tourist Destination Choice: A Case Study of Nigeria. *Tourism Geographies, 6*(3), 303–330.

Blazey, M. A. (1987). The differences between participants and non-participants in a senior travel program. *Journal of Travel Research, 26*(1), 7–12.

Chen, Y.-C., Shang, R.-A., & Li, M.-J. (2014). The effects of perceived relevance of travel blogs' content on the behavioral intention to visit a tourist destination. *Computers in Human Behavior, 30*, 787–799.

Creswell, J. W. (2014). *Research Design: Qualitative, Quantitative and Mixed Methods Approaches.* California: SAGE Publication.

Crouch, G. I. (2011). Destination Competitiveness: An Analysis of Determinant Attributes. *Journal of Travel Research, 50*(1), 27–45.

Dellaert, B., Arantze, T., & Horeni, O. (2013). Tourists' Mental Representations of Complex Travel Decision Problems. *Journal of Travel Research, 53*(1), 3–11.

Dellaert, B. G., Ettema, D. F., & Lindh, C. (1998). Multi-faceted tourist travel decisions: a constraint-based conceptual framework to describe tourists' sequential choices of travel components. *Tourism Management, 19*(4), 313–320.

Doran, A., & Pomfret, G. (2019). Exploring efficacy in personal constraint negotiation: An ethnography of mountaineering tourists. *Tourist Studies, 19*(4), 475–495.
Eby, D. W., & Molnar, L. J. (2002). Importance of scenic byways in route choice: a survey of driving tourists in the United States. *Transportation Research Part A: Policy and Practice, 36*(2), 95–106.
Eusebio, C., & Vieira, A. L. (2011). Destination Attributes' Evaluation, Satisfaction and Behavioural Intentions: a Structural Modelling Approach. *International Journal of Tourism Research, 15*(1), 66–80.
Fleischer, A., & Pizam, A. (2002). Tourism Constraints among Israeli Seniors. *Annals of Tourism Research, 29*(1), 106–123.
Gilbert, D., & Hudson, S. (2000). Tourism demand constraints: A skiing participation. *Annals of Tourism Research, 27*(4), 906–925.
Gossling, S., Bredberg, M., Randow, A., Sandstrom, E., & Svensson, P. (2006). Tourist Perceptions of Climate Change: A Study of International Tourists in Zanzibar. *Current Issues in Tourism, 9*(4–5), 419–435.
Handy, S. (2005). Planning for Accessibility: In Theory and in Practice. *Access to Destinations*, (pp. 131–147).
Huang, S., & Hsu, C. H. (2009). Effects of Travel Motivation, Past Experience, Perceived Constraint, and Attitude on Revisit Intention. *Journal of Travel Research, 48*(1), 29–44.
Jani, D., & Han, H. (2014). Personality, satisfaction, image, ambience, and loyalty: Testing their relationships in the hotel industry. *International Journal of Hospitality Management, 37*(1), 11–20.
Lattman, K., Olsson, L. E., & Friman, M. (2016). Perceived Accessibility of Public Transport as a Potential Indicator of Social Inclusion. *Journal of Transport Geography, 4*(3), 36–45.
Lohmann, M., & Kaim, E. (1999). Weather and holiday destination preferences image, attitude and experience. *The Tourist Review, 54*(2), 54–64.
Manstead, A. S., & Van Eekelen, S. A. (1998). Distinguishing Between Perceived Behavioral Control and Self-Efficacy in the Domain of Academic Achievement Intentions and Behaviors. *Journal of Applied Social Psychology, 28*(15), 1375–1392.
McGuire, F., Dottavio, D., & O'Leary, J. (1986). Constraints to participation in outdoor recreation across the life span: A nationwide study of limitors and prohibitors. *The Gerontologist, 26*(5), 538–544.
McKercher, B. (2008). The Implicit Effect of Distance on Tourist Behavior: a Comparison of Short and Long Haul Pleasure Tourists to Hong Kong. *Journal of Travel & Tourism Marketing, 25*(3–4), 367–381.
Nuraeni, L. (2017). Kemenarikan Daya Tarik Wisata Kawasan Pantai Ujung Genteng, Kecamatan Ciracap, Kabupaten Sukabumi.
Nyaupene, G. P., & Andereck, K. L. (2007). Understanding Travel Constraints: Application and extension of a leisure constraints model. *Journal of Travel Research, 46*(4), 433–439.
Pennington-Gray, L., & Kerstetter, D. (2002). Testing a constraints model within the context of nature-based tourism. *Journal of Travel Research, 40*(4), 416–423.
Prideaux, B., & Carson, D. B. (2010). *Drive Tourism: Trends and Emergin Markets.* Oxfordshire: Routledge.
Qiu, H., Hsu, C., Li, M., & Shu, B. (2018). Self-drive tourism attributes: influences on satisfaction and behavioural intention. *Asia Pacific Journal of Tourism Research, 23*(4), 395–407.
Rahmafitria, F., Nurazizah, G. R., & Riswandi, A. (2016). In S. M. Radzi, M. H. Hanafiah, N. Sumarjan, Z. Mohi, D. Sukyadi, K. Suryadi, & P. Purnawarman, *Heritage, Culture and Society: Research agenda and best practices in the hospitality and tourism industry* (pp. 293–298). London: CRC Press.
Rahmafitria, F., Purboyo, H., & Rosyidie, A. (2019). Agglomeration in tourism: the case of SEZs in regional development goals. *MIMBAR, 35*(2), 342–351.
Ramseook-Munhurrun, P., Seebaluck, V. N., & Naidoo, P. (2015). Examining the Structural Relationships of Destination Image, Perceived Value, Tourist Satisfaction and Loyalty: Case of Mauritius. *Procedia - Social and Behavioral Sciences, 175*, 252–259.
Ritchie, J. B., & Crouch, G. I. (2010). A model of destination competitiveness/sustainability: Brazilian perspectives. *Revista de Administração Pública, 44*(5), 1049–1066.
Rutty, M., Gössling, S., Scott, D., & Hall, C. (2015). the global effects and impacts of tourism: An overview. In C. Hall, D. Scott, S. Gössling, C. Hall, D. Scott, & S. Gössling (Eds.), *Handbook of Tourism and Sustainability* (pp. 36–33). London: Routledge.
Seetanah, B. (2011). Assessing the dynamic economic impact of tourism for island economies. *Annals of Tourism Research, 38*(1), 291–308.
Sugiyono. (2012). *Metode Penelitian Kuantitatif Kualitatif dan R&D.* Bandung: Alfabeta.

Promoting Creative Tourism: Current Issues in Tourism Research – Kusumah et al. (Eds)
© 2021 Taylor & Francis Group, London, ISBN 978-0-367-55862-8

Push and pull factors in visiting a remote nature-based destination

R. Ameliana, L. Somantri, F. Rahmafitria & F.A. Karim
Universitas Pendidikan Indonesia, Bandung, Indonesia

ABSTRACT: Tourists' decision to travel to a destination is arguably influenced by push and pull factors. Thus, this paper aims to analyse the dominant push and pull factors of tourists' visits in Curug Malela, Bandung Barat, Indonesia. Descriptive quantitative methodology was carried out with questionnaires as the research instrument. Based on factor analysis, five push and five pull factors were identified. According to the *Component Transformation Matrix*, the dominant push factor was *adventure and exploration* while *recreation and facilities* is the most dominant pull factor. By identifying these factors, tourists' needs and wants can be identified precisely to drive the focus of development and management of a destination.

Keywords: tourist motivation, pushing & pulling factor, remote nature-based destination

1 INTRODUCTION

One's decision to travel for tourist is influenced by pushing and pulling factors. These factors are generally indicated by two separated relationships between two decisions (Klenosky 2002). Someone travels because of being pushed by internal factor such as motivation and at the same time being pulled by external factors from the destination and its attractions (Cha et al. 1995; Rahmafitria & Marwa 2018).

According to a previous research study, Dianty (2015) identified nine factors from 24 variable indicators of pushing & pulling factors in traveling. The pushing factors included (1) Novelty & Knowledge-seeking, (2) Rest & Relaxation, (3) Fulfilling Dream, (4) Adventure & Exploration and (5) Prestige. While the emerging pulling factors are (1) Safety, Cleanliness & Variety-Seeking, (2) Events & Activities, (3) Affordable Price and (4) Travel Arrangement.

Pulling and pushing factors theory is commonly discussed, however there is still a lack of analysis about the interaction level between pushing and pulling factors, especially with a remote nature-based destination with relatively hard access. Since the number of tourists to remote nature destination are increase, then some questions arise. Does pulling factors attract more than pushing ones? And why remote and hard accessibility influence their motivation? Therefore, this research will discuss tourist motivation within the pushing and pulling factors concept.

The research questions are identified as follows, 1) What are the pushing and pulling factors influencing tourists to go to a remote nature-based destination?; 2) What are the most dominant pushing and pulling factors influencing tourists to go to a remote nature-based destination?

2 LITERATURE REVIEW

2.1 *Pushing factors in tourism*

Ryan in Pitana and Gayatri (2005) stated that there are few pushing factors for someone to travel, such as; escape, relaxation, play, strengthening family bonds, prestige, social interaction, romance, educational opportunity, self-fulfilment and wish fulfilment.

 DOI 10.1201/9781003095484-41

Table 1. Pushing factors theories.

Writers	Pushing Factors
Uysal and Jurowski (1994)	• Desire to escape Rest and relaxation • Prestige • Social interaction • Health and fitness • Adventure
Yuan and McDonald (1990)	• Escape • Relaxation/hobbies • Enhancement of kinship relationships • novelty

Source: Researcher's notes (2019).

Kim (2003) said that at the beginning, tourists are pushed by internal wants or motivation factors like escape motive, relaxation, prestige, family and friend togetherness, knowledge and enjoying natural resources (Table 1).

2.2 *Pulling factors in tourism*

Besides pushing factors, pulling factors also influence tourists to travel which are needed to make traveling happen. These factors came from the destination that attracts tourists to visit (Suwena & Widyatmaja 2017).

Pulling factors, according to Chrouch in Kassean (2013) contribute to tourists' experience in giving positive or negative perception toward a destination. These factors arise as the result of tourists' perception of a destination and these are factors that motivate them in holiday planning.

At the beginning, tourists are drawn by a destination attribute such as historical attraction, environment and weather attraction, expenditure and low travel cost factor and sport and outdoor activities (Goossens 2000). Uysal and Hagan in Zeng (2015) identified several factors that can draw tourists to visit a destination, as follows: Natural and historic attractions, food, people, recreation facilities and marketed image of the destination. Culture is also one of the pulling factors of a destination. The awareness of tourism toward culture is hoped to make every tourism stakeholder to realize about the existing cultural values that need preserving (Andriani et al. 2019).

2.3 *Tourist visiting intention and decision making*

Interest is a pushing factor that attracts someone's attention to another person or object (Widagdyo 2017). The visit intention theory is analogized as a purchasing intention as stated in the research by Albarq (2014).

Tourist's visiting decision is essential to develop a marketing strategy. Tourist behaviour is supported by the general assumption on how a decision was made (Salusu 1996). As a purchasing decision is analogized as visiting decision, the theory about purchasing decision is also used in the visiting decision theory (Jalilvand & Samiei 2012). According to Somantri et al. (2018) and Rahmafitria et al. (2016), other geographical factors that influence a visiting decision include location, accessibility, area differentiation and use value.

3 METHODOLOGY

Curug Malela, Indonesia, is chosen to be the location of the study due to the suitable characteristics of the far distance, hard accessibility and the minimum tourism facilities, and yet has an increasing visit every year (see Table 2). This destination has multiple waterfalls in different altitudes as its unique selling point.

Table 2. Curug Malela tourist visits 2014–2018.

Visitor Number				
2014	2015	2016	2017	2018
1.400	3.000	1.800	4.500	10.000

Source: KBB Cultural & Tourism Board.

This research employed a quantitative approach with multivariate statistic, using factor analysis technique to find correlate factors among independent indicators. The sample was 100 respondents with an accidental sampling method. The indicators in this research are adopted from Pitana and Gayatri (2005); Kim (2003); Uysal and Jurowski (1994); Yuan and McDonald (1990) for pushing factors, and from Goossens (2000); Zeng (2015); Klenosky (2002) for pulling factors.

4 RESULTS AND DISCUSSION

4.1 *Pushing factors analysis*

Pushing factors analysis was done by employing 20 indicators (to be called research variables). After the KMO test and Bartlett's test, the research found that the KMO value is $0{,}811 > 0{,}5$. Meanwhile, Bartlett's test showed a significance value for $0{,}000 < 0{,}5$ ($\alpha = 5\%$). Thus, these variables could be used for the next analysis and prediction.

The communalities table showed that the average communal value from 20 variables is 0,5, with pushing factor X1.13 with 0,807 communal value as the highest (Table 3).

The number of values formed can be seen on the total variance explained table. Variables resulting in >1 eigenvalue are the formed pushing factors. It can be concluded that there were five new factors formed that influence tourists to visit Curug Malela, which were escape and relaxation, novelty and knowledge, prestige, fulfilling dream and adventure and exploration. The value of each factors can be identified by exploring factor loading value on component matrix table. From these new five factors, the next step was deciding whether the new formed variable belongs to the group, which is determined by the highest loading factor value. Based on the component transformation matrix, it is found that the fifth factor; adventure and exploration was the most dominant factor with the highest correlation value of 0,807.

The new 5 factors as finding in this research, hag4 confirm some of previous research studies' findings. There are some similarities of factors finding with Dianty (2015), which are novelty & knowledge-seeking, rest & relaxation, fulfilling dream, adventure & exploration and prestige. Therefore, this research show that basic motivation factors are the main push factors for tourists while visiting remote nature destination.

4.2 *Pulling factors analysis*

The same thing as the pushing factors was applied to pulling factor analysis. After the KMO and Bartett's test, it was found that the KMO value was $0{,}820 > 0{,}5$ and Bartlett's test score showed a significance value of $0{,}000 < 0{,}5$ ($\alpha = 5\%$). Therefore, these variables could be used for the analysis. The communalities table showed 0,5 average score from 20 variables. The biggest score in this analysis was variable X2.14 with 0,784 (Table 4).

The formed pulling factors had a value of >1 that can be seen on total variance explained table. It was arguably that these are the new five factors that influence tourist visits to Curug Malela. The details of the variables can be seen on the component matrix table by analyzing the factor loading value.

From these new five factors, the next step was deciding whether the new formed variable belongs to the group, which is determined by the highest loading factor value. The formed five factors were

Table 3. New pushing loading factors.

No.	Factors		Description	*Loading* Variance	% Variance
1	*Escape and Relaxation*	X1.1	Willingness to escape from daily routine	0,739	36,871%
		X1.2	Finding new interest things	0,714	
		X1.3	Experiencing new life style	0,615	
		X1.4	Doing outdoor activities	0,765	
		X1.5	Enjoy and fun travelling	0,642	
		X1.6	Relaxation	0,591	
		X1.14	Visiting new places	0,570	
2	*Novelty and Knowledge*	X1.7	Enriching knowledge of a destination	0,572	10,172%
		X1.8	Visiting historical places	0,640	
		X1.10	Meeting new people	0,507	
		X1.11	Learning local culture and traditional life style	0,832	
		X1.17	Exploring cultural site	0,700	
3	*Prestige*	X1.9	Sharing travelling experience	0,740	7,418%
		X1.18	Proud for visiting new destination	0,775	
		X1.19	Proud for reaching remote destination	0,771	
4	*Fulfilling Dream*	X1.15	Visiting popular places	0,557	6,091%
		X1.16	Achieving to visit dream places	0,814	
		X1.20	Experiencing different image of visiting destination	0,492	
5	*Adventure and Exploration*	X1.12	Spending time with friends	0,669	5,296%
		X1.13	Adventuring nature based destination	0,743	

Source: Researcher's notes (2019).

Table 4. Pulling factors loading.

No.	Factors		Description	*Loading* Variance	% Variance
1	*Recreation Facilities*	X2.12	Road quality is good	0,826	33,828%
		X2.13	Easiness to find public transport	0,841	
		X2.14	Availability of recreation facilities	0,877	
		X2.15	Availability of information about destination	0,788	
		X2.16	Destination cleanness	0,771	
		X2.17	Destination safety	0,855	
		X2.18	Enjoying local culinary	0,628	
2	*Natural Attraction*	X2.1	Enjoying nature landscape and scenery	0,809	16,433%
		X2.2	Enjoying nature and clean environment	0,713	
		X2.3	Visiting new destination	0,645	
		X2.6	Enjoying fresh air	0,567	
		X2.7	Doing hobby in visiting nature destination	0,623	
		X2.9	Exploring unique water fall	0,667	
3	*Leisure and Culture Activities*	X2.5	Good climate	0,594	6,266%
		X2.8	Interest to destination myth and history	0,752	
		X2.10	Close distance from origin to destination	0,525	
		X2.11	Experiencing recreation activities	0,650	
4	*Affordable Price*	X2.19	Affordable price	0,664	5,467%
		X2.20	Shopping local souvenir	0,728	
5	*Hunting*	X2.4	Finding new photo spots	0,804	5,083%

Source: Researcher's notes (2019).

recreation facilities, natural attraction, leisure & culture activities, affordable price and hunting. The recreation facilities were indicated to be the most dominant factor with the highest correlation value of 0,770. This finding are different with the result of Dianty (2015). She found tour arrangement and natural attractiveness as dominant pulling factors of a destination.

5 CONCLUSION

In the comparison, pulling factors are found to be the strongest factor in attracting tourists to visit remote destinations. Adventure and exploration have a strong motive for tourists, as they try to experience the beauty of nature destination and challenging its physical landscape. The result also showed that distance and access was not a main constraint for tourists to visit. Furthermore, the tourism facilities and attraction were the most dominant pulling factor and tourists' tendency to consider natural attraction first rather than the accessibility. This finding is supported by the adventure and exploration as the dominant pushing factor that explains the hard accessibility and far distance being seen as an attraction and a challenge in the adventure instead of a constrain. This implicates the destination planning and management to turn the accessibility and the distance of the destination as part of the experience for tourists while ensuring their comfort, safety and the development of supporting facilities.

REFERENCES

Albarq, N. (2014). Measuring The Impacts of Online Word of Mouth on Tourists Attitude and Intentions to Visit Jordan: An Empirical Study. *International Business Research*.

Andriani, R., Brahmanto, E., & Purba, B. C. (2019). Value Tari Sigale-Gale dalam Meningkatkan Wisata Budaya di Desa Tomok Kabupaten Samosir. *Journal of Indonesian Tourism, Hospitality and Recreation, 2*.

Cha, S., McCleary, K. W., & Uysal, M. (1995). Travel Motivations of Japanese Overseas Travelers: A Factor-Cluster Segmentation Approach. *Journal of Travel Research, 34*, 33–39.

Dianty, O. (2015). Faktor Pendorong dan Faktor Penarik Wisatawan Surabaya Melakukan Perjalanan Wisata ke Penang, Malaysia. *Program Manajemen Kepariwisataan Program Studi Manajemen Fakultas Eknonomi*.

Goossens, C. (2000). Tourism Information and Pleasure Motivating. *Annals of Tourism Research, 27*, 301–321.

Jalilvand, M. R., & Samiei, N. (2012). The Effect of Word of Mouth on Inbound Tourist Decision for Traveling to Islamic Destinations (The Case of Ishafan as a Tourist Destination in Iran). *Journal of Islamic Marketing*.

Kassean, H. (2013). Exploring Tourist Push and Pull Motivations to Visit Mauritius as a Tourist Destination. *African Journal of Hospitality, Tourism and Leisure, 2*.

Kim, S. L. (2003). The Influence of Push and Pull Factors at Korean National Parks. *Tourism Management*, 169–180.

Klenosky, D. B. (2002). The "Pull" of Tourism Destinations: Ameans-End Investigation. *Journal Travel Research, 40*, 385–395.

Pitana, I. G., & Gayatri, P. G. (2005). *Sosiologi Pariwisata*. Yogyakarta: ANDI.

Rahmafitria, F., & Marwa, S. (2018). A Factor Analysis of Visitors' Motivation in Visiting the Geology Museum of Bandung. *IOP Conference Series: Earth and Environmental Science*.

Rahmafitria, F., Nurazizah, G. R., & Riswandi, A. (2016). Attraction and Destination Readiness Towards Tourists' Intention to Visit Solar Eclipse Phenomenon in Indonesia. *Heritage, Culture and Society: Research Agenda and Best Practices in The Hospitality and Tourism Industry*, 299.

Salusu. (1996). *Pengambilan Keputusan Strategik Untuk Organisasi Publik dan Organisasi Nonprofit*. Jakarta: PT Gramedia.

Somantri, L., Fadhillah, G., & Jupri. (2018). Evaluasi Rute Transportasi Angkutan Kota dengan Menggunakan Sistem Informasi Geografis. *Jurnal Pendidikan Geografi*, 165.

Suwena, I. K., & Widyatmaja, I. N. (2017). *Pengetahuan Dasar Ilmu Pariwisata*. Denpasar: Pustaka Larasan.

Uysal, M., & Jurowski, C. (1994). Testing the Push and Pull Factors. *Annals of Tourism Research*, 844–846.

Widagdyo, K. G. (2017). Pemasaran, Daya Tarik Ekowisata dan Minat Berkunjung Wisatawan. *Jurnal Bisnis dan Manajemen, 7*, 263.

Yuan, S., & McDonald, C. (1990). Motivational Determinates of International Pleasure Time. *Journal Travel Research*, 42–44.

Zeng, G. (2015). *Tourism and Hospitality Development Between China and EU*. Guangzhou: Springer.

Promoting Creative Tourism: Current Issues in Tourism Research – Kusumah et al. (Eds)
 ISBN 978-0-367-55862-8

Can marine debris pollution cause the loss of tourism revenue in Indonesia? An empirical study

P. Hindayani, A. Khosihan & A.R. Pratama
Universitas Pendidikan Indonesia, Bandung, Indonesia

ABSTRACT: As the world's largest archipelagic nation, Indonesia had thousands of islands with beautiful marine resources which had the potential to increase the regional economy. On the contrary, Indonesia was ranked second on the list of the world's biggest plastic waste and marine debris producers. At the moment, marine debris was an emerging global issue which threatened marine tourism in Indonesia. Moreover, some studies indicated that marine debris could harm the tourism sector. Based on literature reviews, publications about marine debris related to the loss of tourism revenue, economic losses and economic impacts in the tourism sector in Indonesia had not been made. In this present paper, we presented an empirical study of the reviews of marine debris in Indonesia related to tourism and identified how to estimate its economic impacts on tourism. We also suggested more comprehensive future research on the economic impact of marine debris on the tourism sector.

Keywords: Marine debris, Loss of revenue, Tourism, Economic impact, Economic Losses

1 INTRODUCTION

Unmanaged waste will end and be carried into rivers or sewage which flows to the sea. It becomes ocean garbage called marine litter or marine debris (MD). One of the dominant global MD compositions is plastics (between 60 and 80%) (Derraik 2002). It is triggered by one of the world's greatest innovations in industries (Agamuthu et al. 2019) that have dominantly increased in the consumer marketplace in the 1930s and 1940s (Jambeck et al. 2015). In the 1950s, the amount of plastics in the environment has increased exponentially (Löhr et al. 2017; UNEP 2016). Moreover, stated by Thevenon and Carroll (2015), the global production of plastics is more than 300 million every year, and approximately eight million tons of plastic end up in our oceans every year. According to Jambeck et al. (2015), 275 million metric tons (MMT) of plastic waste was generated in 192 coastal countries in 2010, with 4.8 to 12.7 MMT entering the ocean. Currently, the issues of MD have become a serious problem in the world's oceans (UNEP 2016). Furthermore, MD has direct and indirect harmful impacts on ocean and coastal ecosystems; such as a threat to the coral reefs (Patterson Edward et al. 2020), mangroves (Debrot et al. 2013), marine organisms of at least 690 species or 17% of the IUCN Red List species (Gall & Thompson 2015), fisheries and aquaculture (Campbell et al. 2017; Hinojosa & Thiel 2009; McIlgorm et al. 2011), human health (Thompson et al. 2009), and financial loss (Jang et al. 2014; Krelling et al. 2017; McIlgorm et al. 2011; Ofiara & Brown 1999; Ofiara & Seneca 2006; Qiang et al. 2020).

Furthermore, MD impacts can harm the entire economic sectors; such as aquaculture and fisheries, transportation, tourism, and public health. Moreover, the estimated economic impacts of plastic pollution in the ocean were eight billion per year in the consumer product sector (UNEP 2016). One of the potential economic losses of MD which is neglected is tourism industries. The presence of marine debris has an impact on tourism industries that can reduce tourists and attractiveness of beaches and shorelines for recreational purposes (Ballance et al. 2000; Jang et al. 2014; Krelling et al. 2017; Ofiara & Brown 1999; Ofiara & Seneca 2006; Qiang et al. 2020). Also,

DOI 10.1201/9781003095484-42

there have been few studies on the relationship between economic losses and MD in Asia Pacific (McIlgorm et al. 2011; McIlgorm et al. 2020), USA (Kahn et al. 1989; NOAA 2019; Ofiara & Brown 1999; Ofiara & Seneca 2006; Swanson & Zimmer 1990; Swanson et al. 1978), the Cape Metropolitan, South Africa (Ballance et al. 2000), in specific sites in Geoje Island, South Korea (Jang et al. 2014), and in the coast of Paraná State, Southern subtropical coast of Brazil (Krelling et al. 2017). Indonesia as the world's largest archipelagic nation, with 17,504 islands and a coastline that extends >54,716 km (Purba et al. 2019), has the enormously threatening potential of economic losses on marine tourism industries due to MD. However, In Indonesia, there had been no studies of economic losses associated with MD in terms of tourism industries.

This paper aimed to present this literature review in terms of the loss of tourism revenue, economic losses or economic impacts as the effects of MD on marine tourism industries. Then, selected published manuscripts from previous studies deeply identified that must be addressed to the loss of tourism revenue, economic losses or economic impacts of MD on tourism industries. Therefore, this study provided insight into how to calculate the loss of tourism revenue, economic losses or economic impacts of MD in Indonesia for tourism industries especially marine tourism. Indeed, this study as a basis strategic approach to determine policy, to mitigate and to minimize economic impacts of MD. Thus, the effect of economic losses due to MD on the marine tourism sector will be reduced.

2 METHODS

This present paper qualitatively reflected on desk studies with references related to the potential economic losses in tourism caused by MD pollution. The data were obtained by examining the existing literature, especially papers, articles, books, government reports, and others. We also looked for literature databases: Google Scholar (http://scholar.google.com), Scopus (http://scopus.com), and the Indonesian scientific literature "Sinta" (http://sinta.dikti.go.id). The manuscripts were searched using the keywords "MD or marine litter", "tourism", and "economic". This study used a qualitative approach to the empirical study of an in-depth review of the potential economic loss and economic impacts in the tourism sector due to MD pollution in Indonesia.

3 LITERATURE REVIEW

3.1 *What is Marine Debris (MD)? An Overview*

MD, or marine debris, ocean garbage, or marine litter, was "any persistent, manufactured, or processed solid material discarded, disposed of, or abandoned in the marine and coastal environment" (Agamuthu et al. 2019; Löhr et al. 2017; NOAA 2019; Purba et al. 2019; UNEP 2005, 2016). The source of MD was waste as a result of human activities that had been dumped into coastal or marine environments, resulted from land-based activities or coastal or sea activities; such as fishing, aquaculture, tourism, shipping (Krelling et al. 2017; Löhr et al. 2017; UNEP 2005). MD was either deliberately or accidentally discharged directly to the beach and sea or indirectly discharged into rivers or sewage which flows towards the sea. Based on UNEP (2005), the source of MD in the world was approximately 80% from the ocean from discharged land-based MD (Michelle et al. 2012; UNEP 2005), including anthropogenic activity, landfills, untreated municipal sewage, industries, and tourism. Remaining 20%, the sources of MD were sea-based like fishing gears, fishing vessels, aquaculture, military fleets, and offshore oil and gas platforms (Michelle et al. 2012; UNEP 2005).

By material types, MD could be divided into several categories: plastic, metal, glass, processed timber, paper and cardboard, rubber, clothing, and textiles (UNEP 2016). Reportedly, one of the material types most commonly found in the world's ocean was plastics (Derraik 2002; Jambeck et al. 2015; Löhr et al. 2017) that was around 60–80% of total global MD and can reach 90 to

95% of the total amount of marine debris (Derraik 2002). According to Zhou et al. (2016), MD could be classified into three categories: 1. MD carried out into the beaches was called deposited marine debris, 2. MD floating on the inshore and offshore ocean was called floating MD, and 3. MD in the seawater was called submerged marine debris. By sizes, MD could be classified into five categories (Lippiatt et al. 2013) i.e. megadebris (>1 m), macrodebris (>2.5 cm), mesodebris (5 mm–2.5 cm), micro-debris (≤5 mm in length), and nanodebris (≤1 μm in length).

4 FINDING AND DISCUSSION

4.1 *Marine Debris (MD) problems in the world*

MD was an environmental and economic problem worldwide. According to UNEP 2005, marine debris worldwide was around 6.4 million metric tones (MMT) per year in 1997 and there were over 13,000 pieces of plastic debris floating on every km^2 of the ocean. Calculated by Jambeck et al. (2015), one of the major sources of MD was mismanaged solid waste on land between 4.8 and 12.7 MMT/yr. Based on Thevenon and Carroll (2015) and UNEP (2016), MD approximately 8 million tons of plastic every year entered the ocean. Furthermore, stated by Parker (2015), in the ocean were 5.25 trillion pieces or 269,000 tons of plastic debris floating on the surface and other some four billion plastic microfibers per square kilometer littered in the deep sea. Meanwhile, recorded by Lebreton et al. (2017), unmanaged waste entering the ocean from rivers was between 1.15 and 2.41 MMT of plastic. A study by Lebreton and Andrady (2019), in 2015, mismanaged plastic waste produced globally was between 60 and 99 MMT. It was predicted by business as usual scenario up to 155–265 MMT per year by 2060 (Lebreton & Andrady 2019).

4.2 *Marine Debris (MD) problems and MD studies related to tourism in Indonesia*

There had been many studies of MD in Indonesia. Indonesia was currently ranked the second on the list of the world's biggest plastic MD around 0.48–1.29 MMT per year and mismanaged plastic waste in Indonesia was 3.22 MMT (Harsono 2019; Jambeck et al. 2015; Purba et al. 2019). Furthermore, Indonesia was also one of the countries with the largest population in the world where waste produced was 0.52 kg per person per day (Jambeck et al. 2015). Stated by Michelle et al. (2012), floating MD in Indonesia consisted of more than four items in every m^2. Seafloor debris or submerged MD in Indonesia was about 690,000 items/km^2 and one of the highest quantity of shoreline debris in the world was reported in Indonesia which was up to 29.1 items per m.

In general, the visible deposited MD in the beaches and shorelines was macrodebris that disturbed the aesthetics in terms of tourism purposes. Selected studies of MD in some tourist areas in Indonesia were already published. The famous tourism areas for both local tourists and foreign tourist included Bali, a world-famous tourist destination (Attamimi et al. 2015; Husrin et al. 2017) which suffered from MD almost every year; Pangandaran Beach, a well-known tourist spot in West Java (Purba et al. 2018); and Pulau Seribu at Jakarta Bay as tourism destinations (Uneputty & Evans 1997; Willoughby 1986) (Table 1).

Previous studies were discussing MD in tourism areas but the studies of MD related to the loss of tourism revenue, economic losses and economic impacts in Indonesia were still not published. Indonesia was facing serious marine pollution due to MD which threatened the tourism sector. As an archipelagic nation with the second-longest of coastline in the world, Indonesia was very vulnerable to be affected economically, especially in the marine tourism sector.

In addition, stated in Presidential Regulation Presidential Regulation No. 97/2017 and No. 83/2018, the Indonesian government has a commitment to reduce waste by 30%, to manage waste properly by 70% of total waste generation, and to reduce 70% in plastic waste entering the sea in 2025. Although Indonesia had a lot of planning in dealing with MD management but there were still many things that needed to be improved. However, in comparison with ASEAN countries (e.g., Singapore, Malaysia, and Thailand) which had higher travel & tourism competitiveness, Indonesia had more marine and coastal areas but less valuable marine and coastal tourism (World Economic

Table 1. Selected studies of Marine Debris at tourism areas in Indonesia.

No	Tourism area	Geographical type of area	Methodology	Dominant debris	By size types	References
1	Pulau Seribu, Jakarta	Beach and ecosystem	Composition, distribution and originating, and effects on tourism	Polyethylene bags	Macrodebris	(Willoughby 1986)
	Pulau Seribu, Jakarta	Beach and ecosystem	Accumulated MD since Willoughby's study (1986)	Bottles	Macrodebris	(Uneputty & Evans 1997)
2	Kuta Beach, Bali	Beach	UNEP/IOC Guideline on Survey and Monitoring of Marine Litter	Wood and plastic	Macrodebris	(Attamimi et al. 2015)
3	Kuta Beach, Bali	Beach, seabed near the shoreline, inshore surface	Abundance, propagation, occurrences, numerical simulations, and environment impact	Plastic and woods	Macrodebris	(Husrin et al. 2017)
4	Pangandaran, West Java	Beach	Using Ocean Conservancy (OC) form to find information about distribution, abundance, and types.	Cigarette butts and rope	Macrodebris	(Purba et al 2018)

Forum 2019). In terms of environmental sustainability and supporting condition policy, Indonesia was the second-lowest country in ASEAN (World Economic Forum 2019). Therefore, Indonesia's government and stakeholder should pay more attention to national tourism development targets on these issues to develop strategies which mitigate potential risks including MD.

4.3 *Lost of tourism revenue due to Marine Debris (MD) pollution*

MD polluted beaches and could reduce the aesthetic quality of beaches for tourist destinations. The visible deposited MD in the beaches was macrodebris that disturbed the comfortable aesthetics for tourism purposes along coastline and beaches. In some cases, MD could affect tourist's perception of the quality of the environment and which led to the loss of tourism. At selected beaches for recreation purposes, MD was one of the vital aspects considered by tourist decisions (Jarvis et al. 2016; Roca et al. 2008; Santos et al. 2005). Moreover, MD might impair the cleanliness of the beaches (Krelling et al. 2017; Qiang et al. 2020; Schuhmann 2019). Even visitors traveled more distant to cleaner beaches (Leggett et al. 2014) and most tourists chose a clean beach as an alternative (Krelling et al. 2017). Tourists would re-visit a beach for recreational purposes if the beach had increased the quality of the natural environment or had been cleaned (Jarvis et al. 2016; Leggett et al. 2014; Qiang et al. 2020; Schuhmann 2019).

These conditions would cause economic impacts on the tourism sector. Reported by McIlgorm et al. (2020), the total cost of direct damage due to MD on the marine economy in the APEC region in 2015 was $10.8 billion where the total cost of direct damage on the marine tourism sector was up to 59.2% or $6.41 billion. Some studies of the economic impacts on the tourism sector had already been published in several places. Most of the studies indicated a decline in tourist numbers and the loss of tourism revenue associated with marine debris. These studies were using the economic valuation approach to investigation.

This study highlighted an empirical investigation of how to calculate the loss of tourism revenue, economic losses and economic impacts on the tourism sector due to MD. This paper was important as it assessed the loss of tourism revenue, economic losses and economic impacts on the tourism sector in Indonesia. Then, this study provided case studies illustrating some economic valuations to assess (see Table 2). Also, any research about the economic valuation of MD in Indonesia had not been published (Purba et al. 2019). The economic valuation might help the government decide on mitigation and solutions. Based on Purba et al. (2019), the control of debris by several parties

Table 2. Some studies of the economic impact due to marine debris on the tourism sector.

No	Location	Type of environment	Market effect	Economic valuation methods	Negative economic impacts	References
1	1976, Long Island, New York, United States	Beach	• Decrease in beach use • Loss of tourists up to 50–60%	Lost expenditures (based on loss in the number of visitors)	$15–25 million	(Swanson et al. 1978)
2	1988, New York Bight, United States	Beach	• Decrease in beach use (30 and 90 million users) • Loss of beach fees	Benefit transfer	$708 to $2399 million $854 to $3429 million (in 2002 values)	(Kahn et al 1989; Ofiara & Brown 1999;
3	1988, New Jersey, United States	Beach	• Decrease in beach use (6.7–37 million users) • Loss of tourists up to 9–44%	Benefit transfer	$209 to $1020 million $397 to $1943 million (in 2002 values)	Ofiara & Seneca 2006; Swanson & Zimmer 1990)
4	1994–1995, Cape Peninsula in South Africa.	Beach	• Decrease in beach use • Loss of tourist revenue • Beach cleaning • Loss of tourists up to 52%	Travel Cost Method (TCM) Beach cleaning cost	R3–R23 million from tourism Expenditure R3 million from beach cleaning during 1994–95	(Ballance et al. 2000)
5	2008 and 2020 Asia-Pacific region	Beach and Ocean.	• Shoreline cleaning • Derelict fishing gear • Damage to transport • Damage to fishing vessels • Decrease in marine tourism	Direct and indirect economic cost	$1.26 billion per year in 2008 terms $10.8 billion per year in 2020 terms (From tourism sector $6.41 billion per year) $216 billion in 2050 (Total)	(McIlgorm et al. 2011; McIlgorm et al. 2020)
6	2011, Geoje Island, South Korea	Beach	• Decrease in beach use • Loss of average expenditure on daytime activities • Loss of lodging • Loss of tourist	Loss of tourists revenue by economic effects	US$29–37 million (29–37 billion KRW).	(Jang et al. 2014)
7	2017, Southern subtropical coast of Brazil	Beach	• Decrease in beach use • Loss of tourists from two Brazilian users • Loss of average expenditure on daytime activities	Loss of tourists revenue by economic effects	$3.2 to US$8.5 million	(Krelling et al. 2017)
8	2019, Alabama, Delaware/ Maryland, California, and Ohio, United States	Beach	• Decrease in beach use • Loss of tourists	Benefit function transfer	$4.5 million in Alabama, $24.0 million in Delaware and Maryland $27.1 million in Orange County, California $8.2 million in Lake Erie beaches in Ohio	(NOAA 2019)

at the national, provincial, and municipal levels had been regarding MD management. However, Indonesia should improve travel and tourism competitiveness, mainly environmental sustainability policies that include MD risks. Therefore, research on the matters will become more comprehensive future research related to an integrated policy approach to reducing the economic impacts.

5 CONCLUSION

Despite its reputation as an archipelagic country which had enormous economic potential in the marine tourism sector, Indonesia was ranked second on the list of the world's biggest plastic waste and marine debris producers. This paper had deeply examined several studies related to the loss of tourism revenue, economic losses and economic impacts due to MD on tourism industries which had been ignored. Based on the literature reviews, published manuscripts associated with economic losses and economic impacts on tourism due to MD were not found. Therefore, this study insighted study how to asses the effects of MD on tourism in the context of economic impacts. However, MD could and had resulted in sizable economic effects and losses. Most MD studies could reduce beach attendances and loss of tourism revenue. Tourists would revisit a beach if it had been cleaned. Based on findings, the empirical studies of economic impact due to MD had been quantified by many economic valuation methods. On the other hand, policies linked with MD management in Indonesia should pay more attention. These conditions gave the awareness and idea to mitigate MD to minimize the economic impacts. Overall, These economic valuation methods as assessments in Indonesia were one of the tools for an integrated policy approach to mitigation and solutions.

REFERENCES

Agamuthu, P., Mehran, S. B., Norkhairah, A., & Norkhairiyah, A. (2019). Marine debris: A review of impacts and global initiatives. *Waste Management & Research, 37*(10), 987–1002. https://doi.org/10.1177/0734242X19845041

Attamimi, A., Purba, N. P., Anggraini, S. R., Harahap, S. A., & Husrin, S. (2015). Investigation of Marine Debris in Kuta Beach, Bali. *The 1st Young Scientist International Conference of Water Resources Development and Environment Protection, June*, 1–7.

Ballance, A., Ryan, P. G., & Turpie, J. K. (2000). How much is a clean beach worth? The impact of litter on beach users in the Cape Peninsula, South Africa. *South African Journal of Science, 96*(5), 210–213.

Campbell, M. L., King, S., Heppenstall, L. D., van Gool, E., Martin, R., & Hewitt, C. L. (2017). Aquaculture and urban marine structures facilitate native and non-indigenous species transfer through generation and accumulation of marine debris. *Marine Pollution Bulletin, 123*(1–2), 304–312. https://doi.org/10.1016/j.marpolbul.2017.08.040

Debrot, A. O., Meesters, H. W. G., Bron, P. S., & de León, R. (2013). Marine debris in mangroves and on the seabed: Largely-neglected litter problems. *Marine Pollution Bulletin, 72*(1), 1. https://doi.org/10.1016/j.marpolbul.2013.03.023

Derraik, J. G. B. (2002). *The pollution of the marine environment by plastic debris: a review. 44*, 842–852.

Gall, S. C., & Thompson, R. C. (2015). The impact of debris on marine life. *Marine Pollution Bulletin, 92*(1–2), 170–179.

Harsono, N. (2019). Government: Indonesia produces up to 0.59 million tons of marine debris per year. *The Jakarta Post.* https://www.thejakartapost.com/news/2019/12/12/government-indonesia-produces-up-to-0-59-million-tons-of-marine-debris-per-year.html

Hinojosa, I. A., & Thiel, M. (2009). Floating marine debris in fjords, gulfs and channels of southern Chile. *Marine Pollution Bulletin, 58*(3), 341–350. https://doi.org/10.1016/j.marpolbul.2008.10.020

Husrin, S., Wisha, U. J., Prasetyo, R., Putra, A., & Attamimi, A. (2017). Characteristics of Marine Litters in the West Coast of Bali. *Jurnal Segara, 13*(2), 129–140. https://doi.org/10.15578/segara.v13i2.6449

Jambeck, J. R., Geyer, R., Wilcox, C., Siegler, T. R., Perryman, M., Andrady, A., Narayan, R., & Law, K. L. (2015). Plastic waste inputs from land into the ocean. *Science, 347, 768*(January).

Jang, Y. C., Hong, S., Lee, J., Lee, M. J., & Shim, W. J. (2014). Estimation of lost tourism revenue in Geoje Island from the 2011 marine debris pollution event in South Korea. *Marine Pollution Bulletin, 81*(1), 49–54. https://doi.org/10.1016/j.marpolbul.2014.02.021

Jarvis, D., Stoeckl, N., & Liu, H. B. (2016). The impact of economic, social and environmental factors on trip satisfaction and the likelihood of visitors returning. *Tourism Management, 52*, 1–18. https://doi.org/10.1016/j.tourman.2015.06.003

Kahn, J., Ofiara, D., & McCay, B. (1989). Economic measures of beach closures, economic measures of toxic seafoods, economic measures of pathogens in shellfish, economic measures of commercial navigation and recreational boatingfloatable hazard. In *Use Impairments and Ecosystem Impacts of the New York Bight.* Waste Management Institute.

Krelling, A. P., Williams, A. T., & Turra, A. (2017). Differences in perception and reaction of tourist groups to beach marine debris that can influence a loss of tourism revenue in coastal areas. *Marine Policy, 85*(September), 87–99. https://doi.org/10.1016/j.marpol.2017.08.021

Lebreton, L., & Andrady, A. (2019). Future scenarios of global plastic waste generation and disposal. *Palgrave Communications, 5*(1), 1–11. https://doi.org/10.1057/s41599-018-0212-7

Lebreton, L. C. M., Van Der Zwet, J., Damsteeg, J. W., Slat, B., Andrady, A., & Reisser, J. (2017). River plastic emissions to the world's oceans. *Nature Communications, 8*, 1–10. https://doi.org/10.1038/ncomms15611

Leggett, C., Scherer, N., Curry, M., & Bailey, R. (2014). Assessing the economic benefits of reductions in marine debris: a pilot study of beach recreation in Orange County, California. In *NOAA Marine Debris Program & Industrial Economics, Inc.* http://marinedebris.noaa.gov/sites/default/files/MarineDebrisEconomicStudy.pdf

Lippiatt, S., Opfer, S., & Arthur, C. (2013). Marine Debris Monitoring and Assessment. In *NOAA Technical Memorandum* (Issue NOS-OR&R-46). http://marinedebris.noaa.gov/sites/default/files/Lippiatt_et_al_2013.pdf

Löhr, A., Savelli, H., Beunen, R., Kalz, M., Ragas, A., & Van Belleghem, F. (2017). Solutions for global marine litter pollution. *Current Opinion in Environmental Sustainability, 28*, 90–99. https://doi.org/10.1016/j.cosust.2017.08.009

McIlgorm, Alistair, Campbell, H. F., & Rule, M. J. (2011). The economic cost and control of marine debris damage in the Asia-Pacific region. *Ocean and Coastal Management, 54*(9), 643–651. https://doi.org/10.1016/j.ocecoaman.2011.05.007

McIlgorm, A., Raubenheimer, K., & McIlgorm, D. E. (2020). *Update of 2009 Apec Report on Economic Costs of Marine Debris To Apec Economies Penultimate Version – in Confidence Update of 2009 Apec Report on Economic Costs of Marine* (Issue March). https://www.apec.org/Publications/2020/03/Update-of-2009-APEC-Report-on-Economic-Costs-of-Marine-Debris-to-APEC-Economies

Michelle, A., Adam, W., David, S., & Paul, J. (2012). *Plastic Debris in the World's Oceans*.

NOAA. (2019). *The Effects of Marine Debris on Beach Recreation and Regional Economies in Four Coastal Communities: A Regional Pilot Study Final Report* (Issue July 2019).

Ofiara, D. D., & Brown, B. (1999). Assessment of economic losses to recreational activities from 1988 marine pollution events and assessment of economic losses from long-term contamination of fish within the New York Bight to New Jersey. *Marine Pollution Bulletin, 38*(11), 990–1004. https://doi.org/10.1016/S0025-326X(99)00123-X

Ofiara, D. D., & Seneca, J. J. (2006). Biological effects and subsequent economic effects and losses from marine pollution and degradations in marine environments: Implications from the literature. *Marine Pollution Bulletin, 52*(8), 844–864. https://doi.org/10.1016/j.marpolbul.2006.02.022

Parker, L. (2015). Ocean Trash: 5.25 Trillion Pieces and Counting, but Big Questions Remain. *NATIONAL GEOGRAPHIC*. There are 5.25 trillion pieces of plastic debris in the ocean. Of that mass, 269,000 tons float on the surface, while some four billion plastic microfibers per square kilometer litter the deep sea.

Patterson Edward, J. K., Mathews, G., Raj, K. D., Laju, R. L., Bharath, M. S., Kumar, P. D., Arasamuthu, A., & Grimsditch, G. (2020). Marine debris — An emerging threat to the reef areas of Gulf of Mannar, India. *Marine Pollution Bulletin, 151*(November), 110793. https://doi.org/10.1016/j.marpolbul.2019.110793

Purba, N. P., Apriliani, I. M., Dewanti, L. P., Herawati, H., & Faizal, I. (2018). Distribution of Macro Debris at Pangandaran Beach, Indonesia. *International Scientific Journal, 103*(7), 144–156. www.worldscientificnews.com

Purba, N. P., Handyman, D. I. W., Pribadi, T. D., Syakti, A. D., Pranowo, W. S., Harvey, A., & Ihsan, Y. N. (2019). Marine debris in Indonesia: A review of research and status. *Marine Pollution Bulletin, 146*(June), 134–144. https://doi.org/10.1016/j.marpolbul.2019.05.057

Qiang, M., Shen, M., & Xie, H. (2020). Loss of tourism revenue induced by coastal environmental pollution: a length-of-stay perspective. *Journal of Sustainable Tourism, 28*(4), 550–567. https://doi.org/10.1080/09669582.2019.1684931

Roca, E., Riera, C., Villares, M., Fragell, R., & Junyent, R. (2008). A combined assessment of beach occupancy and public perceptions of beach quality: A case study in the Costa Brava, Spain. *Ocean and Coastal Management, 51*(12), 839–846. https://doi.org/10.1016/j.ocecoaman.2008.08.005

Santos, I. R., Friedrich, A. C., Wallner-Kersanach, M., & Fillmann, G. (2005). Influence of socio-economic characteristics of beach users on litter generation. *Ocean and Coastal Management, 48*(9–10), 742–752. https://doi.org/10.1016/j.ocecoaman.2005.08.006

Schuhmann, P. W. (2019). *Tourist Perceptions of Beach Cleanliness in Barbados: Implications for Return Visitation Tourist Perceptions of Beach Cleanliness in Barbados: Implications for Return Visitation. May*, 0–13.

Swanson, R. L., & Zimmer, R. L. (1990). Meteorological conditions leading to the 1987 and 1988 washups of floatable wastes on New York and New Jersey beaches and comparison of these conditions with the historical record. *Estuarine, Coastal and Shelf Science, 30*(1), 59–78. https://doi.org/10.1016/0272-7714(90)90077-5

Swanson, R. L., Stanford, H. M., & O'Connor, J. S. (1978). June 1976 pollution of Long Island ocean beaches. *Journal of the Environmental Engineering Division, ASCE, 104*(EE6), 1067–1085.

Thevenon, F., & Carroll, C. (2015). Plastic debris in the ocean: the characterization of marine plastics and their environmental impacts, situation analysis report. In *Plastic debris in the ocean: the characterization of marine plastics and their environmental impacts, situation analysis report*. https://doi.org/10.2305/iucn.ch.2014.03.en

Thompson, R. C., Moore, C. J., Saal, F. S. V., & Swan, S. H. (2009). Plastics, the environment and human health: Current consensus and future trends. *Philosophical Transactions of the Royal Society B: Biological Sciences, 364*(1526), 2153–2166. https://doi.org/10.1098/rstb.2009.0053

UNEP. (2005). *Marine Litter An analytical overview*.

UNEP. (2016). Recommended citation: Acknowledgements: In *Marine plastic debris and microplastics – Global lessons and research to inspire action and guide policy change. United Nations Environment Programme, Nairobi*.

Uneputty, P. A., & Evans, S. M. (1997). Accumulation of beach litter on islands of the Pulau Seribu Archipelago, Indonesia. *Marine Pollution Bulletin, 34*(8), 652–655. https://doi.org/10.1016/S0025-326X(97)00006-4

Willoughby, N. G. (1986). Man-made Litter on the Shores of the Thousand Island Archipelago , Java. *Marine Pollution Bulletin, 17*(5), 224–228.

World Economic Forum. (2019). *The Travel and Tourism Competitiveness Report 2019*. http://www3.weforum.org/docs/WEF_TTCR_2019.pdf

Zhou, C., Liu, X., Wang, Z., Yang, T., Shi, L., Wang, L., You, S., Li, M., & Zhang, C. (2016). Assessment of marine debris in beaches or seawaters around the China Seas and coastal provinces. *Waste Management, 48*, 652–660. https://doi.org/10.1016/j.wasman.2015.11.010

Promoting Creative Tourism: Current Issues in Tourism Research – Kusumah et al. (Eds)
© 2021 Taylor & Francis Group, London, ISBN 978-0-367-55862-8

How risky is liveaboard diving in Indonesia? An empirical investigation on the divers' perceived risk and oceanic geomorphology

R.M. Wirakusuma
Universitas Pendidikan Indonesia, Bandung, Indonesia

M. Lück & H. Schänzel
Auckland University of Technology, Auckland, New Zealand

M.A. Widiawaty & G.P. Pramulatsih
Universitas Pendidikan Indonesia, Bandung, Indonesia

M. Dede
Universitas Padjadjaran, Bandung, Indonesia

E. Dasipah
Universitas Winaya Mukti, Bandung, Indonesia

ABSTRACT: As the home for liveaboard diving, the Komodo National Park becomes the primary destination promoted by the Indonesia Ministry of Tourism. However, the number of foreign tourists have gradually decreased and many diving accidents were reported recently. The purpose of this research is to investigate the liveaboad diver's perceived risk and the dive sites' oceanic geomorphology where the accidents occurred. Fristly, to measure perceived risk, the data were collected from 60 respondents who wrote reviews on online travel agents, online media publications, and Indonesian diving forums. Furthermore, the data is divided into seven types of perceived risks such as physical, social, financial, performance, psychological, equipment, and time risk. Secondly, to obtain accurate 3D geomorphology analysis, this study used data from the Indonesia Geospatial Bureau and interpolated with the official zoning and maps from meteorology, climate, and geophysics bureau. The preliminary result showed that there is another type of risk such as equipment and weather; furthermore, Batu Bolong was considered to be the most dangerous diving site in Indonesia.

Keywords: liveaboard diving, scuba diving, Komodo National Park, marine tourism

1 INTRODUCTION

Liveaboard is the boat equivalent to a person's home and maybe moored permanently at a site or is transient. Thus, the experiences are relatively different, because sailing with live odds is always dealing with the unpredictable environment. Furthermore, the liveaboard is commonly installed with luxurious facilities and could travel islands-hopping to visit unique geographical features (Bowles 2016; Stoeckl et al. 2010). Indonesia has an abundance of marine biodiversities such as Raja Ampat, Komodo National Park, Derawan Islands, and Tulamben Bali, which were crowned by CNN Travel as 4 of the 10 most spectacular dive site destinations in the world (Vera 2017). Hence, that achievement attracts various levels of diversity across the globe and promotes marine

DOI 10.1201/9781003095484-43

tourism significantly in Indonesia. Furthermore, it also captivated international exposure through several exhibitions, underwater pictures, and word of mouth among divers in the multiplatform online forum (LongLiveIndonesia 2019).

Komodo National Park, which is not only recognized for the largest lizard habitat on earth but also as one of the marine tourism destinations, contributed an income of 32 billion rupiahs from the admission only, in 2018 (Lewokeda 2019). Furthermore, the park was declared as 1 of 10 primary destinations promoted by the Indonesia Ministry of Tourism (Soeriaatmadja 2017). The generated tourism impacts a wide range of hospitality operators starting from homestay, hotels, and resorts, and now the spotlight is on a liveaboard. The reason for increased tourist numbers is their rising demand for adventure activities from trekking to scuba diving in one single trip. These tourists also are likely to seek coral conservation provided by the local community (Wirakusuma et al. 2019). Most of the divers agreed that a considerable amount of marine biodiversities in Komodo National Park attract many recreational divers from all levels. Especially during April until October, the destination is in peak season because of summer holidays and the favorable weather conditions (Flores Komodo Expedition 2018).

As a tourist, the diving experience can differ, based on where the diver begins their adventure. Many experienced SCUBA divers choose self-sustained accommodations, such as liveaboard as a first option to fulfill their needs (Fang 2020). Liveaboard is the boat equivalent to a person's home and maybe moored permanently at a site or is transient (Jennings 2007). Thus, the experiences are relatively different, because sailing on a liveaboard is always multi-day, and involves dealing with the unpredictable environment. The diver can experience motion sickness and a lack of physical comfort due to bad weather, which can decrease the diver's level of satisfaction. However, liveaboard commonly offers luxurious facilities and can travel between islands to visit unique geographical features. It can also move to underwater scenery, such as coral reef walls, and the operator will ensure that there is a boat on the surface to pick the divers up at the end, and the dive instructor will carry a brightly colored inflatable marker buoy to ensure the group is visible when it ascends (Buckley 2011).

It is effortless to check the liveaboard cabin availability through www.liveaboard.com, and nowadays, 101 operators have listed their schedules (Liveaboard 2019). The operators offer the trip based on price, vessel, or connectivity with other destinations. The website acts as an online travel agent, and for booking, one simply has to select the date, vessel, number of cabins, and the payment method. However, once the tourists select a specific vessel and pay in advance by credit card, they are reluctant to spend additional money on other expenses, and even local products, because everything is provided onboard. Subsequently, whenever a hidden cost exists, it becomes a financial risk. Furthermore, the liveaboard operator offers various dive spots and few dive in caves or areas with strong currents. Multi-day liveaboard dive charters, which take clients on, for example, open-ocean night dives to submerged pinnacles, will generally only accept experienced clients and check their skills at various less-risky dive sites first. Where they take clients on drift dives, it is considered a physical risk for many SCUBA divers.

The research on liveaboard was very limited; only Fang (2020) has research on passenger motivation. The liveaboard operators are selling holidays with safety and security, and since the demand has been rising, accidents do sometimes occur. Most of the victims were stranded because of unpredictable currents and equipment malfunctions. The danger is inevitable on the open-water, and getting worse because the Komodo Dragons are excellent swimmers (Pet et al. 2005). There were cases where the diver had to save lives from encountering the Komodo Dragon on the surface (Edwards 2008). Thus, the danger will emerge as equipment risk. Many accidents occur and are published in a number of media, ensuring that the concern about the risk posed by tourists will be even more significant. The risk will happen because buying a trip from a third party is vulnerable to uncertainty. This study seeks to conduct spatial analysis related to underwater and geomorphological conditions that cause diving accidents. Hence, the main research question is to measure how aware liveaboard divers are to the risks that can occur, and how was the image of ocean geomorphology when the diving accident happened?

2 LITERATURE REVIEW

2.1 *Perceived risk in tourism*

There is a plethora of studies on perceived risk, affecting the selection of destinations from different perspectives. Lim (2003, p. 219), examined 18 studies in the literature and stated that the perceived risks had been identified in nine types of risk, as financial risk, performance risk, social risk, physical risk, psychological risk, time loss risk, personal risk, privacy risk, and resource risk. However, in these studies analyzed by Lim and in other studies in the literature, it has been observed that these nine dimensions are not entirely present, and some of these types are derived from nomenclature differences. For example, Roehl and Fesenmaier (1992) listed the most common risk types pronounced for the tourism sector as financial risk, functional risk (or performance risk), physical risk, social risk, psychological risk, satisfaction risk, time loss risk, health, political uncertainty, and terrorism. Sharifpour et al. (2014) examined the perceived risk associated with destinations under three headings: psychological risk, physical risk, and performance risk. Tsiros and Heilman (2005, p. 117), in their literature review, indicated that perceived risk was examined under six types, as functional risk, physical risk, financial risk, social risk, psychological risk, and performance risk.

Another study from Rittichainuwat and Chakraborty (2009) claimed there are six dimensions of perceived travel risks such as (1) increase of travel costs, (2) disease, (3) terrorism, (4) lack of novelty, (5) deterioration of tourist attractions, and (6) travel inconvenience. This study emphasizes the tourist's point of view from significant issues when they visit the destination, regarding disease and increase of terrorism.

The risk types, which have an effect on destination selection, are examined in this study as follows: (1) Physical risk: risks associated with personal safety, such as illness or injury, physical danger during vacation (Roehl & Fesenmaier 1992; Tsiros & Heilman 2005). (2) Social risk: the risk of embarrassment to other people or loss of social status due to a failed choice (Tsiros & Heilman 2005). (3) Financial risk: the risk of money loss due to the purchased service (Roehl & Fesenmaier 1992). (4) Performance risk: the risk is that the purchased product does not function properly, that it can only be used for a short period (Lim 2003), that the service characteristics do not respond to the needs of the consumer. (5) Psychological risk: the risk of experiencing disappointment (Lim 2003), and the loss of ego (Tsiros & Heilman 2005). (6) Time loss risk: possibility that a vacation will be a waste of time (Antony & Thomas 2010; Roehl & Fesenmaier 1992). (7) Equipment risk: the possibility that mechanical or equipment failures can occur during the trip (Antony & Thomas 2010).

2.2 *Ocean geomorphology*

Many geomorphologists study why ecosystems are created, understand the past of landforms and forecast changes by a combination of field studies, physical experiments, and computational modeling. This type of study is often used by tourism scholar to bridge interpretation of natural attraction, such as tourist-environmental maps. The government produces these maps, and they can effectively contribute to improving the tourist's knowledge, utilization, and appraisal of the environment of protected areas (Castaldini et al. 2005). Furthermore, the study of seafloor topography (bathymetry) was initially used to help navigation by mapping shifting sand shoals and treacherous reefs. Recent advances in both sonar resolution and GPS positioning now allow the researcher to study seafloor morphology in ever-sharper detail (McAdoo et al. 2000). Bathymetric data are commonly shown as digital shaded relief views illustrating seafloor morphology as clearly as if the oceans had been drained away. Finally, the bathymetric map would be beneficial for tourism stakeholders to create policy and other decision-making outputs.

3 METHODS

All of the respondents' data were collected from major online travel agent reviews, such as liveaboard.com and Tripadvisor.com. Pre-screening reviews were limited from October 2015 to October

2019. Afterwards, the data were divided into seven types of perceived risk based on keywords. The data in the form of comments and reviews were also re-confirmed by the experienced divers to ensure the clarity of the information.

Dive site data coordinates were obtained from various validated online sources. Afterwards, it was calibrated by overlaying dive site locations on high-resolution satellite imagery using ArcGIS software. The software output was a dive site map, and the reported diving accidents were added. This research used the Kernel Density Estimation method to determine the concentration level of diving accidents per 500 meters distance. It is a technique for extrapolating a distribution point through its two-dimensional probability function (Widiawaty 2019). The result is a clustered dangerous dive site location map with red color gradation.

4 RESULTS AND DISCUSSION

The results indicate that there are 692 diving spots in Indonesia with various levels of difficulty. Most of them are offering coral wall and slope, but there are also many diving spots that are located in a strait tending to have tremendous current. Table 1 displays diving accident data from 2008 to 2018 in Indonesia with the sources being obtained from Indonesia diving forums and online news reports. The data were recapitulated based on location and the number of accidents. Furthermore, Table 2 shows the causes of diving accidents in Indonesia. Thus, the destination with the most diving accidents in Indonesia is in the Komodo National Park, with a total of 10 cases recorded.

Table 2 shows that most of the diving accidents were drifted away caused by unpredictable ocean currents, and the second reason is decompression. From the reported dive accidents, there are many varied causes ranging from being swept by the unpredictable down current, unable to control buoyancy, injured from corals reef, to human error. It is hard to extract the information because, in the ocean, the shreds of evidence are nearly impossible to find. The sad fact was when the body was

Table 1. Number of cases diving accident in Indonesia (death fatality) from 2008–2018.

Location	Number of Cases	Location	Number of Cases
Komodo Nat. Park	10	Flores	1
Bali	6	North Minahasa	1
Raja Ampat	3	Wakatobi	1
Bunaken	2	Nias	1
Tanjung Bira	2	Derawan Archipelago	1
Gorontalo	1	Manokwari	1
Kep. Seribu	1	Sumba	1
Sangalaki Island	1	Lombok	1

Table 2. The cause of a diving accident.

Causes	Total
Drifted by Ocean Current	10
Decompression	8
Unknown	6
Human Error	5
Equipment malfunction	2
Crocodile Attack	1
Injured by Ship	1
Injured by coral rocks	1

Table 3. Respondents' demography.

Characteristics	Percentage (%)	Mean
Respondents		
Region		
European	42	
American	23	
Asian	18	
Australian & New Zealand	8	
Others	8	
Divers Experience Level		
High	90	>50 dive logs
Medium to Low	10	10–20 dive logs

found afloat far away from where previously seen. The factors of oceanic geomorphology, current flow prediction, and dive site spatial analysis will be investigated in this research.

4.1 *The perceived risk with liveaboard diving*

This study tried to do an inventory of the reviews of 60 liveaboard divers who wrote about their perceived risks online. Table 3 shows the respondents' demography from their nationality to divers' experience level.

Regarding physical risks, many reviews wrote about the ocean current, such as "But be careful, the Komodo National Park is due to the strong and partly swift currents are only suitable for experienced divers (+80 dives) who are also physically fit." It tells the diver to be more concerned about their dive license. The social risk was related to the person or group on the liveaboard such as "uncomfortable with the other guest in my cabin, and there were some minor shortcomings because the main guide who was not too friendly." The financial risk appeared when the hidden cost was not adequately communicated from the beginning. This review, for instance: "Oh, also be prepared for significant extra park fees not listed in advance. You get what you pay for, bearing in mind elevated Komodo prices compared with other regions. The boats are not worth spending $1000 for two people for two nights. Very expensive compared to other places in Indonesia." The performance risk was related to the facility inside the vessel such as: "No hot water, the ceiling is leaking, AC is down, noisy engine and the fumes entered the room and persistent toilet smell." The psychological risk was concerned with the service, the meals, the boat comfort, or overall service, such as: "I could not recommend this boat enough in terms of comfort and service." The equipment risk related to the worn-out diving gear onboard: "On the boat, we received old equipment. My regulator gauge did not work and stopped at 170 bar, another guest's regulator lost air, and her BCD jacket did not have working quick air release toggles." The last, perceived time risk emerged when the itinerary took longer than the estimated schedule, such as: "I felt the itinerary could have changed to include a checkout dive on the first day and travelled more on the last day before flying." Another perceived risk was the weather risk; according to the reviews, a rainy day could cause underwater visibility to plummet.

4.2 *Geomorphology of Komodo National Park*

This research has collected and analyzed areas of dive accidents with particular danger levels. Figure 1 shows the map of the kernel density estimation (KDE) analysis with a bandwidth of 500 meters. It indicates almost 30 diving spots that are in high accident level and 9 diving spots which are in low accident level. The selection of the KDE method is based on the emergence of the diving accident reports; the more accident points, the denser it will become (Hashimoto et al.

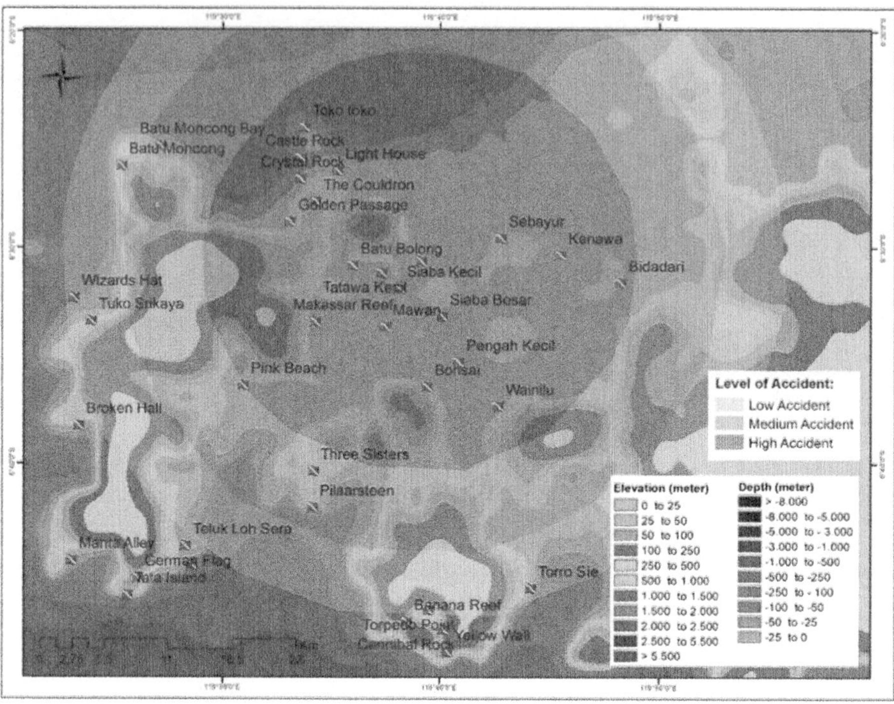

Figure 1. Diving accident map of Komodo National Park.

2016; Setiawan et al. 2019; Widiawaty 2019). The dark red zone is concentrated in the north to the middle of the region, and the center is Batu Bolong Diving spot. Meanwhile, the southern and western regions tend to be safer and marked with pink to no color, indicating a relatively safe zone.

In order to find out the depth level in Batu Bolong, elevation data were interpolated by the ArcGIS Kriging Model. Meanwhile, data on the speed and current direction were obtained from the re-interpolated Geospatial Information Agency website. It is located at the coordinate 119° 36′ 49,85″ BT–119° 36′ 51,07″ BT and 8° 32′ 11,63″ LS–8° 32′ 12,90″ LS. Batu Bolong was one of the most famous diving spots, which was formerly referred to as the Lesser Sunda Islands (Rahardiawan & Purwanto 2014). Figure 2 shows the 3D perspective of Batu Bolong with a depth of approximately 55 meters. It can be seen that most divers only reach the depth of 20 meters to enjoy the underwater scenery, which is indicated by the blue color (top cone).

5 CONCLUSION

After all, the research methodology was limited to the online survey. Thus, further research for the direct survey on location is recommended. The conclusion refers to two particular preliminary results. The result showed that there was another type of risk, such as equipment and weather risks. There is a possibility to find another type of risk on different market segmentation. Furthermore, Batu Bolong was considered to be the most dangerous diving site in Indonesia. Therefore, the liveaboard operator must be well aware of guiding on this dive site. Liveaboard diving is a unique experience yet vulnerable to the risks because of the aspect of adventure in an unpredictable environment. Subsequently, further research is needed to eliminate those risks.

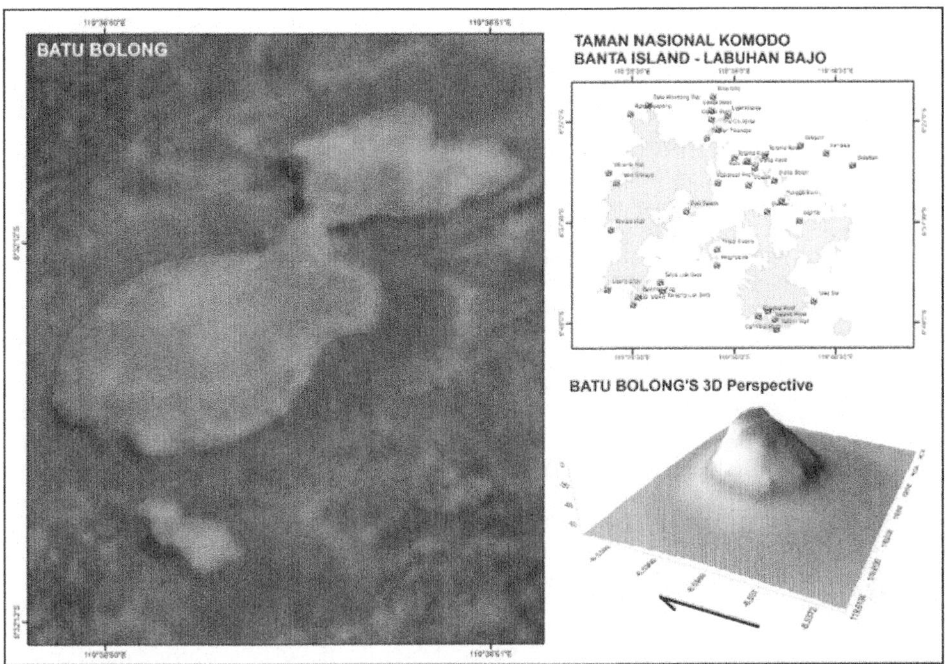

Figure 2. Batu Bolong diving spot.

ACKNOWLEDGEMENT

This research would like to thank Kireina Phinisi Liveaboard and the official Komodo National Park Agency for providing many useful data, and also Universitas Pendidikan Indonesia in partially funding the research.

REFERENCES

Antony, J. K., & Thomas, J. (2010). INFLUENCE OF PERCEIVED RISKS ON THE DESTINATION CHOICE PROCESS: AN INDIAN PERSPECTIVE. In B. Varghese (Ed.), *Evolving Paradigms in Tourism and Hospitality in Developing Countries* (pp. 87–115). Cambridge University Press. https://doi.org/10.1088/1751-8113/44/8/085201

Ardi. (2018). *Best Travel Time for Komodo and Flores*. https://www.floreskomodoexpedition.com/travel-advice/best-travel-time

Bowles, B. (2016). *'Time Is Like a Soup' Boat Time and the Temporal Experience of London's Liveaboard Boaters. 34*(September 2013), 100–112. https://doi.org/10.3167/ca.2016.340110

Buckley, R. (2011). Adventure tourism management. In *Adventure Tourism Management*. https://doi.org/10.4324/9781856178358

Castaldini, D., Valdati, J., & Ilies, D. C. (2005). The contribution of geomorphological mapping to environmental tourism in protected areas: examples from the Apennines of Modena (Northern Italy). *Revista de Geomorfologie, 2004*.

Edwards, R. (2008). *Stranded divers had to fight off Komodo dragons to survive*. https://www.telegraph.co.uk/news/worldnews/asia/indonesia/2095835/Stranded-divers-had-to-fight-off-Komodo-dragons-to-survive.html

Fang, S. (2020). Why Do Divers Want to Liveaboard? Measuring SCUBA Divers' Motivations of Liveaboard Using a Cruise Perspective. *Journal of Tourism and Hospitality Management, 8*(1). https://doi.org/10.17265/2328-2169/2020.01.003

Hashimoto, S., Yoshiki, S., Saeki, R., Mimura, Y., Ando, R., & Nanba, S. (2016). Development and application of traffic accident density estimation models using kernel density estimation. *Journal of Traffic and Transportation Engineering, 3*(3), 263–270.

Jennings, G. (2007). Motorboating. In *Water-Based Tourism, Sport, Leisure, and Recreation Experiences* (pp. 46–63). Elsevier.

Lewokeda, A., & Fardah. (2019). *Komodo national park earned Rp32 billion in 2018*. https://en.antaranews.com/news/121858/komodo-national-park-earned-rp32-billion-in-2018

Lim, N. (2003). Consumers' perceived risk: sources versus consequences. *Electronic Commerce Research and Applications, 2*(3), 216–228. https://doi.org/10.1016/S1567-4223(03)00025-5

Liveaboard. (2019). https://www.liveaboard.com/search/indonesia

LongLiveIndonesia. (2019). *INDONESIA WON 4 AWARDS AT THE ASIA DIVE EXPO*. https://longliveindonesia.com/indonesia-won-4-awards-at-the-asia-dive-expo/

McAdoo, B. G., Pratson, L. F., & Orange, D. L. (2000). Submarine landslide geomorphology, US continental slope. *Marine Geology, 169*, 103–106.

Pet, J. S., Mous, P. J., Muljadi, A. H., Sadovy, Y. J., & Squire, L. (2005). Aggregations of Plectropomus areolatus and Epinephelus fuscoguttatus (groupers, Serranidae) in the Komodo National Park, Indonesia: Monitoring and Implications for Management. *Environmental Biology of Fishes, 74*(2), 209–218. https://doi.org/10.1007/s10641-005-8528-8

Rahardiawan, R., & Purwanto, C. (2014). Struktur Geologi Laut Flores, Nusa Tenggara Timur. *Jurnal Geologi Kelautan, 12*(3), 165–178. https://doi.org/10.32693/jgk.12.3.2014.256

Rittichainuwat, B. N., & Chakraborty, G. (2009). Perceived travel risks regarding terrorism and disease: The case of Thailand. *Tourism Management, 30*(3), 410–418. https://doi.org/10.1016/j.tourman.2008.08.001

Roehl, W. S., & Fesenmaier, D. R. (1992). Risk Perceptions and Pleasure Travel: An Exploratory Analysis. *Journal of Travel Research, 30*(4), 17–26. https://doi.org/10.1177/004728759203000403

Setiawan, I., Dede, M., Sugandi, D., & Agung Widiawaty, M. (2019). Investigating Urban Crime Pattern and Accessibility Using Geographic Information System in Bandung City. *KnE Social Sciences, 2019*, 535–548. https://doi.org/10.18502/kss.v3i21.4993

Sharifpour, M., Walters, G., Ritchie, B. W., & Winter, C. (2014). Investigating the Role of Prior Knowledge in Tourist Decision Making: A Structural Equation Model of Risk Perceptions and Information Search. *Journal of Travel Research, 53*(3), 307–322. https://doi.org/10.1177/0047287513500390

Soeriaatmadja, W. (2017). *Asia*. Www.Straitstime.Com. https://www.straitstimes.com/asia/jokowi-plans-to-replicate-balis-success-in-10-other-indonesian-spots

Stoeckl., et al. (2010). *Live-aboard Dive Boats in the Great Barrier Reef: Regional Economic Impact and the Relative Values of Their Target Marine Species. 16*(4), 995–1018. https://doi.org/10.5367/te.2010.0005

Tsiros, M., & Heilman, C. M. (2005). The Effect of Expiration Dates and Perceived Risk on Purchasing Behavior in Grocery Store Perishable Categories. *Journal of Marketing, 69*(2), 114–129. https://doi.org/10.1509/jmkg.69.2.114.60762

Vera, T. (2017). *The 10 best dive sites in Asia*. CNN Travel. https://edition.cnn.com/travel/article/asia-best-dive-sites/index.html

Widiawaty, M. A. (2019). *Mari Mengenal Sains Informasi Geografis*. Aria Mandiri Group.

Wirakusuma, R. M., Sukirman, O., Waliyudin, R. T., & Putra, R. R. (2019). DESIGNING CORAL REEF TRANSPLANTATION PROGRAM WITH LOCAL COMMUNITY IN FORM OF MARINE ECO-TOURISM TOUR PACKAGE. *Journal of Indonesian Tourism, Hospitality and Recreation, 2*(2), 185–196. https://doi.org/10.17509/jithor.v2i2.20999

Promoting Creative Tourism: Current Issues in Tourism Research – Kusumah et al. (Eds)
© 2021 Taylor & Francis Group, London, ISBN 978-0-367-55862-8

Challenges in sustainable design practices through the lenses of local event organizers

M. Intason
Department of Tourism, Naresuan University, Phitsanulok, Thailand

ABSTRACT: This study aims to investigate the challenges of sustainable event design in practices in Thailand through the lenses of local event organizers. Event design is the process that brings improvement and development in every step of event management. It is a significant process to make events run successfully. To achieve the study aim, in-depth interviews with local event organizers in Bangkok and Chiang Mai were conducted. Thematic analysis was used for data analysis. The findings were reported into three key themes that represent key challenges to achieve sustainable concepts in event design, which are challenges in the selection of alternative resources, challenges in using digital technologies, and challenges in the practices. The results of this study contribute to the understanding of opportunities and limitations of sustainable design in practice based on the local perspective. Additionally, the practical contribution would be the guideline for event professionals to increase the possibility of sustainability in the event design process.

Keywords: Event tourism; Event design; Sustainable practices, Event organizers

1 INTRODUCTION

Event design is an essential process that offers a prediction of successful events. This process allows event organizers to design forms of events and the ways to facilitate the sensory aspects of events that bring positive impressions and experiences to the audiences. Yeoman, *et al.* (2012) suggest that "design brings to the improvement and development in every event level as it is the essence of an event's success." The event design process often involves building up physical objects, decorating the events atmosphere and services in the host location as these activities lead to large consumption of resources either natural or infrastructure. Additionally, the events generate economic advantages to local businesses in a host destination and offer career opportunities for local people. However, organizing an event normally consumes a huge amount of resources to supply the demands of a group of attendees, for example, water, electricity, food, and space. Subsequently, a huge amount of waste was generated event activities were completed.

To sustain a long-term benefit of events to local businesses and local people in a host destination, sustainable event design should be practically conducted, not only a concept. The concept of sustainability provides different ways to respond to the challenges of local environmental, social and cultural and economic dimensions existing in a host destination. The characteristics of the sustainability process provide the framework for maintaining a critical acceptable level of event organization. In addition, it is an evidence that sustainable events can create management perspectives that offer an autonomous cyclical process that can be maintained over time. Sustainability has been a mega-trend in most industries that support the nation's economy since the early 1990s. In the business events industry, sustainable practices have gathered more influence as organizers deal with diverse stakeholders. The principles of event design are the basic practices of creating successful events as these principles are involved in every step of event design such as stage and scenic design, environment design, and experience design. Likewise, event design is the process of using even design principles and techniques to create conceptual, and develop an event for positive

DOI 10.1201/9781003095484-44

and meaningful experiences to audiences (Brown & James, 2004). However, Mokhtar and Deng (2014) noted some limitations in implementing sustainable event design, for instance, paying for sustainability is not preferable cost for event hosts and event attendees because the "green events" in event sponsor perception is somewhat cheaper or less exclusive than non-green events. This can link to a lack of knowledge and information about sustainable practice in events among event stakeholders. Also, event organizers may have difficulties to find event facilities to put sustainability in practices. Thus, this research aims to investigate challenges in sustainable practices of event design from the lenses of local event organizers' perspectives to have a better understanding about practical limitations of the event design process.

2 METHOD

The research methodology employed qualitative approach to achieve the research aim. In-depth interviews were conducted to local 15 local event organizers in Bangkok and Chiang Mai, which are the main event destinations in Thailand. The research respondents were selected through purposive sampling technique. The in-depth interview allows the researcher to deeply explore the research respondents' perspectives on a phenomenon being studied (Guion *et al.*, 2001). Likewise, in-depth interview offers the way to have insight and get in-depth answers about the issues being investigated (Guest *et al.*, 2013).

Following Braun and Clarke (2012) and Walters (2016), the researchers read the transcripts several times to be familiar with the data and identify initial codes based on their similarities and develop basic themes. The author then combined the basic themes into organizing themes and derived the final themes related to the research aim and questions. After this, the developed themes within the final thematic structure were defined and described. The results were categorized into different three crucial themes, which represented the challenges in making sustainability into event design process.

3 RESULTS

The results are reported into three crucial themes that represent the current challenges of putting sustainable concept into action through event design process.

3.1 *Challenges in alternative resources selection*

Various resources are required in organizing an event process, for example, decoration material and equipment, water, food, services, and electricity. Whatever the purposes of gathering people in an event, resources are used, waste is created, and people, environment and economies are affected. This suggests that the challenges/limitations of creating a sustainable event would be found during the process of event design because the event organizers need to decide which resources and what is managing protocol would bring sustainability into action. A local event organizer in Chiang Mai commented on the difficulties that normally occurred in the event design process is that "it is quite difficult to find an alternative material to use for decorating an event venue that offers less environmental impacts with low cost. Additionally, a challenge that usually occurred is we rarely to reuse the used material from previous events for others events. The used material, especially decorating material at the event venue normally became waste after the event finished." Likewise, an local event organizer in Bangkok commented that "alternative resources with low environmental impacts, for example, plywood and compressed paper are not a good idea for using at outdoor events because they cannot retain their forms if it is raining, then they will be damaged and not usable supplies through a period of an event." This suggests that event locations and a life cycle product, which including the possibility to (re)use a product/equipment until the end of its useful life can be challenges in event design process.

Indoor and outdoor locations also bring limitations to the selection of event material, especially the stage of decoration and set-up event venues. A local event organizer in Bangkok commented that "sometimes, we used local resources that available in a host destination for decorating an event venue. For example, using basketworks to decorate an indoor venue and setup basket chairs in outdoor space for event attendees. Likewise, an local event organizer in Bangkok commented that "alternative resources with low environmental impacts, for example, plywood and compressed paper are not a good idea for using at outdoor events because they cannot retain their forms if it is raining, then they will be damaged and not usable supplies through a period of an event." However, it depends on event themes and suitability of event attendees characteristics." This suggests that costs and event themes would be considerable factors that the event organizers should consider as the criteria of event (decorating) material to create a preferable cost. Also, an event theme is an important factor that transfers the sense of an event to event attendees, thus the event organizers should be aware that sustainability will not mislead the event themes and concepts.

3.2 *Challenges in using digital technologies*

Digital devices are useful equipment that offers a sustainable way to organize an event, which has the potential to be reused in different types of events. A local event organizer in Bangkok commented that "digital technology considered as useful equipment for different purposes in an event. For example, we used digital signage for providing direction to event attendees. Also, this equipment can be reused in different events. However, we are facing a difficulty to decompose them after their useful life." Likewise, a local event organizer in Chiang Mai commented that "Digital devices have the potential to be an alternative source that offers low-environmental impact to a host destination because they can be reused."

However, the challenge of using digital technologies is the way to recycle and reduce environmental impacts after the end of their useful life. A local event organizer in Bangkok commented that "every type of digital device has its timeline of useful life, which relates to new trends of technology development. We still challenging with the way to minimize the environmental impacts after disposing of those electronic waste because it generates a toxic chemical that brings damage to the atmosphere." In this case, digital devices would be considered as a potential alternative resource for sustainable event concept. However, managing the electronic product life cycle needs to be considered.

3.3 *Challenges in practices*

An event accommodates a large group of people at the same time and consume a lot of resources such as food, water, and electricity to complete the events' goals successfully, also generate a huge amount of waste each time. To organize an event in a sustainable way that brings less impact to the environment seems to be a significant challenge for event planners/organizers. A local event organizer in Bangkok commented that "Sustainability is still a challenge for us because of the event requires a huge amount of resources for the huge consumption of event attendees. Subsequently, there will be a lot of waste need to be managed after the event finished." Likewise, a local event organizer in Chiang Mai commented that "full practice in sustainable management is quite challenging because it needs every event stakeholders (local people, event organizer, event attendees, local government, private sectors, and service providers) to collaborate in the same way. I think it is difficult to organize the events sustainably if other event stakeholders could not go along with the sustainable concept. Additionally, the cost of using alternative low-environmental impact material is quite high, so I considered this as another challenge in a practical way." This suggests that costs, the collaboration between event stakeholders, and huge consumption are key factors that limit the way to manage the event in an environmental-friendly form.

4 DISCUSSION

To put sustainable concept into practices is challenging to event organizer regarding the way to design and manage an event in a practical way. This research explored the challenges in sustainable event design through the lenses of local event organizers in two main event destinations in Thailand namely, Bangkok and Chiang Mai. The following provides a discussion of key themes related to the research aim.

The first key theme challenges in alternative resource selection, which representing the challenges/difficulties to find alternative material with low-environmental impacts in organizing an event. There are two main factors need to be considered in the selection of event resources. Firstly, event locations can be indoor and outdoor spaces, which influence the event resource selection criteria, for example, the timing of using the material and maintenance requirement during the events. In reality, there are some challenges when applying the sustainable concept in event design, for example, weather (an uncontrollable factor) maintenance requirement, and costs.

Secondly, the product/material life cycle refers to a product that offers less impact on the environment and ability to be reused at the end of its useful life. Likewise, the sustainability of a product life cycle refers to the total life cycle of a product that offers opportunities to minimize waste and pollution as well as contributing to reuse, or recycling of the product at the end of its useful life (Johnson, 1997; Kasarda *et al.*, 2007). However, the findings have shown the high costs of using less-environmental impact products/material are a crucial factor that may slow the development of sustainability of an event. The quality and costs of developing products to be environmentally friendly can lead to additional costs (Kasarda *et al.*, 2007), which subsequently increasing an event budget.

The second key theme is challenges in using digital technologies in an event for a sustainable form. The findings present that digital devices would be the potential alternative equipment for a sustainable event. They are reusable through the end of its useful life, but the proper way to manage electronic waste is a challenge. New technologies are rapidly developed superseding analog appliances. However, managing electronic waste and reduce pollution problems and impacts on the environment is considered a worldwide problem (Kiddee *et al.*, 2013). Likewise, e-waste contains hazardous elements that may negatively impact the environment and human health if not properly managed (Nnorom & Osibanjo, 2008). Additionally, innovation in events can be either ideas that heighten audiences' expectations or using technology to create live experiences which is a popular requirement currently. Digital devices can be reused until their useful lifetime; however, the pollution after the process of damaging themselves is a big challenge for sustainable event management that event professionals and event stakeholders need to consider.

The third key theme challenges in practices. Sustainable event management is the principle of event management that is applied to every step, for example, planning, onsite, and ending. The findings show that collaboration between event stakeholders and the huge consumption of event attendees would be challenges for sustainable event design. Engagements and awareness in sustainable activities of event stakeholders create opportunities for sustainable event design (Kasarda *et al.*, 2007). Likewise, sustainable tourism and events development require cooperation between government bodies and private sectors such as tourism and hospitality sectors, and local communities to achieve the development goal (Timur & Getz, 2008). Additionally, the huge consumption of resources, for example, food, water, and electricity is another challenge for event organizers to consider in sustainable event design. This relates to the nature of events that they are the occasions for a group of people gathering together for specific purposes. As there is a big number of people to attend events each time, so the number of used resources and waste will be large and this matter can impact the environment, social and economic in various aspects (Jones, 2014). Sometimes, there is a big gap between sustainable ideas and required event tasks, for example, opening ceremony and a big show during welcome dinner which needs large consumption of resources, which bringing negative impacts on the environment behind. Therefore, event organizers can offer ideas about sustainability as a theme and concept or putting the idea in creating multi-functional event supplies, for example, digital signage which uses for the main purpose as an event direction, also it can be turned into information screen for advertising and sponsorship showcases. Dealing business with customers by setting up sustainability as a core value of event design and operation would be a possibility to put sustainability in action. For example, using existing event elements in the host destination (local products, and building/spaces), and not destroying the existing environment for setting up an event supply. These activities can educate customers and attendees to attend an event with the responsibility to local society and culture and avoiding activities that might damage the local environment. Also, generating business cooperation with local business sectors such as food and beverage service, local guides, and local transportation can be one of sustainable event management that provides career opportunities to local people and stimulates local economic growth.

These factors are about sustainable detail that event organizers should consider when starting to organize an events, for instance how to use resources efficiently, the degree of event objectives was met, and the possible anticipated outcomes and negative outcomes. Therefore, it is evident from respondents that sustainable events do not necessarily mean consuming fewer resources and using only energy-saving and recycled supplies, but it is about how well you consume resources and reach event objectives effectively. However, some challenges limit the opportunity to bring sustainability into practice, for example, the difficulties to control the amount of pollution and gas emission that might be generated in events and sometimes carbon footprint does not practical for some events. Lack of knowledge about sustainability among people in the industry and the clients is a challenging factor that limits the opportunity to bring sustainability into practice appropriately.

5 CONCLUSION

The results of this study identified key challenges in putting sustainability into event design practices. Sustainability might be considered as a trend that many business sectors pay attention to and try to merge with their management principles. On the other hand, it can be a special concept in an event that offer opportunities to participate in an event with less environmental impact and responsibility to value society and well-being of local people in a host destination. Innovation in events means innovative event ideas that will impress clients and create a meaningful moment for event audiences. However, a limited budget can be a challenge in sustainable event design regarding alternative event resources with less-environment impact and managing electronic waste. Additionally, sustainable events should be touchable and understandable by event attendees. Therefore, sustainable practice in event design is the process of creating ideas with the efficiency of consuming resources and concerning event outcomes, but the ideas should answer event objectives and reach event goals effectively.

The results of this study can contribute to the understandings of opportunities and limitations of sustainability in event design in the sustainable event management. Also, the event professionals can consider the research results as the ideas to develop the event design process and event management that increase the possibility to achieve the sustainable concept in practices. Additionally, future research may investigate alternative event management strategies that provide opportunities to minimize the negative impacts of the environment. Also, examine the opportunities and limitations to practice the virtual event for sustainable event concept, which can maintain the event atmosphere to the audiences.

REFERENCES

Braun, V., & Clarke, V. (2012). Thematic analysis. In Cooper, H. (Ed.), APA Handbook of Research Methods in Psychology (pp. 55–71) Washington, DC: American Psychological Association.

Brown, S., & James, J. (2004). Event design and management: Ritual sacrifice. *Festivals and Events Management*, 53–64.

Guest, G., Namey, E., & Mitchell, M. (2013). In-depth interviews. Collecting Qualitative Data: *A Field Manual for Applied Research*, 113–171.

Guion, L. A., Diehl, D. C., & McDonald, D. (2001). *Conducting an in-depth interview*. University of Florida Cooperative Extension Service, Institute of Food and Agricultural Sciences, EDIS.
Johnson, H. D. (1997). *Green plans: Greenprint for sustainability* (Vol. 7). London: University of Nebraska Press.
Jones, M. (2014). *Sustainable event management: A practical guide*. Oxon: Routledge.
Kasarda, M. E., Terpenny, J. P., Inman, D., Precoda, K. R., Jelesko, J., Sahin, A., & Park, J. (2007). Design for adaptability (DFAD)—a new concept for achieving sustainable design. *Robotics and Computer-Integrated Manufacturing*, 23(6), 727–734.
Kiddee, P., Naidu, R., & Wong, M. H. (2013). Electronic waste management approaches: An overview. *Waste management*, 33(5), 1237–1250.
Mokhtar, S., & Deng, Y. (2014). Sustainable Design in Event Design: Opportunities and Limitations. *Journal of Clean Energy Technologies*, 2(2), 163–167.
Nnorom, I. C., & Osibanjo, O. (2008). Overview of electronic waste (e-waste) management practices and legislations, and their poor applications in the developing countries. *Resources, conservation and recycling*, 52(6), 843–858.
Timur, S., & Getz, D. (2008). A network perspective on managing stakeholders for sustainable urban tourism. *International Journal of Contemporary Hospitality Management*.
Walters, T. (2016). Using thematic analysis in tourism research. *Tourism Analysis*, 21(1), 107–116.
Yeoman, I., Robertson, M., Ali-Knight, J., Drummond, S., & McMahon-Beattie, U. (2012). *Festival and events management*. London: Routledge.

Promoting Creative Tourism: Current Issues in Tourism Research – Kusumah et al. (Eds)
© 2021 Taylor & Francis Group, London, ISBN 978-0-367-55862-8

Rural tourism: The state-of-the-art

A.H.G. Kusumah
Universitas Pendidikan Indonesia, Bandung, Indonesia

ABSTRACT: The concept of a tourist village as an alternative tourist attraction has experienced rapid progress in the past decade in Indonesia. Research and community service regarding tourist villages also experienced significant developments in the last two decades. This paper maps the advancement of research related to tourist villages from the five highest-ranking journals of the Scimago version of 2020. A literature search is carried out through a concise analysis of abstract papers using the MAXQDA 2020 software. This literature study finds that there are three central themes of the rural tourism study, namely destination development, entrepreneurship, and local-community related. This study shows that there are still many areas that have not been much discussed about rural tourism. This paper has implications on the importance of the focus and attention of researchers to conduct studies related to tourism villages, especially with the center of the study on tourism village marketing, traditional accommodation management, and rural tourism management.

1 INTRODUCTION

The tourism village is the concept of developing tourist attractions, which became the flagship program of the Ministry of Tourism and initiated in 2018. This rural tourism program has grown rapidly in several regions in Indonesia since a decade ago. This program was initiated because the majority of Indonesia's territory consists of rural areas. Tourism villages scattered throughout Indonesia are now the center of agricultural tourism activities. Tourism academics are also interested in contributing academically, both through scientific research and community service. On a global scale, research related to rural tourism has been carried out since a few decades ago. Some of the most notable studies on rural tourism were related to social disruption (M. Park & Stokowski 2009), community-based-tourism (Kline et al. 2014), and social entrepreneurs (Mottiar et al. 2018). However, little is known about the state-of-the-art research on rural tourism as a whole. This paper aims to provide an overview of the extent to which research on rural tourism has developed to date.

2 METHODS

For this study, five journals were selected as data sources, namely Tourism Management, Annuals of Tourism Research, Journal of Travel Research, International Journal of Hospitality Management, and International Journal of Contemporary Hospitality Management. These five journals are ranked as the top five in the Scimago ranking for the Journal of Tourism and Hospitality. Papers related to rural tourism were traced through the publisher's website of each journal, there are; sciencedirect.com, journals.sagepub.com, and emerald.com. Two primary keywords, rural tourism and tourism village were used as search-keywords on each publisher's website for each journal. The search-keywords were set to be listed in the title, keyword, or abstract of the article to ensure the searching accuracy process.

DOI 10.1201/9781003095484-45

Figure 1. Word-cloud of rural tourism research topics

Article data extraction was done using Mendeley web importer software. Through this search, a total of 312 articles were found. The article was then manually scanned to ensure the compatibility of the article with the scope of the study. In total, 284 articles were identified after the manual scanning process. The paper elements that obtained through the extraction process were the titles, authors, the year of publication, the name of the journal, and the abstract of the article. All data is then exported into a Research Information Systems (RIS) file. The data is then analyzed using the MAXQDA 2020 software. The type of analysis carried out is to calculate the frequency of words and frequency of combinations of words that appear in all abstracts. The Keyword-in-context feature was used to get the context of using words and word combinations. All of the word-combinations that have been extracted were then categorized into the same group of clumps.

3 FINDINGS AND DISCUSSIONS

Procedure for calculating word frequency using the MAXQDA 2020 software shows the combination of words that appear frequently. As many as 50 combinations of words that appear most frequently were extracted. The term' economic development' is the most common combination of words. They were followed by the phrases 'social capital', 'resident attitude', 'resident perception', and 'market segment'. These five phrases are the most common word combinations that appear in 284 articles that are the subject of this article. The combination of words in the form of word clouds can be seen in Figure. 1

The whole combination of words forms three significant clusters, namely destination development, entrepreneurship, and local-community are related. These three clusters show three big topics that most often appear in previous rural tourism studies.

3.1 *Destination development*

The first cluster formed was the Destination Development cluster, which is also the largest of the two other clusters. In this cluster, two dominant sub-clusters form the cluster, namely economic development and development strategy were closely observed. Economic development studies related to rural tourism focus on poverty reduction (Andereck & Vogt 2000), traditional tourist accommodations (Confalonieri 2011), cost-benefit of tourism (Fleischer & Felsenstein 2000), and resident perceptions of tourism development (Johnson et al. 1994). Sub-cluster development strategy was formed through studies about nature-based tourism promotion (Place 1991), resident support for tourism development (Andereck & Vogt 2000), and rural festival (Chhabra et al. 2003).

3.2 *Entrepreneurship*

The second cluster found in this study is the entrepreneurship-study cluster. Three main sub-clusters formed the entrepreneurship cluster, namely social entrepreneur, market segment, and service quality. Research on social entrepreneurs in a tourism village includes a study of the role of social entrepreneurs in the development of rural tourism destinations (Mottiar et al. 2018) and the social entrepreneur system (Peng & Lin 2016). The market segment sub-cluster discusses the motivations of tourists segment and profile (D. B. Park & Yoon 2009; Rid et al. 2014), vacation-taking behavior (Fleischer & Pizam 2002) and senior traveler profile (Horneman et al. 2002). The third sub-cluster found in this study is the service quality sub-cluster. Its range from a model of perceived service quality (Reichel et al. 2000), small business strategy for service quality (Hernández-Maestro et al. 2009), business consequences of service quality (Melo et al. 2017), and service quality assessment of farm tourism (Rozman et al. 2009).

3.3 *Local community*

The last cluster of studies on rural tourism is local-community related studies. This cluster comprised of three sub-clusters, namely community attitude, community-based-tourism, and social capital. The community attitude studies discussed resident attitudes to tourism development (Johnson et al. 1994; Ma et al. 2020; Madrigal 1993; Mason & Cheyne 2000), non-host community residents' perceptions (Deccio & Baloglu 2002), and place attachment and empowerment (Strzelecka et al. 2017). Local-community related sub-cluster focused on community-based tourism management (Mbaiwa 2011), stakeholder partnership (Hwang et al. 2012), traditional village revitalization (Gao & Wu 2017), and the roles of culture (MacDonald & Jolliffe 2003). The last sub-cluster discussed the roles of social capital to entrepreneurship (Zhao et al. 2011; Zhou et al. 2017) and community conflict (D. B. Park et al. 2012).

4 CONCLUSION

The combination of extracted words shows several large clusters in previous studies. From the results of the clustering, we may conclude there are still many other research topics related to tourism villages that have not been much discussed. Studies related to marketing of tourist villages, traditional accommodation management, to rural tourism management have not been widely explored in previous studies. Besides, a study related to youth travelers visit a special interest attraction (Sari et al. 2018) and homestay guest experience (Aprillea et al. 2018) are also of importance. Researchers must focus on these topics because of the vast growing development of rural tourism. The development of village tourism is becoming increasingly important considering that in the pandemic Covid-19, the future tourism development will be more focused on local tourism so that that village tourism can become an alternative destination for local tourists.

REFERENCES

Andereck, K. L., & Vogt, C. A. (2000). The relationship between residents' attitudes toward tourism and tourism development options. *Journal of Travel Research, 39*(1), 27–36. https://doi.org/10.1177/004728750003900104

Aprillea, R., Kusumah, A. H. G., & Wirakusuma, R. M. (2018). FACTOR ANALYSIS THAT FORMS EXPECTATION OF GUEST HOUSE GUEST EXPERIENCE. *Journal of Indonesian Tourism, Hospitality and Recreation, 1*(2). https://doi.org/10.17509/jithor.v1i2.13763

Chhabra, D., Sills, E., & Cubbage, F. W. (2003). The significance of festivals to rural economies: Estimating the economic impacts of Scottish highland games in North Carolina. *Journal of Travel Research, 41*(4), 421–427. https://doi.org/10.1177/0047287503041004012

Confalonieri, M. (2011). A typical Italian phenomenon: The "albergo diffuso." *Tourism Management, 32*(3), 685–687. https://doi.org/10.1016/j.tourman.2010.05.022

Deccio, C., & Baloglu, S. (2002). Nonhost community resident reactions to the 2002 winter olympics: The spillover impacts. *Journal of Travel Research, 41*(1), 46–56. https://doi.org/10.1177/004728750204100106
Fleischer, A., & Felsenstein, D. (2000). Support for rural tourism: Does it make a difference? *Annals of Tourism Research, 27*(4), 1007–1024. https://doi.org/10.1016/S0160-7383(99)00126-7
Fleischer, A., & Pizam, A. (2002). Tourism constraints among Israeli seniors. *Annals of Tourism Research, 29*(1), 106–123. https://doi.org/10.1016/S0160-7383(01)00026-3
Gao, J., & Wu, B. (2017). Revitalizing traditional villages through rural tourism: A case study of Yuanjia Village, Shaanxi Province, China. *Tourism Management, 63*, 223–233. https://doi.org/10.1016/j.tourman.2017.04.003
Hernández-Maestro, R. M., Muñoz-Gallego, P. A., & Santos-Requejo, L. (2009). Small-business owners' knowledge and rural tourism establishment performance in Spain. *Journal of Travel Research, 48*(1), 58–77. https://doi.org/10.1177/0047287508328794
Horneman, L., Carter, R. W., Wei, S., & Ruys, H. (2002). Profiling the senior traveler: An Australian perspective. *Journal of Travel Research, 41*(1), 23–37. https://doi.org/10.1177/004728750204100104
Hwang, D., Stewart, W. P., & Ko, D. wan. (2012). Community behavior and sustainable rural tourism development. *Journal of Travel Research, 51*(3), 328–341. https://doi.org/10.1177/0047287511410350
Johnson, J. D., Snepenger, D. J., & Akis, S. (1994). Residents' perceptions of tourism development. *Annals of Tourism Research, 21*(3), 629–642. https://doi.org/10.1016/0160-7383(94)90124-4
Kline, C. S., Greenwood, J. B., Swanson, J., & Cárdenas, D. (2014). Paddler market segments: Expanding experience use history segmentation. *Journal of Destination Marketing and Management, 2*(4), 228–240. https://doi.org/10.1016/j.jdmm.2013.10.004
Ma, X. L., Dai, M. L., & Fan, D. X. F. (2020). Land expropriation in tourism development: Residents' attitudinal change and its influencing mechanism. *Tourism Management, 76*. https://doi.org/10.1016/j.tourman.2019.103957
MacDonald, R., & Jolliffe, L. (2003). Cultural rural tourism: Evidence from Canada. *Annals of Tourism Research, 30*(2), 307–322. https://doi.org/10.1016/S0160-7383(02)00061-0
Madrigal, R. (1993). A tale of tourism in two cities. *Annals of Tourism Research, 20*(2), 336–353. https://doi.org/10.1016/0160-7383(93)90059-C
Mason, P., & Cheyne, J. (2000). Residents' attitudes to proposed tourism development. *Annals of Tourism Research, 27*(2), 391–411. https://doi.org/10.1016/S0160-7383(99)00084-5
Mbaiwa, J. E. (2011). Changes on traditional livelihood activities and lifestyles caused by tourism development in the Okavango Delta, Botswana. *Tourism Management, 32*(5), 1050–1060. https://doi.org/10.1016/j.tourman.2010.09.002
Melo, A. J. D. V. T., Hernández-Maestro, R. M., & Muñoz-Gallego, P. A. (2017). Service Quality Perceptions, Online Visibility, and Business Performance in Rural Lodging Establishments. *Journal of Travel Research, 56*(2), 250–262. https://doi.org/10.1177/0047287516635822
Mottiar, Z., Boluk, K., & Kline, C. (2018). The roles of social entrepreneurs in rural destination development. *Annals of Tourism Research, 68*, 77–88. https://doi.org/10.1016/j.annals.2017.12.001
Park, D. B., Lee, K. W., Choi, H. S., & Yoon, Y. (2012). Factors influencing social capital in rural tourism communities in South Korea. *Tourism Management, 33*(6), 1511–1520. https://doi.org/10.1016/j.tourman.2012.02.005
Park, D. B., & Yoon, Y. S. (2009). Segmentation by motivation in rural tourism: A Korean case study. *Tourism Management, 30*(1), 99–108. https://doi.org/10.1016/j.tourman.2008.03.011
Park, M., & Stokowski, P. A. (2009). Social disruption theory and crime in rural communities: Comparisons across three levels of tourism growth. *Tourism Management, 30*(6), 905–915. https://doi.org/10.1016/j.tourman.2008.11.015
Peng, K. L., & Lin, P. M. C. (2016). Social entrepreneurs: innovating rural tourism through the activism of service science. *International Journal of Contemporary Hospitality Management, 28*(6), 1225–1244. https://doi.org/10.1108/IJCHM-12-2014-0611
Place, S. E. (1991). Nature tourism and rural development in Tortuguero. *Annals of Tourism Research, 18*(2), 186–201. https://doi.org/10.1016/0160-7383(91)90003-T
Reichel, A., Lowengart, O., & Milman, A. (2000). Rural tourism in Israel: Service quality and orientation. *Tourism Management, 21*(5), 451–459. https://doi.org/10.1016/S0261-5177(99)00099-0
Rid, W., Ezeuduji, I. O., & Pröbstl-Haider, U. (2014). Segmentation by motivation for rural tourism activities in The Gambia. *Tourism Management, 40*, 102–116. https://doi.org/10.1016/j.tourman.2013.05.006
Rozman, Č., Potočnik, M., Pažek, K., Borec, A., Majkovic, D., & Bohanec, M. (2009). A multi-criteria assessment of tourist farm service quality. *Tourism Management, 30*(5), 629–637. https://doi.org/10.1016/j.tourman.2008.11.008

Sari, D., Kusumah, A. H. G., & Marhanah, S. (2018). Analisis Faktor Motivasi Wisatawan Muda Dalam Mengunjungi Destinasi Wisata Minat Khusus. *Journal of Indonesian Tourism, Hospitality and Recreation, 1*(2). https://doi.org/10.17509/jithor.v1i2.13762

Strzelecka, M., Boley, B. B., & Woosnam, K. M. (2017). Place attachment and empowerment: Do residents need to be attached to be empowered? *Annals of Tourism Research, 66*, 61–73. https://doi.org/10.1016/j.annals.2017.06.002

Zhao, W., Ritchie, J. R. B., & Echtner, C. M. (2011). Social capital and tourism entrepreneurship. *Annals of Tourism Research, 38*(4), 1570–1593. https://doi.org/10.1016/j.annals.2011.02.006

Zhou, L., Chan, E., & Song, H. (2017). Social capital and entrepreneurial mobility in early-stage tourism development: A case from rural China. *Tourism Management, 63*, 338–350. https://doi.org/10.1016/j.tourman.2017.06.027

Promoting Creative Tourism: Current Issues in Tourism Research – Kusumah et al. (Eds)

Hand sign method in playing *angklung* as tourists' involvement in creative tourism: A case study on *Saung Angklung Udjo*

N. Riana & K. Fajri
Sekolah Tinggi Ilmu Ekonomi Pariwisata Yapari, Bandung, Indonesia

ABSTRACT: Creative Tourism offers opportunities to the tourists to develop their creative potencies through an active participation, and digging the learning-by-doing experiences, which are the characteristics of Creative Tourism. Tourists gain the amusement and experience from the chosen activities they had taken. The challenge was, how to build the interaction and communication effectively thus the tourists would feel involved and experience the amusement during the performance. This research aims to describe how the Saung Angklung Udjo design the involvement of the tourists in their performances, and share the cultural knowledge transfer. The research method implemented in this paper is analytical descriptive with qualitative approach, whilst the method of data collection done by field observation and literature studies. The results showed that the concept of simple hand signs and good communication delivered by the *Angklung* Conductor had been attracted the tourists and got their involvement easily. The tourists were asked to play the *Angklung* instrument together, using the Kodály method. This method met and brought the enjoyable experience for the tourists.

Keywords: Tourists' involvement, hand signs, *Kodály* method, creative tourism, *angklung* ensemble

1 INTRODUCTION

Creative Industry is an industry that utilising creativity, ideas, skill to create something that has value in social, cultural, and economy (Purwanto 2019). Stoneman (2010) mentioned some fields that can be categorised in creative industry, namely audio-visual (film, TV, new media, and al), books and publishers, heritage (museum library, and historical environment), entertainment, sport, tourism, and visual art. While Raymond and Richards (2000) defined creative tourism as "Tourism which offers visitors the opportunity to develop their creative potential through active participation in courses and learning experiences, which are characteristic of the holiday destination where they are taken."

Referring from the above understandings, can be concluded that creative tourism is a kind of tourism management that offering enjoyable experience to tourists. Tourists are not persuaded to just come and enjoy the nature or performance, but they asked to be involved and join the activities, together with the performer and other tourists.

The above figure is a result of previous research by Poetry (2011) about attractions in Saung Angklung Udjo. From some attraction aspects of Saung Angklung Udjo, that research showed that the most attractive one is the *angklung* performance (Poetry 2011). Base on that finding, the writer interested to explore further, how the Saung Angklung Udjo set up the Angklung Performance as the most attractive object. The focus of this research especially about the using of hand sign in playing *angklung* together as tourists involvement. This method of hand sign known as Kodály method, a method that created by Hungarian musician, Zoltán Kodály, that is widely used by music teachers in musical ensemble classes.

 DOI 10.1201/9781003095484-46

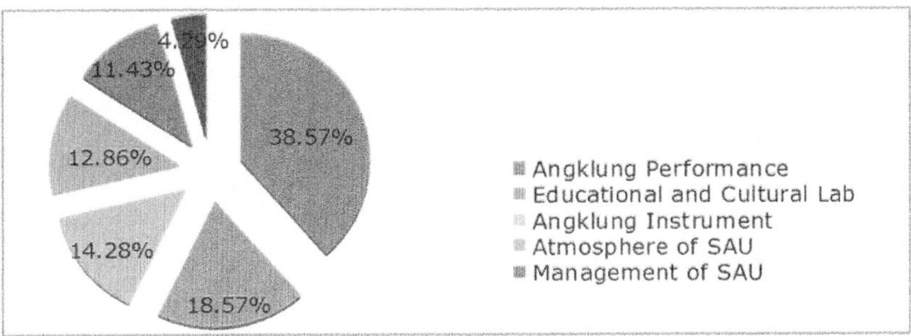

Figure 1. Attraction level in saung angklung udjo.
Source: Poetry (2011).

2 METHOD

This research used a qualitative approach to obtain deeper and more accurate results in the context of analysing the tourists involvement model. In this research, the writer analyse the design of gaining the tourists' satisfaction on Angklung Performance Art through the Tourists Involvement Model. To obtain the data, field observation was conducted directly in the Saung Angklung Udjo, Bandung. While the affirmation gathered from some interviews with tourists. Additional data collection was done through digital observation on Saung Udjo's official social media in Instagram.

3 RESULTS AND DISCUSSION

Saung Angklung Udjo is one of cultural destination in Bandung, especially the Angklung Performance Art. In every performance, Saung Angklung Udjo (SAU) has a unique style, where the Master of Ceremony always involves the tourists in the performance, by playing songs with the angklung instruments. This concept of "learning angklung together" became a special moment for tourists, and appears as an attraction in tourism management (Riana et al. 2016).

In managing the creative tourism, Saung Angklung Udjo create an innovative way by involving the tourists. With this uniqueness, Saung Angklung Udjo now is not only a cultural tourist destination, but it become one of the Creative Tourism destination in Bandung, since it met the transfer of knowledge in art and culture. Saung Angklung Udjo involved the tourists to play angklung together, conducted by a conductor, to play some popular songs. This concept is very attractive and provide happiness for the tourists, especially foreign tourists who never knew about angklung instruments.

Angklung is a traditional Sundanese musical instrument made from bamboo. Every angklung instrument has its own musical tone: Do-Re-Mi-Fa-Sol-La-Ti. In order to get the music harmony, the conductor used a Kodály method of hand sign. The Kodály method is a learning technique on music by converting the notation function to hand sign. The tone from "Do" to "Ti" were presented by some easy forms of hand. (Hidayatullah 2019).

Bermain Angklung Bersama (Learning Angklung Together) session is one of eight sessions in The Program of Pertunjukan Bambu Petang (Afternoon Bamboo Performance). It is a daily activity. This consistency of the performance schedule shows that Saung Angklung Udjo is well managed and consistent in performing the cultural performances. The Program of Pertunjukan Bambu Petang consisted of 8 (eight) sessions below:

1. Demonstrasi Wayang Golek
2. Helaran
3. Arumba
4. Tari Topeng
5. Angklung Mini

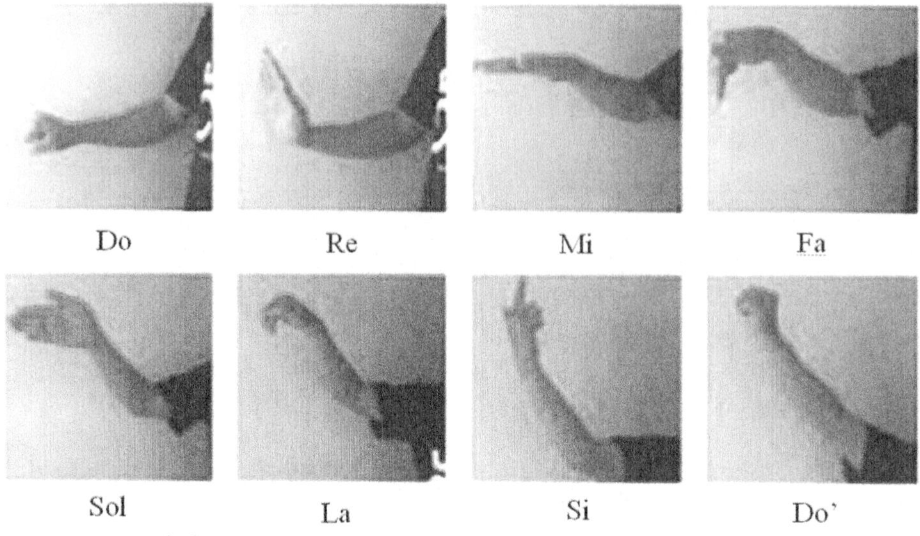

Source: semanticscholar.org

Figure 2. Kodály hand signs of interactive angklung.

6. Angklung Padaeng
7. Bermain Angklung Bersama
8. Menari Bersama

From the above 8 (eight) sessions, the session 7 (seven) is activity with tourists involvement, which is the focus of this research. Base on some interviews, the session of "Bermain Angklung Bersama" in The Bambu Petang Performance is the most attractive on tourists perspective. In this session, tourists had given an angklung, distributed by angklung children performers. Then, the tourists got a short learning session to play angklung. They start with "how to handle" the angklung properly, recognise the tone, and understand the Kodály hand sign. This hand sign is important, not only as a universal symbol, but also to ease from written symbols (do-re-mi notations), since not everyone could read or understand the musical notations. Referring to the interviews result, some tourists admitted that they could not understand or read the musical notations, although in the form of numeric notation. This is due to the beats aspect in written musical notations. While in Kodály hand sign method, all the audiences need was just following the hand signs from the Conductor. They did not need to think about beats. So everybody can easily play the angklung instrument. Some tourists affirmed in the interview that this method of playing angklung was a very interesting session and provided satisfaction for the tourists. Most of the tourists feel happy and satisfied, since they felt that they can play the angklung musical instrument.

Another important thing is, to make sure that all the instruction can be understood by the audience/tourists, the MC/Conductor gave the instruction in multi-language. Their language proficiency and communication skill was very good, so they could build a very nice interaction with the tourists. The Conductor even able to speak in some language. They could greet the tourists in some major languages in the world, such as English, French, Germany, Japanese, Netherlands, etc.

Not only multi languages, but their communication skill were also very good. Refer to some interview results, the tourists felt happy and entertained by the friendly and professional way of the conductor. The multi languages and good skill in communication of the conductor also ease the tourists in understanding the instructions in playing angklung.

Source: @angklungudjo, Instagram official

Figure 3. Tourists involvement in playing angklung.

The result shows that the Saung Angklung Udjo applied the Kodály method of hand sign to ease the tourists in learning playing angklung. The concept of tourists involvement provided a different experience that the tourists not only watching and enjoy the performances, but they also became the performers. This tourists involvement concept is also as a transfer of knowledge about Sundanese culture in The Program Bambu Petang, especially knowledge about angklung musical instrument in Bermain Angklung Bersama (Playing Angklung Together).

4 CONCLUSION

The design of gaining the tourists' satisfaction is delivered by the concept of Tourists' Involvement on the Angklung Performance (playing together session). As the implementation, Saung Angklung Udjo used the Kodály method of hand sign to conduct the angklung ensemble, as this method is easily understand by the tourists. In order to gain the harmony, the conductor gave a simple training at the beginning of the session. The method of simple hand sign delivered with a good communication skill by the Angklung Conductor is attractive and could easily got the tourists' involvement. Tourists Involvement Concept provides an enjoyable experience and smoothing the cultural transfer knowledge to the tourists.

5 RECOMMENDATION

Considering that the tourists came from different countries and the range of age, it would be nice if the management of Saung Angklung Udjo get some more new songs so the millennial tourists would be more familiar with the songs they are playing.

REFERENCES

Bellani, Novika. Gitasiswhara. 2012. *Pengaruh Creative Tourism Saung Angklung Udjo Terhadap Keputusan Berkunjung Wisatawan (Survei pada Pengambil Keputusan Kelompok Rombongan Sekolah Untuk Berkunjung ke Saung Angklung Udjo)*. Tourism and Hospitality Essentials (THE) Journal, Vol. II, No.2, 2012–425.

Bonetti, Enrico. Simoni, Michele. Cercola, Raffaele. 2014. Creative Tourism and Cultural Heritage: A New Perspective. IGI-Global.

Helpiastuti, Selfi Budi. 2019. *Pengembangan Destinasi Pariwisata Kreatif Melalui Pasar Lumpur; Analisis Wacana Grand Opening "Pasar Lumpur" Kawasan Wisata Lumpur, Kecamatan Ledokombo, Kabupaten Jember. Journal of Tourism and Creativity.* Vol 2 No 1. https://jurnal.unej.ac.id/index.php/tourismjournal/article/view/13837.

Hidayatullah, Riyan. 2019. Bahasa Musik dalam Pembelajaran: Metode Kodaily sebagai Alat untuk Berkomunikasi dalam Ansambel. AKSARA Jurnal Bahasa dan Sastra Vol. 20, No. 1, Hal. 25–34, April 2019. http://jurnal.fkip.unila.ac.id/index.php/aksara

Poerwanto. SK. Pembangunan Masyarakat Berbasis Pariwisata: Reorientasi dari Wisata Rekreatif ke Wisata Kreatif. Journal of Tourism and Creativity, [S.l.], v. 1, n. 2, sep. 2019. ISSN 2716-5159. https://jurnal.unej.ac.id/index.php/tourismjournal/article/view/13831

Poetry, RAH. 2011.—-repository.upi.edu

Raymond, C and Richards, G. 2000. http://www.creativetourismnetwork.org/about/

Riana, Nova. Kartika, Titing. Fajri, Khoirul. 2016. Pengemasan Pertunjukan Angklung Sebagai Atraksi Wisata Budaya Di Saung Angklung Udjo Bandung. UGM. Prosiding.

Saung Angklung Udjo, @angklungudjo, Instagram, Official Social Media

Stoneman, Paul. 2010. Soft Innovation: Economics, Product Aesthetics, and the Creative Industries. Oxford University Press.

Widyaiswaratika, Tika Dama (2016) Pengaruh Creative Tourism Terhadap Behavioral Intention Dengan Variabel Moderator Motivasi Wisatawan: (Survei Pada Wisatawan Nusantara Kampung Kreatif Kota Bandung). Other Thesis, Universitas Pendidikan Indonesia. http://repository.upi.edu/23233/

www.kemenparekraf.go.id

www.semanticscholar.org

Promoting Creative Tourism: Current Issues in Tourism Research – Kusumah et al. (Eds)
© 2021 Taylor & Francis Group, London, ISBN 978-0-367-55862-8

Potential of domestic tourist loyalty in Indonesia: A spatial analysis

A.R. Pratama, A. Khosihan & P. Hindayani
Universitas Pendidikan Indonesia, Bandung, Indonesia

ABSTRACT: PP No. 50 of 2011, which contains the Tourism Ministry master plan, states that synergistic, superior, and responsible marketing is one of the efforts that must be made to realize tourism development goals in Indonesia. Marketing activities must be right on target, meaning the government needs to understand how the characteristics of the target market will be targeted. The diversity of tourists in Indonesia causes the need for a typology of tourist character so that marketing efforts are carried out on target. The typology of tourists in Indonesia based on the level of travels and travel duration indicate the type of tourist based on the level of potential and the level of loyalty to tourism activities in an area. Based on the results of the analysis using the cross tabulation method and GIS, it could be described that the spatial typology of tourists in Indonesia was divided into seven, namely, Low Potential Inertia, Low Potential Latent, Low Potential Premium, Middle Potential Inertia, Middle Potential Latent, Middle Potential Premium, and High Potential Latent of Tourist.

Keywords: Potential, loyalty, domestic tourists, tourism marketing, spatial

1 INTRODUCTION

Indonesia, a country with various resources potentially becoming tourist destinations, has great tourism potential. Government Regulation No. 50 of 2011 states that synergistic, superior, and responsible marketing is one of the efforts that must be made to realize the tourism development goals of Indonesia (The Republik Indonesia 2011). As tourists are the consumers in the tourism industry, we need to understand their characteristics to increase their level of travel to a tourist destination. Previous researchers were interested in the characteristics and made a classification in a socio-demographic point of view (Ferrer-Rosell et al. 2015), a classification based on the types of tourism activities such as cultural, sports, and natural tourism activities (Seyidov & Adomaitienė 2017) and a classification based on the distance between the tourists' domiciles and the tourist destination (Abuamoud et al. 2014).

Meanwhile, tourist loyalty is closely linked to the level of tourist satisfaction, so their satisfaction will encourage them to travel to a tourist destination again (Abou-Shouk et al. 2018; Brown et al. 2016; Dayour & Adongo 2015). Therefore, we can say that tourist loyalty has a significant impact on financial performance (Anðelkoviæ et al. 2018). It indicates the need for a potential of loyalty to increase the service quality in a tourist destination. This article discusses the potential of domestic tourist loyalty in Indonesia based on the level of tourist travel, travel duration, and the level of domestic tourist expenditure in Indonesia. The three indicators showed different information to which we could refer while determining efficient tourism development and marketing strategies.

2 RESEARCH METHODS

The research was quantitative research using secondary data published by the Ministry of Tourism of the Republic of Indonesia. The indicators used were the level of tourist travel, travel duration, and the level of expenditure. The level of tourist travel could measure what potentials of domestic tourism in Indonesia were based on the level of tourist travel to a destination. The two other

DOI 10.1201/9781003095484-47

indicators were travel duration and the level of expenditures on tourism activities. The indicators could describe the types of domestic tourism in Indonesia.

Data collected were processed using a multilevel classification method in each indicator. Then, we made a cross-tabulation analysis to identify the typology of domestic tourists in each province in Indonesia. Using GIS, the data was presented in the form of spatial distribution.

3 RESULTS AND ANALYSIS

3.1 *Level of tourist travel*

With diverse characteristics of its provinces, Indonesia has diverse tourism potentials and tourist attractions. In other words, the diversity gives numerous tourist destinations to domestic tourists (Suttikun et al. 2018). Spatially, the level of tourist travel was dominated by tourists from provinces in Java; such as the Special Region of Yogyakarta, Central Java, and West Java. Another province with a high level of tourist travel was Bangka Belitung. Moreover, a moderate level of tourist travel was achieved by tourists from the western part of Indonesia, notably provinces in Sumatera; whereas a low level of tourist travel was dominated by tourists from the eastern part of Indonesia, such as West Papua, Papua, Maluku, and others. The detailed information of these levels of tourist travel is depicted in Figure 1.

Meanwhile, geographic and cultural factors were two factors causing the difference between spatial distribution and the level of tourist travel in Indonesia. It was in line with Zhang et al. (2011), who argued that geographic factors and neighboring effects greatly contributed to tourist distribution at a tourist destination. In other words, geographic factors and the availability of core activities in a spatial structure affected an individual tendency to do tourism activities because the spatial core activities would cause the difference in cultural, educational, and welfare levels. As a result, the more distant the area, the lower the educational and welfare levels. In terms of our cases, the spatial core activities were located in the western part of Indonesia, especially Java. It caused a low level of travel of the tourists from the eastern part of Indonesia despite various and attractive tourist destinations there (Kock et al. 2018; Liu et al. 2018; Nicolau 2008; Tóth & Dávid 2010).

3.2 *Travel duration*

The second indicator was the travel duration. In general, a tourism activity was a travel activity to a tourist destination with the aim of recreation in a temporary period (The Republik Indonesia 2009). By the definition, a tourism activity comes with a temporal dimension or consists of the measure of travel intention and the travel duration. Herington et al. (2013) indicate two classifications of tourists based on their travel duration (i.e., short holiday and long holiday). Besides, there is a one-night tourism activity since it needs one night only to do a certain tourism activity (Pang 2014).

Referring to that parameter, based on travel duration, tourists in Indonesia were classified into three, which were overnight tourists, short tourists, and long tourists. Our findings indicated that long tourists were mostly from the eastern part of Indonesia, such as West Papua, Papua, and Maluku. Meanwhile, short tourists were those from Sumatera (Aceh, South Sumatera, and Riau) and Kalimantan (East Kalimantan and North Kalimantan). Furthermore, most overnight tourists were from Java, especially West Java, East Java, Central Java, Special Region of Yogyakarta, and Bali. The spatial information of tourist origin distribution by travel duration is illustrated in Figure 2. Proportionally, overnight tourists were dominant with a percentage of 55.88%; while short tourists and long tourists were 35.29% and 8.82%, respectively.

The travel duration was the opposite of the level of tourist travel. Tourists from Java as the center of activities in Indonesia were more likely to spend a short time to travel. It was probably due to socio-demographic characteristics of each area in Indonesia. Most people living in the eastern part of Indonesia made a living from agricultural sectors; while those living in Java mostly concentrated on either industrial and office sectors. It became one of the factors causing a difference in travel

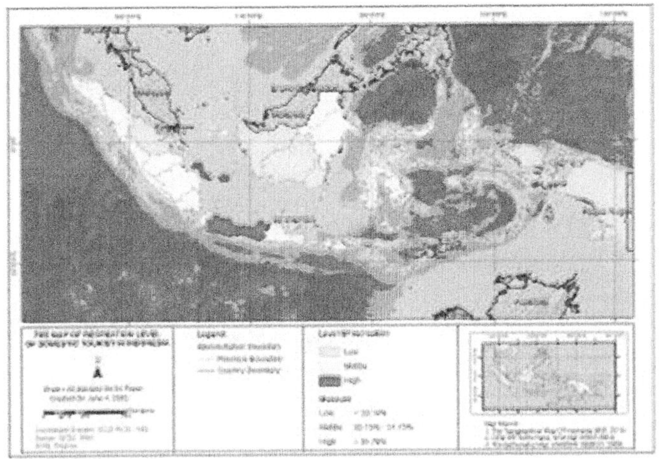

Figure 1. The map of recreation level of domestic tourist in indonesia.
Source: Analysis (2020).

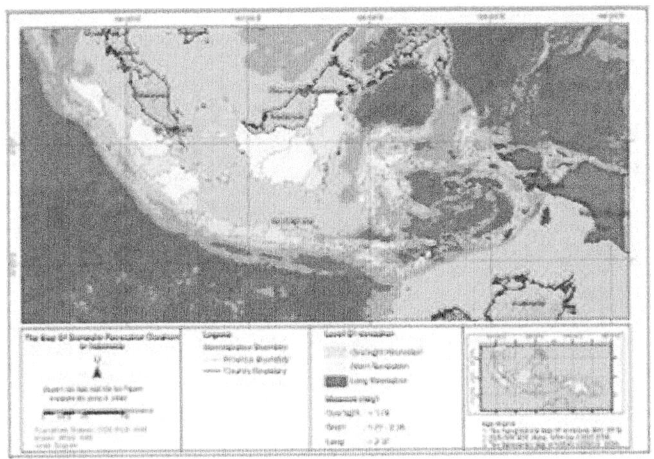

Figure 2. The map of domestic recreation duration in indonesia.
Source: Analysis (2020).

duration. Nicolau et al. (2018) confirmed that there was a relationship among distance, travel duration, and the first travel experience. By their characteristics, people living in Java had a high level of activities, giving them relatively short travel duration. It made them prefer short travel duration and similar destinations visited on weekends. It was different from the people living in the eastern part of Indonesia, particularly Papua, Maluku, and West Papua, whose activities were determined by their farming activities. Their travel duration was relatively long, but they had a low travel intention. It gave them longer travel duration and made them prefer to visit a new place they had never visited. To put it simply, these sort of tourists would opt to visit a destination far from their domiciles and thus had a long travel duration. As a result, they would have a great first-hand experience.

3.3 *Level of expenditure*

In general, the expenditure on tourism activities consisted of accommodation costs including money spent on accommodation, transportation, shopping, food, and beverages (Mudarra-Fernández et al. 2019; Smolčić et al. 2016). Based on our analysis, the level of expenditure on tourism activities

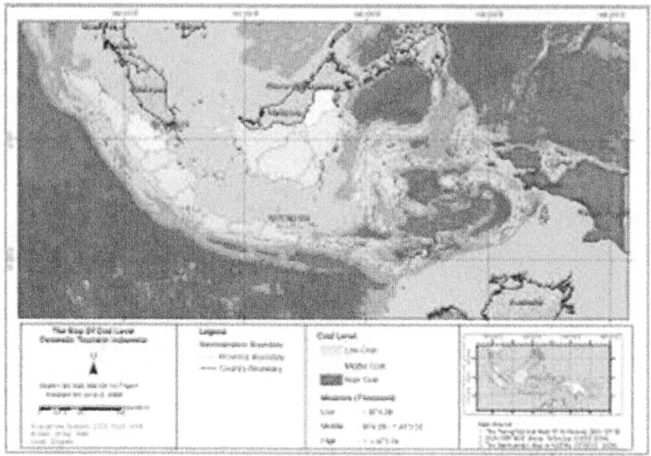

Figure 3. The map of cost level of domestic tourist in indonesia.
Source: Analysis (2020).

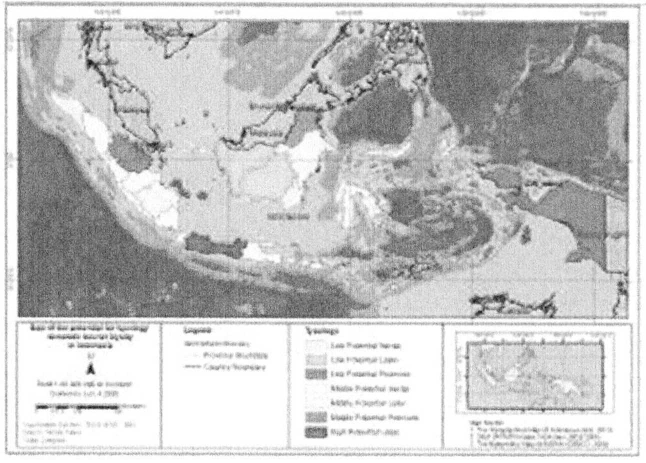

Figure 4. Map of the potential for typology domestic tourist loyalty in indonesia.
Source: Analysis (2020).

varied. However, after classified, a low level of expenditure dominated with a percentage of 82.35%, followed by a moderate level and high level with a percentage of 11.76% and 5.88%, respectively. Based on the distribution, a moderate level of expenditure was achieved by tourists from Riau, Riau Islands, Jakarta, and North Kalimantan. Moreover, West Papua and Papua had a high level of expenditure, while other provinces had a low one. The spatial information on the level of expenditure in Indonesia is indicated in Figure 4.

The average expenditure of tourists from Papua and West Papua was IDR2,072,990.00 and IDR1,643,590.00, respectively, while the average expenditure of domestic tourists in Indonesia was IDR670,110.00 (BPS 2018). It indicated that the level of expenditure of tourists from those two provinces was three times higher than the level of national expenditure.

Wang and Tian (2014) stated that the level of expenditure was one of the standards to determine economic impacts. The level of expenditure was related to travel duration. The longer the travel duration, the higher the expenditure. Following the statement, tourists from Papua and West Papua were premium tourists.

Table 1. The typology of domestic tourists in indonesia based on their potential of loyalty.

Typology	Areas	Description
Low Potential Inertia	Bali, Gorontalo, South Kalimantan, West Sulawesi	The travel potential and potential of loyalty were low.
Low Potential Latent	West Kalimantan, Central Kalimantan, East Nusa Tenggara, Central Sulawesi	The travel potential and potential of loyalty were moderate.
Low Potential Premium	West Papua, Maluku, Papua, Riau	The travel potential and potential of loyalty were high.
Middle Potential Inertia	Banten, Bengkulu, Jambi, East Java, Lampung, West Nusa Tenggara, South Sulawesi, North Sulawesi	The travel potential and potential of loyalty were low.
Middle Potential Latent	Aceh, East Kalimantan, North Maluku Utara, Southeast Sulawesi, South Sumatera	The travel potential and potential of loyalty were moderate.
Middle Potential Premium	Jakarta, North Kalimantan, Riau Islands	The travel potential and potential of loyalty were high.
High Potential Latent	Bangka Belitung, West Java, Central Java, Yogyakarta	The travel potential and potential of loyalty were low.

Source. Analysis (2020).

Meanwhile, there were some rationales of why people spent either low or high cost for tourism activities. The people might want to have nostalgic experiences or new experiences, promote their social status, or support environment acts (Kazeminia et al. 2016; Salem & Salem 2018).

3.4 *Typology of tourist*

Decision making, or policy making, is important to realize development planning. Therefore, making an effective decision is one of the keys to successful development. Meanwhile, a spatial approach defines that each area has different characteristics, so any development in the areas must suit the characteristics. Categorizing areas (typology) helps us identify their characteristics and thus enables us to select alternative development strategies to be implemented.

Based on our findings, in general, referring to their potential of loyalty, the typology of domestic tourists in Indonesia could be divided into seven categories: Low Potential Inertia, Low Potential Latent, Low Potential Premium, Middle Potential Inertia, Middle Potential Latent, Middle Potential Premium, and High Potential Latent of Tourist. The typology indicated the potential of domestic tourist loyalty in Indonesia based on the level of tourist travel, travel duration, and the level of expenditure. Several factors that built the typology were livelihoods, geographic conditions, socio-cultural conditions, and distance from the city center (Nicolau et al. 2018; Suttikun et al. 2018; Zhang et al. 2011). The distribution of the categories of the potential of tourist loyalty is presented in Figure 5.

In Figure 5, we can see that most areas in Indonesia belonged to the category of Middle Potential Inertia. It indicated that tourists from the areas had moderate travel potential and a low potential of loyalty. In other words, they spared a short time and made a low transaction for tourism sectors. Therefore, in general, Indonesian people were not potential loyal tourists, implying that their potential for repeated visits was low. We had identified two factors causing the phenomena which were geographic factors and the characteristics of Indonesia as an agrarian nation.

Furthermore, we also found two potential categories which could contribute to the local revenue: the categories of low potential premium and high potential latent. The first category indicated that tourists from these areas rarely did any tourism activities. However, once they did, they would likely spend a long time and a high amount of money. It certainly contributed to the local revenue from the tourism sectors. Moreover, the second category indicated that tourists from these areas had a high potential of loyalty. Although their time and money allocation for tourism activities were low, their potential for repeated visits was high. It certainly promoted the economy in terms of tourism sectors. The typology of domestic tourists in Indonesia based on their potential of loyalty is shown in Table 1.

4 CONCLUSION

We had explained the indicators of the level of tourist travel, travel duration, and the level of expenditures. The three indicators gave different information and could be used to identify the potential of domestic tourist loyalty in Indonesia. Based on our findings, the typology of the potential of loyalty was divided into seven categories: Low Potential Inertia, Low Potential Latent, Low Potential Premium, Middle Potential Inertia, Middle Potential Latent, Middle Potential Premium, and High Potential Latent of Tourist. Of the seven categories, two were potential categories which could highly contribute to the local revenue. The two categories were Low Potential Premium and High Potential Latent. In other words, tourists from the areas belonging to the category of high potential latent would likely have a high potential of loyalty and thus a high potential for repeated visits.

ACKNOWLEGMENT

Thanks to Universitas Pendidikan Indonesia that gave the researcher a chance to write in this conference and to the Pesona Indonesia Instagram account as the subject of this study.

REFERENCES

Abou-Shouk, M. A., Zoair, N., El-Barbary, M. N., & Hewedi, M. M. (2018). Sense of place relationship with tourist satisfaction and intentional revisit: Evidence from Egypt. *International Journal of Tourism Research, 20*(2), 172–181. https://doi.org/10.1002/jtr.2170

Abuamoud, I. N., Libbin, J., Green, J., & Rousan, R. A. L. (2014). Factors affecting the willingness of tourists to visit cultural heritage sites in Jordan. *Journal of Heritage Tourism, 9*(2), 148–165. https://doi.org/10.1080/1743873X.2013.874429

Anðelkoviæ, D., Vujiæ, M., Liberakos, A., & Zubac, D. (2018). The impact of relationship marketing with customers on the financial performance of the sunflower oil manufacturers in Serbia. *Ekonomika Poljoprivrede, 65*(1), 93–109. https://doi.org/10.5937/ekopolj1801093a

BPS. (2018). Retrieved from www.bps.go.id

Brown, G., Smith, A., & Assaker, G. (2016). Revisiting the host city: An empirical examination of sport involvement, place attachment, event satisfaction and spectator intentions at the London Olympics. *Tourism Management, 55*, 160–172. https://doi.org/10.1016/j.tourman.2016.02.010

Dayour, F., & Adongo, C. A. (2015). Why They Go There?: International Tourists' Motivations and Revisit Intention to Northern Ghana. *Tourism Management 2015, 4*(1), 7–17. https://doi.org/10.5923/j.tourism.20150401.02

Ferrer-Rosell, B., Coenders, G., & Martínez-Garcia, E. (2015). Determinants in Tourist Expenditure Composition — The Role of Airline Types. *Tourism Economics, 21*(1), 9–32. https://doi.org/10.5367/te.2014.0434

Herington, C., Merrilees, B., & Wilkins, H. (2013). Preferences for destination attributes. *Journal of Vacation Marketing, 19*(2), 149–163. https://doi.org/10.1177/1356766712463718

Kazeminia, A., Hultman, M., & Mostaghel, R. (2016). Why pay more for sustainable services? The case of ecotourism. *Journal of Business Research, 69*(11), 4992–4997. https://doi.org/10.1016/j.jbusres.2016.04.069

Kock, F., Josiassen, A., & Assaf, A. G. (2018). On the origin of tourist behavior. *Annals of Tourism Research, 73*(April), 180–183. https://doi.org/10.1016/j.annals.2018.04.002

Liu, H., Li, X. (Robert), Cárdenas, D. A., & Yang, Y. (2018). Perceived cultural distance and international destination choice: The role of destination familiarity, geographic distance, and cultural motivation. *Journal of Destination Marketing and Management, 9*(February), 300–309. https://doi.org/10.1016/j.jdmm.2018.03.002

Mudarra-Fernández, A. B., Carrillo-Hidalgo, I., & Pulido-Fernández, J. I. (2019). Factors influencing tourist expenditure by tourism typologies: a systematic review. *Anatolia*, *30*(1), 18–34. https://doi.org/10.1080/13032917.2018.1495086
Nicolau, J. L. (2008). Characterizing Tourist Sensitivity to Distance. *Journal of Travel Research*, *47*(1), 43–52. https://doi.org/10.1177/0047287507312414
Nicolau, J. L., Zach, F. J., & Tussyadiah, I. P. (2018). Effects of Distance and First-Time Visitation on Tourists' Length of Stay. *Journal of Hospitality and Tourism Research*, *42*(7), 1023–1038. https://doi.org/10.1177/1096348016654972
Pang, K. P. (2014). Impact of socio-demographics, fuel price and airfare on the domestic overnight travel decision in Australia. *Tourism Economics*, *20*(6), 1297–1318. https://doi.org/10.5367/te.2013.0341
Salem, S. F., & Salem, S. O. (2018). Self-identity and social identity as drivers of consumers' purchase intention towards luxury fashion goods and willingness to pay premium price. *Asian Academy of Management Journal*, *23*(2), 161–184. https://doi.org/10.21315/aamj2018.23.2.8
Seyidov, J., & Adomaitienė*, R. (2017). Factors Influencing Local Tourists' Decision-making on Choosing a Destination: a Case of Azerbaijan. *Ekonomika*, *95*(3), 112–127. https://doi.org/10.15388/ekon.2016.3.10332
Smolèiæ, D., Daniela, J., & Frleta, S. (2016). Factors Affecting the Expenditure of Domestic and Foreign Tourists-the Evidence From Rijeka and Opatija, Croatia. *Tourism & Hospitality Industry*, *September*, 418.
Suttikun, C., Chang, H. J., Acho, C. S., Ubi, M., Bicksler, H., Komolsevin, R., & Chongsithiphol, S. (2018). Sociodemographic and travel characteristics affecting the purpose of selecting Bangkok as a tourist destination. *Tourism and Hospitality Research*, *18*(2), 152–162. https://doi.org/10.1177/1467358416637254
The Republik Indonesia. (2009). *Undang-Undang Republik Indonesia*. Indonesia. https://jdih.kemenkeu.go.id/fullText/2009/10TAHUN2009UU.HTM
The Republik Indonesia. (2011). *PP No. 50 Tahun 2011 tentang Rencana Induk Pembangunan Kepariwisataan Nasional Tahun 2010 2025 [JDIH BPK RI]*. Indonesia. https://peraturan.bpk.go.id/Home/Details/5183/pp-no-50-tahun-2011
Tóth, G., & Dávid, L. (2010). Tourism and accessibility: An integrated approach. *Applied Geography*, *30*(4), 666–677. https://doi.org/10.1016/j.apgeog.2010.01.008
Wang, L., & Tian, M. (2014). A discussion on the development model of earthquake relic geopark-a case study of the Qingchuan Earthquake Relic Geopark in Sichuan Province, China. *Journal of Cultural Heritage*, *15*(5), 459–469. https://doi.org/10.1016/j.culher.2013.11.007
Zhang, Y., Xu, J. H., & Zhuang, P. J. (2011). The spatial relationship of tourist distribution in Chinese cities. *Tourism Geographies*, *13*(1), 75–90. https://doi.org/10.1080/14616688.2010.529931

Tourism and education

Promoting Creative Tourism: Current Issues in Tourism Research – Kusumah et al. (Eds)
 ISBN 978-0-367-55862-8

The tourism academic traveler

A.H.G. Kusumah
Universitas Pendidikan Indonesia

Khrisnamurti
Universitas Negeri Jakarta

M. Kristanti
Universitas Kristen Petra

ABSTRACT: Tourist typology and its characteristics are popular topics in tourism studies. Tourism scholars have focused on identifying tourist typologies to understand what tourists want when they travel. This paper aims to understand the typologies and characteristics of travelers who have a background as tourism educators and are considered experts in the field of tourism. As a group who believes in having expertise in the field of tourism, the travel patterns and decision making of this traveler group when traveling are undoubtedly different from tourists in general. The sample in this study was 22 tourism academics from Indonesia who participated in a three-weeks tourism short-course in Hong Kong. Data obtained through in-depth interviews with all short-course participants and content analyzed. Studies show that there are three academic-traveler typologies with each specific characteristic, namely, story-seeker, awe-explorer, and cognitively saturated traveler. This research has implications on how destination management organizations manage a destination and create an experience for their target market.

Keywords: Tourism, academic, traveler.

1 INTRODUCTION

Tourism academic is a profession that can combine elements of work and leisure at the same time. As an academic whose job is teaching students about tourism, traveling is an opportunity to enrich one's knowledge. The newly gained experience and knowledge can be used as learning content in the classroom. Previous research on tourist typology discusses traveling for work (Ojong 2013), tourist shopping typology (Sundström et al. 2011), green tourists (Bergin-Seers & Mair 2009), cultural tourists (McKercher & du Cros 2003; Niemczyk 2013), or millennial tourists (Cavagnaro et al. 2018; King & Gardiner 2015; Sari et al. 2018). However, little is known about the typology of tourism academic travelers. Therefore, it is essential to know the relationship between the background of the traveler as a tourism academic with travel characteristics because it can have an impact on improving the quality of learning in the field of tourism. This study aims to understand the typology of an academic traveler and its characteristics.

2 LITERATURE REVIEW

2.1 *Motivation*

When someone visits a place away from home, there are always driving motives. Motivation in the tourism field refers to the individual's intrinsic and extrinsic travel needs and wants that lead

DOI 10.1201/9781003095484-48

a person to adopt a particular behavior or tourist activity (Correia et al. 2008; Dayour & Adongo 2015; Zhang & Peng 2014). An academic traveler travels for work related to academics, such as international conferences, seminars, workshops, sabbatical leave, research, a courteous visit to partner, and lecturer exchange. The driving motives for these academic travelers can range from networking, interacting and exchanging knowledge, visiting new countries, meeting local people, visiting exciting places, or learning something new.

The driving motives of academic traveler can be developed based on some previous researches discussing tourists' motivation to travel. Correia et al. (2008) concluded there are six motives: knowledge, leisure, socialization, facilities, core attractions, and landscape features. Knowledge is related to the desire to learn about new places, do different things, and increase awareness. Leisure mainly pertains to physical relaxation, escape from the daily routine, and stress relief. Socialization encompasses people's need to develop close friendships, share travel experiences, and go to places where their friends have not been. Zhang and Peng (2014) used the travel career patterns (TCP) model that groups the motives into three layers: core layer, middle layer, and the outer layer. The core layer includes the need to escape and relax, to experience novelty, and to build relationships. The middle layer indicates that tourists with more travel experiences tend to seek close contact with the host community and the local environment as well as striving to fulfill self-development and self-actualization needs. Outer layer motives include seeking romance and looking for isolation. Dayour and Adongo (2015) argued that there are four driving motives to travel, which are culture such as enjoying local food and interacting with a different ethnic group; destination's attractions including local food, cultural artifacts, and religious sites; social contact such as making friends with local people; and adventure/novelty that covers getting close to nature, doing something challenging, experiencing unfamiliar destination, and participating in events. Yoo et al. (2018) found that there are six travel motivations, namely scenery and exotic experience, culture, relaxation, self-actualization, physical refreshment, and pleasure-seeking/fantasy. The scenery and unusual experience talk about natural aesthetics, unique environmental feature, the experience of natural habitat, and novelty. Culture mentions about local people, social interaction, knowledge of new places, cultural sites, and educational value. Relaxation is about feeling emotionally/physically refreshed and having fun. Self-actualization discussed self-actualization and self-fulfillment. Physical refreshment talks about mingling with fellow tourists and being active, and pleasure-seeking discusses escaping from daily life and seeking adventure.

Based on Pearce's travel career ladder (TLC), there was a model called Faculty and Student Travel Career Ladder (FAST-CL) that describes motivations for faculty and student joint travel (Williams & McNeil 2011). There are five levels of FAST-CL, (1) survival needs, (2) safety/security needs, (3) relationship needs, (4) self-esteem needs, and (5) self-actualization needs.

The first level, survival needs, is the motivation to demonstrate the currency and relevance of knowledge (teaching and learning), including undertaking targeted professional development as appropriate. It provides an underlying sense of surviving in terms of maintaining academic qualifications. The second level, safety/security needs, is related to the motivation to demonstrate safe travel competency and administration that execute the travel in a manner consistent with the policies and procedures of the university. The third level, relationship needs, relates to social development and building a relationship through meeting and interacting with people. The fourth level, self-esteem needs, is the motivation of leadership development, and the fifth level, fulfillment motivation.

2.2 *Typology*

Since there are numerous studies about tourist typology that classified tourist behavior in every perspective, only a few studies focus on typology of academic traveler relations over the years (Lichy & McLeay 2018; Ojong 2013; Williams & McNeil 2011). Tourism academics believe that most of their travel and its motivation is something "serious," and not only for leisure. Academics travel to places to discover new knowledge or obtain inspiration while attending academic gatherings such as conferences, seminars, or workshops that could enhance recognition among fellow academicians (Lichy & McLeay 2018). They share their intangible experience, newfound knowledge, and inspiration from the place they visited when they went back to their educational institute. Nevertheless, this "serious" travel has similar intrinsic and extrinsic factors with leisure travel, such as escape from routine, social interaction with like-minded people, visiting places of interest, immersing the local culture, or enjoying local food and beverages (Ojong 2013).

Tourists classified into clusters or groupings with similar characteristics that have the same travel demand since travel motivation explore the socio-physiological reasons on the needs why people tend to travel and also the factors that influence people to visit a tourist destination. Cohen, in his initial studies about typology, combines the demands of tourists for novelty with familiarity (Cohen & Cohen 2019). On the other hand, Plog designed more advanced on a similar model. He points out that the purpose is to explain the type of tourism destination along with the attributes such as accessibilities and amenities that are preferred to a different kind of segmentation of people of its psychographic characteristics (Plog 2001). Later, the model was developed, and it is considered a useful tool for tourism marketing to design their strategies such as segmentation, targeting, and pricing (Decrop & Snelders 2005).

3 METHOD

The data in this qualitative study were collected through in-depth interviews with 22 Indonesian tourism academics who work as a lecturer at universities or colleges in Indonesia. Respondents' working experience as academics ranging between 3 to 25 years of experience. All the interviewees participated in the short-course program initiated by the Indonesian Ministry of Research and Technology to enhance the academic capacity of participants. The short-course runs for 23 days, and the data collection process was carried out during the academic short-course in Hong Kong in 2017. The interview process for each interviewee lasted between 45 to 60 minutes and was conducted during the last week that short-course activity took place. The entire interview process was recorded using a smartphone. The recording was then transcribed verbatim. All transcription results were content analyzed by three researchers separately, and the analysis results were then verified together. Three investigators in this study were also participants of the short-course, and each researcher informally made observations and wrote notes on the activities of all the participants. The results of the observation are considered as one of the elements when conducting the analysis.

4 FINDINGS AND DISCUSSION

The research findings show that there are three types of academic travelers, and each type has unique characteristics. The three types of academic travelers are story-seeker, awe-explorer, and cognitively saturated traveler.

4.1 *Story-seeker*

Story-seeker is someone who is looking for a story in a destination. They tend to find or make a story and shared it with others. This type of tourism academic traveler connects the experience they get with the tourism concepts or theories they can recall. Story-seekers sometimes browse historical stories, characters, anecdotes, or fables related to destinations in order to find compelling stories or perspectives about destinations before going on a trip to form an unforgettable story.

Apart from liking stories and narratives, this type of academic traveler tends to share the collected stories with others through their academic point-of-view. The stories obtained—or created—also become shared stories in classrooms or discussions with colleagues. The stories are saved in a personal memory of the traveler and ready to be recalled to reminisce. In some cases, this personal memory may be converted into a memorable cognitive experience and often internalized into a new value held by the person concerned. This present study supports previous studies that suggest that knowledge-seeking attitude is one of the primary motivators for travel (Babolian Hendijani & Boo 2020; Mangwane et al. 2019). However, although the experiences and stories gathered from traveling are subjects to be recalled, story-seeker travelers tend not to have a particular motivation to share their experiences through social media. This behavior differs from Nurdianisa et al.'s (2018) study that found tourists' tendency to share their experience through social media.

Academics who fall into this category have a particular interest in tourist attractions or destinations that have a story or value related to the social life of the community. This particular interest in the social life of this community can lead to an interest in conducting research or community service regarding phenomena that appear to the community in the visited destination. In addition to liking the story, they are also interested in things that involve the cognitive role, including activities to understand the phenomena in a destination and the rationalization behind the phenomenon. These cognitive activities and the making of the destination stories become the source of their memorable experience.

4.2 *Awe-explorer*

The awe-explorer is the dominant typology of the group. Awe is defined as positive emotion from the overall feeling of humans that is produced from human encounters toward the environment (Chirico 2020). This tourist type likes to discover the highlights of a tourist destination and, if possible, to get every inch of all of the details using their five-sensory human senses to experience the vibe/atmosphere of the place. As stated by Lu et al. (2017), that tourist experience could be increased by awe emotion and could react when a particular aspect of an individual's original frame of reference is confronted by vast perceptual stimuli. Their enthusiasm for novelty experience creates a thrilling feeling that escalates their desire to discover more on the destination and have the tendency to revisit the place to have a nostalgic moment or even searching something that has not yet been discovered. They believe that every moment offers a unique experience. Their behavior to visit tourist attractions is linked to a "globetrotter" idea where the person is profoundly passionate about traveling. They feel an enormous pleasure if the place is undiscovered or least visited by the tourists. This type of academics is flexible in traveling.

In some cases, they could stay in a hostel but also crave for a Michelin restaurant and would not mind spending their expenses on an annual festival or special event, always living for the moment. They are easily overwhelmed and fascinated by every aspect of traveling, such as famous landmarks, stunning natural landscapes, host/local people within their culture, or even their journey to get to the destination. Because of these characteristics, they have a strong tendency to share their experience from a light coffee shop discussion to an academic forum, from talking directly to people to posting their stories online. Through traveling, the awe-explorer has the chance to view reality and his perspective on tourism. For awe-explorers, traveling is a new form of obtaining knowledge through experiences as well as an inspiration for their intellectual endeavor. In supporting the typology, Coghlan, Buckley, and Weaver (Lu et al. 2017) discovered that awe experience in tourism experience has three components, namely, physiological responses, comparative uniqueness, and schema-changing for the future such inspiration. This present study confirmed the previous study in which the findings have similarities. In conclusion, the awe-explorer is the type of traveler that took the most out of the experience of the journey and embraced it that later became their inspiration.

4.3 *Cognitively-saturated traveler*

The third typology of academic traveler is a cognitively saturated traveler. This type of traveler is interested in exploring the world but has some considerations in choosing destinations. They maintain their comfort zone and take cost-value consideration. These considerations will be reflected in choosing accommodation, transportation, touristic destination, and the academic-related event they would joint. They will not mind spending money and time or travel far for something they like if it has value and benefits from their point of view. It is supported by Correia et al. (2008)

who mentioned knowledge and facilities as the driving motives for tourists to travel. When this type of academic traveler has visited or has prior knowledge of a place, they tend not to visit those places. This tendency is mainly formed because they believe the place will give them the same experience with a previous attraction or destination they visited before. They believe they would not find something interesting to discover more.

As a cognitively saturated traveler, an academician has an interest in famous tourist attractions or destinations that could confirm their knowledge and offer values. They tend not to create a story from it, and they will be delighted to go to that place even though they do not explore it. The confirmed knowledge will be brought into the class or discussions with colleagues. Others, these academics, will bring this confirmed knowledge in doing research, publication, and community service. In choosing the place for research and community service, this type of academic traveler will determine a place that fits their comfort zone.

The cognitively saturated traveler has something in common with general tourist travelers in terms of the motivation for traveling. The common motivation of travel such as leisure, socialization, self-esteem, self-actualization, knowledge (Correia et al. 2008; Dayour & Adongo 2015; Williams & McNeil, n.d.; Yoo et al. 2018; Zhang & Peng 2014) is also found in the cognitively saturated traveler characteristics. The differences between the two are more to the purpose of traveling, how to spend time, ideas while seeing touristic destinations, and meeting people. Academic travelers' pattern of traveling is mixing between scholarly and touristic experiences in order to get a package of ideas about other forms of teaching and learning (Ojong 2013).

5 CONCLUSION

The question of how tourism academics utilize this privilege is found in the proposed tourism academic travelers' typology. Three types of travelers found in this study have specific characteristics related to the utilization of their tourism knowledge to gain traveling experience. Besides, each type of traveler also has specific characteristics related to how they utilize their traveling experience in academic activities. If tourism academics can recognize and understand their particular types and characteristics, they can optimize their travel trips into a knowledge enriching process that they can use in their academic obligations, such as in research and learning.

In short, this paper proposes a new typology model of academic traveler specified as tourism educators that are considered an expert in the field of tourism. It describes the travel patterns and decision making of a traveler group that were involved in the same short-course. Tourism academic travelers have professional privileges because they can combine tourism and working activities. Other academic professions do not own this privilege. Therefore, tourism academics must be able to make the best use of this privilege to develop tourism science and improve the quality of tourism education.

The question of how tourism academics utilize this privilege is found in the proposed tourism academic travelers' typology. Three types of travelers found in this study have specific characteristics related to the utilization of their tourism knowledge to gain traveling experience. Besides, each type of traveler also has specific characteristics related to how they utilize their traveling experience in academic activities. If tourism academics can recognize and understand their particular types and characteristics, they can optimize their travel trips into a knowledge-enriching process that they can use in their academic obligations, such as in research and learning.

In short, this paper proposes a new typology model of academic traveler specified as tourism educators that are considered an expert in the field of tourism. It describes the travel patterns and decision making of a traveler group that were involved in the same short-course. Tourism academic travelers have professional privileges because they can combine tourism and working activities. Other academic professions do not own this privilege. Therefore, tourism academics must be able to make the best use of this privilege to develop tourism science and improve the quality of tourism education.

REFERENCES

Babolian Hendijani, R., & Boo, H. C. (2020). Profiling Gastronomes from their Food Experience Journey. *Journal of Hospitality & Tourism Research*, 109634802091774. https://doi.org/10.1177/1096348020917741

Bergin-Seers, S., & Mair, J. (2009). Emerging green tourists in Australia: Their behaviours and attitudes. *Tourism and Hospitality Research*, *9*(2), 109–119. https://doi.org/10.1057/thr.2009.5

Cavagnaro, E., Staffieri, S., & Postma, A. (2018). Understanding millennials' tourism experience: values and meaning to travel as a key for identifying target clusters for youth (sustainable) tourism. *Journal of Tourism Futures*, *4*(1), 31–42. https://doi.org/10.1108/JTF-12-2017-0058

Chirico, A. (2020). Awe. In *The Palgrave Encyclopedia of the Possible* (pp. 1–9). Springer International Publishing. https://doi.org/10.1007/978-3-319-98390-5_30-1

Cohen, S. A., & Cohen, E. (2019). New directions in the sociology of tourism. *Current Issues in Tourism*, *22*(2), 153–172. https://doi.org/10.1080/13683500.2017.1347151

Correia, A., Silva, J. A., & Moço, C. (2008). Portuguese Charter Tourists to Long-Haul Destinations: A Travel Motive Segmentation. *Journal of Hospitality & Tourism Research*, *32*(2), 169–186. https://doi.org/10.1177/1096348007313262

Dayour, F., & Adongo, C. A. (2015). Why They Go There?: International Tourists' Motivations and Revisit Intention to Northern Ghana. *Tourism Management 2015*, *4*(1), 7–17. https://doi.org/10.5923/j.tourism.20150401.02

Decrop, A., & Snelders, D. (2005). A grounded typology of vacation decision-making. *Tourism Management*, *26*(2), 121–132. https://doi.org/10.1016/j.tourman.2003.11.011

King, B., & Gardiner, S. (2015). Chinese International Students. An Avant-Garde of Independent Travellers? *International Journal of Tourism Research*, *17*(2), 130–139. https://doi.org/10.1002/jtr.1971

Lichy, J., & McLeay, F. (2018). Bleisure: motivations and typologies. *Journal of Travel & Tourism Marketing*, *35*(4), 517–530. https://doi.org/10.1080/10548408.2017.1364206

Lu, D., Liu, Y., Lai, I., & Yang, L. (2017). Awe: An Important Emotional Experience in Sustainable Tourism. *Sustainability*, *9*(12), 2189. https://doi.org/10.3390/su9122189

Mangwane, J., Hermann, U. P., & Lenhard, A. I. (2019). Who visits the apartheid museum and why? An exploratory study of the motivations to visit a dark tourism site in South Africa. *International Journal of Culture, Tourism and Hospitality Research*, *13*(3), 273–287. https://doi.org/10.1108/IJCTHR-03-2018-0037

McKercher, B., & du Cros, H. (2003). Testing a cultural tourism typology. *International Journal of Tourism Research*, *5*(1), 45–58. https://doi.org/10.1002/jtr.417

Niemczyk, A. (2013). Cultural tourists: "An attempt to classify them." *Tourism Management Perspectives*, *5*, 24–30. https://doi.org/10.1016/j.tmp.2012.09.006

Nurdianisa, L., Kusumah, A. H. G., & Marhanah, S. (2018). Analisis motivasi wisatawan dalam berbagi pengalaman wisata melalui media sosial Instagram. *Journal of Indonesian Tourism, Hospitality and Recreation*, *1*(1). https://doi.org/10.17509/jithor.v1i1.13291

Ojong, V. B. (2013). Academic Travel: Travelling for Work. *Journal of Human Ecology*, *43*(1), 83–91. https://doi.org/10.1080/09709274.2013.11906614

Plog, S. (2001). Why Destination Areas Rise and Fall in Popularity: An Update of a Cornell Quarterly Classic. *The Cornell Hotel and Restaurant Administration Quarterly*, *42*(3), 13–24. https://doi.org/10.1177/0010880401423001

Sari, D., Kusumah, A. H. G., & Marhanah, S. (2018). Analisis faktor motivasi wisatawan muda dalam mengunjungi destinasi Wisata Minat Khusus. *Journal of Indonesian Tourism, Hospitality and Recreation*, *1*(2). https://doi.org/10.17509/jithor.v1i2.13762

Sundström, M., Lundberg, C., & Giannakis, S. (2011). Tourist shopping motivation: go with the flow or follow the plan. *International Journal of Quality and Service Sciences*, *3*(2), 211–224. https://doi.org/10.1108/17566691111146104

Williams, J., & McNeil, K. (2011). A modified travel career ladder model for understanding academic travel behaviors. *Journal of Behavioral Studies in Business*, 1–10. http://www.aabri.com/manuscripts/09409.pdf

Yoo, C.-K., Yoon, D., & Park, E. (2018). Tourist motivation: an integral approach to destination choices. *Tourism Review*, *73*(2), 169–185. https://doi.org/10.1108/TR-04-2017-0085

Zhang, Y., & Peng, Y. (2014). Understanding travel motivations of Chinese tourists visiting Cairns, Australia. *Journal of Hospitality and Tourism Management*, *21*, 44–53. https://doi.org/10.1016/j.jhtm.2014.07.001

Promoting Creative Tourism: Current Issues in Tourism Research – Kusumah et al. (Eds)
© 2021 Taylor & Francis Group, London, ISBN 978-0-367-55862-8

Competency development problems in tourism and hospitality students' internship in Indonesia

Rosita
Universitas Pendidikan Indonesia, Bandung, Indonesia

ABSTRACT: This study aims to find the problems related to competency development in tourism and hospitality student internships in Indonesia. Qualitative approach was employed in this study. Participants were chosen purposively based on their position and knowledge related to the understanding of the curriculum. They were the head of the study programs and/or the internship program coordinators in the study program. The study programs that participated in this study came from six tertiary institutions. Qualitative data were collected through focus interviews with open-ended questions. The results show that there are four main problems occured during internship program. Collaboration between educators and industry practitioners is vital for problems to be solved, primarily to ensure the suitability of the competencies that will be obtained by trainees.

Keywords: Internship, tourism & hospitality education, competencies, curriculum

1 INTRODUCTION

Research in tourism and hospitality education related to internship programs in Indonesia is still very minimal in number. Certainly, all tourism and hospitality education institutions from the high school to tertiary level have an internship program and many issues that need to be investigated. Now, with the 'Merdeka Belajar Kampus Merdeka' (MBKM) program, a new policy of the Ministry of Education and Culture of the Republic of Indonesia, the vital role of the internship program in developing student competencies is in the spotlight.

The main principles of the MBKM program are written in Regulation of the Minister of Education and Culture No. 3, Article 18 of 2020 concerning National Standards for Higher Education. The two main messages are: firstly, to obtain learning outcomes, students can fully take courses in the study program where he or she is studying, and secondly, students can take some courses from outside the study program where they study, within the institutions or in other tertiary institutions including internships. In this program, students will be given as many as two semesters or equivalent to 40 credits to take courses in the same study program or different study programs within the same tertiary institution and/ or carry out internship programs. Internship program aims at gaining competency deepening and increasing real learning experiences in the community and employment.

In the implementation of this MBKM program, the initial challenge that must be faced by tourism and hospitality education in Indonesia is to identify appropriate competencies. This challenge is indeed, something that will always appear first when competency-based education will be developed (Spady 1978). Then proceed to the next more specific challenge, namely, determining the relevance and balance of each competency (Clement, et.al. 2010) linking it with other competencies, programs in the curriculum, lecture content, and also assessment methods (Griffith 2007; Perlin 2011). This preliminary study aims to find what problems actually occur in developing students' competencies during the internship program.

2 LITERATURE REVIEW

Outcome is the expected learning outcomes or achievements of students that lead to the peak demonstration (performance). Outcome occurs at or after the end of a significant learning experience. It means that the outcome is not a collection or average of previous learning experiences, but is a manifestation of what students can do after they have and complete all of their learning experiences. It also means that outcomes are what students can do from what they know and understand (Spady 1994). The term learning outcomes is often used interchangeably with competence, although both have different understandings of the approach's scope. Competence is a form of learning achievement, so its nature is limited. Achievement is usually stated with competent or incompetent, not in the form of rank (grade). While learning outcomes can be achieved in the form of various levels, even in various ways, and the results can be measured in multiple ways as well. Learning outcomes show learning progress described vertically from one level to another and documented in a qualification framework. Learning outcomes must be accompanied by appropriate assessment criteria that can be used to assess that the expected learning outcomes have been achieved. Competence comes from the Latin 'competere', which means conformity. Competence is generally referenced as suitability for a particular job (Boyatzis 2008; Byars & Rue 2004; Robbins 2003). In vocational education and training, a person is declared competent if he can consistently apply his knowledge and expertise to the performance standards required at work. Competence achieved by a person is the result of structured and tiered learning achieved within a specified period.

Gonczi in Velde (1999) distinguishes competencies into three basic concepts, namely: 1) the 'behaviorist' where competencies are conceptualized in terms of discrete behavior associated with the completion of various tasks. 2) the 'generic' concentrates on attributes such as critical thinking capacity, and 3) the 'integrated' which is a combination of the 'behaviorist' and the 'generic' approaches. Competence, according to Ellstrom in Nilsson & Ekberg (2013), is an attribute of individual/human capital in the form of abilities generated from all knowledge that has been acquired by someone (knowledge, affective and social skills). Competence can also be expressed as a broad concept that can be transformed into productivity and is an attribute of a job, individual potential, or task requirements (qualifications). The combination of the two is the competence that is actually used in the workplace, which is the interaction between individuals and work.

The terms 'competency' and 'main competency' are more widely used by tourism and hospitality education experts than 'basic skills' or 'technical skills' to describe tourism graduates' abilities. Competence is a combination of observable and applied knowledge, skills, and behavior that creates a competitive advantage for an organization. Competence focuses on how an employee creates value and what is achieved (Jauhari 2006). Zehrer and Mossenlechner (2009; in Su 2015) mention four types of competencies, they are professional competencies, methodological competencies, social competencies, and leadership competencies, whereas Lopez Bononilla and Lopez-Bonilla (2012; in Su 2015) consider that the content of different competencies can be developed depending on whether someone is learning to know, learn to do, or learn to develop attitudes.

What competencies are needed by the industrial world and what competencies should be developed for students during the internship program are becoming an exciting discussion theme in tourism and hospitality education to date. Most of the researchers revealed their findings that the industry preferred prospective students to have "soft skills" (Brownell 2008; Fournier & Ineson 2010; Jaykumar et al. 2014; Kim et al. 2017; Ocampoa, et al. 2020; Sisson & Adams 2013; Yang, et al. 2014), and students also agreed that they obtained "soft skills" from their internship experience in addition to the "hard skills" they learned in the classroom (Losekoot, et al. 2018). The internship program is also considered to contribute to the development of student management competencies. Through an internship program, students can develop competencies in several generic management fields, including leadership, human resources, oral and written communication, interpersonal communication, problem-solving, teamwork, planning, and decision making (Losekoot, et al. 2018). Nevertheless, there are still some gaps between the competencies taught with the competencies needed in the hospitality industrial world (Millar et al. 2010).

3 METHOD

This study employed a qualitative approach. Participants were chosen purposively based on their position and knowledge related to the understanding of the curriculum. They were the head of the study programs and/or the internship program coordinators in the study program. The study programs that participated in this study came from six tertiary institutions: Trisakti Institute of Tourism, Bandung Institute of Tourism and Economy (STIEPAR), Telkom University, Indonesia University of Education, Gajah Mada University Vocational School, Udayana University. Qualitative data was collected through focus interviews with open-ended questions. In focus interviews, questions are not written strictly in the interview guidelines. The questions asked can differ in both sentences and their order according to the interviewee's condition. However, the entire question will still refer to the topic under study. Qualitative data analysis is complicated because it is not formulated with certainty like the quantitative analysis. In this problem, the three components (data reduction, data display, drawing conclusions) of qualitative data analysis from Miles and Huberman (1994) would be the researchers' references. Firstly, collection of raw data in form of written notes and transcription was organized into catagories. Secondly, the data was summarized to look for themes and interpreting the meanings of themes in form of written description.

4 RESULTS AND DISCUSSION

All study programs have determined in their curriculum what competencies students should be able to achieve when carrying out apprenticeship activities in the industry. However, the reality in the field of what is to be made is often not achieved. The study program has set industry criteria that students can choose as the internship program location, but it is challenging to apply because it depends on the industry that will accept the student concerned, while students receive more decisions from the recipient industry and consideration from the school. Case in point is the placement of D3 vocational hospitality students with the concentration of Housekeeping that should be in the Housekeeping department but placed by the HRD of the hotel in the F&B department. The school has difficulty intervening in hotel HRD decisions, and students tend to accept industry decisions rather than having to refuse and return from the beginning to find another location.

Some companies already have manager training that regulates the activities of the trainees from the beginning to the end of the internship program and what competencies they have to obtain during that time. However, there are not a few locations of other internship programs that do not have programs for the interns at all. They are only entrusted to the staff and given responsibility for being able to help their work. It is regrettable both by the school and students themselves. In this case, the schools and students also do not have any choice because it is not all, even relatively few, the intended industry has an organized program for interns. However, so far, the school has tried to anticipate it so that the intended competencies can be achieved.

Another problem that influences the acquisition of competencies is the problem of adaptation. The school considers this problem to be a problem of the development of each student who is influenced by various variables outside of education at school, such as family background, student character, motivation, and the like. In this case, the school usually anticipates providing supplies that can support students in the early adaptation periods in the industry, so the briefing is not related to competency skills but rather on motivating and forming attitudes. The adaptation problem becomes more severe than the issue of competency suitability in the early days, or even until the end of the internship. The culture of the working environment is very different from the culture of education in schools. In this case, schools that can implement a culture of behavior with 3G (Greeting, Grooming, Gesture) can at least reduce the burden of student adaptation in the early days. The industry also considers this problem is a problem that almost always occurs in every student internship, the impact is felt from the declining work motivation, discipline, to violations of work rules that end returning students to the school.

Some D3 level vocational study programs have assessed the final results of the apprenticeship program through certification exams. However, the assessment of the final result in most study programs is still general and tends to prioritize attitude criteria such as discipline, courtesy, adaptation, and cooperation. It is based on the importance of assessment criteria from the industry perspective, prioritizing the assessment of attitude rather than technical skills since the technical skills are vary between types of work. However, the effect on students and their knowledge and skills during classroom learning and internship are less appreciated. In this case, the authors propose the school to refer the Indonesian National Work Competency Standards (SKKNI) and the ASEAN Qualification References Framework (AQRF) in tourism and hospitality to identify what competencies students need to master that cover aspect of knowledge, skills and work attitudes.

5 CONCLUSION

An internship program can be a way to bridge the theoretical knowledge learned in the classroom with practical experience in the real world. It also contributes positively to students' academic and career development. However, problems hindering these goals can occur. The results of this study show there are four main problems related to students' competencies development occured during internship program, namely 1) students' placement in the industry, 2) training program management, 3) students' adaptation, and 4) final assessment. In order for the problems to be resolved, the collaboration between educators and industry practitioners is vital, primarily to ensure the suitability of the competencies that will be obtained by the interns.

REFERENCES

Boyatzis, R. E. 2008. Competencies in the 21st century. *Journal of Management Development* 27(1): 5–12.

Brownell, J. 2008. Leading on land and sea: Competencies and context. *International Journal of Hospitality Management* 27:137–150.

Byars, L. L., Rue, L.W. 2004. Human resources management. McGraw-Hill.

Clement, D. G., Hall, R. S., O'Connor, S. J., Qu, H., Stefl, M. E., & White, A. W. 2010. Competency development and validation: A collaborative approach among four graduate programs. *Journal of Health Administration Education* 27(3): 151–173.

Fournier, H., Ineson E. M. 2010. Closing the gap between education and industry: Skills' and competencies' requirements for food service internships in Switzerland. *Journal of Hospitality & Tourism Education* 22(4): 33–42.

Griffith, J. R. 2007. Improving preparation for senior management in healthcare. *Journal of Health Administration Education* 24(1): 11–32.

Jauhari, V. 2006. Competencies for a career in the hospitality industry: An Indian perspective. *International Journal of Contemporary Hospitality Management* 18(2):123–134.

Jaykumar, V., Fukeya, L. N., Balasubramanian, K. 2014. 5th Asia Euro Conference 2014, Hotel managers perspective of managerial competency among graduating students of hotel management programme. *Procedia - Social and Behavioral Sciences* 144: 328 – 342.

Kim, N., Park, J., Choi, J. 2017. Perceptual differences in core competencies between tourism industry practitioners and students using Analytic Hierarchy Process (AHP). *Journal of Hospitality, Leisure, Sport & Tourism Education* 20: 76–86.

Losekoot, E., Lasten, E., Lawson, A., Chen, B. 2018. The development of soft skills during internships: The hospitality student's voice. *Research in Hospitality Management* 8(2): 155–159.

Miles, M. B., Huberman, A. M. 1994. Qualitative data analysis. Thousand Oaks.

Millar, M., Zhenxing, M., Patrick, M. 2010. Hospitality & Tourism Educators vs. The Industry: A competency assessment. *Journal of Hospitality & Tourism Education* 22(2): 38–50.

Nilsson, S., Ekberg, K. 2013. Employability and work ability: returning to the labour market after long-term absence. *A Journal of Prevention, Assessment and rehabilitation* 44(4): 449–457.

Ocampoa, A. C. G. et. al. 2020. The role of internship participation and conscientiousness in developing career adaptability: A five-wave growth mixture model analysis. *Journal of Vocational Behavior* 120: 103–426.

Perlin, M. S. 2011. Curriculum mapping for program evaluation and CAHME accreditation. *Journal of Health Administration Education* 28(1): 27–47.
Robbins, S. P. 2003. Organizational behavior. Prentice Hall.
Sandwith, P. 1993. A hierarchy of management training requirements: The competency domain model. *Public Personnel Management* 22(1); 43–62.
Sisson L. G. & Adams A. R. 2013. Essential hospitality management competencies: The importance of soft skills. *Journal of Hospitality & Tourism Education* 25(3): 131–145.
Spady, W. G. 1978. The concept and implications of competency-based education. Educational Leadership.
Spady, W. G. 1994. Outcome-based Education: Critical Issues and Answers. American Association of School Administrators.
Su, Y. 2015. Lifelong learning in tourism education. In Dredge, D. Airey, D. & Gross, M. J. (Eds.) The Routledge Handbook of Tourism and Hospitality Education, London: Routledge.
Velde, C. 1999. An alternative conception of competence: implications for vocational education. *Journal of Vocational Education and Training* 51(3): 437–447.
Yang L., Partlow C. G., Anand, J., Shukla, V. 2014. Assessing the competencies needed by hospitality management graduates in India. *Journal of Hospitality & Tourism Education* 26(4): 153–165.

Promoting Creative Tourism: Current Issues in Tourism Research – Kusumah et al. (Eds)

Integrative teaching materials for Indonesian Speakers of Other Languages based on Sundanese gastronomy text

M.W. Rizkyanfi, Syihabuddin, F.N. Utorodewo, V.S. Damaianti & D. Turgarini
Universitas Pendidikan Indonesia, Bandung, Indonesia

ABSTRACT: This research aims to develop teaching material with integrative model of Indonesian for Speakers of Other Languages (ISOL) based on Sundanese gastronomy. ISOL teaching materials through an integrative model are compiled as an alternative teaching material addressed to international students who are learning Bahasa Indonesia. The tasks in the teaching materials are thoroughly compiled considering four aspects of language skills: leading, speaking, listening, and writing skills. The research began with the process of formulating the concept of Sundanese gastronomy from several references. Furthermore, the concept of Sundanese gastronomy was included as teaching materials in four language skills by referring to several aspects, namely, (1) content eligibility, (2) presentation, (3) readability, along with (4) ethnic, race, religion, and gender background. The research found that the experts moderately assessed all aspects of the teaching materials by giving 5 points on the Likert scale. The research findings consequence from the guideline was that whenever the point was less than 5, the teaching materials would continue to be improved until discovering the ideal form of ISOL teaching materials. As a recommendation, the development of Sundanese gastronomy-based ISOL teaching materials is significant to be exposed to international students, especially for students who are learning Bahasa Indonesia both as compulsory material in the Indonesian tourism department and as supplementary material in different majors.

Keywords: Teaching material, integrative model, ISOL, Sundanese gastronomy texts

1 INTRODUCTION

The number of Bahasa Indonesia speakers is ranked the 9th as the language being one of the most widely used language on Earth. Based on the large number of users, it can reach more than 159 million people (Hasan 2019). Bahasa Indonesia, along with Malay, is used in several countries including Indonesia, Malaysia, and Brunei Darussalam. According to the Indonesian Ministry of Foreign Affairs data in 2011, Bahasa Indonesia has a high number of native speakers spread abroad by 4,463,950 people (Muliastuti 2015 p.1). Besides, the data obtained from the field show that there are more and more enthusiasts to start learning Bahasa Indonesia. The Indonesian Ministry of Education and Culture (2017) reveals that the Indonesian Darmasiswa program, a scholarship program offered to all international students to study in Indonesia, had attracted many interests in the past years. Overall, since the program started in 1974, this program has produced about 7,215 alumni from 117 countries. In the 2017/2018 academic year, BPKLN (Planning and Cooperation of Foreign Affairs) has managed to attract 1,087 registries from 130 countries. The latest data in 2019 showed that the number of Darmasiswa scholarship participants amounted to 579 people from 101 countries (Kemdikbud [The Indonesian Ministry of Education and Culture], 2019). The awarding of Darmasiswa scholarship is aimed to be given to foreigners who are interested in learning a Bahasa Indonesia course, in addition to tourism, arts, culture, and culinary courses. The data displayed show the depth of concentration in improving the rank position of Bahasa Indonesia among other languages. Therefore, the improvement of the Indonesian human resources quality is the main requirement to progress the Indonesian position as a nation in the global life order. In order to

DOI 10.1201/9781003095484-50

realize the improvement of human resources, improving the quality of Bahasa Indonesia teaching and learning is required since it is one of the portals to the science and technology mastery (Sugono 2001).

Teaching Bahasa Indonesia to speakers of other languages (ISOL) is one of the embodiments of Bahasa Indonesia enthusiasts to internationalize the language. Based on some progress that has been achieved, Indonesia as a nation is able to place itself in the association of the nations that have an affect on the emergence of assumptions from other nations regarding the importance of Bahasa Indonesia as a language with the high number of speakers that keeps gradually increasing.

The data on ISOL development can be proved by one of the language institutions from the Indonesian Ministry of Education and Culture (2012) that suggests that Bahasa Indonesia has been taught to foreigners in various institutions, both domestically and internationally. Domestically, there are approximately 45 institutions that taught Bahasa Indonesia to foreign speakers, in formal as well as informal education. Internationally, teaching Bahasa Indonesia to speakers of other languages (ISOL) has been conducted in approximately 36 countries in 130 institutions around the world, including colleges, foreign cultural centers, the Indonesian Embassy, as well as many other informal educational courses.

The high number of foreigners' interest to learn Bahasa Indonesia is in contrast to the insufficient teaching materials that are suitable to foreigners' wants of how they want to learn Bahasa Indonesia. This situation emerged as a result of the books of ISOL teaching materials scarcity that are available in the bookstore. Teaching materials developed by ISOL teachers can be taken from various sources, not only from the prescribed textbook that has been provided by the official institution.

In addition, the ISOL curriculum or syllabus is remained unplanned by the government, creating the curriculum for every different level of education. The lack of curriculum and syllabus affects every institution that conducts ISOL to create different learning objectives. However, The Indonesian Ministry of Education and Culture has established a ministerial regulation that set graduate competence standard for Bahasa Indonesia courses or training to speakers of other languages in 2017. The standard can be used as a benchmark in creating an ISOL syllabus or curriculum for every institution conducting the program. The researchers of the current study classify seven levels in the graduate competence standard into three different levels; Level 1 and Level 2 ranked in the beginner level, Level 3 up to Level 4 including in the intermediate level, and Level 5 along with level 7 are classified into the advance level. The Researchers have been adopted those tiers from the Common European Framework of Reference for Languages (CEFR). The researchers further determined the intermediate level as a standard to develop current teaching material in this study. The intermediate level selected considering the learners' ability in constructing the elements of the language they acquired in the previous beginner level, but have not acquired the ability to communicate fluently the advanced level ISOL learning.

The teaching materials developed in this study were compiled using integrative models. It is in line with the one developed by Muliastuti (2015) based on her experience that an integrative model can be a great alternative to ISOL learning. The integrative model emphasises the combination of learning-speaking skills and implements on teaching activities that required students to practice integrated language skills. Rizkyanfi (2010), in his research, mentioned that in the integrative model of ISOL teaching materials, the learners could enjoy the process of Bahasa Indonesia teaching and learning in an integrated approach. Rizkyanfi (2010) further suggested that in the integrated approach, the learners are able to practice four basic language skills simultaneously, for instance, during learning to read, the learners can also learn to speak. Besides, the integrative ISOL teaching materials are beneficial for the ISOL program due to the integrated ISOL teaching materials, the richness of Indonesian identity and culture as the nation can be introduced.

The teaching materials developed in the study were based on Sundanese gastronomy, the gastronomy that is often associated with culinary tourism by Indonesian. The development of these teaching materials is materialised due to the shortage of research conducted on gastronomy-related ISOL teaching materials. Considerably, Hartono et al. (2019) explained that the standardized ISOL teaching materials that are created based on culinary tourism could also foster the curiosity of ISOL students on culinary culture across Indonesia. The significant number of foreigners who are interested in studying in Indonesia because of the culture opened the opportunities to introduce Indonesia as a country along with the national language. Consequently, the archipelago of culinary tourism which is integrated into ISOL teaching materials can become Bahasa Indonesia diplomacy media at the international level. Santich (1996) suggested that gastronomy is a guide on a variety of ways involving everything regarding foods and drinks in a highly interdisciplinary manner related to the reflection of a historical, cultural impact, and environmental conditions. Meanwhile, culinary is defined as a discipline related to the art and skill of preparing, crafting, cooking, and serving foods. Gastronomic tours are food-related trips to a place or area with recreational purposes which attempt to achieve a balance between travel benefits.

In another study, Turgarini (2018) revealed that the results of the gastronomic study and all related factors and variations of the food would provide information in the form of the culture. On the other hand, Cousins (as cited in Turgarini 2018) explains that gastronomic is an art or science of a good eating dish related to the enjoyment of eating and drinking. In other words, Fossali (as cited in Turgarini 2018) mentions gastronomy as a study of the relationship between culture and food that learns various cultural components with food at its center (Culinary arts). Cultural and gastronomic relations are formed since gastronomic is a product of cultivation in agricultural activities. Consequently, the food color, aroma, and flavor can be traced from the source of where the raw materials are produced. Therefore, it can be concluded that gastronomic is very much related to culture.

Hartono et al. (2019) explained that the ISOL learners have various reasons in learning Bahasa Indonesia. One of the reasons was the interest in the richness of Indonesian culture diversity. Indonesian cultural richness is in the form of material culture/physical and nonmaterial culture/non-physical culture. One form of Indonesian material/physical culture is in the form of a culinary tourism.

Culinary is one of the elements in cultural tourism that can demonstrate the diversity of culture. Due to the fact that culinary has a very complex cultural form including the idea (recipe innovation), action (the process of cooking), and the product (food and beverages), and in it also presents a variety of aesthetic expressions from the cooks (Suteja & Wahyuningsih 2018). Indonesian cuisine possesses a history and cultural tradition that is a representation of local culture, and for that reason, Indonesians need to introduce it to the international world. How to introduce this culinary can be done by integrating ISOL teaching materials with culinary content that has been standardized by The Indonesian Ministry of Education and Culture. As a result, the process of teaching and learning ISOL will be interesting and attracting foreigners to visit Indonesia for studying a language along with culinary cultures.

In the current modern era, an online platform like YouTube can be a media for foreigners, especially Koreans, who have a YouTube account with culinary content and Indonesian gastronomic content. The YouTube accounts like Hari Jisun, Bandung Oppa, Noona Rossa, Rosakis, Korea Reomit, and many more created contents experiencing the sensation of having Indonesian cuisines, both in Korea and in Indonesia. In addition, one YouTube channel called BNay Channel regularly cooks Indonesian cuisine to be introduced to his friends from various countries including Canada. All YouTube channels with those types of food vlogger gained many subscribers for about a hundred thousand subscribers, some of them even gained more than 1 million subscribers. ISOL teachers should welcome interest of foreigners to Indonesian cuisine because it can also be utilized as additional material for teaching materials.

Based on the aforementioned facts, the author will formulate Bahasa Indonesia teaching materials for the ISOL based on Sundanese gastronomy. The integrative Sundanese-gastronomy-based ISOL teaching materials developed as a result of the ISOL teaching materials shortage, especially related to the implementation of Sundanese gastronomy. Sundanese, together with its idiosyncrasy gastronomy, can be implemented as the framework of ISOL teaching materials for international students who want to learn Bahasa Indonesia that mainly focus on gastronomy. As a part of Indonesian diversity and also as researcher ethnic background, Sundanese gastronomy is one of the areas that can be incorporated into teaching materials. Integrating Bahasa Indonesia for Speakers of Other Languages teaching materials indicated the significant numbers of foreigners who became university students in Indonesia, specifically in West Java Province as the Sundanese epicentre. On the point of the Indonesian Ministry of Education and Culture explanation (2019) as cited in the List of Higher Education Institutions Organising of Darmasiswa Scholarship Program, West Java Province has 11 universities that provide scholarships for international students, including Universitas Pendidikan Indonesian, Universitas Padjadjaran, Universitas Pasundan, Institut Pertanian Bogor. The data shows that many international students are interested in studying in West Java rather than other cities in Indonesia. It is influenced by the fact that there are many foreigners, international tourists, or prospective trainees who want to come to Indonesia to study or to enjoy the diversity of culture and gastronomic in Indonesia.

2 METHOD

The research method used in this study is a descriptive study of qualitative research. The data were collected from relevant literatures: folk stories provided by informants, Internet-lore, films, and literary works by some modern Indonesian writers.

2.1 *ISOL teaching material*

Bistok (1994 p.17) suggests several theories regarding ISOL teaching materials.

- Any writing of ISOL learning materials must be based on a linguistic theoretical basis. The primary parameter of the linguistic theoretical is used as a guideline on the selection of materials such as in the selection of languages (standard, formal, informal) and medium selection (oral or written).
- Other parameters that need to be considered are ISOL learning materials that should be oriented not only on the language (linguistic case) but also on the principle of identity. Bahasa Indonesia should be seen as the subject. Besides, ISOL learning materials should pay attention to the level of cognitive development of the adherent and readability levels.

In addition, according to Widharyanto (2003), there are three issues that become polemics debate about teaching materials, namely:

- teaching materials strictly only created by and produced by ISOL teachers;
- teaching materials derived from the materials in the daily communication and modification as necessary by the teacher;
- the teaching materials that exist in the daily communication without any modifications.

Based on these issues, ISOL materials can be classified as teacher-made teaching materials, and adopted teaching materials and authentic teaching materials. The implementation of those type of teaching materials is adjusted to the level of learners. For the basic level of learners, the teachers can use their own teaching materials at the earliest lesson meetings.

2.2 *Intermediate level language proficiency rating based on Common European Framework of Reference for languages*

CEFR divide foreign language learning skills into three basic levels, namely, A, B, C. Furthermore, each level is divided into two more levels as A1, A2, B1, B2, C1, and C2. The following are the intermediate description and competency tables (B1 and B2) of the CEFR.

2.3 *The Webbed Integrative Model on ISOL teaching materials*

In this study, ISOL teaching materials developed by researchers implemented an integrative model of spider webs, well known as the Webbed Model. The implementation of the model is supported by Sugino (1996), who focused on adult learners who study Bahasa Indonesia as a foreign language.

Table 1. Levels and competencies description.

Level	Description
B1 Intermediate 1	a. Able to understand the main topics of a particular subject, such as work, education, and leisure, to which extent using a standard foreign language is easy to understand; b. Able to cope with almost all the linguistic issues when in foreign language speaking countries; c. Able to describe easily particular subjects or the likes with all related matters; d. Able to share the experiences, events, dreams, hopes, and life goals as well as able to provide a brief explanation of the specific plan/thinking.
B2 Intermediate 2	a. Able to understand the essence of complicated text, either with concrete or abstract themes and able to understand the discussion with certain subjects if appropriate with the field of competence; b. Able to spontaneously and smoothly communicate with Native-speakers to create a reasonable and seamless conversation is that meaningful in two ways communication; c. Able to express clear and detailed opinions on a variety of subjects, and also able to discuss one of the subjects of the actual theme, and can describe both the good and bad decisions are taken.

Sugino (1996) claimed that the communicative-integrative-thematic approach implementing the Webbed Model would provide more opportunities to develop potentials and language skills mastery, guided by the appropriate research methodologies.

According to the 10 models suggested by Fogarty (1991), the researcher selected the webbed model as the foundation of the development of Sundanese gastronomy-based ISOL teaching material. In a similar vein, Aisyah (2007) stated that the term spider web (webbed) is used due to the main design that formed a netlike or a weblike created by a spider, with the discussion theme being the center or as a spider. Based on the theme, sub-themes were determined to show more of the central theme by using aspects of basic skill development. The thematic approach is used by this webbed model to combine several subjects with one big theme covered to analyze the harmonization of concepts, topics, and ideas. This model has the characteristic to develop a curriculum that can be started with a single theme, in which the researcher in this research made the "Sundanese gastronomy" theme as the basis for developing teaching materials.

Fogarty (1991) described the strengths and weaknesses of the webbed model as follows. On the one hand, the strengths of this model are (1) selecting themes that are in line with the desires of learners and will encourage students to actively involved in learning; (2) can be easily implementing by a less-experienced educators/instructors; (3) helping to make it easier to plan teamwork for developing themes in all areas of teaching material; (4) encouraging learners to learn due to the thematic approach; (5) offering learners an easiness in different activities and related ideas within the learning. On the other hand, there are the drawbacks from this model, including (1) class conditions will be a little crowded and teachers will have difficulty dealing with students in the classroom; (2) in learning activities, learners focus more on learning activities rather than developing concepts; (3) requires synchronization between activities and developed teaching materials.

2.4 *The Webbed Integrative Model on ISOL teaching materials*

Turgarini (2018) claims that understanding on gastronomic is not only focused on culinary art or cooking, but also on human behavior, including how to choose raw materials, tasting, sensing, and serving foods; furthermore, experiencing food consumption, searching, studying, researching, and writing about food. Moreover, all matters related to human ethics, etiquette, and nutrition in every nation and country. Gastronomy is a form of art and science, and even an appreciation of the cross-culture, nation, race, group, religion, gender, and culture by studying in detail on eating, food, and drinking culture in a variety of conditions and situations.

The food as the primary source of gastronomic is everything considered biological sources, including agricultural products, plantations, forestry, fisheries, farms, irrigation, and water. Whether it is processed or untreated, that intended as food or beverage for human consumption, including food additives, food raw materials, and other materials used in the preparation, processing, and manufacturing of food or beverage. Thus, another dimension of gastronomic sciences is its nature that relates to food, food and the environment, both physical, biological, and cultural. For example, the cultivation of agriculture has a significant relation with the food aroma, taste, color, and even at the point of origin or location from the environment of raw materials, and the behavior of humans and nationality of the cooks (Soeroso 2014a, 2014b).

Sundanese gastronomy is part of the Sundanese culture, which apart from having a function (benefit) to fulfill basic human needs, for instance, to maintain the life that contains social behavior and norms found in society in general (Macionis & Gerber 2011). Sundanese gastronomy can also be categorized as an art and science related to the culture of eating good food from the Sundanese ethnic group and can be used as a social means to show the identity of the nation, region, and state. In the interest area of ISOL, cultural phenomenon, especially those related to Sundanese gastronomy, will be transmitted through ISOL teaching materials and are expected to provide new experiences in the form of the joy of traveling in Indonesia. The gastronomy components presented by Turgarini (2018) appear in the text of ISOL teaching materials by having the essences of Sundanese, and experts indeed have assessed the suitability. The researchers take the topic in the text of teaching materials from various references that refer to Sundanese gastronomic characteristics, such as Sundanese signature cuisines, Sundanese food ingredients, and Sundanese food philosophy.

3 RESEARCH METHOD

The method used in this research is R&D (Research and Development). This type of method is a used to produce a certain product and analyze the effectiveness of the results. In general, this research is conducted based on a model proposed by Dick and Carey (2009), with the following steps:

- The identification of learning objectives,
- The learning analysis,
- The learner and context analysis,
- The determination of learning objectives,
- The development of assessment instruments,
- The development of learning strategies,
- The development and selection of teaching materials,
- The design and implementation of formative evaluation,
- The revision,
- The design and implementation of summative evaluation.

Furthermore, the development teaching material stages from Dick and Carey (2009) are divided into four steps as follows:

- Needs Analysis Stage,
- Product Design Stage,
- Validation and Evaluation Stage,
- Final Product Stage.

4 RESULT AND DISCUSSION

Based on the results of theoretical studies and assessment of research instruments, the research conducted by 10 experts assessed teaching materials that have previously received a 100% score and undergone two assessments. Thenceforth, teaching materials assessment assisted by ISOL teachers, distributing questionnaires to students in several institutions, as well as interviews with intermediate level ISOL teachers. Ultimately, the researcher obtained a formulation regarding the design of ISOL teaching materials based on Sundanese gastronomy intended for intermediate level, the formulation which will be presented in the next paragraph.

Overall, the ISOL teaching materials developed by researchers have met the eligibility. This can be proven from the summative evaluation given by the experts in the assessment of draft 1, which is 96% that is categorised as very good. However, the materials developed still need improvement, hence, the summative evaluation stage two improved to 100%, which means also categorized as very good. In addition, researchers categorized feasible based on several aspects of consideration, namely:

- content eligibility

This aspect of content eligibility consists of several criteria consisting of the following criteria:

- the suitability of the material with the curriculum,
- the accuracy of the material, also
- supporting ISOL teaching materials based on Sundanese gastronomy text.

The researchers adapted all the feasibility of the content for the intermediate level learners from the intermediate level ISOL curriculum. Thus, the indicators that must appear in intermediate-level ISOL teaching materials are that students must be able to do the following objectives:

- Able to express experiences, hopes, goals, and plans simply and continuously in the form of a simple narrative or description with arguments in the daily work activities and tasks context
- Able to make observation report on events and express ideas on topics according to their scientific background, both concrete and abstract spontaneously and fluently, communicate without any significant obstacles that can damage the understanding of the interlocutor

Concurrently, to fulfill the gastronomy elements material which in this study was narrowed down based on the Sundanese gastronomic text, the researcher had aligned all of the teaching materials according to gastronomic components according to the theory from Turgarini (2018). Researchers developed nine components of the gastronomy by distributing them into nine different topics. From these components or gastronomic elements, the researchers made nine topics of teaching materials, which can be described as follows:

- Topic 1 *Geco*: Authentic Sundanese Cuisine
- Topic 2 *Oncom*: Sundanese Culinary Ingredients
- Topic 3 *Ngariung*: How to Taste the Culinary in Sundanese Culture
- Topic 4 How to Serve Typical Sundanese Food
- Topic 5 Knowing Unique Sundanese Culinary Abbreviations
- Topic 6 Unique Dining Experiences at Bancakan Restaurant, Bandung
- Topic 7 Nutritional Content of *Karedok*
- Topic 8 Philosophy and History of *Nasi Tutug Oncom*
- Topic 9 Eating Ethics and Etiquette in Sundanese Culture

From the revised teaching materials assessment, all experts in their respective fields gave a formative evaluation for draft 1 of 97% accuracy, which is categorized as very good. However, due to the improvement needs, 100% accuracy is acquired, which is an improved category in teaching materials that have been revised. For this reason, the teaching materials compiled by researchers have accurately contained material according to Sundanese gastronomic elements and still implement intermediate-level ISOL materials. Besides, the teachers being interviewed agreed that the material in the teaching materials developed by researchers was appropriate to be given to learners. The result of the ISOL teachers' interview was discovering that they averagely agree that this teaching material is suitable for learners in an intermediate level. They suggested that all teaching materials are necessary to be given to intermediate level ISOL students, especially materials related to improving the four language skills of the learners. In another interview, students also agree that this teaching material can be used as intermediate-level ISOL teaching materials, due to the more than 80% answers that this teaching material can practice their four basic language skills.

– *Presentation*

The aspect of presenting this teaching material is developed by considering several things, which consist of the following criteria:

- the objectives/indicators of presenting learning are clearly stated,
  - visual presentation of ISOL teaching materials with an integrative model based on Sundanese gastronomy,
  - presentation of teaching materials encourages active and creative learners.

Based on the consideration of 10 experts of the assessment, formative evaluation for the presentation aspect according to these criteria is given a 97% assessment, which means very good, but still needs improvement in the assessment of draft 1. Simultaneously, the value changes to 100%, which means very good in the next assessment after the teaching materials being revised.

– *Readibility*

The readibility aspect of the teaching materials used refers to the following three dimensions:

- the linguistic dimension,
- topographic dimensions, also
- layout dimensions.

Based on these dimensions, the formative evaluation of ISOL teaching materials developed by researchers was given a score of 96% by 10 experts, which means very good, but again still needs improvement. However, after going through the improvement process, the value for this readability aspect improved to 100%, which means very good.

– SARAG

SARAG aspect referred to *suku, ras, agama dan gender* (ethnic, race, religion, and gender) background. The formative evaluation obtained by researchers from 10 experts on draft 1 was 96%, which means it is very good and needs improvement. However, after undergoing a process of improvement, the value given by the expert is 100%, which means very good.

In addition, there are several other things that the researchers consider as additional information for the intermediate ISOL teaching materials design implementing integrative model based on Sundanese gastronomic text. Bahasa Indonesia must be seen as a correlated unit as a system. Thus, grammar teaching materials are integrated with other aspects of teaching materials as well as the writing system (spelling). Aspects of learning spoken language (listening and speaking) and aspects of learning written language (reading and writing) are also carried out in an integrated manner. Reading skills are always used as the beginning of learning because reading skills are considered as an opener to find out topics that will be developed. All this matters because the topic of Sundanese gastronomy may be considered as new by intermediate ISOL students, both studying in Indonesia and other countries. Besides, reading skills are considered as a gateway to knowledge about Indonesia as well as being able to introduce the character and identity of Indonesia as a nation.

Based on the results of this research, covering expert assessments, interviews with ISOL teachers and students, and the style of composing intermediate-level ISOL teaching materials through an integrative model based on Sundanese gastronomy, the materials are considered as appropriate for ISOL students in intermediate level. There are only a small amount of revisions to each section that are considered very difficult for intermediate level ISOL learners.

The design of ISOL teaching materials using an integrative model based on Sundanese gastronomic texts developed by researchers can be seen in the following figure.

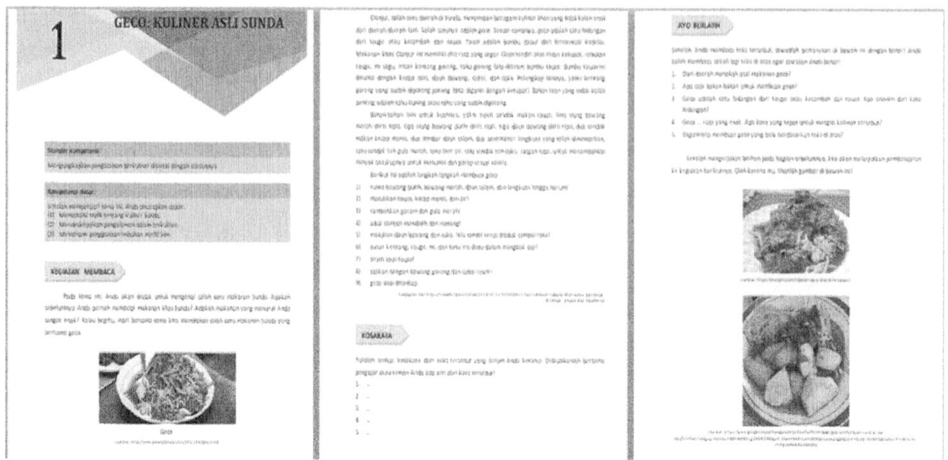

Figure 1. *Layout arrangement of integrative ISOL teaching materials based on Sundanese gastronomy text for intermediate learners.*

5 CONCLUSION AND RECOMMENDATION

Based on the need analysis of teaching materials and reality check results, the ISOL teaching materials based on the Sundanese gastronomic text for intermediate level discovered that there was a gap between the desires of ISOL students and teachers and the real classroom conditions. This gap can be seen in the availability of syllabus and teaching materials that do not match the needs and the curriculum of intermediate level ISOL. Therefore, a syllabus, materials, and teaching materials are required to fulfill the learner's curiosity about Indonesian gastronomy, especially Sundanese gastronomy.

In order to achieve the identifiable needs, researchers developed a syllabus and teaching material models according to the needs of Sundanese gastronomy-based learners. Syllabus was developed from the intermediate level ISOL curriculum. All topics developed in this teaching material are based on Sundanese gastronomy. The Sundanese gastronomy theme is used as a guideline for learners' language competence, which consists of four language skills and grammar aspects. Concurrently, the integrative model used is the webbed model. This type of integrative model is highly suitable for teaching materials with a thematic approach, in line with the big theme in this teaching material, specifically Sundanese gastronomy. Besides, this integrative model also combines studies in the fields of culture, sociology, history, and tourism.

During *the product design stage*, topics were selected for teaching materials based on Sundanese gastronomic components with competency standards for intermediate level ISOL students, covering (1) Geco: Original Sundanese Culinary, (2) Oncom: Sundanese Culinary Raw Materials, (3) Ngariung: How to Taste Culinary in Sundanese Culture, (4) How to Serve Sundanese Specialties, (5) Get to Know Unique Sundanese Culinary Abbreviations, (6) Unique Dining Experiences at Bancakan Restaurant, Bandung, (7) Karedok Nutritional Content, (8) Philosophy and the History of Nasi Tutug Oncom, (9) Ethics and Eating Etiquette in Sundanese Culture. All of these topics are equipped with materials, vocabulary mastery, language knowledge, exercises, and final projects that are suitable for intermediate-level ISOL learners and that integrate four language skills, the fields of culture, history, sociology, and tourism.

In the evaluation and validation stage, information was obtained based on the consideration of 10 experts' assessment of teaching materials that had to receive a 100% score after conducting two assessments. Overall, the integrative ISOL teaching materials based on Sundanese gastronomy text developed by this researcher have met the feasibility. These findings can be proven from the summative evaluation given by the experts in the assessment of draft 1 achieved 96%, while the summative evaluation stage two changes conducted to achieve 100%. It means that there is a 4% increase from draft 1 to the revised draft. Besides, because the development of this teaching material is accomplished with an integrative model, it means that all language skills are presented in an integrated manner, starting with reading language skills, then simultaneously followed by other skills.

The research is still categorized as an early stage in the preparation of Bahasa Indonesia materials for intermediate-level speakers of other languages through the Sundanese gastronomic-based integrative model. Therefore, it is essential to do advanced research that can complement the weakness of the results of this research. In addition to compiling intermediate ISOL teaching materials through this Sundanese gastronomic-based integrative model, it is recommended to be prepared and develop teaching materials for basic as well as advanced levels. The teaching materials for various levels of learners are necessary for the development of the ISOL program, especially for learning in the Catering Industry Management Study Program, which generally can be reused as supplements for other teaching and learning processes in ISOL.

REFERENCES

Aisyah, Siti. 2007. *Pembelajaran Terpadu*. Jakarta: Open University.

A. S., Bistok. 1994. "Beberapa Parameter dalam Pengembangan Bahan Pelajaran Bahasa Indonesia untuk Penutur Asing (ISOL): Suatu Kajian Buku-buku Pelajaran ISOL yang Digunakan di Australia, Amerika dan Eropa", in Satya Wacana Christian University. 1994, KIPISOL. Salatiga: Satya Wacana Christian University.

Common European Framework of Reference for Languages. [Online]. Retreived from: https://en.wikipedia.org/wiki/Common_European_Framework_of_Reference_for_Languages [20 Juni 2020]

Dick, Walter, Lou Carey, and James O. Carey. (2009). *The Systematic Design of Instruction*. New Jersey: Pearson.

Fogarty, Robin. 1991. *How to Integrated the Curricula*. Palatine, Ilinois: IRI/ Skylight Publishing, Inc.

Hartono, Didik, et al. 2019. "Wisata Kuliner Nusantara: Diplomasi Budaya Melalui Standardisasi Materi Ajar ISOL di Era Milenial" in Prosiding Konferensi Internasional Pengajaran Bahasa Indonesia bagi Penutur Asing (KIPBIPA) XI 2019. [Online]. Retreived from: http://kipbipa.appbipa.or.id/?cat=-1 [15 Februari 2020]

Hasan, Rizki Akbar. 2019. "10 Bahasa dengan Penutur Terbanyak di Dunia, Indonesia Urutan Berapa?" [Online]. Retreived from: https://www.liputan6.com/global/read/4063214/10-bahasa-dengan-penutur-terbanyak-di-dunia-indonesia-urutan-berapa [21 Februari 2020]

Kemdikbud. 2017. "Kemendikbud Dorong Alumni Darmasiswa Jadi Duta Indonesia". [Online]. Retreived from: https://www.kemdikbud.go.id/main/blog/2017/05/kemendikbud-dorong-alumni-darmasiswa-jadi-duta-indonesia [17 Januari 2020]

Kemdikbud. 2017. "579 Peserta Program Darmasiswa Ikuti Lokakarya Pengenalan Budaya Indonesia". [Online]. Retreived from: https://www.kemdikbud.go.id/main/blog/2019/09/579-peserta-program-darmasiswa-ikuti-lokakarya-pengenalan-budaya-indonesia [17 Januari 2020]

Kemdikbud. 2019. "*List of University Organizing of Darmasiswa Scholarship Program Academic Years 2020/2021*". [Online]. Retreived from: https://darmasiswa.kemdikbud.go.id/list-of-darmasiswa-university/ [21 Januari 2019]

Macionis, J.J. and L.M. Gerber. 2011. *Sociology*. Toronto, Canada: Pearson Prentice Hall.

Muliastuti, Liliana. 2015. Model Materi Ajar Bahasa Indonesia bagi Penutur Asing Berbasis Common European Framework of Reference for Languages (CEFR) dan Pendekatan Integratif: Sebuah Studi Pengembangan pada Program BIPA-UNJ. Dissertation at Doctoral Program Universitas Negeri Jakarta: Unpublished.

Rizkyanfi, Mochamad Whilky. 2010. Model Integratif Bahan Ajar Bahasa Indonesia bagi Penutur Asing (BIPA) Tingkat Menengah. Thesis at Master program of Indonesia University of Education: Unpublished.

Santich, B. 1996. *Looking for Glavour*. Kent Town, Sout Australia: Wakefield Press.

Soeroso. 2014a. Foodscape, Cultural Landscape, dan Arkeologi: Sebuah Upaya Pelestarian Cagar Budaya dan Pembangunan Ekonomi Indonesia. Paper at the IAAI Congress (Ikatan Ahli Arkeologi Indonesia) on June 24, 2014 at Fort Vredeburg, Yogyakarta.

Soeroso. 2014b. Quo Vadis Gastronomi Indonesia. Paper as an introduction to the FGD of the Indonesian Gastronomy Academy on April 25, 2014 at Founding Father's House, South Jakarta.

Sugino, S. 1996. "Pendekatan Komunikatif-Integratif-Tematis dalam Pengembangan Bahan dan Metodologi Pengajaran BIPA di Indonesia", in The Teaching of Bahasa Indonesia for Speakers of Other Languages proceding. Depok: Literature Faculty of Indonesia University.

Sugono, Dendy. 2001. *Kebijakan Umum Pengajaran Bahasa Indonesia untuk Penutur Asing*. [Online]. Retrieved from: http://www.ialf.edu/kipbipa/papers/DendySugono.doc [30 April 2010].

Suteja, I Wayan dan Sri Wahyuningsih. (2018). "Potensi Kuliner Lokal dalam Menunjang *Cullinary Tourism* di Kawasan Ekonomi Khusus Mandalika Kabupaten Lombok Tengah" in *Media Bina Ilmiah Journal*, Vol.12, No.11 Juli 2018. [Online]. Retrieved from: http://webcache.googleusercontent.com/search?q=cache:aaSQWoCvjxwJ:ejurnal.binawakya.or.id/index.php/MBI/article/download/125/pdf+&cd=4&hl=id&ct=clnk&gl=id [5 Februari 2020]

Turgarini, Dewi. 2018. Gastronomi Sunda sebagai Daya Tarik Wisata Kota Bandung. Dissertation in Doctoral Program at Pascasarjana Universitas Gadjah Mada: Unpublished.

Widharyanto, B. 2003. "Dimensi Autensitas dalam Pembelajaran BIPA" in *Prosiding Konferensi Internasional Pengajaran Bahasa Indonesia bagi Penutur Asing (KIPBIPA) IV*. Denpasar: IALF Bali.

Promoting Creative Tourism: Current Issues in Tourism Research – Kusumah et al. (Eds)
© 2021 Taylor & Francis Group, London, ISBN 978-0-367-55862-8

CIPP model: Curriculum evaluation of the Indonesian gastronomy courses

W. Priantini, I. Abdulhak, D. Wahyudin & A.H.G. Kusumah
Universitas Pendidikan Indonesia, Bandung, Indonesia

ABSTRACT: The diversity of Indonesian gastronomy is an attraction for tourists. Indonesian gastronomy course is an effort to study, record, and document gastronomy to preserve Indonesia's intangible heritage. Indonesian gastronomy was explored from three main aspects: history, culture, and food itself. This study aims to evaluate the Indonesian gastronomy curriculum to ensure that the curriculum is designed and developed according to the needs of the community and science development. The purpose of evaluating Indonesian gastronomic curriculum is to find out the effectiveness and efficiency of the curriculum system. This study uses a qualitative approach, by using the CIPP (Context, Input, Process and Product) curriculum evaluation model. The data were collected from observation, interview, and literature study. The CIPP model was used to measure each component of the curriculum: objectives, content or material, methods and evaluation. The context evaluation under study was lecture facilities and lecturers. Input evaluation is related to plans and strategies in achieving lecturing goals. Process evaluation is related to the implementation of lectures; lecturer skills in managing and method of lectures. Product evaluation is related to the assessment of student learning processes and outcomes. The results of the curriculum evaluation become the basis for reconstructing the curriculum of the Indonesian gastronomy course, in accordance with the established curriculum component.

Keywords: curriculum evaluation, CIPP, Indonesian gastronomy

1 INTRODUCTION

Indonesia has a rich culture and a natural beauty, manifested in tourist attractions. One of the cultural riches is the diversity of traditional foods, which have delicious taste, unique appearance, and the story behind the processing. Tourism is a potential realm for theorizing broader issues of culture and taste (Stringfellow et al. 2013). Traditional food is one of the advantages in attracting tourists. Studies show that there is a correlation between the increased levels of food tourism interest and the retention and development of regional identity, the enhancement of environmental awareness and sustainability, an increase in social and cultural diversity celebrating the production of local food, and the conservation of traditional heritage, skills, and ways of life (Everett & Aitchison 2010).

Gastronomy is a tourist attraction. Along the same line, food history, the influence of acculturation of several cultures and philosophical values of food offerings are subject to study. It is therefore appropriate that hospitality education and training include a gastronomy component in order to give students a greater understanding of the history and culture of food and beverage (Santich 2004). Learning Indonesian gastronomy is not just learning to process food. There is other knowledge that must be learned such as the history, knowledge of food ingredients, and cultural influences on these foods. Food symbolizes the values of local wisdom of an area.

At present, the Catering Industry Management Program teaches the Indonesian Gastronomy course at the level of processing practice. The deepening of historical and cultural aspects has not yet been fully carried out. Indonesian gastronomy is seen from three aspects: food, historical, and culture (UNWTO 2017). Promoting Indonesian culinary tourism should not be merely exposing the

DOI 10.1201/9781003095484-51

sample varieties of the traditional food that Indonesia has, but more importantly, telling the tourists the stories about the socio-cultural values behind the food itself (Wijaya 2019). Courses related to the Indonesian gastronomy must accommodate this effort. In order to respond to current trends in tourism, it is important that hospitality education include a significant and relevant gastronomy component, in addition to practical and business or management courses, so that students develop an understanding of the history and culture of food and drink, and in particular, the history, culture, and traditions of the products of their particular region or country. This might take the form of a dedicated course or program of study, or could be integrated into existing syllabuses (Santich 2004).

Therefore, it is necessary to evaluate the curriculum related to program improvement and ongoing processes. Curriculum evaluation is the process of delineating, obtaining, and providing useful information in making curriculum decisions and judgments (Print 1993). Evaluation work is to provide various information regarding the activities carried out in the process of curriculum development, curriculum implementation, and curriculum evaluation (Hasan 2014). The curriculum evaluation process is to determine the level of success of the curriculum (Rusman 2012). The curriculum evaluation model consists of several types. One of them is the CIPP Model: Context, Input, Process, and Product. The selection of the CIPP model can evaluate all curriculum components, which consist of objectives, content or material, methods and evaluation. The study program must prepare a curriculum that contains a number of topics leading to the preservation of traditional food, industry needs, and development of science, in this case, the existence of gastronomic tourism as one of the attractions of cultural tourism. Evaluating the quality of hospitality and tourism education program accurately has become increasingly important for the growth of hospitality and tourism industry (Shen et al., 2015).

2 METHOD

2.1 *Research approach*

This study evaluated the implementation of curriculum using a qualitative approach. The data collection was carried out through in-depth interviews with people who had links with institutions, documents, and field observations. The program being evaluated is the implementation of the curriculum for the Indonesian Gastronomy Course. The curriculum evaluation approach was carried out by internal evaluation. The evaluator recognizes the characteristics to be evaluated. Therefore, when the evaluator formulates evaluation questions, he can determine which questions have a high priority and high level of relevance to the observed characteristics. Evaluation questions that are formulated with such knowledge will be very useful for the subsequent evaluation process, and the results of the evaluation will provide higher effectiveness (Hasan 2014). According to Hasan (2014), there are four main points of curriculum evaluation with qualitative procedures: (1) determining the focus of the evaluation; (2) problem formulation and data collection; (3) data processing; and (4) determining program improvements and changes.

2.2 *Participants and research location*

Participants involved in this research are people who are directly involved in the research. Participants include lecturers, students, the Ministry of Tourism, Ministry of Education and Culture, local government, local community leaders who understand traditional food, and local culinary entrepreneurs. Participants in this study consisted of lecturers and students of the Catering Industry Management Study Program, Universitas Pendidikan Indonesia.

2.3 *Research instruments*

In this study, researchers used interviews and observations. Interviews were conducted with direct question and answer, between researchers and data sources. Participants interviewed included lecturers from the Indonesian gastronomy and curriculum experts. The data in question included curriculum development procedures, objectives, content or material, methods and evaluation of the gastronomic lectures of the archipelago. In collecting the data, the researchers made observations to obtain information about such aspects as the lecture environment, the condition of the community, lecturers, lecture processes and facilities, local culinary tourism destinations. The data needed includes a curriculum component consisting of objectives, content or material, methods and evaluation of lectures.

Observation allows evaluators to capture the atmosphere that occurs directly in the observed process. Observations were made during the course of the Indonesian Gastronomy lecture. The researcher observed the implementation of the Indonesian Gastronomy lecture from the aspect of the lecture's objectives: material, methods and evaluations carried out just before, during, and after the lecture. Researchers do their own data retrieval so that the data collected is in accordance with required facts in the field.

3 FINDINGS AND DISCUSSION

The curriculum has a central position in the entire educational process as the curriculum directs all forms of educational activities in order to achieve the goals of education (Sukmadinata 2016). The curriculum is dynamic and always developing. Therefore, curriculum changes become common. According to Melrose (1998), the critical paradigm of curriculum evaluation is probably a reaction to earlier paradigms where evaluation was carried out by "expert" educational specialists and evaluators and not by teachers. The purpose of curriculum evaluation according to Zainal (2017) is to find out the effectiveness and efficiency of the curriculum system, both concerning the objectives, content/material, strategy, media, learning resources, environment, and the assessment system itself. Curriculum evaluation becomes a recommendation for improving education quality. According to Hasan (2014), curriculum evaluation is a public policy. Therefore, it must be accountable and prioritize the community's interest, wereas the evaluation according to Ornstein and Hunkins (2014) is a process whereby people gather data in order to make decisions. Evaluating the quality of hospitality and tourism education program accurately has become increasingly important for the growth of the hospitality and tourism industry (Shen et al., 2015). Evaluation is to provide various information regarding activities carried out in the process of curriculum development, curriculum implementation, and curriculum evaluation (Hasan 2014).

3.1 *Higher education curriculum*

Higher education curriculum is a program to produce university graduates. The curriculum should ensure that graduates have qualifications that are equivalent to the qualifications approved in the Indonesian National Qualification Framework. The formulation of capabilities described in Indonesian National Qualification Framework is referred to as learning outcomes, where competencies are included or are part of learning outcomes. Indonesian Gastronomy curriculum is arranged in the following sequence: formulate course learning outcomes; develop semester learning plans; determine the learning process and learning assessment. The developed curriculum tools consist of semester learning plans, lecture program units, knowledge assessment format, skills and attitudes.

From The Guidelines of Higher Education Curriculum Development by the Ministry of Research, Technology and Higher Education (Kemenristek 2018), the determination of learning outcomes is the basis for determining courses. Learning outcomes are usually based on the level of education. Learning outcomes for the Indonesian gastronomy course are still imprecise. What competency will be achieved after the student follows gastronomic recovery is one amongst the uncertainty. Knowledge and skills that students achieve should be in accordance with the needs of the industrial world and keep up-to-date. During this time, the material delivered is still limited to cooking traditional food from several regions. Meanwhile, studying gastronomy isn't just about cooking.

In the UNWTO report, Indonesian gastronomy can be studied from three aspects: food, history, and culture. Much can be extracted from the peculiarities of food from an area in Indonesia. This is a consideration for study programs to reorganize the gastronomic curriculum of the archipelago. The Indonesian gastronomy course is expected to be able to record and document regional food. Thus, it will be one of the efforts to preserve traditional food. It is necessary to evaluate the gastronomic curriculum to be better, in line with the world of work and the development of science.

3.2 *CIPP model curriculum evaluation*

The researcher chooses the CIPP model with the consideration that: (1) the CIPP model can use qualitative and quantitative approaches, so that the data obtained can balance each other; (2) the CIPP model can be done separately. However, it would be better if the study were conducted simultaneously to place the four focuses of evaluations; (3) it can be used to measure each component of the curriculum, namely, objectives, content or material, methods and evaluation; and (4) it is comprehensive and involves all parties involved in the study, namely, lecturers, students, the community.

The curriculum is formed by four components: objectives, curriculum content, methods or strategies, and an evaluation component. As a system, each component must be related to one another. In developing, implementing, and evaluating the curriculum, it must refer to curriculum components. The research framework for the CIPP Model evaluation of the Indonesian gastronomic curriculum can be seen in Figure 1.

Context evaluation is an evaluation of the environment in which the curriculum was developed and will be implemented. In this case, evaluation of facilities, working conditions, number of teachers, teaching tools, physical condition, and learning resources (Hasan 2014). In this study, the context evaluation under study was lecture facilities and lecturers supporting the Indonesian Gastronomy course.

Input evaluation is related to plans and strategies for achieving goals. In this case, the input is translated as a lecture plan. The lecture plan is a curriculum document that contains the objectives, content, learning process, and assessment of learning outcomes (Hasan 2014). Evaluation of documents is in the form of Semester Learning Plans in accordance with The Guidelines of Higher Education Curriculum Development (Kemenristekdikti 2018). Semester Learning Plans contain study program identity; subject identity; course learning outcomes; study materials; method; time; learning; job description; assessment and reference criteria, and indicators. The assessment of student learning processes and outcomes include assessment principles; assessment techniques and instruments; assessment mechanisms and procedures; assessment; appraisal reporting; and student

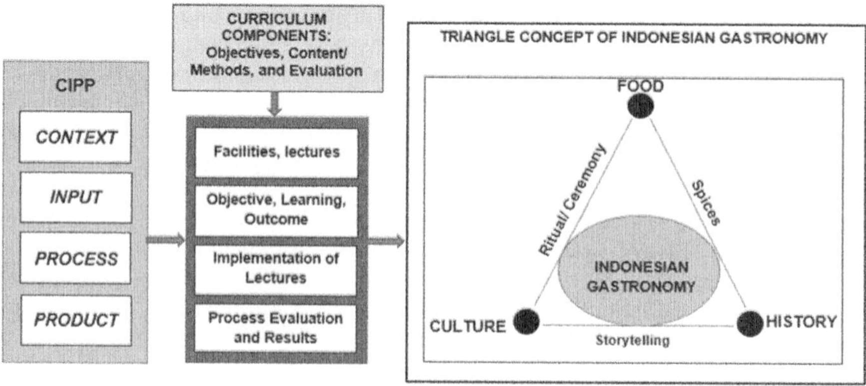

Figure 1. CIPP model: curriculum evaluation of Indonesian gastronomy.

graduation. Assessment should be able to reach important indicators related to honesty, discipline, communication, decisiveness, and confidence that students must have.

Process evaluation is related to the implementation of lectures that have been outlined in the Semester Learning Plans and interaction and communication among lecturers, students, and learning resources. Factors affect learning such as class atmosphere, number of students, learning facilities, schedules, lecturers' workload, and students' workload (Hasan 2014). The ability of lecturers in managing lectures and other factors can hinder the implementation of the learning process.

Product evaluation is related to learning outcomes from cognitive, affective, and psychomotor aspects. Cognitive aspects relate to the ability to think in application, analysis, synthesis, and evaluation. Affective aspects with reference to the behavior of values, attitudes, morals, conscience. Meanwhile, the psychomotor aspects of body movement are combined with cognitive and affective abilities (Hasan 2014). The assessment of student learning processes and outcomes includes assessment principles; assessment techniques and instruments; assessment mechanisms and procedures; conducting assessment; appraisal reporting; and student graduation. The assessment should be able to reach important indicators related to honesty, discipline, communication, decisiveness, and confidence that students must have.

3.3 *Indonesian gastronomy*

The UNWTO world tourism organization from the United Nations revealed that Indonesia has become a gastronomic tourism development. Triangle Concept is a philosophy of pivoting gastronomy on three triangular furnaces; food, culture, and history. This concept explains that between the three there is an inseparable relationship with each other (Setyanti 2017). Imelda (2015) stated that there are at least two reasons, namely, cultural heritage and identity. Indonesia's traditional food heritage contains elements of local wisdom. Conservation can also underlie the development of Indonesia's food globalization program.

In relation to the gastronomic curriculum, the triangle concept becomes one of the considerations in compiling the curriculum. Food and history are connected by the presence of spices. It is undeniable that Indonesia's history in the past is known as a world-famous spice-producing country. The diversity of this spice if used properly will produce a variety of unique foods to Indonesia. Meanwhile, history and culture are connected by a story or folklore, whereas culture and food are associated with various traditional ceremonies held by each area in Indonesia (Setyanti 2017). According to Grew (2018), the historian finds food is tied to economics, technology, commerce, and religion. It particularly satisfies evidence of how ordinary, daily activities are related to larger historical trends.

Studying food, how it is processed, starting from the preparation of raw materials, the tools used, how to cook, and serving methods are the main activities in gastronomic curriculum. Cuisine are the rules which are brought together by ingredients, spices, cooking techniques, and presentation (Brolatte & Giovine 2016).

History is related to the story that accompanies how the traditional food is processed. It covers such aspects as discussing the history of how the food initially became a regional dish, the raw materials obtained, herbs, spices, cooking processes, time spent processing food, and even some foods having rituals in the processing. Many things become storytelling in a traditional Indonesian food.

Culture is related to the influence of several regions or countries on traditional food. Cultural acculturation, geographical location, and ethnicity enrich the characteristics of local food treasures. Culture enriches food, history, and the value of local wisdom. Indonesia boasts a long history with diverse influences from different cultures. This diversity has brought significant influences that have supported the establishment of various unique exotic cuisines in the country (Wijaya 2019). The triangle concept of Indonesian gastronomy can be seen in Figure 2.

The tourists' interest will increase with the growth of culinary tourism destinations from an area or country. Indonesian government has given a great support for the development of culinary tourism

Figure 2. Triangle concept of Indonesian gastronomy (UNWTO survey on gastronomy tourism 2016).

as one special interest in the tourism sector that is promoted extensively to the international market. Promoting Indonesian culinary tourism should not be merely exposing the sample varieties of the traditional food that Indonesia has, but more importantly, telling the market about the socio-cultural values behind the food itself (Wijaya 2019). The role of universities is very large to develop, study, collect, and research everything related to traditional food. Higher education is also required to be able to accommodate the development of science, including the diversity of traditional Indonesian food.

4 CONCLUSIONS

The CIPP model is a tool for evaluating the Indonesian gastronomy course. The use of this model can evaluate all curriculum components. The context evaluation under the study was lecture facilities and supporting lecturers. Inputs is related to plans and strategies in achieving lecturing goals. Process evaluation is related to the implementation of lectures; lecturer skills in managing lectures, choosing the form and method of lectures. Product evaluation is related to the assessment of student learning processes and outcomes. The development of the Indonesian gastronomy curriculum used the triangle concept of Indonesian Gastronomy: food, history, and culture.

REFERENCES

Al-Shanawani, H.M. 2019. *Evaluation of Self-Learning Curriculum for Kindergarten Using Stufflebeam's CIPP Model*, SAGE Open January-March 2019: 1–13, DOI: 10.1177/2158244018822380journals.sagepub.com/home/sgo

Brolatte, R.R. & Giovine, M.A.D. 2016. *Edible Identities: Food as Culture Heritage*, Rouledge NY

Everett, S dan Aitchison C. 2010. The Role of Food Tourism in Sustaining Regional Identity: A Case Study of Cornwall, South West England. *Journal of Sustainable Tourism*, 16: 2, 150–167, doi: 10.2167/jost696.0

Grew, R. 2018. *Food in Global History*, Routledge, NY

Hasan, S.H. 2014. *Evaluasi Kurikulum*. Bandung: PT Remaja Rosdakarya.

Imelda, J. 2015. *Dialog Gastronomi Nusantara*. Akademi Gastronomi Nusantara

Kemenristekdikti, 2018. Panduan Penyusunan Kurkulum Perguruan Tinggi di Era 4.0, Kementerian Riset Teknologi dan Pendidikan Tinggi, Jakarta

Melrose, M. 1988. *Quality in Higher Education* Volume 4, 1998 – Issue 1, Exploring Paradigms of Curriculum Evaluation and Concepts of Quality https://doi.org/10.1080/1353832980040105

Ornstein, A.C. & Hunkins F.P. 2014. *Curriculum: Foundations, Principles, and Issues*. England: Pearson Education Limited.

Palupi, S. & dan Abdilah, F., 2020. *Local Cuisine as a Tourism Signature: Indonesian Culinary and Tourism Ecosystem*, Delivering Tourism Intelegent, Emerald UK

Print, M. 1993. *Curriculum Developmen and Design*, Allen & Unwin, NSW Publisher
Rusman. 2012. *Manajemen Kurikulum*. Jakarta: PT Rajagrafindo Persada
Santich, B. 2004. The Study Of Gastronomy And Its Relevance To Hospitality Education And Training, Hospitality Management 23, 15–24, https://doi.org/10.1016/S0278-4319(03)00069–0 accessed 2/9/2019
Setyanti, C.A. 2017. *Tiga Kunci Populerkan Wisata Kuliner Indonesia ke Dunia*. https://www.cnnindonesia.com/gaya-hidup/20170520 004113-262-216059/tiga-kunci-populerkan-wisata-kuliner-indonesia-ke-dunia
Shen, H., Luo J.M., & Lam, C.F. 2015. *International Journal Of Hospitality Management* 34(1):92–98. September 2013, The Impact Of Urbanization On Hotel Development: Evidence From Guangdong Province In China, Doi: 10.1016/J.Ijhm.2013.02.013
Stringfellow, L. et al. 2013, Conceptualizing taste: Food, culture and celebrities, *Tourism Management* – volume 37 August 2013, Pages 77–85, https://doi.org/10.1016/j.tourman.2012.12.016, accessed 23 Juni 2020
Sukmadinata, N.S. (2016). *Pengembangan Kurikulum: Teori dan Praktik*. Bandung: PT Remaja Rosdakarya.
UNWTO, 2017. Second Global Report on Gastronomy Tourism, Published by the World Tourism Organization (UNWTO), Madrid, Spain
Wijaya, S. 2019. *Journal of Ethnic Foods*, Indonesian Food Culture Mapping: A Starter Contribution To Promote Indonesian Culinary Tourism, https://doi.org/10.1186/s42779-019-0009-3
Zainal, A. (2017). *Konsep dan Model Pengembangan Kurikulum*. Bandung: PT Remaja Rosda

Promoting Creative Tourism: Current Issues in Tourism Research – Kusumah et al. (Eds)
 ISBN 978-0-367-55862-8

Situation analysis of tourism education in the city of Bandung

A. Suwandi, E. Fitriyani, N. Fajria & S.R.P. Wulung
Universitas Pendidikan Indonesia, Bandung, Indonesia

ABSTRACT: The implementation of tourism education is one of the factors that can affect the sustainability of a region's tourism. The success of a tourism destination depends on the quality of service and experience offered by the tourism workforce to tourists. Bandung as a tourism destination and center of higher education must be able to answer the needs of tourism human resources through the implementation of comprehensive tourism education. The purpose of this study is to analyze the situation of organizing tourism in the city of Bandung. The research method uses a qualitative approach with secondary data collection techniques such as articles, journals, reports, website pages of educational institutions, and policy documents related to tourism and human resource issues. Also, data collection in the form of a tourism education curriculum becomes a source of data in primary data collection techniques. The analytical method uses content analysis, exploration, and qualitative descriptive. The findings are expected to be able to answer the objectives of this study, which include identification of the distribution of tourism education programs and the existing conditions of affordability, accessibility, and accountability of tourism education programs in the city of Bandung.

1 INTRODUCTION

Stimulating changes in sustainable development, especially economies that contribute to job creation. The tourism industry demands professionals, workers, and employers to be responsive in managerial, technical, and vocational skills (Baum 2015). On the other hand, tourism education is considered as an applied discipline of higher education, which aims to develop the potential of individuals as leaders with critical thinking and accompanied by managerial skills relevant to industry needs. Sheldon et al. (2011) explained the five main values in the implementation of tourism education, which include ethical values, cooperation, managerial, insight, and professionalism. The application of these values in the tourism education curriculum can improve the quality of tourism human resources (Thapa & Panta 2019).

The implementation of tourism education is currently influenced by various stakeholders with different priorities and interests. It affects tourism education providers that struggle with each other and compete to produce graduates who are responsive to changing industrial needs. One of the external factors that can influence tourism education is the increase in tourist arrivals in the future. This is a trigger to produce skilled human resources and response to changes in managing and maintaining the stability of tourist visits.

Bandung city is one of the tourism destinations in West Java and is well known as the city of education that must be able to answer the needs of tourism human resources through the implementation of comprehensive tourism education. Various issues related to the lack of tourism workforce skills become a challenge for tourism development in the city of Bandung. Given the success of tourism destinations depends on the quality of service and experience offered by the tourism workforce to tourists.

 DOI 10.1201/9781003095484-52

2 METHOD

The approach in this study uses qualitative methods to test and develop an understanding of a phenomenon that is little known, and to enable the discovery of in-depth information about the research subject. Qualitative methodology is based on the form of inductive logic. This involves gathering rich information about a small number of people to understand their roles and behavior or phenomena. This requires a subjective relationship between the researcher and the participants who represent the data. Concepts and variables emerge from the interviewee rather than prior research to produce patterns or theories to explain the phenomenon. This research was conducted for 8 months starting from August 2019 to March 2020 and carried out in the administrative area of Bandung. Determination of research subjects was by using purposive sampling. The sample of this qualitative research subject is a representative sub-group chosen purposively based on criterion sampling (the sample is chosen based on certain criteria). Respondents of this study organizing tourism education in the city of Bandung include secondary, diploma, undergraduate, and postgraduate levels (master and doctor). This is to obtain objective and comprehensive information.

Data collection methods use secondary data such as journal articles, reports, organizational web pages, and policy documents that are reviewed to get a broader perspective of the issue of tourism and human resource development. The data collection is then categorized based on indicators in the development of tourism human resources, which include availability, affordability, accessibility, and accountability. Further data needs were in the form of a tourism education curriculum whose information is obtained through the website of the tourism education providers and direct communication. Curriculum selection is limited based on education levels ranging from diploma, S1, S2, and S3.

3 RESULT

Bandung is one of the largest education providers in Indonesia. Various excellent schools and the best universities are in the city of Bandung. Schools and colleges are scattered in 30 districts in the city of Bandung, but not all districts have schools and colleges.

Based on data from Kopertis in 2019, it can be seen that 80.4% are in senior high schools and 19.6 private, for public vocational high schools at 87.3% with 12.7% private, state diplomas get 18.2% compared to private ones at 81.8% and state universities in Bandung, only 8.6%, while the remaining private universities reach 91.4%, so the distribution of education from the public and private schools is still not proportional. The distribution of high school and vocational schools was still dominated by high schools by 52.3% and vocational high schools by 47.4% and seen in the distribution based on districts in the city of Bandung dominated by schools and education in the city of Bandung are the districts of Lengkong, Cicendo, Andir, and Coblong. By looking at various perspectives, the distribution of public/private, senior/vocational, and educational levels in Bandung has not been proportional.

Bandung is one of the tourist destinations that are much in demand by tourists and should have competent tourism human resources from their attitude, skills, and knowledge. Tourism schools and tertiary education in Bandung are a place to produce competent human resources in the field of tourism, but in producing competent tourism human resources there are far fewer tourism education facilities compared to non-tourism education. Table 1 shows the high school/vocational tourism, that there are 11.5% compared to non-tourism at 88.5%, for the diploma level of tourism 25.8% for non-tourism 74.2% and tourism higher education 11.4% while non-tourism at 88.6%.

For the clarity of the school and higher education gap in tourism and non-tourism, researchers illustrate this in Figure 1.

Tourism education providers in the city of Bandung offer a variety of educational programs both focused on vocational and managerial. At the vocational school (SMK) level of education, the overall focus is on vocational education with five main programs, namely, 1) hospitality accommodation, 2) tourism travel business, 3) catering, 4) beauty, and 5) fashion with a total overall reached

Table 1. Number of schools and tourism education in Bandung.

No.	Education Level	Tourism Education	Non-Education Tourism	Total
1	Senior High/Vocational School	19	146	165
2	Diploma	8	23	31
3	Higher Education	8	62	70

Source: Kopertis IV.

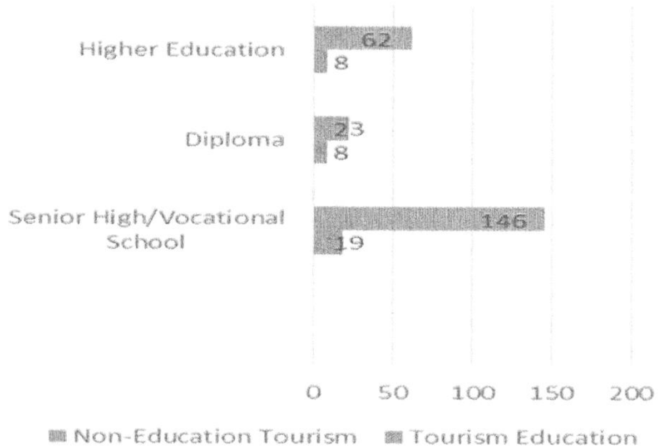

Figure 1. Comparison of tourism education with non-tourism in the city of Bandung. Source: analysis results (2019).

Figure 2. Information on expertise program at SMKN 15 Bandung. Source: survey results (2019).

of 38 education programs. The hotel accommodation and culinary accommodation programs are the most numerous in the city of Bandung at the level of vocational education, while the travel, fashion, and beauty services of other tourism education programs are found in Bandung (Figure 1).

At the tertiary level of education (diploma, bachelor, and master), educational programs are found in vocational and managerial education. As many as 18 tourism education programs are offered at diploma level education (D1, D2, D3, and D4); the number is dominated by hospitality accommodation programs and as many as 11 educational institutions that run it. Furthermore, there

Figure 3. Condition of Telkom Tourism Vocational School in Bandung. Source: survey results (2019).

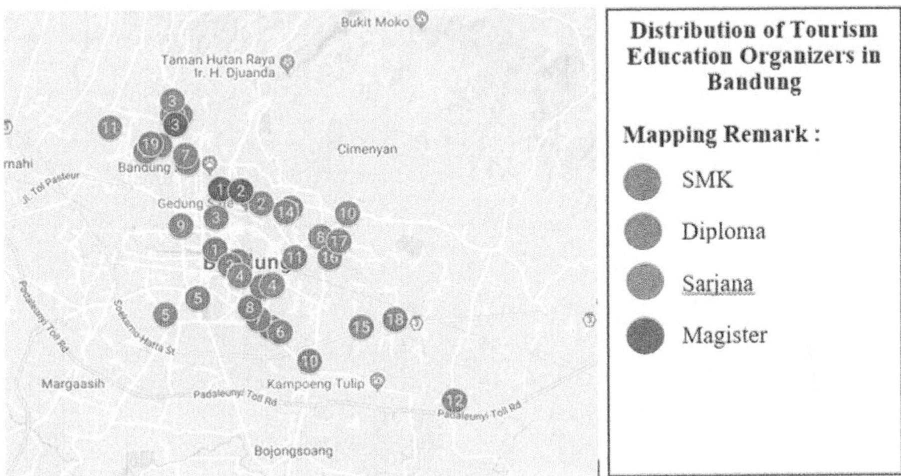

Figure 4. Distribution of tourism education organizer in Bandung. Source: survey results (2019).

are two educational programs at the diploma level that focus on managerial education in the city of Bandung.

4 CONCLUSION

The implementation of tourism education in the city of Bandung is carried out based on the needs of the tourism industry, both in the city of Bandung in particular and in Indonesia in general. Tourism education programs offered in the city of Bandung are classified in vocational and managerial with four stages of education, namely the vocational high school (SMK), diploma, bachelor, and master. The distribution of tourism education that focuses on vocational education tends to be at the level of vocational and diploma education, while for scholars and masters it tends to focus on managerial education.

The affordability of education costs in each tourism education organization in the city of Bandung has two stages, namely, the entrance fee and the semester fee, with differences based on ownership, namely, public and private. Accessibility in the organization of tourism education in the city of Bandung is classified into online accessibility and transportation accessibility. Online accessibility is in the form of websites as every level of tourism education providers in the city of Bandung, in general, has a website. Meanwhile, transportation accessibility, the overall implementation of tourism education, is easily accessed using public transportation and online transportation. Furthermore, the accreditation of the implementation of education in the city of Bandung is on average A and B accredited, which makes its own advantages for the credibility of education providers in the city of Bandung.

REFERENCES

Baum, T. (2015). Human resources in tourism: Still waiting for change? – A 2015 reprise. Tourism Management, 50, 204–212.

Sheldon, P. J., Fesenmaier, D. R., & Tribe, J. (2011). The tourism education futures initiative (TEFI): Activating change in tourism education. Journal of Teaching in Travel & Tourism, 11(1), 2–23.

Thapa, B., & Panta, S. K. (2019). Situation Analysis of Tourism and Hospitality Management Education in Nepal. In Tourism Education and Asia (pp. 49–62). Springer, Singapore.

Promoting Creative Tourism: Current Issues in Tourism Research – Kusumah et al. (Eds)
© 2021 Taylor & Francis Group, London, ISBN 978-0-367-55862-8

Analysis of online learning during the Covid-19 pandemic in tourism education

A. Suwandi, E. Fitriyani & A. Gumelar
Universitas Pendidikan Indonesia, Bandung, Indonesia

ABSTRACT: The online learning process is considered as a new challenge in the era of the industrial revolution 4.0, especially in a pandemic like today. This relates to the current condition of Indonesia amid the Covid-19 virus outbreak where students are required to study at home. Digital-based learning models have been massively maximized almost throughout Indonesia at the time of the pandemic. The purpose of this research is to analyze digital-based learning models due to the Covid-19 pandemic in the online Tourism Vocational Class in Bandung. The research method used is a quantitative analysis of descriptive data analysis in the form of a questionnaire. The results of the study are expected to be able to answer the problem of this study, which covers the implementation of online learning during the Covid-19 virus pandemic.

1 INTRODUCTION

The spread of the COVID-19 virus is the highest cause of death in various countries of the world today. Many victims have died. Even much medical personnel who were victims then died. This is a problem that must be faced by the world today, to carry out a variety of policies including in the country of Indonesia itself. Indonesia also feels the impact of the spread of this virus. More and more it quickly spread to many regions in Indonesia (Velavan & Meyer 2020).

As a result of the COVID-19 pandemic, it has led to the implementation of various policies to break the chain of the spread of the COVID-19 virus in Indonesia. One of the efforts made by the government in Indonesia is to implement an appeal to the community to carry out physical distancing, namely an appeal to keep a distance between the people, avoid activities in all forms of crowds, gatherings, and avoid meetings that involve many people (Ana 2020). These efforts are aimed at the community so that it can be done to break the current spread of the COVID-19 pandemic (Sari 2020).

The government implemented a policy that is Work From Home (WFH). This policy is an effort that is applied to the community to complete all work at home.

Education in Indonesia has become one of the areas affected by the COVID-19 pandemic (Dickinson & Gronseth 2020). With the limitation of interaction, the Ministry of Education in Indonesia also issued a policy that is to dismiss schools and replace the Teaching and Learning process by using a system in the network (Nurhalimah 2020).

The online learning process is considered as a new challenge in the era of the industrial revolution 4.0, especially in a pandemic like today. That is because this year, Indonesia officially just issued a higher education policy specifically responding to the demands of the industrial revolution 4.0, with a policy called an independent campus. This program opens a very wide space for students to determine their areas of learning that are their focus and interest. This program can also encourages students not only to learn in the classroom but also in the community and involve broad agencies (Puncreobutr 2016).

At a time like now, the digital-based learning model has been massively maximized almost throughout Indonesia, although this model also has not been comprehensively reaching the lower social strata in society. Because this learning model also has requirements that must be fulfilled. Namely. access to digital information, the digital learning process can help students in the learning system at home to increase the quality of learning (Elyas 2018).

This online learning system sometimes appears as various problems faced by students and teachers, such as subject matter that has not been completed delivered by the teacher and then the teacher replaces it with other assignments. This becomes a complaint about students because the tasks given by the teacher are more than during the learning system in the classroom (Jamaluddin et al. 2020). Another problem with the existence of an online learning system is access to information that is constrained by signals that cause slow access to in-formation. Students are sometimes left behind with information due to inadequate signals. As a result, they are late in collecting an assignment given by the teacher (Hashim et al. 2020). This is a reference to conduct research related to Digital-Based Learning Analysis at the Covid-19 Pandemic in Tourism Education.

2 METHOD

In this study, researchers used quantitative research methods by taking an approach that focused on the phenomenon of the object under study (phenomenological). Retrieval of data was using a questionnaire with data verification. Thus, the results of this study obtain data in the form of online learning analysis using applications for vocational students. The analysis is aimed at digital-based Learning Programs implemented by Tourism Education, especially Tourism Vocational Schools in West Java.

The objectives to be achieved in this study provide an explanation of the analysis of online learning at the time of the Covid-19 virus pandemic in tourism education, especially at the Tourism Vocational School.

3 RESULT

Since a decree from the Minister of Education and Culture was issued regarding the Corona virus's prevention and dissemination efforts, all conventional learning activities began to be temporarily closed. Educational activities feel locked down. The conventional learning system implemented by some teachers is slowly being eroded and replaced by various online learning applications that can provide a space for direct interaction between teachers and students without having to meet face to face (Kemdikbud 2020).

Teachers, students, and even parents are forced to adapt quickly to this method. Indeed, in a situation like this, the online method is considered the most appropriate solution to do. Even though schools are closed, demands in the learning process can still be accomplished and reached. However, if under normal conditions, many gaps are lacking from this online method (Hikmat et al. 2020).

The lack of technological knowledge of teachers, students, and parents is one of the problems in the application of this online method. Although a teacher must always enrich and upgrade science, being asked to adapt and master various applications that support online learning quickly is not easy (Pangondian et al. 2019).

This is a reference to discuss online learning analysis at the time of the Covid-19 pandemic of the West Java tourism school with the following questions:

- Do you understand online learning?
- Is online learning effective when the Covid-19 pandemic hit Indonesia?
- Is online learning less effective when the Covid-19 pandemic hit Indonesia?
- Are Internet sources an obstacle to online learning?
- Do quotas become a capital for online learning?

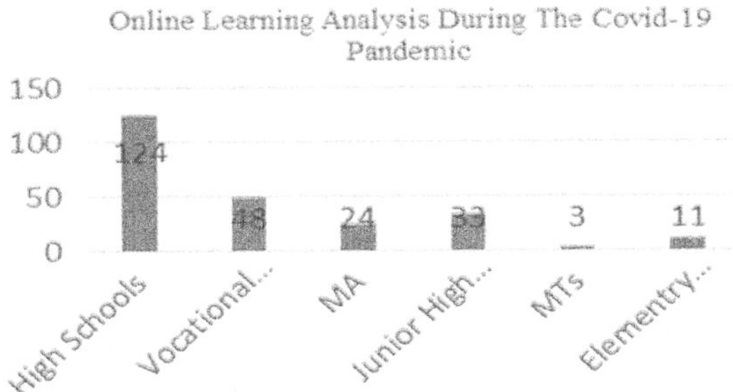

Figure 1. Online learning analysis.

Analysis of online learning analysis during the Covid-19 pandemic at the Tourism Vocational School in quantitative descriptive methods. The survey was conducted with a multistage random sampling technique. The data collection technique uses a questionnaire given using the Google forms application as 246 respondents.

The purpose of the survey is to determine students' perceptions of the implementation of online learning. Respondents by Region After Online Learning was carried out for four weeks with a research instrument in the form of a questionnaire. The results of the questionnaire about online learning analysis during the Covid-19 pandemic can be seen in Figure 1, which came from 124 (50.4%) high schools, 48 (19.5%) vocational schools, and 24 (9.8%) MA. Furthermore, there are 33 (13.4%) junior high schools, only 3 (1.2%) MTS, and 11 cases (4.5%) elementary schools.

A total of 19.5% Tourism Vocational School students in West Java stated that students understand about online learning, online learning is not effectively implemented with various aspects, namely weak Internet network is an obstacle that is often experienced by students. This is especially true for students who live in rural or remote areas, and it will be very difficult to get Internet access. This is one of the important factors for the implementation of online learning (Kemdikbud 2020).

4 CONCLUSION

Online learning is considered to be a solution to teaching and learning activities continue in the middle of the Corona virus pandemic. Although it has been agreed, this method has generated controversy. For teachers, online learning systems are only effective for assignments. They consider that to make students understand the material, the way online is considered difficult.

Besides, the technological and economic capabilities of each student are different. Not all students have facilities that support this distance learning activity. Slow connections, incompetent devices, and expensive Internet quotas are real obstacles. Although many aspects need to be considered when applying online methods, this method also has several advantages. Among these are that teachers and students will be increasingly technology literate and keep abreast of the times, learning activities are not limited to time and place, learning resources are also not limited to teachers but from other sources, students' creativity and criticism will increasingly come out, teachers will be more creative in combining various kinds of online teaching media, teachers are no longer burdened by manually correcting student assignments, and the use of paper will be reduced because it is diverted through online applications.

In the end, in each learning method, there are advantages and disadvantages. However, it is the teacher's job to determine the teaching methods, styles, or techniques that are appropriate to the characteristics of the students being accompanied. Because again, teaching is an art for the formation of character, creativity, criticism, and the nature of student care, not only focused on the delivery of knowledge.

REFERENCES

Ana, A. 2020. Trends in expert system development: A practicum content analysis in vocational education for over grow pandemic learning problems. *Indonesian Journal of Science and Technology*, *5*(2), 71–85. https://doi.org/10.17509/ijost.v5i2.24616

Dickinson, K. J., & Gronseth, S. L. 2020. Application of Universal Design for Learning (UDL) Principles to Surgical Education during the COVID-19 Pandemic. *Journal of Surgical Education, In Press*. https://doi.org/10.1016/j.jsurg.2020.06.005

Elyas, A. H. 2018. Penggunaan Model Pembelajaran E-Learning Dalam Meningkatkan Kualitas Pembelajaran. *Jurnal Warta*.

Hashim, S., Masek, A., Abdullah, N. S., Paimin, A. N., & Muda, W. H. N. W. 2020. Students' intention to share information via social media: A case study of COVID-19 pandemic. *Indonesian Journal of Science and Technology*, *5*(2), 61–70. https://doi.org/10.17509/ijost.v5i2.24586

Hikmat, Hermawan, E., Aldim, & Irwandi. 2020. Efektivitas Pembalajaran Daring Selama Masa Pandemi Covid-19: Sebuah Survey Online. *Digital Library, UIN SUnan Gung Djati, Bandung*.

Jamaluddin, D., Ratnasih, T., Gunawan, H., & Paujiah, E. 2020. Pembelajaran Daring Masa Pandemik Covid-19 Pada Calon Guru: Hambatan, Solusi dan Proyeksi. *Karya Tulis Ilmiah UIN Sunan Gunung Djjati Bandung*.

Kemdikbud, pengelola web. 2020. *Kemendikbud Dorong Pembelajaran Daring Bagi Kampus Di Wilayah Terdampak Covid-19*. Www.Kemdikbud.Go.Id.

Nurhalimah, N. 2020. UPAYA BELA NEGARA MELALUI SOSIAL DISTANCING DAN LOCKDOWN UNTUK MENGATASI WABAH COVID-19 (Efforts to Defend the Country Through Social Distancing and Lockdown to Overcome the COVID-19 plague). *SSRN Electronic Journal*. https://doi.org/10.2139/ssrn.3576405

Pangondian, R. A., Santosa, P. I., & Nugroho, E. 2019. Faktor – Faktor Yang Mempengaruhi Kesuksesan Pembelajaran Daring Dalam Revolusi Industri 4.0. *Seminar Nasional Teknologi Komputer & Sains (SAINTEKS)*.

Puncreobutr, V. 2016. Education 4.0: New Challenge of Learning. *Humanitarian and Socio-Economic Sciences*.

Sari, Y. I. 2020. Sisi Terang Pandemi COVID-19. *Jurnal Ilmiah Hubungan Internasional*. https://doi.org/10.26593/JIHI.V0I0.3878.89–94.

Velavan, T. P., & Meyer, C. G. 2020. The COVID-19 epidemic. In *Tropical Medicine and International Health*. https://doi.org/10.1111/tmi.13383

Promoting Creative Tourism: Current Issues in Tourism Research – Kusumah et al. (Eds)
© 2021 Taylor & Francis Group, London, ISBN 978-0-367-55862-8

Development of a friendly character and working characters for tourism guide practices of SMK's students in Bali

R. Munawar & M. Rahmat
Universitas Pendidikan Indonesia, Bandung, Indonesia

ABSTRACT: Bali is the province with the most beautiful tourist destinations in Indonesia. Foreign and local tourists always visited Bali. For this reason, Bali needs a lot of professional tour guides. The SMK tourism study program is a secondary level educational institution providing tour guides. The research aims to find the key characters of tour guides for internship SMK students in Bali Province. We researched in 2018 at the best SMK in the city of Denpasar: Qualitative research methods, research techniques, interviews with the headmaster, heads of study programs, internships' students, SMK's alumni, and several tour guides. Research results show friendliness is the most important character for tour guides in Bali. They must be friendly to each guest, both to international and local guests, to guests who give big or small tips, to guests who give tips or don't give tips. The most important work character is to have communication skills, can speak English, master tourist destinations, know culinary centers and souvenirs, and understand Balinese culture. The implication is that professional tour guides must be friendly and have a working character related to tourism skilled in communicating, can speak English, know the culinary and souvenir centers, and understand the culture around tourist destinations.

1 INTRODUCTION

Bali is the belle of Indonesia's tourism, which is already well known throughout the world. Besides its natural beauty, especially its beaches, Bali is also famous for its unique and interesting arts and culture (Wikipedia 2020). Bali Island is part of the Lesser Sunda Islands along 153 km and 112 km wide around 3.2 km from Java Island. The total area of Bali Province is 5,636.66 km^2 or 0.29% of the territory of the Unitary Republic of Indonesia. Administratively, Bali Province is divided into 8 regencies, 1 municipality, 55 sub-districts, and 701 villages/wards. The capital of Bali is Denpasar (Wikipedia 2020).

Local or domestic and foreign tourists always visit Bali. Data on domestic tourist visits recorded by the Central Statistics Agency of Bali Province every year is increasing. During the 5 years from 2014 to 2018, there was an increase as follows. In 2014, 6,394,307 to 7,147,100 in 2015, up 11.77%. In 2016, it became 8,643,680, or an increase of 20.94%. In 2017, it became 8,735,633, or an increase of 1.06%; and in 2018 it was 9,757,991, or an increase by 11.70% (BPS 2020a). Foreign tourists under the recommendations of the United Nations World Tourism Organization (UNWTO) are any people who visit a country outside their place of residence, encouraged by one or several needs without intending to earn income at the place visited, and the duration of the visit is not over 12 months (BPS 2020b). Data on tourist arrivals every year is increasing. In 2014, it was 3,766,638, an increase of 14.89% from the previous year. In 2015 it became 4,001,835, or up 6.24%. In 2016, it became 4,927,937, or an increase of 23.14%. In 2017, it became 5,697,739, or increased by 15.62; and in 2018 was 6,070,473, or up 6.54% (BPS 2020c).

An important tourist spot in Bali is Ubud as a center for arts and rest, in Gianyar Regency. Nusa Lembongan is one of the dive sites in Klungkung Regency. Kuta, Seminyak, Jimbaran, and Nusa Dua are some places that become the main destinations for tourism, both beach tourism, and resorts, spas, etc., in Badung Regency. The tourism industry is centered in South Bali and in several other

DOI 10.1201/9781003095484-54

regions. The primary tourist sites are Kuta and surrounding areas such as Legian and Seminyak; eastern areas of the city such as Sanur; city centers like Ubud; and in southern areas such as Jimbaran, Nusa Dua, and Pecatu. Bali as a complete and integrated tourist destination that has a lot of interesting tourist attractions, including Kuta Beach, Tanah Lot Temple, Padang—Padang Beach, Lake Beratan Bedugul, Garuda Wisnu Kencana (GWK), Lovina Beach with its Lumba Lumbanya, Pura Besakih, Uluwatu, Ubud, Munduk, Kintamani, Amed, Tulamben, Menjangan Island, and many others. Now, Bali also has several tourist centers full of education for children such as zoos, three-dimensional museums, water playgrounds, and turtle breeding grounds (Wikipedia 2020).

Along with the development of tourism, the tourism business and services also increased. To compensate, Bali's tourism education has also increased. Tourism secondary education is engaged in the tourism industry, Balinese arts, hospitality services, and tour guides. The education of tour guides in Bali develops the core and work characteristics that are needed for the convenience of tourists. The question is, what kind of character does it develop?

2 METHODS

The research uses a qualitative approach. Data collection techniques are in the form of interviews and observations. Research respondents are headmaster, heads of tourism study programs, several students taking part in practical work, and several alumni of SMK Negeri 2 Denpasar Bali and several Bali tour guides. We chose SMK because this school has the best tourism study program and 100% of its alumni are absorbed in the business world. We conducted research in 2018.

3 FINDINGS

3.1 *Development of a friendly character for tourism guide practices of SMK's students in Bali*

The dominant factors of tourist attraction are tourist destinations and facilities. As for the involved humans (tour guides, hotel employees, and surrounding communities), they are secondary supporting factors. However, the friendliness of the community affects the attractiveness of tourism (Vengesavi et al. 2009).

"Friendly" is the key character of a tour guide in Bali. The head of the Vocational School and the head of the tourism study program explained that friendliness appeared from several characteristics: acceptance, courtesy, respect, and non-discrimination toward tourists. This meaning is in line with the characteristics of tour guides (Sinaga & Utomo 2014). The politeness of the language is important in providing tourism services (Susanthi & Warmadewi 2020).

They further explained, tourists come to Bali to enjoy the natural beauty and distinctive culture, especially tourism destinations. By coincidence, the Balinese also support tourism. The Balinese with its distinctive culture is enough to strengthen the tourist attraction. Every day there are religious and traditional ceremonies that attract tourists. Seven traditional ceremonies are quite enjoyed by tourists: melasti, cremation, mekare-kare, saraswati, napping, galungan, and mepandes (Traveloka 2020).

Tour guides should not damage this tourist attraction factor, stressed a tour guide (research respondent). The hospitality of the tour guide turns out to be stored in the memory of tourists. Respondents shared their experiences, a family of tourists from Europe who visited Bali in December 2001. In October 2002, a Bali Bomb incident devastated Bali tourism. We have been unemployed for over six months. We surrender by tightening expenses. God's power of tourists from Europe called us and asked for a bank account number. He said, "We share our condolences for the Bali Bombing. As a thank you for your service, we transfer some money, hopefully, we can ease your burden!" It was surprising he turned out to be transferring very large money, Rp. 10 million. This shows how satisfied tourists are with performing tour guides. The results of research in Shanghai, the performance of tour guides, was found to have a significant direct effect on tourist satisfaction (Huang et al. 2009).

In contrast is his friend, who was laid off because he was not friendly and was disappointing tourists. He was full of hope that he could tip Rp. 100,000 It turned out that the tourist gave a tip of Rp. 20,000. It was so disappointing he put the money on his forehead by saying, "Ah, only 20,000!" Tourists see the unpleasant events, then report to the management.

Other respondents justified the danger of unpleasant behavior. The head of the vocational school and the head of the tourism study program said that we always emphasize the importance of students practicing friendliness. Students practice saying that we always practice friendly. The alumni also stressed the importance of tour guides to have a friendly character. A tour guide respondent added that a tour guide should not expect tourist tips. They gave a big tip, so you should be grateful. Give a slight tip to be grateful too. Not even giving a tip must be grateful too, because the tour guide has already received an adequate salary/wage.

The Head of Vocational School explained the tour guide education program that we do is to produce graduates who are professional and at the same time have a superior character as tour guides. We always contact leading travel companies and five-star hotels about what tourists like and dislike for tour guides. We also entrust entirely too well-known tourism companies as an internship for our students. The results are satisfying. Our alumni are almost all absorbed in tourism companies in Bali, in other regions in Indonesia and abroad.

We must see acceptance from eye contact and body language to the courtesy and choice of words that tourists need. An apprentice student gives an example of how to start a conversation with tourists as follows.

Tour guide : Hello, good morning Sir. Welcome to Bali. My name is Nyoman and I will be your guide.
Tourist : Hello, good morning. My name is Richard.
Tour guide : Hi Mr. Richard, where are you from?
Tourist : I am from London, England.
Tour guide : Wow, so far away from here.

Tourists often ask to be escorted to famous tourist destinations. But the tour guide can also offer the places in question. The tour guide may offer the first-time visiting Besakih Temple, as the unique tourist destination and religious tourism in Bali. For example, tourists agree.

Tour guide : Our first destination today is Besakih Temple.
Tourist : Okay!
Arriving at Besakih Temple,
Tour guide : Okay Sir. So, welcome to Besakih Temple.
Tourist : Can you tell me more about this temple, Made?
Tour guide : Besakih Temple is one of the famous tourism objects in Bali and it is the biggest Hindu temple in Bali. We can see the wide nature panorama from the top of the temple.
Tourist : Wow, it sounds awesome. Please take me there later on.
Tour guide : Sure, Mr. Richard.
Tourist : I see there are a lot of temples in here. Which one is the main temple?
Tour guide : The main temple is called Penataran Agung Temple. It is the largest temple in here.
Tourist : So, when do Balinese people come to pray? Because I see no one is praying now.
Tour guide : Each temple has its own ceremony celebration. The Balinese people usually come to pray on the full moon or "Sasih Kedasa."
Tourist : Oh, I see. What a unique culture, Made. Okay, let's move to another place.
Tour guide : Okay, sir. Our second destination today is Kuta beach.
Tourist : Kuta beach?
Tour guide : Kuta beach is actually a nice spot for snorkeling, surfing, riding a boat, and swimming. There are places to rent equipment for snorkeling, surfing boards, and some street vendors.
Tourist : I can't wait for that.
…etc.

The sense of comfort of tourists needs to be fully awake. In certain tourist destinations, hawkers often offer their merchandise by force. The behavior of such traders does not make tourists comfortable. The tour guide is how to keep traders away from tourists! Before visiting the tourist destination, tour guides need to remind tourists. If you have arrived at the destination, don't ask about merchandise or prices if you don't intend to buy them. If you ask, the trader will force tourists to buy it. In Maoz's terms, tour guides must be professional hosts (Maoz 2006).

3.2 *Development of working character for tourism guide practices of SMK's students in Bali*

The Head of Vocational School explained, the characteristic of work for tourism study program students is having communication skills, having knowledge about Bali tourist destinations, self-confidence, and attractive appearance. Character of communication skill means is good at talking, is fun, and convincing to tourists. Know how to talk during the initial meeting. Know the right time to talk so it doesn't bother tourists. The tour guide dialogue with Richard's tourist above is an example of a tour guide's communication skills. Research results reinforce the need for tour guides to have these characters. Almost all respondents of local and foreign tourists rated very well of Bali tour guides (Putra et al. 2017).

Tour guides need to master the characteristics, description, details, and history of each tourist destination: Pura Besakih, Kuta Beach, Sanur Beach, etc. Per Holloway, a tour guide must be a provider of information and knowledge sources (Holloway 1981), also as a translator of local culture (Gelbman & Maoz 2012; Salazar 2005). Or per Gelbman and Collins-Kreiner (2016), as a cultural moderator.

Prospective tour guides need to have knowledge about Balinese Hindu religious ceremonies because it is very attractive to tourists. The tour guide must master at least seven religious and traditional ceremonies. First, the Melasti ceremony takes place every year as part of a series of Nyepi Holidays. We intend this ceremony as a sanctification for Hindu residents. This ceremony takes place three to four days before Nyepi Day. Residents come to several sources of sacred springs such as lakes, springs, and seas, which are believed to save the springs of life and purify themselves by taking tirta amertha (immortality). Second, the Ngaben ceremony is the most famous religious ceremony. Ngaben is a special ritual to deliver the bodies of family and relatives whom he loves by burning. This ceremony was held grandly, in procession, and with an atmosphere of fun, because the family delivered the people they love to Nirvana. But in this ceremony, family and relatives are forbidden to not cry. Third, the Mekare-kare ceremony, also known as the "*perang daun pandan*" ("Pandan leaves war"). The players in this traditional ceremony are men as a place to show their ability to fight using sharp, thorny pandan leaves. This ritual is a tribute to Indra, who is famous as the god of war. After finishing, the elder will treat and pray to the participants so that the pain of being stabbed soon disappears.

Fourth, they held the Saraswati Ceremony to celebrate science. Through this traditional ceremony in Bali, Hindus hold a ritual of worshiping the Goddess Saraswati, who is believed to bring knowledge to the earth so that humans become educated. They pray all things related to science such as books and books in the Saraswati ceremony. Not only that, in this series of traditional ceremonies in Bali, but you can also watch dance performances and read stories all night long.

Fifth is Ngerupuk Traditional Ceremony. Still included in the series of Nyepi Day events, they held the Ngerupuk ceremony to expel the Bhuta Kala so as not to interfere with human life while conducting a retreat. They hold this traditional Balinese ceremony exactly the day before the Nyepi day arrives and the community must make offerings to Bhuta Kala. The ritual begins with tearing the house, spraying the house and yard with gunpowder, and hitting objects to make noise. After this traditional ritual in Bali is finished, you can watch the ogoh-ogoh parade paraded with torches around the residents' living area.

Sixth is the Galungan Ceremony. The term galungan comes from the Old Javanese language and means "win." As the name implies, they aim the traditional ceremony in Bali at celebrating the victory against evil. In addition, they also held the Galungan ceremony to commemorate the creation of the universe and its contents. This traditional Balinese ceremony is held every six months in the calculation of the Balinese calendar and is carried out for 10 consecutive days.

Seventh is the Mepandes Ceremony. Also known as Metatah or Mesuguh, the traditional ceremony of Mepandes is performed when a child enters adolescence. They held the teeth cutting ceremony to eliminate dangerous passions such as greed, jealousy, anger, and so on. Both boys whose voices have become heavy and girls who have had their first menstruation will perform religious rituals before they cut their teeth off by the traditional elder as a symbol of maturity (Traveloka 2020).

Other work characteristics for tour guides are attractive in appearance, which is clean, neatly dressed, and pleasing to the eye. Research results support this character, that tourists are very satisfied with the appearance of a tour guide. Finally, the character of confidence, namely confidence in his appearance (which pleases tourists), his mastery of Bali tourist destinations, Balinese customs that are worth the tour, and his ability to guide tourists. In medical research, self-esteem is like the building foundation for a skyscraper (Bailey 2003).

4 CONCLUSION

Destinations and tourist facilities are the fundamental factors of tourist attraction. As for humans, especially tour guides, they are secondary supporting factors. However, the typical character of a tour guide affects the attractiveness of the tourism. The key characteristic of a tour guide is friendly. This character has several characteristics, namely, an attitude of acceptance, courtesy, respect, and is non-discriminatory toward tourists. They must be friendly to foreign and local tourists, to tourists who give tips big or small, even to tourists who do not give any tips, they must remain friendly. They implement this key character in education at SMK Bali. The most important work character is to have communication skills, can speak English, master tourist destinations, know culinary centers and souvenirs, and understand Balinese culture.

Students of SMKs need to master the knowledge of Bali's excellence, tourist destinations, and the unique customs of Balinese Hinduism, which are highly favored by tourists. Coupled with a clean appearance, a tour guide will have confidence as the best tour guide.

The implication is that professional tour guides must be friendly and have a working character related to tourism skilled in communicating, can speak English, know the culinary and souvenir centers, and understand the culture around tourist destinations.

ACKNOWLEDGMENT

No potential conflict of interest was reported by the authors.

REFERENCES

Bailey, Joseph A. 2003. The foundation of self-esteem. *Journal of the National Medical Association*, 95(5): 388–393.

BPS. 2020a. Kunjungan Wisatawan Domestik ke Bali. *https://bali.bps.go.id/statictable/2018/02/09/29/kunjungan-wisatawan-domestik-ke-bali-per-bulan-2004-2018.html*, accessed on 20 Mei 2020.

BPS. 2020b. Konsep dan Definisi Statistik Kunjungan Wisatawan Mancanegara. *https://bali.bps.go.id/subject/16/pariwisata.html*, accessed on 20 Mei 2020.

BPS. 2020c. Kunjungan Wisatawan Asing ke Bali. *https://bali.bps.go.id/statictable/2018/02/09/21/jumlah-wisatawan-asing-ke-bali-menurut-bulan-1982-2019.html*, accessed on 20 Mei 2020.

Gelbman, Alon & Collins-Kreiner, Noga. 2016. Cultural and behavioral differences: tour guides gazing at tourists. *Journal of Tourism and Cultural Change*, Volume 16, 2018 – Issue 2: 155–172. DOI: 10.1080/14766825.2016.1240686

Holloway, J. 1981. The guided tour: A sociological approach. *Annals of Tourism Research*, 8: 377–402.

Huang, Songshan (Sam); Hsu, Cathy H. C.; and Chan, Andrew. 2009. Tour Guide Performance and Tourist Satisfaction: a Study of the Package Tours in Shanghai. *Journal of Hospitality and Tourism Research* 34(1): 3–33. DOI: 10.1177/1096348009349815.

Maoz, D. 2006. The mutual gaze. *Annals of Tourism Research*, 33(1): 221–239.

Putra, Ida Bagus Putu Saskara; Negara, I Made Kusuma; and Wijaya, Ni Made Sofia. 2017. Persepsi Wisatawan Terhadap Kualitas Layanan Pramuwisata di Bali. *Jurnal IPTA*, Vol. 5 No. 1, 2017: 29–34.

Salazar, N. 2005. Tourism and glocalization: 'Local' tour guiding. *Annals of Tourism Research*, 32(3): 628–646.

Sinaga, Endang Komesty & Utomo, Bambang Sapto. 2014. Kualitas Pelayanan Pemanduan Ekowisata di Taman Nasional Tanjung Puting Kabupaten Kotawaringin Barat Kalimantan Tengah. *Jurnal Manajemen Resort & Leisure*, Vol. 11, No. 1, April 2014: 7–23.

Susanthi, I Gusti Ayu Agung Dian & Warmadewi, Anak Agung Istri Manik. 2020. Kesantunan dalam Percakapan Pemandu Wisata di Ubud Bali. *Kulturistik: Jurnal Bahasa dan Budaya*, Vol. 4, No. 1: 22–27. Doi: 10.22225/kulturistik.4.1.155.

Traveloka, 2020. *https://www.traveloka.com/id-id/explore/destination/ritual-upacara-adat-di-bali-acc/30491*, accessed on 25 Mei 2020.

Vengesayi, Sebastian; Mavondo, Felix; and Reisinger, Yvette. 2009. Tourism Destination Attractiveness: Attractions, Facilities, and People as Predictors. *Tourism Analysis* 14(5): 621–636. DOI: 10.3727/108354209X12597959359211.

Wikipedia. 2020. Bali. *https://id.wikipedia.org/wiki/Bali*, accessed on 20 Mei 2020.

Promoting Creative Tourism: Current Issues in Tourism Research – Kusumah et al. (Eds)
© 2021 Taylor & Francis Group, London, ISBN 978-0-367-55862-8

Edutourism: Learning to be the Indonesian society

R.W.A. Rozak, A. Kosasih, M.D. Kembara, N. Budiyanti & V.A. Hadian
Universitas Pendidikan Indonesia, Bandung, Indonesia

ABSTRACT: The tourism sector nowadays leads to the concept of a tourism village. Every village with special characteristics and potential can be promoted to become a tourist destination. The study was conducted to analyze the potentials of Kampung Kasepuhan Babakan Lama to be promoted as a cultural tourism village. In-depth interviews were conducted with an informant who was a prominent figure in the village for three months (April to June 2020). The second party is an individual who grows up and lives in Kampung Kasepuhan Babakan Lama. This was done in order to maintain health during the Covid-19 pandemic. The interview resulted in the finding that Kasepuhan Babakan Lama Village was very representative to become a commercial edutourism. Many aspects can be seen as an edutourism potential, including social interaction, culture, architecture, natural scenery, to interaction with nature. These aspects can be used as the concept of edutourism and useful for the tourism development. Tourists can learn and gain local wisdom, character, and social identity that belong to the people of Kampung Kasepuhan Babakan Lama. Thus, tourists can relax and learn to experience Indonesia by modeling from the community.

Keywords: edutourism, cultured village, learning to become Indonesia, tourism village, tourism potential

1 INTRODUCTION

An increasingly advanced civilization has given people a bunch of information about the lives of various countries. This is often without strong filtration from the community so it tends to follow current trends (Kembara *et al.,* 2019). The times can damage the cultural identity of the community (Blapp & Mitas 2017). Education plays a role to restore the cultural identity of the community (Boyle *et al.*, 2015). Edutourism can provide a different experience in learning local values (Camilleri 2018; Park *et al.* 2019; Wu *et al.* 2020). The community as tourists can be a part or directly witness the local values contained in the tourist area naturally (Mihalic *et al.* 2015; Sie *et al.* 2015; Zhang & Smith 2019). The local wisdom values of a community in the village can be transferred explicitly and implicitly to tourists (Nagy & Segui 2010). The pattern of social interaction can be an attraction in edutourism (Dukić & Volić 2020). Communities in the village (edutourism) can become role models for local characters to foreign tourists, especially domestic tourists who are starting to lose their local identity. Tourists can learn various things, such as philosophy of life, social systems, cultural values, architectural concepts, etc. (Ritchie 2003). This can attract the tourist attention to visit, although the market segmentation is still limited (McGlaggery & Lubbe 2017). Tourism will develop if community participation is well connected (Kanwal *et al.* 2020).

The development of tourism is being intensively carried out by the Indonesian government. This is raised in the slogan Wonderful Indonesia. The Indonesian government is very supportive in developing rural tourism for the sake of regional and national economic growth. Tourism development must be directed toward edutourism, because it is beneficial for educational values and nuances of recreation (Ritchie 2003). The concept of edutourism is divided into two categories, namely, 1) cultural heritage tourism; and 2) natural environment tourism. The concept of edutourism can be applied in Kampung Kasepuhan Babakan Lama, Banten. The village has a number of local intelligence needed by the millennial community nowadays, namely the character and national identity. In addition, there are many aspects that can be assessed and transferred through edutourism, for example social aspect, spiritual aspect, cultural aspect, environmental aspect, etc. (Barkathunnisa *et al.* 2018). However, planning and developing the concept of edutourism must be careful to maintain social preservation and environmental tourism in a sustainable manner (Cascante 2020; Kachniewska 2015; Lin 2019; Luo *et al.* 2016; Min 2017). The planned conception of edutourism can increase the economic income of the community and the country (Gunter *et al.* 2016; Matahir 2017; Rehman *et al.* 2020; Sharma & Sarmah 2019; Tang 2019). In addition, cooperation with educational institutions needs to be held, especially universities for training and development of edutourism (Baiquni & Dzulkifli 2019; Pitman *et al.* 2016).

2 RESEARCH METHOD

Research conducted during the Covid-19 pandemic spread in various regions in Indonesia. Thus, it needs to be circumvented so that the research can possibly be continued. The research lasted for three months, namely April to June 2020. The researcher asked for help from a second party, a resident who is a native descendant of Kampung Kasepuhan Babakan Lama, which is the location of the study. The second party conducted the interview using the interview instrument prepared by the researcher. Interviews were conducted with traditional elders who were leaders in Kampung Kasepuhan Babakan Lama. In addition, interviews were also conducted with village elders. This is intended to get the validity of information.

In-depth interviews conducted will produce the following data: 1) the viewpoint of people's lives from social, cultural and religious aspects; 2) the philosophy of architecture in Kampung Kasepuhan Babakan Lama; and 3) how to interact with the natural surroundings. In addition to interview data, researchers also obtained data in the form of photos of conditions in Kampung Kasepuhan Babakan Lama. The data obtained are then analyzed and synthesized with concepts and theories related to edutourism. In addition, the data is juxtaposed with character concepts and theories. This serves to bring up the character values that are owned by the people of Kampung Kasepuhan so that they are worthy of being used as a model of character learning. The results of the analysis and synthesis are made into the edutourism village design framework with character.

3 COMMUNITY LIFE VALUES OF KAMPUNG KASEPUHAN BABAKAN LAMA

Kampung Kasepuhan Babakan Lama consists of 60 families. This village is a representation of Indonesian society from social and cultural aspects. The culture passed down through generations is still well preserved. This is reflected in various aspects held by the community based on local culture. This is illustrated in Figure 1. Each aspect overlaps with one another. The culture that is preserved in Kampung Kasepuhan Babakan Lama covers all aspects of the lives of the people. Culture becomes the philosophy of local wisdom, which is the identity of the people.

The development of the era cannot disrupt the local identity of the community. This is a joint agreement from every citizen who lives in Kampung Kasepuhan. They do not reject the development but accept the development of the times with the rules of local cultural wisdom. Thus, the nuances of Indonesia are still preserved in this village. The Indonesian nuance in question is hospitality in social interaction, upholding the values of norms, promoting harmony, caring for one another and helping, and utilizing nature wisely. The nuance is shifting and fading in Indonesia because of the rapid development of the era. Local culture is often competed with global culture, so it is considered inappropriate if it is still applied.

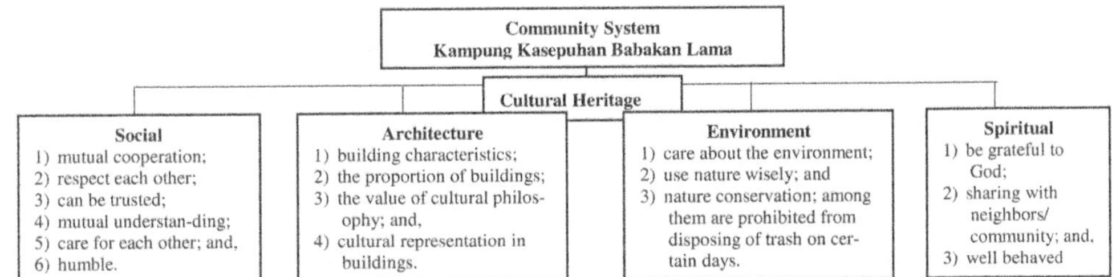

Figure 1. Community base of Kampung Kasepuhan Babakan Lama.

Figure 2. The society work together to renovate one's homes.

3.1 *Social wisdom*

The social wisdom belongs to the people of Kampung Kasepuhan Babakan Lama is mutual assistance between communities. This wisdom is not limited to the surrounding community, but to people outside the village. Kinship is deeply intertwined with harmony among neighbors and residents. Even when there are community members who are struggling, residents will gather, led by community leaders and traditional leaders. The purpose of gathering is to deliberate in producing solutions to problems experienced by residents. This is an endangered Indonesian philosophy, namely, consensus agreement. Every citizen protects and helps each other. This is the manifestation of mutual cooperation that is identical to the way of life of the Indonesian people. In addition, the community often helps other residents who experience difficulties, for example, experiencing disaster. The people of Kampung Kasepuhan will not hesitate to provide temporary shelter at their homes. This is done without being paid by money. Society helps voluntarily and without putting any negative thoughts. This is an entity of character that is owned by exploring local wisdom. This kind of thing would be very good if disseminated because that is the social wisdom that should be owned by the Indonesian society.

Social wisdom formed in the community is a representation of cultural values taught through generations from the ancestors. The implication is that community life in the village is peaceful and harmonious. Another example is when the society works together to renovate one's house (see Figure 2).

The interesting and special thing is that the people who help do not provide fees for the services they provide, but the services are paid for by the hospitality of the host by serving food and drinks. Of course, there are also workers who are hired and paid properly. This behavior is a local identity that is owned by the people by absorbing mutual values.

Figure 3. Houses in Kampung Kasepuhan made of woven bamboo and a roof made from palm fiber and *leuit* saving crops.

3.2 *Architecture and view of community life*

The architecture of the building in Kampung Kasepuhan Babakan Lama has similarities with traditional villages, namely walls made of woven bamboo (bilik), roofs from combined palm fibers, and a pedestal that is not directly hit the ground. This has been maintained until now when building or renovating houses. However, there are also residents' houses that use a wall on condition that it must be matched with the woven bamboo parts, so it is not entirely made from a brick wall. That is, traditional values are maintained strictly to uphold and respect ancestral culture. Thus, architecture is built on a strong cultural foundation. This was done for a reason, when people's houses were modernized, then social inequality would appear. Then, in the event of a natural disaster, such as an earthquake, modern concept buildings tend to be easily damaged and cause material and immaterial losses, such as fatalities. Buildings made of brick walls can cause casualties if they happen to residents. In addition, by using the traditional concept, the house feels cooler and closer to nature.

It can be understood that the concept of traditional architecture in Kampung Kasepuhan is an effort to save people from the social inequalities that are felt in the current 4.0 industrial era. In addition, traditional home building design provides many benefits for the community in terms of social, cultural, and economic. This traditional building concept was also applied to build a place called *leuit* or warehouse for food supplies (rice) owned by each person (see Figure 3). Every citizen in Kasepuhan Village has at least one *leuit* for one family. This is adjusted to the ability of the crops obtained by the community. During the Covid-19 pandemic, the people of Kampung Kasepuhan Babakan Lama were not affected by food shortages because they had stocks for the next few years which were stored in *leuits*. Residents and *leuits* are not located in one location, meaning that the two places are made separately. The *leuit* location is 100 meters from residents' houses.

3.3 *Interaction with nature*

Not only being friendly to humans, citizens are friendly to nature. It can be said that people live side by side with nature and use it wisely. This is an ecovillage concept that was initiated by many researchers to utilize nature. In fact, the people of Kampung Kasepuhan Babakan Lama have applied it long ago. This shows that the character and local identity intelligence of the community are the ideals of many researchers in today's modern times. Utilization of nature includes all aspects that support for people's lives. Clear water is used for bathing, cooking, washing, etc. Local wisdoms of the community also control the community in behaving toward the environment, including being prohibited from cutting down trees without the permission of the customary leader, prohibited from littering on Tuesday, and harvesting once a year. This was carried out obediently by the community and who not question the underlying reasons. This community system can be used as environmental

Figure 4. Clear water because the environment is well maintained and a public place for taking a shower.

Figure 5. Together with women-men separated on banana leaves and *seren taun* activities as gratitude to god.

conservation, especially in Kampung Kasepuhan Babakan Lama. This can be seen from the clear flow of water that flows into the village and is often used as drinking water directly (see Figure 4).

The water channel was made into several channels by the residents, which were distributed to the shelter and then distributed to each house. In addition, water channels are channeled directly to several public places for bathing and washing. The place can be used by all residents by paying attention to its cleanliness. This public bathing place is insulated by using bamboo matting or cubicles (see Figure 4).

3.4 *The way of life of Kasepuhan Babakan Lama villages*

Another interesting thing that can be transferred to the community as tourists is a high social life and always thanking God for the ease and gifts given. Gratitude is often applied when eating together. The community independently prepared dishes to be eaten together. This shows a high social spirit, which is when getting fortune, then share with neighbors or residents in the village. This is shown in Figure 5.

This is evidenced in each phase of activities, for example in agriculture, starting from planting rice to harvesting rice. In the process of planting rice seeds, it is known as *nyarita* (storytelling) activities. Narrative activities are not in the form of storytelling, but are filled with praise and prayers to God, and asking for permission from the ancestors. Some activities are always carried out in every phase of farming to harvest. *Seren Taun* activities are carried out as a thanksgiving to God for the abundance of sustenance and harvest; see Figure 5. The *seren taun* activity ended with a meal together at the house of the adat leader. It is filled with praise and prayers to God. In addition, it is an implementation of family social ties that are always maintained in Kampung Kasepuhan Babakan Lama. These eating activities still pay attention to the norms of decency, namely, that women and men do not mix. This form of local intelligence is very representative to be transferred to other communities.

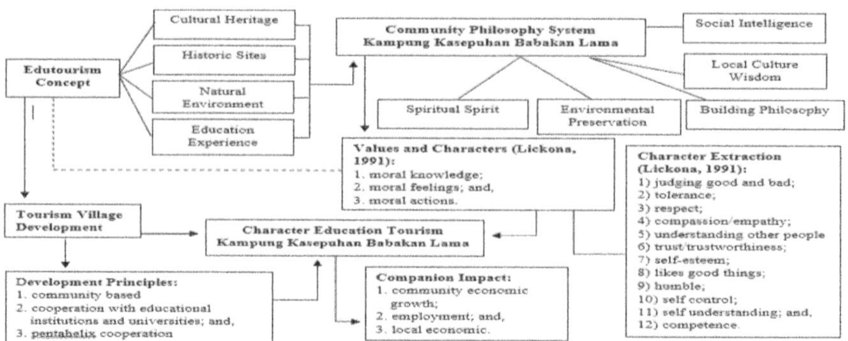

Figure 6. Design of development of values and character education tourism villages in Kampung Kasepuhan Babakan Lama.

4 DISCUSSION

The Kasepuhan Kampung Babakan Lama community is a regional community that still upholds its ancestral heritage. Inheritance is not only in the form of property, but also aspects of life. This aspect is based on a strong cultural philosophy and is trusted by the community. Each aspect does not stand alone, but complements each other and becomes the identity and local intelligence possessed. Based on a high socio-cultural spirit, the people of Kampung Kasepuhan have proven to be able to withstand the domination of foreign cultures that have infected Indonesia. The community is not disturbed by the hegemony of globalization and technological development in the 21st century. Without realizing it, actually the people of Kampung Kasepuhan can become role models for society at large. The local identity of the people of Kampung Kasepuhan is an Indonesian identity that is starting to disappear and be forgotten. Not only that, the social soul also embodies the gratitude of the people to God. Gratitude for the various kinds of goodness is given by God to the community, including a peaceful life and abundance of blessings from agricultural products/gardens. Gratitude is manifested by routine activities held in Kampung Kasepuhan Babakan Lama, including *nyarita* and *seren taun* filled with praise to God. The climax of each ritual is *babacak* or joint meals prepared independently and community cooperation (see Figure 5).

Local intelligence is very helpful in difficult times like now, namely, the spread of the plague Covid-19. When the Indonesian people were widely troubled and confused in facing this outbreak in terms of food security, the people of Kampung Kasepuhan Babakan Lama were not affected by this. This is caused by food security owned by the Kasepuhan community, as they have a reserve food source stored in *leuit* (see Figure 3). Food reserves in the form of agricultural products (rice) are stored for a long period of time, able to last up to one to two years. Thus, it can be understood that the local wisdom of the people of Kampung Kasepuhan Babakan Lama is the intelligence of thought initiated by the ancestors and passed down until now. The intelligence of thinking of the predecessors of the village is an asset worth promoting in the concept of edutourism with character values. The concept of edutourism and local wisdom of the Kampung Kasepuhan community can be made in the design of values and character education. This linkage is visualized in Figure 6.

The concept of edutourism on cultural heritage and the natural environment can be categorized into a number of tourist models, including historic sites, nature reserve parks, cultural heritage, etc., which provides educational experience. Referring to this opinion, character education tours can be designed to be implemented in Kampung Kasepuhan Babakan Lama. In that area, there are at least five aspects that form the basis of community life as the preservation of ancestral cultural heritage. These five aspects are the characteristics of the people of Kampung Kasepuhan Babakan Lama. This is consistent with the concept of character education (Lickona 2019), namely, 1) moral knowledge; 2) moral feelings; and, 3) moral actions. These three characters, if examined, are into 12 character values. The Kampung Kasepuhan Babakan Lama community system is closely related and matches the Lickona character values (see Figure 1). This is something positive and must be disseminated to the wider community. The pattern of distribution can be in the form of destination tourism character education.

Cooperation with educational institutions can be a profitable promotional bridge for the development of edutourism locations. Educational institutions can make Kampung Kasepuhan Babakan Lama a location for research and scientific studies relating to the social sciences and science. The results of his research can be used as written learning material related to the character development of students. The transfer of local wisdom values of the people of Kampung Kasepuhan can be used as a basis for realizing local identity and character that is threatening to disappear in the current era of globalization. This can be combined with technological developments that require intelligence to think, people can behave locally with global insight.

5 CONCLUSION

Kampung Kasepuhan Babakan Lama has a number of local heritages which are loaded with educational content as a representation of local intelligence in addressing the flow of life. Some aspects of Kampung Kasepuhan Babakan Lama community can be transferred to visitors, namely social, cultural, environmental, spiritual, architectural, and philosophical values. The aspects of local intelligence possessed by the Kampung Kasepuhan community are local identities that reflect the character of the Indonesian people. These aspects must be transmitted to the (young) people who visit because they are indicated to have lost the character of the nation (based on various research results). In addition, the adult community (parents) who visit can take advantage of the local intelligence of Kampung Kasepuhan as a nostalgic medium with its former life, which is identical and loaded with wisdom values. The design of tourism villages can be carried out in various regions that have the potential to be studied and promoted as tourism villages. This does not rule out the possibility that with many similar studies it will arouse the village community to develop their village.

REFERENCES

Baiquni M, Dzulkifli M. 2019. Implementing community-based tourism. *Delivering Tourism Intelligence.* (11):61–75. https://doi.org/10.1108/S2042144320190000011006

Barkathunnisa, AB, Lee D, Price A, Wilson E. 2018. Towards a spirituality-based platform in tourism higher education. *Journal Current Issues in Tourism* 22(17):2140–56. https://doi.org/10.1080/13683500.2018.1424810

Blapp M, Mitas O. 2017. Creative tourism in Balinese rural communities. *Journal Current Issues in Tourism.* 21(11): 1285–1311. https://doi.org/10.1080/13683500.2017.1358701

Boyle A, Wilson E, Dimmock K. 2015. Transformative education and sustainable tourism: The influence of a lecturer's worldwide. *Journal of Teaching in Travel & Tourism.* 15(3): 252–63. https://doi.org/10.1080/15313220.2015.1059303

Camilleri MA. 2018. The Planning and Development of the Tourism Product. Camilleri MA, editors.Tourism Planning and Destination Marketing. United Kingdom: Emerald Publishing Limited; 2018. p. 1–23.

Cascante, DM, Suess C. 2020. Natural amenities-driven migration and tourism entrepreneurship: Within business social dynamics conducive to positive social change. *Tourism Management* 81:p. 1. https://doi.org/10.1016/j.tourman.2020.104140.

Dukić V, Volić I. 2020. The importance of documenting and including traditional wisdom in community-based ecotourism planning: A case study of the nature park ponjavica in the village of Omoljica (Serbia). *Sage Journal* 7(1):1–11. https://doi.org/10.1177%2F2158244016681048

Gunter U, Ceddia MG, Troster B. 2016. International ecotourism and economic development in Central America and the Carribean. *Journal of Sustainable Tourism.* 2016; 25(1): 43–60. https://doi.org/10.1080/09669582.2016.1173043

Kachniewska MA. 2015. Tourism development as a determinant of quality of life in rural areas. *Worldwide Hospitality and Tourism Themes.* (5): 500–15. https://doi.org/10.1108/WHATT-06-2015-0028

Kanwal S, Rasheed MI, Pitafi AH, Pitafi A, Ren M. 2020. Road and transport infrastructure development and community support for tourism: The role of perceived benefits and community satisfaction. *Tourism Management* 77: p.1. https://doi.org/10.1016/j.tourman.2019.104014

Kembara MD, Rozak RWA, Hadian VA. 2019. Research-based lectures to improve students' 4c (communication, collaboration, critical thinking, and creativity) skills. *Proceedings ISSEH.*; 306:22–6. https://doi.org/10.2991/isseh-18.2019.6.

Lickona T. 2019. Educating for character: How our schools can teach respect and responsibility. Wahyudi U, Suryani S, editors. Jakarta: Bumi Aksara.

Lin, CL. 2019. Establishing environment sustentation strategies for urban dan rural/town tourism based on a hybrid MCDM approach. *Journal Current Issues in Tourism.* p.1. https://doi.org/10.1080/13683500.2019.1642308

Luo JM, Qiu H, Lam CF. 2016. Urbanization impacts on regional tourism development: A case study in China. *Journal Current Issues in Tourism.* 19 (3): 282–95. https://doi.org/10.1080/13683500.2015.1033385

Matahir H, Tang CF. 2017. Educational tourism and its implication on economic growth in Malaysia. *Asia Pacific Journal of Tourism Research* 22(11): 1110–23. https://doi.org/10.1080/10941665.2017.1373684

McGladdery CA, Lubbe BA. 2017. Rethinking educational tourism: Proposing a new model and future directions. *Tourism Review.* 2017;72 (3): 319–29. https://doi.org/10.1108/TR-03-2017-0055

Mihalic T, Liburd JJ, Guia J.Values in Tourism: The Case of EMTM. 2015. In P. Sheldon, & C. Hsu (Eds.), *Tourism Education.* 21: 41–59. https://doi.org/10.1108/S1571504320150000021010

Min W. 2017. How to construc system guarantee for economic development of eco-tourism resources based on value compensation. *Journal The Anthropologist.* 22(1): 101–12. https://doi.org/10.1080/09720073.2015.11891861

Nagy KXH, Segui AE. 2010. Experiences of community-based tourism in Romania: Chances and challenges. *Journal of Tourism Analysis.* 2010; p. 1. https://doi.org/10.1108/JTA-08-20190033

Park E, Choi BK, Lee TJ. 2019. The role and dimensions of authenticity in heritage tourism. *Tourism Management.* 74: 99–109. https://doi.org/10.1016/j.tourman.2019.03.001

Pitman T, Broomhall S, Majocha E. 2016. Teaching ethics beyond the academy: Educational tourism, life-long learning, and phronesis. *Journal Studies in the Education of Adults.* 43(1): 4–17. https://doi.org/10.1080/02660830.2011.11661600

Rehman A, Ma H, Irfan M, Ahmad M, Traore O. 2020. Investigating the influence of international tourism in Pakistan and its linkage to economic growth: Evidence from ARDL approach. *Sage Journal.* 2020; 10(2): 1–10. https://doi.org/10.1177/2158244020932525

Ritchie, B. W. 2003. Managing Educational Tourism. Channel View Publication: Toronto & Sydney

Sharma N, Sarmah B. 2019. Consumer engagement in village eco-tourism: A case of the cleanest village in Asia-Mawlynnong. *Journal of Global Scholars of Marketing Science.* 2019; 29(2): 248–65. https://doi.org/10.1080/21639159.2019.1577692

Sie L, Patterson I, Pegg S. 2015. Towards an understanding of older adult educational tourism through the development of a three-phase integrated framework. *Journal Current Issues in Tourism.* 19(2):100–36. https://doi.org/10.1080/13683500.2015.1021303

Tang CF. 2019. The treshold effects of educational tourism on economic growth. *Journal Current Issues in Tourism.* 2019; p. 1. https://doi.org/10.1080/13683500.2019.1694869

Wu TC, Lin YE, Wall G, Xie PF. 2020. A spectrum of indigenous tourism experiences as revealed through means-end chain analysis. *Tourism* Management. 2020;76:p. 1. https://doi.org/10.1016/j.tourman.2019.103969

Zhang R, Smith L. 2019. Bonding and dissonance: Rethinking the interrelations among stakeholders in heritage tourism. *Tourism Management* 74: 212–23. https://doi.org/10.1016/j.tourman.2019.03.004

Promoting Creative Tourism: Current Issues in Tourism Research – Kusumah et al. (Eds)
© 2021 Taylor & Francis Group, London, ISBN 978-0-367-55862-8

Interculture language learning: Literacy level determines the development of tourist numbers in Indonesia?

D. Hadianto, V.S. Damaianti, Y. Mulyati, A. Sastromiharjo & N. Budiyanti
Universitas Pendidikan Indonesia, Bandung, Indonesia

ABSTRACT: This study aims to investigate correlation knowledge competencies about Indonesia and the number of Indonesian tourist attractions visited. This research used the correlation method by linking knowledge competencies about Indonesia and the number of Indonesian tourist attractions visited. The data was also supported by interviewing to find several factors that made them compelled to come to Indonesia. The participants is a student from an Indonesian language learner for foreign speaker class or called BIPA. The participants take from BIPA learners academic years 2018–2020 at one of the private Indonesian language courses in Indonesia. The results show that all of tourists from various countries who have a high level of literacy about Indonesia visit more tourist attractions than BIPA students who lack knowledge about tourist attractions in Indonesia. Knowledge about tourist attractions is obtained from BIPA learning and other media. The higher literacy about Indonesia makes their curiosity higher. This implies that language and culture learning for foreigners plays an important role in developing business in the tourism sector.

Keywords: literacy level, development tourist, Indonesian culture

1 INTRODUCTION

The development of literacy is one of the main goals of language in planning education and has grown increasingly significant in the last few decades. In particular, there has been a strong focus on planning to develop the literacy capabilities of large sectors of the population as part of mass education both by governments and non-government organizations (NGOs) such as UNESCO, the IMF, and the World Bank. In part, the emphasis on literacy has grown from the perception that literacy is fundamental to contemporary economic systems and that economic development depends on providing an adequate level of literacy for a broader segment of the population (Achinstein 2012; Walker 2014). Responding to goals such as literacy, language planning has traditionally taken the form of determining how to instill literacy in the broadest possible population segment. The key planning dimension of literacy is to determine the best way to provide as much literacy training as possible in the resources available to the government. Literacy must be recognized as a tool that will change society significantly by achieving extra-linguistic goals of economic development, social improvement, or democratization, although the ability of literacy to achieve these goals is debatable (Fulford 2009; Grimm 2008).

Recently, and especially in the context of social and economic globalization, questions around planning for literacy have become more complex than those related to program formation. Specifically, the nature of literacy has naturally changed as a result of changes in communicative practices, and now there is new literacy that has arisen due to changes in communication, both of which contrast with and add to the old literacy associated with more traditional communicative practices. This means that language planning for literacy is no longer just a matter of planning for improvement in pedagogical methods to teach a stable version of print literacy but instead, it needs to be involved

DOI 10.1201/9781003095484-56

with understanding and conceptualization that emerges and develops and the definition of what literacy means, how, where, and what context and what is the modality. One literacy that can be improved to help the economy is literacy in the tourism sector, both literacy on the part of tourists and the literacy of domestic activists in the tourism sector (Mckercher 2009; Shi et al. 2017). Literacy that can be targeted in developing tourism is the literacy of foreign students, namely, literacy about culture and tourist attractions in Indonesia. Foreign students who are easily reached in the literacy process, one of them is BIPA (Indonesian Language for Foreign Speakers) learners. This foreign student is certainly an illustration for us as what kind of tourism sector is in the outside world. This description can be used as feedback for Indonesia to continue to improve its promotion strategy in order to increase the literacy of outsiders about Indonesia so that they come to Indonesia. This can be an income for the economy in Indonesia, especially in the field of tourism.

In this paper will be raised questions about the role of literacy in the world of tourism. The role of literacy in the world of tourism is currently being developed, especially in the field of promotion. This promotion certainly takes into account the literacy aspect, so researchers ask a number of research questions to uncover the role of literacy in developing the tourism sector. The research questions raised in this article is how does the correlation between the literacy level of BIPA students and the number of tourist attractions visited? What drives BIPA students to visit tourist attractions in Indonesia? What are the sources of media literacy that are BIPA learners' means of obtaining information on tourism development in Indonesia? The results of this study are expected to contribute to the development of tourism in Indonesia in terms of what strategies can be done to increase the amount of tourism in Indonesia.

2 LITERATURE REVIEW

Literacy is the basis for someone to learn other things throughout his life. Literacy plays an important role in human life for its own well-being and social life (Liddicoat 2004; Ollerhead 2019). Initially, literacy is meant as literacy. However, the meaning is the wrong meaning. The literal and technical term literacy only talks about reading and writing. So, literacy is more suitable to be interpreted as literacy. Literacy implies the ability possessed by someone and is the result of language skills, namely reading and writing so as to produce and manipulate science, linguistic, and grammatical analysis that produce oral texts and written texts and from the impact of human history toward the philosophical and social development of western education (Bozkurt et al. 2020; Hillman & Moore 2013). Literacy is deeply influenced by context. Each person and each use of literacy is situated in a world that is interactional, has certain ideologies, and in which changes occur as the context changes (Branch 2014; Vander Zanden 2015).

Even changes and developments in human life from time to time can be said to be the implications of literacy. Literacy in this study focuses on reading skills. Reading is an element of the main activity in literacy besides writing. The concept of reading undergoes a paradigm shift (Hillman & Moore 2013; Loring 2017). Reading is an activity that involves a complex process because the reader goes through several stages during reading until finally the reader gets the information, understands, and interprets the information. As stated by Caldwell (2008), "reading is an extremely complex and multifaceted process." The reader passes a simultaneous process during reading. First, the reader interprets the language codes perceptually and conceptually. After that, the reader collaborates ideas that are interpreted with the schemata they already have with the aim of interpreting in detail the information contained in the reading text, so that in the end the reader can build a new concept from the reading results or can be called "The mental representation that the reader constructs from the text" (Liddicoat 2004; Ollerhead 2019).

As studies in the field of literacy prove, that the better the literacy of a person, the interest in an object increases. Literacy makes someone more knowledgeable about an object, the greater one's interest in the object. Literacy is also able to make it easier for someone to understand information coming from various media (Gillovic et al. 2018; Hall 2015). This theory is the basis of the author to analyze the correlation of literacy with the number of tourist attractions visited by a tourist.

It cannot be denied that the tourism sector is one source of income to improve the economy of a country, including the country of Indonesia (Gillovic et al. 2018; Goethals 2016). Therefore, we must continuously develop strategies to attract tourists from outside. The tourism sector cannot be separated from the language in the target country. When tourists visit a tourist spot in a country, consider the aspect of language. Tourists will definitely prefer to target countries that are more mastered regarding language and literacy aspects about the country (Achinstein 2012; Goethals 2016).

3 RESEARCH METHOD

Correlation method is used in research by linking literacy competencies about tourist attractions in Indonesian BIPA students and the number of tourist attractions in Indonesia (Lee & Lawson 1996). The correlation be-tween these two variables can be used as an illustration for researchers to take further action in developing the economy in tourism. Participants were selected from BIPA students starting from the class of 2018–2020 who studied in one of the private courses in Indonesia. The participants was selected from BIPA students on the grounds that BIPA students who came to Indonesia had the potential to become quite large tourists during the academic year. BIPA learners from 2018–2020 numbered 100 people with 70% of them being female. BI-PA learners come from various countries. The following is the distribution of participant data.

The age of participants is in the range of 20–45 years, dominated by BIPA students who have goals for academic and tourism purposes. The data was obtained through a reading test of understanding about Indonesia, ranging from culture, tourist attractions, and Indonesian language. From the test results the researchers categorized BIPA students into three levels of literacy about Indonesia. The results of the author's analysis correlate with the number of Indonesian tourist attractions they have visited. The data is then supported by interview data regarding BIPA learners' motivation to learn Indonesian.

The first step of the researcher is to record the level of literacy of BIPA students at the beginning of the semester about tourist attractions in Indonesia. The results of this data collection are categorized into high, medium, and low literacy. The results of these data will be correlated with the results of data collection on the number of tourist attractions that have been visited in Indonesia. In addition, the data is supported by in-depth interviews with BIPA students to obtain information on their motivation to visit tourist attractions in Indonesia. Data processing researchers used SPSS by correlating the two data to find a significant level supported by interviews with BIPA learners. Data processing is accompanied not only by correlation data but also by other processing results, which illustrate the relationship between the two variables.

Table 1. Countries and number of participants.

Country	Number
Australia	6
Denmark	4
Japan	10
German	5
South Korea	20
Qatar	5
Uzbekistan	3
Thailand	10
Philiphines	7
Laos	10
Vietnam	10
Myanmar	10

4 RESULT AND DISCUSSION

Based on data processing the following results can be obtained. Data processing from the output of SPSS results is explained in this study to strengthen research arguments and data. Research questions on the correlation can be answered with the processing data below.

From the results ofthe SPSS processing above, Pearson correlation value reached 0.943 or 94.3%. This value can be interpreted that the correlation between literacy level of BIPA learners and the number of tourist attractions visited by BIPA learners can be classified into a very high correlation. This is reinforced by the results of other SPSS processing. The correlation data above shows that the literacy level and the number of BIPA learners visiting tourist attractions are interrelated. BIPA learners who have knowledge of tourist attractions in Indonesia are low in number of visits, and are slightly different from BIPA learners who have good knowledge of tourist attractions in Indonesia. This is in line with several theories and previous studies on literacy. Someone's literacy is a very determining attitude and action to be taken after receiving information to strengthen the data. To strengthen the hypothesis of research "Literacy level really determines the number of tourist attractions visited by BIPA students," the authors display the ANOVA data table SPSS processing results. Anova's results also showed the same conclusion. For more details see Table 3.

Significance test of the regression line equation obtained from the Regression line, namely Fhit (b/a) = 759,164 and P-value 0,000 < 0.05 or H0 rejected. Thus, the regression of Y (Number of Tourist Attractions) over X (Literacy level) is significant or Literacy level is very influential on Number of Tourist Attractions. The significance level above shows that literacy level has an important role in attracting the number of foreign tourists to visit Indonesia. A summary of the overall data can be seen in the next table, Table 4.

The significance test of the correlation coefficient is obtained from the Summary Model table. Seen in the first column the correlation coefficient (rxy) = 0.943 and Fhit = (Fchange) = 759,164, with p-value = 0,000 < 0.05. It can be concluded that (Ho) is rejected or literacy level has a very strong correlation with the number of tourist attractions visited by BIPA students. Of the four tables (Tables 1–4) described above, the four tables are aligned to lead to the result that the literacy level

Table 2. Correlation between literacy level and number of tourist attractions.

		Literacy Level	Number of Tourist Attractions
Literacy Level	Pearson Correlation	1	.943**
	Sig. (2-tailed)		.000
	N	96	96
Number of Tourist Attractions	Pearson Correlation	.943**	1
	Sig. (2-tailed)	.000	
	N	96	96

**. Correlation is significant at the 0.01 level (2-tailed).

Table 3. Processing results of ANOVA[a].

Model	Sum of Squares	Df	Mean Square	F	Sig.
Regression	8033.164	1	8033.164	759.164	.000[b]
Residual	994.669	94	10.582		
Total	9027.833	95			

a. Dependent Variable: Number of Tourist Attractions
b. Predictors: (Constant), Literacy Level

Table 4. Table of model summary.

					Change Statistics				
Model	R	R Square	Adjusted R Square	Std. Error of the Estimate	R Square Change	F Change	df1	df2	Sig. F Change
1	.943[a]	.890	.889	3.253	.890	759.164	1	94	.000

a. Predictors: (Constant), Literacy Level
b. Dependent Variable: Number of Tourist Attractions

of BIPA learners greatly influences BIPA learners' decisions when visiting tourist attractions in Indonesia (Carden 2012; Köseoglu & King 2019).

Based on the above data presentation, the literacy ability of foreign students in capturing information about tourism greatly affects them to travel to tourist attractions in Indonesia. This is evident from the analysis of data processing using SPSS. Based on the results of in-depth interviews, the main factor that drives BIPA students to go to Indonesia as tourists and learners is to draw a picture of the culture and nature of Indonesia that they obtain from various media. The more intense these BIPA learners get information about tourist at-tractions in Indonesia, the more it makes them interested in visiting tourist attractions in Indonesia. The description of culture and tourist attractions in Indonesia that they get through learning becomes a factor that greatly affects these learners visiting tourist attractions in Indonesia. It can be said that attractive packaging in promoting tourist attractions in Indonesia greatly influences their decision to visit tourist attractions in Indonesia. This attractive packaging certainly considers various aspects, namely design, language, and other multi-modal (Goethals 2016; Hall 2015).

In terms of culture, their motivation to visit Indonesia is unique, diverse, and friendly to Indonesian culture in their view. Such depictions of culture in the tourism marketing strategy are very influential, whereas from the aspect of tourist attractions, BIPA learners are interested in visiting because of the exposure of language aspects that are very attractive to tourists, and often these BIPA learners intersect with other texts or media that contain information about the place tourism in Indonesia (Drozdzewski 2011; Hall 2015). From these two data we can conclude that language in conducting marketing strategies plays an important role. Language can make someone's view stronger or change depending on the style of exposure (source) because one's literacy ability can be said to be the process of digesting information by absorbing new information and associating it with knowledge or images that already exist in the schema. Some research on literacy ability to be able to change the view of an object has been done. Therefore, we can make literacy in BIPA learners targeted in marketing.

In addition, Indonesian language is one of their motivations to come to Indonesia. According to them, Indonesian has become one of the languages that is taken into account quite internationally because of the increasing number of speakers outside. Literacy media sources about tourism are electronic media reaching 75%, print media 25%. Electronic media that are included in the category are the official web and social media, while the print media obtained are pamphlets, posters, BIPA learning books, and other print media. The media plays an important role in increasing one's literacy; the media can change one's mindset by directing one's mind by choosing the right language in accordance with the wishes of the author (source). BIPA learning is also loaded with culture and tourist attractions material to be one of the means to attract BIPA students to visit these tourist attractions. Based on data from the results of this study, researchers highlighted the means in developing the literacy skills of BIPA learners. This tool is a strategy of promoting tourist attractions that use Indonesian and English as well as marketing design that greatly influences the attraction of BIPA students to go to tourist attractions. Several studies that have proven the language in promoting tourist attractions greatly affects the number of tourist visits in a country (Mckercher 2009; Pudliner 2007; Shi & Chen 2017).

From an educational point of view, we as actors in the world of education have a very important role in increasing the number of foreign tourists visiting Indonesia. Based on the data obtained, foreign tourists, especially BIPA students, are interested in coming to various Indonesian tourist attractions because of the very at-tractive promotional strategy packaging. Therefore, we as educators must highlight the aspects of literacy in conducting promotional strategies. Text and other aspects or multimodal texts, as well as the delivery of material about the contents of tourist attractions to BIPA learners greatly determine their decision to visit these at-tractions (James-Burdumy et al. 2012; Taboada & Buehl 2012). This level of literacy can be increased through the factor of intensive tourist contact with an object. In the realm of BIPA (Indonesian for Foreign Speakers) learning, instructors can do learning by promoting intercultural language learning. Through this intercultural language learning, BIPA learners can learn Indonesian while receiving information on various cultures and tourist attractions in Indonesia that will make BIPA learners have the potential to visit these tourist attractions. Of course, this can be done by BIPA teachers through learning styles and exposure to material through text or other multimodal texts to improve the literacy ability of BIPA learners (Achinstein 2012; Pudliner 2007)

Interculture language learning can be done in the process of learning a second language. Second language learning through interculture language learning, in addition to being able to introduce the second language being studied, learners can also get to know the culture of second language speakers (Achinstein 2012; James-Burdumy et al. 2012; Leung et al. 2014). Through this strategy, BIPA learners' literacy in Indonesian culture will increase, so that in the end they will be interested in things or objects related to Indonesia, such as tourist attractions or Indonesian culture. This interest will make BIPA learners come to Indonesian tourist sites to find out more about these tourist attractions. Interculture language learning in the process of learning a second language must always be absorbed in every second language learning material, so that learners do not feel bored. The delivery strategy must also be modified as attractive as possible by integrating the material and other aspects of the culture of the target language speakers being studied.

5 CONCLUSION

This study aims to evaluate the relationship between BIPA learners' literacy skills and the number of student visits to tourist attractions. Using the correlation method, the relationship between BIPA learners' literacy skills and the number of student visits to tourist attractions has a very strong correlation. This indicates that the literacy ability of BIPA learners used as participants in this study greatly influences their decision to visit a tourist spot in Indonesia. The motivating factors for BIPA students visiting tourist attractions in Indonesia also lead to the interesting packaging of information they get from various media. Based on these data, from the aspect of education what can be done to help the development of the tourism sector is to make literacy levels a target that must be increased by education practitioners. In this case, the language used in the promotion strategy of a tourist place greatly influences the decisions of foreign tourists. In addition, a friendly service attitude toward foreign tourists also influences whether or not a tourist visits a tourist attraction.

The ability of tourist literacy can be enhanced through the factor of intensive tourists intersecting with an object. In the realm of BIPA (Indonesian for Foreign Speakers) learning, instructors can do learning by promoting intercultural language learning. Through this intercultural language learning, BIPA learners can learn Indonesian while receiving information on various cultures and tourist attractions in Indonesia that will make BIPA learners have the potential to visit these tourist attractions. Of course, this can be done by BIPA teachers through learning styles and exposure to material through text or other multimodal texts to improve the literacy ability of BIPA learners.

REFERENCES

Achinstein, B. (2012). Teachers and Teaching: theory and practice New teacher and mentor political literacy: reading, navigating and transforming induction contexts, (January 2015), 37–41. https://doi.org/10.1080/13450600500467290

Branch, B. D. (2014). Libraries and Spatial Literacy: Toward Next-Generation Education. *College and Undergraduate Libraries*, *21*(1), 109–114. https://doi.org/10.1080/10691316.2014.877745
Carden, S. (2012). Making space for tourists with minority languages: The case of Belfast's Gaeltacht Quarter. *Journal of Tourism and Cultural Change*, *10*(1), 51–64. https://doi.org/10.1080/14766825.2011.653360
Drozdzewski, D. (2011). Language tourism in poland. *Tourism Geographies*, *13*(2), 165–186. https://doi.org/10.1080/14616688.2011.569569
Fulford, A. J. (2009). Cavell, literacy and what it means to read. *Ethics and Education*, *4*(1), 43–55. https://doi.org/10.1080/17449640902860689
Gillovic, B., McIntosh, A., Darcy, S., & Cockburn-Wootten, C. (2018). Enabling the language of accessible tourism. *Journal of Sustainable Tourism*, *26*(4), 615–630. https://doi.org/10.1080/09669582.2017.1377209
Goethals, P. (2016). Multilingualism and international tourism: a content- and discourse-based approach to language-related judgments in web 2.0 hotel reviews. *Language and Intercultural Communication*, *16*(2), 235–253. https://doi.org/10.1080/14708477.2015.1103249
Grimm, K. J. (2008). Longitudinal associations between reading and mathematics achievement. *Developmental Neuropsychology*, *33*(3), 410–426. https://doi.org/10.1080/87565640801982486
Hall, C. M. (2015). On the mobility of tourism mobilities. *Current Issues in Tourism*, *18*(1), 7–10. https://doi.org/10.1080/13683500.2014.971719
Hillman, M., & Moore, T. J. (2013). The web and early literacy. *Web-Based Learning in K-12 Classrooms: Opportunities and Challenges*, 15–21.
James-Burdumy, S., Deke, J., Gersten, R., Lugo-Gil, J., Newman-Gonchar, R., Dimino, J., … Liu, A. Y. H. (2012). Effectiveness of Four Supplemental Reading Comprehension Interventions. *Journal of Research on Educational Effectiveness*, *5*(4), 345–383. https://doi.org/10.1080/19345747.2012.698374
Kato, K., & Horita, Y. (2018). Tourism Research on Japan–Overview on Major Trends: Japanese and English-language Materials. *Tourism Planning and Development*, *15*(1), 3–25. https://doi.org/10.1080/21568316.2017.1325392
Köseoglu, M. A., & King, B. (2019). Authorship Structures and Collaboration Networks in Tourism Journals. *Journal of Hospitality and Tourism Education*, *0*(0), 1–9. https://doi.org/10.1080/10963758.2019.1655433
Lee, C., & Lawson, C. (1996). Numeracy through literacy. *International Journal of Phytoremediation*, *21*(1), 59–72. https://doi.org/10.1080/0965079960040106
Leung, D., Li, G., Fong, L. H. N., Law, R., & Lo, A. (2014). Current state of China tourism research. *Current Issues in Tourism*, *17*(8), 679–704. https://doi.org/10.1080/13683500.2013.804497
Liddicoat, A. J. (2004). Language Planning for Literacy: Issues and Implications. *Current Issues in Language Planning*, *5*(1), 1–17. https://doi.org/10.1080/14664200408669076
Loring, A. (2017). Literacy in Citizenship Preparatory Classes. *Journal of Language, Identity and Education*, *16*(3), 172–188. https://doi.org/10.1080/15348458.2017.1306377
Mckercher, B. (2009). The State of Tourism Research: A Personal Reflection. *Tourism Recreation Research*, *34*(2), 135–142. https://doi.org/10.1080/02508281.2009.11081585
Ollerhead, S. (2019). "The pre-service teacher tango": pairing literacy and science in multilingual Australian classrooms. *International Journal of Science Education*, *0*(0), 1–20. https://doi.org/10.1080/09500693.2019.1634852
Pudliner, B. A. (2007). Alternative literature and tourist experience: Travel and tourist weblogs. *Journal of Tourism and Cultural Change*, *5*(1), 46–59. https://doi.org/10.2167/jtcc051.0
Shi, B., Zhao, J., & Chen, P. J. (2017). Exploring urban tourism crowding in Shanghai via crowdsourcing geospatial data. *Current Issues in Tourism*, *20*(11), 1186–1209. https://doi.org/10.1080/13683500.2016.1224820
Taboada, A., & Buehl, M. M. (2012). Teachers conceptions of reading comprehension and motivation to read. *Teachers and Teaching: Theory and Practice*, *18*(1), 101–122. https://doi.org/10.1080/13540602.2011.622559
Vander Zanden, S. (2015). Productive taboos: cultivating spatialized literacy practices. *Pedagogies*, *10*(2), 177–191. https://doi.org/10.1080/1554480X.2014.985299
Walker, T. R. (2014). Historical Literacy: Reading History through Film, (December), 37–41. https://doi.org/10.3200/TSSS.97.1.30-34

Promoting Creative Tourism: Current Issues in Tourism Research – Kusumah et al. (Eds)
© 2021 Taylor & Francis Group, London, ISBN 978-0-367-55862-8

CATC implementation to strengthen the industrial-based tourism competency of vocational schools

D. Sunarja
STIPAR Yapari, Bandung, Indonesia

O.D. Maharani
SMK Pariwisata Metland School, Bogor, Indonesia

ABSTRACT: CATC is prepared based on the ASEAN Common Competency Standards on Tourism Professionals which aims to prepare vocational students to have international standard competency in tourism, as well as a ground work in the application of the ASEAN Economic Community. Qualitative research methodology was employed in this study by using a case study method: interviews, observation, and documentary. CATC is intended for the tourism industry with hospitality expertise. CATC training is carried out by hospitality instructors and is applied in the learning process in the classroom. Implementation of CATC creation and development of human resources with ASEAN tourism standards and encourages life learning for hospitality workers. A sample of this research are 52 students for the main research. Results of the evaluation of Learning Material Activities showed that the students' scores in the unit of Work in a Socially Diversity Environment is 85.8 with the passing scores of 100%.

Keywords: CATC, tourism education, vocational school

1 INTRODUCTION

ASEAN would be an open market and unity-based production and mobility, the flow of goods, services, investment, capital, and skilled labor would move freely. Meanwhile, Indonesia has undergone concerns that human resources are not competitive compared to other countries in ASEAN. In terms of population, energy purposes, and human resources, the challenge in the face of the liberalization of trade and also ASEAN economic community by Alisjahbana (2014), there are three aspects, namely, (1) maintain the current demographic, (2) increase labor force participation, (3) increase productivity and labor.

This makes the education sector as top quality to the human resources. However, the education sector still faced an amount of unemployment. In the last year, unemployment in 60,000 people was different from unemployment level that fell to 4.99% in February 2020. Seen from education level, unemployment level of vocational high school still is the highest of others, which is 8.49 percent (Central Bureau Statistics 2020).

Vocational high schools as one of the creators of graduates in tourism. Curriculum, facilities, and infrastructure, teachers and strategies of learning in high school have to adjust to the needs of the tourism industry that are thriving now (Figure 1). The Ministry of Education and Culture with the Ministry of Tourism in Indonesia, the Ministry of Employment, and National Board Certification Profession perform together to use a curriculum ASEAN standard called CATC (Common ASEAN Tourism Curriculum). ASEAN CATC are based on common competency standard on tourism professional (ACCSTP), whose aim is to prepare students to have competence for tourism, which an international standard was also the asset in the practice of ASEAN economic community.

CATC is the most simple curriculum that is able to provide solutions to face the employment market challenges of the ASEAN economic community. The application of CATC in vocational

 DOI 10.1201/9781003095484-57

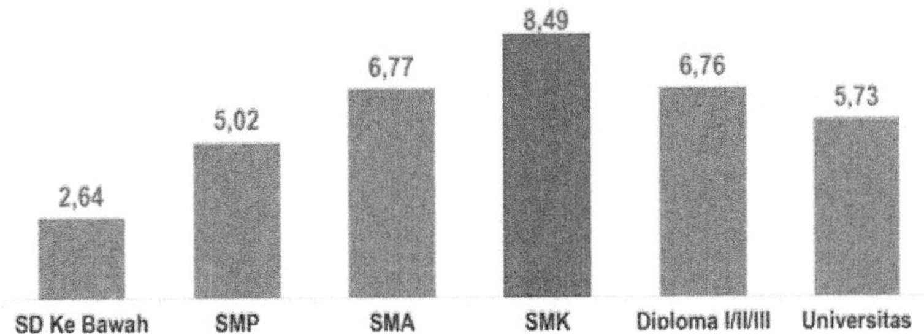

Figure 1. Level of unemployment.

school is expected to improve the quality of human resources for tourism that could support Indonesia tourism. CATC (Common ASEAN Tourism Curriculum) is tourism curriculum approved by AMS (ASEAN Member States) that focuses on competency-based training (CBT). CATC provide resources skills training, knowledge, and attitude needed to professional tourism. Curriculum CATC is based on a series of competency standards arranged in the form of building blocks that include services for hotel-hospitality restaurant, services for hotel-front office and housekeeping, and tour and travel. Competence that is aligned with CATC is divided into six, including Food Production, Food and Beverage Services, Tour Operation, Travel Agencies, Front Office, Housekeeping (ATPRS 2013).

Increased education curriculum and skill of tourism and formulated standard are indispensable in recognition that it leads to the skills and qualifications in the ASEAN (MRA 2013). This means, CATC is a set international curriculum that can be used in vocational high school in Indonesia. Tourism has a particular purpose, a curriculum that is to make the quality of tourism work better. In several countries ASEAN was using the curriculum like of this to develop the quality of their work. One example is Thailand's CATC, having used the curriculum to generally their integrated curriculum.

CATC is applied so that vocational high school graduates should absorbed to the maximum according to ASEAN standards by harmonizing students' competencies with the skills of workers in the international industrial world. CATC is considered to be able to meet the basic capabilities of the tourism sector, especially hospitality, which have the characteristics of an industrial culture, and fully understood the standard operating procedures (SOP) or occupational, health and safety.

CATC is very important to implement as there is a certification system that will guarantee graduates and at the same time recognize the competency achievements of these vocational high school graduates according to international hospitality competency standards. To achieve this, BNSP (Tourism Professional Certification) in Indonesia together with the Ministry of Tourism developed a certification scheme used to certify graduates of vocational high school Tourism in Indonesia that vocational high school graduates are worthy and should be absorbed in the international industry.

In Indonesia itself, there are only 21 vocational high schools implementing CATC. Metland Tourism Vocational School is one of them. Metland Vocational School is a part of an educational institution in Indonesia using the curriculum in 2013 aligned with CATC curriculum with various competency skills such as Hotel Accommodation, Hotel Management, and Culinary but in this research will focus on learning skills in Hotel Accommodation at Work in a Socially Diverse Environment-Trainer Guide. It is linked to learning skills and knowledge that are required to work effectively in a socially diverse environment in various environments within the context of work in the hotel and travel.

Formulation problems related to it are focused on how the CATC with competency standard on a work effectively in socially diverse environment in various environment-trainer in accommodation hospitality skills in Metland vocational school and how is the enhancement increased to the quality of students? Main purpose of the study is to know the impact of the quality of students' development after the application of CATC in schools. In general, the results of research are expected to give contribution as a pilot project related to parties' vocational generation especially schools planning to learn CATC.

2 LITERATURE REVIEW

2.1 *CATC*

CATC (Common ASEAN Tourism Curriculum) is tourism curriculum agreed by AMS focusing on CBT (competency-based training). CATC provides skill training, resources knowledge, and attitudes needed in professional tourism. CATC is based on a series of standards arranged in the form of building blocks, Hotel-Hospitality Restaurant Services, Covering: Hotel-Front Office and Housekeeping Services, Tour and Travel. Competence in CATC curriculum is divided into six divisions labor:

- Food Production (FP)
- Food and Beverage Services (FB)
- Tour Operation (TO)
- Travel Agencies (TA)
- Front Office (FO)
- Housekeeping (HK)

CATC toolbox consists of three main elements: Training Manual, Assessor Manual, Trainer Guide and summary of slides in a coach to PowerPoint of that is used as a handle of training and explained the materials in class activity. There are also clear Standard Competency Skills and knowledge that is required to do a task or activity based on standards at a certain level. There are 242 competency standards in the tourism industry in regional ASEAN that have been developed, which includes knowledge, skill, and the requirement of working on the sixth (ATPRS 2013).

This research focuses on competency standard in the work in a socially diverse environment-trainer guide as follows.

Items in Training Manual CATC are as follows:

- Unit descriptor, explained about the unit: unit code, nominal hours, element, and performance criteria.
- Assessment matrix, explained about the assessment activities which assess the performance of skills, written test, and oral question
- Glossary, collection of terms used in the unit
- Presentation written work

Table 1. Competency standard in the work in a socially diverse environment-trainer guide as follows.

Element 1: Communicate with customers and colleagues from diverse backgrounds	
1.1	Value customers and colleagues from different cultural groups and treat them with respect and sensitivity
1.2	Take into consideration cultural differences in all verbal and non-verbal communication
1.3	Attempt to overcome language barriers
1.4	Obtain assistance from colleagues, reference books or outside organizations when required
Element 2: Deal with cross-cultural misunderstandings	
2.1	Identify issues which may cause conflict or misunderstanding in the workplace
2.2	Address difficulties with the appropriate people and seek assistance from team leaders or others where required
2.3	Consider possible cultural differences when difficulties or misunderstandings occur
2.4	Make efforts to resolve misunderstandings, taking account of cultural considerations
2.5	Refer issues and problems to the appropriate team leader/supervisor for follow-up

- Trainee evaluation sheet, evaluation form contained the unit and performance
- Trainee self-assessment checklist, performance checklist which is filled by the trainee as a form of completion to follow the certification

2.2 *Competency*

Competency showed that skill or knowledge characterized by professionalism in a particular field as the most important as these areas (Suhartini 2012). Welcoming ASEAN economic market vocational school will be more meaningful and effective if the curriculum used international curriculum (CATC) ranging from the preparation of the education program, the implementation, and evaluation (Sunaryo 2002). One important aspect that must not be forgotten in vocational school is relevance competency given to students with the demands of the workforce. The graduates of vocational schools have to have high competence which the global market's need.

Student competency can be measured through cognitive indicators affective and psychomotor (Bloom 1956; Fuad & Ahmad 2009) or in other words knowledge, namely, skill and attitude (Garcia-Barbero 1998). But referring to the CATC competency, students are judged on the standard of competency (CATC) that focuses on industry. Different standards of competency units work in a socially diverse environment. Competency is expected ias follows.

- Students are able to characterize (understand the characters) customers and colleagues from various cultural groups and treat them with respect and sensitivity
- Students are able to consider differences in verbal and non-verbal communication
- Students are able to try to overcome language difficulties
- Students are able to easily get help from colleagues, reference books, or outside organizations as needed
- Students are able to solve problems that can cause conflict or misunderstanding at work
- Students are able to overcome difficulties with the right people and seek help from leaders or others if needed
- Students are able to consider cultural differences as difficulties or misunderstandings occur
- Students are able to achieve success to resolve misunderstandings, taking into consideration cultural considerations
- Students are able to answer problems and problems to the team leader and provide appropriate solutions/follow-up

3 METHOD

Qualitative research methodology is used in this study using the case study method. Interviews, observations, and documentary analysis are used to collect primary data. Qualitative research methodology is used in this study using the case study method. Interviews, observations, and documentary analysis are used to collect primary data about improving student competency after applying the CATC curriculum.

Data collection techniques are through interviews, observation, and study documentation. In-depth interviews are intentional interactions in which the researcher aims to learn what other people know about the topic, to uncover and record what the person is experiencing, and the significance and meaning that someone might have in a social setting (Arthur 2012). Marshall and Rossman (2011) claim that the wealth of data from the field cannot be achieved effectively unless complementary observation is used. Documentation studies will be carried out by analyzing.

- Association of Southeast Asian Nation Mutual Regulations Arrangement on Work in a Socially Diverse Environment-Trainer Guide (104 pages)
- Mutual Regulation Arrangement on Tourism Professional (MRA-TP) Implementation Report (15 pages)
- Agreement on MRA-TP (14 pages)

Table 2. Collecting data technique.

No	Data	Technique	Analysis Data Technique
1	Literature	Documentation	Quality
2	Data to evaluate the effectiveness of teaching materials	Questionnaire and test	Quality and Quantity
3	data on improving students' skills on the standard Work in a socially diverse environment-Trainer Guide	Test	Quality and Quantity

Table 3. Lesson plan.

<table>
<tr><td>Unit title: work in a socially diverse environment</td><td>nominal hours: 25 hours</td></tr>
<tr><td colspan="2">Unit number: D1.HRS.CL1.19; D1.HOT.CL1.02; D2.TCC.CL1.02</td></tr>
<tr><td colspan="2">Unit descriptor: This unit deals with the skills and knowledge required to work effectively in a socially diverse environment in a range of settings within the hotel and travel industries workplace context.</td></tr>
<tr><td>Elements and Performance Criteria</td><td>Assessment Guide</td></tr>
<tr><td>Element 1: Communicate with customers and colleagues from diverse backgrounds
1.1 Value customers and colleagues from different cultural groups and treat them with respect and sensitivity
1.2 Take into consideration cultural differences in all verbal and non-verbal communication
1.3 Attempt to overcome language barriers
1.4 Obtain assistance from colleagues, reference books, or outside organizations when required
Element 2: Deal with cross-cultural misunderstandings
2.1 Identify issues which may cause conflict or misunderstanding in the workplace
2.2 Address difficulties with the appropriate people and seek assistance from team leaders or others where required
2.3 Consider possible cultural differences when difficulties or misunderstandings occur
2.4 Make efforts to resolve misunderstandings, taking account of cultural considerations
2.5 Refer issues and problems to the appropriate team leader/supervisor for follow-up</td><td>The following skills and knowledge must be assessed as part of this unit:
• Knowledge of the principles of effective communication skills
• Ability to use active listening, feedback techniques, and team building techniques to build and maintain interpersonal relationships with customers and colleagues from diverse backgrounds
• Ability to identify the need(s) and concerns of others
• Ability to communicate effectively with a range of people from diverse backgrounds relevant to position and role
• Ability to appropriately deal with cross-cultural misunderstandings.
Context of Assessment
This unit may be assessed on or off the job:
• Assessment should include practical demonstration either in the workplace or through a simulation activity, supported by a range of methods to assess underpinning knowledge.
• Assessment must relate to the individual's work area or area of responsibility.</td></tr>
</table>

The data needed in this study are: (1) literature; (2) data to evaluate the effectiveness of teaching materials; (3) data on improving students' skills on the standard Work in a Socially Diverse Environment-Trainer Guide Data were analyzed qualitatively then described in descriptive form (Basrowi & Suwandi 2008). The study was conducted competently. The phenomena that appear are analyzed, interpreted, and drawn conclusions. This study takes the subject in class X students in hospitality accommodation at Metland Vocational School. The study was conducted in a period of six months. The total sample of this research are 52 students for the main research and 10 students for the limited research.

Table 4. Results of evaluation of the effectiveness of teaching materials.

Indicator	Student average score every item question	
	limited research	main research
People are all different. What are three characteristics that make one person different from another?	86	89
What are two ways you can improve your knowledge about different cultural requirements.	75	80
What are two things you can do to help verbal communication with a customer who has a little understanding of your language?	86	90
What are two points for using appropriate verbal and non-verbal communication when dealing with people from another culture?	81	83
When attempting to overcome language barriers, it is useful to learn a few simple words in a foreign language. What are three topics in which you could learn simple words or phrases?	86	88
Using gestures is an effective way to communicate with someone who does not speak your language. What are three ways you can use gestures to help a customer understand?	79	85
What are two examples of written communication you can use to collect information relating to cultural communication?	83	88
Difficulties or conflict often arise out of misunderstandings. What is this caused by?	85	83
Difficulties or conflict often arise out of misunderstandings. What is this caused by?	82	84
How can you identify conflict in the workplace?	86	89
What are some tips when addressing conflict and misunderstandings?	82	88
What is a cultural difference that a Hindu customer would have in relation to food?	81	85
What are actions you can take to help resolve misunderstandings?	80	86
Total	1072	1029
Average	82.5	85.8

Source: processed data

4 RESULT AND DISCUSSION

The study explored the implementation of the Common ASEAN Tourism Curriculum in Metland vocational School skill competence hotel accommodation. It also sought to understand how this implementation affected Metland Vocational School tourism and hospitality training. A qualitative methodology was employed in which embedded multiple case study methods were adopted as a research strategy.

The findings of CATC implementation and its effect in the cases of Metland Vocational School: The study found this school had implemented the CATC since 2017. Themes emerging from the data analysis suggested that international standard curriculum, student career mobility, school recognition, engaging the relationship with the hospitality industry, teacher skill, and quality development were the factors for Metland Vocational School to adopt and implement the CATC.

As follows is the lesson plan used in Work in a Socially Diverse Environment-Trainer Guide.

Evaluation of the effectiveness of teaching materials is done through limited testing and open testing. Through the question items, the average score of student learning outcomes in the Work in a Socially Diverse Environment-Trainer Guide unit was obtained. The following will present the results of the evaluation of the effectiveness of the teaching materials for each question item.

The results obtained indicate that the material in CATC is appropriate and can be easily understood by students. It can be seen from the assessment results that the expected value is as expected. From the table that has been shown, it shows that students' learning completeness in the Competency Standard in the Work in a socially Diverse Environment-Trainer Guide unit is 100% (Tables 1–4).

By using CATC curriculum, teachers expect the students to be mobilized to a hospitality career in ASEAN region due to CATC may enhance the level of quality and skills in international degree.

The opportunity for students to mobilize for a career in the hospitality industry was a major reason for adopting and implementing the CATC in the school. As a result, Metland Vocational School expected that their graduates would not only be recognized by local employers but also by ASEAN employers across the region.

To implement CATC in Metland Vocational School, they adapted the curriculum to fit with its context, especially the nature of students and school resources, the MRA-TP handbook (ASEAN 2013). The findings suggest that CATC Curriculum is the right and best choice to adopt as a school's curriculum, therefore generating graduates with global competencies that match the industry.

In general, ASEAN tourism economic cooperation and the need for standardization of tourism training were the major motivators driving CATC adoption and implementation. To improve the training program, the development of teacher skills and the enhancement of student learning quality were the reported motivations. Metland Vocational School recognition was found to be one of the key drivers in adopting CATC to improve regional perspective and student's English communications, international (regional) opportunities.

REFERENCES

Alisjahbana, A. S. (2014). Menteri PPN/Kepala Bappenas, tantangan kependudukan, ketenagakerjaan, dan SDM Indonesia menghadapi globalisasi khususnya masyarakat ekonomi Asean: "Arah kebijakan dan program di bidang kependudukan, ketenagakerjaan dan sumber daya manusia menghadapai glob. In Makalah. Jakarta.

ASEAN. (2013). *ASEAN MRA on Tourism Professionals Handbook.* Jakarta: ASEAN Secretariat tourism unit-infrastructure division.

ATPRS. (ASEAN *tourism professional registration system*). (2013). Overview of ASEAN. http://180.250.78.211:2203/atprs/dynamic?request=15. Diakses 10 Juni 2020.

Basrowi dan Suwandi. (2008). Memahami Penelitian Kualitatif. Jakarta: PT. Rineka Cipta.

Bloom, B. S. (1956). Taxonomy of Educational Objectives: The Classification of Educational Goals. London: David McKay Company, Inc.

Central Bureau Statistics. (2020). Statistik Daerah Kecamatan Umbulharjo 2012: Badan Pusat Statistik Kota Yogyakarta.

Fuad, N., & Ahmad, G. (2009). Integrated HRD: Human Resources Development. Jakarta: Grasindo.

Marshall, C., & Rossman, G.B. (2011). Primary Data Collection Methods Designing Qualitative Research. Los Angeles, CA: SAGE.

Suhartini, E. (2012). Motivasi, Kepuasan Kerja dan Kinerja. Makassar: Alauddin University Press.

Sunaryo, D. (2002). Sejarah pendidikan teknik dan kejuruan di Indonesia, membangun manusia produktif. Jakarta: Direktorat Pendidikan Menengah Kejuruan Departemen Pendidikan Nasional RI.

Promoting Creative Tourism: Current Issues in Tourism Research – Kusumah et al. (Eds)
© 2021 Taylor & Francis Group, London, ISBN 978-0-367-55862-8

An exploratory study on Singapore Polytechnic Hospitality and Tourism students' perception toward the use of virtual learning environments (E-learning)

J.M. Pang
Royal Melbourne Institute of Technology University, Ho Chi Minh City, Vietnam

ABSTRACT: With the advancement of technology and the exponential growth of students in the hospitality and tourism field, the use of e-learning has become more prolific amongst higher education institutions. However, the hospitality and tourism line has always been known to be one which requires manual labor with personalized service. In educational institutions, especially at the polytechnic level in Singapore, the training and learning provided to the students have mirrored the industry, with a "hand-on" approach, such as role-plays, simulations, tutorials, and so on. This has led to a dissonance in the pedagogy of the hospitality teaching. This study delves into the perceptions of Singapore hospitality and tourism students toward e-learning and the use of virtual learning environments (VLEs). From the data and analysis obtained, most students who have prior online competencies tend to accept online teaching, with those lacking in the necessary computing skill disliking e-learning. Most students feel that the ability to remote learn from any location was an incentive. The students also noted that the level of satisfaction derived from e-learning is dependent on the subject content and the application and teaching abilities of the lecturer.

Keywords: Singapore, polytechnic, students, online learning, hospitality, tourism

1 INTRODUCTION

1.1 *Technology in education*

The development in technology in recent years has had a tremendous impact on the education process, transforming education curriculum, learning materials, and practices. This worldwide practice has had a major impact on how higher education is currently facilitated. According to Dale and Lane (2004), higher education has traditionally utilized face-to-face delivery or through a paper-based distance learning. However, this change in teaching pedagogy has augmented the teaching landscape and has been adapted by many institutions of higher learning with the use of virtual learning environments (VLEs) (Dale 2003). These VLEs have been designed to enhance students' learning, through the encouragement of discussions, giving feedback, establishing online tasks, and the creation of formative and summative assessments (Dale 2003).

1.2 *Technology education in Singapore*

The Singapore government started the use of technology in education provision in the 1970s (Koh & Lee 2008). In the 1970s, televisions sets, film projectors, and overhead projectors were used frequently in classes to facilitate student learning. The use of these forms of technology, especially the use of videos and televisions, was due to the lack of expertise and competencies amongst graduate teachers, thus foreign imported films were used to fill this gap. In the 1980s, the Singapore Ministry of Education conceptualized a masterplan to provide the infrastructure and teaching capabilities of

DOI 10.1201/9781003095484-58

all educational institutions from the primary school level to the tertiary level (Koh & Lee 2008). Phase 1 from 1997 to 2002 was to set up a basic infrastructure in schools and trained teachers at the cost of $6 billion over six years. Phase 2 from 2003 to 2008 concentrated on integrating information and communications technology into the lessons at the cost of $470 million over three years. The current and final phase from 2009 to the present is targeted at developing better interactive environments to strengthen students' thinking and will look at upgrading all institutions' infrastructures to keep up to date with technology. Given this directive by the ministry, all institutions must incorporate e-learning into their curriculum. At the polytechnic level, the incorporation of e-learning is taken both at national education level and at an individual institution level. At a national level, an online e-learning platform called Polymall is available to all Singapore citizens across all polytechnics. At an institutional level, e-learning is available as a module to students within the framework of their studies. For example, at Temasek Polytechnic, one of the Institutes of Higher Learning in Singapore, the emphasis on e-learning is anchored to the ministry's mandate where students are given ample opportunities for e-learning. Moreover, in the School of Business, each diploma is mandated to have one subject which is facilitated entirely on an e-learning mode. Currently, the School of Business offers nine diplomas, of which two diplomas are specific to the hospitality and tourism line, namely the Diploma in Hospitality and Tourism and the Diploma in Culinary and Catering Management.

Given that the hospitality industry is one which is manually intensive, skilled-based with a fair amount interaction between people, and having had a late adaption to technology (Bull 1995) might have had a significant impact on current students' perception on technology and its relationship to the hospitality industry. Therefore, this study attempts to delve into the perceptions of Singapore polytechnic hospitality and tourism students toward their online hospitality subjects delivered through VLE systems.

2 LITERATURE REVIEW

2.1 *E-learning*

According to the European Union, e-learning can be defined as "the use of new multimedia technologies and the internet to improve the quality of learning by facilitating access to resources and services as well as remote exchanges and collaboration" (Commission of the European Communities 2001). However, there are many ways by which e-learning can be termed. According to Kathawala and Wilgen (2004), e-learning encompasses more than just the use of the Internet, as they defined e-learning as the delivery of learning materials, packages, or content through various forms of electronic media or technology. This includes online learning or web learning as well as older computer-based training, which uses earlier multimedia technologies such as CD-ROMS and videos. This has also been supported by Everley (2011), who purported that e-learning can either be in the form of a computer-based training (CBT) or through a web-based training (WBT).

In the context of the academic institutions, most institutions employ the use of VLE systems to help in the assistance of their teaching (Harrington et al. 2004). Even though there are many types and models available such as Blackboard and Canvas, most VLEs have the following common functions: course content availability, communications and student assessment tools, grade book, and the ability to manage course materials and activities. The course content allows faculty to upload content materials and syllabi. The communication tool includes emails, discussion boards, and chat rooms. In terms of assessments, the VLE allows the faculty to upload and administer quizzes and examinations, which allows the students to take these tests at any location. The system also allows the student to evaluate his/her performance, and information pertaining to the evaluation can be posted up on the gradebook. This information can be done from virtually anywhere and at any time. Student access is similar to the posting and uploading process, where students can log into VLE from any location or at any time (Costen 2009).

2.2 *Students: millennial generation*

The current students who are studying for a polytechnic diploma are aged between 17–20 years old. They are usually born between the years 1981–2004 and are considered the millennial generation. The characteristics of the millennial generation according to Raines (2002) are as follows. They are generally sociable, optimistic, talented, well educated, collaborative, open-minded, and are achievement oriented. They are also known as the Internet Generation, Echo Boomers, Nexters, or the Gen Yers. In the context of the coined classification term given (Internet Generation), Oblinger (2003) supports the notion, as this generation of people have been exposed to technology from a very young age and are adept in the use of technology.

Given their exposure to the latest global issues, such as multiculturism, globalization, and the Internet revolution, the millennial generation displays unique learning styles that embody these themes (Raines 2002). The preferred learning medium by these generations are as follows: teamwork, experiential activities, structure, and the use of technology. Emails and instant messaging are means by which these people communicate amongst themselves to the other generations. As cited by Howe (2003), they are risk adverse and would like to have the best technology available both at home and at their schools. This has been supported by Levine and Arafeh (2002), who cited that students' expectations are generally very high, because they have been exposed to technology all their lives. Given this, Jonas-Dwyer and Pospisil (2004) postulated that academic staff teaching this generation of students will have to increase their own personal technological skills-base, design teaching and learning activities to meet the students' needs and expectations, to be able to communicate with the student through a range of media, and to provide instantaneous feedback and response within 24/7 time frame.

3 PURPOSE OF THE STUDY

3.1 *Purpose*

Given that the hospitality and tourism industry is generally associated with skills, craftsmanship, and team-based activity, hospitality schools often have aligned their lessons to mirror reality, where students are put through simulations, role-plays, and activities. These forms of lessons are often done through traditional forms of pedagogy. Examples of such would be cooking classes, mastery of a reception desk, etc. However, when lessons are taught through a VLE, there seems to be a disconnection with reality as the teaching is done through a medium. The aim of this study is to obtain the Singapore hospitality and tourism polytechnic students' perception of e-learning and evaluate their perceptions on the relationship between what is taught with the industry practices.

4 METHODOLOGY

To better understand how hospitality students viewed e-learning, a hybrid-mode of quantitative and qualitative questions were administered to a cohort of 187 hospitality students who had recently completed an online hospitality subject. The 15 quantitative questions used a 5-point Likert scale, based on Strongly Agree, Agree, Neutral, Disagree, and Strongly Disagree. The 15 quantitative questions were clustered under two different subheaders, in relation to (1) their inclination toward the use of the VLE and (2) the applicability of the VLE in terms of its content and its relationship to the hospitality industry. The qualitative section, which is based on three questions, enabled students to give insights on the use of VLEs in terms of its relationship to the industry and the likes and dislikes in engaging and optimizing the VLE.

A total of 108 students participated in the survey from a cohort of 187 students (58% of the cohort). The online survey was administered to senior year hospitality and tourism students, who are well aware of the hospitality trade, having had two years of academic grounding followed by a 20-week internship. Therefore, this survey is relevant and valid, given the students' understanding of the needs of the industry. The results are shown in Table 1.

Table 1. 5-point Likert scale rating of the use of VLE by Singapore hospitality and tourism students.

Sampling Size = 108	5-point Likert Scale				
	Strongly Agree	Agree	Disagree	Strongly Disagree	Neutral
– Have a clear understanding of the aims of the subject through the VLE	30 (27.8%)	70 (64.8%)	8 (7.4%)	0 (0%)	0 (0%)
– Able to understand the VLE subject contents	32 (29.6%)	74 (68.5%)	2 (1.9%)	0 (0%)	0 (0%)
– The e-learning (VLE) topics were organized to help the understanding of the subject matter	34 (31.5%)	64 (59.3%)	8 (7.4%)	2 (1.9%)	0 (0%)
– The use of the VLE system was manageable	34 (31.5%)	66 (61.1%)	8 (7.4%)	0 (0%)	0 (0%)
– The e-learning subject is able to align practical realism	32 (29.6%)	48 (44.4%)	18 (16.7%)	8 (7.4%)	2 (1.9%)
– The content of the VLE was presented clearly	34 (31.5%)	70 (64.8%)	4 (3.7%)	0 (0%)	0 (0%)
– Assessing the VLE from a geographical distance from campus was an added convenience	58 (53.7%)	40 (37.0%)	8 (7.4%)	0 (%)	2 (1.9%)
– The VLE platform allowed the use of examples and illustrations	38 (35.2%)	68 (63.0%)	2 (1.9%)	0 (0%)	0 (0%)
– In the VLE platform, there were opportunities to participate in activities (blogs, discussions, etc.)	36 (33.3%)	70 (64.80%)	2 (1.9%)	0 (0%)	0 (0%)
– There is a preference to have face-to-face instructions instead of using a VLE	40 (37.0%)	22 (20.4%)	34 (31.5%)	8 (7.4%)	4 (3.7%)
– The learning resources and presentation on the VLE was useful	44 (40.7%)	58 (53.7%)	6 (5.6%)	0 (0%)	0 (0%)
– The assessments were easy to use and submit on the VLE platform	38 (35.2%)	68 (63.0%)	2 (1.9%)	0 (0%)	0 (0%)
– The use of the VLE platform gave granted greater independence to learning	54 (50.0%)	50 (46.3%)	4 (3.7%)	0 (0%)	0 (0%)
– The use of the VLE platform allowed greater critical thinking	42 (38.9%)	60 (55.6%)	4 (3.7%)	0 (0%)	2 (0%)
– The use of the VLE platform allowed good teamwork with fellow students	36 (33.3%)	56 (51.9%)	12 (11.1%)	4 (3.7%)	0 (0%)

5 RESULTS AND FINDINGS

5.1 *Quantitative questions pertaining to VLEs*

There was a total of 15 questions that were related to the use of the VLE. These questions attempted to gauge the perceived receptiveness of the use of VLE. Students rated the VLEs' ability to transmit the subject's content, its effectiveness as a teaching and assessment tool, and the ability to engage and stimulate interaction amongst the students. In terms of its manageability by the students (this implies whether they can navigate and circumvent the system), about 92% were able to manage the system with minimal supervision. About 90% of the students also agreed that the ability to remotely learn from locations other than in the polytechnic's computer laboratories was favorable (there were more than 50% of the students who strongly agreed on this). Most students also felt the VLE can support the illustration of good practices and examples from the hospitality line. They were also able to engage in a wide variety of activities, ranging from the use of blogs, quizzes, discussions, voice-over videos, and bulletin boards. They were also able to effectively obtain information from the VLE, such as downloading information and files for their perusal. For the above three areas, approximately 98% of the students agreed to the benefits of using a VLE. However, despite the above, many students preferred a personal face-to-face format of instruction. In terms of the challenges of studying online, only 57% of the students felt that they did not face difficulties in learning from an online source. Almost 39% of the cohort did not like the online format, with 3.7% with a neutral view on the matter. In terms of the ability to use and submit their assignments online, most students (about 94%) were able to navigate the process of using the plagiarism software Turnitin and submit their assignments. In terms of the learning aspects, most students were highly adaptable. A total of 97% of the students felt that the online subject gave them the ability to learn independently. This e-learning process also reinforced students' ability to develop thinking skills and apply problem-based learning, with 93% of the students agreeing to this. The ability to work in teams was slightly lower with 84% agreeing to its usefulness.

5.2 *Qualitative questions pertaining to VLEs*

From the answers given and derived, keywords findings reflected that the students generally found the content topic relating to the industry manageable, and they were able to cope with the use of technology. Examples of the answers cited were "*All topics were manageable,*" "*Topics were relatively easy to digest,*" and "*Can relate to my internship venue and thus allow me to generate thinking skills...,*" etc. Most students were able to download materials from the VLEs and were able to view the media content, such as the imbedded hyperlinks leading to youtube.com videos. The students were also able to find their class and project groupings in the VLE and participate in the blogs and forum discussions with minimal instructions.

From the answers, the most frequently cited value-add, that is, "likes," from the online lessons was the ability of being able to participate in the lessons without being physically in class. Some students also felt that the use of technology gave them the ability to share and participate in a discussion without being over-shadowed by their classmates.

In terms of the dislikes garnered from the qualitative answers, some student felt that the use of a VLE was dry and difficult to read as compared to use of traditional delivery process (whiteboards and note-pads). Some students highlighted that it was laborious having to type the answers instead of having an actual conversation. Some students highlighted certain difficulties. Examples from citations from students include "*Quite challenging to comprehend,*" "*the whole module was in fact quite dry,*" "*The online forum has its limitations to clarify several questions per topic/ lecture,*" etc. The most frequently highlighted drawback was the fact that the subject was taught on an online mode and had minimal personal contact, which resulted in many students disliking this mode of delivery. Some of the answers given are as follows: "*I don't like the blog. Too confusing and it is difficult for me to track everyone's messages,*" "*Can we change the subject to a face-to-face tutorial/lecture?,*" "*There is a lack of staff-student relationship,*" "*The subject needs to be more hands-on, hard to understand the concept (through a media platform),*" and others.

6 DISCUSSION

From the above data collected, there seems to a mixed reaction from the Singapore polytechnic hospitality and tourism students, also known as the millennial generation, on the use of technology through the VLEs in teaching hospitality students. According to Lim and Theng (2011), Singapore youths are very much equipped and confident in media and computer skills. Their exposure levels are exceptional high due to several reasons, namely the high Internet penetration rate at 85% (Infocomm Development Authority of Singapore 2018), the high self-learning efficacy in computer literacy at 82% (Lim & Theng 2011), and the government's initiative and policy to increase computer usage amongst the lower secondary school students. However, given the technological exposure to e-learning, some of the Singapore polytechnic hospitality students are still uncomfortable with the e-learning. Up to 39% of the participants of the survey felt that they would rather have a face-to-face interaction with their lecturers. According to a study conducted by Allen et al. (2002), students who had face-to-face interaction with educators scored higher in terms of their satisfaction as compared to online learning. This seemed to run against the study done by Schutte (1996) in the literature review section. On another note, this survey gave clear indication that the students were conversant with the VLE and found it easy to manipulate and navigate the computer-aided tool. One of the possible reasons for this dissonance could be due to the nature of the students' studies. The hospitality and tourism industry is often associated with soft skills and service. Given this, many students perceive the hospitality industry as not being associated with technology, leading to the aversion to the use of the computers. Another reason could be due to the curriculum setup for the hospitality students. In most polytechnics offering hospitality and tourism courses in Singapore, the emphasis is on the traditional classroom and lecture sessions, with minimal e-learning subjects. An example would be the Diploma in Hospitality and Tourism Management at Temasek Polytechnic, Singapore, where the total credit units for graduation is 123 points. However, the e-learning subjects total only between 6–12 points. This lack of emphasis in online learning could have had a negative impact on the students' inclination toward technology. Another reason that could have led to this dissonance could be due to the reason that at the secondary school level, there are very few subjects that are related to the use of the computing studies. As extracted from the Singapore Ministry of Education webpage (2019), out of the total number of 76 subjects offered at the GCE 'O' levels, only 4 subjects (Design and Technology, Creative 3D Animation, Media Studies, and Computer Studies) seemed to have a major emphasis on the use of technology. This number (the four subjects) could be a generalized one, as online learning could be imbedded in the individual subject's curriculum. This lack of emphasis on online learning at an educational level prior to the students' entry to the polytechnic could have major impacts and ramifications on the students' outlook toward e-learning.

From the data gathered, it is postulated that the students who felt that e-learning was beneficial were the ones who had the necessary technological competency. A research conducted by Biscomb and Devonport (2008) reflected similar results. In this current study, students who did not appreciate the e-learning subject could be due to several reasons. They could either be disinclined toward the use of technology (known as computer phobic) or they might not be interested in the content-matter of this subject or the industry. As stated earlier, there were a high percentage of responses that indicated that this online subject was not useful for future employment. This response could be indicative that since the students were final-year students, especially after completing their internship, they usually displayed high negativity toward joining the hospitality industry (Chen et al. 2011) or they did not appreciate the technical nature of this subject, which is a facilities management subject (entitled Facilities Management for the Hospitality and Tourism Industry).

7 CONCLUSION

With the advancement in technology, leading to the ease of accessibility and convenience to education, more institutions are embracing the use of technology and VLEs to facilitate learning and internationalization (Radovic-Markovic 2010). This initiative is very prevalent with institutions such as the University of Phoenix offering distance e-learning and using VLEs as a medium of communication with their students. Globalization has also led to an increase in the hospitality workforce, with more students joining and professionalizing the industry. The two factors have led to an increase in student intake at the polytechnic levels and the use of technology in Singapore. From the results gathered, it can be seen that technology through a VLE is a "two-edged sword." It does enable learning to be more effective and efficient in cost and time savings, through the use of new offerings in content, delivery, and assessments. Students who have the competency and prior knowledge of the subject content tend to benefit from this, and this enables them to learn better through decision-making and problem-solving skills, which has been cited by Cho and Schmelzer

(2000). However, on the flip side, some students have given negative feedback on its delivery, such as having a lack of personal touch. Adding to the above, other factors such as the content material and the lecturer's experience and ability to facilitate does have an important impact (McDowall & Lin 2007). In order to cater to the changing needs of the technology and students' expectation, some recommended suggestions would be to better train lecturers in delivery techniques, orienting students to the VLE, or having a blended approach of traditional and e-learning lessons and to keep up with technology and the knowledge of the industry.

REFERENCES

Allen, M., Bourhis, J., Burrell, N., & Malby, E. 2002. Comparing student satisfaction with distance education to traditional classrooms in higher education: A metal analysis. The American Journal of Distance Education, 16(2), 83–97.

Biscomb, K., Devonport, T.J., & Lane, A.M. 2008. Evaluating the use of computer-aided assessment in higher education. Journal of Hospitality, Leisure, Sports and Tourism Education, 7(1), 82–88.

Bull, A. 1995. The economies of travel and tourism (2nd ed.). Melbourne: Pitman.

Chen, C., Hu, J. W., & Chen, C.F. 2011. A study of the effect of internship experiences on the behavioral intention of college students majoring in leisure management in Taiwan. Journal of Hospitality, Leisure, Sports and Tourism Education, 10(2), 61–73.

Cho, W., & Schmelzer, C. 2000. Just-in-time education: Tools for hospitality managers of the future? International Journal of Contemporary Hospitality Management, 12, 31–37.

Commission of the European Communities. 2001. The e-learning action plan: Design Tomorrow's education. Communication from the Commission to the Council and the European Parliament. Brussels: COM.

Costen, W. 2009. The value of staying connected with technology: An analysis exploring the impact of using a course management system on student learning. Journal of Hospitality, Leisure, Sports and Tourism Education, 8(2), 47–59.

Dale, C. 2003. Carry on talking: The use of online discussion groups as a learning tool. LINK8-E-Learning. LTSN for Hospitality, Leisure, Sports and Tourism.

Dale, C., & Lane, A. 2004. Carry on talking: Developing ways to enhance students' use of online discussion forums. Journal of Hospitality, Leisure, Sports and Tourism Education, 3., 52–59

Everley, M. 2011. Training methods. The RoSPA Occupational Safety and Health Journal, Sep, 29–32.

Harrington, C., Gordon, S., & Schibik, T. 2004. Course management systems utilization and implications for practice: A national survey of department chairpersons. Online Journal of Distance Learning Administration, 7(4), 1–13.

Howe, N. 2003. Presidents Institute: Understanding the millennial generation. Retrieved from The Council of Independent Colleges: http://www.cic.org/publications/independent/online/archive/winterspring2003/PI2003_millennial.html

Infocomm Development Authority of Singapore. 2018. Infocomm Usage – Households and Individuals. Retrieved from Infocomm Development Authority of Singapore: https://www.imda.gov.sg/industry-development/facts-and-figures/infocomm-usage-households-and-individuals#3

Jonas-Dwyer, D., & Pospisil, R. 2004. The millennial effect: Implications for academic development. In transforming knowledge into wisdom: Holistic approaches to teaching and learning. HERDSA Conference proceedings.

Kathawala, Y., & A., Wilgen. 2004. E-learning: Evaluation from an organization's perspective. Training and Development Methods, 18, 5.01–5.13.

Koh, T., & Lee, S. 2008. Information communication technology in education: Singapore's ICT masterplan, 1997–2008. Singapore: World Scientific Pub.

Levine, D., & Arafeh, S. 2002. The digital disconnect: The widening gap between internet-savvy students and their schools. Retrieved from http://www.pewinternet.org/reports/pdfs/PIP_schools_internet_Report.pdf

Lim, L., & Theng, Y. 2011. Are you today media literate? A Singapore study on youth's awareness and perceived confidence in media literacy skills. Proceedings of the American Society for Information Science and Technology, 48(1), 1–4.

McDowall, S., & Lin, L. 2007. A comparison of students' attitudes towards two teaching methods: Traditional versus distance learning. Journal of Hospitality and Tourism Education, 19(1), 20–26.

Oblinger, D. 2003. Boomers and gen-xers, millenials: Understanding the 'new students'. Retrieved from EDUCAUSE Review, July/August: http://net.educause.edu/ir/library/pdf/erm0342.pdf

Radovic-Makovic, M. 2010. Advantages and disadvantages of e-learning in comparison to traditional forms of learning. Annals of the University of Petrosani, Economics, 10(2), 289–298.
Raines, C. 2002. Managing millenials. Retrieved from www.generationsatwork.com/articles/millenials.htm
Schutte, J. 1996. Virtual teaching in higher education: The new intellectual superhighway or just another traffic jam? Retrieved from California State University: www.csun.edu/sociology/virexp.htm
Singapore Ministry of Education Singapore. 2019. Joint Admissions Exercise 2019. Retrieved from Ministry of Education, Singapore: https://www.moe.gov.sg/docs/default-source/document/education/admissions/jae/files/booklet.pdf

Promoting Creative Tourism: Current Issues in Tourism Research – Kusumah et al. (Eds)
 ISBN 978-0-367-55862-8

The challenge in disruptive times in tourism education: Toward a redesigned curriculum for new normal from conventional to creative tourism

P.R.M. Tayko
Graduate School of Business Assumption University, Bangkok, Thailand

Foedjiawati
Petra Christian University, Surabaya, Indonesia

ABSTRACT: This reflective analysis, an evidence-based exploration, is a case assessment of tourism education for relevance and responsiveness to emerging needs of the industry in fast-changing times. The results and outcomes become the basis for redesigning the curriculum from conventional to a creative tourism. Considering the tremendous challenges from external environment due to the pandemic impact of Covid-19, it is expedient to review, shift, and redesign the curriculum in terms of context, content, and processes reflective of new perspectives and purpose of the program for future intakes. The methodology uses triangulation, a combination of qualitative and quantitative analysis of content, context, process, perspective around core purpose (2C2PCP). This comprehensive framework of defining, detailing, differentiating, and connecting the essence of each curriculum dimension around the core purpose becomes the map to integrate the findings to make changes in the current curriculum. Survey data from secondary students of selected feeder schools in the area are used to complement and support the analysis. Observation and feedback from the ground support the need to redesign both curriculum and the program thrust as urgent factors to update the university tourism education, a program of choice for prospective entrants. Based on findings, a new design of tourism education is envisioned as the new program for the new normal. The implications of the study will provide new initiatives in program promotion, and implementation for effective instructional strategies and learning processes.

Keywords: redesigning curriculum, creative tourism, tourism education

1 INTRODUCTION

Tourism is one of the leisure activities related to the two motivational forces of escaping from routine activities and seeking recreational opportunities in a form of traveling. However, in a fast-changing pace of the world, especially when the environment becomes more volatile and uncertain with the crises of the Covid-19 pandemic, human mobility and economic activity are curtailed, where the impact affects all sectors specifically in tourism business and education. Tourism education is the way tourists' destinations prepare the human resources to work professionally in developing the tourism sectors (Malihah & Setiyorini 2014).

Although tourism education was first established in a form of vocational schools in Europe, which focused more on training in core competencies such as hospitality, hotel management, and related business skills (Morgan 2004), the interest and demand from public and private sectors were quite high. Therefore, there have been discussions on the balance between vocational and academic focus in order to help students to be broadly educated and responsible in tourism development as well as functional occupation in tourism (Lewis 2005). Previous research by Morgan (2004) suggests tourism educators develop courses which can accommodate the needs of the students and

DOI 10.1201/9781003095484-59

practitioners and in addition, serve the industry's need where educators deliver courses to students to make them ready to function in the industry in whatever capacity, condition, and situation they encounter upon graduation.

Consequently, in this disruptive era of Covid-19, tourism education will not anymore become an attractive major to pursue due to the tremendous challenges with the tourism sectors that are affected. Therefore, in future they will need a new skills set in keeping up with new trends enabling tourism professionals to be updated. There is a need to make a rapid transformation in a new way of learning process and redesigning its curriculum to be in sync with the new normal practice of tourism education.

This paper is an essay with a reflective analysis as a case assessment of tourism education for relevance and responsiveness to emerging needs of the industry in fast-changing times and a survey of prospective students' interest in tourism courses. It is an attempt to link and match the curriculum and adjust the modules to the current challenges.

The purpose of this research is to draw information from senior high school students' perception or interests in program areas and from the assessment of the internal and external factors that have changed in the recent past. The outcome of the research generates insight toward a redesigned tourism education curriculum from conventional tourism packages to a creative tourism curriculum package since tourism education should maintain and sustain the program of tourism education at the university level.

2 METHODS

The approach chosen to conduct this research is a comparative, connective, and reflective process by triangulating the key points of analysis in five areas in curriculum design. It is a process that combines the qualitative and quantitative analysis of 1) CONTEXT—the conditions of the external environment, the circumstances of the industry affected by emerging events; 2) CONTENT—the subject matter or topics of interest that relate to needed competencies to provide services in the industry; 3) PROCESS—the teaching-learning processes with the advent of digital technology that enhance or support learning and instruction; 4) PERSPECTIVES—of global and local proportions considering both growth and fixed requirements; and 5) CORE PURPOSE of the program as well as of the university as integral of its higher education development (2C2PCP) (Tayko 2017). This is a comprehensive framework of defining, detailing, differentiating, determining, and connecting the essence of each dimensions of the curriculum around the core purpose of the program and integrating the findings as a basis for the design of the new curriculum for the new normal, and complemented by quantitative data support from the survey of prospective students from the secondary schools as feeders to the university entrants. The underlying framework of analysis of issues in the areas of context, content, perspectives, and process uses the criteria of consistency, creativity, activity/results, and affinity/relations as relevant to the core purpose of the program. Recommendations for the redesign of the new normal curriculum are the result of the qualitative analysis complemented by a quantitative support from the student's survey.

3 RESULTS AND DISCUSSION

A curriculum and instruction process of any program in the university by design needs to consider five key points, namely, as specified in the methodology. The following Figure 1 on the program design and develop covers 1) context of the industry; 2) the content of the curriculum to develop competencies to work or do business as tourism entrepreneurs; 3) the perspectives of requirements fulfilling quality standards, creative, and innovative features of the program; and 4) the instructional (teaching-learning) processes to develop the competencies of learners in the program to achieve the desired results, and 5) the core purpose of the program around which the elements are designed to

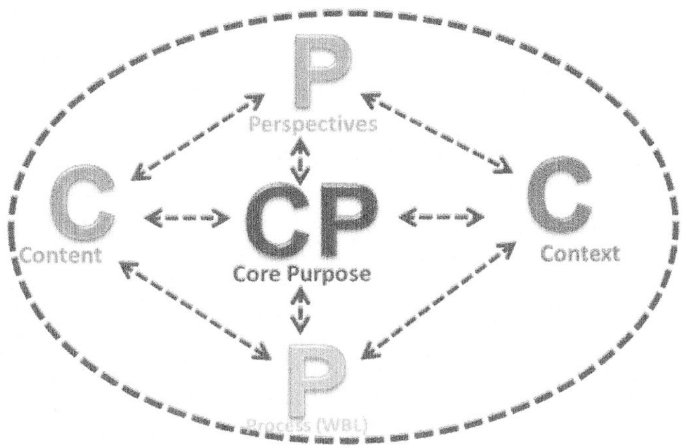

Figure 1. Program design and development framework (Tayko 2015, 2017).

fulfill the purpose of the institution through this program. Each key point is connected to each other by matching the requirements from a holistic criteria framework that fulfills effective, efficient, excellent, elegant, and ethical considerations of services rendered.

The frame/flow as illustrated above is designed for a holistic application of the "brain mapping" processes of the four-brain model of the "high business performance brain" (Lynch 1984, 1993). This is based on the extensive split-brain research on "human information processing skills set" which identifies information distinctively and differently from an internal set of criteria, be it in content or subject matter or context or circumstances in the situations. This process framework was utilized to analyze 43 action researches in organization development (OD) in terms of content, context, process, perspectives, and core purpose (Fernando & Tayko 2013). In a subsequent research on the applicability of whole brain literacy framework in various processes of research projects, the findings showed that the framework provided 1) a way of thinking in holistic perspectives of systems–individual, group, and organization; 2) a technique to broaden and deepen one's understanding of issues; and 3) a way to connect past learning by shifting mindsets inclusive of variations and diversities (Tayko et al. 2017c).

3.1 *Context*

Context analysis is done by identifying the conditions in the industry with the external factors of the environment that create opportunities/challenges for the industry to flourish or decline. This includes in particular the effects of the Covid-19 pandemic in every sector and service in society as well as the internal factors in the university as being impacted by the global changes due to pandemic conditions. In once normal condition, tourism education was also challenged by the threats of a tight competition among institutions both local and overseas, which offer the same program with different benefits to the students. An interview with one head of the leading Tourism Education in Surabaya, East Java, Indonesia (2020), revealed the shortage of the re-sources, networking for collaborations, link and match the curriculum, which have been the challenges that exist internally to most of tourism education. Therefore, tourism education, which has an effort of developing human resources to be better persons, needs to enhance the education. This means colleges and universities are expected to develop a successful education for stakeholders such as students, private sectors, public sectors, as well as community in whatever condition the world is changing. The pressure being felt in the tourism context is likewise brought about by technology and globalization. In the conventional curriculum the inclusion of topics or subjects in the tourism curriculum was mainly based on the "destination" points of interest in terms of location features much more than from the focus of interests of the tourists themselves for personal development.

3.2 *Content*

Content analysis is done by reviewing the scope and coverage of courses of the program that are relevant and related to best practices in the industry. Curriculum mapping is done on the conventional curriculum as well as on the emerging needs for competencies and topics of interests to be considered in the redesign of the conventional curriculum to the new creative curriculum in tourism. Tourism Education was at first taught in the form of a vocation school, and it has evolved to be a study under management that means tourism education has been developed to balance the tourism development demand for fulfilling the broader perspective of managing tourism (Airey & Tribe 2005). Therefore, lots of universities run the tourism education in bachelor degree and continue to post-graduate as well as PhD programs; unlike the early tourism education, which was delivered only in a diploma program and designed with an objective of emphasizing more on vocational training by teaching various practical skills and contemporary subjects related to the tourism industry so that students would be ready to fulfil the human resources vacancy at the operational level of tourism businesses once they graduated.

The conventional curriculum of the bachelor degree program in tourism education is designed by emphasizing the concept of learning in an academic perspective. The combined curriculum was between tourism and management subjects, which focused more on conceptual framework of management on how to solve the tourism management problems in the society. Students need to pursue their studies for four years instead of three years. Henceforth, the subjects have been expanded to respond more on the managerial and multi-discipline approaches rather than skills. Furthermore, the tourism industry needs more resources that are able to manage the work in the tourism business rather than those who are skilled graduates. To enrich and enhance the curriculum, related tourism practitioners and organizations are engaged, some collaborations with overseas universities and tourism industries are also developed in some activities, such as student exchange, immersion program, as well as internships. An applied course is also available for students to gain enrichment in professional development programs designed by inviting some top management and outstanding speakers from various tourism industries to share their knowledge with the students. Those courses are in accordance to Bodger (1998), who defined edutourism as any programs in which participants travel to "location as a group with primary purpose of engaging in learning experience directly related to the location." The activities considered as edutourism cover ecotourism, heritage tourism, rural tourism, and student exchanges between educational institutions. They are incorporated to the learning modality and curriculum.

3.3 *Process*

Process analysis includes identifying learning and instructional engagements for learners to develop the competencies of the program and the industry. The "once normal" modality of F2F or classroom setting for learning may now be reviewed with an alternative of a blended modality of learning making use of online mode of instruction and learning. The shifting of learning from school to home due to Covid-19 has forced all schools, teachers, students, and parents to use the technology to be able to connect. Chou et al. (2010) have defined active interaction in online learning activities including the types of interaction: the learner-self, learner-learner, learner-instructor, learner-content, and learner interface. The learning activities in the course are a combination of forms of interaction between the subjects involved in the teaching, and learning activities include student-content, student-instructor, and student-student interaction (Gradel & Edson 2010).

3.4 *Perspectives*

Both perspectives and process analysis are done to identify fixed and growth mindsets that fulfill a comprehensive curriculum as well as identify the modalities most responsive to the changes for the new normal. To change behavior from once normal to new normal requires, at a deeper level, understanding mindsets as factors that influence perspectives, process, and behavior. Dweck (2006) did an extensive research on fixed and growth mindsets that matter in the way a person views the world and acts on it. It is an orientation that influences the approach and outcome of behavior. Fixed mindset relates to set or fixed expectations with set standards as a basis for action. Growth mindset relates to the process of learning, growing, and becoming, which makes one open to unfolding outcomes. Taking a balanced combination of fixed and growth mindset requirements was opted for effective, efficient, ethical, elegant, and excellent program offering. As explained on the content above, the tourism curriculum consists of more subjects on the knowledge and skills related to the fixed mindset. Some of the samples of the subjects are classified as follows: tourism policy management and productivity, which reflects the fixed mindset; whereas project development, creative engagement, and connectivity with soft skills reflect the growth mindset.

In addition, educational tourism is also offered as an exposure since it is an educational mobility ranges within the duration of 24 hours to 12 months and have learning activities taken or completed within the period. Such activities are excursions and student exchange (Maga & Nicolau 2018). Based on the informal discussion with Yo, Head Program (2020), most adopted subjects are more on the Tourism Policy Management and Productivity, which shows the content of the subjects are taught more on using precision thinking. Therefore, in the pandemic of Covid-19, there is a need to shift the curriculum from the fixed to the growth mindset. Tourism education needs to redesign their curriculum, which will enable their graduates to work in the creative tourism industry. This means that the graduates are ready to work not only in a tour and travel business but also in other fields related to the creative industry of tourism.

A survey of new tourism subjects was conducted in order to have some insights on what the candidate tourism students are keen on learning for subjects that refer to the growth mindset. Based on some definitions of what so-called "creative tourism," the authors classify new clusters of the subjects. Creative tourism as mentioned recently by the Ministry of Tourism & Creative Economy in Indonesia is tourism that offers visitors a creative experience and opportunity to develop their creative potential through active participation in courses and learning experiences. There are four areas in applying creativity in tourism: 1) involving creative person in creative activities; 2) using creative tourism products as attractions; 3) utilizing creative process in providing tourists with creative activities; and 4) using creative environment for visiting creative clusters. Hence, the new four classifications are named as follows: Digital Tourism, Edu Tourism, Leisure Activities, and Conference Events Management and Interpersonal Relations (IPR) soft skills. All the new clusters tend to reflect the growth mindset. In addition, Ernawati (2003) implied that in designing the tourism curriculum, educators should understand the industrial needs of expertise. The scholars need to understand and experience more to the situation. Therefore, edutourism activities should become a point to be considered too.

The survey was distributed to senior high school students through the Google form within one week. There were 101 respondents who filled up the questionnaire. The respondents were asked to rank the first to the last three subjects which they prefer to learn in each of the new cluster.

Here are the results of the survey: There were 41.6% students of year 10, 30.7% students of year 11, and 27.7% students of year 12 (Table 1). They were from 12 private schools and 2 government schools. For the first cluster of Digital Tourism, respondents have chosen subjects on Social Media as the first choice, Tourism Content Creator as the second choice, and Multimedia & Journalists as the third option. The second cluster of Edu Tourism, Introduction to F&B ranks the most in demand, followed by Tourist Destination and Travel Innovation as the last preferred choice. The third cluster of Leisure Activities, respondents are keen on learning Special Interest of Tourism, second option is Arts Management, and the last is Leisure Studies. The fourth cluster of Soft Skill has indicated that respondents are mostly interested in learning Business Communication, followed by Collaboration & Conferencing and Psychology of Business as the last subject they would like to learn. An additional question of Education Tourism is also asked in order to find out whether students are still eager to have traveling activities when the situation and conditions are back to normal. Most students are keen on having a short field trip. This might be when they filled up the

Table 1. Future creative tourism curriculum.

Digital Tourism	Edu Tourism
Social Media Tourism Content Creator Multimedia & Journalist Digital Leadership	Introduction to F&B Business Tourist Destination Transport Management Travel Innovation Medical Tourism Sustainable Tourism Urban Tourism Eco Tourism Agri Cultural Tourism
Leisure Activities	**Soft Skill**
Arts Management Spa Management Leisure Studies Sports Special Interest of Tourism (Cookery, painting, photography, crafts and arts) Tourism Production	Cross Culture Studies Psychology of Business Business Communication & Collaboration

questionnaire, it was in time of Covid-19; therefore, most of them have chosen short field trip and the second chosen activity is internship, and last options chosen are Student Exchange as well as Outbound Tour since those two activities have the same score given by the respondents.

3.5 *Core purpose*

Central to this curriculum review for a redesign is a reflective revisit and review of the core purpose of the program, whether it still reflects the reasons for its existence and the relevance of it to the emerging needs of the stakeholders and determine whether it encompass the systemic purpose of the program, the profession and the industry contributing to the country's development. In going through this research, the faculty involved in the program had to ask themselves the question on relevance and responsiveness of the program. This is one reason why this study is done. In this process, the purpose of the program and that of the university is affirmed at the core for a greater and wider relevant and responsive service to the tourism industry.

4 CONCLUSION

Having gone through the reflective, comparative and connective processes on the five key points, it is clear and expedient that the current conventional curriculum of the tourism program must, need, and should be transformed to respond to the "VUCA" (volatile, uncertain, changing/chaotic/agile/ambiguous) world and be relevant and responsive to the emerging needs of clients in disruptive times. The trend of changes is exponential that adaptation and even transformation of existing models and processes need to happen. Therefore, the current curriculum of the Tourism Management Program needs revision and restructuring in the light of the changes in context and the preferences of prospective entrants in the programs Specifically, the following specific recommendations include 1) re-cluster the courses that reflect both perspectives of requirements of fixed and growth mindset requiring a combination of precision and possibility thinking skills; 2) a redesign of content topics that reflect the emerging changes in the context of creative tourism; 3) a blended approach in the teaching/learning or instructional processes and tapping into the Internet technology as a main modality of learning; 4) development integration of a spectrum of tourism activities to consider the environment and the ecosystem, as these are affected by pandemic issues on public health and economic viability; and 5) affirmation of the core purpose of the program for relevance, responsiveness, and excellence contributing to national development.

REFERENCES

Airey, D., & Tribe, J. (eds), *An International Handbook of Tourism Education*, Oxford, Elsevier, p. 13–24.

Bodger, D. (1998). Leisure, learning, and travel. *Journal of Physical Education Recreation Dance*, 69 (4), 28–31.

Chou, C., Peng, H., & Chang, C. Y. (2010). The technical framework of interactive functions for course management systems: Students' perceptions, uses, and evaluations. Computers and Education, 55(3), 1004–1017.

Dweck, C. S. (2006). *Mindset: The New Psychology of Success*. New York: Random House Publishing Group.

Ernawati, D. B. (2003) Stakeholders' Views on Higher Tourism Education. *Annals of Tourism Research*, Vol. 30(1), 255–258.

Fernando, M. & P.R M. Tayko, (2013). "*Navigating through the First wave of Change: A synthesis of the Action Research Dissertations of the PhD OD Cohort 1-5*. Graduate School of Business Assumption University Commemorative Issue" ABAC ODI Journal Vision.Action.Outcome. Vol. 1, No. 1, November 2013,

Gradel, K., & Edson, A. J. (2010). Cooperative Learning: Smart Pedagogy and Tools for Online and Hybrid Courses, 39(2), 193–212.

Lewis, A. (2005). Rationalising a Tourism Curriculum for Sustainable Tourism Development in Small Island States: A Stakeholder Perspective. *Journal of Hospitality, Leisure, Sports and Tourism Education*, Vol. 4(2), pp. 4–15.

Lynch, D. (1984). *Your High Performance Business Brain: An Operator's Manual –How to fine tune the management mind to increase productivity and profits*. New Jersey: Englewoods Cliffs, Prentice Hall, Inc.

Lynch, D. (1993). *Your Dolphin High Performance Business Brain: 21at Thinking Skills for Ambitious People under Challenge or under fire*. Lakewood, Colorado: Brain Technologies Corporation.

Maga, A., & Nicolau, P. (2018, May). Conceptualizing Educational Tourism and the Educational Tourism Potential (evidence from ASEAN countries). In *International Scientific Conference "Competitive, Sustainable and Secure Development of the Regional Economy: Response to Global Challenges" (CSSDRE 2018)*. Atlantis Press

Malihah, E., & Setiyorini, H. P. D. (2014). Tourism education and edutourism development: sustainable tourism development perspective in education. *Proceding-1st ISOT Eco-Resort and Destination Sustainability: Planning, Impact, & Developmnet*

Morgan, M. (2004). From Production Line to Drama School: Higher Education for the Future of Tourism. *International Journal of Contemporary Hospitality Management*, Vol. 16(2).

Tayko, P. R. M. (2015). OD in Asia in Practicing Organizational Development. USA: John Wiley & Sons, Inc.

Tayko, P. R. M. with Voranit V. (2015) *A Butterfly and Collective Effect: Up Close and Personal.Professional. Positional Engagements in I-I Connection: Bangkok, Thailand: Graduate School of Business, Assumption University, Printed by Bookplus,Pub. Inc.*

Tayko, P. R. M, et al. (2017). *Butterflies in the world: OD Learners/Leaders making a difference in the world.* Bangkok, Thailand, Bookplus Co., ltd.

Promoting Creative Tourism: Current Issues in Tourism Research – Kusumah et al. (Eds)

The suitability of TOEFL-ITP as a tourism industry employment requirement for Indonesian university graduates

G. Ginanjar
Monash University, Melbourne, Australia

M.W. Rizkyanfi
Universitas Pendidikan Indonesia, Bandung, Indonesia

ABSTRACT: Test of English as a Foreign Language (TOEFL) is the most taken English proficiency test with approximately 23 million test-takers worldwide. In Indonesia, TOEFL is one of the pivotal requirements in employment and is seen as a social and economic mobilizer for most Indonesians. For Indonesian university graduates on the tourism and hospitality program, mastering English is an essential requirement for getting a job in the industry. Consequently, it requires them taking TOEFL to fulfill various academic and employability standards. English language skills required by the hospitality and tourism industry should focus on accentuating graduates' good English communication skills. This study aims to discover how accurate paper-based TOEFL test or TOEFL-ITP is in assessing graduates' English proficiency level and English communication skill to be applied in workplace environment. This paper draws on a descriptive qualitative method involving three higher education graduates from the hospitality and tourism industry program that barely begin their employment journey. The data were obtained from the Likert-scale questionnaire and in-depth interviews that identified participants' perceptions regarding the suitability of TOEFL-ITP on tourism industry employment purposes. The study found that the participants shared doubts about the TOEFL-ITP as communicative English skills measurement in a real-life workplace environment. The findings of the study concerned the validity, authenticity, and generalizability of the test. As a recommendation, the findings can help to advance the understanding toward another type of English proficiency test like IELTS, TOEIC, or even TOEFL iBT (Internet-based providing speaking test) that might be more suitable than TOEFL-ITP as a tool of employment requirement, especially for Indonesian University graduates of the hospitality and tourism program.

Keywords: English proficiency test, paper-based TOEFL, tourism industry, university graduates, employment requirement

1 INTRODUCTION

In 2014, the Indonesian Ministry of Research, Technology and Higher Education (2014) mandated that higher education graduates require a competency certificate following their expertise or apart from their study program as an employment requirement. Consequently, the English proficiency certificate emerged as one of the most popular certificate competences for higher education graduates as supplementary to their diploma degree certificate. The English proficiency test measures the level of graduates' English ability that can be utilized in the workplace. There are many English proficiency tests such as TOEFL, IELTS, TOEIC, GRE, and SAT; however, the most widely taken English proficiency test in Indonesia is TOEFL.

TOEFL is considered as an essential test since it is assumed to be described as one's actual English ability and also stands as the "policy tool" for making selection decisions in academic employment aspects (Hamid 2015; Memon & Umrani 2016; Murray et al. 2013; Shih 2010).

DOI 10.1201/9781003095484-60

In EFL countries like Indonesia, previous studies investigate issues surrounding the TOEFL test, including the suitability of the test as an academic and employment requirement (Barnes 2016; Bùi 2016; Fahmi et al. 2016; Saukah 2016). However, there is a gap among the studies. There is no research available yet regarding the suitability of TOEFL ITP in tourism and hospitality industry, even though the TOEFL Institutional Testing Program (ITP) is the most popular paper-based English proficiency test due to its affordable cost as well as widely being used as a job requirement required by many institutions in Indonesia. More than 500,000 ITP tests were taken worldwide every year, and it is showing the significant utilization of the test (Taufiq et al. 2018).

One particular party that had a direct impact from the significance use of TOEFL ITP as employability requirements in Indonesia are university graduates (Nugraheni 2014). Especially in the hospitality and tourism industry, competent human resources are urgently needed in line with Presidential Instruction No. 9 of 2016 that was stipulated to revitalize the tourism industry started by encouraging youth to be a high-quality graduate with international competencies (Muti'ah 2019). Working in the tourism and hospitality industry is closely related to the interaction with foreigners, tourists who came from different countries and cultures. Conclusively, it can be argued that utilizing the TOEFL ITP test to measure the English proficiency level of tourism and hospitality university graduates is crucial as a need for identifying work-ready competence.

However, TOEFL ITP only tested two basic language skills, including listening and reading along with structure and written expression (grammar). The Indonesian government has formulated varied approaches to bolster Indonesian tourism numbers and economy in order to make Indonesia "a world-class tourist destination" (The Indonesian Ministry of Tourism's Strategic Plan 2019). One of the approaches that was targeted by the government is to produce competent employees or workers in the industry, especially English competence to be able to communicate with international tourists with the excellent qualification of spoken and written English. Productive skills covering speaking and writing in the hospitality industry courtesy in every different setting are pivotal since it encourages the process of communication between the employee and the guests or tourists. Consequently, it is essential to discover the suitability of TOEFL ITP among tourism students, since there are no productive skills tested within TOEFL ITP.

Based on the aforementioned facts, the author will investigate the suitability of the TOEFL ITP test for higher education graduates in the tourism and hospitality program. The tourism and hospitality industry demands prospective workers with English communication competence. It is interesting to find out whether or not TOEFL ITP can be utilized as a tool to measure English communication competence. Furthermore, within this research, university graduates' perception and the fundamental concept of the language proficiency test will be used to analyze the data gained from the interview and questionnaire to support empirical research of the interesting issue.

2 LITERATURE REVIEW

2.1 *Test of English as a foreign language in Indonesia*

The language proficiency test is required in order to measure the level of English mastery. One of the language proficiency tests is Test of English as a foreign language, well known as TOEFL. The test is highly perceived as a standard language proficiency for English skills, and it has been widely recognized and accepted by many national and international institutions in the world (Poorsoti & Asl 2015). The TOEFL assessment is mostly used as an indicator for education and employment requirements. In terms of academic, the university always asks the applicants to submit their TOEFL certificate as an indicator of English proficiency level for graduates who want to continue their study to post-graduate programs. Equivalently, when graduates are applying for a job, most institutions require TOEFL as a reliable and valid instrument to measure applicants' English proficiency level to prepare them for an English-speaking work environment.

TOEFL test was originally created by Educational Testing Service (ETS) in the United States, an official institution to produce a certificate for TOEFL Test (Alderson 2009). There are several types of TOEFL tests, TOEFL iBT, TOEFL PBT, TOEFL for junior, and TOEFL ITP (Ananda 2016; Taufiq et al. 2018). Many academic and employment institutions that require TOEFL as an entrance requirement may also allow a TOEFL prediction test, TOEFL-like, or a TOEFL-equivalent test. In developing country like Indonesia, due to the economy factor, many institutions employed more affordable types of TOEFL called TOEFL ITP (Institutional Testing Program) as the employability selection requirements. It is employed in many institutions in different fields of interest, including the hospitality and tourism industry (Taufiq et al. 2018). The TOEFL ITP is conducted in the form of a paper-based test in the printed media; consequently, this research will focus only on paper-based TOEFL as a subject.

In Indonesia, TOEFL ITP is highly recognized by every university graduate that wants to apply for a job. The test is also used in the work area as a medium for getting a promotion to a higher level of the salary in their current job. Due to only testing two different receptive language skills (listening and reading), many people find the test did not fulfill criteria of being valid, authentic, and reliable (Collins & Miller 2018; Taufiq et al. 2018).

2.2 *The fundamental concept of a language proficiency test*

Bachman's (1990) framework focuses on users' communicative ability of a language, which divided into two aspects that are 1) communicative competence and 2) the capacity for producing or implementing those aspects appropriately. Bachman (1990) described the essence of language proficiency test as divided into two approaches, specifically, real-life (RL) approach and interactional ability (IA); both approaches are primarily concerned with the authenticity, validity, and generalizability. The Real-Life approach fixates a characteristic of the test that the proficiency tests should mirror the "reality" of language utilization (Bachman 1990). This approach then raised some issues related to what aspects of language competence to measure from the natural language domain, including in an assessment. Only making some small interaction does not guarantee that the language involved is sufficiently describing the communicative function of the language or suitable in a real-life context (Bachman 1990).

Similarly, the interactional approach (IA) is based on the individual language user/test taker's characteristics, which include topical knowledge, affective aspect, personal characteristics, and language ability. Language ability covered two sub-competencies: strategic competence and language knowledge. Furthermore, language knowledge also has two sub-components: organizational and pragmatic knowledge (Bachman & Palmer 1996).

Furthermore, Bachman's (1990) concept of language proficiency test is supported by recent researchers (Amiryousefi & Tavakoli 2011; Tavakoli 2009), who conducted an empirical study that found a language proficiency test is not the only thing that influences their test performance. Several categories affect test takers' performance, such as motivation, anxiety, and ambiguity. These categories will be explained further in the following paragraph.

Motivation refers to a desire or an emotion that inflames and surrounds the test takers to prepare and learn before taking the test (Amiryousefi & Tavakoli 2011). Test anxiety refers to the test takers' nervousness during the test that arises from their sense of being afraid and then failing to perform their very best on the test, since the process of English proficiency test is in the form of a high standard test that decides his/her future (Rezaabadi 2016). Ambiguity can cause role strain for the test takers on their first time taking the test due to the lack of understanding or misunderstanding on the form of the test, corresponding to expect the uncertainty (Chiang 2016).

3 METHODOLOGY

3.1 *Participants*

Participants in this study are three university graduates from the hospitality and tourism program in Indonesia who used TOEFL ITP as one of the job requirements they fulfilled to get their current job. Furthermore, all three participants had the same reasons and goals on why they are taking the TOEFL ITP: for work requirements. As indicated by Fraenkel and Wallen (2009), contextual investigations include systematic demand for specific social units, for instance, one or a couple of people. Likewise, study cases additionally accentuate the issues experienced by the individual/social units at the research time (Bhattacharya 2017).

For this situation, it included three participants, Sammy, Rami, and Harry (pseudonyms), who had just embarked on their initial hospitality and tourism career after graduating from university.

3.2 *The data collection and analysis*

There are two data collection instruments used by the researcher in this study:

Questionnaire

This questionnaire elicited information on TOEFL ITP that participants took to get a job in the tourism industry. This questionnaire was theoretically conceptualized according to the Concept of Language Proficiency Test proposed by Bachman (1990) and supported by Tavakoli (2009) and Amiryousefi and Tavakoli (2011) whose concept is to capture participants' external factors' possibility on influencing the test result. Each item of the questionnaire used a five-point Likert scale response type from "strongly disagree" to "strongly agree."

In-depth interviews

This type of interview was utilized in order to get more in-depth and detailed information about graduates' perceptions about the suitability of TOEFL ITP as a job requirement. In this in-depth interview, the researcher uses the type of open-ended interview in which the researcher only prepares some guiding questions (Creswell 2013). Guiding questions in in-depth interviews make the process easier for the researcher to explore further information obtained from the participants because of the nature of the questions that are open and flexible to develop.

4 FINDINGS AND DISCUSSIONS

After analysing the questionnaire data, it was revealed that the TOEFL ITP was not considered suitable in measuring English ability for university graduates on the hospitality and tourism program. The data was obtained from the perspective of the graduates after they embarked on their employability journey. There are three aspects to prove why TOEFL ITP was not suitable including authenticity, validity, and generability that will be explored further below.

4.1 *Language proficiency test fundamental concept*

In terms of language proficiency test concept, it was discovered that the university graduates have seen TOEFL ITP did not measure their English ability to be utilized in their workplace: the test has bias results that can be influenced by external factors such as motivation, ambiguity, anxiety, and personality.

4.1.1 *Authenticity*

The terms authenticity in language usage are characterized in real-life and interactional ability that can be analyzed in language test performance (Bachman 1990). In the current study case, all participants Sammy, Rami, and Harry perceived that there was a discrepancy between their personal belief on their real English ability and the TEOFL ITP test. This situation leads to the uncertainty on test authenticity, whether the test is capable of measuring their language ability required in their workplace or not.

The data from the questionnaire was confirmed by the result of the in-depth interview as all participants asserted that the TOEFL ITP test does not represent their actual ability in producing

English as a language (written and orally). They do not understand how valid the TOEFL ITP test is due to the fact that their ability is depicted by number (score) and they might experience difficulty in proving their English in a real-life environment, especially in a workplace environment. Moreover, Sammy even admitted that he joined the TOEFL preparation course before taking the real test in order to get the highest score possible to fulfill a job requirement.

The test setting, situation, and procedure also influence the authenticity of the language proficiency test (Bachman 1990). In the questionnaire responses, the participants concentrated their belief on TOEFL ITP that did not constitute and was not compatible with their real-life workplace situation. TOEFL ITP only covered two language skills plus grammar section. Based on that fact, all three participants that have hospitality and tourism program background clearly cannot apply their prior knowledge they had from their time at university or that is workplace-related. In terms of setting, TOEFL ITP is more identical to a placement test or any other high-stakes standard test that solely focuses on doing the test to get a good score with the allocated time (Taufiq et al. 2018).

Participants expressed that the English proficiency test should have various types of questions and situations, not only multiple choices, and the minimum interview-like test is sufficient to represent real-life or workplace situation. Furthermore, the participants also highlight the skills tested should not be only limited to written language but also should be able to measure how to communicate verbally correctly. All participants pointed out that English has become a demand that generally applies in the world of work, so university graduates should have a standard of ability that can be used as an asset to face the world of work. Harry, who took the TOEFL ITP test for his work requirements of waitress, added that the current TOEFL ITP is too general and was not suitable for his daily interaction in his own work area.

Real-life (RL) approach is defined as the essential authenticity of the language proficiency test, where the test can replicate and measure test-takers' real language ability (Bachman 1990). Overall, TOEFL ITP is not fulfilling an RL approach that includes workplace setting and communicative competence, since this approach applied in the test mirrors the "reality" of non-test language use only.

4.1.2 *Validity*

The term validity in language usage can be analyzed from the test score, or it can be analyzed better in the process of test with test performance (Bachman 1990). Four types of validity will be investigated in this study related to TOEFL ITP: construct validity, content validity, criterion-related validity, and face validity. The participants' perspective toward the validity of TOEFL ITP can be determined from the responses on the questionnaire.

The construct validity can be analyzed through the ability of interaction with other abilities (constructs) (Bachman 1990). The TOEFL ITP test fulfilled validity if the test score reflects the area(s) of language ability they want to measure (Bachman & Palmer 1996). The participants responded that their workplace required them to perform communicative competence in English that can be assessed from productive language skills (writing and speaking). However, the TOEFL ITP only covered receptive language skills, which are listening and reading plus grammar skill. In terms of participants' background on hospitality and tourism workplace, according to construct validity, it can be claimed that TOEFL ITP is not valid for tourism industry employability requirement.

The content validity is the extent to which the domain is being investigated and evaluated by comparing the test specifications to the test contents (Hughes 1989). The participants agreed that TOEFL ITP did not measure their prior English knowledge accurately from their time learning English as tourism program student at the university. It is confirmed by the interview that pointed participants perspectives on TOEFL ITP is more suitable to assess general English rather than specifically to English skills they need for working in their field of interest.

The criterion-related and the face validity can be measured from the test result or score (Hughes 1989). In terms of the result or score of the test, TOEFL ITP employed numbers ranging from 337 to 677 based on the Common European Framework of Reference for Languages (CEFR) (Hernández Montalvo 2014). The participants responded through the questionnaire that the score result of TOEFL ITP might represent their English ability. However, Rami and Harry through the interview emphasized that, regarding the standards of their current workplaces, the score of their TOEFL ITP test did not reflect their English ability since currently, their productive skills are more performed compared to the receptive skills that TOEFL ITP assessed.

4.1.3 *Generability*

Generability refers to external factors that are influenced by the test. There are four external categories analyzed through this study, including motivation, anxiety, and ambiguity. All participants agree that they were motivated during the test and capable of doing well on the test TOEFL ITP. Confirmed in the in-depth interview, their motivation came from their preparation before taking the test; all of them admitted that they had taken the prediction test for paper-based TOEFL. Consequently, they all already are familiar with the instruction, procedure, setting, and situation.

Related to the TOEFL ITP test taken by the participants, test anxiety refers to the test takers' nervousness during the test that arises from their sense of being afraid and then failing to perform their very best on the test, since the process of English proficiency test is in the form of a high standard test that decides their future (Rezaabadi 2016). All the participants agreed on what they experienced during the test, due to the setting and situation of the test. However, in the interview, they suggested that they overcame the anxiety by previously being familiarized with the TOEFL prediction test.

In terms of ambiguity, the participants found that all options on multiple choices implicitly have similar meaning that can make them confused. From the interview, it can be confirmed that the score of TOEFL ITP can be invalid because all participants emphasized that they answer some questions by guessing and hang their score up to luck.

5 CONCLUSIONS

Based on the aforementioned research findings, it is strengthening the concept of English proficiency test proposed by Bachman (1990) that TOEFL ITP should have been measuring the level of university graduates' communicative competence. English proficiency test that focuses on being communicative is considered as more suitable for the needs of tourism graduates to face a real-life environment, which is why paper-based TOEFL tests like TOEFL ITP raise some concerns, particularly related to its authenticity, validity, and generalizability. The results of the questionnaire and an in-depth interview indicate that participants generally agree that the TOEFL test is vital to measure their English ability. However, they also shared doubts about the test as a measurement of communicative competence that can describe their English ability that can be applied in a RL environment. The results could help to advance the understanding that there might be another type of English proficiency test that is more suitable than TOEFL ITP as a tool of selection requirement for university graduates in the hospitality and tourism program.

REFERENCES

Alderson, J. C. (2009). Test review: Test of English as a foreign language™: Internet-based test (TOEFL iBT®). *Language Testing*, *26*(4), 621–631.

Amiryousefi, M. & Tavakoli, M. (2011). The relationship between test anxiety, motivation and the TOEFL iBT reading, listening and writing scores. *Procedia-Social and Behavioral Sciences, 15*, 210–214.

Ananda, R. (2016). Problems with section two ITP TOEFL test. *Studies in English Language and Education*, *3*(1), 35–49.

Bachman, L. F. (1990). *Fundamental considerations in language testing*. New York: Longman.

Bachman, L. F. & Palmer, A. S. (1996). *Language testing in practice: Designing and developing useful language tests (Vol. 1)*. Oxford University Press.

Barnes, M. (2016). The Washback of the TOEFL iBT in Vietnam. *Australian Journal of Teacher Education*, *41*(7), 158–174.

Bhattacharya, K. (2017). *Fundamentals of qualitative research: A practical guide*. Taylor & Francis.

Bùi, T. S. (2016). *The Test Usefulness of the Vietnam's college English Entrance Exam.* Korea University, Seoul.

Chiang, H. H. (2016). A Study of Interactions among Ambiguity Tolerance, Classroom Work Styles, and English Proficiency. *English Language Teaching*, *9*(6), 61–75.

Creswell, J. W. (2013). Qualitative Inquiry & Research Design: Choosing among Five Approaches (3rd ed.). Thousand Oaks, CA: SAGE.

Fahmi, M. R., Lusiana, T. & Mujayana, M. (2016). Rancang bangun aplikasi Test of English as A Foreign Language (TOEFL) prediction pada self access centre (sac) universitas islam negeri sunan ampel surabaya. *JSIKA, 5*(12).

Fraenkel, J. R. & Wallen, N. E. (2009). The nature of qualitative research. *How to design and evaluate research in education, seventh edition. Boston: McGraw-Hill, 420.*

Hamid, M. O. (2016). Policies of global English tests: test-takers' perspectives on the IELTS retake policy. *Discourse: Studies in the Cultural Politics of Education, 37*(3), 472–487.

Hughes, A. (1989). Testing for Language Teachers. *Cambridge: CUP, 28.*

Memon, N. & Umrani, S. (2016). The Impact of IELTS on the Test Preparation Industry of Pakistan. *International Journal of Humanities and Social Science 6(4).*

Murray, J., Cross, J. & Cruickshank, K. (2014). Stakeholder perceptions of IELTS as a gateway to the professional workplace: the case of employers of overseas trained teachers. *IELTS Research Reports, 1.* 1–78.

Muti'ah, A. (2019). Pendekatan Content and Language Integrated Learning (Clil) Dalam Pembelajaran Bahasa Indonesia Di Sekolah Menengah Kejuruan (Smk): Belajar Berbahasa Melalui Materi Peminatan. *FKIP e-PROCEEDING*, 1–16.

Nugraheni, H. (2014). Persepsi Alumni Dan Stakeholder Politeknik Kesehatan Semarang Terhadap Kesesuaian Kemampuan Bahasa Inggris Sebagai Bekal Menghadapi Kebutuhan Dunia Kerja. *Jurnal Kesehatan Gigi, 1*(1), 40–46.

Poorsoti, S. & Asl, H. D. (2015). Iranian Candidates' Attitudes toward TOEFL iBT. *Journal of Applied Linguistics and Language Research, 2*(8).

Rezaabadi, O. T. (2016). The relationships between social class, listening Test anxiety and test scores. *Advances in Language and Literary Studies*, *7*(5), 147–156.

Saukah, A. (2016). The English proficiency of the academics of the teacher training and education institutions. *Jurnal Ilmu Pendidikan*, *7*(1).

Shih, C. M. (2010). The Washback of the General English Proficiency Test on University Policies: A Taiwan Case Study. *Language Assessment Quarterly, 7*(3), 234–254.

Taufiq, W., Santoso, D. R. & Fediyanto, N. (2018, January). Critical Analysis on TOEFL ITP as A Language Assessment. In *1st International Conference on Intellectuals' Global Responsibility (ICIGR 2017).* Atlantis Press.

Tavakoli, M. (2009). The role of motivation in ESP reading comprehension test performance. *TELL*, 9 (3).

Woolfolk, A. & Margetts, K. (2016). *Educational psychology.* (4th ed.). Melbourne, VIC: Pearson Australia.

Zhou, Y., Jindal-Snape, D., Topping, K. & Todman, J. (2008). Theoretical models of culture shock and adaptation in international students in higher education. *Studies in higher education*, *33*(1), 63–75.

Tourism gastronomy

Promoting Creative Tourism: Current Issues in Tourism Research – Kusumah et al. (Eds)
© 2021 Taylor & Francis Group, London, ISBN 978-0-367-55862-8

Canna fettucine: Commodifying culinary Italian Indonesian

S.S. Wachyuni
Universitas Gadjah Mada, Yogyakarta, Indonesia
Sahid Polytechnic, Jakarta, Indonesia

K. Wiweka
Université Angers, Angers, France

R.M. Wirakusuma
Universitas Pendidikan Indonesia, Bandung, Indonesia

ABSTRACT: Pasta has been one of the most popular foods around the world for a long time. Pasta was introduced in the 13th century in Italy by Marco Polo after returning from his trip to China. The essential pasta ingredients include wheat cereal flour and water. Fettuccine is one of many varieties of pasta. Many restaurants or hotels all around the world, including Indonesia, provide this particular Italian food as their main menu. However, some of them have substituted several ingredients according to the local resources. The purpose of this research is to examine the substitution formulation of canna starch (Canna Edulis Ker) in the essential ingredients of fettuccine. Canna is a bulb that grows fertile in Indonesia. This ingredient can be processed as a canna starch, which contains natural carbohydrates and has a high viscosity. The research method used is experimental research based on a completely randomized design (CRD) of one independent variable. The independent variable is the substitution of canna starch in five levels of percentage, 15%, 30%, 45%, 60%, and 75%, that will be tested through organoleptic tests (preference and sensory quality test) by 30 trained panelists. The data was analyzed with ANOVA to identify the differences between several formulations and obtain a suitable fettuccine formula. According to the organoleptic test, the study finds that the use of canna starch has a significant influence on the substitution of wheat flour in making fettuccine. There was a substantial difference between the original fettuccine (control sample) and the canna starch fettuccine. The best formulation is 15%, followed by a 30% substitution of canna starch, which has characteristics of yellow color, pleasant aroma, standard taste, and very springy texture. The implication of this experiment is to find an alternative ingredient, mainly a local resource suitable to be used as international gastronomy products. This commodification will provide an unusual tourists' experience in visiting Indonesia. Whereas the practical implication is to encourage the gastronomist to develop another attractive and tasty product based on local ingredients.

Keywords: Pasta, fettucini, canna, gastronomy tourism, experiment

1 INTRODUCTION

Food issues that are a concern are generally related to basic needs. "Culture" of Indonesian people consuming carbohydrates, on the other hand, can cause food insecurity problems (Pangesthi 2009). This phenomenon tends to encourage higher imports of raw materials. As a raw material for making flour, from 2010 to 2012, wheat imports always increase consistently. APTINDO (Indonesian Flour Producers Association) noted that Indonesia's total wheat imports in 2010 amounted to 4,669,475 tons matrix and an increase of 5,475,148 tons matrices in 2011. Likewise, in 2012 that increased to 6,250,489 tons matrices.

DOI 10.1201/9781003095484-61

Apart from being the main ingredient in making bread, wheat flour is also the main ingredient in making Pasta. Traditionally, pasta is the result of processed hard durum or semolina flour made from hard wheat. Dry Pasta (pasta secca) is made from a mixture of flour and water (Fathullah 2013; Fatmawati 2013). This type of food is also often a substitute for rice for the Indonesia people.

Although diversification of this staple food is a smart solution, the problem that arises then is the dependence on imported raw materials. Therefore, diversification can be optimal if followed by the commodification of local raw materials (Lestari 2020), in other words, utilization of local commodities as commercial products. This phenomenon is known as "Glocalization" or "McDonaldization" (Ikejima & Hisano 2012; Mardatillah et al. 2019). This trend has been implemented in various countries, including the existence of Italian gastronomy in America (Cinotto 2004; Federici & Andrea 2016) or gastronomy as a tourist attraction (Fox 2007).

At present, many palawija plants or tubers are processed into flour, such as taro tubers, which are transformed into taro flour, corn processed into corn flour, and canna that is processed into canna flour. Canna is a non-rice food that has high nutrition, mainly because of the content of phosphorus, calcium, and carbohydrates. Canna nutritional content per 100 grams consists of calories (95.00 cal); protein (1.00 g); fat (0.11 g); carbohydrates (22.60 g); calcium (21.00 g); phosphorus (70.00 g); iron (1.90 mg); vitamin B1 (0.10 mg); vitamin C (10.00 mg); water (75.00 g); and the edible portion is 65.00% (Fathullah 2013; Harmayani et al. 2011; Noriko & Pambudi 2015; Pangesthi 2009; Richana & Sunarti 2004). Whereas, the nutritional content of canna flour (per 100 g) includes water (14 g); protein (0.7 gr); fat (0.2 gr); carbohydrates (85.2 gr); calcium (8 mg); phosphorus (22); iron (1.5 mg); vitamin A (0 Ui); B vitamins (0.09 mg); and vitamin C (0 mg). Canna flour can also be used as raw material for various processed food products (Fathullah 2013).

Canna starch has an amylose content of 38.50% and amylopectin at 31.63%. Based on the comparison of these contents, canna starch has the potential to be used as vermicelli, noodles, rambak, and cendol (Fatmawati 2013; Harmayani et al. 2011; Indrianti et al. 2013). Therefore, the characteristics and nutrition content of canna flour can be used to replace wheat flour in the making of pasta. This strategy is also essential to gradually reduce dependence on imported commodities and to more empower local agricultural products.

Based on this phenomenon, this study focuses on experimenting with the use of canna flour as an additive or substitute for making pasta. This research has several objectives, including identifying the level of consumer preference for fettucine pasta with canna starch substitution. Besides, this study also analyzes the differences in sensory quality of color, aroma, taste, and texture. The ultimate goal is to find the best substitute formula for canna starch in making pasta fettucine.

2 LITERATURE REVIEW

2.1 *Canna*

Canna or Ganyong in Indonesian is one of varieties of tuber. This tuber has not been used optimally, although its availability is quite high. Generally, several techniques are used, including boiling, burning, and frying (Fathullah 2013; Harmayani et al. 2011; Pangesthi 2009; Purwaningsih 2008; Richana & Sunarti 2004). This potential needs to be developed and utilized as an industrial raw material, mainly flour. This product can be categorized as a more durable and low cost in the distribution process (Widowati 2009). Although canna is one of the non-rice food ingredients that have quite high nutrition, especially the content of calcium, phosphorus, and carbohydrates, this raw material is not as popular as sweet potatoes or cassava.

2.2 *Canna flour and starch*

Flour-making techniques in Indonesia generally still use the traditional method with several stages, including selection and cleaning of canna, grating or collision and separation of starch, sedimentation, drying, and re-collision (Richana & Sunarti 2004). Whereas, starch is a source of carbohydrates in food, canna starch is the main processed product from canna tuber, which has a natural carbohydrate content and has a high concentration (Harmayani et al. 2011; Noriko & Pambudi 2015; Richana & Sunarti 2004; Slamet 2010). This starch was obtained from canna tubers, which were 8–12 months old (Harmayani et al. 2011; Indrianti et al. 2013; Richana & Sunarti 2004; Widowati 2009). Nutritional content of canna starch, including water (10.66% bb); ash (0.53% bk); fat (0.29% bk); protein (1.42% bk); carbohydrates (99.06% bk); iron (4.53 mg/100g); calcium (13 mg/100g); phosphorus (35.83 mg/100g); and zinc (0.47 mg/100g).

2.3 *Food processing, food commodification, and food fusion*

Food processing is more than just preparing and cooking the raw products, but it can be proposed to preserve the raw material itself. There are many food processing techniques, such as heating, cooling, drying, salting, filtering, and pickling/fermentation (Azam Ali & Dufour 2008). Processing the tuber into flour through drying and grinding can extend the tuber's shelf life. Meanwhile, to increase the economic value of local products, creativity is needed in processing, among others, by modifying food (food modification) and combining several basic ingredients from different culinary traditions (food fusion). Food modification is an activity to make new variations of food so that it is more interesting (Purnamasari 2019). Yuristrianti and Kuntjoro (2003) added that food modification was also done as a way to improve food taste. Food fusion is an innovation in the field of food that combines elements of different culinary traditions from two or more countries. The innovation will bring out new unique flavors and create new experiences when tasting it (Devina & Jonathan 2020).

2.4 *Pasta*

Pasta is an Italian processed food made from a mixture of flour, water, eggs, and salt. This dough can then be formed into a variety of sizes and shapes (Fathullah 2013; Fatmawati 2013).

3 METHOD

The research method carried out in this study is an experimental method based on a completely randomized design (CRD) of one independent variable. The independent variable in this study is the substitution of canna starch with five levels/treatment, that is 15%, 30%, 45%, 60%, and 75%. The choice of canna starch as a substitution in making fettucine is based on the results of preliminary studies, resulting in the fettucine made from canna starch having a better acceptance compared to fettucine made from canna flour. The dependent variable is the organoleptic quality of canna fettuccine. According to Usman (2009), CRD is a method of manipulation or treating independent variables, then observing, measuring, and analyzing the influence of these manipulations. Quality measurements carried out by organoleptic tests include hedonic tests and hedonic quality to measure product acceptance. Organoleptic test measurements used scoring techniques (Amerine et al. 2013). The hedonic scale can be stretched according to the desired scale (Suryono et al. 2018). The scale of hedonic and hedonic quality used is 1–6 on the parameters of color, aroma, taste, and texture. Organoleptic tests are also intended to test the differences between the five levels of percentage substitution. According to (Faridah 2009), the differentiation test usually uses 15–30 trained panelists, so that the panelists who tested the product in this study numbered 30 trained panelists. Data analysis techniques used ANOVA (analysis of variations) and continued with the least significant difference (LSD) test. The best sample in this study will be carried out as a proximate nutrition test to determine the nutritional content of canna fettuccine.

4 RESULT

In a preliminary study, researchers conducted trials making fettuccine using the basic ingredients of canna flour and canna starch. Fettuccine, which is made from canna starch, has better sensory

Table 1. The average hedonic canna fettuccine test value.

	Substitution				
Parameter	15%	30%	45%	60%	75%
Color	4.25	3.71	3.45	3.05	2.80
Flavor	4.12	3.74	3.46	3.50	3.33
Taste	4.37	3.88	3.58	3.30	3.23
Texture	4.10	3.50	3.05	2.80	2.50
General acceptance	4.21	3.71	3.38	3.16	3.62

Table 2. Color hedonic ANOVA test results.

Source	Type III Sum of Squares	df	Mean Square	F	Sig.
Corrected Model	38.358[a]	4	9.590	9.399	.000
Intercept	1787.031	1	1787.031	1751.514	.000
product	38.358	4	9.590	9.399	.000
Error	147.940	145	1.020		
Total	1973.330	150			
Corrected Total	186.298	149			

results compared to fettuccine made from canna flour. Canna flour has a less pleasant aroma than canna starch, and in terms of texture, canna flour and canna starch can substitute up to 75% flour.

4.1 *Hedonic test*

The hedonic test aims to test the level of parameters in terms of color, aroma, taste, and texture of all canna fettuccine samples tested in this experiment. The average hedonic test values can be seen in Table 1.

The hedonic test tested panelists' preference for five canna fettuccine samples. Each stage of the experiment was repeated three times to reduce the error rate of the test. A total of 30 trained panelists tested the sample in this study. In Table 1, it can be seen that the panelists' preference for the parameters of color, aroma, taste, and texture of fettuccine decreased. The result shows that the sample of 15% canna starch substitution is the most preferred by consumers.

The ANOVA test was used to determine the differences in each preference. As can be seen in Table 1, the ANOVA statistical results of the color likeness test stated a significance value of $0.000 < 0.05$ and <0.01; this means that the degree of color preference was significantly different in fettuccine paste with different percentage of canna starch substitution. After the ANOVA test, the Duncan test is performed to find out which sample has a color difference, and it is known that the error level $\alpha = 0.05$, the fettuccine sample with 15% substitution of canna starch has a significant difference with 30%, 45%, 60%, and 75% samples. Whereas at $\alpha = 0.01$, fettuccine samples with 15% substitution were only significantly different from the 75% sample but not different from 30%, 45%, and 60% samples. Duncan's color hedonic test table can be seen in Table 2.

Furthermore, the hedonic test of aroma, texture, and taste of canna fettuccine was also tested using ANOVA and Duncan tests. The hedonic ANOVA aroma test showed a significance value of $0.122 > 0.05$ and > 0.01. It can be interpreted that the level of aroma preference is not significantly different in the fettucini paste substitution of canna starch with different concentrations. This interpretation means that the respondents' preference level for the whole sample is no different. The results of the average value showed that respondents said they liked the scent of canna fettuccine. However, it tended to decrease along with the increase in the percentage of canna starch substitution.

Table 3. Color hedonic ANOVA test results.

		Notation	
		$\alpha = 0.01$	$\alpha = 0.05$
75%	2.8003	C	d
60%	3.0557	Bc	cd
45%	3.4550	Bc	Bc
30%	3.7117	Ab	B
15%	4.2453	A	A

Table 4. The average value of hedonic canna fettuccine quality test.

	Substitute				
Parameter	15%	30%	45%	60%	75%
Colour	4.03	3.50	3.05	2.80	2.50
Aroma	4.25	3.71	3.45	3.05	2.80
Taste	4.12	3.74	3.46	3.50	3.33
Texture	4.37	3.88	3.58	3.30	3.23

Note: scale 1–6
color = not very yellow – very yellow
aroma = very unpleasant – very pleasant
rasa = very bitter – not very bitter
texture = not very springy – very springy

For example, in the substitution of 75% canna starch, the respondent stated rather like the scent of fettuccine.

The hedonic ANOVA test results showed a significance value of 0.000 <0.05 and <0.01 so that the level of respondents' preference for the texture of the canna fettuccine was significantly different. The tendency of respondents' preference level toward texture decreases with an increasing percentage of canna starch substitution. Duncan's test results are shown with letter notation. Samples that are marked with different letter notations means that they have different qualities, while samples that are marked with the same letter notation mean they do not have quality differences. Duncan's test results at the error level $\alpha = 0.05$, 15% sample have a difference in preference by 60% and 75%. Whereas at $\alpha = 0.01$, the 15% sample only had a difference in preference level with the 75% sample of canna starch substitution. Then, a hedonic taste test was carried out on the canna fettuccine sample. Taste is defined as the taste received by the tongue in the mouth. ANOVA test results show that the significance value of 0.000 < 0.05 and <0.01, then there is a difference in fettuccine with a different percentage of canna starch substitution. Duncan test results, the level of $\alpha = 0.05$ shows that the level of preference for the taste of the sample 15% is significantly different from the sample 75%. While the level of $\alpha = 0.01$ indicates that the sample 15% has a difference with 45%, 60%, and 75%. Nevertheless, samples 45%, 60%, and 75% did not differ from each other.

4.2 *Hedonic quality test*

A quality test or hedonic quality test is carried out to determine the difference in the quality of each sample tested. The results of the average value of the hedonic quality test on the parameters of color, aroma, taste, and texture of the canna fettuccine can be seen in Table 4.

Table 5. Quality hedonic taste ANOVA test result.

Source	Type III Sum of Squares	df	Mean Square	F	Sig.
Corrected Model	11.720[a]	4	2.930	1.853	.122
Intercept	1978.060	1	1978.060	1250.890	.000
PRODUCT	11.720	4	2.930	1.853	.122
Error	229.292	145	1.581		
Total	2219.072	150			
Corrected Total	241.012	149			

a. R Squared = .049 (Adjusted R Squared = .022)
Alpha: 0.05 = .000

Based on the results of the average hedonic canna fettuccine quality test in Table 4, all parameters tested, namely color, aroma, taste, and texture, have decreased quality. The hedonic quality value, which has the highest average value, is a sample of 15%, followed by 30%, 45%, 60%, and 75%. As for the color parameters, a 15% sample has yellow characteristics. The yellow color fades with an increasing percentage of canna starch substitution. Likewise, with the aroma parameters, the increasing substitution of canna starch decreases the quality of the aroma and becomes unpleasant. Furthermore, canna starch generally has a bitter taste. The 15% sample, which is the lowest substitute canna starch, has no bitter taste. In taste testing, the sample is not given any additional seasoning to find out the original flavor characteristics of the canna fettuccine. For the texture, the 15% sample has a chewy texture, and along with the increase of canna starch substitution, there is a decrease in springiness. ANOVA analysis was carried out to determine whether there were statistical differences in the hedonic quality of all samples in the parameters of color, aroma, taste, and texture.

The ANOVA results of color hedonic quality have a significance value of $0.000 < 0.05$ and <0.01, which means the color quality is significantly different from all canna fettuccine samples. The Duncan test results state that in the $\alpha = 0.05$ level and the $\alpha = 0.01$ level, the 15% sample has a different color than the 45%, 60%, and 75% samples. Likewise, with the hedonic aroma quality test results, which showed a noticeable difference from all samples. Duncan's test results show that the 15% sample has a difference with sample 30%, 60%, and 75% at the level of $\alpha = 0.05$, while at the level of $\alpha = 0.01$, the sample 15% shows only a difference with the sample of 75%. It means that the other samples do not show differences in the quality of the aroma.

The hedonic texture quality test results also showed differences in all samples. The 15% sample is different from 45%, 60%, and 75%, but does not differ from the 30% sample at the level of $\alpha = 0.05$ and the level of $\alpha = 0.01$. While on the taste parameter, the significance value is $0.122 > 0.05$ and >0.01. It means that the hedonic quality of taste does not differ in all canna fettuccine samples tested.

The results of the ANOVA quality hedonic taste test results can be seen in Table 5. The Duncan test results show that at the level of $\alpha = 0.05$, the 15% sample has a difference with the 75% sample. However, at the level of $\alpha = 0.01$, the taste of all samples was not different. The best samples in this study were fettuccine with canna starch substitution of 15%. Then the sample was tested for proximate nutrition at PT Saraswanti Indo Genetech, Bogor. Based on the results of nutrition tests, in 100 grams of canna fettuccine, it has a total energy of 2574.63 kcal, energy from fat 163.8, moisture 30%, ash 1.37%, total fat 83.6%, protein 9.51%, and total carbohydrates as much 30.67%. From these results, canna fettuccine has a high nutritional value and can be used as an alternative food to rice.

5 DISCUSSION

Based on this research, canna starch can be used as an alternative to flour in making fettuccine with a substitution percentage of up to 75%. The commodification of Italian cuisine with local ingredients can provide a unique gastronomic experience for tourists visiting Indonesia as well as a useful marketing tool to attract tourists visiting Indonesia. Some studies state that canna starch is used to make pastry product substitute for wheat flour by 20% (Stiti 2016) and dried noodles with substitution by 10% (Budiarsih et al. 2010). Listiyani (2016) also examined that canna starch can be the basic ingredient in making pasta with a substitution of 30%. Canna is a group of tubers that have not been widely used even though it has many advantages, including nutritional value and high fiber content, and is a source of carbohydrates that do not contain gluten.

6 CONCLUSION AND IMPLICATIONS

Starch is a source of carbohydrates that have the characteristics of being easy to expand if exposed to heat. The best formulation, was a sample with a 15% substitution of canna starch. This best sample has the characteristics of yellow, pleasant aroma, bitter taste, and chewy texture. Whereas, the implication of this research includes the use of local food ingredients as substitutes for pasta products is expected to increase the economic value of local ingredients, and to encourage the gastronomist to develop another attractive and tasty product based on local ingredients.

REFERENCES

Amerine, M. A., Pangborn, R. M., & Roessler, E. B. (2013). *Principles of sensory evaluation of food.* Elsevier.

Azam Ali, S., & Dufour, C. (2008). *Home-based fruit and vegetable processing in Afghanistan: A manual for field workers and trainers.*

Budiarsih, K., & Fauza. (2010). *Kajian Penggunaan Tepung Ganyong (Canna Edulis Kerr) Sebagai Substitusi Tepung Terigu Pada Pembuatan Mie Kering.* Jurnal Teknologi Hasil Pertanian, Vol. III, No. 2, Agustus 2010.

Cinotto, S. (2004). *Now that's Italian! Representations of Italian food in American popular magazines, 1950–2000.*

Devina, H., & Jonathan, P. C. (2020). Indonesian Fusion Foods. *Demedia Pustaka.* http://demediapustaka.com/product/indonesian-fusion-foods/

Dewan Ketahanan Pangan. (2006). Kebijakan Umum Ketahanan Pangan 2006–2009. *Jurnal Gizi Dan Pangan, 1*(1), 57–63.

Faridah, A. (2009). Pentingnya Uji Sensori dalam Pengolahan Pangan. *Jurnal Pendidikan Dan Keluarga, 1*(3), 34–48.

Fathullah, A. (2013). *Perbedaan brownies tepung ganyong dengan brownies tepung terigu ditinjau dari kualitas inderawi dan kandungan gizi* [PhD Thesis]. Universitas Negeri Semarang.

Fatmawati, H. (2013). *Pengetahuan Bahan Makanan I.* Jakarta: Kementrian Pendidikan dan Kebudayaan DIrektorat Pembinaan Sekolah Menengah Kejuruan.

Federici, E., & Andrea, B. (2016). *Selling Italian Food in the USA: Pride, History and Tradition.*

Fox, R. (2007). Reinventing the gastronomic identity of Croatian tourist destinations. *International Journal of Hospitality Management, 26*(3), 546–559.

Harmayani, E., Murdiati, A., & Griyaningsih, G. (2011). Karakterisasi pati ganyong (Canna edulis) dan pemanfaatannya sebagai bahan pembuatan cookies dan cendol. *AgriTECH, 31*(4).

Ikejima, Y., & Hisano, S. (2012). *Commodification of Local Resources and its Paradox: A case of traditional vegetables in Kyoto.*

Indrianti, N., Kumalasari, R., Ekafitri, R., & Darmajana, D. A. (2013). Pengaruh penggunaan pati ganyong, tapioka, dan mocaf sebagai bahan substitusi terhadap sifat fisik mie jagung instan. *Agritech, 33*(4), 391–398.

Lestari, N. S. (2020). Roti gambang, acculturation bread from Betawi. *Journal of Indonesian Tourism, Hospitality and Recreation, 3*(1), 40–48.

Listiyani, K. (2016). *Formulasi Pasta Berbasis Pati Ganyong (Canna edulis Kerr) sebagai Pangan Alternatif untuk Anak dengan Sindrom Autisme.* http://repository.ipb.ac.id/handle/123456789/86895

Mardatillah, A., Hermanto, B., & Herawaty, T. (2019). Riau Malay food culture in Pekanbaru, Riau Indonesia: Commodification, authenticity, and sustainability in a global business era. *Journal of Ethnic Foods, 6*(1), 3.

Noriko, N., & Pambudi, A. (2015). Diversifikasi pangan sumber karbohidrat canna edulis Kerr.(Ganyong). *Jurnal Al-Azhar Indonesia Seri Sains Dan Teknologi, 2*(4), 248–252.

Pangesthi, L. T. (2009). Pemanfaatan Pati Ganyong (Canna Edulis) Pada Pembuatan Mie Segar Sebagai Upaya Penganekaragaman Pangan Non Beras. *Media Pendidikan, Gizi, Dan Kuliner, 1*(1).

Purnamasari, N. A. (2019). *Pengaruh modifikasi makan pokok terhadap daya terima balita di taman penitipan anak (TPA) cinta kota palangka raya* [Tugas Akhir Mahasiswa]. Prodi Diploma IV Gizi. http://repo.poltekkes-palangkaraya.ac.id/108/

Purwaningsih, Y. (2008). Ketahanan pangan: Situasi, permasalahan, kebijakan, dan pemberdayaan masyarakat. *Jurnal Ekonomi Pembangunan: Kajian Masalah Ekonomi Dan Pembangunan, 9*(1), 1–27.

Richana, N., & Sunarti, T. C. (2004). Karakterisasi sifat fisikokimia tepung umbi dan tepung pati dari umbi ganyong, suweg, ubi kelapa, dan gembili. *Jurnal Pascapanen, 1*(1), 29–37.

Saliem, H. P., & Ariani, M. (2016). *Ketahanan pangan, konsep, pengukuran dan strategi.*

Slamet, A. (2010). Pengaruh perlakuan pendahuluan pada pembuatan tepung ganyong (Canna edulis) terhadap sifat fisik dan amilografi tepung yang dihasilkan. *Agrointek, 4*(2), 100–103.

Suryono, C., Ningrum, L., & Dewi, T. R. (2018). Uji Kesukaan dan Organoleptik Terhadap 5 Kemasan Dan Produk Kepulauan Seribu Secara Deskriptif. *Jurnal Pariwisata, 5*(2), 95–106.

Usman. (2009). *Metode Penelitian.* Jakarta: Bumi Aksara.

Widowati, S. (2009). Tepung Aneka Umbi Sebuah Solusi Ketahanan Pangan. *Tabloid Sinar Tani, 6.*

Yuristrianti, N., & Kuntjoro, T. (2003). *Pengaruh pelatihan tenaga penjamah makanan tentang sistem pengolahan dan penyajian makanan terhadap mutu makanan pasien di RSUP Prof. Dr. Margono Soekarjo Purwokerto* [PhD Thesis]. [Yogyakarta]: Universitas Gadjah Mada.

Promoting Creative Tourism: Current Issues in Tourism Research – Kusumah et al. (Eds)
© 2021 Taylor & Francis Group, London, ISBN 978-0-367-55862-8

The Salapan Cinyusu (Nona Helix) as a "creativepreneurship" support model for gastronomy tourism in Bandung city

D. Turgarini
Universitas Pendidikan Indonesia, Bandung, Indonesia

ABSTRACT: Bandung since the Dutch colonial era has been known as a vacation destination, so it is known as Parijs Van Java. The local food in this city has become a Sundanese ethnic icon. Unfortunately, now many tourists are less interested in local food because of its conventional-traditional appearance. Apart from that, there is no creativepreunership think tank model where local food is worthy of being part of gastronomic tourism. Therefore, the aim of the study is to identify an interaction model between stakeholders in the gastronomic sector in Bandung Municipality so that it can induce ideas that support entrepreneurship in gastro-tourism. The research method used is descriptive qualitative, using 100 informants as sources. Primary data were collected through in-depth interviews guided by structured questionnaires, discussions, brainstorming, expert meetings, and focus group discussions. Secondary data were taken from publications. Data analysis was carried out through field data collection, then continued by the coding process, data reduction, data display, and finally drawing conclusions. The result showed nine elements of the Salapan Cinyusu or Nona Helix (nine helices) model, which consists of gastronomic businesses that accommodate creativepreneurs; government, workers, suppliers, experts, observers, connoisseurs, non-government organizations, and information technology as bridges must synergize and collaborate to strengthen the existence of tourism gastronomy in Bandung. All elements in the interacting model help to encourage entrepreneurs to come up with creative ideas for managing local Sundanese food.

Keywords: Salapan Cinyusu, Nona Helix, Gastronomi, Creativepreunership, Bandung

1 INTRODUCTION

Bandung has been known as a tourist city for a long time. A nickname, Kota Kembang, was given, due to a lot of beautiful women or a personification as Parijs van Java, like Paris in Java Island, as an indication that since colonial era Bandung has been an exotic tourist destination. The long history has created tourism in Bandung that is different from other regions. Bandung has various iconic places, such as Gedung Sate, Braga road—a venue for Bragastone used to perform, Asia-Afrika road—a historical road where countries from Asia and Africa united to emerge their forces and also a place where young people perform their artistic cosplay costuming along the road (Turgarini 2018). Furthermore, the people of this city are famous for their multitalents in creating ideas and creativities. Bandung is a city where many automotive fan communities are established, one of which is SOG (scooter owners group). It is the first scooter fans' club in Indonesia whose membership is spreading to overseas.

Moreover, Bandung, with its "regional branding" as a culture tourist destination, has an attraction which combines gastronomy and Sundanese ethnic with urban landscape element. The elements combination creates relatively different uniqueness compared to other cities in Java, such as Jakarta, Yogyakarta, Sema-rang, Surabaya, and Malang. Tourist arrivals in Bandung will be continued by enjoying delicious typical local food of Priangan. Some restaurants such as Ampera, Bancakan,

Sindang Siret, and others provide typical Sundanese food menu. Unfortunately, that tourism phenomenon has not become an attractive tourist attraction, especially in term of Sundanese gastronomy handling. So far, the emphasis of food has only been on sources of meeting physical needs. In fact, food, in its form, character, or style of presentation, requires innovation and continuous improvement. The absence of innovation in food makes buyers feel bored. As a result, consumer loyalty has decreased, then switched to other types of food that are considered to have a new variety of attributes, no longer to local cuisine. Meanwhile in Bandung, there is no think tank model, a kind of expert body that can contribute to creative ideas continuously, and channels of communication and brainstorming among stakeholders with the aim of advancing entrepreneurs engaged in the gastronomic tourism sector. Here it needs a role model to induce ideas to be hot, trending support for gastro-tourism in Bandung. Gastronomy tourists, who make their travels to destinations, in this case to Bandung, attempt to taste and to seek experiences on authentic local food (Pullphothong & Sopha 2013). For this reason, it is necessary to look for the form of a creativepreunership support model for gastro-tourism in Bandung. Thus, the purpose of this study is to identify a model of interaction between stakeholders in the gastronomic sector so that it can induce ideas that support creativity from gastro-tourism in Bandung.

2 LITERATURE REVIEW

There are some stakeholder management models that have been developed by many experts through involving various supporting elements. Two models that can become a basic is triple helix which later becomes penta helix for further development.

2.1 *Triple helix*

Triple helix model, initiated by Etzkowitz and Levesdorff (1995), Leydesdorff (2012), shows interdependence between industry and the public sector with universities as a source of knowledge. Public sector necessity encourages technology innovation in universities. Sales result is benefited by all involved sectors, including public ones, in the form of tax. Synergy among universities as the center of Research and Development (R&D) and entrepreneurs in industries and society that is represented to the government shows a collaboration of actors in implementing knowledge, production, and market need fulfilment practice.

The development of the triple helix model that regards consumer behavior is important to be considered furthermore to encourage a new source called Quadruple Helix (Carayannis et al. 2012; Cavallini et al. 2016). Knowledge obtained from consumers will raise knowledge and more solid innovation on products needed by each market segment and target. Quadraple Helix can be as an approach to encourage economic growth (Cavallini et al. 2016). Using Quadruple Helix model, ordinary people can be motivated to develop their resources through education and mentally engaging experiences that they can share with attractive people.

2.2 *Penta helix: ABCGM*

Penta helix is the development of triple helix concept. Five elements involved in the model are Academics, Business, Community, Government and Media. Penta helix model implements a synergy of various sources to actualize innovation (Sturesson et al. 2009; Lindmark et al. 2009).

Business sector is a leverage place for products to become more valuable as well as distribute them to consumers, while Education Institution is a place for the academics to become innovative which means that change theory and the concept to become real products and later are to be updated in the business sector to achieve competitive advantage. Local community participates in business development. They make their businesses as spirits for supporting their daily lives. Government as a regulator and a facilitator will receive a tax return from business activities. Then, media provides information technology as a business development.

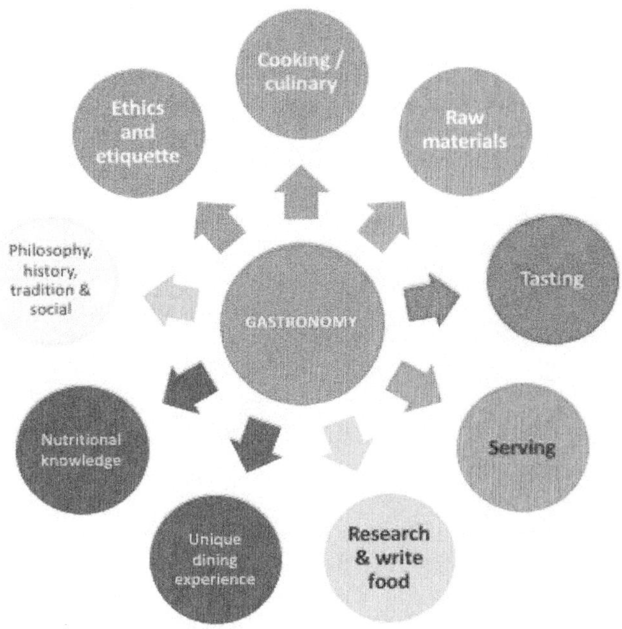

Figure 1. Concept of gastronomy. Source: Turgarini, 2018.

Partnership between industries and universities as well as government and civil society plus non-government organizations (NGO) (REPEC 2012) generates an economic development model to achieve sustainable innovation with a social control application called Penta Helix. As to triple helix, all actors will benefit the result in accordance with each role.

2.3 *Gastronomy*

Soeroso (2014), Soeroso and Susilo (2013), Lilholt (2015), Manolis (2010), Santich (2010) and Pullphothong and Sopha (2013) state that gastronomy has a major difference from culinary (the art of cooking in the kitchen). Gastronomy is said to reflect human behavior in choosing raw material, then in tasting, in feeling, in food serving, and in the experience of consuming as well as in seeking, learning, researching, and writing about food and everything related to ethics, etiquette, and nourishment (Figure 1).

Therefore, it can be stated that culinary is a part of gastronomy. The culinary activity is only about cooking food in the kitchen, while gastronomy includes culture in consuming food.

2.4 *Creativepreneurship*

Creative entrepreneurship or creativepreneurship is a concept initiated by Howkins (2001). Creative-preneurship is creativity in entrepreneurship. A creative entrepreneur is someone who is intelligent, has ex-traordinary skills and talents, and appreciates independence in work. They have a unique concept and vision in accordance with their passions so that they have the potential to open jobs and own wealth through exploi-tation of intellectual rights.

Creativepreneur generates ideas and ways of thinking that are 'out of the box." For that, they make continuous innovations on the products they produce. Creativepreneurs tend to operate off of passion and their personal mission to make an impact in their own creative ways. They think creatively and in an innovative way. They drive off of coming up with new and creative projects or ideas (becomingnatasha 2019). A creativepreneur is an entrepreneur who builds a business around a personal mission, passion, or purpose and runs it from creative principles, using digital teams, tools, and tribes to expand income and influence (Ross 2018). Ross (2018) also explained that creative entrepreneurs have three clear identifiers, namely, (1) creating a business because of passion, mission, or purpose (maybe all three); (2) working in the digital space, hiring virtual help when we need to expand, and (3) understanding creative principles and applying them to work done in business.

Howkins (2001) lists rules for successful creative entrepreneurs. These rules include invent themselves, prioritize ideas over data, be nomadic, learn endlessly, and, most importantly, have fun. A creative income earner takes risks to build assets, in the hope of yielding a profit, but with some key differences. They are the type of people who can calculate risk by creating products that have a clear target market, are good at changing the mindset, have deadlines for achieving goals, don't give up easily, and are persistent in introducing the products offered. The basis of creativepreneur work is creativity, which guides business concepts and executes them into reality.

3 MATERIAL AND METHODS

This research took Bandung Municipality as the object with gastronomy tourism as the focus. This research uses a qualitative descriptive method to describe the phenomena, circumstances, or symptoms that appear on the research object. This study used (1) primary and (2) secondary data. Primary data was taken directly by researchers from in-depth interviews with a structured questionnaire reference to informants, discussions, brainstorming, expert meetings, and FGD (focus group discussion), altogether 100 peoples. The informants were involved from various elements of tourism stakeholders in the city of Bandung, varied from academics, government, business actors, entrepreneur, public figures, and community leaders who are involved in the gastronomy field. Meanwhile, the secondary data were taken from publications.

The data analysis process was carried out in stages. After the field data was collected, data coding and reduction process were carried out (selecting and sorting of relevant data), then displaying the data. The last stage is drawing conclusions. For data validity checking in this research, triangulation technique and sources were used by combining data from various collecting techniques and availability data sources (Sugiyono 2013; Salkind 2010; Pelto 2017). Triangulation technique was conducted through participative observation, in-depth interviews, and documentation on similar data sources simultaneously. Meanwhile, source triangula-tion was conducted through data mining from different sources with identical technique.

4 RESULTS AND DISCUSSION

4.1 *Informant profiles*

The researcher attempted to obtain information from numbers of informants either directly related or not to tourism activity, especially to the catering business (Table 1). Below are 100 informant profiles as the sources of in-formation:

4.2 *Result synthesis*

4.2.1 *Development principles*

The synthesis of research results shows that in developing Bandung as a gastronomy city, aside from requirements determined by UNESCO, it is necessary to re-think and re-engineer about food (Turgarini 2018). This is in line with the spirit of Bandung as a UNESCO creative city, that is involving civil society to improve public space, including the design (UNESCO 2020).

It is important to re-think reminding that the majority of tourists coming to Bandung are gastronomy tourists. As stated by Turgarini (2018), their major motivation to travel is to taste and feel

Table 1. Informant profiles.

No.	Informant	Amount	%
1	Academics	16	16
2	Bureaucrats	13	13
3	Catering entrepreneur	15	15
4	Tourism business actors	17	17
5	Food material suppliers	13	13
6	NGOs	11	11
7	Food observers	15	15
Total	100		

the sensation as well as to experience from consuming local food with typical taste and aroma of Sundanese, yet internationally accepted equal to Shabu-Shabu, Tom Yam Gung, pizza, hamburger, and so on. Therefore, assessment is necessary so that all aspects of gastronomy products start from cultivation, preparation for raw material, and food processing, and then serving style of bringing food from the kitchen to the dining table serving, or the place of taking products as souvenirs, either gesture or costume, must personify Sundanese nuance. Besides, the food produced must fulfill hygiene and sanitation requirements.

The aim is to create Sundanese atmosphere totally so that consumers can feel not only are they purchasing elementary product of food, which is limited to goods and services only, or considering it as merely existence product for commercial and transactional, but they are also accepting it as an integrated gastronomy product involving benefit, pleasure, as well as social value as a form of art created with special skill and technique, nutrition, and even become lifestyle. Therefore, consumers feel that by consuming Sundanese food they will have added value.

The development of Bandung as gastronomy city has fulfilled the criteria determined by UNESCO (2020), that is availability of cooking skill from people and community. There are a lot of restaurants and/or Sundanese chefs that exist in the city having a high level of cooking skill. Local raw material is dominating although there are some imported from overseas. Some food that has "gone international" can still be enjoyed in traditional outlets. Apart from that, awards, contests and wide-supporting facilities like nationally and internationally foodscape have been available. Food observers in Bandung respect the environment by not using endangered animals or plants as food material and always promote sustainable local products. Besides that, cooking schools have included knowledge about the importance of conserving biodiversity in their curriculum (Turgarini 2018).

Next, re-engineering is necessary to make up the products presented to consumers. Food products which are created need to be packaged attractively to have a pleasant and evergreen look (Figure 2). One of which is packing utilization, except giving protection to main products from any disturbance (dropping, crushed, and so on), as well as can be benefited as a means of presenting, introducing, and promoting both food and the city of Bandung.

Unique packaging design can change common food, or easy-to-find food in Bandung (for example Surabi) becomes extraordinary because the packaging can attract many consumers to buy. Therefore, packaging becomes one of the factors for product marketing success, ensuring that the products are safe to the shelves in stores and to consumers' hands.

Four essential factors that need to be notified in food product packaging are, first, raw material relates to flexibility of shape and dimension, capability to protect and to endure crush, safety during delivery process and in transportation, as well as to promote and to introduce core products. Second, the business refers to efficiency (expense efficiency), sustainable (recycle), flexible, and obedience to law/rules/regulation. Third, supply chain relates to endurance, obedience, duration, shelf life, labeling and coding, retailer and consumer needs, and transportation requirement (dimension, shape, size). Last, brand and marketing relate to display shelves readiness in stores or outlets,

Figure 2. Examples of unique and creative food packaging. Source: Businesswire (2020), Shushudesign (2020), Bowman (2019).

design consistency, informative labeling and coding, easy to use, as well as brand representation as a mark and consumers' need.

4.2.2 *Salapan Cinyusu*

Bandung is a city that has many talents in the field of gastronomy—one of which is Andrian Ihsak, a molecular gastronomy chef who has successfully penetrated the boundaries of serving food in general. His exploratory abilities and passion have pushed the boundaries of culinary innovation by blending art with science. Ihsak brings a multi-sensory dining experience to the legendary dishes of Bandung. The local food is served in a molecular gastronomy style so that the appearance and the way to enjoy it are much different than usual, without losing the original taste quality. Ihsak's famous performance is serving shake noodles "Bang Dadeng." This shake noodle uses 27 kinds of spices, and before you enjoy it, you must shake it first. But if usually, the noodle shaker is the seller, Ishak invites the connoisseurs to join together to shake the ingredients. Ishak also made "Arum manis" as a dessert in the shape of a banknote styled to resemble a money tree. Then, the satay dish is made to resemble eggplant and ice pudding that is shaped like a creepy eye ball.

If usually local foods such as Patties, Kupat Tahu (steamed diamond-shaped rice, served with tofu and sweet sauce made from coconut sugar), Lotek (raw vegetable salad served with spicy peanut sauce), and others are only displayed simply in village stalls, so that they are less attractive to buyers, now the food is packed luxuriously. To give a unique impression, consumers are given the attraction of heavy metal music, which is played as the background for serving the local food. Another sensation to accompany the presentation of the local food, Ishak visually presents the atmosphere of eating under the heavy rain complete with the roar of thunder (Andriyawan 2019). However, this way of handling food cannot be done by all entrepreneurs. To produce creative ideas, it needs help and training from many other more experienced gastronomy actors.

The results of interviews, discussions, brainstorming, expert meetings, and FGDs with informants provided direction that it is necessary to rethink and re-engineer local food products to support Bandung as a gastronomic city. The root of the problem is because local Sundanese food sellers are usually not tech-savvy. They are also very conventional in the presentation of food. They do not really care about their formal education (which will be needed when facing competition in the market). The concept of their local food business is unclear, the menu layout is not right, and the food presenters are also perfunctory in grooming. The way of serving the main course with side dishes is out of sync. The place where they put side dishes to keep it looking beautiful and blending with other elements is not taken seriously. The thickness of each food component is also not maintained, so that it seems to be stacked, irregular. The appearance of the food is not special or unique, even though it should be able to increase the customer's appetite. The appearance of food that is interesting, unique, or funny and fun when displayed through social media makes many people willing to visit a restaurant for this. There is no sensation that makes consumers get new experiences when consuming the dishes that are served.

Therefore, the informants saw the need for a creativepreneur forum that could encourage sellers to handle local Sundanese food innovatively. However, this can be done well only if the gastronomic stakeholders in Bandung City synergize or work together. In gastronomic business, besides

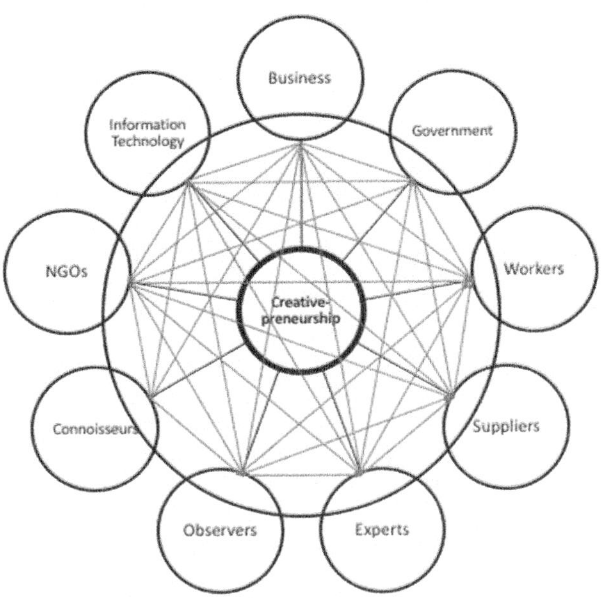

Figure 3. Salapan Cinyusu. Source: Turgarini's Interpretation (2020).

the entrepreneurs who run the industries along with their workers, government as a beneficiary receiver, universities as kitchen of innovation, and NGOs that carry out social control, there is also an involvement of suppliers including farmers who dominate foodshed, intermediary traders, and so on, experts as food practitioners, observers, and civil society as food connoisseurs. Collaboration among those elements, nowadays, is bridged by information technology. The relationship or reciprocal interaction between the elements of the Nona Helix of gastronomy creativepreunership is shown using arrows in Figure 3. The results of this interaction model are expected to encourage the emergence of new creative ideas continuously, which support entrepreneurship in the gastronomy field. The results of these creations can be applied, ranging from foodsheds (for example: food cultivation techniques) to foodscapes (example: serving food, grooming, sales online) in the city of Bandung. See Figure 3.

Those nine-synergizing elements are more than just ABCGM elements in Penta Helix—Academics, Business, Community, Government, and Media (Amrial et al. 2017; Muhyi et al. 2017). The author calls this concept in Sundanese terminology as Salapan Cinyusu (nine springs). It is a form of Nona Helix (nine helices) that connotes a spring which is always dreamt of by living things so that it must not recede because they are interrelated to strengthen the spirit of gastronomy. The absence of one of those elements may cause imbalance of the Sundanese gastronomy constellation in Bandung.

Entrepreneurs (including business associations) cooperate with their workers, who consist of artisan and other human resources, to develop a base for Sundanese food business, while government through the Industrial and Trade Department, Tourism Department, Public Work Department, Education and Culture Department and other related departments become facilitators for that activity. Experts, both academics and practitioners in the field of Sundanese gastronomy, are engaged to give recommendations and ideas in creating innovation for the business. Observers will contribute constructive critics, writings, or reviews for continuous improvement toward resulted products. Connoisseurs are capital owners because they are pure consumers of gastronomy who become partners for entrepreneurs. As for NGO, for example "Hijau Lestari,", "Ketahanan Pangan Mandiri,", "Pemuda Tani Organik Adisa," and "Aku Cinta Masakan Indonesia (ACMI)" can be as control, either at input point in farming origin point, during the process of delivering or processing to become ready food or at the end point on the dining table in Bandung city. Meanwhile, information technology serves as promoting media or a bridge between producers and consumers, or even becomes an expert system for design tools of various supports for gastronomic business like packaging, online shop, etc. Information technology can also be used as an impetus for creative efforts to boost the appearance of food in order to get added value. So, by noticing the people's talent in Bandung, gastronomy city development with salapan cinyusu or Nona Helix concept focuses on gastronomic entrepreneurship based on creativity entrepreneurship or creativepreneurship. Entrepreneurs outpour their creative ideas to start their businesses with added value, namely, art and knowledge. They can produce innovative and out-of-the box products.

5 CONCLUSION

The development of Sundanese gastronomy in Bandung requires the Salapan Cinyusu or Nona Helix model as a think tank forum to support creative-based entrepreneurship or creativepreneurship. The interaction among elements in the model is expected to help in finding creative ideas that are needed by food entrepreneurs in Bandung to develop their business. Gastronomists in Bandung can collaborate and work together to maintain the taste of local Sundanese cuisine while providing an extraordinary experience to consumers. Salapan cinyusu will tell who is playing a role, what is the role, and how that role should be played.

The results showed nine elements of the Salapan Cinyusu or Nona Helix (nine helix) model, which consisted of gastronomic business to accommodate creativepreneurs to include government, workers, suppliers, experts, observers, connoisseurs, NGOs and information technology as bridges who must work together and collaborate to strengthen the existence of tourism gastronomy in Bandung municipality. The failure of one element to meet the performance will cause the forum to be lame. The impetus for the emergence of creative ideas needed by entrepreneurs can be hampered.

The recommendation of this research is that local food entrepreneurs in Bandung need to become members of the Salapan Cinyusu forum, in order to benefit from various elements of creativepreuership, quickly and measurably. The government assists as a facilitator in the formation of Salapan Cinyusu. The interaction of all elements in the forum model must be two-way—giving and receiving input.

However, the model developed by the researcher is still far from perfect and open for further research to make it better. It is because there are elements that should be independently extracted, for example, farmer's behavior and orientation as a cultivator will surely be different from intermediary traders and entrepreneurs as a food renderer.

REFERENCES

Amrial, A., Muhammad, E., and Adrian, A.M. (2017). Penta helix model: A sustainable development solution through the industrial sector. *Conference Paper. Conference: 14th Hokkaido Indonesian Student Association Scientific Meeting,* At Sapporo, Japan in November 2017

Andriyawan, D. (2019). *Andrian Ishak Destruction of 17 Bandung's Legendary Culinary Style of Molecular Gastronomy (Andrian Ishak Destruksi 17 Kuliner Legendaris Bandung ala Molecular Gestronomy).* https://traveling.bisnis.com/read/20191130/223/1176122/andrian-ishak-destruksi-17-kuliner-legendaris- bandung-ala-molecular-gestronomy

Becomingnatasha. (2019). *What Does It Mean to Be A Creativepreneur?* https://www.becomingnatasha.com/life/what-does-it-mean-to-be-a-creativepreneur

Bowman, A. (2020). *11 Eye-Catching Packaging Design Trends for 2019.* https://www.crowdspring.com/blog/packaging-design-trends-2019/

Businesswire. (2020). *ProAmpac Announces Four New Sustainable Packaging Product Groups.* https://www.businesswire.com/news/home/20190304006018/en/ProAmpac-Announces-New-Sustainable-Packaging-Product-Groups

Carayannis, E. G; Barth, T. D; Campbell, D.F.J (2012). The Quintuple Helix innovation model: global warm-ing as a challenge and driver for innovation. *Journal of Innovation and Entrepreneurship,* 1(1): 2. doi:10.1186/2192-5372-1-2

Cavallini, S., Soldi, R., Friedl, J., and Volpe, M. (2016). Using the Qadruple Helix Approach to accelerate the transfer of research and innovation results to regional growth. *European Union Committee of the Regions*
Etzkowitz, H. and Leydesdorff, L. (1995). *The Triple Helix – University-Industry-Government Relations: A Laboratory for Knowledge Based Economic Development*. Rochester, NY
Howkins, J. (2001). *The Creative Economy: How People Make Money from Ideas*. Penguin.
Leydesdorff, L. (2012). *The Knowledge-Based Economy and the Triple Helix Model*. Amsterdam: Amsterdam School of Communications Research, University of Amsterdam
Lilholt, A. (2015). *Entomological Gastronomy: A Gastronomical Approach to Entomophagy*. Addison Lillholt Publisher.
Lindmark, A., Nilsson-Roos, M. and Sturesson, E. (2009). *Difficulties of Collaboration for Innovation: A study in the Öresund region*. Thesis. Sweden: School of Economics and Management, Lund University
Manolis. 2010. *Culinary tourism destination marketing and the food element: a market overview*. https://aboutourism.wordpress.com/tag/culinary-tourism/
Muhyi, H.A., Chan, A., Sukoco, I., and Herawaty, T. (2017). The Penta Helix Collaboration Model in Developing Centers of Flagship Industry in Bandung City. *Review of Integrative Business & Economics Research*, Vol. 6(1): 412–417.
Pelto, P. J. (2017). *Mixed methods in ethnographic research: Historical perspectives*. New York & London: Routledge.
Pullphothong, L. and Sopha, C. (2013). *Gastronomic Tourism in Ayutthaya, Thailand*. Bangkok, Thailand: School of Culinary Art, Suan Dusit Rajabhat University
REPEC. (2012). *Application of Penta Helix Model in economic development*. UK: repec.org. Retrieved from http://ftp://ftp.repec. org/opt/ReDIF/RePEc/osi/eecytt/PDF/EconomyofeasternCroatiayesterdaytoday tomor-row04/eecytt0437.pdf
Ross, O. (2018a). *Creativepreuner*. https://www.urbandictionary.com/define.php?term=Creativepreneur
Salkind, N. (2010). Encyclopedia of Research Design. *SAGE Publication*. doi: http://dx.doi.org/ 10.4135/9781412961288.n469
Santich, B. (1996). *Looking for Glavour*. Kent Town, South Australia: Wakefield Press
Shushudesign (2020). *Package Design*. https://shushudesign.com/portfolio/package-design/
Soeroso, (2014). Foodscape, Cultural Landscape and Archeology: An Effort to Preserve Indonesian Cultural Heritage and Economic Development. *Paper presented at the IAAI Congress (Indonesian Archaeological Association)* on June 25, 2014 at Fort of Vredeburg, Yogyakarta City
Soeroso, A. and Susilo, Y.S. (2013). Traditional Indonesian Gastronomy as a Cultural Tourist Attraction. *Journal Applied Economics for Developing Countries (JAEDC)*, 1: 1-28
Sturesson, E., Lindmark, A. and Nilsson-Roos, M. (2009). *Collaboration for Innovation: A Study in the O¨resund Region*. Paper. Sweden: Lund University Libraries
Sugiyono. (2017). *Qualitative and Quantitative Research Methods and R & D (Metode Penelitian Kuantitatif Kualitatif dan R&D)*. Bandung: Alfabeta Publisher
Turgarini, D. (2018). *Sundanese Gastronomy as a Tourist Attraction in Bandung Municipality*. Unpublished Dissertation. Yogyakarta: Gadjah Mada University
UNESCO (2020). *The Creative Cities Network: A Global Platform for Local Endeavour*. http://www.unesco. org/new/fileadmin/MULTIMEDIA/HQ/CLT/pdf/Creative_cities_brochure_en. pdf

Promoting Creative Tourism: Current Issues in Tourism Research – Kusumah et al. (Eds)
 ISBN 978-0-367-55862-8

The effect of gastronomic festival attributes on behavioral intention at Wisata Kuliner Tjeplak Purwakarta

Rr.M. Vania T., Gitasiswhara & Y. Yuniawati
Universitas Pendidikan Indonesia, Bandung, Indonesia

ABSTRACT: This study aimed to examine the influence of gastronomic festival attributes on behavioral intention. The research method used is descriptive and verification study. The data were collected through questionnaire. The population from this research was visitors from different regions visiting Wisata Kuliner Tjeplak Purwakarta. The technique for choosing the samples is systematic random sampling to get 400 respondents. The statistical analysis used is multiple linear regression using SPSS 20 for Windows. The result of the test shows that gastronomic festival attributes have a significant influence on behavioral intention. Moreover, all dimensions of gastronomic festivals attributes have a significant influence on behavioral intention.

Keywords: gastronomic festival attributes, behavioral intention, Wisata Kuliner Tjeplak Purwakarta

1 INTRODUCTION

The service sector industry is currently trying to keep on at the forefront of the market by offering quality services to their customers (Singh 2015). The more important the role played by the service sector in the global economy, the more research attention should be focused on increasing understanding of the key factors that influence behavioral intention in the service environment (Parvin et al. 2017). One of the businesses that is developing now and has a promising prospect, both regionally and internationally is the tourism industry. The tourism industry is one of the strategic sectors in Indonesia development (Gitasiswhara et al. 2017). In a competitive tourism business environment, satisfying tourists might be enough for short-term business survival, but not for sustainable business performance and success (Dean & Suhartanto 2019). Understanding customer behavior is very important to develop sustainable strategies for a tourist attraction (Dodds et al. 2017).

Behavioral intention is a predictor of the profitability and revenue of service companies (Singh 2015) and can be a good reference for companies in predicting consumer behavior in the future (Tsaur et al. 2015). Visitors who have good behavioral intentions to revisit, not only care about changes in the price (Hallak et al. 2017), but they also tend to offer word-of-mouth to others (Wang 2017). The behavioral intention study in the food and beverage industry illustrates the possibility of returning to a food and beverage provider and the desire to recommend to family, friends and others in the future (Bujisic et al. 2014).

The demand for tourists in the world has changed rapidly. At present, tourists are very keen to find unique experiences by not only visiting historical, cultural and natural sites, but also exploring a destination with something new like gastronomy (Corigliano 2003). Some researchers have shown that culinary tourism can add value to tourist experience (Dimitrovski 2016; Sims 2009), therefore culinary tourism has become a new trend that can contribute to economic and social development in rural and urban areas (Dimitrovski 2016; Yun et al. 2011).

Over the past two decades, festival research has become a major topic in the broader event management and tourism management research literature (Vesci & Botti 2019; Wan & Chan 2013).

 DOI 10.1201/9781003095484-63

Table 1. Number of visits to Wisata Kuliner Tjeplak Purwakarta.

Year	Amount	Percentage
2015	341.695	
2016	533.176	35,91
2017	517.652	−2,91
2018	489.120	−5,83

Source: Dinas Kepemudaan, Olahraga, Pariwisata dan Kebudayaan Kabupaten Purwakarta, 2019

The festival provides the community to celebrate the uniqueness of its tradition and attract tourists and local visitors. This is very important for destination marketing and also for promoting tourism (Chang 2006; Tanford et al. 2017).

Purwakarta is one of the regencies in West Java with an amazing natural beauty, neatly arranged city development and systematic local and regional government. With the charm of its natural, artificial, cultural tourism, culinary centers, Purwakarta now has its own attraction for local and foreign tourists. Seeing this potential, Purwakarta Regency government held an activity called Wisata Kuliner Tjeplak Purwakarta. The activity is a culinary festival which is held every week which aims to move the Purwakarta community to utilize the potential of tourism in Purwakarta, so that it can eliminate the nickname Purwakarta as a "Retired Regency". Wisata Kuliner Tjeplak Purwakarta is also the official weekly event to promote culinary and culture to tourists who come to visit. The following table presents the data on tourist visits to Wisata Kuliner Tjeplak Purwakarta.

Based on the data in Table 1, the number of visits to Wisata Kuliner Tjeplak Purwakarta from 2015 to 2018 was very volatile. In 2016 it increased by 35.91% from the previous year. But in 2017 it decreased by −2.91% from 2016. This happened also in 2018 the next amounted to −5.83%. Overall, Wisata Kuliner Tjeplak Purwakarta has experienced fluctuations in the last 3 years. The surge in the number of visits from 2015 to 2016 was caused by the opening of Sribaduga Fountain which is located near Wisata Kuliner Tjeplak Purwakarta, so that most tourists who came to the Fountain also visited Wisata Kuliner Tjeplak Purwakarta and caused a very high surge of visits in the Wisata Kuliner Tjeplak Purwakarta area from inside and outside Purwakarta Regency.

Based on the observation to Wisata Kuliner Tjeplak Purwakarta, the lack of behavioral intention in the future was caused by the hectic location at the time of the activity, and caused a full parking lot so visitors had to park on the roadside, coupled with congested traffic around the site. It has made visitors less comfortable (Axelsen & Swan 2010; Kruger et al. 2013; Tanford & Jung 2017; Wan & Chan 2013). The lack of satisfaction of visitors is also caused by the lack of facilities provided, such as only provided two portable toilets that are not routinely located on site because it is owned by third party service (Dimitrovski 2016; Wan & Chan 2013). Although there are toilets in the Museum of Diorama and Fountain of Sri Baduga, the location is quite far, making it difficult for visitors.

Culinary events can increase tourist spending in a destination (Everett et al. 2010; Hoon et al. 2011) and create important images that are important to ensure their satisfaction and influence their behavioral intentions towards that goal (Dimitrovski 2016). Therefore, identification of the determinants that satisfy visitors of culinary festivals is very important for festival organizers and industry partners to increase the positive impact of the festival itself (Jang & Yuan 2008; Smith & Costello 2009). This method can be used to popularize certain destinations in many cities (Nilsson et al. 2011). The culinary festival offers not only focuses on delicious dishes, but also on pleasant entertainment (Wong et al. 2014). The culinary festival is very festive, with music and dance being an addition to complementing food-related activities available (Organ et al. 2015).

Food quality in the context of festivals always applies, the elements of other culinary festivals are usually forgotten (Dimitrovski 2016; Yuan et al. 2014). Culinary festivals not only provide a dining experiences, they can also offer a full tourist experience. In Dimitrovski's research, the festival was built by several elements; location and signage, content programs, perceived crowdedness, price, comfort, and entertainment.

Maintaining visitor behavioral intention is a challenge for organizers of culinary festivals. Learn and understand the positive and negative experiences felt by visitors during a visit to Wisata Kuliner Tjeplak Purwakarta with the gastronomic festival attributes using its attributes (location and signage, content programs, perceived crowdedness, price, comfort, entertainment). It is a key to provide and increase visitor satisfaction so that it will result in an increase in behavioral intention.

Based on these problems and phenomena, it is necessary to identify how far the influence of the gastronomic festival attributes on the behavioral intention of visitors who come to Wisata Kuliner Tjeplak Purwakarta.

2 LITERATURE REVIEW

The growing development of the tourism industry is ensured that competition will occur between the tourism industries to be able to attract tourists in order to be able to revisit the attractions offered. Consumer behavior is a study of the processes involved when individuals or groups choose, buy, use, or dispose of products, services, ideas, or experiences to satisfy the needs and desires of customers (Solomon 2017: 28). Behavioral Intention is a consumer behavior, which is often associated with satisfaction from consumers, because behavioral intention occurs when satisfaction has been created so that it is often associated with customer satisfaction. Behavioral intention is post-visit behavior as intention to return, and recommend the company to others (Jang et al. 2011; Ryu 2008; Wu 2013). The behavioral intention model explains that consumers actively think ahead and plan future behavior. The concept of intentions assumes that consumers have a linear time-sensory; they think in terms of past (past), present (present), and future (future) (Solomon 2017: 301).

There are indicators to measure behavioral intention using several dimensions, such as revisit intention and word-of-mouth (Bujisic et al. 2014; Dedeoglu et al. 2018; Jin et al. 2015; Ryu & Han 2010). Revisit intention is defined as someone's intention to return to the same product, brand, business or purpose in the future (Dedeoglu et al. 2018; Kim et al. 2010; Zeithaml et al. 1996). While word-of-mouth is casual communication between people who have consumed goods or services with evaluations that they suggest the goods and services to other people who are interested in experiencing these goods or services (Jalilvand et al. 2017).

Food and beverage management is an important part of the hospitality industry and the overall economy. Like industries whose main food and beverage operations are characterized by diversity (Davis et al. 2018). Gastronomic tourism plays an important role in expanding the attractiveness of destinations such as increasing the number of visits; enhance visitor experience; strengthen regional identity; and stimulate growth in other sectors (Dixit 2019). Gastronomic tourism is an experience journey to a gastronomic region, for recreational or entertainment purposes, which includes visits to primary and secondary food producers, gastronomic festivals, food fairs, events, farmers' markets, cooking shows and demonstrations, tasting quality food products, or what tourism activities even those related to food (Hall et al. 2003). While the festival is full of entertainment rituals, spectacle and memory. Most people participate to enjoy something different (Cudny 2016). Attributes are a source of destination both basic resources and supporting resources. The main source is the attribute can be used as the strongest force to attract, motivate tourists to travel (Smith & Costello 2009b). There are six dimensions of the gastronomic festival attributes according to (Dimitrovski 2016), namely location and signage, content programs, perceived crowdedness, price, comfort and entertainment.

Location and Signage is very important for visitors in many ways (Dimitrovski 2016; Kim et al. 2010; Wan & Chan 2013). If the festival location is close to other tourist attractions, the tourists will be more interested in being involved in the festival activities. Similarly if there is a proper signboard, visitors will find the festival site convenient (Dimitrovski 2016; Wan & Chan 2013). Culinary festivals differ significantly because of the content of the program, with the variety and quality of the program making it possible to influence the level of visitors' satisfaction (Axelsen & Swan 2010; Dimitrovski 2016; Kruger et al. 2013; Wan & Chan 2013). Crowdedness is when the number of people in a limited space limits or interferes with other individual activities, they will feel that the environment is crowded (Dimitrovski 2016). Prices for food and beverages at the culinary program should be seen from among those who visit culinary events. High prices will make visitors dissatisfied and make it difficult to produce loyal visitors (Dimitrovski 2016). Comfort are attributes that directly affect the level of satisfaction (Axelsen & Swan 2010; Kruger et al. 2013; Tanford & Jung 2017; Wan & Chan 2013). If the festival site is comfortable, clean, or adequate in terms of space or size, the perception of the gastronomic festival will be more positive. Culinary festivals are not only about variety and quality of food, but some types of entertainment are needed, especially if the festival is held in an urban environment. (Dimitrovski 2016; Kim et al. 2010; Wan & Chan 2013). It is usually in the form of music entertainment, but it can be adjusted to visitors' request or organizers (Kim et al. 2010; Smith & Costello 2009; Wan & Chan 2013).

3 RESEARCH METHOD

This study aims to analyze the effect of gastronomic festival attributes on the behavioral intention at Wisata Kuliner Tjeplak Purwakarta. The object of the study consists of two variables, namely variable X (independent variable) and variable Y (Dependent variable). This study has two variables, namely the gastronomic festival attributes (Variable X) consisting of location and signage (X1), program content (X2), perceived crowdedness (X3), price (X4), comfort (X5), entertainment (X6). Whereas the dependent variable or dependent variable is behavioral intention (variable Y).

Respondents in this study were visitors to Wisata Kuliner Tjeplak Purwakarta coming from outside Purwakarta Regency area. The method in this study is a cross sectional method. The type of this research is quantitative descriptive research and verification carried out in data collection in the field, the research method that was explanatory survey method. It is in the form of a list of questions such as questionnaires or interviews that will be submitted to respondents from some of the populations studied. The population in this study were 489,120 people in Wisata Kuliner Tjeplak Purwakarta in 2018. The data were collected from Dinas Kepemudaan, Olahraga, Pariwisata dan Kebudayaan Kabupaten Purwakarta in 2019. The determination of the sample using the Slovin formula, the sample size was obtained (n) as many as 400 respondents. This study using a systematic random sampling technique. The types of data collected in this study are primary and secondary data. The data collection techniques used in this study were interviews, observation, literature studies, and questionnaires.

The results of the tests carried out include validity and reliability tests showing that on 24 items of questions for the gastronomic festival variable attributes and 6 items of questions for behavioral intention variables were declared valid and reliable so that they could be used as the correct measuring instrument.

The data analysis technique used in this study is multiple linear regression. The variables analyzed are independent variables (X), namely the gastronomic festival attributes. While the dependent variable (Y) is behavioral intention. The multiple regression equation is formulated as follows:

$$Y = a + b\_1X\_1 + b\_2X\_2 + b\_3X\_(3) + b\_4X\_4$$

Information:
Y = Dependent variable (behavioral intention)
a = Constant Value
b = Number of directions or regression coefficients
X = independent variable (gastronomic festival attributes) location and signage (X1), program content (X2), perceived crowdedness (X3), price (X4), comfort (X5), entertainment (X6).

4 RESEARCH RESULT AND DISCUSSION

Based on the results of the recapitulation of the responses of 400 respondents, the gastronomic festival attributes in Wisata Kuliner Tjeplak Purwakarta can be measured by calculating the overall score from the comparison of the total score and the average score of the gastronomic festival attributes. The total score for the gastronomic festival attributes is equal to 35551 which is classified as high and lying at intervals 32640–40320. The sub variables or dimensions that have the highest rating of visitors to Wisata Kuliner Tjeplak Purwakarta are program content by 21.76%. This is because a large number of respondents felt that the program organized could attract visitors to spend their Saturday night at Wisata Kuliner Tjeplak Purwakarta. Whereas the lowest rating is in the comfort dimension with a percentage of 17.37%.

The results showed that the behavioral intention variable in Wisata Kuliner Tjeplak Purwakarta can be measured by calculating the overall score from the comparison of the total score and the average score of the behavioral intention variable. The total score for behavioral intention is equal to 9318 which is classified as high and lying at intervals of 9120–11520. In this variable, the sub-variable or dimension that has the best response is word-of-mouth with 52.33%. This means that visitors who have come to Wisata Kuliner Tjeplak Purwakarta feel positive things when they are in Wisata Kuliner Tjeplak Purwakarta so they intend to do word-of-mouth in the form of recommendations or solicitation to their friends, colleagues and family so they can be interested in visiting Wisata Kuliner Tjeplak Purwakarta. While for the lower assessment, there is a dimension of revisit intention with a percentage of 47.67%.

Testing the hypothesis obtained regarding the influence of gastronomic festival attributes on the behavioral intention in Wisata Kuliner Tjeplak Purwakarta simultaneously has a total influence of 0.292 or 29.2%. This shows that each dimension of the gastronomic festival attributes (X) contributes 29.2% to the behavioral intention (Y) variable, while the rest is a contribution from other factors not examined in this study. Partially, there are four dimensions that have a significant influence on behavioral intention; location and signage, program content, perceived crowdedness and entertainment. While the other two dimensions namely price and comfort do not have a significant effect on behavioral intention by using a significant level of two-party test of 0.025 or 2.5%.

5 CONCLUSIONS AND RECOMMENDATIONS

5.1 *Conclusions*

In this study, the overall responses of respondents regarding to the implementation of the gastronomic festival attributes in Wisata Kuliner Tjeplak Purwakarta consisting of six dimensions; location and signage, content programs, perceived crowdedness, price, comfort, and entertainment had good ratings from respondents and were in high category. This means that the six dimensions can be used as a strategy to overcome the problem of visits at Wisata Kuliner Tjeplak Purwakarta. The dimensions of the content program get the highest rating. The respondents feel that the program run by Wisata Kuliner Tjeplak Purwakarta is very interesting. While comfort gets the lowest score. This is because visitors consider supporting facilities to be inadequate, for example there are only a few trash cans and toilets that are far from the Wisata Kuliner Tjeplak Purwakarta location and also lack of security in the area, so this reduces the comfort element.

Respondents about the behavioral intention at Wisata Kuliner Tjeplak Purwakarta get a very good rating according to the continuum line. In this behavioral intention study has two indicators, namely revisit intention and word-of-mouth. The highest rating is word-of-mouth, such as willingness to talk about positive things about Wisata Kuliner Tjeplak Purwakarta to friends and relatives. Due to the experiences that visitors feel through culinary diversity and events in Wisata Kuliner Tjeplak Purwakarta are positive and memorable, visitors want to share them with others. While revisit intention gets a lower score because visitors from outside Purwakarta feels they have similar food festival in their respective regions.

The results show that there is a significant influence between gastronomic festivals attributes which consist of location and signage, content programs, perceived crowdedness, price, comfort and entertainment on the behavioral intention at Wisata Kuliner Tjeplak Purwakarta. There is partially a significant influence between gastronomic festival attributes which consist of location and signage, program content, perceived crowdedness, and entertainment towards behavioral intention.

However there is no significant influence between the gastronomic festivals attributes which consist of price and comfort towards behavioral intention. Thus it can be concluded that the gastronomic festival attributes in Wisata Kuliner Tjeplak Purwakarta as a food and beverage festival have a positive value that has an influence on behavioral intention.

5.2 *Reccommendation*

Based on the results of the research that the writer have done, it would be better for the Wisata Kuliner Tjeplak Purwakarta organizers to further enhance the comfort dimension, by adding public toilets and portable toilets around the Wisata Kuliner Tjeplak Purwakarta area, and adding many trash bins in several corners so that visitors feel more comfortable

Based on the results of the research that has been done, the author recommends that the organizers be able to attract visitors to come and come back by further enhancing Wisata Kuliner Tjeplak Purwakarta promotion by maximizing the use of internet media especially social media so as to increase public interest in visiting and returning to visit Wisata Kuliner Tjeplak Purwakarta.

The author suggests that it is good for the organizers to continue to pay attention and increase visitor interest to conduct behavioral intention to Wisata Kuliner Tjeplak Purwakarta by strengthening the gastronomic elements of the festival attributes. The manager can provide innovation and improve marketing so that it will attract new visitors to come and attract visitors who have visited to return to Wisata Kuliner Tjeplak Purwakarta.

The author realizes that in this study still has many shortcomings, limitations and far from perfect, so for further research the authors recommend to conduct further research on the influence of gastronomic festival attributes on other variables such as customer satisfaction, customer loyalty, purchase decision and festival image by adding other indicators such as using the latest dimensions and theories as well as different research methods so that research the discussion can be further developed.

REFERENCES

Axelsen, M., & Swan, T. (2010). Designing festival experiences to influence visitor perceptions: The case of a wine and food festival. *Journal of Travel Research, 49*(4), 436–450.

Bujisic, M., Hutchinson, J., & Parsa, H. G. (2014). The effects of restaurant quality attributes on customer behavioral intentions. *International Journal of Contemporary Hospitality Management, 26*(8), 1270–1291. https://doi.org/10.1108/IJCHM-04-2013-0162

Chang, J. (2006). Segmenting tourists to aboriginal cultural festivals: An example in the Rukai tribal area, Taiwan, *27*, 1224–1234. https://doi.org/10.1016/j.tourman.2005.05.019

Corigliano, M. A. (2003). The route to quality: Italian gastronomy networks in operation.

Cudny, W. (2016). *Festivalisation of Urban Spaces.*

Davis, B., Lockwood, A., Management, H., Management, T., Alcott, P., Tutor, S., ...Management, H. (2018). *Food and Beverage Management.*

Dean, D., & Suhartanto, D. (2019). The formation of visitor behavioral intention to creative tourism: the role of push – Pull motivation. *Asia Pacific Journal of Tourism Research, 0*(0), 1–11. https://doi.org/10.1080/10941665.2019.1572631

Dedeoglu, B. B., Bilgihan, A., Ye, B. H., Buonincontri, P., & Okumus, F. (2018). The impact of servicescape on hedonic value and behavioral intentions: The importance of previous experience. International Journal of Hospitality Management, 72, 10–20. doi:10.1016/j.ijhm.2017.12.007

Dimitrovski, D. (2016, October). Urban gastronomic festivals—Non-food related attributes and food quality in satisfaction construct: A pilot study. In *Journal of Convention & Event Tourism* (Vol. 17, No. 4, pp. 247–265). Routledge.

Dixit, S. K. (2019). *The Routledge Handbook of Gastronomic Tourism. Annals of Tourism Research* (Vol. 50). https://doi.org/10.1016/j.annals.2014.11.004

Dodds, R., Jolliffe, L., & Creating, E. T. (2017). The Handbook of Managing and Marketing Tourism Experiences Article information.

Everett, S., Aitchison, C., Everett, S., & Aitchison, C. (2010). The Role of Food Tourism in Sustaining Regional Identity: A Case Study of Cornwall, South West England The Role of Food Tourism in Sustaining Regional Identity: A Case Study of Cornwall, South West England, (August 2014), 37–41. https://doi.org/10.2167/jost696.0

Gitasiswhara., Yuniawati, Yeni., Fasa, Febi Kusumadewi. (2017). Pengaruh Special Event terhadap Keputusan Berkunjung di Trans Studio Bandung

Hall, C. M., Sharples, L., Mitchell, R., Macionis, N., & Cambourne, B. (2003). Food Tourism Around the World: Development, Management and Markets. *Elsevier*, 1–390. https://doi.org/10.1016/B978-0-7506-5503-3.50005-1

Hallak, R., Assaker, G., & El-haddad, R. (2017). Re-examining the relationships among perceived quality, value, satisfaction, and destination loyalty: A higher-order structural model. https://doi.org/10.1177/1356766717690572

Hoon, Y., Kim, M., & Goh, B. K. (2011). An examination of food tourist's behavior: Using the modified theory of reasoned action. *Tourism Management, 32*(5), 1159–1165. https://doi.org/10.1016/j.tourman.2010.10.006

Jalilvand, M. R., Salimipour, S., Elyasi, M., & Mohammadi, M. (2017). Factors influencing word of mouth behaviour in the restaurant industry. *Marketing Intelligence & Planning, 35*(1), 81–110. https://doi.org/10.1108/MIP-02-2016-0024

Jang, S. I., Lillehoj, H. S., Lee, S. H., Lee, K. W., Lillehoj, E. P., Bertrand, F., …Deville, S. (2011). Mucosal immunity against Eimeria acervulina infection in broiler chickens following oral immunization with profilin in Montanide…adjuvants. Experimental Parasitology, 129(1), 36–41. doi:10.1016/j.exppara.2011.05.021

Jang & Yuan. (2008). The Effects of Quality and Satisfaction on Awareness and Behavioral Intentions: Exploring the Role of a Wine Festival. https://doi.org/10.1177/0047287507308322

Jin, N. (Paul), Goh, B., Huffman, L., & Yuan, J. J. (2015). Predictors and Outcomes of Perceived Image of Restaurant Innovativeness in Fine-Dining Restaurants. *Journal of Hospitality Marketing and Management, 24*(5), 457–485. https://doi.org/10.1080/19368623.2014.915781

Kim, Y. G., Suh, B. W., & Eves, A. (2010). The relationships between food-related personality traits, satisfaction, and loyalty among visitors attending food events and festivals. *International journal of hospitality management, 29*(2), 216–226.

Kruger, S., Rootenberg, C., & Ellis, S. (2013). Examining the influence of the wine festival experience on tourists' quality of life. *Social indicators research, 111*(2), 435–452.

Nilsson, J. H., Svärd, A.-C., Widarsson, Å, & Wirell, T. (2011). 'Cittáslow' eco-gastronomic heritage as a tool for destination development. *Current Issues in Tourism, 14*(4), 373–386. https://doi.org/10.1080/13683500.2010.511709

Organ, K., Koenig-lewis, N., Palmer, A., & Probert, J. (2015). Festivals as agents for behaviour change: A study of food festival engagement and subsequent food choices. *Tourism Management, 48*, 84–99. https://doi.org/10.1016/j.tourman.2014.10.021

Parvin, S., Wang, P. Z., & Uddin, J. (2017). Assessing two consumer behavioural intention models in a service environment. *Asia Pacific Journal of Marketing and Logistics, 29*(3), 653-668. https://doi.org/10.1108/APJML-06-2016-0100

Ryu, K., & Han, H. (2010). *Influence of the Quality of Food, Service, and Physical Environment on Customer Satisfaction and Behavioral Intention in Quick-Casual Restaurants: Moderating Role of Perceived Price. Journal of Hospitality & Tourism Research, 34(3), 310-329.* doi:10.1177/1096348009350624

Singh, A. K. (2015). Modeling passengers' future behavioral intentions in airline industry using SEM. *Journal of Advances in Management Research, 12*(2), 107–127. https://doi.org/10.1108/JAMR-06-2014-0033

Smith, S., & Costello, C. (2009a). Culinary tourism: Satisfaction with a culinary event utilizing importance-performance grid analysis, *15*(2), 99–110. https://doi.org/10.1177/1356766708100818

Smith, S., & Costello, C. (2009b). Segmenting Visitors to a Culinary Event: Motivations, Travel Behavior, and Expenditures. *Journal of Hospitality Marketing & Management, 18*(August 2014), 44–67. https://doi.org/10.1080/19368620801989022

Solomon, M. R. (2017). *Consumer Behavior*.

Tanford, S., Ph, D., & Jung, S. (2017). Festival Attributes and Perceptions: A Meta-Analysis of Relationships with Satisfaction and Loyalty. *Tourism Management, 61*, 209–220. https://doi.org/10.1016/j.tourman.2017.02.005

Tsaur, S. H., Luoh, H. F., & Syue, S. S. (2015). Positive emotions and behavioral intentions of customers in full-service restaurants: Does aesthetic labor matter? *International Journal of Hospitality Management, 51*, 115–126. https://doi.org/10.1016/j.ijhm.2015.08.015

Vesci, M., & Botti, A. (2019). Journal of Hospitality and Tourism Management Festival quality, theory of planned behavior and revisiting intention: Evidence from local and small Italian culinary festivals. *Journal of Hospitality and Tourism Management, 38*(October 2018), 5–15. https://doi.org/10.1016/j.jhtm.2018.10.003

Wan, Y. K. P., & Chan, S. H. J. (2013). Factors that affect the levels of tourists' satisfaction and loyalty towards food festivals: A case study of Macau. *International journal of tourism research, 15*(3), 226–240.

Wang, T.-L. (2017). Destination perceived quality, tourist satisfaction and word-of-mouth.

Wong, J., Wu, H., & Cheng, C. (2014). An Empirical Analysis of Synthesizing the Effects of Festival Quality, Emotion, Festival Image and Festival Satisfaction on Festival Loyalty: A Case Study of Macau Food Festival. https://doi.org/10.1002/jtr

Wu, H.-C. (2013). An Empirical Study of the Effects of Service Quality, Perceived Value, Corporate Image, and Customer Satisfaction on Behavioral Intentions in the Taiwan Quick Service Restaurant Industry. *Journal of Quality Assurance in Hospitality & Tourism, 14*(4), 364–390. https://doi.org/10.1080/1528008X.2013.802581

Yuan, J. J., Cai, L. A., Morrison, A. M., & Linton, S. (2014). Journal of Vacation Marketing. https://doi.org/10.1177/1356766705050842

Yun, D., Edward, P., Hennessey, S. M., Macdonald, R., & Macdonald, R. (2011). Understanding Culinary Tourists: Segmentations based on Past Culinary Experiences and Attitudes toward Food-related Behaviour.

Zeithaml, V. A., Berry, L. L., & Parasuraman, A. (1996). The Behavioral Consequences of Service Quality. *Journal of Marketing Development and Competitiveness*, 31–46. https://doi.org/10.2307/1251929

Promoting Creative Tourism: Current Issues in Tourism Research – Kusumah et al. (Eds)

Culinary experience toward behavioral intention (survey of consumer fusion food on street food in Bandung city)

M.R. Perdana, L.A. Wibowo & Gitasiswhara
Universitas Pendidikan Indonesia, Bandung, Indonesia

ABSTRACT: This study aims to analyze the effect of culinary experience, which consists of fusion food culture, uniqueness, experience, aesthetics, and dining at street food toward behavioral intention on Street Food in Bandung. The research method used is an explanatory survey with a cross-sectional method approach with a total sample of 245 costumers. The data analysis technique used was the path analysis technique. The implementation of culinary experience is in the high category and behavioral intention is in the very high category. Culinary experience, which consists of fusion food culture, uniqueness of the food, new food experience, dining aesthetics, and dining at street food has a positive effect toward behavioral intention of enjoying street food. In addition, sub-variables have the highest influence. In addition, there should be an evaluation and input to improve performance, service quality, and product quality to be even better. The results showed that there was a significant influence between culinary experiences and the behavioral intention.

Keywords: behavioral intention, culinary experience, fusion food, street food

1 INTRODUCTION

Behavioral intention remains an interesting discussion in the field of marketing and tourism (Gupta 2018; Kim & Shim 2019). Some research on behavioral intention defines behavioral intention as the possibility of reusing the product or service, indicating that consumers will remain loyal or not (Namin 2017). An understanding of consumer behavior will allow companies to organize and develop strategies to meet the needs and desires of consumers (Friga et al. 2016). Relevant studies put consumer behavior into a development strategy planning the food and beverage industry (Platania et al. 2016). Behavioral intention plays an important role in the food and beverage business continuity itself (Friga et al. 2016).

Food and beverage business operators must understand the behavioral intention as a measure of business continuity (Bray & Bray, nd). In recent years, behavioral intention is something important because of its connection with comfort that determines the future of consumer choices (Fathema et al. 2015). Until now, the problem is still relevant behavioral intention to be researched in the field of food and beverage (Kristanti et al. 2019; Shin et al. 2018; Young et al. 2018). The importance of understanding behavioral intention in the food and beverage industry has been stated by many experts including research (Muscat et al. 2019; Setiyorini et al. 2019; Young et al. 2018). Behavioral intention itself has benefits for the sustainability of the food and beverage industry contributing to the further development of the industry in relation to tourism (Young et al. 2018).

Food and beverage industry is an industry that covers restaurants, bars, and cafes (Eka 2012: 99), which is also divided into semi-formal, formal, and informal types. One type of informal food and beverage industry is street food (Alfiero 2017b; Dal et al. 2015; Hassan et al. 2016; Leong-Salobir 2019); although this type of food and beverage industry is often overlooked in tourism studies, it has an important role as a form of food service and visitor attraction (Henderson, nd). Some researchers admit that culinary tourism can support the local economy as a source of income culinary (Governance 2019; Malasan 2019; Testa et al. 2019).

 DOI 10.1201/9781003095484-64

Research on street food in the United States is on the kind of food truck. According to IBISWorld, there are at least 117,000 food trucks that operate and their income increased by 9.3% from 2010 to 2015 (Okumus & Sonmez 2018). Street food growth also occurred in China with an estimated current number of 20 million vendors (Governance 2019). These are also trends in Indonesia as research conducted in Surabaya and Malang (Wijaya et al. 2018), Solo (Kristanti et al. 2019), and Bandung (Chan 2017; Komaladewi 2017).

Street food in Bandung named Sudirman Street Day and Night Market and in Cibadak street named Cibadak Culinary Night as the culinary tourist site's most crowded street food in Bandung (Malasan 2017a). Nevertheless, high competition poses and will adversely affect the company, which includes a decrease in traffic, a decrease in sales to liquidation (Liu 2018). There are 74 booths of food vendors in Sudirman Street Day and Night Market, while at Cibadak Culinary Night has 108 active stands. The number of seats and the rate of visits per week can be calculated as many as 4,106 tourists visiting the Sudirman Street Day and Night Market and 4,290 tourists visited Cibadak Culinary Night. Within a period of one year, if the number of visits is calculated on the basis of weekly visits, 217.618 tourists visited the Sudirman Street Day and Night Market and 227.370 tourists visited Cibadak Culinary Night.

The shrinkage of behavioral intention on Street Food in Bandung must be handled immediately because it will cause an impact on visiting level and business continuity (Friga et al. 2016). The impact of the low level of the visit is the loss of consumer confidence in the products and services provider (Ewerhard et al. 2019) and loss of credibility of the company (Jha & Kemper 2019). Most consumers will be looking for new restaurants and only a small part will visit back to the same place (Ewerhard et al. 2019).

The approach used in order to solve the problem of behavioral intention in this research is the theory of consumer decision making. Lodziana-Graboska (2015a) states that there are three stages: consumer decision-making input, process, and output. The previous study states that consumption value consists of emotional value, epistemic value, health value, prestige value, taste and quality value, price value, and the interaction value (Young et al. 2018), consumer attitude (Gupta 2018), food choice (Wang 2018), food image (Setiyorini et al. 2019), perceived quality, authenticity, and price (Muscat et al. 2019) and culinary experience on food and beverage industry (Kristanti et al. 2019). On previous research, food and beverage industry in the form of street food states that culinary experience had a positive impact on the behavioral intention. Culinary experience appeals to consumers to come back (Omar 2017) so that impacts continuing intake (Testa et al. 2019) and is considered important for the company (Björk & Kauppinen-Raisanen 2014). Strategy of culinary experience at On Street Food in Bandung is one of which is identical to the study (Kristanti et al. 2019) consisting of local food culture, uniqueness of the food, new food experiences, dining aesthetic, and dining at restaurants. The phenomenon of fusion food at street food attracts buyers because it is easily accepted socially and culturally, it is cheap and convinient (Leong-Salobir 2019), and irresistible to buyers (Gupta 2018). In addition, the seller can embrace all walks of life and also tourists (Henderson 2017).

This study wanted to see the effect of culinary experience on behavioral intention on street food in Bandung. This study aims to determine the factors that influence the culinary experience visiting and making a fusion food purchase in On Street Food in Bandung.

2 LITERATURE REVIEW

2.1 *Culinary experience*

Culinary is one component of the tourism product that plays an important role in creating a memorable tourist experience including pre-trip, during the trip, and post-trip experience (Sotiriadis 2015), Björk and Kauppinen-Raisanen (2019) define that culinary experience refers to a process that covers prior, during, and after the experience (Kristanti et al. 2019). There are five dimensions of culinary experience used in this study taken from Björk and Kauppinen-Raisanen (2016), encompassing fusion food culture, uniqueness of the food, new food experience, aesthetics of dining, and dining at street food.

2.2 *Behavioral intention*

In the consumer decision making (Lodziana-Graboska 2015a), the process of behavioral intention is illustrated with models of consumer behavior. Consumer behavior is influenced by several factors, ranging from economic, cultural, political, and technological. At the stage of the purchase, the decision process of psychological (motivation, perception, learning, personality, and attitudes) plays an important role in the actual purchase decisions. Furthermore, at the stage of post-purchase, consumers will determine the subsequent purchase behavior as trying a new bid or make repeat purchases. Consumers usually go through three stages, namely, revisit intention, word of mouth, and a willingness to pay more (Young et al. 2018). Muscat et al. (2019) define behavioral intention as the tendency of consumer behavior which was adopted in the future based on their own perceptions. Author receipts indicators of the behavioral intention of Namin (2017), namely, revisit intention and word of mouth to take one dimension of Lim (2019), the willingness to pay more in this research.

2.3 *Fusion food*

The food industry continues to grow with a wide range of new innovations. The term fusion food refers to the result of a mix of regional cuisine with classic French cuisine named Escoffier, but today the term can be defined as an integrated food fusion of cultural elements that are diverse, united in one dish at a dose of balance (Vilmar et al. 2015). Fusion food was divided into three types; sub-regional, regional and continental fusion. Sub-regional fusion is a combination of cuisines from regions such as provinces or cities in a country to create new dishes, regional fusion is a combination of cooking among many countries on the same continent, and Continental fusion (also known as the West Asian fusion food) is the most common type. Sassatelli in the journal (Sassatelli, nd) defines fusion food as food made from components derived from a variety of areas used as a dish without basic identity. Fusion food has become a trend in the culinary world, even fusion food is in vogue with a trend rating.

2.4 *Framework*

Consumer behavior is a study that illustrates how consumers make the decision to use a product or service. A marketer must fully understand consumer buying behavior (Kotler & Keller 2016). There is a five-purchase decision process according to Kotler and Keller (2016), namely, the introduction of the problem, information search, evaluate alternative, purchasing decisions, and post-purchase behavior. Lodziana-Graboska (2015a) states that consumer decisions in determining the purchase is the influence from outside, making purchase decisions and stages after purchase. Knowing consumer behavior can be perceived as marketers in the post-purchase phase. Understanding the behavior of consumer behavior is beneficial for marketers to learn the desires, perceptions, preferences, as well as the purchase behavior of the target market (Serpico et al. 2015).

Culinary is one component of the tourism product that plays an important role in creating a memorable travel experience. Sotiriadis (2015) suggested a food and beverage provider must enable himself in the visitor experience, enriched and differentiating offering and communication or promotion of the products it sells. Besides the local food that has been widely researched, innovative foods such as fusion food as mentioned to provide a unique culinary experience and specialties as well as increase the variety of culinary culture that exists based on previous research relevant to the problem as well as the object of this research (Kristanti et al. 2019). Eventually, researchers summarize the five dimensions of culinary experience, that is, fusion food culture, uniqueness of the food, new food experiences, aesthetics of dining, and dining at street food.

2.5 *Hypothesis*

Because the hypothetical character is still weak, there should be verification by empirical data to test the truth. Researcher in formulating hypotheses about the relationship between behavioral intention culinary experiences is supported by the following premises:

- Experience with food is found to have an effect on behavior. Life experience can create memories that influence in determining the subsequent behavior (Björk & Kauppinen-Raisanen 2014).
- Culinary experience significantly affects culinary experience satisfaction and impacts on behavioral intention. Therefore, culinary experience has an impact on consumer intentions in determining future behavior (Kristanti et al. 2019).

Based on the premises that have been described, the researchers were able to put forward a hypothesis for this study. The author will attempt to choose fusion in this study. Based on the hypothesis of this study is the research, "Culinary Experience Influence on Behavioral Intention."

3 RESEARCH METHODS

This research was conducted in a period of less than one year. The method used is the cross-sectional method. This type of research is descriptive and verification. Research carried out a survey of consumer verification on the Sudirman Street Day and Night Market and Cibadak Culinary Night. The population in this study is consumers On Street Food in Bandung, respectively, as many as 217.618 and 227.370 tourists. Out of which, the average number of 222.494 tourists was taken into account. From the average total number, 245 samples were used for this study based on the formula of Issac and Michael. The sampling technique used in this research is systematic random sampling.

4 RESULTS AND DISCUSSION

4.1 *Visitor responses to the overview of culinary experience on street food in Bandung*

Culinary experience variable comprised of five indicators, that is, fusion food culture, uniqueness of the food, new food experience, aesthetics of dining, and dining at street food. Based on the results of data processing of questionnaires that have been distributed, the results of visitor response in On Street Food in Bandung as is shown in Table 1.

Table 1 illustrates that general visitors have a tendency to try eating with a different atmosphere, in this case, the sub-variables dining at street food with a percentage of 23.8%. This happens because On Street Food in Bandung have interesting products and low prices offered as well as a different dining atmosphere. Sub-variable aesthetics of dining get the second-highest score with a percentage of 23.2%. It is because the experience of eating street food has a distinct aesthetic.

Table 1. Description of visitors to culinary experience.

No.	Dimension	Total Score	Number of Questions	The Average Score	%
1	Fusion food culture	4046	4	1011.5	19.9
2	Uniqueness of food	2761	3	920.333	13.6
3	New food experience	3973	4	993.25	19.5
4	Aesthetics of dining	4729	5	945.8	23.2
5	Dining at street food	4844	5	968.8	23.8
	TOTAL	20 353	21	4839.68	100

Table 2. Description of visitors to behavioral intention.

No.	Dimension	Total Score	Number of Questions	The Average Score	%
1	Revisit intention	2105	2	1052.5	29.3
2	Word of mouth	2892	3	994	41.6
3	Willingness to pay more	2098	3	1044.5	29.1
TOTAL		7176	7	3091	100

Table 3. Path coefficient test results.

Variables	Direct Impact	Indirect Influence X_1	X_2	X_3	X_5	R2Y X1, X2, X3, X5	T_{value}	Sig.	Decision
X1	0.027889	–	0.007093	−0.00061	0.018424	0.052795	2,508	,013	H0 is rejected
X2	0.020449	0.007093	–	0.000118	0.010425	0.038085	2,292	,023	H0 is rejected
X3	0.013924	−0.00061	0.000118	–	−0.00275	0.010678	2,021	,044	H0 is rejected
X5	0.059049	0.018424	0.010425	−0.00275		0.085145	3.629	,000	H0 is rejected
			R2			0.186703			

Sub-variable uniqueness of the food gets the lowest score with a percentage of 13.6%; this is due to the perception that it is unusual to see or buy fusion food, so even though they are not too unique to ignore. According to Yoshida (2017), people tend to hesitate when trying new culinary experiences.

4.2 *Visitor responses to the overview of behavioral intention on street food in Bandung*

Behavioral intention consists of three indicators, namely, revisit intention, word of mouth, and a willingness to pay more. Based on the data processing of distributed questionnaires, it can be seen that the results of visitor response in On Street Food in Bandung are as shown in Table 2.

Data processing results in Table 2 show that visitors have a tendency to revisit On Street Food in Bandung. The visitors assume that On Street Food in Bandung has interesting products and low prices offered as well as a different dining atmosphere, which is influenced by external factors, one of which is an attractive offer of an affordable price.

4.3 *Culinary experience influence against behavioral intention on street food in Bandung*

Simultaneous hypothesis testing (Test F/ANOVA) showed that model 1 obtained a value of Fcount 11.211 with significant value is 0.000, as well as in model 2, which obtained value of Fcount 13.736 with significant value is 0.000 where if Fcount compared with Ftable is 2.44, Fcount has a greater value than the Ftable (Fcount > Ftable).

Based on Table 3, the coefficient of determination (R square) is approximately 0.186, or 18.6%, R square shows that every culinary experience sub-variable (X) contributed 18.6% of the behavioral intention and the remaining 81.4% as a contribution from other factors not examined in the study.

5 CONCLUSION AND RECOMMENDATION

5.1 *Conclusion*

By distributing questionnaire to 245 visitors in Street Food in Bandung through descriptive and verification analysis, using path analysis with the aim to analyze the influence of the culinary experience that consists of fusion food culture, uniqueness of the food, new food experience, aesthetics of dining, and dining at street food to behavioral intention, the researchers presented some conclusions as follows:

- The visitors' feedback regarding the culinary experience in Street Food in Bandung gets higher ratings, and the majority of visitors assume that dining is high; it is because they have interesting products and low prices offered and a different dining atmosphere.
- Visitors' response on behavioral intention in Street Food in Bandung were graded very high, where visitors give highest ratings on revisit intention. This happens because visitors assume that this culinary experience offers a different dining atmosphere with restaurants in general and have a concept of street food but well laid out and packed with erasing the ugly image of the street food.
- The findings of the study empirically show that the influence of culinary experience to behavioral intention simultaneously has a positive and significant impact.

5.2 *Recommendation*

The author recommends a few things regarding the implementation of influence culinary experience toward behavioral intention as follows:

- The recommendation to family and those closest to them and visitors will be willing to pay a heavy price if they feel have a quality experience after visiting On Street Food in Bandung
- Behavioral intention in On Street Food in Bandung is considered to be a good overall indicator. Willingness to pay more is an indicator that has a lower valuation compared with other indicators. Therefore, recommendations is by reanalyzing target market and adjusting prices suitable for their visitors' capabilities.
- The effect of culinary experience on behavioral intention simultaneously has a significant impact and the influence is categorized high, but there is some space for improvement; uniqueness of the food, which has a score of at least influence on behavioral intention. The influence of the four dimensions of culinary experience to the purchase decision is explained as follows:
  - Dining at street food dimensions needs to be maintained and improved by paying more attention to the seller friendliness, speed of service, dining convenience, and hygiene.
  - Fusion food culture dimensions need to be improved with regard to offers or promotions of fusion food so that consumers are more interested in buying, not just for tasting.
  - New food experience dimensions need to be improved with regard to consumer interest in fusion food.
- Recommendations are to improve the behavioral intention through culinary experience offered of improving visitor experience by increasing the types of food, improving quality of service, increasing customer satisfaction On Street Food in Bandung.

REFERENCES

Björk, P., & Kauppinen-Raisanen, H. 2014. *Culinary-Gastronomic Tourism – A Search for Local Food Experiences*. Https://Doi.Org/10.1108/NFS-12-2013-0142

Bray, J., & Bray, J. (ND). Consumer Behaviour Theory *Approaches 26 Models*, 1–33. Https://Doi.Org/10.1509/Jimk.16.1.39

Chan, A. 2017. *Experiential Value Of Bandung Food Tourism*, 6 (1), 184–190.

Dal, R., Cortese, M., Veiros, MB, Feldman, C., & Cavalli, SB. 2015. Food Safety and Hygiene Practices of Vendors During The Chain of Street Food Production In Florianopolis, Brazil: A Cross-Sectional Study. *Food Control*. Https://Doi.Org/10.1016/J.Foodcont.2015.10.027

Ewerhard, A., Sisovsky, K., & Johansson, U. 2019. Consumer Decision-Making of Slow Moving Consumer Goods In The Age of Multi-Channels. *The International Review of Retail, Distribution and Consumer Research*, 00 (00), 1–22. Https://Doi.Org/10.1080/09593969.2018.1537191

Fathema, N., Shannon, D., & Ross, M. 2015. Expanding the Technology Acceptance Model (TAM) Faculty to Examine Use of Learning Management Systems (LMSs) In *Higher Education Institutions*, 11 (2), 210–232.

Governance, MOF, Food, OFS, and In, V. 2019. Modes of Governance of Street Food Vending, (29).

Gupta, V. & Khanna, K.. 2018. A Study on the Street Food Dimensions and its Effects on Consumer Attitude and Behavioral Intentions. Https://Doi.Org/10.1108/TR-03-2018-0033

Hassan, S., Arshad, S., Naru, MH, Tahir, SH, Afzal, L., & Iqbal, MS 2016. Assessment of Hygienic Practices of Street Food Vendors Serving. *Lahore*, 30 (1), 7–13.

Henderson, JC (ND). Street Food and Tourism. *A Southeast Asian Perspective*, 45–57.

Kim, H., & Shim, J. 2019. The Effects of Quality Factors on Customer Satisfaction, Trust and Behavioral Intention In Chicken Restaurants 치킨 전문점 의 품질 요인 이 고객 만족 . 신뢰 와 행동 의 도 에 미치는 영향, 10, 43–56.

Komaladewi, R.2017. The Representation of Culinary Experience as the Future of Indonesian Tourism Cases in Bandung City, West Java:, 2 (5), 268–275. Https://Doi.Org/10.24088/IJBEA-2017-25001

Kristanti, M., Jokom, R., & Widjaja, DC. 2019. Culinary Experience of Domestic Tourists in Tourist Destinations, 69 (Teams 2018), 132–135.

Leong-Salobir, C. 2019. Urban Food Culture. *Urban Food Culture*, 137–164. Https://Doi.Org/10.1057/978-1-137-51691-6

Liu, HY, And HH. 2018. The Impact of Purchase-Decision Involvement on Purchasing Intention: The mediating Effect of Customer Perceived Value, 53 (ICEM 2017), 850–854.

Muscat, B., Hortnagl, T., Prayag, G., & Wagner, S. 2019. Perceived Quality, Authenticity, and Price In Tourists' Dining Experiences: Testing Competing Models of Satisfaction and Behavioral Intentions. *Journal of Vacation Marketing*. Https://Doi.Org/10.1177/1356766718822675

Namin, A. 2017. Revisiting Customers' Perception of Service Quality In Fast Food Restaurants. *Journal of Retailing and Consumer Services*, 34 (September 2016), 70–81. Https://Doi.Org/10.1016/J.Jretconser.2016.09.008

Okumus, B., & Sonmez, S. 2018. An Analysis on the Current Food Regulations For Inspection and Challenges of Street Food: Case of Florida. *Journal of Culinary Science & Technology*, 00 (00), 1–15. Https://Doi.Org/10.1080/15428052.2018.1428707

Omar, MB. 2017. Tourist fulfillment antecedent and revisit Intention of Culinary Experience (July), 63–66.

Platania, M., Platania, S., & Santisi, G. 2016. Author's Accepted Manuscript. *Wine Economics and Policy*. Https://Doi.Org/10.1016/J.Wep.2016.10.001

Serpico, E., Aquilani, B., Ruggieri, A., & Silvestri, C. 2015. Customer Centric Marketing Strategies. In *Marketing and Consumer Behavior* (pp. 666-708). Https://Doi.Org/10.4018/978-1-4666-7357-1.Ch030

Setiyorini, HPD, Abdullah, T., & Ariandani, W. 2019. Does Food Affect Customer Image Intention To Buy Food?, 259 (*Isot 2018*), 318-319. Https://Doi.Org/10.2991/Isot-18.2019.70

Shin, YH, Kim, H., & Severt, K. 2018Antecedents Of Consumers' Intention To Visit Food Trucks. *Journal of Foodservice Business Research*, 21 (3), 239–256. Https://Doi.Org/10.1080/15378020.2017.1368810

Sotiriadis, MD . 2015. Culinary Tourism Assets and Events: Suggesting A Strategic Planning Tool. *International Journal of Contemporary Hospitality Management*, 27 (6), 1214–1232. Https://Doi.Org/10.1108/IJCHM-11-2013-0519

Testa, R., Galati, A., Schifani, G., Maria, A., Trapani, D., & Migliore, G. 2019. Culinary Tourism Experiences In Agri-Tourism Destinations And Sustainable Consumption – Understanding Italian Tourists' Motivations, 2017, 1–17.

Vilmar, F., Da, B., Guilherme, S., Caron, M., Ladeira De Souza Junior, L., Perez de Castro, M., Nunes, A. 2015. Faculdade de Ciencias Da Educação e Saúde FACES Superior de Tecnologia Em Gastronomia.

Wang, S. 2018. Creature of Habit or Embracer of Change? Contrasting Consumer Behavior Daily Food with *The Tourism Scenario*, XX (X), 1–22. Https://Doi.Org/10.1177/1096348018817586

Wijaya, S., King, B., Morrison, A., & Nguyen, T. 2018. Destination Encounters with Local Food: The Experience of International Visitors In Indonesia, 17, 79–91.

Yoshida, M. 2017. Consumer Experience Quality: A Review and Extension of the Sport Management Literature. Sport Management Review, 20 (5), 427–442. Https://Doi.Org/10.1016/J.Smr.2017.01.002

Young, J., Choe, J., & Sam, S. 2018. International Journal of Hospitality Management Effects of Tourists' Local Food Consumption Value on Attitude, Food Destination Image, and Behavioral Intention. *International Journal of Hospitality Management*, 71 (November 2017), 1–10. Https://Doi.Org/10.1016/J.Ijhm.2017.11.007

Promoting Creative Tourism: Current Issues in Tourism Research – Kusumah et al. (Eds)
© 2021 Taylor & Francis Group, London, ISBN 978-0-367-55862-8

Implications of food delivery services for recognition of traditional foods by millennials in Bandung

A. Sudono
Universitas Pendidikan Indonesia, Bandung, Indonesia

ABSTRACT: This study aims to identify the extent to which food delivery service (Go-Food) has an influence on the level of recognition among millennials in Bandung on the traditional food of West Java. A further implication is expected to be the identification of distribution channels for traditional food to reach its market. If these pathways are established, then efforts to preserve traditional food will be easier to do. Research method used is quantitative descriptive, and researchers inventory the availability of traditional foods contained in the Go-Food service in Bandung. The next step is to survey a number of millennial respondents who use Go-Food services. The sampling technique used is snowball sampling, where the number of samples will continue to grow until the data stability is obtained. The data collected is processed using Microsoft Excel to see various trends related to millennial behavior. The results showed in general the types of food that are usually ordered by millennials through the Go-Food service are fast foods that do not include traditional foods. However, 59.1% of respondents realized that there are traditional foods that can be ordered. Some types of traditional foods that are most often ordered include serabi, lotek, balok cake, rengginang, and wajit. Meanwhile 39.8% of respondents felt that they gained new knowledge about traditional food in West Java, while the rest did not feel confident and hesitant.

Keywords: millennials, traditional food, Bandung

1 INTRODUCTION

Prior to the Covid-19 pandemic, 2019 recorded good tourism growth. UNWTO data show that there were 1.5 billion world tourist arrivals (World Travel Organization 2019). The growth achieved was 4%, still lower compared to previous years, which reached 5% to 6%. Even so, in 2019 it was considered a good year for world tourism. Indonesia is a country that is actively developing the tourism sector and recorded a growth of foreign tourists by 1.88% (Badan Pusat Statistik 2020) in 2019 with a total of 16.11 million visits. This growth rate is far below the average growth of world tourists (4%). As a country that has great potential in the field of tourism, this is a big question. So far, Indonesia is known to rely heavily on natural tourism as a mainstay of attraction for tourists. However, statistically this strategy has not succeeded in boosting the level of foreign tourist arrivals.

Further observations related to the attractiveness of Indonesia as a tourist destination country are actually not only on natural attractions. One type of tourism that might be developed is culinary tourism. Basically, almost all regions in Indonesia have unique and diverse culinary potential. West Java is a province in Indonesia which has the most interesting culinary variety in Indonesia. West Java has traditionally inherited strong culinary traditions in the form of traditional food. Bandung as the capital of West Java is a culinary tourism center that represents the culinary diversity throughout West Java. However, the data on the existence of traditional foods as tourism attraction still needs to be improved. This is due to the declining popularity and availability of traditional food itself. Nowadays, it is no longer always easy to get traditional foods in West Java. This condition is considered a weakness, considering food is basically a very important supporting component of

DOI 10.1201/9781003095484-65

tourism. According to Everett and Aitchison (2008), food is one of the potential tools that influence the level of expenditure and the stay of tourists.

The ability and availability of distribution facilities are factors that are thought to greatly affect the existence of traditional foods. In terms of distribution, currently food distribution technology has developed even further. There are companies that specialize in food distribution, including traditional foods. Go-Jek through the Go-Food service is one company that provides online food purchasing services. Through the Go-Food application, consumers can order food from sellers directly with delivery services from Go-Food. This facility is a significant convenience for consumers to obtain the desired food without having to go to the store, as well as for food entrepreneurs, as it is very easy to sell and distribute the food they produce.

This study focuses on describing the extent of the implications of Go-Food services for the recognition of traditional food in West Java by millennials in Bandung. Millennial generation was chosen because they are the majority users of Go-Food services. Another reason is the diminishing recognition of these people for traditional foods. They are generally more familiar with modern food. Yet to develop traditional culinary-based tourism requires a good level of recognition from millennials as the main players in the tourism industry.

2 LITERATURE REVIEW

In tourism, there are at least three important roles of traditional food, namely, extending the stay of tourists, increasing tourist spending, and re-examining the typology of food tourism within a framework of sustainability (Everett & Aitchison 2010) The three roles above are in line with the traditional culinary conditions of West Java, especially in Bandung. As the capital of West Java province, Bandung is a favorite tourist destination with one of the main attractions in food tourism.

Meanwhile Cheung (2013) revealed how Chinese food spread throughout the world. The popularity of Chinese food is so high that it does not only exist, but has become a channel for the spread of broader Chinese culture. This case shows the success of China in preserving and at the same time reviving its food industry throughout the world. In the many examples that can be found, many Chinese immigrants rely on traditional food that they produce and sell. This is a perfect example of how eating culture is no longer limited as a tourist attraction, but has become a way of life that integrates with people of Chinese descent that makes them able to survive wherever they spread their culture.

Bessiere (1998) uncovered unique facts in rural France in food perspective as a cultural heritage as well as local village identity. Basically, every village in France can be identified from traditional food products as well as the cooking skills of its inhabitants. The traditional food specialties become a unique geographical marker, which at the same time can be a promotional tool for local agricultural products.

The PEW Research Center defines millennial as a human being born between 1981 and 1996 (PEW 2010). But the millennial categorization approach is not always based on age; there are also criteria that are based on certain habits of a person. For example, a person who spends more than three hours in front of a cellular phone also makes someone said to enter the millennial community. So that the relevance of this research is maintained, where the millennials referred to in this study are the younger generation, the researchers determine millennial circles are those aged 17 to 40 years old.

3 RESEARCH METHODOLOGY

The method used in this research is quantitative descriptive. Basically, this research is carried out simply. Researchers devised a series of instruments to measure the extent of millennial generation's recognition of traditional food in West Java. This research was conducted in Bandung as the capital of West Java, which is considered to be able to represent the characteristics of West Java in general. There were 200 millennials chosen as respondents. The sampling technique used is snowball sampling, where the sample will continue to grow according to the needs and achieve data stability (Baltar & Brunet 2012).

The instruments were arranged in the form of an online questionnaire (Google form) and distributed to respondents who were accustomed to using Go-Food services. Most of the data collected is obtained in the form of a Likert scale. Then the data is tabulated using Microsoft Excel to obtain certain trends related to millennials' recognition to traditional foods of West Java.

4 RESULT AND DISCUSSION

The tabulated data shows that 47% of respondents use Go-Food services at least once a month, while another 31% more than three times using the same service. The most interesting point is the fact that more than 11% of respondents use Go-Food more than 10 times a month. These statistics illustrate how millennials are very accustomed to using Go-Food services. Some of them are suspected of having some kind of dependency, where they can experience difficulties when not using Go-Food services. According to Everett and Slocum (2013), partnership is one of the most effective ways in marketing a product. In this case, Go-Food shows its ability to partner with consumers and producers at the same time.

The availability of traditional food in Go-Food services has not yet been fully realized by millennials. Only 59% noted the existence of traditional foods in Go-Food services. On the other side, all respondents know better about non-traditional foods. This condition shows the inferiority of traditional foods compared to modern foods. However, it is still better to refer to statistics which show that almost 40% of respondents experienced an increase in knowledge about traditional foods after finding it in Go-Food services.

Even though the level of millennial awareness of traditional foods in Go-Food services cannot be said to be high, 96% of respondents said that they know about the traditional foods they find in Go-Food services. Only about 4% said they only knew certain types of traditional foods after seeing them on Go-Food services. It can be concluded that actually millennials are already familiar with the majority of traditional foods in West Java. In fact, the existence of Go-Food is a kind of reminder of the existence of traditional foods that are getting forgotten.

The most common types of traditional food ordered through Go-Food are seblak, lotek, serabi, balok cake, tutug oncom rice, wajit, and rengginang. There is a quite interesting fact where millennials consider seblak as a traditional food of West Java. Based on a search on Wikipedia (2014) seblak is thought to originate from food in the Sumpiuh area of Central Java. In the 1940s, popular in the area was a type of food called Kerupuk Godog crackers, which in taste, how to cook, and appearance are very similar to seblak. Meanwhile, Lotek, Serabi, Kue Balok, and Nasi Tutug Oncom are traditional food groups whose popularity is quite good in West Java.

In the aspect of millennial interest in ordering traditional food from West Java, data shows that more than 50% expressed interest in ordering traditional food through the Go-Food service—41% of them are still considering, and 9% are not interested. Although the availability of traditional food in the Go-Food service is considered quite complete, 82% of respondents are more interested in buying traditional food directly in the store. This is due to certain traditional foods consumers expect more than just food products, but they also expect the atmosphere of a shop or restaurant that sells these traditional foods. Nguyen et al. (2019) stated that the store atmosphere is an important thing that consumers consider when buying products and enjoying the atmosphere. In connection with traditional foods, atmospheric factors are felt to be more important than modern foods.

5 CONCLUSION

The existence of Go-Food services has succeeded in increasing millennial recognition of traditional food in West Java. This is marked by increasing millennial awareness of the existence of traditional food menus in Go-Food services, increasing knowledge of traditional foods, and drawing high enough interest from millennials to buy traditional foods. Nevertheless, Go-Food services are not completely able to replace the role of offline stores in selling traditional food. This is due to the loss of atmospheric factors in the Go-Food service, which is one of the important factors for consumers to buy and enjoy traditional food. The Go-Food service in this case acts as a catalyst and reminder that is very easy to touch the millennials to stay acquainted with and increase knowledge of traditional foods. The response to the description of perceived authenticity of Sundanese ethnic restaurants in West Bandung Regency is in the high category (good), where the authenticity of atmospherics dimension gets the highest value, while the employee authenticity dimension gets the lowest value.

6 RECOMMENDATION

Distribution is the most important function of the Go-Food service to deliver food from producers to consumers, including traditional food. There are conveniences that previously could not be obtained before the Go-Food service. Since its role is very important, Go-Food should not go alone in increasing access to traditional foods. The government through related agencies, academics, cultural conservationists, traditional food conservationists, and producers should be able to formulate strategies to better reach and educate markets. The life of the traditional food industry is a strong signal that provides great opportunities for traditional food to be sustainable. Tourism as a sector that is very concerned with this will more easily utilize traditional food as a culture that exists and is sustainable.

REFERENCES

Badan Pusat Statistik. 2020. "Perkembangan Pariwisata Dan Transportasi Nasional Desember 2019." *Berita Resmi Statistik* (13): 1–16.

Baltar, Fabiola, and Ignasi Brunet. 2012. "Virtual Snowball Sampling Method Using Facebook." 22(1): 57–74.

Bessiere, Jacinthe. 1998. "Local Development and Heritage: Traditional Food and Cuisine as Tourist Attractions in Rural Areas." *Sociologia Ruralis* 38(1): 21–34. http://doi.wiley.com/10.1111/1467-9523.00061.

Cheung, Sidney C.H. 2013. "From Foodways to Intangible Heritage: A Case Study of Chinese Culinary Resource, Retail and Recipe in Hong Kong." *International Journal of Heritage Studies* 19(4): 353–64.

Everett, Sally, and Cara Aitchison. 2008. "The Role of Food Tourism in Sustaining Regional Identity: A Case Study of Cornwall, South West England." *Journal of Sustainable Tourism* 16(2): 150–67. http://www.tandfonline.com/doi/abs/10.2167/jost696.0 (May 9, 2018).

Everett, Sally, and Cara Aitchison. 2010. "The Role of Food Tourism in Sustaining Regional Identity: A Case Study of Cornwall, South West England." *Taylor & Francis*. http://www.tandfonline.com/doi/abs/10.2167/jost696.0 (May 9, 2018).

Everett, Sally, and Susan L. Slocum. 2013. "Food and Tourism: An Effective Partnership? A UK-Based Review." *Journal of Sustainable Tourism* 21(6): 789–809.

Nguyen, Thuy Van Thi, Haoying Han, Noman Sahito, and Tram Ngoc Lam. 2019. "The Bookstore-Café: Emergence of a New Lifestyle as a 'Third Place' in Hangzhou, China." *Space and Culture* 22(2): 216–33.

PEW, Research Center. 2010. "A Portrait of Generation Next." (February).

Wikipedia, Indonesia. 2014. "Seblak – Wikipedia Bahasa Indonesia, Ensiklopedia Bebas." https://id.wikipedia.org/wiki/Seblak (June 30, 2020).

World Travel Organization. 2019. "International Tourism Highlights International Tourism Continues to Outpace the Global Economy." *Unwto*: 1–24.

Promoting Creative Tourism: Current Issues in Tourism Research – Kusumah et al. (Eds)

Gastronomy tourism as a media to strengthen national identity

R. Fitria, A. Supriatna, K.A. Hakam, S. Nurbayani & Warlim
Universitas Pendidikan Indonesia, Bandung, Indonesia

ABSTRACT: Indonesia is a country that has a diversity of food types. This indicates that everyone's taste for food can be different and can be the same. Food as a concrete result of social activities manifested in the form of ideas and tastes is able to describe the philosophical values that underlie its manufacture. This research aims to describe the development of culinary as a gastronomic tour, one of which is lead rice in the Punclut area. The method used is qualitative with in-depth interview data processing techniques for six months, observation, and literature study. The results show that lead rice culinary in the Punclut area not only shows the potential of tourist attractions but can also be developed as a gastronomic tour. Tasting lead rice is not only just enjoying the taste but further introduction of local wisdom passed down from generation to generation so that it forms into self-identity. The introduction of lead rice by involving the community will provide economic, collective, health, and cultural value.

Keywords: gastronomy, national cultural heritage, national identity

1 INTRODUCTION

Indonesia has a variety of dishes that reflect the diversity of cultures and traditions that play an important role as one of the nation's ancestral heritages. There is a connection between the source of the acquisition of food ingredients, spices, procedures for serving, the tradition of making or cooking, the use of herbs, the nutritional content of each food preparation, and others that bind it into a unique tradition.

In this globalization era, traditional food has started to be abandoned and increasingly unpopular. This is because it has begun to be defeated by foreign food originating from other countries. In Indonesia, there are still many local people who do not really know and like the traditional food of their country. They prefer Chinese food, Japanese food, Western food, especially young people who only enjoy fast food and junk food and consider Indonesian food to be old-fashioned and not modern.

The lack of recognition and appreciation of the traditions and history of the culinary culture of the archipelago is a major factor causing dislikes of traditional foods. The development of information and public interest, as well as lifestyle changes toward modern, consumptive, and selective societies, demand the availability of new, innovative, and standard-compliant facilities, moreover, the characteristics of Indonesian urban communities who like to try new things and experiences, including in the culinary field.

On the other hand, in the midst of the community's own disinterest, foreign tourists who visit are very interested in Indonesian local culture, especially in terms of culinary, which makes foreign tourists addicted and they even wish to learn the history and traditions of Indonesian culinary culture. As one of the attractions of tourism and one of the sub-sectors of the creative economy, culinary plays an important role in the promotion of Indonesian tourism and the culture of the archipelago. According to Hall and Mitchell (2001) in Rahma et al. (2017), food tourism in general can be defined as visits to food producers, food festivals, restaurants, and specific locations for tasting food and/or enjoying/learning the production. Thus food, food production, and special

DOI 10.1201/9781003095484-66

attributes of regional food become the basis and the main driving factors in tourist travel. Culinary tourism is related to local food from the tourist destination. Local food is indispensable in the tourist experience when it can present both cultural and entertainment activities (Hall et al. 2003; Hjalegar & Ricchard 2002; Rahma et al. 2017).

Based on 2016 statistical data, in West Java alone the manufacturing industry business (Category C) and the business of accommodation and provision of food and beverage (Category I) also had a large contribution, each totaling 860,312 businesses (18.93 percent) and more than 600,720 businesses (more than 13.12 percent).

Characteristics of gastronomic tourism include: 1) Gastronomy as an element and indicator of globalization; in particular, the affirmation of regional competition throughout the world, 2) Tourists play a role in the evolution of gastronomic tourism, 3) Tourism as a revealer of regional or local gastronomic potentials and as a contributor to developing or renewing national and subnational identities, 4) Gastronomic tourism as a means of introducing culinary products as a culinary product cultural products, 5) The evolution of gastronomic tourism provides direction for tourism development, 6) Gastronomy as a constructive element in shaping the image of a tourist destination, 7) Gastronomy as a travel destination, 8) Gastronomy as an element of heritage with a tourist dimension (Anton Clave & Knafou 2012).

In the context of developing food-based tourism, the concept of culinary tourism becomes a major part. Culinary tourism can be interpreted as a food-tasting activity somewhere. Culinary according to Pitana (2019) is one of the cultural resources that can be developed into a tourist attraction. Watching firsthand starting from the preparation, how to process, serve, and eat special food is a cultural attraction that is able to attract tourists to visit an area. The development of tourism in the culinary field is known as gastronomy, "culinary tourism, also referred to as gastronomic" (Hjalanger & Richards 2002).

Gastronomic tourism is another way to enjoy tourism objects while also preserving culture through efforts to preserve culture in the field of food and drink or commonly called culinary tourism (Brillat-Savarin 1994). Food-tasting activities are not just tasting food, but there is also an emphasis on cultural aspects as well as studying the history of the food itself, so it can be said as a gastronomic tour.

Gastronomic tourism aspects are very useful and become an "effective weapon" in the context of the introduction of the richness and diversity of cultures and culinary tastes of a nation in the social world constellation.

This study describes the attitude of culinary culture connoisseurs through gastronomic tourism, so that the analysis takes the form of a description of the phenomenon of reality that occurs in the relationship between gastronomic tourism and culture, then the social community of the environment so that it forms a national identity. The analysis will lead to a more concrete presentation on tourism development while strengthening culinary culture that is applicable and useful and can be applied especially in the city of Bandung.

2 STUDY OF LITERATURE

In UURI No.9 / 2010 Article 1 paragraph 6 states that gastronomic tourism can cover one or more administrative areas while the local gastronomy is consumed and the people associated with it become a tourist attraction. There are two benefits that will be obtained by tourists (Turgarini 2018), namely to relieve fatigue after a long time of wrestling with the safe work and to develop themselves.

Ricards (2015) and Castells (2009) in Turgarini (2018) state the relationship between gastronomy and tourism has shifted in the last decade, initially gastronomy only as a set of destinations is now turned into the main reason for tourists to visit a tourist destination, which is the cause of the shift in meaning. is a change in gastronomic and economic position besides gastronomy, which is able to connect between foodscape, cultural identity, and tourism, and this role is increasingly significant in network society.

2.1 *Sundanese gastronomy*

Hjalanger and Richard (2002), Bonow and Rytkonen (2012), and Bassiere (1998) in Turgarini said that Sundanese gastronomy is inseparable from the Pasundan people's creations, tastes, intentions, and works. This local gastronomy plays a different role in tourism, for example, as a complement to tourist destinations, and as a tourism gourmet or as a symbol of local uniqueness, and when tourists choose local gastronomy, they also feel an element of the local character of the area they visit. Based on the explanation, consuming Sundanese gastronomy means absorbing, manifesting, and enjoying intrinsically tourism with the aura of legendary locality.

Bandung city according to Turgarini (2018) has regional branding through a unique blend of Sundanese gastronomic flavors with a nuanced presentation of natural atmosphere as an attraction for visitors. Regional branding is basically used as a unifying tool for a region with the aim of encouraging economic growth, image development, and regional introduction, especially tourism.

According to Cohen and Avieli (2004) and Turgarini (2018), gastronomic tourism will make tourists more deeply involved and gain more experience with the destination environment (tourist destinations) they visit. Gastronomy is part of the social and cultural heritage of the community that reflects the lifestyle of people who live in different geographical areas from tourists and have cooking skills that are rooted in their own culture and traditions.

Bartela (2011) in Turgarini (2018) suggests that gastronomy has so far increased regional income from the tourism sector. Gastronomic tourism is labor intensive, and can create jobs, contributing to the regional economy, thereby strengthening economic aspects, maintaining cultural heritage and the local environment, strengthening the identity and sense of belonging to the local community, extending tourist stays in the tourist season, so that no new investment is needed. It is large, and able to be a bridge to the past, stimulating local agriculture and food production, reducing economic leakage. Based on this, it can be seen that gastronomic tourism provides many advantages in addition to tourists as well as people who are in the area of tourist visits, not only limited to economic needs but can also preserve the social and cultural environment of the local community.

2.2 *Portrait of Punclut tourism region*

The Sunda tourism area, especially Puncrut, is located in the hilly areas of Bandung, offering unspoiled and beautiful natural panoramas. The visitors who came to Puncrut came from various regions around West Java and outside Java. In general, they are interested and want to feel and enjoy the panoramic atmosphere of the hills of Puncrut, which is very beautiful, both during the day and at night, and offers a view of the city of Bandung, which is decorated with sparkling, beautiful city lights visible from the height of the hill Puncrut.

In tourist sites, they generally prefer to walk along the hills to visit interesting and beautiful spots found around the Puncrut area while occasionally capturing moments of their activities through photos. They really enjoy the natural atmosphere that still looks natural and can freely breathe fresh mountain air that is very difficult for them to get in the hustle and bustle of the urban environment.

After the tourists are tired of walking around the Puncrut tourist area, in general they then stop at semi-permanent *lesehan* stalls selling Sundanese special food, namely, lead rice and various vegetables and fresh drinks that are typical such as coconut ice. While relieving, they eat culinary lead rice and side dishes that are appetizing, such as grilled chicken, satay, vegetables, and chili paste/tomato as a complement and fresh drinks such as young coconut.

Lead rice culinary is a characteristic that is very identical to the Sundanese, where this snack becomes something that is almost mandatory and always adorns the lives of Sundanese people. One of the interesting things is that when Sundanese want to travel on an assignment, be it to work in the fields/gardens or want to travel far outside the area, they almost always carry rice stock wrapped in banana leaves and side dishes complete (which is then referred to as rice lead) for them to enjoy on the sidelines of the trip or when they feel hungry so they don't have to bother looking for rice stalls that sell food, and more practically that they can enjoy it anywhere and anytime.

The philosophy of lead rice is a description of the "simplicity" and the "sense of kinship" that is tight between Sundanese people. The concept of simplicity can be seen in the composition of the ingredients forming the lead rice itself such as white/red rice wrapped in banana leaves, side dishes, vegetables, and special chili sauce with a delicious taste, and the manufacturing process is relatively short. While the concept of kinship is reflected in the behavior of Sundanese people, for example when eating leaded rice it is often done in a joint/ lively manner involving at least two or more people, so they generally feel that their sense of kinship is getting closer to the sharing concept.

But on the other hand, culinary lead rice seems to be an irony for the Sundanese people themselves. This is illustrated from the fact that this culinary is felt so closely and populist among the people but its existence is not necessarily made as a very important and meaningful thing as a characteristic, culinary identity, and wealth of a community group, so that its existence is only limited to routine and not necessarily became a truly visible icon and considered a rich culinary cultural heritage for the Sundanese people.

2.3 *Gastronomy and national identity*

The concept of gastronomy is closely related to the national identity of a nation, where through the wealth and diversity of forms, types, and culinary tastes and characteristics of food contained in each region of a country will be able to become the characteristics, character, and pride for the country or nation concerned to then to synergize into a sense of love and a sense of belonging to the country and nationalism.

In this regard, says Hara (2000), nationalism covers a broader context, namely the equality of membership and citizenship of all ethnic and cultural groups in a nation. In the framework of nationalism, pride is also needed to display its identity as a nation. Pride itself is a process that is born because it is learned and not a legacy passed down from one generation to the next.

The diversity and richness of aspects of gastronomic tourism is a very valuable and useful asset as well as being an "effective weapon" in the context of the introduction of wealth and cultural diversity and culinary tastes of a nation in the social world constellation.

3 RESEARCH METHODS

This research uses a qualitative method, that is, by presenting data in descriptive form, words, expressions, opinions, ideas collected by researchers from various sources which are then grouped based on needs with interpretive approaches to the subject, which will then be analyzed (Denzin & Lincoln 2009).

The study was conducted in several stages, which included observing the research location, observing the patterns of life and culture of the community around the research site, observing the patterns and behavioral tendencies of tourists who came to visit, conducting interview sessions with predetermined informants, and studying various sites and documents related to the research theme.

The duration of the research was carried out intensively for six months, including observation, data collection with in-depth interviews, and documentation studies relating to various sides of the information needed.

Data processing is carried out in several stages, namely data analysis, drawing conclusions and verification, and narrative results of data analysis. The data analysis phase is done by analyzing the data that has been obtained from the field and grouping the data. This is done to make it easier to interpret the data so as to minimize errors in determining conclusions. The conclusion and verification phase is carried out by making conclusions from the entire research data, then verifying the data to check the validity and accuracy of the data so that it remains relevant and fulfills the life cycle. Hold narration of the results of data analysis is the last thing to do in which the researcher narratively presents important findings and results of analysis of research data.

4 RESEARCH RESULT

Based on interviews conducted with several people who visited the West Java region, they came to the place to enjoy the cool and fresh atmosphere, one of the main objectives being to enjoy the lead in every direction of the tourist sites. As SY revealed, that he and his family came to Puncrut to exercise while enjoying the fresh air, and after being tired they would usually sit around while enjoying the warm lead along with other foods. Neither the PH, TS, ZS, and BT came from Jakarta, who deliberately came to Puncrut just to walk and sit while enjoying hot water, warm lead rice, fried chicken, vegetables, spicy chili, and other traditional foods. Indeed, one of the characteristics of this tourist spot is the lead rice which is served with white rice or red rice, and this is very reflective of Sundanese customs.

When asked about their philosophical meaning, on average they did not know what it meant; for them, lead rice had existed from time immemorial and had become a Sundanese custom, especially for people who like farming would bring lead rice along with side dishes and vegetables.

Lead philosophy is closely related to the geographical conditions of Indonesia. Most Sundanese livelihoods are farming. Supported by fertile natural conditions and a tropical climate is very appropriate if the West Java region is used for farming. Most farmers grow rice, vegetables, fruits, and raise livestock such as chickens, ducks, goats, sheep, cattle, or buffalo. So almost all of their living needs are obtained from nature around them.

Lead rice is usually carried by farmers or cultivators who want to clear land for agriculture or care for plants. Usually they wrap rice using banana leaves. Banana leaf was chosen because it can make long-lasting rice with a lot of volume. The large amount of rice is pressed using banana leaves until it is very small, and after that, the new banana leaf used is wrapped into rice that has shrunk. In addition, rice wrapped in banana leaves will emit certain special aromas that can arouse appetite. In using banana leaves for wrapping materials, banana leaves cannot be used arbitrarily. Banana leaves must be wiped first and then heated on fire until the banana leaves wither and are no longer hard, so it is very easy to be formed in accordance with the wishes of the rice maker.

Initially, the Sundanese ancestors did not know the advantages of banana leaves; for them, banana leaves are very economical and practical. Banana leaves are said to be economical because to get the leaves one does not need to spend money, instead simply by taking it from their yard or from their garden. Practically in the processing of leaves is only by heating it on the fire, then the leaves will follow whatever shape is the desired shape, and they usually heat the banana leaves the edge of the fireplace or the leaves are stored above the place to cook rice, so that in making it does not take much time. After the rice is cooked, banana leaves are ready to use, the term "once paddle two or three islands exceeded."

The types of banana plants are diverse and can be seen from the bananas produced. The diverse types of bananas also make the leaves have different shapes and textures. Banana leaves are most often used in Indonesia to wrap food before further processing is from the type of banana stone. Banana leaves are thicker, produce a fragrant aroma, and do not cause discoloration in the food wrapped. Therefore, many banana stone trees are cultivated with the aim to take and use only the leaves. The ambon banana leaf is used to wrap processed foods as well, but it is rarer because even though it gives a fragrant aroma to food, usually these banana leaves cause the color of the wrapped food to change after cooking. Based on research results, banana leaves contain polyphenols, which act as antioxidants that are good for health and can reduce the risk of heart disease, blood vessel problems, and cancer. Based on research conducted by Sahaa et al. (2013), it is known that banana leaf (Musa sapientum var. Sylveteris) has the potential to be used in the medical field because of the presence of antimicrobial and antioxidant activity. Research on the flavor of banana leaves is still minimal but from the research of Sahaa et al., it can also be seen that banana leaf extract contains gallic acid, which is a type of catechin. Catechins are included in the polyphenol group and are one of the sources of aroma-producing compounds.

The banana leaf packaging used comes from nature. The pre-modern society utilizes what exists in nature for its survival. The use of natural ingredients in traditional packaging has elements that are not found in modern packaging that use artificial ingredients. The natural cycle of appearance on traditional packaging looks natural, starting from the color, texture, and shape. The aroma of traditional packaging gives a distinctive taste and odor arising from the nature of the ingredients that can affect the product inside. Traditional packaging construction using natural materials has its own strength and elasticity, which cannot be obtained from packaging materials in modern packaging.

Banana leaves as food packaging function as a protector that protects the product, both from external and internal influences. The packaging is able to protect the product from the sun, moisture, and so on and protect from the influence of improper handling and is able to provide protection for packaged food so that it can be transported from the final consumer in good condition, and consumers who feel the product will have the same taste in consuming it even though it is in a different place.

In the packaging of this product, there is something lacking in the area, that is, packaging that is capable of causing traction. If it is only wrapped in banana leaves and put in plastic, it is certainly very unattractive and less appetizing. Lead rice and other snacks should be put together in a special container with a layout that is made as attractive as possible. As it is known, the packaging of a product can be a free advertising/covert promotion if displayed on a display case or at the time of distribution. The more interesting the concept of packaging design and the layout/display, it will further increase the added value of packaged food.

Packaging is a medium to stick the brand image with consumers so that consumers find it easy to remember, fantastic, and attractive. According to the WTO, packaging is an integrated system for preserving and preparing products until they are ready to be distributed to end consumers in an easy and efficient way.

Related to the problem of setting and lay out, it is inseparable from discussing the concept of packaging a product; for centuries, packaging is a functional concept limited to protecting goods or making it easier to carry goods and still seem modest. So that along with the development of increasingly advanced and complex technology, then there is the addition of functional values, especially in the present competition in the business world, which is increasingly sharp, and among producers competing to give attention to prospective consumers, thus the functional concept of packaging has occurred an important part that must cover the entire process of marketing from the conception of the product to the user lastly.

The strength of the brand is very important in marketing, moreover, most consumers will buy goods repeatedly and even regularly buy, especially consumer goods. In the market, people are faced with a large selection of goods, and certainly a lot of devotees are brands that stand out, are known, or glimpsed.

The gastronomic aspect, in addition to being seen from the side of tourism and marketing, is also very closely related to the national identity of a nation, where gastronomy is in direct contact with the background and history of making culinary, which also has its own values and symbols of local wisdom and that later on become a uniqueness and characteristic that is only owned by the region or country.

With so many cultural performances or traditional culinary festivals held regularly and continuously, it will contribute directly to the preservation and introduction of culinary wealth, which has the potential to be an attraction for tourists. Preservation and inheritance of culinary gastronomic cultural values is very important, especially for the young generation so that they can later continue and preserve the existence of their own culinary regions as an inseparable part of local wisdom and regional characteristics.

5 CONCLUSION

Motivation of tourist visits on gastronomic tourism is not only enjoying the panorama provided by nature but also being able to enjoy typical foods of an area and basically is looking for experiences by involving, discovering, feeling, experiencing, researching, understanding, and enjoying what food is in an area with combined local wealth and cultural attractions of the culinary origin.

The support of various parties in the introduction, development, and preservation of gastronomic tourism is an important and absolute thing to do as an effort to maintain the existence of the culinary cultural wealth of an area, so that its existence is not just a routine that is underappreciated but instead will transform into a sacredness of local wisdom and at the same time become a power attraction for tourists who want to share the sensation of being part of a unique and proud cultural and culinary connoisseur.

REFERENCES

Anton Clave, S., & Knafou, R. (2012). Gastronomy tourism and globalization. Paris: Universitat Rovirai Virgili Tarragona, Universite Paris1.

Brillat-Savarin, J.-A. (1994). The Physiology of Taste. (Penerjemah:A. Drayton, Harmondsworth:Penguin.

Denzin, Norman K. dan Yvonna S. Lincoln (eds.). 2009. Handbook of Qualitative Research. Terj. Dariyatno dkk. Jogjakarta: Pustaka Pelajar.

Hara, AE. Kebanggan Berbangsa Indonesia. Kompas, 17 Agustus 2000.

Hjalager, A., & Richards, G. (2002). Tourism and Gastronomic. London: Routledge.

Pitana, I G dan Diarta, S. 2009. Pengantar Ilmu Pariwisata. Jakarta: Andi.

Rahma, Neila., Indah Susilowati, Evi Yulia Purwanti.2017. Minat Wisatawan terhadap Makanan Lokal Kota Semarang. (Tourists' Interest to Local Food In Semarang). Jurnal ekonomi dan pembangunan Indonesia. Vol. 18 No.1 Juli 2017:53–76.

Sahaa, R.K., Srijan A., Syed Sohidul H.S., Priyanka R., 2013, Medicinal activities of the leaves of Musa sapientum var. sylvesteris in vitro, Asian Pacific Journal of Tropical Biomedicine, 3 (6), 476–482.

Turgarini 2018. Gastronomi Sunda sebagai Daya Tarik Wisata Kota Bandung. Yogyakarta. Universitas Gajah Mada.

Promoting Creative Tourism: Current Issues in Tourism Research – Kusumah et al. (Eds)
© 2021 Taylor & Francis Group, London, ISBN 978-0-367-55862-8

Digitalization activities in gastronomy tourism

D. Turgarini & I.I. Pratiwi
Universitas Pendidikan Indonesia, Bandung, Indonesia

T.K. Priyambodo
Universitas Gadjah Mada, Yogyakarta, Indonesia

ABSTRACT: Digital technology is changing the tourism industry. This work aims to describe the development of digital technology activities in gastronomy tourism. The literature review method was used with narrative descriptive approach. Through this approach, both the digitalization phenomenon in gastronomy and the description of the terms in digital gastronomy could be explained. This work concluded that there were human activities not only fulfilling their food needs using digital technology, based on gastronomy tourism concept. There were many activities in digitalization gastronomy, that is, recognizing local food history, tradition and philosophy, finding from local food and drink, finding the origin and distribution of raw material, understanding how to make, how to serve, to enjoy the learning process and getting experience, understanding nutrient content, as well as ethics and etiquette. In addition, the most important thing was that gastronomy digitalization should be used for conservation and education development of gastronomy tourism in the future.

Keywords: digitalization, gastronomy, tourism, education, conservation

1 INTRODUCTION

Gastronomy iFood topic often becomes a subject that appears on the Internet, with interesting food pictures, how to cook, food reviews, and so on. Digital platforms like websites, mobile applications, and social networking become an interaction place for digital generation. These platforms provide different types of information like how food location can be found, how is the eating culture, the health and nutrition in the food, and environment impact from the food (Schneider 2018).

The development of the digital platform also influences human consumption behavior (Davies et al. 2017), and also the experience of consumers or tourists when buying local food. A long time ago, the process of buying food was by choosing food ingredients from the physical characteristic in the market. Nowadays, by digital technology, people buy food based on digital characteristics, for example, from photos, videos, text, and numbers data. The Covid-19 outbreak has also influenced the speed of digital technology adoption in gastronomy. The process of food and drink supply at home is now through the digital process (Oncini et al. 2020).

Digital technology gives a space for new ways of producing, distributing, and consuming food. The relationship between producer and consumer changes. An intermediary who has been monopolizing food systems all this time can be minimized and value margin obtained by producers and consumers will increase. The change in digital technology, which influences a series of gastronomy tourism activities, will be narrated in this research until reaching a conclusion about what digitalization is in gastronomy tourism.

 DOI 10.1201/9781003095484-67

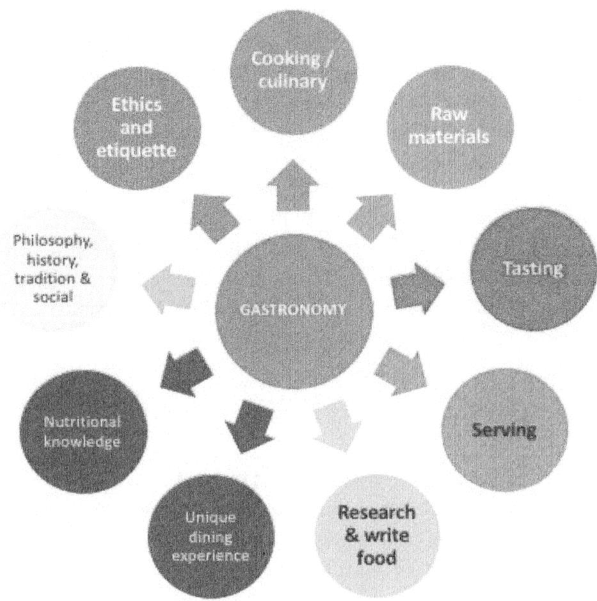

Figure 1. Conceptual of gastronomy. Source: Turgarini, 2018.

2 LITERATURE REVIEW

2.1 *Gastronomy*

Turgarini (2018) explained her understanding of gastronomy not only focusing on culinary arts or cooking methods, but also on human behavior, including choosing raw materials, then tasting, serving food, and experiencing the consuming and searching, studying, researching, and writing about food and all matters relating to ethics, etiquette, and human nutrition in every nation and country (Manolis 2010; Pullphothong & Sopha 2013; Santich 2010). Gastronomy is the art and science, even appreciation that is cross-ethnic, national, racial, group, religion, gender, and culture by studying in detail eating, food, and drink for use in various conditions and situations.

Turgarini (2018) explained that gastronomy emphasizes that food is a core component of every culture. As part of culture, authentic eating (authenticity) is a combination of pleasure, utility, and social (Eagleton 1997), while food is a dictionary of moods and sensations (Ellmann 1993), so the relationship between humans and their food is typical. So, the choice of type and how food is consumed will be a marker of identity and difference (Richards 2002). This perspective shows that food is not just fuel (energy), but also has broad meaning as a means to improve quality of life, a vehicle for socializing, enriching experience, being able to express identity or social status, and even becoming a conflict prevention, and a tool to protect land water. Therefore, regardless of presentation, both local residents and tourists visiting a tourist destination consistently make gastronomic choices according to their social class identity, which in turn will show power and control in a socioeconomic hierarchy (Everett 2009).

Based on the description of Turgarini (2018), local gastronomy is a part of their culture, which besides having a utility function (benefit) to fulfill basic human needs, namely, to maintain life, also contains behaviors and social norms found in human society in general (Macionis & Gerber 2011). Local gastronomy can also be said as art and science related to the culture of eating good food from the local ethnicity and can be used as a social tool to show the identity of the nation, region, and state. In the treasury of tourism, cultural phenomena, especially those related to local gastronomy, will be transmitted through social learning in the community, and are expected to provide new experiences in the excitement of traveling.

This is in line with the explanation of the United Nation World Tourism Organization that describes gastronomy as a value chain of a series of activities starting from food production (including farming, fishery, and food industry), food distribution activity (traditional market, supermarket, and retail), to an activity of how the food arrives home and is processed. It is different from culinary; gastronomy is not only about food served on the plate but also finding how to create food on the plate and a series of activities behind it. Those processes create a value chain and a gastronomy landscape involving history, culture, heritage, and other elements such as policy, regulation, infrastructure, human skill, and the accompanying research.

2.2 *Gastronomy tourism*

Gastronomy tourism as a part of cultural tourism (NIOS 2018) is a type of tourism or travel, which is designed with the primary purpose of making food and drinks as the main motivating factor for someone to make the trip. Gastronomic tourism is defined as "pursuing an eating and drinking experience that is unique and easy to remember (the pursuit of unique and memorable eating and drinking experiences)" (Manolis 2010). Thus, a gastronomic tourist is someone who is willing to travel to other places in an effort to taste and experience authentic local cuisine in a destination (Pullphothong & Sopha 2013).

Turgarini (2018) states that gastronomy tourism destinations can cover one or more administrative regions while the local gastronomic consumed and the people who are related to it become a tourist attraction. Gastronomic tourism means tourists get to learn and appreciate a variety of different cultures; the scope is broader than just learning cooking skills (Santich 2010). Tourists have the motivation to do recreation with the main aim of gaining experience in learning the uniqueness of food in an area (Santich 2010; Travel-Industry-Dictionary 2014). Gastronomic tourism is one of the tourist attractions utilizing cultural diversity.

According to Richards (2015) in Turgarini (2018), gastronomy is very useful for tourism activities because it is the gateway to local culture that can make tourists and local residents enjoy cultural experiences together. Gastronomy is a central part of the tourist experience that will provide a memorable and meaningful experience for tourists. In addition, gastronomy can be a distinctive element of the brand image of the place (Richards 2012), especially tourist destinations, so that it can be a source of identity, even creating a new identity that is hybrid (Scarpato 2002).

In this context, according to Turgarini (2018), local gastronomy cannot be separated from the creations, tastes, intentions, and works of the community land. As a cultural heritage, local gastronomy can relate to the symbiosis of mutualism with tourism. Hjalanger and Richards (2002) say that local gastronomy can play a different role in tourism, for example, as a complementary product for tourist destinations and as a gourmet tourism (Hall et al. 2003) or can be viewed as a symbol of local uniqueness (Bonow & Rytkönen 2012). When tourists choose local gastronomy, they also feel the elements of the local character they visit (Bessière 1998). So, consuming and understanding gastronomy means absorbing and manifesting and enjoying intrinsic tourism with an aura of locality that is even legendary. Usually in a tourist area, local gastronomies are handled by local artisans so that it is seen as an important part of local heritage and culture.

Many cities in many countries have been known to have places to enjoy gastronomy, both those sold at malls, traditional shops, and street-halls. The spectrum of types of food and beverages, how to process raw materials, how to deliver to consumers and consumer behavior is relatively broad. Another gastronomic element that is displayed in several outlets is the way of presentation that has local characteristics.

3 MATERIAL AND METHODS

To explain digitalization in gastronomy, of course, one needs to understand the definition of digital. There are some words with which the definition often overlaps with each other, namely, understanding the word digital, digitalization, and digitations. The word digital in the Indonesian dictionary is something related to numbers or numbering for certain calculation systems. Digitation is a process to change analog data (text, pictures, videos) into digital form, wereas digitalization is digital technology utilization by organizations, industries, or other entities.

Technology development changes human lifestyle and habits, which have influence on social order inside it. Studies that develop include digital behavior change technologies (BCT). Digital technology change in gastronomy is explained in human food interaction (HFI) phenomenon. Human behavior change in adopting technology development and how to react, socially, creates a community group classified as digital community. This community group lives and acts according to information in digital reality. Human interaction in digital world will also influence a decision-making pattern, including in food consumption pattern.

Activities including in value chain, gastronomy definition in this research will contribute in tourism since conservation activity for food as DNA from gastronomy becomes a base for consumers or tourists to visit gastronomy tourism destination. Conservation activity is important in the whole series of activity in gastronomy study because it relates to information education process and knowledge evolution that enrich gastronomy development. Gastronomic activity is included in gastronomy digitalization activity. The first step is to understand and notice the gastronomy concept principle that the digitalization of gastronomy can attract the tourist or consumer if they can access digitalization of the history, tradition, and philosophy about local food and drink, finding the origin and ingredient distribution, understanding how to make, serve, enjoy, learning process and obtaining experience, understanding nutrition, ethics and etiquette.

The next is to do digitalization while doing food production process either when obtaining food from the farm, fishery, food industry, food supply network, restaurant, or even food production using technology like 3D printing.

The third step is gastronomy digitalization in distribution activity like digital food marketing, digital food market, digital food delivery system, food digital media. Digitalization in food distribution, including food marketing, is part of digitalization in gastronomy. This include digital food markets such as websites or application or simple application for food delivery system.

The fourth step is digital food consumption activity, digital money needs for food transaction, availability of food data that can be accessed through online transaction, publication about food behavior, "Mukbang" behavior (eating in huge portion), gastronomy TV program, gastronomy youtube channel, food consumption behavior digital data, and Diet Social Network.

The last is digital gastronomy in food conservation activities such as ebooks, ejournal, and digital archives. Some ebooks include gastronomy textbook material, recipe collection, and popular books related to diet and nutrition and food culture. These kinds of digital gastronomy conservation are parts of gastronomy digital literacy and also parts of gastronomy education. Gastronomy tour information that goes along a digital platform is also part of food conservation in digital gastronomy. You can see the concept model of digitalization gastronomy in Figure 2.

4 RESULTS AND DISCUSSION

Gastronomy knowledge in digital forms will be easily accessed and archived because it can be done by cross region, language, even time. This digital literacy activity plays an important role to provide information and notes for gastronomy development and the next generation. In addition to gastronomy e-book, some digital gastronomy magazines are published as well. Some of them are Spenser Magazine, The Foodie Bugle, Honest Cooking, Easy Eats, College & Cook, also FoodieCrush (Daily 2014). Spenser Magazine publishes twice a month by providing daily food needs information for the readers. The information is about where to buy, how to store as well

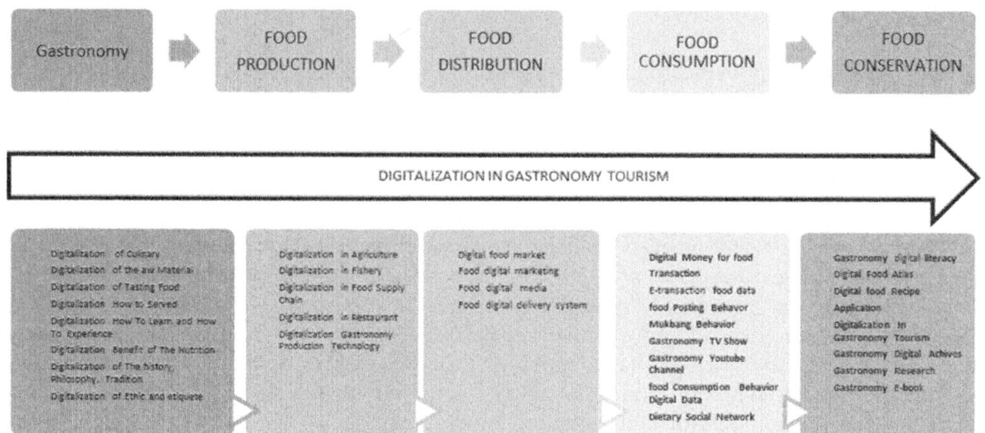

Figure 2. Digital gastronomy tourism conservation.

as how to grow and to treat food material. The Foodie Bugle is a digital magazine that provides information about sources, how to prepare and well enjoy food and drink. Honest Cooking is an online digital magazine that wishes to change the face of food on online media. The topics raised include debate associated with culinary to food recipe. Easy Eats is a digital magazine that specializes on gluten free food information based on personal stories, recipes tested by the community, and advice from diet specialists. College & Cook is a digital magazine intended for students who are associated with food. Foodie Crush is an online magazine intended for food bloggers to share stories, recipes, and food pictures.

Another world of gastronomy digital literacy is blogs and websites. Some contents in food blogs are reviewing restaurants, cooking books, recipe blogs, and blogs for a particular community such as vegetarian, gluten free, halal, baking, and so on. Some blogs and websites are famous enough, including Sweet as Honey, Sprouted Kitchen, Cookie and Kate, Pinch of Yum, I Am a Food Blog, Serious Eats, Recipe Girl, Add a Pinch, and Love & Lemons (Monika 2020). For websites, the content is around restaurants, food producers, distribution business and food material markets, recipes, as well as websites, which become video portals and video streaming like youtube, Twitter, and Reddit. Google searching for the key words food blog shows 7,420,000,000 results in 0.61 second of searching time, while for food website it shows 8,170,000,000 results in 0.53 second of searching time.

Digital technology development includes food social media development as well. Social media accounts spread food pictures, recipes, and tips associated with gastronomy. Some social media accounts managed for gastronomy digital literacy include Andrews Crivani, Julies Kitchen, and Feed Your Soul.

Applications developed by a software developer either in mobile phones or computers have a role in the digital gastronomy world. This application role is quite huge, not only for gastronomy digital literacy but also effects on lots of aspects in gastronomy activities. The development applications are as food material markets, delivery services, restaurants that enable digital transaction, restaurant management, as well as an application that enables users to share information about gastronomy. Those applications include Instagram, Pinterest, and Gojek. One of the most interesting things is the Android application as Information System of Sundanese Gastronomic for Tourism Attraction in Bandung City (Turgarini 2019).

An approach was conducted for education related to eating habits by using serious games in a form of digital novel. This digital game is not applied for entertainment context but has motivational characteristics for learning and changing habits in a more entertaining way (Holzmann et al. 2019). The game is used by schools in Germany to deliver nutrition learning material at elementary and secondary schools. Some computer games used include "The Quest to Lava Mountain," "Squire's Quest," and the "Fit, Food, Fun." Other conservation activity in digital gastronomy is gastronomy virtual tour. The development information in the digital world is related to gastronomy tourism as it contributes in digital gastronomy literacy due to strength in storytelling, to provide information about gastronomy as well as opportunity to tourists to understand and to experience gastronomy products. Some virtual tours with the theme of gastronomy are developing in the Covid-19 pandemic period; some of them are Betawi Culture and Culinary, Cirebon Culinary Virtual Tour, Cook Along Virtual Events, Trip Inspirations Virtual Tour, Sip and Learn, and Virtual Spanish Wine Challenge. Some of the consistent organizers are Atourin, Outing.id, and Devour Tours.

5 CONCLUSION

This work concludes that the definition of gastronomy digitalization is some cases related to human activities to fulfill their food needs using digital technology. A series of gastronomy activities includes not only production, distribution, and consumption about that which is done, but also the activities to reveal history, tradition, and philosophy of local food and drink; to find the origin and raw material distribution; to understand how to make, to serve, to enjoy the learning process and to gain experience; to understand nutrient content; as well as ethics and etiquette, also including food conservation. Digital gastronomy activity can develop along with accompanying technology development. This study is considered as a preliminary study for advanced research to enrich the map of the phenomenon in gastronomy digitalization that may contribute to gastronomic tourism.

REFERENCES

Bessière, J. (1998). Local Development and Heritage: Traditional Food and Cuisine as Tourist Attractions in Rural Areas. *Sociologia Ruralis*, 38, 21–34.

Bonow, M., & P. Rytkönen, (2012). Gastronomy and tourism as a regional development tool: The case of Jämtland. *Food, Hospitality and Tourism,* Vol. 2 (1): 2–10.

Daily, S (Editor). (2014). *6 Great Online Food Magazines You May Not Have Heard of Yet*. Saydaily.com.

Davies, A.R., Edwards, F., Marovelli, B., Morrow, O., Rut, M., & Weymes, M., (2017). Making Visible: Interro-gating the performance of food sharing across 100 urban areas. *Geoforum* 86, 136–149.

Eagleton, T. (1997). *Edible ecriture. The Times*, 24 October, 1997, p. 25. https://www.timeshighereducation.com/features/edible-ecriture/104281. article. Accesed 7 July 2020.

Ellmann, M. (1993). *The Hunger Artists: Starving, Writing, and Imprisonment.* Cambridge, Mass: Harvard University Press.

Everett (2009). Vernacular health moralities and culinary tourism in Newfoundland and Labrador. *Journal of American Folklore*, Vol. 122: 28–52.

Hall, C.M., L. Sharples, R. Mitchell, N. Macionis & B. Cambourne, (2003). *Food Tourism Around the World: Development, Management and Markets*. London: Routledge

Hjalanger, A.M., & Richards, G. (2002). *Tourism and Gastronomy*. Routledge, London

Holzmann, Sophie Laura, Hans Hauner, Christina Holzaphel Hanna Scafer, Georg Groh, David Alexander Plecher, Gudrun Klinker, & Gunther Scahuberger. (2019). Short-term Effects of the Serious Game "Fit, Food, Fun" on Nutritional Knowledge: A Pilot Study among Children and Adolescents. *Nutrients* 2019, 11, 2031.

Macionis JJ & Gerber LM. (2011). *Sociology.* Canada: Pearson Education Canada.

Manolis (2010). *Culinary tourism destination marketing and the food element: a market overview.* https://aboutourism.wordpress.com/tag/culinary-tourism/. Accesed 7 July 2020.

Monika. (2020). *14 Best Food Blogs and Blogger You Need to Follow*. https://blogonyourown.com/best-food-blogs/.

National Institute of Open Schooling. (2018). *Forms of Tourism*. http://oer.nios.ac.in/wiki/index.php/Forms_of_Tourism. Accesed 7 July 2020.

Oncini, Filipo, Emanuala Bozzini, Francesca Forno, & Natalia Magnani. (2020). Towards Food Platforms? An Analysis of Online Food Provisioning Services in Italy. *Geoforum. Elsevier*.

Pullphothong & Sopha (2013). *Gastronomic Tourism in Ayutthaya, Thailand.* Bangkok, Thailand: School of Culinary Art, Suan Dusit Rajabhat University.

Richards, G (2002). *Food and the tourism experience: major findings and policy orientations. In Dodd, D. (ed.) Food and the Tourism Experience*. (pp. 13–46) OECD, Paris

Santich, B. (2010). *Looking for Glavour*. Kent Town, South Australia: Wakefield Press.

Scarpato, R. (2002). Sustainable gastronomy as a tourist product. In Hjalager, A-M. and Richards, G. (Eds) *Tourism and Gastronomy*, (pp. 132–152) London: Routledge.

Schneider, T., Eli, K., Dolan, C., & Ulijaszek, S., (2018). *Digital Food Activism*. London: Routledge.

Travel Industry Dictionary (2014). *Gastro Tourism*. http://www.travel-industry-dictionary.com/gastro-tourism.html. Accesed 7 July 2020.

Turgarini, D. (2018). *Sundanese gastronomy as Tourism Attraction in Bandung City*. Yogyakarta. University Gadjah Mada.

Turgarini, D, Fajri, I & Ridwanudin. (2019). *Android Application as a Information System of Sunda Gastronomic for Tourism Attraction in Bandung City.* Bandung. National Tourism Forum.

Promoting Creative Tourism: Current Issues in Tourism Research – Kusumah et al. (Eds)
© 2021 Taylor & Francis Group, London, ISBN 978-0-367-55862-8

Tape Kareueut Teh Bohay: Students' gastronomic tourism capital Universitas Pendidikan Indonesia

Fahrudin, N.N. Afidah, F. Azis, S. Hamidah & M.W. Rizkyanfi
Universitas Pendidikan Indonesia, Bandung, Indonesia

ABSTRACT: *Tape Kareeut Teh Bohai* is a food brand made from glutinous rice. The processed rice is known as *tapai* (Indonesian) or *tape* (Sundanese). The production process of *Tape Kareeut Teh Bohai* can be used as an entrepreneurial training model for students. With entrepreneurship training through tape production, it can become a gastronomic tourism capital for students of Universitas Pendidikan Indonesia as a solution to the fact that the open unemployment rate of college graduates has increased significantly. In February 2016 the open unemployment rate (TPT) of college graduates reached 8.39% or equivalent to 144,500 people from 80,416 people (4.71%). Open unemployment is a workforce that has absolutely no job. The increase is inseparable from the economic conditions that occurred throughout 2015. The economic slowdown caused the absorption of labor from the industrial sector to fall. The problems in this article are (1) the production process of the *Tape Kareeut Teh Bohai* in Mandalamekar Village, Cimenyan District, and (2) the entrepreneurship training model for the *Tape Kareeut Teh Bohai* production as a capital for the gastronomic tourism of Universitas Pendidikan Indonesia students. Entrepreneurship training on *tape* production was carried out directly at the home industry center of *Tape Kareeut Teh Bohai* Mandalamekar Village, Cimenyan District. The method is said to be a method based on local wisdom. The results of his study show that there is a quality management of glutinous rice and the production process of glutinous rice into the *Tape Kareeut Teh Bohai* in Mandalamekar Village, Cimenyan District. The existence of entrepreneurship training models through the production of *Tape Kareeut Teh Bohai* serves as a gastronomic tourism capital for Universitas Pendidikan Indonesia students in an effort to minimize the numbers unemployed graduates of tertiary institutions in Indonesia.

Keywords: Gastronomic tourism, entrepreneurship training, and *tape kareeut teh bohai*

1 INTRODUCTION

Tape Kareeut Teh Bohai is a food brand made from glutinous rice. The processed rice is known as tapai (Indonesian) or *tape* (Sundanese). The production process of *Tape Kareeut Teh Bohai* can be used as an entrepreneurial training model for students. With entrepreneurship training through tape production, it can become a gastronomic tourism capital for students of Universitas Pendidikan Indonesia. Given the open unemployment rate of college graduates has increased significantly. "In February 2016 the open unemployment rate (TPT) of college graduates reached 8.39% or equivalent to 144,500 people from 80,416 people (4.71%). Open unemployment is a workforce that has absolutely no job. The increase is inseparable from the economic conditions that occurred throughout 2015" (Kasumaningrum 2016). The economic slowdown caused the absorption of labor from the industrial sector to fall.

The problems in this article are (1) the production process of the *Tape Kareeut Teh Bohai* in Mandalamekar Village, Cimenyan District, and (2) the entrepreneurship training model for the *Tape Kareeut Teh Bohai* production as a capital for the gastronomic tourism tour of Universitas Pendidikan Indonesia students.

DOI 10.1201/9781003095484-68

2 RESEARCH METHOD

The methodology used in this community service program is to use the action research method based on local wisdom. Action research is one of research designs that can be done in the classroom and in the community. "Action is an activity to improve something that is planned, implemented, and assessed systematically so that its validity and reliability reach the expected level of the research. Action research is also a process that includes a cycle of action, which is based on reflection, feedback, evidence and evaluation of previous actions and the present situation" (Gunawan 2006). Action research aims to contribute to practical problem solving in fundamental problematic situations and to the achievement of social science goals through joint collaboration in an ethical framework that is mutually adequate and interrelated (Rapoport 1969).

The research process is from time to time between "finding" at the time of research and "action learning". Thus, action connects theory with practice. Action research in the traditional view is a problem solving research framework, in which collaboration between researchers and clients occurs in achieving goals. "Action research as a research method, is set up on the assumption that theory and practice can be closely integrated with learning from the results of planned interventions after a detailed diagnosis of the context of the problem" (Lewin 1990). "Action is divided based on the character of the model (iterative, reflective or linear), structure (rigid and dynamic), the aim of developing organization, system design or scientific knowledge and forms of researcher involvement (collaboration, facilitative or expert)" (Baskerville 1999).

"Action research aims to obtain knowledge for specific situations or targets rather than scientifically generalized knowledge. In general, action research is aimed at achieving three things: 1) improvement of practice, 2) improvement, professional development, understanding practice, 3) improvement of the situation in which practice is implemented" (Madya 2006).

Action research occurs in real situations where problem solving is urgently needed and the results are immediately applied / practiced in related situations. In addition, it appears that, in action research, researchers conduct inquiry, management, empowerment, and development of a product or model.

3 RESULTS AND DISCUSSION

3.1 *Production process of Tape Kareeut Teh Bohai in Mandalamekar village, Cimenyan district*

Tapai or tape is one of the most popular traditional foods in Indonesia until now. *Tape* is made by the fermentation process of ingredients that contain carbohydrates namely, glutinous rice. *Tape* can last at room temperature for 2-3 days (not more than 20 degrees to 25 degrees Celsius).

Tape is made from sticky rice, both from white and black sticky rice. Black sticky rice *tape* became a common treat by the people of West Java during Lebaran. Black sticky rice *tape* has a sweet, sour and fresh taste suitable for hot weather especially when mixed with ice cream. Aside from being a complement to eating ice cream, black sticky *tape* can also be an ingredient in other food innovations. For example, ingredients for making black sticky rice *tape* ice cream, black sticky rice *tape* pudding, *dodol*, and so on. However, In addition to black sticky rice *tape,* white and green sticky rice *tape* can also be a complement to eating ice cream, eating *rangginang*, uli (*ulen*), *opak*, *es doger*, mixed ice, and others.

Here's how to make sweet, fresh and easy sticky rice *tape*. Sticky rice *tape* recipes is practical and simple.

Ingredients:

1000 grams of black or white glutinous rice
1/2 piece of yeast *tape*

How to make black sticky *tape* practical and simple: (1) wash black sticky rice thoroughly; (2) soak the black or white sticky rice for 30–60 minutes, if necessary add natural dyes from *katuk* leaves, *suji* leaves, or food coloring; (3) rinse the sticky rice thoroughly, then drain it briefly; (4) steam sticky rice with presto for 15 minutes, store it in a container; (5) pour 700 ml of clean water on the sticky rice evenly and leave it for 10–15 minutes; (6) steam sticky rice again for 15 minutes; (7) remove and drainit into a container that has been covered with clean banana leaves, put sticky rice on it; (8) let stand until cool; (9) after the sticky rice is cold, sprinkle with the yeast that has been crushed / mashed, flatten it to the entire surface of the sticky rice. Size ½ grain yeast for 1kg glutinous rice; (10) stir the sticky rice so that the yeast is mixed evenly, use clean plastic gloves or clean spoon to stir so that the sticky rice does not stale quickly after cooking; (11) stick the sticky rice in a clean container (jar, cup, plastic clip (standing pounch) and tightly closed; (12) leave the sticky rice at room temperature for three days and do not open the container until the fermentation period is complete; and (13) after three days the sticky rice *tape* is ready. You can move it to small cup containers or wrap it into small pieces using banana leaves or guava leaves, store the *tape* in the refrigerator for more pleasure.

3.2 *Entrepreneurship training model of tape Kareeut Teh Bohai production*

3.2.1 *Programs that provide students with training in entrepreneurship through the production of the Tape Kareeut Teh Bohai as their gastronomic tourism capital*

Workshop on risk ability capability

The willingness and ability to take risks is one of the main values in entrepreneurship. Entrepreneurs who don't want to take risks will find it difficult to start or take the initiative. "An entrepreneur who dares to take risks is a person who always wants to be a winner and win in a good way" (Vellas & Becherel 2008). Entrepreneur is a person who prefers more challenging businesses to achieve success or failure than less challenging businesses. Therefore, entrepreneurs do not like risks that are too low or too high. Courage to take risks should be full of calculations and realistic. Great satisfaction is obtained if the effort is successful in carrying out their duties realistically. Entrepreneurs avoid low risk situations because there are no challenges, and avoid high risk situations because they want to succeed. The choice of risk depends largely on entrepreneurial mentality.

The apprenticeship was attended by eleven students participating in the apprenticeship (students from several study programs at the Faculty of Social Sciences Education), with the hope that students would have an entrepreneurial mentality even though the number of students was minor but it was expected they would later be more motivated to become entrepreneurs. In this workshop, a simulation was made as such how to make sticky rice *tape* with selected raw sticky rice. Participants are challenged to produce more valuable and worth selling products. Next, participants were divided into three groups; the first group acted as sellers of the *Tape Kareeut Teh Bohai*, the second group acted as funders, and the third group acted as marketing agents. From these three parties, a negotiation was carried out to reach an agreement.

Workshop abstinence ability

The training was given in the form of characteristics of an entrepreneur. The following characteristics of good entrepreneurs must be developed in order to find resilient attitude, among others: (a) tenacious and disciplined; (b) independent and realistic; (c) high achievement and commitment; (d) learning from experience; (e) thinking positively and responsibly; (f) taking into account business risks; (g) finding solution; (h) planning before acting; (i) creative and innovative; and (j) effective and efficient. In addition, there are several benefits of being unyielding and resilient, namely: (1) encouraging in doing business; (2) increasing business power; (3) supporting business success; and (4) eliminating despair.

Entrepreneurial motivation workshop

"Motivation in entrepreneurship includes motivation directed at achieving entrepreneurial goals, such as goals that involve the introduction and exploitation of business opportunities" (Baum et al. 2007). Motivation to develop new businesses is needed not only by their confidence in their ability

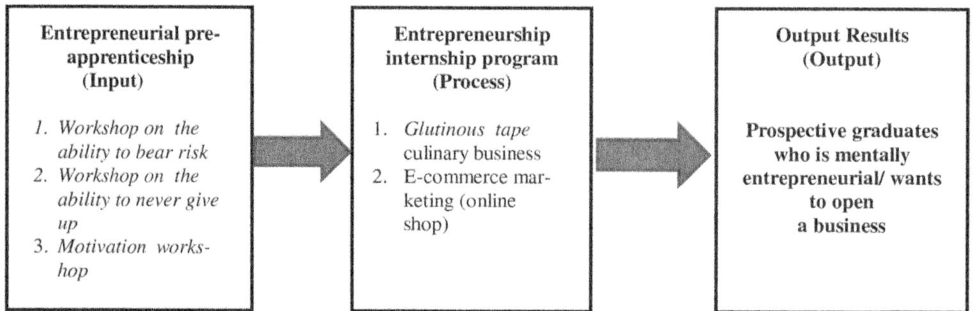

Figure 1. Form of entrepreneurship internship training model.

to succeed, but also by their ability to access information about entrepreneurial opportunities. In expectancy theory it is revealed that specific and periodic information about entrepreneurial opportunities might increase individual expectations that entrepreneurial efforts will produce the expected results, thereby increasing motivation (Figure 1).

3.2.2 *Entrepreneurship internship training process*

This entrepreneurship training begins with a workshop, in the form of providing training with material to bear the risk, the ability to never give up, and providing motivation. From this activity, it is expected that students' thinking patterns change by having an entrepreneurial mentality and ideas. After this activity, the students will intern at the center of the *Tape Kareeut Teh Bohai* home industry. In addition, students will be offered e-commerce marketing programs or online product marketing which are often known as online shops in the hope that after they take part in an internship program, their desire to become entrepreneurs will be stronger. Furthermore, all participants who take part in the internship program will see their interest in entrepreneurship and will be competed so that the students will be evaluated based on their interest in opening a business. At this stage, it is absolutely necessary that there are investors who will lend their funds for entrepreneurship, such as from the University or the banking sector. The expected outputs from this activity will produce students who have an entrepreneurial mind and are willing to open a business. The following results are expected from the internship training process:

- Produce graduates who have an entrepreneurial mind and are willing to open their own businesses;
- Produce cooperation between business partners and graduates, so that later cooperation will be established between universities, business partners, and graduates who are interested in entrepreneurship; and
- Produced an entrepreneurship training model that could be implemented as a preparation for students to become entrepreneurs.

4 CONCLUSION

The production process of the *Tape Kareeut Teh Bohai* in Mandalamekar Village is not easy. There are certain techniques that must be prioritized in the process; the level of cleanliness of glutinous rice when washing, the level of maturity of glutinous rice when cooking, the cleanliness of each container used to make sticky *tape*, and the quality of the used yeast *tape*.

Entrepreneurship programs offered in this training include: the ability to bear risk, the ability to never give up, and the ability to motivate entrepreneurship. The entrepreneurship internship training process begin with the provision of entrepreneurship skills workshops, it is then traded on the home industry *tape*. The expected outcome of this apprenticeship training activity is to produce graduates who have an entrepreneurial mind and are willing to open a business.

REFERENCES

Baum JR, Frese M, & Baron RA. 2007. *The Psychology of Entrepreneurship*. London: Routledge.

Baskerville, R.L. 1999. Investigating Information Systems with Action Research, *Communiucation of the Assosiation for Information Systems*, Volume 2, Article 19 October 1999. http://www.cis.gsu/~rbaskerv/CAIS_2_19/CAIS_2_19.html.

Gunawan. 2006. "Penelitian Tindakan Kelas Proposal, Analisis Data, Monitoring". Makalah disampaikan pada *wokshop*.

Kasumaningrum, Y. 2016. "Sarjana Pengangguran Terus Bertambah". https://www.pikiran-rakyat.com/ekonomi/pr-01255353/sarjana-pengangguran-terus-bertambah-368379.

Lewin, K. 1990. *Action Research and Minorty Problems The Action Research Reader*. (3rd ed.). (Victoria: Deakin Universty).

Madya, S. 2006. *Teori dan Taktik Penelitian Tindakan.* Yogyakarta: Alfabeta.

Rapoport, A. 1969. *House Form and Culture*. Englewood Cliffs, N.J.: PrenticeHall. 47.

Vellas, F. & L. Becherel. 2008. *Pemasaran Pariwisata Internasional: Sebuah Pendekatan Strategi*. Jakarta: Yayasan Obor Indonesia.

Promoting Creative Tourism: Current Issues in Tourism Research – Kusumah et al. (Eds)
© 2021 Taylor & Francis Group, London, ISBN 978-0-367-55862-8

Gastronomy tourism development model on a tourist village

C. Ningsih, D. Turgarini, I.I. Pratiwi & R. Fitrianty
Universitas Pendidikan Indonesia, Bandung, Indonesia

ABSTRACT: Recently, tourist villages become one unique tourism destination. One of the potential tourist villages is Cisambeng, located in Majalengka, West Java. Cisambeng has several Sentra of soya and fermented food production, such as *tempe*, *tofu*, *oncom*, and ketchup; skilled people in gastronomy; and natural tourism destination. The objective of this study is to identify and preserve the traditional Cisambeng food (gastronomy) and analyze its potential to become a tourist village destination. The method of this research is quantitative with mini-regression and qualitative method, through observation, in-depth interviews, documentation, and focus group discussion with the selected residents and gastronomy experts. The result of the study is an inventory of Cisambeng gastronomy product and destination, traveling routes, and tourist package. A design program of gastronomy tourism involved a variety stakeholder, such as some residents, government, scholars, and local and foreign tourists. Using a mini-regression analysis of 47 respondents, the results indicate that gastronomy tourism has a positive significant effect to construct the tourist perception, while tourist attraction has positive effect insignificantly. These results suggest that the curiosity about the food and its culture became the prominent perception for the tourist, with the highest score about the taste of food.

Keywords: gastronomy tourism, development model, tourist village

1 INTRODUCTION

Gastronomic tourism is part of special interest in tourism. Gastronomy is a journey to enjoy food and drinks as a prominent factor in determining the decision of a place visit Turgarini (2013). Cisambeng village, located in Majalengka, West Java, has a potential as a gastronomy tourism destination. Cisambeng village is a part of the administrative area of Palasah Sub-District, Majalengka Regency with an area of 309,529 hectares, that consists of two villages. By geography, Cisambeng village is a region with a height of 42 meters above sea level. The main local food product of this village is tofu, *tempe*, and *oncom* as a kind of fermented food. Cisambeng village has been surrounded by several potential authentic natural tourism destinations.

2 LITERATURE REVIEW

2.1 *Gastronomy tourism*

Gastronomy itself is not only focus on culinary art or cooking methods, but also on human behavior, including choosing raw materials, tasting, serving food, and experiencing, consuming, searching, studying, researching, and writing about food and everything related to ethics, etiquette, and human nutrition in every nation and country (Lilholt 2015; Manolis 2010; Pullphothong & Sopha 2013; Santich 2010; Shenoy 2005; Soeroso 2014a; Soeroso & Susilo 2013). Gastronomy is the art and science that appreciation is cross-ethnic, national, racial, group, religion, gender, and culture. It is reviewing the detail of food to be used in various conditions and situations.

 DOI 10.1201/9781003095484-69

Tourism gastronomy is relatively one of the new tourism types (Chaney & Ryan 2012). Gastronomy tourism as part of a tourism culture (NIOS 2018) that is a type of tourism or travel tour, which was designed with the principal purpose of food and drinks making as a main motivation factor by someone to do the journey. Gastronomy is often identified as food tourism, tasting tourism, or culinary tourism. According to the International Culinary Tourism Association, gastronomy tourism is defined as "the pursuit of unique and memorable eating and drinking experiences" (Manolis 2010). Thus, a gastronomy tourist is someone who is willing to travel to another place in an attempt to taste and look for the experience of the authentic local food in the area of the destination (Pullphothong & Sopha 2013).

2.2 *A gastronomy tourism route*

A gastronomy tourism route has been applied in Europe to develop the potential of agriculture areas, such as grapes, olives, palm oil plantations. According to Murgando (2013), the general activities of gastronomy tourism include visiting the area of a plantation, agriculture/livestock, production area, product processing area, museums or places providing information about the product, such as exhibition and festivals; tasting and buying products in local food area; and living in areas that relate to the route.

Preparation of lane travel or route should consider the local community, such as the owner of the culture, government area, the owner of the business, tourist, tour operators, and investors (Corigliano 2001). Gastronomy is based on attractiveness of local culture (Bruwer 2003).

2.3 *A gastronomy tourism package*

The planning results of a route travel can be applied in a package tour that provides economy value for the local community and government (Nuriata 2014). Travel package is a nontangible product combining the various services provided by different providers, thus the package is a risky decision (Cetin & Yarcan 2017; in Akova & Cetin 2017).

3 RESEARCH METHODOLOGY

3.1 *Methodology*

The method of this research is qualitative and quantitative. The quantitative approach is used in the following stages: (1) data Inventory and identification, (2) field survey, and (3) focus group discussion (FGD). It was also applied in survey and analysis of regression to measure the effect of destination attractiveness to the tourist perception.

Data inventory and identification were obtained from extracting internal/institutional secondary data. The results of FGD and observation resulted in Sundanese cuisine inventory, destinations, attractions, and local culinary product data that exist in the selected location. Data gathered was input into a data base that provides systematic and informative data about the characteristic of the products and its pictures.

3.2 *Sample*

Survey was done to check the information that was obtained from FGD. A survey was done by interviewing and documenting tourist culinary object, as well as the local culinary products that exist in Cisambeng village. The main targets of the survey are tofu, *tempe*, and *oncom* factories, home industries, and other small businesses related to local food uniqueness.

Interviews used in this research are open interviews. Researchers asked directly to the selected informant to provide an overview and information that is used to answer the problems that exist in the research. Distribution of tourist perception questionnaires could only be done to 47 respondents.

Because of pandemic Covid-19, then the deployment of questionnaires was limited and complied with the required standard health protocol.

4 RESULTS AND DISCUSSION

Cisambeng village is located in Palasah District, Majalengka, West Java, with a population of approximately 4,593 people. Most of the residents' livelihoods in this village are *tempe*, tofu, and *oncom* makers. The village has 120 tofu, *tempe*, and *oncom* factories and many other small home industries in the village that make tofu, *tempe*, and *oncom*.

Soybeans, which are the raw material for making tofu, *tempe*, and *oncom* come from US imports at a price of Rp 7,100/kg. The tofu, *tempe*, and *oncom* processors said that by using this American soybean, the tofu, *tempe*, and *oncom* were better when compared to using local soybeans. They believe that the structure of soil and water in Indonesia is not suitable for growing soybeans, and that it can change the taste of processed tofu, *tempe* and *oncom*. So, it can be concluded that there is need for further research related to local soybeans as the main ingredient in making tofu, *tempe*, and *oncom* by taking into account the structure of soil and water in Indonesia. It is important to know that the people of Cisambeng village spend hundreds of tons of soybeans every day for processing tofu, *tempe*, and *oncom*,

Community service in Cisambeng village was welcomed by tofu, *tempe*, and *oncom* businesses as well as Cisambeng village leaders. The villagers were very enthusiastic in following the event after it was held for two days. Presenters and coordinators of the Cisambeng village PKM were invited to tour to see the factory and the process of making tofu, *tempe*, and *oncom*. The village leaders had the idea that the Cisambeng village's tofu, *tempe*, and *oncom* products needed the latest innovation because they saw the reality on the ground that the residents of other villages located around the Cisambeng village also had begun to pioneer the tofu, *tempe*, and *oncom* production business and compete with the villagers of Cisambeng. If innovation is not carried out on these products, then one day the identity of the village of Cisambeng as a tofu, *tempe*, and *oncom* processing village will disappear and lose its competitiveness with other villages in the vicinity. This encourages village leaders to create mini-tofu, *tempe*, and *oncom* laboratories where the laboratory functions to develop product innovations starting from the way of selecting materials, processing, to packaging that is unique and different.

The PKM presenters and coordinators of course visited this mini-laboratory and tried to know what was produced. The taste and texture of the tofu is really delicious and soft when compared to ordinary tofu. However, because the laboratory was not long built, the product that was just being developed was tofu. *Tempe* and *oncom* are other products that they will develop as soon as this tofu product is successful. Cisambeng village leaders also want to make product innovations from tofu, *tempe*, and *oncom* ingredients such as brownies, pudding, and other types of food that can help the community's economy. They also took the initiative to stabilize the prices of tofu, *tempe*, and *oncom* so that every citizen competed healthily in marketing their products. Cisambeng village is also expected to become one of the gastronomic tourist destinations with its mainstay products of tofu, *tempe*, and *oncom* as well as other innovative development product.

The questionnaire involved 46 respondents, which consisted of 25 men and 21 women, in which respondents aged average 17 to 24 years as much as 74.5%, 8.5% aged 25 to 32 years, and the remaining 30 years until more from 41 years old. Respondents, 85% came from Majalengka; the rest reside in Cimahi, Sukabumi, Bandung, and outside the country, Japan. The respondents were identified to have different education, 65.9% graduated from college, and 39.1% graduated SMA/SMK. The respondents' occupational backgrounds are identified as 45.7% students, the other 28.3%, 15.2% private employees and the rest of civil servants and entrepreneurs. The frequency of visits of respondents are as much as one time 45.7%, more than four times as much as 30.4%, and 23.9% had been as much as two to three times. The average reason for visiting is for a tour, or just buying tofu and *tempe* and to visit family.

Based on the data of respondents, the results of the regression as follows.

Table 1. Inventory potential of tourism gastronomy typical village Cisambeng.

Potency	Item
Food Local Food	– White raw tofu – Yellow raw tofu – Sumedang fried tofu – Oncom longish – Oncom kapes – Bobotoh knows – Bobotoh Dage – Surabi soy sauce pulp – Surabi oncom – Sticky coconut sangria – Kecimpring – Opak – Sauteed soy sauce pulp
Food Innovation	Pudding tofu
Things to culture	The culture of eating sissy
Homestay	Eat lunch rice liwet bancakan
Things to home industry	– 120 tofu, *tempe*, and *oncom* factories – Home industry Opak, Kicimpring, Rangginang, Crackers Inul
Cooking class	– Package menu breakfast, Rice Lengko Majalengka, consisting of rice white, sliced cucumber, onion red cooking, pie fried, tempeh fried, know fried bean sprouts fried, soy typical Majalengka – Package menu eating lunch, Rice Liwet Bancakan, the food is placed in the upper leaves of the banana, consisting of see white fried cisambeng, *tempe* fried cisambeng, sautéed kale, salad leaves the distance, fish salty, chili paste, rice liwet, jengkol fries, salad cucumber. – Package menu eat dinner, consisting of Rice Liwet , pedesan Entog, sauteed Dage (Tofu Dregs) Cisambeng, Oncom Gondrong Santan (Cisambeng), sauteed Tofu White Leaves Katel Cisambeng

Table 2. Summary model.

Model	R	R Square	Adjusted R Square	Std. Error of the Estimate
1	.827[a]	683	669	3.64427

a. Predictors: (Constant), Travel gastronomy (X2), Power Pull Tourism (X1)

The values prove that X1 and X2 significantly influence Y is at 0.000 < 0.05 and value of F observed 46 405 > F table 3.20 so it can be concluded there is the effect of X1 and X2 simultaneously to Y.

By using an error rate of 0.05, if the value of sig < 0.05 or t arithmetic < t table then there is the effect of variable X on variable Y.

- Known value of Sig. for the effect of X1 on Y is 0.115 > 0.05 and the value of t is 1.610 < t table 1.678, so it can be concluded that X1 has no significant effect on Y.
- Known value of Sig. for the effect of X2 on Y amounted to 0,000 < 0.05 and the value of t count 4.691 > t table 1.678, so it can be concluded that there is an influence between X2 on Y.

Table 3. ANOVA.

Model		Sum of Squares	df	Mean Square	F	Sig.
1	Regression	1232,583	2	616,292	46,405	.000[b]
	Residual	571,069	43	13281		
	Total	1803,652	45			

a. Dependent Variable: Perception Travelers (Y)
b. Predictors: (Constant), Travel gastronomy (X2), Power Pull Tourism (X1)

Table 4. Coefficients.

		Unstandardized Coefficients		Standardized Coefficients		
Model		B	Std. Error	Beta	t	Sig.
1	(Constant)	5,595	3,564		1,570	.124
	Power Pull Tourism (X1)	.247	.154	.221	1,610	.115
	Gastronomic Tourism (X2)	.399	.085	643	4,691	.000

a. Dependent Variable: Perception Travelers (Y)

4.1 *Gastronomic travel route in Cisambeng village*

Gastronomic tour routes in Cisambeng village are using a decentralized pattern. The activity is centered in Cisambeng village, which is because of the following (Tables 1–4):

- Having easy accessibility because it is close to the toll road access. Likewise, the village has good road conditions.
- There are a large number of tofu, *tempe*, and *oncom* home industries, totaling as many as 120 factories.
- The potential for outlets, gastronomic tourism centers in the Cisambeng village is managed by Bumdes.
- The existence of a mini-lab owned by the Tau Factory community that can become a gastronomic educational tourism center.
- The existence of an active and committed community in developing gastronomic tourism in the Cisam-beng village.
- There are facilities and infrastructure that can be developed into a homestay.
- The presence of PKK figures who have expertise in processing typical foods of Cisambeng Village
- Cisambeng village can become a center of gastronomic tourism and also has potentials that it can be developed in the future.
- The existence of affordable natural tourism borrows from Cisambeng village such as the Panyaweuyan Terasering valley tour, Cipasung Tea Plantation, Paragliding Sports Tourism in Mount Panten, Curun Argapura Waterfall, Cipetey Waterfall, Cipadung, Talaga Nila, Talaga Herang, and Ciboer Pass.

There is a gastronomic tourism education tour package that can be done in Cisambeng village, which is a two-day and a night package.

5 CONCLUSION

Cisambeng village is located in Majalengka, West Java. Cisambeng has several Sentra of soya and fermented food production, such as *tempe*, tofu, *oncom*, and ketchup; skilled people in gastronomy;

and is a natural tourism destination. The objective of this study is to identify and preserve the traditional Cisambeng food (gastronomy), then analyze its potential to become a tourist village destination. The method of this research is quantitative with mini-regression and qualitative method, through observation, in-depth interviews, documentation and focus group discussion with the selected residents and gastronomy experts. The results of the study is an inventory of Cisambeng gastronomy product and destination, traveling route, and tourist packages. A design program of gastronomy tourism involved a variety of stakeholders, such as some residents, government, scholars, and local and foreign tourists. Using a mini-regression analysis of 47 respondents, the results indicate that gastronomy tourism has positive significant effect to construct the tourist perception, while tourist attraction has positive effect insignificantly. These results suggest that the curiosity about the food and its culture became the prominent perception for the tourist, especially the taste of food.

REFERENCES

Dewi Turgarini. 2018. Gastronomi Sunda Sebagai Daya Tarik Wisata di Kota Bandung. Yogyakarta. Universitas Gadjah Mada.

Dewi Turgarini, Andreas Suwandi, Ilma Prastiwi, Oce Ridwanudin. 2019. Yogyakarta. Prosiding Seminar Keunggulan Kuliner Indonesia Universitas Gadjah Mada 4–5 September 2019.

Eugen, Pauli. 1979. *Classical cooking The Modern Way.* Boston, Massachuetts USA: CBI Publihsing Company Inc.

Hiebing Jr, Roman G and Scot W. Cooper. 2004. *The One Day Marketing Plan 3rd Edition: Organizing and Completing Plan That Works.* Mc Graw Hill

Kennedy, Daniel. 1991. *The Ultimate Marketing Plan.* Holbrook: Bob Adams, Inc.

Kotler, Philip and Gary Amstrong. 2012. Prinsip-prinsip Pemasaran. Edisi 13. Jakarta: Erlangga

Laundberg, Donal. 1992. *Understanding Baking.* New York: Van Norstand Reinhold

Louw, Alice & Michelle Kimber. 2007. The Fower Of Packaging. Jurnal Ilmiah. www.insglobal.com/assets/files/The_power_of_packaging_pdf (diakses 18 Juni 2020)

Luther, William M. 2001. *The Marketing Plan: How to Prepare and Implement It.* New York: Amacom Books

McDonald, Malcolm and Adrian Payne. 2006. *Marketing Planning for Services 2nd Edition.* Oxford: Elsevier

Menteri Kesehatan Republik Indonesia. 2011. *Peraturan Menteri Kesehatan Republik Indonesia Nomor 1096/MENKES/PER/VI/2011 Tentang Higiene Sanitasi Jasaboga.* Jakarta (ID): Menteri Kesehatan Republik Indonesia.

Nykiel, Ronald A. 2003. *Marketing Your Business: A Guide to Developing A Strategic Marketing Plan.* New York: Best Business Books

Parantika A, Hurdawaty R. 2018. *Sanitasi, Hygiene dan Keselamatan Kerja.* Jakarta (ID): Penerbit Erlangga

Peraturan Pemerintah Republik Indonesia. 1999. *Peraturan Pemerintah Republik Indonesia Nomor 69 Tahun 1999* Tentang Label Dan Iklan Pangan. Jakarta (ID): Peraturan Pemerintah Republik Indonesia.

Sundoko, L. 2007. PUIR: Keripik & Kerupuk Tahu – Tempe. Jakarta: Gramedia Pustaka Utama

Susuwi S. 2009. *Sanitation Standard Operating Procedures.* Bandung (ID): Universitas Pendidikan Indonesia.

Stratus, Ralf E. 2008. *Marketing Planning by Design: Systematic Planning for Successful Marketing Strategy.* Cornwall: John Wiley & Sons.

Tumpal, Daniel. 2000. *Konsep Dasar Perencanaan Pemasaran dan Proses Penyusunannya.* Modul Perencanaan Pemasaran. Universitas Terbuka

Untari, D. 2013. Bukan Sekedar Tahu Goreng Biasa. Jakarta: Gramedia Pustaka Utama

Westwood, John. 2006. *How to Write A Marketing Plan 3rd Edition.* Cornwall: MPG Books Ltd.

Wirya, Iwan. 1999. Kemasan yang Menjual. Jakarta : PT. Gramedia Pustaka Utama

Promoting Creative Tourism: Current Issues in Tourism Research – Kusumah et al. (Eds)

Exploring the expectation of youth purchasing intention for street food as gastronomy tourism in Bangsaen, Thailand

P. Nakpathom, K. Chinnapha & P. Lakanavisid
Burapha University International College, Thailand

M.R. Putra
Burapha University International College, Thailand
Trisakti School of Tourism, Jakarta, Indonesia

A. Wongla, A. Kowarattanakul, N. Pangket, P. Thanuthep & S.H. Rui
Burapha University International College, Thailand

ABSTRACT: This study aims to explore the expectation of youth purchasing intention for street food as a gastronomy tourism in Bangsaen, Thailand. The samples were 400 participants who purchased street food in Bangsean, Thailand, from March to April 2020. The data collection method was based on online questionnaires. For data analysis, the statistical methods rely mainly on descriptive statistics, exploratory factor analysis, and linear regression analysis. There are four factors of youth purchasing expectation: (1) value of money and infrastructure management, (2) hygiene and cleanliness, (3) integrated brand promotion, and (4) recommendation. Moreover, the linear regression model was developed to predict the youth purchasing intention for street food as gastronomy tourism in Bangsaen, Thailand. The results show that the hypothesized antecedent variables were statistically significant to customer loyalty, particularly the value of money and infrastructure management as ($\beta = 0.370$, $p = 0.000$) and recommendation as ($\beta = 0.290$, $p = 0.000$).

Keywords: purchasing expectation, purchasing intention, street food, youth behavior, gastronomy tourism

1 INTRODUCTION

Thai food has become one of the world-famous cuisines for its succulent coloring and unique taste. The renowned accumulation of sweet, sour, and salty makes Thai food distinctive from others. Currently, Thai street food is very popular and famous for its being low-priced as well as being highly demanded by the local Thai people. This is because the majority of the people who are in industrialized forces prefer to have their daily meals outside their home. Street food can be found commonly in most cities and towns in many developing countries. Aside from ready-made instant pre-prepared-cooked meals at relatively inexpensive prices, local people are attracted to street foods because of its gustatory attributes or tastes. These attributes are linked to the culinary expertise of the vendors (Alimi 2016). Street food in Thailand brings together various offerings of ready-to-eat meals, snacks, fruits, and drinks sold by hawkers or street vendors at food stalls or food carts along street sides throughout Thailand. Tasting of Thai street food is a popular activity for visitors, as it offers a taste of Thai cooking traditions. Bangkok is often mentioned as one of the best places for street food. Many areas in Bangkok that provide plenty of street food include various areas, such as Yaowarat, Talat Noi, Wat Traimit and Chaloem Buri, Nang Loeng, Sam Phraeng, Pratu Phi, Bang Lamphu, Yotse, Sam Yan, Tha Din Daeng, Wongwian Yai, Wang Lang, and Talat Phlu, are among popular examples (Wiens 2020)

 DOI 10.1201/9781003095484-70

In the eastern region of Thailand, there are many well-known tourist destinations, especially famous beaches in various provinces of Chonburi, Chanthaburi, and Rayong. Bangsaen beach is located in Chonburi province and is well known to both Thais and foreigners as one of most popular beach destinations in the eastern region of Thailand. With many tourist destinations to see, visitors do not only travel to these places, but they also wish to explore local dishes and tastes. Various dish items served in Bangsean are mainly based on using seafood ingredients, as being recognizable by tourists for its multiplicity, freshness, deliciousness, and fragrance (Gulasirima et al. 2019). Moreover, Bangsaen's dishes are adapted between local food ingredients mixed with other foreign dishes that produce fusion dishes satisfying the needs of the Thai people. There are many factors that affect customers' expectations toward street food purchasing intention. These can be verified as cleanliness and hygiene, freshness of ingredients, tastes, reasonable priced, and food quality (Hsu & Pham 2015).

Today, there are many places where street food can be found in Bangsaen. Bangsaen has captured the hearts of many teenagers with its convenient location, fantastic nightlife, and beautiful scenery of tourist attractions. The food also plays an important part that attract teenagers. The new generation are always full of curiosity. They like to experience new things. Also, variety of food items becomes a powerful tool that attracts visitors because they enjoy a variety of tastes from all sources of different food selection as well as experiencing Bangsaen's lifestyle and values from local communities and tourist destinations. Overall, this study aims to explore the expectation of youth purchasing intention for street food as a gastronomy tourism in Bangsaen, Thailand.

2 LITERATURE REVIEW

2.1 *Purchasing expectation*

Customer purchasing expectation is the probability of expectation to purchase, or willingness to obtain a good or service in the future. This also refers to a repetitious purchase or returns to uses the goods or services to a reference brand name (Hsu & Pham 2015). Customer expectations are belief about service delivery that serves as standards or reference points against which performance is judged (Nguyen 2017). Nowadays, customers have a wide range of selection, whether it is service or tangible products to fulfill their needs. The expectations will influence the purchasing decision where customers will make their decision to return to the offers or not. If the expectation is low, the business can satisfy the current customers, yet fail to attract more customers who have higher expectations. On the contrary, the too-high expectations will lead to disappointed customers. Customer expectation is one of the keys that affect the satisfaction as well as developing and managing customer relationship.

2.2 *Purchasing intention*

Purchasing intention refers to an individual's readiness and willingness to purchase a certain product or service, and this can influence the purchasing decision of consumers in the future. This can be considered as one of the mechanisms of consumer cognitive behavior on how a consumer intends to purchase a certain product. Additionally, purchase intention is the direct originator of behavior. Purchase intention is the state where an individual is willing to perform a behavior. Defined actual purchase behavior is the individual's willingness to purchase specific manufactured goods or services. Purchase intention was the exchange after customers assessing the whole products such as the states of mind toward purchase behaviors. From all of the theories of purchase intention mentioned above, the researcher can conclude that purchase intention is a process that involves motivation and evaluation before deciding to purchase a certain item or product (Hsu & Pham 2015).

2.3 *Factors on consumer's attributions of purchasing intention*

Nowadays, there are many factors that affect the intention of customers in purchasing street food, including a trigger, which means any that stimulates a consumer to buy a product of a particular brand. This may be an attractive advertising, some special packaging, or any particular attribute of the product which attracts the customer toward the product (Shahid et al. 2017).

Price sensitive consumers are attracted more toward the cheaper products, but the consumers prefer quality over the price; they are less likely to buy the low-cost product regardless of its quality. These consumers think that only expensive products are high quality. An attractive package also attracts the consumers toward the product. A well-packed and advertised product will always be preferred over a poorly packed product (Shahid et al. 2017). Price plays two different roles in consumers' assessments of product alternatives, namely, as a measure of sacrifice and an informational cue. Price may also cause an evaluation of the price image of the item such as prestige street food and lower price (Yanisa & Suwaree 2019). The factor of price has a significant influence on consumers with regards to undertaking price judgments, evaluating the attractiveness of advertised promotions to improve or reduce price search and comparison behaviors (Hsu & Pham 2017).

The perceived quality refers to the consumer's assessment of the performance or superiority of a product instead of the product quality itself. Including the reputation, brand image, cleanliness, and pricing, while the intrinsic characteristics refer to the physical characteristics of the product, such as product characteristics, taste, size, shape, ingredient used, and mouthwatering food items (Mohamed et al. 2019). There are many elements that affect customers' purchase of food, including physical environmental hygiene, food quality, service quality, price and value, lifestyle, emotion, and satisfaction. Product factors include taste, attractiveness, smell, hygienic, variety, freshness, texture, juiciness, and service temperature. To a large extent, customers' decisions and purchasing behaviors depend on the satisfaction evaluation of the overall emotional experience and performance of products and services. According to the taste, appearance, and variety of food are the basis to attract customers to purchase. In addition to the taste and quality of food, hygienic issue is also an important criterion when customers purchase food (Xu 2019).

Places and location are critical aspects to the success of street food. With the diversifications of street food types in urban areas, locations of street foods have become more complex than in the past. At the same time, with rapid urbanization and improved accessibility and comfortability in different regions, transportation becomes a key factor affecting locations of street food.

Vendors' characteristics are pointed out that consumer's purchase decision is affected by their previous experience with the food, which can be perceived by sensory organs. If the seller creates a satisfying experience for consumers, it would be easy to have return customers. Dimensions of service quality include the speed and efficiency a service presents, their friendliness, and the tendency to help. According to Theeramunkong et al. (2017), personal hygiene of street vendors, as well as the environment in which food is being prepared, is the key to the cleanliness of food. Although some street food vendors are aware of personal hygiene, they do not understand key aspects of personal hygiene, such as cleaning food surfaces and controlling the temperature at which food is cooked. These will affect the health aspect of food.

Another important factor that affects the purchase intention is recommendation, which refers to a recommendation from the people using "a word of mouth" as a tool for creating positive or negative attributes as well as other reliable sources of recommendation that can help customers to make decisions on purchasing street food.

2.4 *Street food*

The definition of street food refers to ready food or drink sold along the street or other public places, such as markets or fairs. Sellers often sell food from mobile food booths or food trucks. Street food is a variety of ready-to-eat food and beverages and sometimes is prepared in public places, especially on the street, including food from whole grains and fruits, cooked meat, and drinks. This is often sold in busy public areas such as sidewalks, streets, markets, schools, bus and train stations, beaches, parks, and other public spaces, including land or water, especially on the street. Food is served in the smallest amount for one person with a take-home container. The materials for food containers are made from variety of substances such as disposable plastic, paper, or foam that come in numerous sizes and shapes of crockery, glassware, and eating utensils. The idea is to serve and prepare as quickly as possible. Street food is usually eaten by hand or can be eaten while walking (Rana & Ahirrao 2016).

A wide variety of street food has become part of Thai consumers' daily diet, especially those that belong in student populations. Preparing and cooking at home can be very laborious and time-consuming. In addition, those in the working-class population are competing against each other to have better living conditions. Most people struggle to succeed all the time and without enough energy left to do other activities including preparing their own daily meals at home. Moreover, many people consider the time to study more valuable than time for cooking (Thanh 2014). Consequently, the knowledge and ability to prepare and cook Thai cuisine are decreasing. Furthermore, many households in urban areas have limited space or no kitchen area to prepare and cook food; 20% of the households in Bangsaen do not have a kitchen. As a result, purchasing prepared food or take-out food is usually considered as the best solution for this scenario.

One main reason why people prefer to eat street food is because street food is the foundation of many modern dishes that we have today. Buying street food is perfectly idealistic for gaining a true experience that relates a country's identity and culture. Street food can be found everywhere throughout Thailand. You can literally find street food in every corner of the street. Experiences deriving from buying street food can be fun, cheap, and can be regarded as exotic adventures. It is the best way to experience truly traditional Thai food. Examples of famous street food dishes are Pad Thai, papaya salad, and barbecued seafood (Campusintern 2019). Houses with no kitchen facility and cooking equipment or having issues with buying raw ingredients to prepare and cook food at home can be overcome by purchasing ready-prepared-cooked street food. For foreign visitors to Thailand, your priority is to travel and gain experience on Thai culture or enjoy Thai street food. From noodles to curries, soups to salads, dumplings to spring rolls, and roti to sticky rice, you could spend weeks to taste and experience numerous samples of Thai cuisine. Some of the top street food can be found in Bangkok or elsewhere in other parts of Thailand. This includes Pad Thai, Pad See Ew, Massaman curry, papaya salad, banana roti, and mango sticky rice (Rhodes 2018). Therefore, off-home-premise food consumption or dining out from home has become a common popular interest at any level of Thai society. The term take-out or take away covers food that is eaten elsewhere or food that is bought and then eaten at home. Traditionally, during lunch time, food is bought and consumed outside, whereas for dinner, food is served at home with the family (Khongtong 2016). Eating out on the street is more convenient than eating inside a restaurant, as street food is available at any time of the day and there is a broad range of variety to select from (Rishad 2018).

2.5 *Youth behavior*

The category of the young consumer is a relatively new concept, which originated during the period of marketing development and resulted from the need to define target markets for enterprises' products. An important element of the term "young consumer" is the definition of the age bracket that is categorized into market segments. Regardless of their income level, Bangkok residents consume street food on a regular basis and, therefore, the extra expenditure they would incur in its absence could have a detrimental impact on their expenditure and consumption pattern, particularly in the case of low-income households. The United Nations (UN) defines youth as those aged between 15–24; hence these age groups are the research's primary focus (Ansell 2017). Street foods are widely consumed and produced in almost every country around the world, as a result of nutritional trends in urban areas. The history of ready-to-eat food, those that are offered on the street, can be stretched back to ancient times. It is becoming popular nowadays not only in cities and towns in developing countries but also in many places of European and North American agglomerations too. Street food is preferred by consumers, especially students, because it is cheap, very easy to eat, and is easily obtainable throughout the cities from kiosks or vans that are located along street markets, festivals, and especially in popular, well-known destinations with a high volume of tourists or visitors (Bellia et al. 2016). Today, more and more young consumers in Thailand prefer to eat street food rather than eating at any restaurant.

2.6 *Bangsaen*

Bangsaen is known as a beach destination close to Bangkok, approximately 100 kilometers or 1.30 hours by car from Bangkok. Main tourist attractions are beaches, mountains, temples, cafés, and seafood restaurants by the beach. Moreover, Bangsaen has many street food outlets that attract teenagers throughout the day, especially during early evening to late at night. In recent years, Bangsaen has re-emerged as a trendy district driven by demands of the young generation who gather around the Burapha University (Jansuttipan 2015). Hence, street food has become part of the people and young consumers' lifestyle while gaining easy access to the street food vendors. It becomes a common norm and cheapest way to explore a variety of food and different street food outlets. This is an excellent option for seeking street food experiences in Bangsaen. Bangsaen offers hundreds of Thai street food varieties and vendors that are ready and willing to prepare whatever your preferred taste of desired expectation. Along the beach, many street food venders offer a variety of food such as grilled chicken, grilled squid, roasted whole shrimp, and papaya salad (Wiens 2020).

3 RESEARCH METHODOLOGY

The purpose of this study is to explore the expectation of youth purchasing intention for street food as a gastronomy tourism in Bangsaen, Thailand. This is a quantitative study which focuses on an exploratory factor analysis (EFA) and linear regression analysis.

Researchers developed a list of questionnaires to be reliable and valid. The useful questions from previous studies as stated and cited in a literature review were adapted to the questionnaire. The developed questionnaire was validated by three specialists in hospitality and tourism industries, humanity, and social science and psychology. The questionnaire consists of five sections. The first section focuses on demographic information, which consists of six questions. The second section is about youth behavior. This part consists of three questions. The third section is based on youth purchasing expectation attributes that consists of 32 questions. The fourth section is on the youth purchasing intention that contains three questions. The fifth section is related to research recommendation. All items were measured by a 5-point Likert scale ranging from 1 to 5; 5-Very high expectation, 4-High expectation, 3-Moderate, 2-Low expectation and 1-Very low expectation.

The samples were collected from 400 participants who were randomly selected. The data collection was completed within two months, from the beginning of March to the end of April 2020. The EFA was used while statistical technique was applied for data collection, where an Eigenvalue greater than 1 was accepted and only attributes with factor loading higher than or equal to 0.7 in the final structure of analysis. The last analytical methodology was the use of linear regression analysis that identifies the youth perceived to forecast the trend in the future.

4 DATA ANALYSIS AND RESULTS

There were 400 copies of the questionnaire completed by a group of young people or youth in Bangsaen who were randomly selected, ranging from age between 15 until 24 years old; educational level, religion, income, and region of hometown. The result of the sample's demographic characteristic was analyzed in a percentage as shown in Table 1.

From Table 1, the sample demographic shows 42.7% males and 57.3% females, noting that females are more than males. The average age of the sample was between 15 to 24 years old, noting

Table 1. Demographic information.

Demographic Information		Frequency	Percent
Gender	Male	171	42.7
	Female	229	57.3
Age	15	3	0.8
	16	4	1.0
	17	5	1.3
	18	17	4.3
	19	40	10.0
	20	102	25.5
	21	123	30.8
	22	58	14.5
	23	33	8.3
	24	15	3.8
Education Level	Less than Bachelor's degree	109	27.3
	Bachelor's degree	288	72.0
	Higher than Bachelor's degree	3	0.8
Religion	Buddhism	340	85.0
	Christianity	27	6.8
	Muslim	12	3.0
	Other	21	5.3
Income	Lower than $99	115	28.8
	$100–$299	148	37.0
	$300–$499	76	19.0
	$500–$699	36	9.0
	More than $700	25	6.3
Region of hometown	North	9	2.3
	Northeast	36	9.0
	West	12	3.0
	Central	114	28.5
	East	213	53.3
	South	16	4.0
	Total	**400**	**100.00**

that the age of the sample ranged from highest to lowest as follows: 21, 20, 22, 19, 23, 18, 24, 17, 16, and 15 years old, with 30.8%, 25.5%, 14.5%, 10.0%, 8.3%, 4.3%, 3.8%, 1.3%, 1.0%, and 0.8%, respectively. The highest number of education level is a bachelor's degree (72.0%), less than a bachelor's degree (27.3%), and higher than a bachelor's degree (0.8%). Most of participants respect Buddhism at 85.0%. Highest income is between $100–$299 at 37.0%. The hometown of 216 participants is located in the Eastern region at 53.3%.

From Table 2, youth behavior in Bangsaen shows the characteristics of 400 sampled as follows: the average frequency on how often you purchase street food per week of sample three to four days per week at 43%. Most sample the time that they do purchase the street food at dinner at 73.3% and the sample group they pay per time at the street food around 50–100 baht at 57.8%.

After the researcher summarized the third part of the questionnaire by using EFA with Varimax Rotation, the result was conducted to create correlated variable composites from the original 28 attributes and identify a smaller set of factors, which explained most of the variances between the 32 attributes. The factors were retained if they had Eigenvalues greater than or equal to 1.0, and attributes were retained only if the factor loading was greater than or equal to 0.7. The Eigenvalues

Table 2. Youth behavior.

Youth Behavior		Frequency	Percentage
How often do they purchase street food per week	1–2 days per week	132	33.0
	3–4 days per week	174	43.5
	5–6 days per week	42	10.5
What time do you purchase street food	Every day	52	13.0
	Breakast	25	6.3
	Lunch	69	17.3
	Dinner	293	73.3
	Other	13	3.3
How much do you pay for street food per purchasing	Less than 50 baht	49	12.3
	50–100 baht	231	57.8
	101–150 baht	89	22.3
	More than 151 baht	31	7.8

Table 3. Factor of purchasing intention.

Factors	Loading
Component 1: Purchasing decision (Eigenvalues = 13.736 and Variance = 49.056)	
Street food being easy to accessibility	0.849
Street food being easy for availability	0.846
Street food are convenient to transport	0.836
Variety of street food	0.755
Cheap price	0.741
Reasonable price	0.722
Texture of street food products	0.700
Component 2: Hygiene and Cleanliness (Eigenvalues = 3.282 and Variance = 11.720)	
Physical environmental hygiene of street food	0.823
Food quality of street food	0.817
Cleanliness of street food	0.789
Hygienic street food	0.780
Ingredients used for street food	0.753
Component 3: Attractive (Eigenvalues = 1.609 and Variance = 5.747)	
Attractive packaging of street food	0.801
Attractive advertising	0.799
Advertise promotions of street food	0.776
Component 4: Recommendation (Eigenvalues = 1.043 and Variance = 3.725)	
Online Recommendation (Bloggers, Youtuber)	0.837
Social media recommendation	0.812
Personal recommendation	0.709

suggested that the four-factor solution with 28 attributes explained 70.249% of the overall variance was appropriated. This is summarized in Table 3.

Factor of purchasing decision is the most important for youth purchasing intention with 13.736 of Eigenvalues and variance at 49.056. The attributes of this factor are 1) Street food being easy accessibility; 2) Street food being easy availability; 3) Street food is convenient to transport; 4) Variety of street food; 5) Cheap price; 6) Reasonable price; and 7) Texture of street food products that are presented at the factor loading at 0.849, 0.846, 0.836, 0.755, 0.741, 0.722, and 0.700, respectively. *Factor of hygiene and cleanliness* has Eigenvalues at 3.282 and variance at 11.720. The attributes in this point are 1) Physical environmental hygiene of street food; 2) Food quality

Table 4. The Result of linear regression about the youth expectation toward youth purchasing intention.

	Unstandardized	Coefficients Std. Error	Standardized Coefficients Beta	P
Purchasing decision—purchasing intention	0.386	0.049	0.370	0.000**
Hygiene and cleanliness—purchasing intention	0.026	0.046	0.032	0.578
Attractive—purchasing intention	0.052	0.043	0.065	0.234
Recommendation—purchasing intention	0.231	0.041	0.290	0.000**

of street food; 3) Cleanliness of street food; 4) Hygienic street food; and 5) Ingredients used for street food that show the factor loading at 0.823, 0.817, 0.789, 0.780, and 0.753, respectively. *Factor of attractive* has Eigenvalues at 1.609 and variance at 5.747. The attributes of this factor are 1) Attractive packaging of street food; 2) Attractive advertising; and 3) Advertise promotions of street food that show the factor loading at 0.801, 0.799, and 0.776, respectively. *Factor of recommendation* has Eigenvalues 1.043 and variance at 3.725. The attributes of this factor are 1) Online recommendation (Bloggers, Youtuber); 2) Social media recommendation; and 3) Personal recommendations that show the factor loading at 0.837, 0.812, and 0.709, respectively.

Table 4 showed the results of linear regression and the relationship of coefficients youth expectation toward youth purchasing intention. Coefficients youth expectation toward youth purchasing intention was answered by obtaining the linear regression weights and the p-values for these weights. In Table 5, the results of the linear regression analysis are displayed. The estimated linear regression weights, standard errors, and p-values for all the predictors are given. Safety and Taste were significant predictors of purchasing decision ($\beta = 0.370$, $p = 0.000$) and Recommendation ($\beta = 0.290$, $p = 0.000$), respectively. These two significant variables have positively affected customer satisfaction. However, Hygiene and Cleanliness ($\beta = 0.032$, $p = 0.578$) and Attractive ($\beta = 0.065$, $p = 0.234$) were not significant as predictors of customer satisfaction.

5 DISCUSSION AND CONCLUSION

This study explored the affected factor between youth purchasing expectation and purchasing intention on street food in Bangsaen. This investigation provided some interesting and revealing insights in relation to understanding the satisfaction and expectations of customers to support the street food in Bangsaen. These outcomes were to improve and develop a better quality of street food in Bangsean.

From this research, we found that there are four important factors affecting youth purchasing expectation toward purchasing intention on street food in Bangsaen. All four factors, as shown in Table 3, include factors of purchasing decision, factors of hygiene and cleanliness, factors of attraction, and factors of recommendation. This also agrees with the research results of Xu (2019) that the customers' decisions and purchasing behaviors depend on the satisfaction evaluation of the overall emotional experience and performance of products and services. According to taste, appearance and variety of food are the bases to attract customers to purchase. In addition to the taste and quality of food, hygienic is also an important criterion when customers purchase food. This finding has the same results as that which indicates an important factor that affects the purchase intention is recommendation. This refers to a recommendation from the people or a word of mouth and a reliable source that may influence customer decision making to purchasing street food.

Moreover, the result of linear regression was about the affecting toward youth purchasing decisions. For four factors, Purchasing decision, Hygiene and Cleanliness, Attractive, and Recommendation were significant predictors of the youth purchasing intention, respectively. These four significant variables have positively affected the youth purchasing intention. However, Hygiene and Cleanliness and Attractive are not as significant as predictors of the youth purchasing intention.

The results of this study provide essential information for understanding youth purchasing expectations toward purchasing intention on street food. Secondly, the results of this study will help to improve and can be defined as a supportive written tool to develop better street food in Bangsaen, Thailand. Finally, this research offers the essential elements for a better understanding of customers' purchasing of street food.

REFERENCES

Alimi, B. 2016. Risk factors in street food practices in developing countries. *Food Science and Human Wellness, 5*: 141-148.

Ansell, N. 2017. *Children, Youth and Development*. New York, NY: Routledge.

Bellia, C., Pilato, M., and Seraphin, H. 2016. *Street food and food safety: a driver for tourism?* Retrieved from https://www.researchgate. net/publication/300042432_Street_food_and_food_safety_A_driverfortourism

Campusintern. 2019. *Explore: 7 Must Eat Street-Food In Phuket, Thailand.* Retrieved from http://scoop.coop/explore-7-must-eat-street-food-in-phuket-thailand/

Gulasirima, R., Rasamipiboonb, N., Pichaiyongvongdeeb, S., Chantrarasanamb, C. and Yambunjongc, P. 2019. Developing Potential of Restaurant Entrepreneurs in Pattaya Using Process of Knowledge Management of Local Wisdom on Foods. *Journal of Multidisciplinary in Social Sciences, 15*(1): 32-42.

Hsu, Y. and Pham, H. 2015. *Effects of Reference Pricing on Customer Purchasing Intention, 4*(4): 1156.

Jansuttipan, M. 2015. *Why Bangsaen is now a hip getaway. BK.* Retrieved from https://bk.asiacity.com/travel/news/bangsaen-hip-quick-getaway-thailand

Mohamed, R., Muhamad, B., Borhan, H., Osman, I., Kamaralzaman, S. 2019. The determinant factors of supply chain management on purchase intention of an international braded apparels status quo. *Int. J Sup. Chain. Mgt, 8*(3): 677-684.

Nguyen, A. 2017. *Customers expectation and current offers from food trucks in Helsinki* (Bachelor's Thesis). Finland: Haaga-Helia University of Applied Sciences.

Rana, S. V. and Ahirrao, M. A. 2016. A Study on Street Food Preparation Practices in Jalgaon City. *IBMRD's Journal of Management & Research, 5*(2): 27-29.

Rhodes, E. 2018. *Tips for eating street food in Thailand.* Retrieved from https://www.stuff.co.nz/travel/themes/food/108043914/tips-for-eating- street-food-in-thailand

Rishad, R. 2018. The Conceptual Framework of Determinant Factors of Food Emotional Experience And Outcomes of International Tourist Satisfaction: Empirical Study on Malaysian Street Food, *International Journal of Business, Economics and Law, 15*(2): 44-50.

Shahid, Z., Hussam, T. and Azafar, F. 2017. The impact of Brand awareness on the customers' purchase intention. *Journal of marketing and consumer research, 33*: 36-37.

Theeramunkong, T., Kongkachandra, R., Ketcham, M., Hnoohom, N., Songmuang, P., Supnithi, T., and Hashimoto, K. 2017. *Advances in Intelligent Informatics, Smart Technology and Natural Language Processing.* Switzerland: Springer Nature Switzerland AG.

Wiens, M. 2020. *An Epic Thai Feast at Bangsaen Beach in Chonburi*. Retrieved from: https://migrationology.com/an-epic-thai-feast-at-bang-saen-beach-in-chonburi/

Xu, J. 2019. *Factors affecting consumers' purchase decisions of street food in Bangkok. All Rights Reserved.* Retrived from http://dspace.bu.ac.th/jspui/bitstream/123456789/4231/1/FACTORS%20AFFECTING%20CONSUMERS'%20PURCHASE%20DECISIONS%20OF%20STREET%20FOOD%20IN%20BANGKOK.pdf

Yanisa, M., and Suwaree, A. 2019. Thai street food and brand image development: An investigation from tourists' perspective. *Journal of Tourism, Hospitality & Culinary Arts, 11*(2): 69-82.

Promoting Creative Tourism: Current Issues in Tourism Research – Kusumah et al. (Eds)
 ISBN 978-0-367-55862-8

iDabao during Covid-19: Online-to-offline (O2O) food delivery service and the digitalization of Hawker (street) food during a crisis

E. Tan
Murdoch University, Singapore

ABSTRACT: The 2019 novel coronavirus (Covid-19) health pandemic has triggered unprecedented shocks to the global tourism and hospitality industry. Whilst there has been heightened discourse and rapid investigations analyzing the impacts of this extraordinary crisis on multiple aspects of the industry, the specific evaluation of Covid-19 on hawker (street) food has been lacking. This study investigates the impact of Covid-19 on hawkers in Singapore, and the digitalization of hawker food through online-to-offline (O2O) services such as food delivery apps (FDA). Extant studies on O2O services and related FDA have largely focused on the technological features, informational characteristics and user consumption behavior. In contrast, this study explores the role and digitalization value of O2O services in supporting hawkers and safeguarding Singapore's hawker culture. Preliminary findings of key challenges and opportunities of digitizing hawker food are also presented.

Keywords: iDabao, covid-19, online-to-offline (O2O), food delivery service, digitalization, food crisis.

1 INTRODUCTION

1.1 *Singapore's love affair with hawker food*

Singapore's love affair with its local hawker (street) food is decidedly palpable. The city state is passionately celebrated as a foodie's paradise where scrumptious culinary delights, from Michelin-starred restaurants to local hawker favorites, are savored and shared (Singapore Tourism Board 2020). For many, feeding one's food obsession is more than just satisfying the taste-buds. It is a national pastime and an integral part of the Singaporean identity.

Singapore's distinct local food culture is epitomized by its hawker centres (Henderson 2016). These represent the quintessential spaces for eating and socializing and are central to the everyday life of local residents (Henderson et al. 2012; Tarulevicz 2017). Intrinsically, hawker centres serve as vibrant "community dining rooms" where family, friends and neighbors can socialize and bond over local dishes they love (National Heritage Board 2020). Cuisines served by cooked food hawkers are often depicted as affordable and authentic representations of local culinary heritage, traditions and foodways (Henderson et al. 2012).

When one is ordering food and beverage at a local establishment, it is not uncommon to be asked, "eat here (sic), or dabao?". Colloquially, the term dabao (Cantonese term, meaning to wrap or pack), refers to a takeaway. Whether for eat-in or takeaway, hawker centres are highly patronized, with 83% of local residents indicating that they eat at or buy takeaways from hawker centres at least once a week (National Environment Agency 2019). Although hawker centres are patronized largely by locals, they are also increasingly promoted in Singapore's tourism and heritage narratives (Henderson et al. 2012). Thus, hawker centres and hawker food represent significant national, economic and social value in Singapore society (Tarulevicz 2017). However, the onset of the Covid-19 pandemic has triggered unprecedented impacts to the sector.

DOI 10.1201/9781003095484-71

1.2 *Covid-19 and impacts of the Singapore "Circuit Breaker" on hawkers*

The Covid-19 health pandemic has made its mark on nations, economies and communities across the globe (Nicola et al. 2020). Singapore is no exception. Whilst there has been heightened rhetoric examining the impacts of Covid-19 on the tourism and hospitality industry, the specific evaluations on hawker food and hawkers has been lacking. Particularly absent is the exploration of cooked food hawking within the context of online-to-offline (O2O) services such as food delivery apps (FDA).

This study investigates the impact of Covid-19 on hawkers in Singapore and the digital transformation of hawker food through O2O food delivery platforms. Extant studies on O2O services and related FDA have predominantly focused on the technological features, informational characteristics and user consumption behavior (Alalwan 2020; Ray et al. 2019; Roh & Park 2019). In contrast, this study examines the transformative value of O2O services in supporting and safeguarding Singapore's hawker culture and culinary heritage. Explicitly, it examines such digital transformations accelerated in response to crisis events such as the Covid-19. Preliminary findings of the key challenges and opportunities of digitizing hawker food are also presented.

The Covid-19 pandemic is described as an impactful crisis event due to its restrictions on human contact and mobility resulting from border closures and social distancing measures (Lee & Ong 2020; Nicola et al. 2020). Singapore instituted its early mitigation strategies for Covid-19 in January 2020 and a number of government advisories and measures have progressively been implemented since. On 7th February, the Ministry of Health (MOH) (2020) increased Singapore's Disease Outbreak Response System Condition (DORSCON) alert to Orange, its second highest level. Consequently, safe distancing and crowd limiting measures were elevated across multiple sectors, including food and beverage. As infections escalated, the Multi-Ministry Taskforce implemented the Singapore "Circuit Breaker" on 7th April, instituting a partial lockdown, heightened safe distancing measures and movement restrictions across the city state (Lee & Ong 2020; MOH 2020).

Whilst restaurants, hawker centres and other food and beverage outlets may remain open for takeaways or deliveries during this period, the impact of Covid-19 and the Circuit Breaker for the sector has been immense. It is estimated that hawkers are witnessing a plunge in business by up to 50% due to Covid-19 (Loke & Awang 2020), with many struggling to stay afloat during these challenging times. Adopting the adage of "adapt or die", numerous hawkers and food stallholders, many of whom have never had a website, social media or e-commerce system, are scrambling to digitalize by going online or adopting O2O food delivery platforms to survive (Phoon 2002). Thus, this exploratory study is timely, given the volatility of Covid-19 impacts and nascent discourse on the digital transformation of hawker food through O2O food delivery platforms.

2 LITERATURE REVIEW

2.1 *Digital transformation of food services amid COVID-19*

The Covid-19 pandemic has caused widespread detrimental impacts on global healthcare and economic systems, triggering ripple effects on every aspect of human life and society (Kim 2020; Nicola et al. 2020). Particularly for the food services sector, the Covid-19 crisis continues to exert profound impacts (Gursoy & Chi 2020). In response to these unprecedented challenges, the industry has witnessed a wave of digital transformations and technology-enhanced service innovations.

The drive to innovate and transform through digitalization is not new (Heavin & Power 2018). According to Ponsignon, Kleinhans, and Bressolles (2019), "digital transformation involves leveraging modern digital technologies to radically transform products and services, processes and people and improve performance" (p.S17); particularly when these occur in response to prevalent shifts in consumer consumption behavior and experiences within disrupted industries. Furthermore, these digital transformations can be accelerated in times of a crisis, such as the Covid-19, when fundamental structural changes occur abruptly (Kim 2020).

The velocity of advancements in digital data storage, computing capabilities and internet connectivity have exponentially increased enterprise applications, collaborative social networks, the internet of things (IoT), artificial intelligence (AI) applications and analytics technologies, etc. (Heavin & Power 2018; Ponsignon et al. 2019). In turn, these have radically transformed production, consumption and engagement strategies within the food services sector (Khan 2020). Whilst these service innovations and digital transformations may have been prevalent in the formal hospitality and restaurant sectors, the pace of transformation for the hawker food scene has been significantly slower. However, the onset of the Covid-19 pandemic has changed this.

Disruptive events, like Covid-19, can potentially disrupt an industry's modus operandi to the extent that it is compelled to redesign its strategies in order to survive and sustain the structural change (Khan 2020). Within the context of the food industry, producers and retailers have increasingly begun to harness the opportunities from digitalization, e-commerce and online service platforms to increase competitiveness and better engage with customers (Yeo et al. 2017). Concurrently, Kim's (2020) study on the impact of the Covid-19 on consumers' digital transformations and online consumptive behaviors suggest that whilst the pandemic may be an accelerator of structural changes in digital transformations within the marketplace, many of these disruptions may potentially have long-lasting impacts post-Covid. Thus, this posits a strong implication for the value of technology integration and digital transformation of essential business operations, service and delivery in food service (Gursoy & Chi 2020), wherein these Covid-induced digital transformations can be harnessed to fortify future opportunities. The next section reviews these within the context of O2O services and FDA.

2.2 *Online-to-offline (O2O) services and food delivery apps (FDA)*

Rapid advancements in technology integration, digital transformation and internet connectivity have dramatically changed consumer behavior in recent years. In turn, smart technologies, mobile devices, and related mobile applications (apps) have become an extensive and integral part of daily life (Alalwan 2020; Roh & Park 2019). Consumers today are able to concurrently utilize multiple channels to fulfill their search, comparison and purchase needs. Within the context of food service, the convenience, increased penetration and ubiquity of smart devices and mobile apps have popularized the adoption of online-to-offline (O2O) services and food delivery apps (FDA) within the e-commerce space (Ray et al. 2019; Xu & Huang 2019; Yeo et al. 2017).

Fundamentally, O2O services offer customers the convenience of online services (i.e., to search, compare and order products and services) to an offline reality (i.e., to consume and utilize in an offline venue) (Roh & Park 2019). Consequently, FDA are popularly adopted as innovative channels to engage with customers and provide high-quality food service experiences (Alalwan 2020). It is estimated that the global online on-demand food delivery services market will grow by over US$44billion between 2020-2024, with almost 60% of this growth from Asia Pacific markets (Technavio 2020). Despite the anticipated negative market impacts from Covid-19, compared to previous years' growth forecasts, the market momentum is expected to continue due to consumer preference and demand for convenience and easy accessibility.

Whilst O2O food delivery services may not be the only means of food purchase and ordering, its rising popularity is reflective of the exponential growth in online retailing and services, resulting from changes to consumer lifestyles and society (Kim 2020; Ray et al. 2019). From a supply-side perspective, it also allows food service operators to gain new revenue streams and market development opportunities without expanding physical capacity (Xu & Huang 2019). Given the enormous developments in this field, it is unsurprising that there is growing interest and discourse on O2O services, FDA and its related enterprise applications in recent years.

Broadly, extant research on O2O food delivery services have predominantly examined thematic concepts within (1) technology-driven attributes, (2) consumer-driven attributes, (3) content-driven attributes, and (4) food order-related attributes (e.g., Alalwan 2020; Roh & Park 2019; Ray et al. 2019; Xu & Huang 2019; Yeo et al. 2017). These dimensions are further expanded and outlined in Figure 1 below.

However, despite the growing body of work in this field, there has yet to be an investigation of the contributions and transformative value of O2O services and FDA within the context of supporting

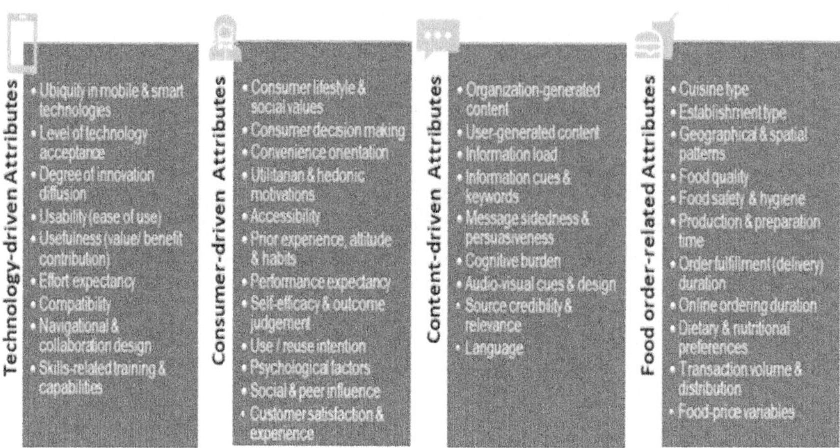

Figure 1. Thematic dimensions of O2O food delivery service.

and safeguarding hawker (street food) culture and culinary heritage. Additionally, this study is beneficial as it examines such digital transformations accelerated in response to crisis events such as the Covid-19 pandemic.

3 METHODOLOGY

This exploratory study aims to investigate the impact of Covid-19 on hawkers in Singapore and the digital transformation of cooked food hawking through O2O services and FDA. The findings and discussions are derived from content analysis of data from news article searches, carried out between April to June 2020, from Google News (https://news.google.com/). Within the context of this study, keywords were combined with boolean operators to narrow or broaden the data collection and expedite the return of adequate congruent results. The boolean phrases and keywords eventually used in the final news article search were as follows: [Covid-19 OR Coronavirus] AND [Singapore Hawkers OR Hawker Centres] AND [Delivery OR Delivery Apps]. In line with the research objectives, search results pertaining to restaurants, fast food, or franchised/chain stores were excluded. A return of 421 results were obtained in the initial search and a final total of 41 articles were retained for analysis, following inspection for content and relevance to the research parameters. The resultant preliminary findings and analysis are discussed in the following sections.

4 FINDINGS AND DISCUSSION

4.1 *iDabao during Covid-19: Hope through digitalization and O2O food delivery services?*

The food services sector represents a key social and economic role in Singapore, contributing about 180,000 jobs (or 5% of total workforce) at over 7,000 establishments nationally (Enterprise Singapore 2020).Within this sector, cooked food hawking has an established history and hawker centers are an iconic feature in contemporary Singapore (Henderson et al. 2012). However, in recent years, this beloved sector is facing challenges and an uncertain future, with aging chefs and food stallholders, rising costs, demanding working conditions and the struggle to preserve traditional recipes and local food heritage (Su 2016; Tarulevicz 2017). The onset of the Covid-19 and Singapore's Circuit Breaker has further intensified the hawker crisis and risks to its legacy.

In 2016, the Singapore government launched the Food Services Industry Transformation Map and Industry Digital Plan, with the aim of transforming the sector and accelerate its digitalization

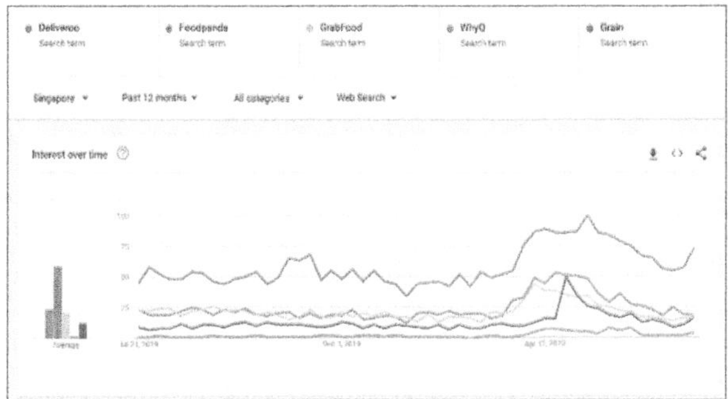

Figure 2. Singapore's top 5 FDA: Interest over time. (12 months) (Source: www.trends.google.com/trends)

journey. (Infocomm Media Development Authority 2020). Subsequently, cooked food hawking in Singapore witnessed an evolution, with a revitalized image and structural changes, such as Michelin-star rankings and the birth of the hawkerpreneurs, i.e., hawker entrepreneurs. These next generation hawkerpreneurs are reinventing the hawker trade and its culinary infrastructure through autonomy, skilled artisan labor, innovation and a passion to preserve their family's food legacies and Singapore's culinary heritage (Tarulevicz 2017). Many have also broken the mold of the hawker stereotype, sharing their food philosophies, artisanal creations and stories through digitalization, service innovations and social media.

With the arrival of the Covid-19 and its resulting Circuit Breaker mitigation measures, the authorities introduced additional support for the sector. The Singapore government allocated over S$500million within its S$33billion Fortitude Budget in May 2020 to support businesses in their digital transformation amid Covid-19 (Ang 2020; Ministry of Finance 2020). Explicitly for hawker centers, stallholders and the food services sector, digital transformations and innovations are leveraged through technology and digital tools to facilitate digitalization, e-commerce, and sustainable digital resilience. This includes a further S$250million set aside specifically to help these businesses accelerate digitalization, build partnerships with digital platform solution providers and augment O2O food service models (Ang 2020).

Concurrently, the Food Delivery Booster Package was introduced to help the food services sector cope with the transition to O2O services and FDA (Enterprise Singapore 2020). Covid-19 implications aside, the ubiquity and prevalence of smart devices, digital tools and advancements in ICTs have dramatically altered consumers' relationship with food and food consumption behavior (Alalwan 2020; Roh & Park 2019). As discussed, whilst the pace of digital transformation and technology-driven service innovations may be widespread in the hospitality and restaurant sectors, its acceptance and adoption within the hawker scene has been sluggish. However, the onset of the Covid-19 pandemic may spur momentum toward digitalization within the sector.

In light of stricter Covid-19 mitigation measures, there has been heightened demand for O2O food delivery services. A search comparison analysis of Singapore's top five FDA and the interest over time (www.trends.google.com/trends), within a 12-month period, indicate a spike in interest during Covid-19; and in particular during the Circuit Breaker period (Figure 2). As observed in similar studies on O2O services and FDA, this trend is indicative of the influencing impact of crisis events such as Covid-19 on consumer demand, consumption culture and behavior (Kim 2020). From a supply-perspective however, the response from hawkers has been mixed. Whilst some indicated that they are grateful for the government grants and other financial assistance, such as rental waivers, others shared that commercial O2O service platforms and commission-based FDA remain unaffordable and unsustainable for small food stallholders over the long run. Nonetheless, a considerable number have digitalized through social media, e-commerce, O2O services and FDA.

A preliminary analysis of narratives articulated in the findings reveal that the Covid-19 and Circuit Breaker measures have had devastating impacts on the majority of hawkers. With a drastic decline in footfalls and patronage at hawker centres, stallholders are experiencing an average loss of between 20–50% in sales; with those in the central business district struggling with up to 90% in lost revenue, as office workers in non-essential services work from home. Thus, whilst some hawkers have shuttered their businesses temporarily or permanently, many others have remained resilient and soldiered on. When queried as to the reasons for their determined resilience to endure this crisis, the top three most common reasons expressed were: (1) preservation or survival of family business/legacy, (2) love of being a hawker and/or dedication to serving hawker food, and (3) safeguarding local hawker culture and community pride. To do so, many hawkers have also harnessed the popularity of social media to augment O2O services and FDA.

As observed in Technavio's (2020) report, social media is a significant marketing and communications platform through which O2O service providers may directly connect and engage with consumers. This was similarly noted in the study. In response to public advocacy to support Singapore hawkers, safeguard its hawker culture and food heritage, a number of ground-up initiatives have emerged on social media. Industry stakeholders, community groups and the general public have publicized social media campaigns and hashtags, such as: #Savefnbsg, #SupportSingpaorehawkers, #HawkersUnited2020, #saveourhawkers, #SupportLocal, etc. For example, *Hawkers United – Dabao 2020*, a Facebook group created by second-generation hawker Melvin Chew at the start of the Circuit Breaker, has become a bustling platform for hawkers and local restaurants to promote their cuisines online and take food orders (Tan 2020). At the time of writing, this vibrant community has grown to over 270,000 members.

Concurrently, others have decided to use this crisis for good. Whilst it is certainly not "business as usual" for many, and despite their own struggles to survive during the Circuit Breaker, some hawkers are using their resources to support the community and those in need. From pledging free meals for the needy and vulnerable elderly, to sponsoring meals for frontline and healthcare workers, these hawkers have taken to social media to offer help to those in need. As Jason Chua, co-founder of *Beng Who Cares Foundation*, shares, *"as long as we don't die, we carry everyone (sic), so everyone survives together"* (Yip 2020). Additionally, some are also using these same channels to "pay it forward", by creating non-profit O2O collaborative platforms (e.g., *SG Dabao*) to directly connect customers with food stallholders and out-of-work delivery drivers/riders.

These notions of unity, solidarity and community dedication in supporting each other and preserving Singapore's beloved hawker culture have been similarly observed across other social media platforms. Hence, the preliminary findings suggest that digitalization and food service transformations through social media and collaborative consumption on O2O platforms have helped to reinforce and revitalize the support for Singapore hawkers and their legacy as custodians of its unique hawker culture and local food heritage. As affirmed avidly on SG Dabao (2020), *"(In) these uncertain times, let's all stand united as one. Our food is our culture. Our food is what makes us Singaporean. Let's all put our best effort in saving our food culture, especially now".*

4.2 *Hawker food gone digital: Opportunities and challenges*

As observed in the previous section, whilst there has been a positive evolution towards digitalization and technology-enhanced service innovations within the hawker trade, this digital transformation journey has not been enthusiastically adopted or satisfactory for all. From the narratives analyzed, it is evident that there is no one-size-fits-all solution to effective digitalization. As Heavin and Power (2018) suggest, a sound digital transformation strategy must go beyond a mere technology-focused approach. While technological innovations and digital tools are integral to transformation strategies, the role of humans (customers, users and managers), organizational culture and strategic planning remain at the heart of successful digitalization. Likewise, not all hawkers and food stallholders share the same business, operational or personal circumstances.

Moreover, to be efficacious, digital platforms must be designed and utilized in a manner that is compatible and comparable to other applications, as well as consumer's values, lifestyle and consumption behavior (Alawan, 2020; Roh & Park, 2019). In this regard, Kim (2020) advises that successful digital transition onto the virtual realm requires: (1) concise information delivery,

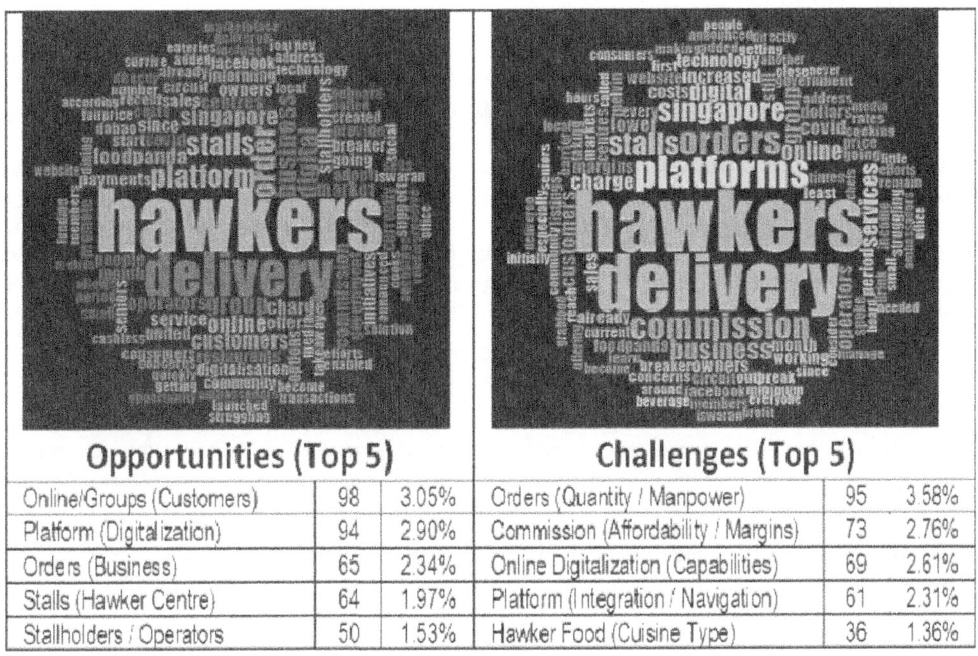

Opportunities (Top 5)			Challenges (Top 5)		
Online/Groups (Customers)	98	3.05%	Orders (Quantity / Manpower)	95	3.58%
Platform (Digitalization)	94	2.90%	Commission (Affordability / Margins)	73	2.76%
Orders (Business)	65	2.34%	Online Digitalization (Capabilities)	69	2.61%
Stalls (Hawker Centre)	64	1.97%	Platform (Integration / Navigation)	61	2.31%
Stallholders / Operators	50	1.53%	Hawker Food (Cuisine Type)	36	1.36%

Figure 3. Opportunities and challenges of digitalization.

(2) interactive content, (3) enhanced audio-visual cues, (4) emotive connections, (5) personalization, (6) trust, (7) honest user reviews, (8) market intelligence from consumer insights, and (9) customer-centric post-purchase relationships.

Within the context of O2O services, FDA and the digitalization of hawker food, the preliminary analysis of narratives suggest that whilst digital transformation opportunities may provide a much needed lifeline for many hawkers during this time of unprecedented crisis, it may not be an affordable, practical nor sustainable solution for all; particularly given the diverse demographics, capabilities, operational characteristics and cuisine types within the hawker community. For example, whilst they may acknowledge the potential benefits of digitalization and connecting with more customers, many aging hawkers or sole proprietors shared that they do not have the technical know-how, language capabilities, or manpower resources to do so. Yet others have indicated that the combination of smaller value orders or quantities and lower margins of some hawker items, versus the high cost of O2O services and commissions-based FDA, make it a poor trade-off. Moreover, most hawkers felt that O2O services and FDA are not their primary business. Figure 3 below summarizes the key opportunities and challenges (top five clusters and word frequency analysis) of digitalization expressed in the narratives.

5 CONCULSION

This research-in-progress presents the preliminary findings examining the impacts of the Covid-19 on hawkers in Singapore and the digitalization of hawker food through O2O services and FDA. Dimensions concerning the digitalization of hawker food and its contributions in safeguarding Singapore hawker culture were also discussed. The findings suggest that digital transformations, social media and technology-enhanced service innovations have stimulated strong communal efforts and solidarity in supporting local hawker food and vulnerable small businesses.

Undoubtedly, there are distinct challenges impeding the digitalization progress of hawker food. These should be addressed in order to ensure future forward momentum and sustainable digital resilience for the sector. However, there is also firm potential for these digital transformations to advocate stronger support, interest and visibility for Singapore's hawker culture, its hawkers and their food heritage narratives.

It is acknowledged that the current study into the digitalization of hawker food and its applications of O2O services and FDA is still embryonic. Thus, while it offers useful preliminary insights, it is not intended as a generalization of all food services sector archetypes. Future research will entail the collection and analysis of empirical data from stakeholders within the sector to yield further in-depth insights for more robust development of future digital transformation agenda and strategies.

REFERENCES

Alalwan, A.A. (2020). Mobile food ordering apps: An empirical study of the factors affecting customer e-satisfaction and continued intention to reuse. *International Journal of Information Management, 50*, 28–44.

Ang, H.M. (2020, May 26). Fortitude budget: More than S$500m allocated to support digital transformation of businesses amid COVID-19 outbreak. *Channel News Asia.* Retrieved from: https://www.channelnewsasia.com/news/singapore/covid-19-fortitude-budget-500m-digital-transformation-business-12770406

Enterprise Singapore (2020). *Food Services.* Retrieved from: https://www.enterprisesg.gov.sg/industries/type/food-services/industry-profile

Gursoy, D. & Chi, C.G. (2020) Effects of COVID-19 pandemic on hospitality industry: Review of the current situations and a research agenda. *Journal of Hospitality Marketing & Management, 29*(5), 527–529.

Heavin, C., & Power, D.J. (2018). Challenges for digital transformation-towards a conceptual decision support guide for managers. *Journal of Decision Systems, 27*(S1), 38–45.

Henderson, J.C. (2016). Foodservice in Singapore: Retaining a place for hawkers. *Journal of Foodservice Business Research*, 1993, 272–286.

Henderson, J.C., Yun, O.S., Poon, P., & Biwei, X. (2012). Hawker centres as tourist attractions: The case of Singapore. International *Journal of Hospitality Management, 31*(3), 849–855.

Infocomm Media Development Authority (2020). *Food Services Industry Digital Plan.* Retrieved from: https://www.imda.gov.sg/-/media/Imda/Files/Programme/SMEs-Go-Digital/Industry-Digital-Plans/Food-Services-IDP-20181219.pdf?la=en

Khan, M.A. (2020). Technological disruptions in restaurant services: Impact of innovations and delivery services. *Journal of Hospitality & Tourism Research, 44*(5), 715–732.

Kim, R.Y. (2020). The impact of Covid-19 on consumers: Preparing for digital sales. *IEEE Engineering Management Review.* DOI: 10.1109/EMR.2020.2990115.

Lee, W.C., & Ong, C.Y. (2020). Overview of rapid mitigating strategies in Singapore during COVID-19 pandemic. *Public Health, 185*, 15–17.

Loke, L., & Awang, N. (2020, February 22). 'We'd try to endure': Hawkers see business plunge by up to 50% due to Covid-19 outbreak. *Today.* Retrieved from: https://www.todayonline.com/singapore/wed-try-endure-hawkers-see-business-plunge-50-due-covid-19-outbreak

Ministry of Finance (2020). *Budget 2020: Fortitude Budget.* Retrieved from: https://www.singaporebudget.gov.sg/budget_2020/fortitude-budget

Ministry of Health, Singapore (2020). *Updates on Covid-19 (Coronavirus disease 2019) local situation.* Retrieved from: https://www.moh.gov.sg/covid-19

National Environment Agency (2019). *High majority of patrons satisfied with hawker centres.* Retrieved from: https://www.nea.gov.sg/media/news/news/index/high-majority-of-patrons-satisfied-with-hawker-centres

National Heritage Board (2020). *Hawker culture in Singapore.* Retrieved from: https://www.oursgheritage.sg/hawker-culture-in-singapore/

Nicola, M., Alsafi, Z., Sohrabi, C., Kerwan, A., Al-Jabir, A., Iosifidis, C., … & Agha, R. (2020). The socio-economic implications of the coronavirus pandemic (COVID-19): A review. *International Journal of Surgery, 78*, 185–193.

Phoon, A. (2020, April 30). 20 famous hawker, zi char stalls with new island wide delivery for circuit breaker. *8 Days.* Retrieved from: https://www.8days.sg/eatanddrink/newsandopening/20-famous-hawker-zi-char-stalls-with-new-islandwide-delivery-for-12647612

Ponsignon, F., Kleinhans, S., & Bressolles, G. (2019). The contribution of quality management to an organization's digital transformation: A qualitative study. *Total Quality Management & Business Excellence, 30*(S1), S17–S34.
Ray, A., Dhir, A., Bala, P.K., & Kaur, P. (2019). Why do people use food delivery apps (FDA)? A uses and gratification theory perspective. *Journal of Retailing & Consumer Services, 51*, 221–230.
Roh, M., & Park, K. (2019). Adoption of O2O food delivery services in South Korea: The moderating role of moral obligation in meal preparation. *International Journal of Information Management, 47*, 262–273.
SG Dabao (2020). *About Us: SG Dabao #saveourhawkers.* Retrieved from: https://sgdabao.com/about-us/
Singapore Tourism Board (STB) (2020). *How to feed your food obsession.* Retrieved from: https://www.visitsingapore.com/editorials/did-you-know-foodies/
Su, E. (2016, October 21). Eating street in Singapore. *Reuters: The Wider Image.* Retrieved from: https://widerimage.reuters.com/story/eating-street-in-singapore
Tan, A. (2020, April 21). Covid-19: Hope for S'pore hawkers comes in the form of a booming Facebook community. *Vulcan Post.* Retrieved from: https://vulcanpost.com/ 696286/hawkers-united-facebook-group-singapore/
Tarulevicz, N. (2018). Hawkerpreneurs: Hawkers, entrepreneurship, and reinventing street food in Singapore. *Journal of Business Management, 58*(3), 291–302.
Technavio (2020). *Online on-demand Food delivery Services Market.* Retrieved from: https://www.technavio.com/report/online-on-demand-food-delivery-services-market-size-industry-analysis
Xu, X., & Huang, Y. (2019). Restaurant information cues, diners' expectations, and need for cognition: Experimental studies of online-to-offline mobile food ordering. *Journal of Retailing & Consumer Services, 51*, 231–241.
Yeo, V.C.S., Goh, S.K., & Rezaei, S. (2017). Consumer experiences, attitude and behavioral intention toward online food delivery (OFD) services. *Journal of Retailing & Consumer Services, 35*, 150–162.
Yip, C. (2020, April 10). 'I'm losing money, might as well do good': The hawkers helping the helpless. *Channel News Asia.* Retrieved from: https://www.channelnewsasia.com/news/cnainsider/covid-19-singapore-hawkers-feeding-the-needy-12628574

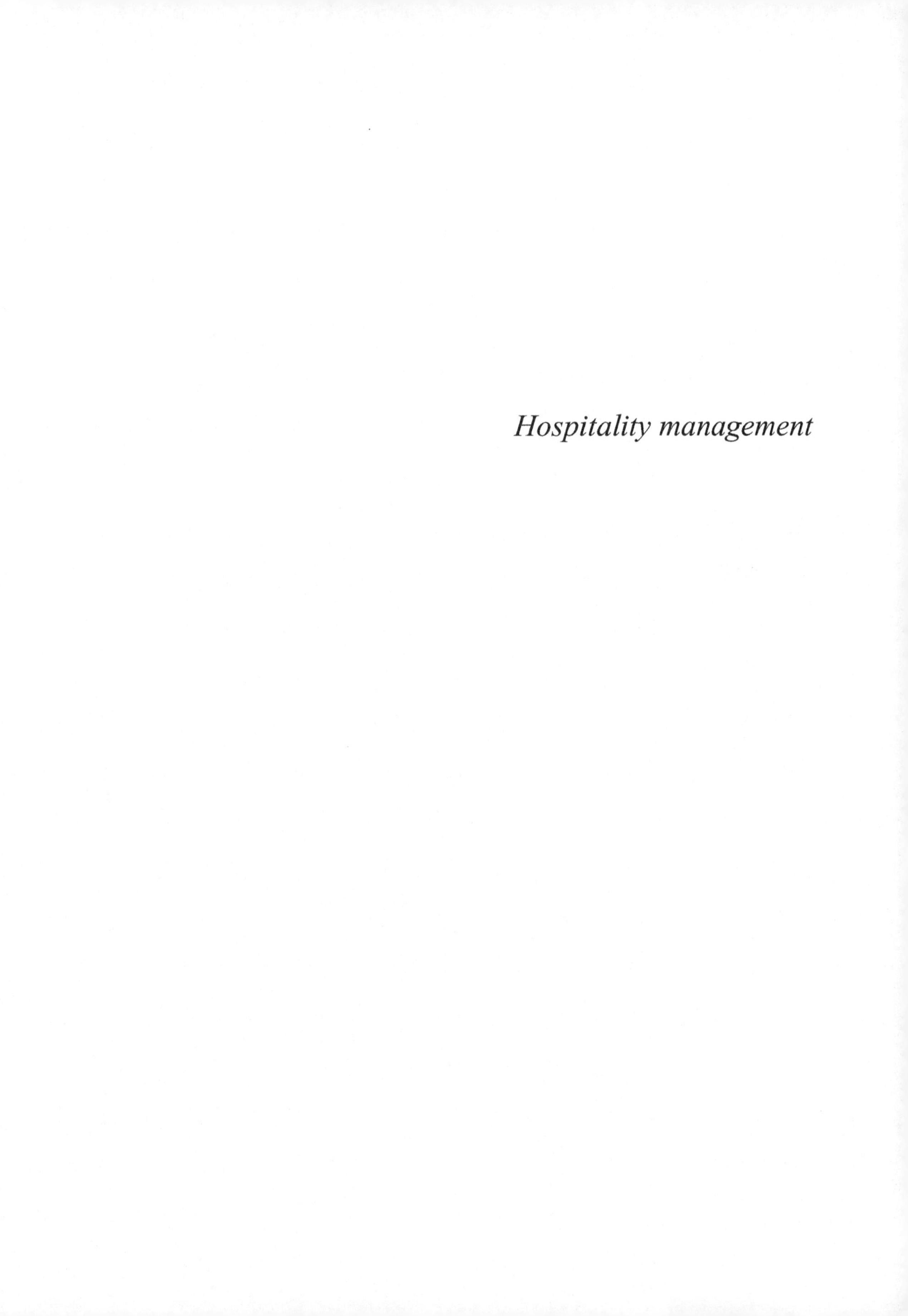

Hospitality management

Promoting Creative Tourism: Current Issues in Tourism Research – Kusumah et al. (Eds)

Re-examining sensory experience on highland nature-based resort rooms

N.H.A. Rahman
International Islamic University Malaysia, Johor, Malaysia

R.M. Wirakusuma
Universitas Pendidikan Indonesia, Bandung, Indonesia

E. Dasipah
Universitas Winaya Mukti, Bandung, Indonesia

ABSTRACT: Tourists are more likely to seek hotel and resort reviews before making a reservation. However, the advertised products are sometimes not in accordance with their actual staying experience. Experience is conceptualized as a combination of senses, influence, and cognition. This research aims to find out the sensory experience by the Resort's guests based on the online reviews as the secondary data. The reviews were gathered from online travel agents, namely Traveloka, Agoda, and TripAdvisor.com. The collected data were analyzed sentence by sentence to discover the sensory experience that has been felt by the guests while staying in the resorts. It can be concluded that the sight experience was related to the superiority of a good view from nature, the serenity as the primary factor for the hearing experience but some rooms in the resorts were not soundproof, and eventually, it had influenced the unpleasant experience to the guests. Furthermore, the un-pleasant touch and smell experience of these resorts were the nuisances from the insects, stiff mattress, and the uncomfortable bathroom. But, the breeze from nature has completed the whole relaxation experience. Hence, this research provides implication on the improvement of guest's experience in the context of five senses in the hospitality and tourism industry.

Keywords: sensory experience, senses, unique accommodation, highland nature-based resort, online travel agents

1 INTRODUCTION

In this highly competitive era, a business must be able to present something different with a unique selling point to attract the attention of people, especially in terms of business and tourism. The accommodation sector is one of the important business in the tourism and hospitality industry. Hence, the accommodation business should be able to create a memorable experience to their guest due to the increasing competitiveness in the hospitality business. An experience felt by a guest is inseparable from the senses possessed by the guest.

The experiential marketing which has been highlighted by Pine and Gilmore (1998), and Schmitt (1999) as cited in Mehraliyev et al. (2020), also proposed the experiences are conceptualized as a blend of senses, affect, and cognition. From the branding perspective, experiences are the takeaway impressions created in the minds of customers because of their interactions with brands (Iglesias et al. 2019). While Brakus et al. (2009) have emphasized that these interactions can be direct or indirect. Direct interaction generally occurs when customer purchase, consume, or use the brand's goods or services, whereas indirect interaction mainly takes place when customer experience the

DOI 10.1201/9781003095484-72

brand's advertising, marketing communication, word-of-mouth recommendations, news report and reviews. Some evidence most likely that the five human senses are very important for individual experiences in the purchasing process, the frequency of purchases, and different consumption (Nandagopal 2015). It is through the human senses that customers can distinguish one brand from the same brand. Brand experience influences the formation of consumer loyalty (Manthiou et al. 2016). By applying the sensory marketing, hotels could treat their customers in an intimate and personal way; the ultimate goal is to emotionally attach the customers to the hotel and to win over their Loyalty (Kim & Perdue 2013)

People in this digitalization era are more likely to see the hotel reviews on mobile applications or through online websites because it is very efficient, especially in providing various choices to the potential customers. The guests used ICT mobile application to find the information, subsequently affect their activity and overall experience (Nurazizah & Marhanah 2020). However, the main issue is the room that has been advertised in mobile applications or websites did not match the actual guest expectation and experience. Consequently, the hotel company most likely receive undesirable reviews. Currently, there was a small study in this field and there are open gaps for further research. Hence, this study aims to explore the sensory experience from the resort guests based on online reviews.

2 LITERATURE REVIEW

2.1 *Understanding the sensory experience*

Several definitions of sensory experience have been raised in previous literature. A commonly adopted definition is that "customer experience originates from a set of interactions between a customer and a product, a company, or part of its organization, which provoke a reaction" (Gentile et al. 2007). Agapito et al. (2017) articulated the physical, intellectual, emotional, social, and sensory dimensions of experience and argued that the literature appears to agree on the central role of sensory dimension among others.

According to Mehraliyev et al. (2020), the terminology of "sensory experience" is not a new term since it has been widely used in other research fields such as in neurosciences (de Lafuente & Romo 2005) and health sciences (Ayo-Yusuf & Agaku 2015). The research on "sensory experience" has been carried out in the field of tourism such as the research that has been done by Agapito (2020), Agapito et al. (2014), Lee et al. (2018), McFarlane et al. (2017), and Agapito et al. (2017), but relatively minimal in the context of hospitality perspective such as the research done by Liu et al. (2016).

Previous research on sensory experience has been done by several academics and experts. In general, the sensory experience has been discussed in two perspectives of tourism geography and tourism marketing. However, most of the research on sensory experience is related to marketing. For instance, the research was done by Satti et al. (2019) has explored the mediating role of sensory marketing in relation to the visual marketing, gustative marketing, auditory marketing, olfactory marketing, and tactile marketing, towards the customer satisfaction.

The research by Satti et al. (2019) is inlined with the research done by Moreira et al. (2017), in which the sensory marketing is seen as the marketing technique to communicate with customers through their five senses. Several studies stated that sensory stimuli would influence the consumers such as flavours and colours (Compeau et al. 1998), touch (Peck & Childers 2006) and fragrance in stores (Spangenberg et al. 2006).

Some previous hotel choice studies have included the sensory attributes, together with cognitive attributes (Callan & Bowman 2000; Devi Juwaheer 2004), including the attractiveness of interior and exterior design, cleanliness of the facilities, bed comfort, and noise. In hospitality, the products and services mostly offer the essence of creating a memorable experience, which means that all the products and services that are served must be felt by the guests to create an experience.

3 METHOD

In this study, the five dimensions of sensory experiences were received through sound, sight, smell, touch, and taste to represent the five sensory experience, respectively. Moreover, this research has utilized online reviews from the highland nature-based resort guests in West Bandung, Indonesia, as the secondary data and analyze them. This research analyzed the online reviews from 120 users in two online booking agents, namely, 1) Traveloka, and 2) Agoda; and the small amount of additional reviews from a travel agent website, tripAdvisor.com. Afterwards, this study combined all the words that have been categorized from various online travel agents and analysed the actual sensory experience by the guests of the resort. All of the data have been analysed qualitatively through narrative description. The following are the stages of the methodology of this research:

4 RESULT

Based on the reviews from 120 users that have experienced the five resorts, the data gathered were categorized into five senses. The research found that 33 users had reviewed that 'the view of the resort from inside the room or from the resort itself', in which to portray the visual experience (Figure 1). The average reviews in relation to the 'resort's view' are very satisfying. Hence, the result has shown that the guests are satisfied with the views of the resorts in West Bandung. Furthermore, there were ten users reviewed the 'noise heard while staying at the resort room' as the auditory experience. Additionally, the mattress had become the focal point on olfactory experience in which, there were 18 users who reviewed the 'smell of the mattress (bed)', and 20 users reviewed the 'smell from food that served by the resort'. There were reviews about the 'cool air' from 29 users in relation to 'the perceived air', and there were ten users who reviewed 'the atmosphere' felt when they stay at a resort in West Bandung regency. Hence, these experiences had depicted the actual tactile experience by the Resort's guests. Moreover, in relation to the sense of touch, the cool climate has influenced the tactile experience because the resorts are located in the highlands, and therefore, it had provided the relaxation experience to the guest. As a result, the guests had reviewed their willingness to visit Bandung again due to its excellent weather.

From the sense of hearing, the silence in the resort area was very influential as it provides the calmness experience that had been felt by the guests. However, the rooms are not soundproof, and therefore, the guests felt uncomfortable. This experience was very influential on the guest's reviews as they had described the bad experience of rooms since they are not soundproof. For the sense of taste experience, the average findings had shown that the guests were less satisfied with the food menu due to its taste and the limited choices of food in the menu provided, in which there was only Indonesian food, and even some resorts only served Sundanese specialties. Hence, it had affected the culinary satisfaction experience for the guests since they need to find food from the outside of the resorts. In relation to the smell experience in the Resort's room, the average findings from the guest's reviews had complained about the unpleasant smell in the room. It is due to the humid conditions of the room due to the weather and the building materials that had been made from wood, which these factors had resulted in the water to enter the room. Hence, these factors had affected the sleeping experience that is bad for guests.

5 CONCLUSION

In conclusion, the research has found the five dimensions of sensory experience that had been perceived and experienced by 120 users in five highland nature-based Resort in West Bandung, Indonesia through sound, sight, smell, touch, and taste to represent the auditory, visual, olfactory, tactile and gustatory experience. Based on the reviews from the three sources, it can be seen that the beautiful scenery had affected the satisfaction for the sense of sight since the guests had experienced and pampered their eyes by seeing the nature. Meanwhile, the uncomfortable hearing

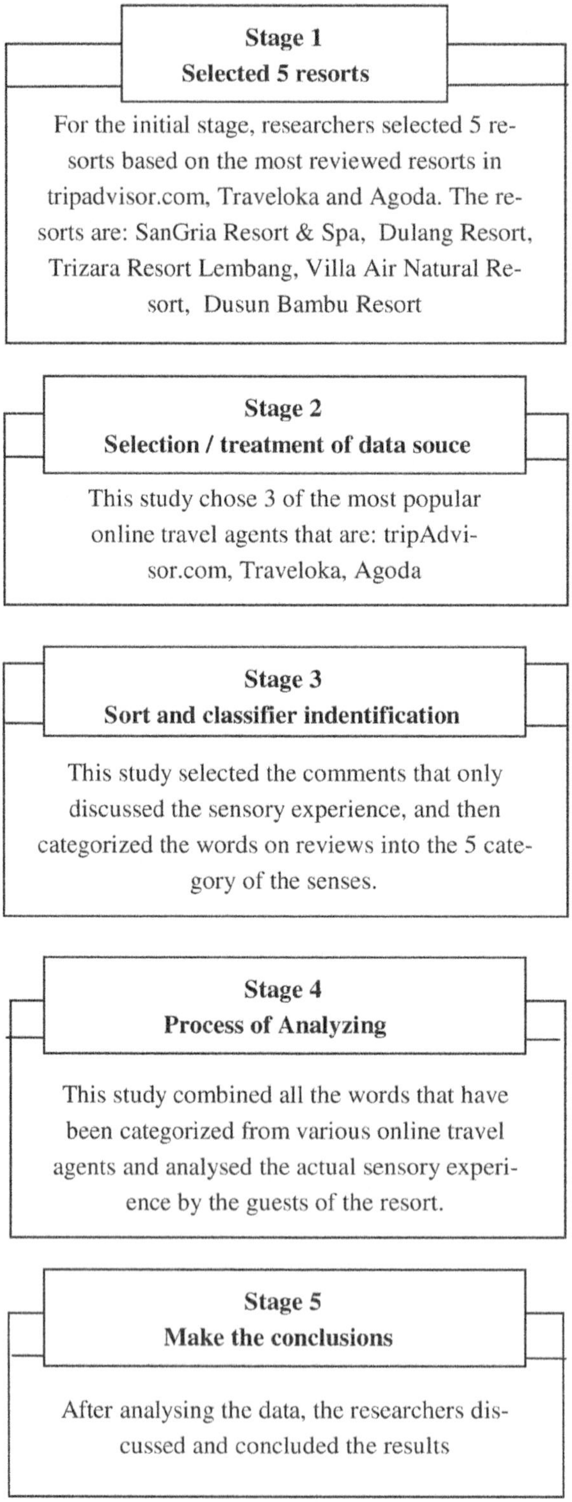

Stage 1
Selected 5 resorts

For the initial stage, researchers selected 5 resorts based on the most reviewed resorts in tripadvisor.com, Traveloka and Agoda. The resorts are: SanGria Resort & Spa, Dulang Resort, Trizara Resort Lembang, Villa Air Natural Resort, Dusun Bambu Resort

Stage 2
Selection / treatment of data souce

This study chose 3 of the most popular online travel agents that are: tripAdvisor.com, Traveloka, Agoda

Stage 3
Sort and classifier indentification

This study selected the comments that only discussed the sensory experience, and then categorized the words on reviews into the 5 category of the senses.

Stage 4
Process of Analyzing

This study combined all the words that have been categorized from various online travel agents and analysed the actual sensory experience by the guests of the resort.

Stage 5
Make the conclusions

After analysing the data, the researchers discussed and concluded the results

Figure 1. Stages of the method.

experience was a result from the bad soundproof in the rooms, unsatisfactory taste experience due to the limitation in serving the different types of food and the unpleasant smell in the room had influenced the guest's smell experience.

This research is vital to the hospitality industry, especially in the accommodation sectors, to discover the sensory experience by their guests. Hence, the accommodation sectors would be able to improve the service quality and to ensure that they can increase the business competitiveness.

REFERENCES

Agapito, D., Pinto, P., & Mendes, J. (2017). Tourists' memories, sensory impressions and Loyalty: In loco and post-visit study in Southwest Portugal. *Tourism Management*, *58*, 108–118. https://doi.org/10.1016/j.tourman.2016.10.015

Agapito, D., Valle, P., & Mendes, J. (2014). The sensory dimension of tourist experiences: Capturing meaningful sensory-informed themes in Southwest Portugal. *Tourism Management*, *42*, 224–237. https://doi.org/10.1016/j.tourman.2013.11.011

Ayo-Yusuf, O. A., & Agaku, I. T. (2015). The Association Between Smokers' Perceived Importance of the Appearance of Cigarettes/Cigarette Packs and Smoking Sensory Experience: A Structural Equation Model. *Nicotine & Tobacco Research*, *17*(1), 91–97. https://doi.org/10.1093/ntr/ntu135

Brakus, J. J., Schmitt, B. H., & Zarantonello, L. (2009). Brand Experience: What is It? How is it Measured? Does it Affect Loyalty? *Journal of Marketing*, *73*(3), 52–68. https://doi.org/10.1509/jmkg.73.3.052

Callan, R. J., & Bowman, L. (2000). Selecting a hotel and determining salient quality attributes: a preliminary study of mature british travellers. *International Journal of Tourism Research*, *2*(2), 97–118. https://doi.org/10.1002/(SICI)1522-1970(200003/04)2:2<97::AID-JTR190>3.0.CO;2-1

Compeau, L. D., Grewal, D., & Monroe, K. B. (1998). Role of Prior Affect and Sensory Cues on Consumers' Affective and Cognitive Responses and Overall Perceptions of Quality. *Journal of Business Research*, *42*(3), 295–308. https://doi.org/10.1016/S0148-2963(97)00126-4

de Lafuente, V., & Romo, R. (2005). Neuronal correlates of subjective sensory experience. *Nature Neuroscience*, *8*(12), 1698–1703. https://doi.org/10.1038/nn1587

Devi Juwaheer, T. (2004). Exploring international tourists' perceptions of hotel operations by using a modified SERVQUAL approach – a case study of Mauritius. *Managing Service Quality: An International Journal*, *14*(5), 350–364. https://doi.org/10.1108/09604520410557967

Gentile, C., Spiller, N., & Noci, G. (2007). How to Sustain the Customer Experience: *European Management Journal*, *25*(5), 395–410. https://doi.org/10.1016/j.emj.2007.08.005

Iglesias, O., Markovic, S., & Rialp, J. (2019). How does sensory brand experience influence brand equity? Considering the roles of customer satisfaction, customer affective commitment, and employee empathy. *Journal of Business Research*, *96*, 343–354. https://doi.org/10.1016/j.jbusres.2018.05.043

Kim, D., & Perdue, R. R. (2013). The effects of cognitive, affective, and sensory attributes on hotel choice. *International Journal of Hospitality Management*, *35*, 246–257. https://doi.org/10.1016/j.ijhm.2013.05.012

Lee, J., Yi, J., Kang, D., & Chu, W. (2018). The effect of travel purpose and self-image congruency on preference toward airline livery design and perceived service quality. *Asia Pacific Journal of Tourism Research*, *23*(6), 532–548. https://doi.org/10.1080/10941665.2018.1483956

Liu, W., Sparks, B., & Coghlan, A. (2016). Measuring customer experience in situ: The link between appraisals, emotions and overall assessments. *International Journal of Hospitality Management*, *59*, 42–49. https://doi.org/10.1016/j.ijhm.2016.09.003

Manthiou, A., Kang, J., Sumarjan, N., & Tang, L. R. (2016). The Incorporation of Consumer Experience into the Branding Process: An Investigation of Name-Brand Hotels. *International Journal of Tourism Research*, *18*(2), 105–115. https://doi.org/10.1002/jtr.2037

McFarlane, J., Grant, B., Blackwell, B., & Mounter, S. (2017). Combining amenity with experience. *Tourism Economics*, *23*(5), 1076–1095. https://doi.org/10.1177/1354816616665754

Mehraliyev, F., Kirilenko, A. P., & Choi, Y. (2020). From measurement scale to sentiment scale: Examining the effect of sensory experiences on online review rating behavior. *Tourism Management*, *79*, 104096. https://doi.org/10.1016/j.tourman.2020.104096

Moreira, A. C., Fortes, N., & Santiago, R. (2017). Influence of Sensory Stimuli on Brand Experience, Brand Equity and Purchase Intention. *Journal of Business Economics and Management*, *18*(1), 68–83. https://doi.org/10.3846/16111699.2016.1252793

Nandagopal R, R. R. (2015). A Study on the Influence of Senses and the Effectiveness of Sensory Branding. *Journal of Psychiatry*, *18*(2). https://doi.org/10.4172/Psychiatry.1000236

Nurazizah, G. R., & Marhanah, S. (2020). Influence of Destination Image and Travel Experience Towards Revisit Intention in Yogyakarta as Tourist Destination. *Journal of Indonesian Tourism, Hospitality and Recreation*, *3*(1), 28–39.

Peck, J., & Childers, T. L. (2006). If I touch it I have to have it: Individual and environmental influences on impulse purchasing. *Journal of Business Research*, *59*(6), 765–769. https://doi.org/10.1016/j.jbusres.2006.01.014

Satti, Z. W., Babar, S. F., & Ahmad, H. M. (2019). Exploring mediating role of service quality in the association between sensory marketing and customer satisfaction. *Total Quality Management & Business Excellence*, 1–18. https://doi.org/10.1080/14783363.2019.1632185

Spangenberg, E. R., Sprott, D. E., Grohmann, B., & Tracy, D. L. (2006). Gender-congruent ambient scent influences on approach and avoidance behaviors in a retail store. *Journal of Business Research*, *59*(12), 1281–1287. https://doi.org/10.1016/j.jbusres.2006.08.006

Promoting Creative Tourism: Current Issues in Tourism Research – Kusumah et al. (Eds)
 ISBN 978-0-367-55862-8

Consumer's complaint behavior between Indonesian and non-Indonesian in the hotel

Y. Machiko, Ivena, M. Kristanti & R. Jokom
Petra Christian University, Surabaya, Indonesia

ABSTRACT: Complaint management is an essential element of hotel business success. Handling customer complaints effectively could result in a future referral or loyalty behavior. Therefore, understanding the customer's complaint behavior will help the hotel operator to create a strategy for maintaining their customers. This paper divides guest's complaint behavior into four categories. They are voice responses, private responses that consist of negative word-of-mouth and exit, third party responses, and taking no action (inertia). These behaviors are related to Hofstede's culture dimension that consists of six dimensions. The authors examined complaint behavior of Indonesian and non-Indonesian, in this case, Asian and Western, and evaluated the significant differences between those three groups of guests. Five-scale Likert questionnaires were distributed to 100 Indonesian, 75 Asian, and 75 Western guests. Then the data was analyzed using One-Way ANOVA. The findings showed that Indonesian and Western guests tend to have voice response complaint behavior, meaning they complained directly to the hotel. Whereas Asian guests were more likely to have negative word-of-mouth behavior, where they will share their bad experience to family and friends. In addition, there is a significant difference for voice, word-of-mouth, and inertia, while there is no significant difference for exit and third-party complaint behavior. The difference in voice responses behavior could be seen in Indonesian and Asian customers, while the difference in word-of-mouth and inertia behaviors were shown between Indonesian and Asian, also Asian and Western groups of customers. This study provides new insight about cross- culture complaint behavior.

Keywords: complaint behavior, Indonesian, non-Indonesian

1 INTRODUCTION

In thehotel industry, maintaining service quality and customer satisfaction is the essential factor. When service failure occurs, inevitably followed by consumer dissatisfaction, the effective handling of complaints becomes central to the recovery of service satisfaction (Jahandideh et al. 2014). Therefore, the ability to understand the customer complaint behavior is critical for hotels to improve customer loyalty. Several studies showed that complaint behaviors vary between customer cultural background (Ekiz & Au 2011; Sann et al. 2020). Hofstede identified six dimensions of culture, which are power distance, masculinity-femininity, individualism-collectivism, uncertainty avoidance, long-term orientation, and indulgence–restraint (Hofstede et al. 2010). Previous studies found that collectivistic societies are less likely to complain than individualistic culture, and that a society which is high in uncertainty avoidance is less interested in engaging in negative word-of-mouth (WOM). For example, Asian customers are more likely to spread negative WOM and tend to take no action due to face issue and conflict avoidance (Kim et al. 2010; Chan et al. 2017). Moreover, the same behavior was reflected in cultures with a higher emphasis on power distance and hierarchy, which are less likely to complain if they feel less powerful than the management (Jahandideh et al. 2014).

DOI 10.1201/9781003095484-73

Indonesian outbound departures increased rapidly from 2013 to 2017. By 2021, the ratio of outbound trips to households in Indonesia was predicted to reach 15.4%, nowhere near that of neighboring Malaysia and Singapore. Moreover, Indonesians spent around 8.3 billion U.S. dollars during their outbound travels in 2017 (Hirschmann 2019). This information proved that Indonesian travelers will become a potential market in the nearest future. However, there is limited study about their complaint behavior compared to the other countries. Therefore, this study aims to identify the complaint behavior of Indonesian guests and compare it with non-Indonesian (Asian and Western guests). This will enrich the cross-cultural understanding of hotel consumer complaint behavior.

2 LITERATURE REVIEW

2.1 *Consumer's complaint behavior*

In the service industries, such as hotels, the staff have a responsibility to provide satisfied service to customers; otherwise, the customers will be dissatisfied and make a complaint. In doing a complaint, the customers have a certain complaint behavior. Complaint behavior can be defined as an action or reaction taken by an individual which involves communicating something negative because of dissatisfaction from unwanted situations during the purchase and the use of a service or good (Istanbulluoglu et al. 2017; Ergun & Kitapci 2018).

Based on Ngai, et al. (2007), Kim et al. (2010), Jahandideh et al. (2014), Istanbulluoglu et al. (2017), and Chan et al. (2017), there are four complaint behaviors, namely, voice, private, third party, and inertia. Voice means the consumer directly reporting their dissatisfaction to the management (Ergun & Kitapci 2018); private responses can be divided into two kinds of responses, exit and negative word-of-mouth. Exit means the consumer stops buying from the company (Ergun & Kitapci 2018). Negative word-of-mouth means that consumers tell their friends and relatives who have not directly encountered the negative experience; third-party responses involve external objects, but these are aimed at organizations that are not directly involved in the dissatisfying transaction, such as media, consumer agencies, or legal firms; and inertia (taking no action), the consumer will not complain to the firm and will not do anything even though they are dissatisfied with the service provided. The consumer has the opinion that it is not worth doing the complaint; there is no value between the effort and the result of complaining.

2.2 *Culture*

Hofstede (in Ergun & Kitapci 2018) defines culture as the various aspects of life in the world such as good and bad, reality and fake, beautiful and ugly, as well as the roles played. Reisinger and Turner (in Jahandideh et al. 2014) said that culture refers to a stable and dominant cultural character of a community shared by most of its individuals and remaining constant for long periods of time

Hofstede et al. (2010) came up with six dimensions of culture, namely power distance, masculinity-femininity, individualism-collectivism, uncertainty avoidance, long-term orientation, and indulgence–restraint. Power distance talks about how the members of a community handle inequalities among people. Masculinity-femininity refers to a society that emphasizes masculine or feminine behavior. The example of masculine behavior is achievement, heroism, firmness, the way of getting money, and material asset; and feminine behavior is helping others, cooperation, modesty, placing relationships with people before money, not showing off, and concern for the quality of life. In individualism-collectivism, individualism is when individuals are expected to make decision by themselves and to take care of only themselves and their nuclear families, the extent to which people feel independent. On the contrary, in a collectivism society, individuals are part of a cohesive group and are expected to work and be rewarded as a group. Uncertainty avoidance is the extent to which people are uncomfortable with uncertain situations; the degree to which the members of a society feel uncomfortable with uncertainty and ambiguity. It has nothing to do with risk avoidance nor with following rules. Long-term orientation deals with change. In a long-term oriented culture, the world is always changing, and preparing for the future is a must. They foster pragmatic values oriented toward rewards, including persistence, saving, and capacity for adaption. In a short-term-oriented culture, society prefer to maintain traditions and norms while viewing changes with suspicion. The values promoted are related to steadiness, respect for tradition, preservation of one's face, reciprocation, and carrying out social obligations. In indulgence-restraint, indulgence is related to enjoying life and having fun. Restraint stands for a society that is regulated by strict social norms.

2.3 *Culture and consumer's complaint behavior*

According to Hofstede et al. (2010), culture as the main factor influences consumer's complaint behavior. It is important for the manager of a company to understand the cultural background of their guests to determine the relationship between a consumer's behavior and the stage of purchasing, and their given cultural characteristics. Ngai et al. (2007) said that consumers in different cultures demonstrate different types of complaint behavior and intentions.

Huang et al. (1996) and Yuksel et al. (2006) said that hotel guests from a country with high power distance have more tolerance to unsatisfactory goods and services and think it is a fact of life and are less prone to complain. It is supported by Ngai et al. (2007) and Chan et al. (2017) that Asian guests, who usually have a higher power distance, rarely complain to hotel management compared with non-Asian guests. On the contrary, Ergun and Kitapci (2018) found that the tendency of individuals to report their dissatisfaction to the hotel management was found to be high in societies where power distance is high, while these societies are also involved in more WOM communication. Consumers at the low end of the hierarchical steps are called "weak consumers," and these consumers do not expect the staff to offer sensitive service to them. As a result, these consumers regard the staff offering a service as strong and specialized in their jobs and, therefore, they choose for no action even if they are dissatisfied. On the contrary, with the power acquired from their hierarchical positions, consumers at the upper end of the hierarchical steps tend to show public action behavior by reporting their dissatisfaction to the hotel management or initiating legal action and reporting their complaints. Asian guests, who are accustomed to a higher power distance, were more likely to complain to hotel management than non-Asian guests.

Huang et al. (1996) and Yuksel et al. (2006) found that hotel guests from a masculinity society want to get things straight, resulting in more complaints to the management and third parties. On the contrary, hotel guests from a feminine society are less likely to complain. For individualistic-collectivist point of view, consumers from an individualistic country are expected to report their dissatisfaction to the hotel or to a third party than individuals from a collectivist society, who are more likely to engage in private responses such as warning friends and relatives. When experiencing dissatisfaction, consumers in an individualistic society are more likely to voice their complaints than those in a collectivist culture; Regarding the individualistic-collectivist society, Ngai et al. (2007) and Chan et al. (2017) said that Asian guests from collectivist cultures think that voicing a complaint is disturbing compared with guests from individualistic cultures or non-Asian guests. Other support comes from Ergun and Kitapci (2018), who say individualist cultures were more willing to voice their dissatisfaction and try to find solutions to their complaints; the individuals adopt an honest and direct communication style.

In addition, Huang et al. (1996) and Yuksel et al. (2006) mentioned that hotel guests from high uncertainty avoidance country might avoid complaining to the hotel or to a third party. On the other hand, guests from a low uncertainty avoidance have a high tolerance for uncertainty and open conflict and should be more likely to complain to the hotel or a third party. These findings are supported by Ngai et al. (2007) and Chan et al. (2017), Asian guests with a higher uncertainty avoidance tend to be more fearful of losing face when making a complaint than non-Asian guests. Asian guests, who have a high tendency to want to avoid uncertainties, tend to be more resistant to change, more fearful of failure, and less likely to take risks. Therefore, they will choose not to complain if they do not know how and where to complain, or to whom they should complain. Consumers included in the uncertainty avoidance dimension prefer private action in the case of dissatisfaction. These guests are also more willing to share their dissatisfaction with others through negative WOM communication and leave the hotel (Ergun & Kitapci 2018). However, most of the non-Asian guests responded that they would try their best to find a way to complain when dissatisfied with a hotel's services, even if they were not familiar with the channels for complaint (Ngai et al. 2007). Based on long-term-oriented dimension, Ergun and Kitapci (2018) said that the long-term-oriented societies were more likely to tolerate bad service than short-term-oriented people. Thus, it can be concluded that the short-term-oriented societies are more likely to complain than the long-term-oriented societies.

Jahandideh et al. (2014) found that Asian consumers respond less actively to dissatisfaction and are less likely to complain to management. They are more likely to engage in private complaining actions such as negative WOM to friends and family members. Asian guests rarely bring complaints to a hotel for fear of losing face. In general, Asian consumers do not complain to the hotel or to the third party since they think that it disrupts the social order, and after having a bad experience in a hotel, they think they should warn their relatives and friends about their experience and the hotel.

Related to Indonesian culture, Hofstede (2020) mentioned that Indonesia has high power distance that makes communication indirect and negative feedback hidden. Indonesia is collectivist and femininity country. Indonesia is less masculine compared to some other Asian countries like Japan, China, and India. Indonesia has low uncertainty avoidance, and Indonesians will not show negative emotion or anger externally. Indonesia has high long-term orientation and restraint culture.

3 METHODS

The survey was held on-site at Juanda International Airport, Surabaya, Indonesia. Potential participants were asked to answer several selection criteria before they started to fill in the questionnaire. The respondents are at least 17 years old and have been staying in hotels in the last one year. A total of 100 Indonesian guests, 75 Asian guests, and 75 Western guests participated in this study.

The questionnaire was prepared in Indonesian and English languages. It was divided into two sections, which are the respondent profile and complaint behavior. Scenario approach was used to measure complaint behavior. Respondents were faced with a case of service failure in a hotel, and they were required to visualize this experience and responded to the questions naturally as if it really happened. The response questions were adapted from Yuksel et al. (2006); Butelli (2007); Kim et al. (2010) and consisted of (1) four indicators as the measurements of voice response, (2) five indicators as the measurements of private response, (3) three indicators as the measurements of third-party responses, (4) five indicators as the measurements of taking no action (Inertia). The grading score of the service failure response was using the f5-Likert scale method, with anchors "strongly disagree" as 1 to "strongly agree" as 5. The data was analyzed using descriptive analysis such as mean and standard deviation to describe their behavior. Moreover, one-way ANOVA was used to examine the behavior differences among those groups, and least-significant difference (LSD) post-hoc testing was conducted to describe the details.

4 RESULT AND DISCUSSION

Most respondents were between 17–25 years old and travel for holiday purpose. In terms of country of origin, there were 100 Indonesian people, 75 Asian people, and 75 Western people. For Asian, most participants were from Malaysia (10.8%), and for Western, most participants were from Australia (10.8%). Most participants were traveling with family (62.4%).

Related to complaint behavior of Indonesian, Asian, and Western people, it can be seen in Table 1 that Indonesian and Western people are more on voice responses, and Asian people are more on private responses, in this case, negative WOM. Based on analysis of variance, there were significant differences for voice, WOM, and inertia responses.

Table 1. Analysis of variance (one-way ANOVA).

Complaint Behavior	Citizenship	Mean	F	Sig	Notes
Voice	Indonesian	3.47	3.32	0.038	Significant
	Asian	3.19			
	Western	3.24			
Word of Mouth	Indonesian	3.05	4.72	0.010	Significant
	Asian	3.36			
	Western	2.94			
Exit	Indonesian	2.78	1.08	0.340	Not Significant
	Asian	2.67			
	Western	2.57			
Third Party	Indonesian	1.79	0.27	0.762	Not Significant
	Asian	1.87			
	Western	1.87			
Inertia	Indonesian	2.08	24.48	0.000	Significant
	Asian	2.92			
	Western	2.11			

Based on Hofstede (2020), Indonesia is high power distance, collectivist, femininity, long-term orientation, restraint, and low uncertainty avoidance societies. By having these characteristics, the complaint behavior of Indonesian people should be indirect and hide negative feedback, maintain good relationship, be adaptive, and not show negative emotion and anger externally. According to Huang et al. (1996), Yuksel et al. (2006), Ngai, et al. (2007), and Chan et al. (2017), a country with larger power distance is less likely to complain to hotel management. However, in contrast to the finding of the research is that Indonesian people are more on voice responses, which means that they will complain directly to hotel management. This finding is supported by Ergun and Kitapci (2018); the societies with high power distance have the tendency to report their dissatisfaction to the hotel management as well as be involved in more WOM communication. Indonesian people are more on voice responses since most respondents are 17–25 years old or in Z generation. The Z generations have the tendency to be individualistic, which means that they more likely to complain and communicate virtually by online platform. By online platform, these generations are free to express their feelings and ideas spontaneously including complaining to hotel management through hotel websites.

It is shown that Asian people are more on negative word-of-mouth responses than voice responses. Mostly, Asian countries are high in power distance, collectivist, and femininity societies. According to Huang et al. (1996) and Yuksel et al. (2006), hotel guests from femininity societies are less likely to complain. Moreover, hotel guests in a collectivist culture are more likely to express private responses compared to individualistic countries that adopt an honest and direct communication style (Ergun & Kitapci 2018). Most of the respondents are Malaysian people that have very high power distance and long-term relationship. Huang, Huang, and Wu (1996), Yuksel, Kilinc, and Yuksel (2006), Ngai, et al. (2007), and Chan, Tang, and Sou (2017) mentioned that Asian guests with high power distance are less likely to complain to hotel management. Jahandideh et al. (2014) said that Asian guests are more likely to engage in private responses such as negative WOM to friends and families. They are less likely to complain to the hotel after having bad experience since they think that it disrupts the social order or harmonious relationship and fear losing face. Related to long-term relationship, Ergun and Kitapci (2018) found that short-term-oriented are more likely to complain than the long-term-oriented societies.

This research found that Western people are more likely to voice responses; they will directly report their dissatisfaction to hotel management. In general, Western people are individualistic, low power distance, masculinity, high in uncertainty avoidance, short-term oriented, and indulgence.

Based on Huang, Huang, and Wu (1996), Yuksel, Kilinc, and Yuksel (2006), countries with low power distance are prone to complain. Others, hotel guests from a high masculinity society, are more likely to get things straight that result in more complaints to the hotel management. Hotel guests from an individualistic country are more expected to complain or voice their complaint to the hotel than individuals from a collectivist country. These findings are supported by Ngai, et al. (2007) and Chan, Tang, and Sou (2017), societies with low power distance are more likely to complain to hotel management than Asian guests. Regarding individualistic and long-term oriented, there is support from Ergun and Kitapci (2018). Individualist societies were more willing to voice their dissatisfaction and try to find solution to their complaints, hotel guests from an individualist country prefer to ask for compensation for bad service rather than admitting it or sharing it with others. Guests from a short-term-oriented country are more likely to complain than from a long-term-oriented country. Related to uncertainty avoidance, Huang, Huang, and Wu (1996) and Yuksel, Kilinc, and Yuksel (2006) said that hotel guests from a high uncertainty avoidance country, mostly Western countries, might avoid complaining to the hotel. This statement is in contrast with the result of the research that Western people are more likely to voice their complaint. It could be because the majority of respondents are from Australia, which is in middle uncertainty avoidance. Australian people are more dominant for low power distance, short-term orientation, individualistic, masculinity, and indulgence. Indulgence means that Australian people travel for holiday purpose, so that they place more importance on leisure time and spend money as they wish (Hofstede 2020); they will complain to hotel management to get a nice holiday and worth their money.

Since there were significant difference for voice, word-of-mouth, and inertia, LSD post-hoc test was done. There was significant difference between Indonesian and Asian people for voice responses since in this case, the Asian people are dominated by Malaysian people that have high power distance, and most Asian people have high uncertainty avoidance compared to Indonesian (Hofstede 2020). These mean Indonesian people are more on voice responses than Asian in general. There was significant difference between Indonesian and Asian, and between Asian and Western for WOM and inertia responses. Based on Hofstede (2020), Western people are dominant in low power distance, individualistic, masculinity, short-term orientation, and indulgence. This means Western people are more on voice responses than WOM and inertia compared to Asian people. It can be seen from Table 1 that mean of inertia of Western people is lower than Asian people. For Indonesian and Asian people, Indonesian people tend to have voice responses compared to Asian people who like to have negative WOM and inertia responses. Asian people choose to tell their friends and families or do nothing since there is no value between the effort and the result of complaining.

5 CONCLUSION AND IMPLICATION

This study offers an essential finding for broadening our understanding on consumers' complaint behavior, which can serve as guidelines for hotel management and academicians. The findings enrich literature on service quality by examining differences in complaining behaviors across distinct cultural backgrounds. Using a scenario about service failure without any recovery action, this study attempted to examine significant differences in behaviors of hotel guests among Indonesian, Asian, and Western. Indonesian guests are more likely to complain directly to hotel management, like Western guests who tend to have voice response, whereas Asian guests are more likely to engage in private responses such as negative WOM to friends and families.

The practical implications of complaint behavior differences across cultures are obvious. Hotel managers can enhance their frontline staff knowledge about how they should customize their service to the guests from different cultural background. Several studies admitted that language barrier plays an important role in the failure of communicating customer's dissatisfaction (Yuksel, Kilinc, & Yuksel 2006; Jahandideh et al. 2014; Chan, Tang & Sou 2017). Therefore, hotels may encourage their employees to pay more attention to the customers' expression of complaint and develop easier ways for these customers to communicate their dissatisfaction with the hotel management rather than elsewhere (Yuksel, Kilinc, & Yuksel 2006). Furthermore, this research has limitations on small sample size and the measurement of complaint behavior. Guests may react differently when they face different cases in the real situation. Therefore, further research could use different approach such as observation or interview to have deeper understanding of customer complaint behavior. Regardless of its limitation, this study confirms that differences in customer complaint behavior exist among different nationalities.

REFERENCES

Butelli, S. 2007. *Consumer complaint behavior (CCB): A literature review*. Northumbria University.

Chan, G.S.H., Tang, I.L.F. & Sou, A.H.K. 2017. An exploration of consumer complaint behavior towards the hotel industry: Case study in Macao. *International Journal of Marketing Studies* 9(5): 56–76.

Ekiz, E.H. & Au, N. 2011. Comparing Chinese and American attitudes towards complaining. International *Journal of Contemporary Hospitality Management* 23(3): 327–343.

Ergun, G.S. & Kitapci, O. 2018. The impact of cultural dimensions on customer complaint behaviors: an exploratory study in Antalya/Manavgat tourism region. *International Journal of Culture, Tourism and Hospitality Research* 12(1): 59–79.

Hirschmann, R. 2019. *Number of outbound departures Indonesia 2013–2017*. Retrieved from https://www.statista.com/statistics/726892/number-of-outbound-travelers-indonesia/

Hofstede, G., Hofstede, G.J. & Minkov, M. 2010. *Cultures and organizations: software of the mind, intercultural cooperation and its importance for survival*. United States: McGraw Hill.

Hofstede, G. 2020. *Hofstede insight*. Retrieved July 12, 2020, from www.hofstede-insights.com/country-comparison/

Huang, J., Huang, C. & Wu, S. 1996. National character and response to unsatisfactory hotel service. *International Journal of Hospitality Management* 15(3): 229–243.

Istanbullouglu, D., Leek, S. & Szmigin, I.T. 2017. Beyond exit and voice: Developing an integrated taxonomy of consumer complaining behavior. *European Journal of Marketing* 51(5/6): 1109–1128.

Jahandideh, B., Golmohammadi, A., Meng, F. & O'Gorman, K.D. 2014. Cross-cultural comparison of Chinese and Arab consumer complaint behavior in the hotel context. *International Journal of Hospitality Management* 41: 67–76.

Kim, M.G., Wang, C. & Mattila, A.S. 2010. The relationship between consumer complaining behavior and service recovery: An integrative review. *International Journal of Contemporary Hospitality Management* 22(7): 975–991.

Ngai, E.W.T., Heung, V.C.S., Wong, Y.H. & Chan, F.K.Y. 2007. Consumer complaint behavior of Asians and non-Asians about hotel services: An empirical analysis. *European Journal of Marketing* 41(11/12): 1375–1391.

Sann, R., Lai, P. C., & Liaw, S. Y. 2020. Online complaining behavior: Does cultural background and hotel class matter? *Journal of Hospitality and Tourism Management* 43: 80–90.

Yuksel, A., Kilinc, U.K. & Yuksel, F. 2006. Cross-national analysis of hotel customers' attitudes toward complaining and their complaining behaviors. *Tourism Management* 27: 11–24.

Promoting Creative Tourism: Current Issues in Tourism Research – Kusumah et al. (Eds)
© 2021 Taylor & Francis Group, London, ISBN 978-0-367-55862-8

Exploring factors influencing homestay operators to participate in the homestay program

S. Haminuddin, S.S. Md Sawari & S.A. Abas
International Islamic University Malaysia, Johor, Malaysia

ABSTRACT: Homestay operators can be considered as the local community who is directly involved with tourism development. In homestay lodging, their perception becomes a major concern to ensure the balance impact in tourism activities. Nevertheless, due to the failure in improving homestay standard, there were many homestay operators who gave up continuing their services. Consequently, to understand the issues, this paper will explore the motivational factors of homestay operators to participate in the homestay program. This study employs qualitative research design by using interview as a method to collect the data. There are seven participants involved in the interview. As a result, there are five major factors that motivate homestay operators to be involved in the homestay program. These factors are significant in this research. The factors were classified into two, which are internal factor and external factor. The internal factor comprises willingness, interest, and excitement while the external factor consists of kinship factor and tourist benefit factor. With the expected tourism development and increasing number of tourists, it is time to investigate the level of acceptance of tourists among the residents.

Keywords: homestay, homestay operator, community-based tourism, homestay program

1 INTRODUCTION

Homestay was defined in various ways in various country. By referring to the Ministry of Tourism, Arts and Culture (MOTAC) Homestay Program Guideline, a homestay program in Malaysia is defined as an experience stay with local communities that have been registered under MOTAC. It is aimed to give an opportunity for the tourist to interact directly with local community to experience the culture and tradition of the village. The main concept for homestay program is more to village lifestyle experience. The objective of a homestay program is to encourage the local community involvement in the tourism industry, encourage economy growth and sharing benefit, and to produce the entrepreneur among local community in rural areas. For benefit of tourists, a homestay program provides the unique experience of the village and culture of local community and at the same time produce a tourism product with a competitive price with level international.

The homestay operator acts as the important key as a homestay program provider because they are parties that are struggling to promote the homestay program (Yusnista & Nik Haziva 2016). Activity of the homestay program is supposed to give benefit to the local community, especially the homestay operator. So, when it involves the tourist problem, they have their own perception, and it is more accurate because the homestay operator is the party that directly interacts with tourists (Hanim et al. 2014a).

There were several issues related with the homestay operator found in the literature. According to Yusof et al. (2018), the homestay operator needs to improve their standard of living to achieve a better life for the future because there are certain periods when the homestay operators were unable to interact with tourists who were to stay at their home because not all went well. Due to this, many homestay operators gave up to continuing their service to be a homestay operator. Through the

 DOI 10.1201/9781003095484-74

homestay concept, it is not clear whether the homestay operator benefits fully from participation in the homestay program activities from the socio-economic aspect.

Despite of the emerging number of homestays across the country, yet in certain areas, the participants are decreasing. This is might be related to resident attitude with the stage of tourism development of the homestay program and the impacts (Fernando & Antonia 2015). As to address the problem, the researcher is eager to fill the gap by looking into the motivational factor of homestay operators participating in a homestay program. This study also will be highlighting the socio-economic impact of homestay from the perception of operators.

2 METHODOLOGY

The data gained was from the primary resources, which are the homestay operators that are active in homestay program activities. Qualitative is a scientific method of observation to gather non-numerical data (Sim et al. 2018). Open coding strategies were used for the data analysis. The coding helps to discover theme and pattern of data collected that lead to the theoretical understanding of social life. Open coding was used to classify and all of the coding will be added with explanation and justification. In this study, research wants to explore the motivational factors that influence the homestay operator joining the homestay program. The researcher chose the village because this village is most active in homestay in Muar, Johor, and gets awards from the tourism of Malaysia (Musa 2017). This village well cooperated with local community and supported tourism development. Due to the study focus on homestay operator, the main participant chosen is the direct participant, which is the homestay operator that actively participated in this homestay.

The researcher was able to find seven homestay operators to participate in this interview. According to Corbin (2015) as cited in Sim et al. (2018), the purpose number of participants was at least five participants and should spend one-hour interviews for theoretical saturation in grounded theory studies. So, it means the researcher is valid to continue the research with seven participants.

3 FINDINGS

This section will be presenting the data based on the interviews that have been done by the researcher with the homestay operator. Through the interview, the researcher was to find out several motivational factors of participation as a homestay operator. The motivational factor was themed into two categories, which are internal factor and external factor. Internal factor was divided into three classes: willingness, interest, and excitement. The external factor was divided into kinship factor and tourist benefit factor.

3.1 *Internal factor*

3.1.1 *Willingness to participate homestay program*

The willingness is about how that person is able to do willingly and eagerly. Based on the interview, the researcher found out the participant has their own willingness in joining the homestay program. There is no force by people. The willingness comes because of their own character that loves people. Here a few answers from the participant.

> *Participant 2: Frankly speaking, I am a cancer patient and totally it is sad for me. With tourist around, I can forget all sadness because I have to entertain them and at the same time I entertained myself too. Since my family are chatterbox, so with tourist we are able to interact and share many things. This is such a treatment for me also. So far there are no negative impacts from tourist, only small matter but we still can handle it by the way.*
>
> *Participant 3: I never feel disturbed by tourist. Before this, I have planned to quit because of the age factor. But when I was thinking back, I like to do this, and there is a thing can do than nothing.*

To improve the homestay program, it is importance to understand the tourist motivation to join the homestay program and how it affected their satisfaction (Vigolo et al. 2018). To learn the tourist motivation, it is important for the homestay operator to be willing and committed to learn about the tourist motivation. This is because willingness is influenced by motivation. The homestay operator has motivation to be involved in a homestay program because it is full of potential to profit-making (Vigolo et al. 2018).

The participation in the homestay program gives benefit to the homestay operator, and it leads to the homestay operator willing to learn related to homestay program activities, marketing, and development, including the tourist preferences. The homestay operator always ensures the available room for the tourist so that they are comfortable.

3.1.2 *Interest toward tourist existence*

The key informants spoke about their interest toward tourists such as feel welcoming, feeling free to entertain, tourist behavior tolerance, and empathy toward tourists. As the evidence, this is the response of the participant toward the interview question, which the researcher asked about the point of view regarding the tourist that come to this village which joined the homestay program. Most of them are really welcoming and feel happy with tourists comeing to join the homestay program.

> *Participant 3: I love tourist. There is no problem created with tourist so far. They are satisfied with my services. I really welcome they come to my house and entertain them. They can fill my day so I do not get lonely, and being alone at my house during my children got to work. There are many things we can learn from them and often open my mind.*

Interest is what leads the homestay operator to enjoy their work and gives the best quality of work. If there is no interest, the homestay program will not last. Interests will enable and lead them to learn more and more about their fieldwork with an aim to give great services to the tourist.

According to the results from the interview, the homestay operator has high interest in tourism, especially involving tourist existence. The homestay operator is always welcoming and happy whenever tourists come to their place since they realize their existence will give benefit to them. Furthermore, the interest will be getting expended when they get support from others or discover connections with personal skills, knowledge, and skill (Black 2017).

3.1.3 *Excitement toward tourist of homestay operator*

As the homestay operator, they are eager and excited to fulfill the satisfaction and make the tourist an important subject that should be protected. Even though the participant has already had 15 years to be anhomestay operator (refer Table 9), they never to stop from learning to improvise their services so that they can give the quality services to their guest. Another question to the homestay operator is about how they are handling the problem if any issues arise from a tourist.

One of the participants was told the way for them to improve their work was through a feedback form from the tourist. How they respond whenever a problem comes out from the tourist shows their level of acceptance of tourist behavior. Based on the interview, the participant answering was like the following:

> *Participant 5: We should not judge the tourist immediately if the tourist shows negative attitude such like wearing inappropriate clothes during visiting the homestay. We should and explained to them about our culture so that they understand. I always prepare our culture clothes such as "baju kurung or kain pelikat" for them to wear if there is problem such like this happened. At the same time they can real experience the kampung style while wearing our tradition the clothes.*

Some of the homestay operators are always excited to welcome tourists. They always take care of their facilities to ensure everything is in good condition and maintained. The cleanliness is always a priority for the tourist. This is to ensure the tourists are comfortable. This is based on the interview:

Participant 4: If I get offered to take tourist as foster family I ensure the room is available before I accept. It is depending on the situation. If that time my family has around, I will pass to the offer other homestay operator. It is because if it is too many people in one house, the tourist will not comfortable. Their comfortness is our priority. As I said before, the tourist is our income so we need to take care of them as gold.

This is all the response that was obtained through interview with seven participants. Their reaction to respond to the interview question was based on the answer given. So, to conclude, the homestay operator was passionate toward the homestay program especially involving the tourist existence. Even though there are issues and problems exist, these participants are able to handle it.

To conclude, homestay operators are excited to welcoming tourists since customers are considered as the most important subject in any tourism business. This is because the number of their appearance shows the success of the business. According to Aliman et al. (2014), tourists are the main resource of host destination income. Due to that, it is really crucial for all the host of tourism products to ensure all the quality of products, services, and facilities are well maintained to protect the destination image. Tourism destination image is the focal point of tourist satisfaction because it creates tourist motivation that leads to decision making (Lai & Li 2016).

3.2 *External factor*

3.2.1 *Kinship*

Sometimes the positive vibe from surroundings can influence and encourage someone to be involved in activities. The encouragement of the family somehow is the main factor of the person who does the decision to join something. In this study, it can be said that the family and friends and the communities are supportive and open for development. Most of the homestay operators get support and encouragement from the people surrounding that help them a lot, which makes their work become easier as homestay operator.

Participant 7: All of my family does not have any problem to entertain tourist. They are really supportive and open minded. That is why, I can be a homestay operator until now which is almost 15 years.

The kinship is defined as the family or close friends (Lee et al. 2016). Kinship is the encouragement of the family, friends, or the surrounding as support that involve emotional and financial assistance (Yusup & Mansora 2016). In this context, the support from the family, friends, and community is the main factor of tourist acceptance. It is important to understand the encouragement from the family and friends about involvement in tourism development.

Based on the interview, the participant admitted their family support is the main factor why there still are homestay operators. It is can be seen from how the family also is together interact with the tourists, who always shared information and agreed to use their room as the tourist room if they are not at home. It can be the the significance of this factor.

3.2.2 *Tourist benefit*

The homestay operator has believed the tourist appearance will give benefit to them. That is why most of them are showing willingness, interest, and are excited to be a part of a homestay program because it can give positive impacts in their lives for a long-term period.

Due to that, the researcher was asked about the importance of tourists to them fully in their satisfaction. Here is the response from the interview:

Participant: Tourist gives a chance for us to be more creative like we able to produce creative product like the variety of krepek from various ingredients. Not only that there are some of us are able to produce creative traditional handicraft for souvenir. In the same time we are involved in small business also become food supplier for some of tourist. So from there we are able to generate money.

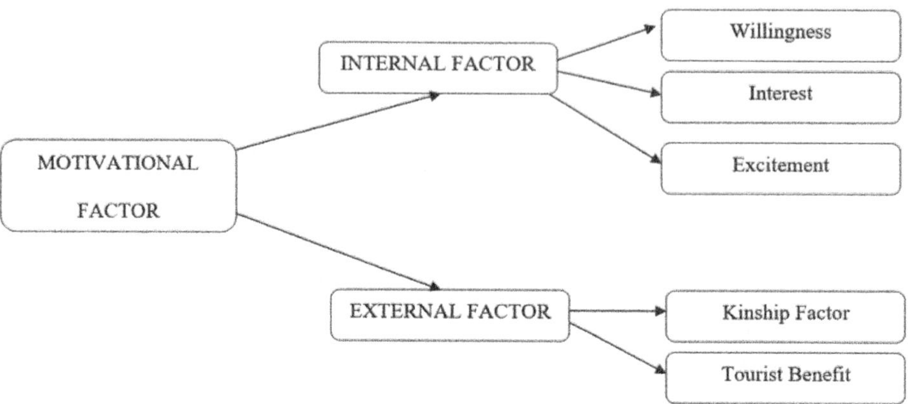

Figure 1. Show the motivational factors framework.

Another benefit is from the environmental side.

> *Participant: We are able to utilize our precious soil for goods. We do cultivation activities like harvesting, planting like pineapple fruits, cassava and etc. It is intended to shos the process of cultivation that also can variety the homestay activities agro-based ventures for tourist. Adding the agro tourism gives opportunity to expand our activities and also to increase our income.*

In this case study, it can be said that tourists give a chance for homestay operators to generate income. In this case study, there are many impacts that are obtained from participating in a homestay program, especially in economy, social, and cultural. The homestay operators are able to be involved in small business, the uses of soil for good purposes, and give good income. The participants told that they are always joining any training or class related to the homestay program just to add knowledge to improve the services. From there, they are able to improve the facilities of the village and also their home, which is the homestay operator or individual that offers their house for a tourist to stay. Not only had that, homestay operators able to gain and learn new knowledge and skills as a trick to attract a tourist's attention to stay in their homestay. To be short, tourism is the main factor of the development of homestay program that could gives benefit to them. So, it is important for homestay operators to preserve and maintain the good services to the tourist to influence their revisit intention since tourists give benefits in the homestay operator's life.

This is all the response that was obtained through interviews with seven participants in response to the first objective, which is to determine the factor that motivates homestay operators to participate in a homestay program. Their reaction ws to respond to the interview question based on the answer given. So, to conclude, the homestay operator was motivated due to internal factor and external factor. For internal factor it is willingness, interest, and excitement while for external factor it comprises kinship factor and tourist benefit. Figure 1 shows the motivational factors framework.

4 CONCLUSION

The findings of this study revealed that there are five major motivational factors in participating homestay programs, which are willingness, interest, excitement, kinship, and tourist benefit. The capability of providing opportunities for the homestay operator to join the homestay program shows the benefits that were gained from it. The expected benefit from it encourages and motivates the participation of local community in homestay program. In this case study, it obviously shows the involvement of a local community to be a homestay operator willingly due to their interest and excitement toward tourism activities. The encouragement from the family and friends is also the factor that motivates to participate in a homestay program. With expected tourism development and increasing numbers of tourists, it is time to investigate the level of acceptance of tourists among residents. This is because residents or local community also indirectly are involved in activities of homestay programs and indicators show the success of a homestay program.

REFERENCES

Aliman, N. K., Hashim, S. M., Wahid, S. D. M., & Haridun, S. (2014). Tourist Expectations, Perceived Quality and Destination Image: Effects on Perceived Value and Satisfaction of Tourists Visiting Langkawi Island, Malaysia. *Asian Journal of Business and Management, 2*(3), 212–222. Retrieved from www.ajouronline.com

Black, S. (2017). Enhance Your Reference Skills by Knowing the Four Phases of Interest Development. ACRL 2017, Baltimore, Maryland, March 22–25, 2017, 200–205.

Fernando, A.G., Antonia, B.V., Rafael, C.M. Resident's Attitude towards the impacts of tourism [J]. Tourism Management Persectives, 2015, 13(1):33–40.

Hanim, N., Salleh, M., Othman, R., Hajar, S., & Idris, M. (2014a). Perception of Homestay Operators towards Homestay Development in Perception of Homestay Operators towards Homestay Development in Malaysia, (December). https://doi.org/10.17576/pengurusan-2014-42-01

Lai, K., & Li, X. (Robert). (2016). Tourism Destination Image: Conceptual Problems and Definitional Solutions. *Journal of Travel Research, 55*(8), 1065–1080. https://doi.org/10.1177/0047287515619693

Lee, Eunju & Choi, Mi Jin & Clarkson-Henderix, Michael, 2016. "Examining needs of informal kinship families: Validating the family needs scale," Children and Youth Services Review, Elsevier, vol. 62(C), pages 97–104.

MOTAC. (2019). Ministry of Tourism, Arts and Culture Malaysia. Retrieved 11 11, 2019, from Homestay Program: http://www.motac.gov.my/en/download/category/10-homestay-kampungstay

Musa, Z. (23 10, 2017). Star Online. Retrieved from Sultan Ibrahim visit awards-winning Bugis village: https;//www.thesatronline.com/new/nation/2017/10/23-Sultan-Ibrahim-visit-awrds-winning-bugis-village

Norhafezah, Y., Ismail, A., Ismail, R., & Aripin', N. (2018). Industry Perspective on Journalism Education Curriculum in Malaysia. *Malaysian Journal of Learning and Instruction 15(1)*, 149–172.

Sim, J., Saunders, B., Waterfield, J., & Kingstone, T. (2018). The sample size debate: response to Norman Blaikie. *International Journal of Social Research Methodology*.

Vigolo, V., Simeoni, F., Cassia, F., & Ugolini, M. (2018). The Effects of Travel Motivation on Satisfaction: The Case of Older Tourists. *International Journal of Business and Social Science, 9*(2), 19–30.

Vigolo, V., Sallaku, R., & Testa, F. (2018). Drivers and Barriers to Clean Cooking: A Systematic Literature Review from a Consumer Behavior Perspective. *Sustainability 10(11)*, 4322.

Yusup, N. B., & Mansora, A. A. (2016). The influence of parents support and its relationship with students achievement in English education. In *International Conference on Education and Regional Development (ICERD)*.

Promoting Creative Tourism: Current Issues in Tourism Research – Kusumah et al. (Eds)
© 2021 Taylor & Francis Group, London, ISBN 978-0-367-55862-8

Analysing the tourist's e-satisfaction of hotel booking website towards online purchase intention in Malaysia

M.M. Jamil & N.H.A. Rahman
International Islamic University Malaysia, Johor, Malaysia

ABSTRACT: Hotel distribution marketing in Malaysia has changed drastically, resulting from the emerging of online hotel booking websites managed by online travel agencies (OTA). However, there has been a limited research conducted in analysing tourist's e-satisfaction on using those websites that influences tourist's purchase intention in Malaysia. This study seeks to identify tourist's e-satisfaction on online hotel booking website in Malaysia, to investigate tourist's online purchase intention on online hotel booking website in Malaysia and to determine the relationship between tourist's e-satisfaction and online purchase intention from online hotel booking in Malaysia. This study applied the quantitative approach with 162 respondents who have experience in using online hotel booking websites operated by OTA in Malaysia. The results have shown that tourists in Malaysia perceived the e-satisfaction of online hotel booking websites and had the intention to make an online purchase from the website. This study also presented that tourist's e-satisfaction was moderately inclined towards online purchase intention on online hotel booking websites. The findings from this research will contribute to online travel agencies in developing hotel booking websites that can enhance tourist's satisfaction and decision to make an online purchase.

Keywords: e-satisfaction, online hotel booking website, online travel agencies, purchase intention

1 INTRODUCTION

The assimilation between tourism with Information Communication Technology (ICT) over the last few years has prompted the invention of e-tourism in which tourists can easily obtain the information or even select tourism products and services using the internet connection (Buhalis & Jun 2011). The impact from the rise of significance in enhancing online hotel booking website experience in getting people's attention and interest to use this technology has led to the creation of indicators that determine hotel website user's excitement level such as e-satisfaction (Abbaspour & Hashim 2015). Moharrer (2013) stated that e-satisfaction is a suitable indicator that can be used in measuring experience perceived by tourists associated with using an online hotel booking website.

In Malaysia, the services offered in online hotel booking websites have received a greater demand among travellers. Bing (2014) had mentioned that a majority of website users visit the online website with the purpose to purchase on travelling products or services such as hotel room reservations. Therefore, various online hotel booking portals have been created to cater the growing number of website users who seek to make online reservation on available hotel deals. An effective hotel booking website's service is demanding a well-planned website that can provide better features, design and quality service to ease the website user's experience on using the website as well as making it interesting (Asraf et al. 2014).

Online travel agency (OTA) is one of the widely known technology media that offers services for tourists to make an online hotel reservation by offering available hotel products from a various range of hotels and deals (Chow 2017). In Malaysia, OTA that focuses on hotel deals business has started to compete in improving website's services and that incorporates giving a satisfying experience to the website's user as a result from the rapid growth of hotel booking website managed

 DOI 10.1201/9781003095484-75

by OTA (Jedin & Ranjini 2017). Despite the increasing development of online hotel booking in Malaysia in the scope of demand among tourist and the number of website's provider, there are still a few drawbacks in terms of analysing tourist's motivation to use the technology and develop their decision to make a purchase from online hotel booking websites.

Therefore, the present study is highly significant in line with the effort of OTA to attract more website consumers because this study is outlining the explicit aspects of online hotel booking websites that contribute towards user's satisfaction and online purchase intention.

2 LITERATURE REVIEW

2.1 *Tourist's online purchase intention*

Online purchase intention is referred to the readiness and willingness of consumers to involve in an online investment deal (Raza et al. 2014). Online purchase intention is often related to the behavior of online consumers in making progress of their buying plan action in which according to Liat and Wuan (2014), online purchase intention will occur when an individual has the motivation to buy particular products or services in the future through an online medium such as e-business website.

Teng (2018) described online purchase intention as a common term that dedicates to consumer's strength of intention to make the purchase of available goods and services through the online website.

After the development of online business websites that have been widely practised and implemented by many consumers and sellers, the reason people make a purchase through the online website becomes a concern, especially among the e-retail business providers. Chen (2012) stated that the satisfied consumers on the online business website would likely influence their purchase intention through the website and a number of previous studies have already achieved the positive results between satisfaction on online tourism business website's influence on tourist's purchase intentions (Chen 2012; Egresi 2017; Teng 2018).

2.2 *E-satisfaction*

E-satisfaction is a terminology developed by Szymanski and Hise (2000) which refer to website user who develops the feeling of contentment and pleasure towards the online portal in which he or she has experienced (Khai & Van 2018). This e-satisfaction also contain the determinants that could be used for better understanding of the quality of online environment perceived by its users, especially in online business portal such as online hotel booking website.

Cetinsoz (2016) mentioned that, e-satisfaction in an online hotel booking website should be discovered widely to emphasise the consumer's satisfaction upon their experience using the website which will greatly contribute towards their intentions to make a purchase through the online medium. Thus, special attention should be emphasised in developing tourist's e-satisfaction in e-tourism business such as online hotel booking website in order to advocate the tourist's intention to make a purchase from the website.

Convenience, site design, product selection, informative, service quality and e-security are the variables of e-satisfaction that have been proposed by Cetinsoz (2016), and it is claimed suitable to be used in identifying tourist's satisfaction on online tourism portal.

3 METHOD

This study is a quantitative research design that used descriptive analysis. Moreover, this study adopted a purposive sampling technique to investigate Malaysian tourists who have experience in using an online hotel booking website managed by OTA as the target population for this study. The survey questionnaire was distributed to target population through distributing the hardcopy of questionnaires and online survey form to be answered by the respondents. This study managed to obtain results from 162 respondents which exceeded the practical minimum size as suggested by Hair, Black, Babin and Andreason (2010).

Each item for both variables of e-satisfaction and online purchase intention had applied the 6-point Likert scale as an indicator of measurement, ranging from 'highly disagreed' (1) to 'highly agreed' (6). Then, the responses from respondents regarding their satisfaction on using online hotel booking from OTA were classified as either positive, neutral or negative based on the mean range for even number-Likert scale from Market Directions (2017).

Moreover, this study implemented the Spearman Correlation to analyze the strength of the relationship between the variables' study. The results from the Spearman Correlation test can be classified in several categories which absolute value of $r = 0.00$ to 0.30 is considered negligible correlation, $r = 0.30$ to 0.50 considered low positive correlation, $r = 0.50$ to 0.70 is moderate positive correlation, $r = 0.70$ to 0.90 is high positive correlation and $r = 0.90$ to 1.00 is very high positive correlation (Mukaka 2012).

4 RESULT

After analysing the data results obtained from 162 respondents of this study, it is found that most of the tourist agreed with their satisfaction on all the variables of e-satisfaction and online purchase intention mentioned in this study. Table 1 shows the mean value for all aspects of e-satisfaction and online purchase intention on online hotel booking was ranged from 4.60 to 5.13 (which $\mu > 3.00$). The highest mean score for tourist's e-satisfaction on online hotel booking was 'Convenience' with 5.13, followed by 'Site Design' that had the mean score of 5.04, continued with the aspects of 'Informative', 'Product Selection' and 'Service Quality' that had the mean score of 4.98, 4.96 and 4.76 respectively while 'E-Security' had the lowest mean score with 4.60. Meanwhile, the average mean value for tourist's online purchase intention on online hotel booking is 4.91.

Hence, it portrays that respondents agreed they satisfied with the convenience, site design, product selection, informative, service quality and e-security of online hotel booking websites also have the intention to make an online purchase from the website they have used as suggested by Cetinsoz (2016).

Among the e-satisfaction aspects perceived in this study, it is found that most of the tourists were satisfied with the convenience aspect on online hotel booking websites from online travel agencies, while e-security was the least satisfying aspect of e-satisfaction perceived by the tourist.

Furthermore, the relationship between tourist's e-satisfaction on online hotel booking towards online purchase intention through online travel agencies in Malaysia is shown in Table 2. The result shows a moderate positive correlation between the tourist's e-satisfaction on online hotel booking towards the online purchase intention. This is based on the Spearman correlation coefficient between e-satisfaction and online purchase intention was $r = 0.668$ in which both variables had moderate positive correlation with the significance value of $p = 0.000$ ($p < 0.05$).

Table 1. Mean score of tourist's e-satisfaction and online purchase intention on the online hotel booking website.

Variables	Mean Score
Convenience	5.13
Site Design	5.04
Informative	4.98
Product Selection	4.96
Service Quality	4.76
E-Security	4.60
Online Purchase Intention	4.91

Table 2. Spearman correlation test between e-satisfaction and online purchase intention.

			E-Satisfaction	Online Purchase Intention
Spearman's rho	E-Satisfaction	Correlation Coefficient	1.000	**.668****
		Sig. (2-tailed)	.	**.000**
		N	162	162
	Online Purchase Intention	Correlation Coefficient	**.668****	1.000
		Sig. (2-tailed)	**.000**	.
		N	162	162

**Correlation is significant at the 0.01 level (2-tailed).

5 CONCLUSION

This study examined the e-satisfaction among 162 tourists on online hotel booking towards their online purchase intention from OTA in Malaysia. This study which adopted e-satisfaction from Cetinsoz (2016) revealed that most of the tourists agreed that they perceived the factors of convenience, site design, product selection, informative, service quality and e-security when utilising online hotel booking website, operated by online travel agencies. The convenience aspect is the most influential factor that has contributed to the e-satisfaction among the Malaysian tourists on online hotel booking. Moreover, most of the tourists are also keen to proceed with online purchases from the hotel booking website managed by online travel agencies as the website is equipped with e-satisfaction aspects previously mentioned. This is due to the research findings showing that e-satisfaction was moderately influenced by the tourist's online purchase intention.

Hence, the findings of this study could be useful for online travel agencies in the hotel industry to provide tourists with a pleasant experience when using online hotel bookings that develop tourist's motivation to make an online purchase through the website.

REFERENCES

Abbaspour, B. & Noor Hazarina Hashim. (2015). The influence of website quality dimensions oncustomer satisfaction in travel website. *International Journal of Science Commerce and Humanities, 3*(5), 6–17.

Amrul Asraf Mohd-Any, Winklhofer, H. & Ennew C. (2014). Measuring Users' Value Experience on a Travel Website (e-Value): What Value Is Cocreated by the User?. *Journal of Travel Research, 54*(4), 496–510.

Bing, W. C. (2014). The Relationship Of Website Quality And Customer E-Satisfaction In Low Cost Carrier. *Universiti Teknologi Malaysia,* 1–207. Retrieve from http://eprints.utm.my/id/eprint/53644/25/Wong ChietBingMFP2014.pdf

Buhalis, D. & Jun, S. H. (2011). E-tourism. *Information technology for strategic tourism management, Prentice Hall, 3.* Retrieved from https://www.goodfellowpublishers.com/free_files/Contemporary-Tourism-Review-Etourism-66769a7ed0935d07653l8203b843a64d.pdf

Çetinsöz, B., C. (2016). The Impact of E-Satisfaction and Confidence Aspects on Web Site Loyalty in Terms of Online Shopping in Tourism: A Study on Domestic Tourists in Turkey. *European Journal of Business and Management, 8*(7), 2222–2839.

Chen, H., Chiang, R., & Storey, V. (2012). Business intelligence and analytics: from big data to big impact. *MIS Quarterly*.

Chow, K., M. (2017). E-service Quality: A Study of Online Hotel Booking Websites in Hong Kong. *Asian Journal of Economics, Business and Accounting, 3*(4), 1–13.

Egresi, I. (2017). Tourists' satisfaction with shopping experience based on reviews on TripAdvisor. *Original scientic paper, 65*(3), 330–345

Hair, J., F,. J., Black, W., C., Babin, B., J., & Anderson, R., E. (2010). *Multivariate data analysis.* Upper Saddle River, NJ: Pearson Education.

Khai, N., T., N. & Van, N., T., X. (2018). The Effects of Website Quality on Customer Satisfaction and Loyalty to Online Travel Agencies in Vietnam. *Journal of Science Ho Chi Minh City Open University, 8*(3), 90–104.
Liat, C., B. & Wuan, Y., S. (2014). Factors Influencing Consumers' Online Purchase Intention: A Study among University Students in Malaysia. *International Journal of Liberal Arts and Social Science, 2*(8), 121–133.
Market Directions. (2017). Discussion Paper on Scales for Measuring Customer Satisfaction. (Discussion Paper No. 800-475-9808). Retrieved from Market Directions for research in discussion on measurement scales in surveys: https://marketdirectionsmr.com/wp-content/uploads/2017/07/survey-scales.pdf
Moharrer, M., Tahayori, H., & Sadeghian, A. (2013). Drivers of Customer Satisfaction in Online Tourism-The Case of European Countries. *Middle-East Journal of Scientific Research, 13*(9), 1172–1179.
Mohd Haniff Jedin & Ranjini, K. (2017). Exploring the Key Factors of Hotel Online Booking through Online Travel Agency. *International Conference on E-Commerce (ICoEC),* 9–14. Retrieve from http://icoec.my/download/icoec_2017/ICoEC2017_106.pdf
Muhammad Ali Raza, Muhammad Abdul Ahad, Muhammad Adeel Shafqat & Muhammad Aurangzaib (2014). The Determinants of Purchase Intention towards Counterfeit Mobile Phones in Pakistan. *Journal of Public Administration and Governance, 4* (3), 1–19.
Mukaka, M., M. (2012). A guide to appropriate use of Correlation coefficient in medical research. *Malawi Medical Journal, 24*(3), 69–71.
Szymanski, D., M. & Hise, R., T. (2000). Evaluation of E-Service Quality through Customer Satisfaction (a Case Study of FBR E-Taxation). *Open Journal of Social Sciences, 5*(9), 309–322.
Teng, T., Y. (2018). Factors Influencing Malaysian Youth Consumers' Online Purchase Intention of Travel Products. *Universiti Tunku Abdul Rahman Faculty of Accountancy and Management,* 1–113. Retrieved from http://eprints.utar.edu.my/2983/1/FACTORS_INFLUENCING_MALAYSIAN_YOUTH_CONSUMERS%E2%80%99_ONLINE_PURCHASE_INTENTION_OF_TRAVEL_PRODUCTS.pdf

Promoting Creative Tourism: Current Issues in Tourism Research – Kusumah et al. (Eds)
 ISBN 978-0-367-55862-8

Sharia-compliant hotel literacy: Profiling the potential Sharia-compliant hotel guest

Y. Rahayu & J. Zuhriatusobah
Nusantara Islamic University, Bandung, Indonesia

ABSTRACT: The current general perception that Sharia-compliant hotels are only meant for Muslims should also be shifted. Even though Muslims are among the largest tourist markets in the world, perceived values of the Sharia-compliant hotel have not been clearly established. This study was measured using two methods. One method used 10 items of 5-point Likert of Sharia-compliant hotel benefit to profile the potential guest and measure literacy level using 15 true-false test questions. Scoring results of true-false test questions were then analyzed with confirmatory factor analysis to test construct validity. Scoring results of 5-point Likert of Sharia-compliant hotel benefit were then analyzed with K-Means clustering analysis used to classify potential guest of sharia-compliant hotels. The results show identifiable characteristics and behavior patterns of potential Sharia-compliant hotel guests. This kind of information can be used in developing effective promotional programs, the feasible marketing strategy, and business sustainability.

Keywords: Sharia-compliant hotel, literacy, potential guest

1 INTRODUCTION

The Muslim market share is a very large market, especially in the ASEAN market with the largest Muslim population today. Therefore, some companies have taken advantage of this concept in their business itself. Not only applied to Islamic banking, Islamic business has also spread in various types of businesses, from start Takaful, Sharia pawn shops, salon Sharia, and Sharia hotels. Problematic issues in using the Sharia label in the hotel business continues to be a source of confusion for owners, managers, and hotel management as well as consumer Sharia. Although Muslims make up one of the largest tourist markets in the world, knowledge related to the different Islamic perspective on tourism (regardless of the very powerful link between the two) is still less represented in the related literature (Eid & El-Gohary 2015). There are still many doubts in the industry about what is called the "Sharia," especially the term "Sharia hotel" (Chan et al., n.d.). A Sharia hotel concept, through its unique value proposition, has rapidly become very much attractive to both Muslim and non-Muslim tourists all over the world. This possibility has taken the Sharia hotel concept to another level of business insight within the marketplace (Putit et al. 2016).

The Indonesian government has made efforts to develop Sharia hotels by issuing Ministerial Regulations on Tourism and Creative Economy on Sharia Hotel Business Guidelines No. 2, 2014. Guidelines for implementing Sharia hotel business as stipulated in the ministerial regulation carry a major contribution to development of Sharia hotels in Indonesia. Therefore, the ministerial regulations are divided into two categories: (1) Hilal Sharia Hotel One, and (2) Hilal Sharia Hotels Two. Ministerial Regulations on Tourism and Creative Economy on Sharia Hotel Business contributed positively to business development in that such regulations have helped the process of hotels undergoing Sharia labels from the beginning of the establishment and those transforming into Sharia hotels (Adirestuty 2019). Sharia according to Ministerial Regulation No. 2 2014 is the principles of Islamic law as to which are arranged in edicts and/or approved by the Indonesian Sharia Council. While understanding of Sharia hotel business is a hotel business, its operation must meet the criteria

DOI 10.1201/9781003095484-76

Table 1. Sharia-compliant hotel requirement.

No alcohol to be served or consumed on the premises
Halal foods (slaughtered in the name of Allah and excluding all pork products and certain other items)
Quran, prayer mats, and arrows indicating the direction of Mecca in every room
Beds and toilets positioned so as not to face the direction of Mecca
Bidets in bathrooms
Prayer rooms
Appropriate entertainment (no nightclubs or adult television channels)
Predominantly Muslim staff
Conservative staff dress
Separate recreational facilities for men and women
All-female floors
Guest dress code
Islamic funding

Source: Henderson, 2010

of Sharia hotel business set out in the Sharia as in this ministerial regulation. The criteria of Sharia hotel business are the formulation of qualifications and/or classification that includes aspects of products, services, and management (Marhanah 2017).

Rosenberg and Choufany (2009) divided Sharia compliant into three sections, which are operations, interior design, and financial. Besides no alcohol and serving halal food, some of the other criteria must also be complied in operation, like majority of staff of Muslim faith, the division of female staff for single female floors, and male staff for single male floors, providing decent television service or channel, Quran and prayer mat in each room. In addition, decent entertainment is allowed with no venue of nightclubs and discotheque, the arts should not depict the human form, and separate wellness facilities for males and females. For the financial part, the hotel must be financed through Islamic financial arrangements and should pay zakat, which is the obligation that a hotel company has to donate a certain proportion of money based on yearly business profit.

However, from a small part of society there are those who think that the Sharia hotel and the non-Sharia (conventional) hotel are the same, the only difference being a "label." Hence, the need for the attention of the Sharia hotel manager today in the increasingly fierce competition, in competing to provide excellent service and quality-oriented to customer satisfaction as the primary goal, but still paying attention to the signs and regulations stipulated by Sharia, so there is a "difference value" between the Sharia and non-Sharia (Marhanah 2017). There are no formal criteria for Sharia-compliance in the hotel sector, but scrutiny of statements made by industry practitioners and analysts reveals broad agreement about a set of attributes that are summarized in Table 1.

The term "Sharia-Compliant" is not clarified enough within the hotel industry. In fact, and not to exaggerate, there are no such hotels in this category and no written rules or classifications currently exist for a Sharia-compliant hotel. Develop and deliver the right Sharia compliant is the biggest challenge. It is because people in the industry only understand the concept and the practices through what is visible to their eyes. The very popular misconception is when there is no alcohol served, and by providing amenities for Muslims, they are already considered as Sharia-compliant. Nevertheless, hotel industries implicitly follow one aspect of Sharia- compliant by having halal certificates in their establishment (Ahmat et al. 2012).

There is still limited study on Sharia-compliant hotel literacy that focuses on an Islamic hotel concept. Literacy is usually associated with knowledge, and it shows that knowledge is one element of things that lead to the behavior of a person (Antara et al. 2016). The authors have a great curiosity about the understanding level of the real concept of Sharia-compliant hotel in the society, especially hotel guests. The objective of this re-search is to carry out a descriptive study on the Sharia-compliant hotel literacy level and determine potential guests in Sharia-compliant hotel, so as to develop the feasible marketing strategy for a Sharia-compliant hotel.

Bandung was the city chosen to be the location of research, which will support the research objectives. Bandung is one of nine Sharia tourism destinations set by the Indonesian government since 2014, with a large number of Sharia hotels compared to other cities. The high number of Muslim tourists from Malaysia and Middle Eastern countries coming to Bandung creates a demand for high Sharia hotel accommodations, which led to the development of Sharia hotels in Bandung instigating the number of hotels that will transform to meet the standards of Sharia compliance. Bandung tourism potential for the Sharia tourism industry is also supported by cultural tourism, historical tourism, culinary tourism, and shopping (Adirestuty 2019).

2 METHODOLOGY

This research was conducted by using two methods, that is factor analysis and cluster analysis. Factor analysis is an analysis by extracting a set of variables so that several factors are formed. Cluster analysis is a multivariate technique whose purpose is to group objects based on their characteristics. Cluster analysis classifies objects so that each object that is closest in common to other objects is in the same cluster.

Factor analysis focuses on grouping statement instruments from variables, to explore the data and give information on how many factors are needed to represent the data, whereas the cluster analysis focused on grouping respondents. This research uses non-hierarchical cluster analysis or K-Means cluster, which is an analysis method where the number of clusters is determined by two and three clusters.

The data for this study are obtained through structured questionnaires and conducted in an online survey. The questionnaire consists of three parts. The first part is socio-economic demographic information of the respondents (gender, age, marital status, education, occupation, and monthly income). The second part is a self-evaluation test regarding the concept of a Sharia-compliant hotel, which consists of design interior, operations, and finance. This test is true-false questions, and the instrument of question was measured using CFA (confirmatory factor analysis). The third part refers to the benefit of Sharia-compliant hotel consisting of novel experience, functional hotel attributes, and the way of life. As measured though a 5-point Likert scale (1—Strongly Disagree to 5—Strongly Agree), then, analyzed using K-Means cluster. This part is to answer the research objective, namely profiling the potential guests of Sharia-compliant hotels.

The samples size used in this study was 227 respondents. There were three clusters identified; these were labeled Hunter Sharia (39% of total respondents), Moderate Sharia (36% of total respondents), and Not Sharia at all (25% of total respondents). The naming of the clusters considered the strength of the clusters' mean scores for segmentation variables. For example, the Hunter Sharia cluster was named based on the respondents' strong attitudes toward and concern they displayed for the halal principles. The hunter sharia group is likely to search out halal products. The moderate sharia cluster is less concerned about halal principles. The cluster of not sharia at all shows negative scores for segmentation variables. In other words, this group is not likely to purchase halal products to make an impression, nor is there concern about the impact products may have on spiritual.

3 RESULTS AND DISCUSSION

Women were the most respondents in this research (51.5%). The average age of respondents was 30–45 years old. They were mostly married (61.2%). They were well educated, 78% of respondents having a college degree. The majority of respondents' occupation were government employees and private employees, and respondents reported an average monthly income between Rp. 2.000.000–Rp. 6.000.000. It is worth to mentioned, 71.4% of the respondents were aware of the existence of Sharia-compliant hotels but most of them never had any experience staying in a Sharia-Compliant hotel.

Table 2. The literacy question items of sharia-compliant hotel concept.

No	Question
Design Interior	
SHL 01	Sharia hotel has a distinctive Exterior and Interior that are different from conventional hotels (there are calligraphy, Islamic music, and Islamic atmosphere)
SHL 02	Sharia hotel must provide praying room for public (both for guests and employee)
SHL 03	Bathrooms in sharia hotels may not face the Qibla
SHL 04	Rooms available of the Quran, prayer mat, and a sign pointing toward the Qibla
SHL 05	In the room, adzan can easily be heard every praying time through the speaker
	Operations
SHL 06	There are Rules of Procedure for guests, i.e., if the guests staying are couples, they must be a muhrim couple (Show ID card); the floor for male, female, and family guests is different; unmarried guest barred from occupying the same bedroom at the same floor
SHL 07	Sharia hotels are not only for Muslims but also for non-Muslims
SHL 08	In organizational structure, must have Sharia Supervisory board
SHL 09	Sharia hotel must have a valid certificate as a Sharia hotel from the National Sharia Council (MUI)
SHL 10	In Sharia hotel, the public facilities such as fitness center, pool, and salon are separated for male and female guests.
SHL 11	Sharia hotel only provide halal food, doesn't provide an alcoholic beverage—neither alcohol nor pork should be served in any of the food and beverage outlets
SHL 12	All employees to be predominantly Muslim
SHL 13	The employee's fashion based on Islamic rules
Financial	
SHL 14	All forms of payment must use Islamic financial institutions
SHL 15	Financial administration arrangement used a Sharia accounting system

Source: Processed by researcher in 2020

Based on the results of calculations POC (proportion of correct answers), POI (proportion of incorrect answers), and POD (proportion of doubtful answers) of each question, it should be noted several of the easiest questions to understood by respondents about the concept of Sharia-Compliant Hotel, respectively, are 2, 4, 6, 11. Conversely, the most difficult questions to understood by respondents are 7, 8, 12, 14, and 15.

Overall, the respondent's literacy level for Sharia-Compliant Hotel is good, at least when measured by the number of respondents in the well literate category (26%) and sufficient literate category (37%). However, respondents whose literacy level is low (37%) should still be the concern of the parties, because it is thought to inhibit the development of Sharia-Compliant Hotels in Indonesia.

Factor analysis is an analysis by extracting a set of factors, so that several factors are formed. The results of using CFA shows that, on the framework of questions about the concept of Sharia-compliant hotel (Table 1), there are seven factors that are reduced or eliminated, because eight factors have been considered to represent 61.89% of variance literacy, from the number of initial indicators of 15. This shows that these eight factors are the most optimal for aspects of literacy that are formed, specifically question numbers 2, 3, 6, 8, 10, 11, 14, and 15 (Table 2).

The concept of Sharia-compliant hotel is very unique, and it needs to be promoted not only toward the Muslim market but also non-Muslims. However, the market needs to be clearly defined in order to effectively promote such a product. There are various market segments for the hotel industry such as leisure travelers and business travelers (Yusof & Muhammad 2013). Based on this condition, profiling guests can help hoteliers to find the proper market positioning for their hotel. It can also be used to select appropriate potential guest groups in target markets to find the marketing opportunities in current potential guest groups and to gain competitive advantages.

Table 3. Summarized the literacy level of sharia-compliant hotel.

Literacy Level	The Number of Respondent	Percentage
Well Literate	60	26%
Sufficient Literate	83	37%
Less Literate	52	23%
Not Literate	32	14%
Total	227	100%

Source: Processed by author in 2020

Table 4. Proposed of grouping potential guest of sharia-compliant hotel.

	Cluster		
	1	2	3
Zscore: Gender	.10930	−.02176	−.10478
Zscore: Age	.71875	.12828	−.88001
Zscore: Marital Status	.70187	.43463	−1.07705
Zscore: Level of Education	.42631	.34282	−.70965
Zscore: Occupation	.26702	.33236	−.52727
Zscore: Salary per Month	.36070	.45720	−.71806
Zscore: Sharia Hotel Benefit: Novel Experience	.65427	−1.00672	−.01046
Zscore: Sharia Hotel Benefit: Functional Hotel Attributes	.69335	−.94734	−.09518
Zscore: Sharia Hotel Benefit: The way of Life	.56300	−.94823	.04867

Source: Processed by researcher in 2020

By cluster analysis author means the partitioning of data into meaningful subgroups, when the number of subgroups and other information about their composition may be unknown. The process of clustering is to divide a set of physical or abstract objects into several groups according to the degree of similarity between them and to constitute groups of similar objects. A cluster is a group of some similar objects, and the objects in different clusters are not similar.

Based on F-test, which is presented in the ANOVA table, it shows that one of the attributes in this study is gender but it is not considered to affect aspects of the grouping potential guest of Sharia-Compliant Hotel (p value = 0,373). The implication for the management of Sharia-Compliant Hotel, gender aspects need not be a priority consideration in formulating marketing strategy, especially those related to determining market segmentation. Based on F-test, the attribute that is the most distinguishing attribute between clusters 1, 2, and 3 is age (F-stat: 105.536) and marital status (F-stat: 214.916).

From the output table above (Table 3), can be defined as follows:

Cluster-1

This cluster, the author named it The Hunter Sharia. This cluster-1 contains respondents who have the value of Sharia-compliant hotel benefits are more than the average studied variables, which consist of novel experiences, functional hotel attributes, and the way of life. This is evident from the positive values found in the Final Cluster Centers table in all variables. Cluster-1 is a grouping of respondents who are predominantly male gender, average age around 43 years, marital status is married, undergraduate education, the majority of occupation is self-employed (entrepreneur), and salary ranges from 6–8 million rupiah.

Cluster-2

Cluster 2, the author named it The Not Sharia at All. Because the characteristics of Sharia-compliant hotel clustered in cluster-2 has a value that is in a position below the average of the studied variables (novel experiences, functional hotel attributes, and the way of life). Cluster-2 is a grouping of respondents, the majority of which have a female gender, average age around 35 years, marital status is married, the majority of education is undergraduate, the majority of occupations are self-employed (entrepreneur), and salary ranges from 6–8 million rupiah.

Cluster-3

Cluster-3, the author named it the Moderate Sharia, because in cluster-3, the value of novel experience and functional hotel attributes has a value below the average of studied variables (negative score). But the value of the way of life shows a positive value. That is, respondents in this group are still concerned with Sharia principles. Cluster-3 is a grouping of respondents, the majority of which have female gender, age 24 years, marital status is single, diploma education, occupation is private employees, and salary ranges from 4–6 million rupiah.

4 CONCLUSION

The result shows that the literacy level of Sharia-compliant hotel is relatively good. Basically, Sharia hotels are hotels as usual, but operations and services have been adjusted to the principles of Sharia or Islamic teaching guidelines, in order to provide a calm, comfortable, healthy, and friendly atmosphere that is needed by guests, both Muslim and non-Muslim. Sharia hotels in general are no different from other hotels, still subject to government regulations, remain open 24 hours, without interruption. Marketers are also open to all groups, both Muslim and non-Muslim.

The key contribution of this study is that it provides interesting and potentially useful information about the potential guest who is interested to stay in a Sharia-compliant hotel. Based on the results of the cluster analysis, Sharia-compliant hoteliers can create the feasible marketing strategy, because each cluster has different socio-economic-demographic characteristics. These different characteristic are regarding the best way to communicate from the selected segments. This information can be used in developing effective promotional programs. The information should be made clear and easily comprehensible to the Muslim and non-Muslim guest.

ACKNOWLEDGEMENT

This research was supported by Faculty of Economics, Nusantara Islamic University (UNINUS). We thank our colleague, Wahdi Suardi, for sharing his pearls of wisdom with us during the course of this research. We are also immensely grateful to all members of the Department of Management-Faculty of Economics, Nusantara Islamic University (UNINUS) for providing assistance and helping in facilitating this research.

REFERENCES

Adirestuty, F. (2019). Customer Perceived Value In Creating Customer Satisfaction And Revisit Intention In Sharia Hotel. *Journal of Islamic Monetary Economics and Finance*, *5*(2). https://doi.org/10.21098/jimf.v5i2.1067

Ahmat, N. C., Ridzuan, A. H. A., & Zahari, M. S. M. (2012). Customer awareness towards Syariah compliant hotel. *ICIMTR 2012 – 2012 International Conference on Innovation, Management and Technology Research, May*, 124–128. https://doi.org/10.1109/ICIMTR.2012.6236373

Antara, P. M., Musa, R., & Hassan, F. (2016). Bridging Islamic Financial Literacy and Halal Literacy: The Way Forward in Halal Ecosystem. *Procedia Economics and Finance*, *37*(16), 196–202. https://doi.org/10.1016/s2212-5671(16)30113-7

Chan, A., Pratami, R. P., & Tresna, W. (n.d.). *Sharia Marketing Analysis In Noor Hotel Bandung.*

Eid, R., & El-Gohary, H. (2015). The role of Islamic religiosity on the relationship between perceived value and tourist satisfaction. *Tourism Management*, *46*, 477–488. https://doi.org/10.1016/j.tourman.2014.08.003

Henderson, J. C. (2010). Sharia-Compliant Hotels. *Tourism and Hospitality Research*, *10*(3), 246–254. https://doi.org/10.1057/thr.2010.3

Marhanah, Sri, Mardhatilla, D. S. (2017). Does Sharia Hotel Meet Its Criteria Study Of Sofyan Hotel Betawi, Jakarta. *People: International Journal of Social Sciences*, *1*(1), 957–966. https://doi.org/10.20319/pijss.2015.s21.957966

Modul 6 analisis cluster. (n.d.). 1–16.

Putit, L., Muda, M., Mahmood, A. N., Ahmad Taufek, N. Z., & Wahib, N. (2016). Linking 'Halal' Friendly Hotel Attributes and Customer Satisfaction:The Islamic Tourism Sector. *Journal of Emerging Economies and Islamic Research*, *4*(4), 43. https://doi.org/10.24191/jeeir.v4i4.9102

Rosenberg, P., & Choufany, H. M. (2009). Spiritual Lodging – the Sharia – Compliant Hotel Concept. *HVS Global Hospitality Services- Dubai*, *April*, 1–7.

Yusof, M. F. M., & Muhammad, M. Z. (2013). Introducing shariah compliant hotels as a new tourism product: The case of Malaysia. *Entrepreneurship Vision 2020: Innovation, Development Sustainability, and Economic Growth – Proceedings of the 20th International Business Information Management Association Conference, IBIMA 2013*, *1*, 1142–1146.

Promoting Creative Tourism: Current Issues in Tourism Research – Kusumah et al. (Eds)
© 2021 Taylor & Francis Group, London, ISBN 978-0-367-55862-8

Environment, food, or employee: Identifying factors in authentic dining experience influencing customer satisfaction

T. Abdullah
University of Otago, Dunedin, New Zealand

N. Latifah & H.P.D. Setiyorini
Universitas Pendidikan Indonesia, Bandung, Indonesia

R.S. Nugraha
Bandung Institute of Tourism, Bandung, Indonesia

ABSTRACT: An authentic dining experience offers a unique and exotic experience to the customers who come to ethnic restaurants. It could become a distinctive feature for ethnic restaurants that can differentiate them from other types of restaurants. Moreover, it is believed that by providing a real authentic dining experience, customers in ethnic restaurants would become more satisfied. The purpose of this study was to analyze the factors of the authentic dining experience that affect customer satisfaction in ethnic restaurants. The correlations between three dimensions of authentic dining experience (i.e., food concern, environmental concern, and employee concern) with customer satisfaction were analyzed. Questionnaires were given to 203 customers of the ethnic restaurants in Bandung, Indonesia. The data were then analyzed using multiple regression analysis. The results show that, simultaneously, there is a significant correlation between the authentic dining experience to customer satisfaction. However, partially, food concern does not significantly influence customer satisfaction; this finding will be further described. Therefore, ethnic restaurants could spend more attention to their employee and environmental dimensions to increase their customers' satisfaction toward authentic dining experience.

Keywords: authentic dining experience, customer satisfaction, ethnic restaurant, food concern, employee concern, environment concern

1 INTRODUCTION

Ethnic restaurants have managed to attract the attention of the market broadly. Two-thirds of the visitors of the restaurants in the United States visit these types of restaurants (Kim et al. 2017). Ethnic restaurants are described as a cultural ambassador for the country or region they from which they originated, and they could even be attractions for tourists (Wood & Muñoz 2007). Ethnic restaurants are generally easy to recognize from their menu and their interior and exterior design. The menu that they offer represents a characteristic of a specific culture (Youn & Kim 2017).

According to Liu and Jang (2009), factors affecting customer satisfaction and behavioral intention in a restaurant are food, service, atmosphere, and authenticity. Authenticity is considered as one of the factors that influence the experience of tourists in visiting a restaurant. It would represent the uniqueness and the original flavor of the food, which is bounded to a specific place or a culture. In culinary tourism, authenticity refers to the environment, time, the layout, the cooking process, the behavior of the surrounding community, customs, and local traditions, which occur in a culinary tourist destination (Kim et al. 2015; Wang 1999). Authenticity became an important criterion used to evaluate an ethnic restaurant, since it can elevate the authentic experience of customers

 DOI 10.1201/9781003095484-77

when consuming food in the restaurant (Lin et al. 2017). The authenticity in this study refers to the food and the environment that can represent the actual flavor and culture of a particular ethnic (Ebster & Guist 2005; Widyakusumastuti 2014). This study believes that employees in an ethnic restaurant also play a crucial role in creating an authentic environment. Therefore, this study placed employees as a separate dimension from environment. Thus, food, environment, and employees were dimensions of authentic dining experience in this study.

The atmosphere created in an ethnic restaurant could describe the culture of where the food originated. Therefore, when customers dine in an ethnic restaurant, they could understand and enjoy other cultures (Van Dijk 1984). The authentic dining experience is one of the factors that could affect the customers' intention to visit an ethnic restaurant (Han et al. 2020). This can be used as a way for an ethnic restaurant to ensure that customers would always be interested in visiting and enjoying the service in the restaurant. Furthermore, any company or organization would always strive to create and maintain customer satisfaction to build customers' loyalty (Pujiastuti et al. 2020). They avoid disappointing customers since it would cause problems (e.g., creating negative word of mouth and declining sales). Ethnic restaurants posit their businesses as representatives of a particular culture, which then establishes a distinctive feature or a uniqueness of their restaurants. Therefore, restaurants of this type could offer an authentic dining experience to their customers. The question is to what extent each of the dimensions of this authentic dining experience can affect customer satisfaction.

Some studies have been conducted regarding authentic dining experiences. Lin et al. (2017) identified important factors that could influence customers' perception of authenticity in the restaurant using content analysis. A survey conducted by Liu and Jang (2009) examined the influence of restaurant attributes on customers' satisfaction and behavioral intentions. They found that food quality, service reliability, and environmental cleanliness are three crucial elements to create satisfied customers and positive post-dining behavioral intentions. The other study was from Tsai and Lu (2012), who analyzed the influence of authentic dining experiences on repurchase intention in ethnic theme restaurants. Some scholars have mentioned that some elements of authentic dining experience could positively influence customer satisfaction (Gaytán 2008; Sukalakamala & Boyce 2007; Wood & Muñoz 2007); however, their studies were not conducted to reveal this correlation. Sukalakamala and Boyce (2007) identified customers' perceptions, acceptance, and expectations regarding the authentic Thai dining experience. Wood and Muñoz (2007) in their study explained how space of consumption could affect the perception of "authentic" culture. Moreover, a study by Gaytán (2008) argues that authenticity is a social construction. Thus, the direct correlation between authentic dining experience and customer satisfaction is still understudied. This study aimed to address this gap; therefore, this study was conducted to analyze the factors of the authentic dining experience that can affect customer satisfaction in ethnic restaurants. This study attempts to answer the question of which dimensions among food, employees, or environment in authentic dining experience could possibly affect customer satisfaction?

This study is imperative because it determines the most important factors between environment, food, or employees in authentic dining experience that could directly influence customer satisfaction. Hence, restaurateurs could decide which dimension should be given attention to in relation to creating customer satisfaction by providing an authentic dining experience.

2 LITERATURE REVIEW

2.1 *Authentic dining experience*

Ethnic restaurants usually represent a specific culture, which allows consumers to experience something new from that culture (Jang et al. 2012; Muñoz & Wood 2009; Youn & Kim 2017). According to Reynold and Hwang (2006), people do not come to an ethnic restaurant only for consuming food; they also look for an exotic dining experience. They stated that customers of this type of restaurant could be viewed as gastronomic tourists. They also believed that the authenticity in ethnic restaurants is imperative because they are a cultural ambassador both for the food and the culture of a particular ethnic.

Lanier (2008) mentioned that in psychology, experience has three components, namely the subject, the object, and the process of experience. The subjects are those who experience and feel the impact of that experience (in this study they are the consumers). The objects are those that give you the experience (in this study, they are ethnic restaurants). The process is an action provided by the objects to the subjects. Furthermore, Lanier explained that experiential experience interface refers to a process by which consumers and providers interact, and marketing experience refers to an experience specifically created and offered for consumers. Lanier also concluded that, nowadays, marketing experience has a vital role in the process of giving experience and creating positive impressions in the minds of consumers.

According to Lanier (2008), there are several types of experiences offered by marketing experience, those are hedonic, leisure, playing, high-risk, and extraordinary experience. "Hedonic experience" is the range of activity with a multi-sensory, fun, and related to some aspects that have sound. Some examples of this experience are enjoying opera, theater, painting, sculpture, and movies. "Leisure experience" or the experience of pleasure, is an activity in which people engage during their free time and is associated with some recreations. Some instances of this experience are dancing, eating, crafting, sewing, and decorating. "Play experience" is the experience associated with hobbies like games and sports. "High-risk experience" is the activity that involves the risk of physical injury and requires physical and mental strength, such as mountain climbing and diving. "Extraordinary experience" is beyond ordinary experience; it has the characteristics of high-level emotional intensity.

This study focused on the "leisure experience" in the social activities of consumers. This experience has a variety of activities that people do during free time (Lanier 2008). According to Canny (2013), dining experience is the total customers' judgment on their overall consumption experience, starting with the quality of food and service to the restaurant environment. Hence, it is not just centered on the menu and beverage alone but also included in the assessment of the overall experience perceived by the consumer. The definition of authentic dining experience is "the level to which the food and dining environment are perceived as original to its type and reflect the concept or themed chosen" (Ishak et al. 2020).

Creating an authentic experience is one of the main goals of ethnic restaurants (Ebster & Guist 2005; Sukalakamala & Boyce 2007). In this study, the author used the dimensions of the authentic dining experience from two studies (Lin et al. 2017; Tsai & Lu 2012). Those are:

a. Environmental concern
   The owners of ethnic restaurants create a different environment with an emphasis on the authentic atmosphere. This environment is related to the authentic atmosphere of a culture which could be felt by consumers in ethnic restaurants. It is considered as an essential component in service management in a restaurant because it can affect the emotional responses of consumers. This environmental concern includes several elements, namely décor, theme, atmosphere, music, cultural feeling, tableware and seating, design, signage language.
b. Food concern
   Customers of ethnic restaurants expect a unique and exotic taste that is different from what they usually eat. Food is an important part of an authentic dining experience; it should represent a culture or a country/region. The elements of food used in this study are authentic food, menu, presentation, flavor, and taste.
c. Employee concern
   Employees are a supporting factor and yet still also imperative for creating an authentic dining experience. The interaction between consumers and service providers can influence consumers' evaluation of the experience during the meal in ethnic restaurants. It would even be considered more authentic if the employees are from the original country. Employee concern consists of the speed of the service, service style, and the appearance of the service staff.

2.2 *Customer satisfaction*

Customer satisfaction is the result of consumers' evaluation of a product and services. It indicates that the product and services are able to meet the needs and the expectations of customers (Zeithaml, Bitner, & Gremler 2013). Furthermore, Kotler and Keller (2016) stated that satisfaction is a feeling of pleasure or disappointment from the comparison between perceived product performance with the expectation that one has. While Kotler and Armstrong (2013) stated that satisfaction is the extent to which the performance of a product's perceived fit with the expectations of buyers. Zeithaml et al. (2013) also explained that customer satisfaction is determined by consumers' internal and external factors, including the product and service features, customer emotion, attributes for service success or failure, perception of equity or fairness, and also other customers, family members, and coworkers. In business, making products that serve customers' hopes and desires will give a positive image for the company (Heung & Gu 2012). Two dimensions of customer satisfaction used in this study are expected and perceived service (Kotler & Keller 2016). There are three hypotheses proposed in this study; those are:

H1: Environment concern in authentic dining experience has a positive influence on customer satisfaction
H2: Food concern in authentic dining experience has a positive influence on customer satisfaction
H3: Employee concern in authentic dining experience has a positive influence on customer satisfaction

3 RESEARCH METHODS

This study was conducted in Bandung city, Indonesia. The local culture in this city is Sundanese; hence, this study analyzed the influence of authentic dining experience on customer satisfaction in three well-known Sundanese ethnic restaurants operated in this city. The participants in this study were 203 customers who had visited the restaurants. There is no specific characteristic of the participants involved in this study; as long as they have been to the restaurants to dine in, they can be included in this research. Data were collected in two months using accidental sampling procedure. Seventy percent of the questionnaires were distributed via Google forms to participants who have been to the restaurants, and 30 % of the questionnaires were distributed directly by approaching the visitors who have just visited the restaurants. After the data were collected, they were analyzed using multiple regression analysis with SPSS software.

4 RESULTS AND DISCUSSION

Bandung is a capital city of West Java Province, Indonesia. It has been one of the tourist destinations which is famous for its cold weather, natural resources, and Sundanese culture. It has also been known as a culinary tourist destination with its Sundanese cuisine and street foods. This study chose three of many Sundanese ethnic restaurants located in Bandung because these restaurants are popular, and they serve a variety of Sundanese food and incorporate all elements to create an authentic dining experience.

Table 1 illustrates the demographic profile of the participants. The participants were 37.9% male and 62.1% female. The majority of the customers were 21–35 years old (70.4%), followed by 36–50 years old (17.2%), 17–20 years old (7.9%), and others (4.4%). Most of them were students (32%), followed by the private employees (29.6%), civil servants (23.6%), entrepreneurs (8.9%), and others (5.9%). Almost all of them came to restaurants with their companions, and most of them knew the information of the restaurants from their family and friends (74.4%), and 17.7% of them knew about the restaurants from the Internet, the rest got the information from mainstream media (2.5%) and others (5.4%).

Table 1. Demographic profile of participants.

Characteristics	Category	Frequency	Per cent
Gender	Male	77	37.90%
	Female	126	62.10%
Age	17–20 years old	16	7.90%
	21–35 years old	143	70.40%
	36–50 years old	35	17.20%
	Others	9	4.40%
Occupation	Student	65	32.00%
	Entrepreneur	18	8.90%
	Civil servant	48	23.60%
	Private employee	60	29.60%
	Others	12	5.90%
Companion	Alone	2	1.00%
	2–3 group	78	38.40%
	3–4 group	64	31.50%
	>5 group	59	29.10%
	Internet	36	17.70%
Source of information	Family/Friends	151	74.40%
	Mainstream Media	5	2.50%
	Others	11	5.40%

Table 2. The results of multiple regression analysis.

	Coefficient Beta	Sig.	Value
(Constant)	23.834	0	
Environmental concern	0.607	0.004	
Food Concern	0.263	0.323	
Employee Concern	1.292	0	
F		0	35.232
R			0.589
R^2			0.347

Based on the results depicted in Table 2, it shows two dimensions significantly influenced customer satisfaction, namely "employee concern" and "environmental concern." The highest influence was from the employee concern (1.292), which was measured by the speed of the service, the authenticity of the service style, and the suitability of uniform. The appearance and greeting of the service personnel are a tangible aspect of a restaurant. Therefore, the employees of the restaurant are also part of the environment as a whole. The other most substantial influence was environmental concern, which in this study was measured by the décor, themes, atmosphere, music, tableware and seating, design, cultural feeling, signage, and language. A study by Liu and Jang (2009) recognized food quality, service reliability, and the environment as the important factors for a restaurant to be successful. However, in this study, the influence of food on customer satisfaction was insignificant.

The findings in this study also slightly contradict a study by Tsai and Lu (2012). In their study, both food concern and environmental concern were crucial for customer repurchase intention, and employee concern was not important. While in this study, employee concern was the dimension that had the strongest influence and was then followed by the environmental factor. This result implies that customers were highly concerned about the courtesy that the restaurant offered. Since

Sundanese people are famous for their hospitality, hence the hospitableness of the staff in these restaurants was highly appreciated. Every attribute related to the employees of these restaurants, including their greeting, gesture, intonation, accent, choice of words, and the politeness of the service personnel should perfectly mimic how a good-mannered Sundanese people usually behave. The type of service and the uniform worn by the staff should also represent the Sundanese culture. These three restaurants have done an excellent work in representing Sundanese culture through their employees, and thus created a great authentic dining experience.

Food concern was the one in which the influence was insignificant. Food concern did not have a significant influence on customer satisfaction because the food served at the restaurants was highly varied. They were not just focused on Sundanese cuisine; instead, they also offered a variety of other cuisines from other ethnics or regions. Based on the findings, two of three hypotheses were proven "accepted," one hypothesis which is hypothesis 2 was not accepted because food concern in authentic dining experience did not influence customer satisfaction.

5 CONCLUSION

This study is meant to be one of the literature sources for those who want to understand more about the phenomenon of the influence of authentic dining experience on customer satisfaction in ethnic restaurants. The results of this study also provide a practical implication, especially for ethnic restaurant managers and owners to support their managerial decisions in creating an authentic dining experience for their restaurants to achieve customer satisfaction. After analyzing some data, it is concluded that in general, there was a significant influence between authentic dining experience on customer satisfaction in Sundanese ethnic restaurants in Bandung. However, if it was analyzed partially, only two sub-variables of authentic dining experience had a substantial impact on customer satisfaction in Sundanese ethnic restaurant in Bandung, namely, employee concern and environmental concern. Food concern did not have a significant influence on customer satisfaction in Sundanese ethnic restaurants in Bandung.

REFERENCES

Boyce, J. M. (2007). Environmental contamination makes an important contribution to hospital infection. *Journal of Hospital Infection Volume 65, Supplement 2*, 50–54.

Canny, I. U. (2013). The role of food quality, service quality, and physical environment on customer satisfaction and future behavioral intentions in casual dining restaurant. *The 7th National Research Management Conference, Sriwijaya University-Palembang, Indonesia (27–28 November 2013).*

Ebster, C., & Guist, I. (2005). The role of authenticity in ethnic theme restaurants. *Journal of Foodservice Business Research, 7*(2), 41–52.

Gaytán, M. S. (2008). From sombreros to sincronizadas: Authenticity, ethnicity, and the Mexican restaurant industry. *Journal of Contemporary Ethnography, 37*(3), 314–341. https://doi.org/10.1177/0891241607309621

Han, G., Akhmedov, A., Li, H., Yu, J., & Hunter, W. C. (2020). An Interpretive Study on Sustainability in the Link between Agriculture and Tourism: Tourist-Stakeholder Satisfaction in Tiantangzhai, China. *Sustainability, 12*(2), 571.

Heung, V. C. S., & Gu, T. (2012). Influence of restaurant atmospherics on patron satisfaction and behavioral intentions. *International Journal of Hospitality Management, 31*(4), 1167–1177.

Ishak, F. A. C., Zainun, N. A., Karim, M. S., Ungku Zainal Abidin, U. F., & Mohamad, S. F. (2020). The Multifaceted of Themed Restaurant: Exploring the Unique and Vulnerable Elements in Staging Authentic Dining Experience. International Journal of Academic Research in Business and Social Sciences, 10(3).

Jang, S. S., Ha, J., & Park, K. (2012). Effects of ethnic authenticity: Investigating Korean restaurant customers in the US. *International Journal of Hospitality Management, 31*(3), 990–1003.

Kim, J. H., & Jang, S. S. (2016). Determinants of authentic experiences. *International Journal of Contemporary Hospitality Management.*

Kim, J.-H., Youn, H., & Rao, Y. (2017). Customer responses to food-related attributes in ethnic restaurants. *International Journal of Hospitality Management, 61*, 129–139.
Kim, Y. H., Duncan, J., & Chung, B. W. (2015). Involvement, Satisfaction, Perceived Value, and Revisit Intention: A Case Study of a Food Festival. *Journal of Culinary Science and Technology, 13*(2), 133–158. https://doi.org/10.1080/15428052.2014.952482
Kotler, P, & Keller, K. L. (2016). *Marketing Management. Global Edition (Vol. 15E).* England: Pearson.
Kotler, Philip, & Armstrong, G. (2013). *Principles of Marketing (16th Global Edition).* Harlow: Pearson.
Lanier Jr, C. D. (2008). *Experiential marketing: Exploring the dimensions, characteristics, and logic of firm-driven experiences.* The University of Nebraska-Lincoln.
Lin, P. M. C., Ren, L., & Chen, C. (2017). Customers' Perception of the Authenticity of a Cantonese Restaurant. *Journal of China Tourism Research, 13*(2), 211–230. https://doi.org/10.1080/19388160.2017.1359721
Liu, Y., & Jang, S. S. (2009). Perceptions of Chinese restaurants in the US: what affects customer satisfaction and behavioral intentions?. *International Journal of Hospitality Management, 28*(3), 338–348.
Muñoz, C. L., & Wood, N. T. (2009). A recipe for success: understanding regional perceptions of authenticity in themed restaurants. *International Journal of Culture, Tourism and Hospitality Research.*
Pujiastuti, E. E., Utomo, H. J. N., & Novamayanti, R. H. (2020). Millennial tourists and revisit intention. *Management Science Letters, 10*, 2889–2896. https://doi.org/10.5267/j.msl.2020.4.018
Reynold, J. S., & Hwang, J. (2006). Influence of age on customer dining experience factors at US Japanese restaurant. *International Journal of Tourism, 2*, 29–43.
Sukalakamala, P., & Boyce, J. B. (2007). Customer perceptions for expectations and acceptance of an authentic dining experience in Thai restaurants. *Journal of Foodservice, 18*(2), 69–75.
Tsai, C.-T. S., & Lu, P.-H. (2012). Authentic dining experiences in ethnic theme restaurants. *International Journal of Hospitality Management, 31*(1), 304–306.
Van Dijk, T. A. (1984). *Prejudice in discourse: An analysis of ethnic prejudice in cognition and conversation.* John Benjamins Publishing.
Wang, N. (1999). Rethinking authenticity in tourism experience. *Annals of Tourism Research, 26*(2), 349–370.
Widyakusumastuti, M. A. (2014). Pengelolaan Keaslian Rasa dan Budaya pada Restoran Etnik Khas Jawa: Analisis Atmospheric Restoran Etnik Khas Jawa di Jakarta Selatan dan Jakarta Pusat. *Humaniora, 5*(2), 977–988.
Wood, N. T., & Muñoz, C. L. (2007). 'No rules, just right' or is it? The role of themed restaurants as cultural ambassadors. *Tourism and Hospitality Research, 7*(3–4), 242–255.
Youn, H., & Kim, J.-H. (2017). Effects of ingredients, names and stories about food origins on perceived authenticity and purchase intentions. *International Journal of Hospitality Management, 63*, 11–21.
Zeithaml, V. A., Bitner, M. J., & Gremler, D. D. (2013). *Services Marketing. Chapter 3: Customer Expectations of Service.* New York: McGraw-Hill Irwin.

Promoting Creative Tourism: Current Issues in Tourism Research – Kusumah et al. (Eds)
© 2021 Taylor & Francis Group, London, ISBN 978-0-367-55862-8

The effect of perceived authenticity on revisit intention of Sundanese restaurants

Y. Yuniawati, T. Abdullah & A.S. Sonjaya
Universitas Pendidikan Indonesia, Bandung, Indonesia

ABSTRACT: This study aimed to investigate the effect of perceived authenticity towards revisit intention of Sundanese ethnic restaurants consumers in West Bandung Regency. The independent variable (X) in this study is perceived authenticity which consists of several sub-dimensions, namely food authenticity, authenticity of atmospherics and employee authenticity and the dependent variable (Y) is revisit intention. The type of this research is explanatory survey. The data were collected through a survey using stratified sampling technique involving 400 respondents consisting 220 consumers of Kampung Daun Culture Gallery & Café and 180 consumers of Sapulidi Resort, Café and Gallery. The data and hypothesis were tested through multivariate regression using IBM SPSS for Windows 20.0 program. The results showed that the perceived authenticity variables had a significant influence on the revisit intention of Sundanese ethnic restaurants consumers in West Bandung Regency. Partially all of perceived authenticity gives significant effect on revisit intention.

Keywords: perceived authenticity, ethnic restaurant, Sundanese

1 INTRODUCTION

There is increasing interest throughout the world on the topics of destination marketing and management because more and more places are competing to get a share of global tourism. This phenomenon is demonstrated through several different phenomena, and one of them is the expansion of professional organizations with a focus on destination marketing and management. Destination management and marketing are two interrelated concepts in tourism. In fact, destination marketing is a function in the broader concept of destination management (Morrison 2018).

Some destination management organizations have developed culinary strategies or plans for their destination products, and this is a good idea because culinary is an attraction that is rich in local cuisine traditions. Although every tourist must eat when they visit tourist destinations, culinary tourism changes food and its preparation are fascinating for tourists. Culinary tourism is about food; explore and discover culture and history through food and food related activities in creating an unforgettable experience. So culinary tourism is not just "exploration and adventure" but also a cultural meeting (Morrison 2018).

Gastronomic tourism in the tourism literature has been referred to as culinary tourism (Cejudo et al. 2019) and is relatively new field of academic research. Gastronomic tourism focuses on the relationship between food and culture. Furthermore, gastronomic tourism acts as a driving factor for visiting tourist destinations and restaurants. Ethnic cuisine is part of gastronomic tourism that tourists seek as cultural experiences through local food (Cejudo et al. 2019). According to the United Nations World Tourism Organization (UNWTO) gastronomy has become a major part of the experience of travel and is an important aspect in the selection of tourist destinations (UNWTO 2017). Thus, the authenticity of a local food and service experience is one of the dominant factors that influence consumer perceptions and trying authentic dishes is the main goal of gastronomic tourism (Dixit 2019).

DOI 10.1201/9781003095484-78

The way consumers perceive a product is influenced by the relationship between a country's cultural heritage and factors that are driven by producers and institutions (Yoo 2018). Consumer cultural experience is a significant factor to bring positive emotions to the authenticity of food or ethnic restaurants (Shawn et al. 2015).

Authenticity is a common word but not yet a very constant concept, especially in the tourism literature. Authenticity is usually defined as a reflection of relationships that are in accordance with the original source (Rudinow 1998). However, various authors provide various meanings and types of authenticity based on their different assumptions. While there are various definitions and perspectives on authenticity, the complex nature of authenticity in tourism is most often classified into several types; objective authenticity, constructive authenticity, existential authenticity, subjective authenticity and type authenticity (Dixit 2019).

From various approaches, constructive/subjective authenticity rejects the idea that authenticity is some-thing that can only be assessed objectively. Subjective authenticity claims that what is considered authentic is relative, negotiable, and depends on the context that cannot be determined objectively. Consumer perceptions of authenticity are formed based on the overall evaluation of consumers on food and beverages served, restaurant design, perceived atmosphere, and servants who interact with consumers (Meng & Kyuhwan 2018; Simon Tsai Chen-Tsang & Lu 2012). According to researchers who use subjective authenticity, authenticity in the context of food services and gastronomic tourism does not need to be objectively authentic because consumers viewe authenticity based on their cultural assumption.

The competitiveness of the restaurant industry has become so fierce that some restaurants cannot survive or grow and are forced to close due to loss of consumers, lack of resources, and bankruptcy. To avoid such impact, restaurants need to adjust to any changes (Petzer & Mackay 2014). As stated earlier,t increasing the value of existing consumers and attracting the revisit behavior of consumer intentions is an effective step rather than having to attract new consumers because with the decline in visits and the low level of revisit intention, the restaurant's income will decrease. Consequently, restaurant management must take the right and effective steps to attract consumers to get more profits efficiently.

Revisit intention has become an important issue that must be considered and improved by restaurant entrepreneurs so that companies can compete and adjust to change. Although many new consumers arrive but if they do not intend to make a repeat visit the restaurant's success is only momentary. It also can lead to a de-crease in the level of consumer visits. Restaurants that do not try to retain existing consumers will be increasingly left behind by other consumers. Increasing and attracting revisit behavior of consumer intention is and effective step rather than having to attract new consumers (Hanai et al. 2008).

Perceived authenticity is an important ethnic restaurant attribute, which is attractive to consumers (J. H. Kim & Jang 2016). Thus, relying on good food or low prices alone cannot guarantee success in this highly competitive market (Y. Liu & Jang 2009). This is reinforced by Parsa et al. (2005) which states that a lack of authenticity is one reason why some hotel and tourism businesses ultimately fail. Particularly, Kim and Baker (2017) explained that consumer perceptions of the authenticity of a restaurant affect revisit intention.

2 METHODOLOGY

This study analyzes perceived authenticity as the independent variable consisting of food authenticity, authenticity of atmospherics, and employee authenticity. And the dependent variable is revisit intention which consists of revisit willingness, revisit propensity, revisit probability in near future, likelihood to recommend and likelihood to be the first choice. Respondents in this study were consumers of Kampung Daun Culture Gallery & Café and Sapulidi Resort, Café and Gallery restaurants. The method in this study is a cross sectional method.

Based on the type of research, this research is descriptive and verification carried out in situ. The method was explanatory survey method. It is in the form of a questions list such as questionnaires or interviews that was submitted to respondents from the observed populations.

The population in this study were consumers of Kampung Daun Culture Gallery & Café and Sapulidi Resort, Café and Gallery as many as 365,800 respondents. Based on the determination of sample using Slovin formula, the sample size (n) was 400 respondents. In choosing the sample, this study using stratified sampling technique. Based on the data, the types of data collected in this study are primary and secondary data. The data collection techniques used in this study were interviews, observation, literature study and questionnaire.

The results of the tests carried out include validity and reliability tests showing that 15 items of statements for perceived authenticity variable and 5 items of statements for revisit intention variable were declared valid and reliable so that they could be used as the correct measuring instrument. There are three dimensions of perceived authenticity which are food authenticity, authenticity of atmospherics, and employee authenticity which are used in this study (S. C. S. Jang et al. 2012; K. Kim & Baker 2017).

The data analysis technique used in this study is multiple linear regression. The variables analyzed were independent variables (X), namely perceived authenticity. While the dependent variable (Y) is revisit intention. The multiple regression equation is formulated as follows:

$$Y = a + b_1X_1 + b_2X_2 + b_3X_3$$

3 RESULT AND DISCUSSION

Based on the results of recapitulation of 400 respondents, perceived authenticity variable in Sundanese ethnic restaurants in West Bandung Regency can be measured by calculating the overall score from the comparison of the total score and the average score of perceived authenticity variable. The total score for perceived authenticity is 23074 which is classified as high and is located at 19040–23520 intervals. The sub variable or dimension that has the highest rating is authenticity of atmospherics with a percentage of 34.26%. This is because consumers get a good and authentic experience while dining at Sundanese ethnic restaurants in West Bandung Regency through the beauty and comfort of a beautiful restaurant atmosphere with natural and Sundanese culture nuances. Whereas, the lowest rating is in the dimension of employee authenticity with a percentage of 32.60%.

The results of the study from responses of 400 respondents, revisit intention variable in Sundanese ethnic restaurants in West Bandung Regency can be measured by calculating the overall score from the comparison of the total score and the average score of revisit intention variable. The total score for revisit intention is 9589 which is classified as high and lies in the 8160–10080 interval. In this variable, the sub-variable or dimension that has the best response is revisit willingness with a percentage of 20.84%. This means that consumers have a high desire to return to Sundanese ethnic restaurants in West Bandung Regency. Whereas the lowest score is in the dimension of revisit probability in near future with a percentage of 18.22%. This is because even though consumers have the desire to return to the restaurant in the future, few will return in the near future.

The hypothesis testing on the effect of perceived authenticity on revisit intention in Sundanese ethnic restaurants in West Bandung Regency simultaneously has proven to have a significant effect. It is shown by the value of F observed (71,241) > F table (2.63) with a significance value of 0,000. Partially, the overall perceived authenticity dimensions which are food authenticity, authenticity of atmospherics, and employee authenticity have a significant influence on revisit intention using a significance level of 0.05.

Based on the results of the calculation of correlation coefficient from perceived authenticity to revisit intention that is equal to 0.592. This value is included in the coefficient interval of 0.400–0.599, indicating that the relationship between perceived authenticity to revisit intention is fairly moderate. Based on the results of calculation, the coefficient of determination is 0.351. This shows that jointly perceived authenticity variable contribute 35.1% to revisit intention, while the rest are influenced by other factors which is not included in the study.

4 CONCLUSION

It can be concluded that the response to the description of perceived authenticity of Sundanese ethnic restaurants in West Bandung Regency is in the high category (good) where the authenticity of atmospherics dimension gets the highest value, while the employee authenticity dimension gets the lowest value. The response to revisit intention of Sundanese ethnic restaurant consumers in West Bandung Regency is in the high category (good), where the revisit willingness dimension gets the highest value while revisit probability in near future gets the lowest value. Simultaneously perceived authenticity has a significant effect on revisit intention and partially all perceived authenticity dimensions which consist of food authenticity, authenticity of atmospherics, and employee authenticity have a significant influence on revisit intention. With employee authenticity gets the highest value of influence, while authenticity of atmospherics gets the lowest value of influence.

The finding is in line with a recent study from Meng dan Kyuhwan (2018) which support the positive effect of perceived authenticity on revisit intention. This study is beneficial for ethnic restaurant managers who wish to increase revisit intention of their consumers.

REFERENCES

Cejudo, B., Patterson, I., & Leeson, G. W. (2019). Senior Foodies: A Developing Niche Market in Gastronomic Tourism. International Journal of Gastronomy and Food Science, 100152. https://doi.org/10.1016/j.ijgfs.2019.100152

Chang, L., Backman, K. F., & Huang, Y. C. (2014). Creative tourism: a preliminary examination of creative tourists' motivation, experience, perceived value and revisit intention. https://doi.org/10.1108/IJCTHR-04-2014-0032

DiPietro, R. B., & Levitt, J. (2017). Restaurant Authenticity: Factors That Influence Perception, Satisfaction and Return Intentions at Regional American-Style Restaurants. International Journal of Hospitality and Tourism Administration, 00(00), 1–27. https://doi.org/10.1080/15256480.2017.1359734

Dixit, S. K. (2019). The Routledge Handbook Of Gastronomic Tourism. (D. S. K. S. E. -, Ed.), The Routledge Handbook Of Gastronomic Tourism. Routledge. https://doi.org/10.4324/9781315147628

Ha, J., & (Shawn) Jang, S. C. (2010). Perceived values, satisfaction, and behavioral intentions: The role of familiarity in Korean restaurants. International Journal of Hospitality Management, 29(1), 2–13. https://doi.org/10.1016/j.ijhm.2009.03.009

Hanai, T., Oguchi, T., Ando, K., & Yamaguchi, K. (2008). Important attributes of lodgings to gain repeat business: A comparison between individual travels and group travels. International Journal of Hospitality Management, 27(2), 268–275. https://doi.org/10.1016/j.ijhm.2007.08.006

Jang, S. (Shawn), Liu, Y., & Namkung, Y. (2011). Effects of authentic atmospherics in ethnic restaurants: investigating Chinese restaurants. International Journal of Contemporary Hospitality Management, 23(5), 662–680. https://doi.org/10.1108/09596111111143395

Jang, S. C. (Shawn), & Ha, J. (2015). The Influence of Cultural Experience: Emotions in Relation to Authenticity at Ethnic Restaurants. Journal of Foodservice Business Research. https://doi.org/10.1080/15378020.2015.1051436

Jang, S. C. S., Ha, J., & Park, K. (2012). Effects of ethnic authenticity: Investigating Korean restaurant customers in the U.S. International Journal of Hospitality Management. https://doi.org/10.1016/j.ijhm.2011.12.003

Kim, J. H., & Jang, S. C. (Shawn). (2016). Determinants of authentic experiences: An extended Gilmore and Pine model for ethnic restaurants. International Journal of Contemporary Hospitality Management. https://doi.org/10.1108/IJCHM-06-2015-0284

Kim, K., & Baker, M. A. (2017). The Impacts of Service Provider Name, Ethnicity, and Menu Information on Perceived Authenticity and Behaviors. Cornell Hospitality Quarterly, 58(3), 312–318. https://doi.org/10.1177/1938965516686107

Kukanja, M., Gomezelj Omerzel, D., & Kodrič, B. (2017). Ensuring restaurant quality and guests' loyalty: an integrative model based on marketing (7P) approach. Total Quality Management and Business Excellence, 28(13–14), 1509–1525. https://doi.org/10.1080/14783363.2016.1150172

Lin, P. M. C., Ren, L., & Chen, C. (2017). Customers' Perception of the Authenticity of a Cantonese Restaurant. Journal of China Tourism Research, 13(2), 211–230. https://doi.org/10.1080/19388160.2017.1359721

Lin, T.-C. (2017). Modeling Consumer Intention to Revisit the Same Restaurant and a Simple Algorithm of Coupon Discount Rate. Theoretical Economics Letters, 07(05), 1179–1188. https://doi.org/10.4236/tel.2017.75079

Lin, T. (2018). Impact of restaurant owners'/managers' handling of customers' unexpected incidents on customers' revisit intention, 15(1), 108–124. https://doi.org/10.1504/IJEBR.2018.10008870

Liu, Hongbo, Li, Hengyun, DiPietro, R. B., & Levitt, J. (2018). the role of authenticity in mainstream ethnic restaurants: evidence from an independent full-service Italian restaurant. International Journal of Contemporary Hospitality Management. https://doi.org/10.1016/j.annals.2015.03.003

Liu, P., & Tse, E. C. Y. (2018). Exploring factors on customers' restaurant choice: an analysis of restaurant attributes. British Food Journal, 120(10), 2289–2303. https://doi.org/10.1108/BFJ-10-2017-0561

Liu, S. Q., & Mattila, A. S. (2015). Ethnic dining: Need to belong, need to be unique, and menu offering. International Journal of Hospitality Management. https://doi.org/10.1016/j.ijhm.2015.04.010

Liu, Y., & Jang, S. C. (Shawn). (2009). Perceptions of Chinese restaurants in the U.S.: What affects customer satisfaction and behavioral intentions? International Journal of Hospitality Management. https://doi.org/10.1016/j.ijhm.2008.10.008

Ma, J., Qu, H., Njite, D., & Chen, S. (2011). Western and Asian customers' perception towards Chinese restaurants in the United States. Journal of Quality Assurance in Hospitality and Tourism, 12(2), 121–139. https://doi.org/10.1080/1528008X.2011.541818

Meng, B., & Kyuhwan, C. (2018). An investigation on customer revisit intention to theme restaurants: the role of servicescape and authentic perception. https://doi.org/http://dx.doi.org/10.1108/JEIM-07-2014-0077

Morrison, A. M. (2018). Marketing and Managing Tourism Destinations. Marketing and Managing Tourism Destinations. https://doi.org/10.4324/9781315178929

Muskat, B., & Wagner, S. (2019). Perceived quality, authenticity, and price in tourists' dining experiences: Testing competing models of satisfaction and behavioral intentions. https://doi.org/10.1177/1356766718822675

Nazriah, W., Nawawi, W., Nor, W., Wan, B., Ghani, A. M., & Adnan, A. M. (2018). Influence of Theme Restaurant Atmospheric Factors Towards Customers' Revisit, (2004), 4–9.

Park, E., Kang, J., Choi, D., & Han, J. (2018). Understanding customers' hotel revisiting behaviour: a sentiment analysis of online feedback reviews. Current Issues in Tourism, 0(0), 1–7. https://doi.org/10.1080/13683500.2018.1549025

Parsa, H. G., Self, J. T., Njite, D., & King, T. (2005). Why restaurants fail. Cornell Hotel and Restaurant Administration Quarterly. https://doi.org/10.1177/0010880405275598

Petzer, D., & Mackay, N. (2014). Dining Atmospherics and Food and Service Quality as Predictors of Customer Satisfaction at Sit-down Restaurants. Journal of Hospitality, Tourism and Leisure, 3(2), 1–14. https://doi.org/10.1017/CBO9781107415324.004

Ramukumba, T. (2018). Tourists revisit intentions based on purpose of visit and preference of the destination . A case study of Tsitsikamma National Park, 7(1), 1–10.

Reichheld, F. F., & Sasser, W. E. (1990). Zero defections: quality comes to services. Harvard Business Review. https://doi.org/10.1016/j.colsurfa.2006.11.029

Rudinow, J. (1998). Race, Ethnicity, Expressive Authenticity: Can White People Sing the Blues? Musical Worlds: New Directions in the Philosophy of Music, 52(1), 159–170. https://doi.org/10.2307/431591

Schiffman, L., & Kanuk, L. L. (2015). Consumer Behavior. PT Indeks.

Shariff, S. N. F. B. A., Omar, M. B., Sulong, S. N. B., Majid, H. A. B. M. A., Ibrahim, H. B. M., Jaafar, Z. B., & Ideris, M. S. K. Bin. (2015). The Influence of Service Quality and Food Quality Towards Customer Fulfillment and Revisit Intention. Canadian Social Science, 11(8), 110–116. https://doi.org/10.3968/7369

Simon Tsai Chen-Tsang, C. T., & Lu, P. H. (2012). Authentic dining experiences in ethnic theme restaurants. International Journal of Hospitality Management, 31(1), 304–306. https://doi.org/10.1016/j.ijhm.2011.04.010

Sthapit, E., & Björk, P. (2017). Relative contributions of souvenirs on memorability of a trip experience and revisit intention: a study of visitors to Rovaniemi, Finland. Scandinavian Journal of Hospitality and Tourism, 2250(August), 1–26. https://doi.org/10.1080/15022250.2017.1354717

Sun, S., Law, R., & Fong, D. K. C. (2018). What Affects the Revisit Intention of Chinese Tourists to Macao? Journal of China Tourism Research, 14(3), 296–309. https://doi.org/10.1080/19388160.2018.1492482

Ting, H., Lau, wee ming, Cheah, J., Yacob, Y., Memon, mumtaz ali, & Lau, E. (2018). Perceived quality and intention to revisit coffee concept shops in Malaysia: a mixed-methods approach. British Food Journal. https://doi.org/http://dx.doi.org/10.1108/MRR-09-2015-0216

UNWTO. (2017). Second Global Report on Gastronomy Tourism. UNWTO.

Wu, H. C., Li, M. Y., & Li, T. (2018). A Study of Experiential Quality, Experiential Value, Experiential Satisfaction, Theme Park Image, and Revisit Intention. Journal of Hospitality and Tourism Research, 42(1), 26–73. https://doi.org/10.1177/1096348014563396

Yan, X., Wang, J., & Chau, M. (2015). Customer revisit intention to restaurants: Evidence from online reviews. Information Systems Frontiers, 17(3), 645–657. https://doi.org/10.1007/s10796-013-9446-5

Yoo, T. (2018). Country of Origin and Diners' Perceptions of a Cuisine: The Moderating Effects of Culinary and Institutional Factors. Journal of Hospitality and Tourism Research, 42(3), 420–444. https://doi.org/10.1177/1096348014565026

Zhang, H., Wu, Y., & Buhalis, D. (2017). Journal of Destination Marketing & Management A model of perceived image, memorable tourism experiences and revisit intention. Journal of Destination Marketing & Management, (June), 1–11. https://doi.org/10.1016/j.jdmm.2017.06.004

Promoting Creative Tourism: Current Issues in Tourism Research – Kusumah et al. (Eds)
© 2021 Taylor & Francis Group, London, ISBN 978-0-367-55862-8

The impact of dining experience towards revisit intention at Mujigae Resto, Bandung

R. Andari, Gitasiswhara & D.A.T. Putri
Universitas Pendidikan Indonesia, Bandung, Indonesia

ABSTRACT: This research is aimed to analyze the impact of dining experience which consists of food, service, environment, and price fairness quality towards the revisit intention on Mujigae Resto at Bandung Outlet (Ciwalk, Festival Citylink and Miko Mall). The independent variable (X) of this research is dining ex-perience among the food quality, service quality, quality of the environment, and price fairness as the sub-dimension of the variable, while the revisit intention is the dependent variable (Y) in this research. The type of research used in this research is a descriptive and verification research. This research also uses the quantitative approach with 400 respondents who were the consumers that doing the purchase at Mujigae Res-to at Bandung Outlet. Service quality gains the highest score while the price fairness gains the lowest score on the research as the sub variable of an independent variable. The revisit intention at Mujigae resto at Bandung Out-let is also the highest score in the category, by the probability to visit again as the highest score and the likeli-hood to be the first choice for the future visit as the lowest one. In conclusion, this research approved that there is a big impact on each other among the dining experience and revisit intention.

Keywords: Dining Experience, Revisit Intention, Food Quality, Service Quality, Quality of Environment, Price Fairness, Mujigae Resto

1 INTRODUCTION

The life of a company just to get customers is not enough. A company must create and get customers to have their revisit intention (Hellier et al. 2003). One principle of marketing that is generally recognized is that regular customers are more profitable than new customers (Yan et al. 2015) because maintaining old customers is cheaper than attracting new customers (Yuen & Chan 2010). Therefore, how to retain existing customers and increase their repeat purchases is an important consideration for practitioners to make a profit (Yan et al. 2015). Revisit intention is considered very valuable because it refers to the possibility of using a product or service again in the future (Azize et al. 2012).

Research on revisit intention has been carried out in several industries including in the retail industry (Jung et al. 2014; Kabadayı & Alan 2012; Shih-Tse Wang & Tsai 2014), health industry (Lee & Kim 2017), banking industry (Shao et al. 2008), transportation industry (Jen & Hu 2003; Lai & Chen 2011; Liu & Lee 2016), tourism industry (Neuvonen et al. 2010; Ramukumba 2018; Wu et al. 2018), and the food and beverage industry (Ting et al. 2018; Yan et al. 2015) (Barnes et al. 2016). Research on revisit intention in restaurant describes consumers' desire to make repeat visits, repurchase, and the desire to recommend to others (Baker & Crompton 2000; Phosaard & Wiriyapinit 2011), the concept of revisit intention has been widely used for research in restaurant industry.

Indonesia is one of the most favoured culinary destinations in the Asia Pacific in addition to Japan, Thailand, and Taiwan which ranks the top. Japanese and Korean tourists pretty much choose Indonesia as their favorite culinary destination and apply to Indonesian tourists who choose their country as their favorite culinary destination based on a survey on 2,700 tourists from nine countries in the Asia Pacific by Markus Schueller, Vice President for Food & Beverage Operations Asia Pacific Hilton Worldwide (source: Travel.kompas.com 2014).

The favorite culinary destination in Indonesia is centered by Bandung. Based on the survey conducted on 3,970 respondents, Bandung received 2,341 or around 59% of the total which outperformed the other 4 cities; Makassar (8%), Surabaya (15%), Padang (14%) and Cirebon (3%). Based on these data, it is not surprising that Bandung has been dominant as a culinary tourism city in Indonesia (Source: travel.detik.com 2014).

As a culinary tourism city, Bandung is a place that has a diverse culinary appeal. It is a choice for tourists visiting and business people to open a culinary business in the city of Bandung. One part of the culinary business that is much in demand by business people today is the restaurant and café business. This is indicated by the number of restaurants and cafes that are increasing every year.

The number of restaurants and cafes in Bandung has caused business people to devise strategies to differentiate themselves from their competitors. As a result, Bandung has a variety of themed restaurants and cafes, one of which is a Korean-themed restaurant and café.

Mujigae Resto is a Korean restaurant that first stood on Ciwalk Bandung since April 2013, in Indonesia. Mujigae Resto became the Top Brand Awards in the Korean Restaurant category, in 2018 by 32.4% (Source: www.topbrand-award.com). In Bandung, the Mujigae Resto entered the top 3 in the search for Tripadvisor in the Korean Restaurant category in Bandung (Source: tripadvisor.co.id), there are 4 outlets of Mujigae Resto in Bandung; at Cihampelas Walk (Ciwalk), Citilink Festival, Miko Mall and at Istana Plaza (recently open).

Mujigae Resto is in the form of ready-to-serve culinary by offering halal and quality food with South Korean special menus. Competition is very clear, especially seen from the increasing number of restaurants and cafes in Bandung and the number of themed restaurants and cafés is the same as Mujigae Resto. According to Mitchell (2001) in Rahmawati (2009), business people must prepare a strategy to be able to please and build consumer enthusiasm into an experience in consuming products and services. Therefore a new paradigm is needed to shift traditional thinking in the foodservice business category, especially restaurants, which previously only provided food (food and drinks) to a modern concept that offers an unforgettable experience (dining experience).

Previous research states that the factors that influence revisit intention are attitude, perceived behavioral control, motivation, and authenticity (Shen 2014), perceived authenticity (DiPietro & Levitt 2017; Kim & Baker 2017; Meng & Kyuhwan 2017), service quality (Lu & Berchoux, 2015) and dining experience (Muskat et al. 2019). Research on dining experience shows that dining experience have a significant positive effect on revisit intention.

Based on the background described above, it is known that dining experience based on theories and research can generate revisit intention. Mujigae Resto is a halal restaurant specializing in Korean food that has four outlets in Bandung, but in this study only examined three outlets; Ciwalk outlets, Citylink Festival, and Miko Mall. Istana Plaza outlets were not examined because the new outlets were recently opened. This restaurant arises because of the phenomenon of the Korean Wave which provides a new and different dining experience from other restaurants. It can be seen that a profitable experience leads to higher customer satisfaction which leads to revisiting intention.

2 LITERATURE REVIEW

An experience can consist of a product, such as a theater game. Experience can also be a complement to the product, such as dinner at a particular restaurant, or experience can be a whole package, making the experience not only a product but also a mental process, a state of mind, one night outside combining eating and seeing a game. The main point is that experience is always more than a product. The essence of the product might be an experience (Darmer & Sundbo 2008)

Marketing experience includes environmental factors such as boundary constraints, themes, atmosphere, performance and security, and includes management factors such as integration, renewal, and provider of experiences that have an impact on the level of participation (active and passive) and relations to the environment (Lanier 2008:22). Nowadays, it cannot be denied that marketing experience has an important role in the process of experience and gives a positive impression on consumers. Of course, consumers will look for special and different offers from a company.

Lanier (2008:18) mentions the type of experience offered by marketing experience, as follows:

a. Hedonic Experience
   Hedonic Experience is defined as those activities that have a multisensory, fantasy, and emotive aspect to them. Examples of the hedonic offering include the performing arts (e.g., opera, ballet, and theatre), the plastic arts (e.g., painting, sculpture, and photography), and the popular arts (e.g., movie, rock concerts, and fashion).
b. Leisure Experiences
   Leisure Experiences are defined as those activities that people engage in during their free time and involve some type of recreation and play. Examples of leisure offerings include social activities (e.g., dancing, dining, and sporting events) and crafts (e.g., carpentry, sewing, and decorating).
c. Play Experiences
   Play Experiences are defined as intrinsically motivating experiences. Examples of play offerings include hobbies, games, sports, and aesthetic appreciation.
d. High-risk Experiences
   High-risk Experience is defined as activities that involve high levels of effort and physical and psychic risk. Examples of high-risk experiential offerings include skydiving, mountain climbing, and scuba diving.
e. Extraordinary Experiences
   Extraordinary Experiences are defined as unusual events that are characterized by high levels of emotional intensity. Examples of extraordinary experiential offerings include river rafting, theme parks, and Renaissance festivals.

This study focuses on Leisure experiences where this experience discusses the social pleasures of consumers. The phenomenon of eating out becomes a social activity in creating a pleasurable experience focusing on the dining experience. The dining experience is the customer's assessment of their overall experience, starting from the quality of food and services for the restaurant environment (Canny 2014).

Dining experiences will be reviewed from four dimensions, including food quality, service quality, quality of the environment, and price fairness (Muskat et al. 2019). The dimension of food quality focuses on 5 indicators namely food presentation, food taste, drink taste, food freshness, and appropriate food temperature. Service quality includes 5 indicators; friendly and courteous employees, prompt service, helpful employees, employees who have knowledge of the products. Quality of environment is measured through 5 indicators, namely interior design and decor, appropriate room temperature, noise level, restaurant cleanliness, and neat and well-dressed employees. While price fairness focuses on two indicators, namely the reasonable price of food and reasonable price of drinks.

3 RESEARCH METHODS

The dining experience consists of 4 dimensions, namely food quality (X1); service quality (X2), quality of environment (X3), and price fairness (X4). This type of research is descriptive research, which as stated by Uma Sekaran & Bougie (2016) as a type of research that describes the main topics of research and is designed to collect data that describes the characteristics of objects, activities, and situations. Through this descriptive research, it can be obtained a description of how dining experience is given by Mujigae Resto Bandung outlet to customers.

The Sampling is based on the population of consumers who have visited the Mujigae Resto Bandung outlet in 2018, which is 152,086 people. The calculation results using the Slovin formula produced a sample of 400 respondents. The sampling technique used in this study is a systematic random sampling technique, where the researcher sets the time for systematic dissemination of consumers, i.e. every lunch hour at three checkpoints alternately, namely Mujigae Resto Cihampelas Walk outlet, Mujigae Resto outlet Citylink Festival, and Mujigae Resto outlet Miko Mall.

4 RESEARCH RESULTS AND DISCUSSION

Based on the responses of 400 respondents, the dining experience variable at Mujigae Resto outlet Bandung can be measured by calculating the overall score from the comparison of the total score and the average score of the dining experience. The dining experience with the highest score is service quality, with a total score of 8,076 and 29.96%. The item which gains the highest score is "The level of politeness of employees in serving consumers" with the acquisition of a total score of 1,709 and a percentage of 21.16%. While the statement item "Level of the speed of employees in serving consumers" has a score of 1,559 with a percentage of 19.30%.

The consumer response to the food quality sub variable at Mujigae Resto outlet Bandung has a total score of 7,764, where the item statement that has the highest value is "The level of attractiveness of the presentation of food and beverages served" with a total score of 1,602 and a percentage of 20,63%. While the statement item "The level of taste of drinks served" has a score of 1,520, this item is the lowest statement with a percentage of 19.58%. This is because consumers think that the most important aspect of food quality is the attractiveness of the presentation of food and beverages, which shows that the attractiveness of serving food and drinks plays an important role to satisfy and retain customers. This is supported by a statement which suggests that the attractiveness of serving food and drinks on food quality is one of the factors used to satisfy and retain customers for restaurants (Yong et al. 2013).

In the sub-variable quality of the environment at Mujigae Resto outlet Bandung, the overall total score is 8.069, where the statement item that has the highest value is "restaurant excrement rate" with the acquisition of a total score of 1,715 and a percentage of 21.25%. While the statement item "Comfort sound level in the restaurant area" has a score of 1,419, this item is the lowest statement with a percentage of 17.59%. This shows cleanliness is an important factor in dining and the results lead to whether consumers will return or not. If the restaurant environment is clean, consumers will feel comfortable and satisfied when dining at the restaurant, and vice versa if the restaurant environment is not clean, consumers are uncomfortable and the impact is that consumers do not want to return to the restaurant. This is supported by a statement which states that customers will not revisit restaurants if the environment is not clean, cleanliness in the restaurant environment is another factor that can affect the dining experience on customer satisfaction and intention to revisit (Yong et al. 2013).

Consumer response to the sub-variable price fairness at Mujigae Resto outlet Bandung has a total score of 3,043, where the statement item that has the highest value is "The level of suitability of food prices with food served" with a total score of 1,538 and a percentage of 50,54%. While the lowest score on this dimension is found in the question item, the level of suitability of food prices with drinks is presented with a score of 1.505 with a percentage of 49.46%. This shows that the suitability of prices on food and beverages is also considered by consumers because it can affect consumer behavior. This is supported by statements (which suggest that prices are one of the determinants of consumers to return to or not to restaurants Muskat et al. 2019).

The dining experience variable that gets the highest rating is service quality with a total score of 8,076 and a percentage of 29.96%. Whereas, the sub-variable with the lowest score is at price fairness with a total score of 3,043 with an average score of 1521.5 and a percentage of 11.29%. This means Mujigae Resto Bandung outlets have implemented service quality such as employee politeness, employee speed, alertness of employees, knowledge of employees, and suitability of the menu at Mujigae Resto outlet Bandung. Whereas the sub-variable that has the lowest score is price fairness, this is due to the prices on food and drinks offered at Mujigae Resto outlet Bandung, for consumers, it is not yet appropriate. Calculations on the continuum line show that dining experience variables belong to the high category, with a score of 26,952.

5 CONCLUSION

In general, respondents' reactions regarding the implementation of a dining experience at Mujigae Resto outlet Bandung, which consists of 4 dimensions, namely food quality, service quality, quality of the environment, and price fairness are in the high category. This means that the application of dining experience is considered good. The service quality dimension gets the highest rating. In the context of the restaurant industry, service quality is seen as one of the core determinants of satisfaction and behavioral intentions because now consumers do not only evaluate food quality but they also consider the service quality that they encounter during their eating experience. Whereas the lowest response is in the dimension of price fairness because for respondents, the suitability of the price of food and drinks is not sufficiently appropriate and the respondent pays attention to it.

Based on the results of this study, it was shown that dining experience consisting of food quality, service quality, quality of the environment, and price fairness had a positive and significant influence on the revisit intention at Mujigae Resto outlet Bandung.

6 RECOMMENDATION

The findings suggest that the management Mujigae Resto Bandung outlets should pay more attention to the taste of drinks, so that consumers feel satisfied with the drinks served to them and pay more attention to the suitability of the price of drinks by increasing the size or by lowering the price.

It is expected that the drinks served are worth the price offered and consumers feel treated fairly. Besides that the management of Mujigae Resto Bandung outlets can add more employees so that when the restaurant is crowded, the needs of consumers are still served quickly.

Furthermore, the Mujigae Resto outlet Bandung must pay attention to the competition of restaurants in Bandung, because there came several competitors or similarly themed restaurants and other themes that are more interesting in the future. In addition to increasing the revisit intention at Mujigae Resto Bandung outlet through the implementation of the dining experience, the management of Mujigae Resto Bandung outlet is expected to be able to control and develop dining experience as a marketing strategy. The recommendation is expected to increase the assessment of likelihood to be the first consumer choice at Mujigae Resto outlet Bandung, so that it can improve the assessment of consumer revisit intention at Mujigae Resto at Bandung outlet.

In this study, there are certainly many shortcomings and limitations. Therefore, for further research, the authors suggest to further find out more about the influence and relationship between dining experience and the revisit intention by using other dimensions so that research and discussion of these two variables, especially with restaurant objects, is getting better and growing. Also, further research needs to be done beyond the variables studied, for example, restaurant attributes, electronic word of mouth, social media marketing, dining service quality, and so on. Likewise with this research is only done to consumers Mujigae Resto Bandung outlets. It is expected that further research can other Mujigae Resto outlets in Jabodetabek.

REFERENCES

Azize, Ş., Cemal, Z., & Hakan, K. ı. (2012). The effects of brand experience and service quality on repurchase intention: The role of brand relationship quality. *African Journal of Business Management*, *6*(45), 11190–11201. https://doi.org/10.5897/AJBM11.2164

Barnes, S. J., Mattsson, J., & Sørensen, F. (2016). Remembered experiences and revisit intentions: A longitudinal study of safari park visitors. *Tourism Management*, *57*, 286–294. https://doi.org/10.1016/j.tourman.2016.06.014

Canny, I. (2014). Measuring the Mediating Role of Dining Experience Attributes on Customer Satisfaction and Its Impact on Behavioral Intentions of Casual Dining Restaurant in Jakarta. *International Journal of Innovation, Management, and Technology*, *5*(1). https://doi.org/10.7763/ijimt.2014.v5.480

Darmer, P., & Sundbo, J. (2008). Creating Experiences ion The Experience Economy. *Business*, 271.

DiPietro, R. B., & Levitt, J. (2017). Restaurant Authenticity: Factors That Influence Perception, Satisfaction and Return Intentions at Regional American-Style Restaurants. *International Journal of Hospitality and Tourism Administration*, *0*(0), 1–27. https://doi.org/10.1080/15256480.2017.1359734

Hellier, P. K., Geursen, G. M., Carr, R. A., & Rickard, J. A. (2003). Customer repurchase intention. *European Journal of Marketing*, *37*(11/12), 1762–1800. https://doi.org/10.1108/03090560310495456

Jen, W., & Hu, K. (2003). Application of perceived value model to identify factors affecting passengers …, 307–327.

Jung, N. Y., Kim, S., & Kim, S. (2014). Influence of consumer attitude toward online brand community on revisit intention and brand trust. *Journal of Retailing and Consumer Services*, *21*(4), 581–589. https://doi.org/10.1016/j.jretconser.2014.04.002

Kabadayı, E. T., & Alan, A. K. (2012). Revisit Intention of Consumer Electronics Retailers: Effects of Customers' Emotion, Technology Orientation, and WOM Influence. *Procedia – Social and Behavioral Sciences*, *41*, 65–73. https://doi.org/10.1016/j.sbspro.2012.04.009

Kim, K., & Baker, M. A. (2017). The Impacts of Service Provider Name, Ethnicity, and Menu Information on Perceived Authenticity and Behaviors. *Cornell Hospitality Quarterly*, *58*(3), 312–318. https://doi.org/10.1177/1938965516686107

Lai, W. T., & Chen, C. F. (2011). Behavioral intentions of public transit passengers-The roles of service quality, perceived value, satisfaction, and involvement. *Transport Policy*, *18*(2), 318–325. https://doi.org/10.1016/j.tranpol.2010.09.003

Lee, S., & Kim, E.-K. (2017). The Effects of Korean Medical Service Quality and Satisfaction on Revisit Intention of the United Arab Emirates Government Sponsored Patients. *Asian Nursing Research*, *11*(2), 142–149. https://doi.org/10.1016/j.anr.2017.05.008

Liu, C. H. S., & Lee, T. (2016). Service quality and price perception of service: Influence on word-of-mouth and revisit intention. *Journal of Air Transport Management*, *52*, 42–54. https://doi.org/10.1016/j.jairtraman.2015.12.007

Lu, C., & Berchoux, C. (2015). Service quality and customer satisfaction: qualitative research implications for luxury hotels. *International Journal of Culture, Tourism, and Hospitality Research*, *9*(2).

Meng, B., & Kyuhwan, C. (2017). An investigation on customer revisit intention to theme restaurants: the role of servicescape and authentic perception. https://doi.org/http://dx.doi.org/10.1108/JEIM-07-2014-0077

Muskat, B., Hörtnagl, T., Prayag, G., & Wagner, S. (2019). Perceived quality, authenticity, and price in tourists' dining experiences: Testing competing models of satisfaction and behavioral intentions. *Journal of Vacation Marketing*. https://doi.org/10.1177/1356766718822675

Neuvonen, M., Pouta, E., & Sievänen, T. (2010). Intention to Revisit a National Park and Its Vicinity. *International Journal of Sociology*, *40*(3), 51–70. https://doi.org/10.2753/IJS0020-7659400303

Ramukumba, T. (2018). Tourists revisit intentions based on purpose of visit and preference of the destination. A case study of Tsitsikamma National Park, *7*(1), 1–10.

Sekaran, U., & Bougie, R. (2016). Research Methods For Business: A Skill Building Approach.

Shao, J. B., Wang, Z., & Long, X. X. (2008). The driving factor of customer retention: Empirical study on bank card. In *2008 International Conference on Management Science and Engineering 15th Annual Conference Proceedings, ICMSE*. https://doi.org/10.1109/ICMSE.2008.4668969

Shen, S. (2014). Intention to revisit traditional folk events: A case study of qinhuai lantern festival, China. *International Journal of Tourism Research*, *16*(5), 513–520. https://doi.org/10.1002/jtr.1949

Shih-Tse Wang, E., & Tsai, B.-K. (2014). Consumer response to retail performance of organic food retailers. *British Food Journal*, *116*(2), 212–227. https://doi.org/10.1108/BFJ-05-2012-0123

Shin, B.-K., Oh, M.-H., Shin, T.-S., Kim, Y.-S., You, S.-M., Roh, G.-Y., & Jung, K.-W. (2014). 한류 문화콘텐츠가 한식 및 한국 제품 구매에 미치는 영향 - 아시아 (중국 , 일본), 미주 , 유럽지역을 중심으로 – The Impact of Korean Wave Cultural Contents on the Purchase of, *29*(3), 250–258.

Ting, H., Lau, wee ming, Cheah, J., Yacob, Y., Memon, mumtaz ali, & Lau, E. (2018). Perceived quality and intention to revisit coffee concept shops in Malaysia: a mixed-methods approach. *British Food Journal*. https://doi.org/http://dx.doi.org/10.1108/MRR-09-2015-0216

Wu, H. C., Li, M. Y., & Li, T. (2018). A Study of Experiential Quality, Experiential Value, Experiential Satisfaction, Theme Park Image, and Revisit Intention. *Journal of Hospitality and Tourism Research*, *42*(1), 26–73. https://doi.org/10.1177/1096348014563396

Yan, X., Wang, J., & Chau, M. (2015). Customer revisit intention to restaurants: Evidence from online reviews. *Information Systems Frontiers*, *17*(3), 645–657. https://doi.org/10.1007/s10796-013-9446-5

Yong, C. K., Siang, D. O. C., Lok, T. W., & Kuan, W. Y. (2013). Factors Influencing Dining Experience On Customer Satisfaction And Revisit Intention Among Undergraduates Towards Fast Food Restaurants, (April).

Yuen, E. F. T., & Chan, S. S. L. (2010). The effect of retail service quality and product quality on customer loyalty. *Journal of Database Marketing & Customer Strategy Management*, *17*(3–4), 222–240. https://doi.org/10.1

Promoting Creative Tourism: Current Issues in Tourism Research – Kusumah et al. (Eds)
© 2021 Taylor & Francis Group, London, ISBN 978-0-367-55862-8

Legal and business sustainability of social enterprises restaurants

N.B. Le
Hong Kong Polytechnic University, Kowloon, Hong Kong

T. Andrianto
Bandung State Polytechnic, Bandung, Indonesia
Hong Kong Polytechnic University, Kowloon, Hong Kong

R. Kwong
Hong Kong Polytechnic University, Kowloon, Hong Kong

ABSTRACT: As part of the tourism industry, there are various kinds of restaurant. One of them is social enterprises that have social goals as added value. This study aims to examine the legal aspects and business models sustainability of social enterprises of restaurants and catering business in Hong Kong, Indonesia, and Vietnam. This study uses a comparative content analysis of documents and interviews script in six selected restaurants. The results confirm that the lack of policy hinders the development of the social enterprise restaurant. The enterprises with unique added values and connections to public funds and corporates tend to be more successful. Current legal structures and hard to withdrawing government subsidies are some of the factors to the lack of social enterprises business sustainability.

Keywords: social enterprise restaurant, legal status, business model, social-enterprise sustainability, unique values

1 INTRODUCTION

Social enterprises are not a new concept. In fact, this type of enterprise has been around in the world for hundreds of years in various forms of self-sufficient charitable organizations (ISEA 2015). However, social enterprises should not be defined by how to get others to make donations or how to teach others to earn a living only–instead, the definition has to focus on how to revolutionize the whole social system to create businesses which are more social-oriented and sustainable (Drayton, as cited in Sheldon & Daniele 2017). Social enterprises may also come in the form of new entrepreneurship, creating a new product or service which meets the needs of a particular group of customers for some social purpose rather than for-profit orientation (Tan et al. 2005). The last few years have seen the growth of social entrepreneurship, especially along with the development of information technology (ISEA 2015). For example, one of the most notable social entrepreneurs is Muhammad Yunus, with his Grameen Bank project in 1983, which was awarded the Nobel Prize in 2006 (no-belprize.org 2006).

Simply put, a business can become a social enterprise when it tries to increase its social effect and benefits, and not only for the merits of its profits (Tan et al. 2005). That being said, a social enterprise can be defined as a business that prioritizes the common benefits of society over the financial benefits of its corporation (Alvord et al. 2004; Baron 2007). However, because of its emphasis on social benefits, it does not always at-tract the most interest in the business world so as to develop structured and established business models (Alegre & Berbegal-Mirabent 2016; Evert 2016). In addition, the legal framework for this type of business has not been well addressed by the law in many countries, especially in developing ones, where, ironically, social enterprises

 DOI 10.1201/9781003095484-80

should be needed the most (Alegre & Berbegal-Mirabent 2016; Defourny & Kim 2011; ISEA 2015, British Council 2007). This paper, as a result, set out in examining the sustainability issues of social enterprises across countries with different levels of economic development in Asia, that is, Hong Kong (China), Indonesia, and Vietnam.

This study aims to examine the legal aspects and business models sustainability of social restaurants in Hong Kong, Indonesia, and Vietnam. In a developed society like that of Hong Kong, social enterprises have for long attracted a great amount of interest–however, ever since the economic crisis in 1997, spending for the social sector has been cut back by the Hong Kong Special Administrative Region (HKSAR) (Ho & Chan 2010). In 2001, efforts were reinitiated by HKSAR to encourage NGOs (non-governmental organizations) to set up social enterprises. The purpose was to reduce poverty, to employ people with disabilities, and to pro-mote social inclusion (Ho & Chan 2010). A good number of social enterprises were created as a result under the form of work-integrated social enterprises (or WISEs) in different lines of business. The restaurant business as adopted for social enterprises is in its maturing stage in Hong Kong with over 80 WISEs in the restaurant and catering business as of 2017 according to iPick (2017). Still, there is very little research literature about social enterprises in the restaurant business in particular and in hospitality and tourism in general (Alegre & Berbegal-Mirabent 2016).

Unlike in Hong Kong, the more immediate problem to social enterprises in Indonesia and Vietnam is on the laws for the creation and operation of this "considered-to-be-new" type of business. While social enterprises in Hong Kong may operate under the various legal status of work-integrated social enterprise or private limited company by guarantee or private limited company with sponsoring organization, etc., there are only one or two types of legal status for social enterprises in Indonesia and Vietnam. The law for "Yayasan," a foundation for social activities, was first adopted in 2001 in Indonesia, and has gone through a number of revisions and regulatory improvements by the Act of 2004 and new regulations in 2008 and 2013 (Undang-Undang RI No.28, 2004; Peraturan Pemerintah No.2, 2013). Another older legal concept is "koperasi," a public union for financial and banking activities to support a social purpose that was passed by the Act of 1992 and was re-modified in 2013 (Undang-Undang RI No.17, 2012). On the other hand, while the legal sections and clauses for social entrepreneurship were drafted and discussed by the Indonesian House of Representatives in 2015, they have not been approved until now (Martaon 2017).

As for Vietnam, social enterprises can only exist under one legal status of "social limited liability company" with policies for tax reduction and refund based on the nature of their business and the thresholds of their profits (British Council 2007). Social enterprises in Vietnam, however, are usually confused with charitable organizations by the general public, and as a result, profits are more than often considered to be unacceptable for this type of business (British Council 2007). In addition, even though both Vietnam and Indonesia now have established their laws for CSR (customer social responsibilities), given the fact that most businesses in their territories are small- to medium-sized and purely for-profit, these laws are often not recognized and their enforcement is rendered ineffective due to the lack of legal tools and practices (ISEA 2015; British Council 2007).

2 LITERATURE REVIEW

In the article "Social enterprise: Through the eyes of the consumer," Allan (2005) identified the following three major characteristics of social enterprises: (1) enterprise orientation, (2) social aims, and (3) social ownership. Of these characteristics, social aims are unique to the individuals or groups who started the social enterprise (Sheldon & Daniele 2017). However, social ownership very much determines the eligible sources of income the social enterprise may facilitate (Osinowo 2014; gatherwell.com 2017). For example, as a benefit corporation or flexible-purpose corporation, the social enterprise will enjoy better profit regulation and tax reduction than a limited partnership or general partnership, which is limited in the amount of earned profits or even only allowed to operate as non-profit. Then, even though the enterprise orientation is usually played down among the three mentioned characteristics of social enterprises because it is all about making profits and involving the production or provision to the market (Allan 2005), it is at the heart of the social enterprise's survival and sustainability (Chahine 2016). In fact, without this characteristic, it will be no longer a social enterprise, since it will become a charitable organization (Ausaid & UTS 2016).

Therefore, it can be said that the sustainability of a social enterprise is, first of all, its market profitability and financial stability (Yunus et al. 2010). However, of course, that is not everything about a social enterprise's sustainability; rather, its social cause or social purposes and public awareness of its social cause is an-other important aspect in the social enterprise's sustainability, since those are the driving force behind (Martin and Osberg 2007). In addition, to sustain its market success and social cause, a social enterprise has to include various groups of stakeholders in its activities so as to maintain the ability to earn income as well as the ability to perform their social functions (Bagnoli & Megali 2011). The hospitality and tourism business as such is always about maintaining a sizable and stable pool of customers and fans or supporters. This is in line with Dickerson and Hassanien (2017), who has put it that a hospitality social enterprise needs to build a community around itself together with the support of that community regarding its sustainability–either environmentally or socially.

Moreover, to retain that support, still, the social enterprise needs a value proposition in a hybrid format for its business model (Parhankangas & Renko 2017). To sum up, Osterwalder and Pigneur (2010) identified nine important factors in the business model: (1) key partner(s), (2) key activities, (3) key resources, (4) value proposition(s), (5) customer relationship, (6) distribution channel, (7) customer segment(s), (8) cost structure, and (9) revenue stream. Performing well on all of these factors, and sustainability is supposed to be more or less ensured. Unfortunately, mistakes and failures will be made during the process of reaching that performance level, and owner and/or managers of social enterprises should keep in mind that failure is always a part of success (Chahine 2016). In addition, contingencies and crises are also identified as inevitable facts in the work of the hospitality and tourism industry, and owners and managers of social enterprises should learn to overcome different types of crisis, occurring to their economic and social purposes (Alegre & Berbegal-Mirabent 2016).

3 METHODOLOGY

A qualitative research methodology is adopted for this study as it examines the processes, structures, and individuals involved in the creation and operation of social enterprises in the restaurant and catering business in Hong Kong, Indonesia, and Vietnam. Since the study is particularly concerned about what works and what does not work in terms of the sustainability for this type of social-enterprises, descriptive, explanatory, and exploratory data would need to be extracted from various sources, rendering the use of quantitative research methods inappropriate compared to those of qualitative research. Qualitative data extraction, in turn, helps provide answers to the social constructionism paradigm of this research, which focuses on socially constructed knowledge on what is sustainability in the restaurant and catering business, how it can be achieved in a social enterprise context, and why it may fail or not in the face of certain crises.

The qualitative research approach for this study is that of narrative on legal documents about social enterprises and of case studies on six different social enterprise restaurants in Hong Kong, Indonesia, and Vietnam (two for each country): "GH" and "C8" from Hong Kong; "KP" and "DCFT" from Indonesia, then "MS" and "BoL" from Vietnam. Codes for names and owners of the restaurant were used to cover the real names and any related information that leads to the person or companies' identity as part of academic purposes and the request of some informants, respectively.

The choice of these social enterprise restaurants came from recommendations of experts on social enterprises in each country as well as from the popularity of these social enterprise restaurants on the Internet and social media. Text and multimedia information about these restaurants, as a result, is relatively abundant. Two types of data collection methods of documents and interviews were adopted, in which past documents were thoroughly examined and in-depth interviews were carried out with the owners and/or managers of these social-enterprise restaurants. Content analysis followed to analyze the actual content of past documents as well as of recorded interview lines before compare-and-contrast data analysis was used for the comparison of one sustainability aspect or situation of one restaurant to that of another. A brief description of each social enterprise in the sample of the study can be found below:

- GH: GH was founded in 2005 by Ms. JM (GH's current CEO) and a group of social workers in an effort to provide employment and life enrichment to retired seniors. Originally self-funded, GH is now reaching out for government grants and donations with its new concept restaurant recently receiving over 1 million Hong Kong dollars in grants. GH is considered to be a success, currently operating up to four restaurants, catering service, food delivery, organic farm, and a career employment service.
- C8: C8 began in 2013 with government subsidies to help employ and train people with physical disabilities or learning disabilities. The TNC is the original sponsoring organization of C8. Another smaller coffee shop called TN also began under this same scheme. C8 is in partnership with the Hong Kong Maritime Museum, which offers its current rental place. C8 mostly serves tourists, tour buses (e.g., Big Bus tours), and nearby offices' workers because of the location.
- KP: This small restaurant in Indonesia is owned by a community, called "KA." The community was founded in 2015 by RH. "KA" is a community that cares about heritage buildings and cultural values of the local Sundanese. KP began in 2015 as the place for members of this community to meet up with one another on a regular basis. Operated and funded by the community (KA) itself until now, KP is not making a profit and has no plan to expand its business.
- DCFT: Privately owned and run by DSA, a 27-year-old female from May 2015, DCFT is the first café in Indonesia to employ deaf people in collaboration with the Deaf Community Movement (Net.Tv 2016). The main social purpose of this enterprise is to help train deaf people become tough and professional at work as well as to inspire them to develop positive thinking in life. Another purpose is to raise awareness of the people who are not deaf to communicate with deaf people (Diansyah 2015). Until now, it has been considered to be a success as they receive all kinds of attention from the general public, the deaf community, the local government, and TV shows (including from some of the top celebrities) (I TV 2016a, 2016b; Net.Tv 2016; Diansyah 2015). Their target market is "everyone who cares about the deaf."
- MS: MS is a restaurant in Ho Chi Minh City, Vietnam, run by two legal owners/directors since 2013. DNDN and Bro. FVH is also assisted by their students/employees as financially disadvantaged youngsters in southern provinces of Vietnam. The actual owner of the MS, however, is a Catholic sect, called the SDB. MS currently has fewer than 100 students/employees, and it offers the three-year program of vocational diploma in Culinary and Restaurant Service from the German Chambers of Commerce Abroad (AHK, for short in German). MS is financially successful and is aiming to build a full-scale vocation school of around 500 students.
- BoL: BoL was founded in 2005 as a bakery store at first, which employs deaf people. Its founders, B and K, are two Americans living in Vietnam, who have the experience of using the sign language and Western food preparation, and would like to pass on those skills to deaf people. The business grew to a full-scale Western restaurant, and was relatively successful with around 20 employees at its peak time. In 2015, its ownership was transferred over to Dr. BJL, a Frenchman, and the business started to decline because of changes in the target customers, products, and locations. Ms. TTTA, the manager of BoL, who has been with the business from its start until now, is trying to recover the business by opening an English Centre for Kids in the same building (Figures 1 and 2).

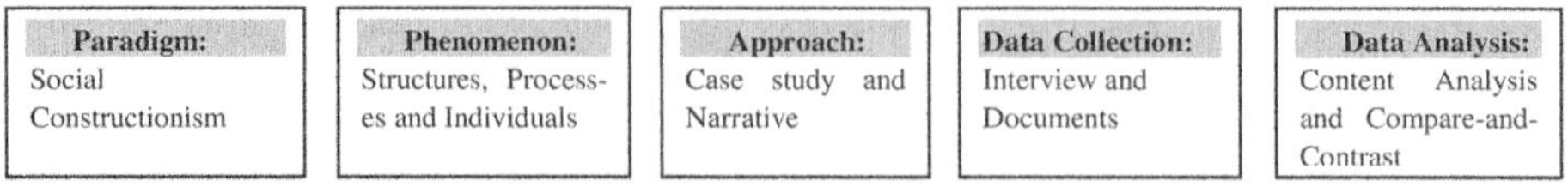

Paradigm:	Phenomenon:	Approach:	Data Collection:	Data Analysis:
Social Constructionism	Structures, Processes and Individuals	Case study and Narrative	Interview and Documents	Content Analysis and Compare-and-Contrast

Figure 1. Researcher's lens for the study of sustainability in social enterprise restaurants.

Hong Kong	Vietnam	Indonesia
C8 Founded in 2013 Mr JM, General Manager	MS Founded in 2013 Bro. DNDM, Director	KP Founded in 2015 Mr RA, Founder (also founder of "KA")
GH Founded in 2005 Ms JM, CEO	The BoL Founded in 2005 Ms TTTA, Manager	DCFT Founded in 2015 Ms DSA, Sole Founder

Figure 2. General information about sampled social enterprise restaurants.

4 FINDINGS AND DISCUSSIONS

No matter what organizational structure or business model a social enterprise may have, a common aspect found of all social enterprises in the study was that they all started because of a social cause or purpose rather than because of some specific business. For example, MS in Vietnam began to help financially disadvantaged youngsters, GH in Hong Kong to support the retired seniors, DCFT in Indonesia to train deaf people. For that reason, there is a sustainability problem of which not each of these social enterprises had a good idea or any experience with its business.

> *"We had no experience of running restaurants prior to starting GH, [we simply wanted to offer help to]... many phone calls [at the Depression Hotline for the Elderly] that expressed boredom and dejection after retirement," said Ms JM, the CEO of GH in the web article by Li (2016).*

As the business progressed and matured, most social enterprise owners and managers would learn to over-come many challenges and obstacles in their line of business. However, there are certain obstacles that they might not always be able to overcome. Those are limitations in the government's laws and policies regarding the creation and operation of social enterprises. Even in Hong Kong, where the laws for social enterprises are relatively established with social enterprises under the legal status of private limited companies by guarantee, receiving tax exemption and government funding, many social enterprises still chose not to adopt that status for the flexibility of their business operation as in the case of GH (Ms JM. Personal communication, November 28, 2017).

At the other end of the spectrum, in Indonesia, social enterprises have not been legally recognized. Most social enterprises like KP or DCFT have to operate like regular businesses with full tax responsibility ("H." Personal Communication, December 12, 2017; Net.Tv 2016). In Vietnam, social enterprises are allowed under the legal status of a social limited liability company with tax reduction, but usually with no government subsidies or funding (Nguyen, N. Personal communication, December 10, 2017). The inadequacy and/or lack of recognition of their legal status usually leads to many hassles in the operational sustainability of many social enterprises.

> *"We were not aware of the tax reduction for social private limited companies at first. But when we found out about it, we chose not register for that status because it involved so much screening and monitoring work by the local government," said Tam Ha, manager of BoL in Vietnam (personal communication, 2017).*

Many social enterprises have found ways to maneuver around to facilitate the nature of their business be-cause of the legal limitations. A common finding of social enterprises' ownership and organizational structure was that they usually established two legal components: one as a foundation or charitable organization and another as the commercial business. The link between these two components helps the latter receive tax reduction or refund through the former while operating like any other regular business. At the same time, the legal status of the former makes it easy in receiving gifts and donations or for educational and training purposes.

For example, GH has one charitable organization called "Everbright Concern Action" and one GH restaurant and catering brand, which currently holds four restaurants, one Hong Kong Kitchen (catering service), one GH Food Delivery, and one LOHAS Organic Farm (Ms JM. Personal communication, November 28, 2017). Or C8 has an NGO called TNC besides C8 and TN coffee shop (Mckinven, J., Personal communication, December 4, 2017). For MS, it has two registrations of MS Social Limited Liability Company for its vocational training programs and MS Restaurant Company for its restaurant business. The same thing was held for KP with one social community called "KA" and its KP, the restaurant business registration ("H." Personal communication, December 12, 2017).

Before digging into the business model of social enterprises, it is important to indicate that most social enterprises do not rely on their business alone for survival and sustainability. Rather, grants and donations are other prevalent sources of income. GH, for example, recently received over 1 million Hong Kong dollars for the startup of its restaurant from the Enhancing Self-Reliance Through District Partnership Programme (ESR) of the Home Affairs Department of Hong Kong (Ms JM. Personal communication, November 28, 2017). An-other example was from MS, which regularly received monetary donations from the SDB and its Catholic subjects (Nguyen, N. Personal communication, December 10, 2017). At the very least example, social enterprises may receive support in their rents or rental place like in the case of KP or DCFT, which initially received their rental places from the "KA" community and the deaf community movement, respectively ("H." Personal communication, December 12, 2017; Net.Tv, 2016). In a rare instance, C8 also received 1.7 million in Hong Kog dollars in a three-year government subsidy granted for its startup (Mckinven, J. Personal communication, December 4, 2017).

> *"Despite what the local and central governments may say about their support for social enterprises, subsidies are hard to get if you are not a state-owned organization. Many times, we have to pay extra 'communication fees' for many of our activities," said Tam Ha, manager of BoL (personal communication, 2017).*

Against all odds, social enterprises in this study strived to make ends meet with their business model. A major difference of the business models of social enterprises from those of regular businesses is that they also need to fulfill certain social purposes (Kim 2007). For example, besides the aim to break even for their restaurant business, MS needs to make use of its restaurant as the formal training place for their in-house students who will ultimately receive the vocational diploma from the AHK (Chambers of Commerce Abroad) of Germany (Nguyen, N. Personal communication, December 10, 2017). BoL and DCFT use their restaurants to teach deaf people the sign language, restaurant skills, and other life skills (Ha, T. Personal communication, December 13, 2017; Diansyah 2015) while GH set out to help the elderly and C8 the disabled people.

The differences between the social purposes and the nature of the adopted business can cause the business model of social enterprises to not be clearly defined. As in the case of KP, since the social purpose is to preserve cultural heritage, historical buildings, and written literature of the "KA" community ("H." Personal communication, December 12, 2017), it is not closely aligned with its restaurant business. For that reason, the business model of KP is not well-defined for future growth and sustainability.

Unlike regular business models, for social enterprises, it is not always the case that their products may determine their customers. However, their customers or employees may define their products from the first place. For instance, the majority of customers of MS are Catholic subjects and foreign expatriates, while BoL are European tourists so that they had to choose Western foods as their specialty. On the other hand, since the employees of GH and C8 are the elderly and disabled people, respectively, who usually work at a slower pace, they can only carry simple refreshments or cuisines (Ms JM & M, J. Personal communication 2017).

Recently, GH even made a move to Vietnamese dishes like rice noodle (pho), fried chicken wing, lemongrass pork-chop rice, etc., which take less time and effort to master the art of cooking and serving (Ms JM. Personal communication, November 28, 2017). Changing the menu constantly from one type of cuisine to another to adapt to the conditions of a certain social enterprise, however, may threaten the sustainability of any restaurant business, as in the case of BoL:

> *"We started out as a bakery, but the profit margin was not good enough..., so, we revise our menu to include American foods,...then, to add more variety, we now also serve Mexican foods,...Still, the number of customers is decreasing," said Tam Ha, the manager of BoL (personal communication, 2017).*

By moving too fast from one type of food to another, BoL lost its original identity as a simple bakery and cafeteria run by the deaf people. However, there were also risks from other changes which led to its currently failing business: change in ownership, change in management, and change in location. Change in ownership and management may disrupt the flow of returned customers who have long been familiar with certain personnel in the restaurant. This risk is also shared by MS and KP as the majority of their customers come from some very specific groups of customers of Catholic subjects, foreign expatriates, and "KA" community members, respectively.

On the other hand, change in location, under certain circumstances, may even put an end to the restaurant business. For example, the GH branch in Jao Tsung-I at Mei Foo, Kowloon, had to close down despite great profits because the Hong Kong Institute for Promotion of Chinese Culture refused to renew the lease (Ms JM. Personal communication, November 28, 2017). In a nutshell, social enterprises share the same or even greater operational risks in running their business when compared to other regular businesses. Hence, before becoming a successful social enterprise, it is important to remind that it needs to successfully function as a regular business first.

Another important aspect of the operation of a social enterprise is its very people. Its employees may not be simply the ordinary employees as they are also its students as in the case of MS, BoL, or DCFT. Its owners and managers may not be simply the business owners or managers as in the case of MS, in which the owner-ship and management is rotated every three to five years by the SDB, a Catholic sect, or the real owner of MS, or in the case of C8 and DCFT, the managers are also the teachers to their disabled employees. Maintenance of a sizable human resource, as a result, is at the heart of a social enterprise's success or failure.

> *"With around 100 students at the present time, we can function well as a restaurant...with almost no turnover. But in order to grow into a full-scale vocational school, we need to aim at around 500 students in the long run..." as commented by Bro. DNDN, the director of MS (personal communication, 2017).*

On the other hand, for smaller restaurants like C8 (with 16 staff with physical or learning disabilities), KP (with 4 staff), and BoL (with 8 staff, including 5 deaf, down from 20), the risk of going out of business is al-ways there if they mishandle the business. As a result, the quality of service is very important in sustaining the business. Quality of service, in turn, depends on the quality of (in-house) training. Except for MS which aims to become a full-scale vocational school with a German diploma being granted (NN. Personal communication, December 10, 2017), the training at the rest of the social enterprises in this study is more or less right on-the-job.

GH tries to add value to its training by including a short orientation for its elderly employees on customer service, team-building, and skills development while disabled employees at C8 used to receive special training courses from the community called "TNC" (MJ. Personal communication, December 4, 2017). Quality training or not, there will always be certain shortcomings in the service being delivered by the elderly, deaf, or disabled employees. For that reason, the ability to handle crisis situations by owners and managers of these social enterprises is another essential determinant of their sustainability. Example from the Director of MS,

> *"... I always tried my best to handle these crises including going to other towns to look for my run-away students, talking continuously to their parents and relatives, maintaining good relationship with government officials, spending my own money for any effort needed..." as recounted by Bro. DNDN, the Director of MS (personal communication, 2017).*

While some crises are very much immediate and critical, others are more on a regular and incremental basis:

> *"The real killer of our business is the ever-increasing rental rate in Hong Kong. While it does not really sound like any crisis, it is always an imminent one. Who knows? Tomorrow, if it rises to certain level that we cannot sustain anymore, then it will be a real crisis as we have to close shop..." commented by Ms. JM, the CEO of GH (personal communication, 2017).*

Indeed, the last but not least essential aspect in ensuring the sustainability of a social enterprise is the marketing and public relations skills of its owners and/or managers. No matter how great or noble the social cause of a social enterprise may be, if no one knows or understands it, sooner or later, that will lead to a complete failure of its business as a social enterprise. For instance, in the case of BoL, they were trying hard for years to help the deaf people, but since they were operating as a for-profit business, the public did not recognize them as an organization for social cause and benefit.

Furthermore, when its ownership was transferred to the new owner, it further confirmed the public perception that this was, by no means, a social enterprise. Or in the case of KP, since the knowledge about its existence and service is only available within the "KA" community, it does not have room for growth so as to attract new customers or to expand its business. If one day when the "KA" community is no longer there to support its business, KP will simply have to stop its existence. Sharing of owners and managers of the social enterprises in this study highlighted unique marketing and public relations approach that should be taken note of:

> *"Besides appealing to the church goers of our sect, we also distribute flyers of our restaurant to people living nearby, educating them of our work and noble cause... While not all of them will become our customers, many will come to appreciate our hard work and values....," commented by Bro. DNDN, the director of MS (personal communication, 2017).*

5 CONCLUSIONS

In conclusion, the study has found that the actual number of successful social enterprises is minimal compared to what many may have expected. In fact, despite all the good things being said to promote social enterprises around the world, current legal structures in many countries do not foster the sustainable development of social enterprises. Government subsidies are usually hard to find even in a developed society like that of Hong Kong, while public suspicion and unfavorable taxation is a reality that social enterprises in a less-developed society like that of Indonesia or Vietnam have to deal with on a daily basis. Therefore, no matter whether the social purpose is to help disabled people or disadvantaged youngsters/elders or any other groups of individuals in trouble, owners and managers of social enterprises need to bear in mind that they have to run a successful and profitable business first before they may realize their social purpose.

As the focus of this paper is on social enterprises in the restaurant and catering business, it was interesting to observe that most social enterprises in reality chose this line of business to start with. One restaurant expert, however, questioned the relevance of choosing the restaurant and catering business for social enterprises. This is not an easy business even for regular for-profit restaurants as the turnover rate is high and success depends on many factors such as chefs, locations, customers, dietary trends, etc. Perhaps it is a common belief that anyone, disabled or not, can learn to cook and the restaurant business does not require as much capital to start with when compared to other types of business. However, such belief may have been wrong, given the lack of sustainable success by social enterprises in the restaurant and catering business.

Social enterprise restaurants may take pride in the fact that they have much lower turnover rate compared to that of regular restaurants because their disadvantaged staff really have no other choice for employment. However, at the same time, it is difficult to provide quality service or to expand the business on the back of such disadvantaged staff. Many social enterprises like BoL or KP or even GH in this study found themselves scaling down the business after years of successful operation for that same reason. Sustainability, thus, becomes a major challenge to social enterprises in the restaurant and catering business.

To attain sustainability, social enterprises in this line of business need not only good operational skills on the part of their managers, but also the ability of the whole organization to deal with contingencies and crises besides the ability of their owners and stakeholders to attract public interest to their social cause or purpose. While this study has discovered various factors affecting the sustainability of social enterprises in the restaurant and catering business, further studies will be required to put them together into some established business models.

ACKNOWLEDGMENTS

All the authors are a final year student of doctoral of Hotel and Tourism Management, School of Hotel and Tourism Management, The Hong Kong Polytechnic University. The second author acknowledges the financial support from Indonesia Endowment Fund for Education (abbreviated as LPDP), Indonesian Ministry of Finance.

REFERENCES

Alegre, I., & Berbegal-Mirabent, J. (2016). Social innovation success factors: hospitality and tourism social enterprises. *International Journal of Contemporary Hospitality Management*, *28*(6), 1155–1176. https://doi.org/10.1108/IJCHM-05-2014-0231

Allan, B. (2005). Social enterprise: through the eyes of the consumer. *Social Enterprise Journal*, *1*(1), 57–77. https://doi.org/10.1108/17508610580000707

Alvord, S. H., Brown, L. D., & Letts, C. W. (2004). Social Entrepreneurship and Societal Transformation. *The Journal of Applied Behavioral Science*, *40*(3), 260–282. https://doi.org/10.1177/0021886304266847

Ausaid and UTS (2016). Private and Social Enterprise business model. Learning Brief. http://enterpriseinwash.info/wp-content/uploads/2016/12/ISF-UTS_Enterprise-in-WASH_Learning-Brief-5-Business-models.pdf

Bagnoli, L., & Megali, C. (2011). Measuring Performance in Social Enterprises. *Nonprofit and Voluntary Sector Quarterly*, *40*(1), 149–165. https://doi.org/10.1177/0899764009351111

Baron, D. P. (2007). Corporate Social Responsibility and Social Entrepreneurship. *Journal of Economics & Management Strategy*, *16*(3), 683–717. https://doi.org/10.1111/j.1530-9134.2007.00154.x

British Council. (2007). Social Enterprise in Vietnam Concept, Context and Policies. *Social Enterprise UK*, 1–70.

Chahine, T. (2016). *Introduction to Social Entrepreneurship. CRC Press*. Boca Raton: Taylor and Francis. https://doi.org/10.2139/ssrn.2071186

Defourny, J., & Kim, S. (2011). Emerging models of social enterprise in Eastern Asia: a cross-country analysis. *Social Enterprise Journal*, *7*(1), 86–111. https://doi.org/10.1108/17508611111130176

Diansyah, Eka (2015, Jul 27). Inspirasi pagi Trans 7 – Cafe Tunarungu FingerTalk [Video File]. Retrieved From URL: https://www.youtube.com/watch?v=slmgxqe_Zec&t=343s

Dickerson, C., & Hassanien, A. (2017). Restaurants' social enterprise business model: Three case studies. *Journal of Quality Assurance in Hospitality and Tourism*, *0*(0), 1–25. https://doi.org/10.1080/1528008X.2017.1363009

Evert. (2016). Literature review on social entrepreneurs as change makers for food security.

Gatherwell.com (2017, December 10). Choosing a legal structure for your Social Enterprise. Retrieved From: http://gatherwell.com/choosing-legal-structure-social-enterprise/

Ho, A. and Chan, K.T. (2010). The social impact of work-integration social enterprise in Hong Kong. *International Social Work, 53*(1), 33–45.

Home Affairs Department of Hong Kong, (2017), Enhancing Self-Reliance through District Partnership Programme [online], Available at: http://www.had.gov.hk/en/public_services/en_self_reli/index.htm [accessed 4 December 2017].

iPick, (2017), SE Restaurants [online], Available at: http://www.ipick.com/hongkong/restaurant/search?page=4&k=social+enterprise+restaurant [accessed 4 December 2017].

ISEA – Institute for Social Entrepreneurship in Asia. (2015, July). Poverty Reduction and Women Economic Leadership: Roles, Potential and Challenges of Social Enterprises in Developing Countries in Asia.

Retrieved From: http://www.isea-group.net/isea_hub/filecontent/ISEA_Oxfam_Poverty_Reduction_and_WEL_Report_pdf.pdf

Kim, A. (2007). Social Enterprise Typology. *Virtue Venture LLC.*

KompasTV (2016a, August 23). Kafe yang memberdayakan Para Difabel - Big Bang Show [Video File]. Retrieved From URL: https://www.youtube.com/watch?v=Nqt6oghnz2Q&t=176s

KompasTV (2016b, August 23). Melatih Kepercayaan Diri dan Keterampilan Difabel – Big Bang Show [Video File]. Retrieved From URL: https://www.youtube.com/watch?v=YIFWrwu3Wr4

Li, Ambrose (2016, September 2). Joyce Mak of Ginko House on social entrepreneurship in Hong Kong. Retrieved From: https://www.timeout.com/hong-kong/blog/joyce-mak-of-gingko-house-on-social-entrepreneurship-in-hong-kong-090216

Martaon, Anggi Tondi (2017, August 31). RUU Kewirausahaan Nasional dibahas di tingkat Pansus. Retrieved From: http://news.metrotvnews.com/peristiwa/Wb7YozaK-ruu-kewirausahaan-nasional-dibahas-di-tingkat-pansus

Martin, R. L., & Osberg, S. (2007). Social entrepreneurship: The case for definition. *Stanford Social Innovation Review*, *5*, 29–39. Retrieved from http://www.ssireview.org/images/articles/2007SP_feature_martinosberg.pdf

Net.Tv (2016, November 4) Cafe Fingertalk, Kafe dengan pramusaji Tunarungu [Video File]. Retrieved From URL: https://www.youtube.com/watch?v=fTXxVNsBtt0&t=78s

Nobelprize.org (2006). The Nobel Prize for 2006. Retrieved From: https://www.nobelprize.org/nobel_prizes/peace/laureates/2006/press.html

Osinowo (2014, January 2). Choosing a legal structure for your Social Enterprise. [Article] Retrieved From: http://www.osinowolaw.com/blog

Osterwalder, A., & Pigneur, Y. (2010). Business Model Generation – Canvas. *Wiley*, 280. Retrieved from http://www.businessmodelgeneration.com/canvas

Parhankangas, A., & Renko, M. (2017). Linguistic style and crowdfunding success among social and commercial entrepreneurs. *Journal of Business Venturing*, *32*(2), 215–236. https://doi.org/10.1016/j.jbusvent.2016.11.001

Tan, W. L., Williams, J., & Tan, T. M. (2005). Defining the "Social" in "Social Entrepreneurship": Altruism and Entrepreneurship. *International Entrepreneurship and Management Journal*, *1*(3), 353–365. https://doi.org/10.1007/s11365-005-2600-x

Undang-Undang RI no.28 (2004). Perubahan atas Undang-undang Nomor 16 Tahun 2001 tentang Yayasan. Retrieved From: http://luk.staff.ugm.ac.id/atur/UU28-2004Yayasan.pdf

Undang-Undang RI No.17 (2012). Perkoperasian. Retrieved From: http://regulasi.kemenperin.go.id/site/download_peraturan/1376

Yunus, M., Moingeon, B., & Lehmann-Ortega, L. (2010). Building social business models: Lessons from the grameen experience. *Long Range Planning*, *43*(2–3), 308–325. https://doi.org/10.1016/j.lrp.2009.12.005

Promoting Creative Tourism: Current Issues in Tourism Research – Kusumah et al. (Eds)
© 2021 Taylor & Francis Group, London, ISBN 978-0-367-55862-8

Restaurant selection of Thai Free Individual Traveler (FIT) by using conjoint analysis approach

K. Pitchayadejanant, L. Dembinski, P. Seesavat, P. Yimsiri & S. Amonpon
Burapha University International College (BUUIC), Thailand

R. Suprina
Trisakti School of Tourism, Jakarta, Indonesia

ABSTRACT: This research is imperative for restaurant owners. The key success of marketing mix is evaluated as in combination concept based on conjoint analysis. The objective of the study is extracting the successful combination for selecting restaurants of Thai Free Individual Traveler (FIT). The sample of study is drawn by using convenience sampling from mobile booking application users aged 21–50 years who traveled individually and made decisions to select the restaurant within the past six months. The questionnaires are distributed through online. There are 400 samples randomly selected in the study. Importantly, four attributes of combination in the study consist of four levels of food, three levels of category, four levels of restaurant, and four levels of price. According to conjoint analysis results, the important level of attributes for selecting the restaurant are type of food, with the importance value about 35.294; price, with the importance value about 28.831; restaurant, with the importance value about 23.124; and category, with the importance value about 17.163, respectively. The best combination of marketing mix to select the restaurant of FIT traveler is Chinese Food, A La Carte, Kiosk, with the price more than 800 baht/person.

Keywords: restaurant selection, Free Individual Traveler, marketing-mix, conjoint analysis

1 INTRODUCTION

Given the rapid development of information communication technology (ICT) and smartphones, smart technologies and mobile application software have become an extensive and integral part of everyday life (Baabdullah et al. 2019; Dwivedi et al. 2016; Ismagilova et al. 2019; Lal & Dwivedi 2009; Malaquias & Hwang 2019; Shareef et al. 2012; Shareef et al. 2016).

An application was designed to provide the product and service to satisfy the people in each diversified needs. In 2019, the application available in Google play is approximately 2,470,000 applications. Google play has the biggest number of applications and is followed with Apple stores' 1,800,000 applications. According to the number of application providers, it can show the various applications designed to serve the people's needs, which are affected by people's behavior. It influences people's attitude and interest at the same time. There are many applications initiated to serve the tourism industry for travelers to be able to explore the interesting destinations, suitable restaurants, hotels, etc. All of these changes affect the travelers' decision and lifestyle to follow the review and comment before making any decision. Presently, the travelers can access information via smartphones easily. Presenting information via application is popular in order to offer products and services' information to travelers.

Currently, travelers can create and design their own trip by comparing products and services according to their preferences without going out to the site. Hence, travelers can plan their trip by using tourism applications by themselves. They do not have to spend money to purchase the trip or book the ticket through a travel agency anymore. They can create and design the trip and select the

DOI 10.1201/9781003095484-81

destinations where they would like to visit by themselves. This type of travelers is known as free individual traveler (FIT).

The study population were free individual traveler in Thailand's tourism industry. The estimated number of in Thailand increased 2.6% in the first quarter of 2019 and was predicted to increase continuously. When FIT decide to travel, they have to contact or book through an application or website, which is called an online travel agency (OTA). Online travel agency is one type of travel agency, but they implement the transaction via online to assist you to manage your booking in hotel, flight, car rental, restaurant, activities, etc. According to current travelers' behavior, the target travelers who book all of traveling facilities and supports via applications are mostly young travelers. Young travelers prefer to explore their traveling experience by themselves, and the trip must be worthy. However, food is one of four basic needs of human living. The income of the tourism industry is generated by accommodation, food, and transportation. Therefore, restaurant business owners should know what feature or product that the FIT want. It is not just grab and go but restaurants with the good service that will attract travelers to obtain their experience during their dining.

Hence, researchers are interested in studying the important attributes for selecting restaurants of FIT based on marketing mix model and also to find out the highest utility of marketing mix combination of choice by adopting conjoint analysis technique. This research aims to recommend the good choice for the restaurant business owner in order to provide the product according to FIT.

2 LITERATURE REVIEW

Yilmaz and Gultekin (2016) mentioned that the information sources in restaurant characteristics that attract the tourists to make decision are to display of a menu, a wide variety of food, a specialty in the menu, looking busy, atmosphere or ambience. In this research, choosing the restaurant selection attributes to be investigated is based on marketing mix to create the concept cards. Marketing strategy is designed to accomplish unfinished desires of people. The goal of marketing strategy appears in two actions: 1) providing good value to customers against their spending, and 2) building strong brand with customers by satisfying customers' needs, resulting in profitability (Nuseir & Madanat 2015). One of the most selected strategies is marketing mix. Marketing mix known as "4Ps" is a conceptual framework that identifies the principal decision making for managers to design their offerings that can satisfy consumers' needs as by-product, price, place, and promotion. Even though marketing mix concept is extended into 7 Ps, which are extended by people, process, and physical evidence, performance is a subjective measure rather than an objective measure. As a consequence, researchers selected 4Ps marketing mix to create the survey for finding the level of satisfaction in each service concept offered by restaurant.

2.1 *Product*

Product is defined as a physical product or service to the consumer for which he is willing to pay. It includes half of the material goods, such as furniture, clothing, and grocery items and intangible products, such as services, which users buy (Singh 2016). Goods and services launched in the market to be consumed by customers to satisfy customers' needs and demands in the market are called "Product" (Armstrong & Kotler 2011). In another opinion, the concept of service as product is based on two aspects. The first is productive service, which is the core benefits, and the second is secondary level service, which consists of tangible and augmented (Hirankitti et al. 2009). Primary aspect of marketing mix is "Product," and every product is different from another product in the basis of its characteristics (Dibb et al. 2005). Borden (1984) mentioned the product characteristics including quality, design, and features. Quality is the major concern of every customer. Design is attractiveness to attract customer, and features enable customers to buy new product and brand name that increase loyalty. However, based on the product criteria studied in this research, the product of the restaurant is the type of the food, which consists of four types: Thai food, Chinese food, Japanese food, and Western food.

2.2 *Price*

Price refers to value charged from the product or service provided to a customer. Price influences essentially on a customer's psychology and helps customers to repurchase decision on a product (Andreasen et al. 2008). To pricing the product depends on many factors, including buying power of customers, cost of product, and cost of product delivery (Parasuraman & Grewal 2000). Price is a major factor on satisfaction and product loyalty if customers receive the most benefit from the product from their spending (Peter & Donnelly 2007). Price is a very important factor in determining how prices are influenced by the cost of the product, marketing strategy, and costs associated with the distribution, advertising costs, or price changes in the nature of the market (Singh 2016). The price criteria as the choice for FIT to select is the price for one person per meal, it is categorized into four categories: 1) not more than 300 baht, 2) 301–500 baht, 3) 501–800 baht, and 4) more than 800 baht.

2.3 *Place*

Regarding the place, place is also known as distribution of the restaurant. It is defined as the process and methods by which products or services reach customers (Martin 2014). The distribution facilities and location are major factors to meet demand and supply of a product on a targeted market (Copley 2004, cited by Nuseir & Madanat 2015). Regarding to the research, place is emphasized on the ambience of restaurant that attracts FIT to select the restaurant. Four types of ambience studied are street food (kiosk), fast dining, casual dining, and fine dining.

2.4 *Promotion*

The promotion concept includes all marketing activities which are used to inform, persuade, and remind the target market about a firm and its products or services, in such a way to create a favorable image in the mind of the customer. Promotion is mentioned to advertisement of a product to sell to customers. It is a process to communicate with customers. Core aspect of promotion is to reduce the communication gap, which occurs between an organization and the customer (Lovelock & Wright 2002). The promotion criteria in this study focuses on the Buffet, Set menu, and A La Carte.

2.5 *Conjoint analysis*

Conjoint analysis is a traditional research technique in assessing consumer preferences that tend to treat each attribute independently and very little information on how consumers are likely to make a favorable or unfavorable buying decision is exposed using old techniques. Consumers do not consider each attribute of a meal experience individually and independently when making decisions. Instead, they consider the bundle range of product attributes at once for their purchase. The conjoint-based approach can help to understand how customers trade off one product attribute against another. Conjoint analysis engages the respondents in a more realistic judgment than another research method. The analysis technique can better predict the overall consumer preference through aggregating the utility scores of all individual product attributes (Levy 1995). It has become a popular method for identifying and understanding the combined effects of product attributes on preferences for a product or service (Hobbs 1996). The incorporation of customized set of attributes for different respondents enables the impact of different product attributes to be analyzed in the context of cues directly relevant to particular market segments (Diamantopoulos et al. 1995).

Researchers must generate the cards with all the combination of the attributes. The number of profile cards affects the burden on respondents; reducing the number of profile cards can decrease the error rates. The minimum number of profile cards and the balance of profile cards are recommended by Ikemoto and Yamaoka (2011, July). They proposed the solution to optimize the problem to minimize the number of cards by the following formula.

$$\text{Number of cards} = \text{Total number of levels} - \text{Number of attributes} + 1 \qquad (1)$$

However, the researchers generate 16 orthogonal profile cards as default by the SPSS program. Orthogonal profile means the attribute, and the profile cards are independent. Two basic assumptions are made in conjoint analysis (Gil & Sánchez 1997). First, a product and service can be described as a combination of levels of a set of attributes. Second, these attribute levels determine consumers' overall judgment of the product or service.

3 RESEARCH METHODOLOGY

The data were collected from two sources: primary and secondary. This research was begun with searching for the secondary data from different journals, websites, and e-books. Electronic journals were obtained from Google Scholar, Science Direct, and other websites.

This study adopts conjoint analysis technique for analyzing the preference of Thai FIT in each service concept offered by the restaurant. The population of interest for this research is FIT who booked any restaurant services in Thailand via mobile booking applications who are age 21–50 years old. Hence, the sample study was drawn via convenience sampling from mobile booking application users aged 21–50 years that have used the service within the past six months and they apply a mobile booking application for the reservation of any restaurant services. However, the population size is undefinable. The sample size is calculated by using Cochran (1997). The sample size in this research is at least 385 samples. Online questionnaire was distributed in Bangkok, Rayong, Chonburi, and Samut Prakan during November 2019-February 2020. Finally, researchers can collect 400 respondents.

The study population were FIT in Thailand. The study deployed convenience sampling method. Restaurant attributes applied in this study according to marketing mix consist of (1) Product has four levels: Chinese food, Western food, Thai food, and Japanese food. (2) Promotion has three levels: Set menu, Buffet, and A La Carte. (3) Place (ambience) has four levels: Fine dining, Fast dining, Casual dining, and Kiosk. (4) Price per person has four levels: lower than 300 baht, 301–500 baht, 501–800 baht, and more than 800 baht (Table 1).

To find the level of interests of FIT in selecting the service from restaurant, researchers use four variables: product, promotion, place (ambience) and price per head. According to four attributes, all possible combination of the cards can be generated into 4*3*4*4=192 concepts. Normally, 192 concepts must be evaluated by comparing as the pairs, 18,336 pairs will be evaluated, which takes time to collect data. However, the statistical program generates 16 orthogonal cards, and the researchers screen out so there is no impossible combination. Then there are 16 orthogonal combinations presenting into the questionnaire as shown in Table 2.

Table 1. Attributes of restaurant.

Restaurant Selection Attributes	Number of Level	Levels
Product	4	Chinese Food, Western Food, Thai Food, Japanese Food
Promotion	3	Set Menu, Buffet, A La Carte
Place (Ambience)	4	Fine Dining, Fast Dining, Casual Dining, Kiosk
Price (per person)	4	Lower than 300, 301–500, 501–800, More than 800

Table 2. Orthogonal combinations presenting into questionnaire.

Card Number	Product	Promotion	Place (Ambience)	Price (per person)
1	Western Food	Set Menu	Fine dining	Lower than 300 baht
2	Thai Food	Buffet	Fine dining	501–800 baht
3	Chinese Food	Set Menu	Fine dining	More than 800 baht
4	Japanese Food	A La Carte	Fine dining	301–500 baht
5	Japanese Food	Set Menu	Fast dining	Lower than 300 baht
6	Chinese Food	Buffet	Casual dining	Lower than 300 baht
7	Thai Food	A La Carte	Kiosk	Lower than 300 baht
8	Western Food	Set Menu	Kiosk	501–800 baht
9	Chinese Food	Set Menu	Kiosk	301–500 baht
10	Western Food	Buffet	Fast dining	301–500 baht
11	Japanese Food	Buffet	Kiosk	More than 800 baht
12	Japanese Food	Set Menu	Casual dining	501–800 baht
13	Thai Food	Set Menu	Fast dining	More than 800 baht
14	Thai Food	Set Menu	Casual dining	301–500 baht
15	Chinese Food	A La Carte	Fast dining	501–800 baht
16	Western Food	A La Carte	Casual dining	More than 800 baht

4 DATA ANALYSIS AND RESULTS

Demographic information of FIT who respond the survey is reported in Table 3. The table shows: female (63.5%), 21–25 years old (36.75%) and more than 40 years old (32.5%). The majority of occupation are private company officer (27.75%) and business owner (26.00%). Salary is between 10,000–50,000 baht per month (61.25%). They mostly travel with family (50.50%) and friends (27.75%).

The top three rank of province where FIT would like to travel in Thailand is shown in Table 4. Chiangmai is the top province where FITs would like to travel. Chiangmai is located in the north of Thailand. This province is outstanding in traditional culture, local product, and northern lifestyle.

The second province is Phuket. It is located in the south of Thailand. This province is the biggest island and its name is the pearl of Andaman Sea because it has many beautiful beaches and nightlife clubs.

The third is Chiangrai, which is located in the top north of Thailand. This province has the border connected with another two countries: Myanmar and Laos, which this area is called golden triangle. In addition, this province has similar culture and lifestyle as Chiangmai.

The fourth is Chonburi province; it is located in the east of Thailand. This province is named as the active beach because Pattaya is located in this province and the province is developed as the new industrial estate of the country (Eastern Economic Corridor: EEC). The province is the strategic area of the country because it is not far from the capital city and has the mega investment of infrastructure: high-speed rail, international airport, and seaport. As a consequence, FITs, especially from Bangkok, select to travel in Chonburi in the top rank.

According to Table 5, product has the highest importance value, the FITs select the restaurant because of the type of food as the first priority (importance value=30.882). Among four types of food, Chinese food has the highest utility estimation (utility estimate = 0.486). Hence, Chinese food is the top priority to select the restaurant. Secondly, the price has the second rank of importance value (importance value = 28.824). Surprisingly, the highest utility value is the price, more than 800 baht/person (utility estimate = 0.557). The result can be implied that although the price is high, FITs are willing to pay for the food; the promotion does not matter for them. For the third and fourth ranks are place (ambience) (importance value = 23.119) and promotion (importance value=17.175), respectively. In addition, the most utility on ambience of FITs to select the restaurant

Table 3. Demographic information.

Demographic Information		Frequency	Percent
Gender	Female	254	63.50
	Male	140	35.00
	LGBT	6	1.50
Age	21–25 Years	147	36.75
	26–30 Years	50	12.50
	31–35 Years	40	10.00
	36–40 Years	33	8.25
	More than 40	130	32.50
Occupation	Private company employees	111	27.75
	Government officer	70	17.50
	Business Owner	104	26.00
	Freelance	34	8.50
	Student	71	17.75
	Other	10	2.50
Salary	Less than 10,000	65	16.25
	10,000–50,000	245	61.25
	50,001–100,000	63	15.75
	More than 100,000	27	6.75
Traveling Partner	Alone	22	5.50
	Family	202	50.50
	Friend	111	27.75
	Couple	65	16.25
	Total	**400**	**100.00**

Table 4. Top three rank of provinces where FIT travelers select.

First Rank		**Second Rank**		**Third Rank**	
Provinces	Frequency	Provinces	Frequency	Provinces	Frequency
Chiangmai	152	Chiangmai	80	Chiangmai	39
Phuket	58	Phuket	62	Phuket	38
Chiangrai	20	Chiangrai	44	Chonburi	25
Nan	17	Krabi	36	Kanchanaburi	24
Krabi	16	Nan	21	Nan	24
Prachuapkhirikhan	16	Kanchanaburi	17	Chiangrai	19

is negative value, which means that FIT do not prefer fine dining (utility estimate = −0.186) but they prefer to select kiosk or street food (utility estimate = 0.156). As a consequence, promotion is not very attractive because it shows the lowest importance value. Moreover, FIT do not prefer Buffet (utility estimate=-0.184) but they prefer A La Carte (utility estimate = 0.130).

5 DISCUSSION

The total utility is quantifiable summation of satisfaction or happiness from the combination of the attributes. The calculation is beneficial for restaurants in order to find out the preference of FIT for selecting the restaurant. The calculation is shown in Table 6.

Table 5. Importance value of each attribute and utilities of each level.

	Utilities			
	Levels	Utility Estimate	Std. Error	Importance Value
Product	1) Western Food	.013	.067	30.882
	2) Japanese Food	−.278	.067	
	3) Thai Food	−.221	.067	
	4) Chinese Food	.486	.067	
Promotion	1) Set Menu	.054	.051	17.175
	2) A La Carte	.130	.060	
	3) Buffet	−.184	.060	
Place (Ambience)	1) Fine dining	−.186	.067	23.119
	2) Casual dining	.046	.067	
	3) Fast dining	−.016	.067	
	4) Kiosk	.156	.067	
Price	1) Less than 300 baht/person	−.586	.067	28.824
	2) 301–500 baht/person	−.179	.067	
	3) 501–800 baht/person	.208	.067	
	4) More than 800 baht/person	.557	.067	
Constant		4.587	.041	

Table 6. Total utility and predicted preference.

Product	Promotion	Place (Ambience)	Price	Total utility
Western Food	Set Menu	Fine dining	Less than 300 baht/person	0.013 + 0.054 − 0.186 − 0.586 + 4.587 = 3.882
Western Food	A La Carte	Fine dining	Less than 300 baht/person	0.013 + 0.130 − 0.186 − 0.586 + 4.587 = 3.958
Western Food	Buffet	Fine dining	Less than 300 baht/person	0.013 − 0.184 − 0.186 − 0.586 + 4.587 = 3.644
Chinese Food	A La Carte	Kiosk	More than 800 baht/person	0.486 + 0.130 + 0.156 + 0.557 + 4.587 = 5.916

According to total utility result, the highest total utility that FIT prefer is 5.916 together with the combination of Chinese Food, A La Carte, Kiosk, with the price more than 800 baht/person. This is the most preferred combination for selecting the restaurant of FIT.

6 CONCLUSION

Chiangmai is the most popular province because it is in the top of FIT travelers' minds. Secondly, Phuket is the second destination. The third is Chiangrai province, and fourth is Chonburi. However, restaurant owners in these provinces can achieve the FIT traveler group by using marketing mix to create the combination of product in restaurant.

The restaurant selection of FIT traveler is based on marketing mix: the first importance attribute is product; the second is price; the third is place (ambience), and the last is promotion. Chinese food has the highest utility estimation. The preferred price of FIT traveler can be more than 800 baht per person. Street food or kiosk is popular for this group of travelers. They like to have A La

Carte rather than the others. As a result, the best combination that FIT preferred is Chinese food with A La Carte. FIT are willing to purchase the food at street food or kiosk with the budget more than 800 baht/person.

REFERENCES

Andreasen, A. R., Kotler, P., & Parker, D. 2008. Strategic marketing for nonprofit organizations (7th Ediction). *Pearson*

Armstrong, G., & Kotler, P. H. 2011. *Marketing an Introduction* (10th ed.). New Jersey: Person Education.

Baabdullah, A. M., Alalwan, A. A., Rana, N. P., Patil, P., & Dwivedi, Y. K. 2019. An integrated model for m-banking adoption in Saudi Arabia. *International Journal of Bank Marketing,* 37(2), 452–478.

Borden, W. S. 1984. *The Pacific alliance: United States foreign economic policy and Japanese trade recovery, 1947–1955*. Madison, Wis.: University of Wisconsin Press.

Dibb, S., Simkin, L., Pride, W. M., & Ferrell, O. C. 2005. *Marketing: Concepts and strategies* (p. 850). Houghton Mifflin

Dwivedi, Y. K., Shareef, M. A., Simintiras, A. C., Lal, B., & Weerakkody, V. 2016. A generalised adoption model for services: A cross-country comparison of mobile health (m-health). *Government Information Quarterly*, *33*(1), 174–187.

Gil, J. M., & Sánchez, M. 1997. Consumer preferences for wine attributes: a conjoint approach. *British Food Journal*, *99*(1), 3–11.

Hirankitti, P., Mechinda, P., & Manjing, S. 2009. Marketing strategies of thai spa operators in Bangkok metropolitan.

Hobbs, J. E. 1996. A transaction cost analysis of quality, traceability and animal welfare issues in UK beef retailing. *British Food Journal*. *98*(6), 16–26

Ikemoto, H., & Yamaoka, T. 2011, July. Conjoint Analysis Method That Minimizes the Number of Profile Cards. In *International Conference on Human-Computer Interaction*. Springer, Berlin, Heidelberg. 23–28.

Ismagilova, E., Hughes, L., Dwivedi, Y. K., & Raman, K. R. 2019. Smart cities: Advances in research - An information systems perspective. *International Journal of Information Management*, *47*, 88–100.

Levy, D. S. 1995. Modern marketing research techniques and the property professional. *Property Management*, *13*, 33-40.

Lovelock, C. H and Lauren, K, Wright, 2002. *Service Marketing and Management, New Jersey*.

Malaquias, R. F., & Hwang, Y. 2019. Mobile banking use: A comparative study with Brazilian and US participants. *International Journal of Information Management*, *44*, 132–140.

Martin, M. Business, *Marketing Marketing Mix*. 2014.

Nuseir, M. T., & Madanat, H. 2015. 4Ps: A strategy to secure customers' loyalty via customer satisfaction. *International Journal of Marketing Studies*, *7*(4), 78.

Parasuraman, A., & Grewal, D. 2000. Serving customers and consumers effectively in the twenty-first century: A conceptual framework and overview. *Journal of the Academy of Marketing Science*, *28*(1), 9–16.

Shareef, M. A., Kumar, V., Dwivedi, Y. K., & Kumar, U. 2016. Service delivery through mobile-government (mGov): Driving factors and cultural impacts. *Information Systems Frontiers*, *18*(2), 315–332.

Singh, M. 2016. Marketing Mix of 4P'S for Competitive Advantage. *IOSR Journal of Business and Management (IOSRJBM)*, *3*(6), 40–45.

Yilmaz, G., & Gultekin, S. 2016. Consumers and tourists' restaurant selections. *Global Issues and Trends in Tourism*, 217–230.

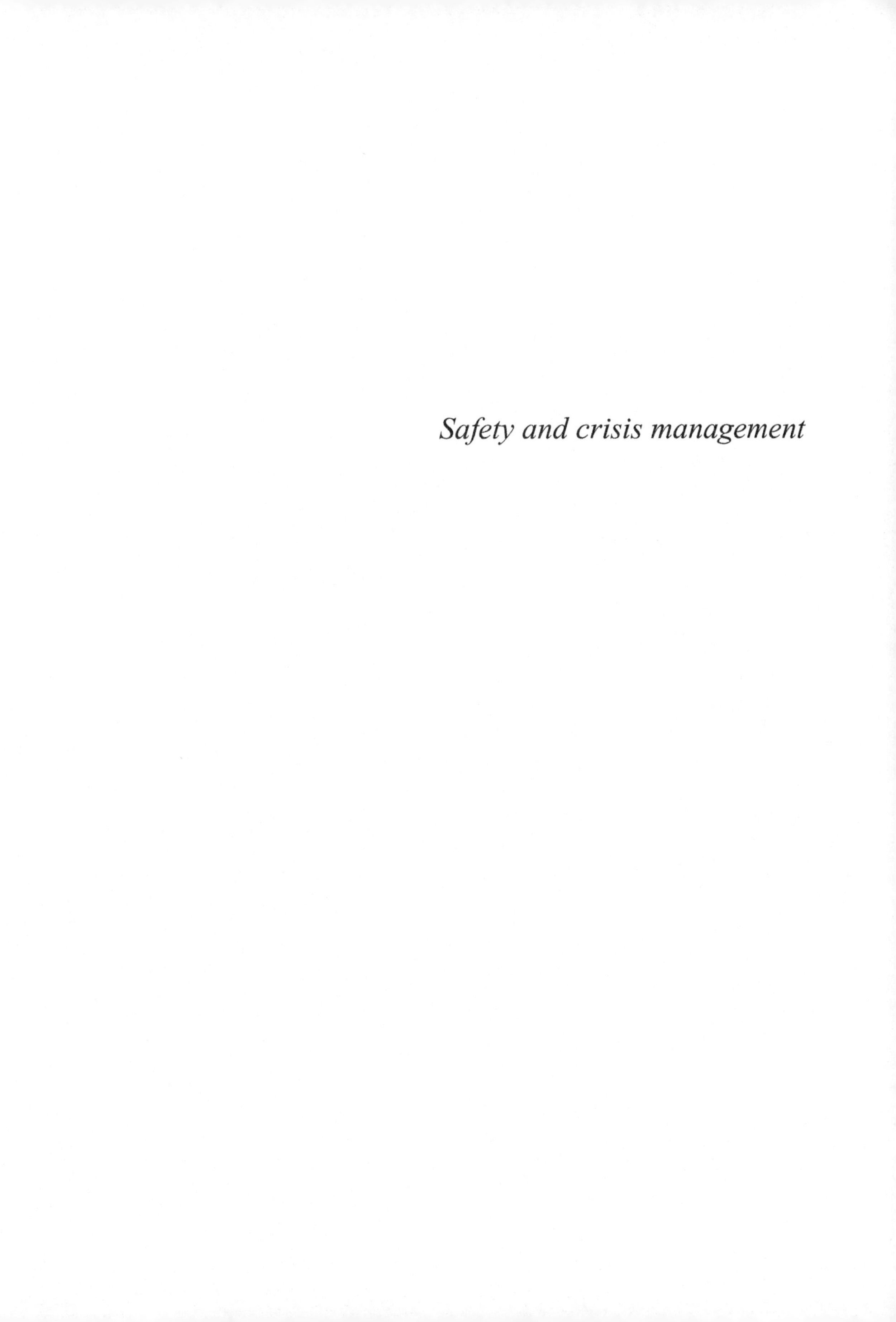

Safety and crisis management

Promoting Creative Tourism: Current Issues in Tourism Research – Kusumah et al. (Eds)

Covid-19 and Indonesian super-priority tourism destinations

S.R.P. Wulung, Y. Yuniawati & R. Andari
Universitas Pendidikan Indonesia, Bandung, Indonesia

ABSTRACT: Covid-19 became a global phenomenon and was declared a global pandemic. Some tourism destinations throughout the world are affected negatively, especially in tourist visits and the cessation of tourism aspect. This study aims to identify the impact of Covid-19 on the tourism aspect in one of Indonesia's super-priority destinations, Borobudur. The research approach used is a qualitative method. The unit of analysis of this research is an indicator of four aspects of tourism. This study uses a case study research type in identifying the Covid-19 pandemic in Borobudur tourism destinations. The data used is secondary data with data collection methods in the form of information obtained by desk study from various policy documents, literature, previous research, and the Internet. The analysis method of this research was using content and descriptive qualitative analysis. To support the government's decision, the manager of the Borobudur tourism destination closed operations from March 20 to May 13, 2020. It aims to minimize the spread of the Covid-19 pandemic. With the closure of Borobudur tourism destinations, 60% of infrastructure development temporarily stopped. Borobudur made a recovery plan for the post-pandemic Covid-19 that will be complemented by the concept of new normal or a new tourist trend based on sanitation systems.

Keywords: Borobudur, Covid-19, destination management

1 INTRODUCTION

Tourism can accelerate global economic growth through job creation and income generation, stimulating infrastructure development, and increasing social welfare in various countries (Dritsakis 2012; Ehigiamusoe 2020; Santamaria & Filis 2019). International tourism travel reached 1.4 billion in 2018, contributing to the world gross domestic product (GDP) of US$ 1.7 trillion and creating jobs for 330 million people (UNWTO 2019; World Economic Forum 2019). In Indonesia, tourism contributed to the foreign exchange of US$ 19.29 million and created 9.7% of jobs (Sakti 2019; WTTC 2020a). Indonesia's tourism development is predicted to continue to increase; foreign exchange receipts from the tourism sector are targeted at US$ 20 million by 2020 (Sakti 2019). Achieving this target will be hampered due to a health disaster at the end of 2019, namely, Corona Virus Disease (Covid-19).

Covid-19 has been declared a global pandemic since March 11, 2020 (WHO 2020). The Covid-19 pandemic was unexpected and an important event for the tourism sector (Higgins-Desbiolles 2020; UNWTO 2020). Covid-19 causes a decrease in the number of international tourist trips by 20%–30% from an estimated growth of 3%–4% in early 2020 (UNWTO 2020). Furthermore, Covid-19 resulted in job losses of 100.8 million people and resulted in a 30% reduction in global GDP from the previous year (WTTC 2020b). In Indonesia, Covid-19 has the potential for the loss of foreign exchange in the tourism sector by US$ 4 billion or equivalent to Rp.54.8 trillion (Nurhanisah 2020). Covid-19 has a negative impact on room occupancy rates of −25% to −50%; room sales price of −10% to −25%; and total hotel revenues ranging from −25% to −50%, a total of 1,266 hotels in 31 provinces in Indonesia were closed due to the COVID-19 pandemic (Kemenparekraf 2020).

DOI 10.1201/9781003095484-82

The high negative impact of the Covid-19 pandemic on tourism was due to the lack of disaster preparedness and restrictions on international, regional, and local travel (Gössling et al. 2020). Covid-19 impacts global tourism destinations (Gössling et al. 2020). This pandemic impacts on decreasing tourist activity and the temporary closure of the tourism industry in all tourism destinations in China (Karim et al. 2020). In Italy, the number of international tourists fell by 63% due to the Covid-19 pandemic (Remuzzi & Remuzzi 2020). In Singapore, the loss of tourists from China is due to the severed accessibility of flights and the decline of the hospitality industry (Pung et al. 2020). Disaster recovery harms people and small and medium businesses in tourism destinations, especially in developing countries that have mortality and economic loss (Kemenparekraf 2020; UNISDR 2009; UNWTO 2020).

Tourism destinations become part of the tourism system and a combination of all products and services for tourists provided by the local community (Buhalis 2000; Gunn & Var 2002; Leiper 1979; Page 2014). Tourism destinations provide products and services for tourists that can be enjoyed/consumed at the same time (Buhalis 2000; Mariani et al. 2014; Neuhofer et al. 2012). Tourism destinations are defined as places visited by tourists and are central to the decision to travel (United Nations 2010). Tourism destinations are places for tourists to travel and to choose to stay temporarily supported by features and characteristics such as tourist attractions (Leiper 1979, 2000; Page 2014). Tourism destinations focus on providing facilities and services designed to meet the needs of tourists (Gunn & Var 2002; Mason 2016).

Most tourism destinations consist of systems that form an ecosystem and are classified in the framework of 6As, including attractions, accessibility, amenities, available packages, activities, and ancillary services (Buhalis 2000; Buhalis & Amaranggana 2013). Morrison (2013) states that there are 10As in the tourism destination system that includes awareness, attractiveness, availability, access, appearance, activities, assurance, appreciation, action, and accountability. The development of tourism destinations requires destination marketing and management to support sustainable tourism development (Mariani et al. 2014; Pike & Page 2014; Sheehan et al. 2007; Volgger & Pechlaner 2014; Wang & Pizam 2011). In Indonesia, tourism development has been determined in Law Number 10 of 2010 concerning Tourism, which covers aspects of tourism destinations, tourism industry, tourism marketing, and tourism institutions. These aspects are identified as core ecosystems in developing tourism destinations (Kemenpar 2018; Mariani et al. 2014; Page 2014). Another important aspect that can support tourism development is tourists. Today's tourists are sensitive and smart in choosing tourism destinations to be visited. This indicates that the tourist factor has an important role in the development of tourism (Beritelli & Laesser 2011; Goffi et al. 2019). The presence of tourists can help to assess the impact of tourism regionally and manage the demand and supply side in maximizing the benefits for all stakeholders (Gunn & Var 2002; Mariani et al. 2014).

Covid-19 has an impact on tourist visits to Indonesia's super-priority tourism destination (Susanto 2020a). One of Indonesia's five super-priority destinations, namely Borobudur, is also negatively affected by the Covid-19 pandemic (Nugroho 2020; Susanto 2020b; Wicaksono 2020). However, the development of tourism destinations continues throughout the pandemic (Taher 2020). Aspects of tourism development are the tourism industry and tourism institutions in Borobudur affected by Covid-19 (Muhyiddin 2020; Riadil 2020; Sugihamretha 2020; Susilawati et al. 2020). Borobudur succeeded in bringing 4,774,000 tourists in 2019; this will not happen in 2020 (Ferri 2020). PT. Taman Wisata Candi (TWC) Borobudur, Prambanan, and Ratu Boko as the main managers of Borobudur tourism destinations explained that in 2020 there will be a decrease in the number of tourists due to the Covid-19 pandemic (Khairunnisa 2020). Four aspects of tourism need to be considered as an effort to increase interest and also facilitate tourists in the post-pandemic (Ningsih 2020; Yawan 2020). The cessation of tourism activities in Borobudur tourism destinations can have a variety of negative impacts, especially on the tourist market and the condition of tourist attractions during and after the pandemic. Various policies and efforts continue to be carried out by the government and managers of Borobudur tourism destinations as an effort to maintain the existence of tourism destinations during and after the COVID-19 pandemic. This study aims to identify the impact of Covid-19 on aspects of tourism in Indonesia's super-priority destination, Borobudur.

2 METHOD

The approach used in this study is a qualitative study with an analysis unit that is an indicator of four aspects of tourism destinations that include tourism destinations and tourist attractions, the tourism industry, tourism marketing, and tourism institutions. This study uses a case study research because the research was carried out by identifying a COVID-19 pandemic toward the management of tourism destination in Borobudur. The type of data used is secondary data with a collection method using information obtained from desk study from various policy documents, literature, previous research, and the Internet related to the COVID-19 pandemic in Borobudur tourism destination. Also, other data obtained was in the form of information materials produced by a social institution, such as magazines, newspapers, bulletins, and statements. This study does not use primary data acquisition such as observations, interviews, and questionnaires because the COVID-19 pandemic and the large-scale social restriction (LSSR) policy are happening. Data analysis used content analysis methods and descriptive qualitative analysis. This analysis is used to consider the form of data and information collected in the form of secondary documents requiring techniques to understand and interpret the data. The descriptive method used aims to link the data that has been obtained based on categories in theory.

3 RESULTS

Tourist arrivals to the Borobudur tourism destination in January 2020 were 235,456 domestic tourists and 10,948 foreign tourists. The enactment of LSSR in February 2020 supports a decrease in tourist arrivals by 30% compared to February 2019, which reached as many as 247,697 domestic tourists and 16,642 foreign tourists (Tosiani 2020). The situation will continue after the Covid-19 pandemic; the target of tourists visiting Borobudur in 2020 will be constrained (Susanto 2020b). Tourists from China are the highest number of tourists who visit, but there is no Covid-19 visit to Borobudur. Since early February, tourist access from China to Indonesia has been closed, resulting in a decrease in the number of tourist visits (Lumbanrau 2020). Foreign tourist arrivals to Borobudur declined by 20-30%, while domestic tourist arrivals were stable and were above those of foreign tourists in February 2020 (Tosiani 2020). The decline in foreign tourist visits is the reason for PT. TWC to target the archipelago tourist market, this is supported by government policies in providing incentive ticket prices to 10 priority tourism destinations (Fauzia 2020). Borobudur tourism destination managers prevent the spread of Covid-19, include the provision of thermal scanners at the entrance to monitor the body temperature of tourists, provide cleaning facilities such as hand sanitizers, provide isolation room facilities, and provide transportation facilities to the referral hospital. The effort aims to educate about Covid-19 precautions to tourists, employees, and the local community. As standard operating procedures during COVID-19, the Borobudur destination manager applies temperature checks to tourists.

3.1 *Impact of Covid-19 on tourist attractions*

The condition of tourist attraction in Borobudur was identified based on three time periods, namely, before, during, and after the COVID-19 pandemic. Before the COVID-19, PT. TWC applies an increase in the price of admission on May 30-June 15, 2019, to Rp.50,000/person for adults (the previous price is Rp.40,000/person) and Rp.25,000/person for children or students (the previous price is Rp.20,000/ person). The price increase coincides with Islamic religious holidays and applies only to domestic tourists. The price increase was accompanied by additional facilities that included facilities for the disabled, 14 transportation units (electric cars and golf carts) for accessibility within the destination, and other attraction facilities such as feeding, taking photos, and witnessing various animals at the Borobudur destination. During COVID-19, the Borobudur Conservation Center (BCC) imposed restrictions on visits to the Borobudur Temple starting on February 13, 2020. Tourists are only allowed to go up to the 8th floor. The policy is an effort to preserve the Borobudur Temple and as a form of sustainable tourism. Since the enactment of LSSR, PT. TWC temporarily closed Borobudur Temple and other tourist attractions in the Borobudur tourism area from March 20 to June 3, 2020 (Idhom 2020; TWC 2020). This was an effort to prevent the spread of the COVID-19 pandemic in Borobudur tourism destination. The temporary closure is accompanied by periodic maintenance of Borobudur destination through the provision of disinfectants and maintaining cleanliness in each tourist attraction.

After the COVID-19 pandemic, the government through the Ministry of Tourism and Creative Economy and PT. TWC plans to prepare tour packages to Borobudur tourism destinations in anticipation of the high interest in travel. The tour package is the first step to attract tourists. Another effort is to educate the COVID-19 pandemic through various media in every tourist attraction in Borobudur tourism destinations. The existence of a planned rebound campaign program in support of MICE tourism is an alternative to increase tourist visits to Borobudur Temple. It is related to the delay of the event in Borobudur tourism destinations during the COVID-19 period. The impact of the COVID-19 pandemic on tourist attractions in Borobudur tourism destinations was identified based on negative and positive. The negative impact is on the absence of events as a tourist attraction, such as Vesak on May 7, 2020. Meanwhile, the positive impact is that the destination manager can carry out regular maintenance of each tourist attraction during the pandemic.

3.2 *Impact of Covid-19 on tourism marketing*

The impact of COVID-19 on tourism marketing in Borobudur tourism destinations was identified based on the period during and after COVID-19. During the pandemic period, destination managers carried out marketing and promotional activities through advertising media, internal sales promotions, public relations, and direct marketing. Destination managers promote Borobudur through a website (http://borobudurpark.com) and social media that includes Twitter (twitter.com/borobudurpark), Facebook (facebook.com/borobudurpark), Youtube (youtube.com/channel/UCKpxRpqgl2uxaXSq22ln1iQ), and Instagram (instagram.com/borobudurpark). On the website, the manager of Borobudur provides official information regarding the temporary closure of destinations and pandemic prevention efforts. At the end of the LSSR period, the manager shares information about the reopening of Borobudur tourism destinations. On Instagram media, PT. TWC regularly shares news and information related to efforts to prevent the COVID-19 pandemic in the destination. There is interaction with netizens through the giveaway program and socializing hashtag *#dirumahaja* with an attractive graphic design. On Twitter media. PT. TWC implements the concept of integrated marketing communication, with words other than information being communicated through Instagram as well as through Twitter and Facebook.

The government plans several marketing strategies after the COVID-19 pandemic, including creating a free video campaign from COVID-19, holding a travel fair, and working with influencers on social media to restore tourism in Indonesia. Meanwhile, for the delayed national tourism events, the government through the Ministry of Tourism and Creative Economy markets and promotes various tour packages to Indonesian tourism destinations. Tourism marketing strategy carried out by the government will directly affect tourist visits to Indonesia, especially to Borobudur destinations. Despite the main focus of tourism marketing for domestic tourists, the government also continues to make foreign tourists a potential tourist market. Foreign tourists are a source of foreign exchange, especially for Borobudur tourism destinations.

3.3 *Impact of Covid-19 on tourism industry*

The tourism industry, such as hotels, restaurants, and tour operators, was affected by the COVID-19 pandemic. The average occupancy rate of hotel rooms around Borobudur destinations has decreased by 40%. The level of tourist arrivals also impacts the restaurant business, where most customers are tourists (Hanoatubun 2020). During the COVID-19 pandemic, several hotels lowered room prices to cover operational costs and temporarily closed their businesses. Restaurants around the Borobudur destination mostly decided to close temporarily due to the drop-in visit rates. Tour operators also have the effect of decreasing orders for tour packages to the attraction of Borobudur. Trips were canceled by tourists, both foreign tourists who planned to visit Indonesia and domestic tourists who travel in their own country.

3.4 *Impact of Covid-19 on tourism institutions*

The impact of COVID-19 on tourism institutions in Borobudur tourism destinations was identified based on efforts by the Ministry of Tourism and Creative Economy, Borobudur Authority Agency, and PT. TWC. Referring to the Circular of the Minister of Tourism and Creative Economy/ Head of Tourism and Creative Economy Agency No. 2 of 2020 regarding Follow-Up Appeal to Prevent the Spread of Coronavirus Disease 2019 (Covid-19), explaining that the Ministry made efforts to address Covid-19 addressed to the Covid-19 department in charge of tourism. The Ministry of Tourism and Creative Economy ensures that the super-priority destination program will continue, even though some physical development has been postponed. This was done as an effort to redesign better and sustainable tourism. The government began to improve tourism development standards so that the area could become a tourist destination (Taher 2020). The preparation of the super-priority destination program is used for the recovery of the post-pandemic tourism sector. Super-priority destination preparations are in the basic sectors, including preparations that are clean, safe, and security in nature. The Ministry of Tourism and Creative Economy is also preparing the development of basic infrastructure related to connectivity in a number of super-priority destinations. This program is one of the incentive packages provided by the government for the tourism sector consisting of airline incentives and business travel tours, joint promotion, tourism promotion activities, and familiarization trips and influencers. While for domestic tourists, the government provides a 30% discount on flights to 10 tourism destinations. The 30% is for 25% quota seats on each flight to 10 tourism destinations. This policy will last for three months and applies to domestic airlines with destinations in Denpasar, Batam, Bintan, Manado, Yogyakarta, Labuan Bajo, Belitung, Lombok, Lake Toba, and Malang.

The Borobudur Authority Implementing Body (BAA) was formed based on Presidential Regulation number 46 of 2017. BAA is a work unit under the Ministry of Tourism and Creative Economy of the Republic of Indonesia. The super-priority National Tourism Strategic Area Management prepared a short-term recovery plan from the crisis due to the COVID-19 pandemic. Infrastructure development in the Borobudur destination stopped temporarily because 60% of funding was for pandemic mitigation. The post-pandemic recovery plan will be complemented by a new normal concept or a new hygiene-based tourism trend where one of the main supporting infrastructure concepts is the sanitation system. Before the COVID-19 pandemic occurred, BAA was developing various features at the Borobodur destination, but the development process was stalled and continued next year. At present, BAA is working with digital content providers, such as digital marketing and assisting local communities to remain productive.

The closure of Taman Wisata Borobudur, Prambanan, and Ratu Boko temples was extended until May 13, 2020, and will be reopened on June 3, 2020. The policy was carried out by monitoring the development of the Covid-19 case in Central Java and Yogyakarta. In addition to closing tourist attractions, PT. TWC also applies for work from home for its employees and prohibits not traveling outside the area during the Covid-19 pandemic disaster emergency, following SE of the Ministry of BUMN RI Number SE4 / MBU / 04/2020 concerning Prohibition of Traveling to Outside the Area and/or Mudik Activities in Preventing the Corona Virus Disease 2019 (Covid-19). PT. TWC explained that the destination manager still carries out maintenance and maintenance of the park and is preparing everything that can provide new experiences and become a tourist attraction if you will visit the destination when conditions are back to normal (Nugroho 2020).

4 CONCLUSIONS

The Covid-19 pandemic in Indonesia has had an impact on foreign tourist arrivals and domestic tourists. Covid-19 impacts hotels, restaurants, and tour operators. PT. TWC conducted different promotions during the Covid-19 Pandemic period. PT. TWC manages websites, Instagram, and Twitter to access official news about Borobudur destinations. After the pandemic, the Ministry of Tourism and Creative Economy plans to prepare attractive tourism packages to increase tourist visits. The government has planned several things after the pandemic, including making video campaigns, holding travel fairs, making promotional tour packages, and much more. This strategy is to restore the tourism sector in Indonesia. This research provides knowledge about the management of tourism destinations as an effort to prevent and reduce Covid-19 risk in Borobudur. Our other super-priority tourism destinations in Indonesia have the opportunity to be identified in Covid-19 risk reduction efforts. It becomes further research-related to destination managers in managing their area during the pandemic and the new-normal era. The research limitations of Covid-19 in Indonesia, especially in tourism destinations, make this research the most recent insight for tourism disaster management science. The research implications provide insights for tourism destination managers in collaborating with stakeholders during disasters, especially non-natural disasters.

REFERENCES

Beritelli, P., Laesser, C., 2011. Power dimensions and influence reputation in tourist destinations: Empirical evidence from a network of actors and stakeholders. Tour. Manag. 32, 1299–1309.

Buhalis, D., 2000. Marketing the competitive destination of the future. Tour. Manag. 21, 97–116.

Buhalis, D., Amaranggana, A., 2013. Smart Tourism Destinations. In: Xiang, Z., Tussyadiah, I. (Eds.), Information and Communication Technologies in Tourism 2014. Springer International Publishing, Cham, pp. 553–564.

Dritsakis, N., 2012. Tourism development and economic growth in seven Mediterranean countries: A panel data approach. Tour. Econ. 18, 801–816.

Ehigiamusoe, K.U., 2020. Tourism, growth and environment: analysis of non-linear and moderating effects. J. Sustain. Tour. 28, 1174–1192.

Fauzia, M., 2020. Pemerintah akan Beri Diskon Tiket Pesawat Hingga 40 Persen ke 10 Destinasi [WWW Document]. Money.kompas.com. URL https://money.kompas.com/read/2020/02/24/194100926/pemerintah-akan-beri- diskon-tiket-pesawat-hingga-40-persen-ke-10-destinasi?page=all (accessed 6.6.20).

Ferri, R.K., 2020. Jumlah Wisatawan Candi Borobudur Baru Mencapai 70 Persen dari Target [WWW Document]. Jogja.tribunnews.com. URL https://jogja.tribunnews.com/2019/12/05/jumlah-wisatawan-candi-borobudur- baru-mencapai-70-persen-dari-target (accessed 6.4.20).

Goffi, G., Cucculelli, M., Masiero, L., 2019. Fostering tourism destination competitiveness in developing countries: The role of sustainability. J. Clean. Prod. 209, 101–115.

Gössling, S., Scott, D., Hall, C.M., 2020. Pandemics, tourism and global change: a rapid assessment of COVID-19. J. Sustain. Tour. 1–20.

Gunn, C.A., Var, T., 2002. Tourism planning: basics, concepts, cases, 4th ed. ed. Routledge, New York.

Hanoatubun, S. (2020). Dampak Covid-19 terhadap Prekonomian Indonesia. EduPsyCouns: Journal of Education, Psychology and Counseling, 2(1), 146–153. Retrieved from https://ummaspul.e-journal.id/Edupsycouns/article/view/423

Higgins-Desbiolles, F., 2020. Socialising tourism for social and ecological justice after COVID-19. Tour. Geogr. 1–14.

Idhom, A.M., 2020. Candi Borobudur-Prambanan akan Dibuka Juni 2020 Usai Tutup 3 Bulan [WWW Document]. Tirto.id. URL https://tirto.id/candi-borobudur-prambanan-akan-dibuka-juni-2020-usai- tutup-3-bulan-fyM4 (accessed 6.10.20).

Karim, W., Haque, A., Anis, Z., Ulfy, M.A., 2020. The Movement Control Order (MCO) for COVID-19 Crisis and its Impact on Tourism and Hospitality Sector in Malaysia. Int. Tour. Hosp. J. 3, 1–7.

Kemenpar, 2018. Sop: Pengelolaan Krisis Kepariwisataan. Kementerian Pariwisata, Jakarta.

Kemenparekraf, 2020. Rencana Mitigasi Sektor Parekraf dalam Menangani Dampak Virus Covid-19.

Khairunnisa, S.N., 2020. Gara-gara Corona, Kunjungan Turis Asing di Taman Wisata Candi Menurun [WWW Document]. Travel.kompas.com. URL https://travel.kompas.com/read/2020/03/11/130100027/gara-gara-corona- kunjungan-turis-asing-di-taman-wisata-candi-menurun?page=all (accessed 6.5.20).
Leiper, N., 1979. The framework of tourism. Towards a definition of tourism, tourist, and the tourist industry. Ann. Tour. Res. 6, 390–407.
Leiper, N., 2000. Are destinations "The Heart of Tourism"? The advantages of an alternative description. Curr. Issues Tour. 3, 364–368.
Lumbanrau, R.E., 2020. Virus corona berpotensi melenyapkan triliunan rupiah devisa: "Seluruh Bali pasti akan kena dampak" [WWW Document]. www.bbc.com. URL https://www.bbc.com/indonesia/indonesia-51324351 (accessed 6.5.20).
Mariani, M.M., Buhalis, D., Longhi, C., Vitouladiti, O., 2014. Managing change in tourism destinations: Key issues and current trends. J. Destin. Mark. Manag. 2, 269–272.
Mason, P., 2016. Tourism impacts, planning and management, Third edit. ed. Routledge, is an imprint of the Taylor & Francis Group, an Informa business, Abingdon, Oxon; New York, NY.
Morrison, A.M., 2013. Marketing and managing tourism destinations, Marketing and Managing Tourism Destinations. Routledge.
Muhyiddin, 2020. Covid-19, New Normal, dan Perencanaan Pembangunan di Indonesia. J. Perenc. Pembang. Indones. J. Dev. Plan. 4, 240–252.
Neuhofer, B., Buhalis, D., Ladkin, A., 2012. Conceptualising technology enhanced destination experiences. J. Destin. Mark. Manag. 1, 36–46.
Ningsih, S.M., 2020. Candi Borobudur Harus Tetap Diprioritaskan [WWW Document]. Suaramerdeka.com. URL https://www.suaramerdeka.com/amp/index.php/smcetak/baca/192328/smtv/ (accessed 6.4.20).
Nugroho, W.S., 2020. Penutupan Akses Wisata di Candi Borobudur, Prambanan, dan Ratu Boko Diperpanjang [WWW Document]. Tribunjogjatravel.tribunnews.com. URL https://tribunjogjatravel.tribunnews.com/2020/04/21/penutupan-akses-wisata-di-candi-borobudur-prambanan-dan-ratu-boko-diperpanjang (accessed 6.4.20).
Nurhanisah, Y., 2020. Dampak COVID-19 terhadap devisa di Indonesi [WWW Document]. Indonesia.go.id. URL https://indonesia.go.id/gallery/pariwisata-indonesia-di-tengah-virus- corona (accessed 6.2.20).
Page, S., 2014. Tourism Management, Fifth edit. ed, Tourism Management. Wiley, Milton, Qld.
Pike, S., Page, S.J., 2014. Destination Marketing Organizations and destination marketing: Anarrative analysis of the literature. Tour. Manag. 41, 202–227.
Pung, R., Chiew, C.J., Young, B.E., Chin, S., Chen, M.I.C., Clapham, H.E., Cook, A.R., Maurer-Stroh, S., Toh, M.P.H.S., Poh, C., Low, M., Lum, J., Koh, V.T.J., Mak, T.M., Cui, L., Lin, R.V.T.P., Heng, D., Leo, Y.S., Lye, D.C., Lee, V.J.M., Kam, K. qian, Kalimuddin, S., Tan, S.Y., Loh, J., Thoon, K.C., Vasoo, S., Khong, W.X., Suhaimi, N.A., Chan, S.J., Zhang, E., Oh, O., Ty, A., Tow, C., Chua, Y.X., Chaw, W.L., Ng, Y., Abdul-Rahman, F., Sahib, S., Zhao, Z., Tang, C., Low, C., Goh, E.H., Lim, G., Hou, Y., Roshan, I., Tan, James, Foo, K., Nandar, K., Kurupatham, L., Chan, P.P., Raj, P., Lin, Y., Said, Z., Lee, A., See, C., Markose, J., Tan, Joanna, Chan, G., See, W., Peh, X., Cai, V., Chen, W.K., Li, Z., Soo, R., Chow, A.L., Wei, W., Farwin, A., Ang, L.W., 2020. Investigation of three clusters of COVID-19 in Singapore: implications for surveillance and response measures. Lancet 395, 1039–1046.
Remuzzi, A., Remuzzi, G., 2020. COVID-19 and Italy: what next? Lancet 395, 1225–1228.
Riadil, I.G., 2020. Tourism Industry Crisis and its Impacts: Investigating the Indonesian Tourism Employees Perspectives' in the Pandemic of COVID-19. J. Kepariwisataan Destin. Hosp. dan Perjalanan 4, 98–108.
Sakti, G., 2019. Siaran pers pariwisata diharapkandorong Indonesia jadi negara maju pada 2045 [WWW Document]. Kemenparekraf. URL http://www.kemenparekraf.go.id/index.php/post/siaran-pers-pariwisata-diharapkan-dorong-indonesia-jadi-negara-maju-pada-2045
Santamaria, D., Filis, G., 2019. Tourism demand and economic growth in Spain: New insights based on the yield curve. Tour. Manag. 75, 447–459.
Sheehan, L., Ritchie, J.R.B., Hudson, S., 2007. The destination promotion triad: Understanding asymmetric stakeholder interdependencies among the city, hotels, and DMO. J. Travel Res. 46, 64–74.
Sugihamretha, I.D.G., 2020. Respon Kebijakan: Mitigasi Dampak Wabah Covid-19 Pada Sektor Pariwisata. J. Perenc. Pembang. Indones. J. Dev. Plan. 4, 191–206.
Susanto, E., 2020a. Data Efek Virus Corona ke Wisata RI per 23 April 2020 [WWW Document]. Travel.Detik.Com. URL https://travel.detik.com/travel-news/d-4928546/data-efek-virus-corona-ke-wisata-ri-per-23-april-2020 (accessed 6.1.20).
Susanto, E., 2020b. Dampak Corona, Wisman di Candi Borobudur Turun 30 Persen [WWW Document]. Travel.detik.com. URL https://travel.detik.com/travel-news/d-4929530/dampak-corona-wisman-di- candi-borobudur-turun-30-persen (accessed 6.4.20).

Susilawati, S., Falefi, R., Purwoko, A., 2020. Impact of COVID-19's Pandemic on the Economy of Indonesia. Budapest Int. Res. Critics Inst. Humanit. Soc. Sci. 3, 1147–1156.
Taher, A.P., 2020. Pembangunan Destinasi Super Prioritas Tetap Jalan Selama COVID-19 [WWW Document]. Tirto.id. URL https://tirto.id/pembangunan-destinasi-super-prioritas-tetap-jalan- selama-covid-19-eNFt (accessed 6.7.20).
Tosiani, 2020. Target Wisatawan ke Tiga Candi Susah Tercapai Terdampak Korona [WWW Document]. Mediaindonesia.com. URL https://mediaindonesia.com/read/detail/295119-target-wisatawan-ke-tiga- candi-susah-tercapai-terdampak-korona (accessed 6.5.20).
TWC, 2020. PT TWC Tutup Sementara Operasional Kawasan Taman Wisata Candi [WWW Document]. Borobudurpark.com. URL http://borobudurpark.com/pt-twc-tutup-sementara-operasional-kawasan- taman-wisata-candi/ (accessed 6.6.20).
UNISDR, 2009. 2009 UNISDR Terminology on Disaster Risk Reduction, International Stratergy for Disaster Reduction (ISDR). the United Nations International Disaster Disaster risk Disaster risk management Strategy for Disaster Reduction (UNISDR) Disaster risk reduction Disaster risk red, Geneva.
United Nations, 2010. International Recommendations for Tourism Statistics 2008 (Chinese version), International Recommendations for Tourism Statistics 2008 (Chinese version). United Nation, New York.
UNWTO, 2019. International Tourism Highlights, 2019 Edition, International Tourism Highlights, 2019 Edition. United Nations World Tourism Organization.
UNWTO, 2020. Tourism and COVID-19. United Nations World Tourism Organization.
Volgger, M., Pechlaner, H., 2014. Requirements for destination management organizations in destination governance: Understanding DMO success. Tour. Manag. 41, 64–75.
Wang, Y., Pizam, A., 2011. Destination marketing and management: Theories and applications, Destination Marketing and Management: Theories and Applications. CABI, Wallington, Oxfordshire, UK; Cambridge, MA.
WHO, 2020. Pernyataan Covid-19 sebagai Pandemi Global [WWW Document]. URL https://www.who.int/emergencies/diseases/novel-coronavirus-2019.
Wicaksono, P., 2020. Kunjungan Wisman ke Candi Borobudur Masih Minim, Tilik Sebabnya [WWW Document]. Travel.tempo.co. URL https://travel.tempo.co/read/1176131/kunjungan-wisman-ke-candi-borobudur- masih-minim-tilik-sebabnya
World Economic Forum, 2019. The Travel and Tourism Competitiveness Report 2019, World Economic Forum. World Economic Forum's Platform for Shaping the Future of Mobility.
WTTC, 2020a. Indonesia 2020 Anual Research: Key Highlights.
WTTC, 2020b. Dampak COVID-19 pada Sektor Pariwisata secara Global [WWW Document]. URL https://wttc.org/COVID-19 (accessed 6.1.20).
Yawan, A.K., 2020. Proyek Infrastruktur Pariwisata Tetap Jalan, Toilet Mewah Didirikan di Borobudur dan Labuan Bojo [WWW Document]. Depok.pikiran-rakyat.com. URL https://depok.pikiran-rakyat.com/ekonomi/ (accessed 6.4.20).

Promoting Creative Tourism: Current Issues in Tourism Research – Kusumah et al. (Eds)

The influence of perceived risk and perceived value toward tourist satisfaction

B. Waluya, O. Ridwanudin & Z.S. Zahirah
Universitas Pendidikan Indonesia, Bandung, Indonesia

ABSTRACT: Satisfaction is the most pivotal factor in the service industry. This study is aimed at finding out the impact of perceived risk and perceived value on tourist satisfaction of Cerme Cave. The dependent variable (X) of the study is perceived risk (X1) consisting of physical risk, performance risk, financial risk, and psychological risk and perceived value (X2) consisting of functional value and emotional value. Meanwhile, the independent variable (Y) is tourist satisfaction. The study applies purposive sampling involving 111 samples. The data is collected through questionnaires, which were then analyzed descriptively and verified through multivariate regression. The result, simultaneously, perceived risk and perceived value, gives significant impact on tourist satisfaction of Cerme Cave. Physical risk and emotional value have the highest score (340; 435,66). It means Cerme Cave succeeded in creating the tourists' satisfaction that was gained from the program, which can help tourists developing their perception on risk and value.

Keywords: perceived risk, perceived value, tourist satisfaction

1 INTRODUCTION

Satisfaction is an important element in service companies. Satisfaction is important in the marketing world because it can influence future purchasing decisions. Researchers in the field of tourism says that marketers need to focus on tourist satisfaction because it can streamline marketing activities (Wu et al. 2018). Satisfaction in tourism marketing literature is one of the most powerful driving factors for tourists to determine future behavior. The research point of view assesses that satisfaction attracts researchers' interest in the tourism industry because satisfaction is the key to evaluating the competitiveness of tourist destinations (Guo et al. 2017).

Research says that understanding satisfaction plays an important role for company managers in the tourism industry because tourist satisfaction can increase loyalty, provide positive word-of-mouth, increase return vis-its, and recommend others (Corte et al. 2015). Research on satisfaction has been carried out in several industries such as the hotel industry (Amin et al. 2014), tour and travel (Chang 2007), and tourism destination industry (Helena et al. 2018). Managing satisfaction effectively has been handled by practitioners and researchers in the field of marketing. Measuring satisfaction is very important for the survival, development, and success of tourist destinations.

Adventure tourism is growing rapidly and has become a favorite choice for tourists who are looking for new experiences in travel. Therefore, the adventure tourist destination must carry out its functions in an integrated manner so that customers such as tourists, destination management staffs, and other visitors can feel satisfied when visiting the destination. Special Region of Yogyakarta Province has a variety of adventure tourism attractions. Cerme Cave is a caving tourism attraction located in Bantul Regency which has ornaments that are still very well preserved.

Perception of risk or perceived risk is a variable that is studied in the marketing literature of tourist destinations and is considered an important variable because risk is one of the factors used as a determinant in choosing tourist destinations and is able to create satisfaction in adventure tourist destinations. Providing a low level of perceived risk is very important because low perceived risk

DOI 10.1201/9781003095484-83

refers to a positive assessment that is manifested in tourist satisfaction, while high perceived risk refers to low tourist satisfaction.

Another factor that influences satisfaction is perceived value. The value of an overall assessment rating of the usefulness of a tourism product is based on the received and given (Kim & Thapa 2017). Tourists capture value in the form of perception called perceived value. Perceived value can create satisfaction for tourists, which will ultimately provide a positive image. Rasoolimanesh et al. (2016) explain that the value received can affect tourist satisfaction at the tourist destination.

Theoretically, the results of this study are expected to contribute scientifically by adding a repertoire of understanding the science of tourism to the industry of the adventure tourist destination by reviewing the perceived risk and perceived value in creating tourists' satisfaction. Empirically, it is expected that this research can build Cerme Cave to be more concerned about tourist satisfaction through the tourist perception, and the results of the research can be an evalution material for Cerme Cave for its marketing strategy.

2 LITERATURE REVIEW

2.1 *Customer satisfaction*

Consumer satisfaction is very crucial to be considered by every company because this is related to the sustainability of the company itself. [9] suggested a method for measuring customer satisfaction called the Customer Satisfaction Index (CSI). This method can be used to measure customer satisfaction both in the form of goods and services. Each respondent assesses the importance of the selected criteria and gives a level of satisfaction using a Likert scale.

$$\mathbf{CSI} = \left(\frac{\mathbf{T}}{\mathbf{5Y}}\right) \mathbf{x\,100\%}$$

Information:
T = Total value of satisfaction
5 = Maximum value on the measurement scale
Y = Total value from the perception column

From these functions the satisfaction level criteria can be seen through the CSI Criteria's of Biesok & Wyród-wróbel (2011). In addition, one of the most widely used satisfaction measurement techniques is Importance Performance Analysis. According to Zagreb (2019), Importance Performance Analysis was first proposed by Martilla and James in 1978. In this technique, respondents were asked to rate the importance of various relevant attributes and the level of company performance (perceived performance) on each of these attributes. Then the average value of the level of importance of attributes and company performance will be analyzed in the Importance Performance Matrix.

2.2 *Perceived risk*

Perceived risk or risk perception was initially spearheaded by Bauer 1960 in a journal of Khan et al. (2017) defined as consumer perceptions of the likelihood that a decision could lead to risks that could influence the level of consumer decisions where the risk was acceptable or not.

Chiu et al. (2019) proposed four dimensions of perceived risk: physical risk, performance risk, financial risk, and psychological risk, which have been applied in academic tourism research. Another dimension used in tourism research is Khan et al. (2017), which proposed five dimensions of perceived risk: physical risk, financial risk, performance risk, socio-psychological risk, and time risk. Hasan et al. (2017) said that equipment risk is pivotal in adventure tourist destination follow by physical risk, financial risk, social risk, and performance risk. This proposal of dimensionality is gaining acceptance among researchers since it encompasses the previous proposals.

2.3 *Perceived value*

Perceived value was originally spearheaded by Zeithaml (1988), which is defined as an overall assessment of consumers of the utility of a product based on perceptions about what is received and what is given. Functional value is the usefulness derived from the quality perceived by the expected product performance. Emotional value is the use that comes from feeling or affective which states that a product is successful.

In the field of nature-based tourism, Kim and Thapa (2017) consider perceived value in four dimensions: quality value, emotional value, price value, and social value. Rasoolimanesh et al. (2016) did research about the influence of perceived value to tourist satisfaction in tourism research, which has two dimensions: functional value and emotional value. We therefore propose the following hypothesis based on Rasoolimanesh et al. (2016) research of the construct: Perceived value has positive impact on tourist satisfaction.

In summary, this paper proposes a model (Figure 3) in which the antecedent role of perceived risk and in the tourism industry perceived by tourists, which consists of four fundamental dimensions (physical risk, performance risk, financial risk, and psychological risk), has positive impact on tourist satisfaction (H1), perceived value composed of two fundamental dimensions: functional value and emotional value has positive impact on tourist satisfaction (H2). In turn, this experience of perceived risk and perceived value happen within, previous to, during, and after their visit in the destination will be decisive in generating greater tourist satisfaction (H3).

3 RESEARCH METHODS

This research uses descriptive quantitative approach. The population in this study were tourists who visited Cerme Cave in 2018, as many as 9,859 people. Based on the determination of the sample using the Tabachnick and Fidel formulas, sample sizes (n) were obtained from 111 respondents. The study used purposive sampling techniques.

Data collection techniques used in this study are interviews, observation, literature study, and questionnaires. The questionnaires were distributed to 111 respondent who visited Cerme Cave. The study of literature in this study were obtained from various sources such as journals and books, printed or electronic media. The other secondary data are collected from Cerme Cave management such as the organization profile and sampling frame (number of participants, name, origins, and email). The data analysis technique used in this study is the multiple linear regression. The data obtained was then analyzed by using SPSS 20.0.

4 DISCUSSION

Profiling the respondents, the majority of them were males (61.2%). This is consistent with several studies that state men have strong participation orientation to do extreme activities (Giddy 2018). Tourists between age of 18–25 dominated the age category (52.2%). Giddy and Webb (2016) said adventure tourist destinations were dominated by young people, usually from the age 18–24 years. Regarding their occupation, an overwhelming majority of them were students (43.2%), which goes in line with the possibility that students have more time to take part in tourism activities as one of the main reasons (Gardiner & Kwek 2017). Respondents from the Special Region of Yogyakarta contributed highest in this research (60.4%). Tourists who got information of Cerme Cave from their friends dominated the information resources category (86.5%). Most respondent (76.6%) have joined the caving program once/first-timer, and the rest (21.6%) have joined the caving program more than twice/repeater. Almost half of respondents' average expenditure (45.0%) are Rp. 100.001–Rp. 150.000.

Table 1. Recapitulation of perceived risk response.

No	Dimension	Total Questions	Total Score	Average Score	Percentage%
1	*Physical Risk*	3	1020	340	27.48
2	*Performance Risk*	3	893	297.66	24.06
3	*Financial Risk*	3	911	303.66	24.54
4	*Psychological Risk*	3	887	295.66	23.90
Total		12	3711	1236.98	100

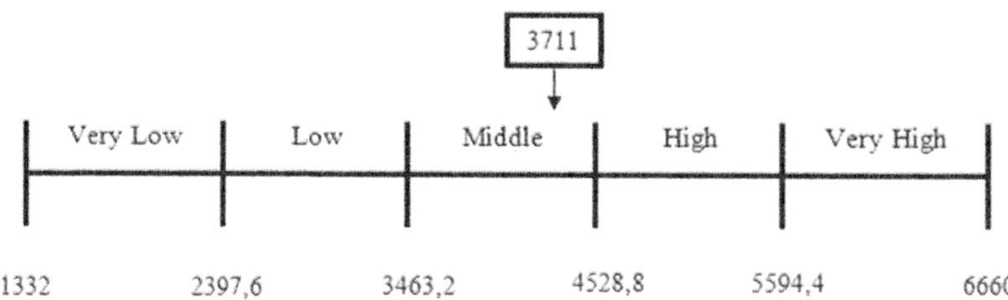

Figure 1. Perceived risk continuum line in Cerme Cave.

Table 2. Recapitulation of perceived value response.

No	Dimension	Total Questions	Total Score	Average Score	Percentage%
1	*Functional value*	3	1242	414	48.72
2	*Emotional value*	3	107	435.66	51.28
Total		6	2549	849.66	100

The results indicated that approximately 74.8% of the respondents had spent 90 minutes in Cerme Cave, while 18.9% of them had spent more than 90 minutes. This was based on the length of the cave, approximately, tourists reach the finish line in 90 minutes.

4.1 *Perceived risk in Cerme Cave*

Four main dimensions of the perceived risk were identified: physical risk, performance risk, financial risk, and psychological risk. The result indicates that more tourists had physical risk (27.48%), financial risk (24.54%), performance risk (24.06%), and psychological risk (23.9%). Specifically, four dimensions of perceived risk were identified in Table 1.

Recapitulation of the tourists' response to the Cerme Cave on the description of perceived risk can be seen in the overall continuum review, in Figure 1.

Based on the continuum line in Figure 1, it showed position of perceived risk value is in the middle category (3711). It means an assessment of the performance of perceived risk in Cerme Cave is middle.

4.2 *Perceived value in Cerme Cave*

Two main dimensions of the perceived value were identified: functional value and emotional value. The result indicates that more tourist had functional value (48.72%) and emotional value (51.28%). Specifically, two dimensions of perceived value were identified in Table 2.

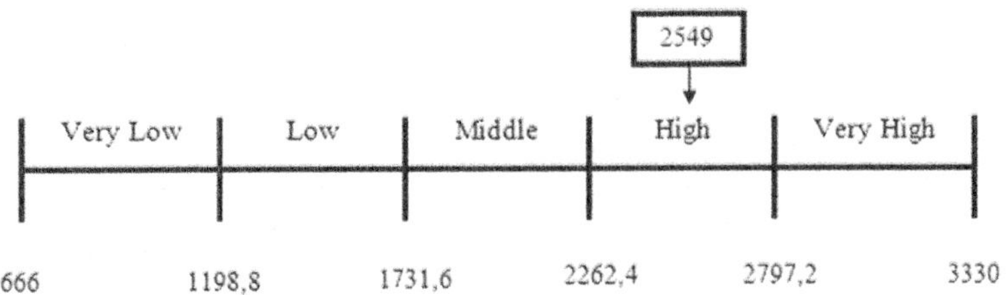

Figure 2. Perceived value continuum line in Cerme Cave.

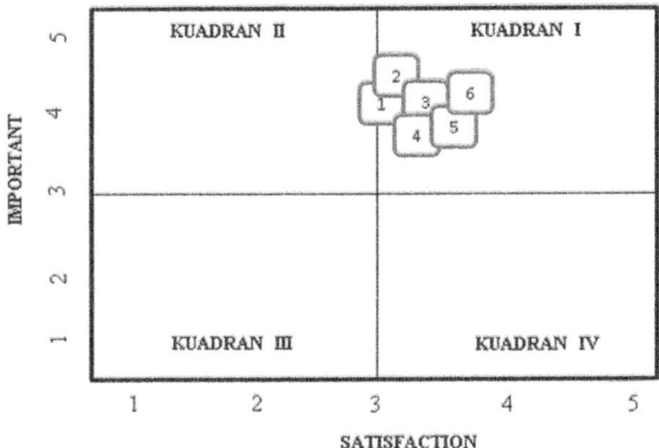

Figure 3. Importance performance analysis result.

Recapitulation of the tourists' response to the Cerme Cave on the description of perceived value can be seen in the overall continuum review, in Figure 2.

Based on the continuum line in Figure 2, it showed position of perceived value is in the high category (2549). It means an assessment of the performance of perceived value in Cerme Cave is high (Figure 3).

4.3 *Customer satisfaction index*

Based on the function of the customer satisfaction index, the results obtained are:

$$\begin{aligned} \text{CSI} &= \frac{210.21}{5(61.53)} \times 100\% \\ &= 68{,}33\% \end{aligned}$$

Based on the calculation results, it can be seen that the tourist satisfaction index of perceived risk and perceived value in the Cerme Cave Tourism Destination has a value of 68.33%, in the category 66%–80.99%, which means that the satisfaction value is in the satisfied category.

Interpretation of tourist satisfaction with perceived risk and perceived value in Cerme Cave uses a Cartesian diagram. After comparing performance with importance, it can be seen that there are six dimensions included in the category of maintaining performance, namely, physical risk (1), performance risk (2), financial risk (3), psychological risk (4), functional value (5), and emotional value (6).

Based on the explanation above, it can be concluded that the distribution of each item from the variable perceived risk and perceived value are all in Quadrant I, which means keep up the good.

This proves that the items included in the quadrant are factors that are considered important and are expected to be a supporting factor for tourist satisfaction, so the company that manages the tourist destination must maintain its performance.

Hypothesis testing obtained about the effect of perceived risk and perceived value on tourist satisfaction in the Cerme Cave Tourism Destination simultaneously has a total effect of 0.653 or 65.3%. This shows that the different dimensions of perceived risk (X1) and perceived value (X2) accounted for 65.3% of the variable tourist satisfaction (Y), while the rest is contributed by other factors that not examined in this study. Partially, the two variables, namely, perceived risk and perceived value, have a significant influence on tourist satisfaction by using a significant level of the two-party test of 0.05 or 5%.

5 CONCLUSIONS

5.1 *Implications*

This study set out to identify the implementation of perceived risk and perceived value in Cerme Cave. Then this research presented some conclusions from perceived risk, which consists of physical risk, performance risk, financial risk, and psychological risk are in a middle category. Physical risk has highest response (27.48%). This happens because physical risk is important for tourists, but the manager has not been able to deal with physical risk concerns that will be faced by tourists. Psychological risk gets the lowest response (23.90%). This happens because the manager of the Cerme Cave has been able to foster self-confidence, minimizing the level of concern and disappointment of tourists when doing activities in the Cerme Cave through the training held every four months by the Bantul District Tourism Office to improve the performance of managers and tour guides in Cerme Cave.

Perceived value which consists of functional value and emotional value is in a high category. Emotional value has the highest response (51.28%). This happens because emotional value is important for tourists. Tourists feel that the emotional value obtained when doing activities in adventure tourist destinations is the main thing compared to other tourist destination categories. Functional value gets the lowest response (48.72%). This happens because the functional value in Cerme Cave Tourism Destinations is not yet maximal. There are other tourist destinations around that can provide better economic value.Overall perceived risk and perceived value at the Cerme Cave have gone quite well.

Understanding perceived risk and perceived value along the six dimensions explored will enable marketers and planners to better design marketing and promotional strategies. In the face of increasing competition, destination marketers are under greater pressure to understand the experiences of travel patrons.

5.2 *Limitations and future research*

The study is not without certain limitations that must be taken into account when evaluating our conclusions. These limitations open up possible future lines of research. The main methodological limitation is the sample, which was a purposive sample and is therefore limited in terms of randomness and representativeness. New research on this topic is therefore needed to guarantee the random and stratified character of the sample analyzed in different cultural and international contexts in order to generalize the results.

A second limitation concerns the use of cross-sectional data, which prevents causal inferences from being drawn. Considering this limitation, future research should study the relationships proposed here using longitudinal data and combining the positivist and interpretative method, as in the case of the ethnographic or phenomenological method (Coviello & Jones 2004). Taking Coviello and Jones as a reference, the reconciliation of positivist and interpretative methodologies would extend knowledge about tourists' perception in their experience of adventure tourist destinations.

Indeed, a longitudinal approach would help to clarify the specific influence of the influential factors in each stage of the process experienced by the tourist (before, during, and after enjoyment).

Our model represents a specific theoretical reference from which to consider new antecedent factors and consequences of the tourist satisfaction valued recently in the specialized literature. In this context, it would be of great interest to complete the analysis of these background factors by assessing the possible interdependencies between them. On the other hand, in relation to the consequential factors, new variables associated with the tourists satisfaction could be included, such as travel experience, destination image, destination attributes, and, especially, the tourism service quality (Chiu et al. 2016; Eid et al. 2019; Hayati & Novitasari 2017). Regarding this last factor, we conjecture that in them, as this improves the visitor's satisfaction. The key is to co-create tourist experiences, which implies involving tourists in environmental practices, as this will contribute to improving the customer experience and, in turn, lead to greater satisfaction.

REFERENCES

Amin, M., Yahya, Z., Faizatul, W., & Ismayatim, A. 2014. *Service Quality Dimension And Customer Satisfaction: An Empirical Study In The Malaysian Hotel Industry*, (December 2014), 37–41. Https://Doi.Org/10.1080/15332969.2013.770665

Biesok, G., & Wyród-Wróbel, J. 2011. Customer Satisfaction – Meaning And Methods Of Measuring Chapter Ii Customer Satisfaction –.

Chang, J.C. 2007. Customer satisfaction with tour leaders' performance: A study of Taiwan's package tours (December 2014), 37–41. https://doi.org/10.1080/10941660500500808

Chiu, L.K., Ting, C., Alananzeh, O.A., & Hua, K. 2019. Perceptions of Risk And Outbound Tourism Travel Intentions Among Young Working Malaysians, *4*(1), 365–379.

Chiu, W., Zeng, S., & Cheng, P.S.T. 2016. The Influence Of Destination Image And Tourist Satisfaction On Tourist Loyalty: A Case Study Of Chinese Tourists In Korea. *International Journal Of Culture, Tourism, And Hospitality Research*, *10*(2), 223–234. Https://Doi.Org/10.1108/Ijcthr-07-2015-0080

Corte, V.D., Sciarelli, M., Cascella, C., & Gaudio, G.D. 2015. Customer Satisfaction In Tourist Destination: The Case Of Tourism Offer In The City Of Naples, *4*, 39–50. Https://Doi.Org/ 10.11648/J.Jim.S.2015040101.16

Coviello, N.E., & Jones, M.V. 2004. Methodological Issues In International Entrepreneurship Research. *Journal Of Business Venturing*, *19*(4), 485–508. Https://Doi.Org/10.1016/J.Jbusvent.2003.06.001

Eid, R., El-Kassrawy, Y.A., & Agag, G. 2019. Integrating Destination Attributes, Political (In)Stability, Destination Image, Tourist Satisfaction, And Intention To Recommend: A Study Of Uae. *Journal Of Hospitality And Tourism Research*, *43*(6), 839–866. Https://Doi.Org/10.1177/1096348019837750

Gardiner, S., & Kwek, A. 2017. Chinese Participation In Adventure Tourism: A Study Of Generation Y International Students' Perceptions. *Journal Of Travel Research*, *56*(4), 496–506. Https://Doi.Org/10.1177/0047287516646221

Giddy, J.K. 2018. A Profile Of Commercial Adventure Tourism Participants In South Africa. *Anatolia*, *29*(1), 40–51. Https://Doi.Org/10.1080/13032917.2017.1366346

Giddy, J.K. & Webb, N.L. 2016. The Influence Of The Environment On Motivations To Participate In Adventure Tourism: The Case Of The Tsitsikamma. *South African Geographical Journal*, *98*(2), 351–366. Https://Doi.Org/10.1080/03736245.2015.1028990

Guo, Y., Barnes, S.J., & Jia, Q. 2017. Mining Meaning From Online Ratings And Reviews: Tourist Satisfaction Analysis Using Latent Dirichlet Allocation. *Tourism Management*, *59*, 467–483. Https://Doi.Org/10.1016/J.Tourman.2016.09.00

Hasan, K., Ismail, A.R., & Islam, F. 2017. Tourist Risk Perceptions And Revisit Intention: A Critical Review Of Literature. *Cogent Business & Management*, *50*(1). Https://Doi.Org/10.1080/23311975.2017.1412874

Hayati, N., & Novitasari, D. 2017. An Analysis Of Tourism Service Quality Toward Customer Satisfaction (Study On Tourists In Indonesia Travel Destinations To Bali). *International Journal Of Marketing And Human Resource Management*, *8*(2), 9–20.

Helena, M., Parreira, A., & Moutinho, L. 2018. Journal Of Destination Marketing & Management Motivations, Emotions And Satisfaction: The Keys To A Tourism Destination Choice. *Journal Of Destination Marketing & Management*, (April), 1–9. Https://Doi.Org/10.1016/J.Jdmm.2018.12.006

Khan, M.J., Chelliah, S., & Ahmed, S. 2017. Factors Influencing Destination Image And Visit Intention Among Young Women Travellers: Role Of Travel Motivation, Perceived Risks, And Travel Constraints. *Asia Pacific Journal Of Tourism Research*, *22*(11), 1139–1155. Https://Doi.Org/10.1080/10941665.2017.1374985.

Kim, M. & Thapa, B. 2017. Perceived Value And Fl Ow Experience: Application In A Nature-Based Tourism Context. *Journal Of Destination Marketing & Management*, (March), 1–12. Https://Doi.Org/10.1016/J.Jdmm.2017.08.002

Rasoolimanesh, S.M., Dahalan, N., & Jaafar, M. 2016. Perceived Value And Satisfaction In A Community-Based Homestay In The Lenggong Valley World Heritage Site. *Journal Of Hospitality And Tourism Management*, *26*, 72–81. Https://Doi.Org/10.1016/J.Jhtm.2016.01.005

Wu, H.C., Li, M.Y., & Li, T. 2018. A Study Of Experiential Quality, Experiential Value, Experiential Satisfaction, Theme Park Image, And Revisit Intention. *Journal Of Hospitality And Tourism Research*, *42*(1), 26–73. Https://Doi.Org/10.1177/1096348014563396

Zagreb, P.I. 2019. Poslovna Izvrsnost Zagreb, God. Xiii (2019) Br. 1 Mikulik J.: Derived-Importance Performance Analysis As A Tool To Identify Priorities …, 77–86.

Zeithaml, V.A. 1988. *Consumer Perceptions Of Price, Quality, And Value: A Means-End*, *52*(July), 2–22.

Promoting Creative Tourism: Current Issues in Tourism Research – Kusumah et al. (Eds)
 ISBN 978-0-367-55862-8

Hospitality industry crisis: How to survive and recovery in the pandemic of COVID-19

E. Fitriyani, D.P. Novalita & Labibatussolihah
Universitas Pendidikan Indonesia, Bandung, Indonesia

ABSTRACT: The purpose of this study is to provide insight into strategies for how to survive and recover in the hospitality industry affected by the Covid-19 pandemic. This study used the content analysis method which was conducted descriptively based on webinars related to the impact of the COVID-19 pandemic on tourism, carried out from April to May 2020. The webinar that the researchers chose was hotel-related, because the hotel industry was one of the industries that had the greatest impact from the COVID-19 pandemic. The results of this study provide tactical steps to survive and recover from the COVID-19 pandemic through operational, financial, human resources, public relations, and recovery processes through collaboration, operation, venture, innovation, and development. Further development of this kit is implemented as a strategic tool for the hotel industry and academia to prepare human resources after the COVID-19 pandemic.

Keywords: Hospitality Industry Crisis, Hotel Industry, Hotel Recovery, Covid-19

1 INTRODUCTION

With most hotels closed or set to experience significantly lower tourism figures, 2020 industry revenue is expected to show a significant decline (e.g. estimated US hotel revenue per available room has a 50.6% decrease in STR, 2020b) (Gössling et al. 2020). The Government of Indonesia works objectively in evaluating the impact of the COVID-19 pandemic in Indonesia (Brahma 2020; Djalante et al. 2020) while the World Health Organization has established protocols to try avoiding the widespread dissemination (Word Health Organization 2020). The tourism industry in Indonesia is developing by providing huge infrastructure and facilities for tourists. The reason why tourists choose Indonesia as a tourist destination for international visitors is because of the way the Indonesian people communicate and warm hospitality.

The COVID-19 pandemic has a disastrous impact on the Indonesian economy (McKibbin & Fernando 2020), in terms of the context of the travel and tourism industry. Among the most significant impacts of this outbreak, which are the declining trend of international visitors entering the most visited destinations, restrictions on people using transportation to travel abroad or domestically, and the Indonesian government has also introduced new regulations to prevent and close transportation in Indonesia for a while. In this context, tourism and hospitality as the third-largest market in the world of beneficiaries and sources of income in the world may have completely worsened with travel fears and the need to maintain physical distance. The examples are: restrictions on visa issuance and closure of international airports, curfews and, applying large-scale social restrictions. Conditions become terrible, catastrophic, and devastating when the whole world becomes locked up or under social pressure, people are afraid to order something before the pain of this disease and virus fading is controlled in the world. From this big phenomenon, it caused several impacts in the tourism and hospitality fields (Riadil 2020).

The COVID-19 case has a significant impact on the hospitality business in Indonesia. 90% of the hotelier predicts that the impact of the COVID-19 outbreak will not last more than 6 months or August. Judging from its impact, the hotelier agrees that co-19 cases are worse than similar

DOI 10.1201/9781003095484-84

Table 1. Webinar Theme.

No.	Date	Webinar Theme	Interviewees	Organizer
1.	9 April 2020	Hotel Crisis Management: Operation Perspective	Director Hotel Corporation	MarkPlus Tourism
2.	14 April 2020	Hotel Crisis Management: HR Perspective	General Manager in Hotel	MarkPlus Tourism
3.	16 April 2020	Hotel Crisis Management: Finance Perspective	General Manager in reputable international chain hotel	MarkPlus Tourism
4.	21 April 2020	Hotel Crisis Management: PR Perspective	General Manager in Hotel	MarkPlus Tourism
5.	5 May 2020	Hotel Industry in 2020 Onward "Impact of Covid-19 and Recovery	Director of Hospitality Management	IHGM and PHRI

Source: Researcher, 2020.

outbreaks (SARS 2003, H5N1 2016). This can be seen from the level of occupancy, average room prices, and total revenue for the first half of 2020 at −25% to −50% when compared to 2019 (PHRI & HTL 2020).

The next agenda is research in the field of work, hospitality that is multidisciplinary in perspective, and multi-method in the tools used, using the most effective way to understand the impact of COVID-19 on the workforce (Baum et al. 2020). It is important to study significant theoretical and/or practical contributions to hospitality theory and practice (Gursoy & Chi 2020). There is a need for a more strategic approach to crisis management in tourism and hospitality, although there are no recommendations to achieve it (Faulkner & Vikulov 2001). This research can provide input in dealing with crises in tourism and hospitality. To fill the problem gap, this paper tries to describe the results of research on how the strategies implemented by the industry to survive and recover from the COVID -19 pandemic, and the results of this study can provide input in dealing with crises in tourism and hospitality from various perspectives.

2 METHOD

This qualitative research aims to explore hotel strategies to recover and survive during a pandemic. This research is a content analysis research conducted descriptively based on webinars related to the impact of the COVID-19 pandemic on tourism carried out from April to May 2020. The webinar that the researchers chose is a webinar related to hospitality because the hotel industry is one of the industries that gets the greatest impact from the Covid-19 pandemic.

Researchers chose webinar topics related to the way the hotel industry survived and restored the situation during the Covid-19 pandemic. The resource persons were from the hotel industry, selected from various perspectives: operational, marketing and public relations finance, and human resources organized by PHRI, IHGM DPD Central Java and MarkPlus Tourism. The theme and identity of the resource persons are as follows (Table 1):

A content analysis conducted by researchers is the result of exposure to webinars from the speakers. The authors concluded to achieve the objectives in this study: to get an overview for hotel industry players to survive and restore the company's business conditions in the period and after the COVID-19 pandemic.

3 RESULTS

3.1 *Hotel crisis management: Operation perspective*

As for anticipation in dealing with crises that must be owned by the company in operational terms, namely:

1) Reserve Funds; 2) Insurance Policies; 3) Emergency SOP's. Some tactical steps of hotel management operationally are:

1) Minimizing costs & expenses.
   High operational costs but not offset by large income will have an impact on losses. One effort that can be done is reducing operational costs. With the decline in the number of guests at the hotel automatically makes the hotel revenue also decline. It is a good idea to avoid spending on goods that are felt to be unnecessary during a pandemic, to reduce electricity, water, and internet usage if the hotel is still operational.
2) Review: Business plan, and financial plan.
   The business plan helps to determine what the company will be like after the pandemic, making a marketing strategy that will be carried out by the company in selling their products or services to post-pandemic consumers. This is done in order to regain customer trust. Rearrange the company's budget in a COVID-19 pandemic situation and plan what costs will be incurred and how much will be incurred.
3) Recovery, restructuring.
   The hotel recovery was carried out when the COVID-19 pandemic began to subside by restructuring the hotel management, staffing structure, and hotel financial recovery. Restructuring is also needed by reducing the number of employees until the hotel finance is deemed stable. As much as possible the company should be able to return to operational activities as before the COVID-19 pandemic (*Hotel Crisis Management: Operation Perspective* 2020).

3.2 *Hotel crisis management: Human resources perspective*

In dealing with the problems that are the effects of the COVI-19 pandemic, there are several things the hotel does from the perspective of Human Resources (HR), namely: Payroll analytic,

1) No revenue: prediction and assumption on how long the cash will continue. Companies must predict how long this pandemic will occur, such as short term (1 month), medium-term (3–4 months), or long term (6–1 years). This is to estimate how long the hotel will be able to survive with the existing costs without any revenue/income.
2) Cut-off salary: Cash flow management and payment schedule. Since there is no revenue, salary deductions are also done as an efficiency measure.
3) Labor Relations:
   a. Final figure and payment plan established (after the analytic payroll is done)
   b. Meeting with a union. In this process, the HR perspective is very important,
   c. General staff meetings,
   d. Outlet and departmental rescheduling: Annual leave, Long leave, and unpaid leave.

Furthermore, there must be Cross support if the hotel is temporarily closed, by ensuring that core security will be available (security & technicians), reopening plans (funds for direct purchases, grounds, payroll needs, training, etc.) If this pandemic is finished, it is not impossible modeling in the hotel industry will change, both from uniforms, cleaner hotel quality, and also changes in guest behavior (*Hotel Crisis Management: HR Perspective* 2020).

3.3 *Hotel crisis management: Finance perspective*

The initial stage of overcoming the tendency to lose the hotel business: Defer merit increase, 4/8/16 days unpaid leave (starts from general manager to supervisory level), annual leave (vacation accrual), freeze hiring & promotion, service & increase, daily worker reduction or stop, utilities cost reduction, food & beverage costs, room operational supplies, food & beverage operational supplies, spa operational supplies, reduce entertainment & activity, cancel business trips & training, sales & marketing promotions, travel agent Commissions, credit cards commission, labor-security contract, labor-reduction contract, maintenance contract discount, procurement savings, and upselling.

It needs a comprehensive strategy not only from the financial side which aims to improve the company's cash flow through cooperation between operations, human resources, and sales. In addition to operations, costs that must be taken into account at the time of the outbreak are electricity costs, while the option to reduce electricity costs is first to reduce temporary power and secondly to temporarily cut costs by passing efficiency in several departments.

The food & beverage department must look at the estimated number of guests present, individuals or groups so that raw materials can be directly processed and not left. Food sales can be made for online sales, introducing more food products to the home market if prices allow for special menus.

Furthermore, if the hotel closes down, there will be a pay cut for general managers, head of department dam supervisors up to 50%, while for rank and file reduction of up to 20% (*Hotel Crisis Management: Finance Perspective* 2020).

3.4 *Hotel crisis management: Public relation perspective*

Seeing the impact of this pandemic, it is estimated that there will be more domestic guests, short-stay guests, more individual segmentation and should be able to implement flexible cancellation policies. Then what should we do now? "All great changes are preceded by chaos", first don't panic, and then create strategies from internal and external.

From external:

a. Maintain connections. Marketing and sales should not be off, even if there are no guests.
b. Community. The repeater community must be maintained because they will later target hotels when reborn apart from the individual market.
c. Empathy to customer needs. When this pandemic ends there will be a change in guest habits, for example, our rooms later will not be able to sell 100% so that it does not occur crowded.

Furthermore, from Internal:

a. Do not get carried away with the state of "unqualified discounting", because it will be difficult to raise prices again later and must balance with the image of the hotel.
b. Focus on marketing initiatives. Must be able to brainstorm so that the existence of hotels, especially on social media, is not lost. It relies on the strength of the sales and marketing team.
c. Maintain marketing efforts. In contrast to these initiatives, there is no need to pay.
d. Maintain service-level beware of customer changing behavior.
e. Don't cut your marketing budget, just evaluated. Don't cut it but it's better to evaluate it from the major operating expenses, marketing expenditure, performance and profitability measure, and other sales expenditure. Once the marketing budget is cut, the hotel is gone.

What is important to do? What is important is the customer and the community how we handle this situation, what they can support us, the community must be the most concerned because they are the royal ones to our hotel. Don't forget that global travel is a lockdown, our opportunity to mix. Move our market to a local (example: promo for people who have Identity in Bandung). Clean your data, prepare for rebirth (*Hotel Crisis Management: PR Perspective* 2020).

3.5 *Hotel industry in 2020 onward "Impact of Covid-19 and recovery"*

Impact/influence COVID-19 on hotels: Disease has an impact on depression individually and companies do not get depression because it will lead to destruction (destruction cannot rise from this destruction). COVID trend in Indonesia cannot be said to go down. This crisis is at the root of the problem of COVID-19 so if the COVID trend has gone down then the business crisis can be said to fall, then the prediction of the end depends on the downward trend of COVID-19 cases, especially in Indonesia.

Recovery: four recovery models can occur after COVID-19 is completed or no more.

a. V-shape recovery, V-shape recovery model like SARS. Looking at the trend of SARS, the trend is over if the vaccine is found, but the recovery business is 6 months after the downward trend. This is the fastest scenario.
b. U-shape, revenue per available room (Revpar), and gross operating profit (GOP) for longer recovery can be said 10 years have not recovered as usual. Revpar is 4 years, but the GOP will be more than 4 years.
c. L-shape, never returns to the position before the crisis because the supply does not match the demand.
d. W-shape, recovery in the form of V-shape, has increased but the second wave crisis occurs again if this happens like a v shape but the longer time can be up to 2 years. After that, who will lead the recovery in 2021? The predicted recovery will be led by Midscale & Economy Hotels.

The crisis must be seen from a different perspective because in every crisis there is an opportunity. In making a recovery strategy three internal looking parts must be adjusted to the condition of the hotel:

1) Income statement, see the condition of the comparison between cost and revenue,
2) Balance sheets see the condition of debt/equity with cash equity.
3) The cash flow statement, see the condition of operating cash flow and financing cash flow.

Broadly speaking the strategy to fight COVID is with COVID: collaborate, operation, venture, innovation, development. The segment that can recover for the first time: domestic market, local market, weekend, corporate, group/postpone an event (*Hotel Industry in 2020 Onward* 2020).

4 DISCUSSION

Our results show that the right strategy needs to be arranged so that the hotel industry can survive, business orientation on high revenue cannot be expected in the current conditions. The most important thing is that hotel operations can continue to run, can pay employees, and the hotel can still exist. Not yet able to speak profit-oriented, but healthy and secure oriented are the points in guest satisfaction at this time. The Crisis Response Phase is followed by the Crisis Recovery stage where the main objective is to bring the organization back to business in its original state. The actions taken at this stage are largely dependent on the magnitude of the impact of the crisis on the organization and the effectiveness of the actions taken at the Crisis Response stage. Although there is a lot of emphasis on recovery, which is often the main focus in organizational recovery after the crisis is reputation and financial, even though it needs the same emphasis for operational recovery (infrastructure and people) (Paraskevas & Quek 2019).

The aviation and hotel industry is most impacted by the coronavirus outbreak hence it is recommended that all service providers involved in this industry, to encourage customers to take precautions to stay safe (Karim et al. 2020). This was started from the results of this study that to carry out operations in this situation required to have an emergency standard operating procedure (SOP) by government regulations, in this case following the health protocol SOP in the prevention of COVID-19 and maintaining the safety and health of employees and guest's hotel. To reduce the spread of this pandemic, all countries have imposed lockdowns, limited international travel, a ban on all foreign visitors; Travel restrictions from various places with confirmed cases. Other restrictions such as suspending all commercial international flights, all travelers subject to 14-day quarantine, and all visa operations are suspended (Bloomberg 2020). Likewise with hotels, to mitigate the spread of this pandemic, many hotels in Indonesia are implementing a temporary close strategy, 1,139 hotels are implementing a temporary close strategy as of April 1, 2020 (Kemenparekraf 2020).

The findings in this study show support for previous research which states that much of the influence of organizational capital is obtained from a manager's dedication to the organization where they work, especially in the team's approach to planning, leadership, and problem-solving.

This is very important for every organization because it measures the success of managers in how they perform during crises caused by external events (Cho & Hambrick 2006), this study provides input for hotel managers to carry out various strategies from the operational, financial, marketing and marketing perspectives. public relations, human resources by looking at the condition of the income statement, balance sheet, and hotel cash flow statement. Through this approach, managers are expected to make decisions by implementing strategies that can overcome the problems of the hotel industry in how to survive during the Covid-19 pandemic and recovery after this pandemic ends.

5 CONCLUSION

Although previous studies have tested the impact of COVID-19 on tourism and hospitality due to social restrictions (Karim et al. 2020), in our study we describe strategies that can be adopted by hotel policymakers in the survival and recovery of the hotel business after the COVID-19 pandemic. We found that the recovery process will be like a hotel re-opening process, where management must make an SOP from checking in to checking out with the Covid-19 protocol and rearranging the budget planning due to New Normal conditions, the hotel cannot get maximum revenue like the conditions before COVID-19.

Recommended research in advancing the knowledge base is to help hotels recover from the COVID-19 pandemic. The suggested research flow is expected to provide actionable insights to promote the development and sustainability of the hotel sector (Jiang & Wen 2020), so it is hoped that this research can be one of the studies suggested in previous studies, namely providing knowledge for the hotel industry to restore, develop sustainability of the hotel sector. Because social media platforms have become an important part of people's lives today (Ahani et al. 2017) a collaborative platform can be relied upon to share information that is useful for managing the impact of unexpected events such as the Coronavirus outbreak on the global economy. Then the next most appropriate strategy to overcome this pandemic is collaboration. The hotel industry can work together for a faster recovery process.

REFERENCES

Ahani, A., Rahim, N. Z. Ab., & Nilashi, M. (2017). Forecasting social CRM adoption in SMEs: A combined SEM-neural network method. Computers in Human Behavior, 75, 560–578. https://doi.org/10.1016/j.chb.2017.05.032

Baum, T., Mooney, S. K. K., Robinson, R. N. S., & Solnet, D. (2020). COVID-19's impact on the hospitality workforce - new crisis or amplification of the norm? International Journal of Contemporary Hospitality Management, 32(9), 2813–2829. https://doi.org/10.1108/IJCHM-04-2020-0314

Bloomberg. (2020). Mapping the Coronavirus Outbreak Across the World.

Brahma, B. (2020). Oncologists and COVID-19 in Indonesia: What can we learn and must do? Indonesian Journal of Cancer, 14(1), 1. https://doi.org/10.33371/ijoc.v14i1.728

Cho, T. S., & Hambrick, D. C. (2006). Attention as the mediator between top management team characteristics and strategic change: The case of airline deregulation. Organization Science, 17(4), 453–469. https://doi.org/10.1287/orsc.1060.0192

Djalante, R., Lassa, J., Setiamarga, D., Sudjatma, A., Indrawan, M., Haryanto, B., Mahfud, C., Sinapoy, M. S., Djalante, S., Rafliana, I., Gunawan, L. A., Surtiari, G. A. K., & Warsilah, H. (2020). Review and analysis of current responses to COVID-19 in Indonesia: Period of January to March 2020. Progress in Disaster Science, 6, 100091. https://doi.org/10.1016/j.pdisas.2020.100091

Faulkner, B., & Vikulov, S. (2001). Katherine, washed out one day, back on track the next: A post-mortem of a tourism disaster. Tourism Management, 22(4), 331–344. https://doi.org/10.1016/S0261-5177(00)00069-8

Gössling, S., Scott, D., & Hall, C. M. (2020). Pandemics, tourism and global change: A rapid assessment of COVID-19. Journal of Sustainable Tourism, 1–20. https://doi.org/10.1080/09669582.2020.1758708

Gursoy, D., & Chi, C. G. (2020). Effects of COVID-19 pandemic on hospitality industry: Review of the current situations and a research agenda. Journal of Hospitality Marketing & Management, 29(5), 527–529. https://doi.org/10.1080/19368623.2020.1788231

Hotel Crisis Management: Finance Perspective. (2020). MarkPlus Tourism.

Hotel Crisis Management: HR Perspective. (2020). MarkPlus Tourism.

Hotel Crisis Management: Operation Perspective. (2020). MarkPlus Tourism.

Hotel Crisis Management: PR Perspective. (2020). MarkPlus Tourism.

Hotel Industry in 2020 Onward. (2020). IHGM & PHRI.

Jiang, Y., & Wen, J. (2020). Effects of COVID-19 on hotel marketing and management: A perspective article. International Journal of Contemporary Hospitality Management, ahead-of-print(ahead-of-print). https://doi.org/10.1108/IJCHM-03-2020-0237

Karim, W., Haque, A., Anis, Z., & Arjie Ulfy, M. (2020). The Movement Control Order (MCO) for COVID-19 crisis and its impact on tourism and hospitality sector in Malaysia. International Tourism and Hospitality Journal. https://doi.org/10.37227/ithj-2020-02-09

Kemenparekraf. (2020). Daftar Hotel Ditutup-Covid-19.

McKibbin, W., & Fernando, R. (2020). The economic impact of COVID-19. Economics in the Time of COVID-19, 45.

Paraskevas, A., & Quek, M. (2019). When Castro seized the Hilton: Risk and crisis management lessons from the past. Tourism Management, 70, 419–429. https://doi.org/10.1016/j.tourman.2018.09.007

PHRI, P., & HTL, H. (2020). Survei Sentimen Pasar Hotel & Restoran di Indonesia Terhadap Wabah Covid-19.

Riadil, I. G. (2020). Tourism industry crisis and its impacts: Investigating the Indonesian tourism employees perspectives' in the Pandemic of COVID-19. Jurnal Kepariwisataan: Destinasi, Hospitalitas Dan Perjalanan, 4(2), 98–108. https://doi.org/10.34013/jk.v4i2.54

Word Health Organization. (2020). Coronavirus disease 2019 (COVID-19) Situation Report – 73.

Promoting Creative Tourism: Current Issues in Tourism Research – Kusumah et al. (Eds)
© 2021 Taylor & Francis Group, London, ISBN 978-0-367-55862-8

From fantasy to reality: Attracting the premium tourists after COVID-19

A.R. Pratama, P. Hindayani & A. Khosihan
Universitas Pendidikan Indonesia, Bandung, Indonesia

ABSTRACT: The COVID-19 pandemic has a great impact on the tourism sector, especially in terms of the number of tourist visits. A surge in the number of tourist visits after the COVID-19 pandemic has been predicted to occur; while on the other hand, due to their economic condition during the pandemic, people will likely prioritize their primary needs so the post-pandemic potential tourists are most likely the premium tourists. Deniz E.G., et al. (2015) argue that the type of tourists will plan their tourism activities. Using the literature study and forum group discussion methods, we reveal the effectiveness of virtual tourism as marketing media which provides spatial knowledge and thus can help the potential tourists choose their tourist destinations. The survey indicates that the concept can influence them. The strategy used is 2S (suggest and safety), which is giving suggestions by the means of virtual tourism and assuring the potential tourists to realize it by the means of safety guarantee against COVID-19.

Keywords: Covid-19, tourism behavior, virtual tourism, tourism marketing

1 INTRODUCTION

The national development goal is to realize a fair and prosperous community, which is mentioned in the preamble of the 1945 Constitution in the fourth paragraph. To realize the goal, the government of Indonesia has formulated a long-term development plan (RPJP) where development principles are stated. The development principles comprise democracy with the principles of togetherness, justice, sustainability, environment, and independence manifested with balancing national progress and unity. Development in all sectors, including in the tourism sector, aims at realizing the development principles. The tourism sector is one of the sectors that greatly contribute to state revenue. Holzner (2011) confirms that the direct impact of the tourism sector on the economic growth is up to 29% of the total sectors. It indicates that the tourism sector is essential in the economy in a certain region so that extraordinary events such as terrorism (Fletcher & Morakabati 2008; Purwomarwanto & Ramachandran 2015), disease outbreak (Kim et al. 2020; Nicula & Oneţiu 2016), and natural disasters (Henderson 2005; Rosselló et al. 2020) can have negative impacts on the tourism sector.

COVID-19 was firstly reported in Wuhan and rapidly transmitted worldwide. Even the World Health Organization has stated it is a global pandemic. It indicates the virus has big impacts. The impacts cause many sectors, one of which is the tourism sector, to suffer from a loss. A high level of virus transmission by humans to humans has made America meet the same fate as Europe and became a virus epicenter (Anderson et al. 2020). The rapid transmission demands tourism management for adapting (Gössling et al. 2020; Haywood 2020). Several experts argue that after the pandemic, a new pattern will be found in the tourism sector which aims at ensuring the safety of the tourists who are visiting a tourist destination.

Indonesia has currently implemented the "new normal" policy, which allows people to do outdoor activities, yet they must strictly pay attention to health protocols. It gives some spaces to the tourism sector to re-open businesses. Meanwhile, the COVID-19 pandemic makes the government implement the "stay-at-home" policy and thus isolates people and makes them inconvenient

 DOI 10.1201/9781003095484-85

(Haywood 2020). Based on the explanation, an outbreak, including the COVID-19 outbreak, gives significant negative impacts on the tourism sector. Through the "new normal" policy, there may be an opportunity to promote the tourism sector. However, potential tourists will be those who have fulfilled their primary needs and thus can afford tourism activities. An identification of tourists' characteristics and habits, especially premium tourists, when planning their tourism activities will help us describe a concept that activates the tourism sector in the new normal era. The concept can be analyzed further in association with the communication strategy used by tourist destination management in the post-pandemic reopening of the tourism sector. Considering the issues, this study of premium tourists is important for the recovery of the tourism sector, especially in Indonesia.

2 LITERATURE REVIEW

Tourism behavior refers to a complex psychological response manifested in the behavior or action when deciding on purchasing products such as tourism from the beginning until the end of the tourism (Hasan 2008). COVID-19 is one of the global challenges in the economic sector, especially in the tourism sector. Nepal (2020) mentions that several countries, including Nepal, have declared that 2020 is the year of tourism visits to Nepal. The strategy is taken to promote tourism in Nepal. Nevertheless, due to the COVID-19 pandemic, all tourism sectors are impacted and deactivated. Besides the regional impacts, there are also individual ones, as policies implemented by the government limit the mobility of communities. The condition has a massive impact and thus disturbs the business mechanism, particularly in the tourism sector (Ioannides & Gyimóthy 2020). In general, tourism behaviors are divided into three stages: pre-visit, on-site or during the trip, and post-visit or after the trip (Cohen et al. 2014). Based on the division, potential tourists will likely perform tourism planning. The planning includes searching for information regarding a tourist destination, access to accommodation, transportation, restaurants, and shopping centers. The pre-visit stage, where potential tourists are making decisions about the tourist destinations offered, is important. The theory of planned behavior (Ajzen 1985) explains that perceived attitudes, social norms, and behavior control influence an individual to behave in a certain way, which leads to actual behaviors. Based on the explanation, the tourism sector greatly relies on the post-COVID-19 pandemic condition. A study of the habits of potential tourists during the COVID-19 pandemic is thus one of the keys to a successful marketing strategy of tourism industries after the pandemic.

3 RESEARCH METHODS

This was qualitative research using a literature study of various relevant references with the habits of premium tourists when determining their tourist destinations and the attitudes to do tourism activities in the destinations. The data were collected by inventorying any available relevant literature, especially articles, books, government reports, and others. The databases used to inventory the literature were Google Scholars, Scopus, and Sinta. The keywords inserted to look for the literature manuscripts to be reviewed were tourism behavior, the impact of COVID-19 on the tourism sector, tourism marketing, and virtual tourism. We interpreted and elaborated on the literature and made a conclusion. The literature study aimed to gain a holistic description of the behaviors showed by premium tourists after the COVID-19 pandemic from various perspectives.

4 FINDING AND DISCUSSION

The tourism sector was the most impacted sector due to the COVID-19 outbreak. It was because of the prohibition to gather or do activities where people would gather despite the COVID-19 outbreak. The government of Indonesia implemented individual-level policies including "stay-at-home," "social-distancing," and "physical-distancing," and a regional-level policy, which was PSBB

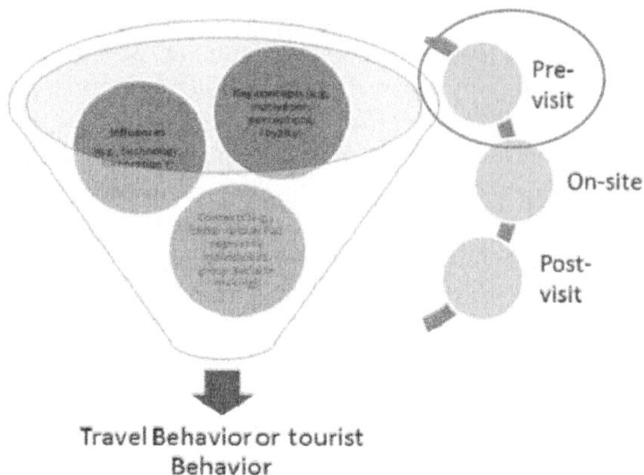

Figure 1. Travel behavior of tourist. (Source: Cohen et al. 2014).

(the Indonesia Large Scale Social Restrictions). The policies limited the community, especially potential tourists, in outdoor tourism activities, so most of them would likely get entertainment from indoor activities: watching TV, playing games, joining virtual tourism activities, and making post-COVID-19 pandemic tourism plan (Tan 2018). The policies would make individuals limit their real social activities, which potentially causes stress. Taylor et al. (2020) argued that COVID-19 might cause stress, indicated by complex symptoms that triggered fear, high vigilance, searching for safety guarantee, and panic buying. It created the type of premium tourists, which was a phenomenon to which the tourist destination management had to pay attention when reopening its destination after the COVID-19 pandemic. Here is a further explanation regarding premium tourists and their tourism behaviors.

4.1 *Tourist typology*

The previous researchers had made tourist classifications using various indicators, that is, a classification based on the type of tourism activities (Seyidov & Adomaitienė 2017), a classification based on socio-demography (Ferrer-Rosell et al. 2015), and a classification based on the distance between the tourist domicile and the tourist destination (Abuamoud et al. 2014). Besides, there was a classification based on the expenditure on tourism. The Government Tourism Office in Bandung Barat (FGD 2020) confirmed that public enthusiasm for tourism might decline after the pandemic because people would prefer to save their money and prioritize their primary needs. As a result, potential tourists would be those who had fulfilled their primary needs. This type of tourists would be willing to spend more to do tourism activities. This type of tourist was called premium tourists. During the COVID-19 pandemic, they were usually planning their tourism activities and looked up information about potential tourist destinations, transportation, accommodation, and the implementation of health protocols (Cohen et al. 2014). The planning activities are depicted in Figure 1. After the Covid-19 pandemic, tourists would prioritize the implementation of health protocols when sorting some potential destinations.

Potential premium tourists became one of the parties who recovered the economic sector, especially in the tourism sector. The management of tourist destinations had to pay special attention to them since their satisfaction level of the implementation of health protocols in the "new normal" era would interest other potential tourists. Besides, their satisfaction would contribute to implementing an effective marketing method, which was "word of mouth." Word of mouth (WOM) was created in the market, and no management could give full control of information in a market (Lapointe 2011; Warrington 2002). Furthermore, the premium tourists could act as influencers who would attract others. A detailed description is indicated in Figure 2.

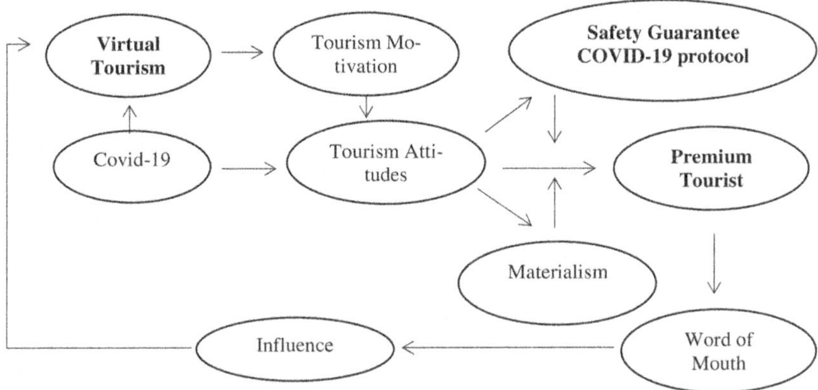

Figure 2. The formation and roles of premium tourists.

4.2 *Attracting the premium tourists after COVID-19*

In general, tourist behaviors were divided into three stages: pre-visit, on-visit, and post-visit (Cohen et al. 2014). The three stages contributed to the behavioral formation of premium tourists, especially after the COVID-19 pandemic. In the pre-visit stage, premium tourists would be making the best tourism plan, starting with planning the destinations and transportations. According to Fiedler (2007), the value of tourism included the desire for or an idea of visiting a destination by considering the visit obstacles. The obstacles could be in terms of financing or the destination worth visiting. Moreover, financing was not a significant obstacle for premium tourists. However, they would need further consideration for some other aspects, such as distance and health and safety insurance. There was a relationship between distance and health insurance, where a long distance was considered riskier than a short one. It motivated premium tourists to choose domestic destinations (Miao & Mattila 2013). It provided a bigger chance for domestic tourism management so they had to convince premium tourism in terms of safety insurance. They could do campaigns of the readiness of tourist destinations to face the new normal. The campaigns could be in the form of videos, pictures, or publications distributed online. The campaigns were expected to stimulate the potential premium tourists to visit the destinations being offered. It was in line with Millie (2014) that tourism-related topics were ranked third as the topics most searched for on Facebook after music and television. It confirmed that the information about access to tourist destinations on social media would be useful for potential tourists. It would be an effective marketing medium for tourism management, especially after the COVID-19 pandemic.

Several premium tourists might be planning to do virtual tourism activities after the pandemic. Kim and Hall (2019) concluded that virtual reality (VR) could stimulate potential tourists to visit a tourist destination. The experience offered by virtual tourism stimulated the cognitive aspect of potential tourists more effectively than their affective aspect. Virtual tourism would give knowledge of a destination through the visual experience being presented. It caused a higher tourist visit percentage in virtual tourist destinations (Huang et al. 2020). Furthermore, the hedonist behavior of premium tourists provided a bigger opportunity to develop tourist destinations after the COVID-19 pandemic because the behavior was related to the sensibility of the use of technology for either traveling or shopping (Kim & Forsythe 2007). The comparison in interests between premium tourists and common tourists after the COVID-19 pandemic is described in Table 1.

Based on Table 1, premium tourists prioritized the assurance of the implementation of health protocols at tourist destinations. Besides, they were also interested in social media and virtual tourism so tourism management could use them as marketing media to attract potential tourists.

Overall, COVID-19 would impact various sectors, especially the tourism sector. The impact triggered stress in the psychological condition of the respective individual. The stress, caused by

Table 1. The comparison of tourism interests between premium tourists and common tourists

Premium Tourists	Common Tourists
– High desire for tourism	– High desire for tourism
– High tourism budget allocation	– Low tourism budget allocation
– High satisfaction standards	– Average satisfaction standards
– Thorough tourism activities planning	– Tourism activities at any time
– High use of technology as an information source	– Average use of technology as an information source
– The priority of the assurance of the implementation of health protocols	– The priority of other matters over the assurance of the implementation of health protocols
– Domestic tourist destination	– Local tourist destination

the policies the government made to break the pandemic chain, had created new phenomena in terms of the tourists' habits in planning their post-COVID-19 pandemic tourism activities. One of the phenomena was the emergence of premium tourists. This type of tourist was the tourists with high access to technology and high standards of satisfaction of services in the tourist destination offered. Reopening a tourist destination whose target market was premium tourists had to engage at least two aspects: technology and safety guarantee. The aspects became the base of marketing, implementation, and evaluation of the post-COVID 19 pandemic tourism activities.

5 CONCLUSION

The tourism sector was the most impacted sector due to COVID-19 pandemic. Some literature predicted a surge in the number of tourists after the pandemic. However, based on study literature and our forum group discussion with the staff of the Government Tourism Office in Bandung Barat, people would likely prioritize their primary needs over tourism activities after the COVID-19 pandemic. By the discussion, the most potential type of tourists as the target of the tourism sector was premium tourists. The premium tourists would plan their tourism activities thoroughly so they needed access to information. The high access to information was strongly related to hedonism behaviors showed by premium tourists. Therefore, effective marketing media in terms of tourism, such as social media and virtual tourism, were needed, where most potential tourists joining virtual tourism would likely do the real tourism. However, there was some major urgency due to the COVID-19 pandemic, that is, health and safety insurance in the form of proper implementation of health protocols at tourist destinations. We suggest tourism management apply a "suggest-and-safety" strategy to attract premium tourists.

REFERENCES

Abuamoud, I. N., Libbin, J., Green, J., & Rousan, R. A. L. (2014). Factors affecting the willingness of tourists to visit cultural heritage sites in Jordan. *Journal of Heritage Tourism*, *9*(2), 148–165. https://doi.org/10.1080/1743873X.2013.874429

Ajzen, Icek. (1985). From intentions to actions: A theory of planned behavior. *Action Control*, 11–39.

Anderson, R. M., Heesterbeek, H., Klinkenberg, D., & Hollingsworth, T. D. (2020). How will country-based mitigation measures influence the course of the COVID-19 epidemic? *The Lancet*, *395*(10228), 931–934. https://doi.org/10.1016/S0140-6736(20)30567-5

Cohen, S. A., Prayag, G., & Moital, M. (2014). Consumer behaviour in tourism: Concepts, influences and opportunities. *Current Issues in Tourism*, *17*(10), 872–909. https://doi.org/10.1080/13683500.2013.850064

Ferrer-Rosell, B., Coenders, G., & Martínez-Garcia, E. (2015). Determinants in tourist expenditure composition – The role of airline types. *Tourism Economics*, *21*(1), 9–32. https://doi.org/10.5367/te.2014.0434

FGD. (2020). *Perencanaan Pariwisata Daerah Pasca COvid-19.*

Fiedler, K. (2007). Construal level theory as an integrative framework for behavioral decision-making research and consumer psychology. *Journal of Consumer Psychology, 17*(2), 101–106. https://doi.org/10.1016/S1057-7408(07)70015-3

Fletcher, J., & Morakabati, Y. (2008). Tourism Activity, terrorism and political the cases of Fiji and Kenya. *International Journal of Tourism Research, 556*(November), 537–556. https://doi.org/DOI: 10.1002/jtr.699

Gössling, S., Scott, D., & Hall, C. M. (2020). Pandemics, tourism and global change: A rapid assessment of COVID-19. *Journal of Sustainable Tourism, 0*(0), 1–20. https://doi.org/10.1080/09669582.2020.1758708

Hasan, Al. (2008). *Marketing*. Media Presindo.

Haywood, K. M. (2020). A post-COVID future: tourism community re-imagined and enabled. *Tourism Geographies, 0*(0), 1–11. https://doi.org/10.1080/14616688.2020.1762120

Henderson, J. C. (2005). Responding to natural disasters: Managing a hotel in the Aftermath of the Indian Ocean Tsunami. *Tourism and Hospitality Research, 6*(1), 89–96. https://doi.org/10.1057/palgrave.thr.6040047

Holzner, M. (2011). Tourism and economic development: The beach disease? *Tourism Management, 32*(4), 922–933. https://doi.org/10.1016/j.tourman.2010.08.007

Huang, C., Wang, Y., Li, X., Ren, L., Zhao, J., Hu, Y., Zhang, L., Fan, G., Xu, J., Gu, X., Cheng, Z., Yu, T., Xia, J., Wei, Y., Wu, W., Xie, X., Yin, W., Li, H., Liu, M., ...Cao, B. (2020). Clinical features of patients infected with 2019 novel coronavirus in Wuhan, China. *The Lancet, 395*(10223), 497–506. https://doi.org/10.1016/S0140-6736(20)30183-5

Ioannides, D., & Gyimóthy, S. (2020). The COVID-19 crisis as an opportunity for escaping the unsustainable global tourism path. *Tourism Geographies, 0*(0), 1–9. https://doi.org/10.1080/14616688.2020.1763445

Kim, Jaewook, Kim, J., Ki Lee, S., & Tang, L. (2020). *Effects of epidemic disease outbreaks on financial performance of restaurants: Event study method approach.* https://doi.org/10.1016/j.jhtm.2020.01.015

Kim, Jiyeon, & Forsythe, S. (2007). Hedonic usage of product virtualization technologies in online apparel shopping. *International Journal of Retail and Distribution Management, 35*(6), 502–514. https://doi.org/10.1108/09590550710750368

Kim, M. J., & Hall, C. M. (2019). A hedonic motivation model in virtual reality tourism: Comparing visitors and non-visitors. *International Journal of Information Management, 46*(July 2018), 236–249. https://doi.org/10.1016/j.ijinfomgt.2018.11.016

Lapointe, P. (2011). The rock in the pond: How online buzz and offline WOM can make a strong message even more powerful. In *Journal of Advertising Research* (Vol. 51, Issue 3, pp. 456–457). Journal of Advertising Research. https://doi.org/10.2501/JAR-51-3-456-457

Miao, L., & Mattila, A. S. (2013). The impact of other customers on customer experiences. *Journal of Hospitality & Tourism Research, 37*(1), 77–99. https://doi.org/10.1177/1096348011425498

Millie, M. B. (2014). *New Research: Topics That Get A Reaction -Facebook vs. Twitter – How Does Engagement Differ On Each?* https://medium.com/@aggiemeesh/new-research-topics-that-get-a-reaction-6e2448d385aa

Nepal, S. K. (2020). Travel and tourism after COVID-19 - business as usual or opportunity to reset? *Tourism Geographies, 0*(0), 1–5. https://doi.org/10.1080/14616688.2020.1760926

Nicula, V., & Onețiu, A.-N. (2016). Dimitrie Cantemir. *Christian University Knowledge Horizons-Economics, 8*(1), 196-200. www.orizonturi.ucdc.ro

Purwomarwanto, Y. L., & Ramachandran, J. (2015). *Performance of tourism sector with regard to the global crisis-a comparative study between Indonesia, Malaysia and Singapore* (Vol. 49, Issue 4).

Rosselló, J., Becken, S., & Santana-Gallego, M. (2020). The effects of natural disasters on international tourism: A global analysis. *Tourism Management, 79*(December 2019). https://doi.org/10.1016/j.tourman.2020.104080

Seyidov, J., & Adomaitienė*, R. (2017). Factors influencing local tourists' decision-making on choosing a destination: A case of Azerbaijan. *Ekonomika, 95*(3), 112–127. https://doi.org/10.15388/ekon.2016.3.10332

Tan, W. K. (2018). From fantasy to reality: a study of pre-trip planning from the perspective of destination image attributes and temporal psychological distance. *Service Business, 12*(1), 65–84. https://doi.org/10.1007/s11628-017-0337-6

Taylor, S., Landry, C. A., Paluszek, M. M., Fergus, T., McKay, D., & Asmundson, G. J. G. (2020). COVID Stress Syndrome: Concept, structure, and correlates. *Depression and Anxiety, May*, 1–9. https://doi.org/10.1002/da.23071

Warrington, T. (2002). The secrets of word of mouth marketing: How to trigger exponential sales through runaway word of mouth. *Journal of Consumer Marketing, 19*(4), 364–366. https://doi.org/10.1108/jcm.2002.19.4.364.4

Promoting Creative Tourism: Current Issues in Tourism Research – Kusumah et al. (Eds)
 ISBN 978-0-367-55862-8

Travel decision-making amid the pandemic

G.R. Nurazizah & Darsiharjo
Universitas Pendidikan Indonesia, Bandung, Indonesia

ABSTRACT: Traveling is a basic human need driven by curiosity to explore new environments and discover new experiences. Comfort and safety are vital in determining one's decisions to travel. Before going on a tour, a prospective tourist performs a mental process to decide the time, duration, place, and transportation mode, and so on. During the Covid-19 pandemic, the global safety factor became threatened and forced changes in all types of people interaction, including in tourism and traveling aspects. Thus, this study aimed to measure the impact of the pandemic on travel decision-making. A total of 382 questionnaires were distributed via Google Forms April–May 2020 to the Indonesians. The study used a simple linear regression test and was assisted by SPSS version 20. The test showed astounding results; for Indonesian people, the influence of this pandemic on the travel decision-making is at a low level.

Keywords: travel decision, pandemic, tourists' behavior

1 INTRODUCTION

Traveling has become a basic human need (Di Ciommo et al. 2018; Tasci & Ko 2017). If these needs are unfulfilled, then physical and psychological health-related problems can arise and threaten a person's quality of life (Lubin & Deka 2012). Through outdoor recreation activities, a person can satisfy needs in gaining new experiences, increasing the insight of a specific destination's attributes, or even only by exposing themselves to the sun and breathing fresh air. The need magnitude depends on the motivation, which is diverse for each individual (Sari et al. 2014). Thus, the decision-making in choosing tourism activities is an individual decision, not the group decision (McCabe 2009).

Besides being pushed by motivation, individual decisions are determined by several factors such as socio-economic level, age, gender (Di Ciommo et al. 2018), destination readiness, and attractiveness (Rahmafitria et al. 2016), security and safety, product quality, and product price (Hsu et al. 2017). Nowadays, the main concepts of how people plan to travel changed steadily to more accurate planning (Di Ciommo et al. 2018). Once there is a restraining factor that arises, the results of decision-making will be changed.

Security and safety are vital in determining travel decisions. This aspect has become a more significant issue in the last two decades and has considerable impacts on the continuation of tourist activities (Khalik 2014). Some researchers believe that terrorism, local conflicts, politics, natural disasters, and global health issues can cause a decline in people's intention to travel (Assaker & O'Connor 2020; Kim 2019). The higher the threats, the more likely people cancel their trip.

Recently, the outbreak of Covid-19 shocked the world, which none have found the vaccine. The disease was rapidly spread cross-continental. As a result, several countries regulated strict travel policies, such as limiting domestic and foreign flights (European Commission 2020; Zaharah 2020). This decision is undoubtedly controversial. The pros saw social restrictions as a mitigation effort, but critics saw it as harm in upholding human freedom. The effectiveness of this regulation has not been proved (Linka et al. 2020).

The social restriction has severely affected the tourism sector. Many countries, such as Indonesia, issued policies to close tourist attractions temporarily. However, this policy is not binding and

 DOI 10.1201/9781003095484-86

does not have proper consequences if violated. It was creating various interpretations for the tour operators and the community. As a result, some travel agents and airlines were still doing their business, even giving a lower price for their customers. Many Indonesians also underestimate and continue to visit various tourist attractions. The complexity of this behavior is interesting to study because it contrasts the theories about the impact of safety issues on tourism. Therefore, this study aimed to assess people's decision-making to travel during the pandemic.

2 LITERATURE REVIEW

Covid-19 is a disease caused by a new coronavirus strain (WHO 2020). The virus transmits through direct contact from the spark of the infected person (coughing or sneezing) that enters through the salivary or tear ducts. The symptoms are similar to colds or coughs. In more severe cases, the infection can cause pneumonia, breathing difficulty, and even death. Unfortunately, the vaccine hasn't been found yet. Although the virus is highly contagious, it can be killed easily using ordinary disinfectants. The ways to prevent and slow the virus spread can be done by maintaining personal hygiene, maintaining ethics in sneezing/coughing, and maintaining physical distance (WHO 2020).

The tourism sectors rely much on social interaction and physical closeness in delivering their services (Minnaert et al. 2009). For instance, when the tour guide accompanies tourists, the waiter offers food, or the hoteliers serve guests. All those activities require close physical interaction. When social restrictions become compulsory, then the tourism sectors should close all the businesses. This limit not only makes people difficult to move for recreation purposes but also business purposes (Benabou 2000). Thus, people are forced to withhold their need to have a direct interaction with other people, nature, or other recreational facilities.

Each individual has his/her different strengths and motivation in obeying the restriction. Some are resistant, but some are weak. Motivation is a "trigger" of the tour process, which can be supported by individual push factors and the destination pull factors (Pitana & Gayatri 2005). The motivation then drives a prospective tourist to decide the travel plan, such as determining the time for travel, duration, location, and how to arrive at a tourist destination. This decision is made in three critical phases (Pitana & Gayatri 2005), namely, the before (when needs arise, information seeking, decision making), during (consumption of accommodation products, transportation, attractions, food), and the after phase (evaluation and determination of subsequent attitudes).

As safety and security factors affect a decision to travel, the pandemic could be one of the factors that ultimately holds someone to have a tour. However, the attitude uniqueness of individuals, indeed, creates different effects on decision-making when deciding to have a trip during a pandemic. It is essential to examine the role of Covid-19's knowledge in affecting travel decision-making. Thus, the hypothesis to be proven is as follows:

H0: knowledge of pandemic does not affect the travel decision-making
H1: knowledge of pandemic affects the travel decision-making

3 RESEARCH METHOD

This study aimed to prove the effect of Covid-19 knowledge on travel decision-making. This study used descriptive methods to assess the knowledge about Covid-19 and its impact on the decision for traveling. A total of 382 questionnaires were distributed via Google Forms from April to May 2020. The respondents were Indonesian, with a minimum age of 17 years old. The questionnaires contained questions about respondents' characteristics, knowledge about Covid-19, and perception in decision-making for traveling amid the pandemic. The researcher analyzed the data with a simple linear regression test with the help of SPSS version 20. The acceptability of the hypothesis was decided by the calculation of t_{value} compared with the t_{table} (1.96). If the t_{value} is higher than the t_{table}, or negative t_{value} is less than the t_{table} (−1.96), then the hypothesis is acceptable.

Table 1. The regression coefficient of attitudes toward Covid-19.

	Unstandardized Coefficients			
	B	Std. Error	t	Sig.
(Constant)	2.798	.254	11.016	.000
Knowledge of Covid-19	.415	.058	7.121	.000

Table 2. The regression coefficient of travel decision-making during the pandemic.

	Unstandardized Coefficients			
	B	Std. Error	t	Sig.
(Constant)	2.441	.273	8.942	.000
Knowledge of Covid-19	−.259	.063	−4.134	.000

4 RESULTS AND DISCUSSION

The selected respondents in this study have various profiles. Resident domiciles spread across 13 provinces in Indonesia, with the majority coming from West Java (30.4%) and varying age ranges from early adolescents to early elderly, dominated by 17–25 years (50%).

The perception of someone's understanding of an issue will affect the attitude toward the problem (Cheng & Wu 2015). In this study, the indicators of disease characteristic understanding, the causes and transmission, and the way to anticipate formed the Covid-19 issue knowledge. Actions measured the attitude during the outbreak, such as following the government's call, maintaining personal hygiene, and supporting the closure of tourist attractions during the outbreak. Regression test results show that the perception of understanding related to Covid-19 has a significant positive effect on a person's attitude. The higher the knowledge, the higher the awareness and attitudes in responding to the outbreak (see Table 1).

In this study, respondents perceive themselves to understand about Covid-19; their understanding is at a very high level (score 4.31). The high value of perceived knowledge is natural because they can access information and news exposure from various mass and electronic media. Moreover, the respondents in the late teens' range have a screen time almost 24/7, and it is natural for them to feel fully aware of this disease. With this understanding, the respondent's attitude toward government instruction and medical direction is considered very concerned (score 4.52). The actions are demonstrated by following the government's call for social distancing, maintaining personal hygiene, and cleanliness. This attitude is also indicated by supporting the policy to close tourist destinations temporarily.

Some theories reveal that traveling decision-making is a very complicated process and influenced by many factors, one of which is in the "before phase," when someone collects information about the overall situation and conditions for traveling (Hamilton & Lau 2005). The results of the analysis show that understanding the Covid-19 issue has a significant negative effect on travel decision-making (see Table 2, $t_{value} < -1.96$, sig. 0.000). The higher the understanding related to the issue is, the lower the tour decisions made during a pandemic, and vice versa.

With the knowledge, respondents tend to decide not to travel during a pandemic. Their decision to have a tour during Covid-19 is at a very low level (score 1.29). Even when cheap ticket promotions are available, the decision remains very low (score 1.55). For them, the economic benefits are not higher than the importance of maintaining health to avoid being exposed to the disease. This

Table 3. Chi-Square Test.

	Value	df	Asymp. Sig. (2-sided)
Pearson Chi-Square	.795[a]	1	.373

Table 4. The regression coefficient of travel decision-making after the pandemic.

	Unstandardized Coefficients			
	B	Std. Error	t	Sig.
4.368	.636	6.869	.000	
Knowledge of Covid-19	−.210	.146	−1.442	.151

finding reinforces the statement of Law (2006) that the perceived risk outweighs the benefit factor, including the economy when making travel decisions.

In this study, the respondents' backgrounds were also identified, specifically related to whether they already had a planned tour before the outbreak or not. Researchers assumed that the experience could affect travel decision-making during a pandemic. The results showed that out of 382 respondents, 54% already had vacation plans long before the epidemic. However, the chi-square test shows that there is no relationship between vacation plans and the decision to continue the plan to travel amid the pandemic (see Table 3). This finding corroborates the statement of Li et al. (2011) that safety and security aspects in the health context have an essential role in travel decision-making.

The decision-making process will be different if world health conditions have improved. Statistical test results show that there is no influence between the understandings of a pandemic related to the decision to travel after the outbreak ends (see Table 4). The regression test shows the t_{value} is smaller than t_{table}, with sig. value $0.151 > 0.05$.

When the pandemic ends, the decision to travel will return to the initial template before the outbreak. Of all respondents, 77% decided to return to travel, while the rest did not intend to travel immediately. This action, of course, can occur for various factors that influence the complexity of decision-making in tourism. These factors provide space for further research related to the decision-making process after a pandemic.

5 CONCLUSION

Understanding the threat of disease will negatively affect the decision to travel. Indonesian people have understood and are aware of the dangers of Covid-19. They also have tried to comply with regulations set by the authorities such as social distancing policies and other health protocols. In making travel decisions, the community is very concerned about matters relating to the transmission of Covid-19. Most people will take a tour after the outbreak ends, but they are reluctant to travel to places that are profoundly affected by Covid-19. Besides, people will better maintain their hygiene if they go on trips after the pandemic ends.

REFERENCES

Assaker, G. & O'Connor, P. 2020. eWOM platforms in moderating the relationships between political and terrorism risk, destination image, and travel Intent: The Case of Lebanon. *Journal of Travel Research*, p.0047287520922317.

Benabou, R. 2000. Unequal societies: Income distribution and the social contract. *American Economic Review* 90(1):96–129.
Cheng, T.M. and Wu, H.C., 2015. How do environmental knowledge, environmental sensitivity, and place attachment affect environmentally responsible behavior? An integrated approach for sustainable island tourism. *Journal of Sustainable Tourism*, 23(4):557–576.
Di Ciommo, F., Pagliara, F. and De Crescenzo, M., 2018. Need-based travel behavior analysis: new potential for mobility survey. *Transportation research procedia* 32: 110–118.
Hamilton, J.M. & Lau, M.A. 2005. The role of climate information in tourist destination choice decision making. *Tourism and global environmental change*:229.
Hsu, S.C., Lin, C.T. & Lee, C. 2017. Measuring the effect of outbound Chinese tourists travel decision-making through tourism destination image and travel safety and security. *Journal of Information and Optimization Sciences* 38(3–4):559–584.
Khalik, W. 2014. Kajian Kenyamanan dan Keamanan Wisatawan di Kawasan Pariwisata Kuta Lombok.
Kim, N. 2019. Q-methodology analysis of Perceived Risks in Tourists and Local residents towards Natural Disaster: The 2016 Gyeongju Earthquake in Republic of Korea.
Law, R. 2006. The perceived impact of risks on travel decisions. *International Journal of Tourism Research* 8(4):289–300.
Li, J., Liu, H., Chen, J., Xue, Q. and Bao, Y., 2011, August. Travel health and health tourism: A perspective of the subject of tourism. In *International Conference on Advances in Education and Management*:56–63. Springer, Berlin, Heidelberg.
Linka, K., Peirlinck, M., Sahli Costabal, F. & Kuhl, E. 2020. Outbreak dynamics of COVID-19 in Europe and the effect of travel restrictions. *Computer Methods in Biomechanics and Biomedical Engineering*:1–8.
Lubin, A. & Deka, D. 2012. Role of public transportation as job access mode: lessons from survey of people with disabilities in New Jersey. *Transportation Research Record: Journal of the Transportation Research Board* 2277:90–97.
McCabe, S. 2009. Who needs a holiday? Evaluating social tourism. *Annals ofTourism Research* 36(4):667–688.
Minnaert, L., Maitland, R. & Miller, G. 2009. Tourism and social policy: The value of social tourism. *Annals of Tourism Research*, 36(2):316–334.
Pitana, I.G. & Gayatri, P.G. 2005. *Sosiologi Pariwisata: Kajian sosiologis terhadap struktur, sistem, dan dampak-dampak pariwisata*. Yogyakarta: Andi.
Rahmafitria, F., Nurazizah, G.R. and Riswandi, A., 2016. Attraction and destination readiness towards tourists' intention to visit solar eclipse phenomenon in Indonesia. Heritage, Culture and Society: Research agenda and best practices in the hospitality and tourism industry, p. 293.
Sari, D., Kusumah, A.H.G. & Marhanah, S. 2014. Analisis Faktor Motivasi Wisatawan Muda Dalam Mengunjungi Destinasi Wisata Minat Khusus. *Journal of Indonesian Tourism, Hospitality and Recreation,* 1(2):11–22.
Tasci. D.A & Ko, Y.J, 2017. Travel needs revisited 1(2):162–183
WHO. (2020). Pesan dan Kegiatan Utama Pencegahan dan Pengendalian Covid-19 di Sekolah. Pesan Dan Kegiatan Utama Pencegahan dan Pengendalian COVID-19 di Sekolah.
Zaharah, Z., Kirilova, G.I. & Windarti, A. 2020. Impact of Corona Virus outbreak towards teaching and learning activities in Indonesia. *SALAM: Jurnal Sosial dan Budaya Syari,* 7(3):269–282.

Promoting Creative Tourism: Current Issues in Tourism Research – Kusumah et al. (Eds)
 ISBN 978-0-367-55862-8

Tourism industry standard operating procedure adaptation preparing Covid-19 new normal in Indonesia

I.I. Pratiwi & A. Mahmudatussa'adah
Universitas Pendidikan Indonesia, Bandung, Indonesia

ABSTRACT: Covid-19 pandemic is a disaster that has an impact on providing challenges to adapt to the tourism industry in the world, including in Indonesia. This study examines how the tourism industry in Indonesia adapts by managing tourism businesses that follow health protocols, one of which is by making changes to standard operating procedures. In-depth interviews were conducted on 23 tourism businesses in Indonesia with a business background in the fields of accommodation, restaurants, tourist attractions, and travel (including tourism transportation businesses). This study found that hygiene and safety factors in standard operating procedures were carried out by the tourism industry in Indonesia by adopting several regulations, including WHO-UNWTO protocols, government policies, and industry agreements.

Keywords: Tourism Management, Standard Operating Procedure, Covid-19, New Normal

1 INTRODUCTION

Since the outbreak of Covid-19 in China at the end of 2019, the world has been rocked and tourism has become the most affected industry (Hoque et al. 2020). Tourism that involves human movement and traveling increases the risk of Covid-19 transmission between countries (Khadka et al. 2020). During the peak of the pandemic in March-May 2020, many countries traveled banning and closed gates to prevent Covid-19 transmission. Indonesia, as a country which is one of the pillars of its economy supported by tourism, issued several policies that hinder tourism growth, one of which is a social distancing policy. Social distancing is the antithesis of all that tourism represents. Countries in the world take this policy to prevent or mitigate the spread of the virus. Quarantine, border closure and travel restrictions, workplace hazard control, and closure of facilities. (Carbone 2020) This policy in Indonesia is called Large-Scale Social Restrictions. The policy is also followed by other policies such as Stay at Home, School From Home, and Work From Home, which one of the points prohibits students from conducting field trip tours and employees are prohibited from traveling on business. The consumption of the tourism industry has declined due to the closure of tourist attractions, malls, and the prohibition of activities that are gathering mass. Some countries even take lockdown policies that close the entrances. (Kumar et al. 2020). In Indonesia, during the Covid-19 pandemic, the decline in foreign tourists reached 64% (Kemenparekraf 2020)

The first peak of the Covid-19 pandemic has passed, but the threat of Covid-19 persists because the vaccine and drug Covid-19 have not been found yet. However, the social restriction policy cannot be continuously taken because it can result in the impact of other snowball crises such as economic crises that lead to social crises. Slowly, several countries took steps to adapt, including in Indonesia. The opening of the activity tap is carried out in stages, and one of them is by allowing the tourism sector to return to operations with the adaptation of new habits. Adaptation of this new habit, in the management of the tourism industry is contained in company documents, one of which is standard operating procedure documents. Adaptation is carried out for example for restaurant businesses that take a new approach to keep operating and serve consumers who stay at home, while air and sea travel stop operations, sporting events, concerts, MICE has been revoked. (Mandabach

DOI 10.1201/9781003095484-87

2020). This study aims to know how the standard operating procedures adapt in tourism industries due to the Covid-19 situation: in accommodation services, in travel services, in restaurant services and also in tourist attractions business.

2 LITERATURE REVIEW

Covid-19 is classified as a health disaster. So far, Covid-19 is an extraordinary event whose impact is felt by almost all countries in the world. Covid-19 transmission is believed to be transmitted between people through droplet transmission primarily when coughing and sneezing when they hold mouth, nose, or eyes. Prevention measures include regular and thorough hand hygiene, social distancing, avoiding touching eyes, nose, and mouth, and good respiratory hygiene. In public areas, cleaning and disinfecting frequently touched objects and surfaces can help reduce the risk of infection (WHO 2020a–2020e).

Post Covid-19, there is a change in demand in the tourism industry by −20% (minus twenty percent) (Bakar & Rosbi 2020). The critical effect of the coronavirus on the human body that is led to severe Pneumonia has grown significant fear among the people (Hoque et al. 2020). The public's tightness to travel, the limitation of travel restrictions, travel restrictions, and quarantine harm the tourism sector including the cancellation of the hotel, flight, cruise ship reservations, museum closures, and tourist attractions. (Carbone 2020). The economic impact of tourism that should arise from the implementation of international transactions, the impact of the economic multiplier from the operation of tourism, increased employment and community income, and investment in tourism, will be disrupted because of the Covid-19 pandemic (Haryanto 2020). This condition will have an impact on the economy which is increasingly declining and will have a social impact because of the many disruptions and new poor people

Policies that need to be taken relating to tourism management in Indonesia must regulate all aspects related to social distancing, especially for religious tourism, adventure tourism, agricultural tourism, and MICE. The transportation system must also be managed by considering social distancing in the origin and destination of tourists, boarding and disembarking space, food service, activities in the accompanying modes of transportation and travel, improving medical facilities in transportation modes, ports, and terminals, and arranging seating arrangements in modes transportation. Travel and entrance restrictions on the number of tourists must be applied to domestic and international destinations. Personal protection equipment will be needed for this type of medical tourism, air travel, and sea travel. Safety and health must be carried out at the destination by implementing periodic controls and controls to control the spread of viruses and diseases. Make indicators to identify the possibility of re-emergence of viruses and further infections. Re-managing work health and safety for residents, workers, and tourists. Carry out mass tests for tourists. Policies related to the crisis must also look at the market demand for tourists, prioritize tourist segmentation, ensure connectivity and cooperation between travel distributors, potential tourists, and destinations, re-start tourist activities to maximize economic, social, l and environmental impacts, and minimize negative impacts on the tourism economy. Tourism events should be reviewed for their implementation especially for sports events, music shows, theaters, concerts, exhibitions and conferences. Accommodation services must also apply social distancing at locations of public facilities such as lobbies, restaurants, food service, and other social activities such as in swimming pools, gyms, and spas. The tourism industry also needs to have other approaches such as the implementation of premium insurance and handling of the possibility of diseases that arise in tourism locations. (Chang et al. 2020)

The policy in the tourism sector carried out by the government of the Republic of Indonesia, namely, the provision of medical facilities and supporting staff for the Covid-19 Referral Hospital handling, coordination with relevant ministries / institutions, proposes various programs that can provide relief for industry and tourism workers and the creative economy sectors, granting subsidies for exemption and tax reduction, empowerment programs for Medium Small Enterprisess in the tourism sector and the creative economy, data collection programs for workers in the tourism sector and creative economy affected by Covid-19, tourism Small Medium Enterprise and regional creative economy programs to prepare food for the affected, and propose a campaign to advance the archipelago coffee, the Tourism Ministry also cooperates with economic actors to strengthen the campaign and communication related to social distancing, online training programs to increase the expertise of the tourism and creative economy actors.

The impact of extraordinary events such as the Covid-19 health disaster will primarily have an economic, social, and environmental impact. In the short term, it will be negative, but in the long run, it will be positive again. Tourism businesses as suppliers and tourists as consumers will adapt in-action, while the interaction of external agencies such as the government, trade unions, and workers associations can also influence the demand and supply of the final product. Intermediate products and factors of production are also needed to formulate new product formulas (Papanikos 2020). This has led to the adoption of new habits in the new normal era of the Covid-19 Pandemic, which will bring novelty to products and travel services.

The tourism industry must continue to position the safety of tourism actors and tourists. Organizers of tourism including hotels, restaurants, travel, and event organizers must have procedures to manage sanitation and safeguard health and safety. Physical collateral is carried out on surfaces that allow contact, such as table surfaces, door handles, toilets, point of sales equipment, keyboards, and ensuring a sanitized environment. Hand washing and the use of hand sanitizers are personal responsibilities to ensure safety and health, especially after touching the surface, holding money, or credit cards. If the employee is not very fit, then he must contact a medical professional, get treatment, and avoid the worksite until he is healthy. (Mandabach 2020). One approach to a recovery strategy for posCOVIDid-19 tourism is to allow people who already have antibodies to Covid-19 to be free to travel with an awareness of their health status. Airlines, hotels, spas, and their tourism businesses must provide these groups with discounts and package offers. (Strielkowski 2020) Management of the consumer database also becomes important during the new normal Covid-19. Some tour operators offer packages for future tours with guaranteed longer time and prices to follow the fixed current rate. (Shresheva 2020)

A Standard Operating Procedure (SOP) is a set of written instructions that document a routine or repetitive activity followed by an organization. The development and use of SOPs are an integral part of a successful quality system as it provides individuals with the information to perform a job properly and facilitates consistency in the quality and integrity of a product or end-result. The term "SOP" may not always be appropriate and terms such as protocols, instructions, worksheets, and laboratory operating procedures may also be used. The SOP is structured to equalize the perceptions and standards of work performed within a company. SOP can minimize miscommunication so that work can be completed effectively and efficiently. SOPs are used not only to produce goods but are also used for service standards. Where often SOPs in service companies are called Service Blueprint.

The Covid-19 condition requires tourism companies to be able to adapt to health protocols issued by WHO. The tourism sector is fully committed to putting people and their well-being first. International cooperation is vital for ensuring the sector can effectively contribute to the containment of COVID-19. UNWTO and WHO is working in close consultation and with other partners to assist States in ensuring that health measures are implemented in ways that minimize unnecessary interference with international traffic and trade. (WHO 2020a–2020e).

3 RESEARCH METHODOLOGY

This study conducts a literature study on Covid-19 policy analysis that has an impact on the implementation of tourism as part of the preparation of standard operating procedures in tourism businesses in Indonesia. In-depth interviews were conducted with 23 tourism business operators in Indonesia during March-May 2020, both in the fields of accommodation services, food supply services, travel services, and tourist attraction services. The aim is to determine the adaptation of standard operating procedures in the tourism business to prepare for the new normal era of Covid-19 (Figure 1).

	Covid-19 Protocol for Accomodation			
	Criteria	No	Indikator	Weight
A	Management Team	1	Action plan	0.167
		2	Mobilisation of resources	0.167
		3	Supervision	0.167
		4	Logbook of actions	0.167
		5	Communication	0.167
		6	Training and information	0.167
B	Reception and Tour Staff	7	Information and communication	0.250
		8	Necessary equipment and medical kit	0.250
		9	Social distancing measures, hand cleaning, and respiratory hygiene	0.250
		10	Monitoring of guests who are possibly ill	0.250
C	Technincal Maintenance	11	Water disinfection	0.250
		12	Dishwashing and Laundry Equipment	0.250
		13	Air-conditioning	0.250
		14	Dispensers	0.250
D	Food Services	15	Information and communication	0.250
		16	Buffets and drinks machines	0.250
		17	Washing dishes, silverware, and table linen	0.250
		18	Table setting	0.250
E	Recreational Area	19	Closing area	0.333
		20	Cleaning and disinfection	0.333
		21	Necessary equipment and medical kit	0.333
F	Cleaning and Housekeeping	22	Cleaning and disinfection	0.250
		23	Monitoring of sick guests	0.250
		24	Availability of Covid-19 supporting materials	0.250
		25	Optional housekeeping programmes	0.250
G	Implementing Enabling Policy	26	Travel Insurance Policy	0.200
		27	Travel Guarantee Policy	0.200
		28	Price Policy	0.200
		29	Visa and Visitor Policy	0.200
		30	Insentif for Industry Policy	0.200
			Total	7.000

Figure 1. Covid protocol for accomodation.

This study makes measurements of criteria and protocol indicators for the implementation of tourism which is compiled based on protocols made by the World Health Organization (WHO), the United Nations World Tourism Organization (UNWTO), and the World Tourism & Travel Council (WTTC). There are 6 criteria, namely Management Team, Front Liner Operational Services, Technical Maintenance, Hospitality Services, Recreational Services, Cleaning Services, and Implementing Enabling Policy. The highest SOP adaptation value will produce number 7. The SOP adaptation classification to the Covid-19 Protocol is carried out by weighting. If the adaptation is good the number> = 6, if the adaptation is quite good the number> = 4, if the adaptation is not good then the number <4. This criterion is used to analyze the entire tourism sector, with different approaches to the indicators. But through weighting, the maximum rating will remain the same as 7.

4 RESULTS

4.1 *Tourism S.O.P Covid-19 adaptation in accommodation services*

The hotel industry is one of the industries that experience the greatest threat from extraordinary events such as pandemics, natural disasters, and terrorist attacks (Chan & Lam 2013; Chen et al. 2011). Different disasters give the industry the consequence of having to adapt to managing threats into challenges. Some hospitality industries utilize Artificial Intelligence (AI) technology and its applications such as robotics. (Zabin 2019). The application of hygiene and cleanliness is important for good hotel operations, this issue becomes a focal point to ensure safety related to the health of tourists and employees (Zhou et al. 2020). Hotels and tourism accommodation establishments are no more susceptible to contagion than other public establishments visited by large numbers of people who interact among themselves and with employees. Nevertheless, they are places where guests stay temporarily in close cohabitation and where there is a high degree of interaction among guests and workers (WHO 2020a–2020e).

Some service protocol approaches at the hotel include opening a health service in the form of a mini-clinic in the hotel, hotel guests and employees who show symptoms and are at risk because of the possibility of contact, and visiting the red zone area is recommended to conduct self-monitoring at body temperature. Guests who show no symptoms are allowed to access public spaces such as restaurants and other hotel facilities. Employees and guests are required to wear masks and keep a personal distance while maintaining hand hygiene. Hotel employees also use Personal Protective Equipment such as face masks and hand gloves (Hoefer et al. 2020).

This study interviewed general managers, supervisors, and front liner staff from hotel and lodging business operators in Indonesia. In-depth interviews were conducted to dig up information about the adoption of SOPs and how SOPs adapted to Covid-19 conditions. The results of interviews processed in the weighting table give a value of 6,033. This value means that the adaptation of accommodation service S.O.P during the Covid-19 period is in a good category.

4.2 *Tourism S.O.P Covid-19 adaptation in restaurant services*

To prevent Pandemic Covid-19 from limiting public interaction, for the restaurant industry some of the policies include limiting restaurant operating hours, prohibiting eating at places, and even prohibiting opening restaurants in the Large-Scale Social Restrictions. A study of fine dining restaurants in Germany gives some of the behavioral changes to cooks, restaurant service employees, restaurant owners, and restaurant visitors. The average restaurant can survive a maximum of around 9.5 weeks and if social restrictions are implemented in the long run, many restaurants will experience financial difficulties and face bankruptcy (Wilekesmann & Wilkesmann 2020).

Standard operating procedures relating to restaurants and food services, which were issued during the Conference for Food Protection, produced several monitoring targets, including those on surfaces, equipment, and machinery that directly or indirectly contact food, both those located in warehouses, kitchens, and restaurants. SOPs related to food will review several things including the applicable regulations, disaster prevention, improve food quality, reduce costs and energy in operations, and improve quality assurance. (Marriott 2020).

This study interviewed managers, supervisors, and front liner staff from restaurant operators in Indonesia. In-depth interviews were conducted to gather information about the application of SOPs and how SOPs adapted to Covid-19 conditions. The results of interviews processed in the weighting table give a value of 3.96, which means that the adaptation of restaurant service S.O.P during the Covid-19 period is not a good category.

4.3 *Tourism S.O.P Covid-19 adaptation in travel services*

The traveling agencies are seen to be impacted by COVID-19 as the lockdown situation is established (Hoque et al. 2020). The Covid-19 condition creates disruption that has an impact on unemployment in countries requiring an adaptation process, especially on the overall stakeholder and tourism supply chain. The transport sector (planes, trains, tolls, ships, taxi), accommodation (hotels, lodges, guest houses, home-stays, and other regular and temporary accommodation), restaurants and other eateries, travel agents, tour operators, and large numbers of small Vendors will provide services to rural and remote tourist destinations throughout the country. (Kumar et al. 2020).

This study interviewed managers, supervisors, and front liner staff from road service providers in Indonesia. In-depth interviews were conducted to gather information about the application of SOPs and how SOPs adapted to Covid-19 conditions. The results of interviews processed in the weighting table give a value of 4.05, which means that the adaptation of accommodation service managers during the Covid-19 period is quite a good category.

4.4 *Tourism S.O.P Covid-19 adaptation in tourist attraction*

The coronavirus provides the condition for the tourism industry facing the greatest threat to the economic slowdown. It faces a tough time which is affecting the countries. The country's image is seen to be more important than earning money. Thus, it is seen that the tour operators' association is stopping their activities for prohibiting the spreading of the coronavirus (Hoque et al. 2020). The Covid-19 pandemic caused many trips, which resulted in a decrease in the number of tourists. The airline industry is also one that experiences the effects of travel bans policies and decreases travel demand. The condition of Covid-19's health crisis is even worse than the period of security crisis caused by terrorism (Lee-Peng et al. 2020).

This study interviewed managers, supervisors, and front liner staff from organizers of tourist attractions in Indonesia. In-depth interviews were conducted to gather information about the application of SOPs and how SOPs adapted to Covid-19 conditions. The results of interviews processed in the weighting table give a value of 2.83 which means that the adaptation of accommodation service managers during the Covid-19 period is not good.

5 CONCLUSION

In this study conducted a study of literature relating to the management of the tourism industry during the pandemic and the new normal period with a review approach to standard operating procedures. In terms of Indonesian government policy, until this research was made, there has not yet been a tourism management protocol published in the framework of adapting Covid-19's new habits. The absence of a protocol has also impacted the absence of legal and policy products that oversee and monitor tourism management in Indonesia to deal with the condition of Covid-19.

The industry moves ahead of the government, where the Covid-19 protocol adaptation is carried out directly by the industry by adapting policies from the Ministry of Health and the Covid-19 task force. The best adaptation is done by industries that serve accommodation, with an average protocol adaptation rate of 6.03. While the lowest adaptation is done by industries that serve tourist attractions with an average number of protocol adaptations of 2.83.

Among the indicators used in this study, the most powerful and widely adopted Covid-19 indicator is the management team indicator, while the least adapted indicator is the indicator relating to company policy that can provide flexibility for visitors. It means the adaptation of S.O.P tourism companies is still limited to forming teams and managing resources. Not yet on using S.O.P policies that can increase the volume of transactions and purchases of tourism service businesses.

This study also has limitations, including generalizations that cannot be made because the number of analytical subjects is still very small (only 22 objects of observation), which allows the final average value of SOP adaptation of tourism management to the Covid-19 protocol to change if a study with a scale is carried out greater. This study can change if there are conditions that also change, for example, the condition of Covid-19 which is getting worse or the occurrence of force majeure which causes the need for more adaptation to the SOP of tourism businesses.

REFERENCES

Bakar, Nashirah Abu, Sofian Rosbi. (2020). *Effect of Coronavirus Disease (Covid-19) to the Tourism Industry.* International Journal of Advanced Engineering Research and Science (IJAERS) Vol-7, Issue-4, Apr-2020. https://dx.doi.org/10.22161/ijaers.74.23

Carbone, Fabio. (2020). *Tourism Destination Management Post Covid-19 Pandemic: A New Humanism for a Human Centred Tourism (Tourism 5.0).* World Tourism, Health Crisis, and Future: Sharing Perspectives.

Chan, E.S., and Lam, D. (2013). *Hotel safety and security systems: Bridging the gap between managers and guests*, International Journal of Hospitality Management, Vol. 32, pp. 202–216.

Chang, Chia-Lin, Michael McAleer, and Vicente Ramos. (2020). *A Charter for Sustainable Tourism after Covid-19.* Sustainability 2020, 12, 3671; DOI:10.3390/su12093671

Haryanto, T., (2020). Editorial: Covid-19 Pandemic and International Tourism Demand. JDE (Journal of Developing Economies), Vol. 5 (1), 1–5.

Hoefer A, Pampaka D, Wagner ER, Herrera AA, Alonso EG-Ramos, Lopez-Perea N, Portero RC, Herrera-Le' on L, Herrera-Le' on S, Gallo DN, Management of a COVID-19 Outbreak in a Hotel in Tenerife, Spain, International Journal of Infectious Diseases (2020), DOI: https://doi.org/10.1016/j.ijid.2020.05.047

Hoque, Ashikul, Farzana Afrin Shikha, Mohammad Waliul Hasanat, Ishtiaque Arif, Abu Bakar Abdul Hamid. (2020) *The Effect of Coronavirus (Covid-19) in the Tourism Industry in China.* Asian Journal of Multidisciplinary Studies Vol. 3, No. 1.

Jiang, Y., & Wen, J. (2020). Effects of COVID-19 on hotel marketing and management: A perspective article. International Journal of contemporary of Hospitality Management. DOI: 10.1108/IJCHM-03-2020-0237

Kemenparekraf (2020). Siaran Pers: Penurunan Kunjungan Wisman ke Indonesia Akibat Pandemi COVID-19 Sesuai Perkiraan. https://www.kemenparekraf.go.id/post/siaran-pers-penurunan-kunjungan- wisman-ke-indonesia-akibat-pandemi-covid-19-sesuai-perkiraan (accessed 29.06.2020)

Khadka, D., Pokhrel, G. P., Thakur, M. S., Magar, P. R., Bhatta, S., Dhamala, M. K., Aryal, P. C., Shi, S., Cui, D., & Bhuju, D. R. (2020). *Impact of COVID-19 On The Tourism Industry in Nepal.* Asian Journal of Arts, Humanities and Social Studies, 3(1), 40–48. Retrieved from http://www.ikprress.org/index.php/AJAHSS/article/view/5089

Kumar, Panka,j and Himanshu B Rout. (2020). *Impact Assessment of Covid-19: In Tourism Perspective.* Dogo Rangsang Research Journal Vol-10 Issue-06 No. 1 June 2020.

Lee-Peng Foo, Mui-Yin Chin, Kim-Leng Tan & Kit-Teng Phuah (2020): The impact of COVID-19 on the tourism industry in Malaysia, Current Issues in Tourism, DOI: 10.1080/13683500.2020.1777951

Listeria Monocytogenes Intervention Committee. (2016). *Sanitation Practices Standard Operating Procedures and Good Retail Practices to Minimize Contamination and Growth of Listeria Monocytogeneses Within Food Establishments.* Conference for Food Protection.

Mandabach KH (2020) *The Worldwide Coronavirus (COVID-19).* J Tourism Hospit 9: e102.

Marriott, Norman G. (1997). *Essentials of Food Sanitation.* International Thomson Publishing.

Marriott, Norman G, and Robert B. Gravani. (2006). *Principles of Food Sanitation.* Springer Science+Business Media, Inc.

Papanikos, Gregory T. (2020). *The Impact of the Covid-19 Pandemic on Greek Tourism.* Athens Journal of Tourism – Volume 7, Issue 2, June 2020, pages 87–100.

Sheresheva MY (2020) Coronavirus and Tourism. Population and Economics 4(2): 72–76. https://doi.org/10.3897/popecon.4.e53574

Strelkowski, Wadim. (2020). *Covid-19 Recovery Strategy for Tourism Industry.* Centre for Tourism Studies, Prague Business School. Prague, Czech Republic.

U.S. Environmental Protection (EPA). (2007). *Guidance for Preparing Standard Operating Procedures (SOPs).* Office of Environmental Information, Washington.

WHO. (2020a). *A Joint Statement on Tourism and Covid-19 – UNWTO and WHO Call for Responsibility and Coordination.* https://www.who.int/news-room/detail/27-02-2020-a-joint-statement-on- tourism-and-covid-19—unwto-and-who-call-for-responsibility-and- coordination (accessed 29.06.2020)

WHO. (2020b). Coronavirus disease 2019 (COVID-19): Situation Report 46. Available at: https://www.who. int/docs/kern-0.5pt /default-source/coronaviruse/situation-reports/20200306-sitrep-46-covid-19.pdf?sfvrsn=96b04adf_2 (accessed 29.06.2020)

WHO (2020c). *Key considerations for repatriation and quarantine of travelers in relation to the outbreak of novel coronavirus 2019-nCoV.* https://www.who.int/news-room/articles-detail/key-considerations-for- repatriation-and-quarantine-of-travellers-in-relation-to-the-outbreak-of- novel-coronavirus-2019-ncov (accessed 29.06.2020)

WHO (2020d). *Operational Considerations for Covid-19 Management in The Accommodation Sector.* Interim Guidance.

WHO (2020e). *WHO recommendations for international traffic in relation to the COVID-19 outbreak.* https://www.who.int/news-room/articles-detail/updated-who- recommendations-for-international-traffic-in-relation-to-covid-19- outbreak (accessed 29.06.2020)

Wilekesmann, Uwe & Maximiliane Wilkesmann. (2020). *(Fine Dining) Restaurants in the Corona Crisis.* Discussion Papers of the Center for Higher Education TU Dortmund University.

Zabin, J. 2019. "Artificial intelligence: Working hand in hand with hotel staff", available at https://hoteltechnologynews.com/2019/07/artificial-intelligence-working- hand-in-hand-with-hotel-staff/ (accessed 29.06.2020)

Zhou, P., Yang, X., Wang, X. et al. A pneumonia outbreak associated with a new coronavirus of probable bat origin. Nature 579, 270–273 (2020). https://doi.org/10.1038/s41586-020-2012-7

Promoting Creative Tourism: Current Issues in Tourism Research – Kusumah et al. (Eds)

Assessing tourist motivation on Tionghoa halal food

C. Ningsih & H. Taufiq A
Universitas Pendidikan Indonesia, Bandung, Indonesia

ABSTRACT: A majority of Tionghoa food is not halal. However, the food has to be adapted in the majority of Muslim countries, such as Indonesia, in order to become a culinary (gastronomy) attraction. The objective of this study examines tourist motivation for having Tionghoa halal food as one unique acculturation culture brand identity for promoting sustainable gastronomy tourism in Bandung, Indonesia. Using a regression analysis of 109 respondents, the results indicate that identity of food, good experience, and exploration have positive significant effect to construct the tourist motivation, while restaurant physical and ambience, food quality and service, halal awareness and certification, and promotion have affected the tourist motivation insignificantly. These results suggest that the curiosity about the food becomes the prominent motivation for the tourist. However, the collected data in this research is limited because of social distancing during the Covid-19 pandemic. Thus, the other large data for this model might be caused by difference results.

Keywords: Tionghoa halal food, gastronomy tourism, tourist motivation

1 INTRODUCTION

The number of Muslim populations has increased by more than 235% in the last 50 years (Ambali & Bakar 2014). Islam is one of the religions that has increased most rapidly, with the number of adherents the second biggest in the world. It is motivation to create halal tourism development in each country, including countries with a majority of non-Muslims, such as Japan, South Korea, Australia, Thailand, and New Zealand. They have provided halal tourism in order to attract the Muslim tourists. Muslim tourists are expected to spend US $ 220 billion by 2020, Muslim-friendly platforms are springing up. Halal tourism is a fast-growing tourism category (Global Muslim Travel Index 2017).

As a country of majority of Muslims, Indonesia should be more intensive in developing halal tourism, within the framework of sustainable tourism. At least, there should be some policies of the Indonesia government to keep its tourism sustainability in the competitive era, including the fulfillment of infrastructure, education, culture, and environmental issue (Ningsih 2016). Indonesia is one of the Asia countries that has developed halal tourism industry in order to compete with other countries. Halal tourism, as one of the tourism-based creative industry and of special interest, should be paid attention to by the stakeholders, and has synergy with the strategy of national development to face a global challenge (Ningsih 2014).

The beginning of the formation of Indonesian food is accultured across various cultural regions and foreign influences, such as Chinese (Tionghoa), Arabic, Indian, and European (Rahman 2018). Bandung, as a capital of the West Java region, has become one of the halal tourism destinations in Indonesia. Bandung has many miscellaneous types of authentic local food and acculturation food from a foreign countries, including from the Chinese (Tionghoa). Most of Tionghoa food are not halal, but along with the interaction of the population of the Chinese who stay longer in Indonesia, and the enhancing of the Tionghoa resident and Muslim populations, the number of home meals and restaurants who offer Tionghoa halal food has been growing. But the existence of

DOI 10.1201/9781003095484-88

the Tionghoa halal restaurants is still limited, so there are still many tourists who don't know about it. By this condition, this research aims to assess the factors that influence the tourist motivation about Tionghoa halal food.

2 LITERATURE REVIEW

2.1 *Gastronomy tourism*

According Gilleisole (2001), gastronomy is an art or food science of good eating. The other explanation mentions that gastronomy is the enjoyment of food and drink. Gastronomy does not only emphasize the skill of cooking but also the observers, lovers, and connoisseurs of food.

Muhilal (1995) mentions that traditional food is a food that has been entrenched in the Indonesian people and has been there since the grandmother ancestral tribes of the archipelago. Traditional food is the food that is closed to the local tradition. Hadisantosa (1993) identified traditional food as food that is consumed by a group of ethnic and a specific region, prepared by recipes which are inherited from one generation to the other. Materials that are used come from local areas and the produced foods are based on the tastes of society.

According to Hall and Michell (2003), summarized by the Ministry of Tourism, gastronomy tourism is the type of tourism in which the tourist has high motivation on food or beverage of the specified area. It is usually associated with expensive restaurants, estate wines, and festivals. Almost all activity tours are associated with interest of it. According to Turgarini (2014), development of gastronomy as an object of travel is recognized as a way to conduct local culinary cultural, stimulate tourism demand, and improve the power competitiveness of destinations. Thus, tourism gastronomy also has emerged as a component that is important. Travelers will get experience more toward local culture, and then the people share their local culture to tourists through the media, in which local residents make representations of food identity. The formation of identity and creation of the image associated with the local food is so it can attract the target market and the beneficial development of tourism gastronomy.

2.2 *A gastronomy tourist motivation factors*

A gastronomy tourist motivation is an impulse that revolves around food, the desire to try new tastes, and an urge to explore the history of food and beverage (gastronomy). Tourist gastronomy has gotten a new form, which has the individual characteristics of and motivation particular in every journey. Here, the factors that influence are (Hall & Sharples 2003; Jacobsen & Hauklend 2002), namely, food identity, views of the environment and culture; experience to try a new taste that has not been found before; exploration of the history and culture of food; physical and atmosphere of the restaurant, food quality, views of flavor, aroma, presentation and color; and quality of service to fulfill the visitor needs.

2.3 *A halal gastronomy*

Gastronomy has been growing rapidly for tourists' destinations in the past few years. The market opportunities are quite large because this tourism is being developed in advanced countries (Sormaz et al. 2016).

Recently, halal is become a universal concept. Halal is the exclusive term that is used in Islam, which means permissible or not. The concept of the virtues of halal, which not only covers the requirements of Sharia but also a concept that has sustainability, hygiene, sanitation, and security aspects, making food that is easily accepted by consumers who care about the food security and healthy lifestyle (Baharuddin 2015). The aim is to make sure that the drinks, foods, and products people make or use are absolutely not harmful, are clean, and safe to human health. Thus, in Islam, the using and consuming halal products are obligatory in serving Allah (SWT). It therefore is worth to note that Muslim communities must be mindful of drinks and food ingredients, handling processes, and packaging of consumable halal products. Processed drinks and foods as well as products are only Halal if the ingredients and used raw materials are Halal, and it is compatible to the Islamic guidelines completely (Zurina 2004)

One of the most important things of halal tourism for Muslim tourist is when the availability of food is halal food. Halal food is already getting attention around the world because it becomes a benchmark to safety, hygiene, and quality guarantee of the product. Products or foods that are produced in accordance with the prescriptions of halal food are easily accepted by Muslims and other religions consumers.

For Muslim consumers, halal food and beverage are products that are required to be in accordance with the Sharia of Islam. For non-Muslims consumers, the halal food and beverages are products that represent a symbol of cleanliness, quality, and strict production that is under the collateral permitted management system.

Based on the description of the above theories, it can be concluded that the halal gastronomy is a gastronomy trip that has culture value itself and covers the Sharia requirements.

2.4 *Gastronomy tourist awareness of halal food*

a. According to Ambali and Bakara (2012), there are some important things that should be noticed in choosing halal food in a restaurant, café, or snacks
b. Relief belief is one of the main determining factors for consuming food
c. Exposure to products promotes a product in various ways
d. Health becomes the based reason for why people realize that it can be consumed
e. Halal certification states the halal status of a product according to Islamic law
f. Halal awareness is the importance of awareness for consumers in consuming food or beverage

3 RESEARCH MODEL AND METHODOLOGY

3.1 *Sample and procedure*

This study conducted an online survey made up 27 questions. The survey was distributed to the consumer of Tionghoa halal food, inside and outside Bandung. In the end, 109 qualified feedbacks from the respondents in total were collected. Thus, to reach them effectively, 30 respondents were placed at the beginning of the survey, to test the validity and reliability of the survey instrument. After that, the survey was spread by social media, such as Facebook and whatsapp, with specific community purpose (Tables 1 and 2).

3.2 *Research model and equation*

Based on analysis of the previous and empirical studied, this research is presented as follow (Figure 1):

The equation of this study as regression model is below:

$$\begin{aligned}\text{Motivation} = {} & \alpha + \beta_1 \cdot \text{Food Id} + \beta_2 \cdot \text{Experience} + \beta_3 \cdot \text{Exploration} + \beta_4 \cdot \text{Resto physical} \\ & + \beta_5 \cdot \text{Resto ambience} + \beta_6 \cdot \text{Food qual} + \beta_7 \cdot \text{ervqual} \\ & + \beta_8 \cdot \text{HalalAwareness} + \beta_9 \cdot \text{Halalcert} + \beta_{10} \cdot \text{Promo}\end{aligned}$$

Table 1. Survey data of halal Chinese restaurants in Bandung.

No	Restaurant Name	Address	Information
1 .	Warung Kopi Purnama	Jl. Alkateri No 22, Braga, Kota Bandung	Halal Non-Halal
2	Kwetiau Sapi Asli A88Gap	Jl. Astana Anyar No.11, Karanganyar, Kec. Astanaanyar, Kota Bandung,	Halal
3	Golden chopstik Restaurant	Jl. Trunojoyo No.32, Citarum, Kec. Bandung Wetan, Bandung City,	Halal
4	Eastern Restaurant	Istana Plaza, Jl. Pasir Kaliki No 121–123, Bandung	Halal
5	Tiga Wonton	Paris Van Java jl Sukajadi no 137–139 Bandung	Halal
6	Sapo Oriental	Istana Plaza Jl Pasir Kaliki N0 121–123, Bandung	Halal
7	Bao Dimsum	Paris Van Java, Jl Sukajadi No 137–139, Bandung	Halal
8	Imperial Kitchen & Dimsum	Transmart Carefour Buah Batu, Jl. Bojongsoang N0 320, Bandung	Halal
9	Furama – El Hotel Royale Bandung	El Hotel Royale Bandung, jl. Merdeka no 2, Bandung	Halal
10 .	Chew by Red Bean	Paris Van Java, Jl. Sukajadi No 137–139, Bandung	Halal
11 .	Ta Wan Restaurant	Jl. Sukajadi no. 131–139	Halal
12	Mulya Cirebon Chinese Food	Jl. Tubagus Ismail Raya, Lebakgede, Kecamatan Coblong, Bandung City	Halal
13	Aliong 88 Chinese Food	Jl. Cibogo No.49, Sukawarna, Kec. Sukajadi, Bandung City	Halal
14	Restoran 499	Jl. Jendral Ahmad Yani No. 687, Padasuka, Kec. Cibeunying Kidul, Bandung City	Halal
15	Bakmie Aloi	Jl. Kebon Jati No.179, Kb. Orange, Kec. Andir, Bandung City	Halal
16	Bubur Ahong	Jl. Kebon Jati No.117, Kb. Orange, Kec. Andir, Bandung City	Halal
17	Wedang Ronde Gardujati	Jl. Gardujati No.52, Kb. Orange, Kec. Andir, Bandung City	Halal

Source: Data processed by the author (2020).

Table 2. Results of multiple linear regression analysis.

		Unstandardized Coefficients		Standardized Coefficients		
Model		B	Std. Error	Beta	T	Sig.
1	(Constant)	1,873	1,043		1,797	.75
	Food_ID	.309	122	.316	2,542	.013
	Experience	109	.069	.195	1,587	.116
	Exploration	.333	.156	.265	2,771	.007
	Resto_Physical	−.007	.72	−1010	−094	.925
	Resto_Ambience	−.137	.74	−.152	−1, 853	.067
	Food_Qual	.016	.059	.032	268	.789
	Service	.444	.079	.051	.556	.580
	Halal_aware	.028	.069	.037	.412	682
	Halal_Cert	.024	.172	.013	.140	.889
	Promo	.025	.121	.017	.210	.834

*Dependent Variable: Motivation.

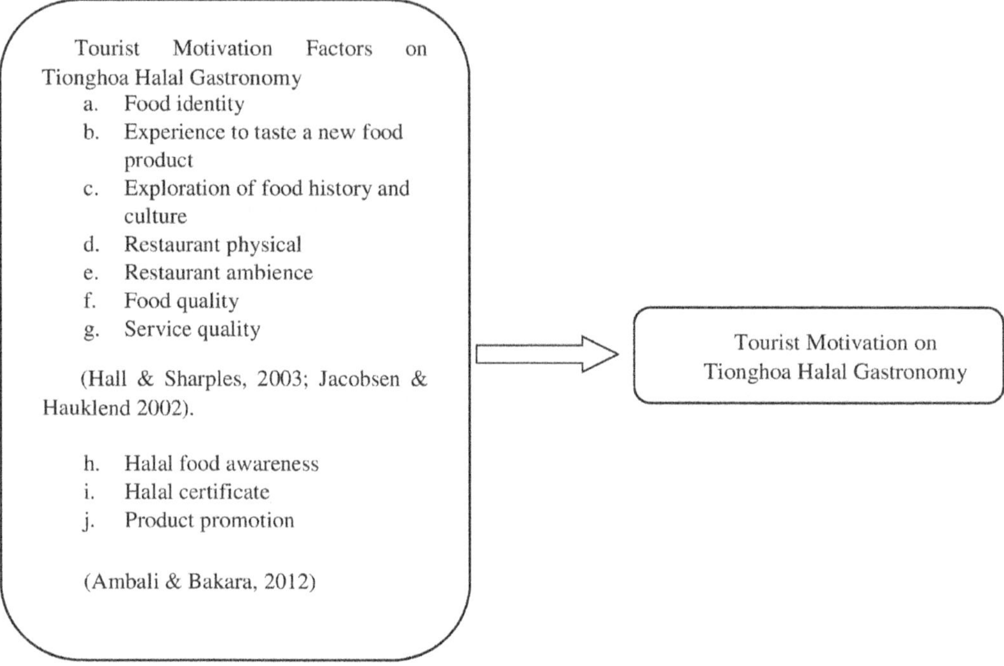

Figure 1. The research model of tourist motivation on Tionghoa halal food.

Table 3. Test results coefficient of determination.

	Summary Model			
Model	R	R Square	Adjusted R Square	Std. Error of the Estimate
1	.687 [a]	.472	.429	1.14907

*Predictors: (Constant), Promo, Food_Qual, Halal_Cert, Service, Halal_aware, Exploration, Food_ID, Experience.

4 RESEARCH RESULTS AND DISCUSSION

4.1 *Results*

The equation regression above shows the relationship between variables independent and variable dependent that can be drawn as follow:

$$\text{Motivation} = 1.873 + 0.309\ \text{Food ID} + 0.109\ \text{Experience} + 0.333\ \text{Exploration}$$
$$-0.007\ \text{Resto Physical} - 0.137\ \text{Resto Ambience} + 0.016\ \text{Food Qual}$$
$$- - +0.444\ \text{Service} + 0.028\ \text{Halal Aware} + 0.024\ \text{Halal Cert} + 0.025\ \text{Promo}$$

Based on Table 3 it can be seen that the coefficient of determination contained in the *R Square* value is 0.472. This means that the ability of the independent variable in explaining the dependent variable is weak, which is equal to 47.2%, and the remaining 52.8 % is explained by other variables not known in this study. The weakness of this model is supposed by the limited data in the Covid-19 pandemic; thus, most of the respondents didn't visit Tionghoa restaurant recently. Thus, the response is only based on their experience in the past.

4.2 *Discussion*

Regression coefficient value of Food Identity is 0.309, which means that if the food identity variable (X1) increases by 1% (ceteris paribus), then the motivation of tourists on Tionghoa halal food increased by 0.309. This shows that the food identity variable provided a positive contribution to the tourist's motivation significantly, so the more food identities provided in Tionghoa halal food in the city of Bandung, the more level of tourist motivation.

Regression coefficient of experiences for trying new foods is 0.109. It mentions that if the experience variable (X2) increases by 1% (ceteris paribus), then the motivation of tourists in Tionghoa halal food will increase by 0.109. This variable has positive impact significantly in the 20% level.

Historical and cultural exploration coefficient value of is 0.333. It means that if the historical and cultural exploration variable (X3) increases by 1% (ceteris paribus), then the motivation of tourists in Tionghoa halal food will increase by 0.333. This historical and cultural exploration variable has positive impact to the tourist's motivation significantly. Thus, the more historical and cultural exploration provided in Tionghoa halal food, the greater increase of the tourist motivation level.

Physical restaurant (X4) and restaurant ambience (X5) coefficient value are -0.007 and -0.137. It notices that physical restaurant and restaurant ambience provided negative impact to the tourist's motivation, although the impact of those variables are not significant.

The regression coefficient value of food (X6) and service quality (X7) are 0.016 and 0.444, which means that food and service quality have positive impact to the tourist's motivation, although the impact of those variables are not significant.

The regression coefficient value of awareness of halal products (X8) and halal certificate (X9) are 0.028 and 0.024, which mentions that awareness of halal products and halal certificate provided a positive contribution to tourist motivation insignificantly.

Regression coefficient of product promotion is 0.025, which means that if the product promotion variable (X10) increases by 1% (ceteris paribus), then the tourist's motivation of Tionghoa halal food will increase by −0.025, although the variable impact is not significant.

5 CONCLUSION

By using a regression analysis of 109 respondents, the results indicate that the identity of food, the good experience in trying new food, and the exploration have positive significant effects to construct the tourism motivation on tourist's motivation of Tionghoa halal food, while restaurants in physical presence and ambience, food quality and service, halal awareness and certification, and promotion have affected the tourist motivation insignificantly. These results suggest that the curiosity about the food becomes the prominent motivation for the tourist.

Even though, the model level is weak, because of the limited data because of the impact of the Covid-19 pandemic.

REFERENCES

Abdul latief, J. (2006). *Manusia, Filsafat, Dan Sejarah*. Jakarta: Bumi Aksara.

Ambali, A. R., & Bakar, A. N. (2014). People's Awareness on Halal Foods and Products: Potential Issues for Policy-makers. *Procedia – Social and Behavioral Sciences*, *121*(March 2014), 3–25. https://doi.org/10.1016/j.sbspro.2014.01.1104

Baharuddin, K., & Kassim, N. A. (2016). Understanding the Halal Concept and the Importance of Information on Halal Food Business Needed by Potential Malaysian Entrepreneurs, (February 2015). https://doi.org/10.6007/IJARBSS/v5-i2/1476

Gilleisole. (2001). *European Gastronomy Into The 21 Century*. Oxford: Oxford University.

Global Muslim Travel Index. (2017)

Hadisantosa. (1993). Makanan Tradisional yang Memiliki Kandungan Gizi dan Keamanan yang Baik. *Makalah Disajikan Dalam Seminar Pengembangan Pangan Tradisional Dalam Rangka Penganekaragaman Pangan.*

Hall, C. (2003). The consumption of experiences or the experience of consumption? An introduction to the tourism of taste. *Food Tourism Around The World*, (December 2003), 1–24. https://doi.org/10.1016/b978-0-7506-5503-3.50004-x

Hall, C. Michael, Sharples, Mitchell, M. and B. C. (2003). *Food Tourism Around The World: Development, Management, and Markets*. UK: Butterworth-Heinemann Elsevier Ltd.

Muhilal. (1995). Makanan Tradisional Sebagai Sumber Zat Gizi dan Non Gizi dalam Meningkatkan Kesehatan Individu dan Masyarakat.

Ningsih, C. (2014). Sinergitas Industri Kreatif Berbasis Pariwisata dengan Strategi Pembangunan Industri Nasional Menuju Globalisasi. Jurnal Manajemen Resort & Leisure, Vol.11. no.1.

Ningsih, C. (2016). Heritage, Culture and Society: Research agenda and best practices in the hospitality and tourism industry.

Rahman, F. (2018). Kuliner sebagai Identitas Keindonesiaan. *Sejarah*, 2(1), 43–62. Retrieved-fromhttp://jurnal.masyarakatsejarawan.or.id/index. php/js/article/download/118/95/

Sormaz, U., Akmese, H., Gunes, E., & Aras, S. (2016). Gastronomy in Tourism. *Procedia Economics and Finance*, 39(November 2015), 725–730. https://doi.org/10.1016/S2212-5671(16)30286-6

Turgarini, D., & Pridia Rukmini Sari, H. (2016). Gastronomy Tourism Attraction in Ternate City, 28(Ictgtd 2016), 90–96. https://doi.org/10.2991/ictgtd-16.2017.17

Zurina, M. B. (2004), Standardization for halal food. Standards and Quality News, 11(4).

Promoting Creative Tourism: Current Issues in Tourism Research – Kusumah et al. (Eds)
© 2021 Taylor & Francis Group, London, ISBN 978-0-367-55862-8

Revenge tourism: Trend or impact post-pandemic Covid-19?

M.N.A. Abdullah
Universitas Pendidikan Indonesia, Bandung, Indonesia

ABSTRACT: The impact of the Corona virus pandemic (Covid-19) for the tourism industry in Indonesia is certainly very influential. Post-pandemic, the number of tourists is predicted to surge. The purpose of this study was to determine the general arguments of the Covid-19 post-pandemic tourism activity that is potentially productive or not. The approach of this study was qualitative method that described and depicted an event naturally. The data collection in this study is using participatory observation for three months in three cities in Indonesia, deep interviews with eight informants, which are divided into six travelers and one expert and one practitioner, documentary studies, and literature studies. The results showed that revenge tourism is very likely to emerge because many people are bored at home. There are two balanced groups where three informants said they would carry out tourism activities after the pandemic, and three other informants said they would travel after the vaccine was found. This study can be implemented in current issues of tourism research, especially the matter of the tourism activity post-pandemic Covid-19's impact.

Keywords: Covid-19, impact, post pandemic, revenge tourism, trend

1 INTRODUCTION

The Corona pandemic resulted in a decline for the tourism sector. The community chose to restrain their desire to travel in the midst of the Corona outbreak. In light of Covid-19, many people had to cancel travel plans they were excited for—be it a weekend getaway or the trip of a lifetime. Either way, recent surveys show that people who can, will want to catch up on missed travel opportunities when the crisis subsides (Heyang & Martin 2020). This has led to the hotel and tourism industry hoping for a wave of "revenge traveling"—people going on extra trips or splurging after the Corona crisis because they were deprived of that possibility for so long (Chinazzi et al. 2020). MarkPlus Tourism's Founder and Chairman Hermawan Kartajaya invites tourism players to remain optimistic in facing Covid-19. After the pandemic ends, tourists will go on revenge tours after months of being at home or what is called revenge tourism. Covid-19 pandemic made many people experience boredom in the midst of the application of social restrictions (Sheresheva 2020). Wuhan's lockdown is over, and travel is opening up around China again. Demand has been returning slowly and gives local and international hoteliers reason for careful optimism (Wells et al. 2020).

The problem in this study is that after the pandemic ends, tourists will go on revenge tours after months of being at home or what is called revenge tourism. After it's over, tourists will go for a walk again, revenge after months at home, or the term *revenge tourism*. That's when tourism actors must use it. Preparations must be made from now on. Some tourism destinations have to continue to promote in different ways. The advantages of these studies was expected to be an investiture of stimulating listing of Indonesia tourism destinations. Because we also believe that Indonesia tourism not only has a wide range of great tourism destinations, but there is deep, inspiring, and meaningful culture behind each destination. It tells about good will, hard work, best quality views, and also local indigeneous people. The purpose of this study is to determine the general arguments of the Covid-19 post-pandemic tourism activity that is potentially productive or not.

DOI 10.1201/9781003095484-89

2 LITERATURE REVIEWS

In China, nearly 85 million domestic tour trips were recorded in the first three days of the Labor Day holiday, according to state media citing China's Ministry of Culture and Tourism (Shen et al. 2018). It seems that now the tourism industry can be given laurels of a time-compressed model of the modern economy: the storm which is yet coming to all areas of economic activity in the collapsed tourism industry immediately, in February 2020, of course, if considering the tourism industry as the activities of a huge network of actors representing a system of related industries (Paget et al. 2010; Sheresheva & Bajo 2014). Tour operators and travel agencies, hotels, sanatoriums, holiday homes and boarding houses, guides, companies providing services for health, recreation and entertainment, booking systems, all types of tourist transportation, catering and souvenir producers—all participants in well-established, often cross-border value chains have found themselves in a dramatic situation of "an aggregate supply shock resulting from contagion containment measures with restrained demand and mobility" (Bénassy-Quéré et al. 2020). The current pandemic is most interesting because it brightly and vividly highlighted all the accumulated challenges of the global economy (Jorda et al. 2020). The idea of the "global village," firstly, was confirmed (all in one boat, and not China alone, as it first seemed to Western analysts). Secondly, it turned out to be its opposite: the global economic system, which for some time has been impersonated as an ideal, in fact has long been a "closed community of the elite, not a global village" (Scott 2001) and is not sustainable at all. According to Jesse Colombo, who predicted the 2008 crisis, the pandemic is dangerous for the world economy not in itself, but as a trigger of a deep crisis, because "we are already very late in the cycle, and coronavirus is basically the one-two punch, but we were already hurtling towards recession before anyone ever heard of coronavirus" (Hall 2020). On the medium-term horizon, the signs of de-globalization, the transition to protectionism and protection of the national priorities by States are becoming increasingly clear. Tourism has shown how fast this is happening: within two months, all tourist destinations were closed, both for organized and independent tourists. The industry has collapsed all over the world. For example, in the United States a drop of 80% in 2020 is predicted (Forum Digital: Will the hotel and tourism industry survive under the conditions of quarantine?) and the tourism industry may lose USD 24 billion (Hirsh 2020). In Turkey, the situation is also catastrophic (Polyanskaya 2020). Logistics chains are crumbling, asset prices are falling, large corporations are in fever, small and medium tourism businesses are stumbling in anticipation of solutions that allow them to survive until better times.

3 METHODS

The approach of the study, Revenge Tourism: Trend or Impact Post-Pandemic Covid-19?, was qualitative method that described and depicted an event naturally. The data collection in this study used participatory observation for three months in three cities in Indonesia, deep interviews with eight informants, which are divided into six travelers and one expert and one practitioner, documentary studies, and literature studies. Qualitative research is defined as follows: Qualitative research is an inquiry process of understanding based on distinct methodological traditions of inquiry that explore a social or human problem. The researcher builds a complex, holistic picture, analysis of words, reports detailed views of informants, and conducts the study in a natural setting.

Researchers create a holistic overview, analyzing, reporting the views of informants in detail, and conduct research in natural situations. Sample technique in this study is defined as snowball sampling. This research used observation, literature study, and in- depth interview data collection techniques.

4 RESULTS

The results showed that revenge tourism is very likely to emerge because many people are bored at home. There are two balanced group where three informants said they would carry out tourism activities after the pandemic, and three other informants said they would travel after the vaccine was found. This study can be implemented in current issues of tourism research, especially the matter of the tourism activity post-pandemic Covid-19's impact.

Acoording to my informant, owner and expertist of Whatravel travel services tourism, the intention to take revenge to the period of quarantine at home with a tour is a natural thing. After limited physical movement at home for weeks, he assessed that everyone naturally wants to enjoy a new atmosphere, from just eating at restaurants, watching films at the cinema, to holidays, to tourist destinations. His opinion was strengthened by the results study that revenge tourism is very likely to arise because many people are bored at home. Going away is one that they want to do now. But after social restrictions are lifted, maybe they will not go to distant destinations. Only next year there will be many trips abroad (Sheresheva 2020).

The trend toward revenge for tourism has also emerged in the United States, according to a poll conducted by Skift Research, a tourism publishing, research, and marketing firm. Their study found that one-third of the US population planned to travel three months after the quarantine of the area was lifted. While in early May, at least 85 million Chinese residents, including in the city of Wuhan, went on a tour after local authorities loosened the "lockdown" policy. As for Seoul, South Korea, on May 10 a new cluster of Covid-19 cases appeared at a nightclub. After 34 people who recently visited the club tested positive for Covid-19, the country was believed to be soon entering the second wave of the pandemic (Gostic et al. 2020).

The question is, how dangerous is the tour after a pandemic? Data from news online study literature, an expert as a spokesman for the Ministry of Tourism and Creative Economy, said that tourism activities can only run normally after the Covid-19 vaccine has been found and can be accessed by the public. Before these normal conditions occur, there must remain a high level of vigilance at the potential for the spread of Covid-19. When the activity is opened in stages, what will be targeted is a tourist destination because people travel to relieve fatigue during work and stay at home. We can no longer behave as before the pandemic. After this health awareness will be higher, he said.

Vaccines, are indeed the only indicators that can ensure that walkers are free from the Covid-19 threat. It is likened it to the yellow fever vaccine for people who will travel to Africa and South America (Currie et al. 2020).

The virus is progressing differently around the world (Bosque-Pérez & Eigenbrode 2011). In a nutshell, China and South Korea seem to have gotten through the storm while Europe is still in the thick of it. In the United States, things are only getting started. This means that travel restrictions for some countries will stay in place for longer while other nations will start easing them soon. Regional differences are also likely as some areas within the same country will recover faster than others and be able to welcome visitors sooner. While travel will become possible again, new travel requirements will probably be imposed. This could include proof of immunity for Covid-19 (or proof of vaccination, once there is one), mandatory testing, and/or quarantining upon arrival (already practiced in countries like Hong Kong and South Korea) (Gostin & Wiley 2020). According to a traveler from Jakarta, a vaccine is indeed the only indicator that can ensure that walkers are free from the threat of Covid-19. But for a tour operator, the vaccine requirements will only be applied to international travel. Domestic tourism would be difficult if it had to be based on a vaccine certificate. Domestic tourism should also need these conditions, but he was not sure everyone feels the need to vaccinate. The number of domestic tourists in Indonesia is very large, and the number of vaccines in the near future may not be that much (Levine et al. 2011).

However, a spokesman for the Ministry of Tourism and Creative Economy said that tourism activities will begin rolling gradually as the government lifts social restrictions. Until now, the Ministry of Tourism and Creative Economy will ask the managers of tourist attractions to ensure the cleanliness, health, and safety of tourists. Right now, in Indonesia, they just can do preparing. Before the vaccine is found, and conditions are not yet normal, there is a possibility of a second or third wave. In 2021, although vaccines have not yet been found, business activities will be reopened.

If at that time the situation is better, even though the Covid-19 case has not been completely resolved, the tourism industry will be ready.

For travelers who want to be safe and comfortable, it turns out the swab test results are not guaranteed because that's the condition at the time of the test. We have to think logically, dare to walk after there is Covid-19's vaccine found and let's don't be silly and die (Hartman & Nickerson 2020).

So, is the urge to reply to the period of self-quarantine immediately on vacation worthy? For people who are desperate, they will still go and ignore the potential for transmission or the second wave (Tang et al. 2020).

Meanwhile, for another traveler from Bandung, her vacation travel will only begin when the Corona case no longer appears in various countries. She will be more confident to take a vacation when the pandemic is over, not because the restrictions are lifted.

The Minister of Tourism and Creative Economy Indonesia, Wishnutama estimates that the number of tourists enjoying Indonesia's tourist destinations in 2020 is only 5 million or down 11 million compared to 2019. This statement is appropriate with conditions on a global scale, that the World Tourism Organization said the tourism sector is experiencing the worst period since 1950 (Niewiadomski 2020).

5 CONCLUSIONS

And in conclusion, again about Indonesia, the Indonesian people are predicted to flood the tourist attractions if the government lifts social restrictions or declares Indonesia to be free of the Covid-19 case. However, the trend referred to by some as revenge tourism is considered to be counterproductive. Tourism is considered to be unable to roll normally if the Covid-19 vaccine has not been found.

REFERENCES

Bénassy-Quéré A, Marimon R, Pisani-Ferry J, Reichlin L, Schoenmaker D, Weder di Mauro B (2020) COVID-19: Europe needs a catastrophe relief plan. VOX CEPR Policy Portal. Research-based policy analysis and commentary from leading economists. https://voxeu.org/article/covid-19-europeneeds-catastrophe-relief-plan [Accessed on 22.04.2020]

Bosque-Pérez, N. A., & Eigenbrode, S. D. (2011). The influence of virus-induced changes in plants on aphid vectors: insights from luteovirus pathosystems. *Virus research*, *159*(2), 201–205.

Chinazzi, M., Davis, J. T., Ajelli, M., Gioannini, C., Litvinova, M., Merler, S., ...& Viboud, C. (2020). The effect of travel restrictions on the spread of the 2019 novel coronavirus (COVID-19) outbreak. *Science*, *368*(6489), 395–400.

Currie, C. S., Fowler, J. W., Kotiadis, K., Monks, T., Onggo, B. S., Robertson, D. A., & Tako, A. A. (2020). How simulation modelling can help reduce the impact of COVID-19. *Journal of Simulation*, 1–15.

Gostic, K., Gomez, A. C., Mummah, R. O., Kucharski, A. J., & Lloyd-Smith, J. O. (2020). Estimated effectiveness of symptom and risk screening to prevent the spread of COVID-19. *Elife*, *9*, e55570.

Gostin, L. O., & Wiley, L. F. (2020). Governmental public health powers during the COVID-19 pandemic: stay-at-home orders, business closures, and travel restrictions. *Jama*, *323*(21), 2137–2138.

Hall R (2020) Analyst who predicted 2008 global financial crash warns another one is on the way – and not just because of coronavirus. Independent. https://www.independent.co.uk/news/world/americas/financial-crisis-2008- coronavirus-donald-trump-economy-stocks-a9392881.html [Accessed on 22.04.2020]

Hartman, G., & Nickerson, N. P. (2020). COVID-19 Impacts on Tourism-Related Businesses: Thoughts and Concerns.

Heyang, T., & Martin, R. (2020). A reimagined world: international tertiary dance education in light of COVID-19. *Research in Dance Education*, 1–15.

Hirsh L (2020) Travel industry could lose $24 billion as coronavirus cripples tourism from outside US. CNBC. https://www.cnbc.com/2020/03/11/coronavirus-travel-industry-could-lose- 24-billion-intourism-from-outside-us.html [Accessed on 22.04.2020]

Jorda O, Singh SR, Taylor AM (2020) Longer-Run Economic Consequences of Pandemics. Federal Reserve Bank of San Francisco Working Paper 2020-09. https://doi.org/10.24148/wp2020-09

Levine, O. S., Bloom, D. E., Cherian, T., De Quadros, C., Sow, S., Wecker, J., …& Greenwood, B. (2011). The future of immunisation policy, implementation, and financing. *The Lancet*, *378*(9789), 439–448.

Niewiadomski, P. (2020). COVID-19: from temporary de-globalisation to a re-discovery of tourism?. *Tourism Geographies*, 1–6.

Polyanskaya A (2020) Man, Hotels in Turkey suspend work. Izvestia. https://iz.ru/991908/aleksandra-polianskaia/vot-zaraza-oteli-turtcii- priostanavlivaiut-rabotu [Accessed on 22.04.2020] (in Russian)

Shen, H., Wang, Q., Ye, C., & Liu, J. S. (2018). The evolution of holiday system in China and its influence on domestic tourism demand. *Journal of Tourism Futures*.

Sheresheva MYu, Bajo R (2014) Network approach in tourist destinations studying: new trends. Initsiativy XXI veka [Initiatives of the XXI century] (2): 58–63. https://www.researchgate.net/publication/281117134_Setevoj_podhod_v_izucenii_turis tskih_destinacij_novye_tendencii [Accessed on 22.04.2020] (in Russian)

Sheresheva, M. Y. (2020). Coronavirus and tourism. *Population and Economics*, 4, 72.

Scott BR (2001) The Great Divide in the Global Village. Foreign Affairs. https://www.foreignaffairs.com/articles/2001-01-01/great-divide-global-village [Accessed on 22.04.2020]

Tang, B., Xia, F., Tang, S., Bragazzi, N. L., Li, Q., Sun, X., …& Wu, J. (2020). The effectiveness of quarantine and isolation determine the trend of the COVID-19 epidemics in the final phase of the current outbreak in China. *International Journal of Infectious Diseases*.

Wells, C. R., Sah, P., Moghadas, S. M., Pandey, A., Shoukat, A., Wang, Y., …& Galvani, A. P. (2020). Impact of international travel and border control measures on the global spread of the novel 2019 coronavirus outbreak. *Proceedings of the National Academy of Sciences*, *117*(13), 7504–7509.

Promoting Creative Tourism: Current Issues in Tourism Research – Kusumah et al. (Eds)
© 2021 Taylor & Francis Group, London, ISBN 978-0-367-55862-8

Spiritual tourism: Study of the experience of fasting on Ramadan during the COVID-19 pandemic in Indonesia

E. Firdaus & M. Rahmat
Universitas Pendidikan Indonesia, Bandung, Indonesia

ABSTRACT: Spiritual tourism in the context of Islamic worship and other religions is in the inner realm of the form of a journey which brings satisfaction and pleasure to the human soul. It is because the comfort of life is not merely based on physical pleasure and happiness. The first and the extraordinary experience of worship and fasting on Ramadan during the lockdown time during the COVID-19 pandemic in securing themselves with the solitude of worship in their respective homes. The methodology is carried out with a case study of the sincerity of individual Islamic worship known only to him/her and his/her Lord alone, without imaging and socializing in the crowd. Seven volunteer participants were recruited for doing the Kurma fasting/date palm fasting on Ramadan (fasting by having pre-dawn meal and breaking the fast only by drinking and having dates for a month). Included were six males, ages of 30, 43, 50, 57, 63 years, and one female aged 61. The findings revealed that they reach the real inner satisfaction as religious spiritual tourism in the God-Human relationship. The higher reduction of someone in his/her relationship to the material shows his/her immaterial qualities and his/her spiritual qualities occur. An absolute in-depth further research supplementation about immaterial and inner worship is recommended.

1 INTRODUCTION

Spiritual tourism in contemporary society had been embraced by many people. There are more modern people leading to secularism, but at the same time, those who represent the sacred journey possess the potency for fulfilling the contemporary tourism. Research had been done in England regarding this phenomenon (Jarratt & Sharpley 2017). In Spain, there was a study about spiritual tourism by heading to the situation of experience in the way of Saint James, by his study on spiritual experiences in multi-religions which interpreted the trend of post-secular (Lopez et al. 2017).

Some researches on that field arose in other parts of the world. Those also have been studies which have some relationships with the contribution on people's welfare. Meanwhile in Australia, study is on spiritual tourism analyzed by the self-determination theory (SDT), which done to check the welfare results on this spiritual tourism by heading to the auto-ethnography (Buzinde 2020). There are also some studies on this, but they are still in progress (Kujawa 2017), along with other researches in Australia (Singleton 2017). Not far from there, research in New Zealand was about spiritual retreat tourism using a semiotic approach. This research is done by exploring the experiences on spiritual retreat tourism that is yoga, which is approaching health tourism (Bone 2013), and research about spiritual tourism guide by seeking personal experiences trough a spiritual, mystics, and divine ones (Parsons et al. 2019).

Research on spiritual tourism in Asia among others in Thailand shows that the driving factors include the novelty, relaxation, transcendent, pride, physical appearance, and seclusion; those factors influence travelers' satisfaction in a spiritual retreat (Ashton 2018). In Japan, by researching on sustainability of spiritual tourism in the future using a critical tourism perspective and full of hope as a platform, it puts the spirituality as a basis for sustainability and advocating by slow involvement with local places and its society (Kato & Progano 2017). In India there is an interesting research

 DOI 10.1201/9781003095484-90

on destination management and satisfaction, but not for business possibilities and for the sense of business "unrealistic of spiritual customization" (Kaur 2016).

One which gives more strength in those researches, they are also investigated with a collaborative research system between countries such as done by US–Australia, with findings about overlapping phenomena, spiritual tourism, which suggests certain motivational themes that might have great uses to help to understand the practice of meditation retreat tourism (Norman et al. 2017); and the collaboration between Australia-Israel on doing the research, which has identified the emergence of binary between the performance of spiritual tourism which is essentially religious and vice versa, as a secular practice (Cheer et al. 2017). The biggest and most interesting research of this spiritual tourism is researching the phenomenon of hajj; although researched from a marketing strategy and its ethnographic perspective, there are academic contributions that focus on hajj as a spiritual journey as the great product and service which is researched in collaboration involving two countries, Australia–Pakistan (Jackson 2009).

Whereas on this occasion, researchers took a different study from the viewpoint of the experiences of fasting in the month of Ramadhan. This is unique as it happened during the COVID-19 pandemic. Determining the case study on worship activities are considered further for inner happiness for those, the doer. They are who maximizing the reducing of material consumed with the intention of bringing his spiritual closer to his/her God, the result is various forms of spiritual happiness. Fasting date palm is a relatively limited Muslim group in the world. The purpose of this research is to investigate how this spiritual tourism takes place and influences religious satisfaction. The satisfaction of worship in Islam as a tour is intended as other worship, just like evening prayers in the month of Ramadan as a kind of tour or taking a rest of Prophet of Muhammad. So, the name of Tarawih (going to a tour/taking a rest) Prayer became popular.

2 METHODS

Study is done to collect the data which planned to be collected from case studies by recruiting seven participants who are fasting date palm in the month of Ramadan as a Muslim obligation. However, this study only observes those who do it voluntarily in a specific way; in the date palm fasting, the subjects only have a pre-dawn meal and break the fast only by drinking and having dates during the month of Ramadan in 2020, at the time of the Covid-19 pandemic. It is not the ordinary fasting as all the Muslim do around the world, who are free to consume all kind of food as long as those are halal. The specificity of this fast is more visible to reduce eating and drinking in moderation by consuming dates among one to seven from dawn to Magrib time in the afternoon. Seven volunteer participants were recruited, six males and one female, males aged 30, 43, 50, 57, 63 years, and one female aged 61, without any coercion. Among them, three participants have experienced fasting like this in the month of Ramadan in the last three years, while others are new volunteers who are doing this kind of fasting for the first time. All participants were willing to be observed, interviewed, and become the object of this study without revealing the confidentiality of the names, identities, and matters relating to their respective privacy, so as not to interfere with the sincerity of their worship of Allah as the Lord who commanded the fasting as an obligation (see Table 1).

The methodology is carried out with a case study on the sincerity of individual Islamic worship, which is only known by himself/herself and his/her God, without imaging and socialization in the crowd. Seven volunteer participants were recruited, those who are taking the date palm fasting on Ramadan (fasting by having pre-dawn meal and breaking the fast only by drinking and having dates for a month). Included were six males, aged 30, 43, 50, 57, 63 years, and one female, aged 61.

All participants were observed for their perceived health and their psychological condition simply and visibly by the observer after their willingness to be the objects of the study with some privacy records that are well maintained. Although the observation was carried out by implementing social distancing during the Covid-19 pandemic in Indonesia. During fasting, they were not found and reported if they were seriously or mildly ill, except the feeling of hunger, thirst, and usual discomfort

Table 1. Participants' identity on study of Ramadan date palm fasting.

No.	Pseudonym	Sex	Age	Date Palm Fasting Experience
1.	Ahmad	M	30	Never
2.	Mahmud	M	43	More than 3 months
3.	Muhammad	M	50	Never
4.	Hamid	M	57	Never
5.	Hammadah	F	61	3 months
6.	Yahmad	M	63	3 months
7.	Zul Hamd	M	43	3 days

Table 2. Implementation of participants in Ramadan date palm fasting.

No.	Pseudonym	Date Palm Fasting Experience	Achievement of Fasting This Year
1.	Ahmad	Never	3 days
2.	Mahmud	More than 3 months	1 month
3.	Muhammad	Never	7 days
4.	Hamid	Never	4 days
5.	Hammadah	3 months	1 month
6.	Yahmad	3 months	1 month
7.	Zul Hamd	3 days	0 day

when one is doing the fasting, refraining from eating, drinking, and having sex in the daytime during the date palm fasting. The differences reported lie in their sincerity and endurance in giving their heart to their Lord by holding back the imagination of the stimuli motivating various mind disorders alone on material physical needs. Sincerity and closeness to the Lord strengthened them in holding on to continue the fasting of these dates palm. But in some cases, ones that have only recently carried out the fasting date, there are those who fail to continue their fast after three days which was experienced by Ahmad and Hamid, so that the success of his fast is only three days for Ahmad and four days for Hamid. The rest of Ramadan like most others do not fast date palm again, but continued to carry out regular fasting by eating at night from dusk until before dawn arrived in the morning.

For Muhammad, who took the date palm fasting for the first time, he succeeded in completing his fasting for seven days and then continued to have general fasting. What is completely different is in the case of Zul Hamd, who was initially able to carry out for three days, but he failed to continue his fasting period on the grounds that he was not strong enough (unable to continue) like he did last year. But for those who have done this fast for at least a month in the past year, this year it turns out they have the power to repeat this fast again in the following year and even in the years to come, this date fast was carried out well by Hamadah and Yahmad, whereas Mahmud had the longest experience of fasting this date so that there was no difficulty for him to do it one month of Ramadan this year (Table 2).

3 FINDINGS

The reasons for participants, why did they do date palm fasting in the year and month of the Covid-19 pandemic? In responding to this question, participants had a variety of answers to each, with various underlying motivations. In general, such as those who carry out fasting in the month of

Table 3. Motivation on Ramadan date palm fasting.

No.	Pseudonym	Description
1.	Ahmad	The advice of the Spiritual Teacher is to be closer to God
2.	Mahmud	Feel the pleasure of worshiping close to God
3.	Muhammad	Modeled the Fasting of the Prophet for submission to God
4.	Hamid	Because Allah ordered the Fast that should be
5.	Hammadah	Wants to be loved by Allah and able to love Him
6.	Yahmad	The true fasting suits to the circumstances of each Prophets in their days, on the most minimal condition on their material life.
7.	Zul Hamd	Trying to fast for having the ability by eating less

Table 4. Date quantity and type consumed per meal.

No.	Pseudonym	Quantity/date	Type/Name of Date	Date Consumed
1.	Ahmad	3–7	Medjol	5 dates
2.	Mahmud	3–5	Ajwa, Sukari	5–7 dates
3.	Muhammad	3–7	Sukari	3 dates
4.	Hamid	3–9	Ajwa	5 dates
5.	Hammadah	1–3	Sukari	1–5 dates
6.	Yahmad	3–5	Tunis, Sukari	3–5 dates
7.	Zul Hamd	3–9	Mazafati	3–7 dates

Ramadan on the reason and statement of Allah SWT had ordered the fasting. But not all people do date palm fasting like you do, so why and what kind of motivation?

The reasons revealed in the motivation according to participants apart from the necessity of any worship, such as the fasting of these dates which are all religiously, they must worship in a sincere state because of Allah SWT. However, the following data unearthed from the field of motivation for fasting dates revealed as in Table 3, (1) As Ahmad said, in a situation that made them not many activities related to others during the Covid-19 pandemic, the role of his spiritual teacher so there are efforts that draw him closer to God; (2) in harmony with Ahmad, Mahmud gets the urge to worship on the basis of being able and often closer to God because of independent quarantine; (3) Can identify fasting in quarantine like the Prophet in the life of the weight of the test for the Prophet Muhammad, who became a guide to human life; (4) The command of God is directed to the full without having to show yourself to people; (5) To be loved by God is more important in this life than just obedience to Sharia; (6) Get motivated by the minimal example of the use/consumption of material; and (7) Only wants to exercise restraint by trying to fast these dates, and therefore the latter is less motivated to be able to continue his fast, which is considered heavy for him (Table 3).

The date palm fasting is by refraining from utilizing the very minimal consumption of food and drinks found in participants. They only consume dates with a limited quantity and are unable to exceed seven dates. As for the other consumption, they fulfill their needs by drinking only water as their complement food. They feel fulfilled by that. In fact, there are certain people who only consumes a date for few days, just as Hammadah did, while others still consume at least three dates to the maximum number of only seven. Three people still consume only three or five seeds, namely Ahmad, Muhammad, and Hamid, while others were in the range of consumption of one to seven (Table 4). On average, they feel comfortable and healthy during fasting, although some of them were facing problems on defecating, but some of them solved it by adding enough fruit to be able to defecate by consuming papaya fruit, or by consuming liquid of Cingcau leaves.

In carrying out this date palm fasting, they obtained the spiritual changes, each in accordance with his hopes to get closer to his Lord. During fasting, they are always bound by their closeness to

Table 5. Experience of hunger and the actions taken.

No.	Pseudonym	Description
1.	Ahmad	Feeling hungry as if in general fasting, two or three times a day
2.	Mahmud	Forgotten the hunger because of the habit of working all day
3.	Muhammad	It is heavy on the first three days, then it gets heavier and could not be strong enough to continue, feeling addicted that cannot be abandoned like drugs
4.	Hamid	No longer able to carry out the date palm fasting after surviving 4 days, because he feels unwell
5.	Hammadah	The first to third day of the first year seem to be hooked, then getting used to up to three times of the month of Ramadan
6.	Yahmad	Had forgotten the feeling of hunger at the beginning of the date palm fasting because gets used to eating dates even in days not fasting too
7.	Zul Hamd	Feel hooked, wants to eat normally and caused his date palm fasting to fail

Table 6. Spiritual experience changes.

No.	Pseudonym	Description
1.	Ahmad	In this date palm fasting could feel closer and always remember Allah, but could not feel the true happier tourism, because only could do three days of date palm fasting
2.	Mahmud	Only could feel the love of God and the love of himself, feel inner pleasure and true happiness
3.	Muhammad	Cannot describe the spiritual change and only could feel happy.
4.	Hamid	More sensitive and have a close sense of remembering God and feel comfortable in worshiping Him
5.	Hammadah	Can have the dialogue as if directly with God closer, because it is the pleasure times of worship by tears and happiness to meet him
6.	Yahmad	Become more diligent in praying and praying at night by spending the night for praying to God
7.	Zul Hamd	It seems like it will change after successfully fasting the dates, doesn't regret and will try again next time to practice the date palm fasting

their Lord, and in addition there are comforts and pleasures that they feel with different intensities. As if everyone is on an adventure in a spendid tour in the spiritual realm. Ahmad found his pleasant tour in three days. Mahmud, as if God came pouring the happiness, his inner happiness. Even though some of them are not aware of the changes like the others, Muhammad felt the changes even though only the feeling of happiness. The pleasure to worship comfortably is felt sensitively by Hamid. Even Hammadah could feel as if he could have a dialogue with his Lord caused by the closeness of his mind, so that he found happiness and joy of worship in the taste and tears of prayer and comfort there. Yahmad experienced an addiction to keep worshiping and praying diligently and keep praying every night, even though he reduced his sleep. He took only for about three to four hours per day to sleep. Whereas the failure of carrying out his fast seemed to Zul Hamd that he was still interested, he still has a desire to try it again to get the experience he heard from others or from his religious/spiritual teacher (Table 6).

The participants of course with changes in the quantity of daily meals will change physically during and after fasting these dates. It was found that no one experienced significant pain due to his fasting, but instead they experienced physical fitness with a lack of weight, possibly including those who were thin as well. But in our observations and interviews it was found that they were healthy with their respective weight loss. Their weight loss occurred between 3 to 11 kg, in the differences in each case as in Table 7. Although some do not weigh themselves, they only feel

Table 7. Physical changes after date palm fasting.

No.	Pseudonym	Description
1.	Ahmad	Do not feel unwell, lose weight 5 kg
2.	Mahmud	Do not feel unwell, lose weight, do not weigh the body, but feels thinner
3.	Muhammad	Do not feel unwell, lose weight 4 kg.
4.	Hamid	Do not feel unwell, lose weight by 3 kg
5.	Hammadah	Do not feel unwell, feel losing weight but do not weigh during this date palm fasting
6.	Yahmad	Do not feel unwell, lose weight in the first week about 11 kg, then his weight is stable until the end of Ramadan
7.	Zul Hamd	Do not feel unwell, only feel hungry, as in the usual fasting of Ramadan, because failed to continue the dates palm fasting

Table 8. The specificity of date palm fasting during the Covid-19 pandemic.

No.	Pseudonym	Description
1.	Ahmad	The sincerity of praying from home alone, without being known or telling others
2.	Mahmud	Suffering caused by the pandemic leads to drawing closer and praying to Allah, moreover, if already owned the feeling of being loved by Allah in this date palm fasting.
3.	Muhammad	The happiness to do the self-quarantine, could pray wholeheartedly
4.	Hamid	The feeling of meeting God can be enhanced without others' presence of knowing the God-man relationship
5.	Hammadah	Days full of happiness in as closer to God inwardly
6.	Yahmad	It must be very intense to worship in quarantine time at home
7.	Zul Hamd	It is a very special moment in solemn and solitary worship opportunities

that their physical scales are reduced, as in the case of Mahmud and Hamadah. Some even felt the stability of their weight during fasting, as in case of Yahmad.

A special characteristic of participants who carry out the fasting of dates is changes in worship patterns. It could be possible because of the lack of business in preparing food before and after mealtime. So that affects many opportunities to carry out additional worship as religious motivation is also influenced by other matters. Because fasting during this pandemic as fasting that does not appear much in physical activities together with other people who tend to lockdown and quarantine in their respective homes, the sincerity of prayer at home becomes special only in the face of his Lord (Ahmad's case). For Mahmud, his love for and from his Lord will continue to increase forged by physical suffering in the real world. In fact, for Muhammad the quarantine and solitude in fasting allowed him to pray fervently. Then in Hamid's experience on the feeling of meeting God could arise, because it is not interfered by the presence of other people that make him to self-promote or create imagery to others. By the quarantine time or staying at home for those who are carrying out the date palm fasting could make him full of happiness of being in love and worship to his Lord in spirit. And Yahmad felt that carrying out the date palm fasting at home ensured that he could do the practice of his worship intensify with his own inner happiness. However, who failed to fast his dates palm, Zul Hamd also recognized the privilege of the dates palm fasting during his confinement in his house (Table 8).

From the interviews and observations of researchers it was found that this date palm fasting was carried out in the case of Muslims in Indonesia. It seemed only a few people were doing it as it is only consuming food and drink in Sharia, only dates and water since sunset in the evening until dawn before dawn prayer, marked by the dawn in the morning. In the afternoon from dawn until dusk they fast. However, those who take date palm fasting and carried them out revealed a desire to approach similarities with the behavior exemplified by the Prophet Muhammad. He (the Prophet) fasted in the perception of the executor of the date palm fasting, they identified with the least consumption of food at night, which according to them is at the time of the Prophet around the 7th century, there was not as much food as in our lives today. Whereas in the time of the Prophet there were only dates, and not many other breads, and fruits were not brought by nomad traders in the middle of the desert, especially in Medina when the fasting order came down (Q.S. 2: 183). We need to imagine the era of the Prophet in the 20th century, where there were not many export-import activities for food commodities, so what happened to any commodity distributed in this world including in Medina where the Prophet lived only with wandering caravans using camel caravan transportation. So that there was very little consumption of the population of Medina except for only dates, and this is what happened, and hereditary for a limited Muslim community exclusively remaining then later as concluded in this study.

Spiritual experience revealed from them is not found in Muslims in general fasting as they do throughout the world. The results of in-depth interviews with participants found that those who carry out the dates palm fasting as physical symbols lacking food intake by avoiding or reducing physicality in the material world could draw closer to activities symbolized by the immaterial. Because God in the teachings of Islam is a Creator who is not material, God can be approached by reducing human dependence entirely on matter.

4 CONCLUSION

Their bowels by carrying out this date palm fasting, they feel closer to their Lord. It can be concluded that God is immaterial, so to approach Him it is necessary to reduce material intake by carrying out the date palm fasting dates. Because this kind of fasting consumes the least amount of consumption of food and drink, including restraining the biological appetite. The meals are only twice, the time after sunset and before dawn by consuming dates and drink water only. The higher reduction in relations with one's material shows the more intense spiritual and spiritual qualities could occur. This date palm fasting is a relatively sufficient means that proves to feel satisfaction and inner happiness during the Covid-19 pandemic of 2020 for those who carry it out. Because it can be said as a spiritual tourism which is meaningful for the solution of human life to the direction of happiness that represents the inner soul.

Any kind of religion in this world has its worship and rules of restraint both physically and non-physically. This activity is a purification of self that aims to get closer or worship to their respective God. Therefore, among the religions in this world they have a common thread of similarity in the physical relationship with the purity of the self from the physical realm into non-physical even to the Lord Himself. The pure worship in these religions is in the non-material realm confined in a space that is less physically but vast and massive in the spiritual realm of immaterial. Worship requires the physical purification of human beings toward their non-physical nature in immateriality. The purity of worship is in the sterility of material from worldly material imagery and the joyful worship in the adventures of the soul is what purifies itself (Q.S. 87:14; 91: 9). Out of happiness comes a heart of sincerity that is only known to man by his Lord alone without the presence of other human eyes who see and judge. Judgment from God is more absolute and correct in the view of religions. So, worship and fasting activities are not in the largest mosque in Makkah or in any mosque which is full of other human values, but at home with their sincerity during the Covid-19 pandemic period which followed the international standard health protocol (WHO, 2020). Finally, happiness could be reach in the aisles and the narrow place of the home for spiritual tourism and do not appear in public places.

DISCLOSURE STATEMENT

No potential conflict of interest was reported by the authors.

REFERENCES

Ashton, A.S. 2018. Spiritual retreat tourism development in the Asia. *Asia Pacific Journal ofTourism Research*, vol. 23, no. 11: 1098–1114.

Bone, K. 2013. Spiritual Retreat Tourism in New Zealand. *Tourism Recreation Research*, vol. 38, no. 3: 295–309.

Buzinde, C.N. 2020. Theoretical linkages between well-being and tourism: The case of self-determination theory and spiritual tourism. *Annals of Tourism Research*.

Cheer, J.M. et al. 2017. The search for spirituality in tourism: Toward a conceptual framework. *Tourism Management Perspectives*, vol. xxx, no. xxx: 1–5.

Jackson, F.H. 2009. Spiritual journey to Hajj: Australian and Pakistani experience and expectations. *Spirituality & Religion, 6:2*, vol. 6, no. 2: 141–156.

Jarratt, D. & Sharpley, R. 2017. Tourists at the seaside: Exploring the spiritual dimension, *Tourist Studies*: 1–20.

Kato, K. & Progano, R.N. 2017. Spiritual (walking) tourism as a foundation for sustainable destination development: Kumano-kodo pilgrimage, Wakayama, Japan. *Tourism Management Perspectives*, vol. xxx, no. xxx: 1–9.

Kaur, G. 2016. Customer Interface in Spiritual Tourism via "Synaptic CRM Gap": An Integrative TechnologyBased Conceptual Model for Relationship Marketing. *Journal of Relationship Marketing*, vol. 15, no. 4: 326–343.

Kujawa, J. 2017. Spiritual tourism as a quest. *Tourism Management Perspectives*, vol. xxx, no. xxx: 1–9.

Norman, A. et al. 2017. Meditation retreats: Spiritual tourism well-being interventions. *Tourism Management Perspectives*, vol. xxx, no. xxx: 1–7.

Lopez, L. et al. 2017. Spiritual tourism on the way of Saint James the current situation. *Tourism Management Perspectives*, vol. xxx, no. xxx: 1–10.

Parsons, H.; Mackenzie, S.H.; & Filep, S. 2019. Facilitating self-development: how tour guides broker spiritual tourist experiences. *Tourism Recreation Research*, vol. xxx, no. xxx: 1–13.

Singleton, A. 2017. The summer of the Spirits: Spiritual tourism to America's foremost village of spirit mediums. *Annals of Tourism Research*, vol. 67, no. xxx: 48–57.

Promoting Creative Tourism: Current Issues in Tourism Research – Kusumah et al. (Eds)
© 2021 Taylor & Francis Group, London, ISBN 978-0-367-55862-8

The new era of tourism: Draw up tourism industry after the pandemic

S. Nurbayani & F.N. Asyahidda
Universitas Pendidikan Indonesia, Bandung, Indonesia

ABSTRACT: The pandemic Covid-19 that occurs throughout the world in fact has had a tremendous impact on all aspects of human needs, from primary to tertiary, including tourism. Tourism sectors experienced a very severe turbulence, considering the industry is focused on pleasure and leisure. However, in the event of a pandemic, all the people of the world are required not to leave their house and carry out programs that have been recommended by the government both nationally and internationally. These conditions have made the tourism industry suspended, and experienced stagnation in the development of the tourism industry in the future. These problems will then be examined in this article about how to redesign a safe tourism industry and promote health protocols. Using a quantitative approach, this research disseminated a questionnaire to the general public about how the tourism industry should be developed after a pandemic. The questionnaire was developed through indicators of how should tourism development after the pandemic. This article maps respondents' answers about the post-covid-19 tourism industry. It is expected that this article can be a valuable input for stakeholders, and it is expected to become a new protocol in the organization of the post-pandemic covid-19 tourism industry.

1 INTRODUCTION

In general, disaster and society are inseparable because a disaster will have a direct or indirect impact on the community. In the narrow sense, disaster is divided into two forms, namely natural disasters and artificial disasters (Mayunga 2007). Natural disasters (such as tsunamis, earthquakes, volcanic eruptions, etc.) are predictable and sometimes unpredictable. Whereas man-made disasters (socio-political/human-made disasters) such as wars, terrorist attacks, politics, or economic crises, are disasters brought or created either intentionally or not by humans (Mulsim et al. 2017; Shiwaku & Shaw 2016).

The covid-19 pandemic disaster is an extraordinary event that attacks all aspects of human life, which can be said to be a socio-political or human-made disaster (French & Monahan 2020). Because the true pandemic is a combination of natural disaster, socio-political crisis, economic crisis, and tourism crisis (Zenker & Kock 2020). Therefore, many researchers about tourism began to talk about how to deal with these unexpected environmental changes, especially in the tourism development sector (Mcgladdery & Lubbe 2017) because the data shows that during this pandemic, 75 million workers in the tourism sector were seriously affected, and also suffered a 2.1 trillion dollar income decline (Sheresheva 2020). The main factor behind this is the recommendation not to travel, gather, and visit crowded places, and it is recommended by the World Health Organization to break the spread of the covid-19 virus (Benjamin et al. 2019). As a result, all borders between countries are heavily guarded, security protocols are improved, and there are many restrictions on entering tourism areas (Yang et al. 2020). The next effect is the disappearance of the trust of prospective tourists to visit tourist areas. The next thing to concern is the stagnation of tourism development, which has an impact on the decline in the level of income of the people around the tourism area (Gössling et al. 2020).

Understandably, many researchers are currently focusing on the continued impact of the Covid-19 disaster because it is feared that the side effects that occur after a pandemic will bring about new

 DOI 10.1201/9781003095484-91

disasters, such as the decline in investment in tourism (Ioannides & Gyimóthy 2020; Li et al. 2020). As a result, the researchers tried to find the main solution to prevent such things from occurring. Therefore, tourism as one of the sectors driving the world economy must develop new indicators of the new normal period (Jennifer et al. 2020), especially in the trust of tourists as in a socio-psychological point of view, as the world is in panic, which is depicted through a series of health protocols issued directly by the World Health Agency. It has then implanted complex thinking about how a new life after the pandemic would be, especially the views of society in the case of tourism in the new normal period. This article will describe the public trust in tourism in the new normal era.

2 METHODS

Covid-19 pandemic is a disaster that attacks all sectors of human life, especially in the economic sector, tourism as one of the driving forces of the world economy. Tourism industry is severely affected because several places of world tourism are forced to close (Hall et al. 2020). However, will this pandemic last forever? Many health protocols are created to prevent the transmission of the Covid-19 virus, especially the importance of implementing health protocols in crowded areas (Sheresheva 2020). In this case, how is the people's trust in the safety and hygiene of post-pandemic tourism area? Several indicators were developed in this article to show how the development of post-pandemic tourism should be developed.

2.1 *Sample and research population*

The data were collected through a questionnaire that was distributed digitally to more than 1,000 people in Asia. The reason for using this technique is to streamline and also follow health protocol standards to keep distance and also comply with social distancing rules.

2.2 *Research indicator*

The indicators in this study were developed from research conducted by Martin Gannon et.al regarding to tourism development. The indicators in the research are community attachment, environmental attitude, cultural attitude, economic gain, involvement, economic perceptions, environmental perceptions, sociocultural perceptions, and support for tourism and resident's perceptions. Furthermore, to emphasize how the development of post-pandemic tourism through the perspective of the community follows, several indicators were taken from articles written by Zanker and Kock, among others: the level of complexity, change in resident behavior, change in the tourism industry, change in destination image, change in tourist behavior, and long-term and indirect effects. Furthermore, before the distribution of the questionnaire was carried out, the questions of the indicator were validated through an expert judgment process beforehand by the tourism experts.

2.3 *Measurement scale*

The measurements in this study iare the Spearman range, where the measurement scale is number 1 to express strongly disagree or something similar, and number 5 to express strongly agree or something similar.

3 RESULTS AND DISCUSSION

Online research has its own obstacles considering that at the time of the pandemic, which lasted for about one semester. Tourism was stagnating, which was quite alarming. Many people considered that post-pandemic tourism would experience a change in patterns. The indicators in the question, namely community attachment, environmental attitude, cultural attitude, economic gain, involvement, economic perceptions, environmental perceptions, sociocultural perceptions, support for tourism, and resident's perceptions will be disrupted by new views about new post-pandemic life (Chang et al. 2020). To find out the various kinds of changes that have occurred post-pandemic to tourism, this article tries to dig deeper with the help of measurements of the level of complexity, change in resident behavior, change in the tourism industry, change in destination image, change in tourist behavior, and long-term and indirect effects (Stankov et al. 2020).

The main focus in this article is formulated as follows: how the community views post-pandemic tourism. This main factor was studied to re-describe how post-pandemic tourism would become, the public's view of the tourism sector should be explored. The analysis has come up with the following results:

Figure 1 shows that people's optimism related to tourism will increase with the end of the pandemic, then interviews were conducted with several respondents to ensure that an increase in tourism would occur after the pandemic. Some respondents answered that this could happen because the community was already feeling very bored with previous activities, lockdown and closure of tourism places. However, several sources gave different opinions regarding the state of tourism after the pandemic. They argued that the growth rate of the tourism sector must also be accompanied by an increase in the economic sector of the community (Gannon et al. 2020). Many people think that the economic recession that occurs in other countries can affect the economic sector in other countries.

Researches focused on handling post-pandemic tourism argue that improving the tourism sector after a pandemic requires extraordinary effort (Zenker & Kock 2020), especially the adaptation of new habits applied in various countries adding to a long list of what forms of tourism are there and must be, to balance these new habits (Li et al. 2020). In addition, an unstable economy can be a big challenge for the tourism industry, because economic stability is one of the keys to developing the tourism sector (Sheresheva 2020).

Benjamin et al. in their article with a focus on the study of tourism equity in the post-pandemic era stated that normal habits before the pandemic could reoccur within the next few years. They realized that the availability of vaccines was the main step so that tourism could again become the driving force of the economy (Benjamin et al. 2019). However, in the French and Monahan article, it was stated that the pandemic was a major disaster for tourism. They explained that the sustainability of tourism would crawl because people would not feel as free as before in doing tourism (French & Monahan 2020).

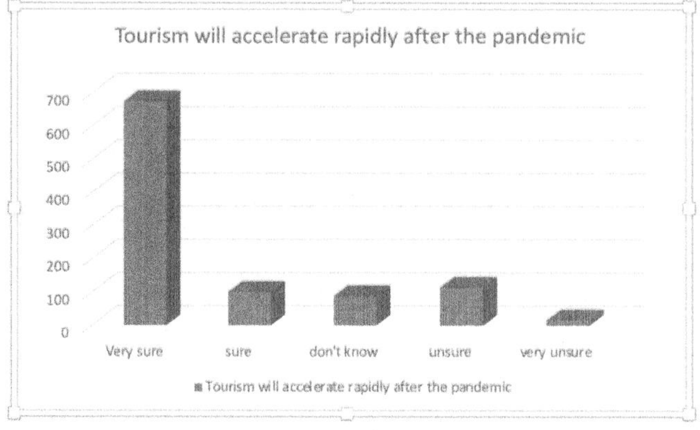

Figure 1. Research result about society perception for tourism accelerating rapidly after the pandemic.

Therefore, in order to re-promote the post-pandemic tourism sector, there needs to be security assurance as well as assurance of good economic progress (Correa-Martínez et al. 2020), because the tourism sector needs support not only from the government but also from the community (Yang et al. 2020). The pandemic that is being fought by all people in the world has succeeded in shifting the tourism sector into a sector that needs to be handled with great care, because this sector really needs communal crowds to be able to continue and also recover (Mostafanezhad et al. 2020).

4 CONCLUSION

The Covid-19 pandemic has had a tremendous impact on the tourism sector. The world community realizes that Indonesia is one of the tourist destinations that must be visited. However, due to this pandemic, tourism in Indonesia and even the world have become suspended. Research was conducted on how the tourism sector is rebuilt with the main focus in this article on the optimism of the community toward increasing post-pandemic tourism which shows that people tend to be optimistic that tourism will again have a central role as one of the advanced sectors after the pandemic. Some notes on the results of research that can be used as a solution to building post-pandemic tourism are: 1) Tourism is very dependent on the stability of the community's economy, and also vice versa; therefore, both the government and those involved in the tourism industry must jointly rebuild the tourism environment. 2) Post-pandemic, community awareness of health has increased; therefore, the post-pandemic tourism environment must be able to convince the public that the area is clean from viruses; and 3) Tourism must be able to become self-healing for the community so that they continue to think positively that a pandemic can be controlled and will end so that the community will again have the trust to carry out tourism activities.

REFERENCES

Benjamin, S., Dillette, A., & Alderman, D. H. 2019. "We can't return to normal": commiting to tourism equity in the post-pandemic age. Hilos Tensados, 1, 1–476. https://doi.org/10.1017/CBO9781107415324.004

Chang, C. L., McAleer, M., & Ramos, V. 2020. A charter for sustainable tourism after COVID-19. Sustainability (Switzerland), 12(9), 10–13. https://doi.org/10.3390/su12093671

Correa-Martínez, C. L., Kampmeier, S., Kümpers, P., Schwierzeck, V., Hennies, M., Hafezi, W., … Mellmann, A. 2020. A Pandemic in Times of Global Tourism: Superspreading and Exportation of COVID-19 Cases from a Ski Area in Austria. Journal of Clinical Microbiology, 58(6), 19–21. https://doi.org/10.1128/JCM.00588-20

French, M., & Monahan, T. 2020. Disease surveillance: how might surveillance studies address covid-19? Surveillance and Society, 18(1), 1–11. https://doi.org/10.24908/ss.v18i1.13985

Gannon, M., Rasoolimanesh, S. M., & Taheri, B. 2020. Assessing the Mediating Role of Residents' Perceptions toward Tourism Development. Journal of Travel Research. https://doi.org/10.1177/0047287519890926

Gössling, S., Scott, D., & Hall, C. M. 2020. Pandemics, tourism and global change: a rapid assessment of COVID-19. Journal of Sustainable Tourism, 1–20. https://doi.org/10.1080/09669582.2020.1758708

Hall, C. M., Scott, D., & Gössling, S. 2020. Pandemics, transformations and tourism: be careful what you wish for. Tourism Geographies, 1–22. https://doi.org/10.1080/14616688.2020.1759131

Ioannides, D., & Gyimóthy, S. 2020. The COVID-19 crisis as an opportunity for escaping the unsustainable global tourism path. Tourism Geographies. https://doi.org/10.1080/14616688.2020.1763445

Jennifer, M., Tilaar, I., Jennifer, M., & Tilaar, I. 2020. The Tourism Industry in A Developing Destination in Time of Crisis The Impact of COVID-19 Pandemic on The Tourism Industry in North Sulawesi, Indonesia Time of Crisis.

Li, J., Hong, T., Nguyen, H., & Coca-stefaniak, J. A. (2020). Coronavirus impacts on post-pandemic planned travel behaviours. Annals of Tourism Research Journal, (January).

Mayunga, J. S. 2007. Understanding and Applying the Concept of Community Disaster Resilience: A capital-based approach. (July), 22–28.

Mcgladdery, C. A., & Lubbe, B. A. 2017. International educational tourism: Does it foster global learning? A survey of South African high school learners. Tourism Management, 62, 292–301. https://doi.org/10.1016/j.tourman.2017.05.004

Mostafanezhad, M., Cheer, J. M., & Sin, H. L. 2020. Geopolitical anxieties of tourism: (Im)mobilities of the COVID-19 pandemic. Dialogues in Human Geography, 1–5. https://doi.org/10.1177/2043820620934206

Mulsim, G. O., Muslim, F. N., Haerani, E., Muslim, D., & Sophian, R. I. 2017. Disaster Awareness Campaign of Indonesian Boy Scout Gerakan Pramuka for Students in Bandung, West Java, 42–47.

Sheresheva, M. Y. 2020. Coronavirus and tourism. Population and Economics, 4(2), 72–76. https://doi.org/10.3897/popecon.4.e53574

Shiwaku, K., & Shaw, R. 2016. Future Perspectives of Disaster Resilience of Education System in Japan and Abroad. Springer Japan.

Stankov, U., Filimonau, V., & Vujičić, M. D. 2020. A mindful shift: an opportunity for mindfulness-driven tourism in a post-pandemic world. Tourism Geographies, 1–10. https://doi.org/10.1080/14616688.2020.1768432

Yang, Y., Zhang, H., & Chen, X. 2020. Coronavirus pandemic and tourism: Dynamic stochastic general equilibrium modeling of infectious disease outbreak Yang. Annals of Tourism Research Journal, (January).

Zenker, S., & Kock, F. 2020. The coronavirus pandemic – A critical discussion of a tourism research agenda. Tourism? Anagement Journal, (January).

Promoting Creative Tourism: Current Issues in Tourism Research – Kusumah et al. (Eds)
© 2021 Taylor & Francis Group, London, ISBN 978-0-367-55862-8

Tourism and tourism crisis management in the COVID-19 pandemic time

E. Edison & T. Kartika
Sekolah Tinggi Ilmu Ekonomi Pariwisata Yapari, Bandung, Indonesia

ABSTRACT: Indonesia's tourism continues to increase from year to year in its competitiveness index. However, Covid-19 has had an impact on tourism throughout Indonesia, including West Java Province. Indonesia, through the Ministry of Tourism has issued Ministerial Regulation Number 10 of 2019 concerning Tourism Crisis Management, but there is no single regulation, theory, or experience that can be used as a reference to anticipate the enormity of the Covid-19 epidemic. Therefore, the authors are interested in examining in-depth about tourism and its relation to tourism crisis management. The methodology used is a qualitative methodology through various literature, legislation, and interviews. From the author's observations, this pandemic has an impact on the tourism sector and many employees have been dismissed and laid off from their jobs. The solution is that the government needs to develop regulations on Tourism Crisis Management that are more comprehensive and can be operationalized during a pandemic. Besides, it is necessary to map the destination areas that can be opened gradually with due observance of strict health protocols.

Keywords: tourism, tourism crisis management, West Java Tourism

1 INTRODUCTION

Tourism is one of the drivers of national development as a source of foreign exchange and is a serious concern of the Indonesian government. Every year Indonesia's tourism growth continues to increase. In the Travel and Tourism Competitiveness Index (TTCI) report issued by the 2019 World Economic Forum, Indonesia is ranked 40th in the world with a score of 4.3. (World Economic Forum 2019). This means that there is a growth of two points when compared to the assessment in 2017.

The tourism sector has become a mainstay of the country's foreign exchange (Kuntadi 2019). The tourism sector has been predicted to be the second largest source of foreign exchange contribution to Indonesia. However, the COVID-19 corona pandemic changed everything (Mutiah 2020). The Covid-19 pandemic globally disrupted domestic supply chains, financial market volatility, shocks to consumer demand, and negative impacts in key sectors such as travel and tourism (Sugihamretha 2020). Even according to UNWTO, Travel, and tourism is among the most affected sectors with airplanes on the ground, closed hotels, and travel restrictions put in place in virtually all countries around the world (UNWTO 2020).

Meanwhile, based on the Indonesian Hotel and Restaurant Association (PHRI) data, the tourism industry has experienced a potential loss of income from foreign tourists of around the US $ 4 billion or the equivalent of Rp. 60 trillion from January 2020 to April 2020 due to the Corona outbreak from the domestic market in the Hotel sector losing potential revenue of around Rp. 30 trillion (Mayasari 2020). It is strengthened by the statement of the Chairman of the Association of Indonesian Hotels and Restaurants (PHRI) of West Java Province, "in West Java as of May there have been 528 hotels with 240 restaurants registered, with 15,050 employees were laid off" (Muchtar 2020).

DOI 10.1201/9781003095484-92

By looking at the background of the problems above, the authors will first examine from the literature aspect on the issue of the global COVID-19 pandemic during the easing period, as a reference in taking an analysis of the area studied. Second, the authors see the legislation related to the handling of the epidemic in terms of its implementation aspect. Third, the authors attempt to build a tourism concept during a pandemic and develop behaviour in new adaptations.

1.1 *Issues of the global COVID-19 pandemic during the easing period*

On Sunday (10/5/2020), the Wuhan Government acknowledged that one person was positive, followed by five others the next day (Rizal 2020). Furthermore, as reported by NBC News, "Wuhan, the epicenter of the Chinese coronavirus outbreak, is conducting nucleic acid testing for its 11 million inhabitants for 10 days after a new cluster of cases raised fears of a second wave of infections." (NBC News 2020). The same thing happened in the tourist area of Itaewon nightclub, shortly after the easing of restrictions was introduced in South Korea. "The Korea Centers for Disease Control and Prevention (KCDC) on Tuesday said at least 102 people had tested positive in connection with cases related to nightclubs and bars" (Cha & Smith 2020). These cases have sparked fears of a second wave after South Korea began relaxing restrictions meant to contain the coronavirus (Tan 2020). This shows that the trend of decreasing cases and slowing down tourism activities during the COVID-19 pandemic can trigger new cases.

However, a conclusion from the results of research conducted by Gössling, Scott, & Hall, States that "COVID-19 provides extraordinary lessons for the tourism industry, policymakers and tourism researchers about the effects of global change. The challenge now is to learn together from this global tragedy to accelerate the transformation of sustainable tourism" (Gössling et al. 2020).

1.2 *Legislation in non-natural disaster mitigation*

If you look at the explanation of Law Number 24 of 2007 concerning Disaster Management, the characteristics of the COVID-19 pandemic are included in the category of non-natural disasters, in which Article 1 (3) states that, "Non-natural disasters are disasters caused by non-natural events or series of events, which include failing technology, failing modernization, epidemics, and disease outbreaks "(Law No. 24/2007). Non-natural factors include a. social / political situation; b. health / infectious disease outbreaks; c. technology; d. environmental pollution; e. economy; and / or f. other non-natural disasters caused by human actions (Ministerial Regulation No.10 of 2009). The framework for Ministerial Regulation Number 10 of 2019 concerning Tourism Crisis Management states that, "The Framework for the MKK consists of phases: a. preparedness and mitigation; b. emergency response; c. recovery; and d. normalization."

Meanwhile, the Covid-19 pandemic was declared a health disaster by the government of the Republic of Indonesia through Presidential Decree No. 11 of 2020 concerning the Determination of the Public Health Emergency for Corona Virus Disease 2019 (COVID-19) on March 31, 2020. This determination is based on the fact that The spread of Corona Virus Disease 2019 (COVID-19) is extraordinary, marked by the increasing and expanding the number of cases and deaths across regions and countries as well as having a broad impact on political, economic, social, cultural, defense and security aspects, as well as the welfare of the community in Indonesia (Feriandi 2020).

To accelerate the handling of COVID-19 as a national health disaster, the President issued Presidential Decree Number 7 of 2020 concerning the Task Force for the Acceleration of Handling Corona Virus Disease2019 (COVID-19). The task force aims to: 1. increase national resilience in the health sector; 2. accelerating the handling of COVID-19 through synergy between ministries and local governments; 3. increase the anticipation of the development of the escalation of the spread of COVID-19; 4. increasing the synergy of operational policy making; and increase readiness and ability to prevent, detect, and respond to COVID-19 (Feriandi 2020). Later the Task Force changed its name to COVID-19 Task Force based on Presidential Regulation No. 82 of 2020.

From the explanation above, it can be concluded that the COVID-19 pandemic is not just a non-natural disaster as described in Ministerial Regulation no. 10 of 2019 concerning Tourism Crisis

Management, but it has entered the category of a national health disaster that needs extraordinary special handling.

2 METHODOLOGY

The research method is a scientific way to get data with specific purposes and uses (Sugiyono 2016). This study used a descriptive qualitative method. The research design was used by conducting interviews, looking at the development of Covid-19 in the mass media, and various kinds of literature, and seeing the effectiveness of existing regulations regarding tourism crisis management. The object of research observation is the province of West Java.

3 RESULT

Learning from Wuhan, China and Itaewon, Korea, the handling of the tourism management crisis, especially easing restrictions needs to be done carefully because this can trigger new cases that get out of control as the spread is through person to person contact, or by equipment that have been touched by someone who is positive for Covid-19. Even recently, WHO stated that Covid-19 can be spread through the air. If you look at the trend in this case, the spread of Covid-19 in Indonesia continues to soar. On the other hand, many tourism destinations in West Java have been closed, employees have been dismissed, and even some employees have been laid off. It can be said that this is a storm that hit West Java and global tourism. This is because there is no theory and / or learning about previous case handling. However, according to the conclusions of research conducted by Higgins-Desbiolles, it is stated that, taking the assumption that the COVID-19 pandemic crisis offers a rare and invaluable opportunity to rethink and reorganize tourism (Higgins-Desbiolles 2020). Moreover, "Returning to pre-pandemic growth patterns will take time and will depend on the depth and extent of the recession triggered by COVID-19" (Prideaux et al. 2020). Therefore, it is necessary to map or regulate safe tourism, so that tourism activities do not completely die.

In addition, the Ministerial Regulation No. 10 of 2019 concerning Tourism Crisis Management, it can be said that this regulation has not been able to anticipate and be operationalized in handling non-natural crises such as the COVID-19 pandemic outbreak. In fact, handling the tourism crisis during a pandemic uses general ad hoc references, such as presidential decrees or presidential regulations. Therefore, it is necessary to reformulate a more comprehensive regulation on crisis management in handling tourism in the case of a pandemic that involves stakeholders. Therefore, the stakeholders (local government, tourism associations, academics, and tourism entrepreneurs) can immediately formulate: safe tourism mapping; procedures based on location and conditions; facility requirements that must be met in the context of health protocols; service and health requirements. The aim is to accelerate the transformation of sustainable tourism. According to Gössling, Daniel Scott, & Hall., "COVID-19 provides great lessons for the tourism industry, policy makers and tourism researchers about the effects of global change. The challenge now is to learn together from this global tragedy to accelerate the transformation of sustainable tourism" (Gössling et al. 2020). In the end, a new normal can be built in tourism activities.

4 CONCLUSION

The predictions for the end of the pandemic are unsettled. While tourism activities are closely related to the economy and the lives of people who depend on tourism activities and their supply chains. Therefore, it is necessary to have a safe tourism mapping, and the government needs to restructure the more comprehensive tourism crisis management, and can technically be operationalized during a pandemic. Experts said that Covid-19 will not disappear and it will always be there. Building a New Normal as stated by the president is a necessity.

REFERENCES

Cha, S., & Smith, J. 2020, May 12. *South Korea investigators comb digital data to trace club coronavirus cluster*. Retrieved from: www.reuters.com: https://www.reuters.com/article/us-health-coronavirus-south korea-idUSKBN22O0OD

Feriandi, Y. 2020. *COVID-19 dan Manajemen Bencana*. Bandung: Pusat Penerbitan Universitas (P2U) Unisba.

Gössling, S., Scott, D., & Hall, C. M. 2020. *Pandemics, tourism, and global change: a rapid assessment of COVID-19*. Retrieved from Journal of Sustainable Tourism: https://doi.org/10.1080/09669582.2020.1758708

Higgins-Desbiolles, F. 2020. Socializing tourism for social and ecological justice after COVID-19. *Tourism Geographies*, 22(3), 610–623. https://www.tandfonline.com/doi/full/10.1080/14616688.2020.1757748

Kuntadi. 2019, Augustus 22. *Kalahkan Migas, Pariwisata Sumbang Devisa Terbesar USD19,2 Miliar*. Retrieved from Okezone.com: https://economy.okezone.com/read/2019/08/22/320/2095457/kalahkan-migas-pariwisata-sumbang-devisa-terbesar-usd19-2-miliar

Mayasari, S. 2020, June 18. *Ini daerah tujuan wisata yang paling terdampak pandemi Covid-19 menurut Kemenparekraf*. Retrieved from Contan.co.id: https://nasional.kontan.co.id/news/ini-daerah-tujuan-wisata-yang-paling-terdampak-pandemi-covid-19-menurut-kemenparekraf

Muchtar, H. 2020, May 12. Ketua PHRI Jawa Barat dan GIPI Kawa Barat. (T. Kartika, Interviewer)

Mutiah, D. 2020, Maret. *Sektor Pariwisata Nyaris Tumbang Akibat Corona Covid-19, Menparekraf Masih Siapkan Solusi*. Retrieved from liputan6.com: https://www.liputan6.com/lifestyle/read/4209455/sektor-pariwisata-nyaris-tumbang-akibat-corona-covid-19-menparekraf-masih-siapkan-solusi

NBC News. 2020, Mei 14. *NBC News*. Retrieved from Wuhan Starts Citywide Testing Again For COVID-19: https://www.youtube.com/watch?v=u-zW15X2BRU

Peraturan Menteri Pariwisata Republik Indonesia nomor10 Tahun 2019 tentang manajemen Krisis Kepariwisataan.

Prideaux, B., Thompson, M., & Pabel, A. 2020. Lessons from COVID-19 can prepare global tourism for the economic transformation needed to combat climate change. *Tourism Geographies*, 22(3), 667–678. https://www.tandfonline.com/doi/full/10.1080/14616688.2020.1762117

Rizal, J. G. 2020, May 12. *Peringatan WHO, Gelombang Kedua Covid-19 dan Pelonggaran Pembatasan*. Retrieved from www.kompas.com: https://www.kompas.com/tren/read/2020/05/12/140300465/peringatan-who-gelombang-kedua-covid-19-dan-pelonggaran-pembatasan?page=all.

Sugihamretha, I. D. 2020. Respon Kebijakan: Mitigasi Dampak Wabah Covid-19 Pada Sektor Pariwisata. *The Indonesian Journal of Development Planning* , 191–206.

Sugiyono. 2016. *Metode Penelitian Kuantitatif, Kualitatif dan R&D*. Bandung: Alfabeta.

Tan, W. 2020, May 12. *https://www.cnbc.com*. Retrieved from Cases in South Korea continue to rise as country tracks new night club cluster: https://www.cnbc.com/2020/05/13/coronavirus-latest-updates-asia-europe.html

UNWTO. 2020, May. *Impact assessment of the COVID-19 outbreak on international tourism*. Retrieved from UNWTO: https://www.unwto.org/impact-assessment-of-the-covid-19-outbreak-on-international-tourism

UU No. 24/2007. (n.d.). *Undang-Undang Republik Indonesia Nomor 24 Tahun 2007 Tentang Penanggulangan Bencana*.

World Economic Forum. 2019, September 4. *The Travel & Tourism Competitiveness Report 2019*. Retrieved from www.weforum.org: https://www.weforum.org/reports/the-travel-tourism-competitiveness-report-2019

Promoting Creative Tourism: Current Issues in Tourism Research – Kusumah et al. (Eds)
© 2021 Taylor & Francis Group, London, ISBN 978-0-367-55862-8

Differences in trust and risk-taking propensity for travelers from Indonesia

A. Njo & F. Andreani
Petra Christian University, Surabaya, Indonesia

ABSTRACT: Traveling is a fun but risky activity depending on the destination. The risks can be reduced by careful planning, especially in the pandemic period. The purpose of this study is to explore trust and risk-taking propensity of Indonesian tourists who travel to other cities or countries. Data collection was carried out by distributing questionnaires online and offline to tourists from Indonesia who did solo or in group traveling and obtained 159 tourists. The results show that there are no differences in trust between solo travelers and group travelers, but there are differences in risk-taking propensity. However, women have more trust than men, and men are more willing to take risks than women. The benefits of this output for tourism practitioners are to create appropriate marketing strategies when offering tourism programs for both groups.

Keywords: trust, risk-taking propensity, traveler

1 INTRODUCTION

Tourism is a fun activity to do individually or in groups after getting tired of doing activities such as work or study. Data from the World Tourism Organization (2019) states that tourist destinations to various regions of the world are leisure, recreation and holiday (56%), visiting relatives or friends, medical treatment or religious activities (27%), and the rest are business activities and others-other. The highest growth in arrivals came from Asia and the Pacific (7%) and Europe (5%) in France, Spain, the United States, China, and Italy as the five highest destination countries. The various tourist destinations are the choice of travelers to do fun activities, because they have certain features or characteristics as points of interest (Buhalis 2000). But after the Covid-19 pandemic, the number of tourists visiting at various destinations has decreased greatly due to the prohibition of arrival in various countries and tourist attractions to reduce the risk of spreading the virus. Developments in the second half of 2020 showed some change in the prohibition of visiting other countries without the right reasons, so this openness makes it possible for tourists to travel again.

Trust is the hope or certainty an individual has. Associated with tourist destination, travelers will take into consideration on the basis of his confidence regarding the intended location. The trust factor is an important antecedent for travelers to travel to these destinations (Mohammed 2016). Collaborative relationships established among different organizations in the tourism industry will reduce risk but at the same time also increase bargaining power in tourism (Wang & Fesenmaier 2007). So, the level of traveler confidence increases and long-term relationships occur with travelers (Fyall et al. 2003; Kim et al. 2009). The main components of trust include honesty, kindness, and competence, so trust has a major successful role in managing tourism destination marketing (Choi et al. 2016). Trust in certain goals influences specific components inherited in personal behavior, such as attitude (Kim et al. 2009, and Sichtmann 2007) and perceptions about risk (Kim et al. 2009; Teo & Liu 2007).

In addition to trust, travelers have considerations about the risks to be faced when choosing a tourist destination. Perceived risk is defined as an individual's perception of uncertainty and negative consequences due to carrying out certain activities (Reisinger & Mavondo 2005), one of

DOI 10.1201/9781003095484-93

which is conducting tourism activities. Risks include organized crime, terrorist activities, economic crises, pandemic, natural disasters, diseases, and other extreme events that increase feelings of fear for travelers. These diverse risk perceptions are a major component of the decision-making process when evaluating goals (Sönmez & Graefe 1998). The most common dimensions of risks perceived by travelers are financial, physical, socio-psychological, health, and performance (Yang et al. 2017). This trust and risk create consideration for the travelers so that they are motivated to decide the best destination.

Swain (1995) introduces the definition of gender in tourism as a starting point for future research. Gender is conceptualized in identities related to men and women, and gender identity is constructed culturally and socially. This study aims to explore the differences expressed between solo travelers and group travelers as well as female and male travelers on the variable of trust and risk. Solo travelers and female travelers face a higher risk and need greater confidence than group travelers and male travelers when choosing travel destinations. This condition is interesting to be investigated further because of the advantages and disadvantages of traveling individually or in groups and based on gender. The benefits of research in the tourism industry for tourism businesses is to conduct reliable strategic planning to overcome the differences in tourism activities so that the forms of promotion and cooperation patterns can be made according to the needs of travelers and which need a very large adjustment in the pandemic and after this pandemic.

2 LITERATURE REVIEW

2.1 *Trust*

Trust not only includes trust in the ability of partner organizations to complete tasks, but also confidence in the good intentions or positive intentions of partners and the perception that partners adhere to acceptable values (Vlaar et al. 2007). Regarding tourism, trust is the result of personality and image in accordance with tourist destinations (Chen & Phou 2013) or the results of the image itself (Loureiro & González 2008). The basis of trust is divided into two domains, namely, affective or cognitive and behavior. The cognitive or affective domain is related to individual beliefs. The behavioral domain relates to individual behavioral tendencies to depend on others to act reliably, emotionally, and honestly (Rotenberg et al. 2005).

Specifically, the components of virtue, honesty, and competence to create trust are attached to people's attitudes; these components also apply to organizations. On the other hand, being honest, kind, and competent in the local population will be the best intermediary at the tourist destination, thereby increasing the level of traveler confidence. Local residents, as part of various public or private institutions at tourist sites, play a key role in the level of travelers' trust in these institutions (Sirdeshmukh et al. 2002). Gender-based trust shows men have independent self-construction, women have independent interdependence. Women are more relation oriented while men are more collective oriented. Gender differences have an impact on the way a person is interdependent with others (Maddux & Brewer 2005).

H1: Women have higher trust than men

2.2 *Risk*

Risk is a consumer's perception of overall negative actions based on the likelihood of evaluating negative results and the likelihood that these results will occur (Mowen & Minor 1998). In tourism literature, personal risks include personal perceptions about pre-trip threats and actual experiences during travel (Tsaur et al. 1997). Risk perception is very important for travel decision making because it is able to change the decision-making process and choice of goals (Sönmez & Graefe 1998; Poon & Adams 2000). Risks that can occur in tourist destinations are crime, terrorism, the spread of disease, and natural disasters (Kozak et al. 2007). Fischhoff et al (2004) found that travelers tend to travel to a destination that is highly predictable at the risk level of the location above or below the traveler's risk tolerance threshold. Hazardous incidents can change risk perceptions and reduce tourist arrivals (Chew * Jahari 2014).

Furthermore, personal risks include social risks, health, financial, and physical (Hajibaba et al. 2015). Some studies find different dimensions of perceived risk such as socio-psychological, physical, financial, and time do not affect tourist visiting intention (Sönmez & Graefe 1998; Qi et al. 2009). Other studies find that physical, financial, and social-psychological risks have negative effects on visit intention and revisit intention (Chew & Jahari 2014). Regarding female travelers, researchers found female travelers would change travel plans if they had an increased risk perception (Kozak et al. 2007). Women also have a higher risk perception than men for certain purposes (Lepp & Gibson 2003). Female travelers pay more attention to safety and security in accommodations and face security threats when walking in remote places (Khoo-Lattimore & Prayag 2015; Khoo-Lattimore & Prayag 2016).

H2: Men tend to dare to take higher risks than women

2.3 *Traveler*

Travelers are also called tourists or those who travel for fun. If done individually, it is called single travelers (Campbell 2009) or solo travelers; traveling with a spouse, parents, children, friends, or relatives or in groups is called group travelers. Tourism activities are influenced by different pre-trip attractions (Bianchi 2016; Jordan 2016) so that these might result in different behavioral patterns. Travel to certain destinations as a group of travelers and solo travelers will create a series of positive or negative experiences, as well as create an impression related to tourist destinations (Walls et al. 2011). A positive experience creates a good destination image so travelers will feel satisfied.

Chhabra (2004) conducted a comparative study between solo and non-solo travelers in Sacramento, California, about travel destinations, type of accommodation, travel planning, length of stay, age, income, and gender. The results found solo travelers were younger than non-solo travelers and on average they stayed longer. Solo travelers spend less money during a visit, despite having almost the same income. Tomaszewski (2003) mentions solo female travelers (backpackers) become stronger spiritually as individuals, are tolerant of risk, and more confident, more independent, and freer during and after the trip. Single female travelers are more concerned with health and safety than solo male travelers (Chiang & Jogaratnam 2006) and they believe that they are more vulnerable to risk (Gibson & Jordan 1998). This study will develop demographic variables, namely, age, education, status, and employment to further deepen the analysis of the traveler.

H3: Solo travelers are more likely to take risks than group travelers

3 RESEARCH METHOD

This research is a comparative study, which is aimed at Indonesian people who like to travel abroad or other regions in the country. These tourism activities can be carried out individually or in groups and are carried out both by females and males. Primary data were collected using questionnaires distributed offline and online to travelers according to the sample criteria. The period of distributing questionnaires was for three months from March-May 2020. Questionnaires could only be collected from 159 respondents, due to pandemic constraints that sufficiently inhibited offline data dissemination and the travel ban during the pandemic period. After that, validity and reliability tests were performed before analyzing data using ANOVA in SPSS program. ANOVA is more appropriate to be used to confirm differences in trust and risk-taking propensity between groups. Table 1 shows the variables and data coding of the variables in this study, consisting of trust and risk, and also demographic data respondents.

Table 1. Research variables.

Variable	Description
Type of Traveler	1 = Solo; 0 = Group
Trust	Likert scale 1–5 (strongly disagree – strongly agree)
Risk Taking Propensity	Likert scale 1–5 (strongly disagree – strongly agree)
Gender	1 = Female; 0 = Male
Age	1 <= 17–25 years; 2 = 26–35 years; 3 = 36–45 years; 4 = >45 years
Education	1 = High school; 2 = Undergraduate; 3 = Postgraduate
Status	1 = Single; 2 = Married
Occupation	1 = Businessman; 2 = Governmental officer; 3 = Private company officer; 4 = Housewife, 5 = Others (Student, Accountant, Architect, Doctor, etc.)

Table 2. Description of respondents.

	Traveler		
Description	Solo	Group	Total
Gender			
Male	42 (26.4%)	24 (15.1%)	66 (41.5%)
Female	41 (25.8%)	52 (32.7%)	93 (58.5%)
Age			
<17–25 years	58 (36.5%)	70 (44.0%)	128 (80.5%)
26–35 years	7 (4.4%)	2 (1.3%)	9 (5.7%)
36–45 years	10 (6.3%)	0 (0.0%)	10 (6.3%)
>45 years	8 (5.0%)	4 (2.5%)	12 (7.5%)
Education			
High school	9 (5.7%)	12 (7.5%)	21 (13.2%)
Undergraduate	61 (38.4%)	60 (37.7%)	121 (76.1%)
Post-graduate	13 (8.2%)	4 (2.5%)	17 (10.7%)
Status			
Single	62 (39.0%)	72 (45.3%)	134 (84.3%)
Married	21 (13.2%)	4 (2.5%)	25 (15.7%)
Occupation			
Businessman	13 (8.2%)	7 (4.4%)	20 (12.6%)
Government officer	3 (1.9%)	0 (0.0%)	3 (1.9%)
Private company officer	15 (9.4%)	8 (5.0%)	23 (14.5%)
Housewife	0 (0.0%)	1 (0.6%)	1 (0.6%)
Others	52 (32.7%)	60 (37.7%)	112 (70.4%)
Total	83 (52.2%)	76 (47.8%)	159 (100%)

4 RESULT AND DISCUSSION

4.1 *Findings*

Questionnaires were distributed online and offline for as many as 159 respondents with the following descriptions in Table 2.

Table 2 shows the respondents who were slightly more dominant on individual tours (solo travelers). Women prefer travel in groups. Respondents predominantly under the age of 17 to 25 years, single status choose to travel individually or in groups. Most respondents have bachelors education and work as professionals (notary, architects, doctors), but there are also some who have not worked

Table 3. Output validity and reliability test.

Code	Description	Pearson Correlation	
		Risk	Trust
Risk1	I like to go camping in the wilderness.	0.675**	–
Risk2	I like to swim far away from the beach or unguarded lake or ocean.	0.672**	–
Risk3	I like to go on vacation to a third-world country without any planned accommodation.	0.536**	–
Risk4	I like to ski beyond my personal abilities.	0.732**	–
Risk5	I like to play white water rafting.	0.749**	–
Risk6	I like to take a sky diving class every weekend.	0.705**	–
Risk7	I like to try bungee jumping off a tall bridge.	0.697**	–
Trust1	In general, do you agree that everybody can be trusted?	–	0.746**
Trust2	Do you agree that most of the time there will be somebody who is willing to help?	–	0.680**
Trust3	Do you think that most people will try to take advantage from you if they have some chance?	–	0.613**
	Cronbach Alpha	0.807	0.584

Description: ** p-value < 0.05; Cronbach Alpha > 0.6

because of being students. Then the validity and reliability tests are performed. Validity test results for risk and trust variables are attached in Table 3.

The test results show all indicators of risk and trust variables are valid because the value below is 0.05. Risk variable is said to be reliable because its Cronbach alpha is above 0.6, but trust variable is said to be quite reliable as its value is less than 0.6. In this study, trust variable is still used. The ANOVA test is then performed to prove the differences in risk and trust in different groups, namely, tourism and gender activities.

The results of Levine test for risk variable is 0.024 < 0.05 and trust variable is 0.000 < 0.05; so, the two variables are declared not to be homogeneous. However, the difference test continues and displays the test results in Table 4 and Table 5 showing the type of tourism, namely solo tourism has a higher risk-taking propensity than group tours. The status of unmarried travelers has a higher risk-taking propensity than married travelers. Interaction test of type of traveling and age as well as type of traveling and educational background shows joint effect on risk. Thus, it proves there are differences in the results of the interaction of the two variables on risk. Trust variable is influenced by age and the interaction between gender and age.

4.2 *Discussions*

Women have more trust also than men, especially those who are at the age of 26–35 and above 45 years. The type of traveling shows no difference in trust. However, the roles of gender and age affect trust, so older women have different beliefs compared to young men. Women tend to depend on others to act reliably, emotionally, and honestly as stated in the research of Rotenberg, et al. (2005). A sense of trust in women increases if supported by honesty, kindness, and competence in the local population of tourist destinations according to the findings of the local people's behavior in the study of Sirdeshmukh et al. (2002).

Research on Indonesian travelers shows that both men and women tend to travel solo at a young age (<17–25 years), especially singles with professional or student backgrounds, while those who choose to travel in groups are young women. This condition is also supported in the results of risk and trust tests. Travelers who are young and single are more willing to take risks do solo traveling; while women who consider safety and health tend to travel in groups. Chhabra (2004) proves that solo travelers are younger. Chiang and Jogaratnam (2006) found that single female travelers were more concerned with health and safety than male solo travelers, and women were more vulnerable to risk (Gibson & Jordan 1998).

5 CONCLUSION AND RECOMMENDATIONS

Regarding trust, women have more trust than men, because women have greater dependence on others. Men are more willing to take risks than women, so they are solo travelers who are willing to take more risk than group travelers. Risk is inherent to young and single travelers, while trust needs to be built from both parties, namely, travelers and those involved in tourism activities in tourist destinations.

Research on tourism in such a pandemic condition is very interesting to be further investigated, because traveling in groups increases health risks, while traveling individually increases safety risks. Financial planning is also needed to realize these tourism activities, because currently tourism funds are increasing quite sharply due to health procedures that must be met. Therefore, to improve the tourism sector, the role of the government and the organizers of tourism activities need to work together to increase travelers' trust by reducing the negative sides that can occur in tourist areas such as pickpocketing, robberies, and kidnappings as well as increasing public facilities that are healthily appropriate. Area tourism that is safe, comfortable, and meets health procedures after a pandemic will increase tourist visits.

REFERENCES

Bianchi, C. 2016. Solo holiday travelers: motivators and drivers of satisfaction and dissatisfaction, *International Journal of Tourism Research*, 18(2), 197–208.

Buhalis, D. 2000. Marketing the competitive destination of the future. *Tourism Management*, 21, 97–116.

Campbell, A. 2009. The importance of being valued: solo 'grey nomads' as volunteers at the national folk festival. *Annals of Leisure Research*, 12(3–4), 277–294.

Chen, C. & Phou, S. 2013. A closer look at destination: image, personality, relationship and loyalty, *Journal of Tourism Management*, 36, 269–278. https://doi.org/10.1016/ j.tourman.2012.11.015

Chew, E.Y.T. & Jahari, S.A. 2014. Destination image as a mediator between perceived risks and revisit intention: a case of post-disaster Japan, *Tourism Management*, 40, 382–393.

Chhabra, D. 2004. Determining spending behavior variations and market attractiveness of solo and nonsolo travelers, *e-Review of Tourism Research*, 2(5), 103–107.

Chiang, C.Y. & Jogaratnam, G. 2006. Why do women travel solo for purposes of leisure?. *Journal of Vacation Marketing*, 12(1), 59–70.

Choi, M., Law, R., & Heo, C. Y. 2016. Shopping destinations and trust-tourist attitudes: Scale development and validation. *Tourism Management*, 54, 490–501.

Fischhoff, B., De Bruin, W. B., Perrin, W., & Downs, J. 2004. Travel risks in a time of terror: Judgments and choices. *Risk Analysis*, 24(5), 1301–1309.

Fyall, A., Callod, C., & Edwards, B. 2003. Relationship marketing: The challenge for destinations. *Annals of Tourism Research, 30*(3), 644–659.

Gibson, H. & Jordan, F. 1998. Traveling solo: a cross-cultural study of British and American women aged 30–50. *Paper presented at the Fourth International Conference of the Leisure Studies Association, Leeds*. 16–20 July

Hajibaba, H., Gretzel, U., Leisch, F., & Dolnicar, S. 2015. Crisis-resistant tourists. *Annals of Tourism Research*, 53, 46–60

Jordan, F. 2016. Tourism and technology: revisiting the experiences of women travelling alone. *Paper presented at the 2nd International Conferences on Information Technology and Business (ICITB), Bandar Lampung*, 15 October.

Khoo-Lattimore, C. & Prayag, G. 2015. The girlfriend getaway market: segmenting accommodation and service preferences. *International Journal of Hospitality Management*, 45, 99–108.

Khoo-Lattimore, C. & Prayag, G. 2016. Accommodation preferences of the girlfriend getaway market in Malaysia: self-image, satisfaction and loyalty. *International Journal of Contemporary Hospitality Management*, 28(12), 2748–2770.

Kim, H. B., Kim, T. T., & Shin, S. W. 2009. Modeling roles of subjective norms and eTrust in customers' acceptance of airline B2C eCommerce websites. *Tourism Management*, 30(2), 266–277.

Kim, T. T., Kim, W. G., & Kim, H. B. 2009. The effects of perceived justice on recovery satisfaction, trust, word-of-mouth, and revisit intention in upscale hotels. *Tourism Management*, 30(1), 51–62.

Kozak, M., Crotts, J. C., & Law, R. 2007. The impact of the perception of risk on international travelers. *International Journal of Tourism Research*, 9(4), 233–242

Lepp, A. & Gibson, H. 2003. Tourist roles, perceived risk and international tourism. *Annals of Tourism Research,* 30(3), 606–624.

Loureiro, S.M.C. & Gonzalez, F.J.M. 2008. The importance of quality, satisfaction, trust, and image in relation to rural tourist loyalty. *Journal of Travel and Tourism Marketing*, 25(2), 117–136. https://doi.org/10.1080/10548400802402321

Maddux, W. W., & Brewer, M. B. 2005. Gender Differences in the Relational and Collective Bases for Trust. *Group Processes & Intergroup Relations, 8*(2), 159–171. https://doi.org/10.1177/1368430205051065

Mohammed, A. 2016. Does eWOM influence destination trust and travel intention: A medical tourism perspective. *Economic Research – Ekonomska Istraživanja*, 29(1), 598–611.

Mowen, J. C., & Minor, M. 1998. *Consumer Behavior* 5th ed. Upper Saddle River, N.J.: Prentice-Hall

Poon, A. & Adams, E. 2000. How the British will travel 2005. Tourism Intelligence International. 29(1), 279–281.

Qi, C.X., Gibson, H.J. & Zhang, J.J. 2009. Perceptions of risk and travel intentions: the case of China and the Beijing Olympic games. *Journal of Sport & Tourism*, 14, 43–67.

Reisinger, Y. & Mavondo, F. 2006. Cultural Differences in Travel Risk Perception. *Journal of Travel & Tourism Marketing*, 20(1), 13–31.

Rotenberg, K. J., Fox, C., Green, S., Ruderman, L., Slater, K., Stevens, K., & Carlo, G. 2005. Construction and validation of a children's interpersonal trust belief scale. *British Journal of Developmental Psychology*, 23(2), 271–292. Doi: 10.1348/026151005X26192.

Sichtmann, C. 2007. An analysis of antecedents and consequences of trust in a corporate brand. *European Journal of Marketing*, 41(9/10), 999–1015.

Sirdeshmukh, D., Singh, J., & Sabol, B. 2002. Consumer trust, value, and loyalty in relational exchanges. *Journal of Marketing, 66*, 15–37.

Sönmez, S. F., & Graefe, A. R. 1998. Influence of terrorism risk on foreign tourism decisions. *Annals of Tourism Research*, 25(1), 112–144.

Swain, M. B. 1995. Gender in tourism, *Annals of Tourism Research*, 22(2), 247–266. https://doi.org/10.1016/0160-7383(94)00095-6.

Teo, T. S., & Liu, J. 2007. Consumer trust in e-commerce in the United States, Singapore and China. *Omega*, 35(1), 22–38.

Tomaszewski, L.E. 2003. Peripheral travelers: how American solo women backpackers participate in two communities of practice. PhD thesis, Texas A&M University, College Station.

Tsaur, S. H., Tzeng, G. H., & Wang, G. C. 1997. The application of AHP and fuzzy MCDM on the evaluation study of tourist risk. *Annals of Tourism Research*, 24(4), 796–812.

Vlaar, P. L., Van den Bosch, F. J., & Volberda, H. W. 2007. On the evolution of trust, distrust, and formal coordination and control in inter-organizational relationships: Toward an integrative framework. *Group & Organization Management*, 32(4), 407–429. Doi: 10.1177/1059601106294215.

Walls, A.R., Okumus, F., Wang, Y.R. & Kwun, D.J.W. 2011. An epistemological view of consumer experiences, *International Journal of Hospitality Management*, 30(1), 10–21.

Wang, Youcheng & Fesenmaier, D.R. 2007. Collaborative destination marketing: A case study of Elkhart County, *Indiana. Tourism Management*. 28, 863–875. Doi: 10.1016/j.tourman.2006.02.007.

World Tourism Organization 2019, International Tourism Highlights, 2019 Ed., UNWTO, Madrid, Doi: https://doi.org/10.18111/9789284421152.

Yang, E.C.L., Khoo-Lattimore, C. & Arcodia, C. 2017. Constructing space and self through risk taking: a case of Asian solo female travelers. *Journal of Travel Research*, 57(2), 260–272.

Promoting Creative Tourism: Current Issues in Tourism Research – Kusumah et al. (Eds)

The impact of travel constraints on travel intention

F. Andreani & A. Njo
Petra Christian University, Surabaya, Indonesia

ABSTRACT: Rapid mass transportation has enabled people to travel from one place to another easily. However, tourists have some constraints that affect their intention to travel, like interpersonal constraints (interaction factors or the relationship between individual' characteristics), intrapersonal constraints (individual psychological states and leisure preferences) and structural constraints (intervening factors between leisure preference and participation). The study is to find the impacts of the trilogy of constraints on travel intention. Factor regression analysis is used to analyze data of 159 respondents. The results of the study show interpersonal constraints have positive but insignificant impact on travel intention; while intrapersonal and structural constraints have negative and significant impacts on travel intention. Thus, it is very useful for tourism and travel industries to accommodate tour packages to minimize the constraints.

Keywords: interpersonal constraints, intrapersonal constraints, structural constraints, travel intention

1 INTRODUCTION

Rapid mass transportation has enabled people to travel from one place to another easily. A lot of people travel from one destination to others. Based on governmental data (Statistik Wisatawan Nusantara 2018 2019), 43.3% out of 100% domestic tourists travel for visiting families/ relatives or friends and 42.9 % for leisure; and the rest are for pilgrimage (4.6%), shopping (2.4%), healthcare (1.8%), business (1.5%), training (0.7%), Meetings Incentives Conferences Exhibitions/ MICE (0.5%), sports (0.3%0 and others (2%). By the end of 2019, domestic tourist movements reached 275 million trips, lower than the number of domestic tourist trips in 2018 which had reached 303.4 million trips. This is due to airline prices that were still high enough (Zuhriyah 2019). In addition, the number of foreign tourists coming to Indonesia in January 2020 increased by 5.85 percent compared to the number of visits in January 2019. However, when compared to December 2019, the number of foreign tourist visits in January 2020 has decreased by 7.62 percent ("Jumlah kunjungan wisman ke Indonesia Januari 2020 mencapai 1.27 juta kunjungan" 2020).

Travelling has become one of the entertaining activities to do. By travelling, tourists learn a lot of from the new surroundings, like culture and nature in the proposed destination. Apart from that, tourists can also enjoy themselves, relax and move out from their day-to-day activities. However, tourists have different constraints that may affect their intention to travel.

Constraints refer to conditions that may hinder tourists to participate in leisure activities. These include lack of time and information, financial conditions, transportation, and others.

Crawford and Godbey (1987) proposed a trilogy of travel constraints, namely interpersonal constraints, intrapersonal constraints and structural constraints. First, interpersonal constraints have something to do with interpersonal interaction or the relationship between individual' characteristics, for example spouse companionship that may affect joint preference for specific leisure activities. Second, intrapersonal constraints involve individual psychological states and leisure preferences, for example stress, anxiety, religiosity and the like. Third, structural constraints refer to intervening factors between leisure preference and participation, for example family life-cycled stage, financial resources, season, climate, availability of opportunity, and others.

 DOI 10.1201/9781003095484-94

Many studies have been taken using the trilogy model to study tourist travelling behavior in different parts of the world. However, there is a little study about travel constraints in relations with travel intention of Indonesian tourists to proposed destinations. So in this study the writers would like to find out the impacts of the three dimensions of travel contraints on travel intention of Indonesian tourists. This study could help managers to accommodate tourists' packages and necessities when tourists are travelling either by themselves or with others.

2 LITERATURE REVIEW

2.1 *Travel constraints*

Constraints are some conditions that may hinder one's freedom, desires and participation. So, travel constraints include factors or barriers that affect individuals or tourist to participate in leisure activities, either locally or internationally. The trilogy of travel constraints, originally conceptualized by Crawford and Godbey (1987) and further developed by Crawford et al. (1991), have made significant contributions to further studies.

The model involves three dimensions of constraints: interpersonal, intrapersonal, and structural constraints (Crawford & Godbey 1987; 1991). Interpersonal constraints occur when individuals have no one to travel with. Thus, it may prevent them to participate in leisure activities as there is no friend, spouse/ mate, or family members to take part with. Intrapersonal constraints refers to individual psychological states or conditions that affect them to participate in the activities of interest. Individuals experience this constraints due to lack of interest, stress, anxiety, depression and religiosity. This kind of constraints are not relatively stable and may change across life stages depending on individual maturity. Structural constraints represent as the intervening factors between leisure preferences and participation. These include lack of time, money, opportunity, climate, information and access (Nyaupane & Andereck 2007; Walker & Virden 2005).

The indicators of interpersonal constraints in this study includes no one to travel with, family and friends not interested, not fun to travel alone. Intrapersonal constraints connsists of indicators like: traveling is risky, not interested at activities in the intended destination and not interested to travel in the intended destination. Then, for the indicators of structural constraints are no money to travel, no time to travel, no sufficient information to travel, unfavourable weather in the intended destination and insufficient transportation in the intended destination.

2.2 *Travel intention*

Behavioral intention reflects individual planned future behaviors. It includes individual positive statements, product or service purchase and even recommendations about product or service being purchased to others. One of these intentions, in leisure and tourism, is the intention to travel or visit a destination (Nunkoo & Ramkissoon 2010). Travel intention is a perceived likelihood of tourists to visit a particular destination in a specific period of time (Ahn et al. 2013)

Jang, Bai, and Hu's (2009) study toward senior travellers in Taiwan suggested that travel intention represents a mental process that leads to travel motivation and transformed into behavior. The behavior to travel is also affected by tourists' attitudes and preferences; and these include travel options like destinations, travel modes and patterns, frequency, companions, duration, and budget (Beerli & Martin 2004).

In addition, Wu (2015) stated that individual behavior to travel is also affected by rational as well as effective product evaluation. Rational evaluation involves the needs which can be fulfilled by the features or environments in the destination; whilst, effective evaluation represents emotions which develop feelings about the destination (Prayag & Ryan 2012). The indicators of travel intention in this study involves being aware of the intended destination, interested in visiting the intended destination and wanting to visit the intended *destination.*

2.3 *The relationship between travel constraints and travel intention*

The trilogy of travel constraints has been adopted by many studies in tourism and leisure. A study towards under-graduate students to join cruise tourism indicates that travel constraints are taken

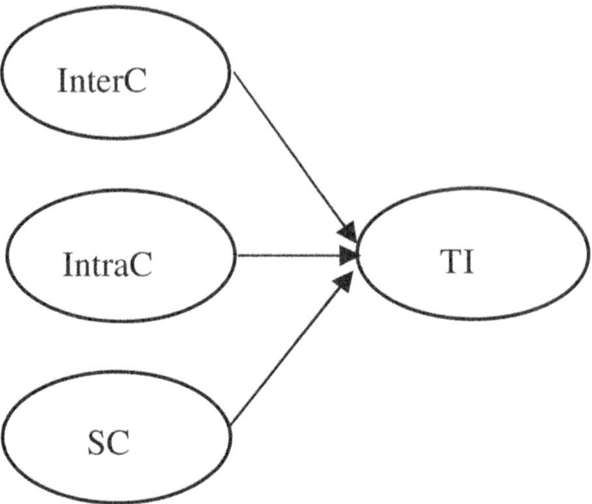

Figure 1. Research model.

as one of variables that may influence individual or tourist decision in join cruises. The results of the study show that travel constraints have negatively influenced tourists travel intention (Hung & Petrick 2012)

Age, income and life stage have significant effect on travel intention (Kattiyapornpong & Miller 2009). Some demographic attributes within structural constraints, like place attributes, lack of time, and lack of money have emerged that prevent tourists to travel. That is why some strategies are needed to overcome the constraints (Nyaupane & Andereck 2007).

Another research finding shows that four determinants shaping the Portuguese south-eastern residents' traveling decisions are travel companion (interpersonal constraints), motivations (intrapersonal constraints), time and money (structural constraints). Travel companion is the strong determinant for travel decision as individuals dislike traveling alone and decide to join a trip only if they have someone to participate with (Silva & Correia 2008).

Moreover, Khan, Chelliah, and Ahmed's study (2019) to Malaysian university students intending to visit India indicates that among three dimensions of travel constraints, interpersonal and intrapersonal constraints have influenced travel intention negatively and significantly; whilst, structural constraints have negative and insignificant impact on travel intention. The findings suggest that tourists having high interpersonal and intrapersonal constraints demonstrated low intention to visit India. While, the insignificant impact of structural constraints on travel intention can be explained by the respondents' profile in this research who are mostly young and educated. So, it's easy for these young respondents to find a lot of information of the proposed destinations, especially climate, travel modes, and places to visit via internet. These were partly similar to the results of previous study (Hung & Petrick 2012) in which vital information about places to visit, climate, modes of transportation, and the like is not a problem for young travellers as this kind of information can be easily searched in the internet.

Based on the previous findings, the writers propose further hypotheses:

H1: There is a negative impact of interpersonal constraints (InterC) on travel intention (TI).
H2: There is a negative impact of intrapersonal constraints (IntraC) on travel intention (TI).
H3: There is a negative impact of structural constraints (SC) on travel intention (TI).

3 RESEARCH METHOD

Online questionnaires were distributed via google forms from March, 12 to April 25, 2020. A five-point Likert scale was used in the questionnaire ranging from 1) strongly disagree, 2) disagree,

3) neutral, 4) agree, and 5) strongly agree. The first part of the questionnaire consists of respondents' demographic data, such as age, gender, education, occupation, and the like. The second part includes 14 items related to travel constraints and intention. The criteria of the respondents are those who have ever travelled at least once within the last one year since March 2020, either by themselves or with others.

The google forms were spread via WhatsApp and LINE groups; but only 159 respondents returned their responses. Due to pandemic Covid-19, it was very difficult to get more respondents as many people didn't do any activities outdoor. Then, data were processed with using descriptive, classic assumption test and regression analysis.

All 14 items in the questionnaires are valid as their correlation significances are 0.00 (less than 0.05). In addition, the values of Cronbach Alpha for interpersonal constraints is 0.319 (low), intrapersonal constraints is 0.560 (moderate), structural constraints is 0.716 (sufficient), and travel intention is 0.906 (perfect). These data have also fulfilled classic assumption tests like normality, heteroscedasticity and multicollinearity tests.

4 FINDINGS AND DISCUSSION

4.1 *Findings*

Most of the 83 respondents (52.2%) have ever traveled alone and the rest 76 respondents (47.8%) have traveled with others. As much as 85.5% respondents have traveled solo once up to twice within the last year; and 47.8% respondents mostly have traveled with others, especially with their families (63.5%), friends (31.4%). They mostly traveled by plane (81.1%) and car (13.2%). The places they visited were Java, Bali, and Asian countries (Singapore, Malaysia, Hongkong, Japan, Thailand, Korean, China, Taiwan), USA, Europe and Australia. Their purposes to travel are to enjoy nature and culinary as well as for fun and leisure, like going to theme parks and shopping.

Furthermore, the mean values of travel constraints can be seen in the following table.

Table 1. The results of mean values.

	M Mean	St Dev
Interpersonal constraints		
No one to travel with	1.89	1.059
Family and friends not interested	1.88	1.052
Not fun to travel alone	3.33	1.395
Average mean	2.37	
Intrapersonal constraints		
traveling is risky	3.48	1.321
not interested at activities in the intended destination	2.02	0.971
not interested to travel in the intended destination	1.87	0.994
Average mean	2.46	
Structural constraints		
no money to travel	2.46	1.184
no time to travel	2.78	1.154
no sufficient information to travel	2.28	1.171
unfavourable weather in the intended destination	2.47	1.030
insufficient transportation in the intended destination	2.18	0.986
Average mean	2.43	
Travel intention		
aware of the intended destination	4.28	0.888
interested at visiting the intended destination	4.35	0.886
wanting to visit the intended destination	4.38	0.832
Average mean	4.34	

The average mean of interpersonal constraints is 2.37. It shows that the respondents do not think that having no one to travel with, having no family or friends interested in traveling, and traveling alone become their constraints. The respondents still want to travel, no matter by themselves or with others. The average mean of intrapersonal constraints is 2.46. This also indicates that the respondents do not feel that this becomes their barriers to travel. Even though traveling is risky, but they are still interested in the activities as well as visiting the intended destination. Moreover, the average mean of structural constraints is 2.43. This represents that respondents do not agree that money, time, sufficient information about climate and transportation in the intended destination become their hindrance. Nowadays people can find any information via internet easily. The average mean of travel intention is 4.34. This shows that respondents are aware, interested and willing to visit the intended destination.

In addition, the value of R is 0.467 showing that the relation among independent variables (interpersonal, intrapersonal, and structural constraints) with dependent variable (travel intention) is good enough. The value of adjusted R square is 0.203 meaning that the three independent variables has influenced travel intention as much as 20.3 %, whereas the rest is influenced by other variables that are not examined in this study.

The regression model is also considered fit as its significance value is 0.000 and F count is higher than F table (14.830 > 2.610). The results of regression analysis can be seen in the following table:

Table 2. The results of regression analysis.

Model	Unstandarized Coefficient	t	Sig.
Constant	−4.118	0.000	1.000
Interpersonal constraints	0.121	1.474	0.143
Intrapersonal constraints	−0.372	−4.560	0.000
Structural constraints	−0.217	−2.495	0.014

So, the multiple regression equation is as follows:

$$TI = -4.118 + 0.121\ \text{InterC} - 0.372\ \text{IntraC} - 0.217\ \text{SC}$$

This equation indicates that the higher constraints the respondents encounter, the less intention they have to visit the intended destination or vice versa.

4.2 *Discussion*

The results in table 2 show that interpersonal constraints have a positive impact (coefficient 0.121) on travel intention insignificantly (sig. 0.143 > 0.05). Thus, the first hypothesis is not supported. The result is not in line with the previous studies. This is due to the fact that most respondents in this study are solo travelers (52.2%), so they do not worry too much if they have no one or friends to travel with. Or even when their families are not interested to travel. They use to travel alone and still have fun. The mean value of interpersonal constraints also indicates that these are not their barriers to travel. With or without anyone or friends and families. the respondents still want to travel and still enjoy themselves.

Furthermore, intrapersonal constraints have negative impact (coefficient −0.372) on travel intention significantly (sig. 0.000). Therefore, the second hypothesis is supported. The respondents realized that traveling is risky. So, the less interests the respondents have in participating in the activities in the intended destination, the higher possibilities they are reluctant to travel. This result provided further support of earlier studies by Silva and Correia (2008) and Khan, et al. (2019) showing that tourists who have high intrapersonal constraints have low intention to visit some destinations.

In addition to this, structural constraints have negatif impact (coefficient −0.217) on travel intention significantly (sig. 0.014). So, the third hypothesis is supported. The respondents who have enough time, money, sufficient information as well as favourable climate and transportation in destinations will have higher intention to travel rather than those who haven't had such information. The higher structural constraints the tourists have, the lower intention they want to travel. This result is also in line with previous studies by Nyaupane and Andereck (2007) as well as Hung and Petrick (2012) demonstrating that the less structural constraints the tourists have, the more intention they have to travel.

Among those three constraints, structural constraints have biggest impact on travel intention. Traveling involves well planned actions especially those related with financial support and quality time. Tourist having enough financial support is not enough, if they do not want to spare some time to enjoy themselves. Even, when they have already had both money and time, it's not enough. They should have sufficient information, weather or climate as well as transportation in the intended destination in order to have fun.

5 CONCLUSION AND RECOMMENDATIONS

Among three hypotheses in this study, only two (the second and third hypotheses) are supported; whilst, the first hypothesis is not supported. Thus, it may provide some feedbacks for tour operators to provide complete and detailed information and pictures in their marketing tools regarding the weather or climate and activities the tourists can participate in. So, they can prepare what clothes to wear to do suitable activities in the intended destination. Besides that, it is also important to provide complete modes of transportation in the destination. By doing so, the tourists can have complete picture to anticipate any emergencies that might appear. For those traveling in groups, it would be advisable to have some pre-departure briefing with all participants to enable them to have questions and answers with tour operators. This also enhances all participant to get to know each other well.

The limitation of this study relies on the values of Cronbach Alpha for interpersonal and intrapersonal constraints which are low to moderate, it is suggested to have more samples for further research. Apart from that some other variables (like travel motivation, financial literacy and perceived risk) can be employed.

REFERENCES

Ahn, T., Ekinci, Y., & Li, G. (2013). Self-congruence, functional congruence, and destination choice. *Journal of Business Research, 66*, 719–723.

Beerli, A., & Martin, J. (2004). Tourists' characteristics and the perceived image of tourist destination: A quantitative analysis—a case study of Lanzarote, Spain. *Tourism Management, 25*(5), 623–636.

Crawford, D., & Godbey, G. (1987). Reconceptualizing Barriers to Family Leisure. *Leisure Sciences, 9*(2), 119–127. doi:https://doi.org/10.1080/01490408709512151

Crawford, D., Jackson, E. L., & Godbey, G. (1991). A hierarchical model of leisure constraints. *Leisure Sciences, 13*(4), 309–320. doi:doi.org/10.1080/01490409109513147

Hung, K., & Petrick, J. (2012). Testing the effects of congruity, travel constraints, and self-efficacy on travel intentions: An alternative decision-making model. *Tourism Management, 33*, 855–867. doi:10.1016/j.tourman.2011.09.007

Jang, S., Bai, B., & Hu, C et al. (2009). Affect, travel motivation and travel intention: A senior market. *Journal of Hospitality & Tourism Research, 33*, 51–73.

Jumlah kunjungan wisman ke Indonesia Januari 2020 mencapai 1.27 juta kunjungan. Badan Pusat Statistik. Retrieved Mei 10, 2020, from https://www.bps.go.id/pressrelease/2020/03/02/1712/jumlah-kunjungan-wisman-ke-indonesia-januari-2020-mencapai-1-27-juta-kunjungan-.html

Kattiyapornpong, U., & Miller, K. (2009). Socio-demographic constraints to travel behavior. *International Journal of Culture, Tourism and Hospitality Research, 3*(1), 81–94. doi:DOI 10.1108/17506180910940360

Khan, M., Chelliah, S., & Ahmed, S. (2019). Intention to visit India among potential travellers: Role of travel motivation,perceived travel risks, and travel constraints. *Tourism and Hospitality Research, 19*(3), 351–367. doi:10.1177/1467358417751025

Nunkoo, R., & Ramkissoon, H. (2010). Gendered Theory of planned behavior and residents' support for tourism. *Current Issues in Tourism, 13*, 525–540.

Nyaupane, G., & Andereck, K. (2007). Understanding travel constraints: Application and extension of a leisure constraints model. *Journal of Travel Research, 20*(40), 1–7. doi:10.1177/0047287507308325

Prayag, G., & Ryan, C. (2012). Antecedents of tourists' loyalty to Mauritius: The role and influence of destination image, plave attachment, personal involvement, and satisfaction. *Journal of Travel research, 51*, 342–356.

Silva, O., & Correia, A. (2008). Facilitators and constraints in leisure travel participation: The case of the southeast of Portugal. *Tourism and Hospitality Research, 2*(1), 25–43.

Statistik Wisatawan Nusantara 2018. (2019). Badan Pusat Statistik. Retrieved May 10, 2020, from https://www.bps.go.id/publication/2019/07/02/5249c2b645e21291b51dfc1a/statistik-wisatawan-nusantara-2018.html

Walker, G., & Virden, R. (2005). Constraints on outdoor recreation. (E. Jackson, Ed.) *Constraints to Leisure*, 201–219.

Wu, C. (2015). Foreign tourists' intentions in visiting leisure farms. (Elsevier, Ed.) *Journal of Business Research, 68*(4), 757–762.

Zuhriyah, D. (2019). *Kemenpar Bidik 275 Juta Pergerakan Wisatawan Nusantara Tahun Ini.* bisnis.com. Retrieved from https://ekonomi.bisnis.com/read/20190910/12/1146568/kemenpar-bidik-275-juta-pergerakan-wisatawan-nusantara-tahun-ini

Promoting Creative Tourism: Current Issues in Tourism Research – Kusumah et al. (Eds)
 ISBN 978-0-367-55862-8

The effect of COVID-19 outbreak to the destination choice and the intention to visit the destination among Thai tourists

M. Worrachananun
Bangkok University, Bangkok, Thailand

N. Srisuksai
University of Phayao, Thailand

ABSTRACT: COVID-19 pandemic is the greatest challenge that mankind has faced since World War II. Governments in different countries attempt to create the responding policies including travel ban, social distancing, and working from home. This affects the tourism industry as both the tourists and the tourism entrepreneurs. This research aims to study the effects of this infectious pandemic to the destination choice and their intention to visit among Thai tourists. The researchers conducted a survey among 400 Thai tourists ages between 18 to 60 years old. The researchers used Pearson correlation method to analyze the data. The results show that Thai tourists will have more concern to travel to domestic destinations over international destinations after the outbreak of the Covid-19 pandemic, and the Covid-19 outbreak has no effect on the intention to visit the destination.

Keywords: Covid-19 outbreak, destination choice, intention to visit

1 INTRODUCTION

In December 2019, a new infectious disease emerged in Wuhan, China, which was later named as Covid-19. The ongoing Covid-19 pandemic is considered as the most significant global public health issue (Huang et al. 2020). According to the situation report from the World Health Organization (2020) at the time of writing (July 4, 2020), the infected confirmed cases are 11,230,357 globally. The pandemic does generally not only affect the public health concern; but it could also be a trigger to the butterfly effect that can create socio-economic crises globally (Chakraborty & Maity 2020). The critical outbreak of the Covid-19 situation can significantly cause the economy recession not only in China or in the United States but for the whole world (Yoo 2020).

This infectious Covid-19 outbreak that was initiated from the seafood wholesale market in Wuhan, China, has caused the pandemic, which creates the health emergency situation around the world (Wang et al. 2020). There are more than 200 countries affected by this outbreak (World Health Organization 2020). Speaking of the pandemic, one of the most possible disease transmission includes travel and tourism activity (Chakraborty &Maity 2020). Therefore, to localize the critical outbreak, many governments exercise crucial enforcement, which is social distancing; some countries may even go to the complete lockdown policy. University classes are cancelled. Academic conferences are postponed or are switched to be online sessions. Shops are shuttered. Many sport leagues are suspended or played behind closed door (Lambert & Hesman Saey 2020).

Travel and the tourism industry is definitely affected by the outbreak. Many international flights are postponed or cancelled, and the other types of transportation are also deferred due to the attempt to restrict the irretrievable damage. People involved in the tourism industry are facing ultimate difficulties among other economic sectors (UNWTO 2020). The output of the tourism industry globally alone decreased approximately 50% to 70% (Organization for Economic Co-operation and Development 2020a).

DOI 10.1201/9781003095484-95

The tourism industry in Thailand created a national revenue of 62 billion USD in 2019, which is impressively increased from 58 billion USD from the previous year (CEIC Data 2020). Unfortunately, the emergence of Covid-19 has affected the tourism industry in Thailand directly. The number of the tourism arrival has decreased approximately 76.4% (CEIC Data 2020). The tourism industry includes many business sectors, such as lodging, restaurant and food service, transportation, and tour operators and travel agencies (Waeoraweewong & VorSittha 2018). The tourism industry has continuously generated income to the country. The researcher aims to study the effect of the Covid-19 pandemic on the destination choice and the intention to visit the destination among the Thai tourists.

2 LITERATURE REVIEW

Despite the limited knowledge of coping with the recent pandemic caused from the Covid-19 virus, its effect have spread vastly, globally in many different industries, including the tourism and hospitality industry (Organization for Economic Co-operation and Development 2020b). As Sommez and Graefe (1998) proposed, the process of destination choice is the sequential process of starting with motivation to travel, followed by a decision to travel. The concept of tourism destination choice suggested by Cohen (1979), Plog (1974), and Pearce (1988) also includes the individuals' mindset and determination to travel.

The decision-making process of the tourism destination is a complicated process. According to Fesenmaier, and Jeng (2000), the decision-making process comprises several sub-decision influenced factors, which are core (time, trip, destination choice, etc.), secondary (tourism attractions and activities), and route decisions (the budget for gifts and shopping stop, etc.). All in all, once the tourists decide to travel, it is called the intention to visit.

However, the tourists' decision on selecting destination or the intention to visit does not only include the tourist inner-self parameter, as the destination itself also plays an important role (Yoo et al. 2018). The intention to visit the destination may be changed or distracted when there is some psychological propaganda among tourists' conscious, such as perceived risks at the destinations, negative stereotypes toward the destination, and negative testimonials from peers (Hughes 2002). Therefore, the authors would like to study about the intention to visit of the Thai tourists toward the destination among this pandemic of Covid-19.

3 RESEARCH OBJECTIVES

- To discover the relationship of Covid-19 effect and destination choice among Thai tourists.
- To discover the relationship of Covid-19 effect and intention to visit the tourism attractions among Thai tourists

4 RESEARCH HYPOTHESIS

H1. COVID-19 pandemic effect has a significant relationship on destination choice among Thai tourists

H2. COVID-19 pandemic effect has a significant relationship with the intention to visit the tourism attractions among Thai tourists

5 RESEARCH METHOD

This research was conducted to determine whether the Covid-19 outbreak affects the destination choice and the intention to visit the destination among Thai tourists. The data were collected via

Table 1. Pearson correlation between the Covid-19 outbreak and the destination decision among Thai tourists

	The effect of Covid-19 outbreak		
Thai tourist decision	Pearson Correlation	Sig. (2- tailed)	Correlation level
International destination decision	−.007	.899	Very low
Domestic destination decision	.041	.426	Very low

Table 2. Pearson correlation between the Covid-19 outbreak and the intention to visit the destination among Thai tourists.

	The effect of Covid-19 outbreak		
Thai tourist decision	Pearson Correlation	Sig. (2-tailed)	Correlation level
Intention to visit the destination	.279**	.000	Very low

self-administered questionnaires composed of four sections, namely, the Covid-19 effect perception of the respondents, desired destination choice of the respondents, intention to visit the destination of the respondents, and demographic information. The researcher designed the questionnaire and sent it to three tourism experts for review. After that, the researcher conducted the pilot test of 50 respondents, to make certain that the questionnaire is comprehended by the general public.

The researcher used G* power program to compute the sample size, which is 370 respondents ages between 18 to 60 years old. At first, the researchers totally collected 400 samples as per guarantee that the valid sample size that the researchers will use is reached and a valid amount. Moreover, the researchers use the Pearson correlation to analyze the correlation of two variables, which includes two tables. Table 1 shows the correlation between the effect of Covid-19 outbreak and Thai tourists' decision on the destination. Table 2 shows the correlation between the Covid-19 outbreak and the intention to visit the destination.

6 RESULTS AND DISCUSSION

The researchers collected 400 samples, which are 103 males and 297 females; ages between 18 to 60 years old, which 81.5% of the sample is at age 21 to 30, and 10.5% of the sample is at age 31 to 40. To reply to the research questions, the researcher conducts correlation coefficient analysis by using Pearson Product Moment Correlation Coefficient, which found that the Covid-19 outbreak has a very low degree of negative correlation with international destination choice decisions among Thai tourists ($r = -.007$). And the Covid-19 outbreak has a very low degree of positive correlation with domestic destination choices among Thai tourists ($r = -.041$). Note that the researcher refers to the conversion of correlation coefficient to five levels from Hinkel (1998) that the correlation coefficient at .00 to .30 equals very low correlation.

And to reply to the H2, the researcher conducts correlation coefficient analysis by using Pearson Product Moment Correlation Coefficient, which found that the Covid-19 outbreak has no correlation with intention to visit the destination among Thai tourists, with statistically significant at .01, as shows in Table 2.

7 DISCUSSION AND IMPLICATIONS

The data from Tables 1 and 2 show that even the Covid-19 outbreak has a very low degree of negative correlation among Thai tourists to make a decision to travel aboard. This is contrary with the decision among Thai tourist to travel domestically, which seems to be more concerned with the positive correlation coefficient. The possible causes could come from many reasons, such as Thai government, which presents the Covid-19 outbreak situation very frequently, and the Thai government restricting the nighttime curfew nationwide. In addition, famous tourism cities, such as, Bangkok and Phuket, have the highest number of Covid-19 cases. The domestic travel is difficult as the Thai government attempts to localize the citizens to stay in their places, to prevent the possibility of carrying the virus.

However, the correlation coefficient among Thai tourists to the Covid-19 outbreak and the domestic destination decision is still in a very low degree. Therefore, to spur the domestic travel, the government could have issued the policy to support the tourism entrepreneur to be able to launch an attractive promotion to the local tourists, as well as ease up the domestic travel requirements to encourage people to travel in the country.

As the intention to visit the destination, Thai tourists seem to be not concerned with the effects of the Covid-19 outbreak, so whenever there is an ease up policy to travel, Thai tourists have a very high possibility to travel wherever they wish to, as the data show that they have no correlation coefficient with the intention to visit. This could be possible from the confidence toward the Thai government that Thai people have, and the hope that the cure of Covid-19 will be here soon.

8 CONCLUSION

Tourism is an industry that has been affected the most by the Covid-19 outbreak as it includes many high touching activities among humans. During the Covid-19 pandemic, people have been directly and indirectly forbidden to travel; however, after the Covid-19 outbreak, Thai tourists have a high possibility to visit the destinations that they wish to; if there is a policy that eases up the social distancing and encourages traveling, the tourism industry may come back in the game again. However, the challenge is to balance the safety and security from the spread of Covid-19 with the encouragement of people to travel safely.

REFERENCES

CEIC Data. (2020). Thailand tourism revenue. CEIC Data. Retrieved May 28, 2020 from https://www.ceicdata.com/en/indicator/thailand/tourism-revenue.

Chakraborty., I. & Maity., P. (2020). COVID-19 Outbreak: Migration, effects on society, global environment and prevention. *Science of the total environment*. 728(2020)138882.

Cohen, E. (1979), "A phenomenology of tourist experiences", Sociology, Vol. 13 No. 2, pp. 179–201.

Fesenmaier, D. P., & Jeng, J. M. (2000). Assessing structure in the pleasure trip planning process. *Tourism Analysis*, 5(1), 13–27.

Hinkle, D.E. (1998). Applied statistics for behavioral sciences. Houghton Miffin College.

Huang., C., Wang, Y., Li, X., Ren, L., Zhao, J., Hu, Y., et al. (2020). Clinical features of patients infected with 2019 novel coronavirus in Wuhan, China. Lancet 395, 497–506. https:// doi.org/10.1016/S0140-6736(20)30183-5.

Hughes., H. (2002). Gay men's holiday destination choice: a case of risk and avoidance. *International journal of tourism research*. 4(2002). p.299–312.

Lambert., J. & Hesman Saey., T. (2020). Social distancing, not travel bans, is crucial to limiting coronavirus'spread. *Science News*. Retrieved May 28, 2020 from https://www.sciencenews.org/article/coronavirus-pandemic-limit-spread-social-distancing-travel-bans

Organization for Economic Co-operation and Development, (2020a). Annual national accounts and OECD calculations. OECD. Retrieved May 28,2020 from http://www.oecd.org/coronavirus/en/

Organization for Economic Co-operation and Development. (2020b). Tourism policy responses to the Coronavirus(COVID-19). OECD. Retrieved August 5,2020 from https://www.oecd.org/coronavirus/policy-responses/tourism-policy-responses-to-the-coronavirus-covid-19-6466aa20/

Pearce, P. (1988), The Ulysses Factor: Evaluating Visitors in Tourist Settings, Springer-Verlag, New York, NY.

Plog, S.C. (1974), "Why destination areas rise and fall in popularity", Cornell Hotel and Restaurant Administration Quarterly, Vol. 14 No. 4, pp. 55–58

Sommez. S,& Graefe, A. (1998). Determining future travel behavior from past travel behavior experience and perceptions of risk and safety. *Journal of travel research*. 37(2):171–177.

UNWTO. (2020). COVID-19: Putting People First. UNWTO. Retrieved May 28, 2020 from https://www.unwto.org/tourism-covid-19.

Waeoraweewong., K. & Vor-Sittha., P. (2018). Economic contribution of transportation on tourism industry in Thailand. *Development Economic Review*. 12(2). P.58–79.

Wang, C., Horby, P.W., Hayden, F.G., Gao, G.F. (2020). A novel coronavirus outbreak of global health concern. Lancet.

World Health Organization. (2020). Coronavirus disease(COVID-19) situation report. World Health Organization. Retrieved May 26,2020. From https://www.who.int/docs/default-source/coronaviruse/situation-reports/20200525-covid-19-sitrep-126.pdf?sfvrsn=887dbd66_2

Yoo.,C., Yoon.,D. & Park.,E. (2018). Tourist motivation: an integral approach to destination choices. *Tourism review*. Vol.73(2). P.169–185.

Yoo., J.H. (2020). The fight against the 2019-nCoV outbreak: an arduous march has just begun. J. Korean Med. Sci. 35–56 https://doi.org/10.3346/jkms.2020.35.56.

Promoting Creative Tourism: Current Issues in Tourism Research – Kusumah et al. (Eds)
© 2021 Taylor & Francis Group, London, ISBN 978-0-367-55862-8

The impacts of COVID-19 at Karangsong Mangrove Centre

D.J. Prihadi & Z. Guanghai
Ocean University of China, Qingdao, China

Khrisnamurti
Universitas Negeri Jakarta, Jakarta, Indonesia

H. Nuraeni
Universitas Padjajaran, Bandung, Indonesia

ABSTRACT: The situation of pandemic Covid-19 creates enormous challenges for the tourism industry, including in Indonesia. Karangsong Mangrove Centre, a nature-based mangrove forest conservation tourist attraction located at Indramayu-West Java, experienced a ceased operation of the place due to strict regulation and policy from the local government. This paper analyzes how the impact of Covid-19 affects Karangsong Mangrove Centre. The unit analysis of this study is the tourist arrival at Karangsong Mangrove Centre. Due to the health protocol in Indonesia, the study was conducted as desk research and collected from various resources related to the topic as secondary data. The data was analyzed using descriptive and content analysis to interpret the data. The study found out that Karangsong Mangrove Centre plans to sustain using the incentive stimulus while waiting for the new regulation to reopen.

Keywords: tourism, Covid-19, tourist arrival, mangrove tourism

1 INTRODUCTION

Tourism is recognized to be fragile. It is influenced by external factors such as political turmoil, natural disasters, and health (Gade & Ankuthi 2016). Several cases have affected tourism in the past years, such as tsunami, SARS, political unrest in several countries, and others (Breiling 2016). Covid-19 pandemic is considered one of the significant threats to human civilization amongst past pandemics such as the Spanish flu, AIDS, H1N1 Swine flu, and others. According to Xinge (2020), Covid-19 was discovered in Wuhan, Hubei, China, in December 2019 and was firstly called 2019 Novel Corona virus (2019-nCoV), and the disease caused by this virus was called Novel Coronavirus Pneumonia (NCP). On February 11, 2020, NCP was officially named Covid-19 (Corona Virus Disease 2019) by the World Health Organization. The Covid-19 pandemic is causing significant damage to the global tourism industry. According to the World Travel and Tourism Council (WTTC 2020) results, the possibility for 50 million people worldwide to lose their jobs is the real effect on the travel and tourism industry due to the outbreak of Covid-19. According to the Indonesian Hotels & Restaurant Association (PHRI), predicted to harm Indonesia's tourism industry due to Covid-19 is reaching 1.5 billion US dollars. The effect of the downturn in tourism on small and medium-sized businesses engaged in food and beverage micro-business has hit 27%. Whereas the effect on small-scale food and beverage industry was 1.77%, and medium-sized businesses were 0.07% (Rahmadi 2020). Given the complex and fast-moving complexity of the situation, it is challenging to predict the effect of COVID-19 on international tourism (UNWTO 2020b). UNWTO reports that international (Rahmadi 2020) tourist arrivals will plunge by 20% to 30% in 2020. It will result in a loss of US$ 300 to US$ 450 billion in foreign tourism receipts (exports)–approximately one-third of the US$ 1.5 trillion globally produced in the worst-case scenario (UNWTO 2020b). The Central

 DOI 10.1201/9781003095484-96

Statistical Agency (BPS 2020) also announced that the number of international tourists visiting the airline decreased from 838,978 in December 2019 to 796,934 in January 2020 and 558,892 in February 2020.

The pandemic also threatens the tourism industry in several districts and regions throughout Indonesia, one of which is the Indramayu district in West Java, which is commonly famous for its nature-based tourism, such as beaches, waterfalls, and mangroves. As one of the most visited tourist attractions, Karangsong Mangrove Centre, situated in the north of the town of Indramayu, with a protected mangrove conservation area of 25 hectares, has discovered a significant decline in tourism activities. It is widely known that mangroves protect coastal towns from storms and erosion and generate income for the community through tourism activities (Azkis 2013; Harahab & Setiawan 2017; Nuddin 2010; Wardhani 2011). The activities include camping and boating–mostly focused on wildlife observation–and fishing. Although other tourists engage in day-to-day or part-day activities, others pursue extended visits for recreational fishing and overnight boating activities. Mangroves may not be the primary drive to select a destination, but they provide a common attraction that can affect the choice of a tourism destination, and its popularity continues to increase (Avau et al. 2011). According to Datta et al. (2012), the main factor affecting the sustainable management of mangrove ecosystems is a cooperation between communities, academic institutions, and governments. These efforts are particularly useful when combined with community education (Dharmawan et al. 2016). As a result, communities play an essential role in mangrove ecotourism and management activities (Masud et al. 2017). Karangsong Mangrove Centre is notable amongst the local population. The uncertain situation means that tourism at the Karangsong Mangrove Center is closed, contributing to a variety of significant negative impacts on the local region. Although tourism in Indramayu plays an essential role in the economic sector, the local government of Indramayu is not going to reopen tourist attractions in any near time, until further notification, due to the pandemic conditions (Rahman 2020). Numerous policies and strategies continue to be undertaken by the local government to improve the tourism sector in Indramayu to endure the impacts of Covid-19. This study seeks to explore the impact of Covid-19 at the Karangsong Mangrove Centre.

2 LITERATURE REVIEW

Mangroves are forest areas found along coastlines, coasts, or along rivers that are affected by seawater. Mangrove forests as natural resources have biodiversity that provides benefits for human life. It can reduce the amount of excess pollution in the soil, thereby reducing the greenhouse gas contribution to global warming by sequestering atmospheric carbon dioxide, while greenhouse gas adds as much as 26% to the greenhouse effect. Some of the resources extracted from the value of mangrove forests are in the form of natural tourism services. Utilization of these products and services has provided additional income and is even a significant income in meeting the needs of people's lives. Mangrove ecosystems have begun to be incorporated in several ways into coastal tourism initiatives, which is why every policy should be a long-range master plan that should be directed at spreading the benefits of tourism (Kustanti 2011; Karma 2015). DasGupta and Shaw (2017) also stated that apart from being a robust buffer zone that protects coastal areas by its thick root network, mangrove forests also act as a tourism attraction that produces significant revenues for communities.

3 METHOD

A desk study conducted this research. All the data collected were from systematic existing secondary data from various resources related to the subject, such as the Management Report of Karangsong Mangrove Centre, news related-issues, literature, and policy documents. The technique applied was due to the Covid-19 pandemic situation that was mandatory to follow for the health and safety protocol. The unit analysis of this study consists of the tourist arrival at Karangsong Mangrove

Centre. The data was analyzed with the descriptive analysis used to describe several data in this study, then content analyzed to identify the impact of Covid-19 at Karangsong Mangrove Centre; the result of the analyses and its interpretation is correspondingly contingent.

4 FINDINGS AND DISCUSSION

4.1 *Tourist arrival*

As shown in Table 1, the total visit to Karangsong Mangrove Centre was 67.127 tourists on 2019. There was one month that had the highest tourists' arrivals in June 2019, which was about 15.674 tourists due to school holidays and Eid Fitr holiday.

Compared to the tourist arrivals in 2019 at Karangsong Mangrove Centre, the impact of Covid-19 has caused a major decline of tourist arrivals visiting Karangsong Mangrove Centre Mangrove area that started in February 2020. The cause was due to government policies that required the community to stay at home and tourist attractions should be closed, in order to break the spread of Covid-19 (Kemenparekraf 2020). On March 2020, the area was temporarily closed along with all tourist destinations across Indonesia. This significant impact has dramatically impacted the revenue of the management of the tourist area. This will cause a great loss on the revenue for Karangsong Mangrove Centre management that will affect numerous employees who relied on it. As UNWTO (2020) stated, tourism faces up to the Covid-19 challenges, and public health is paramount due to the fundamental people-to-people nature of the sector. It is also a sector with a proven capacity to bounce back and multiply recovery to other sectors.Coordinated and substantial mitigationand recovery plansto support the sector can generatemassive returns across the whole economy and jobs (UNWTO 2020b).

Table 1. Total tourist arrivals at Karangsong Mangrove Centre in 2019.

No	Month	Tourists Arrival
1	January	5.605
2	February	3.900
3	March	3.924
4	April	3.888
5	May	2.707
6	June	15.674
7	July	7.917
8	August	4.815
9	September	4.492
10	October	3.485
11	November	4.270
12	December	6.450
	Total	67.127

Table 2. Total tourist arrivals at Karangsong Mangrove Centre in 2020.

No	Month	Tourist Arrivals (person)
1	January	5622
2	February	2551
3	March	1323
Total		9496

4.2 *Tourism recovery*

To handle the situation, the President of the Republic of Indonesia, Joko Widodo, is preparing a social protection program for workers in the tourism sector affected by the new Coronavirus pandemic or Covid-19. Any assessment of the impact of this unparalleled crisis on the tourism sector quickly is surpassed by the fast-changing reality (Pertiwi 2020). The form of social protection program that is expected and much needed is direct assistance in cash (BLT) in the form of fresh money for the next few months. Nevertheless, Indonesia is now more responsive and does various policies such as staying at home, physical/social distancing, requiring people to use masks when traveling, suggesting people to work from home, and arranging public transport. Government policy measures can be followed by the community correctly to break the spread of Covid-19. The management of Karangsong Mangrove Centre is thriving to sustain in order to support the local communities to have a decent income and reduce unemployment. The mangrove ecosystem remains sustainable and alive. As Rahmadi (2020) noted, by taking into account tourism policies that need to strengthen are the food and beverage companies, tourism education subsidies to students studying in state and private tourism tertiary colleges, strengthening Indonesian tourism mitigation SOP, which refers to standardization provided by UNWTO and WHO, is very important when recovery is crucial to early preparation. The aim is to renovate the destination in the comfort of tourist attractions, increase the position of the tourism sector in the tourist field, and improve the legislation on the entry of international visitors. With a visa-free policy for visits from countries that have already been or are vulnerable to outbreaks of disease, all must be checked for the sake of the safety of foreign tourists who intend to visit Indonesia.

5 CONCLUSION

The Covid-19 pandemic has an enormous effect on the global tourism industry, including in Karangsong Mangrove Centre, Indramayu-West Java Indonesia. The decline of the tourist arrivals was due to the national regulation to close every tourism destination and tourist attractions across the country. The Ministry of Tourism and Economy Creative is fully aware of the condition. It aims to recover the tourism industry with several strategies that could withstand and sustain the people who work in the industry. To keep managing Karangsong Mangrove Centre, an incentive stimulus from the government is considered. Due to this pandemic situation, the study's limitation was the data obtained from secondary data, which was limited. Further research of this study is conducting field research to observe the real condition of Karangsong Mangrove Centre directly.

REFERENCES

Avau, J, M Cunha-Lignon, B De Myttenaere, M Godart, and F Dahdouh-Guebas. "The commercial images promoting Caribbean mangroves to tourists: case studies in Jamaica,Guadeloupe and Martinique." *Journal of Coastal Research* 64 (January 2011): 1277–1278.

Azkis, F. "The Compability of the Mangrove Ecosytem and the Ecotourism Development Strategy in the Demak District." Universitas Diponegoro, Semarang, 2013.

BPS. (2020). Retrieved from https://www.bps.go.id/publication/2020/04/29/e9011b3155d45d70823c141f/statistik-indonesia-2020.html

Breiling, M. "The Vulnerability Challenge and Business Continuity Models for ASEAN Countries." *Discussion Paper Series Tourism Supply Chains and Natural Disasters*, 2016.

DasGupta, Rajarshi, and Rajib Shaw. Tokyo: Springer Japan KK, 2017.

Datta, D, R N Chattopadhyay, and P Guha. "Community based mangrove management: A review on status and sustainability." *Journal of Environmental* 107 (2012): 88–85.

Dharmawan, B, M Böcher, and M Krott. "The failure of the mangrove conservation plan in Indonesia: Weak research and an ignorance of grassroots politics." *Ocean and Coastal Management* 130 (2016): 250–259.

Gade, J, and R Ankathi. *Tourism Management Philosophies, Principles and Practices.* Zenon Academic Publication, 2016.

Harahab, N, and Setiawan. "Compatibility Index of Mangrove Ecotourism in Malang District." *ECSOFiM: Journal of Economic and Social of Fisheries and Marine* (Brawijaya University) 04, no. 02 (2017): 153–165.

Karma, K. *Managing Tourist Destination.* New Delhi: Kanishka, 2005.

Kemenparekraf. (2020). Retrieved from https://www.kemenparekraf.go.id/

Kustanti, A. *Management of Mangrove Forest.* Bogo: Institut Pertranian Bogor Press, 2011.

Masud, M M, A M Aldakhij, A A Nassani, and M N Azam. "Community-based ecotourism management for sustainable development of marine protected areas in Malaysia." *Ocean and Coastal Management* 136 (2017): 104–112.

Nuddin, M. *The Economic Valuation of the Mangrove Forest Ecosystem and Its Application in Planning the Coastal Area.* Yogyakarta: Graha Ilmu.

Pertiwi, N L.M. 04 18, 2020. https://amp.kompas.com/travel/read/2020/04/18/230000127/seluruh-pramuwisata-ntt-kehilangan-mata-pencaharian (accessed 05 10, 2020).

Rahmadi, Taufan. 04 07, 2020. https://republika.co.id/berita/q84y62440/7-strategi-pariwisata-indonesia-bangkit-dari-corona (accessed 05 10, 2020).

Rahman, H. 05 29, 2020. https://cirebon.tribunnews.com/2020/05/29/psbb-indramayu-tahap-ketiga-mulai-besok-warga-yang-tak-punya-tujuan-jelas-dilarang-masuk-indramayu (accessed 06 02, 2020).

UNWTO. 03 06, 2020. https://www.unwto.org/news/covid-19-unwto-calls-on-tourism-to-be-part-of-recovery-plans. (accessed 04 13, 2020).

UNWTO. 05 2020. https://www.unwto.org/impact-assessment-of-the-covid-19-outbreak-on-international-tourism (accessed 06 08, 2020).

Wardhani, M K. "Kawasan Konservasi Mangrove: Suatu Potensi ekowisata." *Jurnal Kelautan* 04, no. 01 (April 2011).

WTTC. (2020). *Economic Impact Reports*. Retrieved from World Travel and Tourism Index: https://wttc.org/Research/Economic-Impact

Xinge, M A. *A Handbook for Control and Prevention pf Coronavirus Disease 2019 (COVID-19) on Campus.* Xian Jiatong University Press, 2020.

Promoting Creative Tourism: Current Issues in Tourism Research – Kusumah et al. (Eds)
© 2021 Taylor & Francis Group, London, ISBN 978-0-367-55862-8

Impacts of COVID-19 on national security in Indonesia and the alternative of national policy solutions

A.M. Fawzi & A.T. Nugraha
Universiti Utara Malaysia, Kedah, Malaysia

A.G. Subakti
Bina Nusantara University, Jakarta, Indonesia

ABSTRACT: This research paper is written in the purpose to analyse the issue of COVID-19 pandemic, especially in Indonesia, regarding the influences, challenges, and the alternative of policy solutions toward national security. The data used in this research were the secondary data and based on their characteristics, the data being used were quantitative. The technique used in data collection in this research is the documents and records by viewing and analysing the information available and provided in related websites and online articles regarding the current global issue of COVID-19 pandemic. Then, the type of research analysis technique chosen is descriptive analysis. Specifically, the analysis of this research is mainly concerned with the influence of COVID-19 pandemic on law enforcement, health security, economic security, social security, food security, and supply chain management issue in Indonesia. Also, as the alternative of policy solutions related to COVID-19 issue in Indonesia, the Government can potentially identify both the case size and case location, in which this identification can be done by a combination of Rapid Diagnostic Test (RPD) and Polymerase Chain Reaction Test (PCR), or laboratory tests. Besides, breaking the chain of the coronavirus transmission is the responsibility of all parties and indeed must be properly done with good cooperation between the Government and community in the country.

Keywords: COVID-19, law enforcement, health security, economic security, social security, food security, supply chain management

1 INTRODUCTION

Chinese Research Authorities had declared on 7th of January 2020 that new viruses were identified from the marine food industry in Wuhan city, which is called as 2019-nCOV. During the current spread of this COVID-19 pandemic in 2020, other countries were also confirmed of the COVID-19 cases. For instance, Malaysia on 26th of January, Canada on 27th of January, Sri Lanka, Germany, and Cambodia on 28th of January, United Arab Emirates on 29th of January, Finland, India, and Philippines on 30th of January, Italy on 31st of January, United Kingdom, Sweden, Spain, and Russian Federation on 1st of February, Belgium on 5th of February, and Egypt on 15th of February 2020 of confirmed case of 2019-nCoV (Kumar et al. 2020). The number of cases of COVID-19 in Indonesia continues to increase after months since the first report of case of coronavirus in Indonesia, from the total of 2 cases on March 2nd, to the total of 25,773 cases recently on May 30st, 2020 where the total number of people who have recovered at the same date has reached 7,015 people (Worldometer 2020). In this study, Indonesia is mainly chosen with the consideration as one of the most rapid and fatal spread of the coronavirus along with the critical impacts within the country (Soeriaatmadja 2020).

The role of law in our lives can be illustrated by its vital and possible functions in maintaining social security and order, upholding the ideals of truth and justice, controlling the actions of

DOI 10.1201/9781003095484-97

individuals and community, promoting the growth of social connection, helping community to achieve stability, managing regional and city planning, and ultimately managing the economy. (Syah 2018). Therefore, it is important that we behave responsibly and work with the government to protect our life and other people's lives (Varalakshmi & Swetha 2020). The aspect of health security is referring to both reactive and proactive activities required in order to mitigate the danger and effect of acute public safety and health events that threaten the health of individuals either domestically or globally (Aldis 2008). Countries vary greatly about their ability to deter, identify and respond to the outbreaks (Kandel et al. 2020). Thus, the market balance and economic stabilization should be maintained properly (Wahyono 2017). Consequently, any destruction in the economic cycle results in lower GDP rates and higher rates of unemployment in the country. The research has identified an adversely significant effect of higher loss of job and unemployment concerning with suffer of mental health, for instance the stress, anxiety, and depression (Mamun & Ullah 2020). The main problem in the maintenance of social security, to achieve social stability, is the maintenance of national unity and integration which is basic to the disruption of racial, ethnic, religious and regional problems, and in addition to problems of injustice in politics and economics within a country. A strong national integration covers aspects of territorial integration, elite mass, and integrative values and behavior (Kagan 2020). In addition to mobility, travel, and social distancing controls to support coping the pandemic, effective case identification and exclusion, stringent touch monitoring, checking and having quarantine, as well as surveillance of community-wide, would have to be enforced and maintained (Yezli & Khan 2020). Food security in which households will be categorized as food secure when they are already living above the poverty line along with no fear of hunger (Ferranti et al. 2019). When it is required for a step backwards-taking, investing in improving the short food supply chains and in development locally might enable the country community to move forward in order to maintain the accessibility of food products (Cappelli & Cini 2020). The supply chain management means managing the goods and services in a good flow where involving the whole process of turning raw materials into finished goods. Supply chain is as well seeking to centrally monitor or connect among the production process, shipping, and distribution (Hayes 2019). The shortages were particularly prominent in western countries that usually depend on global supply chains to procure such product forms from certainly low-cost economies. The success of these activities is crucial to the discovery of viable production methods and underused supply chains (Shokrani et al. 2020).

Based on the aforementioned issues, the purpose of this study is mainly to analyse the critical influence of existence of COVID-19 pandemic on law enforcement, health security, economic security, social security, food security, and supply chain management issue in Indonesia, and also the alternative of national policy solutions that potentially can be undertaken by government to cope with the pandemic efficiently.

2 LITERATURE REVIEW

The data in this research are the secondary data in the form of global and national information related to currently coronavirus issues generated through online supported by some related previous studies done and written concerned with national security, especially in Indonesia. Based on their characteristic, the data being used are quantitative data (Sugiyono 2009). Then, the source of data is taken from websites and online articles which are potentially focusing on coronavirus issues, including the influences, challenges, and alternative of solutions toward national security in Indonesia. The technique used in data collection in this research is the documents and records by looking and analysing the information available and provided in related websites and online articles regarding coronavirus issues in Indonesia (Widayat 2004). The type of research analysis technique chosen is descriptive analysis, a method designed to describe or contribute to objects that are examined through data or samples collected together without any further analysis and make conclusions that can be applied to the public (Sugiyono 2009).

3 RESULTS, CONCLUSION, AND POLICY RECOMMENDATIONS

The President of Indonesia, Joko Widodo (Jokowi), has signed Government Regulation No. 21 of 2020 concerning "Large-scale Social Restrictions". The government restriction must also fulfill two criteria, namely the number of positive cases and the number of deaths increasing and spreading rapidly to several regions (Arnani 2020). The "Civil Emergency Policy" has been regulated in a Government Regulation in lieu of Law No. 23 of 1959 concerning the State of Danger. Government had well implemented applicable measurement of control, in which specifically concerned with holiday extension, hospitalization, quarantine, travel restrictions, and fully city lockdown to mitigate significantly and efficiently on decreasing the pandemic's impacts (Lin et al. 2020). Based on Law No. 6 of 2018 concerning Health Quarantine, as stipulated in Article 1 paragraph 2, states that public health emergencies are extraordinary public health events assessed by widely spread of infectious diseases which presenting health threats and massively expanding across regions or countries (Sadikin 2020). Then according to Worldometer (2020), regarding the health issue of COVID-19 cases development in Indonesia, up until June 27th, 2020, it has been identified, with recently addition of 1,385 confirmed cases, that the total confirmed case of COVID-19 in Indonesia has reached 52,812 cases along with total of people who have fully recovered as many as 21,909 people and total of death case as many as 2,720 cases.

The coronavirus not only has a health impact, but also its impact on the Indonesian economy is not small as well. Quoted from the CNN Indonesia, the Minister of Finance of the Republic of Indonesia said that the projection of Indonesia's economic growth would only reach 2.3%, even in fact in the worst situation, the growth of the economy can only reach 0.4%. The causes of this include the decline in consumption and investment, both within the scope of the household and the government (Hidayati 2020). Moreover, the spread of coronavirus in Indonesia also has resulted high fluctuation of exchange rate of Rupiah against US Dollar during the period of January 1st until May 30th, 2020. From IDR13,963 per US$ on the 1st of January, Rupiah has weakened to IDR16,518 per US$ on the 23rd of March, then eventually it has strengthened to IDR14,575 per US$ on the 30th of May, 2020 (Guild 2020).

The "Lock Down" solution is one of the solutions that is real and efficient in breaking the chain of spread of the corona virus in the country. As a result, this has caused many activities to stop at school, offices, tourism, industry, and so on. In other words, everything that becomes a place for public activity must be stopped (Yusnadi 2020). Then, related to food security, it is a primarily concern for governments in Indonesia is one of developing countries that is still struggling with hunger at a significant level. Food shortages or food inflation would affect the population, especially the vulnerable poor, who may spend up to 60% of their income on food even on regular days. However, there is no question that the corona crisis brings other dynamic issues to food security, especially in terms of global cooperation and commerce (Amanta 2020). And then, in the process of supply chain management during the war against the issue of coronavirus, Indonesia often imports from China, in which China is one of Indonesia's largest trading partners. The presence of the COVID-19 pandemic in China has caused Chinese trade to deteriorate. Based on data from the Central Statistics Agency (BPS) of Indonesia, oil and gas and non-oil exports have decreased due to recently China being the largest importer of crude oil (Azizah 2020).

It can be concluded that the result of analysis in this research paper is that firstly, COVID-19 influences the public policy in Indonesia by the impact of the existence of Government Regulation No. 21 of 2020 concerning Large-Scale Social Restrictions. It is the limitation of certain activities of residents in an area suspected of being infected with COVID-19. Also, the Civil Emergency Policy has been regulated in a Government Regulation in lieu of Law No. 23 of 1959 concerning the State of Danger. Secondly, COVID-19 influences the health security in Indonesia by the impact of the existence of Public Health Emergency status in all regions within the country. It is concerned with the interpretation of extraordinary public health events marked by the spread of infectious diseases that pose health hazards and have the potential to spread across regions or countries. Thirdly, COVID-19 influences the economic security in Indonesia by the impact of the existence of potential decreasing trend of the economic growth by reaching only 2.3% of the economic growth and projected to later only reach economic growth of 0.4%. This event is also supported by the decrease of GDP growth and increase of unemployment rate in Indonesia. Next, COVID-19 influences the social security in Indonesia by the impact of the existence of "Lock Down" policy of action taken by the Indonesia Government which has effects toward the stop of any public activities, especially in the schools and workplaces. This is intended to break the chain of the spread of coronavirus within the country. Then, COVID-19 influences the daily consumption security in Indonesia by the impact of the existence of potential spread of coronavirus done by infected food packaging workers who do not wash their hands after being contaminated. Although this way of virus transmission is not significantly proven yet, since it has validly confirmed that the virus itself spreads through droplets of bodily fluids from someone infected with the corona virus and is generally transmitted to others through coughing or sneezing. And eventually, COVID-19 influences the supply chain management issue in Indonesia by becoming a disruption that occurred along with the supply value chain which are mainly being a disturbance in material supply, scarcity of material stock, and overstock due to weak market demand. Thus, it is necessary to assess realistic requests from customers to anticipate demand shortage, optimize production and distribution capacity, identify the safest logistics distribution transportation options, and eventually to conduct good cash management to address liquidity and financial distress issues.

Eventually, related to alternative of national policy solutions, within a short-term period, there are actually some applicable policies that can be immediately taken and done by the Government of Indonesia, in which firstly the Government is supposed to initiate accelerating the distribution of social assistance and as well simultaneously complement the recipient data by combining between Government data and the community data. Secondly, the Government is able to initiate integrating unemployment data and recipients of social assistance data that have been owned by various Government and non-Government institutions. Thirdly, the Government can take immediate action and policy to adjust the Employment Card assistance scheme by prioritizing the underprivileged unemployed people, especially for those who have been affected by COVID-19, with the intention to meet their basic needs efficiently. Next, the Government is supposed to encourage all businesses through providing potential incentives so they are able to optimize all the alternative of solutions to certainly retain their workforce rather than to layoffs. And eventually, the Government should fully optimize the social assistance that has a wider impact on the community's economy in every region in Indonesia.

Moreover, Indonesia Government should be able to implement the following strategies as policy recommendations that can also be taken by the Government themselves, as follows:

Case Identification

The Government can identify the case size and case location. This identification can be done by a combination of Rapid Diagnostic Test (RPD) and PCR tests or laboratory tests. The RPD test will basically find antibodies that Covid-19 positive patients have. However, false negatives often occur because antibodies could not be detected if the patient has no symptoms. This means that transmission can still occur. It is also recommended to the government to improve PCR test (laboratory test) facilities. So that if the detection is done early, mild cases will be easier to find.

Decrease of Death Case

It has been recognized that one of the causes of the high mortality rate due to Covid-19 is due to the lack of facilities in referral hospitals. By increasing Personal Protective Equipment (PPE), medications (symptomatic healers), ventilators, and ICUs can help reducing the risk of Covid-19 patients related to the death case.

Termination of Transmission Chain

Breaking the chain of transmission is the responsibility of all parties and indeed must be done both by the government and the community. It has been noticed that the efforts to break the chain of transmission of this new virus do really require community contributions to carry out social distancing with either proper self-quarantine or regional-quarantine. An example of a country that has successfully carried out the regional-quarantine is China and its success has occurred with 0 cases of transmission issue per day nowadays.

REFERENCES

Aldis, W. (2008). Health Security as A Public Health Concept: A Critical Analysis. *Journal of Health Policy and Systems Research – Oxford Academic*, Vol. 23 Issue 6, pp. 369–375.

Amanta, F. A. (2020, April 4). *Preventing Global Food Crisis Caused by COVID-19*. Retrieved from The-JakartaPost: https://www.thejakartapost.com/academia/2020/04/04/preventing-global-food-crisis-caused-by-covid-19.html

Arnani, M. (2020, April 4). *Corona Virus Update in Indonesia: Details of Covid-19 Cases in 32 Provinces* Retrieved from KOMPAS: https://www.kompas.com/tren/read/2020/04/04/174720065/2092-orang-terinfeksi-corona-ini-rincian-kasus-di-32-provinsi-di-indonesia

Azizah, M. (2020, March 12). *Impact of Corona Virus on the Global Economy, Especially in Indonesia*. Retrieved from DUTA: https://duta.co/dampak-virus-corona-terhadap-perekonomian-global-khususnya-di-indonesia

Cappelli, A., & Cini, E. (2020). Will the COVID-19 Pandemic Make Us Reconsider the Relevance of Short Food Supply Chains and Local Productions? *Trends in Food Science & Technology – ELSEVIER*, pp. 566–567.

Ferranti, Berry, P., & Anderson, E. (2019). The Concept of Food Security. *Encyclopedia of Food Security and Sustainability – ELSEVIER*, Vol. 2, pp. 1–7.

Guild, J. (2020, May 21). *In Indonesia, Will COVID-19 Trigger Another Asian Financial Crisis?* Retrieved from THE DIPLOMAT: https://thediplomat.com/2020/05/in-indonesia-will-covid-19-trigger-another-asian-financial-crisis/

Hayes, A. (2019, August 11). *Supply Chain Management (SCM)*. Retrieved from Investopedia: https://www.investopedia.com/terms/s/scm.asp

Hidayati, K. F. (2020, April 2). *The Corona Virus' Impacts on the Indonesian Economy*. Retrieved from Glints: https://glints.com/id/lowongan/dampak-virus-corona-bagi-perekonomian/#.Xo27DHJS82w

Kagan, J. (2020, April 27). *Social Security*. Retrieved from Investopedia: https://www.investopedia.com/terms/s/socialsecurity.asp

Kandel, N., Chungong, S., Omaar, A., & Xing, J. (2020). Health Security Capacities in the Context of COVID-19 Outbreak: An Analysis of International Health Regulations Annual Report Data from 182 Countries. *World Health Organization – ELSEVIER*, Vol. 395.

Kumar, D., Malviya, R., & Sharma, P. K. (2020). Corona Virus: A Review of COVID-19. *EJMO*, Vol. 4 No. 1, pp. 8–25.

Lin, Q., Zhao, S., Gao, D., Lou, Y., & Yang, S. (2020). A Conceptual Model for the Coronavirus Disease 2019 (COVID-19) Outbreak in Wuhan, China with Individual Reaction and Governmental Action. *International Journal of Infectious Diseases – ELSEVIER*, pp. 211–216.

Mamun, M. A., & Ullah, I. (2020). COVID-19 Suicides in Pakistan, Dying Off not COVID-19 Fear but Poverty? – The Forthcoming Economic Challenges for A Developing Country. *Brain, Behavior, and Immunity – ELSEVIER*

Sadikin, R. A. (2020, April 1). *What Is the Public Health Emergency Status Amid Corona Pandemic?* Retrieved from SUARA: https://www.suara.com/news/2020/04/01/111456/apa-itu-status-kedaruratan-kesehatan-masyarakat-di-tengah-pandemi-corona

Shokrani, A., Loukaides, E., Elias, E., & Lunt, A. (2020). Exploration of Alternative Supply Chains and Distributed Manufacturing in Response to COVID-19: A Case Study of Medical Face Shields. *Materials & Design – ELSEVIER*, Vol. 192.

Soeriaatmadja, W. (2020, April 7). *Indonesia Ranks among World's Worst in Coronavirus Testing Rate*. Retrieved from TheJakartaPost: https://www.thejakartapost.com/news/2020/04/07/indonesia-ranks-among-worlds-worst-in-coronavirus-testing-rate.html

Sugiyono. (2009). *Metode Penelitian Kuantitatif, Kualitatif, dan R&D.* Bandung: Alfabeta.

Syah, N. H. (2018, September 29). *Role and Function of Law*. Retrieved from Kompasiana: https://www.kompasiana.com/nurhadiansyah0650/5baf1e96bde575344f392483/peran-dan-fungsi-hukum?page=all

Varalakshmi, R., & Swetha, R. (2020). COVID-19 Lock Down: People Psychology due to Law Enforcement. *Asian Journal of Psychiatry – ELSEVIER*

Wahyono, W. (2017). National Security in A New Perspective. *Journal of National Defense – University of Gadjah Mada*, Vol. 5 No. 1, pp. 19–34.

Widayat. (2004). *Metode Penelitian Pemasaran Edisi Pertama: Cetakan Pertama.* Malang: UMM Press.

Worldometer. (2020, June 28). *Indonesia – Coronavirus Cases.* Retrieved from Worldometer: https://www.worldometers.info/coronavirus/country/indonesia/

Yezli, S., & Khan, A. (2020). COVID-19 Social Distancing in the Kingdom of Saudi Arabia: Bold Measures in the Face of Political, Economic, Social, and Religious Challenges. *Travel Medicine and Infectious Disease – ELSEVIER*

Yusnadi, A. (2020, March 16). *Corona and Social Impacts.* Retrieved from AnteroAceh: https://anteroaceh.com/news/corona-dan-dampak-sosial/index.html

Promoting Creative Tourism: Current Issues in Tourism Research – Kusumah et al. (Eds)
© 2021 Taylor & Francis Group, London, ISBN 978-0-367-55862-8

Covid-19, technology and tourism: The future of virtual tour?

T. Andrianto
The Hong Kong Polytechnic University, Hong Kong

A.H.G. Kusumah
Universitas Pendidikan Indonesia, Indonesia

N.A. Md Rashid, A.G. Buja & M.A. Arshad
Universiti Teknologi MARA Melaka, Malaysia

ABSTRACT: The Covid-19 pandemic requires everyone to stay at home except for important and urgent needs. One last activity to do outside the house is a trip. This situation encourages new ideas and opportunities for tourism practitioners to take an advantage of online technology platforms. The current trend in Indonesia is the emergence of Virtual Tour (VT) activities amidst community. This paper aims to elaborate on the Virtual Tour phenomenon which is widespread during a pandemic and its challenges after the pandemic is over. This paper also discusses the existing virtual tour concept and its future predictions in Indonesia. From this paper, it is expected that practitioners, academics, government and communities recognize the implementation and future of the Virtual Tour.

1 INTRODUCTION

The covid-19 pandemic has left the world tourism sector paralyzed. Flights among countries have almost stopped completely for some time. All non-essential travels are recommended to be avoided, including tourism activities. 2020 is then considered as the year for the break for tourism. However, some academics and tourism practitioners believe that 2020 will also be a year for rethinking and restarting tourism to be even better.

The Covid-19 pandemic provides an excellent opportunity to rebuild tourism from ground zero and at the same time, generate ideas and opportunities for existing and new online platforms to emerge. In Indonesia, one of the trends is by utilizing the live conference platform called the Virtual Tour (VT). Many people use VT as an alternative activity to spend their time at home. There are also those who professionally provide paid VTs to visit favorite tourist destinations. Some even introduce new concept of travelling and tourist destinations.

But what is the real virtual tour concept like? What would its future be after the pandemic covid-19 is over? This paper aims to elaborate the Virtual Tour phenomenon that during the pandemic and its challenges after the pandemic is over. This paper discusses the existing VT concept and explores its implementation in Indonesia including discussing the VT predictions and the tourism after the Covid-19 pandemic

2 VIRTUAL TOUR

2.1 *Definitions and terminology*

The term Virtual tour is more widely used in the real estate industry. Virtual tour is an activity carried out by prospective house or property buyers who cannot see directly the property they are

DOI 10.1201/9781003095484-98

interested in due to various factors. Through the virtual tour platform, the prospective buyers can see the full condition of the property they are interested in. Although the term virtual tour is more widely used in the real estate industry, the use of 360 photos and videos at tourist attractions and destinations has been increasing in the last 10 years. World-class tourist attractions such as Machu Picchu or the pyramids of Giza also take advantage of this technology. Virtual tours in the real estate industry and tourist attractions promotion use 360 photos or videos of the property to create a complete "presence" for potential buyers.

The sense of presence is an important factor in creating new forms of experience for potential tourists. Tussyadiah et al (2018) in their research related to the use of virtual reality technology in tourist attractions, argue that the feeling of being in the virtual environment increases enjoyment and heighten the feeling of being in a destination. Research related to virtual presence through various media has also been carried out by several researchers from various fields, for example related to child restraints in rural communities (Swanson et al. 2020), virtual field trips in education (Han 2020), virtual environment (Hofer et al. 2020) including in the tourism area such as culture heritage tour (Argyriou et al. 2020), virtual reality in hotel (Bogicevic et al. 2019).

Steuer (1992) in Lee et al. (2020) stated that Telepresence is a feeling experienced by someone who is in a virtual environment. Li et al. (2002) in Lee et al. (2020) added that Telepresence indirectly provides virtual experiences through the website. This means that through technology, a person, in this case a potential tourist, can indirectly bump into his future experience through media that provides telepresence such as a smartphone or computer. Until now, the development of VT in tourism is closely related to the use of VR technology. The development of virtual reality (VR) allows viewers to see the real form of tourist attractions, hotels and restaurants (Tussyadiah et al 2018). In its development, viewers can use VR to dig up information on 3d images, visuals and audio (Tussyadiah et al. 2018). In the use of VR technology, content quality, system quality and vividness are three important keys that must be taken into account in the participant's experience so that it can encourage them to actually visit their destination (Lee et al. 2020). In line with this statement, Bogicevic et al. (2019) stated that image quality is important in VR experiences besides the quality of information and the system used. This is including a good image of a tourist destination in order to attract people to come (Nurazizah & Marhanah 2020).

3 COVID-19 PANDEMIC AND VIRTUAL TOUR IMPLEMENTATION

The Covid-19 pandemic has halted almost all tourism activities; however, at the same time, alternative options for recreation in the work-from-home era appear in various forms such as Google Earth VR, Travel World VR, etc. (Rogres 2020). In Indonesia, one alternative recreational activity that has emerged is a "virtual tour" to visit a tourist destination. This is in accordance with what was conveyed by Rahayu (Aditya 2020) which states that virtual tours provide an opportunity for tourism in Indonesia during the Covid-19 pandemic to keep moving, even though it cannot replace the actual visiting experience.

Although the term used is "virtual tour", the term is not exactly the same as the definition we discussed earlier. In the previous section, the term "virtual tour" refers to a platform that presents a sense of presence to guests during the pandemic covid-19 period. However, in Indonesia, the term virtual tour during the Covid-19 pandemic refers to online guiding tour activities, which is an activity where a tour guide guides "tourists" to a destination or tourist attraction online through online meeting platforms such as zoom or google meet. Thus, the "tourists" are given information and stories about tourist attractions or destinations that they have never visited before. The media used by the "tour guides" vary; starting from using presentation slides containing photos of tourist attractions, to using a spherical view from Google Street as a medium for conducting "tours".

The themes offered through the online guiding tour activities are very diverse, with different approaches. Some of the online guiding tour activities are carried out as an effort to promote a destination to potential tourists who have the desire to visit after the Covid-19 pandemic. Others are more of an activity to share stories, knowledge and experiences of guides when they visit or

Table 1. List of online tour guiding in Indonesia April – June 2020.

Program Name	Areal/Destination/Attraction	Theme	Date
Virtual Community Trip	Jakarta	Coffee Tourism	08 May 2020
Virtual Tour Lawang Sanga	Lawang Sanga	Heritage Tourism	09 May 2020
Virtual Tour Pulau Sumba	Sumba Island	Island Tour	08 May 2020
Ngabubutrip – Tana Toraja	Tana Toraja	Cultural Tour	26 April 2020
Ngabubutrip – Nglanggeran	Desa Wisata Nglanggeran	Rural Tourism	03 May 2020
Ngabubutrip – Tambora	Tambora Mountain	Volcano Tourism	10 May 2020
Ngabubutrip – Sawahlunto	Sawahlunto Mine	Mine Tourism	17 May 2020
Virtual Tour Pulau Belitong	Belitong Island	Island Tour	26 May 2020
Virtual Tour Pulau Sumba	Sumba Island	Island Tour	27 May 2020
Virtual Tour Banjarmasin	Banjarmasin	Urban Tourism	28 May 2020
Virtual Tour Yogjakarta	Yogyakarta	Cultural Tourism	29 May 2020
Virtual Tour Poso	Poso	Urban Tourism	30 May 2020
Virtual Tour Natuna	Natuna	Nature	31 May 2020
Virtual Tour Labuan Bajo	Labuan Bajo	Nature	2 June 2020
Virtual Tour Tanjung Puting	Tanjung Puting	Nature	3 June 2020
Virtual Tour Gunung Rinjani	Gunung Rinjani	Mountain Tourism	4 June 2020
Virtual Tour Danau Toba	Danau Toba, North Sumatera	Lake Tourism	5 June 2020
Virtual Tour Desa Adat Baduy	Baduy, Banten	Cultural Tourism	6 June 2020
Virtual Tour Raja Ampat	Raja Ampat, Papua	Nature	7 June 2020

take tourists to a tourist destination. The various themes, media and activities offered through this online guiding tour are presented in the following table.

4 THE FUTURE OF TOURISM POST-COVID 19

The uncertainty of the future of tourism has surfaced during this Covid-19 pandemic. Analyses from academics and practitioners are emerging about what will happen to the tourism industry after Covid-19. These analyses can be grouped into at least three post-Pan-Covid-19 tourism scenarios, namely the business-as-usual scenario, moderate new-normal, and extreme new-normal.

4.1 *Business as usual*

The first scenario is where the tourism industry will return to normal as before without any significant changes. There is no change in government policies or tourist behavior when traveling. In traveling, there was little or no change in policy in the tourism sector. Policy changes that may arise in this scenario include the obligation to comply with hygiene standard protocols in the tourism industry or in tourist destinations.

This scenario may occur when the Covid-19 pandemic does not last long, or a Covid-19 vaccine is found soon. If this happens, there maybe not much change from the management of the tourism industry, including the tourist consumption pattern. The technology and digital platforms used by tourists in preparing for trips, or enhancing their experiences as long as they are in a destination will not change. Many emerging technology platforms such as 360 virtual tours will remain a tertiary necessity for the tourism sector.

4.2 *Moderate "new normal"*

The second scenario was a moderate "new normal" scenario. This scenario may occur when the Covid-19 pandemic is not over for some time, but is still in a controllable situation. This condition may be described as the condition that occurred as of August 2020. If this scenario occurs, there will be quite a lot of changes in the tourism industry as well as in tourist behavior. Tourists will be very selective in choosing a destination and accommodation when traveling in the desired destinations.

In this scenario, the role of technology becomes increasingly important. Review website platforms like tripadvisor are not sufficient for making decisions. The need for new technology platforms such as 360 virtual tours and virtual reality will be even stronger.

4.3 *Extreme new normal*

The third scenario is Extreme New Normal. In the extreme new normal conditions, all or at least most of the covid-19 pandemic control plan did not work as expected. When this condition occurs, all customs and procedures under normal conditions become obsolete. There will be many changes that occur, including in the tourism industry and tourist behavior.

The extreme point is when the tour becomes very limited to the point that a traveler cannot travel too far from his place of origin. When this scenario occurs, the tourism industry will experience very basic changes. Tourists will stop their long-distance travel and tend to travel in short or medium distances that can be reached using private vehicles. When this happens, the role of new technology platforms such as 360 virtual tours becomes indispensable. New business models that monetize the virtual tour concept will emerge.

4.4 *Virtual tour business model sustainability*

Predictions about how the business model of virtual tours will sustain, both in the form of 360 spherical views and in the form of online guiding tours, are widely discussed by tourism academics. Opinions that doubt the sustainability of the business model for 360 spherical views generally argue that this type of virtual tour is high in interaction cost, but moderate in usefulness. Poor tour design will make visitors confused, loss of direction, and uncomfortable. To get a good level of experience, other information media are needed as the main information provider for visitors. Meanwhile, the 360 spherical view feature on the virtual tour only functions as a supporting feature in creating a sense of space. Even though the 360 feature can increase interaction and experience as applied to cultural heritage tours in Greece (Argyriou et al. 2020).

Meanwhile, doubts regarding the sustainability of the virtual tour model with an online guiding approach are more to the extent to which the program is able to provide real and unforgettable experiences for visitors. And to achieve this unforgettable experience, there are at least three influencing factors, namely the stability of the internet connection from both the tour guide and visitors side, the supporting information media prepared by the tour guide, and the ability of the tour guide in telling stories and compiling a memorable narrative. The stability of the internet connection is very important because it plays a key role in the process of delivering information and stories from the tour guide to visitors. In addition, the media used by tour guides in illustrating guiding narratives also plays an important role in creating a sense of presence for visitors. Mistakes in choosing media such as images, audio or video, can reduce the quality of the experience of virtual visitors. The last factor that plays a role in a virtual tour with an online tour guiding approach is the ability and capacity of the tour guide in telling stories. Based on the author's personal experience following eight tour programs of this type, there are very few virtual tour activities that are able to provide unforgettable experiences for visitors. And one of the factors considered to be the main differentiator between virtual tours is the role and ability of the tour guide in telling stories. Positive responses from virtual visitors can also be seen from several indicators such as actively answering questions, responding to greetings and interactions made by the guide, so that they stay until the end of the tour.

5 CONCLUSION

Nothing is certain when we try to predict the future. Likewise, when we try to predict the future of tourism after the Covid-19 pandemic. It may even be possible that there has never been a "post" pandemic era, and we must make peace and dance with the COVID-19 virus for good. The future of new technology platforms such as virtual tours which are considered as the new big thing in the tourism sector and are considered as alternative solutions during this pandemic are still unpredictable. It is still too early for us to know which direction the future development of world tourism would be, including the development of a sustainable business model for virtual tour activities.

What is certain is that the emergence of new borderless technology platforms can attract new prospective customer (Jaafar et al. 2011). For example, virtual tour has contributed to providing new tone for future tourism development. Although there is the possibility of no meaningful development as a business model, a virtual tour platform can become a new standard in the promotion of a tourist attraction or destination such as a website promotion or social media platform for a destination that is currently considered a common thing and standard in the tourist destinations management

REFERENCES

Aditya, Nicholas Ryan (2020, May 09). Virtual Tour, Peluang Baru PAriwisata di Era New Normal. Retrieved From: https://travel.kompas.com/read/2020/05/09/210800427/virtual-tour-peluang-baru-pariwisata-di-era-new-normal?page=all.

Argyriou, L., Economou, D., & Bouki, V. (2020). Design methodology for 360° immersive video applications: the case study of a cultural heritage virtual tour. *Personal and Ubiquitous Computing*. https://doi.org/10.1007/s00779-020-01373-8

Bogicevic, Vanja, Seo, Soobin, Kandampully, Jay A, Liu, Stephanie Q, & Rudd, Nancy A. (2019). Virtual reality presence as a preamble of tourism experience: The role of mental imagery. *Tourism Management (1982)*, *74*, 55–64.

Han, I. (2020), Immersive virtual field trips in education: A mixed-methods study on elementary students' presence and perceived learning. Br J Educ Technol, 51: 420–435. https://doi.org/10.1111/bjet.12842

Hofer M, Hartmann T, Eden A,Ratan R and Hahn L. (2020). The Roleof Plausibility in the Experience ofSpatial Presence in VirtualEnvironments. Front. Virtual Real. 1:2. https://doi.org/10.3389/frvir.2020.00002

Jaafar, M., Abdul-Aziz, A. R., Maideen, S. A., & Mohd, S. Z. (2011). Entrepreneurship in the tourism industry: Issues in developing countries. *International Journal of Hospitality Management*, *30*(4), 827–835. https://doi.org/10.1016/j.ijhm.2011.01.003

Lee, Minwoo, Lee, Seonjeong Ally, Jeong, Miyoung, & Oh, Haemoon. (2020). Quality of virtual reality and its impacts on behavioral intention. *International Journal of Hospitality Management*, *90*, 102595.

Nurazizah, G.R and Marhanah. S (2020) Influence of destination image and travel experience towards revisit intention in Yogyakarta as tourist destination. Journal of Indonesian Tourism, Hospitality and Recreation.

Rogres, Sol (2020, Mar 18). How Virtual Reality Could Help The Travel & Tourism Industry In The Aftermath OfThe Coronavirus Outbreak. Retrieved From: https://www.forbes.com/sites/solrogers/2020/03/18/virtual-reality-and-tourism-whats-already-happening-is-it-the-future/#b334d1a28a6a

Swanson, M., MacKay, M., Yu, S., Kagiliery, A., Bloom, K., & Schwebel, D. C. (2020). Supporting Caregiver Use of Child Restraints in Rural Communities via Interactive Virtual Presence. Health Education & Behavior, 47(2), 264–271. https://doi.org/10.1177/1090198119889101

Tussyadiah, Iis P, Wang, Dan, Jung, Timothy H, & Tom Dieck, M.Claudia. (2018). Virtual reality, presence, and attitude change: Empirical evidence from tourism. *Tourism Management (1982)*, *66*, 140–154.

Promoting Creative Tourism: Current Issues in Tourism Research – Kusumah et al. (Eds)
© 2021 Taylor & Francis Group, London, ISBN 978-0-367-55862-8

Virtual tour as one of education tourism solutions in Covid-19 pandemic

R. Khaerani
Universitas Pendidikan Indonesia, Bandung, Indonesia

ABSTRACT: Indonesian government policies in the field of education with the temporary closure of several schools, colleges, and universities is an effort to increase the speed of the Covid-19 pandemic transfer. There is disruption of the learning process between students and teachers regarding the decline in interest in learning and the desire to travel to a destination. The problem is a challenge for the government as well as stakeholders to provide educational solutions that can be enjoyed by the public, especially tourists. One of the solutions provided is virtual. The Minister of Education and Culture stated that the community could learn and be able to travel in a unique and new way and add an experience with virtual tour.

Keywords: Covid-19, virtual tour, tourism education

1 INTRODUCTION

1.1 *Background*

The year 2020 is a very tough year for the global community and also the tourism industry in the world, including tourism in Indonesia. This is the year when a virus called Covid-19 (Corona Virus Desease 19) attacked almost the entire world. The Covid-19 virus not only attacks humans but also attacks the economy in the world, especially in the education sector with all the problems that arise from the chain effects it causes.

Many countries decided to close schools, colleges, and universities. Governments in various countries including Indonesia must make a bitter decision to close schools to reduce people's contact (Syah 2020). The learning process in schools is the best public policy tool as an effort to increase knowledge and skills (Caroline 1979). In addition, many students think that schools offer a very fun activity, because students can interact with one another. Schools can improve students' social skills and social class awareness. In addition, the school as a whole is a medium of interaction between students and teachers to increase the ability of integrity, skills, and affection between them. But now the school activities have to be stopped suddenly because of the Covid-19 interference. To what extent is the impact on the learning process in schools? Especially for Indonesia, there is a lot of evidence when schools greatly affect productivity and economic growth (Baharin 2020).

With the sudden arrival of the Covid-19 pandemic, the agency of education in Indonesia needs to follow the path that helps schools in an emergency. Schools need to force themselves to use online media. However, the use of technology is not without problems. There are many variances of problems that hinder the effective implementation of learning with online methods, including: (1) limited mastery of information technology by teachers and students; (2) inadequate facilities and infrastructure; (3) limited Internet access; and (4) inadequate provision of budget.

At present, the Minister of Education and Culture (Mendikbud) Nadiem Anwar Makarim expresses the approach of learning from home as the government's first strategic step in preventing the massive spread of Covid-19. The Minister of Education and Culture (Mendikbud), Nadiem Anwar Makarim, admitted learning in the pandemic was indeed not easy. However, the

 DOI 10.1201/9781003095484-99

Ministry of Education and Culture continues to strive to ensure that learning continues. He appealed to stakeholders to change the cultural paradigm, by exploring the potential of each region, both natural and traditional. Different travel plans will provide a unique experience for tourists.

The closure of several museums under the auspices of the Directorate General of Culture of the Ministry of Education and Culture (Kemendikbud) is part of a series of Ministry efforts to prevent the spread of Coronavirus (Covid-19) in Indonesia. Restricting visits to museums is a choice that must be made in the context of preparedness to deal with the threat of disease and potential health risk factors for Community Health Emergency (KKM) in the public sphere. However, that does not mean the public cannot access museums. The Minister of Education and Culture stated that the public could still access museums and several sites in Indonesia using the technology platform.

Virtual Tour is a part of virtual reality technology in the form of digital content consisting of several series of 360-degree panoramic photos. The 360 Virtual Tour/VR content that is displayed can be used for interactive education, promotion, and presentation needs (indonediavirtual-tour.com).

1.2 *Research question*

Based on the above analysis, there is a research question formulated as follows: What is the solution of tourism education in the Covid-19 pandemic era?

2 LITERATURE REVIEW

2.1 *Education 4.0 era*

Competition in educational services causes education providers to manage their organization like a business, without abandoning their ideological concept, through the concept of marketing. Marketing that only pays attention to product, promotion, distribution, and selling price (4P) aspects is no longer sufficient. Every individual in the organization must be able to see the vision and mission, which is then supported by the formulation of the right strategy (Remiasa 2005).

Education 4.0 era is a phenomenon that responds to the need for the emergence of the fourth industrial revolution where humans and machines are aligned to find solutions to problems and find new possibilities for innovation. The result of the 4.0 industrial revolution is disruptive innovation, which has spread to all sectors including industry, economy, and education.

This phenomenon has also succeeded in shifting people's lifestyles and mindsets. Disruptive innovation can be interpreted as a phenomenon disruption of old industries by new industry players due to the ease of information technology.

2.2 *Education tourism*

Educational tourism is a program that combines elements of tourism activities with an educational content. The program is packaged in such a way as to make annual tourism activities or extracurricular activities of high quality. The materials in the guide have been adjusted to the weight of students and the education curriculum. Every time you visit a tourist attraction, it will be adjusted to the interests of the object and the field of science you are going to study (Akib 2020). Tourism comes from the Sanskrit language, which consists of two syllables, namely the word "Pari," which means touring and the word "Wisata," which means travel (Yoety 2008).

2.3 *Virtual reality technology*

There is a coexistence between real and virtual world results by the use of VR technologies (Milgram & Kishino 1994). The line between the real and virtual is hardly distinguished due to the use of these digital technologies that also offers their users the possibility to increase their level of immersion when living virtual tourism experiences (Jung et al. 2016).

2.4 *Virtual tour*

Virtual Tour is a simulation of a virtual world trip from an existing location, made with several videos and series of still images and other multimedia uses such as sound effects, narrative music, and text. which, according to Ibrahim (2012) that distinguishes the use of live broadcasts for the effect of tele-tourism. The technique of combining photography technology with information technology with the aim of providing space information thoroughly with a 3D and interactive display, which can be turned into an application covering indoor or outdoor which refers to the concept where objects as if they can be explored like the original world, can trace in all directions, looking in all directions, rotating and exploring around (Trimannula 2017).

This virtual tour business has substantial implications for industry travel in terms of marketing strategies and the development and design of visual tourism communities. For tourism organizations, virtual communities have expanded marketing horizons and have a large impact on marketing, sales, product and service development, supplier networks, quality of information, and distribution channels (Wang et al. 2002)

3 RESEARCH METHODS

The method used in this research is the study of literature, whether in the form of scientific journals, books, articles in mass media, or statistical data. The literature will be used to answer the research problems proposed by the author, in this case the tourism education solution in the Covid-19 pandemic.

4 RESULTS AND DISCUSSION

In overcoming problems with the learning process in schools due to the impact of Covid-19, several event organizers have organized virtual tours, including:

- Trans Studio Bandung is one of the places that provides virtual tourism events
- from the role of higher education, tourism students of the University of Education of Indonesia launched a virtual tour project program entitled Bandung 360. This program is also supported by the Bandung City Culture and Tourism Office, and the Bandung Good Guide (BGG). This program is a form of concern for tourism academics at the University of Education of Indonesia toward shifting the conventional tourism mindset toward tourism on a digital platform.
- Atourin, as a local start-up, facilitates tour guides and tour leaders to develop virtual tours. Atourin organizes free online training around virtual tours via the Zoom application.
- PT Mahakreasi Indonesia (Maximum Ulitmate) held a virtual creative educational tour for children in Indonesia with the headline Creative Safari. This virtual event is aimed at introducing three Super Priority Destinations (3DSP) in Indonesia to children, namely Borobudur (Central Java), Lake Toba (North Sumatra), and Labuan Bajo (NTT).
- Travalal market placce dan platform untuk muslim friendly tourism meluncurkan virtual reality tourism se-bagai alternatif pariwisata tanpa mobilitas.
- Travalal provides a wide selection of virtual tourist destinations using video 360 technology and live tours using the video conference application with a wide selection of destinations, both international or domestic, and also religious tourism, namely virtual umrah.

5 CONCLUSION AND RECOMMENDATION

Enjoying a virtual tour turns out to be a challenge, both from the organizers and the audience. Organizers must be able to package material, whether in the form of photos or videos, that can build consumer emotions (virtual tourism lovers). It's not easy to understand these emotions because everyone's taste is different. However, at least this is part of meeting the needs of temporary travel.

From an industrial perspective, virtual tourism can be a business opportunity. When people are still worried about leaving the house but want to enjoy a destination, this can be an alternative and creative collaborative form between tourism and technology.

Sometimes the Internet network becomes an obstacle. Internet instability can interfere with enjoying virtual tours, for example, video buffering suddenly. Another drawback that cannot be denied is the different touch experience between the virtual world and the real one.

Indeed, virtual media cannot replace real-world sensations. In other words, the atmosphere involving the senses of smell, sight, and hearing cannot be completely replaced by sophisticated technology. For example, enjoying culinary tours through virtual media seems unable to provide satisfaction because in it there are aspects of involvement in tasting food and smelling the aroma of culinary dishes.

Virtual tourism can be a part of the recovery toward a new normal era. However, this cannot be used as a complete fulfillment of the needs of its users. This is related to the human sense possessed by the soul of every human being. Experience that is built in real and virtual, of course, will produce different values.

Even so, the presence of virtual tourism can diversify service products for anyone who wants to enjoy them and indirectly become a driving factor for visiting original destinations.

REFERENCES

Akib, E. 2020. Pariwisata Dalam Tinjauan Pendidikan: Studi Menuju Era Revolusi Industri. *PUSAKA (Journal of Tourism, Hospitality, Travel and Business Event)*, *2*(1), 1–7. https://doi.org/10.33649/pusaka.v2i1.40.

Baharin, R., Halal, R., dll. 2020. Impact of Human Resource Investment on Labor Productivity in Indonesia, *Iranian Journal of Management Studies*, 13(1), hal. 139–164.

Caroline Hodges Persell. 1979. *Educations and Inequality, The Roots and Results of Stratification in America's Schools.* United States of America: The Free Press.

Ibrahim. 2012. V*itual Tour Sebagai Simulasi*. Yogyakarta: Graha Ilmu.

Jung TH, tom Dieck MC, Lee H, Chung N. 2016. Effects of virtual reality and augmented reality on visitor experiences in museum. In: Inversini A, Schegg R (eds) Information and communication technologies in tourism. *Springer*, Cham, pp 621–635.

Marcus Remiasa. 2005. Perencanaan Strategis Pemasaran Untuk Menciptakan Sustainable Competitive Advantage (Kasus Pada Program Studi Perhotelan UK Petra di Surabaya). *Jurnal Manajemen Perhotelan*, 1, 14–23.

Milgram P, Khisino F. 1994. A Taxonomy of Mixed reality visual displays. *IEICE Trans Info Syst* 77(12): 1321–1329.

Syah, R. H. 2020. Dampak Covid-19 pada Pendidikan di Indonesia: Sekolah, Keterampilan, dan Proses Pembelajaran. *SALAM: Jurnal Sosial Dan Budaya Syar-I*, *7*(5). https://doi.org/10.15408/sjsbs.v7i5.15314.

Trimanulla, A. 2017. Virtual Tour Berbasis 3D untuk Pengenalan Kampus STIKI Malang. *J-Intech: Jurnal of Information and Technology,* 05(01), 78–82.

Wang, Y., Yu, Q., & Fesenmaier, D. R. 2002. Defining the virtual tourist community: Implications for tourism marketing. *Tourism Management*, *23*(4), 407–417. https://doi.org/10.1016/S0261-5177(01)00093-0.

Yoety, O. A. 2008. *Ekonomi pariwisata: introduksi, informasi, dan aplikasi*. Jakarta: Penerbit Buku Kompas.

Promoting Creative Tourism: Current Issues in Tourism Research – Kusumah et al. (Eds)
© 2021 Taylor & Francis Group, London, ISBN 978-0-367-55862-8

Estimation of short-term economic effect in Geopark Ciletuh-Palabuhanratu tourism due to the coronavirus outbreak

P. Hindayani, A.R. Pratama & A. Khosihan
Universitas Pendidikan Indonesia, Bandung, Indonesia

Z. Anna
Universitas Padjajaran, Bandung, Indonesia

ABSTRACT: Geopark Ciletuh-Palabuhanratu, named as UNESCO Global Geopark (UGG) in 2018 and a new geotourism destination, had an increasing number of tourist visits which was predicted to be more than 1 million in 2020. When Coronavirus (Covid-19) spread throughout the world, tourism industries suffered the most significant negative impacts, and some tourist destinations were closed. In this study, estimation of the short-term economic effect due to Covid-19 in UGG Ciletuh-Palabuhanratu tourism was highlighted. The results showed that the short-term economic effect during the pandemic was IDR1.5–2.02 billion (US$102–137 million) if the Covid-19 outbreak could be ended in December 2020. Considering the optimistic scenario where the government predicted that Covid-19 would end in June 2020, the estimation of economic effect was IDR 605–801 million (US$41–55 million). Meanwhile, based on the moderate scenario where Covid-19 was predicted to end in September 2020, the estimation of economic effect would be IDR1.05–1.42 billion (US$71–96 million). This would help the government prepare applicable policies to plan development strategies for UGG Ciletuh-Palabuhanratu after the Covid-19 outbreak.

Keywords: economic effect, Geopark Ciletuh-Palabuhanratu, tourism, Covid-19

1 INTRODUCTION

Geopark is an outstanding geology area consisting of archeological, ecological, and cultural values (UNESCO 2004; Vdovets et al. 2010). Moreover, Geopark, a new tourism concept for the protection of natural and geological heritage, plays a key role in the development of geotourism (Farsani et al. 2011, 2014; Rosana et al. 2014). One of the geoparks is a new geotourism destination located in Sukabumi, West Java, Indonesia, UNESCO Global Geopark (UGG) Ciletuh-Palabuharatu. UGG Ciletuh-Palabuhanratu has outstanding geological features in its 24 geosites. Each geosite consists of one or more objects (Ikhram et al. 2017). UGG Ciletuh-Palabuhanratu is a new tourist area with an old-rock natural park, the potentials for geological uniqueness, great beautiful natural scenery, the combination of coastal and hilly landscapes, waterfalls, unique geological rocks, diverse unique flora and fauna, and cultural diversity that cannot be found anywhere else (Ikhram et al. 2017; Rosana et al. 2014).

In 2015, Ciletuh-Palabuhanratu Geopark became a national geopark and was stated as UNESCO Global Geoparks (UGG) on April 17, 2018. As a new geotourism destination, the number of visitors, both local and foreign tourists, significantly increased and is predicted to increase more significantly. According to the Ciletuh Geopark Information Center, since its opening in 2012, the number of visitors was 343,910; whereas, in 2017, the number of visitors had reached the 967,311. At the end of December 2019, a novel virus which was officially named Coronavirus disease-2019 (Covid-19) by the World Health Organization (WHO) on February 11, 2020, emerged from Wuhan, China (Wu et al. 2020). The World Health Organization (2020) reported Covid-19 as a global pandemic on March 11, 2020. Prohibition of international travels affected more than 90%

 DOI 10.1201/9781003095484-100

of the world's population. Social distancing, citizen mobility, and almost all of tourism stopped in March 2020 (Gössling et al. 2020). Furthermore, in Indonesia, President Joko Widodo declared the first confirmed two cases of Covid-19 infection on March 2, 2020 (Djalante et al. 2020). Indeed, it had a huge impact on all sectors including economic, social, education, tourism, and others. However, the tourism sector is the most negatively impacted sector and has dramatically deteriorated; there are even no activities in the sector. Meanwhile, UNESCO Global Geoparks (UGG) Ciletuh-Palabuhanratu tourism has been shut down since the Covid-19 outbreak.

There is no research on the economic impact of Covid-19 on tourism at a specific location. However, some studies had estimated the economic impact on the tourism sector due to the SARS pandemic (Siu & Wong 2004; Wilder-Smith 2006; Zeng et al. 2005), the swine flu pandemic (Rassy & Smith 2012), and the MERS pandemic (Al-Tawfiq et al. 2014). Similarly, some studies had calculated the economic impact on the tourism sector due to environmental pollution, including an assessment of the economic effect on the tourism sector due to oil spills (Garza-Gil et al. 2006; Ritchie et al. 2014; Smith et al. 2011), an assessment of tourism sector losses due to air pollution (Anaman & Looi 2000; Sajjad et al. 2014), and estimation of the economic effects on the tourism sector due to marine debris (Jang et al. 2014; Krelling et al. 2017; McIlgorm et al. 2011; Ofiara & Seneca 2006; Qiang et al. 2020).

The purpose of this paper is to present an estimation of economic effects on tourism activities at Geopark Ciletuh-Palabuharatu during the Covid-19 outbreak. The estimated value of economic effect here was simu-lated for the condition during the Covid-19 outbreak. The estimated value of economic effect acts as a basic implication in determining the Geopark Ciletuh-Palabuharatu tourism development strategies after the Coro-navirus outbreak.

2 METHODS

2.1 *Study area*

The study area was UNESCO Global Geopark (UGG) Ciletuh-Palabuhanratu, which has an area of about 126,100 Ha located in Sukabumi, West Java, Indonesia. The research location administratively covered 74 villages and eight sub-districts, namely, Ciemas, Ciracap, Surade, Waluran, Simpenan, Palabuhanratu, Cikrak, and Cisolok (Figure 1).

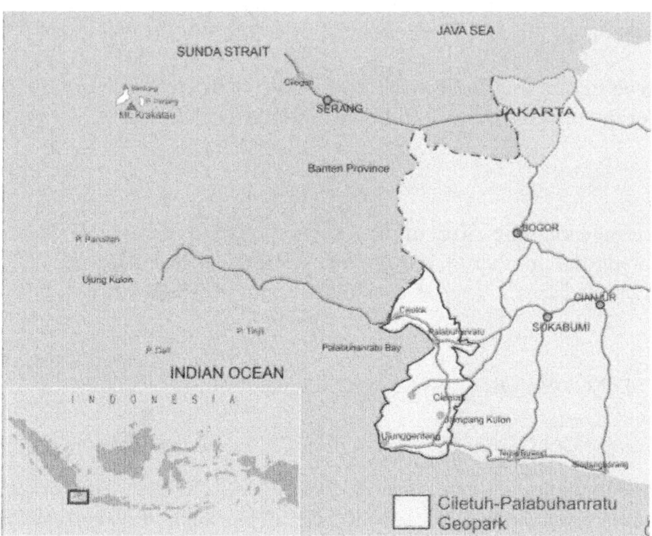

Figure 1. Location of study area. Source: Pusat Penelitian Geopark and Kebencanaan Geologi UNPAD.

Tourist expenditures data were taken from the questionnaires distributed to 295 respondents including 275 local tourists and 20 foreign tourists. These questionnaires were distributed before the Covid-19 outbreak.

Secondary data were collected from the Tourism Office of Sukabumi, Ciletuh Geopark Information Center (GIC), relevant journals, and related documents. The required data were the number of UNESCO Global Ge-opark (UGG) Ciletuh-Palabuharatu tourist visits and the travel cost or tourist expenditures including the over-all costs spent by tourists at Ciletuh-Palabuhanratu UGG, such as the costs of transportation, accommodation, visit duration, food and beverages, souvenirs, attractions, and others.

2.2 *Formulation of the short-term economy effect of the loss of revenue in tourism*

In this paper, the formulation of short-term economic effects of the loss of revenue of tourism due to the Covid-19 outbreak was simple. Firstly, there was a difference between economic losses and economic effects. The definition of economic losses or damages referred to lost economic welfare consisting of lost economic welfare to the consumer (consumer surplus) and the producer (producer surplus). Moreover, in general, economic effects were the changes in economic activities measured by the changes in consumer spending. Meanwhile, the definition of economic effects was "the lost sales of producers" (Jang et al. 2014; Krelling et al. 2017; Ofiara & Seneca 2006). The formulation in this research was modified from the previous studies (Jang et al. 2014; Krelling et al. 2017; Ofiara & Seneca 2006) by multiplying the number of goods by unit price. In other words, the number of goods in this paper constituted the number of tourists; while the unit price was the overall costs spent by consumers, namely, tourist expenditures. The tourist expenditures were obtained from the respondents' answers to the questionnaires.

The number of tourists in 2020 was predicted by time series forecasting. Time series forecasting was usually univariate, which meant a variable based on the past data (Montgomer et al. 2008; Sheldon & Var 1985). The regression for time series models using two types was a linear regression model and exponential regression model, which had the best fit straight line (Montgomer et al. 2008; Sheldon & Var 1985). The statistic tool for time series models was Minitab 17.

In the present study, the number of tourists in 2020 was predicted using hypothetical scenarios with a time series forecasting model. Moreover, the travel cost was the average expenditure or overall costs spent by tour-ists. Furthermore, there were three hypothetical scenarios: 1) Pessimistic scenario, the Covid-19 outbreak would end in 2020, 2) Moderate scenario: the Covid-19 outbreak would end in September 2020, and 3) Op-timistic scenario: the Covid-19 outbreak would end in June 2020.

Thus, this study assumed that there were no tourism activities during the Covid-19 outbreak. The formulation for the present study could be expressed as follows:

$$N = a \times b$$

"N" was the short-term economic effect of the lost revenue of tourism due to the Covid-19 outbreak
"a" was the estimated total number of visitors by hypothetical scenarios
"b" was the average travel cost or tourist expenditures per person

3 RESULT AND DISCUSSION

3.1 *Overview of UNESCO Global Geopark Ciletuh-Palabuhanratu*

Geopark Ciletuh-Palabuhanratu was officially declared as UNESCO Global Geopark (UGG) on April 17, 2018. It was established at the 204th UNESCO Executive Board session, Program and External Relations Commission in Paris, France. According to UNESCO (2006), geoparks had three targets: conservation, educa-tion, and development of the local economy through geotourism.

Reported by Ikhram et al. (2017), UGG Ciletuh-Palabuhanratu as geotourism had an international signifi-cance value and education and tourism applications, such as:

- Cisolok Geyser: the only geyser in Indonesia
- Ciletuh Amphitheatre: the biggest natural amphitheater landscape in Indonesia: Karang Daeu, Mount Badak, Legon Pandan, Batu Naga
- The oldest sedimentary deposits in West Java: Pasir Luhur, Mount Beas, and Sodong Parat

Other offered destinations were Ujung Genteng Beach, Palabuhanratu Beach, Palangpang Beach, Cikadal Beach, Darma Hill, Panenjoan Hill, Awang Waterfall, Cimarinjung Waterfall, Sodong Waterfall, Tengah Wa-terfall, batik village, cultural village, conservation forest, natural tourism park, and others.

3.2 *Socio-demographic characteristics of tourists*

The respondents' age was quite varied. Most respondents, both domestic and foreign tourists, were "41–50 years old" (40.3%); while the rest were "31–40 years old" (34.3%), "21–30 years old" (14.2%), "51–60 years old" (8.1 %), and "under 20 years old" (8.1%).

The education level between domestic and foreign tourists quite differed. Most foreign tourists had a master's degree; while most domestic tourists had a bachelor's degree or higher degree (81.9%), a diploma (13.3 %), and were high school graduates (4.8%).

The respondents' average income significantly differed. The domestic tourists had incomes lower than foreign tourists. The average monthly income level of foreign tourists was US$2,745, or between US$1,694 and US$3,389. Meanwhile, the average monthly income of domestic tourists was approximately US$277, or between US$34 and US$2,372.

3.3 *Tourist expenditures on visiting UNSCO Global Geopark Ciletuh-Palabuhanratu*

UNESCO Global Geopark (UGG) Ciletuh-Palabuhanratu tourism was divided into three zones: 1) Zone 1 consisting of attractions in Palabuhanratu, Kasepuhan, and Cisolok Geysers, 2) Zone 2 consisting of attrac-tions in the Ciletuh area, and 3) Zone 3 comprising attractions in Ujung-genteng and Surade areas. Opened in 2012, UGG Ciletuh-Palabuhanratu was visited by 343,910 tourists. The number increased sharply every year. The increasing number of tourists more than doubled in 2016, which showed a very great interest in UGG Ciletuh-Palabuhanratu as a new destination in West-Java, Indonesia. Until 2017, the number of tourist arrivals reached 967,311, consisting of 802,868 domestic and local visitors and 164,443 foreign tourists. Moreover, during the Eid holiday or peak season, it could reach 300,000 per month. In the following years, it might dramatically mount up, considering that UGG Geopark-Ciletuh had been designated as UNESCO Global Geopark (Figure 2).

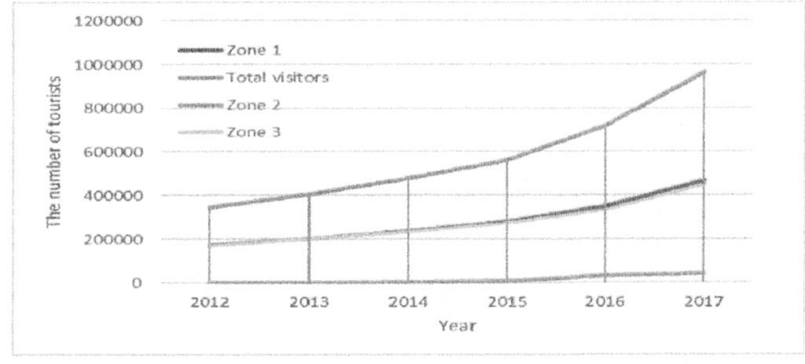

Figure 2. The number of UNESCO Global Geopark Ciletuh-Palabuhanratu tourists (Ciletuh Geopark Information Center).

The majority of domestic tourists including local tourists came from Sukabumi and big cities, such as Jakar-ta, Bogor, Depok, Bekasi, and Bandung, while foreign tourists came from Malaysia, Singapore, Europe, Aus-tralia, and America. The average length of stay for domestic and foreign tourists was three to seven days and one day for local tourists.

The average expenditure of tourists at UGG Ciletuh-Palabuharatu consisted of transportation costs, lodging costs, food and beverage costs, souvenir costs, retribution fees, attractions, and others. Based on the results, the average expenditures of domestic tourists at UGG Ciletuh Palabuhanratu was IDR1,091,040 with a range of IDR10,000–IDR7,300,000 per person; while foreign tourists could spend an average of IDR6,666,500 per person. Thus, the average expenditure of 295 respondents was IDR1,475,554 per visit.

3.4 *Projection of the number of UGG Ciletuh-Palabuhanratu tourists*

The measurement of both Mean Absolute Percentage model Error (MAPE) and determinant coefficient values (R-square) already fitted the criteria for forecasting (Table 1).

The results of the projected number of visitors using statistical analysis to determine the best fit straight line, both linear and exponential regression for time series models had "goodness of fit." Therefore, these models could be used for predicting the number of visitors.

According to Figure 3, the projected number of visitors of Ciletuh-Palabuhanratu UGG was predicted to in-crease in 2020, ranging from 1,230,527–1,650,088 visitors or 102,544–137,507 per month. However, the Covid-19 outbreak hit as a global pandemic, and in Indonesia, the cases were firstly identified on March 2, 2020. By all means, this affected all human activities and mobility, especially the tourism sector. Travel bans to recreational areas as the impact of Covid-19 toppled tourism businesses, including lodging, restaurants, souvenirs, attractions, and others.

This was reinforced by the Indonesian government's policy to begin the temporary closure of direct flights to and from China since February 2020. Flights in January–February 2020, both in domestic and international, were canceled and restricted. Furthermore, there were policy recommendations to "stay at home" for all activities, worship, and work. Besides, in Indonesia, most

Table 1. Performance statistics.

Model	Formulation	R-square	MAPE
Linear	$y = 118488t - 2E + 08$	0.9198	9
Exponential	$y = 2E - 171e^{0,2018t}$	0.9812	4

*95% significance level.

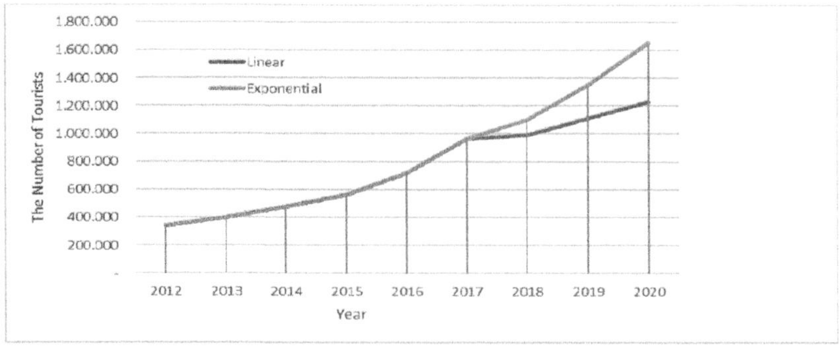

Figure 3. Comparison of the projected number of UNESCO Global Geopark Ciletuh-Palabuhanratu tourist visits using two models.

Table 2. Estimation of economic effects in UGG Ciletuh-Palabuhanratu due to Covid-19 with various scenarios.

Time series model	Optimistic scenario	Moderate scenario	Pessimistic scenario
Linear	IDR605 million (US$41 million)	IDR1.05 billion (US$71 million)	IDR1.51 billion (US$102 million)
Exponential	IDR 811 million (US$ 55 million)	IDR1.42 billion (US$ 96 million)	IDR2.02 billion (US$137 million)

cities dominated by local and domestic tourists of UGG Ciletuh-Palabuhanratu were still large-scale social restrictions (PSBB) so the tourist attractions there were automatically closed. Thus, the projected number of tourists of UGG Ciletuh-Palabuhanratu which should have been a potential profit in 2020 turned out to be a loss of revenue from the tourism sector.

3.5 *Estimation of economic effect in UGG Ciletuh-Palabuhanratu*

This study calculated the short-term economic effect of loss of revenue in UGG Ciletuh-Palabuhanratu dur-ing the Covid-19 pandemic. Hypothetical scenarios for the present study were divided into three categories: 1) Pessimistic scenario: the Covid-19 outbreak would end at the end of 2020, 2) Moderate scenario: the Covid-19 outbreak ended in September 2020, and 3) Optimistic scenario: the Covid-19 outbreak ended in June 2020. The assumptions used in this study were: 1) There had been no tourism visitors since the first case of Covid-19 in Indonesia was confirmed in early March 2020; 2) This study assumed that during the Covid-19 outbreak, there were no tourism activities. Based on Table 2, the estimated short-term economic effect was calculated by multiplying the average tourist expenditures by the hypothetical number of visitor scenarios.

Due to the swine flu pandemic over five months, Mexico lost profits from overseas visitors of around $US2.8 billion (Rassy & Smith 2012). However, the estimated value in this study was much smaller than that in other studies because it was only focused on one attraction, which was UGG Ciletuh-Palabuhanratu. However, this study calculated the short-term economic effects of the loss of tourism revenue as the impact of Covid-19 pandemic, which enormously harmed tourism as a reliable sector. This information is very helpful for the government and stakeholders to organize a new strategy development for geotourism UGG Ciletuh-Palabuhanratu in the "new normal" phase or after the end of the Covid-19 outbreak.

4 CONCLUSION

As a new tourist destination, every year, the tourists of UNESCO Geopark Ciletuh-Palabuhanratu always increased. Based on this present study, the number of tourists, both domestics and foreign in 2020, was forecasted to reach 1,230,527–1,650,088 or 102,544–137,507 per month. However, due to the global pandemic of Covid-19, which was firstly identified in Indonesia in early March 2020, the tourism sector was the most highly impacted sector, evidenced by several tourism sectors being closed. The estimated value of the short-term economic effect of the loss of tourism revenue in UNESCO Geopark Ciletuh-Palabuharatu Indonesia due to Covid-19 was IDR1.51–2.02 billion (US$102–137 million) with the pessimistic scenario of the Covid-19 outbreak ending in late December 2020. An optimistic scenario predicting that Covid-19 would end at the end of June 2020 and that a new normal would be implemented in July had estimated the economic effect of the loss of tourism revenue of IDR605–811 million (US$41–55 million). Meanwhile, a moderate scenario predicting that Covid-19 would end at the end of September 2020 estimated the economic effect of the loss of tourism revenue of IDR1.05–1.42 billion (US$71–96 million). Of course, the assessment in this study was a basic tool to inform and make policies. On the one hand, the government and stakeholders must prepare proper planning and rapid development for UGG Ciletuh-Palabuhanratu after a "new normal" or the Covid-19 ending.

REFERENCES

Al-Tawfiq, J. A., Zumla, A., & Memish, Z. A. (2014). Travel implications of emerging coronaviruses: SARS and MERS-CoV. *Travel Medicine and Infectious Disease*, *12*(5), 422–428. https://doi.org/10.1016/j.tmaid.2014.06.007

Anaman, K. A., & Looi, C. N. (2000). Economic Impact of Haze-Related Air Pollution on the Tourism Industry in Brunei Darussalam. *Economic Analysis and Policy*, *30*(2), 133–143. https://doi.org/10.1016/S0313-5926(00)50016-2

Djalante, R., Lassa, J., Setiamarga, D., Sudjatma, A., Indrawan, M., Haryanto, B., Mahfud, C., Sinapoy, M. S., Djalante, S., Rafliana, I., Gunawan, L. A., Surtiari, G. A. K., & Warsilah, H. (2020). Review and analysis of current responses to COVID-19 in Indonesia: Period of January to March 2020. *Progress in Disaster Science*, *6*(march), 100091. https://doi.org/10.1016/j.pdisas.2020.100091

Farsani, Neda Torabi, Coelho, C., & Costa, C. (2011). Geotourism and geoparks as novel strategies for socio-economic development in rural areas. *International Journal of Tourism Research*, *13*(1), 68–81. https://doi.org/10.1002/jtr.800

Farsani, Neda T., Coelho, C. O. A., Costa, C. M. M., & Amrikazemi, A. (2014). Geo-knowledge Management and Geoconservation via Geoparks and Geotourism. *Geoheritage*, *6*(3), 185–192. https://doi.org/10.1007/s12371-014-0099-7

Garza-Gil, M. D., Prada-Blanco, A., & Vázquez-Rodríguez, M. X. (2006). Estimating the short-term economic damages from the Prestige oil spill in the Galician fisheries and tourism. *Ecological Economics*, *58*(4), 842–849. https://doi.org/10.1016/j.ecolecon.2005.09.009

Gössling, S., Scott, D., & Hall, C. M. (2020). Pandemics, tourism and global change: a rapid assessment of COVID-19. *Journal of Sustainable Tourism*, *0*(0), 1–20. https://doi.org/10.1080/09669582.2020.1758708

Ikhram, R., Rosana, M. F., Agusta, R., & Andriani, S. S. (2017). Study of Significance of Geodiversity in Ciletuh-Palabuhanratu National Geopark, West Java, Indonesia. *International Conference on Earth Sciences and Engineering ICEE*, 1–11.

Jang, Y. C., Hong, S., Lee, J., Lee, M. J., & Shim, W. J. (2014). Estimation of lost tourism revenue in Geoje Island from the 2011 marine debris pollution event in South Korea. *Marine Pollution Bulletin*, *81*(1), 49–54. https://doi.org/10.1016/j.marpolbul.2014.02.021

Krelling, A. P., Williams, A. T., & Turra, A. (2017). Differences in perception and reaction of tourist groups to beach marine debris that can influence a loss of tourism revenue in coastal areas. *Marine Policy*, *85*(September), 87–99. https://doi.org/10.1016/j.marpol.2017.08.021

McIlgorm, A., Campbell, H. F., & Rule, M. J. (2011). The economic cost and control of marine debris damage in the Asia-Pacific region. *Ocean and Coastal Management*, *54*(9), 643–651. https://doi.org/10.1016/j.ocecoaman.2011.05.007

Montgomer, D. C., Jenning, C. L., & Kulahci, M. (2008). *Introduction to Time Series Analysis and Forecasting*. John Wiley & Sons. Inc. http://repositorio.unan.edu.ni/2986/1/5624.pdf

Ofiara, D. D., & Seneca, J. J. (2006). Biological effects and subsequent economic effects and losses from marine pollution and degradations in marine environments: Implications from the literature. *Marine Pollution Bulletin*, *52*(8), 844–864. https://doi.org/10.1016/j.marpolbul.2006.02.022

Qiang, M., Shen, M., & Xie, H. (2020). Loss of tourism revenue induced by coastal environmental pollution: a length-of-stay perspective. *Journal of Sustainable Tourism*, *28*(4), 550–567. https://doi.org/10.1080/09669582.2019.1684931

Rassy, D., & Smith, R. D. (2012). The Economic Impact Of H1N1 on amexico's Tourist and Pork Sectors. *Health Econ.* https://doi.org/10.1002/hec

Ritchie, B. W., Crotts, J. C., Zehrer, A., & Volsky, G. T. (2014). Understanding the Effects of a Tourism Crisis: The Impact of the BP Oil Spill on Regional Lodging Demand. *Journal of Travel Research*, *53*(1), 12–25. https://doi.org/10.1177/0047287513482775

Rosana, M. F., Budiman, H., & Abdurahman, O. (2014). Geology, Geotourism as Definite Factor for Geopark Ciletuh Indonesia. *6th International UNESCO Conference on Global Geoparks*, 80.

Sajjad, F., Noreen, U., & Zaman, K. (2014). Climate change and air pollution jointly creating nightmare for tourism industry. *Environmental Science and Pollution Research*, *21*(21), 12403–12418. https://doi.org/10.1007/s11356-014-3146-7

Sheldon, P. J., & Var, T. (1985). Tourism forecasting: A review of empirical research. *Journal of Forecasting*, *4*(2), 183–195. https://doi.org/10.1002/for.3980040207

Siu, A., & Wong, Y. C. R. (2004). Economic Impact of SARS: The Case of Hong Kong. *Asian Economic Papers*, *3*(1), 62–83. https://doi.org/10.1162/1535351041747996

Smith, L. C., Smith, M., & Ashcroft, P. (2011). Analysis of Environmental and Economic Damages from British Petroleum's Deepwater Horizon Oil Spill. *Albany Law Review*, *74*(1), 563–585. https://doi.org/10.2139/ssrn.1653078

UNESCO. (2004). *United Nations Educational, Scientific and Cultural Organization Organisation des Nations Unies pour l'éducation, la science et la culture Guidelines and Criteria for National Geoparks seeking UNESCO's assistance to join the Global Geoparks Network (GGN)* (Issue January).

UNESCO. (2006). Global Geopark Network. In *Division of Ecological and Earth Sciences Global Earth Observation Section Geoparks Secretariat*.

Vdovets, M. S., Silantiev, V. V., & Mozzherin, V. V. (2010). A national geopark in the republic of tatarstan (Russia): A feasibility study. *Geoheritage*, *2*(1), 25–37. https://doi.org/10.1007/s12371-010-0010-0

Wilder-Smith, A. (2006). The severe acute respiratory syndrome: Impact on travel and tourism. *Travel Medicine and Infectious Disease*, *4*(2), 53–60. https://doi.org/10.1016/j.tmaid.2005.04.004

World Health Organization. (2020). *Critical preparedness , readiness and response actions for COVID-19.* (Issue March).

Wu, Y.-C., Chen, C.-S., & Chan, Y.-J. (2020). The outbreak of COVID-19. *Journal of the Chinese Medical Association*, *83*(3), 217–220. https://doi.org/10.1097/jcma.0000000000000270

Zeng, B., Carter, R. W., & De Lacy, T. (2005). Short-term perturbations and tourism effects: The case of SARS in China. *Current Issues in Tourism*, *8*(4), 306–322. https://doi.org/10.1080/13683500508668220

Tourism marketing

Promoting Creative Tourism: Current Issues in Tourism Research – Kusumah et al. (Eds)
© 2021 Taylor & Francis Group, London, ISBN 978-0-367-55862-8

The effect of tourist satisfaction in the relationship between experiential marketing and revisit intention in Dusun Bambu, Indonesia

R. Khaerani
Universitas Pendidikan Indonesia, Bandung, Indonesia

T. Kartika & B. Basri
Stiepar Yapari, Bandung, Indonesia

ABSTRACT: This study aims to determine the effect of tourist satisfaction in the relationship between experiential marketing and revisit intention in Dusun Bambu, Indonesia. This type of research is quantitative research and sampling technique used was purposive sampling. The data were taken using questionnaire targeting 100 customers in Dusun Bambu. This research concentrates on direct and indirect functions. The analysis tools used are classical assumptions, path analysis, and sobel tests. The analysis was performed using the SPSS 24.0 statistics package. The results of classical assumptions show that all variables consisting of experiential marketing, tourist satisfaction and re-visit intention are in the normal line and have similarity variance. This study also found that tourist satisfaction has been proven to mediate the relationship between experiential marketing and revisit intention. The results of the study show that experiential marketing has a stronger influence on the indirect effects of tourist satisfaction as a mediator than the direct effect on revisit intention. The Sobel test results show that tourist satisfaction has a positive effect in significantly mediating experiential marketing relations with revisit intention.

Keywords: experiential marketing, tourist satisfaction, revisit intention, path analysis

1 INTRODUCTION

In this era of globalization, the activity of tourism becomes increasingly attracted by the community. Even, for some people, a field trip has been regarded as one of the obligation agenda to do every weekend. This is consistent with a statement from Rhenald Kasali about the phenomenon Esteem Economy now come with dragging sweeping changes to community leisure behavior (Kasali 2017).

The presence of varied tourist attractions certainly raises competition between every tourist attraction business provider, so it is important for a tourist destination to not only focus on the products offered but also make an emotional approach to tourists so that the creation of satisfaction when visiting and the desire to re-do repeat visit. The shift in paradigm in this era makes tourists no longer just see the quality of products or services that are good in meeting tourist satisfaction, but tourists want something more in the form of different experiences when visiting a tourist attraction.

People who have vacation usually are happier than those who don't, and happiness will bring positive impact increasing of society's productivity, (Liu as cited in Bintarti & Kurniawan 2017) because from the psycology point of view, by having vacation will increase the happines and also will lower the tress of the society (Cropanzano & Wright 2001). Touring in outdoor environment will increase the quaility of life of a person which will naturally will affect his productivity (Andereck & Nyaupane 2011).

A marketing concept has developed rapidly to an experiential marketing provided by a provider of goods or services, so that consumers have a unique and interesting experience that makes the desire to buy back the product / service increase. Experiential marketing not only provides information and opportunities for customers to gain experience of the benefits, but also evokes emotions and feelings that affect marketing, especially sales (Thejasukmana & Sugiarto 2014).

In the midst of the rapid growth in the numbers of travelers, Dusun Bambu comes with the theme Family Leisure aimed at family tourism on the basis of ecotourism. However, the many interesting places that have sprung up in West Java especially in West Bandung will certainly be an alternative choice for tourists to visit and Dusun Bambu must keep its existence in the midst of such competition.

Based on data from previous studies, from 50 consumers of Dusun Bambu, 30 people made their first visit, 15 people made their second visit and 5 people made their third visit (Khaerunnisa 2016). The results of other studies also state that tourists who make a return visit to Dusun Bambu are only 30% (Amir,2015).

The data shows that the number of interest in repeat visits in the Dusun Bambu is still relatively low. This is very interesting to study where the development of marketing concepts has developed rapidly. In the current era, the concept of marketing is focuses on its consumers instead of its products.

We have found out that there have been various studies about the experiential marketing, satisfaction, and revisit intention relationships while working on this subject (Christian & Dharmayanti 2013; Febriani et al. 2019; Hyunjin 2013; Öztürk 2015).

Dusun Bambu is one of the important objects of eco-tourism in Indonesia, which has a big role to attract domestic and international tourists. For these reasons, this study aims to determine the effects of the experiential marketing of Dusun Bambu on the tourist's satisfaction and revisit intention and determine the interaction between satisfaction and revisit intention. Thus, to take advantage of all the benefits of sustainable tourism practices, the preliminary ideas may occur as what should be done. Based on the analysis above, it is formulated: How influential the experiential marketing on tourist's satisfaction and revisit intention in Dusun Bambu.

2 LITERATURE REVIEW

2.1 *Experiential marketing*

Experiences are events that engage individuals in a personal way (Pine dan Gilmore 1998). Experiential marketing is everywhere. In a variety of industries, companies have moved away from traditional "features-and-benefits" marketing toward creating experiences for their customers. This shift toward experiential marketing has occurred as a result of three simultaneous developments in the broader business environment (Schmitt 1999).

The experiential modules to be managed in Experiential Marketing include sensory experiences (SENSE), affective experiences (FEEL), creative cognitive experiences (THINK), physical experiences, behaviours and lifestyles *(ACT),* and social-identity experiences that result from relating to a reference group or culture (RELATE) (Schmitt 1999) (Figure 1).

2.2 *Tourist satisfaction*

Visitor satisfaction is emotional state spreading positive words from tourists after getting a chance or experience (Mohamad et al. 2014). Tourist satisfaction is considered to be an important element to maintain competitive business in the tourism industry because it affects the choice of destination, and the consumption of products and services (Kozak & Rimmington 2000).

In perspective of 'implied after purchase and use', customer satisfaction as the outcome of purchase and use resulting from the buyer's comparison of the rewards and costs of the purchase relative to the anticipated consequences (Churchill and Surprenant as cited in Alkilani et al. 2012). This study uses 3 measurement concepts according to Shi, Prentice, & He to be used in measuring tourist satisfaction variables; overall satisfaction, confirmation of expectation and performances versus ideal (Shi et al. 2014).

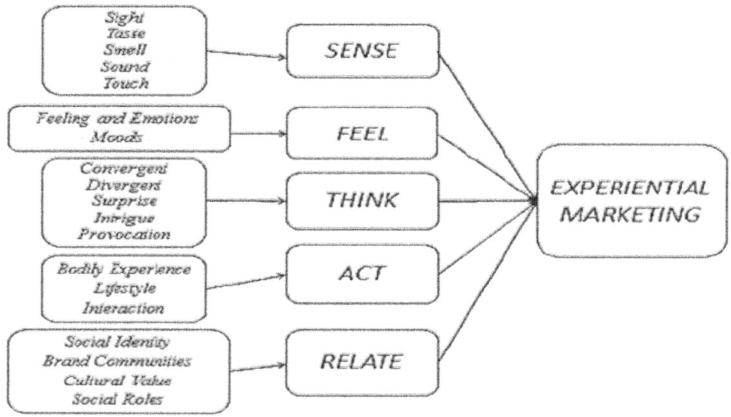

Figure 1. Strategic Experiential Modules (SEMs) by Schmitt 1999.

2.3 *Revisit intention*

The interest in revisiting is the intention to plan certain behaviors. Intention said, when people have strong intentions to engage in behavior, they prefer to conduct recreational behavior in the tourism sector, this takes back on tourism services or recreational services or revisits from destinations or visitors to attractions (Pantouw & Pangemanan 2014). Revisit intention is repeat purchasers continue to buy the same brand even though they don't have an emotional attachment to it (Kotler dan Armstrong s cited in Wulanjani & Derriawan 2017). Revisit intention is to repeat activities to re-visit a destination (Baker dan Crompton as cited Lin 2013). There are two dimensions that issued for revisit intention; intention to recommend and intention to revisit (Baker & Crompton 2000).

3 RESEARCH METHODS

The aim of this study is to determine the relationship among experiential marketing, customer satisfaction, and revisit intention among tourist in West Bandung, whereas intervening variable in this study is customer/tourist satisfaction. Dusun Bambu were selected for this survey. By using Slovin's formula, 100 respondents have been selected for this research. This research is a descriptive and verificative and then tested using path analysis. Data in this study were collected through; questionnaire, interview, observation, and literature study.

4 RESULTS AND DISCUSSION

4.1 *Results of factor analysis*

The survey included several demographic questions (e.g. gender, age, occupation, origin area and frequency of visits). The respondents were predominantly female (65%). The median age group of the respondent was in the age of 21–25 (42%). Most of the respondent works as private-employee (35%). Visitors from outside Bandung accounted for the highest percentages (67%) and the majority of respondents were second-times visitors (80%).

Table 1 gives the results of analyzing the level of experiential marketing of Dusun Bambu. The highest mean is 4.26 which from relate dimension with a very high category. The result of the research indicate that relate marketing implemented by Dusun Bambu is very good. Conversely, for the lowest mean value of 3.76 from act dimensions. Even though it has the lowest mean value,

Table 1. Mean experiential marketing.

Dimension	Mean	Category
Sense	4.07	Very high
Feel	3.95	High
Think	4.12	Very high
Act	3.76	High
Relate	4.26	Very high
Mean	4.02	High

Table 2. Mean tourist satisfaction.

Dimension	Mean	Category
Performances versus Ideal	4.12	High
Overall Satisfaction	3.77	High
Confirmation of Expectation	3.44	High
Mean	3.78	High

Table 3. Mean revisit intention.

Dimension	Mean	Category
Intention to revisit	4.21	Very High
Intention to recommend	4.03	High
Mean	4,12	High

the number is still included in the high category which means that the marketing of Dusun Bambu has also been good.

Tourist satisfaction questionnaire result are reported in Table 2. From the mean data of respondents' answers, it shows the average mean overall was 3.78 with a high category and indicated approval. Then it concluded that visitors were satisfied with the products or services offered by Dusun Bambu.

Table 3 presents the data of respondents' answers which shows mean of 4.12 with a high category. This means that, the respondent in this study mostly had a tendency to revisit Dusun Bambu in the future and be willing to invite and recommend Dusun Bambu to their relatives.

4.2 *Results of data analysis*

The reliability test conducted in this study by looking at the cronbach's alpha value, if the cronbach's alpha value exceeding the recommend threshold level of .70 then the question item can be said to be reliable to use (Widoyoko 2014). In this research, all the question items are acceptable with cronbach's alpha .85, .90 and .82. See Table 4 below:

Table 5 also reports results of testing the discriminant validity of measurements scales. Discriminant validity of the scales is supported because all the items are greater than .306. If the correlation value is less than 0.3 then the question is declared invalid (Sugiyono 2017).

Results from the structural model reported in Figure 2 show that experiential marketing was positively related to tourist satisfaction ($\beta = .254$, $p < .05$) and experiential marketing was positively related to revisit intention ($\beta = .641$, $p < .05$). Results also show that tourist satisfaction was positively related to revisit intention ($\beta = 0{,}344$, $p < .05$). Finally, this study also provided empirical evidence that experiential marketing has a positive effect on tourist satisfaction and revisit intention,

Table 4. Cronbach's alpha and composite validity.

	Cronbach's Alpha	Composite Validity
EM	.85	.63
TS	.90	.86
RI	.82	.91

EM: Experiential marketing, TS: tourist satisfaction, RI: revisit intention.

Table 5. Cronbach's alpha and composite validity.

Direct Effect	Indirect Effect	Total Effect	p-value t-value		
EM→TS	.254	0	.254	.011	2.602
EM→RI	.641	0	.641	.000	8.670
TS→RI	.181	.163	.344	.016	12.741

EM: Experiential marketing, TS: tourist satisfaction, RI: revisit intention.

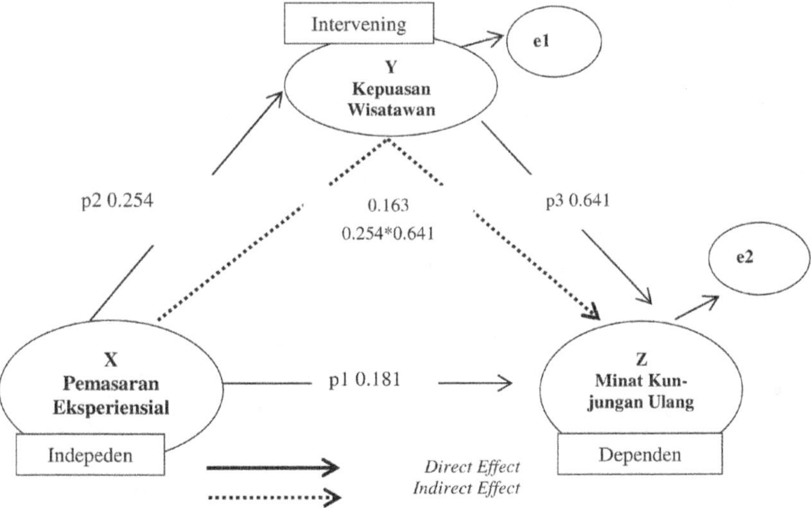

Figure 2. Results from the structural model.

customer satisfaction has a positive effect on revisit intention. R^2 values are also at an acceptable level (see Figure 2). These results showed that all hypotheses were fully supported.

5 CONCLUSION AND RECOMMENDATION

In this study, the relationships among experiential marketing, tourist satisfaction and revisit intention in Dusun Bambu Bandung have been examined. Considering the results of the study, a positive and significant relationship has been determined between experiential marketing and tourist satisfaction. Furthermore, a positive and significant relationship has been determined between experiential marketing and revisit intention. It has been concluded that, experiential marketing practice have importance in explaining tourist satisfaction and tourist's intention to revisit.

Providing an interesting experience to tourists must always be able to bring a sense of satisfaction and make them willing revisit. Therefore, it is necessary for Dusun Bambu to pay attention and follow up on the complaints from the tourists in order to build their repeat intention and loyalty.

REFERENCES

Alkilani, K., Ling, K. C., & Abzakh, A. A. 2012. The impact of experiential marketing and customer satisfaction on customer commitment in the world of social networks. *Asian Social Science*, *9*(1), 262–270. https://doi.org/10.5539/ass.v9n1p262

Amir, F. 2015. *Perancangan Promosi Objek Wisata Dusun Bambu* (pp. 4–42). pp. 4–42. Bandung.

Andereck, K. L., & Nyaupane, G. P. 2011. Exploring the Nature of Tourism and Quality of Life Perceptions among Residents. *Journal of Travel Research*, *50*(3), 248–260. https://doi.org/10.1177/0047287510362918

Baker, D. A., & Crompton, J. L. 2000. Quality, Satisfaction and Behavioral Intentions. *Annals of Tourism Research*, *27*(3), 785–804. https://doi.org/10.1103/PhysRevA.93.032136

Bintarti, S., & Kurniawan, E. N. 2017. A study of revisit intention: Experiential quality and image of Muara Beting tourism site in Bekasi District. *European Research Studies Journal*, *20*(2), 521–537.

Christian, A., & Dharmayanti, D. 2013. Pengaruh Experiential Marketing Terhadap Customer Satisfaction Dan Customer Loyalty the Light Cup Di Surabaya Town Square. *Jurnal Manajemen Pemasaran Petra*, *1*(2), 1–13.

Cropanzano, R., & Wright, T. A. 2001. A Review and Further Refinement of the Happy- Productive Worker Thesis. *Consulting Psychology Journal*, *53*(3), 182–199. https://doi.org/10.1037//106W087.53.3.182

Febriani, I. Y., PA, W. R., & Anwar, M. 2019. Pengaruh Experiential Marketing Terhadap Kepuasan Konsumen Dan Minat Beli Ulang Di Warung Kopi Klotok, Kaliurang, Yogyakarta. *Jurnal Manajemen Bisnis*, *10*(1), 35–54. https://doi.org/10.18196/mb.10167

Hyunjin, J. 2013. The Effect of Experiential Marketing on Customer Satisfaction and Revisit Intention of Beauty Salon Franchise Stores. *Fashion Business*, 17(3), 109–121. https://doi.org/10.12940/jfb.2013.17.3.109

Kasali, R. 2017. *Disruption*. Jakarta: Gramedi Pustaka Utama.

Khaerunnisa, R. M. 2016. *Pengaruh Viral Marketing Terhadap Purchase Decision Involvement Di Dusun Bambu* (pp. 1–10). pp. 1–10.

Kozak, M., & Rimmington, M. 2000. Tourist satisfaction with Mallorca, Spain, as an off-season holiday destination. *Journal of Travel Research*, *38*(3), 260–269. https://doi.org/10.1177/004728750003800308

Lin, C. H. 2013. Determinants of Revisit Intention to a Hot Springs Destination: Evidence from Taiwan. *Asia Pacific Journal of Tourism Research*, *18*(3), 183–204. https://doi.org/10.1080/10941665.2011.640698

Mohamad, M., Ghani, N. I. A., Mamat, M., & Mamat, I. 2014. Satisfaction As A Mediator To The Relationships Between Destination Image and Loyalty. *World Applied Sciences Journal*, *30*(9), 1113–1123. https://doi.org/10.5829/idosi.wasj.2014.30.09.14107

Öztürk, R. 2015. Exploring the Relationships between Experiential Marketing, Customer Satisfaction and Customer Loyalty: An Empirical Examination in Konya. *International Journal of Social, Behavioral, Educational, Economic and Management Engineering*, *9*(8), 2485–2488. Retrieved from www.citeulike.org/user/tilljwinkler/article/10083551.

Pantouw, P., & Pangemanan, S. S. 2014. The Effect Of Destination Image And Tourist Satisfaction On Intention To Revisit In Lembeh Hill Resort. *Jurnal EMBA*, *2*(3), 049–057.

Pine II, B. J., & Gilmore, J. H. 1998. Welcome to the Experience Economy. In *Review Literature And Arts Of The Americas*.

Schmitt, B. 1999. Experiential Marketing. *Journal of Marketing Management*, *15*(1–3), 53–67.

Shi, Y., Prentice, C., & He, W. 2014. Linking Service Quality, Customer Satisfaction and Loyalty In Casinos, Does Membership Matter? *International Journal of Hospitality Management*, *40*, 81–91. https://doi.org/10.1016/j.ijhm.2014.03.013

Sugiyono. 2017. *Metode Penelitian Kuantitatif, Kualitatif dan R&D*. Bandung: Alfabeta.

Thejasukmana, V. A., & Sugiarto, D. S. 2014. Analisis Pengaruh Experiential Marketing Terhadap Pembelian Ulang Konsumen The Vinnette (House of Bovin and Lynette) Surabaya. *Jurnal Manajemen Pemasaran Petra*, *2*(1), 1–14.

Widoyoko, E. P. 2014. *Teknik Penyusunan Instrumen Penelitian*. Yogyakarta: Pustaka Pelajar.

Wulanjani, H., & Derriawan. 2017. Dampak Utilitarian Value Dan Experiential Marketing Terhadap Customer Satisfaction Dan Revisit Intention. *Jurnal Riset Manajemen Dan Bisnis (JRMB) Fakultas Ekonomi UNIAT*, *2*(2), 121–130.

Promoting Creative Tourism: Current Issues in Tourism Research – Kusumah et al. (Eds)
© 2021 Taylor & Francis Group, London, ISBN 978-0-367-55862-8

Virtual public sphere: The overview of Instagram users in responding to the Instagram posts of tourist destinations in Covid-19 pandemic

A. Khosihan, P. Hindayani & A.R. Pratama
Universitas Pendidikan Indonesia, Bandung, Indonesia

ABSTRACT: Technology and information are growing by leaps and bounds, and it caused a major change in the community. The appearance of Instagram as a virtual public sphere raises the question of whether Instagram can carry out the characteristic of the public sphere in the middle of the Covid-19 pandemic as well as it should. This article explained how Instagram features could present the concept of the public sphere virtually through the posts of @pesonaid_travel account in this pandemic. By using qualitative descriptive methods, researcher found that there are 387 comments in 14 posts of the government's tourism promotion official Instagram account (@pesonaid_travel) from March until May 2020. Then, the researcher filtered them into 97 comments to be used as the analysis. The findings of this research show that an Instagram account called Pesona Indonesia has the features that could be the media for the public sphere as well as it should. However, the features that could support this public sphere characteristic do not show comments that led to a significant discussion. Even though, there are some interesting discussions about tourism and pandemic. Practically, there are four characteristic comments related to tourism and the pandemic Covid-19. Those are the comment that only related to the pandemic Covid-19; the comment that only related to the intention of traveling; the comment that related to both the pandemic Covid-19 and the intention of traveling; and the other comment that is not related to the topic presented. This study is useful for presenting the concept of the better version of the virtual public sphere, especially for tourism on Instagram.

Keywords: virtual public sphere, virtual sphere, instagram during Covid-19, instagram post, tourist responses

1 INTRODUCTION

The development of media information is growing fast, including some various types of social media, which is very easy to access. The growth of social media is very useful in tourism; for instance, it could make some expectations for one tourist destination (Narangajavana et al. 2017). Some studies describe social media approaches in tourism, especially in terms of promotion, management, research, and a little bit about product distribution (Leung et al. 2013). One of the most popular social media nowadays is Instagram. NapoleonCat.com stated that Instagram users in Indonesian in April 2020 have reached 65.780.000 accounts and has been the second largest of the largest account users after Facebook (158.160.000 users). The Instagram users in Indonesia, which are dominated by youth (Smith 2018), make this social media interesting to be researched. In tourism, Instagram has an essential role in introducing the power of tourism with some of its tools (Fatanti & Suyadnya 2015), such as hashtags and geotags (Smith 2019). Those facilities on Instagram mean people can interact and communicate with each other using an Instagram feature called comments.

In early 2020, the world was shocked by the explosion of Covid-19 in China, and then the virus spread all over the world quickly, including in Indonesia. There are many sectors affected because of the virus. Each country has its own policy to reduce the spread of Covid-19, for example, lockdown

DOI 10.1201/9781003095484-102

and PSBB (large-scale social restrictions). Studies about Covid-19 have been published, such as the effect of Covid-19 in the tourism industry (Hoque et al. 2020), the effects of misleading media reports about Covid-19 on Chinese tourists' mental health (Zheng et al. 2020), socializing tourism for social and ecological justice after Covid-19 (Higgins-Desbiolles 2020), and pandemics, tourism, and global change, a rapid assessment of Covid-19 (Gössling et al. 2020). In Indonesia, the most affected sector is the tourism sector. Since Covid-19 was announced as a national disaster by BNPB (National Disaster Management Agency) at the end of February and the emergency response until May 29, 2020, tourism in Indonesia is down because tourist destinations are closed. There is no way to travel around. However, in the digital world, activities regarding promoting tourist destinations stay alive. Some national tourism Instagram accounts still upload some photos or videos about tourist destinations in the middle of this pandemic. Facts that the tourism destinations are closed and that tourists cannot visit the tourist attractions are continuously promoted by the government's official tourism promotion account (@pesonaid_travel) make this account a good public space, especially in promoting tourism destinations while campaigning for awareness of the covid-19 outbreak. The researcher wanted to see how these discussions have appeared in the virtual public sphere of Instagram while pandemic Covid-19 is happening.

2 LITERATURE REVIEW

The virtual public sphere was introduced by Habermas to explain the phenomena of modern posts in which the media is a tool of the capitalist and the rulers for propagandizing their own goals. This concept was born to discuss the history of critical discussion as a response to the political reality that happened to the community in the 18th century (Nasrullah 2015). The public sphere has some interpretations, not only sociology perspective but also the point of view of architecture, media activities, and feminism (Prasetyo 2012). The context of this article, the public sphere basic concept that will be discussed, is about the public sphere in the sociology dimension. It is about the main thought that is used as the basis of public sphere theory by Habermas. The purpose of the public sphere by Habermas is to build opinions and will information that has the possibility of generalization and represent the public interest (Gedeona 2006). Nasrullah (2015) stated the public sphere characteristics, status-neglect, freedom of talking about specific topics, and inclusive. Besides those three characteristics, the public sphere has three forms. Those are the public sphere in the form of an arena, the form of public itself (actor), and the form of agency (Jati 2016).

The rapidity of the Internet and technology development finally brings the new public sphere concept. A concept called virtual public sphere was shown as the meeting between cyberspace and conventional public sphere concept. Virtual public space is making the possibility of a new culture in the world of democracy. There are no bourgeoisie-proletariat, men-women, and boundaries (Nasrullah 2015). If the first concept of the public sphere always about politics, then in this modern era, the public sphere, especially the virtual public sphere, becomes more flexible and general. The Internet progress to represent the virtual public sphere finally brings us to the application called Instagram as the new public space. As the promotion tool and the favorite one, Instagram gives a picture of ideology that supports contemporary tourism (Smith 2019). Instagram offers visuals, such as photos and videos; it creates intention to them who want to do traveling (Fatanti & Suyadnya 2015). Because of the photos and videos with the comment feature, so the role of the virtual public sphere can be observed.

3 METHODS

This study used a qualitative approach by using comments in an Instagram post of Pesona Indonesia (@pesonaid_travel) as the official account of the Minister of Tourism and Creative Economy to promote the tourist destination in Indonesia as the analysis content. There are 14 posts from March until May 2020. The researcher processed a total of 378 comments (per May 14, 2020) from 14 posts to map the topic of discussion, which were made by Pesona Indonesia's Instagram account. March to May was chosen because it was a disaster response period determined by BNPB, so the data selection process became more independent.

4 FINDINGS AND DISCUSSION

Based on the analysis of the posts and comments in Pesona Indonesia (@pesonaid_travel), Instagram account, generally, this Instagram account has good connectivity between photos, captions, and comments. There are no inappropriate nor suitable comments at the post made. In this dimension, the Pesona Indonesia Instagram account can fulfill the terms as a public space because it consists of four characteristics of public space by Habermas (Nasrullah 2015). The Instagram facilities in Pesona Indonesia account can be the public space because first, there is no restriction of status social inside the photos or videos that have been posted, the caption that has been made, and the reactions that have been developed. There is no offensive content about economic status or another social status. Admin on the account also never locks the comments column, which shows that the public sphere facility on this account can be used by anyone. Second, the topics in the Instagram posts are in line with the caption. It showed that the criteria of public space that focus on the domain (tourism theme) are not discussed by a specific sphere. Third, the topic used by each post can become a part or a problem together in a common situation. It happened when the Pesona Indonesia Instagram account uploaded a post about Covid-19 and the Day of Silence (Nyepi). The reaction comments that appeared are empathy comments and feeling of silence. The existence of posts related to holidays and pandemics still displays visuals that are eye-catching so that feeds on @pesonaid_travel's Instagram page do not seem lame. This shows that the @pesonaid_travel account does not eliminate its image as a tourism promotion account.

The content that was uploaded consistently by the Instragram account of Pesona Indonesia shows the beauty of tourism destinations in Indonesia. It is in line with a biography written on the profile page of this account. The profile page tells the purpose of the account as media information of official account tourism Indonesia under the Minister of Tourism and Creative Economy of Indonesia. Because of the profile page that explained the aim of this account and consistency in uploading posts related to tourism, it shows one of the principles of the public sphere, which focuses on only one theme. Besides, the Instagram account of Pesona Indonesia also made some posts based on reposts (the post from another account). This media promotion of tourism gives space for the community to create and get involved in promoting tourism in Indonesia. Most of the reposting posts are from the account that also participated in tourism with good quality pictures and videos. Here, the characteristic of the public sphere that does not limit the society to create and promote tourist destinations became indicators that @pesonaid_travel's Instagram account gave access to the virtual public sphere to all the levels of society. Although, not all the posts with hashtag #pesonaIndonesia are reposted by the Pesona Indonesia Instagram account. Because of the open comments feature at the Pesona Indonesia Instagram account, it means anyone can say and write anything in the comment section. Even though the feature on Instagram that means people can reply, the comment gives access for a longer discussion. It shows that the public sphere on Instagram is not exclusive; any post made by any account can be talked about and discussed by other Instagram users. In this context, the post reaction formed in the virtual public space during the pandemic did not insert the condition of the pandemic itself by the community.

Specifically, based on the observation of 14 posts from March until May 2020 (14th May 2020), the researcher processed the data from 387 comments and then analyzed 97 comments that related to tourism. The rest of the comments are with emoticons or the same, and they do not relate to tourism, for example, a comment that promotes medicine. From 97 comments at 14 Instagram posts, the researcher found that, first, most of the comments are about the beauty of tourist destinations, as follows:

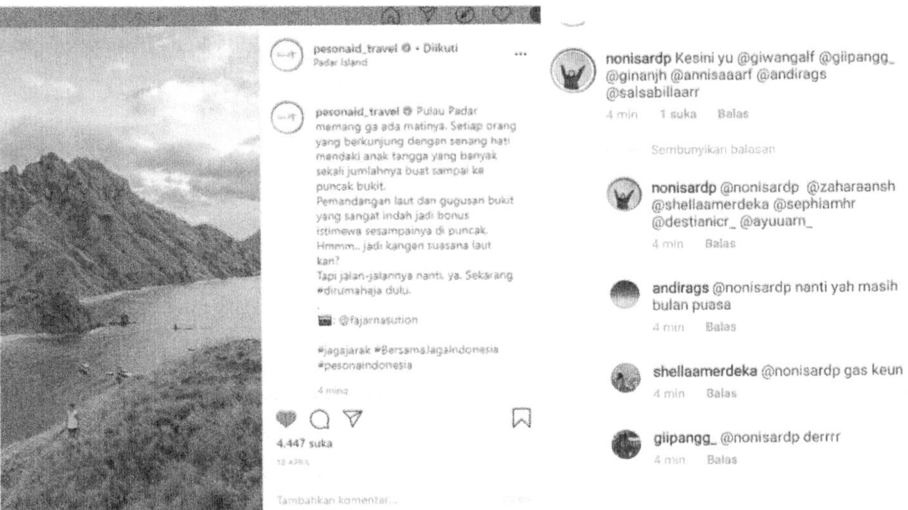

Figure 1. Post and comments about the intention of traveling.

From Figure 1, there is virtual discussion through comment feature provided by Instagram where account @nonisardp mentions her friends to open discussion about going to Padar Island. However, they do not talk about Covid-19 at all. Their problem is just because it was Ramadan so they cannot travel around. The other users only say yes by the comments "gasskeun (let us go)" and "derr (yes)." In this comment, the context seems that the function of Instagram as a virtual public sphere is doing well. However, there are also comments which do not do the discussion. It can be seen at some comments that just say "indahnya (how beautiful)," "keren (cool)," or just emoticon. In the public sphere, there should be a discussion in a group meeting (Nasrullah 2015). Because the context here is tourism, the discussion should be related to tourism. In this context, the post reaction formed in the virtual public space during the pandemic did not insert the condition of the pandemic itself by the community.

The second post of the photo/video tourist destination is connected to pandemic Covid-19 without any intention of the holiday. It can be seen in this picture below:

From Figure 2, it shows that some Instagram users (in this case represented by a comment from @adipratama_putu) have a concern to the pandemic Covid-19. They do not use the sentence with the desire to go traveling, but they pray for the pandemic. The comments about the pandemic do not make the public reaction to discuss it. All the comments about Covid-19 have no response from other users. It shows that the public sphere in this context is just the form of the public itself, where the owner of the account comments as the actor (Jati 2016). This situation can be understandable because in a discussion sphere that most of the posts about a certain topic will be less interesting to be commented on. In this context, comments about Covid-19 are not interesting because the Pesona Indonesia Instagram account is not an account of Covid-19. Pesona Indonesia is an account of tourism. Although, there is a caption that talks about Covid-19 with the hashtag #jagajarak (#keepdistancing) and #stayathome. It happened because the public sphere has a characteristic that focuses on the domain of common concern, which is ignored if the context is not suitable for the usual topic (Nasrullah 2015).

The third is a post from the Pesona Indonesia account since March until May that shows that there are comments related to the desire of traveling and Pandemic Covid-19. It can be seen in this picture (Figure 3):

The picture above explains that Instagram users have wishes to do traveling, but they also have a concern about Covid-19. Those comments indicate that the public sphere on Instagram also depends

Figure 2. Post and comments about pandemic Covid-19.

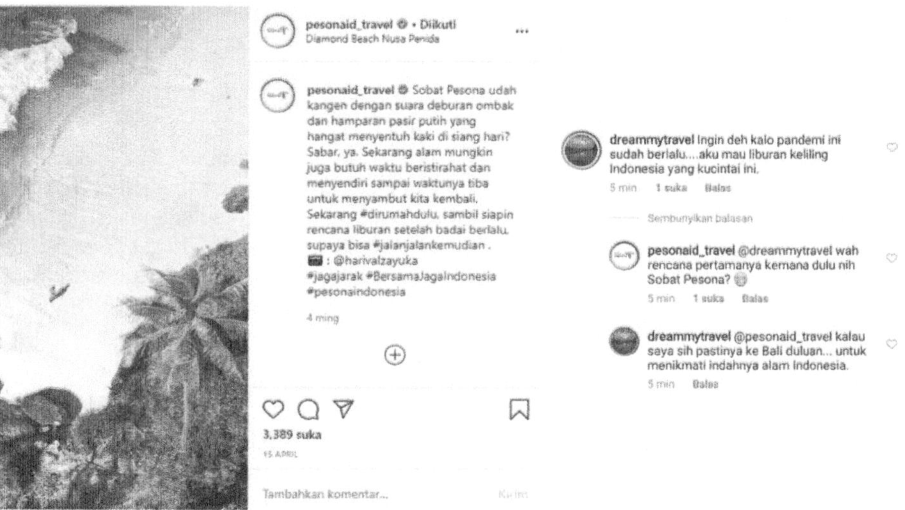

Figure 3. Post and comments about both the intention of traveling and the pandemic Covid-19.

on the stimulus from the caption that consists of some questions. The response of the comments that can be replied will be made with the discussion so the virtual public sphere can be the public sphere as it should be. In this context, there is a rationalist comment that connected the newest situation and the desire to travel. So, Instagram, as a virtual public sphere, can build the public space institution as more real and democratic (Jati 2016).

Fourth, because of the post about the beauty of tourist destinations in this pandemic, there is a comment that has no connection with the tourist destinations and the pandemic. The comment is about another topic. It can be seen in this picture (Figure 4):

This type of comment shows that there are some disadvantages to virtual public space. In this case, the users do not have boundaries to say or write anything in the comments feature. In this context, the characteristic of the public sphere by Habermas cannot be equalized with the condition

Figure 4. Post and comments that do not relate to the purpose of the account.

of the virtual public sphere at the Pesona Indonesia Instagram account. It is because of the contrast of the "domain of common concern," which is the characteristic of the public sphere.

The Pesona Indonesia Instagram account as a media promotion of tourism in Indonesia can be a suitable virtual public space if it has the third main characteristics of public space: those are status-neglected, the freedom to talk about specific topics, and inclusive (Nasrullah 2015). It can be seen by the characteristic of the comment from the Instagram users in the first and third types. Then, it is not suitable for the concept in the fourth comment. It is because there are no boundaries at the comment. The presence of the Pesona Indonesia Instagram account can make Instagram as the media for promoting the tourist destination itself (Fatanti & Suyadnya 2015). However, it also gives duplication to the place; it can be seen from the comment that said the location has similarities with the location at their place. It shows that a public sphere can give the same idea.

5 CONCLUSION

The presence of the Pesona Indonesia Instagram account with 14 posts shows the function of the public sphere in this pandemic, although it is virtual. Based on the Habermas concept, the topic is about politics, but in this account, the topic is about tourism. Instagram account @pesonaid_travel, as a virtual public sphere of tourism in the pandemic of Covid-19, has been doing its function very well. Specifically, it created the fourth characteristic of public regarded by their comments, and those are the comment about tourism, the comment about pandemic Covid-19, the comment about both tourism and pandemic Covid-19, and the comment that has no relation about tourism and pandemic Covid-19.

This study has a limitation of content because it only started from March until May 14, 2020, while the disaster response will continue until 29th May 2020, and the pandemic is still happening. So, it is possible to do other studies in the case of the public sphere that happen because of this pandemic. It includes a map of potential tourism in Indonesia.

ACKNOWLEDGMENT

Thanks to Universitas Pendidikan Indonesia that gave the researcher a chance to write in this conference and to the Pesona Indonesia Instagram account as the subject of this study.

REFERENCES

Fatanti, M. N., & Suyadnya, I. W. (2015). Beyond User Gaze: How Instagram Creates Tourism Destination Brand? *Procedia — Social and Behavioral Sciences, 211*(September), 1089–1095. https://doi.org/10.1016/j.sbspro.2015.11.145

Gedeona T, H. (2006). Masyarakat Multikultural. *Jurnal Ilmu Komunikasi, 2*(34), 33–43.

Gössling, S., Scott, D., & Hall, C. M. (2020). Pandemics, tourism and global change: a rapid assessment of COVID-19. *Journal of Sustainable Tourism, 0*(0), 1–20. https://doi.org/10.1080/09669582.2020.1758708

Higgins-Desbiolles, F. (2020). Socialising tourism for social and ecological justice after COVID-19. *Tourism Geographies, 0*(0), 1–14. https://doi.org/10.1080/14616688.2020.1757748

Hoque, A., Shikha, F. A., Hasanat, M. W., Arif, I., & Abu Bakar Abdul Hamid. (2020). The Effect of Coronavirus (COVID-19) in the Tourism Industry in. *Asian Journal of Multidisciplinary Studies, 3*(1).

Jati, W. R. (2016). *Cyberspace , Internet , dan Ruang Publik Baru: Aktivisme Online Politik Kelas Menengah Indonesia*. it 3(1), 25–35.

Leung, D., Law, R., van Hoof, H., & Buhalis, D. (2013). Social Media in Tourism and Hospitality: A Literature Review. *Journal of Travel and Tourism Marketing, 30*(1–2), 3–22. https://doi.org/10.1080/10548408.2013.750919

Narangajavana, Y., Callarisa Fiol, L. J., Moliner Tena, M. Á., Rodríguez Artola, R. M., & Sánchez García, J. (2017). The influence of social media in creating expectations. An empirical study for a tourist destination. *Annals of Tourism Research, 65*, 60–70. https://doi.org/10.1016/j.annals.2017.05.002

Nasrullah, R. (2015). Komunikator. *Komunikator, 4*(01). http://journal.umy.ac.id/index.php/jkm/article/view/188

Prasetyo, A. (2012). Menuju Demokrasi Rasional: Melacak Pemikiran Jürgen Habermas Tentang Ruang Publik. *Jurnal Ilmu Sosial Dan Ilmu Politik, 16*(2), 37770.

Smith, S. P. (2018). Instagram abroad: Performance, consumption and colonial narrative in tourism. *Postcolonial Studies, 21*(2), 172–191. https://doi.org/10.1080/13688790.2018.1461173

Smith, S. P. (2019). Landscapes for "likes": capitalizing on travel with Instagram. *Social Semiotics, 0*(0), 1–21. https://doi.org/10.1080/10350330.2019.1664579

Zheng, Y., Goh, E., & Wen, J. (2020). The effects of misleading media reports about COVID-19 on Chinese tourists' mental health: a perspective article. *Anatolia, 00*(00), 1–4. https://doi.org/10.1080/13032917.2020.1747208

Promoting Creative Tourism: Current Issues in Tourism Research – Kusumah et al. (Eds)
© 2021 Taylor & Francis Group, London, ISBN 978-0-367-55862-8

Visual ethnography: Tourists' perception of Bandung's destination image

O. Ridwanudin, Y. Yuniawati & V. Gaffar
Universitas Pendidikan Indonesia, Bandung, Indonesia

ABSTRACT: Bandung is known as a culinary city, fashion city, flower city, etc. These various names have an impact to Bandung's brand image. It becomes inconsistence and unclear. This study aims to investigate the tourists' perception of Bandung's destination images visualized from photography. The research method was qualitative with visual ethnography approach. The data were gained from 76 photos taken by tourists and posted through a website. Content analysis was employed to analyze the data. The findings showed that most of the photos were heritage buildings in Bandung. This indicates that tourists tend to perceive Bandung as a Heritage City. Thus, the government should continually maintain the heritage buildings in Bandung as they become the icon of the city.

1 INTRODUCTION

Tourism is inseparable from the tourists' movement visiting various attractions offered in a destination such as sights, monuments, museums, cultural events, shopping centers, and culinary spots. The tourist attraction is one of the motivations for tourists in traveling. Tourists choose a destination as they have emotional bond with that area (Cutler 2015; Ruane et al. 2018). Thus, visiting the intended place is a happy moment that can be enjoyed (Lo et al. 2010).

Photography creates memories of destinations; and it reflects travel experiences (Berger et al. 2007). Many tourists take photos during travel tours as part of their tourism activities (Cederholm 2004). At this point, photos play a symbolic role in creating memories and shaping those memories into a narrative or story (Lo et al. 2010). The photos taken are often shared through social media (Groves 2001). Hence, those who see it can be attracted to visit

Bandung is a famous city providing many interesting tourist attractions. It is often called as 'City of Flowers' 'the City of Parks' 'the City of Culinary Tourism' 'the City of Fashion' or 'Paris Van Java'. All these names are an image of Bandung (Khoiriana & Nurlambang 2017) that is formed as an effort to promote Bandung to the tourist. Destination image is one of the popular city marketing tools nowadays. It refers to the tourists' or society perception of a city that is gained through their experience (Khoiriana & Nurlambang 2017). This perception is then delivered to the other people, then creates a brand image. Brand image must be in accordance with the characteristics of the city as it will influence tourists' satisfaction. If it does not match the tourists' perception, the tourists will not recommend the visited destination to the other, or revisit the destination in the future (Garrod 2009).

The different names owned by Bandung impact its brand image. It causes inconsistency and is unclear. Therefore, this study aims to investigate the tourists' perception of Bandung destination image through photograph created by tourists as Jenkins (2003) mentioned as the most common attribute of tourist behavior. Urry (1990) presents the same view points as photography is a link to the notion of requirement connecting both tourists and tourism provider. The use of photography in scientific research has expanded to encompass a range of emerging areas of interest, such as the ethnographic effects of tourism.

 DOI 10.1201/9781003095484-103

2 LITERATURE REVIEW

Destination image has become a topic that is often discussed in the tourism sector for the past three decades (Pike 2002). It is defined as a cognitive mental representation of an individual consisting of knowledge, ideas, emotional thoughts, and impressions a tourist has for a destination (Baloglu & Mccleary 1999; Fakeye & Crompton 1991; Pan & Li 2011). In other words, it consists of cognitive (beliefs and ideas) and affective components, including impressions and feelings (Baloglu & Brinberg 1993). The combination of these two components will result in the formation of individuals from the overall destination image (Beerli & Martin 2004)

Destination image has a positive impact on tourists' behavior. It influences the decision making process of visiting. It also sets the behavior after decision making (experience at the destination), evaluation (satisfaction) and future behavior consisting of intentions to revisit and willingness to recommend (Chen & Tsai 2007; Chi & Qu 2008).

Many experts state different opinion regarding the dimension of destination image. Echtner and Ricthie (1993) propose eight dimension, such as: comfort/security, interest/adventure, natural state, tour facilities, resort atmosphere/ climate, cultural distance, inexpensiveness, and lack of language barrier. Chen and Tsai (2007) classify the dimension into four categories: (1) destination brand (offers personal safety, a good quality of life, a good name and reputation hospitable and friendly people); (2) entertainment (good night life, a good shopping place, varied gastronomy exotic); and (3) nature and culture (great variety of fauna and flora, spectacular landscape, unusual ways of life and customs); and (4) sun and sand (good weather, good beaches). Meanwhile, various aspects that cover the destination image are Beautiful scenery, Rich culture, Unique sites, Conventional venues, frequently hosted cultural events, good entertainment facilities, high-quality service, high-quality food and beverage, efficiently managed stores, reasonable admission fees, reasonable able food and beverage charges, reasonable merchandise prices (Yeh et al. 2015).

Destination photography taken by both Destination Marketing Organization (DMO) and tourists communicates an image to shape and reshape the tourists' perception of a destination. There are two main components of a photo, namely content and composition (Albers & James 1988). Content refers to the appearance or mark taken in the overall photo. This appearance is interrelated and displayed to others, the combination is a composition of a photograph (Stepchenkova & Zhan 2013).

3 METHODOLOGY

The study used visual ethnography as an approach. The media were photographs. The aim was to assess how tourists' and society's perception of Bandung's destination image through photography. To analyze this phenomenon, two different methods were used: analytical content and photography study/ethnographic photography. A photograph competition was announced, in which tourists were asked to produce photographs that represent the image of Bandung city the most. 155 photos were collected through http://www.mpp.riset.upi.edu, the photos were uploaded from 8 August 2019 to 11 October 2019. Participants also filled in a form containing their personal information such as age, sex, occupation and place of origin. The 155 photos were then assorted based on the location of object photograph which resulted in 76 photos. These photos are then classified into six categories, such as: heritage buildings, tourist transportation, mosque, culture, events, and city parks. These categories are in line with the theme of the study.

4 RESULTS AND DISCUSSION

There are 76 participants whose profile is described demographically and psychographically through gender, age, occupation, and origin.

Table 1. Participant characteristics.

		Frequency			Frequency
Gender	Male	54	Origin	Bandung	42
	Female	22		Cimahi	3
Age	17–25	43		Padalarang	1
	26–35	21		Garut	4
	36–45	23		Sumedang	2
	>45	7		Majalengka	3
Occupation	Students	44		Ciamis	1
	Civil servant	3		Tasikmalaya	1
	Private employee	29		Jakarta	5
	Entrepreneur	15		Bekasi	1
	Lecture	1		Tanggerang	1
	Housewife	2		Lebak	1
				Cilacap	1
				Pekalongan	1
				Tegal	2
				Yogyakarta	1
				Berau	1
				Tebing Tinggi	1
				Pekanbaru	1
				Bima	1
				Lhok Seumawe	1
				Jayapura	1
			Total		76

Table 1 shows that most of the participant in this study are male (76.7% or 54 people). There are only 22 female participants (23.4%). They are mostly 17–25 years old (45.7%). There are only 21 participants who are 26–35 years old; 23 participants who are 36–45 years old, and 7 participants who are above 45 years old. The participants are dominated by students (44 people or 46.8%). There are also private employees (29 people), entrepreneurs (15 people), civil servants (3 people), housewives (2 people), and a lecturer. Regarding the origin, many of them come from Bandung (60 people or 63.8%). There are also participants coming from Jakarta (5 people), Garut (4 people), Majalengka (3 people), Cimahi (3 people), Sumedang (2 people), and Tegal (2 people). In addition, participants also come from Padalarang, Ciamis, Tasikmalaya, Bekasi, Tangerang, Lebak, Cilacap, Pekalongan, and Yogyakarta.

The demographic data of participant indicate that the participants are dominantly by male in which, in this case, tends to have high interest in photography rather than female. In addition, most of them are in a productive age in which they are very active and have great attention to devote moments through picture. This condition is supported by their occupation. Most of them are students who have more free time in taking pictures. They also often shared the result of their shots in their social media or website.

The number of photo in each category can be seen in Figure 1.

Figure 1 shows that most of the photos captured by participants are heritage buildings (30 photos). In this case, participants assume that Bandung is very known by its historical tourism, which is reflected in its heritage buildings. Gedung Sate, Isola Villa, and Gedung Merdeka become favorite places to visualize the image of Bandung. These buildings have beautiful art deco designs and become historic venues for the Asian-African conference.

Next photos are parks (13 photos). Under the government of Ridwan Kamil, Bandung has developed many city parks as a public facility that can be enjoyed by the society. Each park has

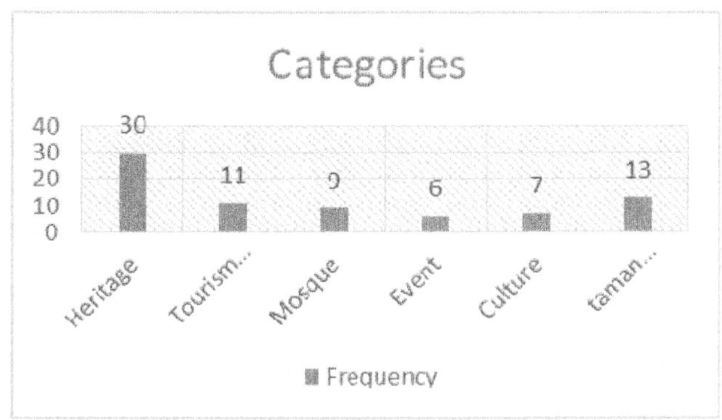

Figure 1. Categories.

its own themes, such as Superhero Park, Lansia Park, Photograph Park, etc. The building of these parks have a positive response from the society. It is proven by the participant who considered it as an image of Bandung.

Then, there are also photos of transportation (11 photos). Nine of eleven photos are Bandung Tour on The Bus (Bandros). It is one of transportation provided by the regional government for those who want to explore Bandung. This transportation explores several attractive places in Bandung. Concerning the photo of Bandros, most of the participant assumes that this transportation is an iconic transportation supporting tourism development in Bandung.

Nine out of 76 photos are those of mosques. Most of the participants take Masjid Agung picture. The construction and renovation of this mosque causes it to be one of the grandest mosques in Bandung. Thus, it becomes participants' targets in visualizing Bandung City.

There are also photos of Wayang Golek (7 photos). Wayang Golek is one of traditional wood puppets in West Java. It is regularly performed at Saung Anglklung Udjo along with traditional dance performance. This traditional puppet is basically almost forgotten by young generation. However, the participants still visualize Wayang Golek as an icon of Bandung. Most of them capture the uniqueness of Wayang Golek creation in Kampung Wanga Urban owned by Mr. Tatang.

The last photos are the events organized by Bandung government (6 photos). The government, in this point, regularly organize annual events especially at the anniversary of the city. The themes are different in each year. One of them is recognizing Sundanese culture by conducting a march around the Bandung's centers.

5 CONCLUSION

The result indicates that participants perceive Bandung as a historical tourism city. It can be shown by many of them taking historical building in visualizing the city. The participants also think that Bandung is a city with a good public space. It provides so many parks that can be enjoyed by the local society or tourists. Meanwhile, the events provided by government cannot be chosen as Bandung's destination image as it is less preferred by the participants. The city of Bandung itself is known as a culinary city, a city of fashion, but from the results of data collection, the participants' photo works were not obtained as their perception of the image of Bandung, which shows that Bandung is a culinary or fashion city. It is assumed that some participants do not know Bandung as a city of fashion or culinary. The rest of the participants recognize the name Bandung as a city of culinary and fashion, but they did not think about visualizing this image, but instead focused on heritage buildings which are one of the icons in Bandung.

REFERENCES

Albers, P. C., & James, W. R. (1988). Travel photography. A methodological approach. Annals of Tourism Research, 15(1), 134–158. https://doi.org/10.1016/0160-7383(88)90076-X

Baloglu, S., & Brinberg, D. (1993). Affective Images of Tourism Destinations. Journal of travel research, 11–15.

Baloglu, S., & Mccleary, K. W. (1999). A Model of Destination Image. Annals of Tourism Research, 26(4), 868–897.

Beerli, A., & Martin, J. D. (2004). Factors Influencing Destination Image. 31(3), 657–681. https://doi.org/10.1016/j.annals.2004.01.010

Berger, H., Denk, M., Dittenbach, M., Pesenhofer, A., & Merkl, D. (2007). Photo-based user profiling for tourism recommender systems.pdf. *E-Commerce and Web Technologies, 8th International Conference, EC-Web 2007, Regensburg, Germany, September 3–7, 2007, Proceedings*, pp. 46–55.

Cederholm, E. A. (2004). The use of photo-elicitation in tourism research – framing the backpacker experience. Scandinavian Journal of Hospitality and Tourism, 4(3), 225–241. https://doi.org/10.1080/15022250410003870

Chen, C., & Tsai, D. (2007). How destination image and evaluative factors affect behavioral intentions? 28, 1115–1122. https://doi.org/10.1016/j.tourman.2006.07.007

Chi, C. G.-Q., & Qu, H. (2008). Examining the structural relationships of destination image, tourist satisfaction and destination loyalty: An integrated approach. *Tourism Management Volume 29, Issue 4*, 624–636.

Crompton, J. L. (n.d.). An Assessment of the Image of Mexico as Vacation Destination and the Influence of Geographical Location Upon That Image.

Cutler, S. Q. (2015). Exploring the Moments and Memory of Tourist Experiences in Peru. Theses and Dissertations (Comprehensive). Retrieved from http://scholars.wlu.ca/etd/1746

Echtner, C. M., & Ritchie, J. B. (1993). The Measurement of Destination Image: An Empirical Assessment. *Journal of Travel Research.*

Fakeye, P. C., & Crompton, J. L. (1991). Image Differences between Prospective, First-Time, and Repeat Visitors to the Lower Rio Grande Valley. *Journal of Travel Research*, 10. https://doi.org/10.1177/004728759103000202

Garrod, B. (2009). Understanding the relationship between tourism destination imagery and tourist photography. Journal of Travel Research, 47(3), 346–358. https://doi.org/10.1177/0047287508322785

Groves, D. L. (2001). Photographic Techniques and the Measurement of Impact and Importance Attributes On Trip Design: A Case Study. Society and Leisure, 24(1), 311–317.

Jenkins, J. (2003). *World Englishes: A resource book for students.*

Khoiriana, R., & Nurlambang, T. (2017). Brand Image Kota Bandung. 8th Industrial Research Workshop and National Seminar.

Lo, I. S., Cheung, C., Law, R., Lo, A., & McKercher, B. (2010). Tourism and online photography. Tourism Management, 32(4), 725–731. https://doi.org/10.1016/j.tourman.2010.06.001

Pan, B., & Li, X. R. (2011). The Long Tail of Destination Image and Online Marketing. Annals of Tourism Research, 38(1), 132–152. https://doi.org/10.1016/j.annals.2010.06.004

Pike, S. (2002). Destination image analysis — a review of 142 papers from 1973 to 2000. 23, 541–549.

Ruane, S. T., Quinn, B., & Flanagan, S. (2018). Tourists' Photographic Constructions of Place in Ireland. 129–144. https://doi.org/10.1108/s1571-504320180000024008

Stepchenkova, S., & Zhan, F. (2013). Visual destination images of Peru: Comparative content analysis of DMO and user-generated photography. Tourism Management, 36, 590–601. https://doi.org/10.1016/j.tourman.2012.08.006

Urry, J. (1990). The 'Consumption' of Tourism.

Yeh, S.-S., Chen, C., & Liu, Y.-C. (2015). Nostalgic Emotion, Experiential Value, Destination Image, and Place Attachment of Cultural Tourists. In Advances in Hospitality and Leisure (Vol. 8). https://doi.org/10.1108/S1745-3542(2012)0000008013

Promoting Creative Tourism: Current Issues in Tourism Research – Kusumah et al. (Eds)
© 2021 Taylor & Francis Group, London, ISBN 978-0-367-55862-8

What makes visitors come again to food festivals? An analysis of a direct influence of culinary festival attributes

T. Abdullah
University of Otago, Dunedin, New Zealand

N.E. Novianti & R. Andari
Universitas Pendidikan Indonesia, Bandung, Indonesia

R.S. Nugraha
Bandung Institute of Tourism, Bandung, Indonesia

ABSTRACT: Along with the growing awareness of the benefits which can be gained from culinary tourism, stakeholders implement some measures to stimulate this type of tourism activity; one of them is by organizing culinary festivals. In order to maintain the continuity of the festival in the future, event organizers need to understand what makes visitors eagerly come again to the next festival. This study was conducted at the Bandung Food Festival, Indonesia, and aims to determine which attributes of a culinary festival that could affect visitors' revisit intention. In order to collect data, questionnaires were distributed among 395 visitors who came to the festival. Gathered data were then analyzed using multiple linear regressions. Even though, in general, culinary festival attributes influenced visitors' revisit intention, this study found that some attributes have insignificant impacts. Hence, this indicates that there were distinctive findings compared to previous studies.

Keywords: culinary festival attributes, food festival, culinary tourism, revisit intention

1 INTRODUCTION

Culinary tourism is one of the promising economic sectors in Indonesia. The total cost incurred by tourists for food and beverages is a third of the total tourist expenditure in global tourism turnover (Obonyo et al. 2012). This statement is in line with the data from the Ministry of Tourism and Creative Economic of the Republic of Indonesia, that in 2010, tourists spent 18–20% of their total traveling cost for food and beverage (Wijaya et al. 2016). Culinary tourism has a positive effect on the development of a tourist destination because of its vital marketing role, hence, many governments have given their focus on developing their culinary potential (Yang et al. 2020). According to Hall et al. (2004) culinary tourism is a form of tourism activity whereas food festivals, restaurants, and unique locations provide opportunities to taste food and/or share experiences on how food is produced as the primary motivation to travel.

Following that definition, visiting a food festival is a form of culinary tourism. The combination of cuisines (the main attraction in culinary tourism) and a pleasant festival atmosphere will develop a satisfying impression for tourists (Lau & Li 2019). This feeling of satisfaction will eventually increase their motivation to revisit a tourist destination (Hussein 2016). Bandung city has many tourism potentials, including in terms of culinary tourism. Besides of its variety and unique local foods, there are many culinary festivals held in Bandung. These festivals have been a distinctive attraction for tourists and local people. One of the famous culinary festivals in this city is Bandung Food Festival. It has been a festival which always been eagerly awaited by the residents and even out-of-town visitors who deliberately come to take part in this festival. Bandung Food Festival provides various ready-to-eat fresh foods directly cooked in the booths available. It is always held in a shopping center in Bandung, hence it is easily accessible for anyone who wants to come. Same with other culinary festivals, Bandung Food Festival has some attributes which attach to it.

Suh and Gartner (2004) stated that tourists are interested to visit a tourist attraction because of its tangible and intangible attributes. The existence of these attributes can affect tourists' satisfaction so that they would willing to come back in the future. This revisit intention is the intention of visitors to visit the same place and recommend it around their circles (Abuthahir & Krishnapillai 2018; Baker & Crompton 2000). Creating and maintaining revisit intention is crucial to maximizing tourists' spending, which would be a preferable goal rather than trying to increase the number of first-time visitors (Pujiastuti et al. 2020). Due to the high promotion cost, companies continue to make innovations to reduce this cost (Kirtiş & Karahan 2011). In the field of marketing, this effort is focused on developing tourists' experience when visiting a destination, hence tourists' revisit intention could increase (Zhang et al. 2018). Nowadays, many tourist destinations have become more concentrated on trying to increase visitors' intention to revisit the destinations (Hasan et al. 2017), followed by many emerging studies which have been conducted to understand the same phenomenon (Li et al. 2018). So far, a developed concept in which good product quality will maximize revisit intention has been an answer for this matter (Liu & Lee 2016). Previous studies identified that visitors' intention to revisit is influenced by tourists' satisfaction (Jang & Feng 2007), destination image (Chew & Jahari 2014), perceived value (Petrick et al. 2001), tourists' previous experience (Huang & Hsu 2009), perceived risk (Çetinsöz & Ege 2013; Chew & Jahari 2014), tourists' motivation (Lee et al. 2014) and attachments (Petrick 2004). According to Kerstetter and Cho (2004), revisit intention generally tends to refer to visitors' past experiences of a destination. Therefore, the experience is an excellent attribute to predict future visitor behavior (Campo-Martinez et al. 2010; Ling & Jabil 2012).

Tourists' perceptions which they get from experiences will directly affect the tendency to make a repeat visit to a tourist destination (Abubakar et al. 2017). Thus, the experience gained by tourists is very influential on revisit intention (Anderson & Sullivan 1993; Baker & Crompton 2000; Brown 1988; Crompton & Love 1995; Ling & Jabil 2012). The key in maximizing tourist satisfaction that will lead to their intention to come again is by creating a pleasant experience which will be remembered by tourists (Yusof et al. 2016).

Festival is certainly not a new thing, studies about festivals have been conducted since many years ago (Lyck et al. 2012). Previous studies examined the attributes of a festival which concluded that the quality of facilities, programs, and entertainment could influence the level of satisfaction, experience, and revisit intention (Cole & Chancellor 2009; Cole & Illum 2006). The Entertainment in an event includes music and other activities such as games which depend on the type of festival. It has a direct influence on visitor satisfaction and leads to visitor intentions to come back (Jung et al. 2015).

Tourists who come to culinary events want to buy food and beverage products and also want to spend money for an unforgettable experience from the supporting attributes provided. The culinary festival attributes are deemed necessary in maximizing revisit intention by analyzing, learning, and understanding experiences based on positive and negative feelings that tourists felt during a visit to Bandung Food Festival. Studies have analyzed the influence of culinary festival attributes towards visitors' experience and satisfaction which then contribute to revisit intention (Cole & Chancellor 2009; Jung et al. 2015). However, this study attempts to reveal the direct influence of the attributes of a culinary festival on visitors' revisit intention. Therefore, this study aims to determine which attributes of a culinary festival that could directly affect visitors' revisit intention.

2 LITERATURE REVIEW

Culinary tourism plays a vital role in the development of the tourism industry. This type of tourism activity has become a new lifestyle where food becomes a primary focus that is inseparable from everyday life (Lopez-Guzman et al. 2014). Many people travel to a tourist destination to try the culinary products from the place they visit, and local food has recently emerged as a means to attract tourists who are looking for culinary experiences and new taste sensations (Gyimóthy & Mykletun 2009; Smith & Costello 2009; Tikkanen 2007). The desire for people to discover where the cuisine is originated and to find new experiences by tasting new foods (Stanley & Stanley 2014) makes tourists eager to attend a food festival.

According to Getz (2008), a festival is a cultural celebration, and it is part of special events (Cudny 2016). A cultural celebration is a celebration activity in which the primary objectives are in relation to cultural goals (Reddy & van Dam 2020). Many cultural festivals are held outside of their countries' origin. For centuries, many communities worldwide, each with their own culture, have recognized the need to put certain times and places for celebratory purposes (Turner 1982). Nurse (2002) and Formica and Uysal (1998) argue that festivals are a significant feature of cultural tourism. In many various types of festivals, food festival is one of the most popular (Wan & Chan 2013). It is increasingly known as an essential element of the cultural tourism market, where food has become a significant focus (Hall & Mitchell 2000; Hjalager & Richards 2002).

Festival attributes are the determining factors to attract visitors to come to a festival (Saleh & Ryan 1993; Smith & Costello 2009). Researchers have examined the effect of environmental factors on consumer behavior (Basera et al. 2013; Gilboa & Rafaeli 2003). Environmental factors play an essential role in shaping customer impressions and furthermore create a destination image (Basera et al. 2013; Bitner 1992; Kotler 1973). Hence, its role is imperative in attracting target markets (Basera et al. 2013).

Smith and Costello (2009) argue that the culinary festival attributes are 'supporting products', which are additional products that add value to the core product, in this case, the core product is food. They become a differentiator of an event, thus one food festival can be distinguished with the others. In this study, we used the dimensions of culinary festival attributes from Tanford and Jung (2017). We chose these dimensions because they are suitable to be used a tool for making an analysis of a culinary festival. We only used five of the six dimensions by excluding the 'concession'. Tanford and Jung explained that concession includes food, beverage and souvenir at the event. We excluded this dimension because we believe that consuming food and beverage is considered as part of activities that visitors enjoy when visiting a festival. Hence the dimensions for culinary festival attributes in this study are activities, authenticity, environment, escape, and socialization. 'Activities' are the main focus of an event, which are things that visitors do at culinary events. It has some elements, such as programs, entertainment, food, and music. 'Authenticity' reflects the unique local content of a culinary festival. It has some elements, namely a unique product, local staff, traditional presentation, and a unique atmosphere (Akhoondnejad 2016). 'Environment' relates to where the event is held. A comfortable environment will attract tourists to come to culinary events. It consists of atmosphere, layout/design, and service (Lee et al. 2014). 'Escape' reflects the emotions and pleasures experienced by visitors. The assessment of pleasure measures enjoyment in a culinary event, happiness, escape, and excitement (Tanford et al. 2012). 'Socialization' is the social aspect between one visitor and other visitors who come to culinary events. Socialization is a warm relationship with other people, a sense of belonging, and a feeling of closeness towards other visitors (Inoue 2016).

The concept of revisit intention originates from behavioral intention (Wu et al. 2018), which furthermore roots from consumer behavior. Consumer behavior is the study of the processes involved when individuals or groups choose, buy, use, or dispose of products, services, ideas, or experiences to satisfy customer needs and desires (Solomon 2017). This study used the indicators of revisit intention by Ramukumba (2018), namely likelihood to visit in the future (the possibility of visitors to visit again in the future), likelihood to recommend it to others (the possibility of visitors to recommend and give word of mouth to others), and likelihood to be the first choice in the future (possibility to make the first choice when intending to visit again).

Table 1. The characteristics of participants (n=395).

Characteristics	Category	Frequency	Per cent
Gender	Male	142	35.9%
	Female	253	64.1%
Age	17–20 years old	101	25.6%
	21–30 years old	275	69.6%
	31–40 years old	13	3.3%
	>40 years old	6	1.5%
Occupation	Student	277	70.1%
	Entrepreneur	17	4.3%
	Civil servant	29	7.3%
	Private employee	49	12.4%
	Others	23	5.8%
Annual Income	<$204	301	76.2%
	$205–$340	43	10.9%
	$340–$476	45	11.4%
	>$476	6	1.5%
Length of Visit	<1 hour	103	26.1%
	1–3 hours	249	63.0%
	4–6 hours	41	10,4%
	>6 hours	2	0.5%
Companion	Alone	31	7.8%
	Family	74	18.7%
	Friends	258	65.3%
	Others	32	8.1%

3 RESEARCH METHODS

This study was conducted at the event of the Bandung Food Festival, which is usually held four times a year. When this study was conducted, this event was held from 19 April until 22 April 2018, on those days, the organizer carried the festival out for four consecutive days on the weekend. We distributed questionnaires between 10.00 until 14.00 for the first and the third days, while on the second and fourth days it was done at 16.00 until 18.00. Questionnaires were handed out in a mall where the event was taken place; they were distributed to visitors who came to the event. We approached visitors who were appeared to be not in a hurry and would spend some time at the festival. Later on, we gave them a brief explanation about the research, and then after they were interested and understood the aims of this study, we asked their consent verbally. Thereafter, we explained the direction to fill the questionnaires. The youngest participants were 17 years old, and the majority of visitors who filled out the questionnaires were 21–30 years old. At the end of the day fourth, 395 participants have filled the questionnaires. The data were then coded and analyzed using multiple linear regression statistical analysis to answer the hypothesis proposed in this study, which is:

H: Culinary festival attributes have a positive influence on visitors' revisit intention.

4 RESULTS AND DISCUSSION

We analyzed the data using multiple linear regression with the help of SPSS version 22.0 for windows. Table 1 shows the demographic profile of the respondents.

The gender distribution was 35.9% male and 64.1% female. The majority of the visitor was 21–30 years old (69.6%), followed by 17–20 years old (25.6%), 31–40 years old (3.3%), and more

Table 2. The results of multiple regression analysis.

	Coefficient Beta	Sig.	Value
(Constant)	4.312	0	
Activities	0.038	0.543	
Authenticity	0.095	0.101	
Environment	0.035	0.275	
Escape	0.342	0	
Socialization	0.417	0	
F			36.333
R			0.564
R^2			0.318

than 40 years old (1.5%). The highest proportion of participants were students (70.1%) followed by the private employees (12.4%), civil servants (7.3%), others (5.8%), and entrepreneurs (4.3%). For the annual income, majority of the participants (76.2%) earned less than $ 204, followed by $ 340–$ 476 (11.4%), $ 204–$ 340 (10.9%) and more than $ 476 (1.5%). In terms of the length of time spent on a visit, the highest frequency was 1–3 hours (63.0%), followed by less than 1 hour (26.1%), 4–6 hours (10.4%), and more than 6 hours (0.5%). Of 65.3% participants visited Bandung Food Festival with their friends, 18.7% with family, 7.8% alone, and the rest with others (8.1%).

Through the results of multiple linear regression analysis depicted in Table 2, in general, this study found that culinary festival attributes influence visitors' intention to come again in the future, hence the hypothesis in this study is accepted. These results confirm previous studies that festival attributes affect visitors' intention to make future visits (Cole & Chancellor 2009; Jung et al. 2015). However, if the results were reviewed based on the influence of each dimension of culinary festival attributes (partially), this study revealed slightly different results than other studies. In this study, the effect of activities, authenticity, and environment towards revisit intention was weak and insignificant (as shown in Table 2, where the significant value of these three dimensions was above 0.05). These results differ from previous studies where activity had the strongest influence on visitor's loyalty (Tanford & Jung 2017). In another study, Jung et al. (2015) found that food quality had the most substantial impact towards visitors' intention to revisit festivals. Moreover, another study also illustrated that entertainment had the strongest influence on revisit intention (Cole & Chancellor 2009). As previously explained, in this study, both food and entertainment are represented as part of activities, hence the results of this study slightly contradict those two studies (Cole & Chancellor 2009; Jung et al. 2015).

The results illustrated that revisit intention was mostly influenced by socialization (0.417). The dimension of socialization is the fulfillment of the social needs of Maslow's Hierarchy of Needs, where there were warm interactions between individuals when visiting culinary festivals, including a sense of acceptance and belonging, as well as closeness among visitors and with food vendors. Socialization also means that by visiting a festival, visitors could make a new relationship or could bring closeness to friends and family with whom they attend the festival. This dimension also includes 'sensation', which is usually included in the excitement when people socialize with others (Smith & Costello 2009). For those who visited and later on want to return for socializing, the existence of culinary events will be highly awaited. Escape had the second strongest influence on revisit intention (0.342). Most of the participants felt that visiting the festival could bring them out of their boredom and exhausting routine (e.g., work or study). In addition, the existence of food and beverage booths and the crowds at the festival kept visitors entertained and for a moment could make them forget their busy life.

5 CONCLUSION

In general, culinary festival attributes positively influence visitors' intention to revisit the festival in the future. Based on the results of this study, we found that three out of five dimensions of culinary festival attributes had an insignificant effect on revisit intention. While the other two dimensions, namely escape and socialization were significantly and strongly influence visitors' revisit intention. These results share a valuable insight which differ from previous studies. In order to increase visitors' revisit intention, organizer might also consider these two dimensions (i.e., escape and socialization) as crucial factors besides others (as suggested by other studies).

REFERENCES

Abubakar, A. M., Ilkan, M., Al-Tal, R. M., & Eluwole, K. K. (2017). eWOM, revisit intention, destination trust and gender. *Journal of Hospitality and Tourism Management*, *31*, 220–227.

Abuthahir, S. B. S., & Krishnapillai, G. (2018). How does the Ambience of Cafe Affect the Revisit Intention among its Patrons? A S on the Cafes in Ipoh, Perak. *MATEC Web of Conferences*, *150*. https://doi.org/10.1051/matecconf/201815005074

Akhoondnejad, A. (2016). Tourist loyalty to a local cultural event: The case of Turkmen handicrafts festival. *Tourism Management*, *52*, 468–477.

Anderson, E. W., & Sullivan, M. W. (1993). The antecedents and consequences of customer satisfaction for firms. *Marketing Science*, *12*(2), 125–143.

Baker, D. A., & Crompton, J. L. (2000). Quality, satisfaction and behavioral intentions. *Annals of Tourism Research*, *27*(3), 785–804. https://doi.org/10.1016/S0160-7383(99)00108-5

Basera, C. H., Mutsikiwa, M., & Dhliwayo, K. (2013). A comparative study on the impact of ambient factors on patronage: A case of three fast foods retail brands in Masvingo, Zimbabwe. *Researchers World*, *4*(1), 24.

Bitner, M. J. (1992). Servicescapes: The impact of physical surroundings on customers and employees. *Journal of Marketing*, *56*(2), 57–71.

Brown, G. (1988). South Asia tourism development. *Tourism Management*, *9*(3), 240–245.

Campo-Martínez, S., Garau-Vadell, J. B., & Martínez-Ruiz, M. P. (2010). Factors influencing repeat visits to a destination: The influence of group composition. *Tourism Management*, 31(6), 862–870.

Çetinsöz, B. C., & Ege, Z. (2013). Impacts of perceived risks on tourists' revisit intentions. *Anatolia*, *24*(2), 173–187. https://doi.org/10.1080/13032917.2012.743921

Chew, E., & Jahari, S. (2014). Destination image as a mediator between perceived risks and revisit intention: A case of post-disaster Japan. *Tourism Management*, *40*, 382–393. https://doi.org/10.1016/j.tourman.2013.07.008

Cole, S. T., & Chancellor, H. C. (2009). Examining the festival attributes that impact visitor experience, satisfaction and re-visit intention. *Journal of Vacation Marketing*, *15*(4), 323–333. https://doi.org/10.1177/1356766709335831

Cole, S. T., & Illum, S. F. (2006). Examining the mediating role of festival visitors' satisfaction in the relationship between service quality and behavioral intentions. *Journal of Vacation Marketing*, *12*(2), 160–173. https://doi.org/10.1177/1356766706062156

Crompton, J. L., & Love, L. L. (1995). The predictive validity of alternative approaches to evaluating quality of a festival. *Journal of Travel Research*, *34*(1), 11–24.

Cudny, W. (2016). *Festivalisation of urban spaces: Factors, processes and effects*. Springer.

Formica, S., & Uysal, M. (1998). Market segmentation of an international cultural-historical event in Italy. *Journal of Travel Research*, *36*(4), 16–24. https://doi.org/10.1177/004728759803600402

Getz, D. (2008). Event tourism: Definition, evolution, and research. *Tourism Management*, *29*(3), 403–428. https://doi.org/10.1016/j.tourman.2007.07.017

Gilboa, S., & Rafaeli, A. (2003). Store environment, emotions and approach behaviour: Applying environmental aesthetics to retailing. *International Review of Retail, Distribution and Consumer Research*, *13*(2), 195–211. https://doi.org/10.1080/0959396032000069568

Gyimóthy, S., & Mykletun, R. J. (2009). Scary food: Commodifying culinary heritage as meal adventures in tourism. *Journal of Vacation Marketing*, *15*(3), 259–273. https://doi.org/10.1177/1356766709104271

Hall, C. M., & Mitchell, R. (2000). Wine tourism in the Mediterranean: A tool for restructuring and development. *Thunderbird International Business Review*, *42*(4), 445–465. https://doi.org/10.1002/1520-6874(200007/08)42:4<445::aid-tie6>3.0.co;2-h

Hall, C. M., Sharples, L., Mitchell, R., Macionis, N., & Cambourne, B. (2004). *Food tourism around the world*. Routledge.

Hasan, M. K., Ismail, A. R., & Islam, M. F. (2017). Tourist risk perceptions and revisit intention: A critical review of literature. *Cogent Business and Management*, *4*(1). https://doi.org/10.1080/23311975.2017.1412874

Hjalager, A.-M., & Richards, G. (2002). 13 Still undigested: research issues in tourism and gastronomy. *Tourism and Gastronomy*, 224–234.

Huang, S., & Hsu, C. H. C. (2009). Effects of travel motivation, past experience, perceived constraint, and attitude on revisit intention. *Journal of Travel Research*, *48*(1), 29–44. https://doi.org/10.1177/0047287508328793

Hussein, A. S. (2016). How Event Awareness, Event Quality and Event Image Creates Visitor Revisit Intention: A Lesson from Car free Day Event. *Procedia Economics and Finance*, *35*(April), 396–400. https://doi.org/10.1016/s2212-5671(16)00049-6

Inoue, Y. (2016). Event-Related Attributes Affecting Donation Intention of Special Event Attendees: A Case Study. *Nonprofit Management and Leadership*, *26*(3), 349–366.

Jang, S. C., & Feng, R. (2007). Temporal destination revisit intention: The effects of novelty seeking and satisfaction. *Tourism Management*, *28*(2), 580–590. https://doi.org/10.1016/j.tourman.2006.04.024

Jung, T., Ineson, E. M., Kim, M., & Yap, M. H. T. (2015). Influence of festival attribute qualities on Slow Food tourists' experience, satisfaction level and revisit intention: The case of the Mold Food and Drink Festival. *Journal of Vacation Marketing*, *21*(3), 277–288. https://doi.org/10.1177/1356766715571389

Kerstetter, D., & Cho, M. H. (2004). Prior knowledge, credibility and information search. *Annals of Tourism Research*, *31*(4), 961–985. https://doi.org/10.1016/j.annals.2004.04.002

Kirtiş, A. K., & Karahan, F. (2011). To Be or not to Be in social media arena as the most cost-efficient marketing strategy after the global recession. *Procedia – Social and Behavioral Sciences*, *24*, 260–268. https://doi.org/10.1016/j.sbspro.2011.09.083

Kotler, P. (1973). Atmospherics as a marketing tool. *Journal of Retailing*, *49*(4), 48–64.

Lau, C., & Li, Y. (2019). Analyzing the effects of an urban food festival: A place theory approach. *Annals of Tourism Research*, 74(October 2018), 43–55. https://doi.org/10.1016/j.annals.2018.10.004

Lee, J., Kao, H.-A., & Yang, S. (2014). Service innovation and smart analytics for industry 4.0 and big data environment. *Procedia Cirp*, *16*(1), 3–8.

Lee, S., Lee, S., & Lee, G. (2014). Ecotourists' Motivation and Revisit Intention: A Case Study of Restored Ecological Parks in South Korea. *Asia Pacific Journal of Tourism Research*, *19*(11), 1327–1344. https://doi.org/10.1080/10941665.2013.852117

Li, F., Wen, J., & Ying, T. (2018). The influence of crisis on tourists' perceived destination image and revisit intention: An exploratory study of Chinese tourists to North Korea. *Journal of Destination Marketing & Management*, *9*, 104–111.

Ling, N., & Jabil, M. (2012). Factors influencing revisit by international tourists: a review. *Geografia. Malaysian Journal of Society and Space*, *8*(3), 1–11.

Liu, C. H. S., & Lee, T. (2016). Service quality and price perception of service: Influence on word-of-mouth and revisit intention. *Journal of Air Transport Management*, *52*, 42–54. https://doi.org/10.1016/j.jairtraman.2015.12.007

Lopez-Guzman, T., Hernandez-Mogollon, J. M., & Di-Clemente, E. (2014). Gastronomic tourism as an engine for local and regional development. *Regional and Sectoral Economic Studies*, *14*(1), 95–104.

Lyck, L., Long, P., & Grige, A. X. (2012). *Tourism, festivals and cultural events in times of crisis*. http://cataleg.udg.edu/record=b1376685~S10*cat

Nurse, K. (2002). Bringing culture into tourism: festival tourism and Reggae Sunsplash in Jamaica. *Social and Economic Studies*, 127–143.

Obonyo, G. O., Ayieko, M. A., & Kambona, O. O. (2012). An importance-performance analysis of food service attributes in gastro-tourism development in Western Tourist Circuit, Kenya. *Tourism and Hospitality Research*, *12*(4), 188–200. https://doi.org/10.1177/1467358413491132

Petrick, J. F. (2004). *Are loyal visitors desired visitors? 25*, 463–470. https://doi.org/10.1016/S0261-5177(03)00116-X

Petrick, J. F., Morais, D. D., & Norman, W. C. (2001). An examination of the determinants of entertainment vacationers' intentions to revisit. *Journal of Travel Research*, *40*(1), 41–48. https://doi.org/10.1177/004728750104000106

Pujiastuti, E. E., Utomo, H. J. N., & Novamayanti, R. H. (2020). Millennial tourists and revisit intention. *Management Science Letters*, *10*, 2889–2896. https://doi.org/10.5267/j.msl.2020.4.018

Ramukumba, T. (2018). Tourists revisit intentions based on purpose of visit and preference of the destination. A case study of Tsitsikamma National Park. *African Journal of Hospitality, Tourism and Leisure*, *7*(1), 1–10.

Reddy, G., & van Dam, R. M. (2020). Food, culture, and identity in multicultural societies: Insights from Singapore. *Appetite*, 149 (September 2019), 104633. https://doi.org/10.1016/j.appet.2020.104633

Saleh, F., & Ryan, C. (1993). Jazz and knitwear. *Tourism Management*, *14*(4), 289–297. https://doi.org/10.1016/0261-5177(93)90063-q

Smith, S., & Costello, C. (2009). Culinary tourism: Satisfaction with a culinary event utilizing importance-performance grid analysis. *Journal of Vacation Marketing*, *15*(2), 99–110. https://doi.org/10.1177/1356766708100818

Solomon, M. R. (2017). Consumer Behavior: Buying, Having, and Being. Harlow. *Pearson Education. in Ergonomics Science*, *8*(1), 1–35.

Stanley, J., & Stanley, L. (2014). *Food tourism: A practical marketing guide*. Cabi.

Suh, Y. K., & Gartner, W. C. (2004). Perceptions in international urban tourism: An analysis of travelers to Seoul, Korea. *Journal of Travel Research*, *43*(1), 39–45. https://doi.org/10.1177/0047287504265511

Tanford, S., & Jung, S. (2017). Festival attributes and perceptions: A meta-analysis of relationships with satisfaction and loyalty. *Tourism Management*, *61*, 209–220.

Tanford, S., Raab, C., & Kim, Y.-S. (2012). Determinants of customer loyalty and purchasing behavior for full-service and limited-service hotels. *International Journal of Hospitality Management*, *31*(2), 319–328.

Tikkanen, Irma (2007). Maslow's Hierarchy and Food Tourism in Finland: Five Cases. *British Food Journal*, *109*(9), 721–734.

Turner, V. (1982). Celebration. *Studies in Festivity and Ritual*.

Wan, Y. K. P., & Chan, S. H. J. (2013). Factors that affect the levels of tourists' satisfaction and loyalty towards food festivals: A case study of Macau. *International Journal of Tourism Research*, *15*(3), 226–240.

Wijaya, S., Morrison, A., Nguyen, T.-H., & King, B. (2016). *Exploration of Culinary Tourism in Indonesia: What Do the International Visitors Expect?* 374–379. https://doi.org/10.2991/atf-16.2016.56

Wu, H.-C., Li, M.-Y., & Li, T. (2018). A study of experiential quality, experiential value, experiential satisfaction, theme park image, and revisit intention. *Journal of Hospitality & Tourism Research*, *42*(1), 26–73.

Yang, F. X., Wong, I. K. A., Tan, X. S., & Wu, D. C. W. (2020). The role of food festivals in branding culinary destinations. *Tourism Management Perspectives*, *34*(October 2019), 100671. https://doi.org/10.1016/j.tmp.2020.100671

Yusof, N. M., Ibrahim, A. A., Muhammad, R., & Ismail, T. A. T. (2016). Determinants of UiTM Students' Revisit Intention to Kopitiam in Penang. *Procedia – Social and Behavioral Sciences*, *222*, 315–323. https://doi.org/10.1016/j.sbspro.2016.05.171

Zhang, H., Wu, Y., & Buhalis, D. (2018). A model of perceived image, memorable tourism experiences and revisit intention. *Journal of Destination Marketing and Management*, *8*(February), 326–336. https://doi.org/10.1016/j.jdmm.2017.06.004

Promoting Creative Tourism: Current Issues in Tourism Research – Kusumah et al. (Eds)
© 2021 Taylor & Francis Group, London, ISBN 978-0-367-55862-8

Women's mountaineering tourism on Instagram: The paradox between gender equality, identity, and objectification

A. Mecca
Universitas Pendidikan Indonesia, Bandung, Indonesia

ABSTRACT: This paper discusses feminist discourse in the practice of mountain tourism through the images of female mountaineers on Instagram. Ideologically, participating female mountaineers on mountain tourism are a manifestation of women's existence to deconstruct gender stereotypes by reconstruction new identity through the images. This paper examines four Instagram sites who have tens of thousands of followers and specifically showing thousands pictures of female mountaineers in high frequency. Data analysis focused on photos posting totaling 9677 posts. Subsequently, data reduction and categorization were carried out to tease up for common themes contained in all photo posts. Textual analysis was used as a method to interpret the visual texts, caption, and comments. The results revealed a paradox condition. On the one hand, pictures were representing gender equality through reconstruct of women's identity image. But on the other hand, the pictures are perceived as objectification women's body for visual pleasure and economic motive.

Keywords: mountaineering tourism, gender, feminism, identity, pictures, Instagram

1 INTRODUCTION

The preference of Mountain Tourism increased in the global context because there occurs a strong relationship between humans and nature (Río-Rama et al. 2019). In Indonesia, nature tourism (including mountain tourism) is the most effective sector to increase in-come to the country, besides oil, gas, and palm. Furthermore, mountain tourism is also being advertised through the social media Instagram. Instagram provides the complete communication feature for tourism branding through photography content (Fatanti & Suyadnya 2015). Branding through photography provides a more trusted and trustworthy image (Thelander & Cassinger 2017) and functionally, photography on Instagram uses as a performative tool that displays a variety of personal and collective/public values (Conti & Lexhagen 2020). In the personal aspect, this has implicated to reach the persuasive purpose of a tourism advertisement because Instagram has a strong personal dimension, which makes a product that is advertised feels so close to its consumers.

Four Instagram accounts @pendakicantik @perempuanpendaki_ @perempuangunung @pendakiperempuan which have tens and hundreds of thousand followers indirectly advertise mountain tour-ism in Indonesia by posting pictures of female mountaineering in each post. Several photos are complemented by the hashtag #wonderfulindonesia #pesonaindonesia #exploreindonesia or #thisisindonesia as a tagline of Indonesia's tourism branding. This fact indicated a transformation gender role from traditional femininity which is associated with the domestic space toward the public space. The involvement of women in the public space through mountaineering activities seemed to confirm ideologically the gender equality that was being fought in society. Women present in a visual representation to reveal the new identity that has rarely to show.

However, this ideological attempt will be confronted with another ideological perception: patriarchal gaze. Aside from personal values, public values that are also contained in Instagram are seen to be full of risks for women to ridicule, hatred, and harassment (Duffy & Hund 2019). Moreover, the narration of mountain climbing in the media is dominated by the discourse of masculinity (Pomfert

& Doran 2015). In the context of tourism consumption, femininity is exploited and manipulated to serve prospective tourists, and this condition implying that patriarchal discourse continues to be embedded in the ad production (Sun 2017). This condition hypothetically implies to a paradoxical condition on gender equality issues which was represented by photo-graphs of female mountaineers in the four Instagram accounts. On the basis of that hypothesis, this paper investigating how the condition represented on Instagram.

In addition to investigation female mountaineer on instagram. Axiologly, this paper is expected to provide awareness for Instagram content creators who make tourism ads to be more sensitive to gender issues. This certainly will give an impact on public perspective about the gender issues through the social media.

2 LITERATURE REVIEW

2.1 *Mountain tourism and gender representation on social media*

Mountaineering Tourism is part of mountain tourism such as hiking, biking, and skiing. Mountain tourism is known as "serious leisure" and "hard adventurous" because it requires mental strength and skill. There are several terms such as hiking, trekking, and mountain climbing to refer to the same thing, but the term mountaineering is considered more popular (Apollo 2014; Marek & Wieczorek 2016; Nepal 2003). Previous researchers assume Mountaineering Tourism as the field that is developing and becoming a potential research topic that is much explored in terms of tourism management and the consumer perspective (Cater & Miller 2019; Mackenzie & Kerr 2012; Pomfert 2016; Vespestad et al. 2019). However, slightly researchers have investigated this issue from the gender and media studies perspective. Therefore, this paper explores these issues to investigate how this female mountaineering is represented on Instagram. In general, tourism is influenced by social change and gender relations. This change can be found in the branding practice of the tourism industry, which is often through social media (Ly & Ly 2020; Rathore et al. 2017). Social media is a platform where users can easily participate, share and create messages in it (Mayfield 2008: 35). The transformation of social activities from the material reality to the virtual space has made social media increasingly recognized as an important source of information on various issues, including tourism.

This new media provide a cheaper marketing and offers many opportunities to involve prospective tourists for visiting tourist destination. Social media as a tourism marketing tool is increasingly convincing marketers that they are an integral part of marketing. At this time, a prospective tourist does not trust advertising that focuses on the advantages and special features of the destinations. Prospective tourist require a personal approach, intelligent, creative, interactive, communication and messages including empathy and emotions (Királ'ová & Pavlíčeka 2015). The aesthetic and functional aspects of social media brands also cannot be ignored. Marketers must ensure that social media technology and interfaces are used not only to focus on their technical aspects, but also to consider their appearance and aesthetic aspects as well (Christou 2015).

To reveal how significant social media is in the context of tourism, several researchers use social media as a primary source of data in their research. For example the comparison of text and images on Flickr affects the views and values of tourists in the mountainous area of Kosciuszko, Australia (Pickering et al. 2020), UCG Analysis to examine photos of tourist destinations on Instagram (Baumann et al. 2018), the characteristics of social media to maintain a sense of belonging when joining online travel communities on the Internet (Kavoura & Stavrianea 2015), the relationship between social media influencers or travel bloggers and marketers in the travel and tourism industry (Stoldt et al. 2019).

In relation to mountain tourism and gender issues there is a study of middle-aged women in Japan creating a new tourism trend of climbing mountains in one to four days of expeditions (Nakata & Momsen 2010) then Pomfert and Doran (2015) reveal media representations that contain masculine narration in mountain climbing tourism and gender experiences in mountain tourism. Then there is also a study Widiastuti (2018) which shows that there are still traditional gender stereotypes inherent in women in Indonesian tourism advertisements.

Several research related to gender construction in social media is also needed to provide a conceptual framework. Bailey et al. (2013) concludes that the stereotypical representation of women as sexual objects seeking the attention of men is commonly found on social networking sites. Likewise Chadha et al. (2020) research that highlights the online harassment of online spaces provides a solution that online space must be open to women, both for matters of principle and fairness and to maximize the use of technology costs. Women should not be pushed out of the online space that contributes to their confidence, develops their critical abilities, enhances their careers, or just gives them pleasure. While research Dwita and Wijayani (2018) concluded the existence of gender equality represented by television media through the portrayal of strong women in the decision making process. Thus, it can be argued that conventional media and new media are a space of ideological contention between feminist discourse in patriarchal construction.

Susan Bordo (1999) says that the industrial context, especially the advertising industry has positioned the sexuality of the female body as a common object in the process of consumption, the naked and near naked female body became an object of mainstream consumption (Bordo 1999: 168). This is also influenced by the psychological assumption that showing a woman's body is more possible than showing a man's body, because the circuit from eyes to brain to the genitals is a quicker trip for men than for women (Bordo 1999: 170). According to Piliang (2010), in a spectacle society, women have a dominant function as forming images and signs of various commodities such as sales promotion girl, cover girl, model girl (Piliang 2010: 331). Hence, that women's bodies are often objectified as visual commodities of the advertising industry.

3 RESEARCH METHODS

The method used is textual analysis by examining the posts of each photo from 4 Instagram accounts (see, Davis 2018). These accounts were selected with consideration the number of followers above 15,000, some even reach 453,000. That means the impact can be quite significant for the audiences. High intensity in posting photos of female mountaineers also to be significant reasoning to decide these accounts as research object, in one day posting could amount to two-four photo or video posts. These four accounts also have in common the visual style of each other, even there are also some photos that are similar to one another.

Actually, these four accounts are not specifically promoting mountain tourism, but commercial accounts that use the User Generated Content (UGC) approach in their marketing strategies. Intended to persuade producers to advertise their product in that account. There are two accounts that use the hashtag #wonderfulin-donesia #pesonaindonesia or #exploreindonesia namely @perempuangunung and @perempuanpendaki_ while two other uses other hashtags, but still related to the context of mountain tourism.

The textual approach used not for the generalized result, but to understanding the meaning and purpose contained in these posts by relying on textual analysis. This approach certainly makes the dominance of the text very strong and this method describe and interpret a situation based on the text as the main analysis tool. The text in this context is visual text and written text.

Until this research was finished, the total posts of the four Instagram totaling 10,077 posts, consisting of 9677 photo posts and 300 video posts, here are the details in table form:

Data analysis focused on photos posting totaling 9677 posts. This amount is calculated from the time all four Instagram accounts were first created until the data collection process is complete. Data was collected from March to June 2020. After the data was collected, data reduction and categorization were carried out to tease up for common themes contained in all photo posts. These themes are as follows: the reconstruction of the identity and objectification of the female body. After this common theme is discovered, an in-depth elaboration of several photos is considered to represent these themes by analyzing the visual, photo caption, and comments. From the results of the reduction and categorization the following is obtained.

Table 1. Account data and number of posts.

Account Name	Followers	Total Posts	Photo	Video
@pendakicantik	453k	3913	3803	110
@perempuangunung	137k	2566	2357	109
@perempuanpendaki_	31.1k	2494	2427	67
@pendakiperempuan	15.1k	1104	1090	14

4 RESULTS AND ANALYSIS

4.1 *Reconstruction identity and various forms of femininity*

In general, the pictures displayed on the four accounts are dominated by the image of beautiful young women wearing hiking clothes and equipment. The background of the picture shows the natural panorama of the mountains. Portrait photography approach is shown frequently rather than snapshot photos. Some pictures also taken a selfie, *wefie*, or the group portrait approach. These pictures shown with a popular aesthetic: *instagramable*. Photos taken not too near distance, the subject and background have proportions that are quite balanced, or even the proportion of subjects smaller than the background. By looking at this fact, it can be argued that the pictures present a reconstruction of the new woman's identity. Woman stereotype that is associated with domestic space is deconstructed, women's activities have moved to public spaces: mountainous nature.

The involvement of women in mountain climbing is unusual for a conventional gender roles and it is implied on some pictures. There are a number of pictures showing a female mountaineering greeting her parents asking for permission to go hiking. The caption is also related to permission from a parent. "*Hayo siapa yang kalo naik gunung ga izin emaknya*?" (who climbed the mountain is not permitted by her mother?). This caption implied that if a woman wants to go to the mountain she will be faced with difficult approval. Because parents as authority holders will not give permit their daughters to go mountain climbing, because climbing mountains are a masculine thing that is considered dangerous. It needs special tricks for female mountaineering can be permitted to go hiking.

Within the framework of this new identity reconstruction, various forms of femininity of women are shown in various representations. There are women who display Islamic religious identity by wearing the niqab, there are those who simple hijab, then there are also women who wear sexy clothes, women who make up, single women, women with children and husband, teenager, students, nurses, female police, security, flag raisers (paskibraka), pregnant women, girls, and then old women.

This picture ideologically represents of gender equality discourse in the mountaineering tourism context. The images on Instagram show that mountain climbing is not only a masculinity privilege, but also for all women, with various femininity and social background. To exist in a space that has been considered a masculine thing, women not have to be masculine. This can be shown by investigating quite a number of posts that showed women's "makeup" on the mountain. Then there are pictures showing makeup tools, or showing the faces of beautiful women with full makeup and mountains in the photo background. Women can reconstruct their identity and showed femininity that is more appropriate with their pleasure and desires as a subject.

However, the ideology contention appears after these posts were produced and responded from the point of view that Mulvey (1999) named as a male gaze. This is triggered because the logic built by these four Instagram accounts is advertising logic that sees women as the object of commodification. Various forms of femininity displayed will be perceived based on masculine logics that tend to be patriarchal. This has also obscured the image of female mountaineering itself. Is it a climber, trekkers or tourists. Because the Instagram account generalizes that anyone who appears in his post can be considered to be a female mountaineer.

4.2 *Objectification and body exploitation*

The women objectification can be identified from the selection of photos by Instagram account holders who prioritize pictures of beautiful women. Most of the photos are dominated by beautiful young women pictures. Even the Instagram account with the most followers is @pendaki_cantik (beautiful mountaineer). This increasingly shows a traditional advertising approach, which exhibit women body for visual pleasure. Where photos consumed by the audience are based on aesthetic visual desires, both in visual composition aspect or the subject / object matter.

This objectification is then emphasized through photo caption and comments. One photo shows a young woman doing gymnastics with trees in the background. On the side of young woman pictures there is a red circle as a pointer to something. The caption is *"Selamat siang gaes, coba fokus titik merah ada apa yoo, yang suka senam di gunung mana suaramu, latih otak dulu eh otot maksudnya biar ga goyang. kali*" (Good afternoon guys, try to focus on the red circle what's up? who likes gymnastics on the mountain where is your voice, train the brain first, I mean it muscle, so as no unstable). If examined carefully, at the red circle nothing is too significant, it is just a sign that leads to a tree. Photo captions only lead the audience to see the breast bulge of the subject of the picture and this was also realized by one spectator, it said *"kita digiring admin untuk fokus ke payudara?*" (we are led by admin to focus on breasts?). Then there are those who answer "mount twin (*gunung kembar*)", "mount gede/big mountain (*gunung gede*)", "round (bulat)", "bulge (*menonjol*)", and other answers that lead to the parable of the breast. But of all those who commented, there are also those who realize that the post is a visual exploitation.

The objectification and exploitation are quite common in other posts. Pictures showing breast protrusions, provoking captions, and audience responses that often equate women's breasts with mountains can be found in the comments column. This is like being something natural and considered normal. Even without pictures that do not show breasts and captions that provoke the audience still responds with a comment sexual contain. This is identified from a picture showing a veiled woman sleeping on prone position while holding a climb stick on a bed with a pink hello kitty bed sheet and beside it was an empty "carrier" bag. The caption is *"Tetap muncak walau #dirumah aja, ada yang bisa nebak mbaknya lagi di gunung mana?"* ("It's still climbing even though #stayathome, can anyone guess where is she?"). Most comments imply to parables of the breast "twin mountains (*gunung kembar*)" such as "inverted mountains (*gunung terbalik*)", "active mountains (*gunung aktif*)". Thus, it can be argued that the main focus of this kind of post and the response aimed at the public is not on the subject's matter activities, but other aspects that have sexual connotations.

Masculine and patriarchal narration can also be identified in a post showing one man flanked by two women with the photo background of Lake Ranu Kumbolo, Mount Semeru. The caption *"Adek mau dimadu apa diracun?*" (Do you want to be poisoned or honey/polygamy)". This photo caption implies the position of men who are superior to women because have the privilege of determining the number of wives. Both honey and poison are not preferring option for women, poison means dead, honey means polygamy. The use of the word "*adek*" (younger sister) also implies the position of men who are stronger than women hierarchically. And in the comments there is a firm statement to refuse of this narration by answering: better poisoned than honey. Similar narration based on religious legitimate also appear in another post who showed three veiled women, with a photo background in the Andong Mountains, Central Java. The photo caption only gives a description of the place. But there is one account that comments "Which of these are ready to polygamy?" then there is another account that answers "everyone". This further strengthens the argument that the production of posts and public response through comments not much focused on the achievements of women in climbing mountains. But it tends to respond to sexual gaze with patriarchal narration.

5 CONCLUSION

This paper has explained how a social media platform, Instagram, becomes a space of ideological contention between the narration of femininity and patriarchal gaze. Thus, it can be concluded that female mountaineering on Instagram pictures were representing gender equality through reconstruction of women's identity image. On the other hand, the pictures are perceived as objectification of women's body for visual pleasure and economic motive. Through the comments, public responses are also in the same way. There are those who appreciate the achievements of these female mountaineering with praise and respect, but there are also those who perceived with a patriarchal ideology by objectifying these pictures with sexual connotations. This condition is revealed because the four Instagram accounts have their own economic motive. That economic interests are more dominant than the necessity to respect and convey women's voices. This matter is often not realized because the advertisements on Instagram reach persuasive goals with the strong personality approach, therefore these forms of objectification look as if they are natural and are just taken for granted. In this case I agreed with Chadha et al. (2020) who stated that online space must be open for women to explore things they want without having to feel disturbed by patriarchal gaze.

REFERENCES

Apollo, M. (2014) Climbing as a kind of human impact on the high mountain environment–based on the selected peaks of Seven Summits, *Journal of Selcuk University Natural and Applied Science* (Special Issue) 2, 1061–71.

Bailey. J, Steeves. V, Burkell.J, Regan.P. (2013). Negotiating With Gender Stereotypes on Social Networking Sites: From "Bicycle Face" to Facebook, *Journal of Communication Inquiry* 37(2) 91–112.

Baumann. F Maria Sofia Lopes. M.S, Lourenço. P, (2018). *Destination image through digital photography*. Instagram as a data collector for UGC analysis. Unpublished

Bordo. S (1999), Beauty (Re)Discovers The Male Body In *"Male Body: A new Look at Men in Public and in Private"*, Farrar, Straus and Giroux, pp. 168–225.

Cater. C & Miller. M.C. (2019) Mountaineering and Trekking Tourism Management: A global perspective, *The Journal of Mountaineering Studies.*

Chadha. K, Steiner. L, Vitak.J. (2020) Women's Responses to Online Harassment, *International Journal of Communication* 14(2020), 239–257.

Christou. E. (2015). Branding Social Media in the Travel Industry, *Procedia – Social and Behavioral Sciences* 175, 607–614.

Conti. E, and Lexhagen. M (2020). Instagramming nature-based tourism experiences: a netnographic study of online photography and value creation, *Tourism Management Perspectives* 34,100650.

Davis. S.E. (2018). Objectification, Sexualization, and Misrepresentation: Social Media and the College Experience. *Social Media + Society* July-September 2018: 1–9.

Duffy. B.E and Hund.E. (2019). Gendered Visibility on Social Media: Navigating Instagram's Authenticity Bind, *International Journal of Communication* 13, 4983–5002.

Dwita. D and Wijayani. I. (2018). Gender Equality In Media Television (Semiotics Analysis of Fair and Lovely Advertisement Issue of Marriage or Master Degree), *Komuniti*, Vol. 10, No. 1.

Fatanti M.N and Suyadnya, I.W. (2015) "Beyond User Gaze: How Instagram Creates Tourism Destination Brand?". *Procedia – Social and Behavioral Sciences* 211, 1089–1095.

Kavoura. A and Stavrianea. A. (2015). Following and Belonging to an Online Travel Community in Social Media, its Shared Characteristics and Gender Differences, *Procedia – Social and Behavioral Sciences* 175, 515–521.

Királ'ová. A and Pavlíčeka. A. (2015) Development of Social Media Strategies in Tourism Destination, *Procedia – Social and Behavioral Sciences* 175, 358–366.

Ly. B and Ly.R (2020) Effect of Social Media in Tourism (Case in Cambodia), Effect of Social Media in Tourism (Case in Cambodia). *J Tourism Hospit* 9:424.

Mackenzie. H.S and Kerr. J. H. (2012) Client experiences in mountaineering tourism and implications for outdoor leaders. *Journal of Outdoor Recreation, Education, and Leadership, 4*(2), 112–115.

Marek, A., & Wieczorek, M. (2015). Tourist Traffic In The Aconcagua Massif Area. *Quaestiones Geographicae* 34(3), 65–76.

Mayfield, A. (2008). What Is Social Media?. iCrossing.

Mulvey. L. (1999) *Visual Pleasure and Narrative Cinema. Film Theory and Criticism: Introductory Readings. Eds. Leo Braudy and Marshall Cohen*. New York: Oxford UP, 1999: 833–44.

Nakata. M and Momsen. J.D. (2010) Gender and tourism: Gender, age and mountain tourism in Japan, *GEOGRAFIA Online Malaysian Journal of Society and Space* 6 issue 2 (63–71).

Nepal, S.K. (2003). Tourism and the environment: Perspectives from the Nepal Himalaya, Innsbruck–Wien–München: Himal Books Patan Dhok & STUDIENVerlag.

Pickering, C Chelsey. W, Barros. A, S Dario Rossi. S.D. (2020) Using social media images and text to examine how tourists view and value the highest mountain in Australia, *Journal of Outdoor Recreation and Tourism* 29 (2020) 100252.

Piliang, Y.A. (2010). *Dunia Yang Dilipat.* Bandung: Pustaka Matahari.

Pomfert. G. (2016) *An exploration of adventure tourism participation and consumption.* Doctoral, Sheffield Hallam University.

Pomfert. G and Doran. A. (2015) Gender and Mountaineering Tourism, Mountaineering Tourism. *Contemporary Geographies of Leisure, Tourism and Mobility*. Routledge, 138–155.

Rathore. A.K, Joshi.U.C, Ilavarasan. P.V. (2017). Social Media Usage for Tourism: A Case of Rajasthan Tourism. *Procedia Computer Science* 122, 751–758.

Río-Rama. M.C, Maldonado-Erazo. A, Durán-Sánchez and J. Álvarez-García, (2019). "Mountain tourism research". A review. *European Journal of Tourism Research* 22, 2019, pp. 130–150.

Stoldt. R , Wellman. M, Ekdale. B,Tully, M. (2019). Professionalizing and Profiting: The Rise of Intermediaries in the Social Media Influencer Industry, *Social Media + Society* January-March 2019: 1–11.

Sun. Z. (2017). Exploiting Femininity in a Patriarchal Postfeminist Way: A Visual Content Analysis of Macau's Tourism Ads, International *Journal of Communication* 11(2017), 2624–2646.

Thelander. A and Cassinger.C. (2017). "Brand New Images? Implications of Instagram Photography for Place Branding". *Media and Communication,* Vol. 5, Issue 4, pp. 6–14.

Vespestad. M.K, Lindberg. F, Mossberg. L. (2019) Value in tourist experiences: How nature-based experiential styles influence value in climbing. *Tourist Studies*. 19(4):453–474.

Widiastuti. A.N, (2018). Perempuan dalam Iklan Pariwisata Indonesia, *MediaTor*, Vol 1 1(2), Desember 2018, 214–226.

Promoting Creative Tourism: Current Issues in Tourism Research – Kusumah et al. (Eds)
© 2021 Taylor & Francis Group, London, ISBN 978-0-367-55862-8

Image of 10 prioritized tourism destinations and its influence on eWOM among tourism students

T.A. Patria, H. Ulinnuha, Y. Maulana, J. Denver & J. Tanika
Bina Nusantara University, Jakarta, Indonesia

ABSTRACT: Generation Z grow up in the internet era and many use social media as a medium of communication. Much of their perception of the world is affected by information and visual forms in the internet, through which they also share their perceptions. This paper examines the influence of image of 10 new prioritized destinations in Indonesia on electronic Word of Mouth (eWOM) among Tourism students in communicating the destinations. Data from 101 Tourism students were obtained using a Google form, which was distributed through WhatsApp groups. Finding of the study shows there is an influence of destination image on eWOM despite the low percentage of influence. This suggests there were other factors aside from destination image indicators that affected the low percentage, which provides opportunities for further studies.

Keywords: Destination image, eWOM, 10 prioritized tourism destinations, Gen Z, tourism students

1 INTRODUCTION

The Indonesian government, through the then Ministry of Tourism (now Ministry of Tourism and Creative Economy), has been launching a program named 10 Prioritized Tourism Destinations (10 PTDs) since 2015 (Kementerian Pariwisata dan Ekonomi Kreatif 2016). The 10 PTDs are Mandalika, West Nusa Tenggara; Morotai Island, North Maluku; Tanjung Kelayang, Bangka Belitung; Lake Toba, North Sumatra; Wakatobi, Southeast Sulawesi; Borobudur, Central Java; Kepulauan Seribu, DKI Jakarta; Tanjung Lesung, Banten; Bromo, East Java; and Labuan Bajo, East Nusa Tenggara. In 2019, the program was focused on Five Super Prioritized Tourism Destinations, including Mandalika, Lake Toba, Borobudur, Labuan Bajo, and the addition of Likupang, North Sulawesi. The goal of the program, also known as the 10 New Balis, is to build new tourism destinations, which will accelerate regional development and contribute to the economic development of the communities who live in and around the destinations.

The Indonesian governments, as well as its agents and various tourism stakeholders, have been making efforts at promoting the program through different kinds of media, including the internet. Use of promotional media through the internet today is inevitable. The former Minister of Tourism, Yahya, stated that almost 70% of tourists today search and share information on tourism destinations through social media (Sukuh.com 2019). Furthermore, digital generations have played a major role of tourism destinations creation and also have become the predominant target market for tourism in Indonesia.

Part of digital generations is Generation Z (Gen Z), a demographic term used to refer to people who were born between 1996 and 2009 (Sladek & Grabinger 2014). One of the characteristics of Gen Z is technology savvy (Wahab et al. 2018). According to Our Parent Survey and Wikia in Sladek and Grabinger (2014), 50% of Gen Z has their own tablet and a 100% is connected online for more than one hour whereas 46% are connected more than 10 hours per day. Additionally, Gen Z is known as a major influencer of brands as well as a generation of demanding consumers due to

 DOI 10.1201/9781003095484-106

their daily exposure to innumerable brands, "If Gen Z has a good or bad experience with a company, they are not going to keep it themselves; they are going to tell the world. They are going to share their thoughts via mobile, social media, and the internet" (Sladek & Grabinger 2014). While this is true within the US context, it generally can be considered more or less true within the global context.

This paper was based on a study that examined the influence of the image of the 10 PTDs built on social media on electronic word of mouth (eWOM) among Tourism students. Little was known about how destination image (DI) affects social media users to share contents about the destination with other fellow users. Thus, a hypothesis was formulated and examined to find if the image of the 10 PTDs has any influence on eWOM among Tourism students.

Selection of Tourism students for the study was based on a consideration that Tourism students are expected to build a sense of awareness of the occurring tourism phenomena as well as to develop critical thinking skills at assessing various aspects of tourism, including tourism-related policies, programs, and activities. Aside from that, current Tourism students are part of Gen Z so it can be assumed that they are exposed to the current tourism occurrences through social media. The majors of students are specifically focused on tourism destination-related ones, such as Tourism Destination Planning and Tourism Destination Management, which are to be differentiated from Hotel Management and Tours and Travel Management majors.

When conducting the study, the authors used theories about the relationship between DI and eWOM. Echtner and Ritchie (1991) describe DI as "impressions of a place" or "perceptions of an area". It is one of factors that influences tourists' perceptions of the destination and intentions, from sharing the image with others to visiting the destination. A positive image of a tourism destination on social media will encourage users, specifically tourists, to do a range of actions, from sharing positive statements about the destination with the others to visiting the destination. On the opposite side, a negative image of a tourism destination will discourage users from sharing positive statements about it or to visit it.

Findings of the study can be used as input to the policy making related to promotion of the 10 PTDs, particularly aimed at Gen Z in general and Tourism students specifically.

2 LITERATURE REVIEW

Studies about relationships between eWOM and DI have been conducted by many scholars. However, influences of DI on eWOM have been underresearched.

Thurau, et al. in Andriani et al. (2019) defined eWOM as any positive or negative statement of potential, actual, and previous customers about a product or a company made accessible to wider audience through the internet. Within the context of conventional goods, which is different from that of tourism and hospitality goods and services, studies about relationships between eWOM and intention to purchase products are numerous (for example, Erkan & Evans 2016; Matute et al. 2016; Yusuf et al. 2018). It was suggested that positive eWOM influences customers' decisions to purchase or repurchase products.

Within the tourism domain, eWOM has been investigated to find its relationships with a tourism DI. One of the most generated research findings is that eWOM, including its attributes or attitudes toward it, has positive and significant influence on DI, tourist attitude, or travel intention (for example, Abubakar 2016; Andriani et al. 2019; Jalilvand 2016; Jalilvand & Heidari 2017; Wang 2015a,b; Zarrad & Debabi 2015). In one of the studies, it was suggested that eWOM has a more powerful effect on DI, attitude, and travel intention than face-to-face eWOM (Jalilvand & Heidari 2017). All these studies indicate that tourists trust the eWOM they read and it is one of the factors that encourages them to build a positive image of a destination and to travel to the destination.

Other scholars who conducted studies about relationship of eWOM and tourism DI but with more diverse sub-variables or using Structural Equation Modeling (SEM) include Assaker & O'Connor 2020, who suggested that use of eWOM platforms as opposed to travel review sites can indirectly enhance DI and revisit intention; Bigne et al. (2019), who found that the interactions between positive (vs. negative) online reviews, specific (vs. general) online reviews, and familiarity with a destination enhance digital DI and intention to visit a tourist destination; and Abubakar et al. (2017), who suggested that eWOM influences intention to revisit and destination trust, and such impacts are higher in men than women. These studies show a current expansion of research themes that are related to eWOM and DI among scholars.

However, studies about influences of DI on eWOM have not been undertaken as much as those of the aforementioned themes. Hunt in Echtner and Ritchie (1991) defined DI as perceptions that visitors or potential visitors have about a destination. Whereas Echtner & Ritchie made an attempt to define and operationalize indicators of a DI. They compiled previous studies about DI, which resulted in a hierarchy of DI based on the levels of tangibility of the components. They suggested that the nature of DI was functional and psychological. Functional characteristics are directly observable or measurable, such as prices and store layout, whereas psychological cannot be directly measured, such as friendliness and atmosphere (p.40).

Few findings under this topic include: heritage destination attachment has a significant positive effect on eWOM intention (Pandey & Sahu 2020); DI has a positive relationship with eWOM, aside from tourist loyalty, intention to visit, and tourist satisfaction (Kanwei et al. 2019); and DI positively and significantly affect eWOM and revisit intention and satisfaction (Prayogo & Kusumawardhani 2016; Prayogo et al. 2016). A study that was not specifically examined DI but familiarity with a destination showed that such familiarity not only influences consumers' cognitive evaluations of the destination but also affects their feelings about it, which translates into their intention to travel to destination (Kim et al. 2019). It can be synthesized that, despite the low number of studies about the influences of destination image and eWOM, there is a positive relationship between DI and eWOM. In other words, it can be suggested that a positive DI can increase tourists' intention to spread positive eWOM about the destination.

These previous studies still provide ample research opportunities to conduct. The authors decided to investigate the influence of DI on eWOM intention of Tourism students as part of Gen Z as it has not been done.

3 RESEARCH METHODS

The study was based on a proposition that a positive image of a tourism destination will encourage positive communication about the destination among internet users. Specific to this context, a positive image of the 10 PTDs will encourage positive eWOM among Tourism students. Based on the goal of the study, the following hypotheses were proposed:

H0: There is no influence of image of the 10 prioritized tourism destinations on eWOM among Tourism students

H1: There is an influence of image of the 10 prioritized tourism destinations on eWOM among Tourism students

Population of the study was Tourism students of Tourism Department, Sekolah Tinggi Pariwisata Bandung (Bandung Institute of Tourism). The number of students in the department was approximately 350, thus, using Slovin's sampling method, a total of at least 78 respondents were needed for the study. However, the researchers decided to involve about 100 students. Questionnaires were distributed among students from five classes of three levels (specifically, they were Sophomores, Juniors, and Seniors representing three study programs, namely Tourism Destination Management, Tourism Business Management, and Tourism Destination Study). The classes were selected conveniently, when the students were on campus on the day when the survey took place. In line with the eWOM concept, respondents include those who had and had not visited any of the 10 PTDS. Questionnaires were distributed electronically and respondents accessed the e-questionnaires using their electronic gadgets, including mobile phones and laptops. A total of 104 questionnaires were gained but only 101 were considered valid.

The indicators of DI and eWOM were adopted from Byon and Zhang (2010) and Goyette et al. (2010) respectively, and were slightly adjusted to suit the actual condition of the destinations and the students. The indicators included in DI were: 1) Attractiveness of tourism/recreational activities; 2) Attractiveness and well maintenance of natural environment; 3) A wealth and a variety of local history and culture; 4) A flourishing and sociable community; 5) Sufficient facilities and infrastructures; 6) Stable economic and political conditions; and 7) Pleasant environment and atmosphere, whereas those in eWOM were: 1) Receipt of information online; 2) Share of information online; 3) Recommendation of destination online; and 4) Pride of destination. The scale used was Likert, with 1 representing Strongly Disagree, 2 Disagree, 3 Neither Disagree nor Agree, 4 Agree, and 5 Strongly Agree. Students marked options that closely represent their perceptions of the indicators. The indicators of the two variables are as follow:

The data obtained were analyzed using a Simple Linear Regression. Transformation of ordinal data into interval data were carried (Successive Interval) to meet the assumptions.

4 RESULTS & DISCUSSION

4.1 *Results*

Profiles of the respondents are presented as follows (Tables 1–3).

For the analysis purpose, the Y variable, namely Image of the 10 PTDs, was defined first before the X variable, which was eWOM. Each score for the Y and X variables was obtained by summing the score of questions for each variable that has been transformed, then divided by the maximum score (45 for Y and 35 for X) and then multiplied by 100%. So, the size of the variables involved was in the percentage of 0–100%.

The data were analyzed using regression equation eWOM = 43,50 + 0,3200.destination. This equation depicts relationship between the two variables, that are DI as independent variable and eWOM as dependent variable.

The equation is interpreted with a value of 43.50 as intercept. Interpretation of this intercept is if the percentage of DI is 0%, the percentage of eWOM is 43.50%. This gives an impression that, without the influence of DI, the percentage of eWOM is quite high (almost 50%). While the DI coefficient in the equation, which is 0.320, indicates that for each increase of the DI for as much as 1%, there will be an increase of percentage of eWOM for as much as 0.32%. This shows that the influence of DI in increasing the percentage of eWOM tends to be very low. To see whether it is significant or not, it can be seen in the following (Table 4).

Interpretation of the above results shows that the DI variable has a significant influence on the eWOM variable even though, descriptively speaking, the influence is very small. It can be seen through the p-value column, which was as much as 0, smaller than whatever alpha used in this examination. The result is used to answer the hypothesis: There is an influence of image of the 10 PTDs on eWOM among Tourism students. Summary of this test for X signification is to reject H0, which means X has significant influence on Y.

Model summary is as follows:

S	R-sq	R-sq(adj)	R-sq(pred)
9,95190	14,08%	13,21%	10,41%

Result in the above table shows the regression model in depicting relationship between DI and eWOM. The indicator used is the number shown in R-sq column. The number 14.08% shows that eWOM variable is able to explain variation of the DI variable for as much as 14.08%, whereas the remaining 85.92% of the information on DI is explained by other factors aside from eWOM. Based on this data, we can recapitulate that the DI factor is quite small in affecting eWOM compared to the other factors.

Table 1. Profile of the respondents.

Gender	Male 44.2% **Female 55.8%**
Age	<18 y.o. 1% 18–19 y.o. 26.9% **20–21y.o. 42.3%** >21 y.o. 29.8%
Place of Origin	**Java 77.9%** Bali & Nusa Tenggara 1% Sumatera 11.5% Kalimantan 1.9% Sulawesi, Maluku & Papua 7.7%
With Whom They Visited	Alone 17.3% Friends 70.2% **Parents/Family 58.7%** School/Campus 51% Travel Community 1% Non-Travel Community 1% Others 3.8%
Purpose of Visit	**Leisure 77.7%** Education 12.6% Other 5.8% Work 3.9%
Types of social media and applications used	**Instagram 93.3%** Youtube 73% Twitter 56.7% Traveloka 56.7% Facebook 33.7% Other 20.2%
Average duration of using social media per day	**3–5 hours 35.6%** 1–3 hours 25% 5–7 hours 23.1% >8 hours 11.5% <1 hour 4.8%

Table 2. Destination image.

Statements	Scales				
	1	2	3	4	5
Offer attractive tourism/recreational activities	1.9	2.9	23.3	**45.9**	22.3
Have attractive and well-maintained natural environment	1	6.8	17.5	**38.8**	35.9
Have a wealth and a variety of local history and culture	1	1.9	15.5	**38.8**	42.7
Community in the destinations is a flourishing and sociable community	2.9	5.8	29.1	**39.8**	22.3
Availability of sufficient facilities and infrastructure at the destinations	3.9	14.6	**33**	27.2	21.4
Stable economic and political conditions at the destination	1	13.6	**42.7**	23.3	15.5
Pleasant environment and atmosphere at the destinations	1	0	13.6	39.8	**45.6**

Table 3. EWOM.

Statements	Scales				
	1	2	3	4	5
I often receive information on the destinations through online media	1	6.8	6.8	33	**52.4**
I always share the information on the destinations received through online media	3.8	23.1	22.1	**28.8**	22.1
I wish to recommend the destinations through online media	1	10.7	21.4	**36.9**	30.1
I am proud of the destinations	1	1.9	8.7	**38.5**	50
I wish to promote the destination through online media	7.7	13.5	**32.7**	26	20.2
I trust the positive information on the destinations on online media	3.8	8.7	30.8	**41.3**	15.4
I trust the negative information on the destinations on online media	6.7	18.3	**42.3**	26	6.7
I want to discuss the quality of destinations on online media	4.8	7.7	29.8	**37.5**	20.2
I want to discuss my experiences in visiting the destinations on online media	3.8	6.7	21.2	**38.5**	29.8

Table 4. Significance of destination image.

Term	Coef	SE Coef	T-Value	P-Value	VIF
Constant	43,50	5,56	7,82	0,000	
Image	0,3200	0,0794	4,03	0,000	1,00

4.2 *Discussion*

The influence of DI on eWOM can be seen from the significance table of DI, particulary in the p-value column. If the p-value is greater than alpha, then H0 is not rejected, whereas if the p-value is less than alpha, then H0 is rejected.

Finding of the study suggested that there is an influence of image of the 10 PTDs on eWOM among Tourism students even though the percentage of the influence is low. This means there are other factors aside from eWOM that affected the low percentage. The other larger percentage may be caused by the major in which the students were involved, which is Tourism, and by the fact that they were part of Gen Z, where the internet and social media have become part of their daily activities. Additionally, finding of the study is in line with those of previous ones, that there is a relationship between DI with eWOM. Specifically, a positive image of a tourist destination influences positive eWOM among internet users, in this case is Tourism students.

5 CONCLUSION, LIMITATION AND FUTURE RESEARCH

The current research constructs the fundamental concept of regression, which explains the causal relationship between DI and eWOM. The result depicts the linear relationship between DI and eWOM. It shows that by increasing the DI by 1%, the eWOM will also increase by 0.32%. To summarize, there is a linear causal relationship between DI and eWOM even though it is relatively small, which could be indicated by the coefficient of determination (14,09%).

The inference of this research indicates that Tourism students potentially can be employed as effective ambassadors of a tourism destination, considering their behavior in actively engaging with social media and educational backgorund in Tourism. In this regard, the eWOM occurences relating to the DI is particularly interesting. Yet, there is 85.92% of the information on eWOM explained by other factors than destination image, which could be a focus for further research.

Limitation of the study was the tendency for the respondents to be biased about the destinations, considering their background as Tourism students. Thus, future research may involve non-Tourism students as well as explore factors in a larger percentage.

ACKNOWLEDGEMENT

This work is supported by Research and Technology Transfer Office, Bina Nusantara University as a part of Penelitian Terapan Binus entitled "Pengaruh Citra 10 Destinasi Pariwisata Prioritas terhadap eWOM di Kalangan Mahasiswa Kepariwisataan" with contract number: No. 026/VR.RTT/IV/2010 and contract date: 6 April 2020.

REFERENCES

Abubakar, A.M. 2016. Does eWOM influence destination trust and travel intention: a medical tourism perspective. *Economic Research – Ekonomska Instrazivanja*, 29(1): 598–611.

Abubakar, A.M., Ilkan, M., Al-Tal, R.M. & Eluwole, K.K. 2017. EWOM, revisit intention, destination trust and gender. *Journal of Hospitality and Tourism Management*, 31, June 2017: 220–227.

Andriani, K., Fitri, A. & Yusri, A. 2019. Analyzing influence of electronic word of mouth (eWOM) towards visit intention with destination image as mediating variable: A study on domestic visitors of Museum Angkut in Batu, Indonesia. *Eurasia: Economics & Business*, 1(19), January 2019: 50–57.

Assaker, G. & O'Connor, P. EWOM platform in moderating the relationship between political and terrorism risk, destination image, and travel intent: the case of Lebanon. *Journal of Travel Research*, 2020.

Bigne, E., Ruiz, C. & Curras-Perez, R. 2019. Destination appeal through digitalized comments. *Journal of Business Research*, 101: 447–453.

Byon, K.K. & Zhang, J.J. (2010). Development of a scale measuring destination image. *Marketing Intelligence & Planning*, 28(4): 508–532.

Echtner, C.M. & Ritchie, J.R.B. 1991. The meaning and measurement of destination image. *The Journal of Tourism Studies*, 14(1): 37–48.

Erkan, I. & Evans, C. 2016. The influence of eWOM in social media on consumers' purchase intentions: An extended approach of information adoption. *Computers in Human Behavior*, 61, August 2016: 47–55.

Goyette, I., Ricard, L., Bergeron, J. & Marticotte, F. 2010. eWOM scale: word-of-mouth measurement scale for e-services context. *Canadian Journal of Administrative Sciences/Revue Canadienne des Sciences de l'Administration,* 27(1): 5–23.

Jalilvand, M.R. 2016. Word-of-mouth vs. mass media: Their contributions to destination image formation. *An International Journal of Tourism and Hospitality Research*, 28(2): 151–162.

Jalilvand, M.R. & Heidari, A. 2017. Comparing face-to-face and electronic word-of-mouth in destination image formation: The case of Iran. *Information Technology & People*, 30(4): 710–735.

Kanwei, S., Lingqiang, Z., Asif, M., Hwang, J., Hussain, A. & Jameel, A. 2019. The influence of destination image on tourist loyalty and intention to visit: Testing a multiple mediation approach. *Sustainability 2019*, 11(22): 6401.

Kementerian Pariwisata dan Ekonomi Kreatif. 2016. Menpar Bersama gubernur dan bupati membahas 10 destinasi pariwisata prioritas. Retrieved July 5, 2020 from https://www.kemenparekraf.go.id/index.php/post/menpar-bersama-gubernur-dan-bupati-membahas-10-destinasi-pariwisata-prioritas

Kim, S., Lehto, X. & Kandampully, J. 2019. The role of familiarity in consumer destination image formation. *Tourism Review*, 74(4): 885–901.

Matute, J., Polo-Redondo, Y. & Utrillas, A. 2016. The influence of eWOM characteristics on online repurchase intention: Mediating roles of trust and perceived usefulness. *Online Information Review*, 40(7): 1090–1110.

Pandey, A. & Sahu, R. 2020. Modeling the relationship between service quality, destination attachment and eWOM intention in heritage tourism. *International Journal of Tourism Cities*, Vol. ahead-of-print No. ahead-of-print.

Prayogo, R.R. & Kusumawardhani, A. 2016. Examining relationships of destination image, service quality, ewom, and revisit intention to Sabang Island, Indonesia. *Asia-Pacific Management and Business Application*.

Prayogo, R.R., Ketaren, F.L.S. & Hati, R.M. (2016). Electronic word of mouth, destination image, and satisfaction toward visit intention: an empirical study in Malioboro Street, Yogyakarta. *Proceedings of the 1st International Conference on Social Media and Political Development* (ICOSOP 2016).

Sladek, S, Grabinger, A (2014) Gen Z. Introducing the first generation of the 21st century. URL (consulted December 2016)

Sukuh.com. 2019. Media sosial tetap jadi andalan promosi pariwisata. Retrieved July 5, 2020 from https://sukuh.com/media-sosial-tetap-jadi-andalan-promosi-wisata/

Wahab, A., Ang, M.C., Jenal, R., Muktar, M., Elias, N.F., Arshad, H., Sahari, N.A. & Shukor, S.A. 2018. Mooc implementation in addressing the needs of generation z towards discrete mathematics learning. *Journal of Theoretical and Applied Information Technology*, 96(21): 7030–7040.

Wang, P. 2015. Exploring the influence of electronic word-of-mouth on tourists' visit intention: A dual process approach. *Journal of Systems and Information Technology*, 17(4): 381–395.

Wang, Y. 2015. A study on the influence of electronic word of mouth and the image of gastronomy tourism on the intentions of tourists visiting Macau. *Tourism: An International Interdisciplinary Journal*, 63(1): 67–80.

Yusuf, A.S., Hussin, A.R.C. & Busalim, A.H. 2018. Influence of e-WOM engagement on consumer purchase intention in social commerce. *Journal of Services Marketing*, 32(4): 493–504.

Zarrad, H. & Debabi, M. 2015. Analyzing the effect of electronic word of mouth on tourists' attitude toward destination and travel intention. *International Research Journal of Social Science*, 4(4): 53–60.

Promoting Creative Tourism: Current Issues in Tourism Research – Kusumah et al. (Eds)
© 2021 Taylor & Francis Group, London, ISBN 978-0-367-55862-8

Nation brand culture tourism to improve the nation image

Wilodati, S. Komariah & N.F. Utami
Universitas Pendidikan Indonesia, Bandung, Indonesia

ABSTRACT: Cultural tourism is essential not only about visiting tourist destination with the purpose of recreation, but it can also relate to cultural potential that become the attraction of tourists to the destinations they visit. This is also the case in the Sasak Sade Tourism Village, Lombok, an area that still preserves and treats Lombok's traditions to become a popular cultural tourism site in, West Nusa Tenggara. The purpose of this research is to uncover the tips of the local community to care for and maintain the traditions of their ancestors, so that in the end it is able to optimize the potential of its territory and lift the image of Indonesia. This research uses a qualitative approach and descriptive method. Data collection techniques employed are observation, interviews, and documentation studies. The data were analyzed through data reduction, data display, and verification and the data validity was tested by using triangulation. The results of the study showed that to realize the elevation of the image of the Indonesian people, the community must continue to preserve the cultural heritage of their ancestors, which also can provide economic benefits for them.

Keywords: Cultural tourism, National image, Sasak Sade

1 INTRODUCTION

Nation brand is the most important indicator to improve the image of the nation in this digital era (Elena 2016; Miazhevich 2018). Through the nation brand, the Indonesian people have strong characteristics in international relations (Macedo 2019; Masango & Naidoo 2018) that is able to form the national image or national identity of Indonesian to make the nation become a positive, unique and different nation from other nations (Qeis 2015).

The image of the Indonesian people who have friendly, open values, respect differences, have strong and humble mutual cooperation, is a social capital that is not owned by any nation (Unayah 2017). These positive values are formed through a long process of history. This differentiation is known by other nations about Indonesia, namely a pluralistic, multicultural, beautiful, natural, yet exotic society. But in reality, these positive values are also accompanied by various negative values. Indonesian people are also known for being feudal, lazy, undisciplined, feeling inferior, uncritical and inferior (Taryoto 2010). However, these negative values reflect the social character of the Indonesian nation. If these negative characters are known by other nations, then it will weaken the image of Indonesia in the eyes of the world.

The importance of building a positive national brand not only strengthens the positive image of the Indonesian nation in international relations, but also will facilitate bilateral and multilateral cooperation with other countries. Increased bilateral and multilateral cooperation that is well established will certainly bring Indonesia more advances in various sectors (Masango & Naidoo 2018). One effort to strengthen the Indonesian national brand can be pursued through the tourism sector (Fitri 2020). Because the Indonesian people have a variety of exotic values as potentials, the most strategic tourism to continue to be developed, innovated and preserved is cultural tourism (Djukardi et al. 2020). Cultural tourism is considered a weak brand power of the Indonesian people (Rumbiak 2020).

 DOI 10.1201/9781003095484-107

Weak cultural tourism can cause a lack of investment, trade and tourism itself. Other countries that do not have potential resources such as Indonesia have seriously developed the tourism sector as one of their nation's brands. These efforts are then proven by increasing the country's soft power reputation (Mudhoffir 2017; Schofield & Eaglen 2011) This step can be emulated by Indonesia. Through a strategy of developing good cultural tourism and sustainable management, various Indonesian tourism can be created internationally. Even better, a positive reputation can be created by itself since many tourists who come to Indonesia can feel the beauty and cultural uniqueness of the Indonesian people. One of the cultural tourism owned by the Indonesian people and has been known by international tourists is the Culture of the Sade Tribe, in West Nusa Tenggara (Febriana et al. 2018).

The Sade culture in West Nusa Tenggara is one of Indonesia's exotic cultural tours. International tourists know the Sade with a variety of unique environment, buildings, communities and arts. The cultural element of the Sade is a strong nation brand because international tourists associate the brand with the identity of the Indonesian people. Based on the background above, it is important to conduct research on Sade Tribe cultural tourism and its relation to efforts to build a strong nation brand. Through this research, a new mechanism and pattern will be obtained, how a culture is perceived and interpreted to be a unique feature of national identity. As well as how cultural tourism of the Sade Tribe can achieve a positive reputation in the eyes of international tourists. So that the findings of this research can formulate various new instruments in formulating important and good policies.

This study aims to reveal various new findings regarding the potential of the Sasak Sade tribe as an indicator in strengthening the nation brand. Furthermore, these potentials can make strategic contributions to many national development agendas.

2 LITERATURE REVIEW

Sasak Sade tribe has been known as one of the famous cultural tourism sites in the world (Widisono 2019). The course of cultural tourism activities in Sade builds various social interactions between individuals, individual groups, and among groups within (Sadede et al. 2019). The process of social interaction with different backgrounds will create a process of exchanging information on values and culture (Canavan 2016; Jain & Thakkar 2019).

The process and patterns of interaction in cultural tourism are also able to build cultural experiences for tourists. Cultural experience is an important motivational element in tourism activities (Richards 2018). In addition to cultural experience, good cultural tourism also gives an authentic experience to tourists so that tourists feel a positive impression and a meaningful experience toward the tourism site. In addition, one of the important factors for tourist arrivals is cultural motivation. Cultural motivation is one of the many motives for one's attraction to visit a place that has a cultural identity. A thick cultural identity of values is a great potential to enhance the brand of a nation (Macedo 2019). Through a strong national brand, a country will easily establish good relations with other countries, both bilateral and multilateral (Macedo 2019). This can be achieved when a nation brand is built very well involving elements of history, politics, economics, and culture.

In addition, a review of the perspective of sociological education on the nation brand and culture of Sasak Sade is part of an ethnopedagogy study. Through this study, culture is seen as having local wisdom values that can be actualized in the learning process. So that implicitly, this research provides various novelty references that can be used as a scientific review of the cultural study.

3 METHOD

The approach that has been used in this research is qualitative approach with the research type as descriptive research method.

The qualitative approach was chosen because it has the ability to discover new mechanisms of a phenomenon. A qualitative approach is also used to make this study more in-depth and able to accommodate various findings related to the research objectives, without being tied to the specified variables.

The data needed in the study were collected through observation, interview, and documentation. In this research, the informants involved are four elements in community, such as Village Head, Sade Village tourism organization manager, community, and the government.

4 RESULTS AND DISCUSSION

In the results and discussion, various characteristics of the Sade Tribe will be described as one of Indonesia's leading cultural tourism destinations. The study precisely investigate how the Sasak Sade Tribe maintains and carries out its tradition as well as how the tradition is packaged into a cultural tourism activity and become one of the nation's Indonesian brands. The findings and discussion were then analyzed with various theoretical points of view.

Sade Village is located in Rembitan Village, Central Lombok District, West Nusa Tenggara Province. In this village lived a group of people called the Sade Tribe. Sade Village and Sade Tribe are two elements that cannot be separated from the establishment of the Sade traditional tradition. The Sade tribe lives for hundreds of years, and the products of life are manifested in a variety of values and cultural artifacts.

The privileges of the Sasak Sade are physical and non-physical. Physical privileges in the form of cultural artifacts as the structure of houses that are still preserved from their ancestors and the village environment that is still preserved. Meanwhile the non-physical features possessed by the Sasak Sade Tribe are in the form of community values contained in one term "awiq-awiq", a collection of norms in the life of the Sasak Sade tribe. The Sasak Sade house building offers a very thick traditional feel. The walls are still made of woven bamboo, the roof uses straw, and the floor is clay which is often mixed with dung dirt. However, it did not cause any unpleasant odor.

The entire construction of the building does not use nails at all. The concept of the Sasak Sade tribe is adapted to the community organization system. For example there is a house called Bale Bonter which is commonly used as a residence for the village head and his staff. There is also Bale Kodong which is used for newly married families or parents who are no longer working. And the third is Bale Tani, used by families who already have children or offspring. Although the concept of a house owned by the Sasak Sade Tribe has various types, in general the building of the house always has three main rooms.

The front room is for men, the living room is used for women and the kitchen, and the back room is for mothers who want to give birth to their children. The composition of the space is also adjusted to the customs of the Sasak Sade Tribe, which of course has certain aims and objectives. The unique building structure owned by the Sasak Sade is a concept of environmental management that is still relevant today. In building houses, the Sasak Sade tribe still pay attention to the concept of environmentally friendly and one with nature.

This is still done by the Sasak Sade Tribe by adapting modern buildings such as ATMs, mini markets to toilets that have a traditional appearance. In addition to various uniqueness in the form of building houses, Sasak Sade tribe also has a rich culture and traditions of the people. Sasak Sade tribe has a variety of arts, typical dances, baleq drums to gamelan. The various arts and traditions are increasingly preserved when tourism activities develop in the village of Sade. The tradition that can also be found in the Sasak Sade Tribe is the marriage tradition. A woman in the Sasak Sade tribe is not allowed to get married when she is unable to weave. In addition to being required to be able to weave, the marriage process will be carried out when a prospective bridegroom is able to abduct the bride-to-be. The uniqueness is a feature that belongs to the Sasak Sade Tribe and continues to be preserved.

Sade Tourism Activities The various cultures owned by the Sasak Sade Tribe are then packaged in a cultural tourism activity. The cultural tourism of the Sasak Sade tribe is well known in foreign countries (Widisono 2019). It is also the impact of the sustainable management of cultural tourism. The existence of cultural tourism has a very positive impact on the Sasak Sade Tribe. The existence of cultural tourism also broadens social interaction between the Sasak Sade Tribe and international tourists. From the process of interaction occurs the process of exchanging information, values to culture (Canavan 2016; Jain & Thakkar 2019; Sadede et al. 2019).

The Sasak Sade tribe has a friendly attitude towards tourists. This friendly attitude is generalized by tourists as a positive attitude of the Indonesian people. In the process of the tour, the Sade community which is part of the Sade Sasak Tribe becomes a local guide to guide tourists around exploring every corner of the village. The presence of this local guide certainly provides a unique experience for tourists. Tourists can experience the cultural climate of the Sasak Sade tribe directly.

This kind of cultural experience and authentic experience are essential in a process of cultural tourism (Richards 2018). Cultural tourism also involves motivation as a pull factor for tourists to come to an area. Motivation is not only single, but can also be a combination of motivation (Mahika 2011). The arrival of tourists to Sade is not only the desire to visit and have fun, but can also be driven by the desire to conduct research and even business. The motivation and pull factor is owned by the Sasak Sade Tribe as one of the cultural motivations of tourists. The process of interaction and the desire of tourists which are driven by various motives, will give birth to mutual understanding and mutual understanding between cultures. Because, indirectly, tourists are encouraged to get to know, know and understand a culture. Various positive values were also obtained by the Sasak Sade Tribe, who obtained various new values from tourists. Based on the above processes, cultural tourism of the Sasak Sade Tribe actually becomes very promising for the formation of a strong nation brand. Among the advances of increasingly competitive times, nation brands such as those of the Sasak Sade Tribe have clearly become a great potential for national identity (Macedo 2019).

To sum up, the concept of a nation brand is to be a country that is unique and different from the others (Macedo 2019). Through the various uniqueness of the Sasak Sade tribe, Indonesia can promote a positive national image. The concept of a nation brand is very appropriate because of its elements. The concept of a nation brand is not only a matter of promotion of special things, but also involves the entire image including history, politic, economic, and culture.

5 CONCLUSION

Sasak Sade tribe is a cultural tourism which is well-known by foreign countries because of its unique privileges. The cultural tourism of the Sasak Sade tribe is one of Indonesia's great potentials to build a strong nation brand that is able to improve the image of the Indonesian nation and facilitate various relations between Indonesia and other countries.

REFERENCES

Canavan, B. (2016). Tourism culture: Nexus, characteristics, context and sustainability. *Tourism Management*, 53, 229–243. https://doi.org/10.1016/j.tourman.2015.10.002.

Djukardi, D. M., Ketut Rachmi, I. G. A., & Sumiarni, E. (2020). Indonesian Government Policy and The Importance of Protection Cultural Heritage for National Identity. *Advances in Social Science, Education and Humanities Research*, 389(Icstcsd 2019), 213–219. https://doi.org/10.2991/icstcsd-19.2020.43.

Elena, R. (2016). Branding the Nation in the Digital Era: Why don't you come over? *Culturesofcommunications*, 1, 1–8.

Febriana, L. L. R., & Pranoto, Kurniawati, Y. S. (2018). The Study of Sasak Baby Care in Sade Village Cental Lombok. *BELIA: Early Childhood Education Papers*, 7(2), 76–81.

Fitri, A. (2020). The Role of Brand Personality on Consumer Behavior and Branding Challenges in Asia. *UCT Journal of Management and Accounting Studies* 2020, 8(01), 15–20.

Jain, R., & Thakkar, J. (2019). Experiencing Craft and Culture: An Emerging Cultural Sustainable Tourism Model in India. *Cultural Sustainable Tourism*, 29–35. https://doi.org/10.1007/978-3-030-10804-53

Macedo, G. B. (2019). Nation Branding as an instrument for public diplomacy. *Politica Internacional y Diplomacia*, 1–35.
Mahika, E. (2011). Current trends in tourist motivation. *Cactus Tourism Journal*, 2(2), 15–24. https://doi.org/10.5772/1421.
Masango, C., & Naidoo, V. (2018). An Analysis of Nation Brand Attractiveness: Evidence from Brand Zimbabwe Cleven. *Journal of Economics and Behavioral Studies*, 7(6), 1–25.
Miazhevich, G. (2018). Nation branding in the post-broadcast era: the case of RT. *European Journal of Cultural Studies*, 7, 1–25.
Mudhoffir, A. M. (2017). Teori Kekuasaan Michel Foucault: Tantangan bagi Sosiologi Politik. MASYARAKAT: *Jurnal Sosiologi*, 18(1). https://doi.org/10.7454/mjs.v18i1.3734.
Qeis, M. I. (2015). Citraan Indonesia dalam Iklan Televisi "Visit Indonesia Year 2008." *Journal.Lppmunindra*.Ac.Id, 1, 63–77.
Richards, G. (2018). Cultural tourism: A review of recent research and trends. *Journal of Chemical Information and Modeling*, 53(9), 1689–1699. https://doi.org/10.1017/CBO9781107415324.004.
Rumbiak, N. R. E. (2020). Peran Asean Tourism Forum dalam Mengembangkan Potensi Pariwisata Indonesia di Era MEA (Masyarakat Ekonomi Asean). *Sociae Polites*, 125–142.
Sadede, M., Boham, A., & Runtuwene, A. (2019). Peranan Komunikasi Dinas Pariwisata Kota Bitung Guna Meningkatkan Pengetahuan Berbahasa Inggris Masyarakat dalam Melayani Wisatawan di Pulau Lembeh. *Acta Diurna Komunikasi*, 4.
Schofield, & Eaglen, M. (2011). Nation in transformation: tourism and national identity in The Kyrgyz Republic. *Sociological Review.*
Taryoto, A. H. (2010). Telaahan Ulang Ciri-Ciri Manusia Indonesia. *Jurnal Penyuluhan Perikanan Dan Kelautan*, 4(2), 68–84. https://doi.org/10.33378/jppik.v4i2.19.
Unayah, N. (2017). Mutual Help Activities as Social capital in Handling of Poverty (Gotong Royong Sebagai Moal Sosial Dalam Penanganan Kemiskinan). *Sosio Informa*, 3(1), 49–58.
Widisono, A. (2019). The Local Wisdom on Sasak Tribe Sade Hamlet Central Lombok Regency. *Local Wisdom: Jurnal Ilmiah Kajian Kearifan Lokal*, 11(1), 42–52. https://doi.org/10.26905/lw.v11i1.2711.

Promoting Creative Tourism: Current Issues in Tourism Research – Kusumah et al. (Eds)
© 2021 Taylor & Francis Group, London, ISBN 978-0-367-55862-8

The role of social media in Generation Z travel decision-making process

Khrisnamurti, R. Fedrina & U. Suhud
Universitas Negeri Jakarta, Jakarta, Indonesia

D.J. Prihadi
Ocean University of China, Qingdao, China

ABSTRACT: As an integral part of their life, social media inspired Generation Z to travel and recommend the destination that suits them. Therefore, social media became an essential role in tourism for marketing the tourism destination and several businesses attached to it. This paper analyzes how Generation Z, particularly the university students majoring in tour and travel operations studying at Universitas Negeri Jakarta, is influenced by social media to travel through a decision-making process. A quantitative methodology used in this study was by using the e-questionnaire to obtain the data. The findings showed that social media is often used by Generation Z, which was represented by students and alumni of Tour and Travel Operations Study Program, Universitas Negeri Jakarta, in every level of the travel decision-making process: pre-phase trip (activities before traveling), during-phase trip (on-site/undergoing activities), and post-phase trip (activities after traveling) as the source of information for traveling.

1 INTRODUCTION

Brought up by technology, Generation Z, or as is widely known as Gen Z, is considered a digital native and has the privilege by the abundance of the information all around them. According to Bloomberg analyses of United Nations data, this type of generation born after 1995 contributes one of the significant portions of the global population chart for about 32%, which means that Gen Z is a potential market. Looking at the fact that this generation is tech-savvy and an addressable market, marketers find one of the effective ways to come close to Gen Z channeling through their infamous communication platform called social media such as Facebook, Twitter, and Instagram, and others (CMO Adobe 2019). Social media and travel-decision process have been extensively studied by scholars (Björk & Jansson 2008; Cox et al. 2009; Dwityas & Briandana 2017; Fotis et al. 2011; Huertas & Marine-Roig 2018; Kim & Fesenmaier 2017). However, there is slightly unexplored research that integrated travel-making decisions with Gen Z, particularly in the Indonesia context, which is state of the art in this research. This study tries to analyze how social media influences the travel decision-making process of Gen Z that will be represented by a specific community, the students, and alumni of the Tour and Travel Operations Study Program of Universitas Negeri Jakarta.

2 LITERATURE REVIEW

2.1 *Gen Z and social media*

Tulgan (2013) defines Gen Z, or postmillennial, as a descendant of Gen X that is entrenched in digital technology cultures, such as the Internet, smartphones, networks, and digital media.

DOI 10.1201/9781003095484-108

Although it could agree that every generation has their own characteristics, there are some common characteristics that share younger generations. Degraffenreid (2008) acknowledges that Millenials and Gen Z are the creators and early adopters of new trends. They are used to new technologies, optimistic, non-linear thinkers, innovative in problem-solving. They make the most of social media to be connected as a new norm to bond between friends and family. Kaplan and Haenlein (2010) define social media as "a group of the Internet-based applications that build on the ideological and technological foundations of Web 2.0, and that allow the creation and exchange of User Generated Content (UGC)." In terms of traveling, Gen Z will follow their predecessor, Millenials, as Gardiner et al. (2014) point out that due to their habits on traveling at a young age, in comparison to the older generation, they have a hankering on traveling, mainly abroad. In the future, they will become the primary consumers in the travel industry. Uysal, Perdue, and Sirgy (2012) note that social media is becoming a reliable source of information for travelers to plan their trip. The primary reason is that social media promptly updates and consists of various integrated links in their search engine results (Huertas & Marine-Roig 2018).

2.2 *Travel decision-making process*

According to Martin and Woodside (2012), travel decision-making is not just one single decision. However, it consists of many sub-decisions and a dynamic process involving a series of seemingly unstructured and unique decisions. The use of social media is a reliable source of information in travel planning stages (de Souza & Machado 2017); with the use of social media, tourists could visualize what kind of experience they will have at the tourist destination to reduce uncertainty and increase expectation (Zivkovic et al. 2014). The research conducted by Dwityas and Briandana (2017) found out how social media's role is essential as a source of information in the travel decision-making process. They create a model of decision making of tourism through uses of social media by creating three phases: pre-trip phase, during trip phase, and post-trip phase. The phases describe how the information from social media is gathered, implied, evaluated, and shared by travelers. As Xiang et al. (2015) point out, that for young people, social media are considered an essential source of information during the planning of a trip and also the same case as during trip phase. Social media in the post-trip phase have a different usage compared to the pre-trip phase, and during-trip phase, most travelers in this phase tend to share their experiences and comments with others rather than search for information (Kim & Fesenmaier 2017) (Figure 1).

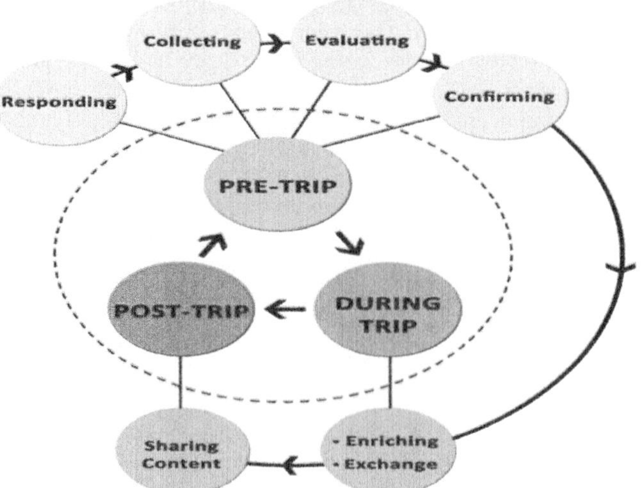

Figure 1. Social media in travel decision-making process by Dwityas and Briandana (2017).

3 METHOD

A quantitative study was conducted in June 2020 using an online questionnaire of consumers known to use social media to gather information when they travel, specifically Gen Z, born after 1995, particularly university students and alumni of Tour and Travel Operations Study Program at the Universitas Negeri Jakarta. The data collected from 153 samples of participants contained three critical sets of multiple-response questions borrowed from Dwityas and Briandana (2017). The samples have the following characteristic: As per age, almost half of the respondent were 18–22 years old (47.7%), while 23–25 years old were 37.3% and above 25 years old were 15%. The vast majority of the respondents were female, 70.6%, and 29.4% male. The online questionnaire was designed with close-ended questions and consisted of three sections. Most of the respondents travel more than four times in a year (39.9%), 22.2% three times, 26.1 twice, and 11.8% once. The recapitulations of the top three reasons for traveling were leisure 86.3%, 52.3% boredom, and culinary 36.6%. The first section described the sample profile, which consists of age, gender, travel frequency, and travel motivation. The second section captured the usage of social media by Gen Z. The last section was divided into three phases: the pre-trip and during-trip used to describe the purposes of social media usage as sources of information. At the same time, post-trip evaluated their experience on social media. This study uses descriptive analyses to describe the feature characteristics and preferences of Gen Z usage of social media in each phase.

4 FINDINGS AND DISCUSSION

4.1 *The usage of social media*

For the intensity of social media usage, most of the respondents use social media more than 4 hours a day (57.5%), 2–4 hours (32%), 1–2 hours (8.5%), and less than an hour (2%). Based on the recapitulations of social media, they frequently used Instagram (91.6%), Youtube (86.9%), and Twitter (86.6%). In terms of traveling, the most social media accounts they follow on social media were the influencers who like to travel (71.2%), Official States Tourism Institution/DMO (53.6%), and Travel TV program (50.3%). Gen Z is investing their duration of time predominantly with social media; however, the likes, comments, and posts could cause addictions, desires, anxieties, and joy (Naidu 2019). Their compulsive behavior on mobile device traveling makes them incredibly dependent on accessing social media (Research Skift 2019) and could lead to the Fear of Missing Out (FOMO) behavior phenomenon. It supports the statement from Cox et al. (2009) that prospective travelers such as Gen Z have established a strong dependency as the sources of online information for travel planning.

4.2 *Travel decision-making process*

Pre-trip phase

The pre-trip phase's findings describe how Gen Z utilizes social media to seek inspiration and information in terms of traveling: 62% of Gen Z strongly agree when they want to travel; they use social media to get inspiration. Most of them agree to be motivated (pull factor) to travel (52%) when they saw content posted about traveling, and 83% of them agree that the reason they travel was from social media. Interestingly, about 48% agree that comments on social media affect their choice to travel; this is also in relation with Hyung-Park et al. (2007) that note that online consumer reviews are often considered more trustworthy and credible than information that is provided by suppliers of products and services, assumedly because consumers considered to provide more factual information.

Table 1 shows that Gen Z predominantly uses Influencer (who likes to travel) social media account as their prime source of information (64.1%), while the tourism state institution, both website (39.2%) and social media account (34.6%), were the secondaries. It indicates that Influencer has

Table 1. Sources of information used in the pre-trip phases.

Source of Information	Pre-Trip
Guidebooks	4,6%
Tripadvisor	28,8%
State Tourism Institution/DMO (website)	39,2%
State Tourism Institution/DMO (social media account)	34,6%
Travel Blogger (website)	15%
Travel Blogger (social media account)	44,4%
Tour Operator of Travel Agent (including OTA)	30,1%
Leaflet or brochure	8,5%
Influencer/celebrity who like travel (social media account)	64,1%
News Portal website	10,5%
TV Program	20,3%

*Note: Recapitulation multiple response questions.

Table 2. The most sought information in the pre-trip using social media.

Searching for Information	Pre-Trip
Tourist attractions	941%
Tour Package	131%
Accommodations	484%
Food and Beverages	588%
Cultural information	314%
Souvenir/shopping Center	98%
Weather	92%
Accessibility (including transportation mode)	595%

*Note: Recapitulation multiple response questions.

a substantial impact on Gen Z, particularly in terms of traveling information (Fotis 2011; Kotler et al. 2010; Litvin et al. 2008). Jacobsen and Munar (2012) found out that the pre-trip phase of a traveler is the most crucial aspect of traveling. It provides an abundance of information that would inspire them to decide their travel pattern. Within such information search processes, consumers rely on other travelers' experiences to increase the exchange utility and decrease uncertainty.

The finding also reveals that the Gen Z top three main interest they seek for information in the pre-trip planning were tourist attractions (94.1%), Food and Beverages (58.8%) and Accommodations (48.4%), which supports the research by Jacobsen and Munar (2012) that found out that the usage of social media is mainly to seek information related to activities, accommodation, and dining (Table 2).

During-trip phase

Huertas and Marine-Roig (2018) stated that searching for information in traveling does not stop at the pre-trip phase. The result showed that 57% always check their social media account relating to the information update of the destination. Social media also creates a high expectation for Gen Z to experience what they imagine before (58%). This generation likes to share everything, including traveling. That is why 81% of them share their experience on social media and felt pleased if somebody responded (52%). This study also has been done by Bangkok Post (2017), that found out over half (56%) of Gen Z frequently shared their pictures on social media while traveling. Patel (2017) also found out that Gen Z tends to post a constant stream of content in order to let other people know how exciting was their trip. Therefore, it was not surprising that most of Gen Z are

Table 3. Sources of information used in during-trip phases.

Source of Information	During-trip
Guidebooks	7,2%
Tripadvisor	34,6%
State Tourism Institution/DMO (website)	35,3%
State Tourism Institution/DMO (social media account)	36,6%
Travel Blogger (website)	20,9%
Travel Blogger (social media account)	45,8%
Tour Operator of Travel Agent (Including OTA)	25,5%
Leaflet or brochure	14,4%
Influencer/celebrity who like travel (social media account)	51%
News Portal website	14,4%
TV Program	14,4%

*Note: Recapitulation multiple response questions

Table 4. The most sought information in during-trip phase.

Searching for Information	During-trip
Tourist Attractions	817%
Tour Package	118%
Accommodations	418%
Food and Beverages	647%
Cultural Information	275%
Souvenir/shopping Center	19%
Weather	85%
Accessibility (including transportation mode)	451%

*Note: Recapitulation multiple response questions.

worried if they do not promptly update their trip experience (57%), and that is why they do not deactivate their social media during their trip (54%).

According to Table 3, the behavior of Gen Z relating to seeking information was slightly different; in comparison to the pre-trip phase, they mainly trust the information from travel bloggers (45.8%) social media account instead of from the Influencer, while social media accounts (36.6%) and websites (35.3) from state tourism institutions still follow similar to the previous phase. It shows that Gen Z tends to trust a reliable, passionate, experienced person and finds a living based on traveling. The travel blogger is depicted as more authentic than the Influencer, and it gives a sincere vibe for the audience that connects to them (Beckford 2019). Nevertheless, this behavior contradicts Conrady and Buck (2009), who stated that social media is a game-changer for young travelers to obtain information. However, they still heavily rely on guidebooks and a travel consultant's expertise that works at a travel agent.

The results for the most sought information in during-trip phase are, to some extent, similar to the pre-trip phase. Gen Z was predominantly searching for information about tourist attractions (81.7%), followed by food and beverages (64.5%) and accessibilities (41.8%) (Table 4).

Post-trip phase

The last phase of the travel decision-making process merely focuses on evaluating the trip produced/shared in social media, unlike the two previous phases that travelers seek for information (Huertas & Marine-Roig 2018). As we acknowledged, Gen Z's characteristics are willing to share and inspire others. It is the ultimate verdict of their entire trip, whether they experienced a great holiday or not (Tulgan 2013). The survey results show that 52% write their experience on social media after the trip, while 61% share photos and videos. The result supports Mcdonald's et al. (2019) statement that one of the primary activities for Gen Z after a holiday is sharing their trip experiences through posting pictures and reviewing their holiday through social media. This finding supported Gen Z's behavior that resulted in how they decide to travel, which is shaped and inspired by other people's travel experiences from social media that they could upload and share text, image, audio-video that adopted the user-generated content (UGC) platform (Timothy 2019).

5 CONCLUSION AND IMPLICATION

This study's primary purpose is to explore the role of social media that influences Gen Z on the travel decision-making process as a source of information. The result showed that Gen Z relied on and utilized social media in the three phases (pre-trip, during-trip, post-trip). They simultaneously use social media to seek travel information and inspirations mostly from travel bloggers or Influencer due to their honest reviews and their horizontal relationship amongst them. After the trip, they often share their trip experiences publicly to others, inspiring people to follow the same activities.

This study's limitation is the small sample of the research that does not generalize findings to large survey samples, mainly Gen Z in Indonesia. Further research is necessary to overcome the limitations and take the research onto a more significant sample across the nation with more in-depth analysis. In terms of implications, Gen Z has disrupted tourism marketing, and it is essential to understand their characteristics with their powerful tool that is the social media, and the strategies need to be reshaped and refocused in order to accommodate and to build a relationship with the potential future travelers.

REFERENCES

Bangkok Post. (2017, february 21). *www.bangkokpost.com.* Retrieved june 25, 2020, from https://www.bangkokpost.com/travel/1751414/social-media-makes-travel-plans-for-gen-z-booking-com-survey

Beckford, G. (2019). *www.packslight.com.* Retrieved 06 25, 2020, from https://www.packslight.com/gen-z-marketing-tips-for-the-social-media-generation-from-an-actual-gen-z/

Björk, P., & Jansson, T. (2008). Travel Decision-making: The Role of Habit. *Tourismos: An International Multidisciplinary Journal of Tourism, 3* (2), 11–34.

CMO Adobe. (2019). Retrieved may 24, 2020, from www.cmo.adobe.com: https://cmo.adobe.com/articles/2019/5/generation-z-travel.html#gs.8rhdsd

Conrady, R., & Buck, M. (2009). *Trends and Issues Global Tourism 2009.* Springer Publishing.

Cox, C., Burgess, S., Sellitto, C. J., & Buultjen, J. (2009). The role of user-generated content in tourist travel planning behaviour. *ournal of Hospitality Marketing & Management, 18* (18), 743–764.

Degraffenreid, S. (2008). *Understanding the Millennial Mind: A Menace or Amazing?* Just Brilliant Services.

de Souza, S. C., & Machado, D. F. (2017). Use and Influence of Social Media on Trip Planning: a quantitative study. *Revista Turismo Em Análise, 28* (2), 254–270.

Dwityas, N. A., & Briandana, R. (2017). Social media in travel decision making process. *International Journal of Humanities & Social Sciences, 7* (7), 193–201.

Fotis, J., Buhalis, D., & Rossides, N. (2011). Social Media Impact on Holiday Travel Planning: The Case of the Russian and FSU Markets. *International Journal of Online Marketing, 1* (4), 1–19.

Gardiner, S., Grace, D., & King, C. (2014). the Generation Effect: The Future of Domestic Tourism in Australia. *Journal of Travel Research, 53* (6), 705–720.

Huertas, A., & Marine-Roig, E. (2018). Searching and Sharing of Information in Social Networks During The Different Stages of a Trip. *Cuadernos de Turismo* (42), 185–212.

Hyung-Park, D., Lee, J., & Han, I. (2007). The Effect of On-line Consumer Reviews on consumer Purchase Intention: The Moderating Role of Involvement. *International Journal of Electronic Commerce, 11* (4), 125–148.

Jacobsen, J. K., & Munar, A. M. (2012). Tourist information search and destination choice in a digital age. *Tourism Management Perspectives, 1*, 29–47.

Jepsen, A. (2006). Information Search in Virtual Communities: Is it Replacing Use of Off-Line Communication?. *Journal of Marketing Communications, 12* (4), 247–261.

Kaplan, A. M., & Haenlein, M. (2010). Users of the World, unite! The Challenges and Opportunities of Social Media. *Business Horizons*, *53* (1), 59–68.

Kim, J. J., & Fesenmaier, D. R. (2017). Sharing Tourism Experiences: The Post-trip Experience. *Journal of Travel Research*, *26* (1), 28–20.

Kotler, P., Bowen, J., & Makens, J. (2010). *MArketing for Hospitality and Tourism.* Pearson.

Litvin, S. W., Goldsmith, R. E., & Bing, P. (2008). Electronic Word-of-mouth in Hospitality and Tourism Management. *Journal of Tourism Management*, *29*, 458–468.

Martin, D., & Woodside, A. G. (2012). Structure and Process Modeling of Seemingly Unstructured Leisure-travel Decisions and Behavior. *international Journal of Contemporary Hospitaity Management*, *24* (6), 855–872.

McDonald, J. S., Bennet Jr., J. R., Merwin, K. A., & Merwin Jr., G. A. (2019). *Cultural Tourism in the Wake of Web Innovation.* Hershey PA, USA: IGI Global.

Naidu, S. (2019). A Study on the influence of Social Mdia on Tourist Psychology. *International Journal of Management, tehcnology and Engineering*, *IX* (1), 537–545.

Patel, D. (2017). Retrieved 06 12, 2020, from www.forbes.com: https://www.forbes.com/sites/deeppatel/2017/08/11/how-companies-are-adjusting-to-gen-zs-shifting-travel-demands/#6668aa916f5c

Research Skift. (2019). Retrieved June 22, 2020, from www.research.skift.com: https://research.skift.com/wp-content/uploads/2019/08/MillennialGenZ_Final.8.12.pdf

Semerádová, T., & Weinlich, P. (2020). *Impacts of Online Advertising on Business Performance.* Hershey PA, USA: IGI Global.

Starcevic, S., & Konjikusic, S. (2018). Why Millenials As Digital Travelers Transformed Marketing Strategy In Tourism Industry. *Tourism in Era of Digital Transformation* (pp. 221–240). Serbia: Univerzitet U Kragujevcu.

Timothy, D. J. (2019). *Handbook of Globalisation and Tourism.* Cheltenhem: Edward edgar Publishing.

Tulgan, B. (2013). *Meet Generation Z: the second generation within the giant 'Millennial' cohort".* Retrieved 06 29, 2020, from Rainmaker Thinking: http://rainmakerthinking.com/assets/uploads/2013/10/Gen-Z-Whitepaper.pdf

Uysal, M., Perdue, R., & Sirgy, J. (2012). *Hanbook of Tourism and Quality-of-life Researc: Enhancing the Lives of Tourists and Residents of The Host Communities.* New York: Springer Science & Business Media.

Xiang, Z., Schwartz, Z., Gerdes, J. H., & Uysal, M. (2015). What Can Big Data and Text Analytics Tell Us About Hotel Guest Experience and Satisfaction? *International Journal of Hospitality Management*, *44*, 120–130.

Zivkovic, R., Gajic, J., & Brdar, I. (2014). The Impact of Social Media on Tourism. *Singidunum Journal of Applied Sciences Supplement*, 758–761.

Promoting Creative Tourism: Current Issues in Tourism Research – Kusumah et al. (Eds)
 ISBN 978-0-367-55862-8

The influence of halal tourism destination attributes on tourist satisfaction in Bandung

N. Wildan & M.N. Della
Universitas Pendidikan Indonesia Kampus Sumedang, Sumedang, Indonesia

O. Sukirman
Universitas Pendidikan Indonesia, Bandung, Indonesia

ABSTRACT: Supported by various facilities, infrastructure, diversity of tourist attractions, core needs, and services such as halal food and Sharia hotels that are the main source in meeting the needs of Muslim tourists make Bandung as one of the cities developing halal tourism destinations in West Java. This study aims to determine the influence of halal tourism destination attributes aspects (i.e., access, communications, environment, and services) on tourist satisfaction. The type of research used was descriptive and verification with quantitative approach. The sample in this study was 203 respondents of foreign Muslim tourists who visited Bandung. The analysis technique used was path analysis. The result of the research shows that halal tourism destination attributes on tourist satisfaction simultaneously and partially influential. In this study, the government, related institutions, as well as business actors in the field of halal tourism in particular must work together in an effort to develop the City of Bandung as a halal tourism destination.

Keywords: halal tourism, destination attributes, tourist satisfaction

1 INTRODUCTION

Indonesia has invested significant resources to improve and promote its tourism industry with a target of 20 million international arrivals by 2019 (Mastercard & CrescentRating 2018). Developing the potential of halal tourism is one of the strategies in achieving this target. The strategic issue was triggered based on global trends in halal tourism. Some non-Muslim countries like Japan, Philippines (Battour 2018), Taiwan, Vietnam, Korea, China (Battour,& Ismail 2016), Thailand, Australia, countries in Europe (Chanin et al. 2015), even prepare the needs of Muslims to support comfort and safety in enjoying visiting time in their countries. Several factors for the development of halal tourism are considered to have bright prospects in improving the economy of the tourism industry sector (KEMENPAR 2015). In 2018, there were 140 million global Muslim visitors. This number is expected to further grow to 230 million Muslim visitors by 2026 and the overall global travel expenditure by the Muslim travel market to hit US$300 billion (Mastercard & CrescentRating 2019).

The seriousness of the government in developing halal tourism destinations is shown by improving Indonesia's position as a Muslim-friendly destination. In the last four years, Indonesia's ranking in the Mastercard–CrescentRating Global Muslim Travel Index has improved significantly: sixth rank in 2015, fourth rank in 2016, third rank in 2017, and second rank in 2018. With all the improvements in the past years, Indonesia has a great opportunity and deserves be the number one position in the next Global Muslim Travel Index Mastercard and CrescentRating (2019).

The concept of halal tourism is growing and starting to generally be accepted by emphasizing service and facilities on sightseeing provided by the tourism industry in meeting the needs of Muslim travelers (Battour 2018; Battour et al. 2011; Napu & Nurhidayat 2019; Sofyan 2017). Hospitality and tourism providers have to consider the importance of going halal to gain a competitive advantage

 DOI 10.1201/9781003095484-109

in the industry (Mohsin et al. 2016). The concept become the model in developing halal tourist destinations in Indonesia, one of which is the city of Bandung, West Java. With diverse tourist attractions and supported by various facilities, core needs and services such as halal food and Sharia hotelsare developing to accommodate the needs of Muslim tourists and increase the level of tourist satisfaction. This is indicated by the improvement data on the percentage of tourist visits. Although there was a decrease in foreign tourist arrivals, in the past three years, the city of Bandung showed an improvement in the percentage of tourist visits: −9.09% in 2017, −6.88% in 2018, and 1.46% BPS (2019).

According to Battour et al. (2014), Eid and El-gohary (2015), and Zailani et al. (2016), halal tourism relates to the satisfaction of tourists. This research aims to know: (1) the overview of halal tourism destination attributes in the city of Bandung; (2) the influence of halal tourism destination attributes on tourist satisfaction.

2 LITERATURE REVIEW

2.1 *Halal tourism destination attributes*

The halal tourism trend has raised the concepts and theories used by Muslim-friendly destinations to become more focused and detailed in accordance with the needs of Muslim tourists and has become part of the strategy (Hall et al. 2020) in developing Muslim-friendly destinations. There are several dimensions in halal tourism destinations such as attributes, that is, availability of Halal and Muslim-friendly hotels, halal food, Muslim-friendly phone applications, Muslim-friendly airports, halal holidays, halal health care facilities and services, halal cruises, halal swimming suits, and halal tourism websites (Battour 2018). In addition to concepts and theories, the national Sharia council (DSN-MUI) provides a fatwa number 08/DSN-MUI/X/2016 regarding the implementation of Sharia tourism with the scope of Sharia hotels, tourists, tourist destinations, spas, saunas and massage, Sharia travel agents, and Sharia tour guides.

Indonesia's strategy in developing Muslim-friendly destinations is through the ACES model adapted from the Global Muslim Travel Index (GMTI). The ACES model looks at four areas: access, communication, environment and services (CrescentRating 2019; Mastercard & CrescentRating 2019). These four areas represent the critical components that destinations need to focus on to improve and become a more Muslim-friendly travel destination. The ability to satisfy each of the four strategic areas is critical for Muslim-friendly travel and will have a different impact on destinations and their Muslim travelers.

Access component considers the following sub-criteria: air access, rail access, sea access and road infrastructure. This component measures the ease of accessibility of a destination via several modes of transportation. Destinations that are not easy to access will not be considered by Muslim travelers.

Communication component considers the following sub-criteria: Muslim visitor guides, stakeholder education, market outreach, language capabilities of tour guides, and digital marketing (i.e., Muslim-friendly phone application and halal tourism websites) (Battour 2018; Stephenson 2014). This component measures the level of awareness and the degree of market outreach of Muslim travelers' needs. It also accounts for the ease of communication between the travelers and the destinations. A destination with poor communication will remain relatively unfamiliar for Muslim travelers.

Environment component considers the following sub-criteria: domestic tourist arrivals, international tourist arrivals, Wi-Fi coverage at airports and commitment to halal tourism (Battour 2018). This component measures the destination's Muslim Travel Climate and the environment's overall safety and comfort. An environment that is hostile and alien to Muslim travelers will generally create an undesirable experience.

Services component considers the following sub-criteria: halal restaurants, mosques, airports, hotels, and attractions. This component measures the faith-based needs services provided by the destination. These services are crucial to enable Muslim travelers to travel freely while still fulfilling and remaining faithful to their religious requirements even when they travel. The importance of service aspects becomes an interesting part to be made into research by several researchers in detail, for example, Muslim-friendly hotels deliver Muslim guests with all services that are compliant with Islamic teachings such as Qibla direction, halal food, halal certificate, alcohol-free beverages, and prayer room with call for prayers, no gambling or prostitution, halal management and operation, design facilities, and appropriate dress code (Battour 2018; Hall et al. 2020; Henderson 2010; Nassar et al. 2015; Sahida et al. 2011; Wardi et al. 2018).

2.2 *Tourist satisfaction*

According to Kotler (2016) satisfaction is defined as the level of one's satisfaction after comparing performance or perceived results compared with expectations (Chahal & Devi 2015) in Yüksel & Yüksel 2007). If the customer is comfortable with what he gets, then the customer is likely to feel satisfied (Saputro et al. 2018). Tourism destinations and service providers must pay even greater attention to customer satisfaction in the modern day environment because of the fas-evolving competitive landscape resultant from recent consumer and technological trends, which make customer satisfaction more important than ever (Wang 2016).

2.3 *Halal tourism destination attributes and tourist satisfaction*

The results showed a positive and significant relationship between halal tourism destination attributes and tourist satisfaction. Battour et al. (2014) have investigated the link between tourism motivation's attributes and satisfaction. They revealed that Islamic tourism attributes have a significant and positive impact on satisfaction of tourists. "Islamic morality" was found to be the most important Islamic attribute, indicating that travel agents could select hotels for Muslims that are located far from red-light districts. Tourist guides should also avoid visiting these places on tour programs. The availability of halal food in hotels and restaurants could represent a high priority for Muslim tourists in destination selection. "Worship facilities" represent important factors identified to satisfy Muslim tourist needs on their vacation. The provision of maps indicating the locations of mosques/prayer facilities could be made available in key areas, such as tourist information centers, airports, hotels, and parks to please and satisfy Muslim tourists. Providing worship facilities for Muslim tourists may encourage them to travel to a specific destination.

Eid and El-gohary (2015) found that customer-perceived value has a significant impact on their satisfaction. The perceived value might be measured in an Islamic specific value. The perceived value on Islamic tourism attributes also has a link to their satisfaction. Zailani et al. (2016) explain that halal practices in the tourism sector will also have an impact on tourists' satisfaction. Research conducted by Saputro et al. (2018) shows that the halal tourism dimensions have a positive and significant impact on the satisfaction of tourists in West Sumatra. Therefore, there are proposed hypotheses:

- H_0: Access has no significant impact on tourist satisfaction
  H_1: Access has a significant impact on tourist satisfaction
- H_0: Communication has no significant impact on tourist satisfaction
  H_1: Communication has a significant impact on tourist satisfaction
- H_0: Environment has no significant impact on tourist satisfaction
  H_1: Environment has a significant impact on tourist satisfaction
- H_0: Services have no significant impact on tourist satisfaction
  H_1: Services have a significant impact on tourist satisfaction

3 METHOD

The type of research used was descriptive and verified with a quantitative approach. The analysis technique used is path analysis. Data was obtained through the dissemination of the questionnaire directly and through the Google Form. The sample in this study was 203 respondents of foreign Muslim tourists who visited Bandung.

The study of literature in this research was obtained from various sources such as journals, annual reports, and books that were printed or electronic media. The other secondary data are collected from MUI (Majelis Ulama Indonesia), BPS (Badan Pusat Statistik), IMTI (Indonesia Muslim Travel Index) 2019, GMTI (Global Muslim Travel Index) 2019, and hotel and restaurant websites such as the number and name of participants and address. The data was analyzed by using SPSS 23.0.

4 FINDINGS

4.1 *Tourist profile*

Profile of respondents was foreign Muslim tourists in Bandung dominated by tourists from Malaysia (43.84%) followed by Singapore (33%) and India (5.91%); 50.74% of Muslim foreign tourists is gen Y (millennials) and female are the most tourists with a percentage of 51.72%. Majority of the surveyed traveled usually with their families (41.38%). This travel behavior is aligned with Islamic values on the importance of family togetherness and maintaining kinship (Mastercard & CrescentRating 2018).

Among the people surveyed, the most common purposes were leisure (69.61%), business (13.73%), spirituality (5.88%), and other (10.78%). The majority spend on average USD$101–$500 per expenditure (72.41%) and the average tourist spends time visiting for one to three days (Table 1).

4.2 *Overview of halal tourism destination attributes in the city of Bandung*

Based on processed data through a questionnaire distributed to 203 respondents of foreign Muslim tourists who visited Bandung, the dimensions of services have the highest average score of 42%, while access has the lowest average score of 14%. Services are an important element needed by foreign Muslim tourists in fulfilling Islamic principles in traveling. This is in line with Ahmad (2018) concept that the availability of food and drink that is guaranteed halal is a primary need for tourists in traveling. Battour and Ismail (2014) mention that the worship facilities, such as mosque, Qiblah direction, and wudhu (ablution) facilities, are an important element for Muslim tourists when they are on a holiday (Table 2).

4.3 *Hypothesis test and simultaneous test (F test)*

$F_{count} = 104.241$ is the value of F_{count} from the correlation of each dimension of halal tourism destination attributes (X) to the variable of tourist satisfaction (Y) simultaneously with the probability level sig. 0.000. $F_{count} = 104.241$, when compared with F_{table} (3.94) then $F_{count} > F_{table}$, shows that each dimension of halal tourism destination attributes (X) simultaneously has a significant influence on tourist satisfaction variables (Y).

Significance value $(0.000) < 0.05$ then H_0 = rejected and H_1 = accepted. Hi: $PYxi \neq 0$ means that there is a significant influence between halal tourism destination attributes consisting of access, communications, environment, and services to tourist satisfaction.

4.4 *Hypothesis test and partial significane testing (t test) (Table 3)*

The data shows that there is a partial effect between halal tourism destination attributes on tourist satisfaction. By comparing t_{count} with t_{table}, it can be explained:

- There is a significant impact between the dimensions of access on tourist satisfaction with a sig. value of $0.072 > 0.050$ and t_{count} $1.449 > t_{table}$ 1.66055 so that H_0 = rejected and H_1 = accepted.

Table 1. Recapitulation of respondent profile.

	Category	Frequency	Percentage %
Nationality	Malaysia	89	43.84%
	Singapore	67	33%
	India	12	5.91%
	Turkey	7	3.45%
	Saudi Arabia	5	2.46%
	others	23	11.33
Generation	Baby Boomers (1946–1964)	7	3.45%
	Gen X (1965–1980)	42	20.69%
	Gen Y (Millennials) (1981–1994)	103	50.74%
	Gen Z (after 1994)	51	25.12%
Gender	Male	98	48.28%
	Female	105	51.72%
Travel Companions	Family	84	41.38%
	Friends	39	19.21%
	Spouse/ Partner	46	22.66%
	Alone	21	10.34%
	Others	13	6.40%
Purpose	Leisure	142	69.61%
	Business	28	13.73%
	Spirituality	12	5.88%
	Others	22	10.78%
Travel Expenditure	<100$	20	9.85%
	101$–500$	147	72.41%
	501$–1500$	27	13.30%
	>1500$	9	4.43%
Visit duration	1–3 days	91	67.49%
	4–6 days	65	21.67%
	>6 days	47	10.84%

Table 2. F test output (ANOVA[a])

Model	Sum of Squares	df	Mean Square	F	Sig.
1 Regression	9154.562	2	3964778	104.241	**0.00[b]**
Residual	6021.079	201	31.271		
Total	15175.641	203			

a. Dependent Variable: Y
b. Predictors: (Constant), X1.4, X1.3, X1.2, X1.1

Table 3. Test result of halal tourism destination attributes path coefficient on tourist satisfaction.

Path coefficient		t_{count}	Sig.	Conclusion
P_{yx1}	0.138	1.449	0.072	H rejected
P_{yx2}	0.149	1.686	0.089	H rejected
P_{yx3}	0.226	2.707	0.054	H rejected
P_{yx4}	0.533	8.114	0.00	H rejected

- There is a significant impact between the dimensions of communication on tourist satisfaction with a sig. value of 0.089 > 0.050 and t_{count} 1.686 > t_{table} 1.66055 so that H_0 = rejected and H_1 = accepted.
- There is a significant impact between the dimensions of environment on tourist satisfaction with a sig. value of 0.054 > 0.050 and t_{count} 2.707 > t_{table} 1.66055 so that H_0 = rejected and H_1 = accepted.
- There is a significant impact between the dimensions of service on tourist satisfaction with a sig. value of 0.000 > 0.050 and t_{count} 8.114 > t_{table} 1.66055 so that H_0 = rejected and H_1 = accepted.

The influence of halal tourism destination attributes on tourist satisfaction is presented in Figure 1. Based on the path diagram of hypothesis in Figure 1, direct and indirect calculations were made between the dimensions of the halal destination attributes to the visiting decisions presented in Table 4.

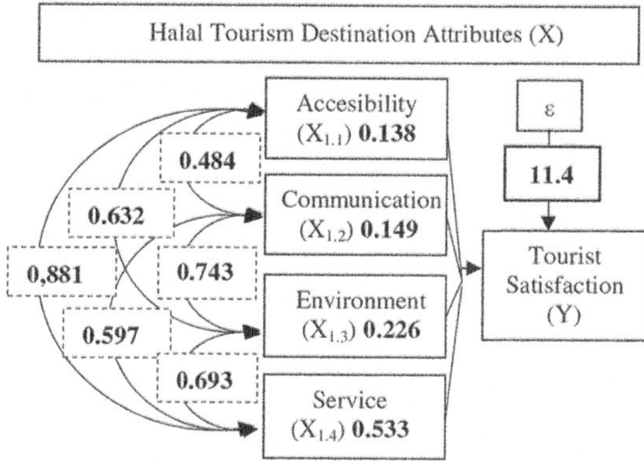

Figure 1. Path diagram of hypothesis testing halal tourism destination attributes on tourist satisfaction.

Table 4. Test result for path coefficients, direct and indirect influence of halal tourism destination attributes on tourist satisfaction.

	Direct Effect on Y	Indirect Effect			
X_1		$X_{1.1}$	$X_{1.2}$	$X_{1.3}$	$X_{1.4}$
$X_{1.1}$	0.019		0.010	0.020	0.065
$X_{1.2}$	0.022	0.010		0.025	0.063
$X_{1.3}$	0.51	0.020	0.025		0.072
$X_{1.4}$	0.533	0.065	0.063	0.072	
X_1	R^2Y $X_{1.1}Y$ $X_{1.2}Y$ $X_{1.3}Y$ $X_{1.4}Y$	Total Effect %		Result	
$X_{1.1}$	0.114	11.4		H_1 accepted	
$X_{1.2}$	0.120	12		H_1 accepted	
$X_{1.3}$	0.168	16.8		H_1 accepted	
$X_{1.4}$	0.484	48.4		H_1 accepted	
R^2		88.6			

Based on the results of the calculation, the influence of the total halal tourism destination attributes on tourist satisfaction was 0.885 or 88.5% and the rest of 11.4% is affected by other factors that are not included in this research. The results of data processing indicate that this research is in line with the research that halal tourism attributes (i.e., Islamic facility, Halalness, general Islamic morality, and free from alcohol and gambling) might have an impact on tourists' satisfaction (Battour et al. 2014; Eid & El-gohary 2015; Saputro et al. 2018; Zailani et al. 2016).

5 CONCLUSION

The result of the research shows that halal tourism destination attributes consisting of access, communications, environment, and services on tourist satisfaction are simultaneously and partially influential. The dimensions of services have the highest average score of 42% while access has the lowest average score of 14%. Services are an important element needed by foreign Muslim tourists in fulfilling Islamic principles in traveling. Based on the responses of respondents regarding the implementation of halal tourism destination attributes in the city of Bandung are considered good, so that it supports the activities of Muslim tourists such as primary needs; Muslim-friendly hotels, halal-certified restaurants, and other supporting facilities have an impact on the satisfaction of Muslim tourists. The government, related institutions, as well as business actors in the field of halal tourism in particular must work together in an effort to develop the City of Bandung as a halal tourism destination. This will certainly affect massively the development of the halal tourism sector.

REFERENCES

Ahmad, Hafizuddin. 2018. "Sosialisasi STP NHI Bandung 11 Jan 2018 V2."

Battour, Mohamed. 2018. "Muslim Travel Behavior in Halal Tourism." *Mobilities, Tourism and Travel Behavior – Contexts and Boundaries*: 1–14.

Battour, M., Battor, M., & Bhatti, M. A. 2014. "Islamic Attributes of Destination: Construct Development and Measurement Validation, and Their Impact on Tourist Satisfaction." *International Journal of Tourism Research* 16(6): 556–64.

Battour, M., & Ismail, M. N. (2014). The Role of Destination Attributes in Islamic Tourism . *SHS Web of Conferences 12*.

Battour, Mohamed, and Mohd Nazari Ismail. 2016. "Halal Tourism: Concepts, Practises, Challenges and Future." *Tourism Management Perspectives* 19: 150–54. http://dx.doi.org/10.1016/j.tmp.2015.12.008.

Battour, Mohamed, Mohd Nazari Ismail, and Moustafa Battor. 2011. "The Impact of Destination Attributes on Muslim Tourist's Choice." *International Journal of Tourism Research* 13(6): 527–40.

BPS. 2019. "STATISTIK Perkembangan Pariwisata Dan Transportasi Nasional 2018." *Berita Resmi Statistik* (110): 1–12.

Chahal, Hardeep, and Asha Devi. 2015. "Destination Attributes and Destination Image Relationship in Volatile Tourist Destination: Role of Perceived Risk." *Metamorphosis*, Vol 14(2), 1–19.

Chanin, Oraphan, Piangpis Sriprasert, Hamzah Abd Rahman, and Mohd Sobri Don. 2015. "Guidelines on Halal Tourism Management in the Andaman Sea Coast of Thailand." 3(8): 8–11.

CrescentRating. 2019. "Indonesia Muslim Travel Index (IMTI) 2019." Singapore: CrescentRating Pte.Ltd. & Mastercard Asia Pacific Pte.Ltd.

Eid, Riyad, and Hatem El-gohary. 2015. "The Role of Islamic Religiosity on the Relationship between Perceived Value and Tourist Satisfaction." *Tourism Management* 46: 477–88. http://dx.doi.org/10.1016/j.tourman.2014.08.003.

Henderson, Joan C. 2010. "Sharia-Compliant Hotels." *Tourism and Hospitality Research* 10(3): 246–54. http://dx.doi.org/10.1057/thr.2010.3.

KEMENPAR. 2015. Laporan Akhir Kajian Pengembangan Wisata Syariah *Laporan Akhir Kajian Pengembangan Wisata Syariah*. Jakarta: Asdep Litbang Kebijakan Kepariwisataan.

Kotler, & Keller. 2016. 39 Pearson Prentice Hall *Marketing Management, 14th Edition*. 14th ed. New Jersey: Pearson Prentice Hall.

Mastercard-CrescentRating. 2018. "Digital Muslim Travel Report 2018." Singapore: CrescentRating Pte.Ltd. & Mastercard Asia Pacific Pte.Ltd.

Mastercard, and CrescentRating. 2018. "Indonesia Muslim Travel Index." Singapore: CrescentRating Pte.Ltd. & Mastercard Asia Pacific Pte.Ltd.

Mastercard-CrescentRating. 2019. "Global Muslim Travel Index 2019." Singapore: CrescentRating Pte.Ltd. & Mastercard Asia Pacific Pte.Ltd.

Mohsin, A., Ramli, N. and Alkhulayfi, B. A. 2016. "Halal Tourism: Emerging Opportunities." *Tourism Management Perspectives*: 137–143.

C. Michael Hall, Nor Hidayatun Abdul Razak, and Girish Prayag. 2020. *The Routledge Handbook of Halal Hospitality and Islamic Tourism*. ed. C. Michael Hall and Girish Prayag. New York: Routledge.

Napu, Della Maghfira, and Wildan Nurhidayat. 2019. "The Effect of Halal Destination Attributes on Visiting Decision." 259 (Isot 2018): 57–61.

Nassar, Mohamed A., Mohamed M. Mostafa, and Yvette Reisinger. 2015. "Factors Influencing Travel to Islamic Destinations: An Empirical Analysis of Kuwaiti Nationals." *International Journal of Culture, Tourism, and Hospitality Research* 9(1): 36–53.

Sahida, Wan, Suhaimi Ab Rahman, Khairil Awang, and Yaakob Che Man. 2011. "The Implementation of Shariah Compliance Concept Hotel: De Palma Hotel Ampang,Malaysia." 17: 138–42.

Saputro, Muhammad Sindhu Danu, Yunia Wardi, and Abror Abror. 2018. "The Effect of Halal Tourism on Customer Satisfaction." (October).

Sofyan, R. 2017. *Prospek Bisnis Pariwisata Syariah*. Jakarta: Buku Beta.

Stephenson, Marcus L. 2014. "Deciphering 'Islamic Hospitality': Developments, Challenges and Opportunities." *Tourism Management* 40: 155–64. http://dx.doi.org/10.1016/j.tourman.2013.05.002.

Wang, Ying. 2016. Griffith Institute for Tourism Research *More Important than Ever: Measuring Tourist Satisfaction*.

Wardi, Yunia, Abror Abror, and Okki Trinanda. 2018. "Halal Tourism: Antecedent of Tourist's Satisfaction and Word of Mouth (WOM)." *Asia Pacific Journal of Tourism Research* 23(5): 463–72.

Yüksel, Atila, and Fisun Yüksel. 2007. "Shopping Risk Perceptions: Effects on Tourists' Emotions, Satisfaction and Expressed Loyalty Intentions." *Tourism Management* 28(3): 703–13.

Zailani, Suhaiza et al. 2016. "Predicting Muslim Medical Tourists' Satisfaction with Malaysian Islamic Friendly Hospitals." *Tourism Management* 57: 159–67. http://dx.doi.org/10.1016/j.tourman.2016.05.009.

Promoting Creative Tourism: Current Issues in Tourism Research – Kusumah et al. (Eds)
 ISBN 978-0-367-55862-8

mGuiding (Mobile Guiding) – using a mobile GIS app for guiding Geopark Ciletuh Palabuhanratu, Indonesia

R. Arrasyid, Darsiharjo, M. Ruhimat, D.S. Logayah, R. Ridwana & H.R.M. Isya
Universitas Pendidikan Indonesia, Bandung, Indonesia

ABSTRACT: The Geopark mGuiding Gateway is an Android application that can perform some functions related to driving and guiding travelers. The development of this application is based on a user assessment designed to be executed on a digital mobile device. User evaluation data are obtained from a cohesion survey consisting of a sample of 30 responses and processed through Black Box. The results revealed the traveler preferences for the travel guidance system. This application is designed according to the tourist preferences. The new features of this application include: (1) an attractive and readable Geopark map; (2) augmented reality (AR) with a camera function and a global positioning system (GPS); (3) user-friendly interface design. This application integrates tracking and location functions using GPS coordinates, multi media guides (beautiful location descriptions, videos, photos and audio presentations), tour and search options suggestion using AR. Geopark mGuiding's Advanced Gateway System offers a friendly user interface that can be used by travelers on their own smart phones for their entire travel activities.

Keywords: driver, application, geographic information system (SIG), Global Positioning System (GPS), ridits analysis and importance-performance analysis

1 INTRODUCTION

In the last few years, a rapid increase in the development and use of SIG technology has led to the development of a number of integrated GIS applications, which can be exploited by various needs including the needs of the tourism industry (Chu et al. 2011; Devlin et al. 2008; Duncan et al. 2009; Duncan & Mummery 2007; Jung et al. 2006). Shoval and Isaacson (2007), for example, used SIG applications and GPS methods to collect data on traveller activity spatially and temporally.

The use of SIG technology and mobile GPS can improve and become a solution in the travel guidance process by offering visual and graphic representation of multimedia information to users (Cheverst et al. 2000; Frank et al. 2004; Kabassi 2010; Zipf 2002). GIS and GPS technologies have been used to investigate individual space and time behaviour in society (Ahas & Mark 2005; Quiroga & Bullock 1998; Shoval & Isaacson 2007). GIS can provide comprehensive access to the spatial database, query features, create theme and lay-out, and report creation (Ardagna et al. 2009; Chen 2007) based on cellular GIS technology functions.

The development of Mobile map application technology began to grow rapidly in the mid-1990s (Sarjakoski & Sarjakoski 2007). With the advent of mobile computing devices and cheap location sensing systems, location information has been an important resource for mobile and computer users (Leonhardt et al. 1996). Mobile GIS has already been implemented for the development of location-based services (LBSs)where someone's location is used to personalize the service (Ardagna et al. 2009). The Kema-Juan technology has dominated mobile devices, such as Smart phones, personal digital assistants (PDAS) and GPS units have made the development of LBS-guiding information services. For example, one can see the success of the project implementation of WebPark (October 2001–September 2004), which gives visitors information about a natural area. The Platform for this system was developed from a combination of smart phones and GPS (WebPark 2010)

 DOI 10.1201/9781003095484-110

In addition to the rapid increase in technology related to smart phones and their mobile application solutions, GIS professionals have experienced an explosion in technology related to mobile GIS.

Software applications for smart phones are becoming increasingly popular and useful. Mobile devices and their respective applications enable GIS professionals to monitor the spatial patterns of visitors to a place, and record and verify information for tourists. There were more than 500,000 applications available for global users to download in 2010. A large number of these applications clearly demonstrate the comprehensive effect cellular technology has for users. In this study, we have chosen the Android platform to develop a new version of the tour guide system. The iOS platform is al-so developed for iPhone users.

This paper provides a prototype of a cellular travel system that integrates GIS technology and cellular GPS. Geopark mGuiding Web Browser Application uses a cellular GIS platform that can operate on Android 8.0.0 android Oreo (iOS version of mGuiding also developed for iPhone users). The main goal is to offer multi media guidance to travelers for places that attract natural landscapes and culture in the Geopark Ciletuh Harbor area. In addition, user evaluation and evaluation from the Geopark mGuiding Port Shield Application system, and the results were used to help us improve the new mGuiding application. This application offers features that include: (1) an attractive and readable Geopark map; (2) augmented reality (AR) with camera functions and global positioning systems (GPS); (3) user-friendly interface design.

2 METHODS

This research is an implementation of the development of the Mobile GIS application for mGuiding, this application offers navigation, interpretive information, digital maps, selection of access guides for users, geopark areas, geographical coverage, biological coverage, cultural diversification, tourist destinations, and travel plans. The target of using this media is guides and tourists at UNESCO Global Geopark Ciletuh Palabuhanratu. The development method for research use system testing using Black Box and quesioner. System testing using Black Box is intended to find errors or flaws in software that have been tested (Arrasyid et al. 2019). Testing intends to know which software to create meets the criteria that are compliant with the software's design objectives. In this research testing conducted on the system IE testing blackbox and questionnaire. Blackbox testing is done on the development side that records all errors and usage problems. Testing of Blackboxes is done in a controlled environment. A testing plan is a test of the functionality contained in the application, whether that functionality is as expected or not.

Stages of validation or application testing are treated qualitatively and quantitatively. This is qualitative as a questionnaire for guiding and tourists about application products using the black box meth-od. This is a quantitative assessment sheet material experts and media experts. At this stage, before conducting the trial, the initial test (design validation) is carried out through tourist. The point is to ensure that the material and appearance of the application are suitable and suitable for use.

This calculation used the likert scale with scale 1–4. The result of the calculation process is presented in the table to obtain the eligibility test score media. The interval is determined by using the equation with a maximum score of 4 and score a minimum of 1. Then by equation (3.2) found the interval class (i) 0.75. Minimum scale is set to 1.00. Then the resulting categories can be seen in

Table 1. Categories interval

Interval	Categories
3,28–4,03	Excellent (E)
2,52–3,27	Good (G)
1,76–2,51	Poorly (P)
1,00–1,75	Bad (B)

Table 2. Tourist's assessment.

Categories	Scale	Frequency	Total	Percentage (%)
Excellent (E)	4	28	112	94,9
Good (G)	3	2	6	5,1
Poorly (P)	2	–	–	–
Bad (B)	1	–	–	–
Total		30	118	100
Average			3,933	
Categories	Excellent			

Table 3. The likert scale was analyzed by calculating the score of each interval from the statement given to the respondent. The following is the result of the assessment of the test against the user with each variable.

3 RESULTS AND DISCUSSION

3.1 *Android app development for mGuiding Geopark Ciletuh Pelabuhanratu*

The rapid development of information and communication technologies, GPS technology, and location based service (LBS) technology led to a significant impact on the world of tourism. In this research, we strive to develop a technology that can help and improve the service of tour guide in Ciletuh Palabuhanratu Geopark, Indonesia. The purpose of this mobile guide system application is to provide information for the guide and information about interesting places in the form of tour packages that can be chosen by the spatial data based tourists. This system is installed on PDAs and is provided for the tourists to use in Ciletuh Pelabuhanratu Geopark. The distribution of questionnaires is done to help evaluate the traveller's experience by using this mguiding application system. Below the assessment table of tourists.

The assessment of the results of the distribution of questionnaires to tourists on the testing of the mGuiding application system resulted that the developed application was rated as Excellent as 94.9% of 30 tourists. With the reason that the interface design of this application system has been considered easy to use and friendly to users, as well as having information related to interesting tourist attractions.

The improvement of the system is to improve the new version of the (mguiding geopark ciletuh Pelabuhanratu application) tour guide system operating on 8.0.0 devices (Android oreo). Users can download the mGuiding application on the Google Play Store via a Wi-Fi network or cellular communication service from their telecommunications company. The hardware used is as follows: (1) Electric Compass, (2) Camera, and (3) GPS / Assisted Global Positioning System (APGS). While the software used is ArcGIS Explorer for Android, ArcGIS Online is integrated into the application.

ArcGIS Explorer 20.2.1 and ArcGIS online devices are used to develop systems for spatial data display, queri functions, data manipulation, and analysis. MGuiding Application systems can operate in Androis 8.0.0 or higher OS. Users can download the app to their personal smart phone and install it for use as a mobile travel-driving service related to Ciletuh Pelabuhanratu Geopark. The above-mentioned devices are used to develop a LBS mobile travel guide application for Ciletuh Pelabuhanratu Geopark.

3.2 *Implementation of the mGuiding app*

mGuiding was designed and developed primarily on Android 8.0.0 Oreo (Veris iOS is also developed for iPhone users). When used for traveller l, the application system has the following functions

Table 3. Menu interface implementation.

<table>
<tr><th>No</th><th>Interface</th><th>Description</th></tr>
<tr><td>1</td><td>Select Main Menu and Choose Menu How to Use

</td><td>The main menu interface is the menu that first appears when accessed by the user. The following displays the interface from the main menu that has been implemented.
Implementation of the main menu interface, there are menu choices that can be chosen according to the information needed.
The interface of how to use is a menu that can be accessed by users to find out how to use the application. The following is an interface display of how to use that has been implemented</td></tr>
<tr><td>2</td><td>Selecting the Exit Button and Selecting the Current Location Button (GPS)
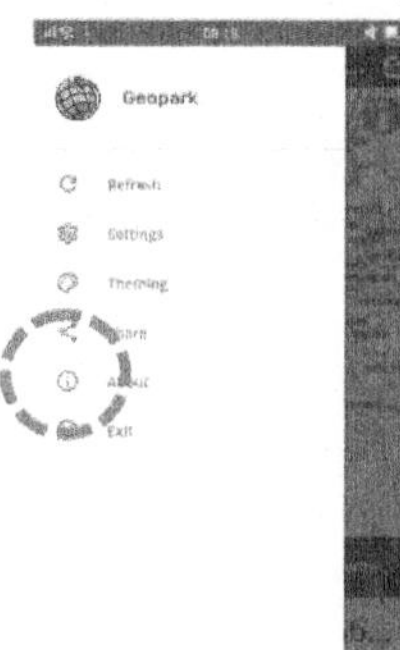

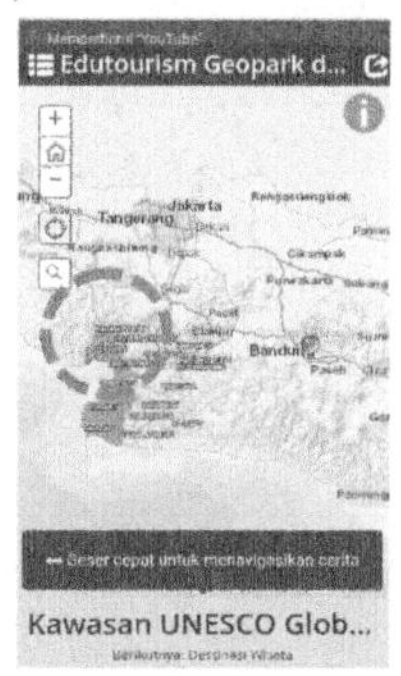
</td><td>Implement the exit menu interface and display the GPS menu to find out the user's location.</td></tr>
<tr><td>3</td><td>Displays location search and displays related object descriptions

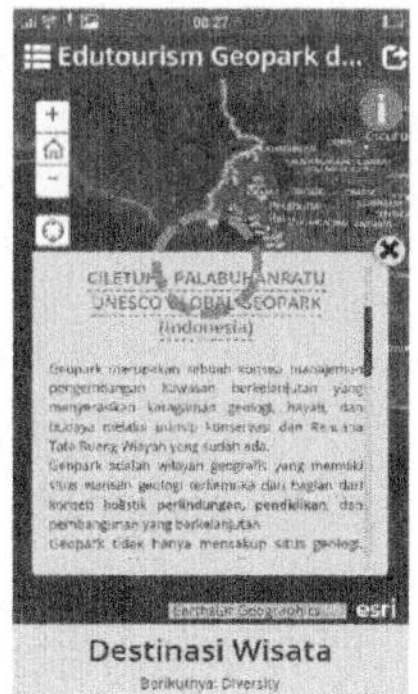
</td><td>Display the location search menu to find the location of the search that has been implemented. Displays the information menu to explain information and descriptions related to the sought object that has been implemented.</td></tr>
</table>

(1) providing five tour packages for tourists to choose from, (2) Providing narrative information about places of interest and beauty, (3) using AR Technology to guide, (4) have a friendly and easy-to-use interload design for the user. The design of each utility and its significance are described in detail below.

Table 4. Objek interface implementation.

No	Interface	Description
1		Displaying Interface related to spatial information on the location of the UNESCO Global Geopark Ciletuh Palabuhanratu that has been implemented Displaying interfaces related to spatial information, distribution, location, and destination attributes in the UNESCO Global Geopark Ciletuh Palabuhanratu region that has been implemented

Table 5. Display testing menu.

No	Components Tested	Test Scenarios	Test Result
1	Menu	Select Menu Using Applications	[v] Accepted [ ] Rejected
		Select Menu How to Use	[v] Accepted [ ] Rejected
		Select the Exit Button	[v] Accepted [ ] Rejected
		Select the existing window according to the desired object and information	[v] Accepted [ ] Rejected

The results of Black Box analysis show that the main assessment of this mGuiding application is on the simplicity of manipulation (easy to use) and the provision of a friendly interface. In addition, we also improve the accuracy and expression of location information. Table 1 shows the function of the system application mGuiding part of the menu display. Table 2 shows the function of the application mGuiding section display objects (Table 4).

3.3 *Display object testing*

The object testing was carried out to determine whether each object displayed has an error or not using Black Box testing, the test results can be seen in the following Tables 5 and 6

Table 6. Display object testing.

No	Components Tested	Test Scenarios	Test Result
1	Object	Unesco Global Geopark Ciletuh Palabuhanratu area	[v] Accepted [ ] Rejected
		Objects / Destinations in the Unesco Global Geopark Area Ciletuh Palabuhanratu	[v] Accepted [ ] Rejected
		Diversity	[v] Accepted [ ] Rejected
		Biodiversity	[v] Accepted [ ] Rejected
		Geodiversity	[v] Accepted [ ] Rejected
		Culturdiversity	[v] Accepted [ ] Rejected
		Community	[v] Accepted [ ] Rejected
		Accessibility and Global Potition System	[v] Accepted [ ] Rejected

4 CONCLUSION

Based on the results of the object's appearance, a menu using Black box testing can be concluded that a functionally constructed application has issued a result that is as expected. While. The results of the questionnaire showed that the design of the interface of the application system has been rated easy to use and user friendly, and has information related to the tourist attractions are interesting.

REFERENCES

Ahas, R., & Mark, Ü. (2005). Location based services – new challenges for planning and public administration? Futures, 37(6), 547–561. doi:10.1016/j.futures.2004.10.012

Ardagna, C.A., Cremonini, M., & Gianini, G. (2009). Landscape-aware location-privacy protection in location-based services. Journal of Systems Architecture, 55(4), 243–254. doi:10.1016/ j.sysarc.2009.01.00

Arrasyid, R., Setiawan, I., & Sugandi, D. (2019). Developing Learning Media Based on Geographic Information System for Geography Subject in Senior High Schools. *JURNAL PENDIDIKAN ILMU SOSIAL, 28*(1), 1–7.

Chen, R.J.C. (2007). Geographic information systems (GIS) applications in retail tourism and teaching curriculum. Journal of Retailing and Consumer Services, 14(4), 289–295. doi:10.1016/ j.jretconser.2006.07.00

Cheverst, K., Davies, N., Mitchell, K., Friday, A., & Efstratiou, C. (2000). Developing a context-aware electronic tourist guide: Some issues and experiences. In T. Turner, G. Szwillus, M. Czerwinski, & F. Paterno (Eds.), Proceedings of the CHI2000: Conference on human factors in computing systems (pp. 17–24). New York, NY: Association for Computing Machinery.

Chu, T.H., Lin, M.L., Chang, C.H., & Chen, C.W. (2011). Developing a tour guiding information system for tourism service using mobile GIS and GPS techniques. Advances in Information Sciences and Service Sciences, 3(6), 49–58.

Devlin, G.J., McDonnell, K., & Ward, S. (2008). Timber haulage routing in Ireland: An analysis using GIS and GPS. Journal of Transport Geography, 16(1), 63–72. doi:10.1016/j.jtrangeo.2007.01.008

Duncan, M.J., Badland, H.M., & Mummery, W.K. (2009). Applying GPS to enhance understanding of transport-related physical activity. Journal of Science and Medicine in Sport, 12(5), 549–556. doi:10.1016/j.jsams.2008.10.010

Duncan, M.J., & Mummery, W.K. (2007). GIS or GPS? A comparison of two methods for assessing route taken during active transport. American Journal of Preventive Medicine, 33(1), 51–53. doi:10.1016/j.amepre.2007.02.042

Frank, C., Caduff, D., & Wuersch, M. (2004). From GIS to LBS, an intelligent mobile GIS. In M. Raubal, A. Sliwinski, & W. Kuhn (Eds.), Proceedings of GIS days 2004 (pp. 277–287). Münster: IfGI Prints.

Jung, H., Lee, K., & Chun, W. (2006). Integration of GIS, GPS, and optimization technologies for the effective control of parcel delivery service. Computers & Industrial Engineering, 51(1), 154–162. doi:10.1016/j.cie.2006.07.00

Kabassi, K. (2010). Personalizing recommendations for tourists. Telematics and Informatics, 27(1), 51–66. doi:10.1016/j.tele.2009.05.003

Leonhardt, U., Magee, J., & Dias, P. (1996). Location service in mobile computing environments. Computers & Graphics, 20(5), 627–632. doi:10.1016/S0097-8493(96)00036-2

Sarjakoski, T., & Sarjakoski, L.T. (2007). A real-time generalisation and map adaptation approach for location-based services. In A.M. William, R. Anne, & L.T. Sarjakoski (Eds.), Generalisation of geographic information (pp. 137–159). Amsterdam: Elsevier Science B.V.

Shoval, N., & Isaacson, M. (2007). Tracking tourists in the digital age. Annals of Tourism Research, 34(1), 141–159. doi:10.1016/j.annals.2006.07.007

Promoting Creative Tourism: Current Issues in Tourism Research – Kusumah et al. (Eds)
© 2021 Taylor & Francis Group, London, ISBN 978-0-367-55862-8

Virtual tour: Tourism opportunities in the new normal era

Labibatussolihah, D.P. Novalita, N. Fathiraini & E. Fitriyani
Universitas Pendidikan Indonesia, Bandung, Indonesia

ABSTRACT: Technological advances in the digital era emerged with new innovations that combined the Internet of Things (IoT) with tourism. At present, tourism is being challenged on how to survive amid a pandemic. Due to the outbreak which occurred around the globe, it makes people aware that they should keep their distance, stop travelling, maintain cleanliness, and avoid crowds. As a result, it decreased visitors' numbers and weakened tourism service providers. In dealing with this condition, virtual tours utilized the Internet network as an answer so that the tourism sector sub-section can survive. A virtual tour realm is not something new, but the pandemic outbreak has triggered virtual tour providers in cyberspace vigorously. This paper aimed to illustrate tourists' necessities, who choose virtual tours, which began to increase from March to June 2020. Netnography is a research method used in analyzing 12 netizen comments and interviews with 8 virtual tourists. Comments were identified into two categories, positive and negative comments. These comments were responses toward virtual tours which were provided by two Instagram accounts. The study's result describe virtual tourism's management in accordance with market share based on attractions, amenities, and activities. Virtual tours do not replace tourist experiences in reality. But it has potential, due to the public enthusiasts, as well as to help economic recovery in the new normal era.

Keywords: Instagram, netnography, new normal, virtual tour

1 INTRODUCTION

Indonesian natural bounty is one key that boosts hospitality economy of the nation. Its tropical climate makes the attractions available for tourists to visit almost all year. This condition helped the influx of visitors, although the number is still lower than other Southeast Asian nations, such as Thailand and Malaysia. In 2017, there were 14,039,799 tourists; this increased to 15,810,305 in 2018 and 16,106,954 tourists in 2019 (Badan Pusat Statistik 2020). This development directly affected the increase in foreign exchange earnings that generated 19.29 billion USD in 2019 (Kementerian Keuangan 2020). From budget structure, there are different characters of foreign and domestic visitors. Most foreigner's expenditure in tourism are food or restaurants (20%), domestic transportation (13%), and accommodation (4%). The rest is for services (guiding and attractions) and souvenirs (LPEM FEB UI 2018). It differs from domestic tourists, which mostly spent for public transportation (37%), restaurants (23%), followed by non-food industrial products (15%) and accommodation (10%) (2018).

The growth of the tourism sector in recent years has fluctuated due to problems faced by the global community. Some of these issues are economic crisis, terrorism, political instability, natural disasters, and plague (Boniface & Cooper 2005; Kementerian Keuangan 2020). The current situation of Covid-19 pandemic restricted social mobility and also hampered the benefit of the sector to the economy conventionally. But, there are some potential aspects especially for agents and providers of the business to adapt a new way of conditions such as using information technology to overcome the gap between demand, supply, and obstacles between them as studied in consumers' behavior that the thoughts, feelings, and actions of each individual or group, as well as the environment influenced what will become of their characteristics (Kotler & Keller 2016).

DOI 10.1201/9781003095484-111

Maditinos and Vassiliadis (2008) stated that one of big loses in sudden global changes such as the pandemic for tourism is the decrease of mass job availability. One of the activities that will help its resistances, keeping the workforce in the sector's line, is through developing virtual tours. As Osman et al. (2009) point out, a virtual tour is a technology that places or simulates audiences or tourists into situations through images and video. Specifically, social media application could be one of the most efficient and influential aspects included in this usage (Abdallah et al. 2007). Instagram, as an example, become an informative instrument to promote this kind of touring. The purpose of this study is to find an alternative way of tourism industry using virtual tour's opportunities as an adaptation in the pandemic. Here, we will review contents from a virtual tour service provider, then analyze the response of users who have experienced it. Through this study, it is expected to contribute by delivering practical advice for managing virtual tours in the new normal era.

2 LITERATURE REVIEW

2.1 *Covid-19 pandemic's impact on tourism industry*

Tourism sector in Indonesia currently is hit hard by the Covid-19 pandemic. Data from the Badan Pusat Statistik (2020) shows that in February 2020 there were 863,960 foreign tourists, then in March it decreased to 470,970 foreign tourists. In April, foreign tourists were only 160,042 people. Some were due to the impacts of lockdown policy, quarantine measures, travel ban, and closure of international air services. This also made thousands of people laydown or were impacted so hard financially. In just three months, the loss reached USD 5.7–6.9 billion and keeps rising (Kompas 2020).

2.2 *Tourist behavior in social media marketing*

In communication, there are eight marketing instruments, one of which is online and social media marketing. Social media is a means for users to share photos, videos, or information content with each other. Social media like Instagram has become a factor in how we relate to others. This media allows people to connect with almost anyone, anywhere, and anytime. In this case, connections and interactions' expansion is the development of various mobile social media applications (Wang et al. 2015). In recent years, the use of social media has become an integral part of most teenagers and adults' lives. The use of social media for young people has developed and become more sophisticated (Kranzler & Bleakley 2019). Social media marketing is defined as a dialogue that is often triggered by consumers circulating among the parties mentioned to drive open communication about some promotional information so that it allows learning from users and experiences with each other, ultimately benefiting all parties involved (Furner et al. 2013). Consumer behavior is the study of how individuals, groups, and organizations choose, buy, use, and determine goods, services, ideas, or experiences to meet their needs and desires (Kotler & Keller 2016a). Thus, it can be emphasized that to fulfill consumers' needs for pleasure, one of which is to utilize social media.

2.3 *Virtual tour maintains the existence*

The virtual tour is an answer as a way to survive so that tourism activity can continue amidst the pandemic. Although this is not new, the Covid-19 pandemic has triggered virtual tours vigorously. A social media platform that mostly promotes virtual tour is Instagram. Instagram (IG) is an image- and video-based social media. Instagram utilization to promote virtual tours makes it further well known.

Virtual tours offer local and international walks without being hindered by distance by viewing video sequences and/or images. The costs are also affordable, and some are free through giveaways. Technical guiding conducted online through communication applications using videos such as Zoom meetings, Google Meet, Jitsi meet, or similar applications. The tourists visualized to gather at a point, then the guide explains the direction and purpose of the tour. In general, virtual tours utilize various applications narrated by guides. Utilized tools used to guide include Google Maps, Google Earth, Google Street View, Youtube, photos, and videos. A proper name for a user who takes a virtual tour is a virtual tourist. Virtual tourists are simply equipped with a stable connection to enjoy virtual tours through a laptop or device screen.

A creative guide can maintain his profession during a pandemic by guiding virtually. Tourism transformation from visiting in person or called offline into online is a way for service sellers to survive amidst the pandemic. Personal photos and video archives can be discussed during the tour. A virtual tour is closely related to online object tourism development. Anggraini (2017) offers six main component characteristics or known as an A framework to analyze tourism destinations, which consist of attractions, amenity, accessibility, available packages, activities, and ancillary services. However, the characteristics that are suitable for virtual tours in the new normal era include attractions, amenities, and activities. Further explained is that the attraction component becomes very important because it relates to tourist attractions that tourists can enjoy. Amenities are a supporting facility such as entertainment or catering services so that the tourists feel comfortable. Activities are activities that involve tourists to gain experience in their visiting areas such as attending ceremonies, festivals, exhibitions, and others.

2.4 *Experience in a virtual tour*

A virtual tour gives different experiences for tourists after joining the tour. Smith (1993) defines a tourist experience as an interaction between tourists with a variety of facilities, services, and attitudes of various companies that exist in tourism sites. The perceived experience is the feeling of "having been" in a tourist area. Although they have not visited the area, yet they already know the place complexity. A memorable experience that can create a good experience for tourists depends on the services they obtain. Based on tourist experience from Buonincontri and Micera (2016), there are three phases of the tourist experience: pre-trip, on-site, and after-trip.

The tourists' process in gaining experience can be observed for the first time from their interest and the way tourists get initial information as a pre-trip phase. Second, the on-site phase is when tourists are in a tour that can be extracted from their story. Third, the after-tour phase is by interviewing tourists regarding expectancy and meanings obtained after doing the virtual tour. The perception of value in the actor's mind is also influenced by their experiences of other destinations (Nasir et al. 2020).

The observed experience from all three phases is considered as a response. This response can be expressed through writing or emote icons that appear on social media providers of virtual tours. If the social media account gets lots of likes and positive comments, it will give a good reputation to the social media account (Price 2014). A positive response will be shown by the tendency to expect a virtual tour, while negative responses are raised from words or comments that indicate dislike, rejection, and a tendency to avoid virtual tours. The comments column can be filled openly by users so they can write a variety of responses that are felt or thought about.

3 RESEARCH METHOD

This study uses a qualitative approach with the netnography method, another name for ethnography. Netnography focuses its study on culture and online communities. Hine (2011) called it a virtual ethnography. Whereas Kozinets (2002) defines netnography as a qualitative research methodology that adapts ethnographic research techniques to study culture and community that is conducted in computer-mediated communications. Netnography research is also carried out to reflect the implications of online communication for culture and vice versa. The investigation was conducted by exploring social relations in social networks, namely, relationships between Instagram users. This research was conducted online by accessing data on the platform of the research subjects.

Collecting data process in netnographic research needs to be understood to establish communication with members of a community (Kozinets 2010). This communication can take many forms that are important for the interaction between researchers and the community. Researchers observed the interaction between virtual tour providers and citizens of the social media @jktgoodguide and @atourin.official as tour operators.

In this study, data collection techniques were carried out with archived data, field notes, and interviews. Archive data was obtained from content, comments, reviews, and virtual tourists' testimonies regarding operator tour's services. Facts and data obtained through archive are then identified into two, namely, positive and negative responses. Responses were written in the field notes. Interviews were conducted with tourists who have taken a virtual tour with in-depth interviews via direct messages on Instagram. There are seven questions asked given to informants, as follows:

1. Why are you interested in virtual tours?
2. How did you find out about virtual tours?
3. Tell us about your experience while taking a virtual tour!
4. In your opinion, what's interesting about a virtual tour?
5. What do you expect for the future of the virtual tour?
6. Are you interested in taking the virtual tour in the future?
7. When the conditions are new normal, what would you choose, virtual tours or do you go directly to tourism site?

Subjects in the study were tourists who had taken a virtual tour during the April–June period. Informants randomly interviewed as many as eight people, which consisted of three men and five women. The following is the informant's profile:

Table 1. Informants' profile.

No.	Name Initial	Sex	Age	Domicile	Job
1.	EM	F	50	Medan	Entrepreneur
2.	SP	F	38	Kota Tentena	Entrepreneur
3.	J	M	33	Surabaya	Private Employees
4.	MPM	F	22	Jakarta	Student
5.	F	F	22	Nagoya	Private Employees
6.	A	M	20	Jakarta	Freelance
7.	PP	M	19	Yogyakarta	Student
8.	AN	F	18	Bogor	Student

The research's object is tourist's responses and netizen's comments on Instagram virtual tour provider's account. The data is then analyzed using the inductive approach to be extracted into its essences (Kozinets 2010). The inductive approach in this study is responses given by tourists regarding the virtual tour. These responses are the basis for developing better virtual tour services in the future.

4 RESULTS

4.1 *Virtual tour service provider*

Virtual tourism becomes a trend to channeling citizens' hobby for wandering during the large-scale social restrictions (PSBB). This research focuses on two Instagram accounts of virtual tour service providers: @atourin.official and @jktgoodguide. Researchers explored the content posted from the two Instagram accounts in the form of pictures and videos. @atourin.official has been conducting a virtual tour since April 2020, which until now there are 50 posts consisting of 49 images and 1 video. The virtual tour was already conducted 15 times with a total of 657 virtual tourists. It is different from @jktgoodguide since the Covid-19 disaster status announced in Jakarta, which effected many walking tours that cannot be carried out and maneuver to virtual tours. The transition began to be promoted since March, and the tour started in April. The number of posts are 45 feeds consisting of 37 images, 7 videos, and 1 IG TV. The tours that have been facilitated more or less 90 times.

The virtual tour's cost is varied. @atourin.official offers for one person Rp. 30,000; two people Rp. 50,000; and five people Rp. 100,000. Whereas @jktgoodguide uses the pay as you wish system as a form of appreciation for guides without any provision for the amount of payment. There are similarities between these service providers that are both using the Zoom Meeting application as a media liaising guide and tourists. When conducting the tour, the guide is supported by photos, videos, youtube, Google Street View and Google Earth.

When observing from the Instagram feed, there are differences in the tourist destinations offered. Virtual tour destinations are from @atourin.official, specifically tourist attractions in Indonesia such as Natuna Island, Nglanggeran Tourism Village, Belitong Island, Sumba Island, Banjarmasin, Yogyakarta, Poso, Labuan Bajo, Tanjung Puting, Mount Rinjani, Lake Toba, Lake Adar Baduy, Raja Amat, the Dieng Plateau, and Malang Raya. Based on the provided tour, they showcase the natural and cultural attractions. Before the tour, there were remarks from the local tourism office or the communication and information department. Tour guides come from the village head, pokdarwis, tour observers, and local guides. This shows that the relationship owned by @atourin.official is very broad. The interesting thing is that they will have a giveaway for netizen. Then, the luckiest netizen will get a free virtual tour. Furthermore, there are always comments on each post.

Another case with Instagram content is from @jktgoodguide for which there is no giveaway. Tours offered are more to cultural and historical tourism. The tour is categorized into four types of city tour: thematic, urban legend, and foreign tourism. First, the city tour tourist destinations include Blok M, Bogor, Cikini, Chinatown, City Center, Jalan Thamrin, Jalan Sabang, Jatinegara, Karet Kuningan X Ereveld, Kebon Raya Bogor, Kemang Timur, Heroes' Tomb, Manggarai, Matraman, Molenvlet, Monas, Old Town, Oranje Boulevard, Pasar Baru, Senayan, Sunter, Senen, Skyscrapers, Tanah Abang, Palmerah, and Weltevreden. Second, Jakarta's thematic tours provided are old mosques, legendary coffee shops, station roaming, and visits to monuments. Third, urban legend visits places that have scary stories like Jeruk Purut, Manggarai Station, Casablanca Tunnel, Mall Klender, and Ancol. Fourth, foreign tours that are visited virtually on weekends are Paris, Rome, Siena, Prague, Budapest, Beijing, Wuhan, New York City, Ho Chi Minh City, and Barcelona.

Viewing from citizens' commentaries and videos on IG TV highlights the challenges faced by both accounts, namely, providing new insights about virtual tours. There are netizens who commented on the account @jktgoodguide "How do I join, sir?" Ad-min replied, "The tour is going to conduct through an application called zoom. It accessibly through a cell phone or laptop. If you register, we will send an email with a link to click when the tour starts." This shows there are still citizens who are not familiar with virtual tours.

4.2 *Tourists' responses regarding the virtual tour*

Buonincontri and Micera (2016) mentioned three phases to see tourists' responses. The pre-tour stage can be observed from @atourin.official's account followers' responses, as many as 1,653, and @jktgoodguide as many as 26.8k. The followers' number is interpreted as citizens' trust in Instagram accounts. These supported by posted likers number that compiled in June 2020. Based on both accounts, @atourin.official posted 15 destination pictures for virtual tours, which reached 814 likers. On the other hand, @jktgoodguide posted 11 destination pictures, which reached likers 559. A possible reason @atourin.official has more likers is due to their giveaway program. The next section discusses the positive response of citizens from @atourin.official's account.

Netizen 1 : *"Sumba is a dream destination ... let's go virtual 💕💕💕"*

Netizen 2 : *"Follow the quiz so that you can join the virtual tour to Banjarmasin for free. Bubur sumsum ikan asin, Assalamualaikum Banjarmasin."*

Netizen 3 : *"woohoo explore the Natuna, desperately want to wander the island that contested by many countries 🙏."*

Netizen 4 : *"Pandemics cannot prevent us from going up the plateau, one way that cannot do but some other way as choice can't go there directly, yet there are virtual ones that can help."*

Netizen 5 : *"Got twice opportunities to go to Kupang, I can only harbour the desire to be able to continue the tour to Labuan Bajo..hiks..hiks .. it's okay for now on we heal the dream through the virtual tour, yeah, yeah.. let's hit the road sister."*

Based on the above response, it appears that netizens have an interest since the pre-travel phase toward virtual tourism. The netizen 6 even shared feelings by posting in Labuan Bajo as follows:

> *"...While I waiting for this pandemic to end, to fulfil my desire for LBJ. I want to take a virtual tour from @ atourin.official, yeah. Hopefully, I will be one of 10 people who get free tickets from @ atourin.official for virtual trips to LBJ. ☐ I also want to invite my friend dear ... to follow the virtual walk to LBJ with @ atourin.official. Because at that time Siti suddenly had to cancel her ticket to LBJ, so I had to leave alone... 😇"*

The comment was given by citizen 6 in line with interview result with F via Instagram DM regarding his opinion on virtual tours by saying that the tour is better than youtube channel as "*we can get stories that maybe if you want to know more you have to research it.*" One comment showed dislike in the @atourin.official account, to compare the service as common as youtube free vlogs. Then the admin replied by emphasizing its social benefit: "*It becomes an alternative media to help tour guides affected by corona, and we also have a special cross-subsidy for all health workers in Indonesia, all for free*"

A similar response shown from an interview with SP said that the advantages as well as disadvantages of the virtual tour, especially its absent of social interactions that we gain from the actual visit to the site: "*the virtual trip was cool, [but] By visit directly, we can ...actually interact with other people.*"

Both responses above indicate an objection toward costs and sensations issue that considered virtual tour as less attractive. Thus, it can be described as a positive and negative response from the first account. For the @jktgoodguide there are positive responses from netizen's comments such as excitement of waiting for the tour, the schedule, or the turn. Virtual tour provider's response by giving clarity to tourists who have the interest to take part on tour. The similar thing also was shown in the interview to A that said, "*A virtual tour is fascinating...to find out an area first before deciding to visit...what is fun...so, even though we only see photos or videos but...[through] the narration...we...can feel and know...as if we are in the site.*"

The @jktgoodguide uses bilingual Bahasa and English. The aim is to attract foreigners to take part in the virtual tour. The admin was also responsive in handling this. Some comment directly asked about English guiding service, of course from non-Indonesian visitors or prospective users. In IG TV, the founder explained that virtual tour expands the scope of market share. Walking tours are usually limited to tourists from Jabodetabek (Jakarta Bogor, Depok, Tangerang and Bekasi). Yet, through a virtual tour, there is tourist diversity such as from Semarang, Surabaya, and Bali. It also attracts tourists from Japan and America. Some of them show preference to take a direct tour, as mentioned by netizen10 and interviews with MPM:

Netizen 10 : *"In my opinion, it is still more exciting to go for a walk, especially if it's the culinary section. Hopefully, the pandemic will pass quickly, miss for wandering together @jktgoodguide.t Stay health for all team of @jktgoodguide 🙏."*

MPM : *"Of course, it will not be the same as the real tour, but several times following the virtual tour coupled with an active and fun guide the sensation is as fun as if we tour in real life."*

Researchers observed that 90% showed a positive response to the virtual tour beginning from the pre-trip phase. Thus, it is in line to the on-site phase, the interviews' result with PP and J, who shared their experiences while taking a virtual tour. However, the interviews' result with tourists after taking a virtual tour as an answer to the after-tour phase indicate the following unfulfilled expectations:

EM : *"Hope that the guide's internet connection guide more stable and please deliver multiply photos/videos at past."*

AN : *"Go to the extreme places. Like to North Korea or Chernobyl 😉😉"*

MPM : *"From several virtual tours, the only thing stumbling block is the availability of a good internet network, understandably living in a remote area. I feel much more pleased with the virtual tour that the tour guide is fun, friendly, and knows the area, so I hope that in the future for each virtual tour inviting tour guides who are as fun as I have followed."*

J : *"Explore more destinations and sound effects"*

Netizen 11 : *"Please create more, or other European countries also would great "if I don't ask too much, please cover Milan ... huhu"*

Netizen 12 : *"Yorkshire and Iceland please 😁"*

Based on the collected responses, it can conclude things that still need to be improved are Internet networks, technology-literate guides, destinations, prices, and offered facilities. Both netizen and virtual tourists prefer direct tours. It's because basically a virtual tour is impossible to replace the real experience. Yet, it can enhance the experience or called pre-tour. A virtual tour has many enthusiasts from tourist's numbers, likers, and comments. Thus, with the consumer's availability it means potential

5 DISCUSSION

5.1 *Strength and weakness of virtual tour*

Researchers identified virtual tour's strengths as follows: firstly, a guide challenged to think creatively, so they can organize tours during a pandemic safely. Secondly, virtual tours are flexible because there are connecting media such as Zoom meetings that can be accessed anywhere. These can be an advantage for foreigners to participate in tourism without geographical limitations. Thirdly, an increasing middle-class has changed a consumption tourism pattern, where tourism becomes a primary need so that even though the pandemic ends, a virtual tour can stay exists as alternative tourism. While a virtual tour weakness is due to the Internet network in Indonesia that has not distributed evenly yet, virtual training for guides still needs to be intensified so that guides can operate the application better.

5.2 *Virtual tour management*

A virtual tours' development in the new normal era refers to Anggraini (2017) by recommending "3As" that consist of attraction, amenity, and activity. Attraction defines a tourist attraction. In this case, the guide needs to make videos, collect images, and make events lists such as traditional ceremonies, dances, religious ceremonies, local traditions, with personal documentation.

Amenity becomes traveling activities supporters by preparing accommodation. Because virtual tours do not require a place to stay, the amenity prepared by guide is replaced by providing limited edition souvenirs for virtual tourists that are sent via a freight forwarding service. It can also be a donation to craftsmen affected by the pandemic. The virtual tour effect is not only felt by guides but also by handcraft and souvenir producers. This method will help the new normal economic recovery.

An activity is tourist activities that directly involved participating in events held by tourism objects. To engage tourists more interactively with guides, they can deliver inquiry regarding visited sites through https://www.sli.do/. By doing so, a developed communication is not only through a question and answer system by default. These efforts are expected to provide experience for tourists.

These three things can become the government's study material, which in the KEM PPKF document of the Ministry of Finance (2020) mentions 2P as the focus of improvement efforts, namely, promotion and participation enhancement of private businesses. A virtual tour can also indirectly help promote various Indonesian tourism objects across countries.

Furthermore, virtual tours also have the potential to develop to be friendly tourism for the disabled community. This potential development can be made possible by an integrative research with skillfull sign language experts who translated the guiding for them.

REFERENCES

Abdallah, A., Rana, N. P., Dwivedi, Y. K., & Algharabat, R. (2017). Telematics and Informatics Social media in marketing : A review and analysis of the existing literature. *Telematics and Informatics*, *34*(7), 1177–1190. https://doi.org/10.1016/j.tele.2017.05.008

Anggraini, D. (2017). *Analisis Hubungan Komplementer Dan Kompetisi Antar Destinasi Pariwisata (Studi Kasus: 10 Destinasi Pariwisata Prioritas Di Indonesia).* (Tesis). Sekolah Pascasarjana, Universitas Indonesia, Jakarta.

Badan Pusat Statistik. (2020). Jumlah Kunjungan Wisman Menurut Kebangsaan dan Bulan Kedatangan Tahun 2017–2020. [Online]. Accessed from: https://www.bps.go.id/dynamictable/2018/07/30/1548/jumlah-kunjungan-wisman-menu rut-kebangsaan-dan-bulan-kedatangan-tahun-2017—2020.html

Boniface, B. & Cooper, C. (2005). The future geography of travel and tourism. Dalam B. Boniface & C. Cooper, *Worldwide destinations; The geography of travel and tourism, (fourth edition)* (hlm. 476-88). Italy: Elsevier Butterworth-Heinemann.

Buonincontri, P., & Micera, R. (2016). The experience co-creation in smart tourism destinations: a multiple case analysis of European destinations. *Information Technology & Tourism 16(3)*, 285–315.

Furner, C. P., Racherla, P., & Babb, J. S. (2013). Social media marketing and advertising. *The Marketing Review*, *13*(2), 103–123. https://doi.org/10.1111/jan.13030

Hine, C. (2011). *Virtual Ethnography*. London: SAGE Publications Ltd.

Kabadayi, S., & Price, K. (2014). Consumer–brand engagement on Facebook: liking and commenting behaviors. *Journal of Research in Interactive Marketing*, 8(3), 203–223.

Kementerian Keuangan RI. (2020). *Kerangka ekonomi makro dan pokok-pokok kebijakan fisikal tahun 2021*. Jakarta: Kemenkeu.

Kompas, 5 May 2020. "Terdampak virus corona, devisa sektor pariwisata bisa turun US$ 6,9 miliar". https://nasional.kontan.co.id/news/terdampak-virus-corona-devisa-sektor-pariwisata-bisa-turun-us-69-miliar

Kotler, P. & Keller, K.L. (2016). *Marketing Management 15 Global Edition*. https://doi.org/10.1080/08911760 903022556

Kozinet, R.V. (2010). *Netnography: Doing Ethnographic Research Online*. London: SAGE Ltd.

Kozinets, R.V. (2002). The Field Behind the Screen: Using Netnography for Marketing Research in Online Communities. (February 2002). *Journal of Marketing Research, 39*(1), 61–72.

Kranzler, E. C. & Bleakley, A. (2019). Youth Social Media Use and Health Outcomes: #diggingdeeper. *Journal of Adolescent Health*, *64*(2), 141–142. https://doi.org/10.1016/j. jadohealth.2018.11.002

LPEM FEB UI. (2018). Retrieved from https://www.lpem.org/id/

Maditinos, Z. & Vassiliadis, C. (2008). Crises and Disasters in Tourism Industry: Happen locally – Affect globally. *In Management of international business and economics systems, MIBES conference* (pp. 67–76)

Nasir, C. S. & Aprilia, C. (2020). *Kebahagiaan turis dan pengalaman perjalanan berkesan.* Aceh: Syiah Kuala University Press

Osman, A., Wahab, N. A., & Ismail, M. H. (2009). Development and Evaluation of an Interactive 360° Virtual Tour for Tourist Destinations. *Journal of Information Technology Impact. 9*(3), 173–182.

Smith, P. (1993). Measuring Human Development. *Asian Economic Journal Volume 7, Issue 1.*

Wang, C., Lee, M. K. O., & Hua, Z. (2015). A theory of social media dependence: Evidence from microblog users. *Decision Support Systems*, *69*, 40–49. https://doi.org/10.1016/j.dss.2014.11.002

Promoting Creative Tourism: Current Issues in Tourism Research – Kusumah et al. (Eds)
© 2021 Taylor & Francis Group, London, ISBN 978-0-367-55862-8

The effect of internet marketing and electronic word of mouth of Sundanese gastronomy tourism on tourist visit motivation to Bandung

D. Valentina, D. Turgarini & I.I. Pratiwi
University Indonesia of Education, Indonesia

ABSTRACT: This study used the quantitative method to see and examine if Internet marketing (X1) and electronic word of mouth (X2) affect the tourist visit motivation to Bandung. The sampling technique used was incidental sampling, with local tourists as the respondent. This study used 100 respondents. Instrument testing technique in this study was the validity and reliability test, while the data analysis technique used the classic assumption test, multiple linear regression analysis, and descriptive analysis. The study result shows that 1) Internet marketing and electronic word of mouth simultaneously affected the motivation of tourism to visit Bandung; 2) Internet marketing didn't affect the tourist visit motivation to Bandung; 3) electronic word of mouth affected the tourism visit motivation to Bandung; and 4) the most known menus of Sundanese food and beverage in Bandung by the respondent are a menu that can be found easily in Bandung.

Keywords: Internet marketing, electronic word of mouth, tourist motivation, Bandung

1 INTRODUCTION

Bandung has a very big gastronomy tourism potency. This is provided by the big variety of food and beverages that also differ from the other regions in Indonesia, ranging from the raw material, processing technique, presentation, and also the historical and philosophy of the food. According to the result of a focus group discussion held by Dr. Dewi Turgarini on January 14, 2018, it says that there are 304 Sundanese gastronomy recipes, consisting of 40 recipes of the main course, 92 recipes of side dishes, 27 recipes of a complete meal, 78 recipes of snacks, 26 recipes of hahampangan (a type of cake), 16 recipes of fresh snacks, 12 recipes of beverages, and 13 recipes of sambal (sauce). The result of this inventory shows that Sundanese gastronomy has a big chance to be the prime gastronomy tourism in Bandung (Turgarini 2018).

In this 4.0 industry revolution, the marketing systems are also developed. Internet marketing becomes popular and widely used by people in their business. According to the data that's released by e-Marketer, the use of the mobile phone is continuously increasing every year. In 2019 2,381 million people were using mobile phones around the world to access information digitally. The collaboration between Sundanese ethnic food in Bandung, Internet media, and the power of online review is highly expected to motivate tourists to visit the city of Bandung.

This study aims at investigating if there is any significant effect between Internet marketing (X1) and electronic word of mouth (X2) on tourism visit motivation to the city of Bandung (Y). This study also analyzes the Sundanese food or beverages that are mostly known by the respondent and the Sundanese gastronomy tourism that are mostly posted at online reviews.

2 LITERATURE STUDY

2.1 *Tourism*

Tourism is a set of activities and businesses of providing services such as tourist attractions, accommodation, transportation, and other services to fulfill the need of people who are on holiday

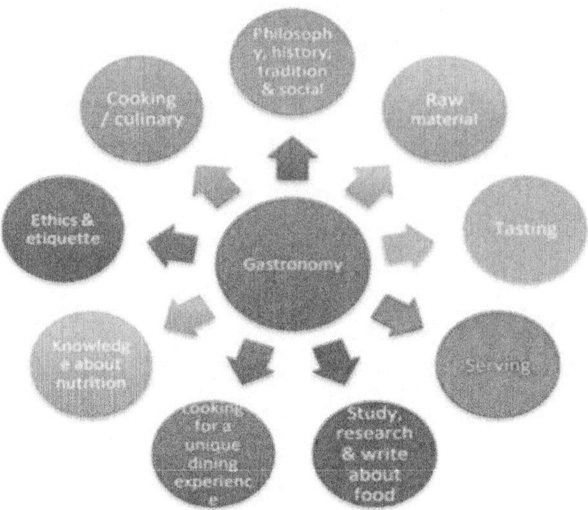

Figure 1. Gastronomy tourism components. Source: Dewi Turgarini (2012).

(Sugiama 2011). Collaboration between the creative industry and tourism industry will be able to encourage national industrial development and overcome the negative impact of world economic liberalization and globalization (Ningsih 2016).

Tourism has its uniqueness compared to the other sectors, its ability to combine both products and services. Tourism has the real products that we can see, touch, buy, and save, but on the other side tourism also has products that can't be seen: that's the services. The service product is a kind of product that we can purchase and we enjoy the benefits when we use it.

2.2 *Gastronomy tourism*

Gastronomy is a study that discusses the relationship between culinary and culture. It focuses on the cultural elements and culinary as the main idea (culinary art) (Taqwani 2012). According to Turgarini (2018), there are nine important components in gastronomy (Figure 1).

Gastronomy tourism is about the journey of people especially to enjoy food and beverages as the main goals to visit someplace. Gastronomy tourism provides tourists an authentic culinary education, greater understanding, and a high appreciation of the cultural richness and local food from each place. By gastronomy consumption, it means that tourists contributing to absorb and realize the intrinsic tourism with local and legendary aura (Turgarini 2018).

2.3 *Marketing communication*

According to Philip Kotler and Kevin Lane Keller (2012), marketing communication is how firms attempt to inform, persuade, and remind consumers–directly or indirectly–about the products and brands they sell. It means that marketing communication is a medium for the company to inform, persuade, and remind the consumer about their product directly or indirectly. The point is marketing communication presented the company thought and their brand. Through the marketing communication there could be the dialog and connection between the company and their consumers.

Another definition of marketing communication is a marketing activity that uses various techniques of communication aimed at providing information to reach the company goals, to strengthen the marketing strategy, and also to expand the market segmentation (Machfoedz 2010).

2.4 *Internet marketing*

According to Rafi A. Mohammed et al. (2003), Internet marketing is a process in building and having a connection with consumers by online activity that enables the product exchange, services, and ideas that are aimed to meet the satisfaction from both sides.

According to Supranto (2006), some indicators can be used to measure the use of Internet marketing: information, service existence, service responsive, process, and usability.

The big use of the Internet in Indonesia is as a big opportunity to attract local or foreign tourist interest to experience the gastronomy tourism in Bandung. There are a lot of benefits of Internet marketing, such as simplicity and practically in products or services marketing, saving cost, the fast-selling time, right target, easiness in changing information, directly communicating with consumers, wider scope, simplicity of management, supporting at the market research, increasing the selling, and actual selling result.

2.5 *Electronic word of mouth*

According to Jalilvand and Samiei (2012), electronic word of mouth is a positive or negative review made by actual consumers, potential consumers, and ex-consumers about a product or a company, which this information can be accessed by everyone in Internet media.

One of electronic word of mouth (WOM) that has an important role nowadays is the online review, which is considered as a provider of important and trusted information. The appearance of online reviews brought a breakthrough in marketing (Kertajaya 2008). According to Jeong and Soocheong (2011), the E-WOM dimension has been reflected through three dimensions: 1) Concern for others, 2) Expressing positive feelings, and 3) Helping the company.

2.6 *Tourist visit motivation*

According to Jeffrey et al. (in Suryani 2008), the motivation process occurs as a response to the unfulfilled needs, desires, and expectations that may cause tension. At a certain level, this tension will turn into an ambition that encourages individuals to carry out certain behaviors to meet their needs, desires, and expectations.

According to Turgarini (2018), there are 10 main factors that influence the tourist visit motivation to Bandung: 1) Tourist special objectives, 2) Affordable prices, 3) Refreshment, 4) Unique restaurant, 5) Culture heritage, 6) Sightseeing, 7) Sundanese gastronomy, 8) Personal, 9) Range of reachability, and 10) Vitality.

3 RESEARCH METHOD

This study used a quantitative method. The instrument testing technique in this study was the validity and reliability test, while the data analysis technique used the classic assumption test, multiple linear regression analysis, and descriptive analysis. This study used questioner to collect data from respondents and used the Likert scale in the questioner. This study used 100 respondents that have been determined using incidental sampling techniques. Questionnaires were distributed online using Google Forms to the domestic tourists in Bandung. Respondents were asked some questions to see their suitability to the criteria first before they fulfill the questioner.

4 RESULTS AND DISCUSSION

4.1 *Instrument testing*

Before the data collection, the questionnaire was tested to see its validity and reliability. Validity test and reliability test were done to 30 respondents, as a minimal requirement for validity and reliability test. Testing was done using SPSS.

Validity Test

To find out whether each item is valid or invalid, it must meet the following conditions:

a. If rcount $\geq$ rtable, the instrument is declared valid.
b. If rtable $\geq$ rcount, the instrument is declared invalid.

Table 1. Validity test result.

Variable	r_{count} average	r_{table}	Result
Internet Marketing	0,691	0,361	VALID
Electronic Word of Mouth	0,72	0,361	VALID
Tourist Visit Motivation	0,66	0,361	VALID

Source: Data processed researchers, 2020.

Table 2. Reliability test result.

Variable	Cronbach's Alpha	N of Items	Result
Internet Marketing	0,883	11	RELIABLE
Electronic Word of Mouth	0,881	9	RELIABLE
Tourist Visit Motivation	0,926	18	RELIABLE

Source: Data processed researchers, 2020.

The rtable value for n = 30 with level of error (α) 5% is 0,361. Based on the table above, it can be seen that the average of each variable's recount value is greater than the retable value. It means that all the statements contained in variable Internet marketing, electronic word of mouth, and tourist visit motivation are valid.

Reliability Test

Reliability testing using SPSS will be conducted using reliability analysis statistics with Cronbach Alpha (α). If the Cronbach alpha value (α) > 0.60, it can be said that the variable is reliable.

The table above shows that Cronbach's alpha value for each variable in this research is greater than 0,60. Thus, it can be concluded that all statements on the three variables are reliable.

4.2 *Classic assumption test*

Whether or not the data distribution is normal is done by looking at the significant value in the Kolmogorov-Smirnov table. If the significance value is greater than 0,05 (Sig > 0,05) then the data is normally distributed.

According to table results on SPSS, the value of Sig. in the section Kolmogorov-Smirnov is 0.075, so it can be concluded that the distribution of data obtained by the researcher is normal because the value of the signal on the table Kolmogorov-Smirnov is greater than 0.05.

4.3 *Multiple linear regression analysis*

Multiple linear regression analysis aims to know the relation between the independent variable and the dependent variable. Based on the result from SPSS, regression formulation can be formulated as follows:

$$Y = 20{,}500 + 0{,}036X1 + 1{,}405X2 + e$$

It means that if there is no influence from Internet marketing (X1) and electronic word of mouth (X2), then the value of Tourist Visit Motivation (Y) is 20.500 units. If Internet marketing (X1) is increased by 1 unit, the value of Tourist Visit Motivation (Y) is increased by 0.036 units. And if the electronic word of mouth (X2) increases by 1 unit, the value of Tourist Visit Motivation (Y) is increased by 1.405 units.

4.4 *F Test*

F test is done to know whether the influence of Internet marketing and electronic word of mouth to tourists visit motivation simultaneously.

Based on the result table on SPSS, the value of sig = 0.000 with an F-count value of 36.765 and F-table value of 3.94. Based on the calculation result shows that F-count (36.765) > F-table (3.94). So, it can be concluded that Internet marketing and electronic word of the mouth simultaneously influence tourist visit motivation. So, the variable Internet marketing and electronic word of mouth jointly affect the variable tourist visit motivation.

4.5 *t-Test*

t-test is done to see the presence or absence of the influence of independent variables in a partial to variable dependent.

According to the t-test result on SPSS, the value of t-count for the variable Internet Marketing (X1) is 0.174, and the value of significance (SIG) 0.862. The value of significance (SIG) is greater than 0.025 and the value of t-count is smaller than 1.98498 then H0 received Ha rejected. It means that Internet marketing (X1) has no partial influence on Tourist Visit Motivation. The value of t-count for the variable electronic word of mouth (X2) is 6.281 and the value of significance (SIG) is 0.000. The value of significance (SIG) is smaller than 0.025 and the value of t-count is greater than 1.98498 then H0 rejected Ha accepted. That is, the electronic word of mouth (X2) has a partial influence on Tourist Visit Motivation.

4.6 *Determination coefficient*

The determination coefficients aren't used to determine how the large variation dependent variable on the independent variable. The coefficient of determination value ranges from zero to one. A value approaching one means the independent variable provides almost all the information needed to predict the variation of the dependent variable.

From the table result on SPSS, the coefficient determination (Adjusted R Square) value is 0.419 or 41.9%, which means that 41.9% of the variation in the Tourist Visit Motivation variable can be explained by both independent variables (Internet marketing and electronic word of mouth), while 58.1% is explained by another variable not studied in this study.

4.7 *Respondent descriptive analysis*

Respondents of this research are the domestic tourists of Bandung who are less than 25 years old to more than 35 years. The majority of respondents in this study were less than 25 years old with a percentage of 84%. Female respondents were more than male respondents with a far-off number, 75% female and 25% male. According to the original area, the majority of respondents came from outside West Java with a percentage of 71%. Based on the number of visits to Bandung, the average respondent has visited Bandung more than five times with a percentage of 55%, and the reason for the most visits is to do sightseeing with a percentage of 39%.

The analysis shows that the majority of respondents use Instagram with a percentage of 80% as a means to search for Sundanese gastronomic tourism information in Bandung, as well as the media to hear or see Sunda gastronomic tourism reviews that are most widely used is Youtube with a percentage of 69%.

Based on the results of the analysis of respondents to the Sundanese food menus in Bandung, the main food menu the most widely known by respondents was Opor Hayam with a percentage of 61%, the companion food menu most widely known by respondents was Lotek with a percentage of 74%. Then, the complete menu that is most widely known by the respondent is Mie Kocok Bandung with a percentage of 74%, the most widely known snack menu is Cireng with a percentage of 82%. Furthermore, the Sundanese snack menu that is most widely known by respondents is Kerupuk Udang, the most widely known fresh snack is Es Cingcau with a percentage of 73%. Menu of special Sundanese sambal in the city of Bandung, the most widely known is Sambal Terasi with a percentage of 73%, and the menu of Sundanese beverage that is most widely known by the respondent is Bandrek with a percentage of 77%.

4.8 *Variable descriptive analysis*

Based on the results of variable descriptive analysis of Internet marketing, the "Usability" indicator has a higher average score of 4.4. This shows that through the Internet media-related information on Sundanese gastronomic tourism in Bandung is more easily obtained by tourists because through the Internet information can be obtained, complete, and in a short time.

The descriptive analysis results of the variable electronic word of mouth show that the "Concern for Others" indicator has the highest average score of 4.26. This proves that the reviews that tourists hear or see on social media or websites affect the knowledge of tourists to Sundanese gastronomic attractions in Bandung.

In the descriptive analysis of the Tourist Visit Motivation variable, the "Special Purpose" indicator has the highest average score, which is 4.69. This proves that the majority of respondents want to visit Bandung to travel with their families

5 CONCLUSION

The following conclusions are based on the results of research and testing:

- Internet marketing has no partial effect on Tourist Visit Motivation to Bandung. This indicates that the information of Sundanese gastronomic tourism on the Internet does not affect the motivation of tourists to visit Bandung.
- There is a partial positive and significant influence between the variable electronic word of mouth with tourist visit motivation to Bandung. This proves that positive experiences shared by tourists through online reviews can influence tourists to visit motivation to Bandung. The more positive opinion of tourists about Sundanese gastronomic tourism in Bandung, the motivation of other tourists to visit Bandung will increase.
- Internet marketing and electronic word of mouth jointly affect Tourist Visit Motivation to Bandung. This shows that Sundanese gastronomic tourism information found on the Internet and positive experiences shared by tourists through online reviews can motivate other travelers to visit Bandung.
- The results of the analysis of the introduction Sundanese food and beverage shows that the most known menus of Sundanese food and beverage in Bandung by the respondent are a menu that can be found easily in Bandung. The menus are Opor Hayam, Lotek, Mie Kocok Bandung, Cireng, Kerupuk Udang, Es Cingcau, Sambal Terasi, and Bandrek.

REFERENCES

Jalilvand, M. R., & Samiei, N. (2012). The Effect of Electronic Word of Mouth on Brand Image ad Purchase Intention: An empirical study in the automobile industry in Iran. Marketing Intelligence and Planning, 30(4).

Jeong, E., & Soocheong, J. (2011). Restaurant Experiences Triggering Positive Electronic Word-of-Mouth (eWOM) Motivations. International Journal of Hospitality Management, 30(356).

Kotler, Philip, & Keller, K. (2012). Marketing Management (12th ed.). Jakarta: Erlangga.

Machfoedz, M. (2010). Modern Marketing Communication. Yogyakarta: Cakra Ilmu.

Mohammed, R. A., Jaworski, B. J., Fisher, R. J., & Paddison, G. J. (2003). Internet Marketing: Building Advantage in a Network Economy (2nd ed.). New York: McGraw-Hill Book Co.

Ningsih, C. (2016). The Synergy of Tourism Based on Creative Industries with the National Industry Development Strategy Towards Globalization. 11 (1), 59–64. https://doi.org/10.17509/jurel.v11i1.2903

Sugiama, A. (2011). Ecotourism: Nature Conservation Based on Tourism Development. Bandung: Guardaya Intimate.

Supranto, J. (2006). Measuring Customer Satisfaction Level: To Increase Market Segmentation. Jakarta: Rineka Cipta.

Suryani, T. (2008). Consumer Behavior: Implications for Marketing Strategies. Yogyakarta: Graha Ilmu.

Taqwani, D. (2012). Culture of Gastronomy and Speech Acts in Pragmatic Studies in Ratatouille Films.

Turgarini, D. (2018). Sundanese Gastronomy as a Tourist Attraction in Bandung. Yogyakarta: Gadjah Mada University.

Author index